STANDARD 7: **CONCEPTS OF WHOLE NUMBER OPERATIONS**

In grades K–4, the mathematics curriculum should include concepts of addition, subtraction, multiplication, and division of whole numbers so that students can—

- develop meaning for the operations by modeling and discussing a rich variety of problem situations;
- relate the mathematical language and symbolism of operations to problem situations and informal language;
- recognize that a wide variety of problem structures can be represented by a single operation;
- develop operation sense.

STANDARD 8: **WHOLE NUMBER COMPUTATION**

In grades K–4, the mathematics curriculum should develop whole number computation so that students can—

- model, explain, and develop reasonable proficiency with basic facts and algorithms;
- use a variety of mental computation and estimation techniques;
- use calculators in appropriate computational situations;
- select and use computation techniques appropriate to specific problems and determine whether the results are reasonable.

STANDARD 9: **GEOMETRY AND SPATIAL SENSE**

In grades K–4, the mathematics curriculum should include two- and three-dimensional geometry so that students can—

- describe, model, draw, and classify shapes;
- investigate and predict the results of combining, subdividing, and changing shapes;
- develop spatial sense;
- relate geometric ideas to number and measurement ideas;
- recognize and appreciate geometry in their world.

STANDARD 10: **MEASUREMENT**

In grades K–4, the mathematics curriculum should include measurement so that students can—

- understand the attributes of length, capacity, weight, area, volume, time, temperature, and angle;
- develop the process of measuring and concepts related to units of measurement;
- make and use estimates of measurement;
- make and use measurements in problem and everyday situations.

STANDARD 11: **STATISTICS AND PROBABILITY**

In grades K–4, the mathematics curriculum should include experiences with data analysis and probability so that students can—

- collect, organize, and describe data;
- construct, read, and interpret displays of data;
- formulate and solve problems that involve collecting and analyzing data;
- explore concepts of chance.

STANDARD 12: **FRACTIONS AND DECIMALS**

In grades K–4, the mathematics curriculum should include fractions and decimals so that students can—

- develop concepts of fractions, mixed numbers, and decimals;
- develop number sense for fractions and decimals;
- use models to relate fractions to decimals and to find equivalent fractions;
- use models to explore operations on fractions and decimals;
- apply fractions and decimals to problem situations.

STANDARD 13: **PATTERNS AND RELATIONSHIPS**

In grades K–4, the mathematics curriculum should include the study of patterns and relationships so that students can—

- recognize, describe, extend, and create a wide variety of patterns;
- represent and describe mathematical relationships;
- explore the use of variables and open sentences to express relationships.

Please refer to the inside back cover for the Curriculum Standards for Grades 5–8.

Mathematics for Elementary Teachers

An Interactive Approach

Second Edition

Mathematics for Elementary Teachers

An Interactive Approach

Second Edition

Thomas Sonnabend
Montgomery College

Saunders College Publishing
Harcourt Brace College Publishers
Fort Worth Philadelphia San Diego New York Orlando Austin
San Antonio Toronto Montreal London Sydney Tokyo

Publisher: John Vondeling
Executive Editor: Bill Hoffman
Product Manager: Nick Agnew
Developmental Editor: Terri Ward
Project Editor: Linda Boyle
Production Manager: Alicia Jackson
Art Directors: Caroline McGowan; Joan Wendt

Cover Credit: Kathleen Flanagan; Joan Wendt

About the Cover

The shape on the cover is a Mobius strip. To find out more about this amazing shape, see the Special Exercise on page 464.

Printed in the United States of America
MATHEMATICS FOR ELEMENTARY SCHOOL TEACHERS: AN INTERACTIVE APPROACH, second edition

0-03-018367-7

Library of Congress Catalog Card Number: 97-70493

7890123456 069 10 987654321

To my mother Norma, and in memory of my stepfather Irving: Thanks for your love, your inspiration, and your sacrifices.

Preface

The NCTM Standards state that "prospective teachers must be taught in a manner similar to how they are to teach—by exploring, conjecturing, communicating, reasoning, and so forth." This necessitates devoting more class time to discussing and discovering ideas and less time to lecturing.

Can a mathematics book actively involve students in developing and explaining mathematical concepts? Yes, by using a carefully organized, interactive lesson format that promotes student involvement and gradually leads the student to a deeper understanding of mathematical ideas. The interactive format also allows for more class discussion and small group work.

To implement the NCTM Standards, one needs a textbook that provides numerous opportunities for investigation and discourse. Rather than having a few special puzzle and investigation problems, this textbook presents a substantial collection of exercises in every lesson and homework set that involve reasoning, investigating, or communicating.

Most people do not enjoy reading mathematics textbooks. Can a mathematics textbook be interesting to read and study? Yes, if the mathematical presentation is straightforward and clear, and the lesson content is enriched with investigations, appropriate uses of technology, humor, history, and interesting applications. A text for this course should also make it clear how the topics in each chapter relate to both the NCTM Standards and the current school mathematics curriculum.

You are looking at the result of my effort to address all these concerns while covering new topics and examining underlying concepts and connections in elementary-school mathematics. This textbook is the culmination of 17 years of work with university, college, and community college students.

Teaching elementary school is one of the most important and challenging professions in our society. It is my hope that this textbook provides material for a more interesting course that produces more competent teachers with a better sense of what elementary-school mathematics is all about.

The second edition of *Mathematics for Elementary Teachers: An Interactive Approach* retains the same goals as the first edition. Changes in the second edition reflect the latest trends in elementary school mathematics, feedback from students and professors, and a thorough review and refinement of the material on nearly every page of the first edition.

DISTINCTIVE FEATURES RETAINED IN THIS EDITION

This textbook places *greater emphasis* on

- discovery, discussion, and explanation of concepts that involve students and deepen their understanding

- investigations and activities that require higher-level thinking
- applications of mathematics that connect mathematical concepts to everyday life
- explaining a concept or solving a problem in more than one way so that future teachers can see a variety of approaches
- how material relates to the current elementary-school curriculum
- difficult concepts in the elementary-school mathematics curriculum that future teachers have trouble explaining
- connecting lessons, homework exercises, and chapter review exercises
- multiple representations of concepts so that students develop a broader understanding
- inductive and deductive reasoning, showing how these two processes are used throughout mathematics
- models for arithmetic operations that establish the connection between an operation and its everyday applications
- common student error patterns in arithmetic and measurement that prepare future teachers for diagnosing student difficulties
- spatial perception and perspective drawing so that students become more adept at three-dimensional geometry
- algebra as a mathematical language and a generalization of arithmetic
- important statistical applications including statistical deceptions, surveys, and standardized tests

In order to make time for these topics, this textbook places less emphasis upon (1) topics more students have studied or will study in other mathematics or education courses and (2) topics that are relatively more remote from the current elementary-school curriculum. Such topics include arithmetic and algebra skills, formal logic, geometry terminology, networks, trigonometry, and isometries.

Other Features Retained in this Edition

- Students use relevant NCTM Standards to review each chapter and increase their awareness of specific standards.
- Charts at the end of each chapter show grade-level coverage of related elementary-school topics.
- Extensive coverage of geometry and measurement topics strengthens students' spatial and analytical abilities.
- Extensive coverage of statistics and probability incorporates many realistic applications.
- Mental computation and estimation with whole numbers, fractions, decimals, and percents make use of properties of operations and complement the use of calculators.
- Humor and historical vignettes humanize the mathematics.

FEATURES NEW TO THIS EDITION

- Many new topics are now covered in lessons, including complement of a set (2.1), Roman numeration (3.1), introduction to exponents (3.1), missing addend (3.2), constructivism (3.4), counting divisors (4.3), Euclidean Algorithm (4.4), colored counter models (5.1), types of rounding (7.1), complementary and supplementary angles (8.1), chord (8.2), relative error (10.1), translating words into equations and tables (11.3), line plot (12.1), stem-and-leaf plot (12.1), mode (12.3), common sampling techniques (12.5), and permutations and combinations (13.3).
- Introductory probability begins with experiments. Concepts are developed from these hands-on activities.
- Estimation now precedes computation in the sections on decimals and percents. This reflects the current practices in elementary and middle school mathematics.
- Updated Technology Coverage also reflects current practices in elementary school. The second edition has expanded coverage of fraction calculators (6.2), spreadsheets (7.2), drawing software (8.2), and graphing calculators (11.2). MINITAB is used to analyze data in Chapter 12. Material (optional) on BASIC has been relegated to the Appendix.
- Updated coverage of elementary school curriculum topics. Examples include precision, surveys, misleading statistics, fair vs. unfair games, independent events, drawing space figures, spreadsheets, algebra as a language, and algebra using graphs/tables/equations.
- Expanded coverage of the history of mathematics and women in mathematics further enriches the course and encourages women to study mathematics.
- Expanded coverage of the van Hiele model familiarizes students with this useful pedagogical model.
- Expanded coverage of compass constructions now includes the constructions that students in certain states need to learn.
- Basic homework exercises are reorganized so the teacher and students can easily recognize an initial set of exercises to work on. These would include the odd-numbered exercises and those exercises that are marked (*). Answers to all these exercises are provided in the back of the textbook.
- More answers and solutions to the Lesson Exercises are provided. This makes more feedback available to students. Teachers and students who desire a more traditional approach with more worked examples can regularly refer to the answers and solutions to Lesson Exercises.
- Selected lesson exercises that are especially suitable for discussion are now marked with the icon . Some Lesson Exercises are classified as more appropriate for whole class discussion ("Class Lesson Exercise") or cooperative work ("Cooperative Lesson Exercise"). These classifications will assist students and teachers in determining the best way to approach lesson exercises.
- Investigations are now labeled with the icon .
- Terminology has been updated to reflect current usage in elementary school mathematics.

- The *Instructor's Manual* has expanded commentaries on each section to help the teacher use the text more effectively. Guidelines for working in small groups or pairs are provided.
- Reading lists at the end of each chapter have been updated.

TEACHING AND LEARNING AIDS

The following features make it easier for students and professors to make more effective use of this book.

- Chapter introductions set the stage for each new chapter.
- Subsection headings help students understand how each lesson is organized.
- Exercises within each lesson actively involve students in developing and reinforcing concepts.
- Lessons gradually lead students to a deeper understanding of difficult concepts in the elementary-school curriculum.
- Answers to most lesson exercises provide students with the opportunity to check their work as they develop new ideas.
- Important definitions, theorems, and properties are boxed with boldfaced headings.
- All mathematical terms are set boldface.
- Color is used to highlight mathematical ideas and add clarity to figures.
- Sample textbook pages show newer topics in elementary-school mathematics, thus increasing student awareness of recent changes.
- A "friendly" writing style as well as cartoons, drawings, and photographs stimulate student interest.
- Homework exercises are categorized as basic, extension, computer, and special so that instructors can select the appropriate level and type of homework exercises.
- Answers to most odd-numbered and all starred even-numbered homework exercises and all chapter review exercises provide students with adequate feedback.
- Calculator and computer exercises are marked.
- Chapter study guides list important concepts and terms by section to help students recall major ideas.
- Chapter review exercises help students prepare for exams.
- Suggested reading lists for each chapter direct students to interesting resource material.

Lesson Organization

Each lesson presents a series of exercises that develop the basic ideas of the section. These exercises should be completed before attempting the related questions in the Homework Exercises. The Lesson Exercises may be done individually, in pairs, in small groups, in class discussion, or they can be completed at home. Answers to nearly all of the Lesson Exercises appear at the end of the lesson or in the lesson.

CHAPTER ORGANIZATION AND FEATURES

Chapter 1 introduces mathematical reasoning processes and problem solving techniques that are used throughout the course. Its broader view of mathematical reasoning includes inductive and deductive reasoning as well as patterns and problem solving.

Chapter 2 covers set concepts that are used later in the course to clarify other concepts. Venn diagrams are used to solve problems in Chapters 1 and 2.

Chapters 3–7 cover the number systems from whole numbers to real numbers. Categories for each whole number operation (e.g., compare and take away) are studied in depth in Chapter 3 and used with other number systems in Chapters 5–7. Mental computation and estimation are also used with each number system. In number theory (Chapter 4), students do proofs and learn two different methods for solving certain problem types. In Chapters 5–7, students learn how to explain difficult procedures in integer, fraction, and decimal arithmetic, using models and realistic applications. Students also compare different representations of real numbers.

Chapters 8–10 cover geometry and measurement. In Chapter 8, the van Hiele model is used to develop categories and definitions of quadrilaterals. Sections on space figures and spatial perception strengthen students' spatial abilities. In Chapter 9, students connect transformation geometry to congruence, symmetry, and similarity, and students analyze a series of constructions using congruence properties. In Chapter 10, students perform a series of activities to learn about the metric system and to develop area formulas in a logical sequence.

Chapter 11, Algebra and Coordinate Geometry, follows and extends arithmetic from Chapters 3–7 and geometry and measurement from Chapters 8–10. Students learn different ways to represent a function or relation. Multiple representations (words, tables, graphs, and formulas) are used in solving realistic application problems. Algebra is studied as a mathematical language.

The final two chapters cover statistics and probability. Chapter 12 emphasizes choosing the most appropriate graph or statistic to summarize results. Choices include the relatively new stem-and-leaf plots and box-and-whisker plots. The extensive applications of statistics include statistical deceptions, surveys, and standardized tests.

In Chapter 13, students learn the connection between theoretical and experimental probabilities and use simulations to generate experimental probabilities. Students see the importance of probability in insurance, drug testing, and gambling games.

COURSE OUTLINES

This textbook provides for some flexibility in organizing the course. The textbook contains more than enough material for two four-semester-hour mathematics courses for preservice elementary teachers. The material in Chapters 1–7 should be studied in sequence; however, a number of sections and parts of sections can be omitted.

In Chapters 1–7 you should cover the following material.

1. Chapter 1
2. In Section 2.1, review set notation and define whole numbers. In Section 2.2, cover "Intersection and Union."
3. In Section 3.1, cover all subsections beginning with "Models for Place Value." Cover Sections 3.2 and 3.3. In Section 3.4, cover the properties. Cover Section 3.5.
4. In Section 4.1, cover "factors." In Section 4.2, cover "Multiples," "Divisibility Tests for 2, 5, and 10," and "Divisibility Tests for 3 and 9." Cover Sections 4.3 and 4.4.
5. Cover Section 5.1 and "Inverses" in Section 5.3.
6. Cover Chapter 6.
7. In Section 7.1, cover "Place Value" and "Exponents." In Section 7.2, cover "Adding and Subtracting Decimals", "Multiplying Decimals", and "Dividing Decimals." Cover Sections 7.3 and 7.4. In Section 7.5, cover "Basic Percent Problems."

After studying the basic matierial in Chapters 1–7, the major ideas of Chapters 8–13 can be studied independently of one another with the following exceptions.

Material	**Prerequisite**
Logo in Chapters 9 and 10	Sections 8.6, 8.7
Chapter 10	Chapter 8
Section 10.6	Section 9.4
Section 11.5	Chapters 8–10
Section 13.4	Section 12.3

One-Semester Course

Design a one-semester course that suits your needs, with the following conditions.

1. Cover the required sections and parts of sections in Chapter 1–7 as already outlined.
2. You can skip optional Section 3.6, investigations, extension exercises, and special exercises.
3. Round out the course with additional material from Chapters 1–7, or select sections from Chapters 8–13. Depending upon the length of your course, the

ability of your students, and the amount of time you spend on extension exercises, you might be able to cover about 5 to 15 more sections. Suggestions for additional material would include Sections 8.1–8.5, 9.1, 10.1–10.3, 12.1–12.3, 13.1 and 13.2.

Two-Semester Course

In two four-semester one-hour courses, one can cover most of the sections in the book, depending upon the number of homework exercises covered. One could skip some of Sections 3.6, 9.2, 10.6, 11.5, 13.4, or 13.5, or optional subsections that are listed in the *Instructor's Manual.*

SUPPLEMENT FOR TEACHERS

The *Instructor's Manual* contains teaching suggestions, overhead transparency masters, and exercise answers that are not given in the textbook. The *Test Bank* contains sample test questions for each chapter. The *Computerized Test Bank,* available in IBM and Macintosh formats, contains the questions from the *Test Bank.*

Manipulatives Kit

This kit will consist of Cuisenaire mathematics manipulatives. The kit includes geometric solids, connecting cubes, and the following materials for the overhead projector: a geoboard, fraction tiles, base-ten blocks, and a tangram puzzle. This kit will be provided free to professors upon adoption of this text.

SUPPLEMENT FOR STUDENTS

The *Student Resource Manual* for the first edition is still available. It provides additional hands-on activities for each chapter. These activities utilize material cards that come with the manual. Also included are activities that focus on calculators, critical thinking, connections, and the NCTM standards.

Saunders College Publishing may provide complimentary instructional aids and supplements or supplement packages to those adopters qualified under our adoption policy. Please contact your sales representative for more information. If as an adopter or potential user you receive supplements you do not need, please return them to your sales representative or send them to

Attn: Returns Department
Troy Warehouse
465 South Lincoln Drive
Troy, MO 63379

ACKNOWLEDGMENTS

First, I would like to thank my wife Celia for her support, encouragement, and patience during the past 9 years, not to mention her many suggestions based upon her own extensive experience as an elementary-school teacher. I am also grateful for the encouragement and advice provided by my mother, my sister, my late stepfather, my friends, Judge Robert Mark, and author Dan Benice.

Next, I would like to thank the editorial staff at Saunders College Publishing, especially Developmental Editor, Terri Ward, Project Editor Linda Boyle, and Editorial Assistant Elisa Krantweiss, who put a great deal of time and effort into the project during a difficult time when the staff was shorthanded. Freelance copyeditor Mary Patton made a significant contribution to the improved exposition in the second edition.

The attractive design of the book is due to the direction of Art Directors Caroline McGowan and Joan Wendt, and the many graphic artists who rendered the figures, with special thanks to the cartoonists at Rolin Graphics. Thanks also to Alicia Jackson and the production staff who kept the book moving along through the production process.

Next, I would like to thank my professors from graduate school, who provided many inspiring ideas that led to the development of this textbook. They included Neil Davidson, Jim Fey, Zal Usiskin, Stephen Willoughby, Jim Henkelman, and Mildred Cole—six outstanding mathematics educators.

Finally, I would like to thank my colleagues, my students, and the many reviewers whose helpful suggestions significantly improved this book. The names and affiliations of principal reviewers for the first two editions are as follows.

Principal Reviewers

James Arnold, *University of Wisconsin (Milwaukee)*
Susan Beal, *Saint Xavier University*
Judy Bergman, *University of Houston*
Peter Braunfeld, *University of Illinois (Urbana)*
Laura Cameron, *University of New Mexico*
Gail Cashman, *former Montgomery College student*
Jane Crane, *Boise State University*
Terry Crites, *Northern Arizona University*
Cynthia Davis, *Truckee Meadows Community College*
Gregory Davis, *University of Wisconsin (Green Bay)*
Donald Dessart, *University of Tennessee (Knoxville)*
Billy Edwards, *University of Tennessee (Chatanooga)*
Ann Farrell, *Northern Arizona State University*
Joseph Ferrar, *Ohio State University*
Iris B. Fetta *Clemson University*
Gail Gallitano, *West Chester University*
Virginia Ellen Hanks, *Western Kentucky University*
Alan Hoffer, *Boston University*
John Koker, *University of Wisconsin (Oshkosh)*
Flora Metz, *Jackson State Community College*
William A. Miller, *Central Michigan University*
Barbara Moses, *Bowling Green State University*

John Neil, *Boise State University*
Pamela Pospisil, *Montgomery College*
Donald Ramirez, *University of Virginia*
Janice Rech, *University of Nebraska (Omaha)*
Alice I. Robold, *Ball State University*
Sharon Ross, *California State University (Chico)*
Elizabeth Shearn, *University of Maryland*
Jane F. Sheilack, *Texas A & M University*
Keith Shuert, *Oakland Community College*
Peter Sin, *University of Florida*
Jim Trudnowski, *Carroll College*
C. Ralph Verno, *West Chester University*
Jeannine Vigerust, *New Mexico State University*
Richard Vinson, *University of South Alabama*
Stephen S. Willoughby, *University of Arizona*
Deon Woodward, *New Mexico State University*
John E. Young, *Southeastern Missouri State University*

Contents Overview

Contents

Introduction

Why Learn Mathematics?

As a teacher, you will introduce the next generation of students to mathematics, teaching them arithmetic, geometry, and statistics. Your students will solve problems, think logically, find patterns, and apply their knowledge to practical situations.

But why does anyone teach children mathematics as part of their education? For one thing, mathematics offers a unique way of looking at the world. Mathematics gives simple, abstract descriptions that illuminate general relationships among quantities or shapes while simplifying or ignoring qualities. Mathematics can show that, ounce for ounce, Brand A is a better buy than Brand B, focusing upon the amount and the price while ignoring other differences in the products and the companies that produce them. Such mathematical information can help one decide which brand to buy.

Furthermore, the logical, objective approach of mathematics is applied to many situations. Mathematics has had an impact on nearly every area of life, from philosophy to economics to art. Philosophers apply the logic of mathematics to ultimate questions. Economists employ quantitative methods to describe financial trends. Artists use geometry when they represent our three-dimensional world on canvas.

Finally, mathematics is the language of technological societies. Businesses use mathematics in record-keeping and analysis. As a consumer and a citizen, each of us needs mathematics to make financial decisions and to interpret statistics in political and economic news.

What have your mathematical experiences been like? Is mathematics one of your favorite subjects? Or have you felt like Peppermint Patty in the following cartoon?

Reprinted by permission of UFS, Inc.

Your professor may ask you to complete the following survey.

Introductory Survey

Name ______________________________

1. For your two most recent mathematics courses, list the course name, when and where you took it, and your final grade.
____
____
2. One of my favorite things about recent mathematics courses has been
____
____
3. One of my least favorite things about recent mathematics courses has been ______________________________
____
4. Do you feel adequately prepared for this class?
(a) Yes (b) No (c) Not sure
5. (a) How many credit hours are you taking this term? ________
(b) How many hours will you work each week? ________
(c) Do you also have children to take care of? ________
6. Rate your overall time schedule this term.
(a) Very busy/may be too much
(b) Busy
(c) Comfortable
7. Would you be interested in meeting with other students in the class
(a) to do homework? Yes No Maybe
(b) to study for exams? Yes No Maybe

1 Mathematical Reasoning

How do we determine what is true in mathematics classes and in everyday life? You have been doing mathematics for many years, yet you may know little about the approaches you use over and over again in solving problems.

The two central processes of mathematics are inductive and deductive reasoning. Becoming aware of these reasoning processes will give you a better understanding of how you and your students learn mathematics. After studying these processes, you will use them throughout this course to make and prove generalizations.

Mathematics is sometimes called the study of patterns. Induction and deduction are used to generalize mathematical patterns, to extend them, and to prove generalizations about them.

In recent years, a greater emphasis has been placed on helping all students develop problem-solving abilities in a structured way. Elementary-school books now teach some version of Polya's four-step problem-solving procedure: Read, plan, solve, and check. When it comes to the second step, making a plan, the elementary-school curriculum now includes specific processes and strategies for problem solving. The most useful ones are induction, deduction, choosing the operation, making a table, and drawing a picture. More specialized strategies include guessing and checking and working backwards. These processes and strategies are introduced in Chapters 1 and 3.

1.1 INDUCTIVE REASONING

How do we form our beliefs? For example, do you believe that your friend Alan is trustworthy?

Alan

Perhaps you would answer "yes" if your previous encounters with Alan suggest that he is trustworthy.

Similarly, many mathematical ideas arise from observing a pattern in a series of examples. The "number magic" in Lesson Exercise 1.1 will lead your entire class to form a generalization based upon a pattern in their individual results.

Class Lesson Exercise 1.1

(a) Follow these instructions.

Pick a number.
Multiply it by 2.
Add 10.
Divide by 2.
Subtract your original number.

(b) What number do you end up with?

(c) So why all the fanfare? Pick a different number and follow the instructions in part (a) again (or find out what everyone else in the class got). What number do you end up with this time? Are you mystified?

(d) Make a generalization based upon your results in part (b) and part (c).

(e) Are you sure that your generalization will work for all starting numbers?

You may wonder how the "magic" in Lesson Exercise 1.1 works. When we return to it later, in Section 1.3, you'll be able to prove that it works for all starting numbers.

The four statements in the cartoon at the beginning of the lesson and the "number magic" in Lesson Exercise 1.1 illustrate inductive reasoning. Inductive reasoning works like this. You might be adding pairs of odd numbers and see a pattern in the results: $1 + 3 = 4$, $3 + 7 = 10$, and $35 + 31 = 66$. Then you make a generalization: The sum of any two odd numbers is an even number. This process of reasoning from the specific to the general is called inductive reasoning.

Definition **Inductive Reasoning**

Inductive reasoning is the process of making a generalization based upon a limited number of observations or examples.

Inductive reasoning is the basis of experimental science and a fundamental part of mathematics. A scientist uses inductive reasoning to formulate a hypothesis based upon a pattern in experimental results. A mathematician uses inductive reasoning to form a reasonable generalization from a pattern in a series of examples.

In addition, many of our beliefs, prejudices, and theories are based upon this process. This lesson will help you make intelligent use of inductive reasoning in mathematics and in everyday life. Use inductive reasoning with care, since your generalizations may turn out to be wrong. For example, your friend Alan may eventually turn out to be untrustworthy.

MAKING GENERALIZATIONS

The essence of inductive reasoning is making generalizations (also called "conjectures") that appear to be true. The following example and exercises will help you become familiar with the process of making generalizations.

EXAMPLE 1.1

Does every number squared equal itself? Use inductive reasoning to make a generalization based upon the examples given.

Examples: $0^2 = 0$ $\quad 1^2 = 1$
Generalization: ______________________

Solution

In both examples, a number squared equals itself. Then, by inductive reasoning, I conclude that every number squared equals itself.

Generalization: Every number squared equals itself. (This will turn out to be a false generalization.) ■

In each of the following exercises, use inductive reasoning to make a generalization based upon the examples given. Write each generalization as a complete sentence.

Lesson Exercise 1.2

The product of two odd numbers is what type of number?

Examples: $5 \cdot 9 = 45$ $3 \cdot 7 = 21$
Generalization: ____________________

Lesson Exercise 1.3

Examples: The two times I went on a picnic, it rained.
Generalization: ____________________

____

(*Note:* The answers to many lesson exercises can be found at the end of the section, just before the homework exercises.)

Figure 1-1 shows how the second-grade *Mathematics Plus* textbook (Harcourt Brace Jovanovich, 1994) asks the child to make a generalization after a series of subtraction problems.

Of course, we would prefer that our generalizations were always true. In everyday life, we settle for generalizations that are true *most* of the time. When we wake up each morning, we assume that the refrigerator will be working and that we will have hot water. Such generalizations are reasonable and usually true. They give our lives order and enable us to plan ahead. However, such generalizations would not meet mathematical standards. In mathematics, we seek generalizations that are *always* true.

Unfortunately, inductive reasoning does not always lead to true generalizations. Although the mathematical generalization you made in Lesson Exercise 1.2 is true, the mathematical generalization in Example 1.1 is false. Often a false generalization results when one selects a small or nonrepresentative set of examples. By observing a larger number of individual events, one can make more reasonable generalizations that are more likely to be true.

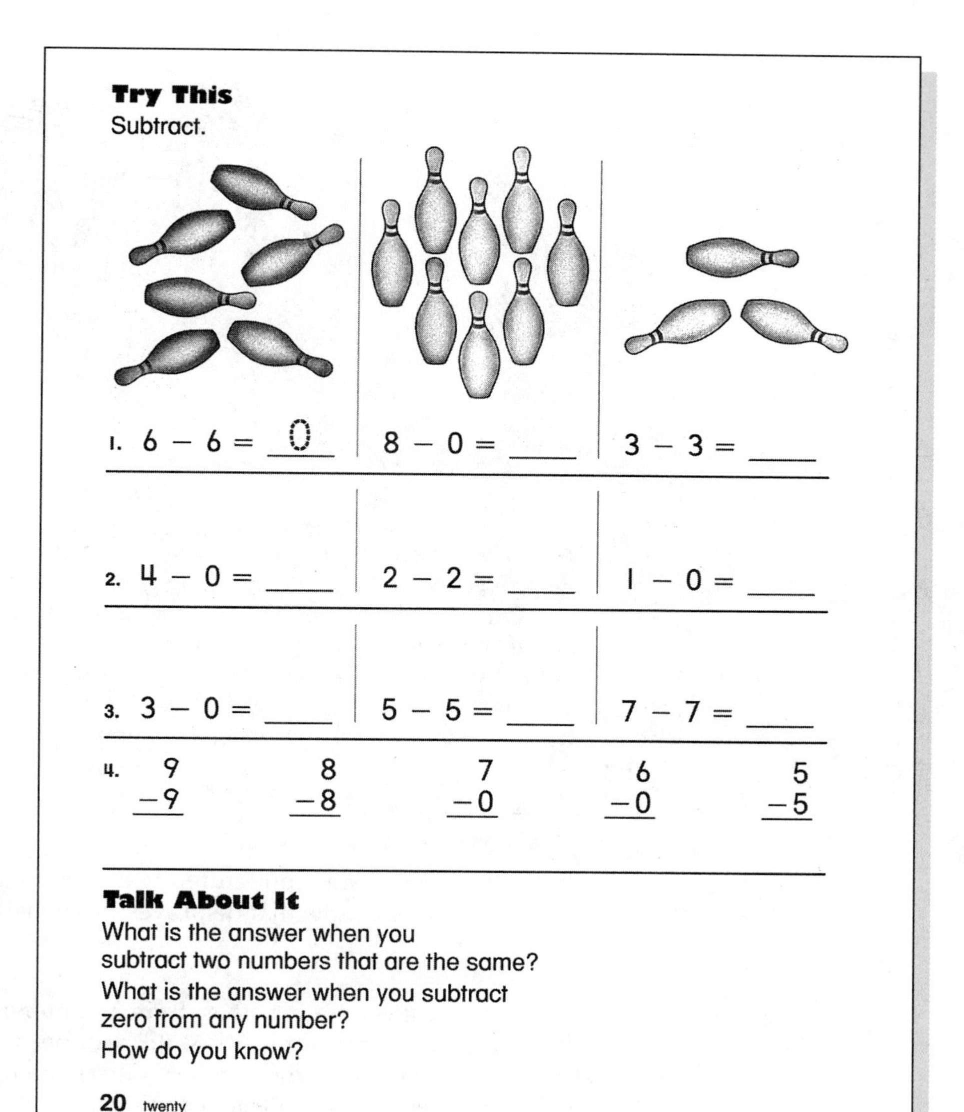

Try This

Subtract.

1. 6 − 6 = 0 | 8 − 0 = ____ | 3 − 3 = ____

2. 4 − 0 = ____ | 2 − 2 = ____ | 1 − 0 = ____

3. 3 − 0 = ____ | 5 − 5 = ____ | 7 − 7 = ____

4. $\begin{array}{r} 9 \\ -9 \\ \hline \end{array}$ $\begin{array}{r} 8 \\ -8 \\ \hline \end{array}$ $\begin{array}{r} 7 \\ -0 \\ \hline \end{array}$ $\begin{array}{r} 6 \\ -0 \\ \hline \end{array}$ $\begin{array}{r} 5 \\ -5 \\ \hline \end{array}$

Talk About It

What is the answer when you subtract two numbers that are the same?

What is the answer when you subtract zero from any number?

How do you know?

20 twenty

From Mathematics Plus, *Grade 2 (San Diego: Harcourt Brace and Company 1994), p. 20.*

Figure 1–1

USING COUNTEREXAMPLES TO DISPROVE FALSE GENERALIZATIONS

How can one recognize false generalizations? Often it is a matter of finding just one exception. For over 1900 years, people accepted Aristotle's (384–322 B.C.) generalization that heavier objects fall faster than lighter objects. To test this generalization, Galileo (1564–1643) is alleged to have dropped two metal objects, one much heavier than the other, from the Leaning Tower of Pisa (Figure

Figure 1–2

1-2). The objects hit the ground simultaneously! This single **counterexample** disproved Aristotle's longstanding generalization.

One usually cannot be sure whether a generalization reached by induction is true, because one cannot examine every single possibility. Although one cannot *prove* a generalization is true by picking examples, one can show a generalization is false by finding just *one* counterexample! Example 1.2 illustrates how one counterexample is used to show that a mathematical generalization is false.

EXAMPLE 1.2

In Example 1.1, it was conjectured that every number squared equals itself. Find a counterexample that disproves this generalization.

Solution

This statement is false for other decimal numbers. For example, $2^2 = 4$, not 2. In making a generalization, it would have been wiser to use more than two examples, including some numbers other than 0 and 1 (since 0 and 1 have special properties for multiplication). ■

In the following Lesson Exercises 1.4 and 1.5, decide whether the generalization is reasonable or false. If possible, try a few other specific examples. If the generalization is false, disprove it by giving a counterexample. (The 👥 in front of an exercise (such as Lesson Exercise 1.5) indicates that it is a good exercise to discuss as a class or in groups. Communicating your ideas orally or in writing helps to clarify them.)

Lesson Exercise 1.4

I take a number and add it to itself. Then I take the same number and multiply it by itself. I obtain the same answer either way in two examples: $0 + 0 = 0 \cdot 0$ and $2 + 2 = 2 \cdot 2$. So I generalize that this will always happen.

Lesson Exercise 1.5

I add the first two, three, and four odd numbers.

$$1 + 3 = 2^2 \qquad 1 + 3 + 5 = 3^2 \qquad 1 + 3 + 5 + 7 = 4^2$$

From this, I generalize that the sum of the first N odd numbers is N^2, where $N = 2, 3, 4, \ldots$. (*Hint:* Write the next two equations that continue the pattern, and see if they are true.)

How can you make more *reasonable* generalizations and detect faulty ones? By checking more examples and a greater variety of them.

APPLICATIONS OF INDUCTIVE REASONING

At its best, inductive reasoning leads to reasonable generalizations. In mathematics, these generalizations are then proved true (if possible) using deductive reasoning. (Deductive reasoning will be discussed in the next section.) In everyday life, however, a reasonable generalization based upon a large number of examples is often the best we can do. It is often impossible to prove that such a generalization is true.

Because nonmathematical generalizations are often impossible to prove, one should be careful about making them. Unfortunately, people are not always careful about making nonmathematical generalizations. Consider prejudice. One example of prejudice is a generalization about a whole group of people. Some people learn prejudices from their family or community, but prejudices can also be developed using inductive reasoning.

After I meet a few people from a certain group and observe a particular common characteristic, I may assume that all people from that group share that characteristic. No matter what generalization I make about a group of people, there are bound to be some exceptions.

In Lesson Exercise 1.6, consider how some superstitions arise from inductive reasoning. (Note the use of the instruction "*Explain*" in the exercise. In this book, *Explain* means you should write an explanation that you might give to an elementary-school student. The explanation can be brief but should proceed step by step from start to finish.)

Lesson Exercise 1.6

Explain how someone could use inductive reasoning to develop a superstition.

Lesson Exercise 1.7

Give an example of how you have used inductive reasoning to form a general idea.

AN INVESTIGATION: SUMS OF CONSECUTIVE WHOLE NUMBERS

An important part of mathematics is exploring new situations and making conjectures about them. The following investigation can be done in class (alone or in groups) or as homework.

Lesson Exercise 1.8

Use inductive reasoning to answer the following questions.

(a) Is the sum of any three consecutive whole numbers divisible by 3?

(b) Is the sum of any four consecutive whole numbers divisible by 4?

(c) Is the sum of any five consecutive whole numbers divisible by 5?

(d) Investigate further examples and complete the following generalization:

The sum of N consecutive whole numbers is divisible by N when $N =$ ________________

ANSWERS TO SELECTED LESSON EXERCISES

1.2 The product of two odd numbers is an odd number.

1.3 Every time I go on a picnic it rains.

1.4 False; $3 + 3 \neq 3 \cdot 3$

1.5 Reasonable

1.6 A person may associate events that have occurred in succession a few times, such as a black cat crossing the person's path and something bad happening soon after. After such experiences, the person generalizes that every time a black cat crosses one's path, something bad will happen soon after.

1.1 HOMEWORK EXERCISES

Basic Exercises

1. Consider the following number trick.

Pick any number at all.
Add 8.
Multiply by 2.
Subtract 10.
Subtract your original number.

(a) Try the trick with two or three different numbers.

(b) What is the general pattern in your results?

2. $3 \cdot 5 = 5 \cdot 3 \qquad 6 \cdot 2 = 2 \cdot 6 \qquad 1 \cdot 9 = 9 \cdot 1$
Write a generalization based upon these results.

3. A Martian met three people from Boston, and they all wore high heels. Now the Martian thinks everyone from Boston wears high heels. Clearly, this generalization is false. Did the Martian use inductive reasoning correctly?

4. Why is it usually impossible to show that a statement is true by checking examples?

5. How can a proposed generalization be disproved?

In Exercises 6–10(a), decide whether the generalization is reasonable or false. If it's false, give a counterexample.

***6.** *Examples:*

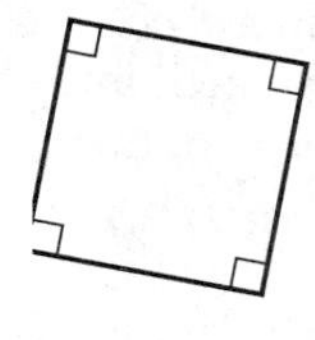

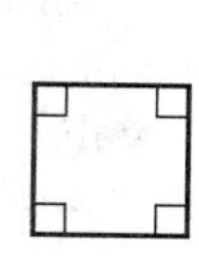

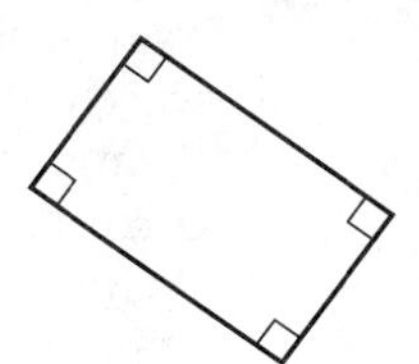

Generalization: All four-sided figures have four right angles.

7. *Examples:* José likes to drink beer and watch football on TV.

Brad likes to drink beer and watch football on TV.

Generalization: All men like to drink beer and watch football on TV.

8. *Examples:* $1 + 2 = \frac{2(3)}{2}$

$$1 + 2 + 3 = \frac{3(4)}{2}$$

$$1 + 2 + 3 + 4 = \frac{4(5)}{2}$$

Generalization: The sum of the first N counting numbers is $N(N + 1)/2$ for $N = 2, 3, 4, \ldots$.

9. *Examples:*

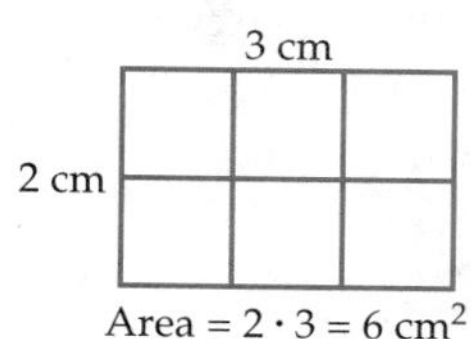

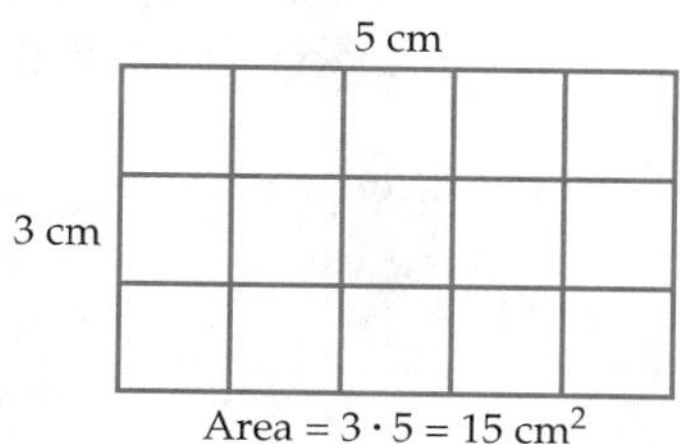

Generalization: The area of a rectangle in square units equals its length times its width.

10. (a) A farmer uses fertilizer for 3 years in a row, and his corn crop yield improves each year. Based upon these results, he generalizes that fertilizer improves corn crop yield.

(b) If the generalization in part (a) is true, what other factors should the farmer consider in deciding whether or not to use this fertilizer to grow corn?

11. Ready for some math magic?

$$112 \times 124 = 13{,}888$$

No, of course that's not the whole trick! Read on.

(a) Reverse the digits in each factor and multiply.
$211 \times 421 =$ ________

(b) What is interesting about the result in part (a)?

(c) $312 \times 221 =$ ________
$213 \times 122 =$ ________

(d) Make a generalization based upon the results to parts (a) and (c).

(e) Try some other examples, and decide whether your generalization is reasonable or false.

12. (a) $12{,}345{,}679 \times 36 =$ ______________

(b) $12{,}345{,}679 \times 45 =$ ______________

(c) Write two more equations that extend this pattern.

(d) *Explain* how this pattern works. (*Hint:* What is $12{,}345{,}679 \times 9$?)

13. Galileo used inductive reasoning to develop hypotheses about the length of a pendulum and the period of a full swing.

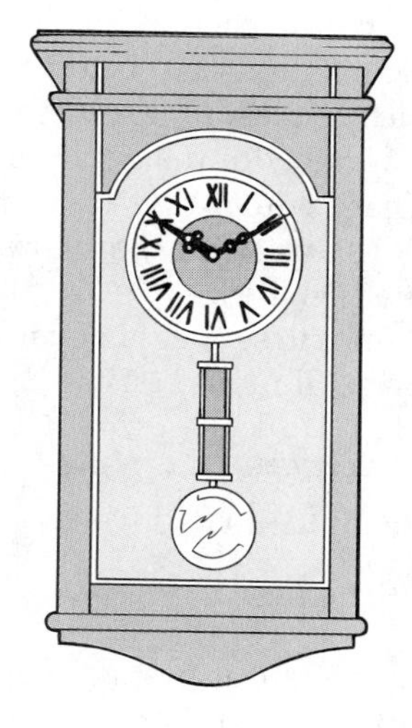

Use the following data to develop your own hypothesis.

L (length of pendulum)	P (period of a full swing)
1 unit	1 second
2 units	1.41 seconds
3 units	1.73 seconds
4 units	2 seconds
9 units	3 seconds

(a) If the pattern continues, what is the *length* of a pendulum that has a *period* of 4 seconds?
(b) What is the general relationship between the length and the period?
(c) Write a formula relating L to P.
(d) Use your formula from part (c) to estimate the period of a pendulum 5 units long.

14. How would one use inductive reasoning to form a prejudice?

15. Give an example showing how each of the following groups might use inductive reasoning.
(a) Students (b) Mathematicians

16. Give an example showing how each of the following groups might use inductive reasoning.
(a) Teachers
(b) College admissions officers

17. (a) *Explain* how each of the following generalizations can be formed using inductive reasoning and (b) whether or not you agree with the generalization.
1. There will always be wars.
2. Boys are better than girls at mathematics.
3. A movie sequel is never as good as the original movie.

Extension Exercises

18.

E (weight on Earth)	100 lb	200 lb	300 lb
M (weight on Mars)	38 lb	76 lb	114 lb

(a) An astronaut in a space suit might weigh about 400 lb on Earth. If the pattern continued, what would be her weight on Mars?
(b) If someone's weight doubled on Earth, what do you think would happen to the person's weight on Mars?
(c) Make a conjecture about the general relationship between weight on Earth and weight on Mars, based upon these data. (Compare each E value to the corresponding M value.)
(d) Write an equation relating E to M, based upon your conjecture.

19. Try some examples, and decide whether each of the following statements is probably true or definitely false.
(a) The product of any three consecutive whole numbers is divisible by 3.
(b) The product of any four consecutive whole numbers is divisible by 4.
(c) The product of any five consecutive whole numbers is divisible by 5.
(d) Investigate further examples, and state a generalization of your results.

20. (a) Is every number that is divisible by 4 also divisible by 3?
(b) Is every number that is divisible by 5 also divisible by 3?
(c) Is every number that is divisible by 6 also divisible by 3?
(d) Investigate further examples, and state a generalization of your results.

21. When people clasp their own hands, do they all prefer to put the right thumb on top?

Special Exercise

22.

Mapmakers often shade adjacent states or countries in different colors. One of the most famous problems in mathematics is the map-coloring problem. It asks: How many different colors are needed to color any map drawn on a flat surface or a sphere so that adjacent regions are different colors?

In 1878, Arthur Cayley posed this problem to the London Mathematical Society. Ninety-eight years later, Wolfgang Haken and Kenneth Appel of the University of Illinois found and proved the answer using over 1000 hours of computer time.

(a) Find the minimum number of different colors needed to color the map of the United States so that no two adjacent regions are the same color. (You don't have to color in every state.)
(b) Based on your results, make a conjecture about the answer to the map-coloring question.

1.2 DEDUCTIVE REASONING

It's the first day of class. You assume that the teacher will be older than most of the students and that the teacher will settle in at the front of the room. A middle-aged woman walks in the door and heads directly to the front of the room. Next, she lays her things on the teacher's desk. What would you conclude about this person?

Or how about this? I'm thinking of a one-digit odd number that is greater than 5 and divisible by 3. Can you deduce what number it is?

In drawing conclusions from information given in the preceding examples, one can use a second important reasoning process: deductive reasoning.

> ***Definition*** **Deductive Reasoning**
>
> **Deductive reasoning** is the process of reaching a necessary conclusion from given facts or hypotheses.

One of the most famous deductive thinkers is Sherlock Holmes. Notice how he uses deductive reasoning in the following scene.

A FAMOUS APPLICATION OF DEDUCTIVE REASONING

The following scene is adapted from *The Adventures of the Dancing Men* by Sir Arthur Conan Doyle. Places! Action!

HOLMES: *(cooly)* So Watson, you were just discussing gold investments.
WATSON: *(astonished)* How on earth did you know that?!

HOLMES: In a few moments, you'll say it's so absurdly simple.
WATSON: *(in a huff)* I certainly will not.
HOLMES: After seeing chalk between your index finger and thumb, I feel sure that you were discussing gold investments.
WATSON: *(still confused)* What is the connection?
HOLMES: Let me explain and diagram my argument using step-by-step deductive reasoning.

1. You have chalk between your index finger and thumb.
2. Therefore, you must have played billiards.
3. You must have played billiards with Thurston, since you always play with him.
4. Thurston had an option to invest in gold no later than today and wanted you to invest with him, so you must have discussed it.

WATSON: How absurdly simple! You've done a proof like I used to do in high-school geometry class.

Curtain (and applause).

In reaching his conclusion, Sherlock Holmes used deductive reasoning. Each statement in his explanation necessarily leads to the next statement when combined with certain assumptions. For example, Holmes combined the fact that Watson had chalk between his index finger and thumb with the assumption that, if Watson had chalk on his fingers, then he must have been playing billiards. The necessary conclusion is that Watson was playing billiards.

Chalk on fingers → Played billiards → Was with Thurston → Discussed gold investment

Deductive reasoning is used by detectives solving crimes, mathematicians drawing conclusions, and philosophers thinking logically. When the assumptions are true, deductive reasoning produces certain results. Now it's your turn to solve a mystery using deductive reasoning.

Lesson Exercise 1.9

The painting "By Numbers" was stolen from Jane Dough's Illinois home last night. Today the painting was found in an alley. Can you find the thief?

The only suspects are Jane Dough (the owner of the house), Jeeves (the butler), Sharky (the pool man), and Fluffy (the pet Doberman pinscher). The police have gathered the following evidence:

1. The painting was stolen between 8:00 and 9:00 P.M. last night.
2. Fluffy can't carry a painting.
3. Jeeves has been visiting Mumsy in Liverpool all week.
4. Sharky's and Jane's fingerprints were on the canvas.
5. Jane was with three friends at the movie "Return of the Stepson of Darth Vader's Cousin" from 7:30 to 10:00 P.M. last night.

DRAWING CONCLUSIONS

The essence of deductive reasoning is drawing a necessary conclusion (from given hypotheses). You used this process repeatedly in the preceding exercise.

Lesson Exercise 1.10

Fill in the conclusion that follows from the given hypotheses.

Hypotheses: The painting was stolen from the house between 8:00 and 9:00 P.M.

Jane was at a movie from 7:30 to 10:00 P.M.

Conclusion: ______________________________

Mathematicians often use this same process. The following lesson exercise illustrates how applying a geometry theorem to a particular figure requires deductive reasoning.

Lesson Exercise 1.11

Fill in a conclusion that follows from the given theorem and hypothesis.

Theorem: The opposite sides of a parallelogram are equal in length.
Hypothesis: $ABCD$ is a parallelogram.

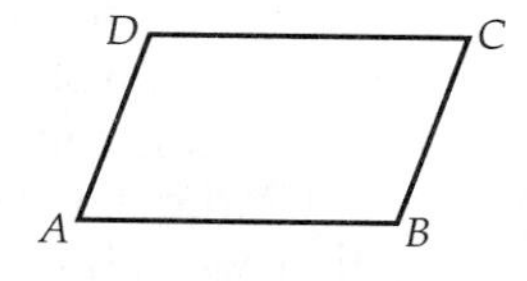

Conclusion: ______________________________

In the preceding exercise, you applied the theorem that the opposite sides of a parallelogram are equal in length in a particular parallelogram $ABCD$. Mathematicians also use deductive reasoning in applying a definition to a particular case, as is done in the following lesson exercise. The following exercise introduces a second format that is commonly used in deductive reasoning: the "if-then" format.

Lesson Exercise 1.12

Fill in a conclusion that follows from the given hypotheses.

If all numbers that are divisible by 2 are even, and 264 is divisible by 2, then ______________________________.

In the preceding exercise, the definition that an even number is divisible by 2 was applied to the number 264. If we know that all numbers that are divisible by 2 are even and that 264 is divisible by 2, then we can conclude that 264 is even. In Lesson Exercise 1.12, the hypotheses are (1) "All numbers that are divisible by 2 are even numbers" and (2) "264 is divisible by 2." The conclusion is "264 is an even number." In the "if-then" statement, the hypotheses come after the word "if," and the conclusion is stated after the word "then."

Lesson Exercise 1.13

Consider the following statement: If $x = 4$ and $y = x + 3$, then $y = 7$.

(a) What are the hypotheses?

(b) What is the conclusion?

The mathematician, like the detective, uses deductive reasoning in a process of elimination. This is illustrated in the following exercise.

Lesson Exercise 1.14

(a) Fill in the blank and (b) identify the hypotheses and conclusion.

Two distinct lines in a plane are either parallel or intersecting. Suppose distinct lines a and b in a plane are not parallel. Therefore, ____________.

Aristotle (384–322 B.C.), the father of deductive reasoning, was the first to study logic systematically. He stated the most famous example of deductive reasoning. It consists of two hypotheses and a necessary conclusion.

Example 1.3 models Aristotle's famous example of deductive reasoning with a Venn diagram (set picture). Aristotle didn't have the luxury of using a Venn diagram. John Venn (1834–1923) invented these diagrams, which are used for illustrating set relationships, more than 2000 years after Aristotle's death.

EXAMPLE 1.3

Deduce the conclusion and represent it with a Venn diagram.

Hypothesis: All people are mortal.
Socrates is a person.

Conclusion: ____________

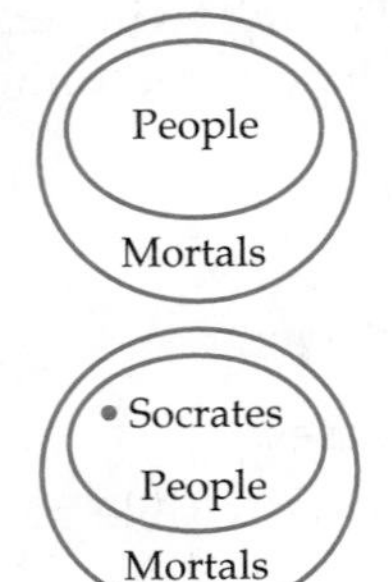

Figure 1–3

Solution

Since Socrates is a person and all people are mortal, we can conclude that Socrates must be mortal. The same conclusion can be confirmed with Venn diagrams, as shown in Figure 1-3. The first hypothesis, "All people are mortal," is shown by drawing a circle or oval that represents the set of "all people" inside a circle or oval that represents the set of "all mortals." This Venn diagram shows that the set of "all mortals" contains the set of "all people." The second hypothesis, "Socrates is a person," is shown by drawing a point that represents Socrates inside the set of "all people," since the set of "all people" contains Socrates. (A point is used to represent a single element.)

Now the point representing Socrates is also contained in the circle representing mortals. So the set of "mortals" contains Socrates. We have our conclusion: Socrates is mortal. ■

Lesson Exercise 1.15

Deduce the conclusion to the following, and represent both hypotheses in a Venn diagram. (*Hint:* For the Venn diagram, identify a larger group [circle or oval] that contains another smaller group [circle or oval].)

Hypotheses: All doctors are college graduates.
All college graduates finish high school.

Conclusion: ______________________________

Figure 1-4, on the next page, shows how the first-grade *Mathematics Plus* textbook asks children to draw a conclusion from a set of pictures.

DOES DEDUCTIVE REASONING ALWAYS WORK?

Are the conclusions that result from deductive reasoning always true? Find out what happens if we make a slight change in the preceding exercise.

Lesson Exercise 1.16

(a) Fill in the blank and (b) identify the hypotheses and conclusion.

If all doctors are college graduates and all college graduates like roller skating, then ______________________________.

Lesson Exercise 1.16 illustrates that valid (correct) deductive reasoning can sometimes lead to a false conclusion.

FALSE HYPOTHESIS —Valid reasoning?→ FALSE CONCLUSION

Lesson Exercise 1.17 contains another example of this type.

Lesson Exercise 1.17

Deduce the conclusion to the following.

Hypotheses: A square is also a rectangle.
All rectangles have six sides.

Conclusion: ______________________________

When deductive reasoning is done correctly, it is called valid deductive reasoning, and the conclusion is called a valid conclusion. In **valid** deductive

From Mathematics Plus, *Grade 1 (San Diego: Harcourt Brace and Company, 1994), p. 289.*

Figure 1–4

reasoning, the conclusion automatically follows from the hypotheses. Deductive reasoning is said to be **invalid** (done incorrectly) when the conclusion does not automatically follow from the hypotheses. The use of the word "valid" or "invalid" does *not* indicate whether any hypothesis or conclusion is true or false.

Lesson Exercises 1.16 and 1.17 are examples of how valid deductive reasoning can lead to a false conclusion. Lesson Exercises 1.16 and 1.17 each have a false conclusion that results from having a false hypothesis. In order to guarantee that the conclusion reached by deductive reasoning is true, all the hypotheses must be true.

True Conclusions from Deductive Reasoning

If the hypotheses are true and the deductive reasoning is valid, then the conclusion must be true.

In analyzing a deductive sequence, check to see whether the hypotheses and conclusion are true or false, and check whether reasoning from the hypotheses to the conclusion is valid (done correctly) or invalid.

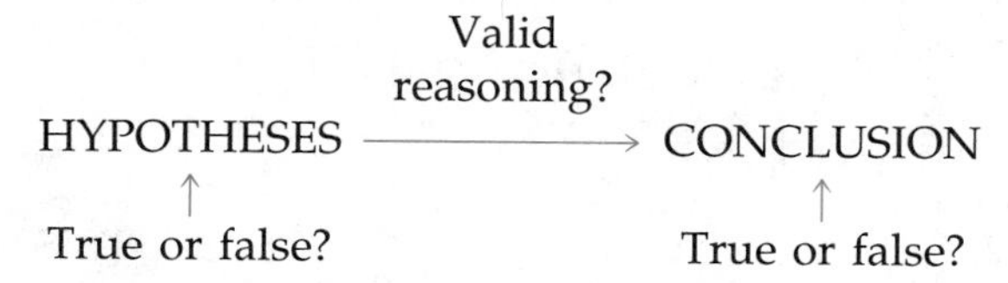

VENN DIAGRAMS AND DEDUCTIVE REASONING

How can you recognize invalid deductive reasoning? It occurs when someone attempts to use deductive reasoning even though it does not apply to a situation. Venn diagrams, which were used earlier to illustrate valid conclusions, are also helpful in detecting invalid deductive reasoning. If deductive reasoning is valid, it is impossible to draw a Venn diagram that satisfies the hypotheses but not the conclusion. If deductive reasoning is invalid, it *is* possible to draw a Venn diagram that satisfies all the hypotheses but not the conclusion.

EXAMPLE 1.4

Is the following conclusion valid? If so, draw a Venn diagram that illustrates it. If not, draw a Venn diagram showing that the conclusion does not follow from the two hypotheses.

Hypotheses: All dogs are animals.
All cats are animals.
Conclusion: All dogs are cats.

Solution

The conclusion does not seem to follow from the two hypotheses. If the conclusion is invalid, there should be a Venn diagram that satisfies the hypotheses but not the conclusion (see Figure 1-5). Such a diagram would be a counterexample to the conclusion. The first hypothesis, "All dogs are animals," is shown by drawing an oval that represents the set of all dogs inside a circle

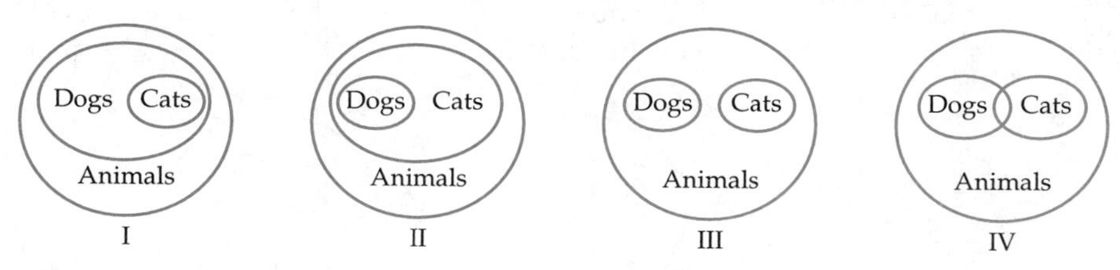

Figure 1–5

that represents the set of all animals, since the group of animals contains all dogs (plus other animals).

Similarly, the hypothesis "All cats are animals" is shown by drawing an oval that represents all cats inside the circle that represents all animals, since the group of all animals contains all cats.

It is possible to draw a diagram (I, III, or IV) that satisfies the hypotheses but contradicts the conclusion. The "Dogs" oval does not have to be inside the "Cats" oval as the conclusion suggests, so the conclusion does not necessarily follow from the hypotheses. ■

The result of Example 1.4 can be illustrated as follows.

$$\underset{\text{(true)}}{\text{HYPOTHESIS}} \xrightarrow[\quad]{\substack{\text{Invalid}\\\text{reasoning}}} \underset{\text{(false)}}{\text{CONCLUSION}}$$

The error in Example 1.4 is obvious. However, advertisers sometimes use the same form of faulty reasoning, and many people do not notice it. Example 1.5 illustrates this technique.

EXAMPLE 1.5

Does the conclusion follow from the two hypotheses? If so, draw a Venn diagram that illustrates it. If not, draw a Venn diagram showing that the conclusion does not follow from the two hypotheses.

Hypotheses: All beautiful women use Coverall lipstick.
Sally uses Coverall lipstick.

Conclusion: Sally is a beautiful woman.

Solution

According to the first hypothesis, Coverall users include all beautiful women. Draw an oval for beautiful women inside a circle for Coverall users. Sally is an individual represented by a point (not an oval) inside the circle for Coverall users (Figure 1-6). The point representing Sally does not have to be inside the "beautiful woman" circle. In fact, Sally may be a child! The Venn diagram in Figure 1-6 satisfies the hypotheses but not the conclusion. Therefore, the deductive reasoning must be invalid. ■

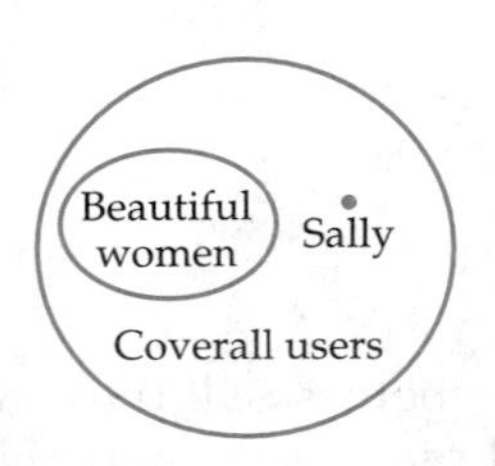

Figure 1–6

Note that Figure 1-6 and diagram III in Figure 1-5 have the same structure. This diagram results from one common form of invalid deductive reasoning. As shown in Figure 1-7, if A is in C and B is in C, then the deduction that B is in A is invalid.

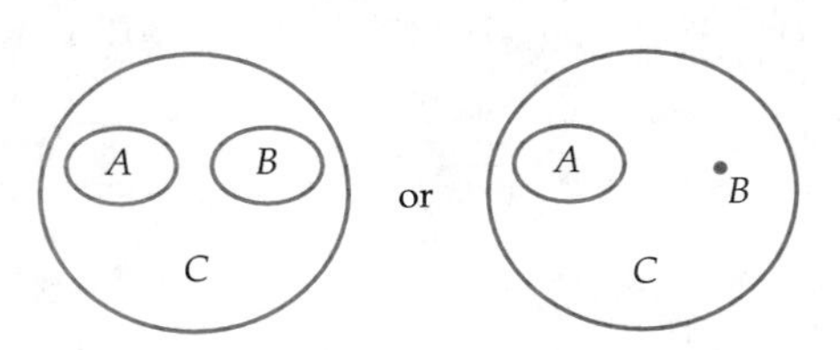

Figure 1–7

In Lesson Exercises 1.18 and 1.19, *assume the first two statements (hypotheses) are true,* and decide whether the third statement (conclusion) can be deduced from the first two. If the conclusion is valid, draw a Venn diagram that illustrates this. If the conclusion is invalid, draw a Venn diagram that disproves it.

Lesson Exercise 1.18

A rectangle is a polygon. A triangle is a polygon. Therefore, a rectangle is a triangle.

Lesson Exercise 1.19

Harold uses Some laundry detergent. People who use Some have clean clothes. Therefore, Harold has clean clothes.

A GAME: PICA-CENTRO

Pica-Centro is a game of deductive reasoning for two players that Douglas Aichele describes in his May 1972 article in *Arithmetic Teacher*. The object of the game is to determine the three-digit number your opponent has secretly selected by using deductive reasoning (and a little luck).

To begin the game, Player A secretly chooses a three-digit number with three different digits. Player B then keeps guessing three-digit numbers until Player B can determine Player A's number. For example, suppose Player A secretly picks the number 807. Player B records guesses as follows.

Guesses by Player B	Responses by Player A	
Digits	Pica: Correct Digit, Wrong Position	Centro: Correct Digit, Correct Position
7 4 2	1	0
4 2 5	0	0
3 8 7	1	1
8 0 7	0	3

For each of Player B's guesses, Player A tells how many digits are correct and whether they are in the correct positions, as shown in the preceding table. This process continues until Player B finds Player A's number. Then the players switch roles.

***Cooperative* Lesson Exercise 1.20**

Find a partner and play a game of Pica-Centro.

ANSWERS TO SELECTED LESSON EXERCISES

1.9 Sharky

1.10 Jane did not steal the painting.

1.11 $AB = CD$ and $AD = BC$.

1.12 264 is an even number.

1.13 (a) $x = 4$ and $y = x + 3$ (b) $y = 7$

1.14 (a) a and b are intersecting.
(b) The first two sentences (about the lines) are the hypotheses, and the conclusion is "a and b are intersecting."

1.15 All doctors finish high school.

1.16 (a) All doctors like roller skating.
(b) The phrase after "If" and before "then" contains the two hypotheses, and the conclusion is "all doctors like roller skating."

1.17 All squares have six sides.

1.18 Invalid

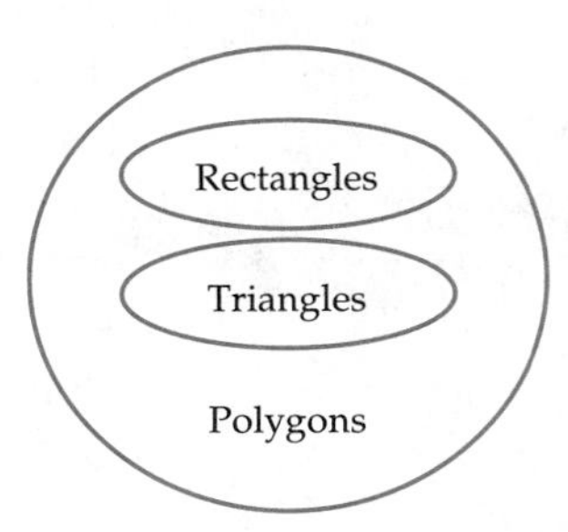

1.19 Valid

1.2 HOMEWORK EXERCISES

Basic Exercises

1. Use deductive reasoning to fill in a conclusion that follows from the given statements, and draw a Venn diagram (set picture).
   All rectangles are parallelograms. All parallelograms are quadrilaterals.
   Conclusion: ______________________

2. Use deductive reasoning to fill in the conclusion.
   Watson was playing billiards today. Watson only plays billiards with Thurston.
   Conclusion: ______________________

3. (a) Fill in the blank and (b) identify the hypotheses and conclusion.

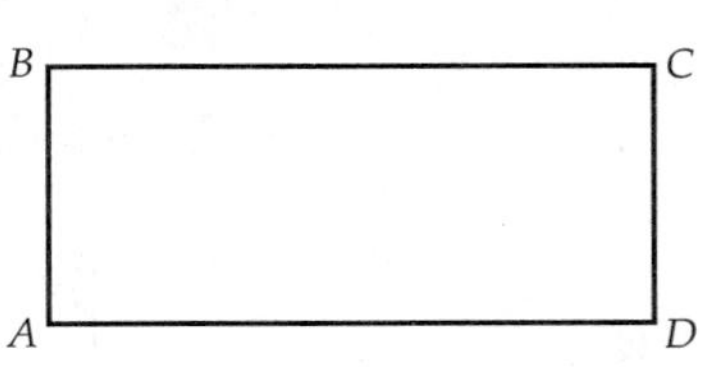

   If $ABCD$ is a rectangle, and the opposite sides of a rectangle are parallel, then ______________________.

4. What will guarantee that the conclusion reached through valid deductive reasoning is true?

5. The first person to study deductive reasoning systematically was
(a) Sherlock Holmes (b) Sly Stallone
(c) my mathematics teacher
(d) Newton (e) Aristotle

In Exercises 6–9, decide whether or not the conclusion can be deduced from the first two hypotheses. If so, draw a Venn diagram that supports your answer. If not, draw a Venn diagram that satisfies the hypotheses but does not support the conclusion.

6. *Hypotheses:* All elephants have legs.
All chairs have legs.
Conclusion: All elephants are chairs.

7. *Hypotheses:* Two is a whole number.
All whole numbers can be written as fractions.
Conclusion: Two can be written as a fraction.

8. *Hypotheses:* All cockroaches are young.
All young things are beautiful.
Conclusion: All cockroaches are beautiful.

9. *Hypotheses:* Models drink diet cola.
I drink diet cola.
Conclusion: I am a model.

In Exercises 10 and 11, decide whether or not the third statement can be deduced from the two hypotheses.

***10.** *Hypotheses:* Martina is taller than Gabriela.
Gabriela is taller than Steffi.
Conclusion: Martina is taller than Steffi.

11. *Hypotheses:* Martina beat Gabriela at tennis.
Gabriela beat Steffi at tennis.
Conclusion: Martina will beat Steffi at tennis.

***12.** Consider the following Pica-Centro Game.

Guesses	Responses	
Digits	Pica: Correct Digit, Wrong Position	Centro: Correct Digit, Correct Position
5 3 2	2	0
4 2 3	1	0

Which digit is definitely part of the secret number? Justify your answer.

13. Consider the following Pica-Centro game.

Guesses	Responses	
Digits	Pica: Correct Digit, Wrong Position	Centro: Correct Digit, Correct Position
5 2 3	2	0
0 1 2	0	0
6 7 8	0	1
9 7 6	0	0

(a) Which digits can be eliminated?
(b) Which digit is in the correct position in the third guess? Explain why.
(c) Which two digits are correct in the first guess?
(d) What are the correct positions of the two correct digits in the first guess? Explain why.
(e) What is the secret number?

14. Consider the following Pica-Centro game.

Guesses	Responses	
Digits	Pica: Correct Digit, Wrong Position	Centro: Correct Digit, Correct Position
3 0 7	0	0
2 6 4	1	1
6 8 4	0	1
1 5 3	1	0

Find the secret number.

In Exercises 15–18, use deductive reasoning to fill in a conclusion that follows from all the given statements.

15. If you have your teeth cleaned twice a year, you will have less tooth decay. If you have less tooth decay, you will lose fewer teeth.
Conclusion: ____________________

16. If my students don't like me, they will complain to the dean. If the dean hears complaints about me, then I won't get a raise in salary. If I give low grades, then my students won't like me.
Conclusion: ____________________

17. *Hypotheses:* Some adults watch television.
People who watch television don't have time to read.
Conclusion: ____________________

18. *Hypotheses:* Some people throw litter on the street.
People who throw litter do not care about their surroundings.
Conclusion: ____________________

19. Fill in the blank with a deduction suggested by the advertiser, and state whether or not it is true.
If I use Listermint mouthwash, then
____________________.

20. Complete the following to create an example of valid deductive reasoning.
Assumptions: All elephants are good dancers.
____
Conclusion: ____________________

21. Add a hypothesis so that the given conclusion follows deductively from the two hypotheses.
(a) *Hypotheses:* Ada is a mathematics teacher.
____
Conclusion: Ada loves mathematics.
(b) *Hypotheses:* You are having fun.
____
Conclusion: Time flies.

22. Joe and Sue are the same age. Joe is younger than Paul. Paul is older than Jane. Is Joe older than Jane, younger than Jane, or is it impossible to tell from the given information?

23. What conclusions can be drawn about Sandy based upon the following diagram?

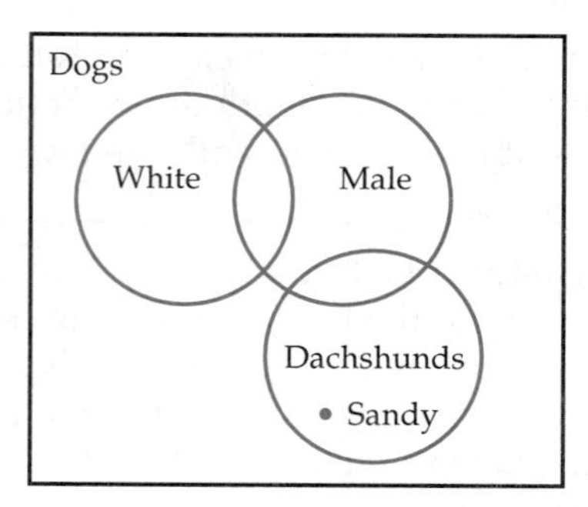

Extension Exercises

24. Assume the following statements are true.

All people are created equal.
People have the right to life, liberty, and the pursuit of happiness.
A government should be supported only as long as it protects these rights.

(a) These statements are taken from a famous document, although the word "men" has been changed to "people." What is the name of this document?
(b) Based upon the above three statements, which of the following statements *must* be true or *must* be false?
1. Men and women are created equal.
2. Everyone has a right to a good job.
3. People should support a government even if it doesn't protect their liberty.

Photo courtesy of Library of Congress

25. Everyone in the Smith family has dressed up for Halloween, but they didn't change their shoes or socks. Match each family member out of costume to the same family member in costume. Write a deductive explanation of your answer.

26. Lewis Carroll (real name: Charles Dodgson), author of *Alice's Adventures in Wonderland*, was also a mathematics professor who devised logic puzzles. The following are from his book *Symbolic Logic*. In each part, draw a valid conclusion that follows from all the given statements.

(a) Everyone who is sane can do logic.
No lunatics are fit to sit on a jury.
None of your sons can do logic.

(b) No ducks waltz.
No officers decline to waltz.
All my poultry are ducks.

(c) No birds except ostriches are 9 feet high.
There are no birds in this aviary that belong to anyone but me.
No ostrich lives on mince pies.
I have no birds less than 9 feet high.

27. (a) Draw a Venn diagram that represents the following statements.

All dolphins are swimmers.
All swimmers wear bathing suits.
No tigers wear bathing suits.

(b) Which of the following statements are confirmed by your diagram?

1. All dolphins wear bathing suits.
2. If you are not a tiger, then you wear a bathing suit.
3. All dolphins are tigers.
4. If you are not a swimmer, then you are not a dolphin.

28. The Three Stooges are a real pain. When you ask one of them a question, they all answer. Furthermore, one of them always lies, and the other two answer truthfully. I asked them who is the smartest. They replied:

CURLY: I am not the smartest.
MOE: Larry is the smartest.
LARRY: Curly is the smartest.

Who is really the smartest, and who is the liar?

29. Raymond Smullyan, a contemporary mathematician, devised the following puzzle:

The Politician Puzzle

One hundred politicians attended a convention. Each politician was either crooked or honest. Also:

1. At least one of the politicians was honest.
2. Given any two politicians, at least one of the two was crooked.

How many politicians were honest, and how many were crooked?

30. You want to determine whether the following statement is true for the cards shown: "If a 1 is printed on one side, then a 2 is printed on the other side." Which of these cards *must* you turn over?

1 2 3

31. Three boxes contain yellow and white tennis balls. One box has all yellow tennis balls; one has all white tennis balls; and one has some yellow and some white balls.

Unfortunately, all three boxes are labeled incorrectly! How can you determine the correct labeling after selecting just one ball?

Y W W and Y

32. Mr. Barber is a barber who shaves every man in town who does not shave himself. Who shaves Mr. Barber?

Special Exercises

33. (a) Fill the numbers in the puzzle. Each number should be used once.

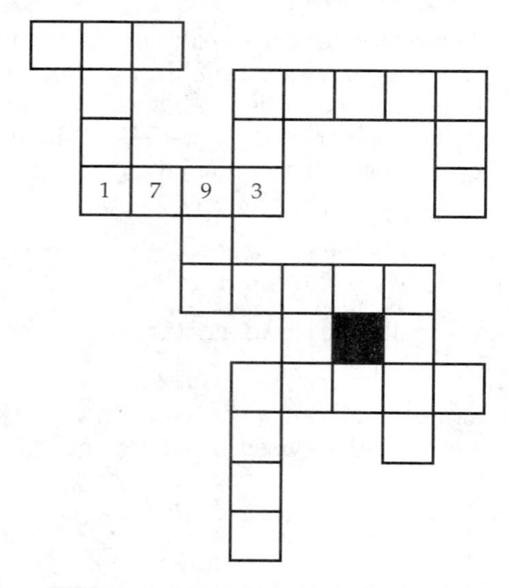

3 Digits	4 Digits	5 Digits
632	1562	12683
683	1793	14623
904	3081	98729
913	8421	
981		

(b) Give an example showing how you used deductive reasoning to fill in the numbers in this puzzle.

34. The following puzzle is adapted from the magazine *Murder Ink*.

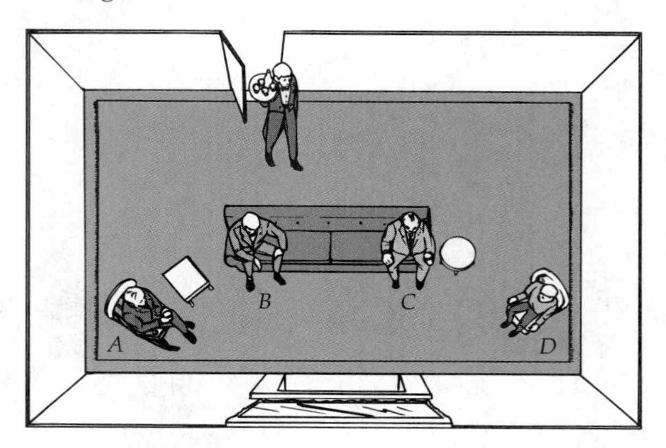

Boggs has been found dead. Four men, seated on the sofa and two chairs in front of the fireplace (as shown), are discussing the murder. Their names are Howell, Scott, Jennings, and Wilton. They are—not necessarily respectively—a general, a schoolmaster, an admiral, and a doctor.

1. The waiter brings a whiskey for Jennings and a beer for Scott.
2. In the mirror over the fireplace, the general sees the door close as the waiter departs. The general then speaks to Wilton, next to him.
3. Neither Howell nor Scott has any sisters.
4. The schoolmaster does not drink alcohol.
5. Howell is the admiral's brother-in-law. Howell is sitting in a chair, and the schoolmaster is next to him on his left.
6. The murderer suddenly moves to drop something in Jennings' whiskey. No one has moved from his seat.

(a) Tell the position of each man, give his profession, and name the murderer.
(b) Give an example showing how you used deductive reasoning to solve this puzzle.

35. Make up your own mystery like the one in the preceding exercise.

1.3 INDUCTIVE AND DEDUCTIVE REASONING

Mathematicians use inductive and deductive reasoning to create new mathematics. First they propose generalizations (using induction), and then they try to prove the logical validity of their generalizations (using deduction).

In this lesson, you'll first learn to tell the difference between induction and deduction. Then you'll see how mathematicians use these two processes to create new mathematics.

INDUCTIVE vs. DEDUCTIVE REASONING

By permission of Johnny Hart and Creators Syndicate, Inc.

Figure 1–8

Lesson Exercise 1.21

Are the men in the cartoon in Figure 1-8 discussing deductive or inductive reasoning?

How can one avoid confusing inductive reasoning and deductive reasoning? Remember, inductive reasoning leads to a generalization from specific examples; deductive reasoning draws a necessary conclusion from given assumptions. Induction involves an "inductive leap" from specific examples to a general idea. Deduction leads to a conclusion that *follows automatically* from the given assumptions.

Can you tell the difference? Find out in the following exercises.

In Lesson Exercises 1.22 through 1.24, identify whether induction or deduction is being used.

Lesson Exercise 1.22

The fingerprints on the wall match yours. I know that everyone has a unique set of fingerprints. I conclude that you touched the wall.

Lesson Exercise 1.23

I meet two people from France who wear berets. I conclude that all people from France wear berets.

Lesson Exercise 1.24

Given: $m = n + 6$ and $n = -8$. I conclude that $m = -2$.

INDUCTIVE AND DEDUCTIVE REASONING

Many mathematical ideas are developed using a two-step process. First inductive reasoning is used to make a generalization about an observed pattern in a variety of examples. Then deductive reasoning demonstrates that the general statement *must* be true if certain assumptions are made.

Remember the mystifying number trick from Class Lesson Exercise 1.1 in Section 1.1? You observed that no matter what number you started with, you ended up with 5. Unfortunately, no matter how many examples you tried, you could not be sure you would *always* end up with 5.

Inductive reasoning carries the risk of an incorrect generalization, so the conclusion it produces lacks the certitude required in mathematics. A mathematician would use deductive reasoning to prove that the number trick always produces a 5.

EXAMPLE 1.6

Prove that the following number trick always results in 5.

Pick a number.
Multiply by 2.
Add 10.
Divide by 2.
Subtract your original number.

Solution

Mathematicians use algebra and deductive reasoning to prove general results about numbers. In the case of the number puzzle, instead of starting with a number such as 3 or -2, start with a variable N that could represent any number. Then go through all the steps in the puzzle, and deduce what happens to the variable along the way.

1. Pick a number. N

2. Multiply by 2. ______ (Deduction?)

What goes in the blank? If you take N and multiply by 2, you obtain $2N$.

Next, how do we deduce the result of step 3 from step 2?

2. Multiply by 2. $2N$
3. Add 10. _______ (Deduction?)

Take $2N$ and add 10, and you obtain $2N + 10$. Continuing this series of steps and deductions yields the following proof.

1. Pick a number.	N
2. Multiply by 2.	$2N$
3. Add 10.	$2N + 10$
4. Divide by 2.	$N + 5$
5. Subtract your original number.	5

Deductive reasoning shows that no matter what N you start with, you will end up with 5. ■

Lesson Exercise 1.25

Consider the following number trick.

Pick any number at all.
Subtract 3.
Multiply by 2.
Subtract your original number.
Add 6.

(a) Try two or three different numbers, and use inductive reasoning to make a conjecture about what will happen with any number.
(b) Use deductive reasoning to prove your generalization from part (a).

Mathematics is our unique attempt to construct a system of ideas based upon precise deductive reasoning. A mathematical system begins with a few simple assumptions and then, through step-by-step logical arguments, arrives at new conclusions. The conclusions of deductive reasoning are irrefutable if the assumptions are correct.

Ancient Greek philosophers required that all their mathematical theorems be based upon deductive reasoning, and it has remained that way for over 2000 years! Mathematicians will not accept any statement as a theorem of mathematics unless it has been proved deductively.

THE CONVERSE OF A STATEMENT

What will happen if we interchange the hypothesis and conclusion in an "if-then" statement? Will the new statement be true?

"If your fingerprints are on the murder weapon, then you must have touched it." This statement is generally true, although Sherlock Holmes once had to disprove it to solve a case. Would the following also be generally true?

"If you touched the murder weapon, then your fingerprints must be on it." This last statement is the converse of the first.

An "if-then" statement can be written as "If A, then B," or $A \rightarrow B$. The **converse** is the new "if-then" statement formed by interchanging the hypothesis and the conclusion. The converse of $A \rightarrow B$ is $B \rightarrow A$. When you have shown an "if-then" statement is true, you may also want to examine its converse.

The following exercise contains another example of a statement and its converse.

Lesson Exercise 1.26

Statement: If a number is divisible by 10, then the number is even.
Converse: If a number is even, then the number is divisible by 10.

The statement is true. Is the converse also true? Try some examples and see whether the converse is probably true or definitely false.

Lesson Exercise 1.27

If $x + 3 = 5$, then $x = 2$.

(a) Is this statement true?

(b) Write its converse.

(c) Is the converse true or false?

Lesson Exercises 1.26 and 1.27 suggest the following.

Statements and Their Converses

The converse of a true "if-then" statement may be either true or false.

The extension (homework) exercises address two other statements that are related to an "if-then" statement: the contrapositive and the inverse.

"IF AND ONLY IF"

It would be simpler if all conditional statements were in an "If A, then B" format, but they are not. Two other common forms are "A if B," as in

$$X = 2 \quad \textit{if} \quad X + 3 = 5$$

and "A only if B," as in

You have mononucleosis *only if* you have a fever and a sore throat.

How are the "A if B" and "A only if B" formats related to the "If A, then B" format? Each of the two preceding statements can be rewritten using the "if-then" format. The first statement

$$X = 2 \quad \text{if} \quad X + 3 = 5$$

is the same as

$$\text{If} \quad X + 3 = 5 \quad \text{then} \quad X = 2$$

Both of these statements mean that whenever $X + 3 = 5$, it is also true that $X = 2$. See if you can rewrite the "only-if" statement in the "If A, then B" format in the following lesson exercise.

Lesson Exercise 1.28

Which of the following are equivalent to the statement "You have mononucleosis only if you have a fever and a sore throat"?

(a) People who have mononucleosis must have a fever and a sore throat.

(b) If you have a fever and a sore throat, then you must have mononucleosis.

(c) If you have mononucleosis, then you have a fever and a sore throat.

The preceding exercise illustrates that "A only if B" means that whenever A is true, B is also true. The preceding example and exercise are special cases of the following.

Definition **"If" Statements and "Only-If" Statements**

"A if B" means "If B, then A."
"A only if B" means "If A, then B."

Use these definitions to translate the following statements.

Lesson Exercise 1.29

Write each of the following statements in an "if-then" format.

(a) You may teach only if you have a college degree.

(b) N is an even number if N is divisible by 2.

Definitions and theorems sometimes contain the more powerful phrase "if and only if," which combines the "if" and "only-if" formats. Consider a compound statement that contains both "if" and "only if"!

$$x = 2 \quad \text{if and only if} \quad x + 3 = 5$$

An "if-and-only-if" statement is actually two statements in one. The statement "$x = 2$ if and only if $x + 3 = 5$" means

1. $x = 2$ if $x + 3 = 5$ and
2. $x = 2$ only if $x + 3 = 5$

These two statements can be rewritten in the "if-then" format as

1. If $x + 3 = 5$, then $x = 2$ and
2. If $x = 2$, then $x + 3 = 5$

In symbols, $x = 2 \leftrightarrow x + 3 = 5$ means "If $x = 2$, then $x + 3 = 5$" *and* "If $x + 3 = 5$, then $x = 2$."

So an "if-and-only-if" statement describes a very strong connection between two conditions. It gives you two statements in one: a statement *and* its converse. It means that the two conditions (separated by "if and only if") are equivalent. For example, to say "$x + 3 = 5$" is equivalent to saying "$x = 2$." Whenever one is true, the other must be true; whenever one is false, the other must be false.

Definition "If-and-Only-If" Statements

"*A* if and only if *B*" means

1. If *A*, then *B*, *and*
2. If *B*, then *A*.

In other words, "*A* if and only if *B*" means that *A* implies *B* and *B* implies *A*. This relationship is often written symbolically as $A \leftrightarrow B$. "If and only if" is a standard phrase used in many mathematical definitions and theorems. Look for it later in this book. Remember: "If and only if" indicates a theorem or definition that satisfies *two* "if-then" statements.

Lesson Exercise 1.30

Write the following statement as two "if-then" statements.

A whole number is divisible by 10 if and only if its ones digit is a 0.

ANSWERS TO LESSON EXERCISES

1.21 Inductive reasoning (generalization based on over 800,000 examples)

1.22 Deduction

1.23 Induction

1.24 Deduction

1.25 (a) I always end up with my original number.
(b) Assume that $N =$ my original number. Then the following deductions must be true.

1. Subtract 3, and I get $N - 3$.
2. Multiply by 2, and I get $2N - 6$.
3. Subtract my original number N, and I get $N - 6$.
4. Add 6, and I end up with N.

No matter what number N that I start out with, I end up with the same N. In short form, $N \rightarrow N - 3 \rightarrow 2N - 6 \rightarrow N - 6 \rightarrow N$.

1.26 No. The number 4 would be a counterexample.

1.27 (a) Yes
(b) If $x = 2$, then $x + 3 = 5$.
(c) True

1.28 (a) and (c)

1.29 (a) If you teach, then you have a college degree.
(b) If N is divisible by 2, then N is even.

1.30 If whole number is divisible by 10, then its ones digit is a 0. If its ones digit is a 0, then a whole number is divisible by 10.

1.3 HOMEWORK EXERCISES

Basic Exercises

1. What kind of reasoning involves a leap from given examples to a general conclusion?

2. Which type of reasoning leads to a true conclusion whenever the given statements are true?

3. Make up an example of valid deductive reasoning that leads to a false conclusion.

In Exercises 4–8, identify whether induction or deduction is being used.

4. My mother and grandmother were homemakers. I conclude that all women are homemakers.

5. My teacher gave a quiz five Thursdays in a row. I conclude that she will give a quiz every Thursday.

6. Given that $AB = CD$ and $AB = 6$ cm, I conclude that $CD = 6$ cm.

7. On the last two Friday the 13th's, I had bad luck. I conclude that Friday the 13th is an unlucky day for me.

8. Since $4 \times 6 = 24$ and $2 \times 8 = 16$, I conclude that the product of two even numbers is an even number.

9. Our courts accept fingerprints as evidence because, after millions of comparisons, no two identical sets of fingerprints have been found. This is an example of ________ reasoning.

10. How do mathematicians use inductive and deductive reasoning in sequence to develop a new idea?

11. Consider the following number trick.

Pick a number.
Multiply by 3.
Subtract your original number.
Add 8.
Divide by 2.

(a) Try two or three different numbers, and use inductive reasoning to make a conjecture about what will happen with any number.
(b) Why don't your results to part (a) *prove* that the number trick adds 4 to any number?
(c) Use deductive reasoning to prove your generalization from part (a).
(d) *Explain* how you used deductive reasoning to derive the result of the fourth step using the result of the third step.

12. Consider the following number trick.

Pick a number.
Subtract 8.
Add your original number.
Divide by 2.
Add 5.

(a) Try two or three different numbers, and use inductive reasoning to make a conjecture about what will happen with any number.

(b) Use deductive reasoning to prove your generalization from part (a).
(c) *Explain* how you used deductive reasoning to derive the result of the third step using the result of the second step.

13. The number trick in Exercise 11 can also be verified with drawings of blocks (such as algebra tiles or algebra lab gear). Represent the number selected with an unknown block ▭, and represent ones with squares ▫. Begin with ▭, and show what blocks you would have after each step in the number puzzle.

14. Show how to verify Example 1.6 with drawings of blocks. (Refer to the preceding exercise.)

15. How do you write the converse of an "if-then" statement?

16. Consider the following statement: If a figure is a square, then the figure has exactly four sides.
(a) Is the statement true?
(b) Write the converse of the statement.
(c) Is the converse true?

17. Consider the following statement: If you have a fever, then you are sick.
(a) Is the statement true?
(b) Write the converse of the statement.
(c) Is the converse true?

18. Write the following statements in an "if-then" format.
(a) You can graduate only if you have taken a writing course.
(b) The number 5 is odd if 4 is even.

19. Write the following statements in an "if-then" format.
(a) You may be successful if you are smart.
(b) A triangle is equilateral only if it is isosceles.

20. Write the following statement as two "if-then" statements: A triangle is a right triangle if and only if it has a right angle.

21. Write the following statement as two "if-then" statements: A triangle has two equal sides if and only if it has two equal angles.

22. Write the following two statements as one "if-and-only-if" statement.

If you are my parent, then I am your child.
If I am your child, then you are my parent.

23. Write the following two statements as one "if-and-only-if" statement.

If $2x = 4$, then $x = 2$.
If $x = 2$, then $2x = 4$.

24. Four friends are trying to decide whether to jump into a cold swimming pool. Greg will jump in only if everyone else does. Suzanne will go in if at least one other person does. Mike will go in only if Suzanne has decided to go in. Gina will jump in regardless of what the others do. How many will jump in the pool?

Extension Exercises

25. Make up a number puzzle in which a person will always end up with the number 2. (Make it complicated enough so that someone else could not easily see how it works.)

26. Make up a number puzzle in which a person will always end up with a number that is 6 more than the person's original number.

27. Assume the following five statements are true.

Canaries like to sing. Blue jays like to sing. Tweety is a canary. Cardinals like to sing. Arnold is a pig.

(a) Use some of the preceding statements to create an example of valid inductive reasoning.
(b) Use some of the preceding statements to create an example of valid deductive reasoning.

28. (a) Try some examples, and answer the following question. What is the sum of two even numbers?
(b) Fill in the blanks in the following deductive proof.
1. Assume your two even numbers are $2M$ and $2N$, where M and N are whole numbers.
2. The sum of the two even numbers is ________.
3. $2M + 2N = 2($________$)$.

4. $2(M + N)$ is an even number because ______________.

5. So the sum of two even numbers is an even number.

29. In this lesson, you studied the converse of a statement. Another related statement of this type is the inverse. To make the **inverse** of a statement, negate the hypothesis and the conclusion of the statement.

Statement: If it is raining, then I shall wear a raincoat.
Inverse: If it is not raining, then I shall not wear a raincoat.

(a) Write the inverse of the following statement: If $x + 3 = 5$, then $x = 2$.
(b) Write the inverse of the following statement: If I confess, then I am guilty.
(c) If possible, write a true "if-then" statement that has a false inverse.

30. Write the inverse of the following statements.
(a) If I drive a truck, then I need a driver's license.
(b) If a woman uses nose powder, then her nose looks beautiful.

31. The **contrapositive** of a statement combines the converse and the inverse. To make the contrapositive, interchange the hypothesis and conclusion *and* negate them both.

Statement: If a number is divisible by 100, then the number is even.
Contrapositive: If a number is not even, then the number is not divisible by 100.

(a) Write the contrapositive of the following statement: If $x + 3 = 5$, then $x = 2$.
(b) Write the contrapositive of the following statement: If I confess, then I am guilty.
(c) If possible, write a true "if-then" statement that has a false contrapositive.

32. Consider the following statement: If you are a U.S. Marine, then you are a real man.
(a) Write the converse.
(b) Write the contrapositive.
(c) Write the inverse.

33. Suppose the following statement is true: If it was raining, then I drove to work.
(a) Write the converse.
(b) Write the contrapositive.
(c) Write the inverse.
(d) Which of the three statements in parts (a), (b), and (c) is (are) true?

34. Suppose the following statement is true: If it rains today, then I will scream. Which of the following must also be true?
(a) If I do not scream today, then it is not raining.
(b) If I scream today, then it is raining.
(c) If it does not rain today, then I will not scream.

35. Suppose the following statement is true: If a four-sided figure is a square, then it is a rectangle. Which of the following must also be true?
(a) If a four-sided figure is a rectangle, then it is a square.
(b) If a four-sided figure is not a square, then it is not a rectangle.
(c) If a four-sided figure is not a rectangle, then it is not a square.

36. (a) Based upon the preceding exercises, whenever an "if-then" statement is false, its contrapositive
(1) is true (2) is false
(3) could be true or false
(b) Based upon the preceding exercises, whenever an "if-then" statement is true, its contrapositive
(1) is true (2) is false
(3) could be true or false

37. Suppose the following advertising claim is true: "If you rent a car from Mary, then you will be able to travel around with ease." Is it then true that "if you do not rent a car from Mary, you will not be able to travel around with ease"?

38. Which of the following are the same as "If A, then B," and which are the same as "If B, then A"?
(a) A implies B. (b) A follows from B.
(c) A is a sufficient condition for B.
(d) A is a necessary condition for B.
(e) If A is false, then B is false.

Special Exercise

39. Fill in the correct numbers in the square, using the following clues.

1. Each number from 1 to 16 is used once.
2. Each row adds up to 34, and each column adds up to 34.
3. K is twice as large as G and three times E.
4. J and N add up to F.
5. H is 5 times C.
6. B and D are two-digit numbers.

A	B	C	D
E	F	G	H
I	J	K	L
M	N	O	P 8

1.4 PATTERNS

Look out the window. At first glance, the world appears complex and confusing, but in reality it is full of patterns. Patterns give order to the world. In mathematics, we find patterns in shapes and quantities.

Human beings are better than other animals at finding mathematical patterns. The ability to find patterns is supposed to be a sign of intelligence.

Experimenters at Tulane University wanted to teach a rat the concept of "two." The rat had to choose among three doors. One door had one mark, one had two marks, and one had three marks. In each trial, the food was behind the door with two marks. How long would it take you to discover such a pattern? It took the rat 1500 trials.

People are better at finding mathematical patterns than rats. (Sometimes we find patterns even where they don't exist, as the cartoon in Figure 1-9 shows.)

Figure 1–9

Class Lesson Exercise 1.31

Describe some patterns you see in your classroom.

The ability to see patterns is important in mathematical problem solving. Working with patterns also develops number sense. In this lesson, you will study three common types of mathematical pattern problems: sequences, finding a rule, and patterns in sums.

Discovering and extending patterns requires inductive and deductive reasoning. First, conjecturing a general pattern based on examples is inductive reasoning. Then, proving the generalization requires deductive reasoning. Writing new examples of a pattern based on a general rule is another use of deductive reasoning.

G. H. Hardy (1877–1947), a British mathematician, said, "A mathematician, like a painter or poet, is a maker of patterns. If his patterns are more permanent than theirs, it is because they are made with ideas. The mathematician's patterns, like the painter's or the poet's, must be beautiful . . . [and] fit together in a harmonious way."

SEQUENCES

The first set of numbers children study is the **counting** (or **natural**) **numbers** {1, 2, 3, . . .}. A 3-year-old child who is 94 cm tall may grow about 6 cm per year until age 9, setting up the following pattern: 94 cm, 100 cm, 106 cm, 112 cm, 118 cm, 124 cm, 130 cm. These are two examples of number sequences. A **sequence** is an ordered arrangement of numbers, figures, or letters.

Lesson Exercise 1.32

(a) What is the fourth term in the sequence 16, 17, 19, 23, 31?

(b) What will the next term in the sequence be?

Many number sequences follow a pattern based upon addition, subtraction, multiplication, division, or a combination of these operations. In each example, *induction* is used to find a general rule. Then the general rule is used to *deduce* the next term.

EXAMPLE 1.7

Find the next term in each of the following sequences.

(a) A 64-g mass of a radioactive substance decays as shown every 5 years (its half-life).

64 g, 32 g, 16 g, 8 g, ________

(b) 5, 18, 70, 278, ________

Solution

(a) I conjecture that each new term is obtained by dividing the preceding term by 2 (*induction*). If so, the next term is 4 g (*deduction*).

(b) I conjecture that each new term is obtained by multiplying the preceding term by 4 and subtracting 2 (*induction*). If so, the next term is 1110 (*deduction*). ■

Try to complete the sequences in Lesson Exercises 1.33 and 1.34.

Daisy (13)

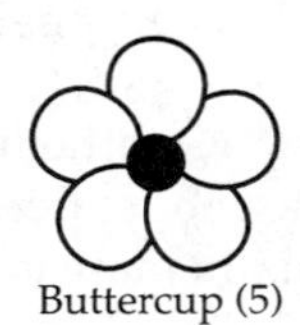

Buttercup (5)

Figure 1–10

Lesson Exercise 1.33

The number of petals in some varieties of flowers (see Figure 1-10), is found in the following sequence (called the **Fibonacci sequence** after a twelfth century mathematician).

1, 1, 2, 3, 5, __, __, __, __, __, __

Lesson Exercise 1.34

1, 2, 5, 14, ________

(a) The next term is ________.

(b) In guessing the general rule for the sequence, you used ________ reasoning.

(c) In applying the general rule to find the next term, you used ________ reasoning.

Figure 1-11, on the next page, shows a sequence problem one would find in the first-grade *Mathematics Plus* textbook.

Some number sequences involve squares, cubes, or squaring or cubing combined with addition or subtraction.

EXAMPLE 1.8

Find the next term in the following sequence: 2, 5, 10, 17, ________.

Solution

Each term is 1 more than a square number (1, 4, 9, 16). I generalize that each term in the sequence will be 1 more than a square number (*induction*). If so, the next term is $25 + 1 = 26$ (*deduction*). Or I note that the differences between successive terms are 3, 5, and 7. I generalize that the difference between terms will increase by 2 each time (*induction*). The 5th term is $17 + 9 = 26$ (*deduction*). ■

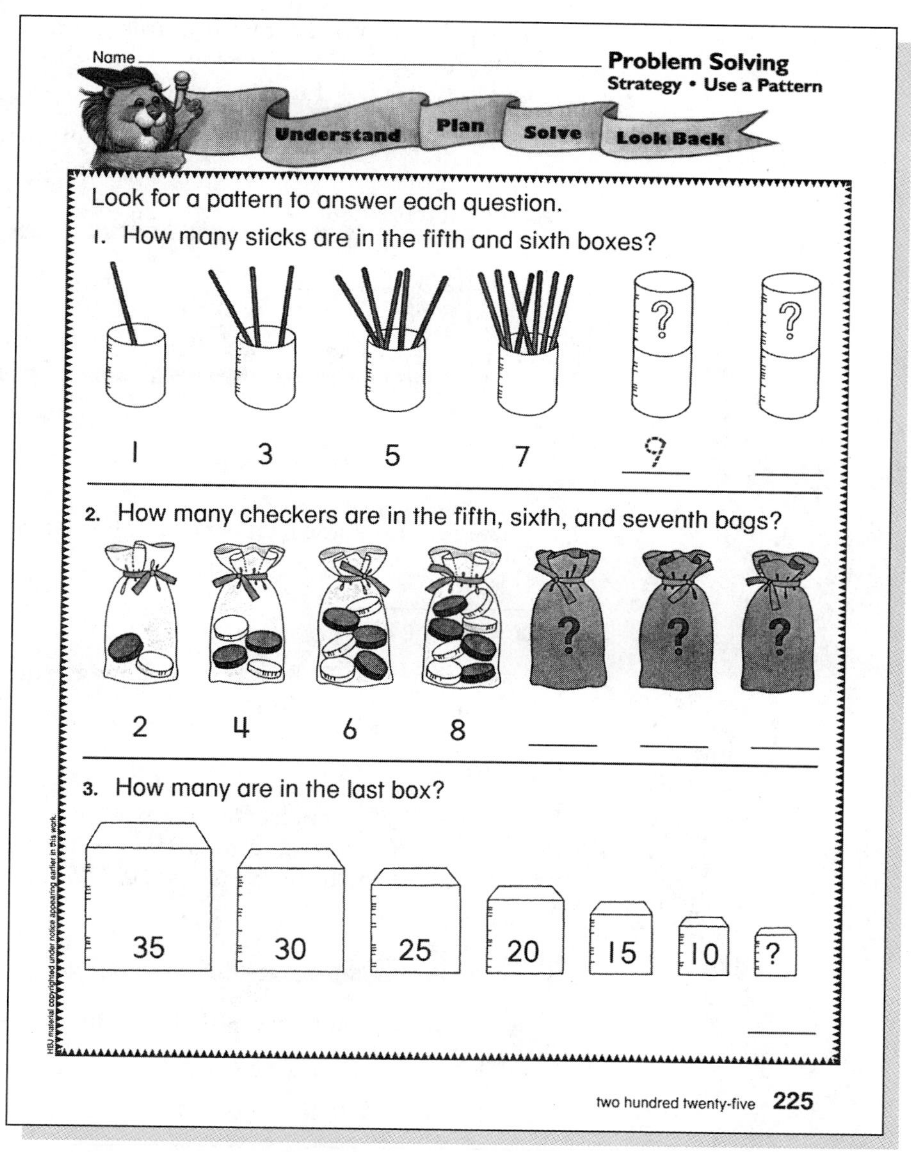

Name ____________________ **Problem Solving**
Strategy • Use a Pattern

Understand Plan Solve Look Back

Look for a pattern to answer each question.

1. How many sticks are in the fifth and sixth boxes?

1 3 5 7 9 ____

2. How many checkers are in the fifth, sixth, and seventh bags?

2 4 6 8 ____ ____ ____

3. How many are in the last box?

35 30 25 20 15 10 ?

____

HBJ material copyrighted under notice appearing earlier in this work.

two hundred twenty-five **225**

From Mathematics Plus, *Grade 1 (San Diego: Harcourt Brace and Company, 1994), p. 225.*

Figure 1–11

Lesson Exercise 1.35

Find the next term in the following sequence: 2, 7, 14, 23, ________.

Finding a rule for a sequence often includes the following steps.

1. See if the sequence is a relatively simple one, involving only one of the following: addition, subtraction, multiplication, division, squares, or cubes.

2. If Step 1 does not uncover the rule, a more complicated sequence may involve multiplication, division, squares, or cubes *combined with* addition or subtraction. Look for this type of rule.

The next two sequences each fit into one of these categories.

Lesson Exercise 1.36

2, 10, 42, 170, ______

Lesson Exercise 1.37

1, 8, 27, 64, ______

The rule for some sequences can be stated algebraically.

EXAMPLE 1.9

The sequence 3, 9, 27, 81 has the following rule: The nth term $= 3^n$. Use the rule to deduce the 7th term.

Solution

In the rule "3^n," n represents the position of the term. For example, the 1st term is 3^1, or 3. The 2nd term is 3^2, or 9. To find the 7th term, substitute 7 for n in the rule. The 7th term is $3^7 = 2187$ (*deduction*). ■

Lesson Exercise 1.38

The rule for a sequence is that the nth term $= 2n + 1$.

(a) What is the 1st term?

(b) What are the next three terms?

The reverse process, finding the formula for the nth term of a given sequence, is more challenging.

EXAMPLE 1.10

Consider the sequence 2, 6, 10, 14, ...

(a) What is the rule for the nth term?

(b) What is the 50th term?

Solution

(a) The rule for the nth term relates each term to its *position* rather than to the preceding term.

Position	1	2	3	4
Number	2	6	10	14

What is a simple rule that relates 1 to 2, 2 to 6, 3 to 10, and 4 to 14? It does not involve a single arithmetic operation or squaring or cubing. Next, try multiplication with addition or subtraction. The number goes up by 4 each time. This suggests that the rule is related to counting by 4, as in 4, 8, 12, 16, ... However, each term in the example is 2 less than the counting-by-4 sequence. So the sequence rule is "4 times the position minus 2," or "The nth term $= 4n - 2$" (*induction*).

(b) The 50th term is $4 \cdot 50 - 2 = 198$ (*deduction*). ■

Lesson Exercise 1.39

Consider the sequence 4, 7, 10, 13, ...

(a) What is the next term?

(b) What is the most basic number sequence children learn that increases by 3 each time?

(c) What is the rule for the nth term?

(d) What is the 60th term?

(e) Does part (c) require inductive or deductive reasoning?

(f) Does part (d) require inductive or deductive reasoning?

(g) Look at Lesson Exercises 1.38 and 1.39 and Example 1.10, and complete the following generalization.

If the difference between each successive pair of terms in a sequence is the same number k, then the nth-term formula includes the term ________.

The next sequence has a different type of nth-term rule.

Lesson Exercise 1.40

Consider the sequence 0, 3, 8, 15, ...

(a) What is the next term?

(b) What is the rule for the nth term?

(c) What is the 100th term?

The following chart summarizes the use of induction and deduction with sequences.

	Induction			*Deduction*	
Numerical examples	⟶	General rule	General rule and given sequence	⟶	Apply rule to extend sequence

FINDING A RULE

In many everyday situations, a rule relates two sets of numbers. For example, when tickets are sold, the total amount of money collected is directly related to the number of tickets sold. One can often express such a rule with a formula.

Lesson Exercise 1.41

Suppose you light a 10-inch candle when the lights go out, and you record the following data.

T (elapsed time in hours)	0	1	2	3
H (candle height in inches)	10	8	6	

(a) Fill in the last value for H.

(b) What is H when $T = 5$?

(c) Describe the relationship between H and T.

(d) Write a formula: $H =$ ________.

(e) Graph each pair of values for T and H.

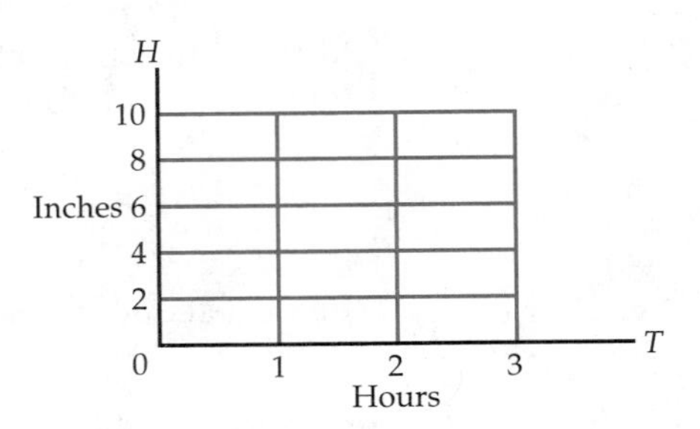

(f) What pattern do you see in the graph?

In "find-a-rule" patterns, the objects come in pairs. The same pattern or rule always relates the first object to the second object. The rules often involve arithmetic, as they did with sequences. There may be more than one possible rule. The object is to find a rule that works for all given pairs of numbers or objects.

Lesson Exercise 1.42

Fill in the blanks by finding a rule that works for every number pair.

Hours of Labor		Charge
1	⟶	\$30
2	⟶	\$40
3	⟶	\$50
4	⟶	___
N	⟶	___

(*Hint:* The charge includes a set fee for travel and an hourly wage.)

In Lesson Exercise 1.43, look for a pattern that relates each X value to its corresponding Y value, rather than looking for a pattern among the Y values.

Lesson Exercise 1.43

X	1	2	3	4	5
Y	2	5	10	17	

(a) Fill in the last value for Y.

(b) A formula relating Y to X is $Y =$ ________

(c) What is Y when X is 8?

(d) Answering part (c) by applying your formula from part (b) is an example of ________ reasoning.

(e) Complete a *number-sequence* problem that is equivalent to part (a): Find the 5th term of ________________.

(f) Write a rule for the nth term of your sequence in part (e).

PATTERNS IN SUMS

Pythagoras and his followers were the first to study numbers for their own sake. The Pythagoreans thought of quantities geometrically. They represented numbers with pebbles.

Lesson Exercise 1.44

What is the common name for numbers such as the following?

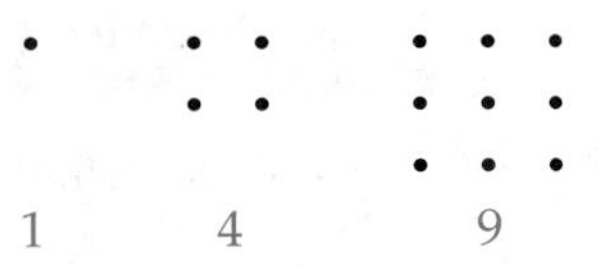

Thinking of numbers geometrically reveals patterns in sums and establishes a connection between arithmetic and geometry.

Lesson Exercise 1.45

$$1 = (\ )^2$$
$$1 + 3 = (\ )^2$$
$$1 + 3 + 5 = (\ )^2$$

(a) Fill in the missing numbers. (Recall that $n^2 = n \cdot n$.)

(b) Draw geometric dot pictures of the three sums that show the pattern. (*Hint:* Use squares.)

(c) What would the next equation be if the pattern continued? Is this equation true?

(d) The sum of the first three odd numbers is ________ squared.

(e) The sum of the first four odd numbers is ________ squared.

(f) Write a generalization for any counting number N, based upon parts (d) and (e).

(g) Part (f) involves ____________ reasoning.

(h) Use your generalization to compute $1 + 3 + 5 + 7 + \cdots + 79$.

Square number patterns are part of a branch of mathematics called number theory, a subject you will study further in Chapter 4. Number theory also includes such familiar topics as factors, multiples, and prime numbers.

Other patterns in sums may be too complicated to show geometrically, but your experience with number sequences will help you find the patterns.

Lesson Exercise 1.46

$$2^2 - 1^2 = 3$$
$$3^2 - 2^2 = 5$$
$$4^2 - 3^2 = 7$$

(a) If the pattern continues, what is the next equation? Is the next equation true?

(b) Complete the following generalization for any counting number c.
$c^2 -$ ________ $=$ ________.

(c) Show that your equation in part (b) is true.

(d) Part (b) involves ________ reasoning, and part (c) involves ________ reasoning.

ANSWERS TO SELECTED LESSON EXERCISES

1.32 (a) 23 (b) 47

1.33 8, 13, 21, 34, 55, 89

1.34 (a) 41 (b) inductive (c) deductive

1.35 34

1.36 682 (times 4 plus 2)

1.37 125 (cubes)

1.38 (a) 3 (b) 5, 7, 9

1.39 (a) 16 (b) 3, 6, 9, . . . (c) $3n + 1$
(d) 181 (e) Inductive
(f) Deductive (g) $k \cdot n$

1.40 (a) 24 (b) $n^2 - 1$ (c) 9999

1.41 (a) 4 (b) 0
(c) H starts at 10 and T at 0. When T increases by 1, H decreases by 2.
(d) Note the pattern in the H values:

$$10 = 10 \qquad 8 = 10 - 2 = 10 - 2(1)$$
$$6 = 10 - 2 - 2 = 10 - 2(2)$$

Therefore, $H = 10 - 2T$.
(f) The points all lie on a straight line.

1.42 \$60; $20 + 10N$ dollars

1.43 (a) 26 (b) $X^2 + 1$
(c) 65 (d) deductive
(e) 2, 5, 10, 17, ... (f) $n^2 + 1$

1.44 Square numbers

1.45 (a) 1; 2; 3
(b) 1 1 + 3 1 + 3 + 5

1^2 2^2 3^2

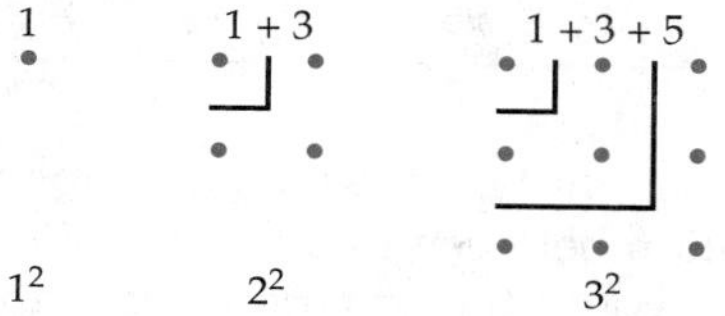

(c) $1 + 3 + 5 + 7 = 4^2$; yes
(d) 3 (e) 4
(f) The sum of the first N odd numbers is N^2.
(g) inductive (h) $(40)^2 = 1600$

1.46 (a) $5^2 - 4^2 = 9$; yes
(b) $c^2 - (c - 1)^2 = (c + c - 1)$ (*Hint:* Look at the pattern as you read an equation from left to right.)
(c)
$$c^2 - (c - 1)^2 \stackrel{?}{=} (c + c - 1)$$
$$c^2 - (c^2 - 2c + 1) \stackrel{?}{=} 2c - 1$$
$$c^2 - c^2 + 2c - 1 \stackrel{?}{=} 2c - 1$$
$$2c - 1 = 2c - 1$$
(d) inductive; deductive

1.4 HOMEWORK EXERCISES

Basic Exercises

In Exercises 1–6, find the next term in each sequence.

1. (a) 1, 16, 81, 256, ____
(b) Explain how you used both induction and deduction in solving Exercise 1(a).

2. 3, 7, 15, 31, ____

3. 10, ____ , 8, 14, 6, 16

4. −1, 2, 7, 14, ____

5. 4, 12, 14, 42, 44, ____

6. Man, 2; lion, 4; cockroach, 6; spider, ____; cat, ____ (The last answer is not 10.)

7. (a) Find two different reasonable answers for the next term in the sequence
1, 2, 4, ____.
(b) What rule did you use to obtain each of your answers?

8. In 1772, an astronomer named Bode found a pattern in the distances of the six known planets from the sun (where the distance from the Earth to the sun is 10 units).

Planet	Actual Distance	Bode's Pattern
Mercury	4	4
Venus	7	$4 + 3 = 7$
Earth	10	$4 + (3 \times 2) = 10$
Mars	15	$4 + (3 \times 2^2) = 16$
??????		$4 + (3 \times 2^3) = 28$
Jupiter	52	$4 + (3 \times 2^4) = 52$
Saturn	96	$4 + (3 \times 2^5) = 100$

(a) What might be an explanation for the extra equation between Mars and Jupiter?
(b) What would the equation after Saturn's be?
(c) In 1781 the next planet, Uranus, was discovered. It is 192 units from the sun. Is this close to Bode's prediction?
(d) In 1801 the asteroid Ceres was discovered 28 units from the sun. How does this relate to Bode's model?
(e) Neptune, the planet after Uranus, is 301 units from the sun. How close is this to Bode's prediction?

9. The Fibonacci sequence begins with two 1's. Each term after that is obtained by adding the preceding two terms.

$$1, 1, 2, 3, 5, 8, \ldots .$$

(a) Write the first ten terms of the sequence.
(b) Compare the sum of the first 3 terms to the 5th term and the sum of the first 4 terms to the 6th term. What pattern do you see?
(c) Write a generalization of your results.
(d) Write another example that supports your generalization.

10. Examine the following pattern based on the Fibonacci sequence (see the preceding exercise).

$$1^2 + 1^2 = 1 \times 2$$
$$1^2 + 1^2 + 2^2 = 2 \times 3$$
$$1^2 + 1^2 + 2^2 + 3^2 = 3 \times 5$$

What would the next equation be if the pattern continued? Is this equation true?

11. All living things contain carbon. Some scientists use radioactive carbon-14 to figure out the age of fossils. Carbon-14 has a half-life of 5600 years, meaning that *half its mass decays every 5600 years.*
(a) Complete the following table.

Time (years)	0	5600			
Fraction of C-14 Left	1	1/2	1/4		

(b) Prehistoric charcoal paintings in Lascaux, France, were found to have about 1/7 of their original C-14. Estimate their age.

12. Strontium-90 is a common waste product of the production of nuclear weapons and nuclear energy. It has a half-life of 28 years. Some scientists estimate that it will be safe when about 1/100 of it is left. About how long will this take?

13. Counting only blood relatives, state the total number of
(a) parents a person has.
(b) grandparents a person has.
(c) great-grandparents a person has.
(d) Great-grandparents are 3 generations back. How many direct ancestors would a person have in the 12th generation back?

14. The rule for a sequence is that the nth term $=$ $n + 4$. Write the first four terms of the sequence.

15. The rule for a sequence is that the nth term $=$ $10 - n$. Write the first five terms of the sequence.

16. What is the 9th term in a sequence with the following five characteristics?
1. Starting with the 3rd term, each odd-numbered term is the sum of the two terms that precede it.
2. The 1st term is 1.
3. The 4th term is 6.

4. The 5th term is 10.
5. Starting with the 4th term, each even-numbered term is double the term two places to its left.

17. (a) Write the even numbers as a number sequence.
(b) Write a rule for the nth term of the sequence.
(c) What rule do you see relating each term to the next one?

***18.** (a) Write the odd numbers as a number sequence.
(b) Write a rule for the nth term of the sequence.
(c) What rule do you see relating each term to the next one?

19. Examine the following design.

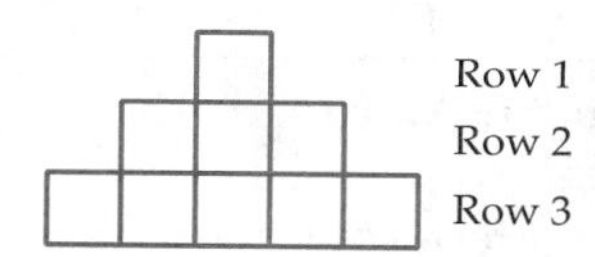

(a) If the design were to continue downward, how many squares would there be in the 50th row?
(b) What would be the total number of small squares in the first 50 rows?

20. Examine the following designs.

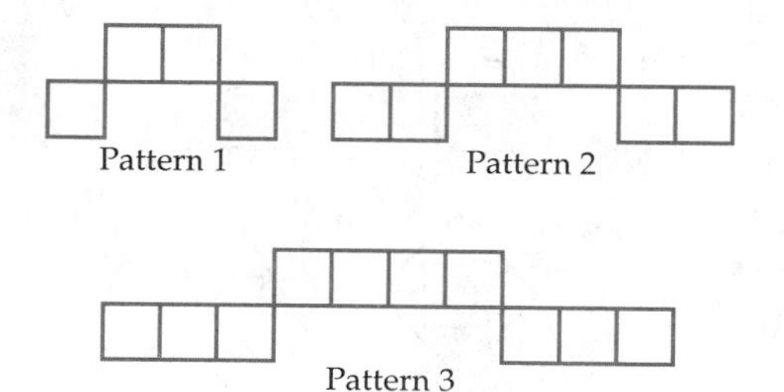

(a) How many squares would there be in Pattern 5?
(b) How many squares are in Pattern N, where $N = 1, 2, 3, \ldots$?

21. Consider the following sequence.

$$11, 13, 15, 17, \ldots$$

(a) What is the next term?
(b) What is the rule for the nth term?
(c) What is the 40th term?
(d) Does part (c) require induction or deduction?

22. Consider the following sequence.

$$2, 6, 12, 20, \ldots$$

(a) What is the next term?
(b) What is the rule for the nth term?
(c) What is the 20th term?

23. Consider the following sequence.

$$3, 6, 11, 18, 27, \ldots$$

(a) What is the next term?
(b) What is the rule for the nth term?
(c) What is the 30th term?

24. Consider the following sequence.

$$50, 49, 48, 47, \ldots$$

(a) What is the rule for the nth term?
(b) What is the 25th term?

25. Consider the following sequence.

$$2, 9, 16, 23, \ldots$$

(a) What is the rule for the nth term?
(b) What is the 100th term?

26. Determine how many numbers are in the following sequence.

$$3, 9, 15, 21, 27, \ldots, 159$$

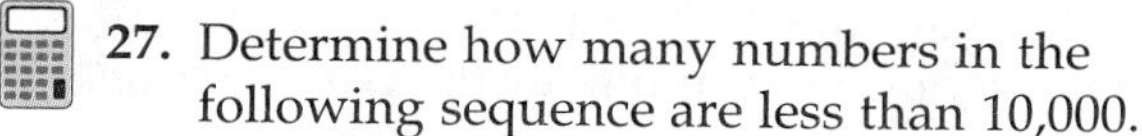

27. Determine how many numbers in the following sequence are less than 10,000.

$$3, 9, 27, 81, \ldots$$

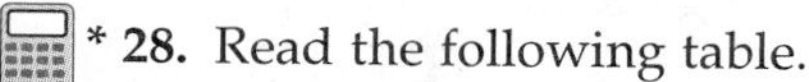

*** 28.** Read the following table.

N (number sold)	1	2	3	4	5
P (profit)	1	3	5	7	

(a) Write a formula that relates each N value to its corresponding P value.
(b) Use your rule to fill in the last value of P.
(c) You found a rule in part (a) by using ____________ reasoning.
(d) Plot each pair of values for N and P on a graph.
(e) What pattern do you see in your graph?
(f) Write a rule for the nth term of

$$1, 3, 5, 7, \ldots$$

29. Read the following table.

L (length)	1	2	3	4	5
V (volume)	1	8	27	64	

(a) A rule that relates L to V is $V =$ ________.
(b) Use your rule to fill in the last value of V.
(c) Write a rule for the nth term of

$$1, 8, 27, 64, \ldots$$

30. Read the following table.

X	1	2	3	4	5
Y	5	12	31	68	

(a) A rule that relates X to Y is $Y =$ ________. (*Hint:* See the preceding exercise.)
(b) Use your rule to fill in the last value of Y.

31. Examine the following designs.

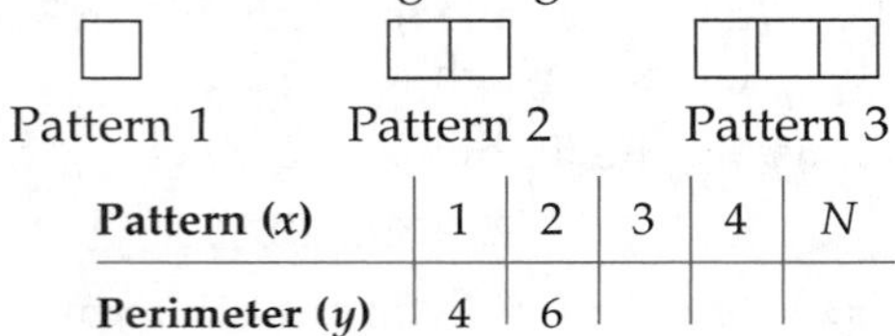

Pattern 1 Pattern 2 Pattern 3

Pattern (x)	1	2	3	4	N
Perimeter (y)	4	6			

(a) Find the perimeter of Pattern 3 and write it in the table.
(b) Draw Pattern 4 and find its perimeter.
(c) Plot each pair of values for x and y.
(d) What pattern do you see in your graph?
(e) Figure out the formula for the perimeter of a Pattern N.

32. Examine the following designs.

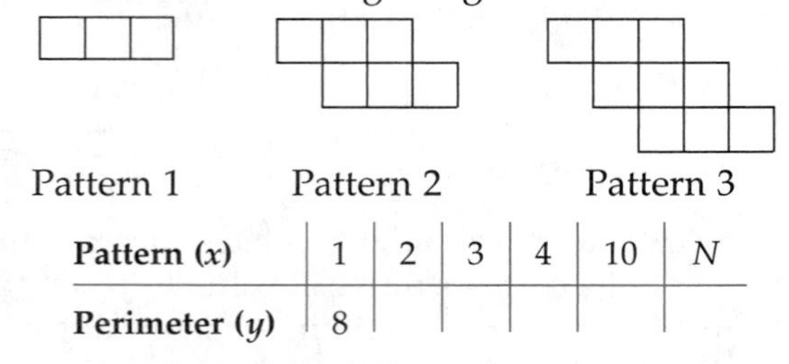

Pattern 1 Pattern 2 Pattern 3

Pattern (x)	1	2	3	4	10	N
Perimeter (y)	8					

(a) Find the perimeters of Patterns 2 and 3 and write them in the table.
(b) Draw Pattern 4 and find its perimeter.
(c) Graph each pair of values for x and y.
(d) What pattern do you see in your graph?
(e) Try to guess the perimeter for Pattern 10.
(f) Figure out a formula for the perimeter of Pattern N.

33. Find a rule that works for all of the following number pairs, and use that rule to fill in the blanks.

Expenses		**Reserves**
0	⟶	8
1	⟶	7
2	⟶	6
3	⟶	—
N	⟶	—

34. If you jump in the air, the length of time you remain in the air is related to how high you jump.

Height (ft)	1	4	9	
Time in the Air (s)	0.5	1	1.5	

(a) If the pattern continues, fill in the next height and time in the table.
(b) A formula relating time T to height H is $T =$ ________.

35.

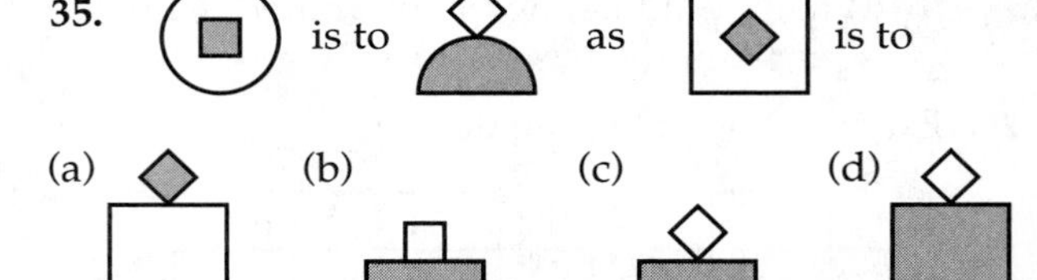

is to as is to

(a) (b) (c) (d)

36.

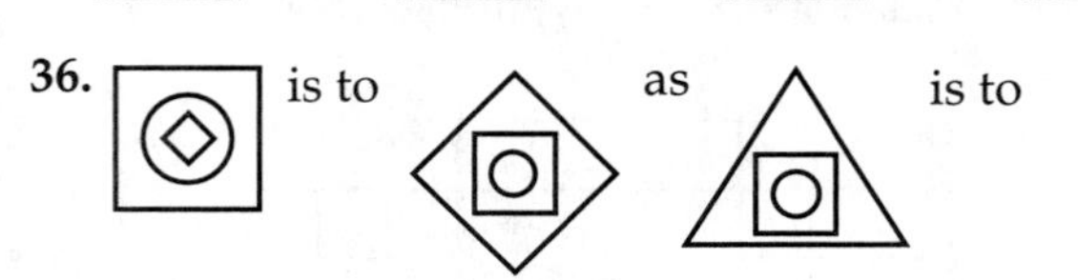

is to as is to

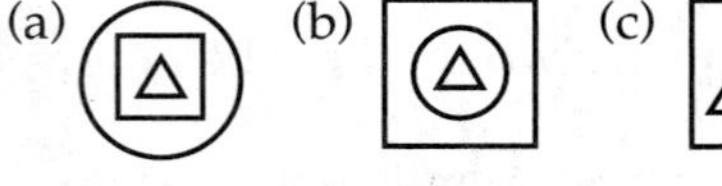

(a) (b) (c) (d)

37. Examine the following pattern.

$$3^2 + 4^2 \qquad = (\ )^2$$
$$3^3 + 4^3 + 5^3 = (\ )^3$$

(a) Fill in the missing numbers.
(b) What would the next equation be if the pattern continued? Is it a correct equation?

38. The Pythagoreans called certain numbers such as 1, 3, and 6 "triangular."

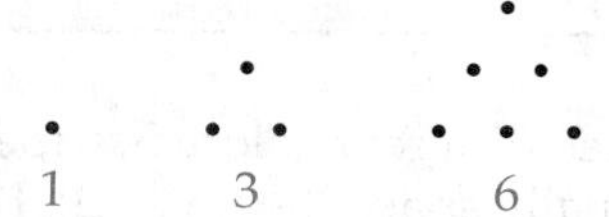

(a) What is the next triangular number?
(b) Each triangular number can be written as the sum of consecutive numbers. Show how this works with a dot drawing.
(c) Make dot drawings that show how 4, 9, and 16 are each the sum of two triangular numbers.
(d) What generalization does part (c) suggest?

39. Examine the following pattern.

$$1 + 8 \cdot 1 = 3^2$$
$$1 + 8 \cdot 3 = 5^2$$
$$1 + 8 \cdot 6 = 7^2$$

(a) What would the next equation be if the pattern continued? Is this equation true?
(b) A general formula for these equations would be $1 + 8\left(\frac{n(n+1)}{2}\right) = $ ________.
(c) Show that the two sides of your equation in part (b) are equal.

40. Examine the following pattern.

$$2^2 - 0^2 = 4$$
$$3^2 - 1^2 = 8$$
$$4^2 - 2^2 = 12$$

(a) What would the next equation be if the pattern continued? Is it a correct equation?
(b) Complete the following generalization. For any counting number c, $c^2 -$ ________ $=$ ________.
(c) Prove that your equation in part (b) is correct by showing that both sides are equal.

41. Examine the following pattern.

$$1^2 + 2^2 + 2^2 = (\ \ )^2$$
$$2^2 + 3^2 + 6^2 = (\ \ )^2$$
$$3^2 + 4^2 + 12^2 = (\ \ )^2$$

(a) Fill in the missing numbers.
(b) What would the next equation be if the pattern continued? Is this equation true?
(c) Complete the following generalization of the pattern. For any counting number N, $N^2 +$ ________ $+$ ________ $= (\qquad)^2$.
(d) Show that the two sides of your equation in part (c) are equal.

Extension Exercises

42. Examine these multiplication facts.

$$2 \times 9 = 18$$
$$3 \times 9 = 27$$
$$4 \times 9 = 36$$
$$5 \times 9 = 45$$
$$6 \times 9 = 54$$

(a) Name two patterns you see in these multiplication facts of 9.
(b) Do the patterns you mentioned in part (a) also work for 7×9, 8×9, and 9×9?

43.

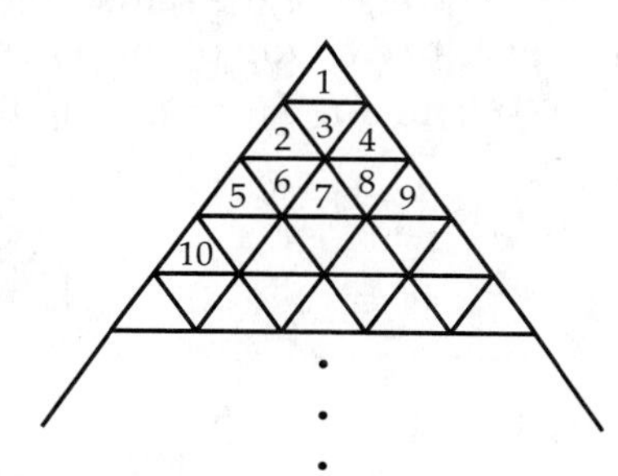

(a) If the pattern continued, where would 125 be located?
(b) Is 125 on a triangle that points up or down?
(c) Where would 457 be located?
(d) Is 457 on a triangle that points up or down?

44. (a) Look at the following table and find an approximate formula for obtaining the flash number N from the ASA film speeds. (*Hint:* Find $\sqrt{S}$.)

ASA Film speed (S)	100	200	400	800
Flash Number (N)	120	170	240	340

(b) Use your formula to find the flash number for an ASA film speed of 64.

45. A finite sum has a definite answer, but infinite sums can be tricky.

(a) The sum of

$$(1 - 1) + (1 - 1) + (1 - 1) + \cdots$$

appears to be ________.

(b) The sum of

$$1 - (1 - 1) - (1 - 1) - (1 - 1) \cdots$$

appears to be ________.

(c) Write

$$(1 - 1) + (1 - 1) + (1 - 1) + \cdots$$

without parentheses.

(d) Write $1 - (1 - 1) - (1 - 1) - \cdots$ without parentheses.

(e) Is $(1 - 1) + (1 - 1) + (1 - 1) + \cdots$ the same as

$$1 - (1 - 1) - (1 - 1) - \cdots ?$$

46. A cubical set of ten blocks has edges of 1 cm, 2 cm, 3 cm, . . . , 10 cm. Is it possible to build two 5-cube "towers" of the same height by stacking cubes one on top of the other?

In Exercises 47–50, find the next term in the sequence.

47. 2, 9, 39, 161, _____

48. 4, 10, 5, 17, 7, 31, _____

49. 7, 9, 40, 74, 1526, _____

50. x, $2x - 4$, $4x - 12$, $8x - 28$, _____

51. Make up two of your own challenging number-sequence exercises.

52. (a) Investigate the rule for the nth term of sequences with a constant difference between successive terms (called **arithmetic sequences**), such as 5, 7, 9, 11, ... Look for examples in the lesson and in previous homework exercises.

(b) Generalize your results from part (a) by completing the following. If the difference between successive terms is d and the first term is f, the rule for the nth term is ________________.

53. (a) Two different colored stripes are painted (equally spaced) on a stick, as shown at the top of the next column. How many different distinguishable sticks can be made by changing the positions of the colors?

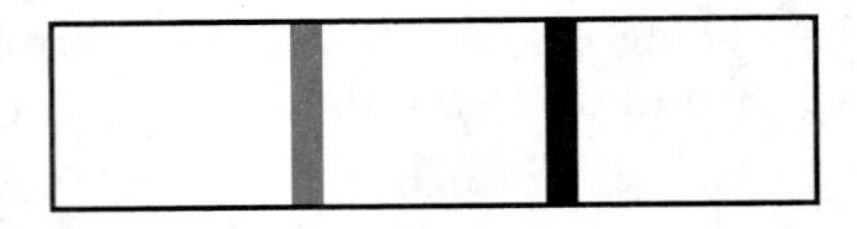

(b) Three different colored stripes are painted (equally spaced) on a stick. How many different distinguishable sticks can be made by changing the positions of the colors?

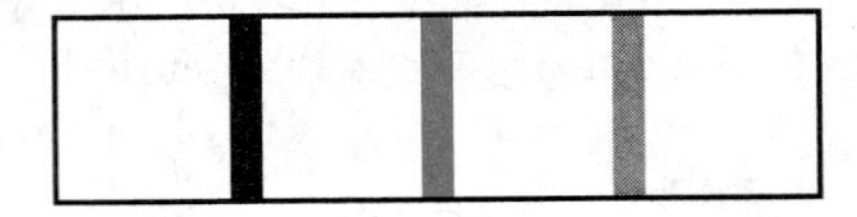

(c) Four different colored stripes are painted (equally spaced) on a stick. How many different distinguishable sticks can be made by changing the positions of the colors?

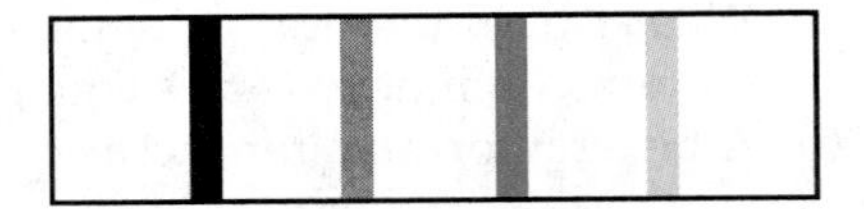

54. (a) Three different colored stripes are painted (equally spaced) on a ring, as shown. How many different distinguishable rings can be made by changing the positions of the colors?

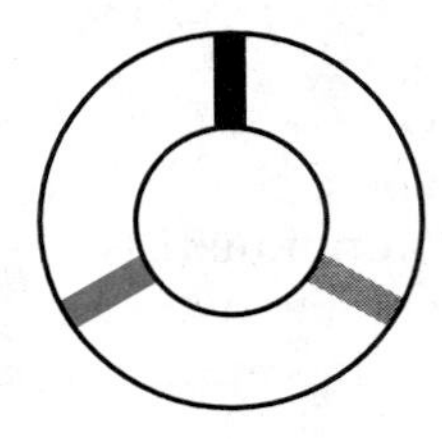

(b) Four different colored stripes are painted (equally spaced) on a ring. How many different distinguishable rings can be made by changing the positions of the colors?

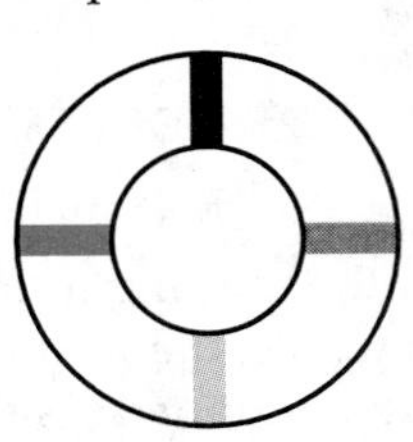

55. In a **geometric sequence**, each term is multiplied by the same number to obtain the next term. For example, in the sequence 3, 6, 12, 24, . . . , each term is multiplied by 2 to obtain the next term.

Find the next term of each geometric sequence.
(a) 2, 8, 32, _____
(b) 8, 4, _____

56. (a) Make up the first five terms of a geometric sequence of your choice.
(b) Compute the difference between each pair of successive terms.
(c) Would these differences form a geometric sequence?
(d) Repeat parts (a), (b), and (c) for two other geometric sequences and write a conclusion.

57. If a fixed number is added to each term of a geometric sequence, is the resulting sequence geometric? Try some examples and decide.

58. If each term of a geometric sequence is multiplied by a fixed number, is the resulting sequence geometric? Try some examples and decide.

59. (a) Drop a ball from different heights and see how high it goes on the rebound. Describe any pattern in your results.
(b) Predict how high the ball will go on the *second* bounce in relation to its initial height. Check your prediction.

60. Select a current elementary-school mathematics textbook. What kind of pattern problems does it have?

Special Exercises

61. Carl Friedrich Gauss (1777–1855) was one of the greatest mathematicians who ever lived (Figure 1-12). According to one story, when Gauss was 10, his teacher, desiring to keep the children occupied, asked them to add a sum like $1 + 2 + 3 + \cdots + 98 + 99 + 100$. Expecting the students to be busy for a half hour or so, the teacher was astonished when Gauss came up with the answer in less than a minute. How did Gauss do it?

Carl Friedrich Gauss
Photo courtesy of Library of Congress.

Figure 1–12

(a) Gauss probably paired off the numbers.

$$1 + 2 + 3 + \cdots + 98 + 99 + 100$$

What do you notice about the pairs?
(b) How many of these pairs are there?
(c) What is the sum?

62. Use the method of the preceding exercise to find the sum of the following:

$$5 + 10 + 15 + \cdots + 290 + 295 + 300$$

63.

x	1	2	3	4	5	6	N
y	1	3	6	10	15		

(a) When $x = 6$, $y =$ _____.
(b) When $x = N$, $y =$ _____.

64. The following diagram shows that

$$1 + 2 = \frac{1}{2}(2 \times 3)$$

(a) Make a similar diagram showing that

$$1 + 2 + 3 = \frac{1}{2}(3 \times 4)$$

(b) Suggest a shortcut for computing

$$1 + 2 + 3 + \cdots + 98 + 99 + 100$$

using the same approach, and see if you obtain the correct sum.

65. A 15-row auditorium seats 15 people in the first row, 16 in the second, 17 in the third, and so on. Use shortcuts to find the total number of seats in the auditorium.

1.5 PROBLEM SOLVING

For too long, school mathematics has emphasized learning isolated facts and skills while devoting little time to mathematical reasoning. The most important part of school mathematics is not specific facts and skills; rather, it is learning how facts and skills are used by citizens analyzing data, consumers deciding what to buy, and workers dealing with technology or finance.

Admittedly, it is easier to teach children to perform routine computations and memorize facts and formulas, but as teachers we must tackle a more significant task: developing children's problem-solving ability. The National Council of Teachers of Mathematics (NCTM) says, "Problem solving must be the focus of school mathematics." You have already studied two of the most important components of mathematical reasoning: induction and deduction. The next two sections will further develop your problem-solving ability.

What is problem solving? In order to understand what problem solving is, it is helpful to know what a problem is. A **problem** has two characteristics: (1) It requires a solution, and (2) the solution is not immediately obvious. The best classroom problems offer a situation of interest to the learner, in which the need for mathematics arises naturally.

TYPES OF PROBLEMS

What kinds of problems are studied in elementary school? Read and solve the following three examples of elementary-school problems.

Lesson Exercise 1.47

Pièrre has 21 oranges. He gives Jane 12. How many oranges does Pièrre have left?

Lesson Exercise 1.48

The members of the Environment Club want to raise $50 by selling apples at $.25 each. So far, they have sold 120 apples. How many more apples must they sell?

Lesson Exercise 1.49

Farmer Laura had 36 cows. She sold all but 10. How many cows does she have left?

Like the preceding three problems, many elementary-school problems fit into one of three categories: (1) one-step translation problems, (2) multi-step translation problems, and (3) puzzle problems. These categories can be used to help organize lesson plans.

A **one-step translation** problem can be solved with a single arithmetic operation. Lesson Exercise 1.47 is an example of a one-step translation problem, the most common type in elementary school. One-step translation problems illustrate common applications of arithmetic and help reinforce arithmetic skills. Some mathematics educators would rather not call these "problems," since they become familiar and routine for most students who practice them.

A **multi-step translation** problem can be solved with two or more arithmetic steps. Lesson Exercise 1.48 is an example of a multi-step translation problem. Multi-step translation problems also illustrate common applications of arithmetic and reinforce arithmetic skills, but they require higher-level thinking than one-step problems. Current elementary-school textbooks contain more multi-step problems than less recent textbooks.

A **puzzle** problem is often solved with some unusual approach or insight. Lesson Exercise 1.49 is an example of a puzzle problem. Puzzle problems are nonroutine problems that develop flexible thinking.

These classifications are not precise; they depend upon a student's background. Lesson Exercise 1.47 might be a routine one-step subtraction problem for a second-grader but nonroutine for a first-grader, who might solve it by counting back.

Read Lesson Exercises 1.50–1.52, and tell whether each would usually be classified as a one-step translation problem, a multi-step translation problem, or a puzzle problem.

Lesson Exercise 1.50

A classroom has 5 rows of desks. There are 6 desks in each row. If all but 2 desks are occupied, how many children are in the class?

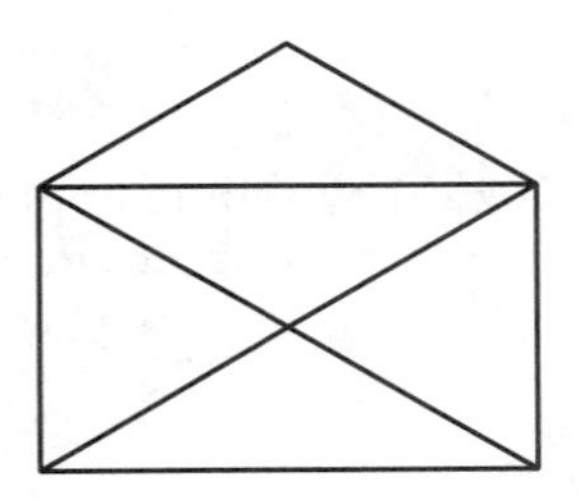

Figure 1–13

Lesson Exercise 1.51

Draw the figure shown in Figure 1-13 without lifting your pencil off the paper or retracing any line segments.

Lesson Exercise 1.52

Sidney has 8 incredibly good chocolate brownies. Four of us are waiting to eat them. If we all eat the same number of brownies, how many will we each eat?

George Polya

Photo courtesy of G.L. Alexanderson, Santa Clara University.

Figure 1–14

PROBLEM SOLVING

Solving problems is the specific achievement of intelligence and intelligence is the specific gift of mankind: solving problems can be regarded as the most characteristically human activity. . . . you can learn it only by imitation and practice.
(George Polya, *Mathematical Discovery*, J. Wiley, 1962, p. ix.)

George Polya (1887–1985; Figure 1-14), a brilliant teacher and mathematician, advocated problem-solving activities in mathematics beginning with his famous book, *How To Solve It* (1945). More recently, organizations such as NCTM have promoted this idea, and it has become a standard topic in the elementary-school mathematics curriculum. Until recently, school mathematics focused upon computational skills; now problem solving is considered to be of central importance in mathematics, and computational skills play a supporting role.

Good problem solvers take time to solve problems. After doing some thinking, they are not afraid to go ahead and try out some way of solving the problem. If one approach does not work, they are flexible and try another approach. They solve lots of problems and learn from their experiences.

Thanks to George Polya, we can learn specific methods for solving problems that increase our chances of being successful. First, Polya described four steps that can be used in analyzing many mathematical and nonmathematical problems:

Four Steps in Problem Solving

1. Understand the problem.
2. Devise a plan.
3. Carry out the plan.
4. Look back.

A mathematician could use these steps to solve a mathematics problem. A doctor could use them to treat an illness. A mechanic could use them to repair a car. Elementary-school mathematics textbooks utilize three-, four-, five-, or six-step variations of this approach.

People frequently solve problems without thinking about each step in Polya's plan. However, if one is having difficulty with a problem, the four-step plan is a useful guide. First, read the problem carefully and see if you understand it. What do you know, and what do you want to figure out? Second, in devising a plan, consider how the unknown is connected to the given in the problem. Also, how does the problem relate to concepts you know or other problems you have solved? Third, carry out the steps of your plan. Finally, look back, reviewing and checking your results. Have you answered the original question? Is there a way to check your answer to see if it is reasonable? Look back over the problem to improve your understanding of it. You can use this knowledge to solve related problems in the future.

Most elementary textbook series introduce the problem-solving steps early on. Figure 1-15 on the next page shows how the second-grade *Mathematics Plus* textbook introduces Polya's steps. Most elementary-school textbooks present these four steps as a sequence:

Understand → **Plan** → **Solve** → **Look Back**

Real problem solving is more dynamic; you may want to move backward and forward in this sequence. For example, the attempt to devise a plan may lead you to go back and try to understand the problem better.

Simply put, the four steps are understand, plan, solve, and check. How are these steps used in solving problems? Try them out as you work on Lesson Exercises 1.53 and 1.54.

Lesson Exercise 1.53

Twenty-two years ago, Jamaal's daughter was 1/4 his age, and his dog Yapper was 2. Today Jamaal's daughter is 1/2 his age. How old are Jamaal and his daughter now?

Understanding the Problem

(a) What are you supposed to figure out?

(b) Do you have enough information to do it?

(c) Is any information given that is not needed?

Devising a Plan

(d) You can solve this problem by using a guess-and-check (trial-and-error) approach or by using algebraic equations. Which will you try? (Guessing and checking is easier for most people.)

(e) Which of the following will you use in solving the problem: a calculator, paper and pencil, or mental computation?

Carrying Out the Plan

(f) Carry out your plan and obtain an answer.

Looking Back

(g) Check your answer.

(h) Make up a similar problem that can be solved in the same way.

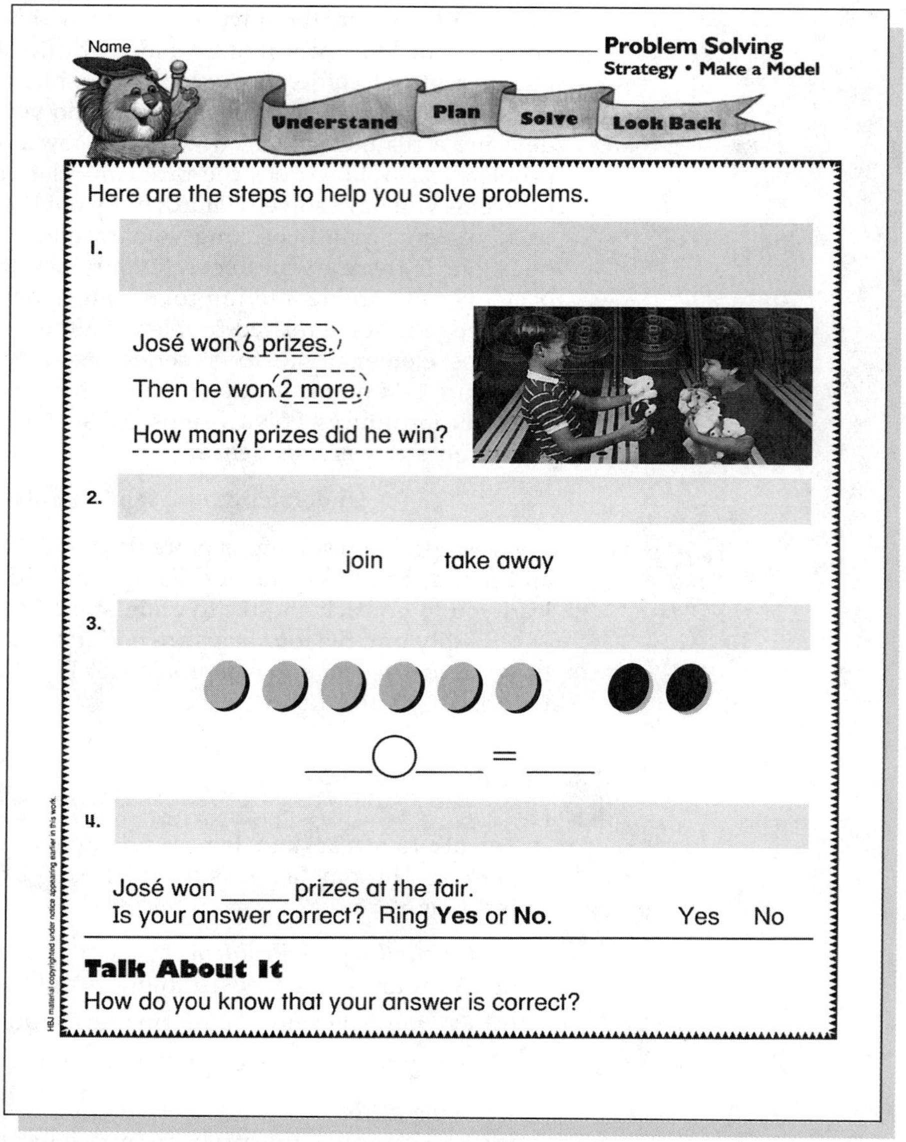

Name ______________________________

Problem Solving
Strategy • Make a Model

Understand Plan Solve Look Back

Here are the steps to help you solve problems.

1\.

José won 6 prizes.
Then he won 2 more.
How many prizes did he win?

2\.

join take away

3\.

___ ◯ ___ = ___

4\.

José won _____ prizes at the fair.
Is your answer correct? Ring **Yes** or **No**. Yes No

Talk About It

How do you know that your answer is correct?

From Mathematics Plus, *Grade 2 (San Diego: Harcourt Brace and Company, 1994), p. 13.*

Figure 1–15

Cooperative Lesson Exercise 1.54

Consider the following problem. Wanda Cash spent 1/4 of her money at a grocery store. Then she sat on a bench for 10 minutes. After that, she spent 1/2 of what was left at a record store. Now she has \$6. How much money did Wanda start out with?

Understanding the Problem

(a) What are you supposed to figure out?

(b) What information do you have?

Devising a Plan
A good strategy for this problem is called **working backward**. Start at the end and work backward step by step until you reach the beginning.

Carrying Out the Plan
(c) Try the working-backward strategy on this problem.

Looking Back
(d) Check your answer.

(e) Make up another problem like this one.

AN INVESTIGATION: COUNTING SQUARES

Cooperative **Lesson Exercise 1.55**

(a) Solve the following problem using Polya's four steps. Make up one appropriate activity or question and answer it for each of the four steps. See Lesson Exercises 1.53 and 1.54 for ideas.

How many different squares (of all sizes) are in the following picture?

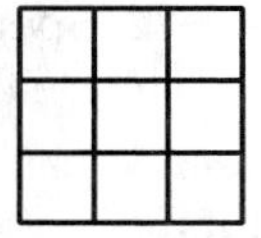

(b) How many different squares are in the following picture?

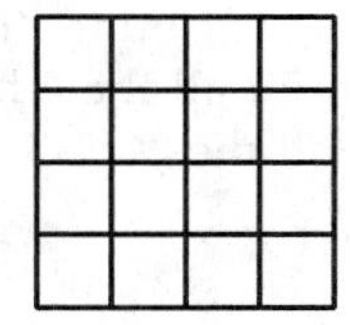

(c) How many different squares are in the following picture?

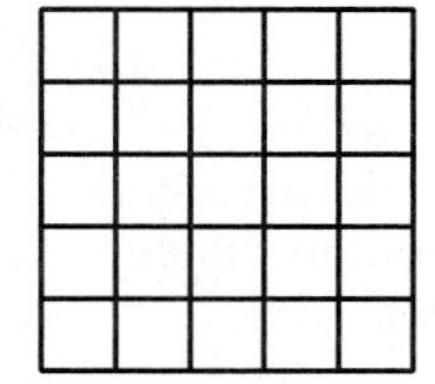

(d) What pattern do you see in the numbers in parts (a), (b), and (c)?

(e) Solve a famous problem. How many different squares are there on an 8-by-8 checkerboard? (Use a shortcut.)

The four-step approach does not guarantee that you will be able to solve a problem, but it does provide some guidance. Previous experience solving problems is also a great help in solving new problems.

ANSWERS TO SELECTED LESSON EXERCISES

1.47 9

1.48 80

1.49 10

1.50 Multi-step translation

1.51 Puzzle

1.52 One-step translation

1.53 (a) The current ages of Jamaal and his daughter
(b) Yes (c) Yapper's age
(f) 66 years; 33 years

1.54 (a) How much money Wanda started out with
(c) She has $6 now. So before she went to the record store, she had $12. And before she went to the grocery store, she had $16.
(d) Start with $16. Spend 1/4 of it at the grocery store. That leaves $12. Spend 1/2 of that at the record store. That leaves $6.

1.55 (a) ***Understanding the Problem*** What are you supposed to find? (The total number of squares of all sizes in the picture.)

Devising a Plan How will you do this? (Group the squares by size. Count all the 1-by-1 squares. Then count all the 2-by-2 squares. Finally, count all the 3-by-3 squares.)

Carrying Out the Plan Carry out your plan and obtain an answer.

Size	Number of Squares
1 by 1	9
2 by 2	4
3 by 3	1
Total	14

Looking Back Make up another problem like this one. How many different squares are in the following picture?

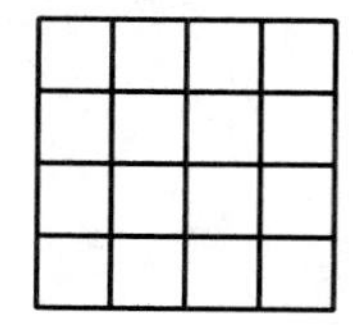

After solving this problem and the preceding one, develop a general solution for solving the same kind of problem with any size square.

(b) 30 (c) 55 (e) 204

1.5 HOMEWORK EXERCISES

Basic Exercises

Tell whether Exercises 1–3 would usually be classified as one-step translation, multi-step translation, or puzzle problems.

1. There are 10 boys and 12 girls going on a field trip. Each car will hold 5 children and 1 parent. How many cars are needed?
2. A classroom has 4 rows with 5 desks in each row. How many desks are there in all?
3. About how many marbles would fit inside a basketball?
4. Describe with one word each of Polya's four steps for solving a problem.
5. Explain how a doctor treating an illness could use Polya's four steps.
6. Consider the following problem. "The area of a square field is 49 m^2. What is the length of a fence that goes around the outside of the field?" Describe a plan for solving this problem.
7. Seth Borgas buys two shirts for $12.98 each and a cap for $8.98. The sales clerk says the total cost is $48.94. Without computing the

exact answer, tell whether the clerk's total seems reasonable. Why or why not?

8. A family is taking a 500-mile car trip. Make up a word problem about the trip that uses multiplication and subtraction.

9. Make up a word problem about the Country Kitchen's menu that involves multiplication and addition.

THE COUNTRY KITCHEN	
Roadside squirrel	$1.50
Twice-baked kelp	$1.05
Scalloped corn	$ .60
Pond water	$ .10

10. Solve the following problem using Polya's four steps. Make up an appropriate question or activity and answer it for each of Polya's four steps. The problem is as follows. "You have 10 black socks, 10 white socks, and 4 red socks in a drawer. What is the greatest number of socks you would have to pull out (without replacement) on a dark morning to be sure of obtaining a matching pair?"

11. Solve the following problem using Polya's four steps. Make up an appropriate question or activity and answer it for each of Polya's four steps. The problem is as follows. "How many cuts does it take to divide a log into five cross-sectional (cylindrical) pieces?"

12. How many cuts does it take to divide a log into
(a) six equal cross-sectional pieces?
(b) seven equal cross-sectional pieces?
(c) N equal cross-sectional pieces?

13. How many squares are in the following picture?

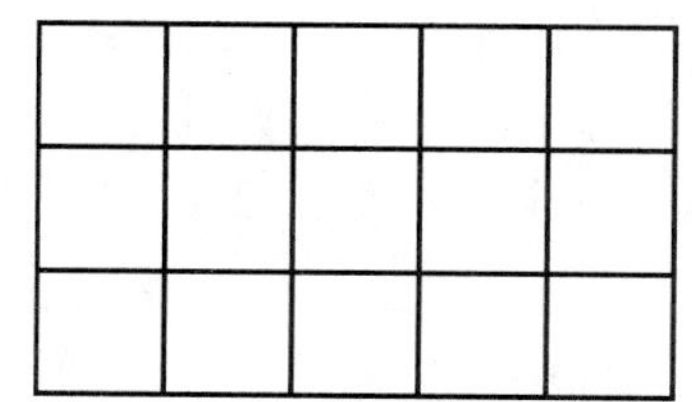

14. (a) A 2-by-2-by-2 cube is built from eight 1-by-1-by-1 cubes. How many cubes of all sizes are there?
(b) A 4-by-4-by-4 cube is built from sixty-four 1-by-1-by-1 cubes. How many cubes of all sizes are there?

15. (a) The Bathula family has 2 sons. Each son has 3 sisters. How many children are there?
(b) The Dulfano family has T sons. Each son has N sisters. How many children are there?

16. A pizza restaurant has 10 different toppings for its cheese-and-tomato pizza: mushrooms, peppers, pepperoni, sausage, onion, anchovies, tuna, pickles, shredded wheat, and celery. How many different kinds of pizza can be made by varying the combination of toppings? (*Hint:* Try the same kind of problem with 1 topping, then 2 toppings, and so on, and look for a pattern.)

17. Fill in the missing numbers.

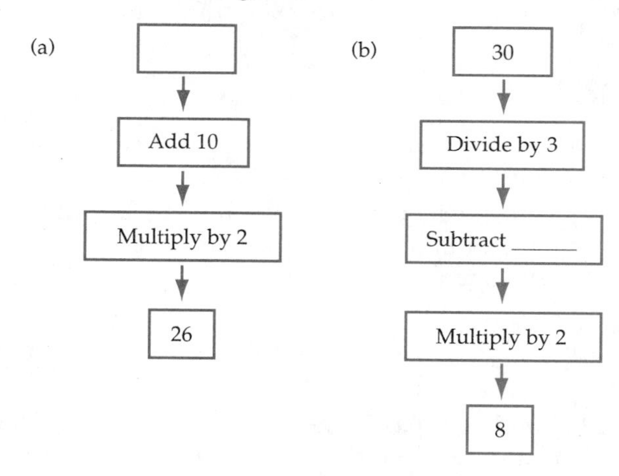

18. Find three numbers whose sum is 10 and whose product is as large as possible.

19. Consider the following problem. Chick N. Little bought 3 dozen eggs, and 2 of the eggs were broken. How many good eggs were there?
(a) Write a problem about 12-year-old Bea Young and her father that involves computing $3 \times 12 - 2$ to find his age.
(b) Write a problem about the total number of people in rows of chairs that uses the same $3 \times 12 - 2$ computation.

20. Remove two toothpicks so that you have two squares of different sizes.

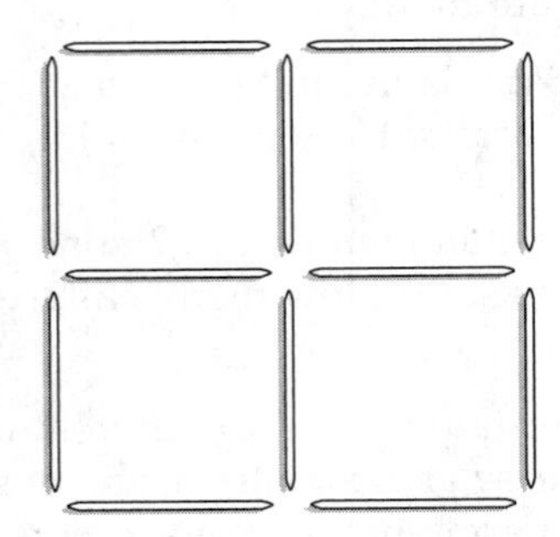

21. Move only three dots to make this design

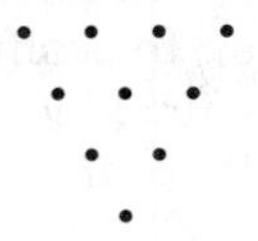

look like this:

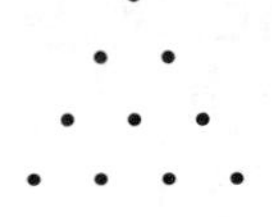

Extension Exercises

22. Connect the nine dots by drawing four lines, without lifting your pencil off the paper or retracing any lines.

• • •
• • •
• • •

23. You have 16 coins and a balance. Fifteen of the coins are regular, and one is a lighter counterfeit coin. Explain how you could identify the counterfeit coin after three weighings.

24. You have 24 coins and a balance. Twenty-three of the coins are regular, and one is a lighter counterfeit coin. Explain how you could identify the counterfeit coin after three weighings.

25. Consider the entire set of problems that are like the preceding two exercises. What is the greatest number of coins you could start with and be sure to identify the counterfeit coin after

(a) three weighings? (b) two weighings?

26. The goal in each part of this exercise is to switch the positions of the coins. A coin may move forward one space or jump over a single coin.

(a) A penny (P) and a nickel (N) are placed as shown:

P		N

How many moves are needed to switch the penny and the nickel?

(b)

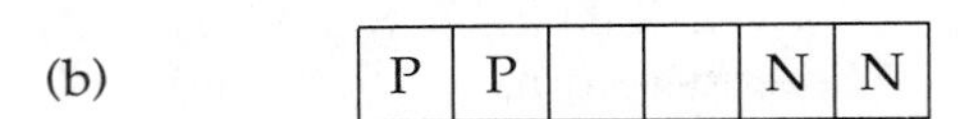

How many moves are needed to switch the pennies and nickels?

27. In word boxes, you can go from box to box as long as you move one step down or to the right. For example, for the word "BE," the word box allows you to complete the chart as follows.

	[1] B
[2] B	E

Number of Letters	Number of Paths Starting at		Total Number of Paths
	Box 1	Box 2	
2	1	1	2

(a) Complete the chart for "ZIP".

		[1] Z
	[2] Z	I
[3] Z	I	P

Number of Letters	Number of Paths Starting at			Total Number of Paths
	Box 1	Box 2	Box 3	
3				

(b) Complete a similar chart for "MATH."
(c) Based upon the results for two-, three-, and four-letter words, guess how many paths there would be for a word that has N letters.

28. (a) A number is divisible by 8. It has three digits that add up to 13. Guess the number.
(b) Find three other possible answers.

29. Write a paragraph describing Polya's four steps for problem solving.

Special Exercise

30. A woman lines up five red checkers and five black checkers.

R	R	R	R	R	B	B	B	B	B
1	2	3	4	5	6	7	8	9	10

She asks: Can you switch the positions of only four checkers so that the black and red checkers alternate?
(a) How would you do it?
(b) If you started with only 1 pair of checkers, how many checkers would you have to move to solve the same problem?
(c) If you started with 3 pairs of checkers, how many checkers would you have to move to solve the same problem?
(d) What is the general pattern in the numbers of checkers that must be moved in parts (a), (b), and (c)?
(e) Does this pattern work for 2 pairs of checkers?
(f) Does it work for 4 pairs of checkers?
(g) Can you think of a different rule that would work for parts (e) and (f)?
(h) Make a conjecture about how many checkers must be moved in the same problem involving 25 pairs of checkers.
(i) Make a conjecture about how many checkers must be moved in the same problem involving 50 pairs of checkers.

1.6 PROBLEM-SOLVING STRATEGIES

Do you ever have difficulty solving mathematics problems? Devising a plan, the second step of Polya's procedure, is often the most difficult step. Elementary-school children now learn specific strategies that they can use to solve a variety of problems. Learning these strategies can help any student become a better problem solver.

Children now do sets of problems that focus upon specific processes or strategies. The following problem-solving strategies are discussed in this book.

Some Problem-Solving Processes and Strategies

1. Using inductive reasoning
2. Using deductive reasoning or logic
3. Guessing and checking
4. Making a table or list
5. Drawing a picture
6. Choosing an operation
7. Working backward
8. Using a graph
9. Using an equation
10. Taking a break and trying again

Problem-solving strategies are now included in most elementary-school textbooks. You have already studied induction and deduction. Choosing an operation and working backward will be introduced in Chapter 3. Chapter 11 discusses using graphs and equations to solve problems.

In this section you will study three useful problem-solving strategies: guessing and checking, making a table, and drawing a picture. Examples of each strategy follow.

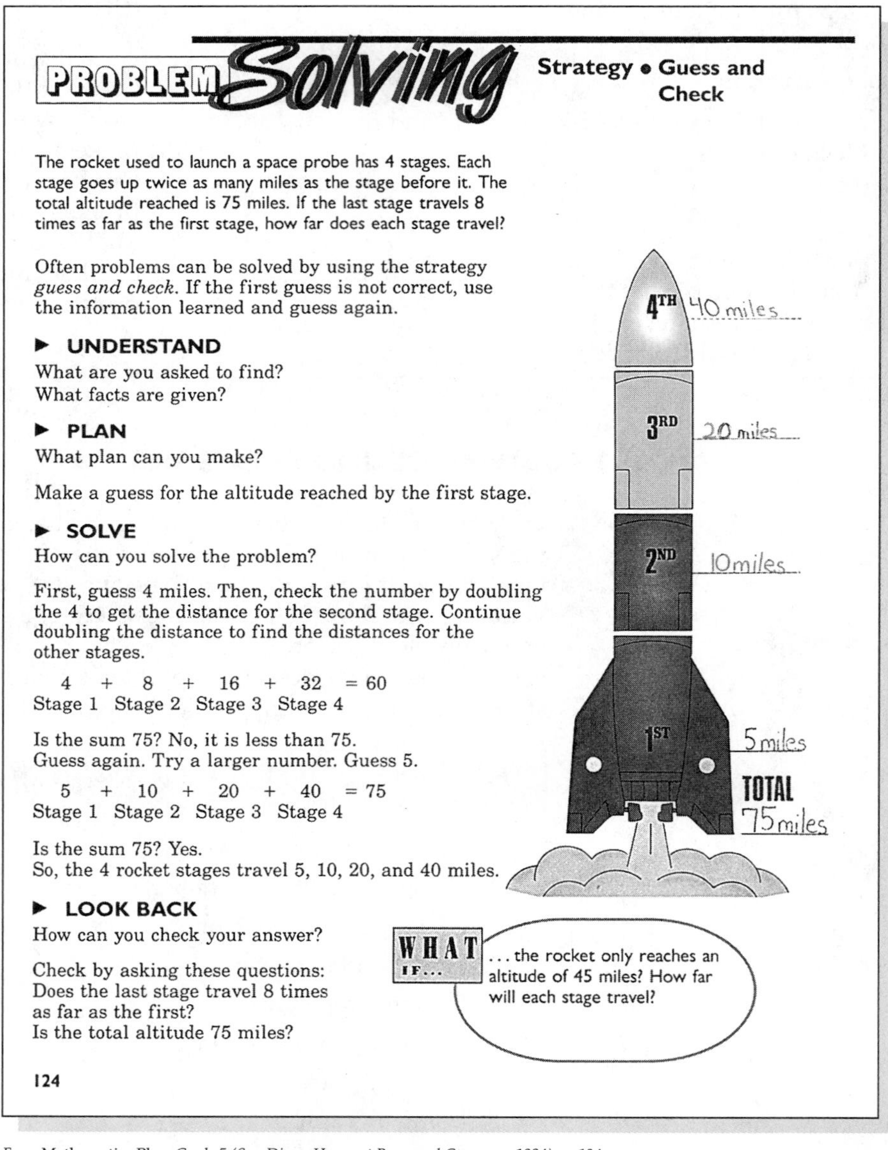
PROBLEM Solving Strategy • Guess and Check

The rocket used to launch a space probe has 4 stages. Each stage goes up twice as many miles as the stage before it. The total altitude reached is 75 miles. If the last stage travels 8 times as far as the first stage, how far does each stage travel?

Often problems can be solved by using the strategy *guess and check*. If the first guess is not correct, use the information learned and guess again.

▶ **UNDERSTAND**

What are you asked to find?
What facts are given?

▶ **PLAN**

What plan can you make?

Make a guess for the altitude reached by the first stage.

▶ **SOLVE**

How can you solve the problem?

First, guess 4 miles. Then, check the number by doubling the 4 to get the distance for the second stage. Continue doubling the distance to find the distances for the other stages.

4 + 8 + 16 + 32 = 60
Stage 1 Stage 2 Stage 3 Stage 4

Is the sum 75? No, it is less than 75.
Guess again. Try a larger number. Guess 5.

5 + 10 + 20 + 40 = 75
Stage 1 Stage 2 Stage 3 Stage 4

Is the sum 75? Yes.
So, the 4 rocket stages travel 5, 10, 20, and 40 miles.

▶ **LOOK BACK**

How can you check your answer?

Check by asking these questions:
Does the last stage travel 8 times as far as the first?
Is the total altitude 75 miles?

124

From Mathematics Plus, *Grade 5 (San Diego: Harcourt Brace and Company, 1994), p. 124.*

Figure 1–16

The best way for you or your students to learn a problem-solving strategy is to practice using it first by itself. After each of the three individual strategies becomes familiar, you can try to solve problems in which you must decide which of the three strategies to use.

GUESSING AND CHECKING

Sometimes you have to be bold in mathematics! The guessing-and-checking strategy requires you to start by making a guess and then checking how close your answer is. Next, based upon this result, revise your guess and try again. Figure 1-16, on the preceding page, shows how the fifth-grade *Mathematics Plus* textbook presents guessing and checking.

Try guessing and checking in the following exercise.

Lesson Exercise 1.56

Sandy bought 18 pieces of fruit (oranges and grapefruits), which cost $4.62. If an orange costs $.19 and a grapefruit costs $.29, how many of each did she buy? (*Guess and check.*)

Consider guessing and checking when you have a question with a limited number of possible answers, and you can check how close a guess is to the correct answer.

MAKING A TABLE OR LIST

Organizing information often makes it easier to solve a problem. You may have used a table in the preceding lesson exercise. A table can also help you solve the problem in Example 1.11.

EXAMPLE 1.11

A cashier wants to make change for $1 using nickels, dimes, and quarters. Furthermore, he will use at least 2 quarters and no more than 8 coins all together. How many ways are there to do this? (*Make a table.*)

Solution

Understanding the Problem The cashier must use 2, 3, or 4 quarters and no more than a total of 8 coins to make $1.

Devising a Plan All the solutions can be organized into a table.

Carrying Out the Plan Fill out the table in an *organized* way.

Quarters	Dimes	Nickels
4	0	0
3	2	1
3	1	3
3	0	5
2	5	0
2	4	2

There are 6 different ways.

Looking Back Check the table to see that all combinations are included. They are. ■

Tables enable one to organize information in a simple, clear way.

Lesson Exercise 1.57

It will take 11 quarts of paint to paint your new apartment. Quarts cost \$3.50 and gallons cost \$8.50. (*Note:* 1 gallon = 4 quarts.)

(a) What are the different ways you can purchase the paint for the job? (*Make a table or list.*)

(b) Which way is the least expensive?

Consider making a table or list when you have a limited number of options that can be listed in an organized way, and you want to examine all those options.

DRAWING A PICTURE

Sometimes a drawing or a diagram can help you analyze a problem. For example, you can use Venn diagrams to help check deductive reasoning. Solve Lesson Exercise 1.58 by *drawing a picture*.

Lesson Exercise 1.58

A well is 30 ft deep. An athletic snail at the bottom climbs up 3 ft each day and slips back 2 ft each night. On what day does the snail reach the top of the well? (*Hint:* Draw a picture. The answer is *not* the 30th day!)

Consider drawing a picture for any problem that involves measurements or geometric shapes. Besides using problem-solving strategies, successful problem solvers tend to (1) approach new problems with confidence, (2) analyze and explore the conditions of the problem, and (3) obtain a lot of experience in solving problems.

AN INVESTIGATION: PROBLEM ANALYSIS

The following lesson exercises are all related to the preceding puzzle problem. By doing this series of problems, you will deepen your understanding of this kind of puzzle problem.

Lesson Exercise 1.59

A well is 30 ft deep. A muscle-bound worm at the bottom climbs up 4 ft each day and slips back 1 ft each night. On what day does it reach the top of the well? (*Draw a picture.*)

Lesson Exercise 1.60

A well is 50 ft deep. A muscle-bound worm at the bottom climbs up 4 ft each day and slips back 2 ft each night. On what day does it reach the top of the well? (*Draw a picture.*)

Lesson Exercise 1.61

A well is 80 ft deep. Complete a worm or snail problem so that the slimy animal reaches the top on the
(a) 20th day. (b) 19th day. (c) 18th day.

ANSWERS TO SELECTED LESSON EXERCISES

1.56 6 oranges and 12 grapefruits

1.57 (a)

Number of Gallons	Number of Quarts	Total Cost
0	11	$38.50
1	7	$33.00
2	3	$27.50
3	0	$25.50

(b) Three gallons would be the least expensive.

1.58 28th day (*Hint:* Where is the snail after 27 days and nights?)

1.59 10th day

1.60 24th day

1.61 *Hint:* Use an "up" number that is 4 more than the "down" number.

1.6 HOMEWORK EXERCISES

Basic Exercises

1. Which step of Polya's four-step scheme involves the consideration of various problem-solving strategies?
2. Find two consecutive even numbers whose product is 50,624. (*Guess and check.*)
3. Craig Chandler is 5 years older than Linda. The product of their ages is 1184. How old are they? (*Guess and check.*)
4. Find $\sqrt{1444}$ without using a calculator $\boxed{\sqrt{\ }}$ key. (*Guess and check.*)
5. Approximate a solution of $(x + 2)(x + 3) = 50$ to one decimal place. (*Guess and check.*)
6. A bank sells the following packets of traveler's checks: five \$20 bills, three \$50 bills, three \$100 bills, and five \$100 bills. How many different ways are there to buy \$500 in traveler's checks? (*Make a table.*)
7. A baseball league has 6 teams: the Bats, the Diamonds, the Flies, the Goose Eggs, the Hot Dogs, and the Relish. If every team plays each of the other teams 4 times, how many games must be scheduled? (*Make a table.*)
8. (a) A square has an area of 81 square units. What is its perimeter? Recall that the perimeter is the distance around the outside. (*Draw a picture and use the correct units.*)
   (b) Computing the perimeter of a square when you know the length of each side involves what type of reasoning, inductive or deductive?
9. A well is 100 ft deep. A muscle-bound worm at the bottom climbs up 5 ft each day and slips back 2 ft each night. On what day does it reach the top of the well? (*Draw a picture.*)
10. A well is 60 ft deep. Complete a worm or snail problem so that the slimy animal reaches the top on the
   (a) 20th day. (b) 19th day. (c) 18th day.

11. A telephone company offers two plans for local calls. Plan A charges \$.09 per call. Plan B charges \$4 for the first 50 calls and \$.06 for each call after that. Determine under what conditions each plan is cheaper. (*Make a table and guess and check.*) The table headings are as follows:

Number of Calls	Cost of Plan A	Cost of Plan B

12. During 6 weeks in the summer, each of 4 workers will take 3 weeks off. No more than 2 workers may be off at one time, and everyone wants at least 2 consecutive weeks off. Complete the following schedule.

	Week 1	Week 2	Week 3	Week 4	Week 5	Week 6
Molly	X			X	X	
Paul	X		X	X		
Wong						
Clio						

13. An office staff includes a manager (M), an assistant manager (AM), two secretaries (S), and two sales agents (SA). Construct a 7-week vacation schedule for them, taking into consideration the following constraints.
   1. The manager is taking weeks 5, 6, and 7 off. Everyone else gets 2 consecutive weeks off.
   2. No more than two people can be on vacation at one time.
   3. At least one secretary and one sales agent must be present in the office each week.
   4. When the manager is away, the assistant manager and both sales agents must be at work.

(table on the next page)

	M	AM	S1	S2	SA1	SA2
Week 1						
Week 2						
Week 3						
Week 4						
Week 5	X					
Week 6	X					
Week 7	X					

14. Make up a word problem that can be solved with each of the following strategies.
(a) Guess and check (b) Draw a picture
(c) Make a table

15. Solve the following problem and tell which of the three strategies you used. An apartment has two adjacent rectangular rooms, each 9 ft by 12 ft. They share a 9-ft-long wall. What is the perimeter of the apartment?

16. Solve the following problem and tell which of the three strategies you used. A 10-lb bag of mixed nuts contains 20% peanuts. How many pounds of peanuts should be added to change the mixture to 80% peanuts?

17. Solve the following problem and tell which of the three strategies you used. After throwing 25 darts at the following target, what is the *highest* score below 100 that it is *impossible* to score?

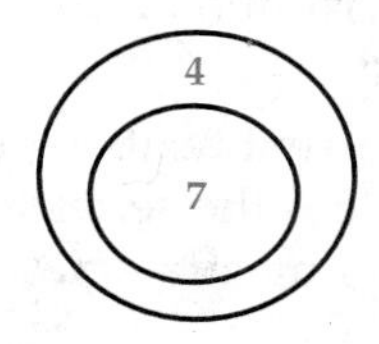

Extension Exercises

18. (a) A woman has 40 yards of fencing for her yard. What is the maximum rectanglar area she can enclose? (*Draw a picture, make a table, and guess and check.*)
(b) A woman has F yards of fencing for her yard. What is the maximum rectangular area she can enclose? (*Draw a picture, make a table, and guess and check.*)

19. (a) A woman wants to enclose 100 yd^2 of field with a rectangular fence. What is the minimum perimeter of fencing she can use?
(b) A woman wants to enclose N yd^2 of field with a rectangular fence. What is the minimum perimeter of fence she can use?

20. A travel agency is planning a one-week trip to Italy that has an initial cost of $2000 plus $1140 per person. At a price of $1600, 15 people will sign up. Past experience suggests that for each $20 decrease in the price, an additional person will sign up. Also, for each $20 increase in the price, one less person will sign up. Approximately what price will maximize the profit? (*Make a table and guess and check.*)

21. A farmer wants to transport a fox, a goose, and a bag of corn across a river in a boat. He

can take only one of the three across on each trip. He cannot leave the fox and the goose alone, since the fox will eat the goose. He cannot leave the goose and the corn alone, since the goose will eat the corn. How will he get the fox, the goose, and the corn across the river? (*Draw a picture and guess and check.*)

22. A man and two small children want to cross the river in a small boat. The boat is only big enough to hold either the man or the two children, both of whom can row. How can all three get across the river in the boat?

23. There are 20 children in a class, including exactly 10 girls and 12 eight-year-olds. How many eight-year-old girls could there be? Give *all* possible answers. (*Draw a picture and guess and check.*)

24. There are 26 children in a class, including exactly 12 girls and 20 eight-year-olds. How many eight-year-old girls could there be?

25. There are C children in a class, including exactly G girls and E eight-year-olds. Assume that $G < E < C$. How many eight-year-old girls could there be?

26. (a) I'm inviting 20 people to a party, and I want to seat them all at one long table. I'm going to put together a series of card tables that seat one person on each side, and form one long table. If I arrange the card tables in a single row, how many card tables will I need?

(b) Give a general solution for seating $2N$ people (N is a whole number greater than 3). How many tables are needed?

27. Another option for the party in the preceding exercise is to use rectangular tables, as shown.

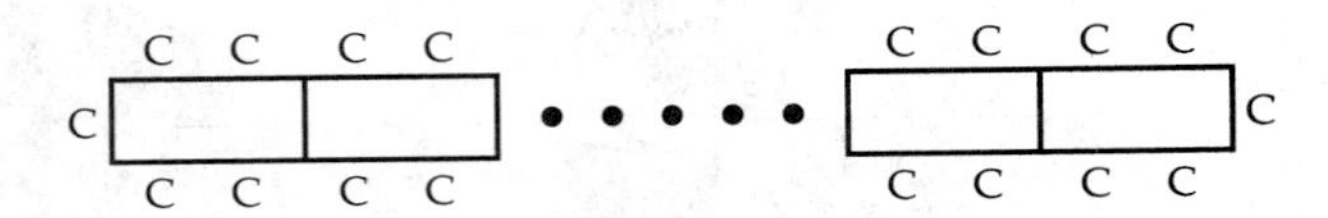

What is a formula that relates the number of chairs (C) to the number of tables (T)?

28. In the preceding exercise, one student gives the answer $C = 2 + 4T$, and another gives the answer $C = 6 + 4(T - 1)$. Are both answers correct? Explain why or why not.

29. Another arrangement of rectangular tables (see the preceding exercise) gives more room on the table for serving platters.

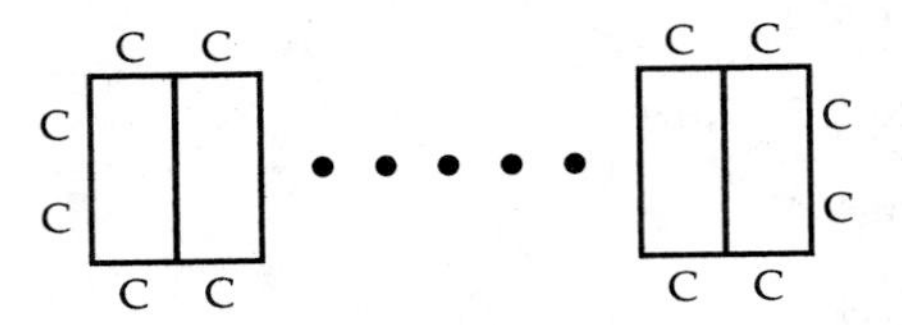

What is the formula that relates the number of chairs to the number of tables?

30. At a party where the guests have never met one another, everyone wants to shake hands with everyone else. How many handshakes will occur if there are

(a) 2 people at the party?
(b) 3 people at the party?
(c) 4 people at the party?
(d) 8 people at the party? Use the pattern from parts (a), (b), and (c).

31. A group of girls are running in a race. How many orders of finish are possible if there are

(a) 2 runners? (b) 3 runners?
(c) 4 runners? (d) N runners?
(e) Making a generalization in part (d) involves ______________ reasoning.

32. (a) A farmer is asked by a passing stranger how many rabbits and chickens she has. The clever farmer replies, "Between the rabbits and the chickens, there are 76 eyes and 126 feet." How many rabbits and chickens does she have?

(b) Repeat part (a) for E eyes and F feet.

33. Write a description of the guess-and-check strategy.

34. Look at a current textbook series for Grades 1–6. What does the series teach about problem-solving strategies in each grade?

SUMMARY

What methods of reasoning do mathematicians use that people also use to analyze problems in everyday life? Two methods commonly used to discover new ideas in mathematics and everyday life are inductive and deductive reasoning. Induction involves making a reasonable generalization from specific examples. Scientists and mathematicians use induction to develop general hypotheses or theorems. People may also develop prejudices and superstitions with induction.

To verify conjectures, one uses deductive reasoning, the process of drawing a necessary conclusion from given assumptions. Detectives use deduction in drawing conclusions, and mathematicians use deduction to prove theorems. Mathematicians often develop new ideas by using induction to make a generalization or state a theorem, followed by deduction to prove that the generalization or theorem must be true.

Both types of reasoning can lead to false conclusions. Induction is not totally reliable, because what happens a few times may not always happen. Superstitions and prejudices are examples of erroneous generalizations. Deductions are sometimes false because valid deductive reasoning based upon false assumptions can lead to a false conclusion.

A superior ability to find patterns distinguishes human beings from other animals and machines. Mathematicians develop new ideas by recognizing and generalizing patterns. Finding a pattern makes use of induction, and extending the pattern based upon a general rule requires deduction. Finding and extending patterns enables people to make predictions about the future and to develop classifications.

Problem solving is now a focus of school mathematics. Whenever possible, mathematical ideas should be developed from problem situations of interest to the learner. Three types of problems children study are one-step translation problems, multi-step translation problems, and puzzle problems. In solving most of these problems, Polya's four steps can be followed: (1) Understand the problem, (2) devise a plan, (3) carry out the plan, and (4) check and assess your results.

Devising a plan or strategy is often the hardest step. Three strategies commonly used to solve mathematics problems are guessing and checking, making a table, and drawing a picture. Arithmetic, algebra, and geometry problems can sometimes be solved more easily with these strategies.

STUDY GUIDE

To review Chapter 1, see what you know about each of the following ideas and terms that you have studied. You can also use this list to generate your own questions about the chapter.

THE NCTM CURRICULUM STANDARDS AND MATHEMATICAL REASONING

The National Council of Teachers of Mathematics (NCTM) *Curriculum and Evaluation Standards for School Mathematics* (1989) is the most important document about the current elementary-school mathematics curriculum. Some states and communities have told textbook publishers they will only use books that follow "the Standards."

At the end of each chapter, you can learn about the Standards and review the chapter by considering how the Standards relate to the material you have just studied. If you have access to elementary-school textbooks, you can also investigate how these materials incorporate the Standards.

The NCTM Curriculum Standards and Rational Numbers

Selected NCTM Curriculum Standards

There are about 50 objectives for students in Grades K–4 and 70 objectives for students in Grades 5–8. The following standards come from these two lists of objectives.

- Recognize and apply inductive and deductive reasoning.
- Use mathematics in their daily lives.
- Recognize, describe, extend, and create a wide variety of patterns.
- Describe and represent relationships with tables, graphs, and rules.
- Develop and apply strategies to solve a wide variety of problems.
- Reflect on and clarify their thinking about mathematical ideas and situations.

1. Describe how each standard listed relates to the material you studied in Chapter 1.
2. Select any current elementary-school mathematics textbook series and describe sample lessons or exercises that illustrate each standard listed.

MATHEMATICAL REASONING IN ELEMENTARY SCHOOL

The following chart shows at what grade levels selected mathematical reasoning topics typically appear in elementary-school mathematics textbooks. Underlined numbers indicate grades in which the most time is spent on the given topic.

Topic	Typical Grade Level in Current Textbooks
Using inductive reasoning	1, 2, 3, 4, 5, 6
Using deductive reasoning	1, 2, 3, $\underline{4}$, $\underline{5}$, $\underline{6}$
Number sequences	1, 2, $\underline{3}$, $\underline{4}$, $\underline{5}$, 6
Finding patterns	1, 2, 3, 4, 5, 6
Problem-solving steps (Polya)	1, 2, $\underline{3}$, $\underline{4}$, $\underline{5}$, $\underline{6}$
Problem-solving strategies	$\underline{1}$, $\underline{2}$, $\underline{3}$, $\underline{4}$, $\underline{5}$, $\underline{6}$

REVIEW EXERCISES

1. What is deductive reasoning?

In Exercises 2–4, identify whether induction or deduction is being used.

2. People have died in the past. I assume everyone will die in the future.
3. I know Alicia is in the bedroom, the kitchen, or the bathroom. I do not find her in the bedroom or the kitchen. I conclude that she is in the bathroom.
4. I know that $x > y$ and $y > z$. I conclude that $x > z$.

5. Explain how an auto mechanic repairing an engine could use Polya's four steps.

6. Does the given conclusion follow from the two hypotheses? If so, draw a Venn diagram that illustrates it. If the conclusion is invalid, draw a Venn diagram that satisfies the hypothesis but does not support the conclusion.

Hypotheses: All squares have four sides. All rhombuses have four sides.
Conclusion: Therefore, all rhombuses are squares.

7. *Explain* how a student might use inductive reasoning.

8. Make up an example of valid deductive reasoning that
(a) leads to a true conclusion.
(b) leads to a false conclusion.

9. Is the sum of any three consecutive counting numbers divisible by 3?

10. (a) Write the converse of the following statement: If it is raining, then the ground gets wet.
(b) Is the converse true?

11. Write the following statements in an "if-then" format.
(a) A rectangle is a square if it has four congruent sides.
(b) I will call only if there is a problem.

12. Consider the following number trick.

Pick a number.
Add 5.
Multiply by 3.
Subtract 9.
Subtract your original number.
Divide by 2.

Prove that you will always end up with three more than you started.

13. (a) Write a formula for the nth term in the following sequence.

$$20, 19, 18, 17, \ldots$$

(b) What is the 40th term?
(c) Make up a chart of x and y values that is analogous to the sequence in part (a).
(d) Does part (a) involve inductive or deductive reasoning?

14. Suppose you have to wait for a train to pass at a railroad crossing. The waiting time is related to the speed of the train.

Speed (mi/hr)	20	30	40	50	60
Waiting time (min)	4.5	3	2.25	1.8	

(a) Fill in the waiting time for 60 miles per hour.
(b) A formula that relates waiting time T to speed S is $T =$ ________.

15. (a) Fill in the missing numbers.

$$1 = (\ \ )^3$$
$$3 + 5 = (\ \ )^3$$
$$7 + 9 + 11 = (\ \ )^3$$

(b) What would the next equation be if the pattern continued? Is the equation true?

16. Examine the following pattern.

$$4^2 - 1^2 = 3 \cdot 5$$
$$5^2 - 2^2 = 3 \cdot 7$$
$$6^2 - 3^2 = 3 \cdot 9$$

(a) Write the next example that follows the pattern.
(b) Complete the following generalization. For any counting number N greater than two, $N^2 -$ ________ $=$ ________.
(c) Prove that your equation in part (b) is true.

17. Make up an example of a multi-step translation problem.

18. How many rectangles are there in the figure?

19. Consider the following problem. You took 11 orders for either a tuna salad sandwich ($1.90) or a buckwheat Newburg ($1.50) but forgot how many of each. If the bill comes to $17.70, how many of each were ordered?
(a) What would be a good strategy for solving this problem?
(b) Solve the problem using your strategy.

20. An employer wishes to select 2 people for a job from Alice, José, Mary, and Reggie. How many ways are there to do this?

21. Solve the following problem using Polya's four steps. Make up an appropriate question and answer it for each of the four steps.

Examine the following figure.

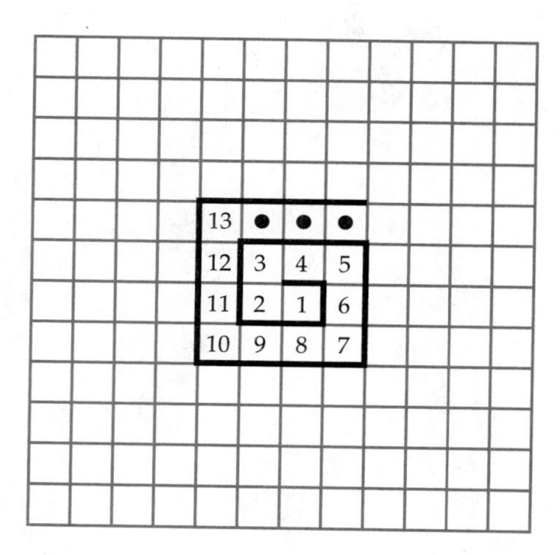

If the pattern in the square continues, what number will appear in the upper right-hand corner?

22. A well is 40 ft deep. A snail climbs up 9 feet each day and slips back 5 feet at night. How long will the snail take to reach the top of the well?

23. A well is 40 ft deep. Make up a snail problem in which the snail reaches the top on the
(a) 18th day. (b) 19th day.

24. (a) Mack, Ahmad, Mona, and Rosie each own a pet. The pets are an ant, a crocodile, a gorilla, and a sea turtle. Determine who owns each pet by using the following facts.
1. Mack is allergic to crocodiles and gorillas.
2. Mona and the owner of the gorilla live in the same apartment building.
3. Ahmad is not allowed to have a pet larger than a goldfish in his building.

(b) Does this problem involve induction or deduction?

25.

$$142857 \times 1 = 142857$$
$$142857 \times 2 = 285714$$
$$142857 \times 3 = 428571$$

(a) Write the next example, continuing the pattern.
(b) Make a generalization about the digits in the result.
(c) How long does the pattern continue? Try 5, 6, and 7 as factors.

26. Write a paragraph describing inductive reasoning and deductive reasoning and how to tell the difference between them.

27. Write a paragraph describing how to find the rule for the nth term of a number sequence.

SUGGESTED READINGS

Brown, S., and M. Walter. *The Art of Problem Posing.* Palo Alto, Calif.: Dale Seymour, 1983.

Burns, M. *The Book of Think.* Boston, Mass.: Little, 1976.

Charles, R. and F. Lester. *Teaching Problem Solving: What, Why, and How.* Palo Alto, Calif.: Dale Seymour, 1982.

Coburn, T., and others. *Patterns.* Reston, Va.: NCTM, 1992.

Equals. *Get It Together.* Berkeley, Calif.: Equals, 1989.

Gardner, Martin. *Aha! Insight.* New York: Freeman, 1978.

Jacobs, Harold. *Mathematics: A Human Endeavor.* 3rd ed. New York: Freeman, 1994.

Mason, J., L. Burton, and K. Stacey. *Thinking Mathematically.* Rev. ed. Reading, Mass.: Addison-Wesley, 1995.

National Council of Teachers of Mathematics. *Curriculum and Evaluation Standards for School Mathematics.* Reston, Va.: NCTM, 1989.

National Council of Teachers of Mathematics. *Problem Solving in School Mathematics.* 1980 Yearbook. Reston, Va.: NCTM, 1980.

National Council of Teachers of Mathematics. *New Directions for Elementary School Mathematics.* 1989 Yearbook. Reston, Va.: NCTM, 1989.

National Council of Teachers of Mathematics. *Professional Development for Teachers of Mathematics.* 1994 Yearbook. Reston, Va.: NCTM, 1994.

National Council of Teachers of Mathematics. *Professional Standards for Teaching Mathematics.* Reston, Va.: NCTM, 1990.

Polya, G. *How To Solve It.* 2nd ed. Princeton, N.J.: Princeton Univ. Press, 1973.

(*Note:* Recent issues of *Teaching Children Mathematics* would be another good source for most of the topics covered in this book.)

2 Sets

Every area of mathematics utilizes sets in some way. For example, Chapters 3, 4, 5, and 7 each concern a particular set of numbers (e.g., whole numbers). You will also study different sets of shapes (e.g., rectangles and parallelograms) and the relationships among them. Graphing involves sets of ordered pairs. In statistics one studies data sets, and in probability one examines sets of possible outcomes. You have already used set pictures (Venn diagrams) to illustrate logical relationships.

This chapter introduces the basic ideas of sets, but this book as a whole does not place great emphasis on formal set theory. Neither does the current elementary-school mathematics curriculum.

2.1 SETS

George Boole
Photo courtesy of Library of Congress

Figure 2–1

George Boole (1815–1864; Figure 2-1) developed set theory. Born to an English working-class family, Boole taught himself six foreign languages and a good deal of mathematics. He believed that the essence of mathematics lies in its deductive organization.

George Cantor (1845–1918; Figure 2-2, on the next page), a German mathematician, extended set theory to include infinite sets, a topic so controversial that it caused one of his former teachers to turn against him. Partly as a result of being ridiculed by that teacher, Cantor suffered a nervous breakdown. He died in a mental hospital.

Today Boole's and Cantor's work with finite and infinite sets is an accepted part of mathematics. In this text, sets will be used to define whole-number addition and multiplication and will serve as a model for each of the four whole-number operations in Chapter 3.

In the 1960s and early 1970s, some elementary-school teachers used sets to explain counting and arithmetic. Some of you may remember studying sets as children. Others may claim to be too young to remember 1970. (Others, like myself, can't believe it isn't still 1970.)

Teaching that was heavily based upon set theory was not well received in elementary schools, and today sets are generally not studied as a separate topic at that level. However, we often talk about sets of numbers or shapes, about

George Cantor
Photo courtesy of Library of Congress

Figure 2–2

subsets, and about the elements two sets have in common. Venn diagrams are introduced in some middle-school programs as an aid to problem solving.

SETS

What education courses must you pass to receive state teacher certification? Who in your class likes to eat pizza? The answers to these questions would each comprise the members of a set.

A **set** is a collection of objects called **members** or **elements**. To define a set, one can list the members and enclose them in braces. For example, the set of primary colors is {red, yellow, blue}.

The expression "$x \in A$" means that x is a member of set A. The expression "$x \notin A$" means that x is *not* a member of set A.

Lesson Exercise 2.1

Fill in each blank with $\in$ or $\notin$.

(a) 4 _____ {2, 4, 6}

(b) Orange juice _____ the set of all junk foods.

A set with no members is called an **empty** or **null set**. For example, the set of all pink elephants in your math class is an empty set. An empty set is denoted by the symbol { } or the Danish letter $\varnothing$.

Lesson Exercise 2.2

Make up another description that would represent an empty set.

People invented numbers so they could count objects. At first, people used only the first few counting numbers. The infinite set of counting or natural numbers is the result of thousands of years of work in expanding this initial set of numbers and developing more efficient notation.

We take the symbol 0 for granted, but at first people thought a symbol for "nothing" was unnecessary. Once 0 was included as a symbol for "nothing" and as a placeholder, numeration systems made significant advances (see Chapter 3).

By putting together 0 and the set of counting numbers, one obtains the set of whole numbers.

Definition Whole Numbers

The set of **whole numbers** $W = \{0, 1, 2, 3, \ldots\}$.

EQUAL, EQUIVALENT, FINITE, AND INFINITE SETS

One might compare two sets to see whether they are identical or whether they contain exactly the same number of elements. Two sets are **equal** if and only if they contain exactly the same elements. For example, if $A = \{2, 3\}$ and $B = \{3, 2\}$, then $A = B$.

To compare two sets, one can try to place their elements into a one-to-one correspondence. A **one-to-one correspondence** pairs the elements of two sets so that, for each element of one set, there is exactly one element of the other. For example, the sets {1, 2} and {Bill, Sue} can be placed into the two different one-to-one correspondences shown in Figure 2-3.

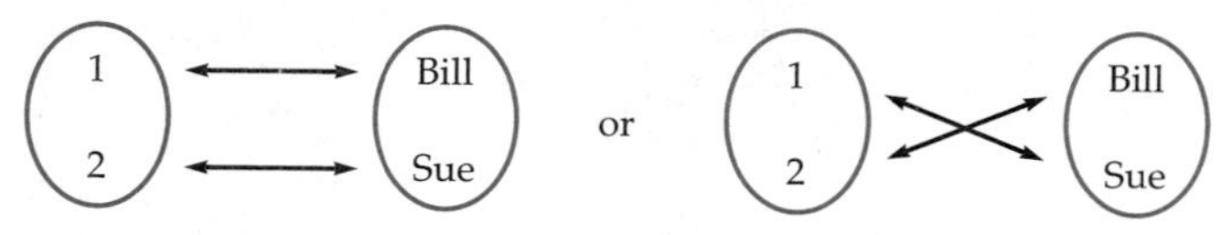

Figure 2–3

Two sets are **equivalent** if and only if there is a one-to-one correspondence between the sets. The following exercise examines the differences between equal and equivalent sets.

Lesson Exercise 2.3

$A = \{1, 2, 3\}$ $B = \{1, 3, 5\}$ $C = \{1, 2, 3, 4\}$

D = {Sara, John, Will} E = {bat, ball, glove, cap} $F = \{1, 3, 2\}$

(a) Which sets are equivalent?

(b) Which sets are equal?

(c) If two sets are equivalent, are they equal?

(d) If two sets are equal, are they equivalent?

(e) Show a one-to-one correspondence between A and D.

A set is a **finite set** if it is empty or if it can be placed into a one-to-one correspondence with a set of the form $\{1, 2, 3, \ldots, N\}$, where N is a whole number. The number of elements in a finite set must be a whole number. The set of days of the week is a finite set containing seven elements. The set of whole numbers is an **infinite set** because it does not contain a finite number of elements. A more formal definition of an infinite set appears in the extension exercises.

Lesson Exercise 2.4

Decide whether each of the following sets is a finite set or an infinite set.

(a) The set of whole numbers less than 6

(b) The set of all the pancakes in Arizona right now

(c) The set of whole numbers greater than 6

UNIVERSAL SETS AND SUBSETS

If you wanted to decide which days to exercise, you would choose from the 7 days of the week. Let

U = {Sunday, Monday, Tuesday, Wednesday, Thursday, Friday, Saturday}

U is the **universal set** that contains all elements being considered in a given situation.

Assume you decide upon a 3-day schedule, either A or B, or a very light exercise schedule, C.

A = {Monday, Wednesday, Friday}

B = {Tuesday, Thursday, Saturday}

$C = \{\ \}$

Under schedule A, the set of days on which you will not exercise is $\overline{A}$ = {Sunday, Tuesday, Thursday, Saturday}, the complement of A.

> ***Definition*** **Complement of a Set**
>
> The **complement** of a set D, written $\overline{D}$, is the set of elements in the universal set that are not in D.

Think of set $\overline{D}$ as "not D."

Lesson Exercise 2.5

(a) Write the set of elements in $\overline{B}$ in the exercise-scheduling example.

(b) What does $\overline{B}$ represent in relation to the days on which you will exercise?

Figure 2-4 shows that set A is an example of a subset of U, since each element of A is contained in U. For the same reason, B and C are also subsets of U.

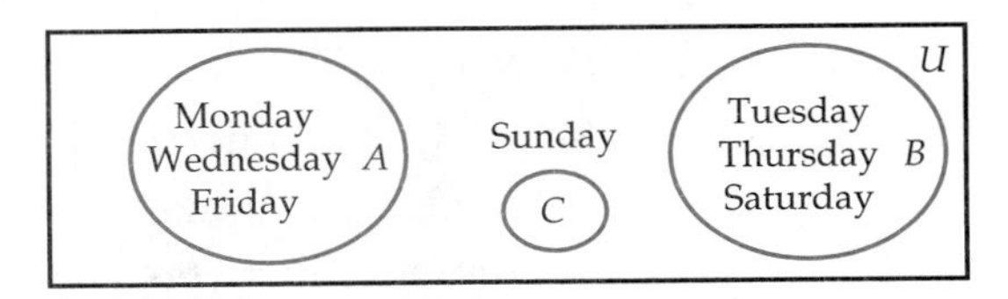

Figure 2–4

The definition of a subset is as follows.

Definition Subset

Set A is a **subset** of B, written "$A \subseteq B$," if and only if every element of A is also an element of B.

When A is not a subset of B, written "$A \not\subseteq B$," it means that A contains an element that is not in B.

Lesson Exercise 2.6

Which of the following sets is a subset of $\{4, 6, 8\}$?

(a) $\{4, 8\}$ (b) $\{3\}$ (c) $\{\ \}$ (d) $\{4, 6, 8\}$ (e) $\{4, 6, 8, 10\}$

You may wonder why $\{\ \}$ in the preceding exercise is a subset of $\{4, 6, 8\}$. In order for $\{\ \}$ to be a subset, every element in $\{\ \}$ must also be in $\{4, 6, 8\}$. Since there are no elements in $\{\ \}$, it is true that every element in $\{\ \}$ is also an element of $\{4, 6, 8\}$!

A subset of $\{4, 6, 8\}$ represents a possible choice. For example, you might choose $\{4, 6\}$ or $\{4, 6, 8\}$ from $\{4, 6, 8\}$. Another choice would be to select no elements, creating the subset $\{\ \}$.

You may be familiar with $\subset$, the notation for a **proper subset**. The expression "$A \subset B$" means that every element of A is also an element of B *and* that B contains at least one element that is *not* in A. So $\{1, 2\}$ is a proper subset of $\{1, 2, 3\}$, but $\{1, 2, 3\}$ is not a proper subset of $\{1, 2, 3\}$. The symbols $\subseteq$ and $\subset$ are analogous to $\leq$ and $<$ for relationships between numbers.

How does one distinguish between $\subseteq$ and $\in$? The symbol $\subseteq$ shows a relationship between *sets*, as in "$\{8\} \subseteq \{8, 10\}$." It shows that a *set* containing 8 can be formed using elements of $\{8, 10\}$. The symbol $\in$ shows that one object is a member of a set, as in "$8 \in \{8, 10\}$" or "$\{8\} \in \{\{8\}, \{10\}\}$."

Lesson Exercise 2.7

Fill in each blank with $\in$ or $\subseteq$.

(a) 12 _____ $\{10, 11, \ldots, 19\}$ (b) $\{2, 4\}$ _____ $\{2, 4, 6\}$

AN INVESTIGATION: SUBSETS

Lesson Exercise 2.8

Is there a pattern in the number of subsets that different-sized sets have?

(a) ***Understanding the Problem*** What are you supposed to do in this investigation?

(b) ***Devising a Plan and Carrying Out the Plan*** Complete the following table. (*Hint:* The empty set is a subset of every set.)

Set	List of Subsets	Number of Subsets
{1}	{ } {1}	2
{1, 2}		
{1, 2, 3}		
{1, 2, 3, 4}		

(c) *Explain* why {1, 2, 3, 4} has twice as many subsets as {1, 2, 3}.

(d) Based upon your response to part (a), a set with N elements appears to have _____ subsets.

(e) ***Looking Back*** In part (d), you used ______________ reasoning.

ANSWERS TO SELECTED LESSON EXERCISES

2.1 (a) $\in$ (b) $\notin$

2.2 The set of all positive numbers less than 0

2.3 (a) A, B, D, and F are equivalent, and C and E are equivalent.
(b) $A = F$ (c) No (d) Yes
(e)

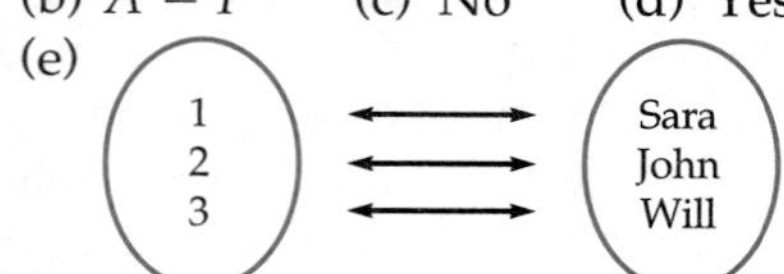

2.4 (a) Finite (b) Finite (c) Infinite

2.5 (a) {Sunday, Monday, Wednesday, Friday}
(b) The days you do not exercise

2.6 (a), (c), (d)

2.7 (a) $\in$ (since 12 is not a set) (b) $\subseteq$

2.1 HOMEWORK EXERCISES

Basic Exercises

1. Name five English words that refer to a set (for example, a coin *collection* or a *herd* of cattle).

2. Name two sets of which you are a member.

3. List the members of the set of whole numbers less than 10 that are even.

***4.** Let A be the set of odd numbers. Are the following true or false?
(a) $11 \in A$ (b) $10 \in A$
(c) $6 \notin A$ (d) $\{3\} \in A$

5. Which of the following would be an empty set?
(a) The set of purple crows
(b) The set of odd numbers that are divisible by 2

6. Which of the following represent equal sets?

$D = \{\text{orange, apple}\}$ $F = \{\text{apple, orange}\}$
$A = \{1, 2\}$ $G = \{1, 2, 3\}$
$E = \{\ \}$ $N = \varnothing$ $H = \{5, 6\}$

7. (a) Which of the following represent equivalent sets?

$D = \{\text{orange, apple}\}$ $F = \{\text{apple, orange}\}$
$A = \{1, 2\}$ $G = \{1, 2, 3\}$
$E = \{\ \}$ $N = \varnothing$ $H = \{5, 6\}$

(b) Show all one-to-one correspondences between D and H.

8. How can a first-grader use the idea of one-to-one correspondence to determine which of two sets of blocks has more blocks?

9. (a) True or false? If two sets are not equal, then they are not equivalent.
(b) If part (a) is true, give an example that supports it. If part (a) is false, give a counterexample.

10. Historical evidence reveals that some hunting tribes counted using equivalent sets. The tribal mathematician placed a rock in a pile for each hunter who left on an expedition. What do you think the mathematician did when the hunters returned (besides eating)?

11. Decide whether each set is a finite set or an infinite set.
(a) The set of whole numbers greater than 6
(b) The set of all the grains of sand on Earth
(c) The set of all perfect square numbers

12. Suppose the universal set $U = \{\text{math, science, English, Spanish, history, art}\}$, and consider two possible schedules, $A = \{\text{math, English, history}\}$ and $B = \{\text{Spanish, art}\}$. List the elements of $\overline{A}$ and $\overline{B}$.

13. Suppose the universal set $U = \{1, 2, 3, 4, 5, 6, 7, 8, 9, 10\}$, $A = \{1, 3, 5, 7, 9\}$, and $B = \{1, 2, 3, 4, 5\}$. List the elements of $\overline{A}$ and $\overline{B}$.

14. The Venn diagram at the top of the next column shows sets A, B, and U. Shade in the interior part of the rectangle that represents $\overline{A}$.

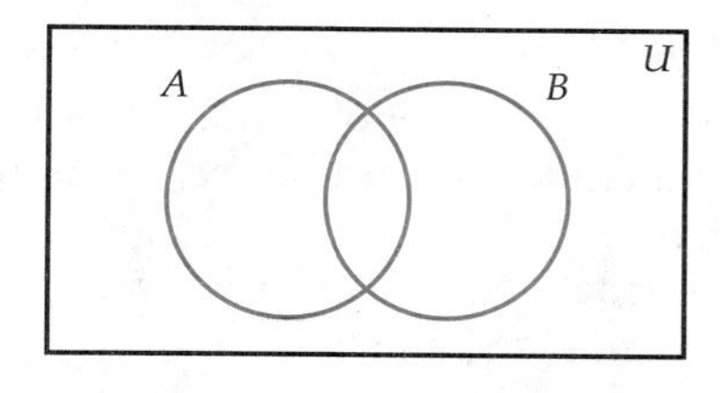

15. Which of the following sets are subsets of $\{1, 3, 5, 7\}$?
(a) $\{1, 3, 5, 7\}$ (b) $\{9, 11, 13, \ldots\}$
(c) $\{\ \}$ (d) $\{5\}$

***16.** Rewrite the following expressions using symbols.
(a) A is a subset of B.
(b) The number 2 is not a member of set T.

17. Let $A = \{3, 4, 5, 6, 7, 8\}$. Are the following true or false?
(a) $2 \notin A$ (b) $\{4\} \in A$
(c) $4 \in A$ (d) $\{4\} \subseteq A$

18. If $A \subseteq \varnothing$, then $A =$ ________.

19. Fill in each blank with $\in$ or $\subseteq$.
(a) $\{\ \}$ ________ $\{1, 3\}$
(b) Jean ________ {Tom, Jean}

20. The United Nations Security Council has 15 members, of which 5 are permanent and 10 are elected for 2-year terms. No measure can pass unless all 5 permanent members (the United States, the Russian Federation, the United Kingdom, France, and China) vote for it. Overall, nine votes are needed to pass a proposal.
(a) How many votes are needed from temporary members?
(b) You may refer to the 10 temporary members as $T_1, T_2, T_3, \ldots, T_{10}$. List five different winning coalitions that each contain exactly nine members.

21. A committee C = {Bobbie, Rupert, Sly, Jenny, Melissa} requires a majority vote to pass any new rules. Each of the following two sets contains a winning coalition with enough voters to pass a new rule. Complete the list of winning coalitions, using just the first letter of each person's name. (Make an organized list.)

{B, R, S} {B, R, S, J}

Extension Exercises

22. Decide which symbol, $\in$, $\notin$, $\subseteq$, or $\not\subseteq$, is equivalent to the underlined word in each sentence.
(a) German shepherds are dogs.
(b) Juanita is Spanish.
(c) Jane is not a skydiver.

23. How many one-to-one correspondences are possible between each of the following pairs of sets?
(a) Two sets, each having 2 members
(b) Two sets, each having 3 members
(c) Two sets, each having 4 members
(d) Two sets, each having N members

24. How can you create a one-to-one correspondence between the points of the triangle and the points of the circle?

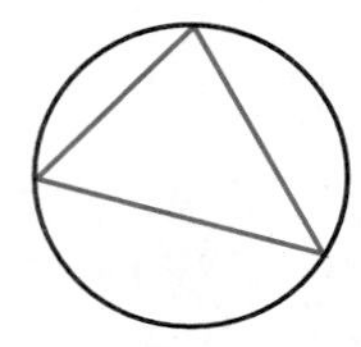

25. (a) A bag contains 2 different colored balls. How many balls must you select to be sure of getting at least 2 of one color?
(b) A bag contains 2 different colored balls. How many balls must you select to be sure of getting at least 3 of one color?
(c) A bag contains 2 different colored balls. How many balls must you select to be sure of getting at least 4 of one color?
(d) A bag contains 2 different colored balls. How many balls must you select to be sure of getting at least N of one color?

26. (a) A bag contains 3 different colored balls. How many balls must you select to be sure of getting at least 2 of one color?
(b) A bag contains 3 different colored balls. How many balls must you select to be sure of getting at least 3 of one color?
(c) A bag contains 3 different colored balls. How many balls must you select to be sure of getting at least 4 of one color?
(d) A bag contains 3 different colored balls. How many balls must you select to be sure of getting at least N of one color?

27. A bag contains K different colored balls. How many balls must you select to be sure of getting at least N balls of one color?

28. Set A is **infinite** if and only if it can be put into a one-to-one correspondence with a proper subset of itself. For example, {0, 1, 2, 3, . . .} is infinite because it can be put into a one-to-one correspondence with its proper subset, {10, 11, 12, 13, . . .}.

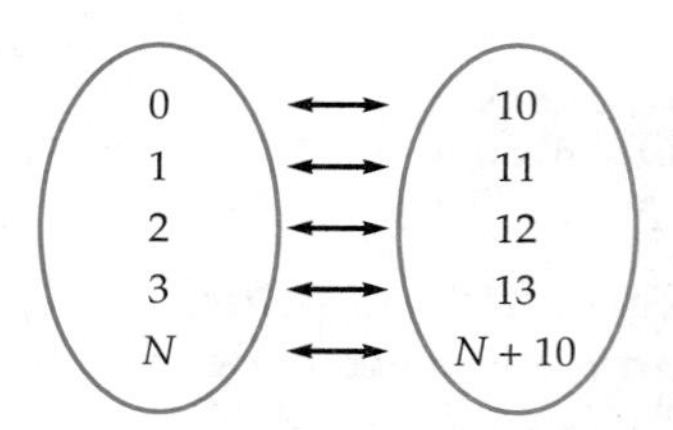

Show that the following sets are infinite.
(a) {0, 2, 4, 6, 8, . . .}
(b) {20, 21, 22, . . .}

29. Georg Cantor was the first to apply set theory to infinite sets. This led to some strange and surprising results.
(a) Make a conjecture about which set has more elements, W = {0, 1, 2, 3, . . .} or E = {0, 2, 4, 6, . . .}.
(b) Cantor reasoned that two equivalent infinite sets (like finite sets) are those that can be put into a one-to-one correspondence. How did he match the members of W and E in a one-to-one correspondence to show that they are equivalent sets?

30. (a) Make up a nonempty set, and list all possible subsets.
(b) How does the number of subsets with an even number of elements compare to the number of subsets with an odd number of elements? (*Note:* 0 is an even number.)
(c) Repeat parts (a) and (b) for a different nonempty set.
(d) Propose a generalization of your results.

31. Write a verbal description of each set.
(a) {4, 8, 12, 16, . . .}
(b) {3, 13, 23, 33, . . .}

2.2 OPERATIONS ON TWO SETS

Finding the complement of a set is an operation on one set that produces another set. Other set operations act on two sets to produce another set, just as the operation of addition on two numbers, such as 2 + 3, results in another number, 5. This section covers three set operations: intersection, union, and Cartesian products.

INTERSECTION AND UNION

Should Pam and Matt get married? Pam likes skydiving, bronco busting, and chess. Matt likes Sumo wrestling, mountain climbing, and chess. What interests do they have in common?

If A is the set of Pam's interests and B is the set of Matt's interests, the answer to the question is the intersection of sets A and B (members of A *and* B), written $A \cap B$. The only member of $A \cap B$ is "chess." "And" is a key word suggesting intersection. In everyday language and in mathematics, "and" indicates that *both* conditions must be true.

Definition **Intersection**

The **intersection** of sets A and B, written $A \cap B$, is the set containing the elements that are in both A and B.

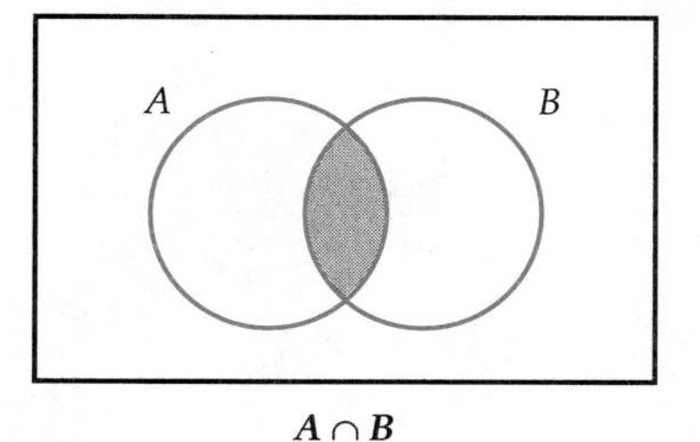

Figure 2–5

The shaded part of the set picture in Figure 2-5 is $A \cap B$. If A and B have no elements in common, then $A \cap B = \{\ \}$ or $\varnothing$.

Lesson Exercise 2.9

The math club membership M = {Joe, Sam, Li, Clara} and the science club membership S = {Juan, Li, Martha}.

(a) What is $M \cap S$?

(b) What characteristic do members of $M \cap S$ have?

(c) Does $S \cap M = M \cap S$?

Lesson Exercise 2.10

$A \cap A =$ ________.

Lesson Exercise 2.11

If $B \cap C = C$, what is the relationship between B and C?

When two or more sets are joined together, the new set they form is called the union of those sets. Getting back to Pam and Matt, Pam has a sports car and an airplane, and she has a bicycle that she and Matt received as an engagement gift. Set $C = \{\text{car, plane, bicycle}\}$. Matt has a motorcycle, a skateboard, and that same bicycle. Set $D = \{\text{motorcycle, skateboard, bicycle}\}$. If Pam and Matt marry, they will combine these vehicles to form the union of items belonging to either Pam *or* Matt *or both*, written $C \cup D$. The set $C \cup D = \{\text{car, plane, bicycle, motorcycle, skateboard}\}$.

Definition Union

The **union** of sets A and B, written $A \cup B$, is the set containing all elements that are either in A or in B or in both A and B.

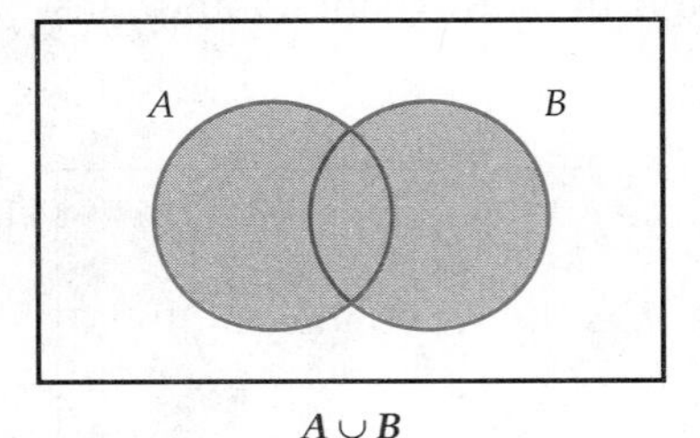

$A \cup B$

Figure 2–6

"Or" is a key word suggesting union. "Or" has a slightly different meaning in mathematics than it sometimes does in everyday speech. In everyday speech, "or" often means "one or the other (but not both)." In mathematics, "A or B" means "A or B *or both*." The shaded area in the set picture in Figure 2-6 is $A \cup B$.

Union and intersection are called **set operations** because they replace two sets with a third set, just as arithmetic operations replace two numbers with a third number. Two properties of set operations are discussed in the homework exercises.

Lesson Exercises 2.12 and 2.13 refer back to Lesson Exercise 2.9.

Lesson Exercise 2.12

(a) In Lesson Exercise 2.9, what is $M \cup S$?

(b) Does $M \cup S = S \cup M$?

Lesson Exercise 2.13

The math club (M) and the science club (S) have a joint activity with all their members. Which set represents the people attending?

(a) $M \cup S$ (b) $M \cap S$ (c) Neither (a) nor (b)

Lesson Exercise 2.14

If $B \cup C = C$, then how are sets B and C related?

Lesson Exercise 2.15

Fill in each blank with $\in$, $\cup$, $\cap$, or $\subseteq$.

(a) $\{3, 5\}$ ____ $\{6\} = \{3, 5, 6\}$

(b) 3 ____ $\{1, 2, 3\}$

(c) $\{3\}$ ____ $\{1, 2, 3\}$

TWO-SET VENN DIAGRAMS

All teachers are college graduates. *No* dirty clothes smell nice. Each of these statements can be illustrated with a two-set Venn diagram, as shown in Figure 2-7.

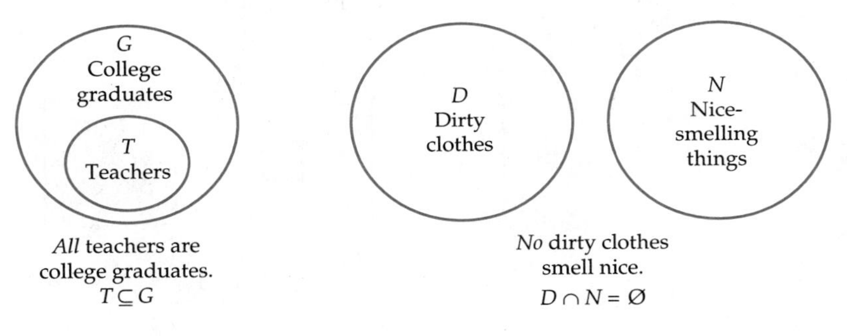

Figure 2–7

Lesson Exercise 2.16

Draw a two-set Venn diagram illustrating each of the following statements, and write the relationship between the two sets, using symbols.

(a) All fish are good swimmers.

(b) No U.S. senators are teenagers.

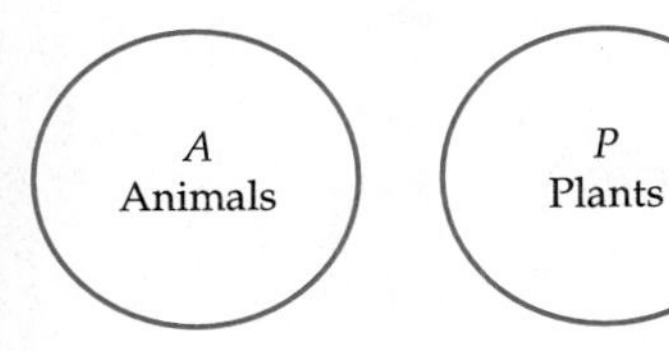

Figure 2–8

Lesson Exercise 2.17

(a) Write a sentence describing the set relationship shown in Figure 2-8.

(b) Use symbols to represent the relationship between sets A and P.

Lesson Exercise 2.18

Let U = {all students at Euclid College}, M = {students taking a math course}, and S = {students taking a science course.}.

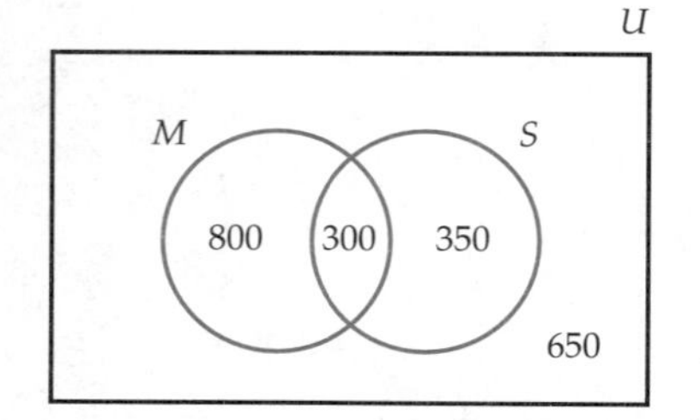

Figure 2–9

The Venn diagram in Figure 2-9 shows student enrollment in mathematics and science courses. How many students are

(a) taking mathematics but not science?
(b) taking mathematics and science?
(c) taking neither mathematics nor science?

ATTRIBUTE BLOCKS

As a teacher, you will sometimes use objects or models to introduce ideas to your students. Attribute blocks are manipulatives (objects) used in elementary school to study shapes, sets, and classification. Attribute blocks are so named because they have a variety of attributes: shape, color, size, and (sometimes) thickness. A typical set of attribute blocks is shown in Figure 2-10. You can buy a set or make one of your own.

Figure 2–10

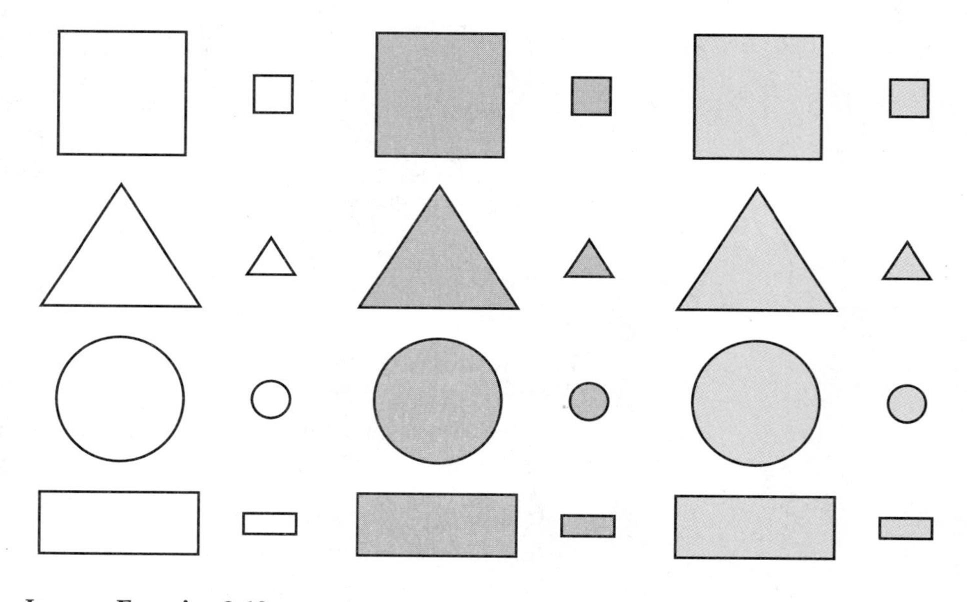

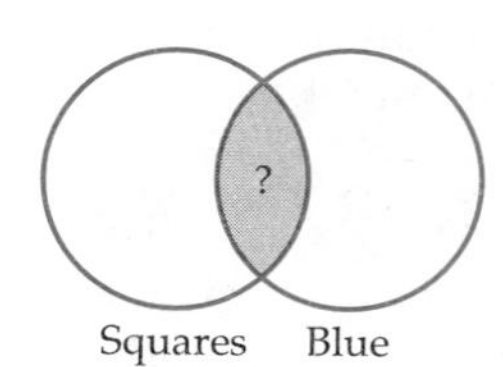

Figure 2–11

Lesson Exercise 2.19

Which attribute blocks in Figure 2-11 belong in the overlapping part?

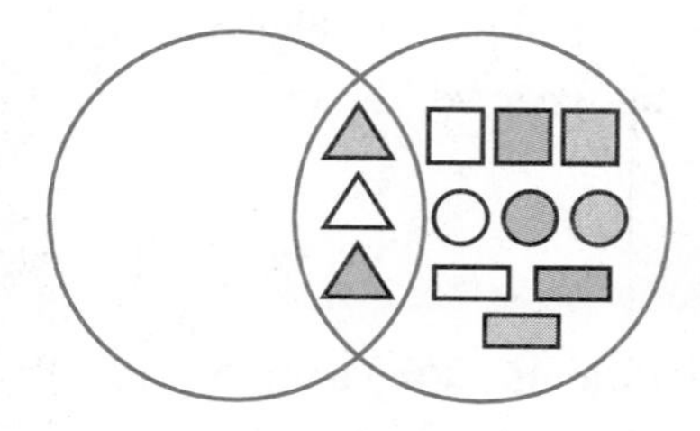

Figure 2–12

Lesson Exercise 2.20

Which attribute blocks in Figure 2-12 belong to the left-hand part of the circle on the left?

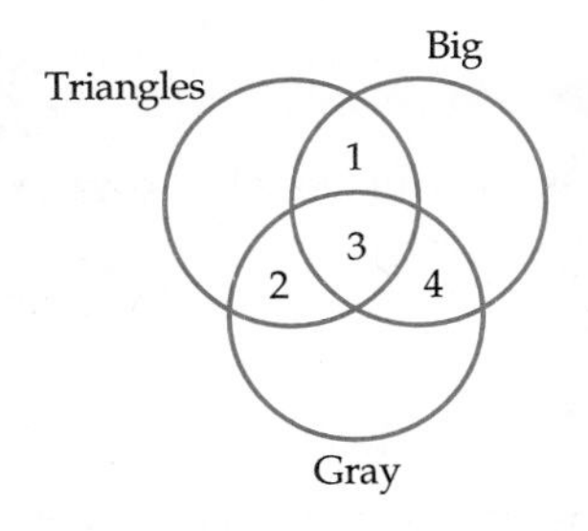

Figure 2–13

Lesson Exercise 2.21

Which attribute blocks in Figure 2-10 belong in Sections 1, 2, 3, and 4 of Figure 2-13?

CARTESIAN PRODUCTS

Suppose you have 3 different shirts and 2 different pairs of pants. If everything matches, how many different outfits can you create? The answer involves the Cartesian product, another set operation. The **Cartesian product** of two sets is the set that shows all the ways one can pair a member of one set with a member of the other set.

Lesson Exercise 2.22

(a) Suppose you have 3 shirts, $S = \{\text{purple, yellow, green}\}$, and 2 pairs of pants, $P = \{\text{red, blue}\}$. How many possible shirt-pant outfits could you create?

(b) Complete the following. The Cartesian product $S \times P$ = {(purple, red), (purple, blue), (,   ),(,   ), (,   ), (,   )}.

The Cartesian product $S \times P$ is a set of *ordered* pairs. The shirt color is written first and the pants color second. The ordered pair "(purple, red)" means "purple shirt and red pants," and it is in $S \times P$, but "(red, purple)," meaning "red shirt and purple pants," is not in $S \times P$.

In the preceding exercise, you saw that if S has 3 elements and P has 2 elements, then $S \times P$ has 6 elements. Consider some more examples and see if you can find a general pattern.

Lesson Exercise 2.23

(a) If A has 3 elements and B has 1 element, then $A \times B$ has ______ elements.

(b) If A has 3 elements and B has 3 elements, then $A \times B$ has ______ elements.

(c) If A has m elements and B has n elements, then $A \times B$ has ______ elements.

As the preceding exercise suggests, if A has m elements and B has n elements, then $A \times B$ has mn elements. Use this principle to do the next two exercises.

Lesson Exercise 2.24

Marissa has 10 shirts and 7 skirts. How many different shirt-skirt outfits can she create?

Lesson Exercise 2.25

Barry has 6 shirts, 5 pairs of pants, and 3 pairs of shoes. How many different shirt-pants-shoes outfits can he create?

ANSWERS TO SELECTED LESSON EXERCISES

2.9 (a) {Li}
(b) They are members of both the math and science clubs.
(c) Yes

2.10 A

2.11 $C \subseteq B$

2.12 (a) {Joe, Sam, Li, Clara, Juan, Martha}
(b) Yes

2.13 (a)

2.14 $B \subseteq C$

2.15 (a) $\cup$ (b) $\in$ (c) $\subseteq$ ({3} is a set)

2.16 (a)

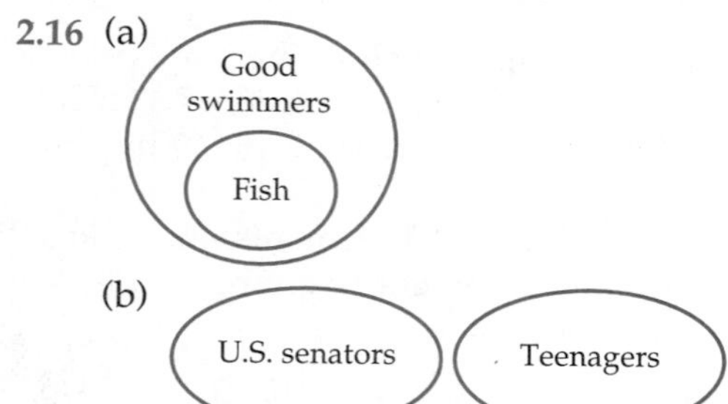

2.17 (a) No animals are plants.
(b) $A \cap P = \varnothing$

2.18 (a) 800 (b) 300 (c) 650

2.19 The small and large blue squares

2.20 The large gray, blue, and white triangles

2.21 1 = {big blue and white triangles},
2 = {small gray triangles},
3 = {big gray triangles},
4 = {big gray squares, circles, and rectangles}

2.22 (a) 6
(b) (yellow, red), (yellow, blue), (green, red), (green, blue)

2.23 (a) 3 (b) 9

2.24 $(10)(7) = 70$

2.25 $(6)(5)(3) = 90$

2.2 HOMEWORK EXERCISES

Basic Exercises

1. $A = \{1, 3, 5, 7, 9, 11\}$ and $B = \{3, 6, 9, 12, 15, 18\}$.
(a) $A \cap B =$ ________
(b) $A \cup B =$ ________
(c) Does $A \cap B = B \cap A$?
(d) Is $A \subseteq B$?

2. U is the set of all months of the year, T is the set of all 30-day months, and S is the set of school-vacation months (June, July, and August). What is
(a) $T \cap S$? (b) $T \cup S$?
(c) In parts (a) and (b), you used sets T and S, set definitions, and ________ reasoning.

3. Explain how an intersection of two streets is like an intersection of two sets.

4. The operation $\cap$ is **commutative** because $A \cap B = B \cap A$ for all sets A and B. Is the operation $\cup$ commutative?

5. (a) The operation $\cap$ is **associative** because $(A \cap B) \cap C = A \cap (B \cap C)$ for all sets A and B. Write a comparable equation for $\cup$ and decide whether the operation $\cup$ is associative.
   (b) Support your conclusion in part (a) by shading two Venn diagrams and comparing them.

*6. A has 3 members, and B has 2 members.
   (a) What is the largest number of members $A \cup B$ could have?
   (b) What is the smallest number of members $A \cup B$ could have?

7. A has a members and B has b members, where $a \geq b$.
   (a) What is the largest number of members $A \cup B$ could have?
   (b) What is the smallest number of members $A \cup B$ could have?

8. In an apartment complex of 30 apartments, 20 apartments receive the *New York Times* and 14 apartments receive the *Sporting News*.
   (a) What is the largest number of apartments that could receive *both* publications?
   (b) What is the smallest number of apartments that could receive *both* publications?

9. A sixth-grade class has 24 students. Four of them are in the chess club, 6 are in the Spanish club, and 2 are in both clubs.
   (a) Shade squares with a pencil to represent the students in the chess club.

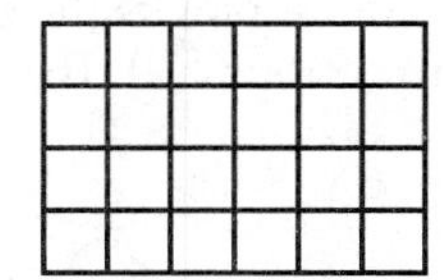

   (b) Shade squares with a pen to represent the students in the Spanish club (and remember that 2 are in both clubs).
   (c) Based on your diagram, how many students are in neither club?

10. In an apartment complex of 40 apartments, 25 have porches and 22 have storage rooms.
   (a) What is the largest number of apartments that could have a porch and a storage room?
   (b) What is the smallest number of apartments that could have a porch and a storage room?

11. Use symbols to represent the following statements.
   (a) E is the intersection of F and G.
   (b) The union of A and E is E.

12. A logician says, "Tomorrow will be rainy or cold." What is the logician possibly predicting for tomorrow?

13. Assume $2 \in (A \cup B)$. Which of the following statements could be true?
   (a) 2 is in A but not in B.
   (b) 2 is in B but not in A.
   (c) 2 is in both A and B.
   (d) 2 is in neither A nor B.

14. Fill in each blank with $\in$, $\cup$, $\cap$, or $\subseteq$.
   (a) $\{3\}$ ____ $\{2, 3, 6\}$
   (b) 8 ____ $\{7, 8, 9\}$

15. Fill in each blank with $\in$, $\cup$, $\cap$, or $\subseteq$.
   (a) 1 ____ $\{1, 2\}$
   (b) $\{1, 3, 5\}$ ____ $\{5\} = \{5\}$
   (c) $\{6, 7\}$ ____ $\{5, 6, 7\}$

16. Make up two sets A and B such that the number of elements in A equals the number of elements in $A \cap B$.

17. If possible, make up two sets such that the number of elements in A plus the number of elements in B is
   (a) less than the number of elements in $A \cup B$.
   (b) equal to the number of elements in $A \cup B$.
   (c) greater than the number of elements in $A \cup B$.

18. Draw a two-set Venn diagram to illustrate each of the following statements, and use symbols to represent the relationship between the two sets.
   (a) All musicians are creative people.
   (b) No square is a triangle.

19. Draw a two-set Venn diagram to illustrate each of the following statements, and use symbols to represent the relationship between the two sets.
(a) All parallelograms are quadrilaterals.
(b) No man is an island.

20. (a) Write a sentence describing the set relationship shown in the figure.

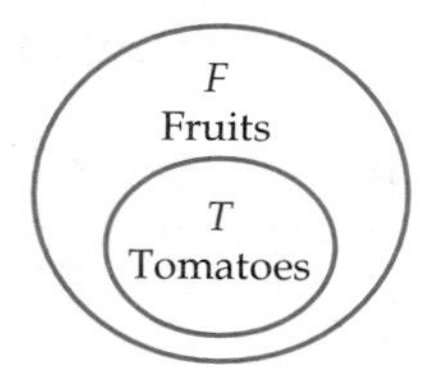

(b) Use symbols to represent the relationship between sets F and T.

21. A survey is taken of 500 adults. The results follow. Let U = {all adults}, T = {adults who watch television}, and R = {adults who read books}.

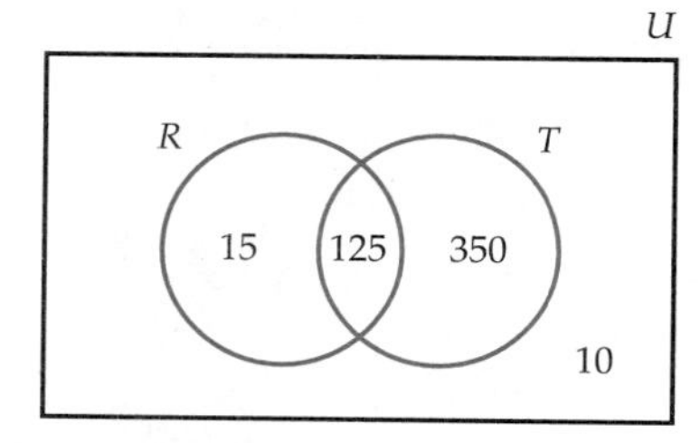

How many adults
(a) watch television and read?
(b) watch television but do not read?
(c) do not watch television and do not read?

22. A survey is taken of 300 college students about whether they eat breakfast and whether they have a job. Two hundred seventy eat breakfast, and 150 have a job. Twenty have a job and do not eat breakfast.
(a) Draw a two-set Venn diagram that displays these results.
(b) How many students eat breakfast but do not have a job?

23. A survey asked 100 students whether they play tennis or swim. Twenty-one play tennis, and 30 swim. Fourteen play tennis and swim.
(a) Draw a two-set Venn diagram that displays these results.
(b) How many students play neither sport?

24. Suppose P = {people who like prunes}, G = {people who like grapes}, and R = {people who like raspberries}. The regions of a Venn diagram of sets P, G, and R are labeled 1–8.

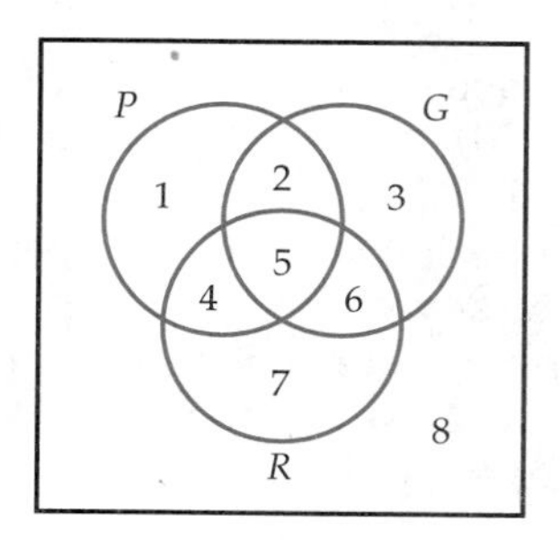

Which region or regions would contain
(a) your name?
(b) people who like raspberries and grapes?
(c) people who like raspberries and grapes but not prunes?
(d) people who like raspberries?
(e) people who like raspberries but not grapes?

25. Consider a set of attribute blocks; T = {triangles} and G = {gray}.
(a) What is $T \cup G$?
(b) What is $T \cap G$?

26. Using the set of attribute blocks described in the lesson, tell which ones would belong in the right-hand section of the circle on the right.

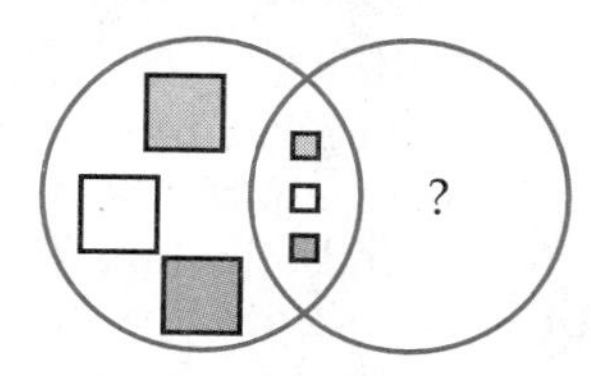

27. Which attribute blocks described in the lesson belong in Sections I, II, III, and IV?

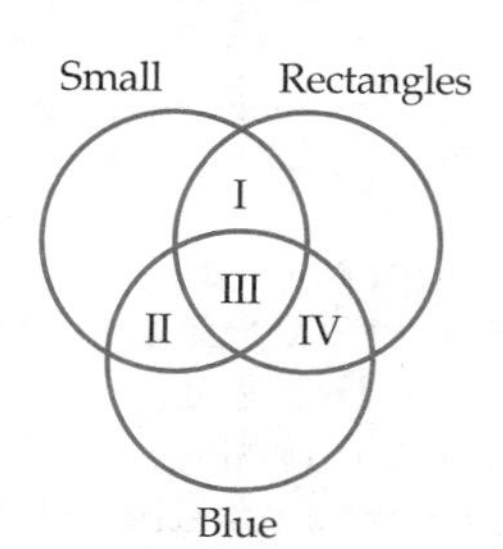

28. 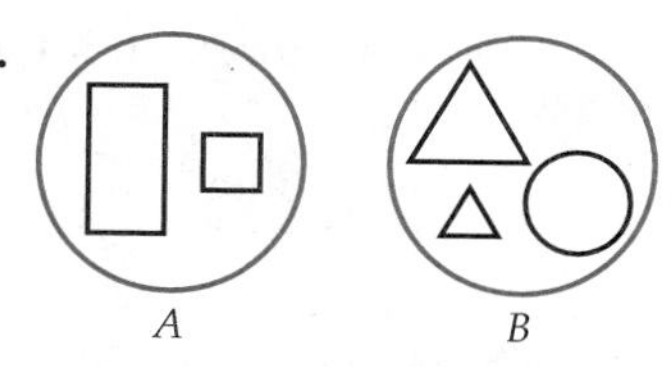

All the attribute pieces in set A have a certain attribute that the shapes in set B do not have. What attribute could it be?

29. The universal set contains all the attribute blocks. Define sets for each Venn diagram so that at least one piece goes in each section.

(a) 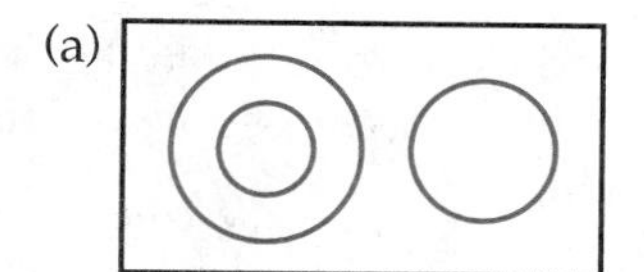

(b) 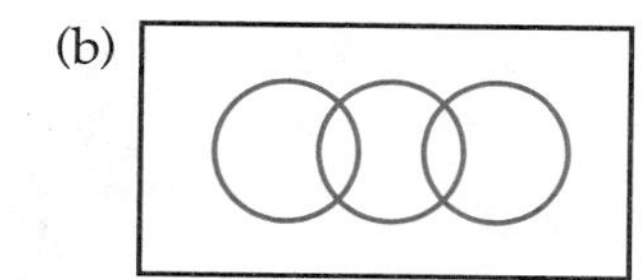

30. Teachers sometimes have children sort keys. Name two different attributes of keys that the children could use to sort them.

31. 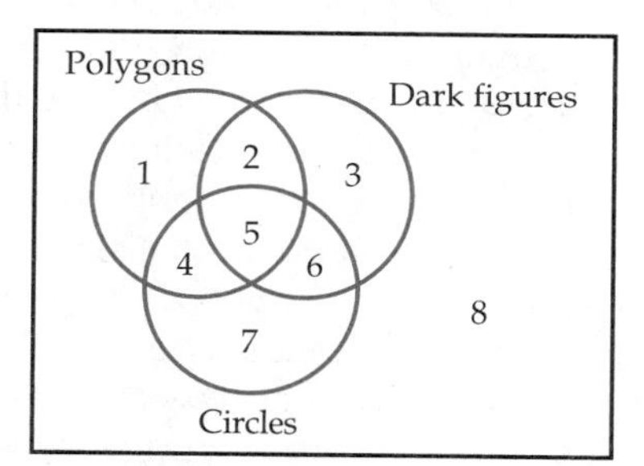

Which region would each figure go in?

(a) 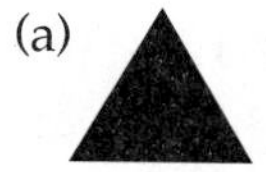 (b) 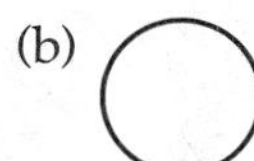 (c) 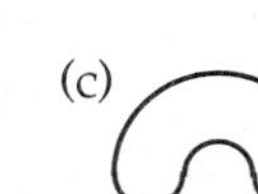 (d) 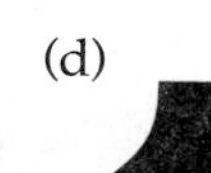

32. Following are words that do or do not qualify for set P.

Yes LATE PIE ROAM
No LOW HOUSE ENTREE

Which of the following is in set P?

(a) EITHER (b) QUIT (c) SHE (d) I

33. Following are figures that do or do not qualify for set P.

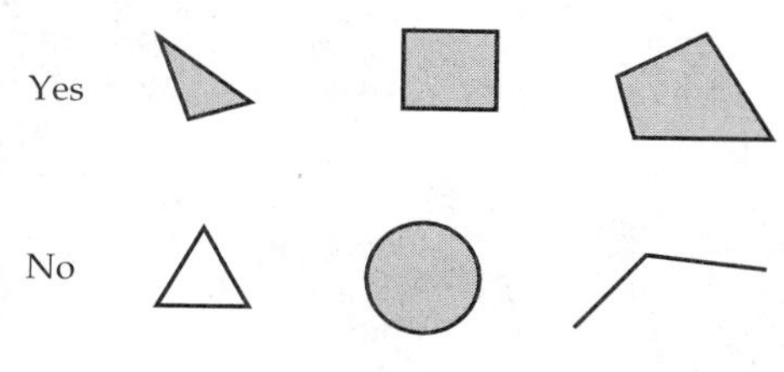

Which of the following is in set P?

(a) 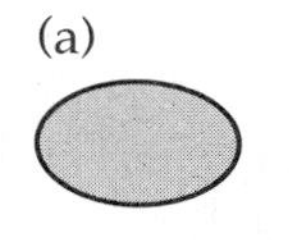 (b) 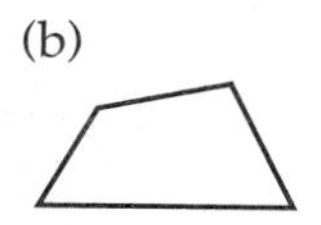 (c) 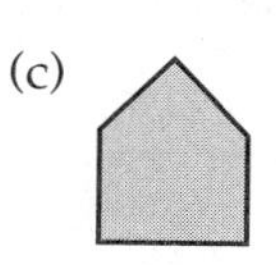 (d) 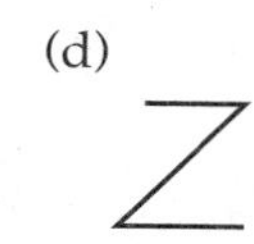

34. True or false?
(a) If $A \subseteq B$, then $B \subseteq A$.
(b) $A \cup \varnothing = \varnothing$
(c) $A \cap \varnothing = \varnothing$
(d) $0 \in \varnothing$

35. $A = \{1, 2\}$ and $B = \{0, 2, 4, 6\}$
(a) $A \times B =$ ________
(b) How many ordered pairs are in $A \times B$?

36. Create sets A and B so that $A \times B$ has 6 members.

37. Lois has 2 different skirts and 6 different shirts. How many possible shirt-skirt outfits can she create?

38. The Cartesian product $A \times B$ is {(Joe, Mary), (Joe, Elisa), (Marco, Mary), (Marco, Elisa)}. What are the elements of sets A and B?

39. Chef Robert is making a dinner with an appetizer, an entree, and a vegetable. He chooses the menu from 3 appetizers, 6 entrees, and 5 vegetables. How many different menus can he create?

Extension Exercises

40. Write a sentence or two telling the difference between the union of two sets and the intersection of two sets.

41.

Reprinted with special permission of NAS, Inc. © Field Enterprises, Inc., 1976.

(a) What do the percentages in the cartoon add up to?
(b) Why can the percentages add up to more than 100% and the data still be correct?

42. W is the set of overweight people, S the set of cigarette smokers, and E the set of people who exercise regularly. Describe the following sets in words.
(a) $W \cap S$ (b) $W \cup S$ (c) $W \cap S \cap E$
(d) $\overline{W}$ (e) $\overline{W} \cap \overline{S}$

43. U is the set of all college students, M the set of male college students, F the set of female college students, E the set of education majors, and Y the set of college students under 22 years of age.
(a) Describe $E \cap F$ in words.
(b) Describe $\overline{E} \cap \overline{Y}$ in words.
(c) $M \cup F =$ ________
(d) $M \cap F =$ ________

44. Following are the results of a survey about preferred alcoholic beverages.

	Don't Drink	Drink Only Beer or Wine	Drink Hard Liquor
Age	**(*N*)**	**(*B*)**	**(*H*)**
18–30 (*Y*)	40	42	18
31–55 (*M*)	26	46	28
Over 55 (*E*)	21	43	36

Using the letters in parentheses from the table to represent each set, tell how many people are in each of the following sets.
(a) Y (b) $E \cap B$ (c) $M \cap N$
(d) $Y \cup M$ (e) N (f) $Y \cup M \cup E$

45. Blood types can be illustrated by a Venn diagram. There are three antigens, A, B, and Rh, that may or may not be present in any human's blood. If you have the A antigen or the B antigen in your blood, that letter appears in your blood type. If you have neither the A nor the B antigen in your blood, the letter O appears in your blood type. If you have the Rh antigen in your blood, a plus sign (+) appears in your blood type. If you don't have Rh, a minus sign (−) appears in your blood type. (You probably wonder how anyone ever made up this system to type blood. I wonder, too.)
(a) The blood types can be shown nicely in a Venn diagram. I started it for you. Fill in the types in the regions that have question marks.

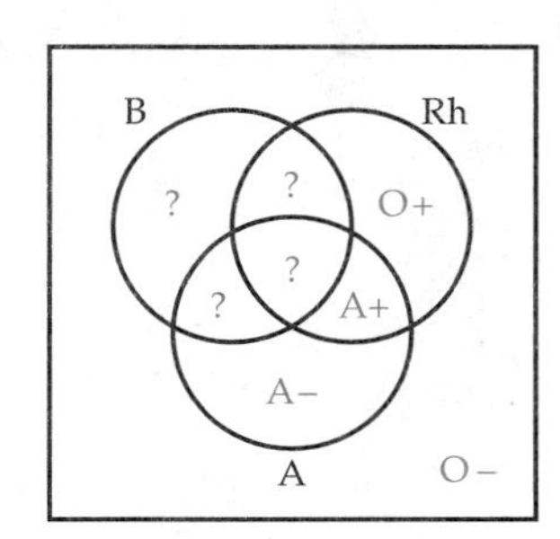

(b) The approximate percents of each blood type in the world are as follows.

	A	O	B	AB
+	37%	32%	11%	5%
−	6%	6.5%	2%	0.5%

Fill in these percents in the appropriate regions of a Venn diagram.

(c) The following chart can be used for transfusions.

Blood Type	Can Receive
O	O
A	A, O
B	B, O
AB	A, B, AB, O

Which blood type is the "universal donor"?

46. (a) At a college, 90% of all first-year students take English and 80% take mathematics. What is the minimum percentage of students who take both subjects?

(b) At the same college, 75% of the first-year students take social science. What is the minimum percentage of students who take all three subjects?

47. The shaded area represents

(a) $B \cap C$

(b) $(A \cap B) \cup (B \cap C)$

(c) $(B \cup A) \cap C$

(d) $(B \cap C) \cap B$

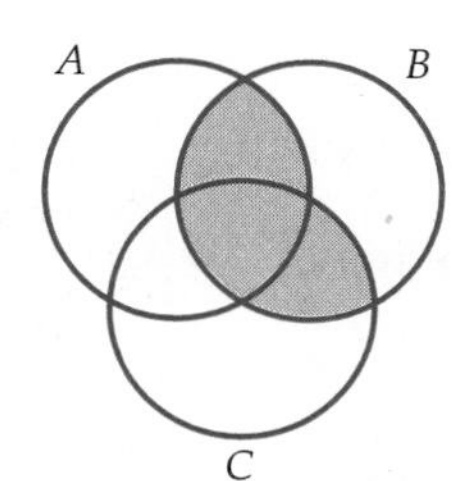

48. C is the interior of the circle, S is the interior of the square, and H is the interior of the hexagon. Shade $(C \cap S) \cup (C \cap H)$.

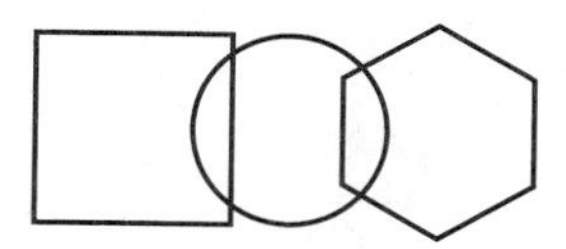

49. How can you create a one-to-one correspondence between the points of $\overline{AC}$ and the points of $\overline{AB} \cup \overline{BC}$?

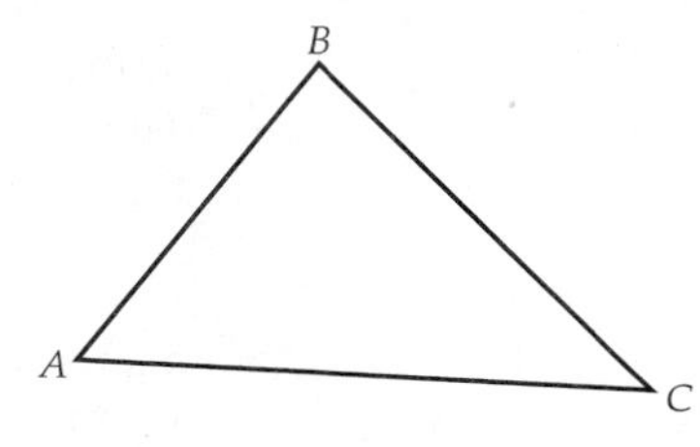

50. Assume that the following information is true.

$A \subseteq B$ $\quad B \subseteq C$ $\quad a \in A$

$b \in B$ $\quad c \in C$ $\quad d \notin B$

Which of the following must be true?

(a) $a \in C$ (b) $b \in A$ (c) $c \notin A$

(d) $A \subseteq C$ (e) $d \notin A$

51. The counting numbers 1, 2, 3, 4, 5, 6, and 7 can be placed in the 7 regions of a three-set Venn diagram so that the 4 numbers in each circle have the same sum. See if you can complete the diagram.

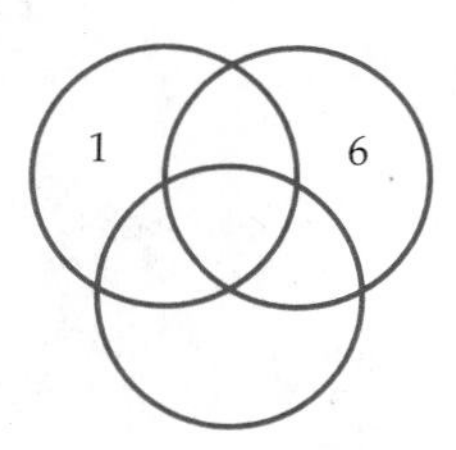

52. Which of the following is equal to $\overline{A \cap B}$?

(a) $\overline{A} \cup \overline{B}$ (b) $\overline{A} \cap \overline{B}$ (c) $A \cap B$

53. Which of the following is equal to $\overline{A \cup B}$? (*Hint:* Make up sets of elements for A, B, and U, or shade some Venn diagrams.)

(a) $\overline{A} \cup \overline{B}$ (b) $\overline{A} \cap \overline{B}$ (c) $A \cap B$

SUMMARY

Sets are an organizing concept in mathematics. In elementary-school arithmetic, one examines the operations and properties of sets of numbers. Geometric shapes can also be grouped into sets according to various characteristics, such as the number of sides. In statistics, data are collected in sets. In probability, one studies sets of possible outcomes.

A set can be represented by a list of elements, a verbal description, or a symbol. In studying two sets, one may want to compare their elements. If one set is contained in the other, then it is a subset of the other set.

One can operate on two sets to create a third set that is related to the other two. The intersection of two sets contains all objects that are in both sets. The union of two sets contains all objects that are in one set or the other set or both. The Cartesian product of two sets creates ordered pairs from the elements of the two sets.

"Children's geometric ideas can be developed by having them sort and classify models of plane and solid figures" (*Curriculum and Evaluation Standards*, NCTM, p. 49), such as attribute blocks. Attribute blocks can also be used within Venn diagrams to give a visual representation of relationships among sets.

STUDY GUIDE

To review Chapter 2, see what you know about each of the following ideas or terms that you have studied. You can also use this list to generate your own questions about the chapter.

The NCTM Curriculum Standards and Sets

Selected NCTM Curriculum Standards

The following standards come from the NCTM document.

- Relate physical materials, pictures, and diagrams to mathematical ideas.
- Relate everyday language to mathematical language and symbols.
- Recognize and apply deductive and inductive reasoning.
- Develop common understandings of mathematical ideas, including the role of definitions.

1. Describe how each standard listed relates to the material you studied in Chapter 2.
2. Select any current elementary-school mathematics textbook series and describe a sample lesson or exercise that illustrates each standard listed.

SETS IN ELEMENTARY SCHOOL

Although set language is used in all areas of mathematics, set theory is not studied in most elementary schools currently as it was in the 1960s and early 1970s. The following chart shows at what grade levels set topics typically appear in elementary-school mathematics textbooks.

Topic	Typical Grade Level in Current Textbooks
One-to-one correspondence	1
Venn diagrams	5, 6 (enrichment topic)

REVIEW EXERCISES

1. Draw a Venn diagram illustrating the following statement, and use symbols to represent the relationship between the two sets: "All elementary-school teachers are college graduates."

2. What attribute blocks belong in Section I?

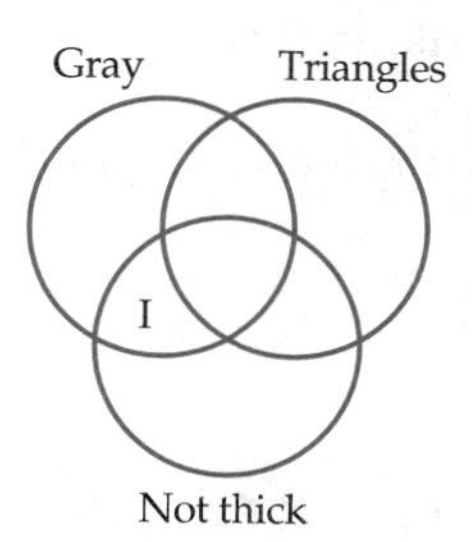

3. Let $R = \{31, 32, 33, 34, 35\}$, and let W be the set of whole numbers. Determine whether the following are true or false.
(a) $R \subseteq W$ (b) $10 \notin R$ (c) $R \cup W = W$

4. Fill in each blank with $\in$, $\cup$, $\cap$, or $\subseteq$.
(a) $\{7, 9\}$ ____ $\{7, 9, 10\} = \{7, 9\}$
(b) 8 ____ $\{3, 5, 8\}$

5. (a) How many subsets does $\{3, 5, 7\}$ have?
(b) If the universal set is $U = \{3, 4, 5, 6, 7\}$, create a set A that is equivalent to $\{3, 5, 7\}$.
(c) Show a one-to-one correspondence between set A and $\{3, 5, 7\}$.
(d) What is $\overline{A}$?

6. This semester, 1000 students registered for classes at a college; 300 of them signed up for math.
(a) What is the smallest number of students who could have registered for a course other than math?
(b) What is the largest number of students who could have registered for an English class?
(c) What is the largest number of students who could have registered for both math and English?

7. In which region does each of the following belong?

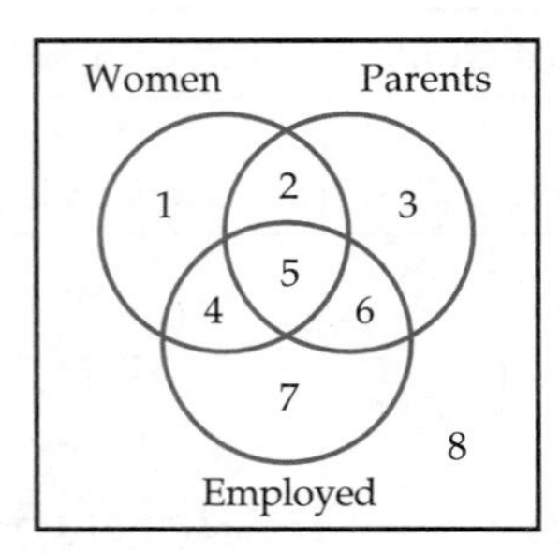

(a) A 7-year-old boy who plays when he's not in school
(b) A man with two young children who works
(c) You

8. Use the following information to determine how many households in a community have dogs.
1. Seven households have only cats.
2. Six households have no pet.
3. Two households have only dogs.
4. There are no pets in the community other than cats and dogs.
5. There are 18 households.

9. A partition of a set divides it into nonempty proper subsets. Two partitions of the set $\{1, 2, 3\}$ are shown.

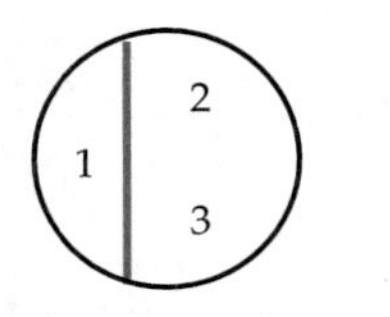

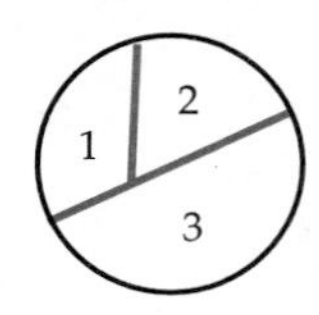

(a) How many different partitions are there of $\{1, 2, 3\}$?
(b) How many different partitions are there of $\{1, 2, 3, 4\}$?

SUGGESTED READINGS

Eves, H. *Great Moments in Mathematics After 1650.* Washington, D.C.: Mathematical Association of America, 1983.

Goodnow, J. *Moving On with Attribute Blocks.* Oak Lawn, Ill.: Creative Publications, 1989.

National Council of Teachers of Mathematics. *Topics in Mathematics for Elementary School Teachers.* 1964 Yearbook. Reston, Va.: NCTM, 1964.

Vilenkin, N. *Stories About Sets.* New York: Academic Press, 1969.

3

Whole Numbers

Many of us take whole numbers for granted, not fully appreciating that whole numbers have many significant applications and that they provide the basis for working with fractions, decimals, and integers. Whole numbers help us locate streets and houses. They help us keep track of how many bananas we have. And in a country with about 270 million people, they help us keep track of the federal budget and the amount of unemployment.

3.1 NUMERATION SYSTEMS

Figure 3–1

Today most countries use the simple, efficient base-ten place-value system. Simple as it appears, it took thousands of years to develop. Using place value and a base, one can express amounts in the hundreds or thousands with only a few digits! Big deal, you say? Try writing a number such as 70 using tally marks (Figure 3-1).

THE HISTORY OF NUMERATION SYSTEMS

People first used numbers to count objects. Before people understood the abstract idea of a number such as "four," they associated the number 4 with sets of objects such as four cows or four stars. A major breakthrough occurred when people began to think of "four" as an abstract quantity that could measure the sizes of a variety of concrete sets.

No one knows for sure, but the first numeration system was probably a tally system. The first nine numbers would have been written as follows.

| || ||| |||| ||||| |||||| ||||||| |||||||| |||||||||

A system such as this, having only one symbol, is very simple, but large numbers are difficult to read and write. Computation with large numbers is also slow.

Five thousands years ago, even before your professors were born, the ancient Egyptians improved the tally system by inventing additional symbols for 10, 100, 1000, and so on. Symbols for numbers are called **numerals**. Most of our knowledge about Egyptian numerals comes from the Moscow Papyrus (1850? B.C.) and the Rhind Papyrus (1650 B.C.).

Egyptian numerals are rather attractive symbols called hieroglyphics (Figure 3-2). Using symbols for groups as well as for single objects was a major advance that made it possible to represent large quantities much more easily. The ancient Egyptian numeration system uses ten as a **base**, whereby a new symbol replaces each group of 10 symbols. The Egyptian system is **additive**, since the value of a number is the sum of the values of the numerals. For instance, 99III represents 100 + 100 + 1 + 1 + 1, or 203.

Egyptian Numerals	Value
/ (staff)	1
∩ (heel bone)	10
9 (scroll or coil of rope)	100
(lotus flower)	1,000
(pointing finger)	10,000
(tadpole or fish)	100,000
(astonished man)	1,000,000

Figure 3–2

Lesson Exercise 3.1

(a) Speculate as to why the Egyptians chose some of the symbols shown in Figure 3-2.

(b) Write 672 using Egyptian numerals.

You can see from Lesson Exercise 3.1(b) that the early Egyptian system requires more symbols than our base-ten system. The Egyptians later shortened their notation by creating a different symbol for each number from 1 to 9 and for each multiple of 10 from 10 to 90, as shown in Figure 3-3. For example, 7 changed from IIIIIII to Z. This change necessitated memorizing more symbols, but it significantly shortened the notation.

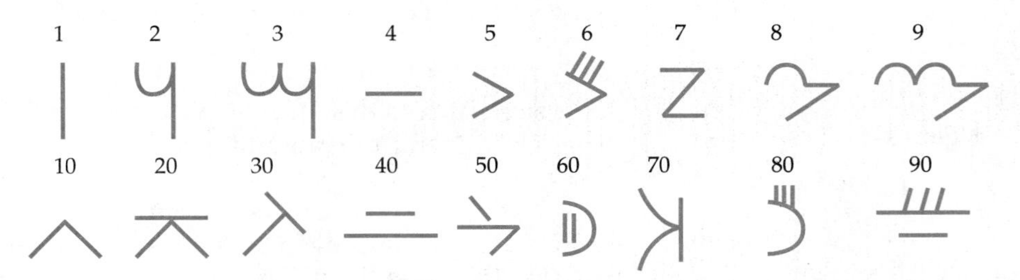

Figure 3–3

Lesson Exercise 3.2

Write 672 using the later Egyptian system.

The Babylonians made another significant improvement by developing a system based on **place value**, in which the value of a numeral changed according to its position. A place-value system reduces the number of different symbols needed.

The Babylonians used place value (for numbers greater than 59), a base of 60, and two symbols: < for 10 and Y for 1. The 3800-year-old tablet in Figure 3-4 shows Babylonian numerals.

The University Museum, University of Pennsylvania (neg. #69434).

Figure 3–4

Lesson Exercise 3.3

To write 672 in Babylonian numerals, use eleven 60's and twelve 1's. (*Note:* $672 \div 60 = 11$ R 12.) Since <Y represents 11 and <YY represents 12, the Babylonian numeral for 672 is <Y<YY. How does the number of symbols needed to write 672 compare with the Egyptian system used in the preceding exercises?

The Babylonians lacked a symbol for zero until 300 B.C. Before that time, it was impossible to distinguish between certain numerals. For example, 83 and 3623 are $1 \cdot 60 + 23$ and $1 \cdot 60^2 + 23$, both written as Y<<YYY. The Babylonians sometimes, but not always, put a space between groupings to indicate an empty place.

Zero serves two purposes: as a placeholder and as a symbol for nothing. Around 300 B.C., the Babylonians invented 𒑱 as a placeholder. The expression $1 \cdot 60^2 + 23$ was then written as 𒁹𒑱<<𒁹𒁹𒁹, where 𒑱 indicates that there are zero 60's. Even then, they did not use 𒑱 on the right, so 1 and 60 were both written as 𒁹.

Lesson Exercise 3.4

What base-ten numeral could represent the same number as <𒑱<<<𒁹?

Between them, the Egyptian and Babylonians developed most of the foundations of our numeration system: a base (ten), unique symbols for one through nine, place value, and a partial symbol for zero. The symbol 𒑱 worked like 0 within a number but was not used at the end of a number.

THE ROMAN NUMERATION SYSTEM

Roman numerals can still be found today in books, outlines, movies, and monuments.

Lesson Exercise 3.5

Where do you find Roman numerals

(a) in a book?

(b) in a movie?

The basic Roman numerals have the following base-ten values.

Roman numeral	I	V	X	L	C	D	M
Base-ten value	1	5	10	50	100	500	1000

To find the value of a Roman numeral, start at the left. Add when the symbols are alike or decrease in value from left to right. Subtract when the value of a symbol is less than the value of the symbol to its right. For example, XI is 11 and IX is 9. The Roman numeration system was developed between 500 B.C. and A.D. 100, but the subtractive principle was not introduced until the Middle Ages.

Lesson Exercise 3.6

(a) Write MCMLIX as a base-ten numeral.

(b) Write 672 (base ten) as a Roman numeral.

OUR BASE-TEN PLACE-VALUE SYSTEM

Between A.D. 200 and 1000, the Hindus and Arabs developed the base-ten numeration system we use today, which is appropriately known as the **Hindu-**

Arabic numeration system. The Hindus (around A.D. 600) and the Mayans (date unknown) were the first to treat 0 not only as a placeholder but also as a separate numeral.

Lesson Exercise 3.7

Which numeration system generally uses the smallest total number of symbols to represent large numbers?

(a) The tally system
(b) The early Egyptian system
(c) The Babylonian system
(d) The Roman system
(e) Our Hindu-Arabic system

The Hindu-Arabic numerals 0, 1, 2, 3, 4, 5, 6, 7, 8, and 9 were developed from 300 B.C. to A.D. 1522. The numerals 6 and 7 were used in 300 B.C., and the numerals 2, 3, 4, and 5 were all invented by A.D. 1530.

We use these numerals in a base-ten place-value (Hindu-Arabic) system to represent any member of the set of whole numbers, $W = \{0, 1, 2, 3, \ldots\}$. The Hindu-Arabic system has three important features: (1) a symbol for zero, (2) a way to represent any whole number using some combination of ten basic symbols (called **digits**), and (3) **base-ten place value**, in which each digit in a numeral, according to its position, is multiplied by a specific power of ten. Each place has ten times the value of the place immediately to its right.

The following diagram shows the first four place values for whole numbers, starting with the "ones" place on the far right.

*T*housands (10^3)	*H*undreds (10^2)	*T*ens (10^1)	*O*nes (1)
3	4	2	6

3426 can be written in expanded notation as

3 thousands, 4 hundreds, 2 tens, and 6 ones

$$\text{or } (3 \times 1000) + (4 \times 100) + (2 \times 10) + 6$$

$$\text{or } (3 \times 10^3) + (4 \times 10^2) + (2 \times 10^1) + 6$$

This **expanded notation** shows the value of each place.

Exponents offer a convenient shorthand. In base ten, $10^1 = 10$, $10^2 = 10 \cdot 10 = 100$, and $10^3 = 10 \cdot 10 \cdot 10 = 1000$. In the numeral 10^3, 10 is the base and 3 is the exponent. René Descartes invented the notation for an exponent in 1637. In an expression of numbers being multiplied such as $10 \cdot 10 \cdot 10$, each 10 is called a factor. (The term "factor" will be defined formally in Section 3.3.)

Definition **Base, Exponent, and a^n**

If a is any number and n is a counting number, then

$$a^n = \underbrace{a \cdot a \cdot \ldots \cdot a}_{n \text{ factors}}$$

where a is the **base** and n is the **exponent**.

Your calculator may have special keys for working with exponents such as $\boxed{x^2}$ (the squaring key), $\boxed{y^x}$ (the y-to-the-xth-power key), or $\boxed{\wedge}$ (the powering key). On some calculators, the powering key is $\boxed{x^y}$.

On a scientific calculator, one could compute 10^2 in the preceding exercise with the squaring key, by pressing $\boxed{10}$ $\boxed{x^2}$. The $\boxed{y^x}$ key can be used to compute powers other than 2. For example, to compute 5^3, one could press $\boxed{5}$ $\boxed{y^x}$ $\boxed{3}$ $\boxed{=}$.

On a graphing calculator, 5^3 is often computed by $\boxed{5}$ $\boxed{\wedge}$ $\boxed{3}$ $\boxed{\text{ENTER}}$. You may want to use the exponent keys on your calculator to do some of the homework exercises.

Lesson Exercise 3.8

A number has digits $\underline{A}\ \underline{B}\ \underline{C}$. Write this numeral in expanded notation.

Lesson Exercise 3.9

What is the difference between a place-value numeration system such as the Hindu-Arabic system and a system such as the Egyptian system that does not have place value?

Abacists vs. Algorists

Figure 3–5

You may be so accustomed to using place value that you don't give it a second thought. The Babylonians made some use of place value in their systems. Our own place-value system was first developed by the Hindus and was significantly improved by the Arabs.

Until the twelfth century, when paper became more readily available, the abacus was the most convenient tool for computations. Paper offered a new medium for doing arithmetic. A battle ensued from 1100 to 1500 between the *abacists*, who wanted to use Roman numerals and an abacus, and the *algorists*, who wanted to use Hindu-Arabic numerals and paper (Figure 3-5). By 1500 the algorists had prevailed.

MODELS FOR PLACE VALUE

To understand and communicate mathematical concepts, one must be able to relate concrete and pictorial models to abstract ideas. For example, children learn about base ten using a variety of concrete materials such as Dienes blocks. Dienes blocks for base ten come in the shapes shown in Figure 3-6.

Block	Name	Value
	Unit	1
	Long	10
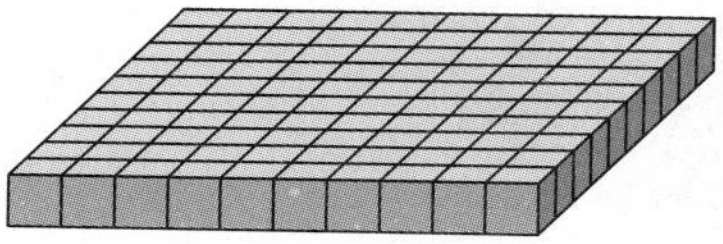	Flat	100
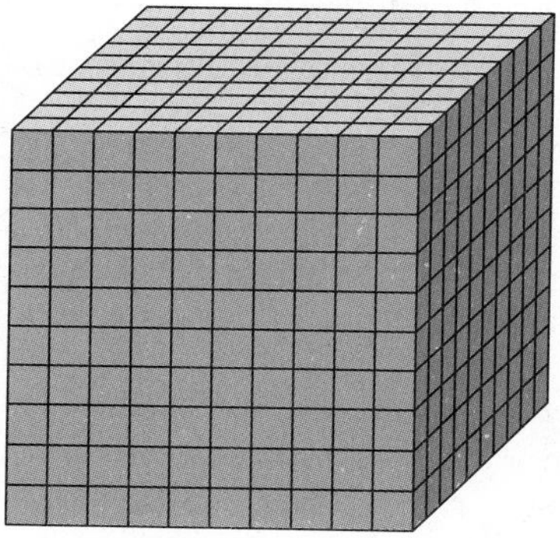	Block	1000

Figure 3–6

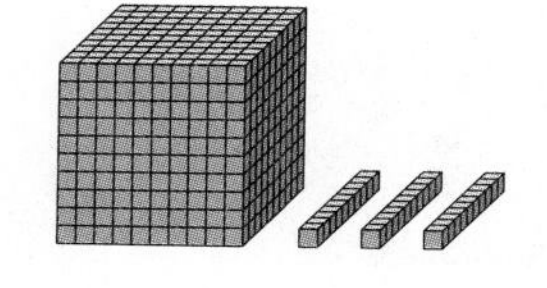

Figure 3–7

Lesson Exercise 3.10

What number is represented in Figure 3-7?

Another widely used place-value model is the abacus. A simplified abacus like the one in Figure 3-8 may be used in elementary school after children have had experience with more concrete models, such as base-ten blocks. The abacus in Figure 3-8 shows 1233.

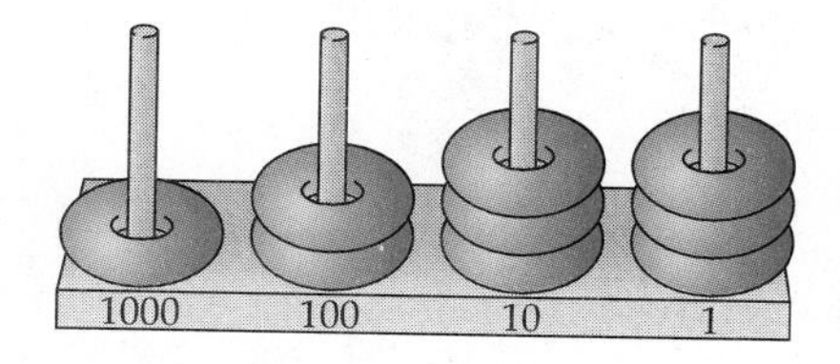

Figure 3–8

Lesson Exercise 3.11

What makes the abacus more abstract than base-ten blocks?

The abacus was developed over 2400 years ago as a calculating device. Today many people still use it for computation.

COUNTS AND COUNTING UNITS

Suppose 386,212 people attended a concert. It would be possible to count the number of people at the concert one by one. The number 386,212 is a count, and "people" is the counting unit. The **count** (or **cardinal number**) of a set is the number of elements in the set. Every count has a counting unit.

Lesson Exercise 3.12

Name the count and counting unit for

(a) the number of letters in your first name.

(b) the number of seats in your classroom.

MEASURES AND MEASURING UNITS

Jackie weighs 108 pounds. The number 108 is a **measure**, and pounds are the unit of measure. Unlike counting units such as people, **units of measure** such as pounds can be split into smaller parts. Other examples of units of measure are hours, meters, and square feet.

Lesson Exercise 3.13

Name the measure and unit of measure for your height in inches.

Lesson Exercise 3.14

Tell whether the number in each part is a count or a measure.

(a) 26 cows (b) 26 seconds (c) 26 miles

ROUNDING

24,000 Attend U2 Concert

If you recently read about attendance at a concert or looked up the population of the world, you probably read an approximate, or rounded, figure. A number is always rounded to a specific place (for example, to the nearest thousand). The following exercise concerns the rounding of whole numbers between 24,000 and 25,000.

Lesson Exercise 3.15

(a) If a whole number between 24,000 and 25,000 is closer to 24,000, what digits could be in the hundreds place?
(b) If a whole number between 24,000 and 25,000 is closer to 25,000, what digits could be in the hundreds place?

As you saw in the preceding exercise, numbers between 24,000 and 24,500 are closer to 24,000 than to 25,000, and they have 0, 1, 2, 3, or 4 in the hundreds place. In rounding to the nearest thousand, these numbers would be rounded *down* to 24,000. However, numbers that are between 24,500 and 25,000, such as 24,589, would be rounded *up* to 25,000 because they are closer to 25,000 than to 24,000. All of these numbers have 5, 6, 7, 8, or 9 in the hundreds place.

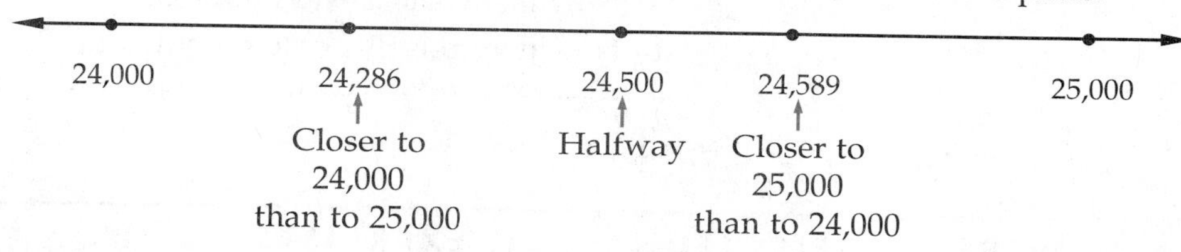

The number 24,500 is exactly halfway between 24,000 and 25,000, so it could be rounded either way. The rule used in our schools is to round up all such numbers that are exactly in the middle. All other whole numbers from 24,501 to 24,599 with 5 in the hundreds place should be rounded up to 25,000. Rounding 24,500 up to 25,000 enables children to learn one rule for all numbers with a 5 in the hundreds place.

Rounding Whole Numbers

The steps for rounding a whole number to a specific place are as follows.

1. Locate the place and then check the digit one place to its right.
2. Round *up* if the digit to the right is 5 or more, and round *down* if it is 4 or less.

Lesson Exercise 3.16

Exactly 386,212 people attended the Earaches' rock concert.

(a) Round this number to the nearest thousand.
(b) Use a number line to show that your answer is correct.

Lesson Exercise 3.17

A newspaper headline says "83,000 SEE KITTENS DEFEAT DUST BUNNIES." If this is a rounded figure, the actual attendance could have been between what two numbers?

Whole-number rounding will be used in estimation later in this chapter.

Rounding is not always done to "the nearest." The other two types of rounding are rounding up and rounding down. In buying supplies, one often has to round up to determine how much to purchase.

Lesson Exercise 3.18

I need 627 address labels for a company mailing. They come in groups of 100.

(a) How many labels will I buy?

(b) How does this illustrate rounding up?

ANSWERS TO SELECTED LESSON EXERCISES

3.1 (a) The Egyptians would have used a staff in traveling. The scroll relates to their development of papyrus. The lotus flower, tadpoles, and fish would have been found in a culture that developed around the Nile River. The astonished man is amazed by the amount of 1,000,000.
(b) 𓍢𓍢𓍢𓍢𓍢𓍢𓎆𓎆𓎆𓎆𓎆𓎆𓎆𓏺𓏺

3.2 𓍢𓍢𓍢𓍢𓍢𓍢ꓘЧ

3.3 It takes fewer distinct symbols and a smaller total number of symbols.

3.4 $10 \cdot 60^2 + 31 = 36{,}031$

3.5 (a) Page numbers in the preface
(b) Copyright date

3.6 (a) 1959 (b) DCLXXII

3.7 (e)

3.8 $(100 \times A) + (10 \times B) + C$

3.9 In a place-value system, the value of a numeral changes depending upon its location. In a system that does not have place value, the value of a numeral is always the same.

3.10 1030

3.11 The abacus distinguishes place value with position but not size. Base-ten blocks do use size.

3.14 (a) Count (b) Measure (c) Measure

3.15 (a) 0, 1, 2, 3, or 4 (b) 5, 6, 7, 8, or 9

3.16 (a) 386,000
(b) 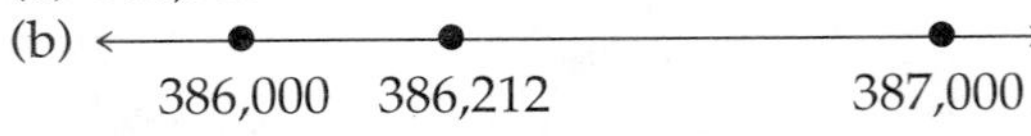

3.17 82,500 and 83,499 (inclusive) if attendance is rounded to the nearest thousand

3.18 (a) 700
(b) 627 is rounded up to the next 100.

3.1 HOMEWORK EXERCISES

Basic Exercises

1. Tell whether each word in quotation marks is used like a numeral (that is, as a symbol) or like a number (an idea).
(a) "Cat" is a three-letter word.
(b) A "cat" is a warm, furry animal.

2. Write the following Egyptian numerals as Hindu-Arabic numerals.
(a) 𓍢𓍢∩∩∩∩∩ (b) 𓆼∩IIII

3. Write 328 as an early Egyptian numeral.

4. (a) Do both of the following Egyptian numerals represent the same number?
∩II I∩I
(b) What difference between Egyptian and Hindu-Arabic numerals is suggested by part (a)?

5. What are three important characteristics of our base-ten numeration system?

6. Write a Roman numeral for
(a) 4 (b) 24 (c) 1998

7. Give the Hindu-Arabic numeral for
(a) XIV (b) XL
(c) MDCXIII (d) MCMLXIV

8. Name three places where you might find Roman numerals.

9. How many symbols would be needed to write 642 using each of the following?
(a) Tallies
(b) Early Egyptian numerals
(c) Roman numerals

10. Write 72 as
(a) an early Egyptian numeral.
(b) a Babylonian numeral.
(c) a Roman numeral.

11. Write the following base-ten numbers using expanded notation.
(a) 407 (b) 3125

12. A base-ten numeral has the digits $\underline{A}\ \underline{A}\ \underline{A}$. The value of the A at the far left is ____ times the value of the A at the far right.

13. A base-ten numeral has the digits $\underline{A}\ \underline{B}\ \underline{C}\ \underline{D}$. Write this numeral using expanded notation.

14. Rewrite each of the following as a base-ten numeral.
(a) $(8 \times 10^3) + (6 \times 10) + 2$
(b) $(2 \times 10^4) + (3 \times 10^3) + (4 \times 10^2) + 5$

15. What number is represented in each part?
(a)

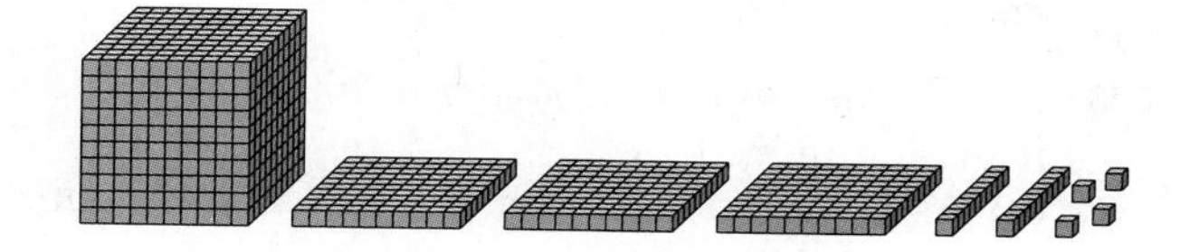

(b) 2 flats and 7 units

16. How would you represent 380 using base-ten blocks?

17. Chip trading is a common place-value model that uses colored chips and a mat. In base ten, the columns on the mat represent powers of ten. For example, 4523 is represented by 4 red, 5 green, 2 blue, and 3 yellow chips.

Red	Green	Blue	Yellow
4	5	2	3

(a) How would you represent 372?
(b) Ten blue chips can be traded in for one ________ chip.
(c) Which is more abstract, chip trading or base-ten blocks? Explain why.

18. What number is represented in the following picture?

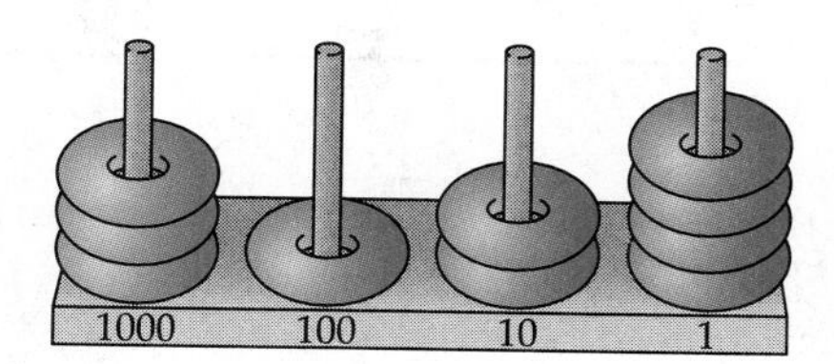

19. One way to represent 127 is with 12 tens and 7 ones. List three other ways.

20. How can you distinguish between counting units and units of measure?

21. Tell whether the number in each part is a count or a measure.
(a) 42 yards (b) 42 pens (c) 42 minutes

***22.** Is money a counting unit or a unit of measure? Consider whether the definition of a unit of measure in the text would apply to 25 dollars and to 25 cents.

23. Round each of the following numbers to the nearest hundred.
(a) $625, the price of lunch for two at La Vielle Chaussure
(b) $36,412, the one-game salary of a superstar athlete

24. The 1996 population of Cleveland was estimated at 2,311,000. Round this number to the nearest million.

25. Round 472
(a) up to the next hundred.
(b) down to the preceding hundred.
(c) to the nearest hundred.

26. About 8000 people attended a baseball game. If this figure has been rounded to the nearest thousand, the actual attendance was between what two numbers?

27. Give the place to which each number has been rounded.

	Original Number	Rounded Number
(a)	418	420
(b)	418	400
(c)	368,272	370,000

28. If the quarterback in the following cartoon is right, what is the greatest number of people that could be there watching?

29. You have $126 in a checking account. The automated cash machine will give you only $10 bills.
(a) What is the most money you can withdraw?
(b) Does this illustrate rounding up, rounding down, or rounding to the nearest?

30. Ball-point pens come in packs of 10. You want 23 pens for your class.
(a) How many pens will you order?
(b) Does this illustrate rounding up, rounding down, or rounding to the nearest?

31. Match the amount in each part with one of the following estimates.

4 40 400 4000
4,000,000 400,000,000

(a) Number of students in a typical elementary school
(b) Population of Chicago
(c) Number of miles an adult can walk in an hour
(d) Distance, in miles, from Boston to San Diego
(e) Population of North America

32. What is the largest number you can enter into your calculator?

Extension Exercises

33. Write a ten-digit numeral in which the first digit tells the number of zeroes in the numeral, the second digit tells the number of 1's, the third digit tells the number of 2's, and so on. (*Hint:* There are no 8's or 9's in the numeral.)

34. Consider the following problem. "A number is called a palindrome if it reads the same forward and backward; 33 and 686 are examples. How many palindromes are there between 1 and 1000 (inclusive)?"

(a) Devise a plan and solve the problem.

(b) Make up a similar problem and solve it.

35. Writing a time using hours, minutes, and seconds is comparable to using what base?

36. The Mayan numeration system used only three symbols.

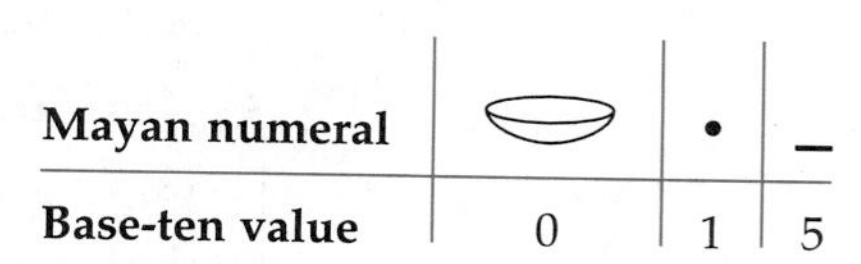

Mayan numeral		•	—
Base-ten value	0	1	5

They used combinations of these three symbols to write numerals for 0 through 19. For example, 8 was written as , and 19 was written as . Notation for whole numbers greater than 19 used vertical place value. The next vertical place value is 20. For example, represents eight 1's and five 20's, or 108, in base ten.

Write each Mayan numeral as a base-ten numeral.

(a) (b) (c)

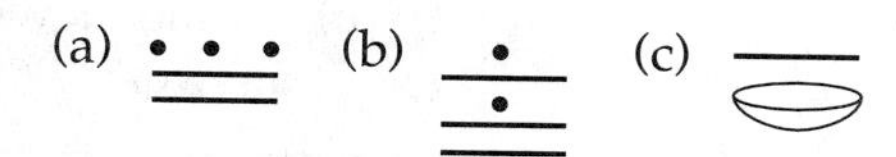

37. In the five-digit zip code, what do the different digits tell the post office?

38. What do the first three digits of the Social Security number indicate?

3.2 ADDITION AND SUBTRACTION OF WHOLE NUMBERS

Addition and subtraction involve more than computing $3 + 6$ or $63 - 27$. Computational skills are not useful unless you know when to use them in solving problems. Children should learn how to recognize situations that call for addition or subtraction.

ADDITION DEFINITION AND CLOSURE

In elementary school, teachers explain the meaning of addition by giving examples. Later, one can formally define important ideas such as addition. The idea of combining sets (that is, union) is used to define addition. An example is combining 2 fresh bananas with 3 older bananas to obtain a total of 5 bananas.

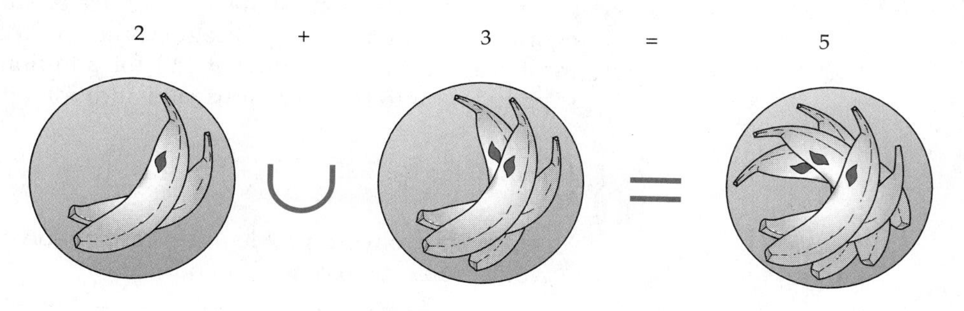

Definition **Whole-Number Addition**

If set A contains a elements, set B contains b elements, and $A \cap B = \emptyset$, then $a + b$ is the number of elements in $A \cup B$.

In the addition equation $a + b = c$, a and b are called **addends**, and c is called the **sum**. The German mathematician Widmann used the + sign for addition (and the − sign for subtraction) in his book published in 1498.

Lesson Exercise 3.19

(a) If two whole numbers are added, is the result always a whole number?

(b) Would a whole-number addition problem ever have more than one answer?

Your responses to the preceding exercise should confirm that the sum of two whole numbers is a unique whole number. For example, 3 + 5 equals 8, a unique whole number. This is called the closure property of addition of whole numbers.

The Closure Property of Addition of Whole Numbers

If a and b are whole numbers, then $a + b$ is a unique whole number.

Closure requires that an operation on two members of a set results in a unique member of the same set, "unique" meaning that the result is the only possibility.

CLASSIFYING ADDITION APPLICATIONS

Research with children indicates that the most difficult aspect of word problems in arithmetic is choosing the correct operation. How do children learn which operation to use?

Consider addition. Many applications of addition fall into one of two categories. If children learn to recognize these two categories, they will recognize most real-world problems that call for addition. The following exercise will introduce you to the classification of addition applications.

Lesson Exercise 3.20

Make up two word problems that can be solved by adding 3 and 5. Compare your problems to others in your class.

Do your word problems fit into one of the following categories? Many addition applications do.

1. **Combine Sets** Combine two nonintersecting sets and find out how many objects are in the new set.

   Example: Tina has 3 cute beagle puppies, and Maddy has 5. How many puppies do they have altogether (Figure 3-9)?

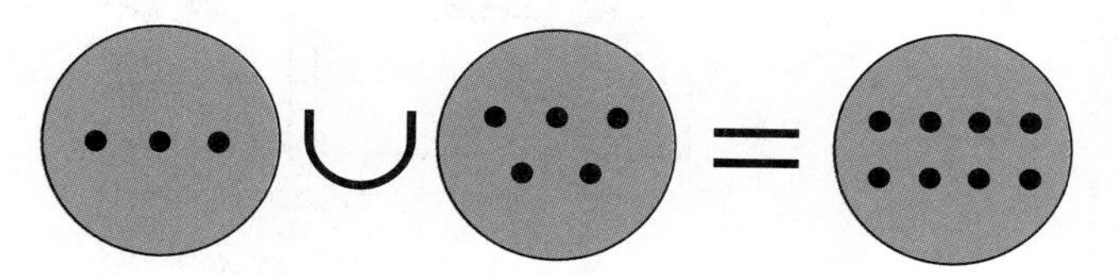

Figure 3–9

2. **Combine Measures** Combine two measures of the same type and find the total measure.

   Example: Laura is going to Paris for 3 days and to London for 5 days. How long is the whole trip (Figure 3-10)?

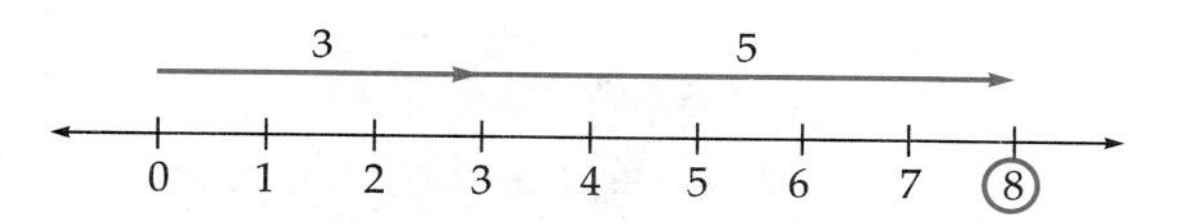

Figure 3–10

Numbers are used to count and to measure. The only difference between categories 1 and 2 is the difference between "sets" and "measures." Set problems involve objects that *can be counted* with whole numbers; objects such as apples, dogs, or people. Measure problems involve measures such as distance, weight, and time, which are not limited to having whole-number values. Most arithmetic applications can be classified as either set problems or measure problems.

Not all addition applications fit into the "combine sets (counts)" or "combine measures" categories, but these two categories identify specific characteristics of addition applications. Children learn to recognize these two addition categories in elementary school. Although your future students may not learn the terms "combine sets" and "combine measures," these categories will help you (the teacher) to identify more specifically what your students do or do not understand. For example, a child may know what $2 + 4$ equals but may not be able to recognize combining sets as an addition situation.

Lesson Exercise 3.21

Classify the addition situation in the cartoon in Figure 3-11.

By permission of Johnny Hart and Creators Syndicate, Inc.

Figure 3–11

Lesson Exercise 3.22

Classify the following addition application. "Hy Climber climbs to an altitude of 3600 ft. If he climbs 400 more feet, what is his new altitude?"

Teachers first model the concepts of combining sets and combining measures for children using manipulatives (objects) and pictures. Manipulatives and pictures establish connections between addition and everyday applications of it.

EXAMPLE 3.1

Make a drawing that represents $2 + 3 = 5$ (Lesson Exercise 3.21) using a set of counters.

Solution

■

Lesson Exercise 3.23

Show how to solve Lesson Exercise 3.22 using a number line, and write the addition equation it illustrates.

As Example 3.1 and Lesson Exercise 3.23 suggest, problems that involve combining sets can be represented with a set of identical objects, and problems that involve combining length measures can be represented with a number line.

A major intellectual breakthrough occurred when people realized that addition applies to all kinds of objects. The equation $3 + 4 = 7$ describes a relationship that applies to sets of people, apples, and rocks. The power of mathematics lies in the way we can apply basic ideas such as addition to such a wide variety of situations.

Using addition with many different kinds of sets also has its drawbacks. Numbers emphasize similarities between situations. People must then make allowances for differences in the *qualities* of the living things or objects being analyzed. Although 3 peaches plus 4 peaches is the same mathematically as 3 bombs plus 4 bombs, peaches and bombs are rather different.

Lesson Exercise 3.24

Angela has 2 apples and buys 3 more. Now she has 5 apples. What information about the applies is ignored in this description?

SUBTRACTION DEFINITION

In school, children learn how subtraction is related to addition.

Lesson Exercise 3.25

Name an addition equation that is equivalent to $6 - 2 = 4$.

In answering Lesson Exercise 3.25, you used the following definition to convert a subtraction equation to an equivalent addition equation.

Definition **Whole-Number Subtraction**

For any whole numbers a and b, $a - b = c$ if and only if $a = b + c$ for a unique whole number c.

In the subtraction equation $a - b = c$, c is called the **difference**. You can illustrate the relationship between addition and subtraction with a diagram.

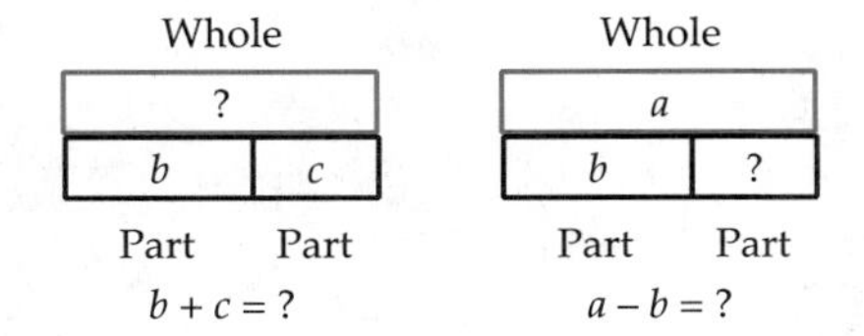

Because of this relationship, subtraction is called the inverse operation of addition. Although subtraction is defined in terms of addition, most children first understand subtraction as the process of removing objects from a set.

CLASSIFYING SUBTRACTION APPLICATIONS

Children can learn to recognize applications that call for subtraction by recognizing the common categories of these applications.

Lesson Exercise 3.26

Make up two word problems that can be solved by subtracting 2 from 6. Try to create two different types of problems. Compare your problems to others in your class.

Do your problems fit into any of the following categories?

1. **Take Away Sets** Take away some objects from a set of objects.

   Example: Dan had 6 pet fish. Two of them died. How many are left (Figure 3-12)?

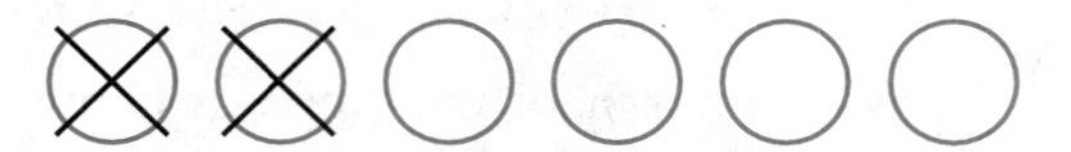

Figure 3–12

2. **Take Away Measures** Take away a certain measure from a given measure.

   Example: A lumberjack has 10 ft of rope. He uses 6 ft of rope to tie some logs. How much rope is left (Figure 3-13)?

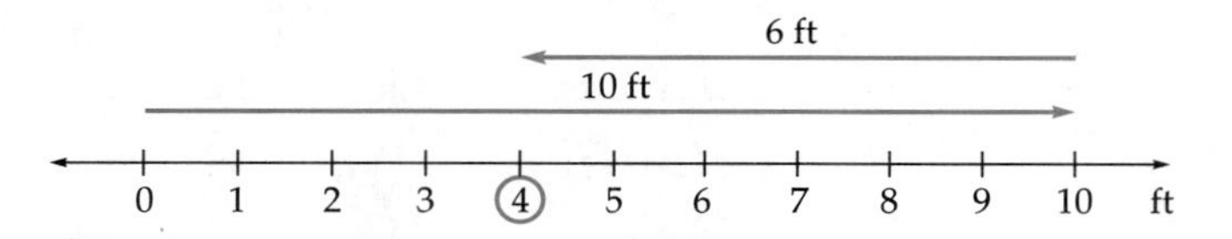

Figure 3–13

3. **Compare Sets** Find how many more objects one set has than another.

   Example: Maria has 6 cuddly kittens and Pete has 2. How many more kittens does Maria have than Pete (Figure 3-14)?

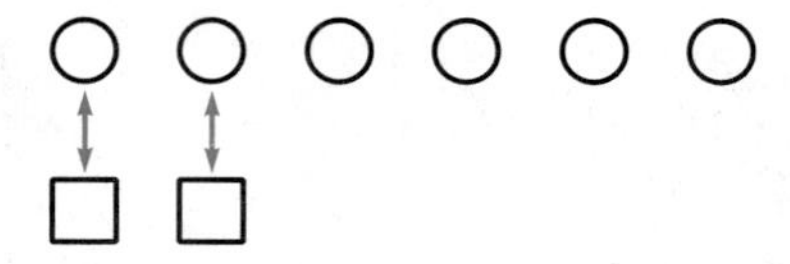

Figure 3–14

4. **Compare Measures** Find how much larger one measure is than another.

   Example: Sue is 56 inches tall. Twelve months ago, she was 51 inches tall. How much has Sue grown in the last year?

56 in.	
51 in.	?

Missing addend problems such as $2 + \square = 6$ are abstract examples of category 3 or 4.

Once they learn to recognize these four types of subtraction situations, children usually know when to use subtraction in solving problems.

In Lesson Exercises 3.27 and 3.28, classify each subtraction application.

Lesson Exercise 3.27

Hy Climber climbs to an altitude of 3600 ft. The summit is at 4000 ft. How much farther does he have to climb?

Lesson Exercise 3.28

The following is a true story. I use to have 3 umbrellas. I no longer have 2 of them. I lost 1 on the subway, and the other suffered an early internal breakdown. How many umbrellas do I have now?

The take away and compare models of subtraction are usually introduced with manipulatives and pictures in elementary school.

Lesson Exercise 3.29

Show how to solve Lesson Exercise 3.27 using a number line, and write the arithmetic equation it illustrates.

AN INVESTIGATION: NUMBER CHAINS

Lesson Exercise 3.30

A **number chain** is created by adding and subtracting. The number in each square in Figure 3-15 is the sum of the numbers that are next to it on both sides.

Figure 3–15

What is the relationship between the far-left-hand and far-right-hand numbers on the number chain in Figure 3-16?

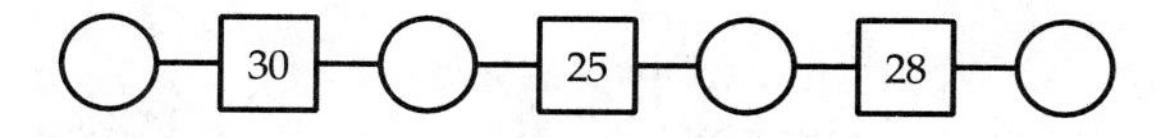

Figure 3–16

(a) ***Understanding the Problem*** If you put 10 in the far-left-hand circle, what number results in the far-right-hand circle?
(b) ***Devising and Carrying Out a Plan*** Try some other numbers in the far-left-hand circle, and fill in the remaining circles. Look for a pattern.
(c) Make a generalization of your results from part (b).
(d) ***Looking Back*** Did you use inductive or deductive reasoning in part (c)?
(e) Start with X in the far-left-hand circle and fill in all the other circles to prove your conjecture in part (c).

Lesson Exercise 3.31

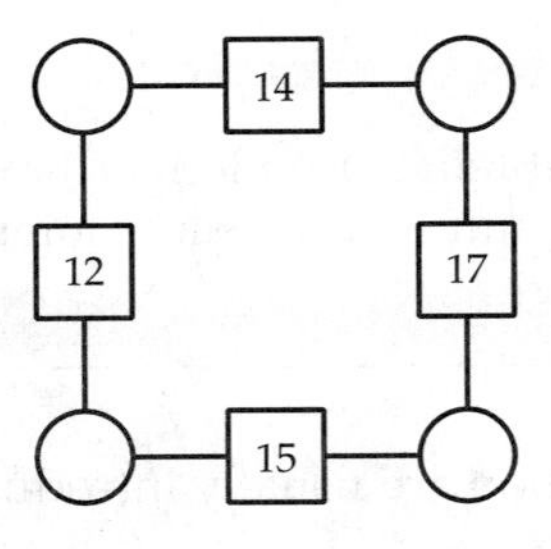

Figure 3–17

Consider the number chain in Figure 3-17.

(a) Fill in the circles with numbers that work.
(b) If possible, find a second solution.
(c) Start with X in the upper left-hand circle and use algebra to fill in all the circles.
(d) What does your answer in part (c) tell you about the relationship between the numbers in opposite corners?

ANSWERS TO SELECTED LESSON EXERCISES

3.19 (a) Yes (b) No

3.20 See the examples that follow Lesson Exercise 3.20.

3.21 Combine sets

3.22 Combine measures

3.23 3600 ft + 400 ft = 4000 ft

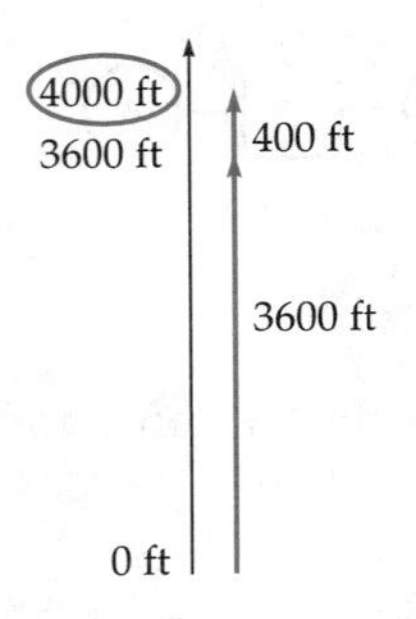

3.24 Are they fresh? What kind of apples are they? How much did she pay? What will she do with them? Were they sprayed with pesticides?

3.25 2 + 4 = 6

3.26 See the examples that follow Lesson Exercise 3.26.

3.27 Compare measures

3.28 Take away sets

3.29 4000 ft − 3600 ft = 400 ft

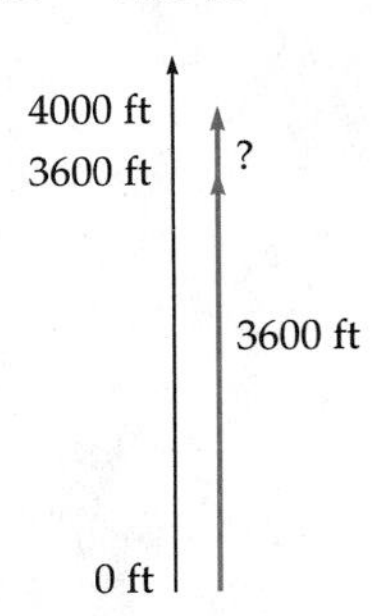

3.30 (a) 23 (d) Inductive

3.2 HOMEWORK EXERCISES

Basic Exercises

1. Which of the following sets are closed under addition?
   (a) {2, 4, 6, 8, . . . }
   (b) {1, 3, 5, 7, . . . }
   (c) {0, 1}
2. Which category of addition, combine sets or combine measures, is illustrated in the following problem? "Mary sold 61 apples yesterday and 48 apples today. How many apples did she sell in the last two days?"
3. The Samuels built a house and lived in it for 8 years. They sold it to the Wangs, who have now lived there for 11 years. How old is the house?
   (a) Which category of addition, combine sets or combine measures, is illustrated?
   (b) Name some other significant information that is not included in the problem.

*4. (a) What addition fact is illustrated in the diagram?

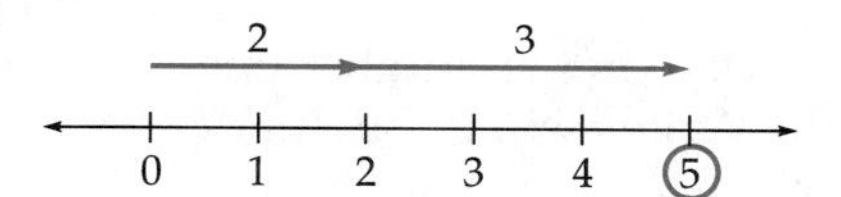

   (b) How would you use a number line to show a child that $4 + 2 = 6$?

5. (a) What addition fact is illustrated by placing the rods together as shown?

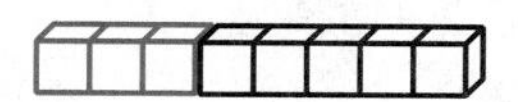

   (b) Which category of addition is illustrated?

6. Use the definition of subtraction to rewrite each of the following subtraction equations as an addition equation.
   (a) $7 - 2 = ?$
   (b) $N - 72 = 37$
   (c) $586 - N = 23$

In Exercises 7 and 8, classify each application as one of the following: take away sets, take away measures, compare sets, or compare measures.

7. Mary is 4 feet, 8 inches tall and Sidney is 4 feet, 3 inches tall. How much taller is Mary?

*8. Kong had 14 bananas. Kong ate 3. How many are left?

9. Show how to solve the preceding exercise using blocks, and tell what arithmetic equation it illustrates.

*10. (a) What subtraction fact is shown?

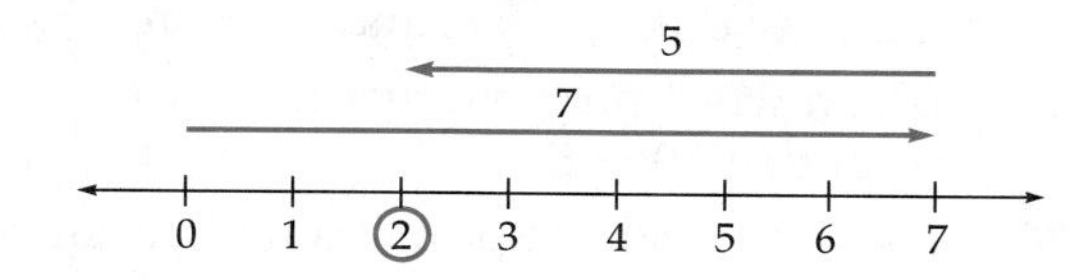

   (b) How would you use a number line to show a child that $5 - 2 = 3$?

11. (a) What subtraction fact is illustrated?

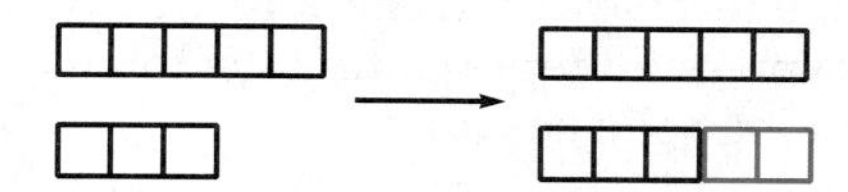

   (b) Which category of subtraction is illustrated?

12. (a) What subtraction fact is illustrated?

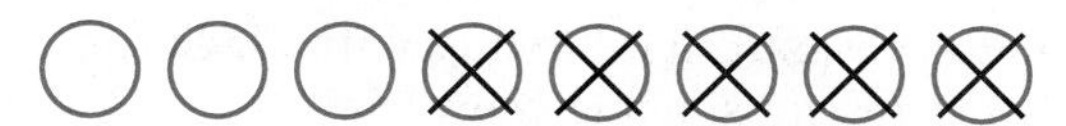

   (b) Which category of subtraction is illustrated?

13. Make a drawing that shows $8 - 2 = 6$, using
   (a) take away sets. (b) compare sets.
   (c) compare measures.

*14. Is whole-number subtraction closed? (*Hint:* Check the two parts of the closure property.)

15. The following two third-grade problems require subtracting 12 from 30. How do they differ in tone?
   (a) Jody Winer had 30 records. She gave 12 to her brother. How many records does she have left?
   (b) Jody Winer has 30 records. Her brother has 12. How many more records does she have than her brother?

16. Explain why you cannot use addition to solve the following problem. "How much is 1 cup of water plus 1 cup of sugar?"

17. What operation and classification are illustrated by the following problem? "A clerk takes 24 cartons of orange juice out of the storeroom. Now 10 cartons are left in the storeroom. How many were there to start with?"

***18.** What operation and classification are illustrated by the following problem? "A rectangular field is 20 ft by 26 ft. What is its perimeter (distance around the border)?"

19. Write a word problem that requires computing $10 - (3 + 5)$.

20. Make up a realistic problem illustrating the compare-sets category of subtraction.

21. Solve the following using addition and subtraction: "Margarita starts off with $\$A$. She buys food for $\$F$ and clothes for $\$C$, and then she receives a paycheck for $\$P$. Write an expression representing the total amount of money she has now."

22. Represent the following with algebraic expressions, using the variable n.
(a) The sum of 4 and a number
(b) 3 less than a number
(c) A number increased by 6

23. Represent the following with algebraic expressions, using the variable x.
(a) The difference between 10 and a number
(b) A number decreased by 2
(c) The sum of a number and 6

24. Solve the following equations mentally.
(a) $\square + 5 = 9$ (b) $x - 6 = 8$
(c) $9 - n = 3$

25. If $a - b = c$, then $a - (b + 1) =$ ________.

26. The political party in power sometimes tries to redraw the boundaries of a voting district in a way that gives it control of a greater number of districts. Find two possible ways to form 4 voting districts, each containing between 135,000 and 160,000 voters, in the region shown at the top of the next column. Each precinct must remain intact, and a district cannot contain a region that does not share a common border with at least one other region in that district. The numbers in the diagram represent the population, in thousands, in each precinct.

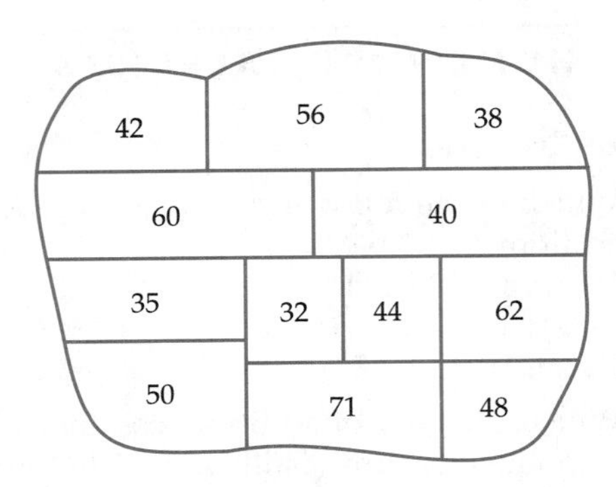

27. You are familiar with the "greater than" (>) symbol. "Greater than" can be defined using addition. For whole numbers a and b, $a > b$ when $a = b + k$ for some counting number k.
(a) Use the definition to show why $8 > 6$.
(b) Write a definition for "less than" (<) using subtraction.

28. "Less than" and "greater than" can be defined in terms of a number line.

0 1 2 3 4 5 6

(a) For whole numbers a and b, $a > b$ means that a is located to the ________ (left, right) of b on the number line.
(b) Write a similar definition of $a < b$.

Extension Exercises

29. (a) If possible, write each of the following numbers as the sum of two or more consecutive counting numbers. For example, $13 = 6 + 7$ and $14 = 2 + 3 + 4 + 5$.

1 = 6 =
2 = 7 =
3 = 8 =
4 = 9 =
5 = 10 =

(b) Propose a hypothesis regarding what kinds of counting numbers can be written as the sum of two or more consecutive counting numbers.

30. Consider the following problem. "How could you measure 1 oz of syrup using only a 4-oz container and a 7-oz container?"
(a) Devise a plan and solve the problem.

(b) Make up a similar problem.

31. Last night, three women checked in at the Quantity Inn and were charged \$30 for their room. Later, the desk clerk realized he had charged them a total of \$5 too much. He gave the bellperson five \$1 bills to take to the women. The bellperson wanted to save them the trouble of splitting \$5 three ways, so she kept \$2 and gave them \$3.

Each woman had originally paid \$10 and received \$1 back. So the room cost them \$9 apiece. This means they spent \$27 plus the \$2 "tip." What happened to the other dollar?

32. It took 600 digits to label the pages of a book starting with page 1. How many pages does the book have?

33. Consider the following problem. "Under what conditions for whole numbers a, b, and c would $(a - b) - c$ be a whole number?"
(a) Devise a plan and solve the problem.
(b) Make up a similar problem.

34. Make up a set containing one element that is closed under addition.

35. Consider the following problem. "Last year, your salary was $\$A$ and your expenses were $\$B$. This year, you received a salary increase of $\$X$, but your expenses increased by $\$Y$. Write an expression showing how much of your salary will be left this year after you have paid all your expenses."
(a) Devise a plan and solve the problem.
(b) Make up a similar problem.

36. (a) Find six pairs of solutions to $x - y = 3$ for whole numbers x and y.
(b) Graph $x - y = 3$ for whole numbers x and y.
(c) Repeat parts (a) and (b) with $x - y = 4$.
(d) Propose a generalization about $x - y = k$ in which x, y, and k are whole numbers.

37. All the pairs of whole numbers that add up to 6 comprise a **fact family**.

$$0 + 6,\quad 1 + 5,\quad 2 + 4,\quad 3 + 3,$$
$$4 + 2,\quad 5 + 1,\quad 6 + 0$$

(a) Plot the points (0, 6), (1, 5), (2, 4), (3, 3), (4, 2), (5, 1), and (6, 0) on an xy graph.
(b) What geometric pattern do you see in the points?
(c) On a separate graph, plot the points for the fact family of 7.
(d) What geometric pattern do you see in the points in part (c)?
(e) Make a generalization based upon your results to parts (b) and (d).
(f) Is part (e) an example of induction or deduction?

38. (a) Divide the clock face into 3 regions such that the sum of the numbers in each region is the same.

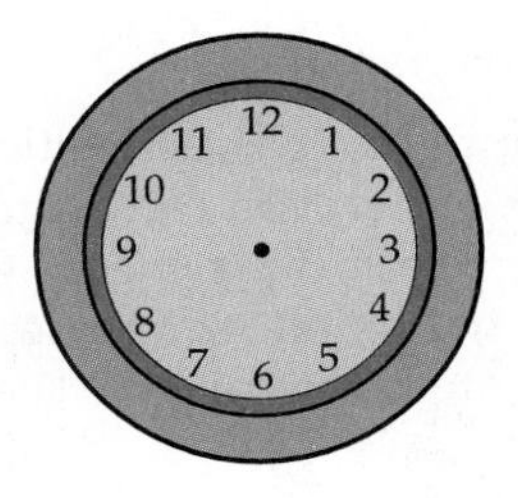

(b) Divide the clock face into 6 regions such that the sum of the numbers in each region is the same.

39. Consider the following problem. "Place the digits 1 to 9 in the squares so that all horizontal, vertical, and diagonal sums of three numbers equal 15."

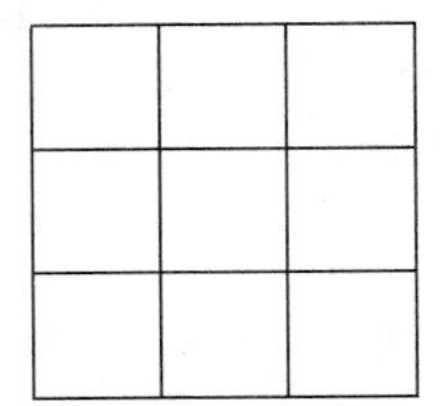

(a) Devise a plan and solve the problem.
(b) Make up a similar problem.

40. Explain why the number 9 could *not* be in one of the corners in the preceding exercise.

41. Each letter represents a digit. Find a possible solution.

$$\begin{array}{r} A \\ A \\ A \\ +\ \ B \\ \hline BA \end{array}$$

42. Fill in the digits from 1 to 9 in the circles so that the sum on each side of the triangle is the same *and* as large as possible.

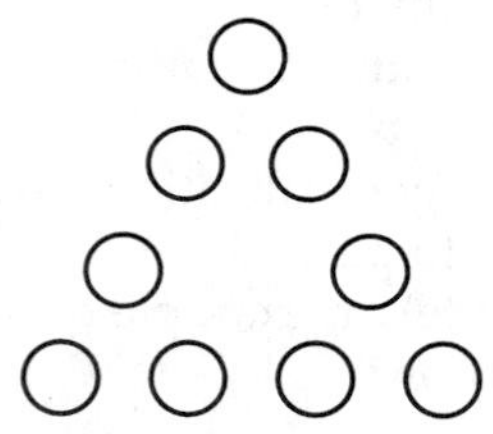

43. In the equation $x + y + z = 3$, can x, y, and z all equal 1?

Special Exercise

44. By quantifying characteristics and adding them together, you can estimate your risk of cardiac arrest (not a cheery topic, but useful).

(a) Fill in the following questionnaire, which is adapted from one by The Executive Health Group.

Risk of Heart Disease

1. Cigarette smoking
   (0) Never (1) None in past year
   (2) Smoked in past year or under 10 per day
   (3) 20 per day
   (5) 40 or more per day
2. Family history (parents, grandparents)
   (0) No heart disease before age 75
   (1) One relative with heart disease between age 60 and 75
   (2) Two relatives with heart disease between age 60 and 75
   (3) One relative with heart disease under age 60
   (4) Two relatives with heart disease under age 60
3. Diabetes (parents, grandparents, siblings)
   (0) None (1) One relative
   (2) Two relatives
   (4) You have diabetes between ages 20 and 60.
4. Exercise
   (Subtract 1) Regular aerobic exercise
   (1) Moderate exercise during and after work
   (3) Sedentary work and moderate recreation
   (5) No exercise
5. Disposition
   (0) Always easy-going
   (2) Frequently impatient
   (4) Constantly hard-driving
6. Age/gender
   (0) Under 45, F
   (1) Under 40, M
   (2) 45–55, F, or 40–50, M
   (3) Over 55, F
   (4) 50–60, M
   (5) Over 60, M
7. Blood pressure
   (0) Low (3) Medium (6) High
8. HDL cholesterol (if known)
   (0) Under 3 (3) 6 to 7.9
   (1) 3 to 4.5 (4) 8 to 9.9
   (2) 4.6 to 5.9 (5) 10 or over
9. Weight
   (0) Not much overweight
   (1) About 10 lb overweight
   (2) About 23 lb overweight
   (3) About 38 lb overweight
   (4) About 53 lb overweight

To obtain your score, add up the numbers of all your choices. The results are as follows: 0–17, low risk to acceptable; 18–22, average; 23–32, above average—try to lower some risk factors; 33 and up, high—consult with a physician.

(b) Which factors in the questionnaire are most under your personal control?

3.3 MULTIPLICATION AND DIVISION OF WHOLE NUMBERS

Well, what were you expecting to study after addition and subtraction, scuba diving? To understand multiplication and division better, you will classify the kinds of situations that call for multiplication or division. Sound familiar?

MULTIPLICATION DEFINITION

Historically, people developed multiplication as a shortcut for certain addition problems, such as $4 + 4 + 4$.

Lesson Exercise 3.32

Write $4 + 4 + 4$ as a multiplication problem.

This repeated-addition model of multiplication can be used to define multiplication of whole numbers.

Definition **Whole-Number Multiplication**

For any whole numbers a and b where $a \neq 0$,

$$a \times b = \underbrace{b + b + \cdots + b}_{a \text{ terms}}.$$

If $a = 0$, then $0 \times b = 0$.

According to the definition, $3 \times 4 = 4 + 4 + 4$. However, because multiplication is commutative (see Section 3.4), $4 \times 3 = 4 + 4 + 4$ is also an acceptable equation in elementary school. In $a \times b = c$, a and b are called **factors** and c is called the **product**.

An English minister, William Oughtred (1574–1660), created over 150 mathematical symbols, only 3 of which are still in use, the most important being the symbol $\times$ for multiplication. Some seventeenth-century mathematicians objected that the symbol $\times$ for multiplication would be confused with the letter x. One of them, Gottfried Wilhelm von Leibnitz (1646–1716), invented the dot symbol $\cdot$ for multiplication in 1698. Multiplication of two variables such as m and n can be written as mn without the use of either multiplication symbol.

Lesson Exercise 3.33

(a) If you multiply any two whole numbers, is the result always a whole number?

(b) Would a whole-number multiplication problem ever have more than one answer?

Your responses to the preceding exercise should suggest that whole-number multiplication is closed.

The Closure Property of Multiplication of Whole Numbers

If a and b are whole numbers, then $a \cdot b$ is a unique whole number.

Classifying Multiplication Applications

Multiplication is used in a variety of applications, which usually fall into one of five categories.

Lesson Exercise 3.34

Make up two word problems that can be solved by computing 3×4. Try to create two different types of problems. Compare your problems to others in the class.

Do your problems fit into any of the following categories?

1. **Repeated Sets** Find the total number of objects, given a certain number of equivalent sets, each of which has the same number of objects.

   Example: I bought 3 packages of tomatoes. Each package contained 4 tomatoes. How many tomatoes did I buy (Figure 3-18)?

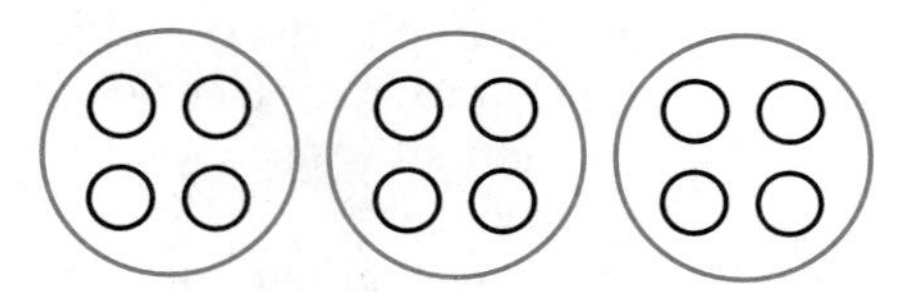

Figure 3–18

2. **Repeated Measures** Find the total measure that results from repeating a given measure a certain number of times.

   Example: On a long car trip, I average 50 miles per hour. How far will I travel in 6 hours (Figure 3-19)?

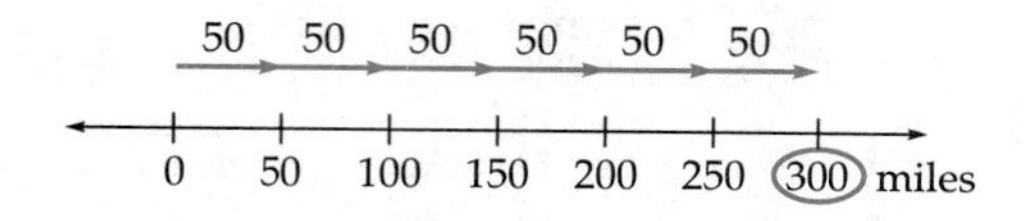

Figure 3–19

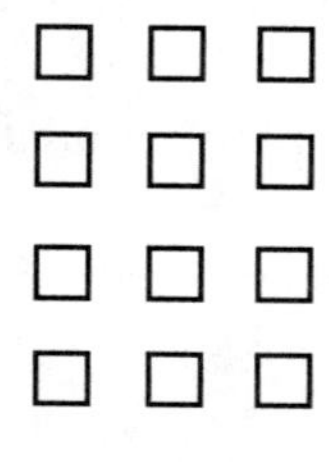

Figure 3–20

3. **Array** Find the total number of objects needed to occupy a given number of rows and columns.

   Example: A class has 4 rows of desks with 3 desks in each row. How many desks are there (Figure 3-20)?

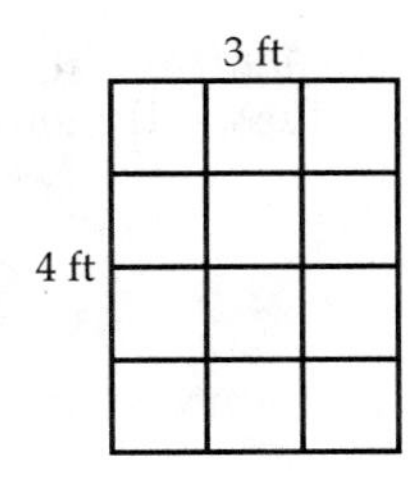

Figure 3–21

4. **Area (Measures)** Find the total measure in square units, given the width (number of rows of squares) and length (number of columns of squares). This category establishes an important connection between multiplication and area.

   Example: A rug is 4 feet by 3 feet. What is its area in square feet (Figure 3-21)?

5. **Cartesian Product (Pairings)** Find the total number of different pairs formed by pairing any object from one set with any object from a second set.

   Example: I have 3 shirts and 4 pairs of pants. How many different shirt-pant outfits can I create (Figure 3-22)? In each case, the first item is chosen from $S = \{S_1, S_2, S_3\}$ and the second item from $P = \{P_1, P_2, P_3, P_4\}$. The set of ordered pairs forms the Cartesian product $S \times P$. The Fundamental Counting Principle (Section 13.3) states that we can count the ordered pairs by multiplying 3 by 4.

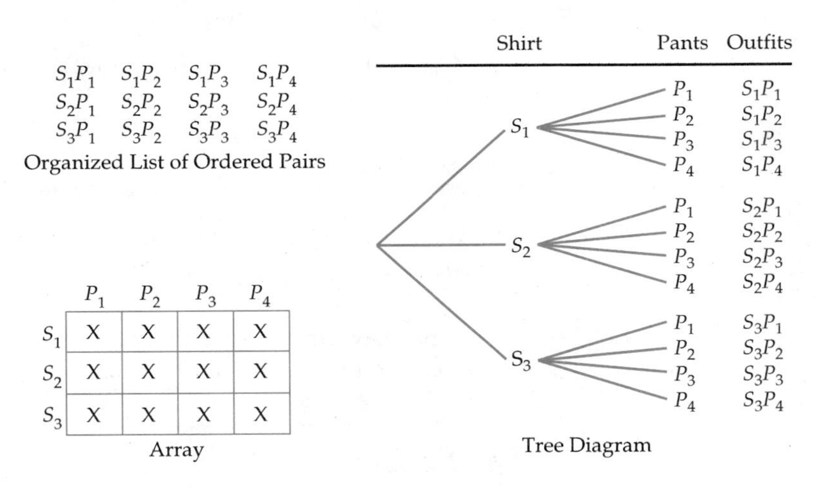

Figure 3–22

Anyone who can recognize these five situations will usually know when to multiply to solve a problem. Lesson Exercises 3.35 and 3.36 will give you practice in recognizing different types of problems that require multiplication.

Lesson Exercise 3.35

Classify the following multiplication application. "A juice pack contains 3 cartons of juice. How many cartons are in 4 juice packs?"

Lesson Exercise 3.36

Classify the following multiplication application. "A rectangular tabletop is 5 feet long and 4 feet wide. What is its area?"

In elementary school, repeated sets, repeated measures, arrays, area, and Cartesian products are usually introduced with manipulatives and pictures.

Lesson Exercise 3.37

Show how to solve Lesson Exercise 3.35 using a set of counters, and write the arithmetic equation it illustrates.

Lesson Exercise 3.38

Show how to solve Lesson Exercise 3.36 using paper squares, and write the arithmetic equation it illustrates.

DIVISION DEFINITION

Just as subtraction is defined in terms of addition, division is defined in terms of multiplication.

Lesson Exercise 3.39

What multiplication equation corresponds to $12 \div 3 = 4$?

In answering Lesson Exercise 3.39, you used the following relationship between multiplication and division.

Definition **Whole-Number Division**

If x, y, and q are whole numbers, and $y \neq 0$, then $x \div y = q$ if and only if $x = y \cdot q$.

In $x \div y = q$, x is called the **dividend**, y is called the **divisor**, and q is called the **quotient**. The division ($\div$) symbol first appeared in print in 1659 in a work by the Swiss mathematician Johann Rahn. The definition of division establishes that division is the inverse operation of multiplication. The connections among whole-number operations are summarized in the following diagram.

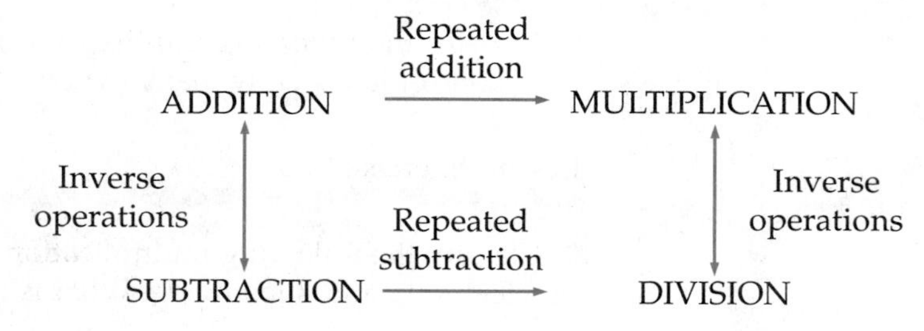

All of these connections have already been discussed, except for division as repeated subtraction. Imagine that a child (who does not know what "division" is) has 8 cookies to distribute. She serves 2 cookies per person, and she wants to figure out how many people she can serve. She could use repeated subtraction to determine the answer.

$8 - 2 = 6$
$6 - 2 = 4$
$4 - 2 = 2$
$2 - 2 = 0$

2 2 2 2
0 2 4 6 8

Whole-number division problems such as $8 \div 2$ can be thought of as asking how many times 2 can be subtracted from 8 until nothing remains. The answer is 4.

The preceding definition of division applies to examples with whole-number quotients and no remainder. For other whole-number division problems with nonzero divisors, one can always find a whole-number quotient and a whole-number remainder. For example, $13 \div 4 = 3$ R 1.

Lesson Exercise 3.40

Using multiplication and addition, write an equation that is equivalent to $13 \div 4 = 3$ R 1.

The preceding exercise illustrates the division algorithm.

The Division Algorithm (for $a \div b$)

If a and b are whole numbers with $b \neq 0$, then there exist unique whole numbers q and r such that $a = bq + r$, in which $0 \leq r < b$.

The division algorithm equation $a = bq + r$ means

$$\text{Dividend} = \text{divisor} \cdot \text{quotient} + \text{remainder}$$

The quotient q is the greatest whole number of b's that are in a. The remainder r is how much more than bq is needed to make a.

CLASSIFYING DIVISION APPLICATIONS

Have you ever needed to divide 12 by 3? What types of situations require computing $12 \div 3$?

Lesson Exercise 3.41

Make up two word problems that can be solved by dividing 12 by 3. Try to create two different types of problems. Compare your results to others in your class.

Do your problems fit into any of the following categories?

1. **Repeated Sets** Find how many sets of a certain size can be made from a group of objects.

   Example: I have a dozen eggs. How many 3-egg omelets can I make (Figure 3-23)?

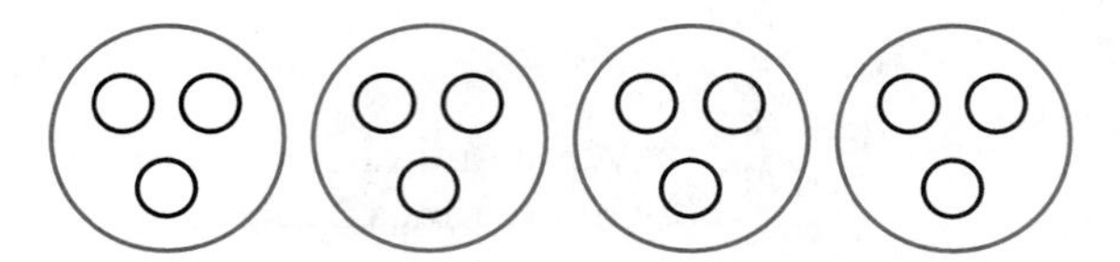

Figure 3–23

2. **Repeated Measures** Find how many measurements of a certain size equal a given measurement.

   Example: A walk is 12 miles long. How long will it take if a walker averages 3 miles per hour (Figure 3-24)?

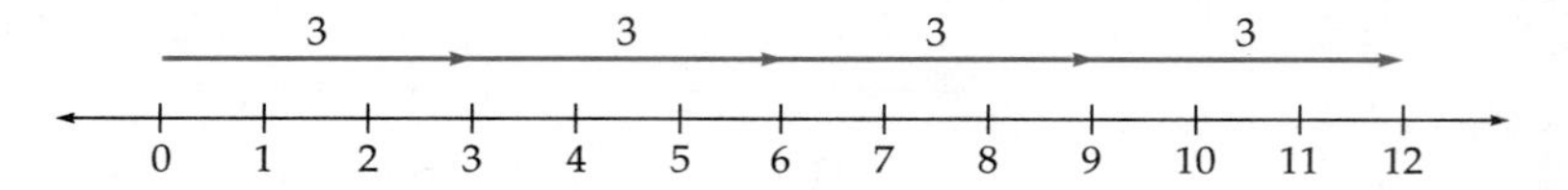

Figure 3–24

3. **Partition Sets** Find how many are in each group when you divide a set of objects equally into a given number of groups.

   Example: We have 12 yummy strawberries for the 3 of us. How many will each person get if we are fair about it (Figure 3-25)?

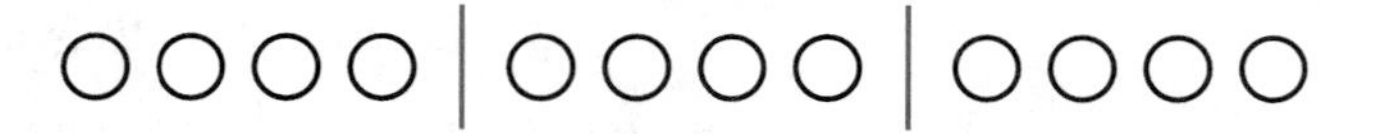

Figure 3–25

4. **Partition Measures** Find the measure of each part when you divide a given measurement into a given number of equal parts.

   Example: Let's divide this delicious 12-inch submarine sandwich equally among the three of us. How long a piece will we each receive (Figure 3-26)?

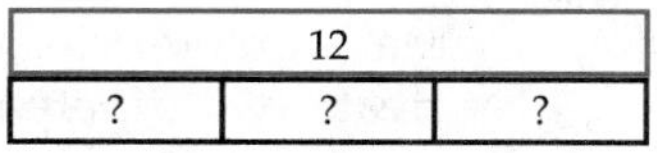

12		
?	?	?

Figure 3–26

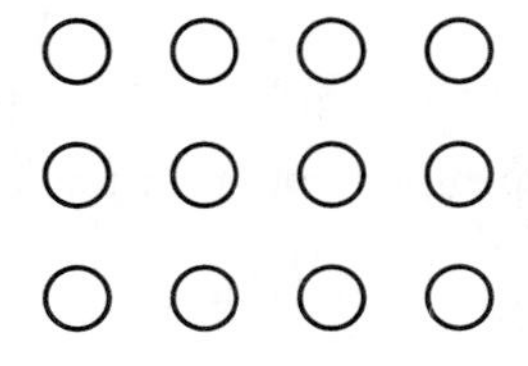
Figure 3–27

5. **Array** Find the number of rows (or columns), given an array of objects and the number of columns (or rows).

   Example: Twelve people are seated in 3 rows. How many people are there in each row (Figure 3-27)?

6. **Area** Find the length (or width) of a rectangle, given its area and its width (or length).

   Example: A rug has an area of 12 m^2 and a length of 3 m. What is its width (Figure 3-28)?

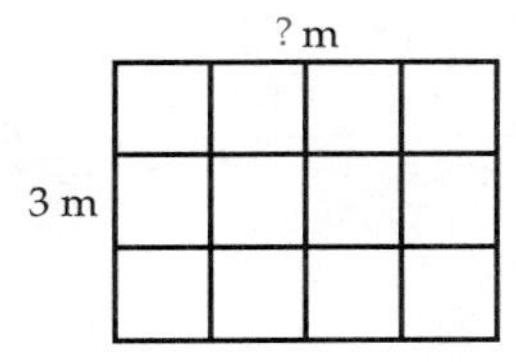

Figure 3–28

Students who recognize these types of situations will usually know when to divide to solve a problem. The hardest distinction for people to make is between the "repeated" categories (1 and 2) and the "partition" categories (3 and 4). For 12 ÷ 3, a "partitioning" problem would say, "Divide 12 into 3 equal parts," and a "repeating" problem would say, "How many 3's make 12?" Note that the partitioning categories (3 and 4) include problems in which something is shared equally.

Division is also related to fractions and ratios. These topics are discussed in Chapters 6 and 7, respectively.

In Lesson Exercises 3.42, 3.43, and 3.44, classify the division applications.

Lesson Exercise 3.42

Judy, Joyce, and Zdanna share a box of 6 pears equally. How many pears does each woman receive?

Lesson Exercise 3.43

The Arcurris drove 1400 km last week. What was the average distance they drove each day?

Lesson Exercise 3.44

My National Motors Gerbil gets 80 miles per gallon. How many gallons would I use on a 400-mile trip?

Repeated sets and measures, partition sets and measures, arrays, and area are usually introduced with manipulatives and pictures in elementary school.

Lesson Exercise 3.45

Show how to solve Lesson Exercise 3.42 using a set of counters, and write the arithmetic equation it illustrates.

Lesson Exercise 3.46

Show how to solve Lesson Exercise 3.44 using a number line, and write the arithmetic equation it illustrates.

The following chart summarizes the most common classifications of arithmetic applications.

Classifying Arithmetic Applications	
Addition	*Subtraction*
1. Combine sets	**1.** Take-away sets
2. Combine measures	**2.** Take-away measures
	3. Compare sets
	4. Compare measures
Multiplication	*Division*
1. Repeated sets	**1.** Repeated sets
2. Repeated measures	**2.** Repeated measures
3. Array	**3.** Partition sets
4. Area	**4.** Partition measures
5. Cartesian product	**5.** Array
	6. Area

DIVISION BY ZERO

Division problems involving zero are a source of confusion. What is $4 \div 0$? $0 \div 0$? $0 \div 4$? According to the definition of division, $4 \div 0$ and $0 \div 0$ are undefined because you cannot have 0 as a divisor. However, $0 \div 4$ is defined. Children wonder why one problem is defined, but the other two are not.

Lesson Exercise 3.47

Why are problems such as $4 \div 0$ undefined?

Expressions involving division by zero are most easily examined by converting the expression to a related multiplication equation. They can also be analyzed using any of the six division categories. The repeated-sets model is usually the easiest to use.

EXAMPLE 3.2

(a) What is $4 \div 0$?

(b) *Explain* why, using multiplication or the division category of your choice.

Solution

Method 1 Using the inverse-of-multiplication model, $4 \div 0 = ?$ would mean $0 \times ? = 4$. There is no solution to $0 \times ? = 4$, so we make $4 \div 0$ undefined.

Method 2 Using repeated sets, $4 \div 0$ would mean "How many sets of 0 will make 4?" No number of sets of 0 will make 4. So we make $4 \div 0$ undefined.

■

Lesson Exercise 3.48

(a) What is $0 \div 4$?

(b) *Explain* why, using multiplication or the division category of your choice.

Lesson Exercise 3.49

(a) What is $0 \div 0$?

(b) *Explain* why, using multiplication or the division category of your choice.

In elementary school, we simply teach children that they cannot divide a number by zero, after doing some examples such as $4 \div 0$ and $2 \div 0$. The difference between $0 \div 0$ and these other examples is not discussed. You are unlikely to find $0 \div 0$ in an elementary school textbook, but a child may ask you about it.

WORKING BACKWARD

In Chapter 1, you studied problem-solving strategies such as induction, guess and check, and drawing a picture. In the previous two sections, you have seen how children can learn to choose the correct operation to solve a problem. Now you can add another strategy, working backward, to your repertoire. Sometimes you can solve a problem by **working backward** from the final result to the starting point. Some working-backward problems involve the four operations of arithmetic you just studied. The following example illustrates this strategy.

EXAMPLE 3.3

Hy Roller got his weekly paycheck. He spent half of it on a gift for his mother. Then he spent \$8 on a pizza. Now he has \$19. How much was his paycheck?

Solution

Understanding the Problem You know that he ended up with $19, and you know how he spent his money. How much did he start with?

Devising a Plan Draw a diagram and then work backward.

Carrying Out the Plan The diagram shows the sequence from the beginning to the end of the problem.

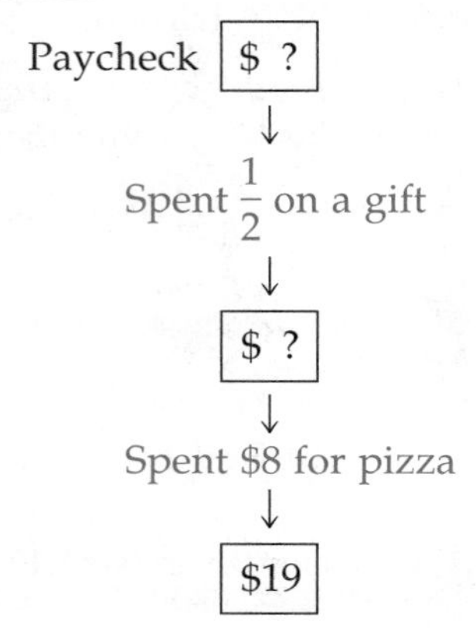

Now work backward and fill in the question marks. He ended up with $19. What did he have before he spent $8 on the pizza? $27.

So Hy had $27 just after he spent 1/2 on a gift. How much did he start with? He started with 2 × $27, or $54.

Looking Back After you obtain an answer, start with $54 and work forward to check the answer. Start with $54. Spend 1/2 on a gift. That leaves $27. Spend $8 on a pizza. That leaves $19. It checks! ■

One could solve the preceding example using inverse operations.

Original Problem	**Same Problem Working Backward**
☐	☐
↓ ÷ 2	↑ × 2
☐	☐
↓ − 8	↑ + 8
$19	$19

Now you try one.

Lesson Exercise 3.50

(a) Hy Roller received a raise in his weekly pay. After he received his first new paycheck, he spent $6 on tacos. He spent 1/2 of what was left on a gift for his father. Now he has $25. How much is his weekly pay? (Draw a diagram and work backward.)

(b) Does working backward involve inductive reasoning or deductive reasoning?

ORDER OF OPERATIONS

What is $7 + 3 \times 6$? When you do a computation, the order in which you add and multiply may affect the answer.

***Class* Lesson Exercise 3.51**

(a) Everyone in the class should do 7 [+] 3 [×] 6 [=] on a calculator. Collect the results.

(b) *Explain* how you can obtain 60 as an answer.

(c) *Explain* how you can obtain 25 as an answer.

(d) Which answer is right?

Because problems such as $7 + 3 \times 6$ are ambiguous, mathematicians have made rules for the order of operations. Normally, we use parentheses to indicate the order, by writing either $(7 + 3) \times 6$ or $7 + (3 \times 6)$. However, when using a calculator, parentheses are often omitted. All computers and most calculators automatically use the following rules for order of operations with and without parentheses.

Order of Operations

1. Compute inside parentheses first.
2. Evaluate all exponents.
3. Multiply and divide, working from left to right.
4. Add and subtract, working from left to right.

These same rules will also apply to negative numbers, fractions, and decimals later in the course. Apply the order of operations in the following exercise.

Lesson Exercise 3.52

Compute the following using the correct order of operations.

(a) $6 - 18 \div 3 + 7 \times (2 + 1)$

(b) $12 - 3^2 + 2 \times 5$

AN INVESTIGATION: OPERATING ON 3's

Lesson Exercise 3.53

Using four 3's and any of the four operations, try to obtain all the numbers from 1 to 10. Here, 0 is done as an example.

$$0 = 3 - 3 + 3 - 3$$

ANSWERS TO SELECTED LESSON EXERCISES

3.32 3×4

3.33 (a) Yes (b) No

3.34 See the examples that follow Lesson Exercise 3.34.

3.35 Repeated sets (4 is repeated 3 times)

3.36 Area

3.37 ○○○ ○○○ ○○○ ○○○

$3 \times 4 = 12$

3.38

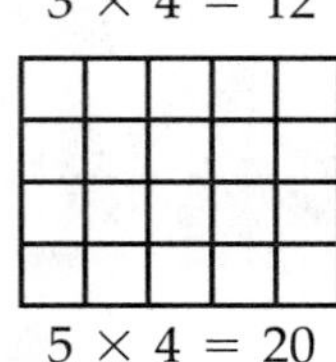

$5 \times 4 = 20$

3.39 $12 = 3 \times 4$

3.40 $4 \times 3 + 1 = 13$

3.41 See the examples that follow Lesson Exercise 3.41.

3.42 Partition sets (since the divisor 3 partitions the 6 pears)

3.43 Partition measures (the divisor 7 partitions the distance)

3.44 Repeated measures (how many 80's make 400)

3.45

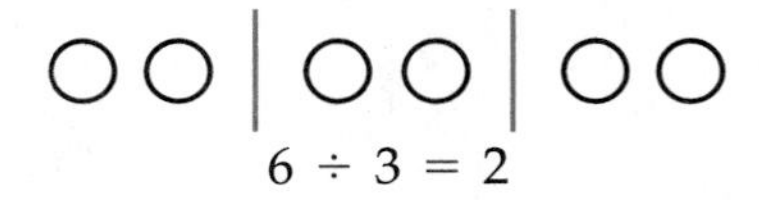

$6 \div 3 = 2$

3.46

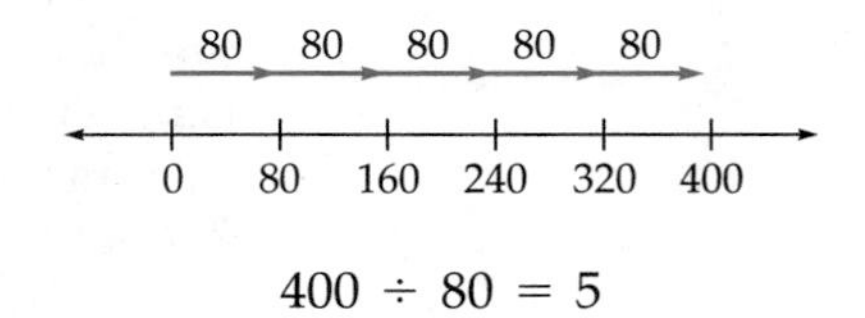

$400 \div 80 = 5$

3.48 (a) $0 \div 4 = 0$
(b) $0 \div 4 = ?$ is the same as $4 \times ? = 0$. So $? = 0$.

3.49 (a) $0 \div 0$ is undefined.
(b) $0 \div 0 = ?$ is the same as $0 \times ? = 0$. So ? could stand for any number! Since each division problem must have one definite answer, we say $0 \div 0$ is undefined.

3.50 (a) \$56 (double \$25 and add \$6)
(b) Deductive

3.52 (a) 21 (b) 13

3.3 HOMEWORK EXERCISES

Basic Exercises

1. Rewrite $7 + 7 + 7$ as a multiplication problem.

2. Explain what it means to say that whole numbers are closed under multiplication.

3. Tell which of the following sets are closed under multiplication. If a set is not closed, show why not.
   (a) $\{1, 2\}$ (b) $\{0, 1\}$ (c) $\{2, 4, 6, 8, \ldots\}$

In Exercises 4–7, classify each application as one of the following: repeated sets, repeated measures, array, area, or Cartesian product.

4. In an election, 5 people are running for president and 3 people are running for vice president. How many different pairs of candidates can be elected to the two offices?

5. Each person in the United States creates about 6 pounds of solid garbage per day. How much solid garbage does a person dispose of in a year?

6. An index card is 3 in. by 5 in. What is its area?

7. Tacos come 10 to a box. How many tacos are in 3 boxes?

*8. How would you explain to a child why the area of the rectangle shown is 6 ft^2?

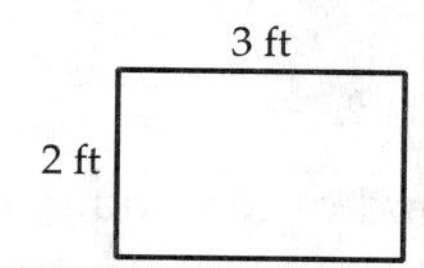

9. The following two word problems require multiplying 8 by 3. Which problem illustrates a more realistic use of mathematics?
   (a) Bill has 8 pencils. Courtney has 3 times as many. How many pencils does Courtney have?
   (b) Concert tickets cost \$8 each. How much would 3 tickets cost?

*10. (a) What multiplication problem is shown in the following diagram?

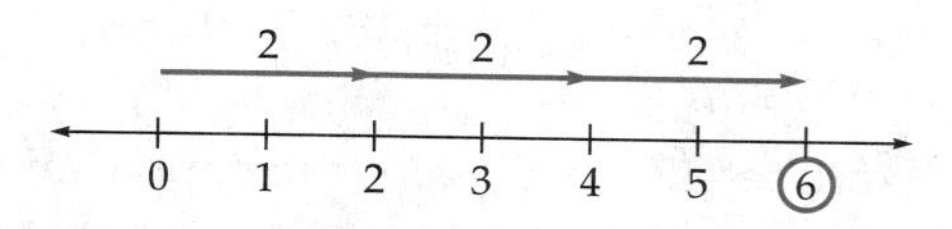

(b) How would you use a number line to show a child that $3 \times 4 = 12$?

11. Make a drawing that shows $2 \times 5 = 10$, using
    (a) repeated sets.
    (b) repeated measures.
    (c) an array.

12. What is the next number in each sentence?
    (a) 7, 15, 31, _____
    (b) 20, 37, 71, 139, _____

13. Fill in the blanks in each part by following the same rule used in the completed examples.

(a)		(b)	
$4 \to$	26	$3 \to$	10
$5 \to$	35	$4 \to$	14
$6 \to$	46	$5 \to$	18
$8 \to$	_____	$10 \to$	_____
$X \to$	_____	$X \to$	_____

(c) The process of making a conjecture about the general formula in the last blank in parts (a) and (b) is an example of ________ reasoning.

14. (a) $11^2 =$ ________ (b) $111^2 =$ ________
    (c) Without a calculator, guess the values of 1111^2 and $11{,}111^2$.
    (d) Use a calculator to check your answers to part (c).

15. (a) Write the number of small squares using an exponent other than 1.

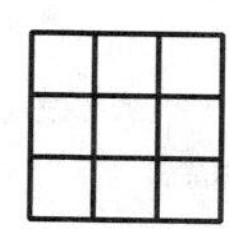

(b) Write the number of small cubes using an exponent other than 1.

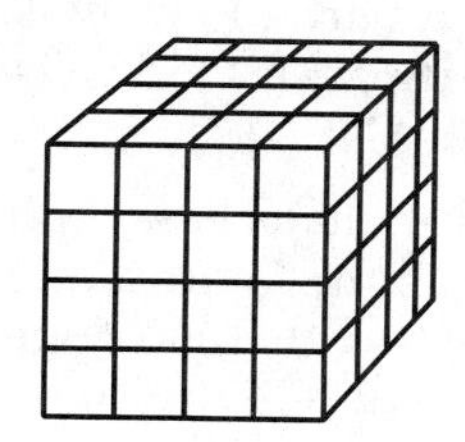

16. Use the definition of division to rewrite each division equation as a multiplication equation.
(a) $40 \div 8 = ?$ (b) $32 \div N = 6$
(c) $0 \div N = C$

In Exercises 17–20, classify each application as one of the following: repeated sets, repeated measures, partition sets, partition measures, array, or area.

17. A car travels 180 miles in 4 hours. What is its average speed?

18. If I deal a deck of 52 cards to 4 people, how many cards does each person receive?

19. A car travels 180 miles, averaging 45 miles per hour. How long does this trip take?

20. A teacher wants to seat 30 students evenly in 5 rows. How many children sit in each row?

21. (a) What division fact is illustrated in the diagram?

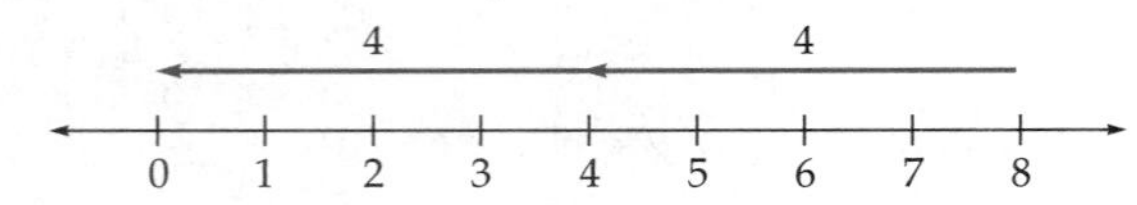

(b) How would you use a number line to show a child that $10 \div 5 = 2$?

***22.** (a) What division fact is illustrated in the diagram?

(b) Use a set-partition picture to show that $6 \div 3 = 2$.

23. Make a drawing that illustrates $10 \div 5 = 2$, using
(a) an array.
(b) partition measures.
(c) repeated sets.

24. (a) When whole numbers are divided by 5, how many different possible remainders are there?
(b) Repeat part (a) for division by 6.
(c) Repeat part (a) for division by a counting number N.

25. Helmer Junghans has 37 photographs to put in a photo album. Write a question about Helmer's task that has an answer of
(a) 10 (b) 1 (c) 9

26. You are sewing buttons on a blouse whose front is 22 in. long. The first button is 2 in. from the top, and the buttons are 3 in. apart. How many buttons do you need?

27. The Hersheys buy a 6-slice pizza for the 4 members of their family. They each eat an equal number of slices and give any leftover slices to their pet goat Dainty.
(a) How many slices does each eat?
(b) How many slices does Dainty get?
(c) Is the answer to part (a) a quotient, a remainder, or neither?
(d) Is the answer to part (b) a quotient, a remainder, or neither?

28. (a) Ninety-seven children want to go on a field trip in school buses. Each bus holds 40 children. How many buses are needed?
(b) Is the answer a quotient, a remainder, or neither?

29. (a) What is $7 \div 0$?
(b) Explain why, using multiplication or the division category of your choice.

30. (a) What is $0 \div 7$?
(b) Explain why, using multiplication or the division category of your choice.

31. Try computing $4 \div 0$ on a calculator. What is the result? What does the result mean?

***32.** (a) Fill in the missing numbers.

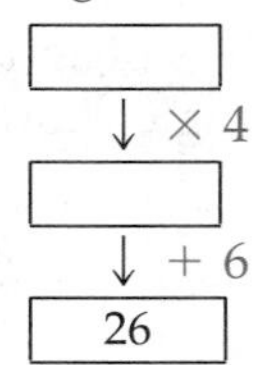

(b) What kind of reasoning, inductive or deductive, do you use to fill in the blanks?

33. Sharon takes a number, adds 2 to it, and then multiplies the result by 3. She ends up with N. What was the original number in terms of N?

34. Carroll Matthews went to a gambling casino. He did not look in his wallet beforehand, but he remembers how he spent his money. He paid \$5 for parking. Then he spent \$3 on an imitation orange juice drink. Then he lost half of his remaining money gambling. Now he has \$14. How much did Carroll start with?

35. Frank picked some daisies. He gave half of them to his wife. Then he divided what was left evenly among his brother, his daughter, and his horse. They each got 6 daisies. How many daisies did he pick?

36. Compute the following.
(a) $10 \times 2 - 6 \div 2 + 1$
(b) $18 - 2 \times (4 + 1)$
(c) $8^2 - (2 + 4 \times 3)$

37. Compute the following.
(a) $(3 + 5) - 4 + 2 \times 3$
(b) $7^2 + 2 \times 3^2$
(c) $8 - (7 - 2) + 3 \times 2$

38. Represent the following expressions using the variable y.
(a) The product of 8 and a number
(b) A number divided by 3
(c) A number decreased by 7

39. Represent the following expressions with the variable t.
(a) Twice a number
(b) 5 less than a number
(c) The product of a number and 5

40. Solve the following problems mentally.
(a) $4 \times \square = 28$ (b) $\frac{8}{n} = 2$
(c) $7(y + 2) = 35$

41. Solve the following problems mentally.
(a) $\frac{r}{5} = 20$ (b) $\frac{35}{t + 2} = 7$
(c) $8(y - 2) = 40$

42. Make up an application problem that illustrates the repeated-measures model of multiplication.

43. Make up an application problem that illustrates the partition-sets model of division.

44. In each part, complete the last two number pairs using the rule from the three completed pairs.
(a) $20 \rightarrow 0$
$30 \rightarrow 1$
$40 \rightarrow 2$
$50 \rightarrow$ __
$N \rightarrow$ __

(b) $12 \rightarrow 7$
$14 \rightarrow 8$
$16 \rightarrow 9$
$30 \rightarrow$ __
$N \rightarrow$ __

45. What is the next number in the sequence 8, 23, 68, 203, . . ?

***46.** Is whole-number division closed? (*Hint:* Check the two parts of the closure property.)

47. A 5×5 square grid has 16 squares on the border.

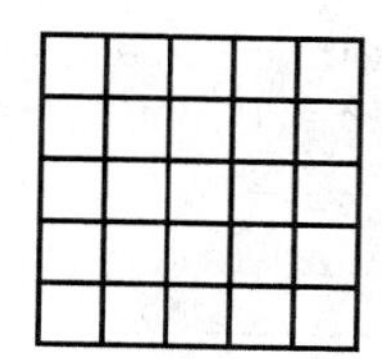

How many squares would an $n \times n$ grid have on the border? (Simplify your answer.)

48. At a restaurant, you can order a chicken, turkey, cheese, or roast beef sandwich on rye, wheat, or white bread. How many different kinds of sandwiches are possible?

49. What two operations and categories are used in each of the following problems?
(a) At a carnival, Puneet tried to knock down 4 rows of 3 pins. On his first throw, he hit 7 pins. How many were left?
(b) Wally ate 2 pears. Juanita ate twice as many. How many pears did the two of them eat all together?

50. What two operations and categories are illustrated in each of the following applications?
(a) How many miles must you travel each day to complete a 3260-mile car trip in 2 weeks?
(b) A National Motors Capon travels 276 miles on 12 gallons of gas. How far does it travel on a full 20-gallon tank of gas?

51. What two operations and categories are illustrated in each of the following applications?
(a) You have 10 library books. Then you return 2 of them and lend 3 to a friend. How many books do you have left?
(b) An auditorium has 20 rows with 28 seats in each row. If 16 seats are empty at a concert, how many people are seated there?

52. Write a realistic word problem that requires computing $(4 \times 8) - 5$.

53. Write a realistic word problem that requires computing $(3 + 6) \times 7$.

*54. Gwen Cyrkiel buys A shirts for \$$B$ each and C skirts for \$$D$ each. What is the total cost of all her purchases?

55. For whole numbers a, b, and c, with a an even number and $b \neq 0$, if $a \div b = c$, then $a \div 2b =$ ________.

56. For whole numbers a, b, and c, if $ab = c$, then $(2a)(3b) =$ ________.

57. (a) The numeral 5^{100} has 70 digits. What is the last digit? (*Hint:* Use inductive reasoning.)
(b) T is a one-digit whole number. If the last digit of T^{100} is T, what are the possible values of T?

58. $3 + 6 = 9$ is to $2^3 \cdot 2^6 = 2^9$ as $5 - 2 = 3$ is to ____________.

59. You need to measure 7 oz of medicine, and all you have are two transparent 2-oz measuring cups and a glass. How can you do it?

60. The square of a number is 9 less than 10 times the number. What is the number? (Guess and check.)

61. Suppose the sum of two numbers is even and their product is odd. What can you deduce about the two numbers?

Extension Exercises

62. $25 \times 25 = 625$
$26 \times 24 = 624$
$27 \times 23 = 621$
(a) Predict the answer to the next multiplication problem that extends the pattern. (*Hint:* Look at the last digit of each factor.)
(b) Check your guess in part (a).
(c) Repeat parts (a) and (b).
(d) Start with $40 \times 40 = 1600$ and repeat parts (a) and (b) three times.
(e) Make a generalization about your results.
(f) Use algebra to explain why your generalization is true. (*Hint:* Use $(x + a)(x - a)$.)

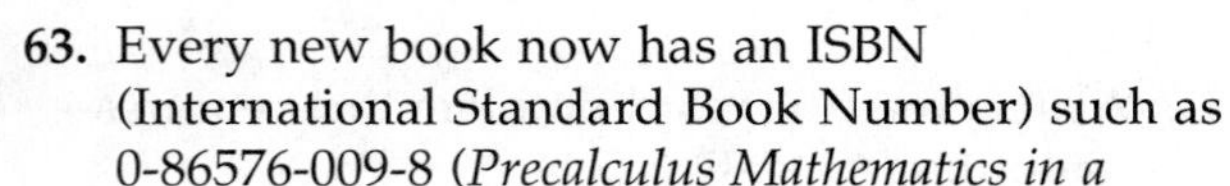

63. Every new book now has an ISBN (International Standard Book Number) such as 0-86576-009-8 (*Precalculus Mathematics in a Nutshell*). The first digit indicates the language of the country in which the book is published (for example, 0 for English). The digits 86576 represent the publisher (William Kaufmann, Inc.). The digits 009 identify the book for the publisher. Finally, the last digit is the **check digit**, which is used to check that the rest of the number is recorded correctly. To obtain an ISBN check digit:

Step 1 Multiply the first nine digits by 10, 9, 8, 7, 6, 5, 4, 3, and 2, respectively. For 0-86576-009-8, $(0 \times 10) + (8 \times 9) + (6 \times 8) + (5 \times 7) + (7 \times 6) + (6 \times 5) + (0 \times 4) + (0 \times 3) + (9 \times 2) = 245$.

Step 2 Divide the sum by 11 and find the remainder.

$$245 \div 11 = 22 \text{ R } 3$$

Step 3 Subtract the remainder from 11 to match the check digit.

$$11 - 3 = 8 \qquad \text{Yes, it checks!}$$

The check digit is used to check ISBN numbers that are copied onto order forms. Check the following ISBN numbers to see if they seem to be correct.
(a) ISBN 0-03-008367-2
(b) ISBN 0-7617-1326-8

64. (a) Write down any two numbers.
(b) Write down a third number that is the sum of the two numbers.
(c) Write down a fourth number that is the sum of the second and third numbers.
(d) Continue this process until you have ten numbers.
(e) Add up all ten numbers.
(f) The mathemagician does part (e) faster by multiplying the seventh number by 11. Try this.
(g) Represent the first two numbers with x and y, and show why parts (e) and (f) yield the same result.

65.

5	10	20	5
4	8	16	4
6	12	24	6
9	18	36	9

(a) Copy the table and circle four numbers, making sure that only one number is in each row and only one number is in each column.
(b) Multiply your four numbers together.
(c) Repeat parts (a) and (b) after picking a different group of four numbers.
(d) Repeat parts (a) and (b) after picking a different group of four numbers.
(e) Explain the pattern in the table.
(f) Make your own three-row-by-three-column table that does the same thing.

66. The Universal Product Code (UPC) appears on many grocery items. The first digit identifies the type of product. The next five digits identify the manufacturer. The next five digits identify the specific product. The 12th digit (not always printed) is the check digit, which will reveal some errors in the first 11 digits.

For the product 0-16300-15114 (a brand of orange juice), the check digit is obtained as follows.

1. Add the digits in positions 1, 3, 5, 7, 9, and 11 and triple the sum.
2. Add the sum of the remaining digits to the result from step 1.
3. Create a check digit that makes the sum resulting from step 2 end in 0.

In this example, $0 + 6 + 0 + 1 + 1 + 4 = 12$ and $12 \times 3 = 36$. Then $1 + 3 + 0 + 5 + 1 + 36 = 46$. Make the check digit 4 so that $46 + 4$ ends in 0.

Find the check digit for
(a) 0-14019-02747 (*Consumer Reports* magazine)
(b) 0-74333-47052 (a brand of peanut butter)
(c) Find a product at home with a UPC and verify that the check digit is correct.

67. In a two-person game of Last Out, the players take turns removing either one or two chips. The person who removes the last chip loses the game. If you go first and the game starts with 11 chips, describe a winning strategy. (Work backward.)

68. A true-false test has 5 questions. How many different ways can the test be completed if each question is answered true or false?

69. (a) Suppose you know how many groups of equal size there are and the number of items in each group. You are asked to find how many items there are in all. What operation and category does this illustrate?
(b) Suppose you know how many items there are in all and how many equal-sized groups there are. You are asked to find the number in each group. What operation and category does this illustrate?

70. Whole-number multiplication can also be defined using the Cartesian product. For whole numbers a and b, if set A contains a elements and set B contains b elements, then $a \cdot b$ is the number of elements in $A \times B$. Use this definition with $A = \{1, 2\}$ and $B = \{1, 2, 3\}$ to explain why $2 \cdot 3 = 6$.

71. Consider the following problem. "Find a two-digit number that is twice the product of its digits."
(a) Devise a plan and solve the problem.
(b) Make up a similar problem.

72. At a Chinese restaurant, A people divide the check evenly. Their dishes cost \$$B$, \$$C$, \$$D$, \$$E$, and \$$F$. They add 20% for tax and tip. How much does each person pay?

3.4 WHOLE NUMBERS: PROPERTIES, ALGORITHMS, AND ERROR PATTERNS

Properties of whole-number operations make it easier to memorize basic facts and do certain computations. If you learn $7 \times 9 = 63$, then you know what 9×7 equals. The sum $(24 + 2) + 8$ is more easily computed as $24 + (2 + 8)$. These properties are also the basis for the efficient procedures we use to add and multiply larger numbers. The properties have names such as "commutative," "associative," "identity," and "distributive." Do you feel your memory being jarred?

THE COMMUTATIVE AND ASSOCIATIVE PROPERTIES

Is a "night light" the same as a "light night"? Is 7×9 the same as 9×7? These questions involve the commutative property. In mathematics, the commutative property says that you can change the *order* of two numbers in certain arithmetic operations and still obtain the same answer. Which of the four whole-number operations are commutative? Investigate this question in Lesson Exercises 3.54 and 3.55.

Lesson Exercise 3.54

Answer the following questions to help determine whether whole-number addition is commutative.

(a) When you combine two sets or two measures, does it matter which is the first and which is the second?

(b) Does $6 + 8 = 8 + 6$?

(c) For any two whole numbers x and y, do you think $x + y = y + x$?

(d) Your answer to part (c) is based upon ______________ reasoning.

Lesson Exercise 3.55

Try some examples and see whether you think the following whole-number operations are commutative. Give a counterexample for any operation that is not commutative.

(a) Subtraction (b) Multiplication (c) Division

By now you should be convinced that whole-number addition and multiplication are commutative.

The Commutative Property of Addition for Whole Numbers

For any whole numbers x and y, $x + y = y + x$.

The Commutative Property of Multiplication for Whole Numbers

For any whole numbers x and y, $xy = yx$.

The term "commutative" was first used by Francois Servois in 1814. The commutative properties can be stated in words as follows: When you add two whole numbers, you may add them in either order, and when you multiply two whole numbers, you may multiply them in either order. Figure 3-29 illustrates these properties using measures.

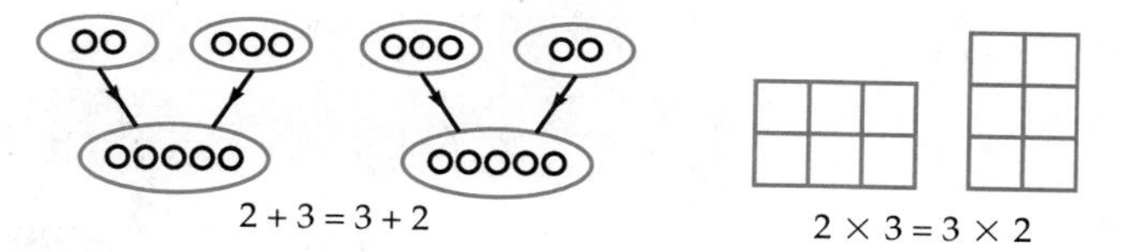

Figure 3–29

Note that the commutative property applies to expressions involving *one* operation, either addition or multiplication. But how do human beings use this property? It makes learning the addition and multiplication tables a lot easier.

Lesson Exercise 3.56

How does the commutative property reduce the number of one-digit addition facts (for example, $3 + 5 = 8$) that must be memorized?

What about the associative (grouping) property? This property says that the grouping of numbers for an arithmetic operation will not change the answer. So which whole-number operations are associative? The following exercises will help you decide.

Lesson Exercise 3.57

Answer the following questions and see whether you think whole-number addition is associative.

(a) Does $(6 + 3) + 5 = 6 + (3 + 5)$?

(b) Does $(24 + 2) + 18 = 24 + (2 + 18)$?

(c) Do you think $(x + y) + z = x + (y + z)$ for all whole numbers?

Lesson Exercise 3.58

Try some examples and see whether you think the following whole-number operations are associative. Give a counterexample for any operation that is not associative.

(a) Subtraction (b) Multiplication (c) Division

Are you now entirely convinced that whole-number addition and multiplication are commutative and associative? They are.

The Associative Property of Addition for Whole Numbers

For any whole numbers x, y, and z, $(x + y) + z = x + (y + z)$.

The Associative Property of Multiplication for Whole Numbers

For any whole numbers x, y, and z, $(xy)z = x(yz)$.

The associative properties can be stated in words as follows: When you add a series of whole numbers, parentheses have no effect on the results, and when you multiply a series of whole numbers, parentheses also have no effect on the results. In these kinds of problems, you can move parentheses around or remove them altogether.

Figure 3-30 illustrates these properties using sets.

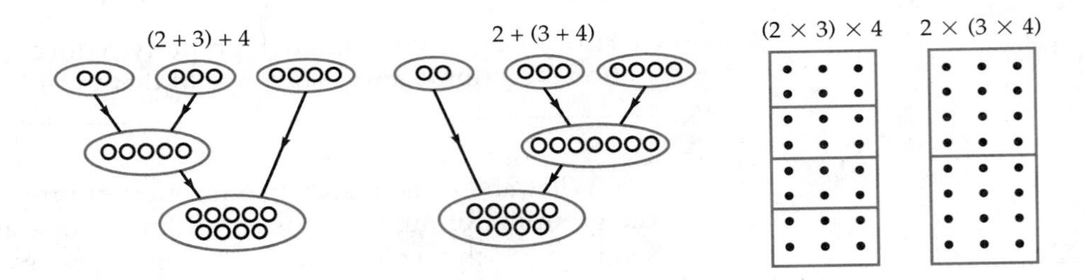

Figure 3–30

Like the commutative property, the associative property applies to expressions involving *one* operation, either addition or multiplication. The commutative property allows you to change the order of the numbers being added (or multiplied). The associative property allows you to change the order in which the operations are performed. We use the associative property in algebra to simplify an expression such as $8 \cdot (3x)$.

Lesson Exercise 3.59

(a) Does $8 \cdot (3 \cdot x) = 8 \cdot 3 \cdot 8 \cdot x$? (If you're not sure, try substituting some numbers for x.)

(b) What property says that, for a whole number x, $8 \cdot (3 \cdot x) = (8 \cdot 3) \cdot x$?

(c) Part (b) confirms that $8 \cdot (3 \cdot x)$ simplifes to ________.

Together, the commutative and associative properties allow us to reorder and regroup numbers in an addition problem in any way we want! We can put the addends in any order, and we can remove or put them in parentheses. For example, we can change $46 + (28 + 4) + 2$ to $(46 + 4) + (28 + 2)$ using the commutative and associative properties to create easier computations. The same reordering and regrouping may be done to any multiplication problem.

Lesson Exercise 3.60

(a) Rewrite $8 \cdot (4 \cdot 7) \cdot 25$ to make it easier to compute mentally.

(b) What properties justify your answer to part (a)?

THE IDENTITY PROPERTY

Some whole-number operations have a unique number called an identity element. The following exercises will guide you through an investigation of identity elements.

Lesson Exercise 3.61

What whole number, if any, can go in both blanks in the following statement? For every whole number w, $w + \_\_\_\_ = \_\_\_\_ + w = w$.

Lesson Exercise 3.62

What whole number, if any, can go in both blanks in the following statement? For every whole number w, $w \cdot \_\_\_\_ = \_\_\_\_ \cdot w = w$.

Lesson Exercise 3.63

What whole number, if any, can go in both blanks in the following statement? For every whole number w, $w - \_\_\_\_ = \_\_\_\_ - w = w$.

Lesson Exercise 3.64

What whole number, if any, can go in both blanks in the following statement? For every whole number w, $w \div \_\_\_\_ = \_\_\_\_ \div w = w$.

In Lesson Exercises 3.61 through 3.64, did you find that only addition and multiplication have a unique whole number that goes in both blanks and works for all whole numbers? These special numbers are called **identity elements**.

Identity Elements for Whole-Number Addition and Multiplication

Zero is the unique additive identity such that, for every whole number w, $w + 0 = 0 + w = w$.

One is the unique multiplicative identity such that, for every whole number w, $w \cdot 1 = 1 \cdot w = w$.

There is no identity element for whole-number subtraction or division. Don't let this upset you too much. Let's return to addition and multiplication and enjoy the fact that these operations do have identities.

Lesson Exercise 3.65

(a) The equation $3 + 5 = 8$ is an example of a one-digit addition fact. What one-digit addition facts use the identity property?

(b) What one-digit multiplication facts use the identity property?

Together, the commutative and identity properties of addition significantly reduce the number of separate addition facts that need to be memorized. In multiplication, the commutative and identity properties, along with the fact that $0 \times a = 0$ for every whole number a, greatly reduce the number of separate multiplication facts that children must memorize. You will learn additional strategies children can use to memorize addition and multiplication facts if you take a course in methods of teaching mathematics.

THE DISTRIBUTIVE PROPERTY

The distributive property plays an important part in the procedure for multiplying numbers such as 34×2. The distributive property is the only property in this lesson that involves two different operations at the same time. The distributive property of multiplication over addition is

$$F \cdot (G + H) = (F \cdot G) + (F \cdot H)$$

***Cooperative* Lesson Exercise 3.66**

(a) Try some examples and see whether you think the distributive property of multiplication over addition holds for all whole numbers.

(b) The conclusion is based upon ____________ reasoning.

Lesson Exercise 3.66 suggests that the whole numbers possess the distributive property of multiplication over addition.

The Distributive Property of Whole-Number Multiplication Over Addition

For any whole numbers x, y, and z, $x(y + z) = (xy) + (xz)$.

Figure 3-31 illustrates the distributive property using arrays.

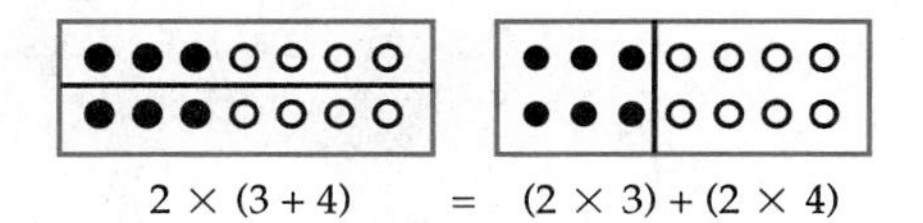

Figure 3–31

Lesson Exercise 3.67

Draw a set diagram showing that $3 \times (2 + 5) = (3 \times 2) + (3 \times 5)$.

People use the distributive property in algebra to combine like terms. Consider the following exercise.

Lesson Exercise 3.68

If x is a whole number, the distributive property of multiplication over addition says that $7 \cdot x + 3 \cdot x =$ ________________.

Do any other distributive properties work for whole-number operations?

Cooperative **Lesson Exercise 3.69**

(a) $F + (G \times H) = (F + G) \times (F + H)$ would be called the distributive property of ________________ over ________________.

(b) Try some whole-number examples and see whether you think the property in part (a) is true.

Cooperative **Lesson Exercise 3.70**

(a) $F \times (G - H) = (F \times G) - (F \times H)$ would be called the distributive property of ________________________.

(b) Try some whole-number examples and see whether you think the property in part (a) is true.

(c) The conclusion in part (b) is based upon ____________ reasoning.

As Lesson Exercise 3.70 suggests, another distributive property holds for whole numbers.

The Distributive Property of Whole-Number Multiplication Over Subtraction

For any whole numbers x, y, and z, $x(y - z) = (xy - xz)$.

Lesson Exercise 3.71

A storekeeper buys 24 televisions for $99 each. A method for mentally multiplying 24×99 is to compute $(24 \times 100) - (24 \times 1)$. Use a distributive property to explain why these two expressions are equal.

The following chart summarizes all the properties of whole-number operations presented in this section.

Properties of Whole-Number Operations

Whole-number addition is commutative and associative.

Whole-number multiplication is commutative and associative.

The additive identity for whole numbers is 0.

The multiplicative identity for whole numbers is 1.

Multiplication is distributive over addition and subtraction for whole numbers.

THE ADDITION ALGORITHM

People have developed procedures to perform paper-and-pencil computations involving large numbers more easily. Despite having such procedures, people in the Middle Ages considered whole-number arithmetic a college-level subject! Today children in elementary school study some of the best of these computational procedures, called **algorithms**. We typically use algorithms to perform computations with two- and three-digit numbers.

"Algorithm" is a term you can use to impress friends at parties when they ask, "What *are* you studying in that math class, anyway?" Tell them, "We're analyzing a few algorithms."

In the most common addition algorithm, children line up digits according to place value and proceed from right to left, adding the digits in each column. This method enables children to compute the answer to a problem such as 32 + 42, in which the sum of the digits in each column is less than 10 (Figure 3-32).

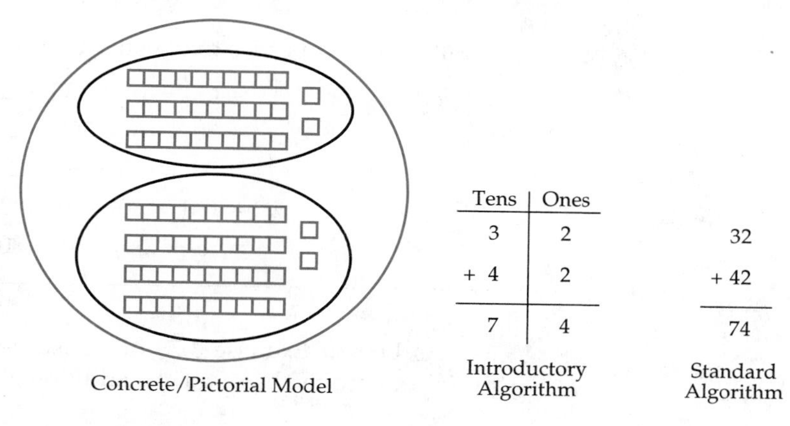

Figure 3–32

If the digits in any column add up to 10 or more, regrouping is needed. Dienes (base-ten) blocks help children understand the addition algorithm. They are especially useful for showing how regrouping works.

Lesson Exercise 3.72

Suppose you want to introduce a child to the procedure for computing

$$\begin{array}{r} 38 \\ +54 \\ \hline \end{array}.$$

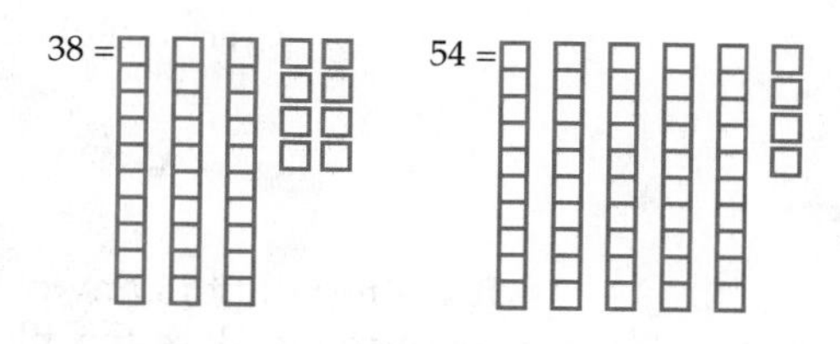

Figure 3–33

Describe how you would get the answer 92 by adding the blocks shown in Figure 3-33 and trading in. Follow the same sequence of steps that one uses in computing $\begin{array}{r} 38 \\ +54 \\ \hline \end{array}$ with paper and pencil.

One can use the properties of whole-number operations to show how the addition algorithm works. For example, in computing $\begin{array}{r}56\\+32\\\hline\end{array}$ why are we allowed to add 2 + 6 and then add 50 + 30 to that result?

Lesson Exercise 3.73

Fill in the properties that justify the last three steps.

$56 + 32 = (50 + 6) + (30 + 2)$	Expanded notation
$= 50 + (6 + 30) + 2$	+ is associative
$= 50 + (30 + 6) + 2$	______________
$= (50 + 30) + (6 + 2)$	______________
$= (6 + 2) + (50 + 30)$	______________

As Lesson Exercise 3.73 demonstrates, the commutative and associative properties are the basis for the addition algorithm.

THE SUBTRACTION ALGORITHM

In studying the subtraction algorithm, children first solve problems that do not require regrouping. Children line up digits according to place value and proceed from right to left, subtracting each digit from the one above it. This process is sufficient to compute the answer to a problem such as 74 − 32 (Figure 3-34).

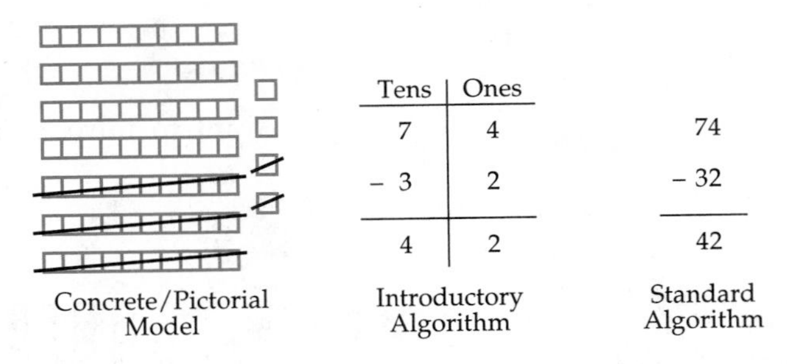

Figure 3–34

If the top digit in any column is less than the bottom digit, the child must use regrouping. Base-ten blocks help demonstrate regrouping in a problem such as 54 − 38.

Lesson Exercise 3.74

Suppose you want to introduce a child to the procedure for computing $\begin{array}{r}54\\-38\\\hline\end{array}$. Using words and pictures, describe how you would get the answer (16) by subtracting and trading in blocks. Start with 54 and take away 38. Follow the same sequence of steps that you would use in computing $\begin{array}{r} 54 \\ -38 \\ \hline \end{array}$ with paper and pencil. Start with the ones.

The **equal-additions algorithm** is another subtraction algorithm that has been used in some U.S. schools in the last 60 years. The equal-additions algorithm is based upon the concept that there is an infinite number of equivalent subtraction problems having a given difference. Many European schools in the fifteenth and sixteenth centuries used this algorithm. See how you think it compares to the standard algorithm after working the next three lesson exercises.

Lesson Exercise 3.75

Suppose you add the same amount to both numbers in a subtraction problem. What happens to the answer? Try the following.

(a) What is 86 − 29?

(b) Add 1 to both numbers in part (a) and subtract. Do you obtain the same answer?

(c) Add 11 to both numbers in part (a) and subtract. Do you obtain the same answer?

Lesson Exercise 3.76

As the preceding exercise suggests, you can add the same amount to both numbers in a subtraction problem without changing the result. For example, this technique can be used to change 53 − 27 into 56 − 30, an easier problem.

(a) Rewrite 44 − 18 as an equivalent but easier subtraction problem.

(b) Rewrite 322 − 96 as an equivalent but easier subtraction problem.

Lesson Exercise 3.77

The property developed in the preceding two exercises is the basis for the equal-additions algorithm. For example, in computing 563 − 249, one needs to add 10 to the 3. To compensate, one adds 10 to 249. Then the subtraction can be done without regrouping.

$$\begin{array}{r} 5\ 6\ 3 \\ -2\ 4\ 9 \\ \hline \end{array} \qquad \begin{array}{r} 5\ 6\ {}^{1}3 \\ -2\ {}^{5}\!\not{4}\ 9 \\ \hline \end{array} \qquad \begin{array}{r} 5\ 6\ {}^{1}3 \\ -2\ {}^{5}\!\not{4}\ 9 \\ \hline 3\ 1\ 4 \end{array}$$

(a) Compute 86 − 29 using the equal-additions algorithm.

(b) How do you think this algorithm compares to the standard algorithm?

Research indicates that children can learn to do subtraction quickly and accurately with either the standard algorithm or the equal-additions algorithm. However, research also confirms that the standard algorithm is easier for children to understand since it is based upon place-value trades that can be more easily illustrated with manipulatives such as base-ten blocks or the abacus.

THE MULTIPLICATION ALGORITHM

In studying the multiplication algorithm, children first multiply a two-digit number by a one-digit number. Each digit in one factor is multiplied by each digit in the other factor (Figure 3-35).

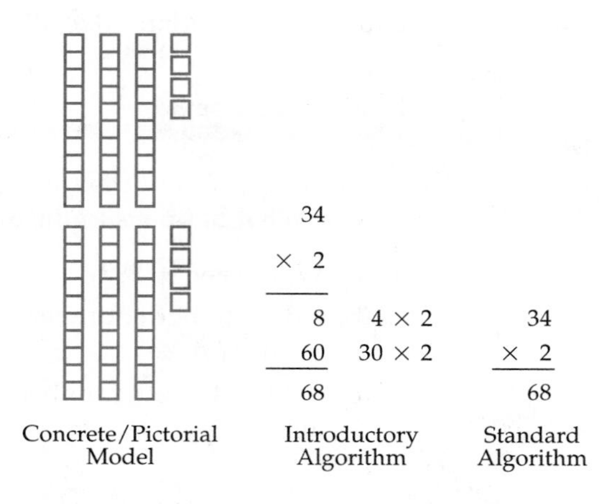

Figure 3–35

Base-ten blocks can help children understand the standard procedure.

Lesson Exercise 3.78

Suppose you want to introduce a child to the procedure for computing $\begin{array}{r} 26 \\ \times\ 3 \\ \hline \end{array}$. Using words along with pictures of base-ten blocks, describe how you would obtain the answer (78) by multiplying and trading in blocks. Follow the same sequence of steps you would use in computing 26×3 using paper and pencil.

Whole-number properties help justify the standard procedure. For example, why are we allowed to multiply 2×4 and 2×30 and add the results together? Try the following exercise.

Lesson Exercise 3.79

Fill in the properties that justify the last two steps.

$34 \times 2 = (30 + 4) \times 2$	Expanded notation
$= (30 \times 2) + (4 \times 2)$	______________
$= (4 \times 2) + (30 \times 2)$	______________

The preceding exercise illustrates how the distributive property is used in the multiplication algorithm. Multiplying larger numbers is merely an extension of this process (Figure 3-36).

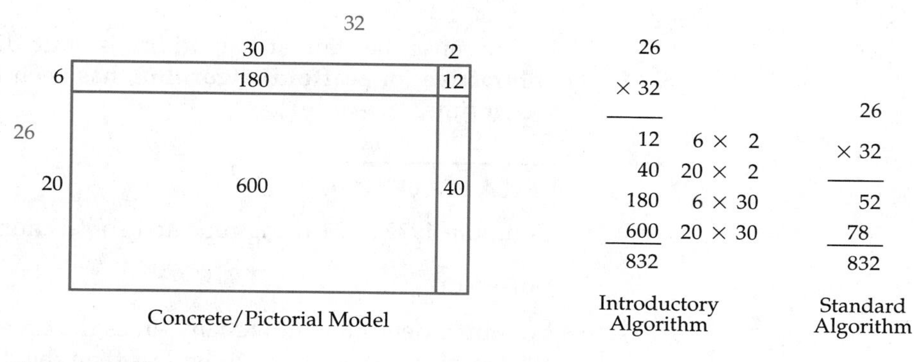

Figure 3–36

THE DIVISION ALGORITHM

The standard long-division algorithm for whole numbers is more complicated than, and somewhat unlike, the algorithms for the other three operations. Instead of lining up the digits of the numbers as in the other operations, one writes the divisor to the left of the dividend. Unlike the other three operations, one finds the quotient from *left to right* (Figure 3-37). And while one simply computes one-digit basic facts to perform the other algorithms, the division algorithm may require rounding the divisor (if it is more than one digit) and the dividend and then using estimation (Section 3.5) to obtain digits in the quotient.

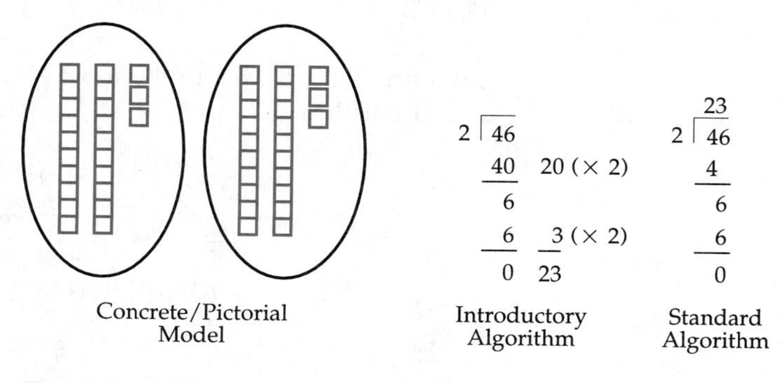

Figure 3–37

Base-ten blocks are often used to develop the standard division algorithm (Figure 3-37).

Lesson Exercise 3.80

Suppose you want to introduce a child to the procedure for computing 54 ÷ 2. Start with 5 tens and 4 ones, and use words and pictures to explain how to divide 54 into 2 equal groups. Follow the same sequence you would use in computing 54 ÷ 2 with paper and pencil.

Because the standard algorithm is quite difficult, a second algorithm, the **subtractive** (or **scaffold) algorithm**, has been taught in a few U.S. schools in recent times.

EXAMPLE 3.4

Compute 4972 ÷ 21 using repeated subtraction and the subtractive algorithm.

Solution

You must determine how many times the divisor 21 can be *subtracted* from 4972 to obtain 0 or a remainder less than the divisor 21. One could start by subtracting groups of one hundred 21's and continue as shown until the remainder is less than 21. The right-hand column is a record of how many 21's have been subtracted. (*Note:* In each step, the groupings of 21's you subtract may be of any size, but it helps to do this efficiently.)

```
21 | 4972
     2100     100 (× 21)
     ----
     2872
     2100     100 (× 21)
     ----
      772
      630      30 (× 21)
     ----
      142
      126       6 (× 21)
     ----     ---
       16     236
```

So 4972 ÷ 21 = 236 R 16, or 4972 = (21)(236) + 16. ■

Compare Example 3.4 to the most efficient subtractive algorithm and the standard algorithm.

```
     236R 16                             236R 16
21 | 4972                           21 | 4972
     42         200 (× 21)               42
     --                                  --
     77                                  77
      63         30 (× 21)                63
      --                                  --
      142                                 142
      126         6 (× 21)                126
      ---       ---                       ---
       16       236                        16
Subtractive algorithm               Standard algorithm
```

Lesson Exercise 3.81

(a) Compute 6874 ÷ 22 using the subtractive algorithm.
(b) Compute 6874 ÷ 22 using the standard algorithm.
(c) What are the advantages and disadvantages of each algorithm?

Most people find that the subtractive algorithm is longer and easier to understand than the standard long-division algorithm. The standard algorithm is simply a more efficient and compact version of the subtractive algorithm.

COMMON ERROR PATTERNS IN ALGORITHMS

The bad news is that not all students learn the algorithms just the way you teach them. Confused students may initially develop their own erroneous procedures. The goods news is that you get to play detective in trying to uncover students' error patterns.

Children's written work provides the first evidence of many learning difficulties. Try your luck at finding the error patterns in the following exercises.

In Lesson Exercises 3.82 through 3.85, (a) complete the last example, repeating the error pattern in the completed examples, and (b) write a description of the error pattern.

Lesson Exercise 3.82

$$\begin{array}{r} 49 \\ +37 \\ \hline \textit{76} \end{array} \qquad \begin{array}{r} 67 \\ +43 \\ \hline \textit{100} \end{array} \qquad \begin{array}{r} 92 \\ +39 \\ \hline \end{array}$$

Lesson Exercise 3.83

$$\begin{array}{r} 92 \\ -39 \\ \hline \textit{67} \end{array} \qquad \begin{array}{r} 408 \\ -322 \\ \hline \textit{126} \end{array} \qquad \begin{array}{r} 921 \\ -376 \\ \hline \end{array}$$

Lesson Exercise 3.84

$$\begin{array}{r} 36 \\ \times\ 8 \\ \hline \textit{568} \end{array} \qquad \begin{array}{r} 42 \\ \times\ 6 \\ \hline \textit{302} \end{array} \qquad \begin{array}{r} 72 \\ \times\ 9 \\ \hline \end{array}$$

Lesson Exercise 3.85

$$\begin{array}{r} \textit{121} \\ 4\overline{)623} \end{array} \qquad \begin{array}{r} \textit{184} \\ 8\overline{)912} \end{array} \qquad \begin{array}{r} \\ 3\overline{)782} \end{array}$$

CHILDREN CREATING ALGORITHMS

The most influential theory behind recent curriculum reform is constructivism. According to **constructivism**, each child must create mathematics in his or her own fashion rather than imitating and internalizing previously developed procedures.

What would happen if first and second graders were not taught the traditional addition and subtraction algorithms but instead were allowed to invent their own algorithms? Research by Rob Madell, Constance Kamii, and others suggests that most children would add or subtract from left to right instead of from right to left. (See the December 1993 issue of *Arithmetic Teacher*.)

Try the following algorithms that were developed by first and second graders.

Lesson Exercise 3.86

(a) Kamii, Lewis, and Livingston describe the following steps in the procedure a child invented for 18 + 17.

$$10 + 10 = 20 \rightarrow 8 + 7 = 15 \rightarrow 20 + 10 = 30 \rightarrow 30 + 5 = 35$$

Write the steps for computing 26 + 15 the same way.

(b) Madell describes the following steps in the procedure a child invented for 53 − 24.

$$50 - 20 = 30 \rightarrow 30 - 4 = 26 \rightarrow 26 + 3 = 29$$

Write the steps for computing 32 − 19 using the same procedure.

ALTERNATIVE LOW-STRESS ALGORITHMS

In the 1976 NCTM yearbook, Barton Hutchins describes alternative algorithms that are easier for some children. Today his subtraction algorithm and a modified version of his addition algorithm are sometimes used with older students who have been unable to master the corresponding standard algorithm. The modified form of Hutchins' low-stress addition algorithm is called **scratch addition**.

EXAMPLE 3.5

Compute 57 + 86 + 39 using scratch addition.

Solution

Scratch addition requires only addition of single digits. Write the example vertically.

57 8$\not{6}_3$ +39	Add the numbers in the units place, starting at the top. If the sum is 10 or more, "scratch" a line through the last digit added and write the number of units just below it.

$$\begin{array}{r} 57 \\ 8\not{6}_3 \\ +3\not{9} \\ \hline 2 \end{array}$$

Continue adding units.

$$\begin{array}{r} {\scriptstyle 1\;2} \\ 5\;7 \\ \not{8}_5\not{6}_3 \\ +\;3\;\not{9} \\ \hline 1\;8\;2 \end{array}$$

Count the number of scratches in the column, and write it at the top of the next column. Repeat the procedure for each successive column. ■

Lesson Exercise 3.87

(a) Compute 38 + 97 + 246 using scratch addition.
(b) What is easier about scratch addition?

Hutchins' **low-stress subtraction** algorithm for regrouping differs from the standard procedure in two ways. First, the renamed minuend (first or top number) is written between the minuend and the subtrahend (second or lower number). Second, all regrouping is done before any subtraction.

EXAMPLE 3.6

Compute 324 − 168 using low-stress subtraction.

Solution

Write the problem vertically and draw a box between the minuend and the subtrahend where the renamed minuend will go.

$$\begin{array}{r} 324 \\ \boxed{} \\ -168 \\ \hline \end{array}$$

First, do all needed regrouping in the minuend. Start with the ones. Since 8 is bigger than 4, regroup from the tens to the ones.

$$\begin{array}{r} 324 \\ \boxed{1\,{}^1 4} \\ -168 \\ \hline \end{array}$$

Next, check the tens. Since 6 is bigger than 1, regroup from the hundreds to the tens.

$$\begin{array}{r} 324 \\ \boxed{2\,{}^1 1\,{}^1 4} \\ -168 \\ \hline \end{array} \quad \xrightarrow{\text{Now subtract.}} \quad \begin{array}{r} 324 \\ \boxed{2\,{}^1 1\,{}^1 4} \\ -168 \\ \hline 156 \end{array}$$

■

Lesson Exercise 3.88

(a) Compute 827 − 469 using low-stress subtraction.
(b) What is easier about low-stress subtraction?

The homework exercises contain examples of older algorithms that are not as efficient or as easy to understand as those we teach.

ANSWERS TO SELECTED LESSON EXERCISES

3.54 (a) No (b) Yes (d) Inductive

3.55 (a) No, $2 - 1 \neq 1 - 2$
(c) No, $2 \div 1 \neq 1 \div 2$

3.56 After memorizing any fact with two different addends (e.g., $3 + 5 = 8$), you will also know a companion fact with the addends reversed (e.g., $5 + 3 = 8$). This reduces the number of basic facts that need to be learned from 100 to 55.

+	0	1	2	3	4	5	6	7	8	9
0	0	1	2	3	4	5	6	7	8	9
1		2	3	4	5	6	7	8	9	10
2			4	5	6	7	8	9	10	11
3				6	7	8	9	10	11	12
4					8	9	10	11	12	13
5						10	11	12	13	14
6							12	13	14	15
7								14	15	16
8									16	17
9										18

3.57 (a) Yes (b) Yes

3.58 (a) No, $(3 - 2) - 1 \neq 3 - (2 - 1)$
(c) No, $(8 \div 4) \div 2 \neq 8 \div (4 \div 2)$

3.59 (a) No
(b) Associative property of ×

3.60 (a) $(4 \cdot 25) \cdot 8 \cdot 7$
(b) Commutative and associative properties

3.61 0

3.62 1

3.63 None

3.64 None

3.65 (a) 0 + any number or any number + 0
(b) 1 × any number or any number × 1

3.66 (b) inductive

3.67

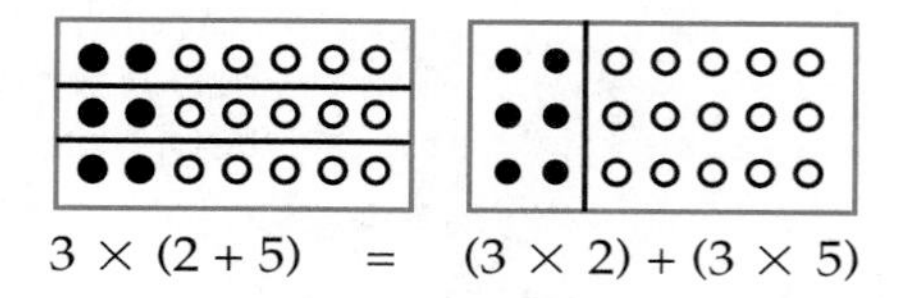

$3 \times (2 + 5) = (3 \times 2) + (3 \times 5)$

3.68 $(7 + 3) \cdot x$

3.69 (a) addition over multiplication
(b) No

3.70 (a) multiplication over subtraction
(b) Yes
(c) inductive

3.71 $24 \times 99 = 24 \times (100 - 1) = (24 \times 100) - (24 \times 1)$

3.72 Start with 3 tens and 8 ones, and 5 tens and 4 ones. Next, combine 8 ones and 4 ones to obtain 12 ones. Trade 10 ones in for 1 ten, leaving 2 ones. Combine 1 ten, 3 tens, and 5 tens to obtain 9 tens. The sum is 9 tens, 2 ones = 92.

3.73 + is commutative; + is associative; + is commutative

3.74 Start with 5 tens and 4 ones. You cannot take away 8 ones from 4 ones. Regroup 5 tens, 4 ones as 4 tens, 14 ones. Now take 8 ones away from 14 ones, leaving 6 ones. Take away 3 tens from 4 tens, leaving 1 ten. The difference is 1 ten, 6 ones = 16.

3.75 (a) 57 (b) Yes (c) Yes

3.76 (a) 46 − 20 (b) 326 − 100

3.77 (a)

```
 1  2
    3 8
    9̸ 4 7̸ 5
 + 2 4 6̸
   3 8 1
```

3.78 3 × 26 means 3 sets of 26. Show this with base-ten blocks.

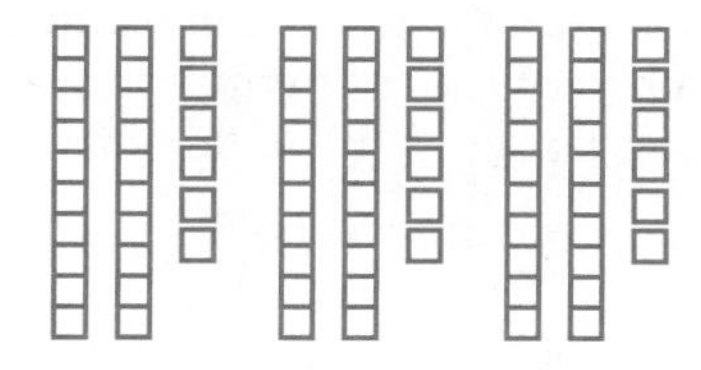

Combine the ones together. You have 18 ones. Trade in 10 ones for 1 ten. Combine the tens together. You end up with 7 tens and 8 ones = 78. So 3 × 26 = 78.

3.79 Distributive; + is commutative

3.80 54 ÷ 2 means to divide 54 into 2 equal groups. Start with 5 tens and 4 ones.

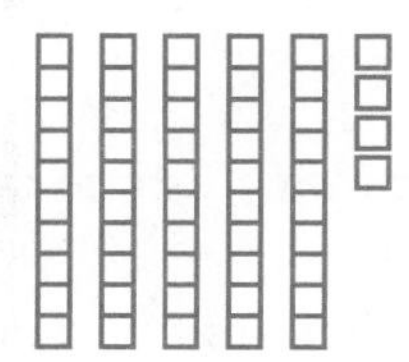

First, put the 5 tens into 2 equal groups. Each group gets 2 tens. Trade the leftover ten for 10 ones. Now, divide the 14 ones into the 2 equal groups that have the other tens. Each group ends up with 2 tens and 7 ones = 27. So 54 ÷ 2 = 27.

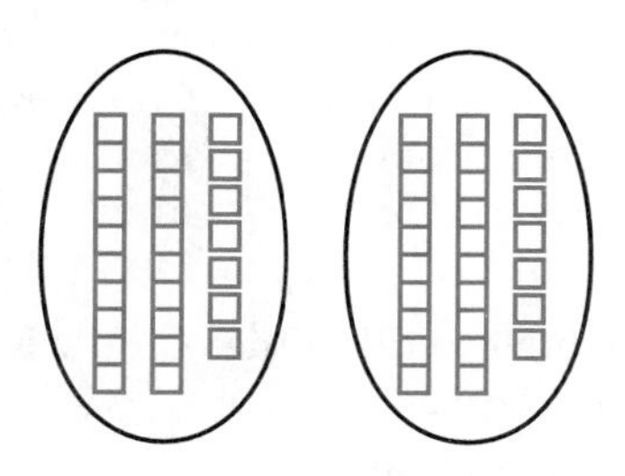

3.81 (a) Possible solution:

```
        312 R 10
  22 ) 6874
       6600     300 (× 22)
        274
        220      10 (× 22)
         54
         44       2 (× 22)
         10     312
```

3.82 (a) 121
(b) The child does not regroup (''carry'') the 1 to the tens column.

3.83 (a) 655
(b) The child subtracts the smaller digit from the larger in each place-value column.

3.84 (a) 728
(b) In regrouping the ones to the tens column, the child adds the number to the tens digit *before* multiplying.

3.85 (a) 221
(b) The child moves from left to right, dividing the smaller digit into the larger and discarding all remainders.

3.86 (a) 20 + 10 = 30 → 6 + 5 = 11 → 30 + 10 = 40 → 40 + 1 = 41
(b) 30 − 10 = 20 → 20 − 9 = 11 → 11 + 2 = 13

3.87 (a)

```
 1  2
    3 8
    9̸ 4 7̸ 5
 + 2 4 6̸
   3 8 1
```

(b) It requires only the addition of one-digit numbers.

3.88 (a)

```
    8 2 7
   [7 ¹1 ¹7]
  − 4 6 9
    3 5 8
```

(b) It requires less switching back and forth between different procedures (regrouping and subtracting).

3.4 HOMEWORK EXERCISES

Basic Exercises

1. Explain all the different ways to compute $3 \cdot 2 \cdot 4$ by multiplying.

2. X and Y are whole numbers. $X + (Y + 3) = (X + Y) + 3$ illustrates the ________ property of ________.

3. (a) Four sets of 3 is the same as 3 sets of 4 because of the ________ property of ________.
(b) Draw a picture that shows that the two amounts in part (a) are the same.

4. (a) What whole-number operations are commutative?
(b) What whole-number operations are associative?

5. Draw arrays showing that $(2 \times 4) \times 5 = 2 \times (4 \times 5)$.

***6.** Some people confuse $8 \div 2$ and $2 \div 8$.
(a) $8 \div 2 \neq 2 \div 8$ is a counterexample disproving what property?
(b) Explain the difference between $8 \div 2$ and $2 \div 8$.

7. Give a counterexample showing that whole-number subtraction is not associative.

8. You buy a simulated Astroturf carpet for $38, a talking wastebasket for $57, and a half-pound chocolate moose for $2. What is the easiest way to compute your total bill?

9. A carpet that is 6 ft by 9 ft costs $5 per square foot. Explain an easy way to compute mentally the total cost of the carpet.

10. Three items cost $246, $38, and $4. Explain an easy way to compute mentally the total cost.

11. How does the commutative property reduce the number of one-digit multiplication facts that must be memorized?

12. Is the operation of putting on your shoes and socks commutative?

13. In English, the meaning of a phrase may change depending upon which words are associated. "Slow-motion picture" and "slow motion picture" have different meanings. Therefore, the phrase "slow-motion picture" is not associative. Which of the following phrases is associative?
(a) Man eating shark
(b) Smart handsome stranger
(c) Hot dog salesperson
(d) High-school student

14. In English, the meaning of a phrase may change depending upon the order of the words. Does changing the order change the meanings of the following phrases?
(a) Hall light (b) Killer shark

15. Do the problems in the first column with a calculator. Then find a pattern you can use to do the problems in the second column without a calculator.

(a) $37 \times 3 =$ ______ $37 \times 12 =$ ______
$37 \times 6 =$ ______ $37 \times 18 =$ ______
$37 \times 9 =$ ______ $37 \times$ ______ $= 888$

(b) $3367 \times 3 =$ ______ $3367 \times 15 =$ ______
$3367 \times 6 =$ ______ $3367 \times 21 =$ ______
$3367 \times 9 =$ ______ $3367 \times$ ______ $= 40{,}404$

16. Without computing their exact values, tell which is greater, 3^{14} or 9^{10}.

17. $6 + 0 = 0 + 6 = 6$
$8 + 0 = 0 + 8 = 8$
These examples illustrate that ______ is the ________ for ________.

***18.** What one-digit multiplication facts use the identity property?

19. Until fourth or fifth grade, many mathematics textbooks refer to some of the properties by alternative names. Match each name in column 1 with its corresponding name in column 2.

Lower Elementary Grades	Upper Elementary Grades
Order property	Associative
Zero property	Commutative
Grouping property	Identity for addition
Property of one	Identity for multiplication

20. Show how to fill in the blank without computing 12×34.

$$(12 \times 34) - 48 = 12 \times \underline{\qquad}$$

21. (a) Find the total area of the yard shown in the following picture.
(b) Find the total area a different way.
(c) If possible, show that your two methods are related by the distributive property.

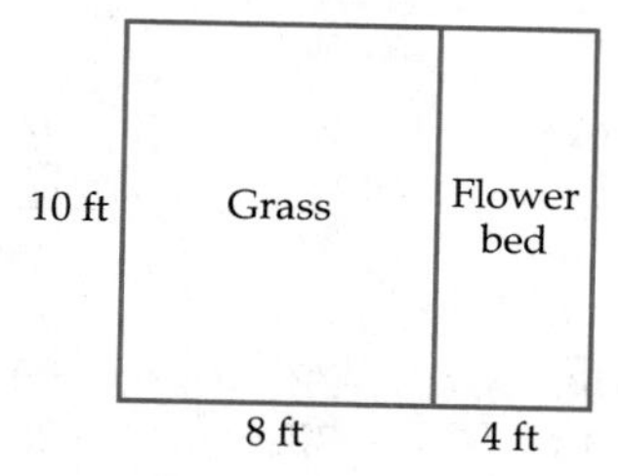

22. You drove your car on two different days for business purposes, going 120 miles and 62 miles. The company will reimburse you at the rate of 20 cents per mile. Show two ways of computing the total cost that illustrate the distributive property.

23. Name the property illustrated.
(a) $3 \cdot (2 \cdot r) = (3 \cdot 2) \cdot r$
(b) $4 \cdot x + 3 \cdot x = (4 + 3)x$
(c) $n \cdot 2 = 2 \cdot n$
(d) $(8 + 2w) + 5w = 8 + (2w + 5w)$

24. Match each property name with an equation.

Associative property of addition
Commutative property of addition
Identity property of addition
Distributive property of multiplication over addition

(a) $a + b = b + a$
(b) $a(b + c) = ab + ac$
(c) $a + 0 = a$
(d) $a + (b + c) = (a + b) + c$

25. If x is a whole number, justify each of the following steps.

$(x + 3)(x + 6) = x(x + 6) + 3(x + 6)$ ______
$= x^2 + 6x + 3x + 18$ ______
$= x^2 + (6 + 3)x + 18$ ______
$= x^2 + 9x + 18$ Basic addition fact

26. Fill in the blank with the property that justifies the step.

$36 + 8 = (30 + 6) + 8$ Expanded notation
$= 30 + (6 + 8)$ ______

27. Describe how to compute 182 + 336 using base-ten blocks. Follow the sequence of steps that one uses in computing 182 + 336 with paper and pencil.

28. When you add 39 + 48 using the standard algorithm, you add 9 + 8 + 30 + 40 and regroup. The following shows why this method is correct. Fill in the blanks with properties you studied in this lesson.

$39 + 48 = (30 + 9) + (40 + 8)$ Expanded notation
$= 30 + (9 + 40) + 8$ ______
$= 30 + (40 + 9) + 8$ ______
$= (30 + 40) + (9 + 8)$ ______
$= (9 + 8) + (30 + 40)$ ______
$= 17 + (30 + 40)$ Addition fact
$= 7 + 10 + (30 + 40)$ Expanded notation
$= 7 + (10 + 30 + 40)$ ______

29. Show how to compute 336 − 182 using base-ten blocks. Follow the sequence of steps that one uses in computing 336 − 182 with paper and pencil.

30. In subtracting 87 from 325, 325 must be regrouped as ____ hundreds, ____ tens, and ____ ones.

31. Compute the following using the equal-additions algorithm.
(a) $\begin{array}{r} 72 \\ -47 \\ \hline \end{array}$ (b) $\begin{array}{r} 821 \\ -376 \\ \hline \end{array}$

32. Use a number line to show that 21 − 7 = 24 − 10.

33. Describe how to compute 24 × 4 using base-ten blocks. Follow the same sequence of steps that is used in computing $\begin{array}{r} 24 \\ \times\ 4 \\ \hline \end{array}$ with paper and pencil.

34. Describe how to compute 47 × 2 using base-ten blocks. Follow the sequence of steps that is used in computing 47 × 2 with paper and pencil.

35. In multiplying 62 × 3, we use the fact that (60 + 2) × 3 = (60 × 3) + (2 × 3). What property does this equation illustrate?

36. Draw a rectangular array with four subdivisions that correspond to the four partial products shown. (*Hint:* Find a helpful drawing in the lesson.)

$$\begin{array}{rl} 34 & \\ \times 23 & \\ \hline 12 & 3 \times 4 \\ 90 & 3 \times 30 \\ 80 & 20 \times 4 \\ 600 & 20 \times 30 \\ \hline 782 & \end{array}$$

37. Before learning the standard multiplication algorithm, children may learn to multiply 74 by 23 as follows.

$$\begin{array}{rl} 74 & \\ \times 23 & \\ \hline 222 & \leftarrow \ 3 \times 74 \\ 1480 & \leftarrow \ 20 \times 74 \\ \hline 1702 & \end{array}$$

Multiply 86 by 42 using this introductory algorithm.

38. The beginning of the division algorithm for $197 \div 3$ follows. Explain the meanings of the $6 \times 3 = 18$ and the 17.

$$\begin{array}{r} 6 \\ 3\overline{)197} \\ 18 \\ \hline 17 \end{array}$$

39. Compute the following using the subtractive algorithm.
(a) $437 \div 7$ (b) $8921 \div 37$

40. Compute the following using whichever algorithm you prefer.
(a) $3498 \div 12$ (b) $25{,}792 \div 32$

41. Describe how you would compute $246 \div 2$ using the partition model with base-ten blocks. Follow the sequence of steps that is used in computing $246 \div 2$ with paper and pencil. Start with the hundreds.

42. Describe how to compute $371 \div 2$ using base-ten blocks. Follow the sequence of steps that is used in computing $371 \div 2$ with paper and pencil.

In Exercises 43–50, (a) complete the last example, repeating the error pattern in the completed examples, and (b) write a description of the error pattern.

43.
$$\begin{array}{r} \mathit{1} \\ 76 \\ +\ \ 6 \\ \hline \mathit{142} \end{array} \qquad \begin{array}{r} \mathit{1} \\ 98 \\ +\ \ 7 \\ \hline \mathit{175} \end{array} \qquad \begin{array}{r} \\ 87 \\ +\ 8 \\ \hline \end{array}$$

44.
$$\begin{array}{r} \mathit{1} \\ 62 \\ +\ 57 \\ \hline \mathit{110} \end{array} \qquad \begin{array}{r} \mathit{1} \\ 52 \\ +84 \\ \hline \mathit{37} \end{array} \qquad \begin{array}{r} \\ 57 \\ +76 \\ \hline \end{array}$$

45.
$$\begin{array}{r} \mathit{1} \\ 86 \\ -48 \\ \hline \mathit{48} \end{array} \qquad \begin{array}{r} \mathit{1} \\ 72 \\ -37 \\ \hline \mathit{45} \end{array} \qquad \begin{array}{r} \\ 93 \\ -28 \\ \hline \end{array}$$

46.
$$\begin{array}{r} 40 \\ -27 \\ \hline \mathit{20} \end{array} \qquad \begin{array}{r} 306 \\ -215 \\ \hline \mathit{101} \end{array} \qquad \begin{array}{r} 809 \\ -763 \\ \hline \end{array}$$

47.
$$\begin{array}{r} 36 \\ \times\ \ 8 \\ \hline \mathit{248} \end{array} \qquad \begin{array}{r} 42 \\ \times\ 6 \\ \hline \mathit{242} \end{array} \qquad \begin{array}{r} 72 \\ \times\ 9 \\ \hline \end{array}$$

48.
$$\begin{array}{r} 82 \\ \times\ 37 \\ \hline \mathit{574} \\ \mathit{246} \\ \hline \mathit{820} \end{array} \qquad \begin{array}{r} 41 \\ \times\ 79 \\ \hline \mathit{369} \\ \mathit{287} \\ \hline \mathit{656} \end{array} \qquad \begin{array}{r} 27 \\ \times 32 \\ \hline \end{array}$$

49.
$$\begin{array}{r} \mathit{36} \\ 6\overline{)378} \\ \mathit{36} \\ \hline \mathit{18} \\ \mathit{18} \\ \hline \end{array} \qquad \begin{array}{r} \mathit{57} \\ 5\overline{)375} \\ \mathit{35} \\ \hline \mathit{25} \\ \mathit{25} \\ \hline \end{array} \qquad \begin{array}{r} \\ 8\overline{)512} \end{array}$$

50.
$$\begin{array}{r} \mathit{26} \\ 4\overline{)824} \end{array} \qquad \begin{array}{r} \mathit{15} \\ 5\overline{)525} \end{array} \qquad \begin{array}{r} \\ 6\overline{)1254} \end{array}$$

51. Describe two errors children might make in computing 27×30.

52. Consider the following steps in the procedure a child invented for $53 - 24$.

$$50 - 20 = 30 \rightarrow 4 - 3 = 1 \rightarrow 30 - 1 = 29$$

Write the steps for computing $32 - 19$ using the same procedure.

53. Consider the following steps in the procedure a child invented for 125×4.

$$100 \times 4 = 400 \rightarrow 20 \times 4 = 80 \rightarrow$$
$$5 \times 4 = 20 \rightarrow 400 + 80 + 20 = 500$$

Write the steps for computing 382×7 the same way.

54. Compute the following using scratch addition.

(a)
```
  386
   97
   58
+  87
```
(b)
```
  4679
   345
+  276
```

55. Compute the following using low-stress subtraction.

(a)
```
  68
 −39
```
(b)
```
  324
 −157
```

56. **Lattice multiplication** was passed along from the early Hindus and Chinese to the Arabs to medieval Europe. It appeared in the earliest known arithmetic book, which was published in Italy in 1478. Today it is sometimes taught as an enrichment topic in the upper elementary grades. The algorithm for 27×34 is shown here.

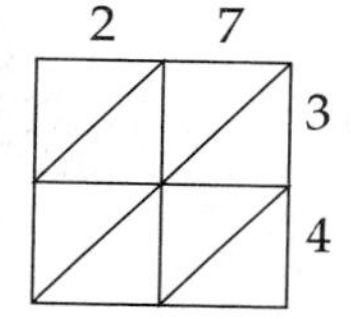

Step 1: Write the numbers.

Step 2: Multiply.

Step 3: Sum the numbers on each diagonal, beginning at the right. On the second diagonal, $1 + 2 + 8 = 11$. Put down the 1 and carry the other 1 to the third diagonal. Then $1 + 2 + 6 = 9$.

So $27 \times 34 = 918$.

(a) Compute 38×74 using lattice multiplication.

(b) Compute 123×35 using lattice multiplication.

(c) How do you think this algorithm compares to the standard ones?

57. The **Russian peasant multiplication algorithm** was used thousands of years ago by the Egyptians and until recently by Russian peasants. The Russian peasant algorithm for multiplying employs halving and doubling. Remainders are ignored when halving. The algorithm for 33×47 is shown here.

	Halving	**Doubling**	
Smaller factor →	33	(47)	← Larger factor
	16	94	
	8	188	
	4	376	
	2	752	
	1	(1504)	

Circle and add all the numbers in the doubling column that are paired with *odd* numbers in the halving column. $47 + 1504 = 1551$, so $33 \times 47 = 1551$.

(a) Compute 28×52 using this algorithm.

(b) Compute 18×127 using this algorithm.

(c) How does it compare to the standard algorithm?

58. A cashier may count off your change using the **cashier's subtraction algorithm.** If you give the cashier \$40 for a \$27 purchase, a cashier using this algorithm would say "\$28, \$29, \$30, \$40" while giving you change. Explain how the algorithm works in this example.

59. Name three consecutive counting numbers whose sum is

(a) 84. (b) 411.

(c) N, a counting number divisible by 3.

60. Write a sentence telling the difference between the commutative property of addition and the associative property of addition for whole numbers.

Extension Exercises

61. Find the quotient and *remainder* of $8569 \div 23$, using a calculator.

62. A, B, and C are whole numbers with $C \neq 0$. The following represent three fairly common algebra errors. For each equation, find one set of numbers that makes the equation true and one set of numbers that makes the equation false.

(a) $(A - B)^2 = A^2 - B^2$

(b) $(A + B)^2 = A^2 + B^2$

(c) $A(BC) = (AB)(AC)$

63. A, B, and C are whole numbers, and $A \neq 0$ and $B \neq 0$. For each equation, find one set of numbers that makes the equation true and one set of numbers that makes the equation false.

(a) $A \div B = B \div A$
(b) $C \div (A \div B) = (C \div A) \div B$
(c) $A - (B - C) = (A - B) - C$
(d) $A - B$ is a whole number.
(e) $A \div B$ is a whole number.
(f) $A \times (C \div B) = (A \times C) \div (A \times B)$
(g) $A - (C \div B) = (A - C) \div (A - B)$

64. (a) When you double both numbers in a subtraction problem, what happens to the difference?
(b) Use the distributive property to prove your answer to part (a).

65. (a) When you double both addends, what happens to the sum?
(b) Use the distributive property to prove your answer to part (a).
(c) What kind of reasoning did you use in part (a)?
(d) What kind of reasoning did you use in part (b)?

66. Show two other ways besides the standard algorithm to compute each of the following.
(a) $41 - 26$ (b) 41×26

67. When you multiply 39×48 using the standard algorithm, you multiply and add partial products $(9 \times 8) + (30 \times 8) + (9 \times 40) + (30 \times 40)$. The following proof shows why this is correct. Fill in the blanks with properties you studied in this lesson.

39×48
$= (30 + 9) \times (40 + 8)$ Expanded notation
$= [(30 + 9) \times 40] + [(30 + 9) \times 8]$ ______________
$= [(30 \times 40) + (9 \times 40)] + [(30 \times 8) + (9 \times 8)]$ ______________
$= [(30 \times 8) + (9 \times 8)] + [(30 \times 40) + (9 \times 40)]$ ______________
$= [(9 \times 8) + (30 \times 8)] + [(9 \times 40) + (30 \times 40)]$ ______________

68. (a) Explain why someone might incorrectly compute $2 \cdot (3 \cdot 4)$ as $2 \cdot 3 \cdot 2 \cdot 4$.
(b) How would you explain the correct answer to this person?

69. (a) Select a three-digit number whose first and third digits are different.
(b) Reverse the digits of your number and subtract the smaller of the two numbers from the larger. (Parts (c)–(f) follow.)
(c) Select another three-digit number and do the same thing.
(d) Select another three-digit number and do the same thing.
(e) What pattern do you see in your answer?
(f) Finding the general pattern from examples involves ______________ reasoning.

70. Fill in each box using each digit from 1 to 9 once.

```
  □□□
- □□□
-----
  □□□
```

71. Fill in the missing digits.

```
     2??
 6 ) ?2??
     ??
     --
      3
      ?
      --
      ??
      ??
      --
       0
```

Special Exercises

72. Consider the set of whole numbers and the operation ∇, which takes the larger of any two numbers as the result. For example, $3 \nabla 6 = 6$ and $8 \nabla 2 = 8$.
(a) Is ∇ commutative for whole numbers?
(b) Is there an identity number for ∇ in W?
(c) Is ∇ associative for whole numbers?

73. Consider the set of numbers $A = \{5, 6, 7, 8\}$ with a made-up operation called $*$. The results for $*$ are shown in the table.

*	5	6	7	8
5	8	7	6	5
6	7	6	5	8
7	6	5	8	7
8	5	8	7	6

(a) Is $*$ commutative for set A?
(b) Is there a number I (an identity element) in A such that $I * a = a * I = a$ for all numbers a in A?
(c) Does $(5 * 6) * 8 = 5 * (6 * 8)$?
(d) Is $*$ associative for set A?

3.5 WHOLE NUMBERS: MENTAL COMPUTATION AND ESTIMATION

Believe it or not, your ancestors used to do all their arithmetic without electronic calculators! In order to cope with this hardship, they developed a repertoire of computational shortcuts. Today, most adults solve complicated arithmetic problems with a calculator. But what happens if you don't have a calculator with you? Would you be limited to doing one-digit arithmetic?

Mental computation is used to obtain an *exact* answer to a computation that can be done easily without paper and pencil or a calculator. **Estimation** (usually done mentally) is used to obtain an *approximate* answer to a more difficult computation.

LEFT-TO-RIGHT ADDITION AND MULTIPLICATION

Do you enjoy doing something differently from the way it's "normally" done? Sometimes that's a good idea! Addition and multiplication can sometimes be done mentally by working from left to right instead of from right to left.

EXAMPLE 3.7

A woman buys a dress for $46 and a coat for $79. How can she mentally compute the (exact) total cost?

Solution

Two break-apart methods can be used: breaking apart one *or* both addends using the place value of the number.

Method 1 Add the entire first number to the tens place of the second number. Then add the ones digit from the second number. To compute 46 + 79, think $46 + $70 = $116 and $116 + $9 = $125.

Method 2 Add the tens, add the ones, and then add the answers together. Think 40 + 70 = 110 and 6 + 9 = 15. The answer is $110 + $15 = $125. ■

Do Lesson Exercise 3.89 mentally, using one of the methods from Example 3.7.

Lesson Exercise 3.89

(a) A man buys a coat for $56 and a cassette player for $39. How can he mentally compute the (exact) total cost?

(b) What property makes $50 + 6 + 30 + 9 = 50 + 30 + 6 + 9$?

(c) Describe a way to compute $56 + 39$ using the chart in Figure 3-38.

50	51	52	53	54	55	56	57	58	59
60	61	62	63	64	65	66	67	68	69
70	71	72	73	74	75	76	77	78	79
80	81	82	83	84	85	86	87	88	89
90	91	92	93	94	95	96	97	98	99

Figure 3–38

Multiplication can also be done from left to right.

EXAMPLE 3.8

A restaurant uses 68 gallons of water per day. Explain how to compute mentally how much the restaurant would use in a week.

Solution

Multiply from left to right. To compute 7×68, think $7 \times 60 = 420$ and $7 \times 8 = 56$. Add $420 + 56 = 476$ gallons. ■

Lesson Exercise 3.90

A car trip in the mountains takes 9 hours of driving at an average of 35 miles per hour. How can you mentally compute the total distance traveled?

ESTIMATION

An estimate may be preferred over an exact answer when (1) an estimate is easier to obtain, (2) an estimate is easier to use in computations, or (3) an overestimate or underestimate includes a safety factor.

Lesson Exercise 3.91

In each situation, tell why an estimate might be preferred over an exact answer.

(a) You want to tell how many people attended a political rally.

(b) You want to decide how much money to take on a summer trip.

(c) You want to figure out how long it will take to drive from Chicago to Cleveland.

Estimation is often used in computations when it is easier and provides sufficient accuracy. In order to estimate, one converts a problem to a similar but easier problem that can be computed mentally. Estimation shows another side of mathematics, in which one seeks a reasonable result rather than an exact one. People use estimation either to obtain an approximate answer without having to use a calculator or to check work done with a calculator or by another person.

Cooperative Lesson Exercise 3.92

Name a situation in which you have used estimation.

The three most useful strategies for whole-number estimation are rounding, the compatible-numbers strategy, and the front-end strategy.

ROUNDING STRATEGY

How many calories should you eat each day? If you are a young adult, as I used to be, you could multiply your weight in pounds by 19 to get a rough calorie estimate.

Lesson Exercise 3.93

May Wong is a 14-year-old who weighs 93 pounds. How would you estimate how many calories she should consume to maintain her weight if the recommended number of calories is 18 times the number of pounds she weighs?

Did you use rounding to estimate the answer to the preceding exercise? You could have estimated 93×18 by rounding the factors to 90 and 20, multiplying 9 by 2 mentally, and adding two zeroes. So the answer is about 1800 calories. In symbols, this is written $93 \times 18 \approx 1800$, in which the symbol $\approx$ means "is approximately equal to."

The rounding strategy involves two steps.

Rounding Strategy

1. Round the numbers to obtain a problem you can compute mentally.
2. Add, subtract, or multiply the rounded numbers to obtain an estimate.

Use the rounding strategy to estimate the answers to Lesson Exercises 3.94 and 3.95.

Lesson Exercise 3.94

A stadium seats 58,921 people. If only 3426 tickets are left for next Sunday's match between the Poodles and the Goulash, *explain* how you would estimate the number of tickets that have been sold.

Lesson Exercise 3.95

In estimating $A - C$, in which $A > C$, you round A up and C down. Your estimate

(a) is too high

(b) is too low

(c) could be too high or too low

THE COMPATIBLE NUMBERS STRATEGY

In division, rounding to the "nearest" does not always yield a simpler problem.

Lesson Exercise 3.96

A group of milking cows produces 712 oz of milk one fine day. How would you estimate the number of quarts produced? (One quart contains 32 oz.)

In the preceding exercise, you needed to estimate $712 \div 32$. You could have rounded it to $700 \div 30$ or $710 \div 30$, but a better choice would be $600 \div 30$, $750 \div 30$, or $700 \div 35$. For example, $712 \div 32 \approx 750 \div 30 = 25$. The average per cow is a little less than 25 quarts.

The computation $750 \div 30$ is an example of **compatible numbers,** a set of numbers whose sum, difference, product, or quotient is easy to compute mentally. Numbers that do not divide evenly, such as $700 \div 30$, are not compatible.

The compatible-numbers strategy involves two steps.

The Compatible-Numbers Strategy

1. Round the numbers to nearby compatible numbers.
2. Perform the computation with the compatible numbers, and use the answer as an estimate.

Use the compatible-numbers strategy to estimate in the following exercises.

Lesson Exercise 3.97

You want to pay for an $8462 National Motors Ulcer in 24 easy, interest-free monthly payments. How could you estimate the cost of a monthly payment?

Lesson Exercise 3.98

A and C are whole numbers. In estimating $A \div C$, you round A up and C down. If you assume that $C \neq 0$, your estimate

(a) is too high
(b) is too low
(c) could be too high or too low

You can also use compatible numbers to estimate when adding three or more numbers, by adding groups of numbers that total approximately 100 or 1000 (or some other round number).

EXAMPLE 3.9

A movie theater had six shows of "Return of the Dodecahedron" on a Saturday. The numbers of tickets sold at the six shows were 64, 59, 32, 41, 27, and 77. How can you estimate the total number of tickets sold, using the compatible-numbers strategy?

Solution

Add compatible numbers.

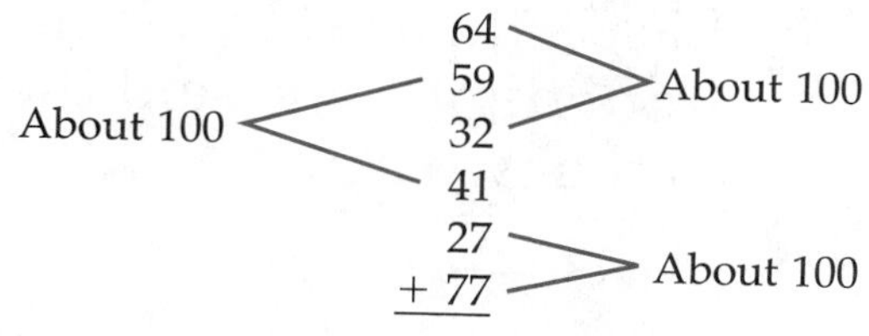

The sum is about 300 tickets. ■

Lesson Exercise 3.99

(a) The next day, the movie theater mentioned in the preceding example had six more shows. The total numbers of tickets sold at the six shows were 36, 52, 51, 98, 71, and 96. Explain how to estimate the total number of tickets sold, using the compatible-numbers strategy.

(b) What properties allow you to reorder and regroup the numbers?

THE FRONT-END STRATEGY

Front-end estimation is especially useful in addition. It is based upon adding a column of numbers from left to right. The estimate is made by adding the digits in the left-hand column and is then adjusted by considering the digits in the next column over to the right.

Figure 3-39 shows an example done with rounding and the front-end strategy from *Mathematics Plus*, Grade 3.

ESTIMATING SUMS

The refrigerator cars on trains keep food at the right temperature. In one car are 183 boxes of lettuce. In another car are 315 boxes of lettuce. About how many boxes of lettuce are in the two cars?

Brian estimated the sum by rounding. He said there are about 500 boxes in the two cars.

$$\begin{array}{r} 183 \\ +315 \end{array} \rightarrow \begin{array}{r} 200 \\ +300 \\ \hline 500 \end{array}$$

Round to the nearest hundred.

Lisa estimated by adding the values of the front digits. She said there are about 400 boxes in the two cars.

$$\begin{array}{r} 183 \\ +315 \end{array} \rightarrow \begin{array}{r} 100 \\ +300 \\ \hline 400 \end{array}$$

Figure 3–39

Estimation is a relatively new topic in school. If you are not familiar with the front-end strategy, study Lisa's method in Figure 3-39 before solving the next exercise.

Lesson Exercise 3.100

A salesperson sells three used cars for \$3793, \$4391, and \$2807. *Explain* how you would use front-end estimation to estimate the salesperson's total sales.

ANSWERS TO SELECTED LESSON EXERCISES

3.89 (a) Compute $56 + 30 = 86$ and $86 + 9 = 95$. Or compute $50 + 30 = 80$ and $6 + 9 = 15$. Then $80 + 15 = 95$.
(b) Commutative property of addition.

3.90 Think $9 \times 30 = 270$ and $9 \times 5 = 45$. Then $270 + 45 = 315$ miles.

3.91 (a) It is not possible to find the exact answer.
(b) You use an overestimate to be safe.
(c) You cannot know the exact time in advance.

3.94 Round the problem to $59{,}000 - 3000 = 56{,}000$ tickets.

3.95 (a) *Hint:* Try $29 - 11$.

3.97 $8462 \div 24 \approx 8000 \div 20 = \$400$

3.98 (a) *Hint:* Try $29 \div 11$.

3.99 (a) $36 + 71$, $52 + 51$, 98 and 96 are each about 100. The sum is about 400.
(b) Commutative and associative properties of addition

3.100 Add $3 + 4 + 2$ thousand $= \$9000$. Then $\$793 + \$391 + \$807$ is about another \$2000. Compute $\$9000 + \$2000 = \$11{,}000$.

3.5 HOMEWORK EXERCISES

Basic Exercises

1. The distance from Washington, D.C., north to Baltimore is 39 miles; the distance from Baltimore north to Philadelphia is 97 miles.
(a) Explain how to compute mentally the total distance from Washington, D.C., to Philadelphia.
(b) What property justifies the expression $30 + 9 + 90 + 7 = 30 + 90 + 9 + 7$?

2. A watermelon costs 39¢ per pound.
(a) Explain how to compute mentally the cost of a 6-pound watermelon.
(b) What property justifies the expression $39 \times 6 = (30 \times 6) + (9 \times 6)$?

Explain how to compute Exercises 3–5 mentally.

3. $(500)^3 =$ ________

4. The distance from Memphis to Albuquerque is 1000 miles. How long would it take to drive this distance, averaging 50 mph?

5. A restaurant serves 4000 people per week. How many does it serve in 300 weeks?

***6.** In each situation, tell why an estimate might be used instead of an exact answer.
(a) You want to predict the U.S. population in 2010.
(b) You want to decide how many people should ride on an elevator that carries up to 2000 pounds.
(c) You want to tell how long it took you to finish a mathematics test.

7. *Explain* how you would use rounding to estimate mentally $5692 + 8091 + 3721$.

8. An auditorium has 56 rows, each seating 23 people.
(a) Explain how you would use rounding to estimate mentally the total number of seats by rounding.
(b) Which category of multiplication is illustrated?

9. The distance from Boston to Buffalo is about 437 miles.
   (a) Explain how you would mentally estimate the time the drive would take if you averaged 55 mph.
   (b) Which category of division is illustrated?
10. A college has 4832 students and 324 teachers.
   (a) Explain how you would mentally estimate the number of students per teacher (called the student-teacher ratio).
   (b) Which category of division is illustrated?
11. Use mental computation to write a realistic multiplication word problem that has an answer of 72,000.

*12. *Explain* how you would mentally compute the following using the compatible-numbers strategy.

$$\begin{array}{r} 59 \\ 32 \\ 42 \\ +\ 97 \\ \hline \end{array}$$

13. A and C are two-digit whole numbers. In estimating $A \div C$, you round A up and C down. Is your estimate too high or too low, or is it impossible to tell?
14. Without computing the products, tell which pairs of factors have the same product.
   (a) 36×22 and 18×11
   (b) 14×22 and 7×44
   (c) 35×66 and 105×22
15. You can mentally divide by 25 if you think of four 25's making 100 (or four quarters making a dollar, if you prefer). In other words, each 100 has four 25's. Try the following.
   (a) $200 \div 25$ (Think of four 25's per 100.)
   (b) $700 \div 25$
   (c) How many quarters in $9?
   (d) $650 \div 25$
   (e) How many quarters in $4.25?
16. Three towns with populations of 3692, 1527, and 4278 make up a voting district. *Explain* how to estimate the total population of the voting district using front-end estimation.
17. The Department of Education spent the following amounts on three projects: $3,462,871, $830,212, and $21,172,806. Explain how to estimate the total expense to the nearest *million* dollars, using the front-end or rounding strategy.
18. Crusty's Pizza sold 2621 pizzas in March, 1522 pizzas in April, and 2218 pizzas in May. Explain how to estimate the total number of pizzas sold for those 3 months, using front-end estimation.
19. Consider the following problem. "A company orders 32 boxes of lightbulbs. Each box contains 48 lightbulbs. What is the total number of lightbulbs?"
   (a) Show how to use one step of front-end estimation to obtain an answer.
   (b) Show how to use rounding to obtain an estimate.
   (c) Which estimate is closer to the exact answer?
20. **Clustering** is a method of estimating a sum when the numbers are all close to one value. For example, $3648 + 4281 + 3791 \approx 3 \cdot (4000) = 12{,}000$. Show how to estimate the following using the clustering strategy.
   (a) $897 + 706 + 823 + 902 + 851 \approx$ ______
   (b) $36{,}421 + 41{,}362 + 40{,}987 + 42{,}621 \approx$ ______
21. Tell whether an estimate or an exact answer is more appropriate in each problem.
   (a) You want to buy two books costing $7.95 and $8.95. You have $20. Is that enough money?
   (b) You want to buy two books costing $7.95 and $8.95. How much change will you receive back from a $20 bill?
22. In each situation, tell whether it makes more sense to compute mentally or to use a calculator. Assume the calculator is stored somewhere nearby.
   (a) You want to estimate the total revenue at a baseball game with 38,542 people who paid an average of about $5 per ticket.
   (b) You just purchased 264 scientific calculators for your school at $18 each and you want to check the total bill.
23. How could you use a calculator *and* mental computation to figure the cost of buying 20 carpets that are 18 ft by 24 ft, if each square foot costs $5?

24. In each exercise, estimate the second factor so that the product falls in the range given. Check your guess on a calculator and revise it as needed.
(a) $300 \times$ _____ = (between 6000 and 6500)
(b) $46 \times$ _____ = (between 700 and 750)
(c) $67 \times$ _____ = (between 2500 and 2600)

25. In each exercise, estimate the divisor so that the quotient falls in the range given. Check your guess on a calculator and revise it as needed.
(a) $463 \div$ _____ = (between 80 and 90)
(b) $3246 \div$ _____ = (between 200 and 300)
(c) $4684 \div$ _____ = (between 65 and 70)

26. (a) Place the digits 4, 5, 6, 8, and 9 in the blanks to obtain an answer that is as close as possible to 50,000.
__ __ __ __ $\times$ __ $\approx$ 50,000
(b) Use a calculator to see how close you came.
(c) Make a second guess and try to get closer to 50,000.
(d) Check your second guess on a calculator.

27. (a) Calculate 15^2, 25^2, and 35^2.
(b) Devise a shortcut for mentally squaring a two-digit counting number that ends in 5.
(c) Try your shortcut on 65^2 and check your guess on a calculator.

28. Solve mentally.
(a) $\square - 10 = 26$ (b) $\dfrac{100}{r + 2} = 5$
(c) $4(x + 8) = 200$

29. Solve mentally.
(a) $34 - n = 20$ (b) $\dfrac{x + 3}{9} = 20$
(c) $30(y - 2) = 150$

30. Select 6 different digits. Use each of the digits once and obtain the largest possible quotient.

□□)□□□□

31. Consider the following problem. "Use each of the digits from 1 to 6 once to obtain the largest possible product."

```
 ???
×???
```

(a) Devise a plan and solve the problem.
(b) Make up a similar problem.

32. Predict which of the following problems have the smallest and largest answers. Then compute the answers with a calculator.
(a) 46×72 (b) 42×76
(c) 24×67 (d) 74×26

33. Use estimation to tell whether the following calculator answers are reasonable. Explain why or why not.
(a) $657 + 542 + 707 =$ [543364]
(b) $26 \times 47 =$ [1222]
(c) $3650 \div 25 =$ [1825]

34. (a) Compute the following using a calculator.

$15{,}873 \times 7 =$ _______
$15{,}873 \times 14 =$ _______
$15{,}873 \times 21 =$ _______

(b) Based upon the pattern in part (a), guess the following products.

$15{,}873 \times 35 =$ _______
$15{,}873 \times 63 =$ _______

35. For each computation, tell which computation method you would use (mental computation, paper and pencil, or calculator) and why.
(a) $87{,}347 \times 144$ (b) $750 + 422 + 250$
(c) $782 - 246$

Extension Exercises

36. **Compensation** is another method of mental computation in which changes made to the numbers in an arithmetic problem balance out. The result is that the overall problem is easier to do, and it still has the same answer! For example:

$$783 + 597 = 780 + 600 = 1380$$
$$352 - 238 = 354 - 240 = 114$$
$$288 \div 36 = 72 \div 9 = 8$$

In parts (a), (b), and (c), change the example to an easier problem and compute the result mentally.
(a) $896 + 364$ (b) $821 - 489$
(c) $3600 \div 72$
(d) Show algebraically why compensation to

$x + y$ using an amount c does not change the result.
(e) Show algebraically why compensation to $x - y$ using an amount c does not change the result.

37. Show how to compute the following mentally, using compensation.
(a) $389 + 895$ (b) $4990 + 3627$
(c) $762 - 89$

38. There is a shortcut for multiplying a whole number by 99. For example, consider 15×99.
(a) Why does $15 \times 99 = (15 \times 100) - (15 \times 1)$?
(b) Compute 15×99 mentally, using the formula in part (a).
(c) Compute 95×99 mentally, using the same method.

39. (a) Develop a shortcut for multiplying 8×999 mentally. (*Hint:* See the preceding exercise.)
(b) Compute 6×999 mentally, using the same shortcut.
(c) Show how this shortcut is based upon the distributive property.

40. (a) Develop a shortcut for multiplying by 25 mentally in a computation such as 24×25.
(b) Compute 44×25 using the same shortcut.

41. (a) Develop a shortcut for multiplying by 15 mentally in a computation such as 24×15.
(b) Compute 42×15 using the same shortcut.

42. $33 \times 37 = 1221$
$46 \times 44 = 2024$
$58 \times 52 = 3016$

(a) Make up another example of this type.
(b) How can these examples be computed mentally?
(c) Prove that the shortcut always works. (*Hint:* Represent the two digits of the two numbers by $\underline{a}\,\underline{b}$ and $\underline{a}\,\underline{10 - b}$, and write them in expanded notation.)

43. Michigan has about 10 million people, each using 70 gallons of water per day. Explain how to compute mentally how much water these people use in a 30-day month.

44. Consider the following problem. "Find four consecutive whole numbers whose product is 303,600." Devise a plan and solve the problem.

45. Estimate the total number of restaurants in the United States.

46. Look at a current series of mathematics textbooks for Grades 1–6. What do they teach about estimation and mental computation with whole numbers in each grade?

3.6 PLACE VALUE AND ALGORITHMS IN OTHER BASES

You are well acquainted with base-ten place value and arithmetic. This familiarity will make it harder for you to understand the difficulties your students will encounter in learning about place value and arithmetic for the first time.

Studying place value and the algorithms in less familiar number systems will enable you to re-experience learning these concepts. Working in other bases will also deepen your understanding of place value and the algorithms.

OTHER NUMERATION SYSTEMS

Today, most countries use base-ten numeration systems. The basis for this is the tendency to count on our fingers.

In fact, any counting number greater than 1 could be used as a base. The Babylonians used base sixty, the Mayans used base twenty, and some ancient tribes used base two. Today computers work with bases two, eight, and sixteen, and some goods, such as eggs and pencils, are grouped in dozens and grosses (groups of 12 and 12^2).

In representing 19 items, one could group them in different ways, as shown in Figure 3-40.

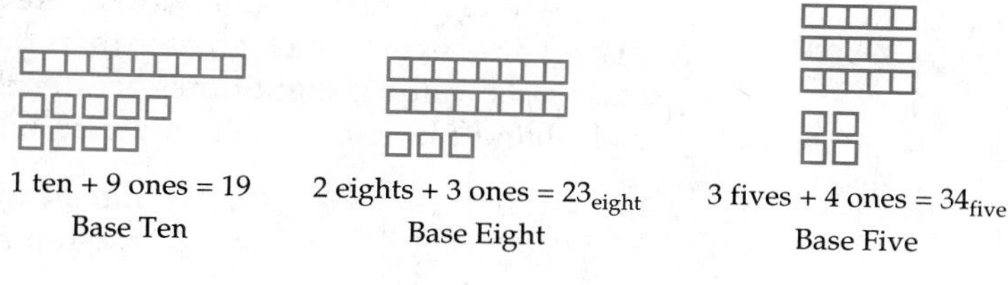

Figure 3–40

Consider base eight, the **octal** system, one base that is used in computers. Just as base ten has ten digits (0, 1, 2, 3, 4, 5, 6, 7, 8, 9), base eight has eight digits (0, 1, 2, 3, 4, 5, 6, 7).

Lesson Exercise 3.101

What digits would base five have?

Counting with the eight digits of base eight goes like this: 0, 1, 2, 3, 4, 5, 6, 7, 10, 11, 12, ... In order to indicate a numeral in a base other than ten, one writes the base as a subscript, as in 12_{eight} (read "one-two, base eight").

Figure 3-41 shows how to represent some base-eight numerals, with base-eight blocks.

Base Eight Symbol	Representation
0_{eight}	
1_{eight}	
2_{eight}	
3_{eight}	
4_{eight}	
5_{eight}	
6_{eight}	
7_{eight}	
10_{eight}	
11_{eight}	
12_{eight}	

Figure 3–41

Lesson Exercise 3.102

What base-ten numeral is the same as 32_{eight}?

Base-eight place value works like base-ten place value. In base ten, each place value is ten times greater than the place value to its right. In base eight, each place value is eight times greater than the place value to its right, with the right-hand place for whole numbers being the ones place.

100	10	1
3	1	4

← **Place values expressed using base-ten numerals** →

64	8	1
3	1	4

$314 = 3(100) + 1(10) + 4(1)$

$$314_{\text{eight}} = 3(64) + 1(8) + 4(1) = 204$$

Thus, 314_{eight} is the same as 204.

Lesson Exercise 3.103

Convert 426_{eight} to base ten.

Lesson Exercise 3.104

Convert 2134_{five} to base ten. (*Hint:* First, write place-value columns for base five.)

One can also convert base-ten numerals to any other base.

EXAMPLE 3.10

Convert 150 to base eight.

Solution

Base-eight numbers use groups of 1, 8, 64, 512, and so on. Start with the largest grouping that is less than or equal to 150—namely 64. How many 64's are there in 150? There are 2.

$$\begin{array}{r} 2 \\ 64\,\overline{)\,150} \\ \underline{128} \\ 22 \end{array}$$

This leaves 22. Proceed to the next lower grouping, 8's. How many 8's are in 22? There are 2.

$$8\overline{)22}\quad 2\text{R}\,6$$

This leaves 6 ones. So $150 = (2 \times 64) + (2 \times 8) + 6$, or 226_{eight}. ■

Lesson Exercise 3.105

(a) Convert 302 to base eight.

(b) Convert 302 to base five.

ALGORITHMS IN BASE FIVE

The base-ten arithmetic algorithms also work in other bases. Studying algorithms in other bases will increase your understanding of them. How does one compute $34_{\text{five}} + 22_{\text{five}}$? How is it similar to base-ten addition?

EXAMPLE 3.11

Compute $\begin{array}{r} 34_{\text{five}} \\ +22_{\text{five}} \\ \hline \end{array}$

Solution

Follow the same steps as in the base-ten algorithm, but remember that the base-five digits are 0, 1, 2, 3, and 4 and that regrouping (carrying) involves groups of 5 rather than groups of 10.

Step 1

$\begin{array}{r} 1 \\ 34_{\text{five}} \\ +22_{\text{five}} \\ \hline 1_{\text{five}} \end{array}$ Add the ones. There are 6 ones. Trade 5 ones for 1 five.

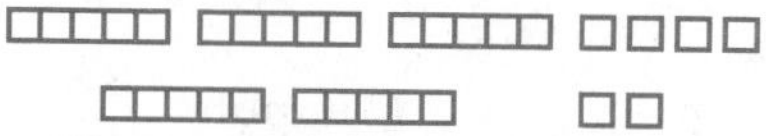

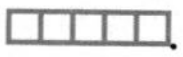

Step 2

$\begin{array}{r} 1 \\ 34_{\text{five}} \\ +\ 22_{\text{five}} \\ \hline 111_{\text{five}} \end{array}$

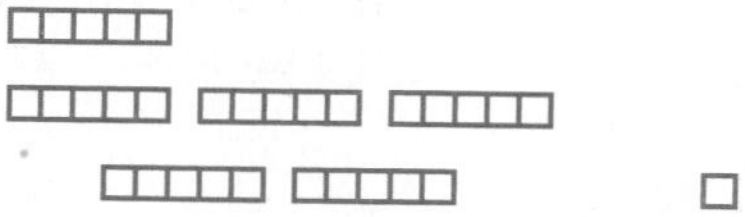

Add the fives. Trade 5 fives for 1 twenty-five.

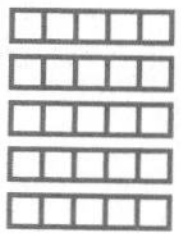

So $34_{\text{five}} + 22_{\text{five}} = 111_{\text{five}}$, or 1 twenty-five, 1 five, 1 one. ■

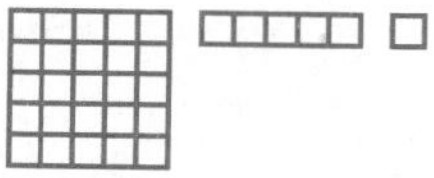

Lesson Exercise 3.106

Compute $24_{five} + 14_{five}$.

Lesson Exercise 3.107

Compute $32_{five} - 14_{five}$. Remember: Regrouping involves 5's, not 10's.

Next, consider the multiplication algorithm. A base-five multiplication table will be helpful.

Lesson Exercise 3.108

Complete the base-five multiplication table.

×	0	1	2	3	4
0	0	0	0	0	0
1	0	1	2	3	4
2	0	2	4	11	13
3					
4					

Example 3.12 utilizes this table to compute $23_{five} \times 14_{five}$.

EXAMPLE 3.12

Compute $\begin{array}{r} 23_{five} \\ \times 14_{five} \\ \hline \end{array}$

Solution

Follow the same steps as in the base-ten algorithm. Refer to the base-five multiplication table to obtain the basic facts. First multiply $4_{five} \times 3_{five}$, which the table says is 22_{five}. Put down the 2 and carry the 2.

$$\begin{array}{r} {\scriptstyle 2} \\ 23_{five} \\ \times 14_{five} \\ \hline 2_{five} \end{array}$$

Then $4_{five} \times 2_{five} = 13_{five}$, and $13_{five} + 2_{five} = 20_{five}$.

$$\begin{array}{r} {\scriptstyle 2} \\ 23_{five} \\ \times\ 14_{five} \\ \hline 202_{five} \end{array}$$

Next, compute $10_{five} \times 23_{five}$ and add the partial products.

$$\begin{array}{r} 23_{five} \\ \times\ 14_{five} \\ \hline 202_{five} \\ 230_{five} \\ \hline 432_{five} \end{array}$$

■

Lesson Exercise 3.109

Compute $44_{five} \times 22_{five}$.

Long division in base five can be done with a long-division algorithm analogous to the base-ten algorithm.

EXAMPLE 3.13

Compute $1442_{five} \div 4_{five}$.

Solution

Using long division, follow the same steps as in base ten. First, how many times can 4_{five} go into 14_{five}? Two times. (You can refer to the base-five multiplication table.) $2_{five} \times 4_{five} = 13_{five}$. Subtract this from 14_{five}, and bring down the next digit in the next dividend.

$$\begin{array}{r} 2\phantom{42_{five}} \\ 4_{five}\ \overline{)\,1442_{five}} \\ \underline{13}\phantom{42_{five}} \\ 14\phantom{2_{five}} \end{array}$$

Next, divide 4_{five} into 14_{five}. It goes in 2 times, as it did before. Continuing the algorithm in this manner yields the following results.

$$\begin{array}{r} 221 \text{ R } 3 \\ 4_{five}\ \overline{)\,1442_{five}} \\ \underline{13}\phantom{42_{five}} \\ 14\phantom{2_{five}} \\ \underline{13}\phantom{2_{five}} \\ 12\phantom{_{five}} \\ \underline{4}\phantom{_{five}} \\ 3\phantom{_{five}} \end{array}$$

■

Lesson Exercise 3.110

Compute $1343_{five} \div 3_{five}$.

The homework exercises include problems involving alternative algorithms in other bases.

ANSWERS TO SELECTED LESSON EXERCISES

3.101 0, 1, 2, 3, 4

3.102 $3(8) + 2 = 26$

3.103 $4(64) + 2(8) + 6 = 278$

3.104 $2(125) + 1(25) + 3(5) + 4 = 294$

3.105 (a) 456_{eight}
(b) 2202_{five}

3.106 43_{five}

3.107 13_{five}

3.108

×	0	1	2	3	4
0	0	0	0	0	0
1	0	1	2	3	4
2	0	2	4	11	13
3	0	3	11	14	22
4	0	4	13	22	31

3.109 2123_{five}

3.110 244_{five} R 1

3.6 HOMEWORK EXERCISES

Basic Exercises

1. How would you group

X X X X X X X X X X X X X

in each of the following bases?
(a) Base eight (b) Base five

2. How many different digits are needed for base twelve?

3. Write the first 12 counting numbers in base three.

4. What base-eight numeral follows 377_{eight}?

5. Convert each of the following to base ten.
(a) 75_{eight} (b) 423_{five} (c) 213_{eight}

***6.** Convert each of the following base-ten numerals to numerals in the indicated base.
(a) 46 to base eight
(b) 26 to base five
(c) 324 to base five

7. (a) Using quarters, nickels, and pennies, what is the minimum number of coins needed to make 86¢?
(b) Write a mathematics problem involving bases that is equivalent to the problem in part (a).

***8.** How can one distinguish even whole numbers from odd ones in the following systems?
(a) Base eight (b) Base five

9. Consider the following problem. "How can one recognize a base-five numeral that is divisible by 5?" Devise a plan and solve the problem.

10. Change 36_{eight} to base five.

11. Which is larger, 1011_{five} or 72_{eight}?

12. Write a numeral for the following set, using the base indicated by the groupings.

XXXXX
XXXXX
XXXXX
XXXXX
XXXXX XXXXX XXXXX X

13. Write a base-five numeral represented by the base-five blocks shown. (Make all possible trades first.)

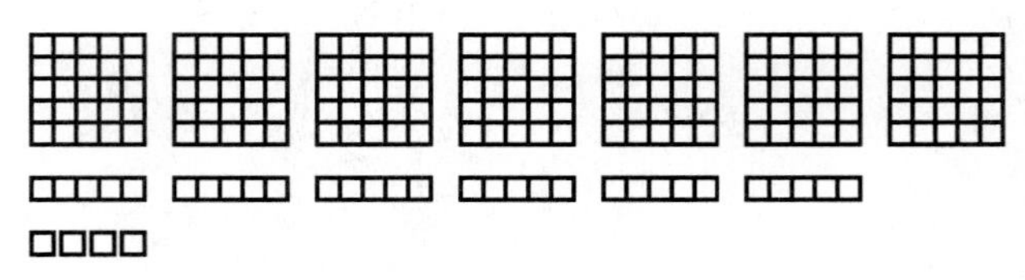

***14.** Perform the following computations.

(a) $\begin{array}{r} 23_{\text{five}} \\ +34_{\text{five}} \\ \hline \end{array}$ (b) $\begin{array}{r} 324_{\text{five}} \\ +132_{\text{five}} \\ \hline \end{array}$ (c) $\begin{array}{r} 432_{\text{five}} \\ +233_{\text{five}} \\ \hline \end{array}$

15. Perform the following computations.

(a) $\begin{array}{r} 41_{\text{five}} \\ -23_{\text{five}} \\ \hline \end{array}$ (b) $\begin{array}{r} 312_{\text{five}} \\ -133_{\text{five}} \\ \hline \end{array}$ (c) $\begin{array}{r} 432_{\text{five}} \\ -143_{\text{five}} \\ \hline \end{array}$

***16.** *Explain* how to compute $31_{\text{five}} + 22_{\text{five}}$ using base-five blocks. Follow the same sequence of steps used in the addition algorithm.

17. *Explain* how to compute $43_{five} - 24_{five}$ using base-five blocks. Follow the same sequence of steps used in the subtraction algorithm.

***18.** Perform the following computations.

(a) $\begin{array}{r} 34_{five} \\ \times 23_{five} \\ \hline \end{array}$ (b) $\begin{array}{r} 412_{five} \\ \times 321_{five} \\ \hline \end{array}$

19. Perform the following computations.

(a) $213_{five} \div 4_{five}$ (b) $4123_{five} \div 3_{five}$

20. Compute $324_{five} \div 4_{five}$ using the following algorithms.

(a) Subtractive algorithm
(b) Standard algorithm
(c) Which algorithm do you prefer? Why?

21. Complete the following base-eight addition and multiplication tables.

+	0 1 2 3 4 5 6 7
0	
1	
2	
3	
4	
5	
6	
7	

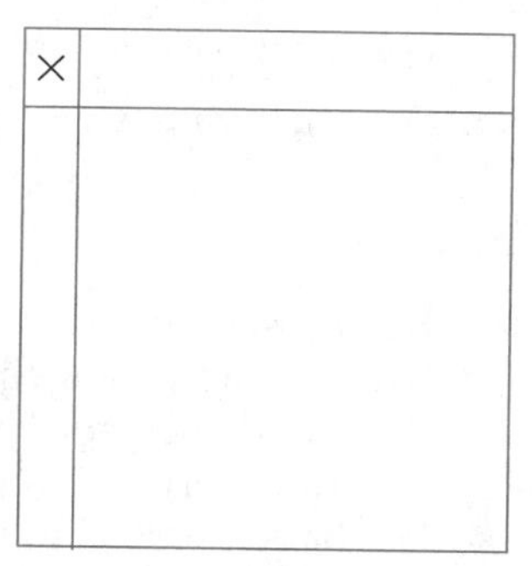

***22.** Perform the following computations.

(a) $\begin{array}{r} 32_{eight} \\ +66_{eight} \\ \hline \end{array}$ (b) $\begin{array}{r} 132_{eight} \\ -\ 66_{eight} \\ \hline \end{array}$ (c) $\begin{array}{r} 24_{eight} \\ \times 35_{eight} \\ \hline \end{array}$

23. Perform the following computations.

(a) $124_{eight} \div 4_{eight}$ (b) $756_{eight} \div 3_{eight}$

24. Fill in the missing digits.

(a) $\begin{array}{r} 1\ 3\ \_\ _{eight} \\ +\ 2\ \_\ 4\ _{eight} \\ \hline \_\ 2\ 1\ _{eight} \end{array}$ (b) $\begin{array}{r} 2\ 0\ \_\ \_\ _{five} \\ -\ 1\ 0\ 2\ 2\ _{five} \\ \hline \_\ \_\ 1\ 3\ _{five} \end{array}$

Extension Exercises

25. Consider the following problem. "An inspector wants to check the accuracy of a scale in weighing any whole-number amount from 1 to 15 pounds. Find the smallest number of weights the inspector needs, and tell how heavy each of these weights is."

(a) What amounts must the inspector be able to weigh?
(b) Devise a plan and solve the problem.
(c) Make up a similar problem.

26. Computers use base two since it contains two digits, 0 and 1, that correspond to electronc switches in the computer being off or on. For example,

Base-Two Place Value

Eights 2^3	Fours 2^2	Twos 2^1	Ones 1
	1	0	1

So $101_{two} = 5$.

(a) Write 1101_{two} as a base-ten numeral.
(b) Write 17 as a base-two numeral.
(c) $1011_{two} + 111_{two} =$ ________ (base two).

27. (a) Construct addition and multiplication tables for base two.

Use your tables to compute the following.

(b) $1101_{two} + 101_{two}$ (c) $1101_{two} \times 101_{two}$

28. Computers sometimes take numbers in base eight or sixteen rather than base ten, because they are easier to convert to base two. Can you figure out why?

(a) Convert 555_{eight} to base two.
(b) Describe an easier way for you or a computer to do part (a). (*Hint:* Work with each digit separately.)
(c) Use the shortcut from part (b) to convert 642_{eight} to base two.

29. In base sixteen (**hexadecimal**), the digits are 0 through 9 and *A*, *B*, *C*, *D*, *E*, and *F* for 10 through 15.

(a) Convert $B6_{sixteen}$ to base ten.
(b) Convert $B6_{sixteen}$ to base two.
(c) How can a computer convert $B6_{sixteen}$ to base two without first converting it to base ten?

30. (a) Convert 11101_{two} to base eight.
(b) Convert 101101_{two} to base sixteen.

31. A magician asks a volunteer to think of a number from 1 to 15. The magician then shows the following four cards and asks the volunteer which cards contain the volunteer's number.

1	3
5	7
9	11
13	15

2	3
6	7
10	11
14	15

4	5
6	7
12	13
14	15

8	9
10	11
12	13
14	15

By adding the numbers in the upper left corner of the selected cards, the magician finds the secret number.

(a) Pick a number from 1 to 15 and show that the trick works.

(b) Explain how the trick works, using base-two place value.

32. In base twelve, the **duodecimal** system, one uses 12 digits: 0, 1, 2, 3, 4, 5, 6, 7, 8, 9, T, and E (T and E represent ten and eleven).

(a) Convert $E6_{\text{twelve}}$ to base ten.

(b) Convert 80 to base twelve.

33. Pencils are often packed by the dozen or by the gross (144).

(a) Inventory shows that a college bookstore has 3 gross, 2 dozen, and 8 pencils. How many pencils does the store have?

(b) Write a mathematics problem involving bases that is equivalent to the problem in part (a).

34. Perform the following computations.

(a) $\begin{array}{r} 321_{\text{four}} \\ +233_{\text{four}} \\ \hline \end{array}$ (b) $\begin{array}{r} 1001_{\text{two}} \\ -\ 110_{\text{two}} \\ \hline \end{array}$

35. Perform the following computations.

(a) $\begin{array}{r} 35_{\text{six}} \\ \times 42_{\text{six}} \\ \hline \end{array}$ (b) $\begin{array}{r} 312_{\text{four}} \\ \times 132_{\text{four}} \\ \hline \end{array}$

36. Perform the following computations.

(a) $123_{\text{six}} \div 3_{\text{six}}$

(b) $123_{\text{seven}} \div 3_{\text{seven}}$

37. Compute $\begin{array}{r} 243_{\text{five}} \\ +231_{\text{five}} \\ \hline \end{array}$ using scratch addition.

38. Compute $\begin{array}{r} 321_{\text{seven}} \\ -255_{\text{seven}} \\ \hline \end{array}$ using low-stress subtraction.

39. Compute $(452_{\text{six}}) \cdot (35_{\text{six}})$ using lattice multiplication.

40. For what base b would $56_b + 44_b = 111_b$?

41. Consider the following problem. "If $37_{\text{eight}} = 133_b$, what base is b?"

(a) Devise a plan and solve the problem.

(b) Make up a similar problem.

42. For what bases a and b would $12_a = 22_b$?

43. For what bases b would $32_b + 25_b = 57_b$?

44. For what values of a and b does $a_b = b_a$?

Special Exercise

45. Businesses that use the ZIP + 4 (nine-digit) bar code on postage-paid reply envelopes use the binary system. The tall bars represent 1's and the short bars represent 0's. The first and last bars are used to mark the beginning and end. Each group of five bars besides the first and last represents a digit. The last group before the end bar is a check digit that is not part of the zip code. The five-bar codes are *not* the binary equivalents of the digits. Find some postage-paid reply envelopes and try to figure out what five-bar code is used for each digit in a zip code. For example, ıııll represents 1.

SUMMARY

Our Hindu-Arabic numeration system took a long time to develop. By contrasting the Hindu-Arabic system with older systems such as ancient Egyptian and Babylonian numeration, one gains a greater appreciation of the time it took to develop a base-ten place-value system that has a symbol for zero. Children today study place value with concrete materials such as the base-ten blocks and the abacus.

One essential component of what it means to understand an operation is recognizing conditions in real-world situations that indicate that the operation would be useful in those situations. Other components include building an awareness of models and the properties of an operation, seeing relationships among operations, and acquiring insights into effects of an operation on a pair of numbers.
(NCTM, *Curriculum and Evaluation Standards*, p. 41).

In order to know when to add, subtract, multiply, or divide in everyday life, one must recognize common application categories for each operation. Although educators do not agree on the exact number or names of the categories, it is clear that each operation has anywhere from two to seven different classifications. These categories can be illustrated with pictures or objects.

Whole-number operations possess properties that simplify certain computations as well as memorization of the addition and multiplication tables. Addition and multiplication are commutative and associative, and the distributive property holds for multiplication over addition and multiplication over subtraction. The whole numbers also possess identity elements—0 for addition and 1 for multiplication.

Algorithms are faster and more efficient methods for paper-and-pencil computation involving larger numbers. Elementary-school children study simple introductory algorithms before learning more efficient procedures for all four operations. Manipulatives such as base-ten blocks clarify the steps in the algorithms. The addition and multiplication algorithms are based upon the commutative, associative, and distributive properties.

Adults frequently use mental computation and estimation. Our school curriculum has not reflected this usage. The elementary-school curriculum is beginning to devote more time to teaching children about whole-number computations that are especially easy to compute mentally. Children are also learning more about estimating with rounding, compatible numbers, and front-end strategies.

One can develop a deeper understanding of place value and algorithms for the four operations by studying other bases. Computers are well suited to base two, with its two digits that can correspond to "on" and "off."

STUDY GUIDE

To review Chapter 3, see what you know about each of the following ideas or terms that you have studied. You can also use this list to generate your own questions about the chapter.

The NCTM Curriculum Standards and Whole Numbers

Selected NCTM Curriculum Standards

The following standards come from the NCTM document.

- Relate physical materials, pictures, and diagrams to mathematical ideas.
- Recognize that a wide variety of situations can be represented by a single operation.
- Relate the mathematical language and symbolism of operations to problem situations and informal language.
- Understand how the basic arithmetic operations are related to one another.
- Develop, analyze, and explain procedures for computation and techniques for estimation.
- Relate various representation of concepts or procedures to one another.

1. Describe how each standard listed relates to the material you studied in Chapter 3.
2. Select any current elementary-school mathematics textbook series, and describe sample lessons or exercises that illustrate each standard listed.

WHOLE NUMBERS IN ELEMENTARY SCHOOL

The following chart shows at what grade levels selected whole-number topics typically appear in elementary-school mathematics textbooks. Underlined numbers indicate grades in which the most time is spent on the given topic.

Topic	Typical Grade Level in Current Textbooks
Place value	1, 2, 3, 4, 5
Addition concepts	1, 2
Addition algorithm	1, 2, 3
Subtraction concepts	1, 2
Subtraction algorithm	1, 2, 3
Multiplication concepts	2, 3, 4
Multiplication algorithm	3, 4, 5
Division concepts	3, 4
Division algorithm	3, 4, 5, 6
Working backward	3, 4, 5, 6
Order of operations	5, 6
Estimation	2, 3, 4, 5, 6

REVIEW EXERCISES

1. (a) Make a set picture illustrating why $3 \times 5 = 5 \times 3$.
   (b) What property is illustrated?

2. Show how to compute $362 - 187$ using the equal-additions algorithm.

3. Find one set of whole numbers A, B, and C, in which $B \neq 0$ and $C \neq 0$, that makes the following equation true; then find one set that makes the equation false.

$$(A \div B) + (A \div C) = A \div (B + C)$$

4. (a) What is an easy way to compute $(8 \times 24) + (8 \times 16)$?
   (b) What property justifies your answer to part (a)?

5. Suppose the number 36,472 is rounded to 36,500. Give the place to which 36,472 is being rounded.

6. Describe the difference between a numeration system that has place value and one that does not.

7. (a) What is $3 \div 0$?
   (b) *Explain* why, using multiplication or the division category of your choice.

8. What is the total number of symbols needed to write 37 with each of the following?
   (a) Tallies
   (b) Late Egyptian numerals
   (c) Roman numerals

9. Make a drawing that shows $7 - 3 = 4$ using compare measures.

10. You are planning a 2270-mile trip. Your car gets 37 miles per gallon.
    (a) Explain how you could mentally estimate how much gas you will need.
    (b) Which category of division is illustrated?

11. "My math class has 36 students, and your class has 27 students. How many more students are there in my class?" What operation and category are illustrated in this problem?

12. "A box holds 10 diskettes. How many boxes does Diane Lorenz need to hold 40 diskettes?" What operation and category are illustrated in this problem?

13. (a) Is whole-number division associative?
    (b) If so, give an example. If not, give a counterexample.

14. "Ira Reed bought 6 cans of tennis balls. Each can contained 3 tennis balls. Since then, he has lost 2 balls. How many tennis balls does he have left?" What two operations and categories are illustrated in this problem?

15. "A man has agreed to transport 1600 lb of boxes. He has already transported 600 lb of them. If each box weighs 40 lb, how many boxes does he have left to transport?" What two operations and categories are illustrated in this problem?

16. In each part, complete the last example, repeating the same error pattern that occurs in the two completed examples.

    (a) $\begin{array}{r} 82 \\ +79 \\ \hline 1511 \end{array} \qquad \begin{array}{r} 46 \\ +37 \\ \hline 713 \end{array} \qquad \begin{array}{r} 89 \\ +74 \\ \hline \end{array}$

    (b) $\begin{array}{r} 37 \\ -10 \\ \hline 20 \end{array} \qquad \begin{array}{r} 426 \\ -302 \\ \hline 104 \end{array} \qquad \begin{array}{r} 371 \\ -130 \\ \hline \end{array}$

    (c) $\begin{array}{r} {\scriptstyle 5} \\ 37 \\ \times 48 \\ \hline 296 \\ 178 \\ \hline 2076 \end{array} \qquad \begin{array}{r} {\scriptstyle 2} \\ 27 \\ \times 93 \\ \hline 81 \\ 203 \\ \hline 2111 \end{array} \qquad \begin{array}{r} 48 \\ \times 57 \\ \hline \end{array}$

17. *Explain* how to compute $326 + 293$ using Dienes (base-ten) blocks. Follow the same sequence of steps as in the standard algorithm.

18. (a) Each letter in the following expression represents a digit. Find a possible solution.

$$\begin{array}{r} AA \\ +BBB \\ \hline ACD \end{array}$$

(b) *Explain* how you solved the problem in part (a).

19. A man earns D dollars per hour. If he works H hours and then spends S dollars for taxes and expenses, how much money does he have left?

20. Joy spent $30 on groceries. Then she spent half of her remaining money on a book. She now has $8. How much money did she start out with?

21. Convert 100 to base six.

22. Compute $1324_{\text{seven}} \div 6_{\text{seven}}$.

23. Find bases a and b such that $11_a = 22_b$.

24. Write a paragraph describing the difference between the repeated-sets and partition-sets classifications for division.

25. Write a sentence or two describing the difference between the rounding strategy for multiplication and the compatible-numbers strategy for division.

SUGGESTED READINGS

Ashlock, R. *Error Patterns in Computation.* 6th ed. Englewood Cliffs, N.J.: Prentice-Hall, 1994.

Burton, G., and others. *Number Sense and Operations.* Reston, Va.: NCTM, 1993.

Edwards, R. *Operation Magic Tricks.* Pacific Grove, Calif.: Critical Thinking Press, 1995.

Eves, H. *Introduction to the History of Mathematics.* 6th ed. Philadelphia: Saunders College Publishing, 1990.

National Council of Teachers of Mathematics. *Historical Topics for the Mathematics Classroom.* Reston, Va.: NCTM, 1989.

National Council of Teachers of Mathematics. *Developing Computational Skills.* 1978 Yearbook. Reston, Va.: NCTM, 1978.

National Council of Teachers of Mathematics. *Estimation and Mental Computation.* 1986 Yearbook. Reston, Va.: NCTM, 1986.

National Council of Teachers of Mathematics. *New Directions for Elementary School Mathematics.* 1989 Yearbook. Reston, Va.: NCTM, 1989.

4 Number Theory

Written records indicate that until the Pythagoreans came along around 500 B.C., people used numbers primarily in practical applications. The Pythagoreans believed that numbers revealed the underlying structure of the universe, so they studied patterns of counting numbers. This was probably the first significant work in the field of mathematics that we now call number theory.

Number theory is useful in certain computations with fractions and in algebra. Divisibility checks are used to verify codes on merchandise and food, identification numbers on tickets and books, signals from compact discs and TV transmitters, and data from space probes. People also study number theory for the excitement of discovering number patterns and the challenge of determining whether or not the patterns apply to all whole or counting numbers.

4.1 FACTORS

Pythagoras was the charismatic leader of a group of 300 called the Pythagoreans, who lived about 2500 years ago. One of the female Pythagoreans was Theano, a talented student of Pythagoras, who later married him. After Pythagoras died, Theano continued his work. The Pythagoreans observed some rather strange rules. They (1) never ate meat or beans except after a religious sacrifice, (2) never walked on a highway, and (3) never let swallows sit on their roofs.

They were also mathematicians. Believing that numbers were the basis of all things, they associated numbers with ideas, just as some people today believe that 7 is lucky or 13 is unlucky. The Pythagoreans associated the number 1 with reason (a consistent whole), 2 with opinion (two sides), 4 with justice (balanced square and product of equals), and 5 with marriage (unity of their first odd or masculine number, 3, and the first even or feminine number).

The Pythagoreans' interest in numbers also led them to investigate factors. They labeled a few special numbers "perfect" and some special pairs of numbers "amicable," depending upon what factors they possessed.

DEFINITION OF A FACTOR

Before finding out which numbers are perfect or amicable, you'll need to review factors. As you learned in Chapter 3, the term "factor" is used in describing the parts of any multiplication problem such as $2 \times 5 = 10$.

$$\begin{array}{ccc} 2 \times 5 & = & 10 \\ \uparrow \quad \uparrow & & \uparrow \\ \text{Factors} & & \text{Product} \end{array}$$

Lesson Exercise 4.1

Write the two division equations that are equivalent to $2 \times 5 = 10$.

In the equation $10 \div 5 = 2$, 5 is the divisor, and in the equation $10 \div 2 = 5$, 2 is the divisor. So 2 and 5 are called either divisors or factors of 10. Although the word "factor" generally refers to multiplication and "divisor" to division, factors and divisors are the same in number theory.

Lesson Exercise 4.2

If someone asked you why 3 is a factor of 21, how would you respond

(a) using division?

(b) using multiplication?

The formal definition of "factor" or "divisor" uses multiplication rather than division.

Definition **Factor or Divisor**

If A and B are whole numbers, with $A \neq 0$, then A is a **factor** or **divisor** of B, written $A \mid B$, if and only if there is a whole number C such that $A \cdot C = B$.

The statement "A is a factor of B" means that A divides B evenly. If C is the quotient for $B \div A$, the relationship between A and B can be shown in the following three ways.

$$A \overset{\displaystyle C}{\overline{\smash{)}B}} \qquad \text{or} \qquad A \cdot C = B \qquad \text{or} \qquad A \mid B$$

If asked to explain why 6 is a factor of 30, you might say that it is because 6 divides 30 evenly ($30 \div 6 = 5$). Using the *definition* of a factor to explain why 6 is a factor of 30 requires using a *multiplication* equation. By the definition, 6 is a factor of 30 because there is a whole number 5 such that $6 \cdot 5 = 30$.

One advantage of defining "factor" using multiplication instead of division is that a multiplication equation is easier to work with in proofs involving factors.

Lesson Exercise 4.3

(a) Use the definition of a factor to explain why 2 is a factor of 40.

(b) If 2 is a factor of R, then there must be a whole number N such that ______________. (Write a multiplication equation.)

(c) Applying the definition to 2 and R in part (b) to fill in the blank involves ______________ reasoning.

The notation $2 \mid 12$ means "2 is a factor of 12" or "2 divides 12." In $a \mid b$, the first number, a, represents the divisor, or factor. To symbolize that 5 is not a factor of 12, one writes $5 \nmid 12$.

Lesson Exercise 4.4

(a) Name all the factors of 12.

(b) True or false? $12 \mid 3$.

(c) Show the factor pairs of 12 using rectangles. The rectangles for 2 and 6 are shown in Figure 4-1.

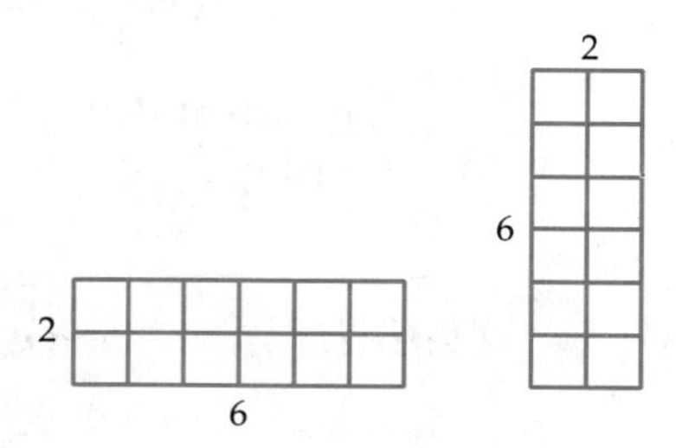

Figure 4–1

(d) The number 12 is divisible by ______________ (list all possibilities).

(e) Use the definition of a factor to explain why $8 \nmid 12$.

AN INVESTIGATION: FACTORS

Lesson Exercise 4.5

(a) Find three examples of counting numbers that have exactly three different factors.

(b) Propose a generalization about counting numbers that have exactly three different factors.

PERFECT AND AMICABLE NUMBERS

What makes a counting number perfect? According to the Pythagoreans, a **perfect number** equals the sum of all its factors that are less than itself. The factors of 6 are 1, 2, 3, and 6. All the factors of 6 that are less than 6 add up to 6: $1 + 2 + 3 = 6$. How perfect!

Lesson Exercise 4.6

Test each number to determine if it is a perfect number.

(a) 20 (b) 28 (c) 38

Before 1950, only 12 perfect numbers were known. Since then, people have used computers to find about 20 more perfect numbers. These perfect numbers have the form $2^{k-1}(2^k - 1)$, where k is a whole number and $2^k - 1$ is prime.

The Pythagoreans also studied amicable numbers. According to the Pythagoreans, two numbers are **amicable** (or **friendly**) if and only if each is the sum of the factors of the other (excluding the numbers themselves as factors). They found only one pair of amicable numbers, 284 and 220. The sum of the factors of 220 that are less than 220 $(1 + 2 + 4 + 5 + 10 + 11 + 20 + 22 + 44 + 55 + 110)$ is 284. The sum of the factors of 284 that are less than 284 $(1 + 2 + 4 + 71 + 142)$ is 220.

If two Pythagoreans were good friends, they wore medallions with amicable numbers written on them.

It was over 2000 years later, in 1636, that Fermat found another pair of amicable numbers: 17,296 and 18,416.

THEOREMS ABOUT FACTORS

Patterns they noted in the factors of particular numbers led the Pythagoreans and other Greek mathematicians to investigate more general questions about factors. For example, if A is a divisor of B, then A must also be a divisor of what other numbers?

In Lesson Exercises 4.7–4.9, try numerical examples and use inductive reasoning to make an educated guess regarding whether each statement is true or false. Remember that in mathematics and logic, a "true" statement must *always* be true. If a mathematical statement is false in even one instance, it is "false." Give an example if you think a statement is true, and give a counterexample if you discover a statement that is false.

Lesson Exercise 4.7

True or false? If A is a factor of an even number, then A is even.

Lesson Exercise 4.8

True or false? If $A \mid 1200$, then $A \mid 2400$.

Lesson Exercise 4.9

In Lesson Exercise 4.8, why isn't $A = 7$ a counterexample?

When you see the phrase "there exist(s)," you need to find only one instance of it, as in the following exercise.

Lesson Exercise 4.10

True or false? There exist counting numbers A, B, and C such that $A \mid (B + C)$ and $A \mid BC$.

You may have guessed that the general statement in Lesson Exercise 4.8 is true. But how can you be sure? In this same situation, the classical Greeks, whenever possible, proved their conjectures using deductive reasoning or found a counterexample to disprove their conjectures. Euclid organized and published many of the Greeks' proofs in his work *Elements*.

Deductive proofs about factors usually employ the definition of a factor. The following exercises will help prepare you for these proofs.

Lesson Exercise 4.11

Suppose A, B, and C are counting numbers and $A \cdot B = C$. What can you deduce about factor relationships among A, B, and C?

Lesson Exercise 4.12

Suppose $A \mid B$. Using the definition of a factor, one can deduce that ______________________________.

EXAMPLE 4.1

Prove that if $A \mid 1200$, then $A \mid 2400$.

Solution

In Lesson Exercise 4.8, you probably found a few different examples that worked. For example, when $A = 2$, $2 \mid 1200$ and $2 \mid 2400$.

To show that the statement is true for any counting number A when $A \mid 1200$, use the definition of a factor on the hypothesis $A \mid 1200$.

1. $A \mid 1200$
2. $A \mid 1200$ means that there is a whole number W such that $A \cdot W = 1200$.
3. ?
4. ?

The last (fifth?) step of the proof is

5. $A \mid 2400$

Now try to work backward from the last step. If step 5 says $A \mid 2400$, how would we show that in step 4? By finding a whole number ? such that $A \cdot ? = 2400$.

1. $A \mid 1200$
2. $A \cdot W = 1200$, in which W is a whole number
3. ?
4. $A \cdot \_\_ = 2400$, in which __ is a whole number
5. $A \mid 2400$

Now, how do you get from step 2 to step 4? To make $A \cdot W = 1200$ look like $A \cdot ? = 2400$, multiply both sides by 2. This is step 3. Since we know that $2W$ is a whole number, this shows that $A \mid 2400$, because $A \cdot 2W = 2400$.

1. $A \mid 1200$
2. $A \cdot W = 1200$, in which W is a whole number
3. $A \cdot 2W = 2400$
4. $A(2W) = 2400$, in which $2W$ is a whole number
5. $A \mid 2400$

■

The preceding proof and the one in the following exercise can be carried out as follows.

1. Write the first (hypothesis) and last (conclusion) steps.
2. Write a multiplication equation for each factor statement to obtain step 2 and the next-to-last step.
3. Use algebra to go from the equation(s) in step 2 to the equation in the next-to-last step.

Try to prove the statement in the following exercise.

Lesson Exercise 4.13

Prove that if $3 \mid A$, then $3 \mid 5A$, in which A is a whole number. (*Hint:* Start by saying $3 \mid A$, and then write an equation that follows from this fact. At the end of the proof, consider what equation will show that $3 \mid 5A$.)

The statements in Lesson Exercises 4.14–4.17 are possible theorems about divisors. Try numerical examples to determine which of these statements might be true. Give a counterexample for any statement that is false. If you believe that a statement is true, prove that it is true with deductive reasoning. In all statements, A, B, and C are whole numbers, with $A \neq 0$.

Lesson Exercise 4.14

True or false? If $A \mid B$, then $B \mid A$.

Lesson Exercise 4.15

True or false? If $A \mid B$ and $A \mid C$, then $A \mid (B + C)$.

Lesson Exercise 4.16

True or false? If $A \mid B$, then $A \mid BC$.

Lesson Exercise 4.17

True or false? If $A \mid BC$, then $A \mid B$ and $A \mid C$.

Were you able to prove the statements in Lesson Exercises 4.15 and 4.16? Lesson Exercise 4.15 is a new theorem called the Divisibility-of-a-Sum Theorem. Lesson Exercise 4.16 is called the Divisibility-of-a-Product Theorem. Both these theorems have the word "divisibility" in them. For counting numbers A and B, A is **divisible** by B if and only if B is a factor of A. The Divisibility-of-a-Sum-Theorem is illustrated in Figure 4-2 and stated as follows.

The Divisibility-of-a-Sum Theorem

For any whole numbers A, B, and C, with $A \neq 0$, if $A \mid B$ and $A \mid C$, then $A \mid (B + C)$.

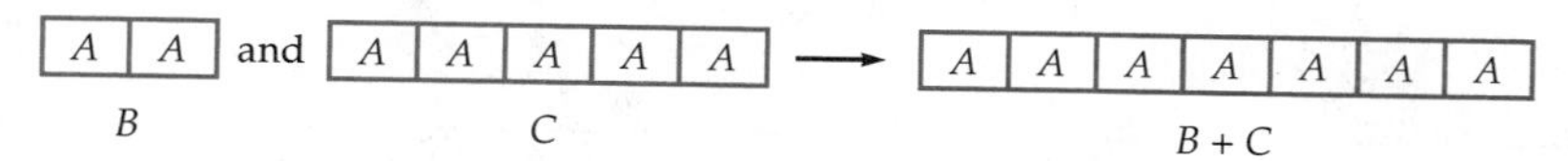

Figure 4–2

This theorem is useful in explaining why the divisibility tests in the next lesson work. See if you can apply the theorem in the following exercise.

Lesson Exercise 4.18

According to the Divisibility-of-a-Sum Theorem, if $5 \mid 4000$ and $5 \mid 200$, then ______________.

A similar theorem works for subtraction. It is modeled in Figure 4-3. The proof is left for the homework exercises.

The Divisibility-of-a-Difference Theorem

For any whole numbers A, B, and C, with $A \neq 0$ and $C > B$, if $A \mid B$ and $A \mid C$, then $A \mid (C - B)$.

A	A

and

A	A	A	A	A

→

A	A	A

B C $C - B$

Figure 4–3

The Divisibility-of-a-Product Theorem is as follows.

The Divisibility-of-a-Product Theorem

A, B, and C are whole numbers with $A \neq 0$. If $A \mid B$, then $A \mid BC$.

See whether you can apply the Divisibility-of-a-Product Theorem in the following exercises.

Lesson Exercise 4.19

According to the Divisibility-of-a-Product Theorem, if A and C are whole numbers, with $A \neq 0$ and $A \mid 3$, then ______________.

Lesson Exercise 4.20

Explain how you could use the Divisibility-of-a-Product Theorem to recognize that $6 \mid (12 \cdot 37)$.

ANSWERS TO SELECTED LESSON EXERCISES

4.1 $10 \div 5 = 2$; $10 \div 2 = 5$

4.3 (a) 2 is a factor of 40 because $2 \cdot 20 = 40$.
(b) $2N = R$ (c) deductive

4.4 (a) 1, 2, 3, 4, 6, 12 (b) False
(c)

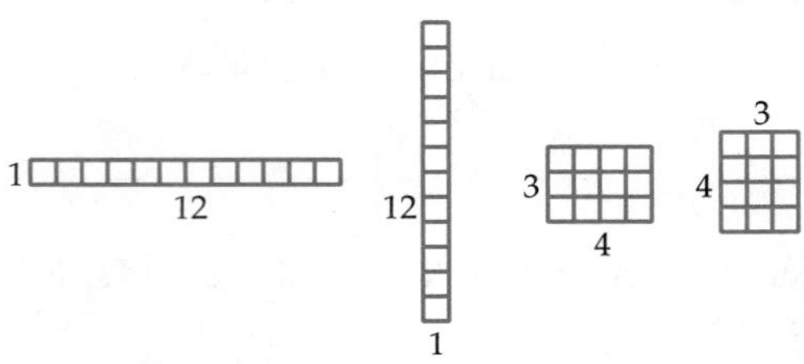

(d) 1, 2, 3, 4, 6, 12
(e) 8 is not a factor of 12 because there is no whole number C such that $8C = 12$.

4.5 (a) 4, 9, 25 (b) They are perfect squares.

4.6 (a) No (b) Yes (c) No

4.7 False for $A = 3$

4.8 True; $A = 2$ is an example.

4.9 Because $7 \nmid 1200$.

4.10 True for $A = 1$ and any B and C

4.11 A and B are factors of C.

4.12 A times some whole number equals B.

4.13
1. $3 \mid A$
2. $3C = A$, in which C is a whole number
3. $3(5C) = 5A$, in which $5C$ is a whole number
4. $3 \mid 5A$

4.14 False; $A = 2$, $B = 4$

4.15 True
1. $A \mid B$ and $A \mid C$
2. $A \cdot D = B$ and $A \cdot E = C$, in which D and E are whole numbers
3. $A \cdot D + A \cdot E = B + C$
4. $A \cdot (D + E) = B + C$, in which $D + E$ is a whole number
5. $A \mid (B + C)$

4.16 True
1. $A \mid B$
2. $AD = B$, in which D is a whole number
3. $A(DC) = BC$, in which DC is a whole number
4. $A \mid BC$

4.17 False; $A = 2$, $B = 3$, $C = 4$

4.18 $5 \mid 4200$

4.19 $A \mid 3C$

4.20 $6 \mid 12$, so 6 is a factor of 12 times any whole number such as $12 \cdot 37$.

4.1 HOMEWORK EXERCISES

Basic Exercises

1. (a) Name all the factors of 28.
(b) Use the definition of a factor to explain why 6 is not a factor of 28.

2. Use the definition of a factor to explain why $11 \mid 0$.

3. True or false? $8 \nmid 2$.

***4.** Test each of the following numbers to see if it is a perfect number.
(a) 120 (b) 315 (c) 496

5. In 1866, a 16-year-old Italian student named Niccolo Paganini found the second-lowest pair of amicable numbers. One of them is 1184. What is the other?

Decide whether the statements in Exercises 6–11 are true or false. Assume that A, B, C, and K are whole numbers, with $A \neq 0$. If a statement is true, explain why or give an example. If it is false, explain why or give a counterexample.

6. True or false? If $2 \nmid B$, then $4 \nmid B$.
7. True or false? If $A \mid B$, then $A \mid KB$.
8. True or false? If $A \mid B$, then $B \mid A$.
9. True or false? If $A \mid B$ and $A \mid C$, then $A \mid (B + C)$.
10. True or false? There exist A, B, and C such that if $A \mid B$, then $AC \mid B$.
11. True or false? There exist A, B, and C such that if $A \mid BC$, then $A \mid B$ and $A \mid C$.
12. If $2 \mid B$, then there exists a whole number C such that ______________.
13. If $3 \cdot A = B$, then _____ is a factor of _____, and _____ is a factor of _____.
14. The statement "If $5 \mid 20$ and $5 \mid 40$, then $5 \mid 60$" is an instance of what theorem?
15. The statement "If $5 \mid 20$, then $5 \mid 200$" is an instance of what theorem?
16. According to the Divisibility-of-a-Sum Theorem, if $10 \mid 1000$ and $10 \mid 20$, then ______________.
17. According to the Divisibility-of-a-Sum Theorem, if C is a whole number such that $8 \mid C$ and $8 \mid 16$, then ______________.
18. According to the Divisibility-of-a-Difference Theorem, if $7 \mid 49$ and $7 \mid 91$, then __________.
19. According to the Divisibility-of-a-Product Theorem, if A and C are whole numbers, with $A \neq 0$, and $A \mid 8$, then ______________.
20. If $3 \mid A$, then $3 \mid 2A$. This is an instance of what theorem?
21. (a) You want to arrange 20 blocks into a rectangular shape. What are all the ways this can be done?
    (b) What mathematical concept does part (a) illustrate?

Extension Exercises

In Exercises 22–30, decide whether each statement is true or false. If it is true, prove it. If it is false, give a counterexample. Assume that A, B, and C are whole numbers, with $A \neq 0$.

22. True or false? $1 \mid B$ for all B. (Read the instructions that precede Exercise 22.)
23. True or false? If $5 \mid B$, then $10 \mid B$. (Read the instructions that precede Exercise 22.)
24. True or false? If $A \mid B$, then $A \mid (B + 1)$.
25. True or false? If $A \mid B$, then $A \mid B^2$.
26. True or false? If B and C are divisible by 2, then BC is divisible by 2.
27. True or false? If $A \neq 0$, $B \neq 0$, and $A \mid B$ and $B \mid C$, then $A \mid C$. (*Hint:* You will have to write equations based on $A \mid B$ and $B \mid C$.)
28. True or false? If $B > C$ and B and C are divisible by A, then $B - C$ is divisible by A.
29. True or false? The product of two even numbers is even. (*Hint:* C is **even** if and only if $C = 2W$ for some whole number W.)
30. True or false? The product of two odd numbers is odd. (*Hint:* C is **odd** if and only if $C = 2W + 1$ for some whole number W.)
31. Use the Divisibility-of-a-Product Theorem to show why, for whole numbers x and y with $x \neq 0$, $x \mid x^2y^2$.
32. Prove the Divisibility-of-a-Difference Theorem.
33. Prove that for any whole number $N \geq 2$, $3 \mid (N^3 - N)$. (*Hint:* Factor $N^3 - N$.)
34. Consider the following problem. "Investigate the result of adding 1 to the product of four consecutive whole numbers."

Understanding the Problem

(a) Give an example of adding 1 to the product of four consecutive whole numbers.

Devising a Plan and Carrying Out the Plan

(b) Repeat part (a) for four different consecutive whole numbers.

(c) Propose a generalization.

Looking Back

(d) Prove your generalization using N as the first whole number.

35. A company asks you to design a box that holds 12 rectangular sponges all laid out flat in the same position. If each sponge is 3 in. by 4 in., give all possible lengths and widths the box could have.

4.2 DIVISIBILITY

Three drugstore owners pool their resources to buy 3456 bottles of aspirin. Can they divide the bottles evenly among themselves?

What is a shortcut for determining if a number such as 3456 is divisible by 3? The shortcut can be discovered by looking at the multiples of 3.

MULTIPLES

Do different numbers that are divisible by 3 have anything in common? Consider 0, 3, 6, 9, 12, 15, 18, 21, 24, 27, ...

In order to study divisibility by 3, you would look at the multiples of 3, {0, 3, 6, 9, 12, 15, 18, 21, 24, 27, . . . }. Multiples are generated when you count by a number starting at 0. The multiples of 3 would be written as $3 \cdot 0$, $3 \cdot 1$, $3 \cdot 2$, $3 \cdot 3$, $3 \cdot 4$, or 3 times any whole number.

The definition of a multiple is based upon the definition of a factor. A number B is a **multiple** of A if and only if A is a factor of B. For example, 20 is a multiple of 5 because 5 is a factor of 20.

Lesson Exercise 4.21

List the multiples of 8.

Lesson Exercise 4.22

According to the definition, 50 is a multiple of 10 because there is a whole number ____ such that ____ · ____ = ____.

Students sometimes confuse factors and multiples because they are closely related ideas. Try the following.

Lesson Exercise 4.23

Fill in each blank with "factor(s)" or "multiple(s)."

(a) 7 is a ____________ of 14.

(b) 30 is a ____________ of 10.

(c) $8x$ is a ____________ of x when x is a whole number.

(d) Every counting number has an infinite set of ____________.

DIVISIBILITY TESTS FOR 2, 5, AND 10

Is 77,846,379,820 divisible by 2, 5, or 10? The longer way to find out is to divide this 11-digit number by 2, 5, and then 10. There is a shorter way.

The divisibility tests for 2, 5, and 10 are all based upon checking the last digit of the number. Do you see the patterns in the last digits of each of the following sets of numbers?

Numbers divisible by 2: {0, 2, 4, 6, 8, 10, 12, 14, 16, 18, . . . }
Numbers divisible by 5: {0, 5, 10, 15, 20, 25, . . . }
Numbers divisible by 10: {0, 10, 20, 30, 40, 50, . . . }

See whether you can describe the divisibility tests in the following exercise.

Lesson Exercise 4.24

Fill in the blanks.

(a) A whole number is divisible by 2 if and only if ______________.
(b) A whole number is divisible by 5 if and only if ______________.
(c) A whole number is divisible by 10 if and only if ______________.

The divisibility tests for 2, 5, and 10 are as follows.

Divisibility Tests for 2, 5, and 10

- A whole number is divisible by 2 if and only if its ones digit is divisible by 2.
- A whole number is divisible by 5 if and only if its ones digit is 0 or 5.
- A whole number is divisible by 10 if and only if its ones digit is 0.

These divisibility tests are grouped together because they all require checking the last digit of the whole number.

Lesson Exercise 4.25

Without dividing, determine whether each number in parts (a)–(c) is divisible by 2, 5, and/or 10.

(a) 8,479,238 (b) 1,046,890 (c) 317,425
(d) Applying the divisibility rules to answer parts (a), (b), and (c) is an example of ______________ reasoning.

Lesson Exercise 4.26

Complete the number so that it is divisible by 2 but not by 5 or 10. Place one digit in each blank.

8 6 3, 1 __ __

Divisibility tests are useful in studying number theory and fractions. Later in this chapter, divisibility tests will be used to factor a number and to test whether a number is prime. In Chapter 6, divisibility tests will be used to simplify fractions and to find common denominators.

DIVISIBILITY TESTS FOR 3 AND 9

Do you know how to tell if a number is divisible by 3 or 9? Both divisibility tests require adding up the digits of the number.

Numbers divisible by 3: {0, 3, 6, 9, 12, 15, 18, . . . }

Numbers divisible by 9: {0, 9, 18, 27, 36, 45, . . . }

Lesson Exercise 4.27

(a) State the divisibility tests for 3 and 9, if you recall them. Use the sets just listed to check your guesses.

(b) What is the relationship between divisibility by 3 and divisibility by 9?

The rules are as follows.

Divisibility Tests for 3 and 9

- A whole number is divisible by 3 if and only if the sum of all its digits is divisible by 3.
- A whole number is divisible by 9 if and only if the sum of all its digits is divisible by 9.

These divisibility tests are grouped together because they both require computing the sum of the digits.

Lesson Exercise 4.28

Without dividing, determine whether each number is divisible by 3 or 9.

(a) 468,172 (b) 32,094

Lesson Exercise 4.29

Complete the number so that it is divisible by 3 and 9. Place one digit in each blank.

1 0, 8 2 1, 7 __ __

Lesson Exercise 4.30

Without dividing, complete the following chart.

Number	Divisible by				
	2	3	5	9	10
8,172	X	X		X	
403,155					
800,002					
68,710					

In Lesson Exercises 4.31–4.33, do the following. (a) Try some examples and decide whether the statements are true or false. (b) If a statement is true, give an example that supports it. If a statement is false, give a counterexample. (c) Draw a Venn diagram showing the relationship between the sets of numbers in each statement.

Lesson Exercise 4.31

True or false? If a number is divisible by 6, then it is divisible by 3.

Lesson Exercise 4.32

True or false? If a number is divisible by 2 and 4, then it is divisible by 8.

Lesson Exercise 4.33

True or false? If a number is not divisible by 2, then it is not divisible by 4.

WHAT MAKES THE DIVISIBILITY TESTS WORK?

Why can we tell that 3640 is divisible by 10 just by looking at the last digit? The numbers of thousands, hundreds, and tens do not matter, because any number of thousands, hundreds, or tens is divisible by 10. Only the ones digit matters!

Lesson Exercise 4.34

(a) In 3640, the 3 represents ______, the 6 represents ______, and the 4 represents ______.

(b) The number 3640 has place values of 3000, 600, 40, and 0. Is each of these place values divisible by 10?

(c) Then 3000 + 600 + 40 + 0 (or 3640) is divisible by 10 because of the ______________ Theorem.

The following example gives a more general proof of the divisibility test for 10.

EXAMPLE 4.2

Prove that if $\underline{A}\,\underline{B}\,\underline{C}\,\underline{0}$ is a four-digit number, then $\underline{A}\,\underline{B}\,\underline{C}\,\underline{0}$ is divisible by 10. (A, B, and C represent the first three digits of the number.) In other words, a four-digit number that ends in 0 is divisible by 10.

Solution

Understanding the Problem Assume that $\underline{A}\,\underline{B}\,\underline{C}\,\underline{0}$ is a four-digit number. Show that it must be divisible by 10.

Devising a Plan The first step is the hypothesis, and the last step is the conclusion. Let's estimate that it will take five steps.

1. $\underline{A}\,\underline{B}\,\underline{C}\,\underline{0}$ is a four-digit number.

2.

3.

4.

5. $\underline{A}\,\underline{B}\,\underline{C}\,\underline{0}$ is divisible by 10.

Use expanded notation and the Divisibility-of-a-Sum Theorem to work from step 1 to step 5.

Carrying Out the Plan

1. $\underline{A}\,\underline{B}\,\underline{C}\,\underline{0}$ is a four-digit number.

2. $\underline{A}\,\underline{B}\,\underline{C}\,\underline{0} = 1000A + 100B + 10C + 0$

3. 1000A, 100B, 10C, and 0 are each divisible by 10.

4. By the Divisibility-of-a-Sum theorem, $1000A + 100B + 10C + 0$ is divisible by 10.

5. $\underline{A}\,\underline{B}\,\underline{C}\,\underline{0}$ is divisible by 10.

Looking Back In a whole number ending in 0, there are no ones. All other place values (tens, hundreds, and so on) are divisible by 10 no matter what digit is in that place. Therefore, all whole numbers ending in 0 are divisible by 10. ■

Now it's your turn.

Lesson Exercise 4.35

Prove that if a three digit number ends in 5 or 0 (denoted $\underline{A}\ \underline{B}\ \underline{0}$ or $\underline{A}\ \underline{B}\ \underline{5}$), then the number is divisible by 5.

Lesson Exercise 4.36

Fill in the steps in the proof of the divisibility test for 3: If $\underline{A}\ \underline{B}\ \underline{C}\ \underline{D}$ is a four-digit number with A + B + C + D divisible by 3, then $\underline{A}\ \underline{B}\ \underline{C}\ \underline{D}$ is divisible by 3.

(a) Write the hypothesis as step 1.

(b) Write $\underline{A}\ \underline{B}\ \underline{C}\ \underline{D}$ in expanded notation as step 2. $\underline{A}\ \underline{B}\ \underline{C}\ \underline{D}$ = ________.

(c) In step 3, regroup the terms in expanded notation so that each term is divisible by 3. For example, 1000A may not be divisible by 3, but 999A is, so pull out a 999A. Fill in the blanks to complete step 3.

$$1000A + 100B + 10C + D = 999A + 99B + 9C + \_\_\_\_\_\_\_\_$$

(d) Step 4 says that 999A, 99B, 9C, and A + B + C + D are each ________.

(e) Step 5 says that 999A + 99B + 9C + A + B + C + D is ________.

(f) Step 6 says, therefore, $\underline{A}\ \underline{B}\ \underline{C}\ \underline{D}$ is divisible by 3.

ANSWERS TO SELECTED LESSON EXERCISES

4.21 0, 8, 16, 24, 32, ...

4.22 5; $10 \cdot 5 = 50$

4.23 (a) factor
(b) multiple
(c) multiple
(d) multiples

4.24 See the box that follows the exercise.

4.25 (a) 2 (b) 2, 5, 10
(c) 5 (d) deductive

4.26 The last digit is 2, 4, 6, or 8. The other digit can be any number.

4.27 (b) Any whole number that is divisible by 9 is also divisible by 3.

4.28 (a) Neither
(b) 3, 9

4.29 Any two digits with a sum of 8

4.30

Number	Divisible by 2	3	5	9	10
8,172	X	X		X	
403,155		X	X	X	
800,002	X				
68,710	X		X		X

4.31 True; 6 is an example

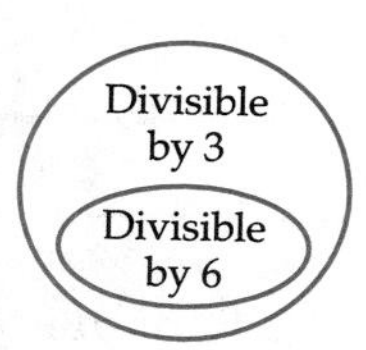

4.32 False for 4

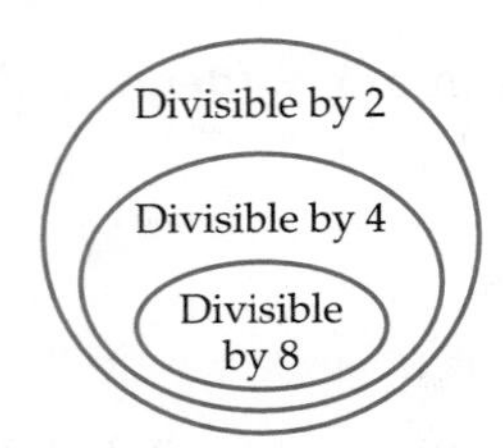

4.33 True; 7 is an example

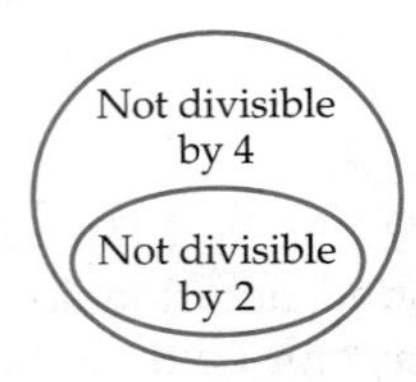

4.34 (a) 3000; 600; 40
(b) Yes (c) Divisibility-of-a-Sum

4.35 1. $\underline{A}\,\underline{B}\,\underline{0}$ and $\underline{A}\,\underline{B}\,\underline{5}$ are three-digit numbers.
2. $\underline{A}\,\underline{B}\,\underline{5} = (A \cdot 100) + (B \cdot 10) + 5$ and $\underline{A}\,\underline{B}\,\underline{0} = (A \cdot 100) + (B \cdot 10)$
3. $(A \cdot 100)$, $(B \cdot 10)$, and 5 are all divisible by 5.
4. $(A \cdot 100) + (B \cdot 10) + 5$ and $(A \cdot 100) + (B \cdot 10)$ are divisible by 5.
5. $\underline{A}\,\underline{B}\,\underline{0}$ and $\underline{A}\,\underline{B}\,\underline{5}$ are divisible by 5.

4.36 (a) $\underline{A}\,\underline{B}\,\underline{C}\,\underline{D}$ is a four-digit number with $A + B + C + D$ divisible by 3.
(b) $(A \cdot 1000) + (B \cdot 100) + (C \cdot 10) + D$
(c) $A + B + C + D$
(d) Divisible by 3
(e) Divisible by 3

4.2 HOMEWORK EXERCISES

Basic Exercises

1. What are the multiples of 7?
2. X is a whole number. What are the multiples of X?
3. Explain why 3 is not a multiple of 6.
4. Name a number that is a factor and a multiple of 15.
5. Fill in each blank with "multiple" or "factor."
   (a) 1 is a ________ of every counting number.
   (b) 3 is a ________ of 12.
   (c) 25 is a ________ of 5.
   (d) $2X$ is a ________ of $8X^3$ when X is a whole number not equal to 0.
6. Without dividing, complete the following chart.

Number	Divisible by 2	3	5	9	10
5,260					
8,197					
345,678					

7. There will be 219 children in next year's third grade. If the school has 9 teachers, can the school assign each teacher the same number of students?
8. Three sisters earn a reward of $37,500 for solving a mathematics problem. Can they divide the money equally?
9. Complete each number so that it is divisible by the indicated number or numbers. Determine all possible answers.
   (a) 4 1, __ 7 2 (by 3)
   (b) 8 2 6, 3 __ __ (by 2 and 5)
   (c) 4 1 7, 2 __ __ (by 2, 3, and 10)
10. A warehouse contains 352 identical boxes of supplies. There is room for up to 10 boxes in a stack. The manager wants the same number of boxes in each stack. What is the greatest number of boxes that can be placed in each stack?
11. (a) Suppose you want to mail a package that requires $1.96 in postage, using $.25 stamps and $.15 stamps. Is it possible to find stamps totaling the exact amount needed?
    (b) Explain the result of part (a) using divisibility.
12. Consider the numbers 0, 1, 4, 8, and 64.
    (a) Tell which number does not belong with the others, and why.

(b) Give a reason why each of the other numbers could be given as the answer to part (a).

In Exercises 13–16, do the following. (a) Tell whether the statement is true or false. (b) If the statement is true, tell why or give an example that supports it; if the statement is false, give a counterexample. (c) Draw a Venn diagram showing the correct relationship among the sets of numbers in each statement.

13. True or false? If a number is divisible by 5, then it is divisible by 10.

14. True or false? If a number is not divisible by 5, then it is not divisible by 10.

15. True or false? If a number is divisible by 6 and 8, then it is also divisible by 48.

16. True or false? If a number is divisible by 8 and 10, then it is also divisible by 40. (Use inductive reasoning.)

17. If $15 \mid N$, then what other counting numbers must be factors of N?

18. If M is a multiple of 20, then M must also be a multiple of what other whole numbers?

19. Fill in the blanks to complete the following proof. A three-digit even number has the form $\underline{A}\ \underline{B}\ \underline{C}$ where $C = 0, 2, 4, 6$, or 8. In expanded notation, the number is ____________.

100A, ________, and ________ are each divisible by 2. Therefore, ____________ is divisible by 2.

20. Prove that if $\underline{A}\ \underline{B}\ \underline{0}$ is a three-digit number, then it is divisible by 10.

Extension Exercises

21. Prove that if the sum of the digits of a four-digit number $\underline{A}\ \underline{B}\ \underline{C}\ \underline{D}$ is divisible by 9, then the number itself is divisible by 9.

22. Prove that if $\underline{A}\ \underline{B}\ \underline{C}$ is a three-digit number and $A + B + C$ is divisible by 3, then $\underline{A}\ \underline{B}\ \underline{C}$ is divisible by 3.

23. Devise a divisibility test for 4. (*Hint:* Look at the last two digits of any number that is divisible by 4.)

24. Devise a divisibility test for 8. (*Hint:* It is an extension of the divisibility test for 4.)

25. (a) What patterns do you see in numbers divisible by 6, such as 6, 12, 18, 24, 30, and 36?
(b) Devise a divisibility test for 6.

26. Devise a divisibility test for 25.

27. (a) What is the divisibility test for 100?
(b) A leap year must be divisible by 4. Furthermore, if a leap year is divisible by 100, then it must also be divisible by 400. Which of the following are leap years?
(1) 1776 (2) 1994 (3) 1996
(4) 2000 (5) 2010

28. Devise a divisibility test for 15. (*Hint:* Look at the divisibility test for 6.)

29. (a) Use the divisibility tests for 6 and 15 to complete the following generalization. If x is the product of two primes a and b, then a whole number y is divisible by x if and only if ____________________.
(b) Try some examples, and conjecture whether the following more general hypothesis is true. If x is the product of two relatively prime numbers a and b, then a whole number y is divisible by x if and only if y is divisible by a and b. (*Hint:* Start with $a = 4$ and $b = 5$.)

30. Consider the following problem. "The ages of a woman and her granddaughter have a surprising property. First of all, they were born on the same day of the year. And for the last 6 years in a row, the grandmother's age has been divisible by her granddaughter's age! How old are they today?"
(a) Devise a plan and solve the problem.
(b) Check your answer.

31. Find a nine-digit number such that the first two digits form a numeral divisible by 2, the first three digits form a numeral divisible by 3, and so on up to the nine-digit numeral, which is divisible by 9.

In Exercises 32–36, if the statement is true, prove it. If the statement is false, give a counterexample. Assume that A and B are whole numbers with $B \neq 0$.

32. True or false? If A is divisible by 2, then $4 + A$ is divisible by 2.

33. True or false? If a whole number is divisible by $A + B$, then it is divisible by A. (Read the directions that precede Exercise 32.)

34. True or false? The sum of any five consecutive whole numbers is divisible by 5.

35. True or false? If A is a multiple of B, then $2A$ is a multiple of B.

36. True or false? If the sum of all the digits of a number is divisible by 6, then the number is divisible by 6.

37. A, B, C, and D are whole numbers. The number $A \cdot B \cdot (C + D)$ is even whenever ________ is even. (Any or all may be correct.)
   (a) A (b) B (c) C (d) D

38. 8! is shorthand for $8 \cdot 7 \cdot 6 \cdot 5 \cdot 4 \cdot 3 \cdot 2 \cdot 1$. Similarly, $10! = 10 \cdot 9 \cdot 8 \cdot 7 \cdot 6 \cdot 5 \cdot 4 \cdot 3 \cdot 2 \cdot 1$.
   (a) Guess how many terminal zeroes the number equal to 8! has. (For example, 100 has two terminal zeroes.) Check your answer on a calculator.
   (b) Repeat part (a) for 10!.
   (c) How can you tell how many terminal zeroes you will get?
   (d) Without computing, tell how many terminal zeroes 6! and 15! have.

39. (a) Choose a two-digit number and reverse the digits. Subtract the smaller number from the larger. Is the difference divisible by 9?
   (b) Repeat part (a) with a different number.
   (c) Prove that the difference will always be divisible by 9.

40. $10^2 - 8^2$ is divisible by 18.
   $20^2 - 3^2$ is divisible by 23.
   (a) Write another statement that fits this pattern, and see if it is true.
   (b) Use algebra to explain why this pattern works for $a^2 - b^2$ in which a and b are counting numbers and $a > b$.

41. Form an eight-digit number by writing a four-digit number twice in succession (for example, 12,341,234).
   (a) Is your number divisible by 73?
   (b) Repeat part (a) with another number of this type.
   (c) Propose a generalization.
   (d) Prove your generalization.

42. Find the remainder when $3^{888,888}$ is divided by
   (a) 4. (b) 5.

43. (a) Devise a divisibility test for 4 in base eight.
   (b) Devise divisibility tests for 2 and 7 in base eight.

44. Write a sentence telling the difference between a factor of x and a multiple of x.

Special Exercise

45. (a) Think of any three-digit number.
   (b) Write it down twice to form a six-digit number (e.g., 382,382).
   (c) Is your number divisible by 7?
   (d) Is your number divisible by 11?
   (e) Is your number divisible by 13?
   (f) Repeat parts (a)–(e) with a new number.
   (g) What is $7 \times 11 \times 13$?
   (h) What is 382×1001?
   (i) Can you explain why your six-digit numbers were divisible by 7, 11, and 13?

4.3 PRIME AND COMPOSITE NUMBERS

The Greeks were the first to study numbers systematically. They discovered a subset of the whole numbers that are the building blocks of all whole numbers greater than 1.

PRIME NUMBERS AND COMPOSITE NUMBERS

The prime numbers are the essential components of counting numbers greater than 1.

Lesson Exercise 4.37

Try to write a definition that tells what property of a counting number makes it a prime number. (*Hint:* Say something about factors.)

Compare your definition to the first part of the following.

> ***Definitions*** **Prime and Composite Numbers**
>
> A counting number greater than 1 is **prime** if it has exactly two distinct factors (divisors): 1 and itself. A counting number is **composite** if it has more than two distinct factors (divisors).

The number 11 is prime since it has exactly two factors: 1 and 11. The number 10 is composite (and not prime) since it has four factors: 1, 2, 5, and 10. A composite number has a factor other than 1 and itself.

Lesson Exercise 4.38

How many factors does 1 have?

Since 1 has only one factor, it is neither prime nor composite. The Pythagoreans, who studied prime numbers 2500 years ago, called 1 the unity that generates all prime and composite numbers. The Pythagoreans studied prime and composite numbers geometrically.

Lesson Exercise 4.39

In this exercise, you will draw all possible rectangles using the given number of squares. For example, with 4 squares you could draw rectangles that are 2 by 2, 1 by 4, and 4 by 1.

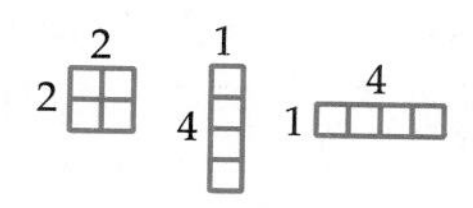

In parts (a)–(d), draw all possible rectangles using

(a) 2 squares.

(b) 6 squares.

(c) 7 squares.

(d) 1 square.

(e) Based upon your results, how is the number of different rectangles related to whether or not the number is prime?

In geometric terms, you can represent a prime number with exactly two rectangles with counting-number dimensions. You can represent a composite number with more than two such rectangles. (Note that 1, which is neither prime nor composite, has exactly one rectangle with counting-number dimensions.)

Lesson Exercise 4.40

Name all prime numbers that are less than 20.

Children use prime numbers to find common denominators and to simplify fractions. This will be discussed in Chapter 6.

THE PRIME NUMBER TEST

It's fairly easy to see that 2, 3, 5, 7, and 11 are primes. But how can you tell if a larger number is prime?

Lesson Exercise 4.41

How would you check to see if 367 is prime?

In the preceding exercise, one would know that 367 is prime if it were not divisible by 2, 3, 4, 5, . . . , 365, 366. However, it is not necessary to try all these numbers as divisors. This procedure can be shortened. Consider the following questions.

Lesson Exercise 4.42

(a) If 2 is not a divisor of a number (such as 367), then what other numbers could not possibly be divisors of the number?

(b) If 3 is not a divisor of a number (such as 367), then what other numbers could not possibly be divisors of the number?

The approach of Lesson Exercise 4.42 would go as follows for 367.

$2 \nmid 367$. Therefore, 4, 6, 8, 10, . . . , 366 are not divisors of 367. (The Divisibility-of-a-Product Theorem says that 4, 6, 8, . . . cannot be divisors unless 2 is a divisor.)

$3 \nmid 367$. Therefore, 6, 9, 12, . . . , 366 are not divisors of 367. The divisor 4 is already eliminated.

$5 \nmid 367$. Therefore, 10, 15, 20, 25, . . . , 365 are not divisors of 367. The divisor 6 is already eliminated.

$7 \nmid 367$. Therefore, 14, 21, 28, 35, . . . , 364 are not divisors of 367. The divisors 8, 9, and 10 are already eliminated.

What numbers do we have to check as divisors? Only primes. If *they* are not divisors, other whole numbers greater than 1 (which are all multiples of primes) could not be divisors either.

Do we need to check all primes less than 367? No. There's an additional shortcut. It results from the fact that every composite number x has a prime factor less than or equal to $\sqrt{x}$. Recall that $\sqrt{x}$ means the nonnegative or **principal square root** of x, which is a number $y \geq 0$ such that $y^2 = x$.

We need to check only primes less than or equal to $\sqrt{x}$. Consider 360. Any larger factor of 360, such as 180, has a smaller companion factor (in this case, 2).

$$\begin{aligned}
360 &= 180 \cdot 2 \\
&= 120 \cdot 3 \\
&= 90 \cdot 4 \\
&= 72 \cdot 5 \\
&= 60 \cdot 6 \\
&= 40 \cdot 9 \\
&= 36 \cdot 10 \\
&= 30 \cdot 12 \\
&= 24 \cdot 15 \\
&= 20 \cdot 18
\end{aligned}$$

In fact, any factor larger than $\sqrt{360} \approx 19$ has a companion factor smaller than $\sqrt{360} \approx 19$. So, if a number does not have a factor less than or equal to its square root, then it is prime. Putting the two ideas together, one needs to check only prime numbers less than or equal to the square root of the number.

The Prime Number Test

To determine if a number N is prime, find out if any prime number less than or equal to $\sqrt{N}$ is a divisor of N.

EXAMPLE 4.3

You want to determine if 367 is prime.

(a) What is the minimum set of numbers you must try as divisors?

(b) Use the Prime Number Test to determine if 367 is prime.

Solution

(a) A calculator shows that $\sqrt{367} \approx 19.2$. So we need to test all prime numbers less than 19.2 as divisors. This would include 2, 3, 5, 7, 11, 13, 17, and 19.

(b) By divisibility tests, $2 \nmid 367$, $3 \nmid 367$, and $5 \nmid 367$.

$$367 \div 7 = 54 \text{ R } 6, \text{ so } 7 \nmid 367$$

$$367 \div 11 = 34 \text{ R } 9, \text{ so } 11 \nmid 367$$

$$367 \div 13 = 29 \text{ R } 6, \text{ so } 13 \nmid 367$$

$$367 \div 17 = 22 \text{ R } 9, \text{ so } 17 \nmid 367$$

$$367 \div 19 = 20 \text{ R } 3, \text{ so } 19 \nmid 367$$

(You could also check the divisors 11, 13, 17, and 19 with a calculator.)

367 is not divisible by any prime number $\leq \sqrt{367}$. This means that 367 cannot have any prime factors less than 367. Therefore, 367 is a prime number. ■

Lesson Exercise 4.43

Use the Prime Number Test to determine if the following numbers are prime.

(a) 347 (b) 253

PRIME FACTORIZATION OF COMPOSITE NUMBERS

Is it possible to factor any composite number using only prime numbers {2, 3, 5, 7, 11, 13, 17, 19, 23, . . .}? For example, $8 = 2 \cdot 2 \cdot 2$ and $264 = 2 \cdot 2 \cdot 2 \cdot 3 \cdot 11$. In other words, is it possible to write any composite number as a product of prime numbers?

Lesson Exercise 4.44

(a) Select a composite number. Can you write it as a product of primes?

(b) Can you find a different set of prime factors for this same number?

(c) Repeat parts (a) and (b) for two other composite numbers.

(d) What generalization can you make from parts (a)–(c)?

(e) Part (d) is an example of ____________ reasoning.

After studying many examples, Greek mathematicians conjectured that every composite number could be written as a product of a unique set of prime numbers. Euclid, a mathematics professor at the University of Alexandria, proved this in Book 9 of his *Elements*.

The prime numbers are the building blocks for all counting numbers greater than 1. Every whole number greater than 1 is prime or can be expressed using prime factors. The prime factorization is like a signature for each composite number. For example, $12 = 2 \cdot 2 \cdot 3 = 2^2 \cdot 3$, and $26 = 2 \cdot 13$.

It is customary to write the prime factors in increasing order and to use exponents as a shorthand for repeated factors. The possibility of writing any composite number as a unique product of numbers is so important that it is called the Fundamental Theorem of Arithmetic.

The Fundamental Theorem of Arithmetic

Every composite number has exactly one prime factorization.

Two methods are commonly used to find prime factorizations. Most middle-school textbooks teach the **factor-tree method**. The other method I call the "prime-divisor" method.

EXAMPLE 4.4

Find the prime factorization of 84 using both the factor-tree and prime-divisor methods.

Solution

The Factor-Tree Method Start with 84 and find any two factors (for example, 2 and 42).

Then continue finding prime factors of these factors until all the factors are prime.

```
      84
      /\
    2 · 42
    |   /\
    2 · 2 · 21
    |   |   /\
    2 · 2 · 3 · 7
```

The result, $2 \cdot 2 \cdot 3 \cdot 7$, is the prime factorization of 84. Repeated factors are usually written with exponents.

$$84 = 2^2 \cdot 3 \cdot 7$$

The Prime-Divisor Method In this method, try all prime numbers in increasing order as divisors, beginning with 2. Use each prime number as a divisor as many times as possible.

Try 2.	⟶	2 ⌊ 84
It works twice.	⟶	2 ⌊ 42
Next, try 3.	⟶	3 ⌊ 21
7 is prime.		7

The divisors and final quotient give the prime factorization of 84.

$$84 = 2 \cdot 2 \cdot 3 \cdot 7$$

■

Try the following exercise using both methods from Example 4.4.

Lesson Exercise 4.45

Find the prime factorization of 120 using both methods.

COUNTING DIVISORS

How can you find out how many different divisors a number such as 162 has by looking at its prime factorization?

Lesson Exercise 4.46

(a) The number $162 = 2 \cdot 3^4$. First, find all the divisors of 2.

(b) Next, find all the divisors of 3^4.

(c) How many different divisors does 162, or $2 \cdot 3^4$, have?

(d) Try the same process for $225 = 3^2 \cdot 5^2$. First, find all divisors of 3^2.

(e) Next, find all divisors of 5^2.

(f) How many different divisors does 225, or $3^2 \cdot 5^2$, have?

(g) Based upon your results in parts (a)–(f), if p and q are prime, how many different divisors does $p^m \cdot q^n$ have?

The results of the preceding exercise can be summarized as follows.

Counting Divisors

If the prime factorization of a number $n = p_1^{m_1} \cdot p_2^{m_2} \cdot \ldots \cdot p_k^{m_k}$, then the number of divisors of n is $(m_1 + 1)(m_2 + 1) \cdot \ldots \cdot (m_k + 1)$.

For example, $225 = 3^2 \cdot 5^2$ has $(2 + 1)(2 + 1)$, or 9, divisors. Apply this shortcut in the following exercise.

Lesson Exercise 4.47

(a) How many divisors does $3^2 \cdot 5^3 \cdot 11^4$ have?

(b) Show how to use the prime factorization to determine how many divisors 48 has.

(c) A child lists the divisors of 72 as 1, 2, 3, 4, 6, 8, 9, 12, 24, 36, and 72. Show how to use the prime factorization of 72 to determine if the list is complete.

(d) What is another way to find the missing divisor of 72 in part (c)?

FAMOUS UNSOLVED PROBLEMS

Some mathematics problems have remained unsolved for centuries. Since number theory deals only with numbers, its unsolved problems are the simplest to describe and the most well known. Do you want to be famous? Just solve one of these problems!

One famous unsolved problem concerns twin primes. **Twin primes** are any two consecutive odd numbers, such as 3 and 5, that are prime.

Lesson Exercise 4.48

(a) Find all twin primes less than 100.

(b) Look at the number between each pair of twin primes greater than 3. What property do all these numbers have?

(c) Do you think there is an infinite number of twin primes?

Euclid proved that there is an infinite number of primes. Number theorists believe that there is also an infinite number of twin primes, but no one has been able to prove it!

Another famous unsolved problem is Goldbach's conjecture. Christian Goldbach (1690–1764) said that every even number greater than 2 is the sum of two primes. No one knows for sure whether this is true or false.

Lesson Exercise 4.49

Show that Goldbach's conjecture is true for the following numbers.

(a) 8 (b) 22 (c) 120

Lesson Exercise 4.50

Goldbach's conjecture has been checked for all even numbers up to 100 million (using a computer), and it works! Why isn't this a sufficient proof?

ANSWERS TO SELECTED LESSON EXERCISES

4.38 One

4.39 (e) A prime number can be represented by exactly two different rectangles with lengths that are counting numbers.

4.40 2, 3, 5, 7, 11, 13, 17, 19

4.42 (a) Multiples of 2 (b) Multiples of 3

4.43 (a) Yes (b) No, divisible by 11

4.44 (a) Yes (b) No (e) inductive

4.45 $120 = 2^3 \cdot 3 \cdot 5$

4.46 (a) 1, 2
(b) 1, 3, 9, 27, 81
(c) 10 (1, 2, 3, 6, 9, 18, 27, 54, 81, 162)
(d) 1, 3, 9
(e) 1, 5, 25
(f) 9 (1, 3, 5, 9, 15, 25, 45, 75, 225)
(g) $(m + 1)(n + 1)$

4.47 (a) $3 \cdot 4 \cdot 5 = 60$
(b) $48 = 2^4 \cdot 3^1$, and $5 \cdot 2 = 10$ divisors
(c) $72 = 2^3 \cdot 3^2$ and $4 \cdot 3 = 12$ divisors. The list is missing one divisor.
(d) Pair off divisors that multiply to make 72. $4 \cdot 18 = 72$, so 18 is missing.

4.48 (a) 3 and 5, 11 and 13, 17 and 19, 29 and 31, 41 and 43, 59 and 61, 71 and 73
(b) They are all divisible by 6.

4.49 (a) $8 = 3 + 5$ (b) $22 = 11 + 11$
(c) $120 = 47 + 73$ or $7 + 113$ or $11 + 109$ or $13 + 107$ or $19 + 101$, and so on

4.50 This is inductive reasoning. Some even number greater than 100 million may turn out to be a counterexample.

4.3 HOMEWORK EXERCISES

Basic Exercises

1. Classify the following numbers as prime or composite, and draw all possible rectangles with counting-number dimensions for each number.
(a) 47 (b) 51

2. Classify the following numbers as prime or composite, and draw all possible rectangles with counting-number dimensions for each number.
(a) 87 (b) 89

3. One can often obtain a prime number by performing the following steps. Begin with each number from 1 through 10, and see how many prime numbers you end up with.
Step 1. Choose a counting number.
Step 2. Multiply it by the next highest counting number.
Step 3. Add 17 to your result.

4. The formula $2^n - 1$ often produces prime numbers when n is a counting number. In 1992 a group of British mathematicians used a computer to show that $2^{756,839} - 1$ is prime.
(a) Compute $2^n - 1$ for $n = 2, 3, 4, 5$, and 6.
(b) Which of your resulting numbers are prime?

5. Many mathematicians have tried to find formulas that produce only prime numbers. Fermat conjectured that $2^{2^n} + 1$ would be prime for $n = 0, 1, 2, 3, \ldots$. The resulting numbers, called **Fermat numbers**, are $2^{2^0} + 1 = 3$, $2^{2^1} + 1 = 5$, $2^{2^2} + 1 = 17$, and so on.
(a) The first five Fermat numbers are prime. Find the fourth Fermat number.
(b) The sixth Fermat number is a composite number divisible by 641. Find this number, and show that it is divisible by 641.

6. In 1978 Hugh Williams discovered the following prime number with 317 ones.
11,111,111, . . . ,111. Explain why this number is not divisible by 2, 3, or 5.

7. Is 1 prime or composite?

8. In order to determine if 817 is prime, what is the minimum set of numbers you must *try* as divisors?

9. In order to determine if 431 is prime, what is the minimum set of numbers you must *try* as divisors?

10. Use the Prime Number Test to classify the following numbers as prime or composite.
(a) 91 (b) 577

11. Use the Prime Number Test to classify the following numbers as prime or composite.
(a) 71 (b) 697

12. The following numbers are prime.

31 331 3331 33,331 333,331
3,333,331 33,333,331

(a) Write a generalization of this pattern.
(b) Does part (a) involve inductive or deductive reasoning?
(c) Use the Prime Number Test to classify 333,333,331 as prime or composite.

13. According to the Fundamental Theorem of Arithmetic, every composite number has ______________________.

14. Consider a property similar to the Fundamental Theorem of Arithmetic. Show that every composite number greater than 1 cannot be *uniquely* expressed as a sum of prime numbers.

15. Find the prime factorization of each number using both methods.
(a) 76 (b) 320

16. Write the prime factorizations of the following numbers.
(a) 90 (b) 3155

17. (a) How many different divisors does $2^5 \cdot 3^2 \cdot 7$ have?
(b) Show how to use the prime factorization to determine how many different factors 148 has.

18. (a) How many different factors does 120 have?
(b) List all the divisors of 84. Use the prime factorization of 84 to confirm that your list is complete.

19. Find all twin primes between 101 and 140.

20. **Triplet primes** are any three consecutive odd numbers that are prime.
(a) Find a set of triplet primes.
(b) Why can't there be any other sets of triplet primes?

21. Show that Goldbach's conjecture is true for the following numbers.
(a) 12 (b) 30 (c) 108

22. In how many ways can 28 be expressed as the sum of two prime numbers?

23. The number 5 can be written as $1 + 2^2$. Find three other prime numbers that have the form $1 + W^2$ for some whole number W.

Extension Exercises

24. Complete the following table for counting numbers from 1 to 25. The numbers 1 through 6 have already been placed in the appropriate columns for their numbers of divisors.

Total Number of Divisors for Counting Numbers

1	2	3	4	5	6	7	8	9	10
1	2	4	6						
	3								
	5								

Describe any pattern you see in the numbers in any particular column.

25. (a) Choose any prime number greater than 3. Square it. Add 17. Find the remainder that results when you divide by 12.
(b) Repeat part (a) for two other prime numbers greater than 3.
(c) Make a table of numbers from 1 to 102 with six numbers in each row. Circle all prime numbers. Where are all prime numbers other than 3 located?
(d) All prime numbers greater than 3 have the form $6n + 1$ or $6n - 1$, in which n is a counting number. Show why the procedure in part (a) always results in a remainder of 6.

26. Here is a way to find five consecutive composite numbers. First, compute $2 \times 3 \times 4 \times 5 \times 6 = 720$. Then, using the Divisibility-of-a-Sum Theorem, fill in the blanks.
(a) $2 \mid 720$ and $2 \mid 2$, so __________.
(b) $3 \mid 720$ and $3 \mid 3$, so __________.
(c) $4 \mid 720$ and __________.
(d) Complete the pattern of parts (a), (b), and (c).

27. Use the method of the preceding exercise to find 10 consecutive composite numbers.

28. Euclid proved that if p is prime and $2^p - 1$ is prime, then $(2^p - 1)2^{p-1}$ is perfect. This formula has been used to discover the 27 known perfect numbers. Use the formula to determine if the following values of p result in perfect numbers.
(a) 2 (b) 3 (c) 5 (d) 6

29.

$$2 + 1 = 3$$
$$2 \times 3 + 1 = 7$$
$$2 \times 3 \times 5 + 1 = 31$$

(a) Write the next two equations that continue the pattern.
(b) The numbers 3, 7, and 31 are prime. Are the numbers on the right sides of your equations in part (a) prime?
(c) Show that the next equation that continues the pattern results in a composite number divisible by a prime number between 50 and 60.

30. (a) Is $3^2 \cdot 2^4$ a factor of $3^4 \cdot 2^7$?
(b) Is $3^2 \cdot 2^5$ a factor of $3 \cdot 2^7$?

31. Suppose that x and y are whole numbers.
(a) Is x^3y^5 a factor of x^2y^8?
(b) Is x^2y^2 a factor of x^3y^5?

32. Sophie Germain (1776–1831) studied divisibility problems like the following.

Suppose we want $x^3 \div p$ to have a remainder of 2 for a counting number x and an odd prime p. For example, $2^3 \div 3$ has a remainder of 2. If possible, in parts (a)–(c), find an x such that:
(a) $x^3 \div 5$ has a remainder of 2.
(b) $x^3 \div 7$ has a remainder of 2.
(c) $x^3 \div 11$ has a remainder of 2.
(d) What pattern do you see in the values of x?

33. Without computing the results, explain why each of the following problems will result in a composite number.
(a) $3 \times 5 \times 7 \times 11 \times 13$
(b) $(3 \times 4 \times 5 \times 6 \times 7 \times 8) + 2$
(c) $(3 \times 4 \times 5 \times 6 \times 7 \times 8) + 5$

Special Exercises

34. Eratosthenes, a Greek mathematician, developed the Sieve of Eratosthenes about 2200 years ago as a method for finding all prime numbers less than a given number. Follow the directions to find all prime numbers less than or equal to 50.

1	2	3	4	5	6	7	8	9	10
11	12	13	14	15	16	17	18	19	20
21	22	23	24	25	26	27	28	29	30
31	32	33	34	35	36	37	38	39	40
41	42	43	44	45	46	47	48	49	50

(a) Copy the list of numbers.
(b) Cross out 1 because 1 is not prime.
(c) Circle 2. Count by 2's from there, and cross out 4, 6, 8, . . . , 50 because all these numbers are divisible by 2 and therefore are not prime.
(d) Circle 3. Count by 3's from there, and cross out all numbers not already crossed out because these numbers are divisible by 3.
(e) Circle the smallest number not yet crossed out. Count by that number, and cross out all numbers that are not already crossed out.
(f) Repeat part (e) until there are no more numbers to circle. The circled numbers are all the prime numbers.
(g) List all the primes between 1 and 50.

35. At Hexagon High, the students play mathematical pranks. There are exactly 400 lockers, numbered 1 to 400. One day, the 400 students filed into school one by one. The first student opened every locker. The second student closed every even-numbered locker. The third student changed (opened or closed) every third locker (3, 6, 9, . . .). The fourth student changed every fourth locker, and so on. After all the students were finished, which lockers were open?

4.4 COMMON FACTORS AND MULTIPLES

After studying factors and multiples of individual whole numbers, it is natural to examine the factors and multiples that two whole numbers have in common. With the exception of 0, any two whole numbers have a largest common factor and a smallest nonzero common multiple.

THE GREATEST COMMON FACTOR (GCF)

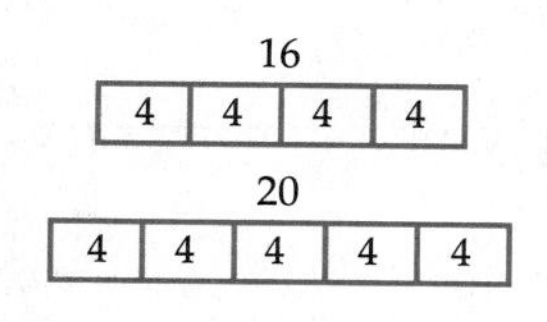

Figure 4–4

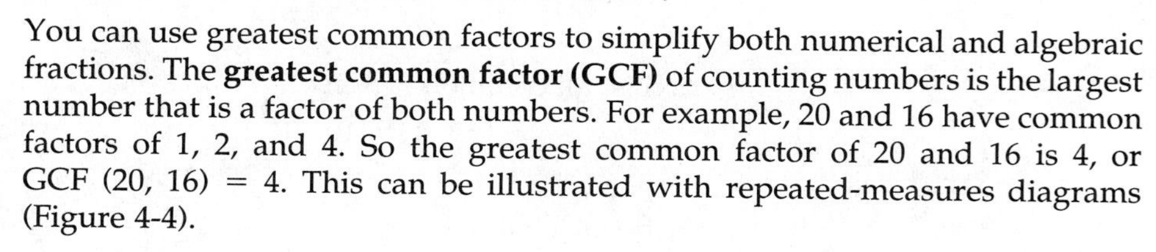
You can use greatest common factors to simplify both numerical and algebraic fractions. The **greatest common factor (GCF)** of counting numbers is the largest number that is a factor of both numbers. For example, 20 and 16 have common factors of 1, 2, and 4. So the greatest common factor of 20 and 16 is 4, or GCF (20, 16) = 4. This can be illustrated with repeated-measures diagrams (Figure 4-4).

Lesson Exercise 4.51

(a) What are the common factors of 20 and 30?

(b) What is the greatest common factor of 20 and 30?

(c) Draw repeated-measures diagrams illustrating your result in part (b).

Note that the definition of the greatest common factor excludes 0. The following exercise concerns what would happen if 0 were included in the definition.

Lesson Exercise 4.52

The definition of GCF is limited to counting numbers. Suppose 0 were included in the definition. What would the greatest common factor of 0 and any counting number be?

What follows are two methods for finding the greatest common factor: the factor-list method and the prime-factorization method. (A third method, the Euclidean algorithm, is presented later in the section.)

EXAMPLE 4.5

Find the greatest common factor of 60 and 140.

Solution

Two methods are as follows.

The Factor-List Method

1. List all factors.

60: 1, 2, 3, 4, 5, 6, 10, 12, 15, 20, 30, 60

The Prime-Factorization Method

1. Write the prime factorizations.

$$60 = 2 \cdot 2 \cdot 3 \cdot 5$$
$$140 = 2 \cdot 2 \cdot 5 \cdot 7$$

The Factor-List Method

140: 1, 2, 4, 5, 7, 10, 14, 20, 28, 35, 70, 140

2. List all common factors.

1, 2, 4, 5, 10, 20

3. So GCF(60, 140) = 20.

The Prime-Factorization Method

(*Note:* Since 2, 2, and 5 are on both lists, 2 is a common factor, $2 \cdot 2$ is a common factor, and $2 \cdot 5$ is a common factor, but we want the greatest common factor.)

2. Multiply all common prime factors.

$$2 \cdot 2 \cdot 5$$

3. So GCF(60, 140) = 20. ■

The factor-list method is less abstract, so children can more easily understand why it works. For this reason it is the first method children study in school. Although both methods work well for small numbers, the prime-factorization method is usually superior for larger numbers that have many factors.

Lesson Exercise 4.53

Compute the GCF of 546 and 650 using whichever method you think is easier.

When two counting numbers have no common factors other than 1, they are said to be **relatively prime**. If the numerator and denominator of a fraction are relatively prime, the fraction is in simplest form. For example, 7/10 is in simplest form because 7 and 10 are relatively prime. In general, A and B are **relatively prime** if and only if GCF(A, B) = 1. For example, 7 and 10 are relatively prime because GCF(10, 7) = 1. Note that A and B need not be prime numbers to be relatively prime.

Lesson Exercise 4.54

Which of the following pairs of numbers are relatively prime?

(a) 8 and 6 (b) 8 and 25 (c) 13 and 21

Lesson Exercise 4.55

(a) If A and B are two different prime numbers, are A and B relatively prime? Try some examples and make an educated guess.

(b) Your guess in part (a) is based upon ____________ reasoning.

EUCLIDEAN ALGORITHM FOR FINDING THE GCF

Suppose you want to find the GCF of two numbers that are difficult to factor, such as 374 and 1173. There is a third method for finding the GCF that is more efficient.

Lesson Exercise 4.56

(a) According to the Divisibility-of-a-Difference Theorem from Section 4.1, if $x \mid 374$ and $x \mid 1173$, then ________________.

(b) If $y \mid 374$ and $y \mid 1173$, then ________________.

The preceding exercise suggests a method of simplifying a GCF problem. Any number (such as x or y in the preceding exercise) that is a common factor of 374 and 1173 is also a common factor of 374 and 799. If x is GCF(374, 1173), then x is also GCF(374, 799).

Continuing this way, x is GCF(374, 799 − 374), or GCF(374, 425). Then x is also GCF(374, 425 − 374), or GCF(374, 51). This simplifies the original problem from GCF(374, 1173) to GCF(374, 51).

Lesson Exercise 4.57

51 is simply the remainder after you select as many 374's as possible from 1173. How can you find 51 from 374 and 1173 more quickly?

The method of simplifying GCF(374, 1173) with division is called the Euclidean algorithm. Once we reach GCF(374, 51), we repeat the process on those two numbers. Here is the method from start to finish.

EXAMPLE 4.6

Find GCF(374, 1173) using the Euclidean algorithm.

Solution

$$\text{GCF}(374, 1173) \rightarrow 374 \overset{\displaystyle 3}{\overline{\big)\,1173}} \qquad 1173 - 1122 = 51$$

$$\text{GCF}(374, 51) \rightarrow 51 \overset{\displaystyle 7}{\overline{\big)\,374}} \qquad 374 - 357 = 17$$

$$\text{GCF}(17, 51) \rightarrow 17 \overset{\displaystyle 3}{\overline{\big)\,51}} \qquad 51 - 51 = 0$$

This shows that GCF(17, 51) = 17. Therefore, GCF(374, 1173) = 17. ■

Try one yourself.

Lesson Exercise 4.58

Use the Euclidean algorithm to find GCF(253, 322).

THE LEAST COMMON MULTIPLE (LCM)

Least common multiples of denominators are the least common denominators used to add and subtract fractions. The **least common multiple (LCM)** of two counting numbers is the smallest counting number that is a multiple of both numbers. For example, the nonzero multiples of 10 are 10, 20, 30, 40, 50, . . . , and the nonzero multiples of 8 are 8, 16, 24, 32, 40, 48, ... So the least common multiple (LCM) of 10 and 8 is 40, or LCM(10, 8) = 40. This can be illustrated with a repeated-measures diagram, as shown in Figure 4-5.

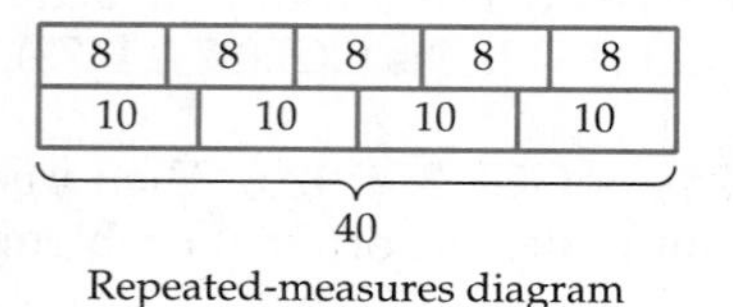

Repeated-measures diagram

Figure 4–5

Note that 40, 80, 120, 160, . . . are all common nonzero multiples of 10 and 8. Any two counting numbers have an infinite set of common multiples.

Lesson Exercise 4.59

(a) Name three common multiples of 10 and 15.

(b) What is the least common multiple of 10 and 15?

(c) Draw a repeated-measures diagram that supports your answer.

Lesson Exercise 4.60

(a) Is 18 a common multiple of 3 and 6?

(b) Is 40 a common multiple of 5 and 8?

(c) Based upon parts (a) and (b), what numbers would you expect to be a common multiple of counting numbers M and N?

(d) What kind of reasoning did you use in part (c)?

(e) Use the definition of a multiple to prove that your answer to part (c) is true.

Lesson Exercise 4.60 verifies the following theorem.

Products as Common Multiples

MN is a common multiple of M and N for all counting numbers M and N.

Computing the product of two counting numbers is an easy way to find a common multiple. However, the result may not be the *least* common multiple. Two methods for finding the LCM are the multiple-list method and the prime-factorization method. Both methods are used in the following example to find the LCM of 10 and 8.

EXAMPLE 4.7

Find the LCM of 10 and 8.

Solution

Two methods are as follows.

The Multiple-List Method

1. List the multiples of each number.

10: 10, 20, 30, 40, 50, . . .
8: 8, 16, 24, 32, 40, 48, . . .

2. Find the first common multiple.

10, 20, 30, 40, 50, . . .
8, 16, 32, 40, 48, . . .

3. So the LCM = 40.

The Prime-Factorization Method

1. Write the prime factorizations.

$$10 = 2 \cdot 5$$
$$8 = 2 \cdot 2 \cdot 2$$

2. Multiply all prime numbers that appear in either factorization. Use each prime number the greatest number of times it appears in either factorization. The LCM of 10 and 8 must contain $2 \cdot 5$ (which is 10) and $2 \cdot 2 \cdot 2$ (which is 8). So it is $2 \cdot 5 \cdot$ ____. In order for the LCM to contain 8, we need two more 2's as factors. $2 \cdot 5 \cdot \underline{2 \cdot 2} = 40$.

3. So the LCM = 40. (*Note:* $2 \cdot 2 \cdot 2 \cdot 5$ contains $2 \cdot 2 \cdot 2$, or 8, and $2 \cdot 5$, or 10, so it is a multiple of both 8 and 10.) ■

The multiple-list method is less abstract, so children can more easily understand why it works. For this reason, it is the first method taught in school. Although both methods work well for smaller numbers, the prime-factorization method is usually superior for larger LCMs.

Lesson Exercise 4.61

Compute the LCM of 120 and 72 using both methods.

Lesson Exercise 4.62

Compute the LCM of 62 and 80 using either method.

AN INVESTIGATION: LEAST COMMON MULTIPLES

Lesson Exercise 4.63

Consider the following problem. "When is MN the least common multiple of counting numbers M and N?"

(a) Restate the question in your own words.

(b) Devise a plan and solve the problem.

ANSWERS TO SELECTED LESSON EXERCISES

4.51 (a) 1, 2, 5, 10 (b) 10
(c)

20

10	10

30

10	10	10

4.52 The counting number

4.53 26

4.54 (b), (c)

4.55 (a) Yes (b) inductive

4.56 (a) $x \mid 799$ (b) $y \mid 799$

4.57 Find the remainder for $1173 \div 374$.

4.58 23

4.59 (a) 30, 60, 90 (b) 30
(c)

10	10	10
15	15	

30

4.60 (a) Yes (b) Yes (c) MN
(d) Inductive
(e) MN is a whole number times M, so it is a multiple of M.

MN is a whole number times N, so it is a multiple of N.

4.61 360

4.62 2480

4.4 HOMEWORK EXERCISES

Basic Exercises

1. (a) Complete the Venn diagram.

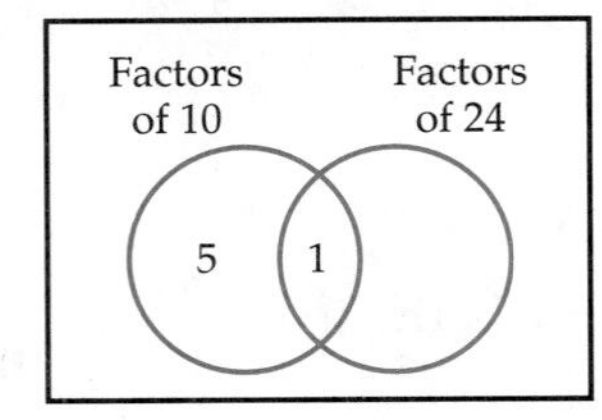

(b) Use the diagram to find the common factors of 10 and 24.

(c) Use the diagram to find the GCF of 10 and 24. (Part d is in the next column.)

(d) Draw repeated-measures diagrams illustrating your result in part (c).

***2.** (a) Compute the GCF of 42 and 120 using factor lists.

(b) Compute the GCF of 42 and 120 using prime factorizations.

3. Compute the GCF of 126 and 270 using prime factorizations.

4. Compute the GCF of 20, 48, and 172.

5. (a) Find the GCF of 4 with each of the following: 5, 6, 7, 8, 9, 10, 11, and 12.

(b) Write a generalization of your results in part (a).

***6.** $a = 3^2 \cdot 5 \cdot 11^4$, and $b = 3 \cdot 5^4 \cdot 7 \cdot 11$. Why isn't 11 the GCF?

7. $a = 2^2 \cdot 3^4 \cdot 5^7$, and $b = 2^3 \cdot 3 \cdot 5^6$. What is GCF($a$, b)? (You may write your answer as a prime factorization.)

8. $a = 5 \cdot 7 \cdot 11^3$, and $b = 2^3 \cdot 5^2 \cdot 7 \cdot 11$. What is GCF($a$, b)?

9. Which of the following pairs of numbers are relatively prime?
(a) 11 and 12 (b) 34 and 51
(c) 157 and 46

10. Which of the following pairs of numbers are relatively prime?
(a) 28 and 33 (b) 14 and 15
(c) 186 and 97

11. Find the GCF for each of the following, using the Euclidean algorithm.
(a) 627 and 665 (b) 851 and 2035
(c) 551 and 609

12. Find the GCF for each of the following, using the Euclidean algorithm.
(a) 583 and 795 (b) 602 and 1330
(c) 1705 and 527

13. (a) Name three common multiples of 10 and 12.
(b) How many common multiples do 10 and 12 have?
(c) What is the LCM of 10 and 12?

14. Suppose 0 were included as a possible multiple in the definition of the least common multiple. What would be the LCM of any two whole numbers?

15. Place the numbers 2, 6, 10, 16, 24, 40, and 48 in the correct regions of the Venn diagram.

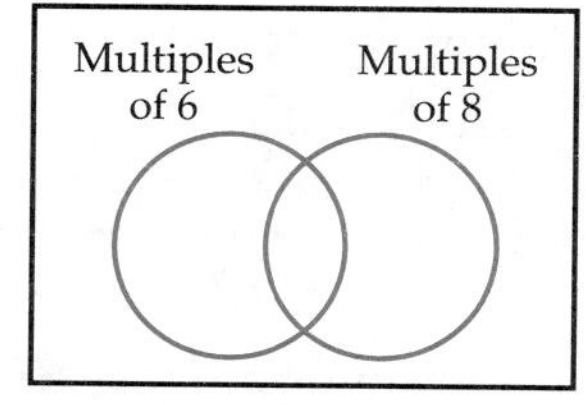

Multiples of 6 and 8

16. Which of the following could describe set A, set B, set C, and set D?

Multiples of 3 Multiples of 10
Divisors of 20 Prime numbers
Odd numbers

(a)

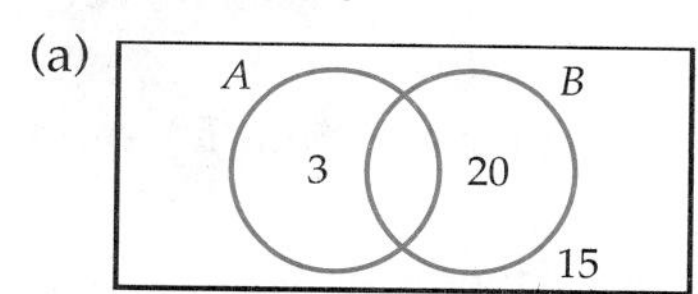

(b)

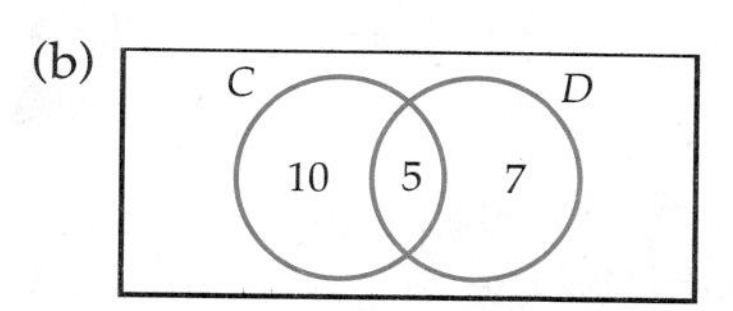

17. (a) Find the LCM of 20 and 32 using multiple lists.
(b) Find the LCM of 20 and 32 using prime factorizations.
(c) Draw a repeated-measures diagram that supports your answer.

18. Two buses leave the terminal at 8 A.M. Bus 36 takes 60 minutes to complete its route; bus 87 takes 75 minutes.
(a) When is the next time the two buses will arrive together at the terminal (if they are on time)?
(b) Draw a repeated-measures diagram that supports your answer.

19. Compute the LCM of 108, 32, and 20.

20. You have a balance and an unlimited supply of 3-oz and 5-oz weights. It is possible to weigh any whole-number amount of ounces using these weights on one or both sides of the balance!
(a) How would you weigh the following amounts of a substance with the 3- and 5-oz weights?
(1) 13 oz (2) 2 oz (3) 1 oz
(b) Find another set of weights (measured in ounces) of two different denominations that you could use to weigh any whole-number amount of ounces.

21. $a = 2 \cdot 3^2 \cdot 5^7 \cdot 7 \cdot 11$, and $b = 2^3 \cdot 3^4 \cdot 5 \cdot 11$.
(a) What is LCM(a, b)?
(b) What is GCF(a, b)?

22. $a = 2^5 \cdot 7 \cdot 11^3$, and $b = 2^3 \cdot 3 \cdot 7^2 \cdot 11$.
(a) What is LCM(a, b)?
(b) What is GCF(a, b)?

23. $a = 2^3 \cdot 5^2 \cdot 7^3$, GCF$(a, b) = 2 \cdot 5^2 \cdot 7$, and LCM$(a, b) = 2^3 \cdot 3^2 \cdot 5^4 \cdot 7^3$. Find b.

24. $a = 2 \cdot 3^2 \cdot 7^3$, and GCF$(a, b) = 2 \cdot 3^2 \cdot 7$. Give two possible values for b.

25. For what counting numbers a and b would $7^a = 5^b$?

26. The lowest common denominator for two fractions is the same as the ________ of the denominators.
(a) LCM (b) GCF (c) Neither

27. In simplifying 30/45, divide the numerator and denominator by the ________ of 30 nd 45.
(a) LCM (b) GCF (c) Neither

28. If A is a counting number, then GCF$(A, A^2) =$ ________.

29. Without multiplying out the bases, tell which number in each pair is larger.
(a) 2^{10} or 4^4 (b) 3^{12} or 10^6

In Exercises 30–32, A and B are counting numbers. If you think a statement is true, give an example. If a statement is false, give a counterexample.

30. True or false? If A and B are relatively prime, then A is prime and B is prime.

31. True or false? If A is a factor of B, then A is a factor of every multiple of B.

32. True or false? The LCM of two prime numbers is their product.

33. Find the greatest common factor of $3x^2y$ and $6y^2$, in which x and y are counting numbers.

34. Find three common multiples of x^2 and xy, in which x and y are counting numbers.

Extension Exercises

35. You want to pack new "Fiber 'n Wood Chips" cereal boxes standing up in a carton. The cereal boxes and all possible cartons are 10 in. high. The cereal boxes are 8 in. long and 3 in. wide. Which of the following cartons could be used to pack them without any wasted space between boxes?

(a) 24" × 10"   (b) 32" × 15"   (c) 30" × 16"

(Top view: All boxes are 10" high.)

36. Suppose that, in the preceding exercise, you want to use a box with a square base (since it would use less material). What are the smallest possible dimensions of the base?

37. Consider the following problem. "John has a cloth that is 30 in. by 48 in. He wants to cut out the largest possible squares of the same size and use all the material. How big can the squares be?"
(a) Select a problem-solving strategy.
(b) Solve the problem.

38. When the entire fifth grade is divided into groups of 4, there is 1 child left over. When the fifth grade is divided into groups of 3, there are 2 children left over. How many children could be in the fifth grade? (Give all possible answers.)

39. Let x, y, and z be counting numbers. Tell whether each of the following is always true, sometimes true, or never true. Give a numerical example or examples to support your answer.
(a) GCF$(x, y) = x$
(b) LCM$(x, y) < x$
(c) If GCF$(x, y) =$ GCF(x, z), then $y = z$.

40. A and B are counting numbers. If LCM$(A, B) = B$, then GCF$(A, B) =$ ___.

41. How are GCF(a, b) and LCM(a, b) related?
(a) Select three different pairs of counting numbers greater than 2, and fill in the table.

a	b	GCF(a, b)	LCM(a, b)	GCF$(a, b) \cdot$ LCM(a, b)

(b) Make a generalization based upon your results in part (a).

42. What is the largest counting number that is a factor of all the following numbers: $(1 \cdot 3 \cdot 5)$, $(3 \cdot 5 \cdot 7)$, $(5 \cdot 7 \cdot 9)$, $(7 \cdot 9 \cdot 11)$, . . . ?

SUMMARY

"Number theory offers many rich opportunities for explorations that are interesting, enjoyable, and useful. These explorations have payoffs in problem solving, in understanding and developing other mathematical concepts, in illustrating the beauty of mathematics, and in understanding the human aspects of the historical development of number" (NCTM, *Curriculum and Evaluation Standards*, p. 91).

The classical Greeks were the first to study patterns in the set of counting numbers in detail, breaking down the counting numbers greater than 1 into their basic components. This analysis led to the study of factors (divisors) and multiples (by reversing factor-number relationships). The Divisibility-of-a-Sum Theorem and the Divisibility-of-a-Product Theorem describe the relationships between factors of different numbers. These theorems can be used with expanded notation to show how divisibility tests work. Elementary-school children learn divisibility tests for 2, 3, 5, 9, and 10.

The crowning achievement of ancient number theory was the discovery of the basic building blocks of counting numbers greater than 1: prime numbers. All composite numbers can be uniquely factored using prime numbers. This result is called the Fundamental Theorem of Arithmetic.

The Greeks also studied the relationship between the factors and multiples of counting numbers. Any two counting numbers have a greatest common factor and a least common (nonzero) multiple. In simplifying fractions, one divides the numerator and denominator by their greatest common factor. In order to find the lowest common denominator for adding or subtracting two fractions, one can compute the least common multiple of the denominators.

Number theory continues to fascinate mathematicians because it contains seemingly simple conjectures that have remained unproved for hundreds of years. People have tested these conjectures for many examples, but no one has been able to prove these statements.

STUDY GUIDE

To review Chapter 4, see what you know about each of the following ideas or terms that you have studied. You can also use this list to generate your own questions about the chapter.

The NCTM Curriculum Standards and Number Theory

Selected NCTM Curriculum Standards

The following standards come from the NCTM document.

- Develop common understanding of mathematical ideas, including the role of definitions.
- Discuss mathematical ideas and make conjectures and convincing arguments.
- Recognize and apply deductive and inductive reasoning.
- Develop and apply number theory concepts (e.g., primes, factors, and multiples) in real-world and mathematical problem situations.

1. Describe how each standard listed relates to the material you studied in Chapter 4.
2. Select any current elementary-school mathematics textbook series, and describe a sample lesson or exercise that illustrates each standard listed.

NUMBER THEORY IN ELEMENTARY SCHOOL

The following chart shows at which grade levels selected topics in number theory typically appear in elementary-school mathematics textbooks. Underlined numbers indicate grades in which the most time is spent on the given topic.

Topic	Typical Grade Level in Current Textbooks
Divisibility tests	5, 6
Factors	4, 5, 6
Prime and composite numbers	5, 6
Multiples	4, 5, 6
GCF and LCM	5, 6

REVIEW EXERCISES

1. Find a three-digit number N for which $N - 8$ is divisible by 8, $N - 7$ is divisible by 7, $N - 6$ is divisible by 6, and $N - 5$ is divisible by 5.

2. By the definition of a multiple, N is a multiple of 8 if and only if ________________.

3. Write the prime factorization of 312.

4. The numbers A and 10 are relatively prime. Give two possible values of A.

5. To simplify 28/40, you divide the numerator and denominator by ____, which is the ________________ of 28 and 40.

6. Use the Divisibility-of-a-Product Theorem to explain why, for any whole number $x > 2$, $(x - 2) \mid (x - 2)(x + 3)$.

In Exercises 7–9, assume that A, B, and C are counting numbers. If a statement is true, give an example that illustrates it. If a statement is false, give a counterexample.

7. True or false? If $A \mid B$ and $A \nmid C$, then $A \nmid (B + C)$.

8. True or false? The GCF of any two prime numbers is 1.

9. True or false? If $A \mid B$, then $A \mid B^2$.

In Exercises 10 and 11, if the statement is true, prove it. If the statement is false, give a counterexample.

10. X, Y, and Z are whole numbers, with $X \neq 0$ and $Y \neq 0$. True or false? If $XY \mid Z$, then $XY \mid (XY + Z)$.

11. True or false? The LCM of any prime number and any composite number is their product.

12. Complete the number so that it is divisible by 3, 5, and 10.

$$3\,2,0\,0\,5, \underline{\quad}\,2\,\underline{\quad}$$

Determine all possible answers.

13. $a = 2 \cdot 5^2 \cdot 11^3$, and $b = 2^3 \cdot 3 \cdot 5^4 \cdot 11^2$.
 (a) What is GCF(a, b)?
 (b) What is LCM(a, b)?

14. Suppose that you want to determine if 577 is prime. Using the Prime Number Test, what is the minimum set of numbers you must try as divisors?

15. Use expanded notation and the Divisibility-of-a-Sum Theorem to prove that if a three-digit numeral ends in 0 or 5 ($\underline{A}\,\underline{B}\,\underline{0}$ or $\underline{A}\,\underline{B}\,\underline{5}$), then the number is divisible by 5.

16. Loanne wants to lay three rows of floor tiles of different sizes and colors. One row will contain white tiles, each 8 cm long. One row will contain blue tiles, each 10 cm long. One row will contain gray tiles, each 24 cm long. At what lengths will the tiles line up across all three rows?

17. Write a paragraph describing how to construct a proof of a theorem about factors.

SUGGESTED READINGS

Brown, S. *Some Prime Comparisons*. Reston, Va.: NCTM, 1978.

Burton, D. *The History of Mathematics*. 3rd ed. Dubuque, Iowa: William C. Brown, 1994.

National Council of Teachers of Mathematics. *Historical Topics for the Mathematics Classroom*. Reston, Va.: NCTM, 1989.

National Council of Teachers of Mathematics. *New Directions for Elementary School Mathematics*. 1989 Yearbook. Reston, Va.: NCTM, 1989.

5 Integers

As far back as 200 B.C., Chinese accountants used black (negative) rods for debts and red (positive) rods for credits. But 1700 years passed before Giralumo Cardano (1501–1576) gave the first detailed description of negative numbers and their properties.

Cardano was a physician who did mathematics and astrology in his spare time. After publishing a horoscope of the life of Jesus Christ, Cardano was imprisoned for heresy. In spite of this scandal, he was later appointed as an astrologer to the papal court! Some sources claim that Cardano killed himself in 1576 so that his prediction of the date of his death would be correct.

Not everyone accepted Cardano's negative numbers. René Descartes called them "false" numbers. However, by the eighteenth century, negative numbers were widely used.

Today, negative numbers retain their importance in accounting. If you have \$80 and spend \$100, you have a net worth of −\$20. People also use negative numbers to measure temperatures, golf scores, and changes in stock prices.

In mathematics, the set of integers results from enlarging the set of whole numbers to create solutions for every whole-number subtraction problem (for example, $6 - 10$). The set of integers retains many properties and patterns of whole-number operations.

In this course, negative whole numbers (integers) are introduced before fractions and decimals to illustrate more clearly the connections among different sets of numbers in elementary school. However, in elementary-school mathematics, most fraction and decimal topics precede most topics involving negative numbers.

5.1 ADDITION AND SUBTRACTION OF INTEGERS

When you add or multiply any two whole numbers, the result is a whole number. However, some whole-number subtraction problems, such as $2 - 5$, do not have whole-number answers. These subtraction problems provided the mathematical motivation for the creation of negative integers.

To compute 2 − 5 on a number line, we would start at 2 and move 5 to the left. The whole-number line goes only 2 units to the left of 2, to reach 0. We mark units to the left of 0 and create negative integers.

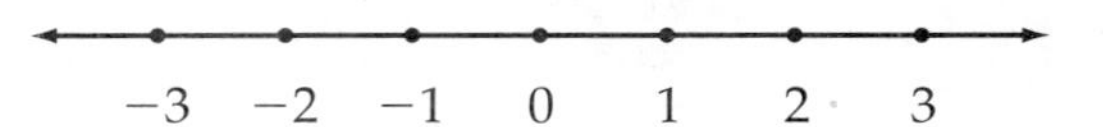

For example, 3 is 3 units to the right of 0, and "negative 3," written −3, is 3 units to the left of 0. For each positive number to the right of 0 on a number line, there is a corresponding negative number the same distance to the left of 0. The set {−1, −2, −3, . . .} is called the set of **negative integers**.

By combining the set of whole numbers with the set of negative integers, one obtains the set of integers.

Definition **Integers**

The union of the set of whole numbers and the set of negative integers is called the **integers**. The set of integers is denoted by $I = \{\ldots -3, -2, -1, 0, 1, 2, 3, \ldots\}$.

The integers are an extension of the whole numbers. With whole numbers, we can only count forward from 0, or move to the right from 0 on a number line. With integers, we can count forward or backward from 0 and move to the right *or* left from 0 on a number line. A greater number is always to the right of a lesser number on the standard integer number line. For integers m and n, $m > n$ if and only if m is located to the right of n on the number line.

Lesson Exercise 5.1

What number is 2 units to the left of −130 on the standard number line?

Using numbers greater than or equal to 0 is sufficient for measurements such as weight, area, and the number of raisins in a bowl of cereal. Such measurements have a lower limit of 0. In some measurement scales, such as temperature, the scale extends above *and* below 0.

Figure 5-1 illustrates common uses of negative numbers. Investors do not like negative numbers, but golfers love them!

Figure 5–1

Lesson Exercise 5.2

Tell in words what the − sign means in

(a) a temperature of −5°C.

(b) a golf score of −4.

(c) an account balance of −$400.

(d) a change in a stock price of −3 points.

As a measure, the number −5 (called "negative 5") can represent a location on a number line or a move of 5 to the left.

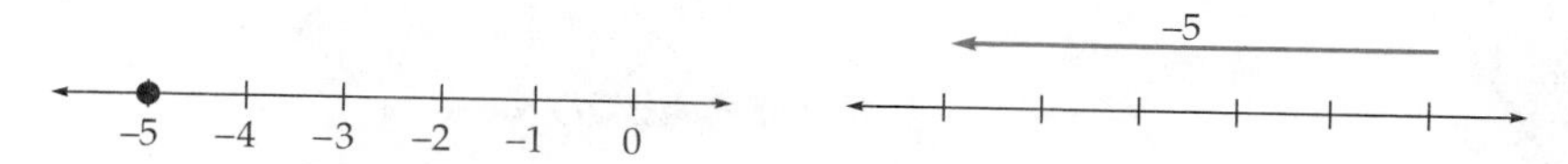

Using a set model, −5 can be represented by 5 black counters from a set of blue (positive) counters and black (negative) counters.

–5

The number −5 could represent a debt of $5. Positive numbers such as 2 would be represented by differently colored (blue) counters.

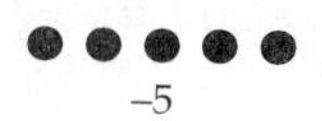

2

Lesson Exercise 5.3

Represent -2 in three ways: as a position on a number line, as a move on a number line, and as a set of colored counters.

The number -5 is also "the opposite of 5." A number and its **opposite** (for example, 5 and -5) are the same distance from 0 but on opposite sides.

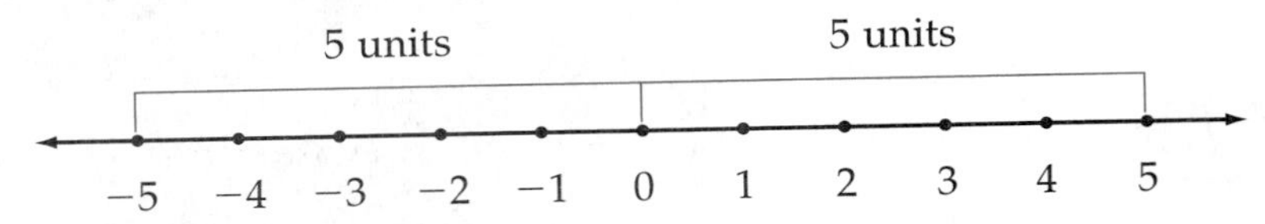

The opposite of -5, written $-(-5)$, is 5. If x is an integer, x can be positive, negative, or zero; consequently, $-x$ can be negative, positive, or zero. Thus, calling $-x$ "the opposite of x" instead of "negative x" may be less confusing. The use of the $-$ sign to indicate "the opposite of" is one of the three uses of the minus sign: as a subtraction symbol, as a negative sign, and as an opposite sign.

The distance between 0 and an integer x on the integer number line is called the **absolute value** of x and written $|x|$. The absolute values of both 5 and -5 are 5. In symbols,

$$|5| = 5 \quad \text{and} \quad |-5| = 5$$

Lesson Exercise 5.4

Using the word the "distance," explain why $|-13| = |13|$.

Integer arithmetic is more abstract than whole-number arithmetic. People are sometimes surprised by the answers to integer arithmetic problems. The remainder of this lesson shows how to explain the results of integer addition and subtraction.

INTEGER ADDITION

If you are \$3 in debt and you receive a bill for \$2, what is your new net worth? If a football player loses 3 yards on one play and 2 yards on the next, what is the player's overall yardage on the two plays? Both these questions are applications of $-3 + (-2)$. (*Note:* Any time a signed number is written after an operation symbol, parentheses are placed around that signed number.)

The first question is best modeled by a set of colored counters. The second question is best modeled by a number line. Example 5.1 shows how to compute $-3 + (-2)$ using colored counters or a number line.

EXAMPLE 5.1

Explain how to compute $-3 + (-2)$ using

(a) colored counters.

(b) a number line.

Solution

(a) Suppose positive numbers are represented by blue counters and negative numbers by black counters. Show -3 as 3 black counters and -2 as 2 black counters.

Combine these counters to obtain 5 black counters, which would represent -5. So $-3 + (-2) = -5$.

(b) Go to -3. Add the second number, -2, by moving 2 to the left from -3.

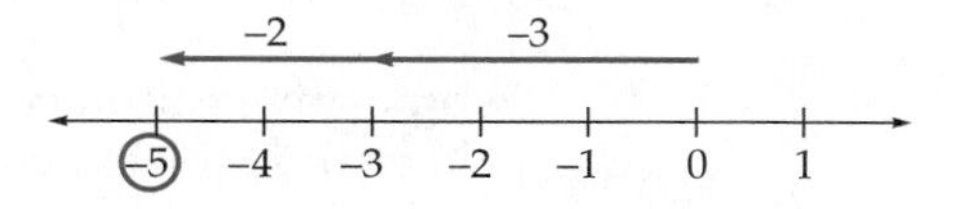

You end up at -5. So $-3 + (-2) = -5$. ■

Next, consider some examples involving a positive number and a negative number. To use the counters, note that a positive counter and a negative counter add up to 0. In other words, 1 positive counter cancels out 1 negative counter just as \$1 cancels out a debt of \$1.

The simplest addition of a positive integer and a negative integer has the form $a + (-a)$.

Lesson Exercise 5.5

(a) Explain how to compute $3 + (-3)$ with a number line.

(b) Explain how to compute $3 + (-3)$ with colored counters.

(c) The general result for a whole number a is $a + (-a) =$ ________.

Next, consider addition of the form $a + (-b)$ for whole numbers a and b, in which $a > b$ or $a < b$. Suppose you reach an intersection and are unsure about whether to turn right or left. So you turn right and drive for 2 miles. Uh oh. This can't be the right way, so you turn around and head 3 miles in the opposite direction. Where are you now in relation to the intersection? This trip is an application of $2 + (-3)$.

EXAMPLE 5.2

(a) Explain how to compute $2 + (-3)$ with a number line.

(b) Explain how to compute $2 + (-3)$ using colored counters.

(c) In adding 2 and -3, why does the answer come out negative?

Solution

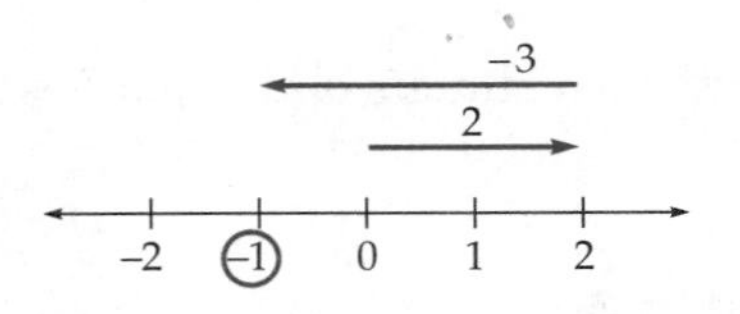

Figure 5–2

(a) Go to 2. Add the second number, -3, by moving 3 to the left from 2 (Figure 5-2). You end up at -1. So $2 + (-3) = -1$. This process corresponds to going 2 miles to the right and 3 miles to the left. You end up 1 mile to the left of where you started.

(b) Represent 2 with 2 blue counters and -3 with 3 black counters. Combine the counters.

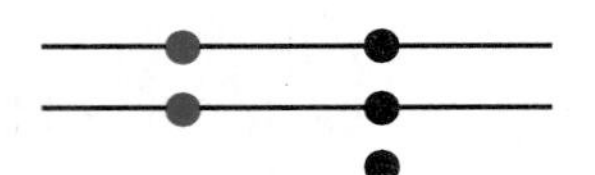

Two positive counters and two negative counters add up to 0, leaving one negative (black) counter. So $2 + (-3) = -1$.

(c) The answer comes out negative because -3 is larger in absolute value than 2. ■

Lesson Exercise 5.6

(a) Explain how to compute $4 + (-1)$ with a number line.

(b) Explain how to compute $4 + (-1)$ with colored counters.

(c) In adding 4 and -1, why does the answer come out positive?

What are the general rules for addition involving negative integers? First consider the sum of two negative integers.

Lesson Exercise 5.7

(a) The general rule for whole numbers a and b is $-a + (-b) =$ ________.

(1) $-(a + b)$ (2) $b - a$ (3) $a + b$ (4) $a - b$

(b) To add two negative integers, add their ________________ and make the result ____________.

It is more difficult to express the rule for adding a positive integer and a negative integer in words.

Lesson Exercise 5.8

(a) In adding a positive integer and a negative integer, the sign of the sum is determined by ______________________.

(b) To add a positive integer and a negative integer, find the ________ of their absolute values. The sum has the sign of the integer with the ____________.

Lesson Exercises 5.7 and 5.8 suggest the rules for addition involving negative integers. Combined with whole-number addition rules, these rules comprise the *definition* of integer addition.

Definition **Addition of Integers**

1. For any integer a, $a + 0 = 0 + a = a$.
2. To add two positive integers, add them as whole numbers.
3. To add two negative integers, add their absolute values and make the result negative.
4. To add a positive integer and a negative integer, compute the larger absolute value minus the smaller absolute value. The sum has the sign of the integer with the larger absolute value. If both integers have the same absolute value, the sum is 0.

Consider the following applications of integer addition.

Lesson Exercise 5.9

(a) In today's mail you will receive a check for $282 and a bill for $405. Write an integer addition equation that gives your overall gain or loss.

(b) What type of addition application is this?

Although most applications fall into the set and measure categories, money does not fit clearly into either category. Money is taught as a measurement in school, since it measures the value of goods and services. Yet money is more easily modeled by counters (sets) in units of dollars or cents. In this text, applications concerning money will be classified as "measures/sets," indicating that money has characteristics of both measures and sets.

Can you write a realistic application of a given integer addition problem? You might use temperature, money, or a journey. Try the following.

Lesson Exercise 5.10

Write an application problem that represents $-10 + (-14)$.

INTEGER SUBTRACTION

In elementary school, a child who is learning whole-number subtraction may ask, "What is $3 - 5$?" The child is asking a question that motivates the creation of negative numbers. The four approaches in Lesson Exercise 5.11 can be used to develop subtraction. See which of them you already know about. Then continue reading to learn about the others.

Lesson Exercise 5.11

Show how to determine the result of $3 - 5$ by

(a) using a number line.

(b) using a temperature or money application.

(c) adding the opposite.

(d) using colored counters.

This section covers all the approaches in the preceding exercise. First, how can you explain the answer to $3 - 5$ by using an application of it? Integer subtraction can be illustrated by situations involving money or temperatures. Using take-away, $3 - 5$ can represent an application such as "I had \$3 and I spent \$5. What is my net worth now?" or "The temperature was 3°C, and it went down 5°. What is the temperature now?" In both cases, the answer is -2 with the appropriate units (−\$2 or −2°C).

Lesson Exercise 5.12

Make up a temperature or money problem for each of the following, and give the result.

(a) $-2 - 1$

(b) $2 - (-3)$ (*Hint:* Use a temperature comparison.)

(c) $-3 - (-2)$ (*Hint:* Use a debt and take away some of it.)

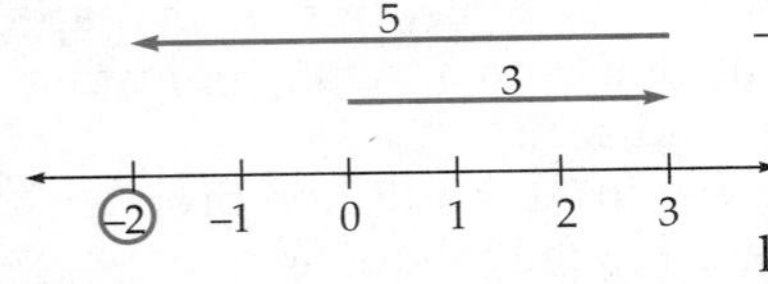

Figure 5–3

Like integer addition, integer subtraction can be illustrated with a number line or with colored counters. First consider a number line. To show $3 - 5$, first go to 3. Then move 5 to the left. You end up at -2. So $3 - 5 = -2$ (Figure 5-3).

The following exercise concerns addition and subtraction on an integer number line.

Lesson Exercise 5.13

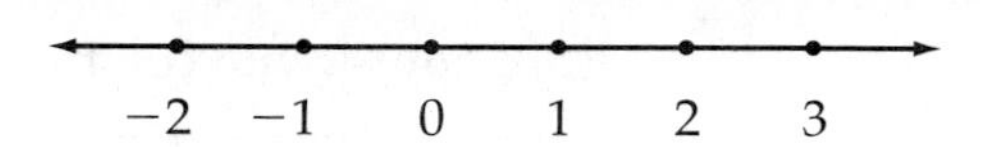

(a) On the standard number line, adding a positive integer is shown as a move to the ________.

(b) On the standard number line, adding a negative integer is shown as a move to the ________.

(c) On the standard number line, subtracting a positive integer is shown as a move to the ________.

(d) Based upon parts (a)–(c), it would make sense for subtraction of a negative integer to be shown as a move to the ________.

Apply the ideas of the preceding exercise to the three numerical examples from Lesson Exercise 5.12.

Lesson Exercise 5.14

Explain how to compute each of the following using a number line.

(a) $-2 - 1$ (b) $2 - (-3)$ (c) $-3 - (-2)$

Many elementary-school textbooks now use colored counters as a model for integer subtraction. The easiest type of computation to work out with colored counters is one like $-3 - (-2)$.

Lesson Exercise 5.15

To compute $-3 - (-2)$ with colored counters, first show -3. Then take away -2 and see what is left. Write a more complete explanation, and include a drawing of the counters.

Many integer subtraction computations are more difficult to work out with counters. They require changing the way you represent the first number in the subtraction problem.

Lesson Exercise 5.16

(a) Which of the following shows 4?

(1) ● ● ● ●   (2) ● ● ● ● ●   (3) ● ● ● ● ●   (4) ● ● ● ● ● ●

(b) Show 4 with colored counters in another way.

EXAMPLE 5.3

Show how to find $3 - 5$ using colored counters.

Solution

First show 3.

● ● ●

Now take away 5. It's not possible. We need to represent 3 in a form in which we can take away 5 positive (blue) counters. Add two 0 pairs, and show 3 as

Now take away 5.

Two *black* chips are left. So $3 - 5 = -2$. ■

Lesson Exercise 5.17

Show how to find the following using colored counters.

(a) $-2 - 1$ (b) $-2 - (-3)$

Using applications, a number line, or colored counters, one finds that $3 - 5 = -2$, $-2 - 1 = -3$, $2 - (-3) = 5$, and $-3 - (-2) = -1$. But there is a shorter way: the add-the-opposite rule. The rule tells how to rewrite an integer subtraction problem as an equivalent integer addition problem. You probably remember this rule, but do you know why it works?

The rule can be explained using money or a number line.

Lesson Exercise 5.18

Suppose you are $10 in debt.

(a) Would you rather lose $5 or gain another $5 debt?

(b) Part (a) shows that subtracting _____ is the same as adding _____.

(c) $-10 - 5 =$ _____ and $-10 + (-5) =$ _____.

(d) Would you rather have someone remove a \$5 debt, or would you rather receive \$5?

(e) Part (d) shows that subtracting _____ is the same as adding _____.

(f) $-10 - (-5) =$ _____ and $-10 + 5 =$ _____.

Lesson Exercise 5.19

(a) On a number line, why is subtracting a positive number (such as 3) the same as adding its opposite (-3)?

(b) On a number line, why is subtracting a negative number (such as -5) the same as adding its opposite (5)?

(c) Make a generalization based upon parts (a) and (b).

The preceding two exercises suggest the add-the-opposite rule for subtracting integers, which will be our definition of integer subtraction.

Definition **Integer Subtraction**

Subtracting an integer is the same as adding its opposite. If x and y are integers, $x - y = x + (-y)$.

For example, $-2 - (+1) = -2 + (-1) = -3$. So any subtraction problem can be rewritten as an addition problem. In order to compensate for changing the subtraction sign to an addition sign, the sign of the number being subtracted must also be changed.

How would you compute $-2 - (+1)$ on a calculator? Most calculators have a [+/−] change-of-sign key that takes the opposite of the number on the display. To enter -2, try [2] [+/−] or [+/−] [2]. (The [−] key is used for subtraction, not for negative signs.) To compute $-2 - (+1)$ on your calculator, try

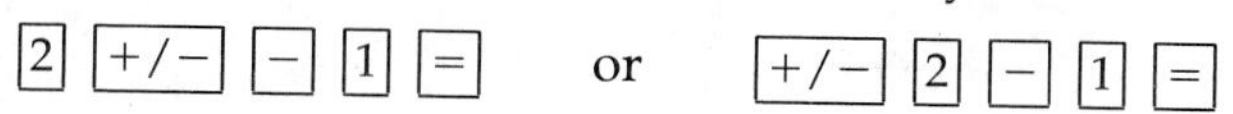

Lesson Exercise 5.20

Use the integer subtraction rule to compute the following. Then check your results with a calculator.

(a) $420 - (-506)$ (b) $-208 - 80$

The familiar categories fit applications of integer arithmetic.

Lesson Exercise 5.21

At 2 P.M., the temperature was 14°C. Since then it has dropped 38°.

(a) Write an integer subtraction equation for this situation.

(b) What is the temperature now?

(c) What subtraction category does this illustrate?

ANSWERS TO SELECTED LESSON EXERCISES

5.1 −132

5.2 (a) 5 degrees below 0 (b) 4 under par
(c) A debt of $400
(d) A decrease of $3/share

5.3

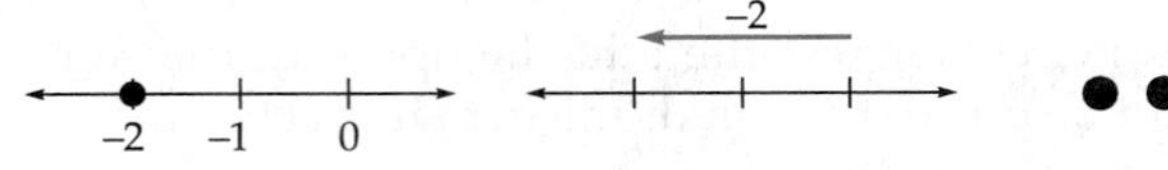

5.4 −13 and 13 are both equal distances (13) from 0.

5.5 (a) Go to 3. To add −3, move 3 to the left.

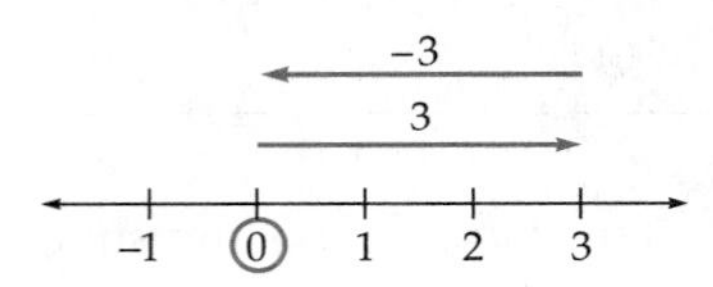

You end up at 0. So 3 + −3 = 0.

(b) Show 3 as 3 blue counters and −3 as 3 black counters. Each blue counter cancels out a black one. You are left with 0. So 3 + (−3) = 0.

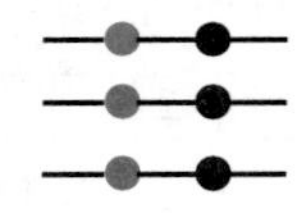

(c) 0

5.6 (a) Go to 4. Then move 1 to the left.

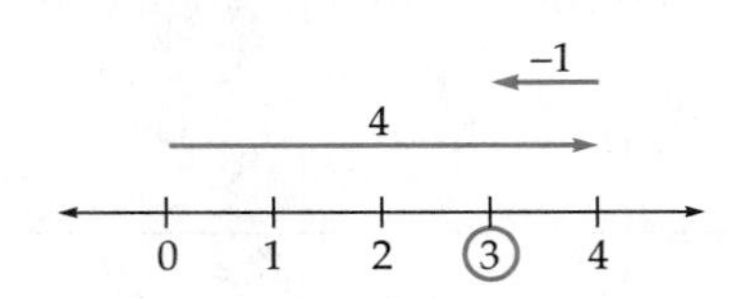

You end up at 3. So 4 + (−1) = 3.

(b) Show 4 as 4 blue counters and −1 as 1 black counter. Cancel 1 blue counter and 1 black counter. This leaves 3 blue counters. So 4 + (−1) = 3.

(c) The answer comes out positive because 4 has a larger absolute value than −1.

5.7 (a) (1) (b) absolute values; negative

5.8 (a) The number with the larger absolute value
(b) difference; larger absolute value

5.9 (a) $282 + (−$405) = −$123
(b) See the discussion that follows the exercise.

5.12 (a) The temperature is −2°C, and it goes down 1°. What is the temperature now? −3°C. So −2 − 1 = −3.
(b) How much higher is 2°C than −3°C? 5°. So 2 − (−3) = 5.
(c) I am $3 in debt, and $2 of the debt is forgiven. What is my debt now? $1. So −3 − (−2) = −1.

5.13 (a) right (b) left
(c) left (d) right

5.14 (a) Go to −2. Move 1 to the left. You end up at −3. So −2 − 1 = −3.

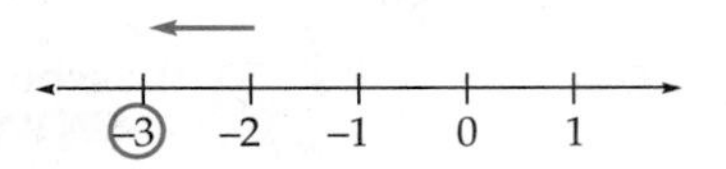

(b) Go to 2. Adding −3 would move to the left so subtracting −3 moves 3 to the right. You end up at 5. So 2 − (−3) = 5.

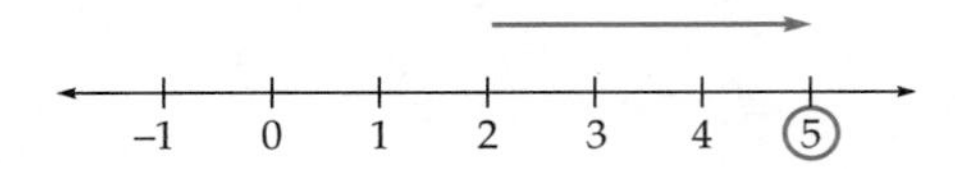

(c) Go to -3. Adding -2 is a move to the left, so subtracting -2 is a move of 2 to the right. You end up at -1. So $-3 - (-2) = -1$.

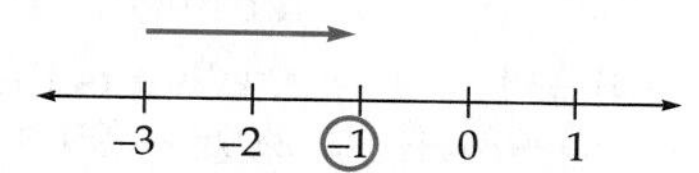

5.15 Show -3 as 3 black counters.

Now take away -2.

One black counter is left. So $-3 - (-2) = -1$.

5.16 (a) (1), (2), (4)

5.17 (a) Show -2 as 2 black counters. To be able to take away 1, we must change -2 to 3 black counters and 1 blue counter.

Now take away 1 blue counter.

Three black counters are left. So $-2 - 1 = -3$.

(b) Show -2 as 2 black counters. To be able to take away -3, we must change -2 to 3 black counters and 1 blue counter.

Now take away 3 black counters.

One blue counter is left. So $-2 - (-3) = 1$.

5.18 (a) No difference
(b) 5; -5
(c) -15; -15
(d) No difference
(e) -5; 5
(f) -5; -5

5.19 (a) In subtracting a positive integer a, you move a units to the left, and in adding $-a$, you move a units to the left.
(b) In subtracting a negative integer $-a$, you move a units to the right, and in adding a, you move a units to the right.

5.20 (a) 926 (b) -288

5.21 (a) $14 - 38 = -24$
(b) -24°C
(c) Take away measures

5.1 HOMEWORK EXERCISES

Basic Exercises

−1. Which of the following are integers?
(a) -4 (b) 0 (c) $\frac{2}{3}$ (d) $\frac{-8}{4}$

0. What number is 3 to the right of -100 on the standard number line?

1. What is the largest negative integer?

2. Write an integer to represent
(a) a debt of \$40. (b) a fever of 102°.
(c) a loss of 4 yards on a football play.

3. Explain the meaning of the − sign in.
(a) making -2 yards on a football play.
(b) a federal budget balance of $-$\$140 billion (called a deficit).
(c) an altitude of -50 ft.

4. Find an example of a place where negative numbers appear in the newspaper.

5. Represent -4 in three ways: on a number line, as a move on a number line, and as a set of colored counters.

6. A newspaper lists the following information on "departure from normal high temperature."

Boston -6 Wash. D.C. $+7$

Explain what these numbers mean.

7. (a) Use colored counters to represent a debt of \$3.
(b) Use a vertical number line to represent a temperature of 4° below 0°C.

8. $a < 0$. Then $-a$ is
(a) positive (b) zero (c) negative

9. Give the opposite idea and the integer that represents it.
(a) 20 seconds after liftoff (b) A loss of \$3
(c) 3 floors higher

10. Using the word "distance," explain why $|-5| = 5$.

11. (a) $|-3| =$ _____
(b) $|-7| =$ _____
(c) If $x < 0$, then $|x| =$ _____.

12.

A 0 B

A and B are 9 units apart. A is twice as far from 0 as B. What are A and B? (Guess and check.)

13. What integer *addition* problem is shown on the number line?

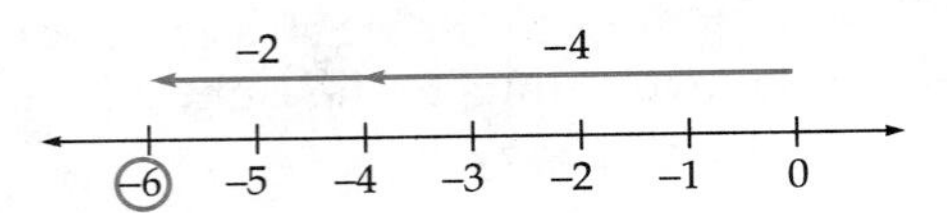

14. Explain how to compute $6 + (-2)$ with
(a) colored counters. (b) a number line.

15. (a) Explain how to compute $-5 + 3$ with a number line.
(b) What addition category is illustrated by part (a)?
(c) Explain how to compute $-5 + 3$ with colored counters.

16. State the rule for adding two negative integers.

17. If a and b are whole numbers, and $a > b$, then a formula relating $-a + b$ to whole-number subtraction is $-a + b =$
(a) $a - b$ (b) $a + b$
(c) $-(a - b)$ (d) $-(a + b)$

18. An ion contains 42 protons, each with a single positive charge, and 44 electrons, each with a single negative charge.
(a) What is the overall positive or negative charge of the ion?
(b) Write an addition equation for this situation.
(c) What addition category is illustrated?

19. In today's mail you will receive a check for \$86, a bill for \$30, and a bill for \$20.
(a) Write an integer addition equation that gives the overall gain or loss.
(b) What application category is this?

20. $-1 + 0 + 1 + 2 = 2$. What is the sum of $-99 + (-98) + \cdots + 98 + 99 + 100 + 101$?

21. Fill in the blanks, continuing the same pattern.

$2 + 2 = 4$
$2 + 1 = 3$
$2 + 0 = 2$
____
____

22. Compute the following without a calculator.
(a) $-23 + 30$ (b) $-8 + (-17)$
(c) $10 + (-25)$

23. Which of the following is a correct way to say $-(-6)$?
(a) "Minus minus 6"
(b) "The opposite of negative 6"
(c) "Minus negative 6"

24. Which of the following are correct ways to say $-3 - (-2)$?
(a) "Minus 3 minus negative 2"
(b) "Negative 3 minus negative 2"
(c) "Minus 3 minus minus 2"
(d) "The difference between minus 3 and minus 2"
(e) "The difference between negative 3 and negative 2"

25. Make up a temperature or money problem for $-6 - 4$, and give the result.

26. Make up a temperature or money problem for $4 - 8$, and give the result.

27. *Explain* how to compute $5 - (-2)$ on a number line.

28. *Explain* how to compute $-6 - (-3)$ on a number line.

29. Show two ways to represent each integer with colored counters.
(a) 3 (b) -2

***30.** Tell what subtraction problem each picture illustrates.
(a) ✖✖✖●● (b) ✖✖✖✖✖ / ●●● (c) ●●●●● / ✖✖

31. *Explain* how to compute the following with colored counters.
(a) $-6 - (-2)$ (b) $2 - 6$

32. *Explain* how to compute the following with colored counters.
(a) $-2 - 3$ (b) $4 - (-2)$

33. Tell whether each of the following is easier to compute with a number line or with colored counters.
(a) $2 - 5$ (b) $-3 - (-2)$
(c) $-1 - 3$ (d) $2 - (-4)$

34. (a) Taking away a debt of \$10 is the same as receiving ____.
(b) Part (a) illustrates that subtracting ____ is the same as adding ____.

35. (a) On a standard number line, subtracting 3 is the same as moving ____ units to the ____.
(b) On a standard number line, adding -3 is the same as moving ____ units to the ____.
(c) What conclusion is suggested by parts (a) and (b)?
(d) Make a broader generalization based upon part (c).

36. Compute the following without a calculator. Then check your result with a calculator.
(a) $21 - 48$ (b) $34 - (-10)$

37. Compute the following without a calculator. Then check your result with a calculator.
(a) $-51 - 22$ (b) $-32 - (-70)$

38. Compute the following without a calculator.
(a) $-3 - 4 + 2$ (b) $-8 + (-2) - (-5)$
(c) $4 - (-10) + (-2)$ (d) $-5 + 7 - (-3)$

39. Solve mentally. (Guess and check.)
(a) $\square + 8 = -6$ (b) $r - 10 = -2$
(c) $4 - x = -6$

40. Solve mentally. (Guess and check.)
(a) $9 - \square = -3$ (b) $r + (-2) = -7$
(c) $10 - y = 14$

41. Suppose that a and b are whole numbers, and $a < b$. Then $-a - b =$
(a) $a + b$ (b) $-a + b$
(c) $-a + (-b)$ (d) $a + (-b)$

42. A football team runs four plays with results of +6 yd, +10 yd, +3 yd, and −5 yd.
(a) What is the net yardage?
(b) How could you use this example to explain why $14 - (-5) = 19$?

43. Suppose that x and y are negative integers, and $x > y$. Then $x - y$ is
(a) positive (b) zero (c) negative

44. True or false? All whole numbers are integers.

45. $I = \{\ldots, -2, -1, 0, 1, 2, \ldots\}$
$P = \{1, 2, 3, \ldots\}$
$N = \{-1, -2, -3, \ldots\}$
$W = \{0, 1, 2, 3, \ldots\}$
(a) $N \cup W =$ ____
(b) $N \cap P =$ ____

46. Euclid was born around 360 B.C.
(a) If he lived for 50 years, when did he die?
(b) Write an integer equation for this situation.
(c) What operation and category does this illustrate?

47. An elevator is at an altitude of −10 ft. The elevator goes down 30 ft.
(a) Write an integer equation for this situation.
(b) What is the new altitude?
(c) What subtraction category does this illustrate?

48.

	Continent	
	North America	*Europe*
Highest point	Mt. McKinley	Mt. Elbrus
Altitude	6194 m	5642 m
Lowest point	Death Valley	Caspian Sea
Altitude	−86 m	−28 m

(a) What is the difference between the highest and lowest elevations in North America?
(b) What is the difference between the highest and lowest elevations in Europe?
(c) What operation and category are illustrated in parts (a) and (b)?

49. Consider the following problem. "A football team gains 5 yards on their first play and loses 12 yards on the second play."
(a) What is their net gain or loss?
(b) What operation and category does this illustrate? (Give two possible answers.)

50.

Wind-Chill Temperature (°F)						
Wind Speed	**Actual Temperature (°F)**					
	50°	*40°*	*30*	*20°*	*10°*	*0°*
10 mph	40	28	16	4	−8	
20 mph	32	18	4	−10	−24	
30 mph	28	13	−2	−17	−32	
40 mph	26	10	−6	−22	−38	

(a) The temperature is 40°F, and there is a 30-mph wind. What is the wind-chill temperature?
(b) The weather report said that the temperature is 10°F, and it feels like −25°F. What is the wind speed?
(c) Use the pattern in each row to fill in the last column of the chart.
(d) The temperature is 50°F, and the wind speed is 15 mph. Estimate the wind-chill temperature.

51. Fill in the chart.

Stock	Monday Price	Tuesday Price	Change
Maytag	$32	$31	−1
Wang	$65	$62	
Texas Oil & Gas	$87		−4
K-Mart		$53	−2

Extension Exercises

52. One can find the results to integer subtraction by extending patterns in whole-number subtraction. For example, to find $3 - 5$, we work our way down from more familiar problems in which a smaller positive number is subtracted from 3.

$$3 - 1 = 2$$
$$3 - 2 = 1$$
$$3 - 3 = 0$$
$$3 - 4 = \_\_\_$$
$$3 - 5 = \_\_\_$$

(a) Fill in the blanks, continuing the pattern.
(b) Find the answer to $3 - (-2)$ by creating a similar pattern, beginning with three whole-number subtraction examples. (*Hint:* Try 3 minus some small positive numbers.)
(c) Find the result of $-2 - 1$ by extending a pattern in three whole-number subtraction examples.

53. Find the result of $4 - 7$ by extending a pattern in three whole-number subtraction examples. (*Hint:* Start with $9 - 7$ or $4 - 2$.)

54. Find the result of $5 - (-2)$ by extending a pattern in three whole-number subtraction examples.

55. Whole-number subtraction such as $6 - 2 = n$ can be rewritten in the form $n + 2 = 6$ using the definition of subtraction. The same definition can be used with integers.
(a) Using this approach, $6 - (-2) = n$ is the same as ________ and $n =$ ________.
(b) Using this approach, $-3 - 2 = n$ is the same as ________ and $n =$ ________.

56. Use the method of the preceding exercise to complete the following.
(a) Using this approach, $7 - 10 = n$ is the same as ________ and $n =$ ________.
(b) Using this approach, $-4 - (-5) = n$ is the same as ________ and $n =$ ________.

57. Which integers can be written as a sum of each of the following?
(a) Two consecutive integers
(b) Three consecutive integers

58. Consider the following subtraction algorithm for $52 - 27$.

$$\begin{array}{rl} 52 & \\ -27 & \\ \hline -5 & (2 - 7) \\ 30 & (50 - 20) \\ \hline 25 & \end{array}$$

Will this algorithm work for all whole-number subtraction problems?

59. The sum of two integers is 4. The difference of the two integers is 10. What are they? (Guess and check.)

60. Consider the following problem. "For what integers x, y, and z does $(x - y) - z = x - (y - z)$?"

(a) Devise a plan and solve the problem.
(b) Make up a similar problem.

61. Decide whether each statement is true or false for all integers x and y. If a statement is true, give an example that supports it. If it is false, give a counterexample.
(a) True or false? $|x - y| = |y - x|$.
(b) True or false? $|x - y| = |x| - |y|$.

62. (a) For what integers x is $|x| < x$?
(b) For what integers x is $|x| = x$?
(c) For what integers x is $|x| > x$?

63. If possible, find integers a, b, and c such that
(a) $|a + b + c| < |a| + |b| + |c|$.
(b) $|a + b + c| = |a| + |b| + |c|$.
(c) $|a + b + c| > |a| + |b| + |c|$.

64. If possible, find integers m and n such that
(a) $|m + n| < |m| + |n|$.
(b) $|m + n| = |m| + |n|$.
(c) $|m + n| > |m| + |n|$.

65. Fill in each of the numbers $-8, -6, -4, -2, 0, 2, 4, 6, 8$ in one square so that every row, column, and diagonal has the same sum.

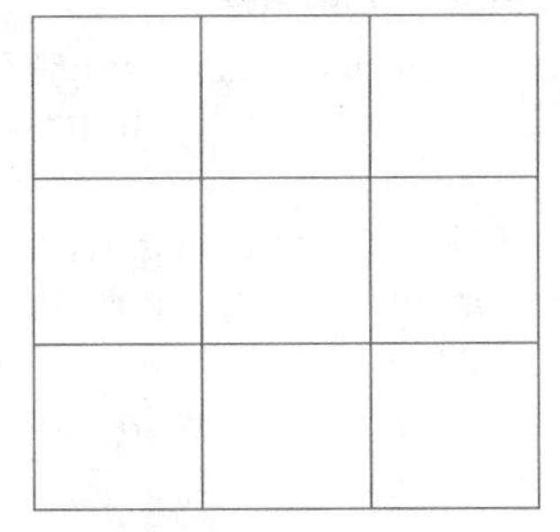

66. Fill in each of the numbers $-15, -12, -9, -6, -3, 0, 3, 6, 9$ in one square so that every row, column, and diagonal has the same sum.

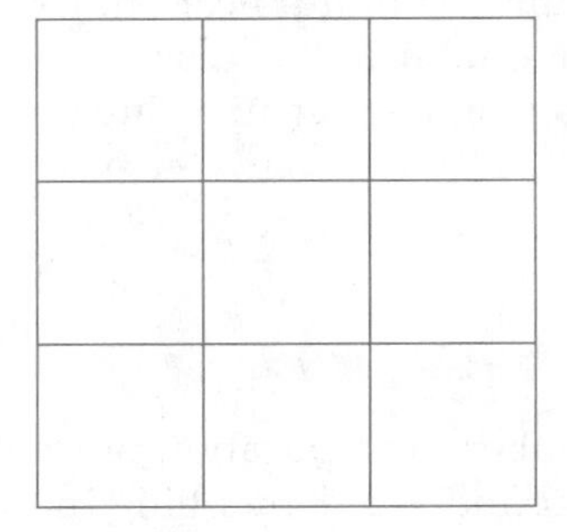

67. Are there more integers than whole numbers? Show that the sets are equivalent by establishing a one-to-one correspondence between the whole numbers and the integers.

Special Exercise

68. How can you increase your chance of living a longer life? The following predictor of life expectancy is adapted from Wallechinsky and Wallace's *People's Almanac #2*.
(a) Start with 72 and add the integers for all applicable descriptors.

Descriptor	
• Male	-3
Female	$+4$
• Urban residence over 2,000,000	-2
Rural residence under 10,000	$+2$
• Job with regular heavy labor	$+3$
Exercise five times per week	$+2$
• Alone for each 10 years since 25	-1
Live with spouse/friend	$+5$
• Easily angered, very aggressive	-3
Easygoing, a follower	$+3$
• Happy	$+1$
Unhappy	-2
• College graduate	$+1$
Graduate degree	$+2$
• One grandparent lived to 85	$+2$
All grandparents lived to 80	$+6$
• Parent died of stroke, heart attack before 50	-4
• Immediate family under 50 has cancer, heart disease, diabetes	-4
• Smoke $>$ 2 packs/day	-8
Smoke 1 to 2 packs	-6
Smoke 1/2 to 1 pack	-3
• Drink at least 1/4 bottle liquor/day	-1
• 50 or more pounds overweight	-8
30 to 49 pounds overweight	-4
10 to 29 pounds overweight	-2
• 30 to 39 years old	$+2$
40 to 49	$+3$
50 to 69	$+4$

The number you obtain is your life expectancy.
(b) According to this scoring system, what are the three most important factors affecting your life expectancy, besides your current age?

5.2 MULTIPLICATION AND DIVISION OF INTEGERS

Integer multiplication and division are similar to whole-number multiplication and division. In multiplying and dividing integers, the one new issue is whether the result is positive or negative. This lesson shows how to explain the sign of an integer product or quotient using patterns, applications, and definitions.

INTEGER MULTIPLICATION

Why would someone want to multiply $3 \times (-4)$ in everyday life? Suppose the temperature drops 4° per hour for 3 hours. How much does it change altogether? $-12°$. This problem suggests that $3 \times (-4) = -12$. Some temperature and money applications are useful illustrations of integer multiplication.

Lesson Exercise 5.22

(a) Write a temperature or money application for $2 \times (-5)$ and give the result. (In your question, ask for the change over time so that the person answering must include the negative sign or a word such as "decreases.")

(b) What category does your application illustrate?

(c) The temperature is now 0°C, and it has been increasing 5° per hour. The temperature 2 hours ago was ____. Going back 2 hours represents -2, so this application represents $-2 \times$ ____ = ____.

(d) Based upon parts (a) and (c), it appears that a negative integer times a positive integer or a positive integer times a negative integer results in a ________________.

(e) Proposing a general rule based upon a pattern in examples requires ______________ reasoning.

Another way to determine the sign of an answer to a multiplication problem is to extend patterns in whole-number multiplication. Extending whole-number multiplication patterns can establish the results of multiplying a positive integer by a negative integer and a negative integer by a positive integer.

Lesson Exercise 5.23

Suppose a child knows whole-number multiplication but not integer multiplication. Show how the child can find $-2 \cdot 3$ by filling in the blanks, continuing the same pattern. (*Note:* Always begin with enough whole-number examples—at least three—to establish a pattern.)

(a)
$$\left.\begin{aligned}2\cdot 3 &= 6\\ 1\cdot 3 &= 3\\ 0\cdot 3 &= 0\end{aligned}\right\}\text{ Whole-number multiplication}$$
$$\begin{aligned}-1\cdot 3 &= \_\_\_\_\\ -2\cdot 3 &= \_\_\_\_\end{aligned}$$

(b) Extend a pattern in whole-number multiplication to determine $2 \cdot (-2)$.

One can use the knowledge that a negative integer times a positive equals a negative to establish the result of a negative integer times a negative. Lesson Exercise 5.24 shows how to do this by extending a pattern.

Lesson Exercise 5.24

Suppose a child knows that a negative times a positive is a negative and wants to use a pattern to determine what $-2 \cdot (-3)$ equals. Fill in the blanks, continuing the same pattern.

(a)
$$\left.\begin{aligned}-2\cdot 3 &= -6\\ -2\cdot 2 &= -4\\ -2\cdot 1 &= -2\end{aligned}\right\}\text{ Negative} \cdot \text{positive}$$
$$\begin{aligned}-2\cdot 0 &= 0\\ -2\cdot(-1) &= \_\_\_\_\\ -2\cdot(-2) &= \_\_\_\_\\ -2\cdot(-3) &= \_\_\_\_\end{aligned}$$

(b) Based upon part (a), a negative integer times a negative integer results in a ____________.

In Lesson Exercise 5.22, you worked with applications of multiplication of positive and negative integers. Next, consider the most difficult integer multiplication situation to model, a negative integer times a negative.

The following example explains how to set up an application of $-3 \times (-2)$.

EXAMPLE 5.4

Describe an application that suggests why $-3 \times (-2) = 6$.

Solution

Temperature, money, and mailing problems are particularly suitable. Whichever one you use, start by saying that the amount is *now* 0. Then describe a change in the negative direction. To end the application, you must ask a question that goes back in time.

	Temperature	Money	Mailing
Now 0	It is now 0°C.	I have $0.	I have no cards left.
−2	The temperature has been dropping 2° per hour.	I have been spending/losing $2 per hour.	I mailed 2 greeting cards each day.
−3 (back 3 in time)	What was the temperature 3 hours ago?	How much did I have 3 hours ago?	How many cards did I have 3 days ago?
Answer	6°C	$6	6
	So $-3 \times (-2) = 6$.	So $-3 \times (-2) = 6$.	So $-3 \times (-2) = 6$.

Now you try one.

Lesson Exercise 5.25

Describe an application problem that suggests why $-5 \times (-6) = 30$.

The results of Lesson Exercises 5.22–5.25 support the rules for multiplication involving negative integers. Each rule relates integer multiplication to whole-number multiplication. These rules comprise the definition of integer multiplication (when combined with the rules for whole-number multiplication).

Definition **Multiplication of Integers**

If a and b are whole numbers, then

$$a \cdot b = +(ab)$$
$$a \cdot (-b) = -(ab)$$
$$-a \cdot b = -(ab)$$
$$-a \cdot (-b) = +(ab)$$

For *nonzero* a and b, this means that

"Positive · positive = positive"
"Positive · negative = negative"
"Negative · positive = negative"
"Negative · negative = positive"

When an integer is multiplied by a positive number, the sign of the product is the same as the sign of the integer. When an integer is multiplied by a negative number, the sign of the product is the opposite of the sign of the integer.

Lesson Exercise 5.26

Use the rules just presented to compute -3×4.

INTEGER DIVISION

When, in everyday life, would someone do division involving negative numbers? Suppose that a town's population drops by 400 in 5 years. What is the average population change per year? -80 people. This problem suggests that $-400 \div 5 = -80$. Population, money, and temperature problems offer useful models of integer division.

Lesson Exercise 5.27

Write a temperature or money problem for $-12 \div 3$, and give the result.

Just as whole-number division is defined as the inverse of whole-number multiplication, integer division is defined as the inverse of integer multiplication.

Definition **Integer Division**

If x, y, and q are integers and $y \neq 0$, then $x \div y = q$ if and only if $x = y \cdot q$.

This definition enables someone who knows integer multiplication to solve integer division problems. Using this definition, one can convert any integer division problem with a nonzero divisor to an equivalent multiplication problem and solve it. For example, $-42 \div 6 = n$ is equivalent to $6 \times n = -42$. Thus, $n = -7$. So $-42 \div 6 = -7$. Use this approach to determine the rules for integer division in the following exercises.

Lesson Exercise 5.28

(a) Use the definition of integer division to rewrite $-8 \div 2 = n$ as a multiplication question and solve it.

(b) The result of part (a) suggests that a negative integer divided by a positive integer results in a ____________________.

Lesson Exercise 5.29

(a) Rewrite $10 \div (-5) = n$ as an equivalent multiplication problem, and solve it.

(b) The result of part (a) suggests that a positive integer divided by a negative integer results in a ________________.

Lesson Exercise 5.30

(a) Rewrite $-12 \div (-3) = n$ as an equivalent multiplication problem, and solve it.

(b) The result of part (a) suggests that a negative integer divided by a negative integer results in a ________________.

The following chart summarizes the rules for division of integers.

Division of Integers

If a and b are whole numbers, with $b \neq 0$, then	For *nonzero* a and b, this means that
$a \div b = +(a \div b)$	"Positive ÷ positive = positive"
$-a \div b = -(a \div b)$	"Negative ÷ positive = negative"
$a \div (-b) = -(a \div b)$	"Positive ÷ negative = negative"
$-a \div (-b) = +(a \div b)$	"Negative ÷ negative = positive"

Lesson Exercise 5.31

Use one of the rules just presented to compute $-24 \div (-8)$.

The introductory applications of division modeled a negative integer divided by a positive. It is more difficult to describe applications of a positive integer divided by a negative and a negative integer divided by a negative.

Lesson Exercise 5.32

The temperature was 8° higher 2 hours ago.

(a) What was the average temperature change per hour?

(b) Write an integer division equation for this application.

(c) What division category does this illustrate?

(d) This problem could be reworded as follows. "The temperature dropped 8° in the last 2 hours. What was the average temperature change per hour?" Write an integer division equation for this application.

The homework exercises include an application that models a negative integer divided by a negative.

COMMON ERROR PATTERNS

As a teacher, you will encounter certain common errors in your students' integer arithmetic. The following problems will give you practice in detecting them.

In Lesson Exercises 5.33–5.35, (a) complete the last two examples, repeating the error pattern in the completed examples, and (b) describe the error pattern.

Lesson Exercise 5.33

$4 - 6 = \underline{2}$ $\quad 8 - 11 = \underline{3}$ $\quad 2 - 7 = \underline{5}$

$6 - 10 =$ ____ $\quad 10 - 12 =$ ____

Lesson Exercise 5.34

$-6 \times (-5) = \underline{-30}$ $\quad -4 \times (-2) = \underline{-8}$ $\quad -3 \times (-2) = \underline{-6}$

$-5 \times (-4) =$ ____ $\quad -3 \times (-6) =$ ____

Lesson Exercise 5.35

$-8 + 5 = \underline{-13}$ $\quad -4 + 4 = \underline{-8}$ $\quad -2 + 6 = \underline{-8}$

$3 + (-6) =$ ____ $\quad -2 + 8 =$ ____

ANSWERS TO SELECTED LESSON EXERCISES

5.22 (a) Min has been spending $5 an hour for 2 hours. What is her overall change in finances? −$10. So $2 \times (-5) = -10$. (*Note:* If you say "How much did she *lose*?" the person will answer "$10" and may think the answer is *positive* 10.)
(b) Repeated sets/measures
(c) −10°C; 5; −10
(d) negative integer
(e) inductive

5.23 (a) −3; −6
(b) $2 \cdot 2 = 4, 2 \cdot 1 = 2, 2 \cdot 0 = 0, 2 \cdot (-1) = -2, 2 \cdot (-2) = -4$

5.24 (a) 2; 4; 6
(b) positive

5.26 −12

5.27 −4

5.28 (a) $2 \times n = -8$, so $n = -4$.
(b) negative

5.29 (a) $-5 \times n = 10$, so $n = -2$.
(b) negative

5.30 (a) $(-3) \times n = -12$, so $n = 4$.
(b) positive

5.31 3

5.32 (a) −4° per hour
(b) $8 \div (-2) = -4$
(c) Partition measures
(d) $-8 \div 2 = -4$

5.33 (a) 4; 2
(b) The child computes the larger number minus the smaller.

5.34 (a) -20; -18
(b) The child thinks a negative number times a negative number is a negative number.

5.35 (a) -9; -10
(b) The child adds a negative number and a positive number by adding their absolute values and placing a negative sign in front of the result.

5.2 HOMEWORK EXERCISES

Basic Exercises

1. Suppose a child knows how to multiply a negative by a positive. Fill in the blanks, continuing the same pattern, to show the result of a negative times a negative.

$$-5 \cdot 2 = -10$$
$$-5 \cdot 1 = -5$$
$$-5 \cdot 0 = 0$$

____

____

2. Show how -2×4 can be solved by extending a pattern in whole-number multiplication.

3. Mike lost 6 pounds each week for 4 weeks.
(a) What was the total change in his weight?
(b) Write an integer equation for this situation.
(c) What application category does this illustrate?

4. A school population has been dropping 15 students per year.
(a) How many more students were at the school 2 years ago?
(b) Write an integer equation for this situation.
(c) What application category does this illustrate?

5. Make up a dieting, money, population-change, or temperature problem for $4 \times (-5)$.

6. Make up an application that suggests why $-3 \times (-5) = 15$.

7. Make up an application that suggests why $-4 \times (-6) = 24$.

8. Write a temperature or money problem for $-20 \div 5$, and give the result.

9. If a and b are integers, what conditions would make each of the following true?
(a) $a \times b = 0$
(b) $a \times b < 0$

10. Consider $(-2)^N$, in which N is a whole number. For what values of N is the result negative?

11. If x is a negative integer, which is larger, $(x + x + x + x)$ or $(x \cdot x \cdot x \cdot x)$?

***12.** Rewrite each problem as an equivalent multiplication problem, and give the solution.
(a) $-54 \div (-6) =$ _____
(b) $32 \div (-4) =$ _____

13. Rewrite each problem as an equivalent multiplication problem, and give the solution.
(a) $0 \div (-3) =$ _____
(b) $-3 \div 0 =$ _____

14. Fill in the blanks, continuing the same pattern.

$$8 \div 2 = 4$$
$$4 \div 2 = 2$$
$$0 \div 2 = 0$$

____

____

15. A store lost \$480,000 last year.
(a) What was the average net change per month?
(b) Write an integer equation for this situation.
(c) What application category does this illustrate?

16. The temperature was 8° lower 2 hours ago.
(a) What was the average temperature change per hour?

(b) Write an integer equation for this situation.
(c) What application category does this illustrate?

17. Make up a dieting, money, or temperature problem for $-40 \div (-5)$, and give the result.

18. Make up a dieting, money, or temperature problem for $-35 \div 7$, and give the result.

19. Compute the following without a calculator.
(a) $3 \times (-8)$
(b) $-5 \times (-8) \times (-2) \times (-3)$
(c) $-24 \div 8$
(d) $(-1)^{20}$

20. Compute the following, using the correct rules for order of operations.
(a) $-2 - (-3)^3$ (b) $-12 + 3 \times (-2)$

21. Compute the following, using the correct rules for order of operations.
(a) $-2^2 - 3$ (b) $-5 + (-4)^2 \times (-2)$

22. Solve mentally.
(a) $-10 \div \square = 2$ (b) $-4 \times \square = -32$
(c) $9(r + 3) = -45$

23. Solve mentally.
(a) $-5t = 0$ (b) $y^2 = 4$ (c) $\frac{30}{n} = -6$

24. If x is a member of $\{-3, -2, -1, 0, 1, 2\}$ and y is a member of $\{-6, -4, -2, 0, 2, 4\}$, find the largest and smallest possible values of each of the following.

(a) $|x + y|$ (b) $x - y$ (c) xy (d) $\frac{x}{y}$

In Exercises 25 and 26, (a) complete the last two examples, repeating the error pattern in the completed examples, and (b) write a description of the error pattern.

25. $-3 + (-4) =$ <u>7</u> $-8 + (-2) =$ <u>10</u>
$-2 + (-3) =$ ______ $-6 + (-1) =$ ______

26. $3 - (-6) =$ <u>−3</u> $4 - (-5) =$ <u>−1</u>
$6 - (-2) =$ ______ $-3 - (-2) =$ ______

27. Why might the error pattern in Exercise 25 occur?

28. Describe two errors a child might make in computing $-28 - 65$.

29. A stock changes as follows for 5 days: $-2, 4, 6, 3, -1$. What is the average daily change in price?

30. If x is a positive integer, $(-x)^3$ is
(a) positive (b) zero (c) negative

31. If y is a negative integer, $(-y)^5$ is
(a) positive (b) zero (c) negative

Extension Exercises

32. Use the following five integers to fill in the blanks in the box. You may use the same number more than once.

$-4 \quad 2 \quad 1 \quad -8 \quad -2$

1	×	____	×	____	=	−8
+		+		×		
____	−	____	−	____	=	4
+		+		×		
____	÷	____	÷	____	=	1
=		=		=		
1		−10		4		

33. The expression $3 \times (-2)$ can be written as $-2 + (-2) + (-2) = -6$ using repeated addition. Show how to compute the following using repeated addition.
(a) $4 \times (-8)$ (b) $2 \times (-5)$

34. Consider the following problem. "If x and y are integers, and $x < y$, what values of x and y would make $x^2 < y^2$?" Devise a plan and solve the problem.

35. If x and y are integers and $x < y$, what values of x and y would make $x^3 < y^3$?

36. Multiplication involving negative integers can also be modeled by motion along a number line.

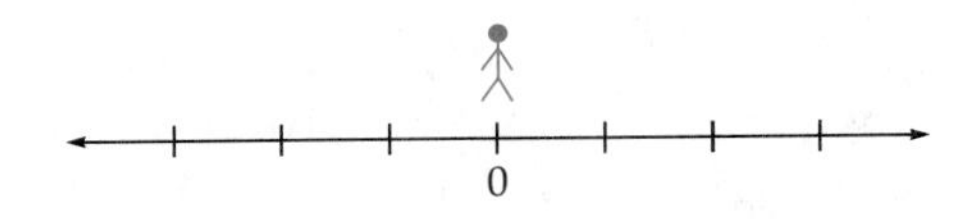

(a) A walker is now at 0 and moves 2 units per hour to the right. Where was the walker 3 hours ago?
(b) Part (a) shows that ____ × ____ = ____.
(c) Write a similar description of a walker to show that $-3 \times (-2) = 6$.

37. Place a check in the appropriate column for each expression.

Expression	Value of X	Value of Expression		
		Negative	*0*	*Positive*
X^4	$X < 0$			
$X^3 - 6$	$X < 0$			
$-3X$	$X < 0$			

38. Place a check in the appropriate column for each expression.

Expression	Values of X and Y	Value of Expression		
		Negative	*0*	*Positive*
$-3X + 2Y$	$X < 0, Y > 0$			
$-3X + 2Y$	$X > 0, Y < 0$			
$2X^2 + Y^2$	$X < 0, Y > 0$			

Special Exercises

39. *Four negative twos.* Using only four -2's and any combination of arithmetic symbols, write expressions equal to 1, 2, 3, . . . 9. The first one has been done for you. Some of them are impossible.

$$-2 \div (-2) + (-2) - (-2) = 1$$

40. Look at two current elementary-school textbooks and report how they explain that the product of two negative numbers is a positive number.

5.3 INTEGER PROPERTIES AND ALGORITHMS

Does $-3 \cdot (4 \cdot 5) = -3 \cdot 4 \cdot 5$, or does $-3 \cdot (4 \cdot 5) = -3 \cdot 4 \cdot (-3) \cdot 5$? Why does $-5x + 2x = -3x$? You can answer these questions if you understand integer properties.

First, let your mind relax. Now, let those whole-number properties you studied in Chapter 3 re-enter your consciousness. Do you recall that whole-number addition is commutative and associative and that it has the identity number 0? Furthermore, do you recall that whole-number multiplication is commutative and associative and that it has the identity number 1? And that whole numbers have the distributive properties of multiplication over addition and multiplication over subtraction?

Why recall these properties at this particular time? Well, wouldn't it be wonderful if integer arithmetic had these properties, too? It does!

WHOLE-NUMBER PROPERTIES RETAINED!

Integer operations retain the same commutative, associative, identity, and distributive properties as whole-number operations. The following list summarizes these properties.

Properties of Integer Operations

1. Integer addition and multiplication are closed. For any integers x and y, $x + y$ is a unique integer and xy is a unique integer.
2. Integer addition and multiplication are commutative. For any integers x and y, $x + y = y + x$ and $xy = yx$.
3. Integer addition and multiplication are associative. For any integers x, y, and z, $(x + y) + z = x + (y + z)$ and $(xy)z = x(yz)$.
4. The unique additive identity for integers is 0, and the unique multiplicative identity for integers is 1. For any integer x, $x + 0 = 0 + x = x$ and $x \cdot 1 = 1 \cdot x = x$.
5. Integer multiplication is distributive over addition, and integer multiplication is distributive over subtraction. For any integers x, y, and z, $x(y + z) = xy + xz$ and $x(y - z) = xy - xz$.

The following exercises utilize these properties.

Lesson Exercise 5.36

(a) What is the easiest way to add $(3 + (-5)) + 5$?

(b) What property does part (a) illustrate?

Lesson Exercise 5.37

What property guarantees that for any integer y, $-2 \cdot (3y) = (-2 \cdot 3)y$?

Lesson Exercise 5.38

I is an integer. According to the distributive property of multiplication over addition, $-5I + 2I =$ ________.

What about integer subtraction and division? Are they commutative or associative?

Lesson Exercise 5.39

(a) Since 2 and 3 are integers, $2 - 3 \neq 3 - 2$ shows that integer subtraction is not ____________________.

(b) Explain why any commutative or associative property that does not hold for all whole numbers could not possibly hold for the set of integers.

INVERSES

So far, the integers have had all the same properties as the whole numbers. Big deal, you say? Well, it is nice to have some consistency. How would you like it if gravity stopped working?

Something different is also nice once in a while. The integers do have one additional property for addition that the whole numbers do not have. This additional property concerns the fact that people use integers to solve problems such as 3 + _____ = 0 and 4 + _____ = 0. The numbers that go in the blanks are the additive inverses of 3 and 4.

> ***Definition*** **Additive Inverse**
>
> The integer y is an **additive inverse** of the integer x if and only if $x + y = y + x = 0$ (the additive identity).

Any integer added to its additive inverse should result in the additive identity, 0. Do all integers have unique additive inverses that are also integers? The results of Lesson Exercises 5.40 and 5.41 will help you decide.

Lesson Exercise 5.40

Fill in each blank with all possible answers.

(a) 3 + ______ = 0, and ______ + 3 = 0. Therefore, ______ is an additive inverse of 3.

(b) −8 + ______ = 0, and ______ + −8 = 0. Therefore, ______ is an additive inverse of −8.

(c) Did each part have a unique (exactly one) answer?

Lesson Exercise 5.41

What is the additive inverse of each of the following?

(a) 7 (b) −2 (c) 0

Lesson Exercises 5.40 and 5.41 should convince you that every integer has a unique additive inverse that is an integer. This is an additional property for integers that does not work for the set of whole numbers.

Additive Inverses for Integers

For each integer x, there is a unique integer $-x$ such that $x + (-x) = -x + x = 0$.

CLOSURE

One reason for creating integers is to provide answers to problems such as $2 - 3$. Do all integer subtraction problems have integer answers?

Lesson Exercise 5.42

If x and y are integers, is $x - y$ always a unique integer?

Lesson Exercise 5.42 may have led you to the following conclusion.

The Closure Property for Integer Subtraction

Integer subtraction is closed. For any two integers x and y, $x - y$ is a unique integer.

ALGORITHMS

All integer arithmetic problems can be converted to whole-number arithmetic problems with positive or negative signs in front of them. For example, $34 - 52 = -(52 - 34)$, and $-86 \times 7 = -(86 \times 7)$. Thus, the whole-number algorithms can also be used in integer problems with larger numbers.

Lesson Exercise 5.43

$-456 + 78 = ?$

(a) $456 + 78$ (b) $456 - 78$ (c) $-(456 - 78)$ (d) $-(456 + 78)$

Lesson Exercise 5.44

$876 \div (-23) = ?$

(a) $876 \div 23$ (b) $-(876 \div 23)$ (c) $876 - 23$ (d) $-(876 - 23)$

AN INVESTIGATION: ORDER IN SUBTRACTION

Integer subtraction is not commutative, but $x - y$ is related to $y - x$ for all integers x and y.

Lesson Exercise 5.45

If $x - y = 2$, then $y - x =$ _____. (*Hint:* Try some numbers for x and y.)

Lesson Exercise 5.46

For integers a and b, how does $a - b$ compare to $b - a$?

(a) Devise a plan.

(b) Carry out the plan.

(c) Make a generalization based upon your results.

(d) What kind of reasoning is used to make a generalization from examples in part (c)?

Lesson Exercise 5.47

For what integer values of m and n does $m - n = n - m$?

ANSWERS TO SELECTED LESSON EXERCISES

5.36 (a) Add $-5 + 5$, and then add 3.
(b) Associative property of addition

5.37 Associative property of multiplication

5.38 $(-5 + 2)I$

5.39 (a) commutative
(b) If a rule does not apply to *all* whole numbers, then it cannot apply to *all* integers, since every whole number is also an integer. Whatever was a counterexample for whole numbers is also a counterexample for the set of integers.

5.40 (a) -3; -3; -3 (b) 8; 8; 8 (c) Yes

5.41 (a) -7 (b) 2 (c) 0

5.42 Yes

5.43 (c)

5.44 (b)

5.45 -2

5.47 When $m = n$

5.3 HOMEWORK EXERCISES

Basic Exercises

1. (a) What integer operations are commutative?
(b) What integer operations are associative?

2. During 3 consecutive years, a man gains 14 pounds, loses 37 pounds, and then loses 14 pounds.
(a) Write an integer *addition* expression that represents his overall change.
(b) What is an easy way to add the numbers?

3. What is an easy way to multiply $-5 \times (7 \times -8)$?

4. What property guarantees that for an integer m, $4 \cdot (-3m) = (4 \cdot -3)m$?

5. What properties guarantee that for an integer n, we know $2n + 8 + n + -6 = (2n + n) + (8 + -6)$?

***6.** In adding a series of integers, it is often easier to add all the negative numbers and positive numbers separately and then add the results together.
(a) Compute $(-6 + 4) + (4 + (-3)) + (-7 + 5)$ mentally, using this method.
(b) Compute $-5 + (-2) + 6 + (-3) + 8 + 7$ mentally, using this method.
(c) What two properties enable you to add integers in a different order and still obtain the same answer?

7. How can you compute -27×6 using a distributive property?

8. The equation $-5 \times (-3) = -3 \times (-5)$ illustrates the ________ property of ________.

9. For an integer n, the distributive property of multiplication over subtraction states that $2n - 5n =$ ________.

10. For an integer n, the distributive property of multiplication over addition states that $-6n + -3n =$ ________.

11. (a) $-12 \times 99 = -12 \times (100 - 1) =$
________ $-$ ________ $=$ ________
(b) Part (a) uses what integer property?
(c) Use the procedure from part (a) to compute -34×99. Do as much of it mentally as you can.

12. Give a counterexample that shows that integer subtraction is not commutative.

13. Explain why a person who has studied whole-number operations would know that integer division is not associative.

14. What is the integer identity element for addition?

15. The examples $-3 \times 1 = -3$ and $-5 \times 1 = -5$ illustrate that ________ is the ________ for ________.

16. Any number added to its additive inverse equals ____.

17. What property guarantees that $6 - 12$ has a unique integer answer?

18. $82 - (-47) = ?$
(a) $82 + 47$ (b) $-(82 + 47)$
(c) $82 - 47$ (d) $-(82 - 47)$

19. $-63 + 78 = ?$
(a) $-(78 - 63)$ (b) $78 - 63$
(c) $78 + 63$

Extension Exercises

20. Fill in the blanks, following the rule from the completed examples.

$-2 \rightarrow 3$
$-4 \rightarrow 5$
$6 \rightarrow -5$
$-5 \rightarrow$ __
$8 \rightarrow$ __
$N \rightarrow$ __
__ $\rightarrow -17$

21. If $a \div b = 10$, then $b \div a =$ ____.

22. Consider the following problem. "For nonzero integers a and b, how does $a \div b$ compare to $b \div a$?"
(a) Devise a plan and solve the problem.
(b) For what integer values of a and b does $a \div b = b \div a$?

23. Suppose you did not know that $-2 \cdot 4 = -8$. You could prove that $-2 \cdot 4 = -8$ using the additive inverse property and whole-number multiplication.
(a) Since $2 \cdot 4$ is the additive inverse of $-2 \cdot 4$, write an addition equation that shows their relationship.
(b) How does this show that $-2 \cdot 4 = -8$?

24. Suppose you did not know that $-2 \cdot (-4) = 8$. How could you use the result of the preceding exercise to prove it?

25. An **even** integer is any integer that can be written in the form $2m$, in which m is an integer. Show that the product of two even integers is even.

26. Show that the sum of two even integers is even.

SUMMARY

"As students reach grade 5, they begin to recognize—in both arithmetic and geometric settings—the need for numbers beyond whole numbers. . . . The integer −1 becomes necessary so that the whole number problem 2 − 3 has a solution, such as when a player loses three points in a game when he or she only has two points. As students expand their mathematical horizons . . . they need to understand both the common ideas underlying these number systems and the differences among them" (NCTM, *Curriculum and Evaluation Standards*, pp. 91–92).

People use negative integers to measure stock prices, golf scores, and altitudes. Integers were also developed as solutions to whole-number subtraction problems, such as $2 - 3$, that have no whole-number solution.

With integers, as with whole numbers, addition and multiplication are defined, and then subtraction is defined as the inverse of addition, and division is defined as the inverse of multiplication. The results of integer arithmetic can be illustrated with definitions, with applications such as temperature and money, or with models such as a number line, or colored counters.

The set of integers retains the commutative, associative, and distributive properties of whole-number operations. Both integers and whole numbers have the same identity elements for addition (0) and multiplication (1). The set of integers has two important additional properties that whole numbers do not have: additive inverses and closure for subtraction.

STUDY GUIDE

To review Chapter 5, see what you know about each of the following ideas or terms that you have studied. You can also use this list to generate your own questions about the chapter.

The NCTM Curriculum Standards and Integers

Selected NCTM Curriculum Standards

The following standards come from the NCTM document.

- Understand and appreciate the need for numbers beyond the whole numbers.
- Formulate problems from mathematical situations.
- Extend their understanding of whole-number operations to integers.
- Understand how basic arithmetic operations are related to one another.
- Develop, analyze, and explain techniques for computation.

1. Describe a topic you studied or an exercise you completed in this chapter that illustrates each standard listed.
2. Select any current elementary-school mathematics textbook series and describe sample lessons or exercises that illustrate each standard listed.

INTEGERS IN ELEMENTARY SCHOOL

The following chart shows at what grade levels selected integer topics typically appear in elementary-school mathematics textbooks.

Topic	Typical Grade Level in Current Textbooks
Integer concepts	6
Adding and subtracting integers	6
Multiplying and dividing integers	6 (enrichment)

REVIEW EXERCISES

1. Give an example showing that integer subtraction is not associative.

2. What integer operations are commutative?

3. Make up a temperature or money problem for $-3 - 6$.

4. Write a realistic application that is solved by computing $-3 \times (-5)$.

5. Explain how to compute $-4 - (-2)$ with
(a) colored counters.
(b) a number line.

6. Suppose a child knows whole-number subtraction but has not yet studied integer subtraction. Show how $4 - (-2)$ can be solved by extending a pattern in whole-number subtraction.

7. (a) An army that loses 300 soldiers is how much better off than an army that loses 800 soldiers?
(b) Write an integer equation for this situation.
(c) What subtraction category does this illustrate?

8. Sandy lost 18 pounds in 9 months. What was the average monthly change in her weight?
   (a) Write an integer equation and a solution for this situation.
   (b) What operation and category does this illustrate?

9. An elevator is at an altitude of -30 ft. If it goes up 20 ft, what is its new altitude?
   (a) Write an integer equation and a solution for this situation.
   (b) What operation and category does this illustrate?

10. Explain how to compute $2 - 4$ with
   (a) a number line.
   (b) colored counters.

11. $-86 - 47 = ?$
   (a) $86 + 47$ (b) $86 - 47$
   (c) $-(86 + 47)$ (d) $-(86 - 47)$

12. If a, b, and c are whole numbers, then $a + (-b) + c = ?$ (Select all correct answers.)
   (a) $a + b - c$ (b) $(a + c) - b$
   (c) $b - (a + c)$ (d) $-(b - a + c)$
   (e) $-(b - a) + c$

13. Name a *property* that the integers have that whole numbers do not.

14. Write a paragraph defining whole numbers and integers and telling how these two sets of numbers are related.

SUGGESTED READINGS

Ashlock, R. *Error Patterns in Computation.* 6th ed. Englewood Cliffs, N.J.: Prentice-Hall, 1994.

National Council of Teachers of Mathematics. *Estimation and Mental Computation.* 1986 Yearbook. Reston, Va.: NCTM, 1986.

6 Rational Numbers as Fractions

The ancient Egyptians were using fractions long before the invention of negative integers. The Rhind Papyrus (1600 B.C.) gives a systematic treatment of unit fractions $\left(\frac{1}{\text{counting number}}\right)$. Teachers follow this historical order when they teach fractions before integers in elementary school.

Practical applications that led to the development of fractions include sharing problems (for example, 4 loaves of bread for 10 people) and measuring problems (for example, $2\frac{1}{2}$ ft). Today, we also talk about 3/4 of a tank of gas or a 2/3 majority to override a veto in Congress.

This chapter examines a subset of the set of fractions called the rational numbers: fractions that have an integer numerator and a nonzero integer denominator. In a mathematical development of number systems, one expands the whole numbers to the integers in order to have answers to all subtraction problems. Still lacking answers to many whole-number and integer division problems (such as $2 \div 3$), one can expand the integers to form the rational number system, which provides answers to all integer division problems with nonzero divisors.

6.1 RATIONAL NUMBERS

The Babylonians and Egyptians used fractions in agriculture and business over 3500 years ago. Today, the increased use of computers has reduced the importance of fractions relative to decimals. However, fractional notation is still used in algebra, geometry, and calculus as well as for certain sharing and measuring applications. In this section, you'll examine the uses of fractions and methods for comparing the sizes of two fractions.

Copyright: *The Fresno Bee*

Figure 6–1

Lesson Exercise 6.1

Why do you think a fraction was used in the speed limit sign in Figure 6-1?

ELEMENTARY FRACTIONS

When you hear the word "fractions," how do you react?

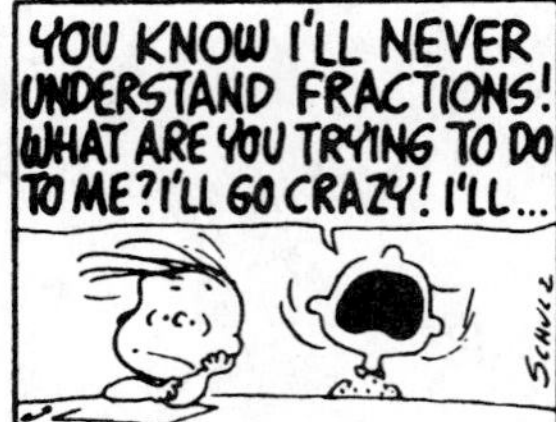

The fractions that children study in elementary school are referred to as elementary fractions. An **elementary fraction** is a number of the form $\frac{a}{b}$ in which a and b are whole numbers and $b \neq 0$. Elementary fractions include $\frac{2}{3}$, $\frac{1}{2}$, and $\frac{11}{4}$ $\left(\text{or } 2\frac{3}{4}\right)$, whereas $\frac{-2}{3}$ and $\frac{\sqrt{2}}{3}$ are not elementary fractions. In a fraction $\frac{a}{b}$, a is called the **numerator**, and b is called the **denominator**.

Why do we use the terms "numerator" and "denominator"? Well, suppose I cut an apple pie into 8 equal pieces. The size of any serving is expressed in eighths (the denomination). The number of pieces that someone eats determines the numerator. If you eat 3 (number) pieces that are eighths (denomination), then you have eaten 3/8 of the pie. (You also must have been rather hungry.)

FOUR MEANINGS OF AN ELEMENTARY FRACTION

Do you realize that elementary fractions have four meanings that children typically learn in elementary school?

Cooperative Lesson Exercise 6.2

What are some different things the fraction $\frac{2}{3}$ can represent in mathematics or in everyday life?

Did you come up with the following uses of $\frac{2}{3}$?

Four Meanings of Elementary Fractions

1. **Part of a whole (or region):** $\frac{2}{3}$ means to count 2 parts out of 3 equal parts.

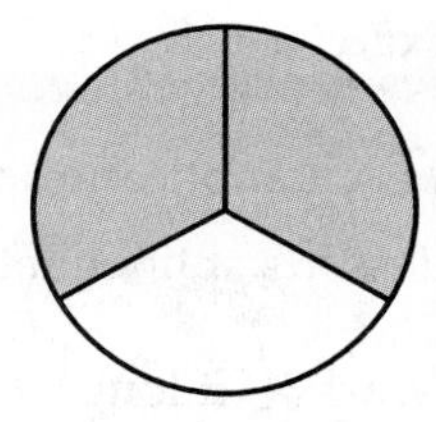
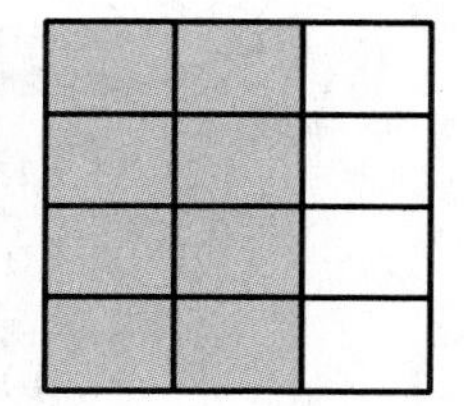

2. **Part of a set (or group):** $\frac{2}{3}$ means to divide a set into 3 equal groups and shade 2 out of every 3.

3. **Location on a number line:** $\frac{2}{3}$ is a number between 0 and 1. Divide the interval from 0 to 1 into 3 equal parts, and count 2 parts over from 0 to 1. In other words, go $\frac{2}{3}$ of the way from 0 to 1.

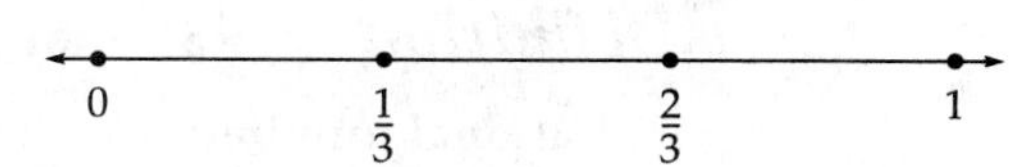

4. **Division:** $\frac{2}{3}$ means $2 \div 3$, the numerator divided by the denominator.

2

Note that $\frac{2}{3}$ can also represent a ratio. This meaning of $\frac{2}{3}$ will be addressed in Chapter 7.

Lesson Exercise 6.3

Describe four common meanings of $\frac{3}{4}$. For each meaning, give its name, make a drawing, and describe how it works for $\frac{3}{4}$.

Everyday applications of elementary fractions match up with these four meanings.

Lesson Exercise 6.4

Match each application with a fraction meaning.

(a) 2 desserts split equally among 4 people	(1) Part of a whole
(b) A scarf $3\frac{1}{2}$ ft long	(2) Location on a number line
(c) 2 slices of an 8-slice pizza	(3) Division
(d) $\frac{3}{5}$ of a group of 20 prefer juice over soda.	(4) Part of a set

---

RATIONAL NUMBERS

Children in elementary school study elementary fractions. In secondary-school mathematics, students also study the negatives of elementary fractions. The union of the set of elementary fractions and their negatives is the rational numbers. Secondary-school mathematics also includes other types of fractions, such as fractions that contain square roots.

This chapter focuses on rational numbers.

> ***Definition*** **Rational Numbers**
>
> **Rational numbers** are all numbers that *can be written* as a quotient (ratio) of two integers $\frac{p}{q}$, in which $q \neq 0$.

The set of rational numbers includes all integers, since any integer can be written in rational form. For example, $-4 = \frac{-4}{1}$. Many decimals such as 0.3

are also rational numbers, since $0.3 = \frac{3}{10}$. Percents such as 42% are rational, since $42\% = \frac{42}{100}$. Decimals and percents will be discussed further in Chapter 7.

What numbers are *not* rational? It is impossible to write $\sqrt{2}$ and $\sqrt{3}$ as an integer over a nonzero integer! This will be proved for $\sqrt{2}$ in Chapter 7.

Lesson Exercise 6.5

Which of the following are rational numbers?

(a) $-3/4$ (b) 5 (c) $\sqrt{2}$ (d) 0 (e) 0.37 (f) $\sqrt{25}$

Both elementary fractions and the set of rational numbers written as fractions are subsets of the set of fractions. A **fraction** is a number of the form $\frac{a}{b}$ in which a is any kind of number and b is any nonzero number (not necessarily integers). Examples of fractions include $\frac{11}{4}$, $\frac{-3}{8}$, $\frac{\sqrt{5}}{7}$, $\frac{\sqrt{-3}}{2}$, and $-\frac{\pi}{2}$.

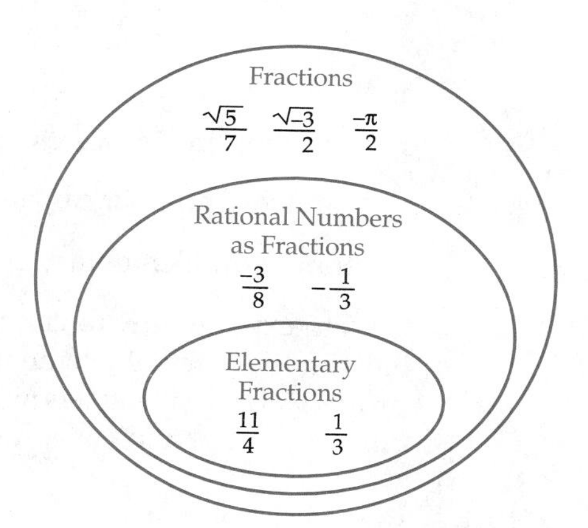

Lesson Exercise 6.6

Where in the preceding Venn diagram would each of the following go?

(a) $\frac{3}{1}$ (b) $\frac{0}{1}$ (c) $\frac{-2}{1}$ (d) $\frac{\sqrt{3}}{2}$

All properties and rules in this chapter will be stated for elementary fractions or rational numbers, although they usually apply to all fractions.

Which of the four meanings of elementary fractions apply to rational numbers less than 0? Try the following.

Lesson Exercise 6.7

Which of the four meanings are easily applied to $\frac{-3}{4}$?

As you may have discovered in the preceding exercise, two of the four meanings, location on a number line and division, work well with negative rational numbers.

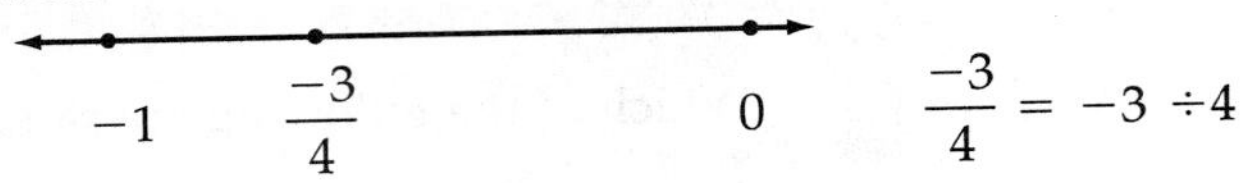

The length of someone's finger might be $2\frac{1}{4}$ inches, a mixed number. A **mixed number** is made up of an integer and a fractional part of an integer. Mixed numbers also can be interpreted using some of the meanings of elementary fractions. A mixed number such as $2\frac{1}{4}$ is a location on a number line.

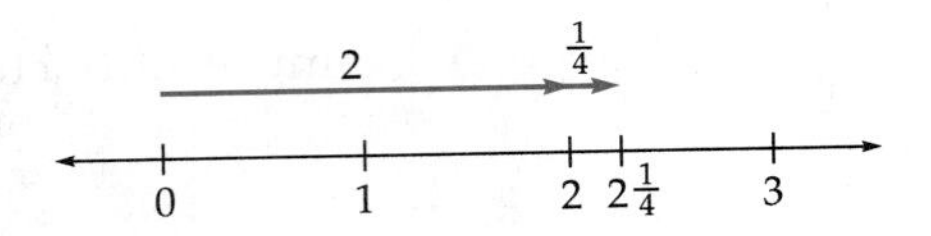

The number line indicates why $2\frac{1}{4} = 2 + \frac{1}{4}$.

A mixed number can also be written as an improper fraction. An **improper fraction** is an elementary fraction $\frac{a}{b}$, in which $a \geq b$. Fraction models can be used to explain how to change improper fractions to mixed numbers. Roughly half of the current elementary-school textbooks do not use the term "improper." They simply show students how to write a mixed number as a "fraction" or a "fraction" as a mixed number.

Lesson Exercise 6.8

Show why $\frac{9}{4} = 2\frac{1}{4}$, using

(a) division.

(b) the part-of-a-whole meaning.

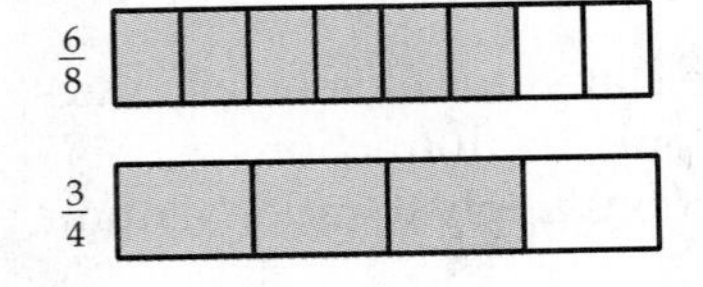

Figure 6–2

EQUIVALENT FRACTIONS FOR RATIONAL NUMBERS

A friend of mine went on a diet. Instead of cutting a cake into 8 equal pieces and eating 6, he cut it into 4 equal pieces and ate only 3. Fractions such as 3/4 and 6/8 that look different may represent the same rational number (Figure 6-2).

Equivalent fractions are two fractions that represent the same number. Equivalent elementary fractions represent the same part of a whole. What is the pattern in the numerators and denominators of equivalent elementary fractions?

Lesson Exercise 6.9

By splitting thirds into equal parts, one can create equivalent fractions.

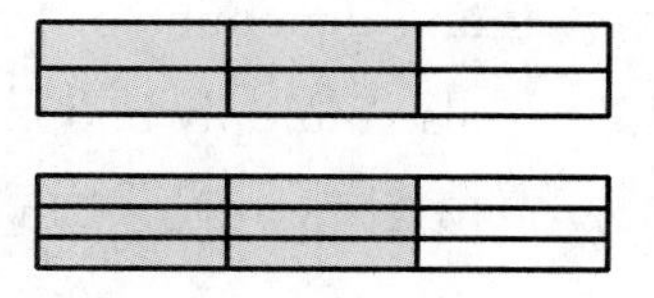

The diagrams show that $\frac{2}{3}$ = ____ = ____.

The pattern could be extended without end: $\frac{2}{3} = \frac{4}{6} = \frac{6}{9} = \frac{8}{12} = \frac{10}{15}$ and so on, indefinitely. In general, we can rename a fraction by multiplying or dividing its numerator and denominator by the same positive whole number.

What about multiplying the numerator and denominator by the same negative integer? Does $\frac{2 \times -2}{3 \times -2} = \frac{-4}{-6}$?

Lesson Exercise 6.10

Use division rules for integers and equivalent elementary fractions to show why $\frac{-4}{-6} = \frac{2}{3}$.

A number-line model can be used to show that the same pattern applies to negative rational numbers:

$$-\frac{2}{3} = -\frac{4}{6} = -\frac{6}{9} = -\frac{8}{12} = -\frac{10}{15} = \ldots$$

-1 $\quad -\frac{2}{3}$ $\quad -\frac{1}{3}$ $\quad 0$

-1 $\quad -\frac{4}{6}$ $\quad 0$

and so on.

The Fundamental Law of Fractions describes the general relationship between equivalent fractions.

The Fundamental Law of Fractions

For any rational number $\frac{a}{b}$ and any integer $c \neq 0$, $\frac{ac}{bc} = \frac{a}{b}$.

The Fundamental Law works in two ways. We can divide the numerator and denominator by the same nonzero integer c to simplify $\frac{ac}{bc}$ to $\frac{a}{b}$. We can also multiply both numerator and denominator by the same nonzero integer c to change $\frac{a}{b}$ to $\frac{ac}{bc}$. In most elementary-school texts, fractions are simplified with division symbols in accordance with the Fundamental Law.

$$\frac{12}{15} = \frac{12 \div 3}{15 \div 3} = \frac{4}{5}$$

Lesson Exercise 6.11

Use the Fundamental Law of Fractions to show why $\frac{6}{8} = \frac{3}{4}$.

Lesson Exercise 6.12

Can you *add* the same counting number to the numerator and denominator of an elementary fraction $\frac{a}{b}$ without changing its value? That is, does $\frac{a}{b} = \frac{a + c}{b + c}$ for counting numbers a, b, and c?

SIMPLIFYING ELEMENTARY FRACTIONS

In simplifying an elementary fraction such as 12/20, one divides the numerator and denominator by the same counting number. The following exercise illustrates this process.

Lesson Exercise 6.13

(a) To write $\frac{12}{20}$ in simplest form in one step, divide the numerator and denominator by ______.

(b) To write $\frac{28}{42}$ in simplest form in one step, divide the numerator and denominator by ____.

(c) In parts (a) and (b), how is the answer related to the numbers in the original numerator and denominator?

(d) Give a general procedure for simplifying an elementary fraction, based upon your response to part (c).

A fraction $\frac{a}{b}$ is in **simplest form** if GCF(a, b) = 1. When the numerator and denominator of an elementary fraction have a greatest common factor (GCF) greater than 1, you can divide the numerator and dominator by the GCF, and the resulting fraction will be in simplest form! Try this method in the following exercise.

Lesson Exercise 6.14

Find the GCF of 148 and 260, and use it to simplify $\frac{148}{260}$.

In simplifying fractions that have relatively large numerators and denominators, such as $\frac{148}{260}$, the GCF may make the work easier. Most fractions children encounter in elementary school $\left(\text{for example, } \frac{8}{16}\right)$ can be simplified just as easily without finding the GCF. But when children end up computing $\frac{8}{16} = \frac{4}{8} = \frac{2}{4} = \frac{1}{2}$ and ask why it took so many steps, you can tell them about the GCF. It always gets the simplification done in *one* step!

VERIFYING THE EQUIVALENCE OF FRACTIONS

In one group, 18 out of 30 people $\left(\frac{18}{30}\right)$ prefer butter to guns, and in a second group, 24 out of 40 people $\left(\frac{24}{40}\right)$ prefer butter to guns. How do the group preferences compare?

Since $\frac{18}{30} = \frac{3}{5}$ and $\frac{24}{40} = \frac{3}{5}$, the same fraction of each group prefers butter.

Lesson Exercise 6.15

(a) Show that $\frac{8}{20} = \frac{6}{15}$ by putting both fractions in simplest form.

(b) Show that $\frac{2}{3} \neq \frac{3}{5}$ by writing both fractions in terms of the least common denominator (which is the LCM of the two denominators).

Writing two fractions in terms of a common denominator is one of the two common ways of showing that two fractions are equivalent. The other method compares cross products. You probably already know that the equation $\frac{a}{b} = \frac{c}{d}$ can be rewritten as $ad = bc$ using "cross multiplication." The following exercise shows why this works.

Lesson Exercise 6.16

Suppose $\frac{a}{b} = \frac{c}{d}$.

(a) Use the Fundamental Law of Fractions to write $\frac{a}{b}$ and $\frac{c}{d}$ in terms of the common denominator bd. Fill in the resulting numerators in the following equation.

$$\frac{\quad}{bd} = \frac{\quad}{bd}$$

(b) Now if $\frac{ad}{bd} = \frac{bc}{bd}$, then ______ = ______.

The preceding exercise shows that $\frac{a}{b} = \frac{c}{d}$ is equivalent to $ad = bc$. Therefore, by comparing cross products ad and bc, one can tell if $\frac{a}{b} = \frac{c}{d}$.

Two Methods for Verifying the Equivalence of Fractions

1. **LCD (Least common denominator).** After putting both fractions in simplest form, write both fractions with the least common denominator (LCM of the denominators) and see if the numerators are equal.

2. **Compare cross products.** $\frac{a}{b} = \frac{c}{d}$ if and only if $ad = bc$.

Use the second method in the following exercise.

Lesson Exercise 6.17

Show that $\frac{8}{20} = \frac{6}{15}$ by comparing cross products.

To solve for a missing numerator or denominator in one of two equivalent fractions, one can use either the Fundamental Law of Fractions or cross products.

Lesson Exercise 6.18

$$\frac{3}{2} = \frac{15}{N}$$

(a) Solve for N by using the Fundamental Law of Fractions. $\left(\textit{Hint:} \text{ What number was multiplied by the numerator and denominator of } \frac{3}{2}?\right)$

(b) Solve for N by using equal cross products.

Lesson Exercise 6.19

Two companies conduct surveys asking people whether they want stricter handgun control. The first company asks 500 people, and the second asks 1000. Make up results for each survey so that the results would be considered equivalent.

UNEQUAL ELEMENTARY FRACTIONS

Someone with a basic understanding of elementary fractions will have little difficulty determining which of two elementary fractions with the same denominator is larger $\left(\text{for example, } \frac{5}{6} > \frac{2}{6}\right)$. In general, for whole numbers a, b, and c with $c \neq 0$, if $a > b$, then $\frac{a}{c} > \frac{b}{c}$. On a standard number line, $\frac{a}{c}$ will be to the right of $\frac{b}{c}$.

When denominators are unequal, it may be more difficult to determine which of two elementary fractions is larger. For example, how do $\frac{7}{12}$ and $\frac{5}{9}$

compare in size? One way to find out is to rename both fractions with the least common denominator and then see which one has a larger numerator.

$$\text{LCM}(9, 12) = 36$$

$$\frac{7}{12} = \frac{7 \cdot 3}{12 \cdot 3} = \frac{21}{36} \quad \text{and} \quad \frac{5}{9} = \frac{5 \cdot 4}{9 \cdot 4} = \frac{20}{36}$$

Since $\frac{21}{36} > \frac{20}{36}$, we know that $\frac{7}{12} > \frac{5}{9}$.

Comparing cross products can also be used to compare the sizes of fractions. The following exercise develops the rule for comparing cross products of unequal elementary fractions.

Lesson Exercise 6.20

Suppose $\frac{a}{b}$ and $\frac{c}{d}$ are elementary fractions, and $\frac{a}{b} > \frac{c}{d}$. Find a common denominator and simplify to derive an equivalent inequality that does not contain fractions.

Exercise 6.20 shows that $\frac{a}{b} > \frac{c}{d}$ and $ad > bc$ are equivalent for whole numbers a and c and counting numbers b and d.

Comparing Cross Products of Unequal Elementary Fractions

If a and c are whole numbers and b and d are counting numbers, then $\frac{a}{b} > \frac{c}{d}$ if and only if $ad > bc$.

EXAMPLE 6.1

How do $\frac{3}{7}$ and $\frac{5}{11}$ compare in size?

Solution

The LCD (Least-Common-Denominator) Method Write both fractions with the least common denominator and compare them. The least common denominator is 77.

$$\frac{3}{7} = \frac{3 \cdot 11}{7 \cdot 11} = \frac{33}{77} \quad \text{and} \quad \frac{5}{11} = \frac{5 \cdot 7}{11 \cdot 7} = \frac{35}{77}$$

So $\frac{5}{11} > \frac{3}{7}$.

The Compare-Cross-Products Method Compare the cross products: $3 \cdot 11 < 5 \cdot 7$. So $\frac{3}{7} < \frac{5}{11}$.

Note that the cross product with the numerator of a fraction corresponds to that fraction in the comparison. ■

Lesson Exercise 6.21

In one class, 6 out of 20 students prefer swimming over soccer. In another class, 10 out of 32 students prefer swimming over soccer. Compare the preferences of the two classes, using the LCD method and the compare-cross-products method.

ANSWERS TO SELECTED LESSON EXERCISES

6.1 To catch people's attention

6.3 $\frac{3}{4}$ represents:

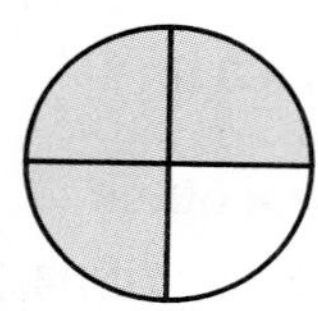

Part of a whole (shade 3 of 4 equal parts)

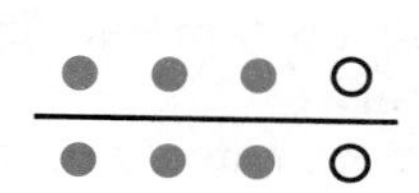

Part of a set (shade 3 out of every 4)

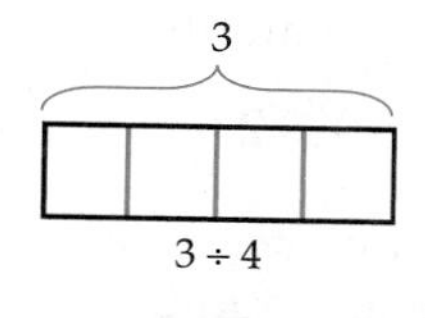

Division (divide the numerator by the denominator)

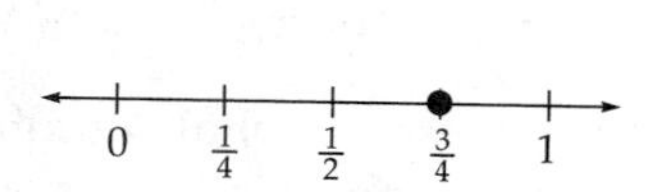

Number line location (go $\frac{3}{4}$ of the way from 0 to 1)

6.4 (a) (3) (b) (2) (c) (1) (d) (4)

6.5 (a), (b), (d), (e), (f)

6.6 (a), (b) In the elementary-fractions oval
(c) In the rational-numbers oval and outside the elementary-fractions oval
(d) In the fractions oval and outside the other two ovals

6.7 Number-line location and division

6.8 (a) $\frac{9}{4} = 9 \div 4 = 2\frac{1}{4}$
(b)

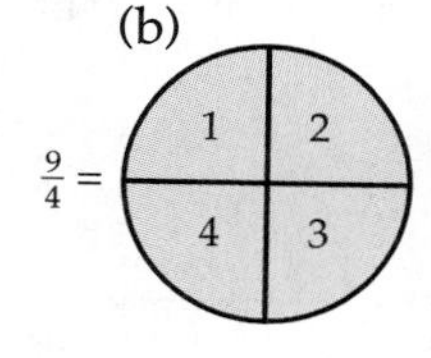

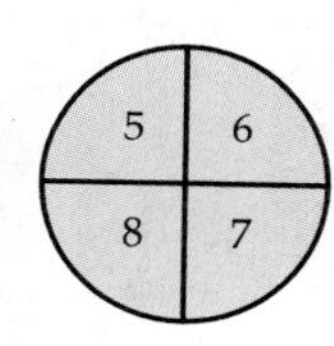

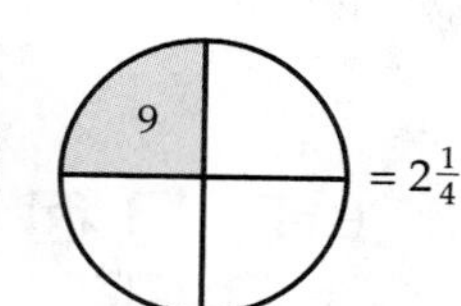

6.9 $\frac{4}{6}; \frac{6}{9}$

6.10 $\frac{-4}{-6} = -4 \div -6 = 4 \div 6 = \frac{4}{6} = \frac{2}{3}$

6.11 $\frac{6}{8} = \frac{6 \div 2}{8 \div 2} = \frac{3}{4}$

6.12 No

6.13 (a) 4 (b) 14
(c) $4 = \text{GCF}(12, 20)$ $14 = \text{GCF}(28, 42)$
(d) The answer follows the exercise.

6.14 GCF = 4; $\frac{148}{260} = \frac{148 \div 4}{260 \div 4} = \frac{37}{65}$

6.15 (a) $\frac{2}{5} = \frac{2}{5}$ (b) $\frac{10}{15} \neq \frac{9}{15}$

6.16 (a) $\frac{ad}{bd} = \frac{bc}{bd}$ (b) ad; bc

6.17 $8 \cdot 15 \stackrel{?}{=} 20 \cdot 6$ Yes, $120 = 120$.

6.18 (a) $2 \cdot 5 = 10 = N$
(b) $3N = 2 \cdot 15$. So $3N = 30$, or $N = 10$.

6.20 $\frac{ad}{bd} > \frac{bc}{bd}$. So $ad > bc$.

6.21 LCD: $\frac{6}{20}$? $\frac{10}{32}$ $\rightarrow$ LCD = 160 $\rightarrow$ $\frac{6}{20} = \frac{48}{160}$ and $\frac{10}{32} = \frac{50}{160}$. So $\frac{6}{20} < \frac{10}{32}$.

Compare cross products: $6 \cdot 32 < 10 \cdot 20$. So $\frac{6}{20} < \frac{10}{32}$.

6.1 HOMEWORK EXERCISES

Basic Exercises

1. Explain how to complete each diagram so that it shows $\frac{3}{5}$.

(a) 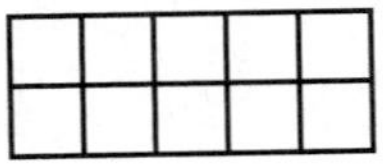 (b) 

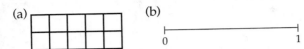

2. Explain how to complete each diagram so that it shows $\frac{2}{3}$.

(a) 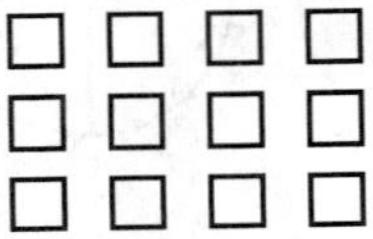 (b) 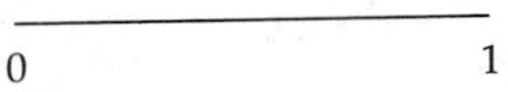

3. A child shows $\frac{4}{5}$ as 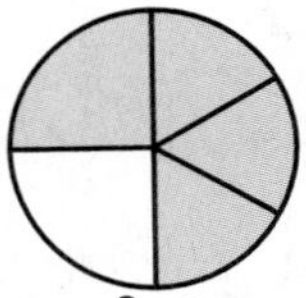. What is wrong with the diagram?

4. Describe four common meanings of $\frac{5}{8}$. For each meaning, give its name, make a drawing, and describe how it works for $\frac{5}{8}$.

5. Describe four common meanings of $\frac{1}{3}$. For each meaning, give its name, make a drawing, and describe how it works for $\frac{1}{3}$.

6. We know that: $\frac{8}{4} = 8 \div 4$ and $\frac{9}{3} = 9 \div 3$. Write an algebraic formula that generalizes this pattern.

7. Match each application with a fraction model.

Application	*Fraction Model*
(a) Your height in inches	(1) Part of a whole
(b) Fraction of those surveyed who jog	(2) Location on a number line
(c) The fraction of the day for which you are asleep	(3) Division
(d) Evenly sharing 3 pizzas among 5 people	(4) Part of a set

8. (a) What fractions can you find in a daily newspaper?
   (b) What denominators appear most often?

9. Show that each number is rational by writing it as a quotient of two integers.
   (a) 3 (b) -3 (c) $4\frac{1}{2}$ (d) -5.6
   (e) 25%

*10. Which of the following are rational numbers?
   (a) -7 (b) $\frac{2}{3}$ (c) $\sqrt{3}$ (d) 0.2

11. W is the set of whole numbers, I is the set of integers, and Q is the set of rational numbers.
   (a) Is $I \subseteq Q$?
   (b) $W \cap Q =$ ____

12. If $\frac{c}{d}$ is an improper fraction, then c __?__ d.
   (a) $\geq$ (b) $=$ (c) $\leq$

13. Show why $\frac{13}{3} = 4\frac{1}{3}$, using
   (a) division.
   (b) the part-of-a-whole meaning.

14. *Explain* why $\frac{1}{3} > \frac{1}{4}$, using the part-of-a-whole meaning.

15. *Explain* why $\frac{0}{5} = 0$.

16. *Explain* why $\frac{0}{0}$ is undefined.

17. Make two different designs by shading half of each figure.

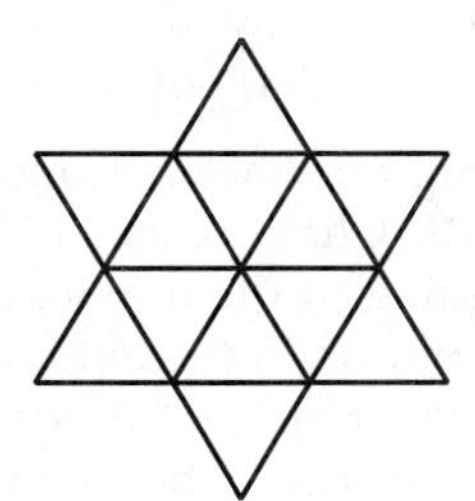
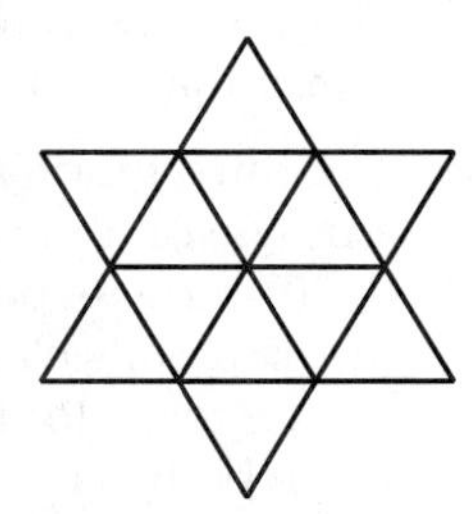

***18.** Use division rules for integers to show why $\frac{-4}{7} = \frac{4}{-7}$.

19. Which of the following are equal to $\frac{-3}{5}$?

(a) $-\frac{3}{5}$ (b) $\frac{-3}{-5}$ (c) $\frac{3}{-5}$ (d) $\frac{3}{5}$

***20.** Show why $\frac{1}{2} = \frac{2}{4}$, using

(a) the part-of-a-whole meaning.
(b) the Fundamental Law of Fractions.

21. Use fractions to explain the error made by the man in this cartoon.

BLAIR
"CUT MY PIZZA INTO FOUR PIECES...
NO WAY I COULD EAT EIGHT."

22. The relationships between the numerators and denominators of equivalent fractions can be studied using points on a graph.

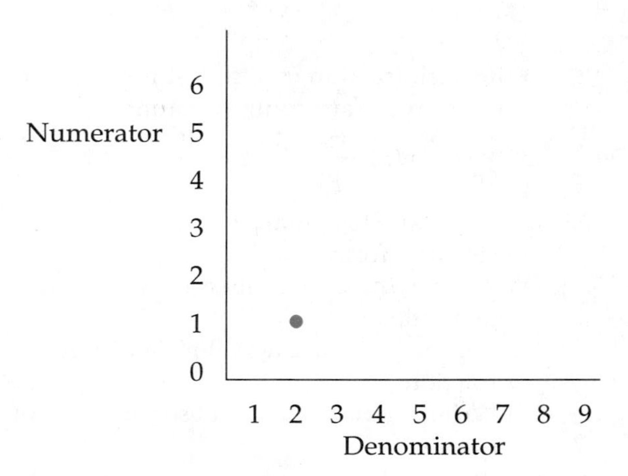

The graph shows the point that represents $\frac{1}{2}$.

(a) Plot points for $\frac{2}{4}$, $\frac{3}{6}$, and $\frac{4}{8}$.
(b) Connect all four points. What patterns do you see?
(c) Plot points for $\frac{2}{3}$, $\frac{4}{6}$, and $\frac{6}{9}$.
(d) Connect the points from part (c). What patterns do you see?
(e) Write a generalization of the results of parts (b) and (d).
(f) Making a generalization based upon the examples of parts (b) and (d) is an instance of ______________ reasoning.

23. If x and y are counting numbers, use the Fundamental law of Fractions to show why $\frac{x}{y} = \frac{2x^2}{2xy}$.

24. Describe an everyday situation in which one would use $\frac{15}{100}$ instead of $\frac{3}{20}$.

25. Name a situation in fraction arithmetic in which one would change

(a) $\frac{12}{15}$ to $\frac{4}{5}$. (b) $\frac{4}{5}$ to $\frac{12}{15}$.

26. Make up a numerical counterexample that shows the algebra is incorrect.

$$\frac{x}{y+x} = \frac{\overset{1}{\cancel{x}}}{y+\underset{1}{\cancel{x}}} = \frac{1}{y+1}$$

27. Write each fraction in simplest form. (Assume that x, y, and z are counting numbers.)
(a) $\frac{168}{464}$ (b) $\frac{xy^2}{xy^3z}$ (c) $\frac{48}{264}$ (d) $\frac{2^8 - 2^7}{2^7 - 2^6}$

28. (a) Write an elementary fraction that is in simplest form.
(b) What is the GCF of its numerator and denominator?
(c) Repeat parts (a) and (b) for a different fraction.
(d) Write a generalization based upon your results.
(e) In part (d), you used ________ reasoning.

29. Suppose that you rewrite $\frac{4}{6}$ as $\frac{2}{3}$. Explain why it is misleading to call this "reducing."

30. By what do you multiply both sides of $\frac{a}{b} = \frac{c}{d}$ to obtain $ad = bc$, assuming $b \neq 0$ and $d \neq 0$?

31. (a) Write $\frac{3}{4}$ and $\frac{7}{10}$ with a common denominator, and tell if they are equivalent.
(b) Compare cross products, and tell if $\frac{3}{4}$ and $\frac{7}{10}$ are equivalent.

32. (a) Write $\frac{9}{12}$ and $\frac{15}{20}$ with a common denominator, and tell if they are equivalent.
(b) Compare cross products, and tell if $\frac{9}{12}$ and $\frac{15}{20}$ are equivalent.

33. Solve for N.
(a) $\frac{N}{2} = \frac{5}{4}$ (b) $\frac{4}{N} = \frac{N}{9}$ (c) $\frac{5}{N} = \frac{6}{M}$

34. Solve for N.
(a) $\frac{55}{90} = \frac{N}{18}$ (b) $\frac{5}{15} = \frac{7}{N}$ (c) $\frac{3}{4} = \frac{N}{M}$

35. Two companies conduct surveys asking people if they favor stronger controls on air pollution. The first company asks 1500 people, and the second asks 2000. In the first group, 1200 say yes. Make up results for the second group that would be considered equivalent.

36. The third grade is voting on whether to go to a movie or a play. In Ms. Chan's class, 12 out of 20 students prefer going to the movie. In Ms. Brussat's class, 15 out of 25 students prefer going to the movie. Explain in what sense both classes equally prefer the movie over the play.

37. Use the LCD method and the compare-cross-products method to show how $\frac{4}{7}$ and $\frac{8}{15}$ compare in size.

38. Which is greater, $-2\frac{1}{4}$ or $-\frac{1}{2}$? (*Hint:* Think of a number line.)

39. In one class, 14 out of 23 students would rather go to an aquarium than a circus. In another class, 17 out of 30 students would rather go to an aquarium than a circus. Use the LCD or compare-cross-products method to compare their preferences.

40. In one fifth-grade class there are 35 students, 16 of whom are boys. In the other class there are 32 students, 14 of whom are boys. Write two multi-step mathematical questions that could be asked about one or both of these classes.

41. If you select a card at random from a regular deck of 52 cards, what fraction of the time would you expect to pick each of the following?
(a) An ace
(b) A picture card (jack, queen, or king)

42. Teachers can use familiar objects to illustrate fractions. Name a common object that is naturally divided into each of the following.
(a) Halves (b) Fourths (c) Twelfths

Extension Exercises

43. Consider the following regions, which represent manipulatives called pattern blocks.

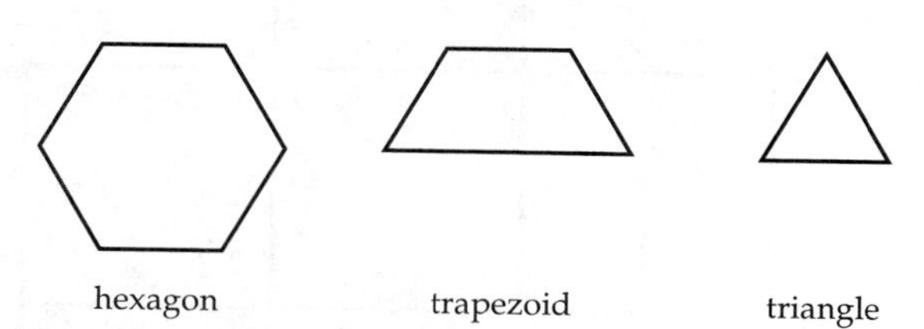

(a) How many triangular regions would it take to cover the hexagonal region?
(b) What fractional part of the hexagonal region is the triangular region?
(c) What fractional part of the hexagonal region is the trapezoidal region?
(d) Two triangular regions would be what fractional part of the trapezoidal region?

44. Cuisenaire® rods are whole-number lengths from 1 to 10 cm long in the colors indicated (the actual widths are 1 cm).

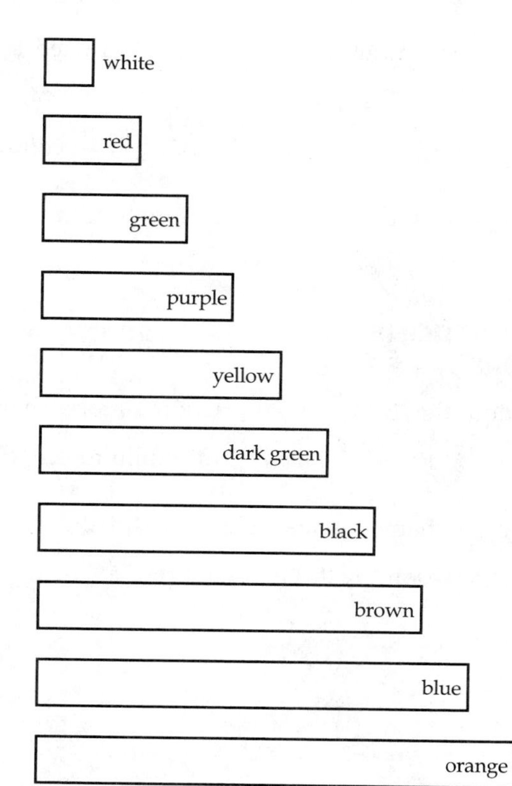

Any Cuisenaire® rod can be used as the unit rod.
(a) If the red rod is a unit, what fraction does the white rod represent?
(b) If the orange rod is a whole, what fraction does the green rod represent?
(c) If the white rod represents 1/4, what does the red rod represent?
(d) If the purple rod is 2/3, what color represents a whole?

45. 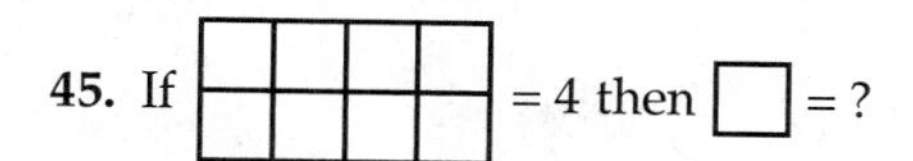

46. If ▦ = 2 then ▯▯ = ?

47. $\dfrac{1+1}{4+3}$ is between $\dfrac{1}{4}$ and $\dfrac{1}{3}$.

(a) Is $\dfrac{2+5}{3+6}$ between $\dfrac{2}{3}$ and $\dfrac{5}{6}$?

(b) If $\dfrac{a}{b}$ and $\dfrac{c}{d}$ are positive and $\dfrac{a}{b} < \dfrac{c}{d}$, is $\dfrac{a+c}{b+d}$ between $\dfrac{a}{b}$ and $\dfrac{c}{d}$? Try more examples and decide whether this is reasonable, or find a counterexample to show that it is false.

48. (a) Which is greater, $-\dfrac{1}{3}$ or $-\dfrac{1}{2}$? (*Hint:* Think of a number line.)

(b) Which is greater, $-\dfrac{1}{4}$ or $-\dfrac{3}{8}$?

(c) Which is greater, $-\dfrac{7}{10}$ or $-\dfrac{5}{8}$? (*Hint:* Find a common denominator.)

49. Prove: If a and b are counting numbers and $a > b$, then $\dfrac{1}{b} > \dfrac{1}{a}$.

50. Using only the digits 1, 3, 7, and 8, replace each question mark with a digit to make a true statement. (Guess and check.)

$$?\,\frac{?}{?} = \frac{?\,?}{?}$$

Special Exercise

51. Copy the picture and cut out all nine squares. Fit the nine pieces together into one large square so that all the edges that touch have equivalent fractions.

1; 0	$1\frac{1}{3}$; $\frac{1}{4}$	$\frac{1}{2}$; $\frac{7}{7}$; $\frac{2}{5}$
$\frac{2}{16}$; $\frac{4}{3}$; $\frac{3}{4}$	$\frac{1}{3}$; $\frac{2}{6}$; $\frac{2}{8}$	$\frac{1}{3}$; $\frac{6}{8}$; $\frac{3}{6}$; $\frac{10}{12}$
$\frac{1}{3}$; $\frac{4}{10}$	$\frac{5}{6}$; $\frac{2}{4}$; $\frac{0}{3}$	$\frac{1}{2}$; $\frac{1}{8}$

6.2 ADDITION AND SUBTRACTION OF RATIONAL NUMBERS

You plan to spend $\frac{3}{4}$ hour doing mathematics homework and $\frac{1}{2}$ hour doing English. What is the total amount of time you will need? You have $\frac{1}{4}$ yd of fabric and you need a total of $\frac{1}{2}$ yd. How much more fabric do you need to purchase?

To solve the first problem, you would add $\frac{3}{4}$ and $\frac{1}{2}$. To solve the second, you would subtract $\frac{1}{4}$ from $\frac{1}{2}$. Adding and subtracting rational numbers are the subjects of this lesson.

ADDING AND SUBTRACTING FRACTIONS THAT HAVE LIKE DENOMINATORS

Children first add and subtract fractions that have the same denominators. They can solve problems like the following. "Pamela walked $\frac{1}{5}$ of a mile to school and then $\frac{3}{5}$ of a mile from school to Stephanie's house. How far did she walk altogether? How much farther was the second walk than the first?"

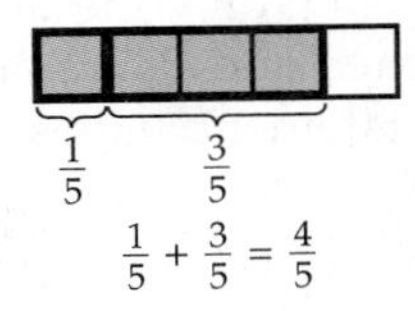

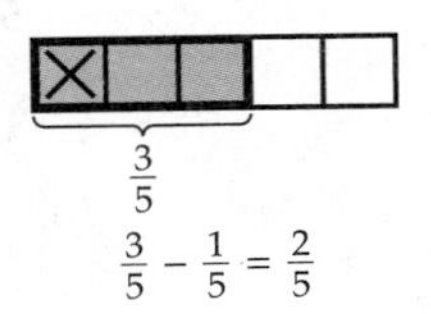

Figure 6–3

Pictures and manipulatives help clarify why one should not add or subtract the denominators (Figure 6-3). The general addition and subtraction rules for like denominators can be given algebraically—by you! Complete the following.

Lesson Exercise 6.22

(a) For rational numbers, $\frac{a}{c} + \frac{b}{c} =$ ________.

(b) For rational numbers, $\frac{d}{f} - \frac{e}{f} =$ ________.

The rules for adding and subtracting rational numbers that have the same denominator are as follows.

Addition of Rational Numbers That Have Like Denominators

$$\frac{a}{c} + \frac{b}{c} = \frac{a + b}{c}$$

Subtraction of Rational Numbers That Have Like Denominators

$$\frac{d}{f} - \frac{e}{f} = \frac{d - e}{f}$$

Use the rules for adding and subtracting rational numbers that have like denominators to compute the following.

Lesson Exercise 6.23

Assume that the following fractions represent rational numbers.

$$\frac{4x}{x + 2} - \frac{3x}{x + 2} = \underline{\qquad\qquad}$$

ADDING ELEMENTARY FRACTIONS THAT HAVE UNLIKE DENOMINATORS

Does $\frac{1}{2} + \frac{1}{3} = \frac{2}{5}$? According to a recent National Assessment of Educational Progress (NAEP), about 30% of the seventh-graders in the United States thought so (40% thought the answer was $\frac{5}{6}$, and 30% got some other answer).

How would you convince someone who thinks $\frac{1}{2} + \frac{1}{3} = \frac{2}{5}$ that $\frac{2}{5}$ is the wrong answer? Pictures of fraction bars are helpful in showing why $\frac{1}{2} + \frac{1}{3} \neq \frac{2}{5}$. Pictures of $\frac{1}{2}$, $\frac{1}{3}$, and $\frac{2}{5}$ are shown in Figure 6-4.

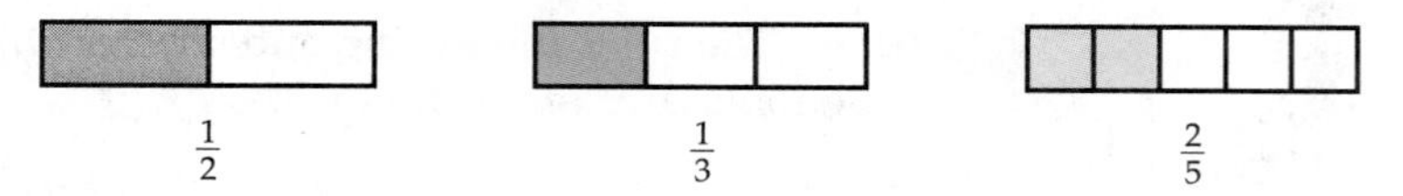

Figure 6–4

Clearly, $\frac{1}{2} + \frac{1}{3} \neq \frac{2}{5}$, because $\frac{1}{2}$ by itself is more than $\frac{2}{5}$! The following example continues from this point. Pictures of fraction bars can show why a common denominator is needed. Then, by trading for fraction bars with the common denominator, one can find the answer.

EXAMPLE 6.2

(a) Draw a fraction bar for each fraction in $\frac{1}{2} + \frac{1}{3}$.

(b) Draw a fraction bar for the sum, and explain why a common denominator is needed.

(c) Draw a fraction bar for each fraction with the least common denominator.

(d) Show how to compute $\frac{1}{2} + \frac{1}{3}$ using a picture.

Solution

(a)

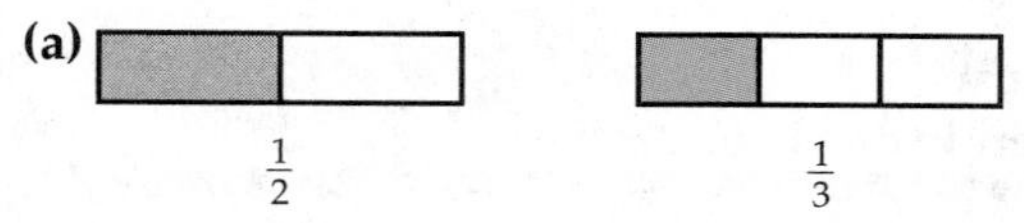

(b) Try to determine the sum of $\frac{1}{2}$ and $\frac{1}{3}$ without a common denominator. What part of the whole is shaded to show the sum in Figure 6-5? There is no name for the sum, since it is made out of two different units (denominations), halves and thirds. The answer cannot be determined without a *common denominator*.

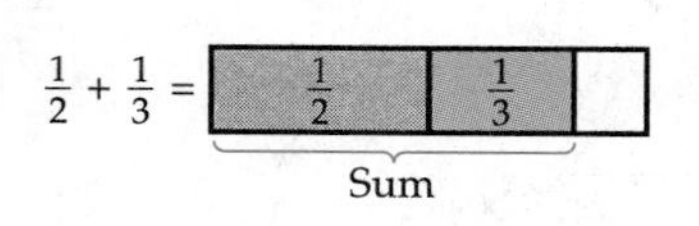

Figure 6–5

(c) 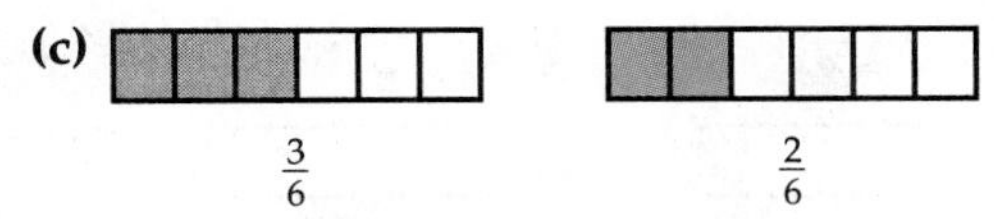

(d) After rewriting both fractions with a common denominator (sixths), one can use a combine-measures pictures to show the sum.

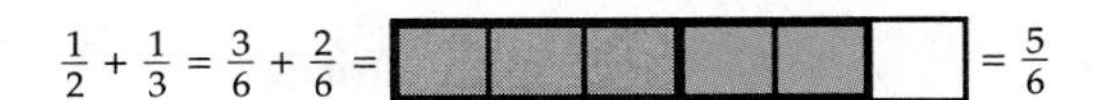

Aren't common denominators wonderful? Without them, we could not name sums or differences of fractions that have unlike denominators.

Lesson Exercise 6.24

(a) Draw a fraction bar for each fraction in $\frac{1}{2} + \frac{1}{4}$.

(b) Draw a fraction bar for the sum, and explain why a common denominator is needed.

(c) Draw a fraction bar for each fraction with the least common denominator.

(d) Show how to compute $\frac{1}{2} + \frac{1}{4}$ using a picture.

SUBTRACTING ELEMENTARY FRACTIONS THAT HAVE UNLIKE DENOMINATORS

Suppose a family spends $\frac{1}{3}$ of their income on taxes and $\frac{1}{4}$ of their income on rent. How much more of their income do they spend on taxes than on rent? This example requires subtracting fractions that have unlike denominators.

Fraction pictures clarify subtraction rules for unlike denominators. The following example explains why one needs a common denominator to compute $\frac{1}{3} - \frac{1}{4}$.

EXAMPLE 6.3

(a) Draw a fraction bar for each fraction in $\frac{1}{3} - \frac{1}{4}$.

(b) Explain why a common denominator is needed.

(c) Draw a fraction bar for each fraction with the least common denominator.

(d) Show how to compute $\frac{1}{3} - \frac{1}{4}$ using a picture.

Solution

(a)

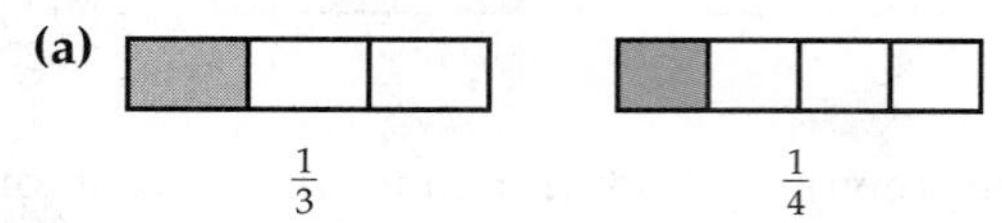

(b) How much is left when we take $\frac{1}{4}$ away from $\frac{1}{3}$? We can't tell. A common denominator is needed.

(c)

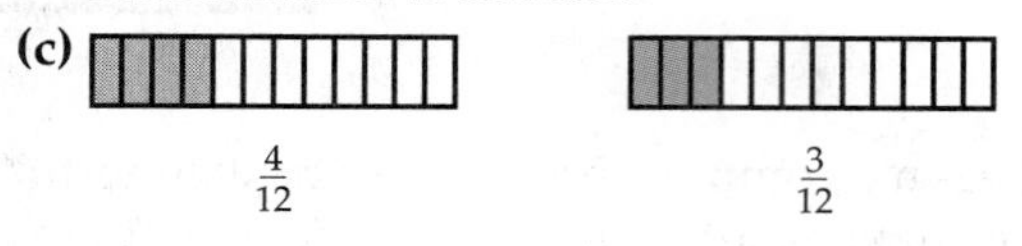

(d) $\frac{1}{3} - \frac{1}{4} = \frac{4}{12} - \frac{3}{12} =$ $= \frac{1}{12}$ ■

Note that one can also use compare-measures subtraction with two separate fraction bars and see how much longer $\frac{4}{12}$ is than $\frac{3}{12}$.

Lesson Exercise 6.25

(a) Draw a fraction bar for each fraction in $\frac{1}{2} - \frac{1}{3}$.

(b) Explain why a common denominator is needed.

(c) Draw a fraction bar for each fraction with the least common denominator.

(d) Show how to compute $\frac{1}{2} - \frac{1}{3}$ using a picture.

RATIONAL NUMBERS THAT HAVE UNLIKE DENOMINATORS

Before you can add or subtract rational numbers that have unlike denominators, you must rewrite them with a common denominator. As you saw in Section 6.1, you can find the least common denominator (LCD) using the least common multiple (LCM) of all the denominators. For example, the LCD for $\frac{1}{6}$ and $\frac{7}{8}$ is the LCM of 6 and 8, which is 24.

$$\frac{1}{6} + \frac{7}{8}$$

$$\text{LCM} = 24$$

$$\frac{1 \cdot 4}{6 \cdot 4} + \frac{7 \cdot 3}{8 \cdot 3} = \frac{4}{24} + \frac{21}{24} = \frac{25}{24} = 1\frac{1}{24}$$

Negative rational numbers can be added and subtracted in the same way. In order to retain the positive denominator, I will always write $\frac{-3}{5}$ rather than $\frac{3}{-5}$.

The general rules for adding and subtracting rational numbers that have unlike denominators are as follows.

Addition of Rational Numbers That Have Unlike Denominators

To add rational numbers $\frac{a}{b} + \frac{c}{d}$ in which $b > 0$, $d > 0$, and $b \neq d$:

1. Rename each fraction with the least common denominator, that is, LCM(b, d).
2. Add the fractions using the addition rule for like denominators.

Subtraction of Rational Numbers That Have Unlike Denominators

To subtract rational numbers $\frac{a}{b} - \frac{c}{d}$ in which $b > 0$, $d > 0$, and $b \neq d$:

1. Rename each fraction with the least common denominator, that is, LCM(b, d).
2. Subtract the fractions using the subtraction rule for like denominators.

Use these rules in Lesson Exercise 6.26.

Lesson Exercise 6.26

(a) Compute $\frac{3}{20} - \frac{1}{12}$.

(b) What is the least common denominator of $\frac{3}{4x}$ and $\frac{5}{6}$?

Are you aware that fractions can be added with any common denominator? However, if you do not use the *least* common denominator, additional simplification will be required at the end.

$$\frac{5}{12} + \frac{1}{6} = \underset{\text{(LCD)}}{\frac{5}{12} + \frac{2}{12}} = \frac{7}{12} \quad \text{or} \quad \frac{5}{12} + \frac{1}{6} = \underset{\left(\substack{\text{Other common}\\ \text{denominator}}\right)}{\frac{30}{72} + \frac{12}{72}} = \underset{\text{(Extra step)}}{\frac{42}{72} = \frac{7}{12}}$$

Until the seventeenth century, most people used the *product* of all the denominators as the common denominator. For the last 300 years, most people have used the least common denominator in textbook examples, and they rarely need prime factorizations to do it.

Children learn addition and subtraction involving negative rational numbers in junior high school. Since they have already studied elementary fraction and integer arithmetic, they can relate any rational-number addition or subtraction problem to elementary fraction and integer arithmetic.

Lesson Exercise 6.27

$-2\frac{3}{4} + 4\frac{5}{8} = ?$

(a) $4\frac{5}{8} - 2\frac{3}{4}$ (b) $2\frac{3}{4} + 4\frac{5}{8}$ (c) $-\left(4\frac{5}{8} - 2\frac{3}{4}\right)$

Do you recall how to change a mixed number such as $1\frac{3}{4}$ into an improper fraction? You can verify the results with addition of fractions.

Lesson Exercise 6.28

(a) What is a shortcut for converting $6\frac{2}{3}$ to an improper fraction?

(b) Verify the shortcut in part (a) by changing $6\frac{2}{3}$ to $\frac{6}{1} + \frac{2}{3}$. Add $\frac{6}{1} + \frac{2}{3}$. Is the result the same as your answer to part (a)?

The classifications for whole-number operations apply to some addition and subtraction applications involving rational numbers.

Lesson Exercise 6.29

What operation and category are illustrated in the following problem? "Last week you worked $37\frac{1}{2}$ hours, and this week you worked 45 hours. How many more hours did you work this week?"

FRACTION CALCULATORS

Calculators such as the TI-Explorer and Explorer Plus, the Casio fx-55 FRACTION MATE, and the Sharp EL-E300 give results of fraction arithmetic in fraction form. To compute $\frac{1}{2} + \frac{3}{8}$ on a TI-Explorer or Explorer Plus, press 1 [/] 2 [+] 3 [/] 8 [=]. The answer will be displayed as 7/8. The fraction key is [b/c] on the Casio and [$\frac{x}{y}$] on the Sharp. The Casio and the Sharp will display $\frac{7}{8}$.

Lesson Exercise 6.30

If you have a fraction calculator, compute $\frac{5}{8} - \frac{1}{2}$.

The [Simp] key can be used to simplify fractions on the TI-Explorer and the Casio FRACTION MATE. Suppose you display 12/16 on the TI-Explorer. To specify what number to divide into the numerator and denominator, type that number between [Simp] and [=]. For example, 12 [/] 16 [Simp] 2 [=] will display 6/8.

Lesson Exercise 6.31

If you have a fraction calculator, enter $\frac{50}{60}$ and see if you can simplify it in

(a) one step. (b) two steps.

All three calculators convert fractions to decimals and convert improper fractions to mixed numerals. If you have a fraction calculator, you can experiment with these features. Refer to your manual for help.

ANSWERS TO SELECTED LESSON EXERCISES

6.22 (a) $\frac{a + b}{c}$ (b) $\frac{d - e}{f}$

6.23 $\frac{x}{x + 2}$

6.24 (a)

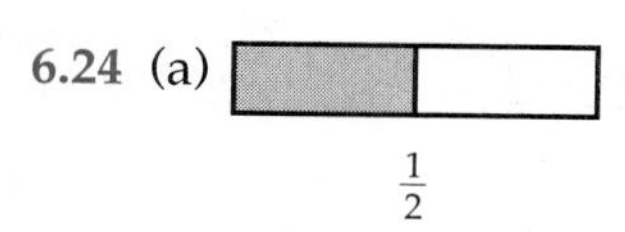

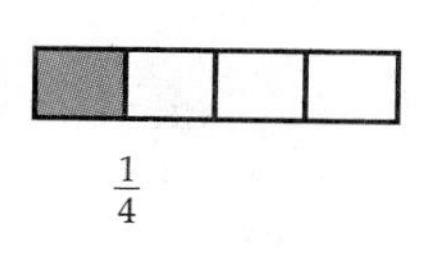

(b) $\frac{1}{2} + \frac{1}{4}$ would equal the shaded region shown,

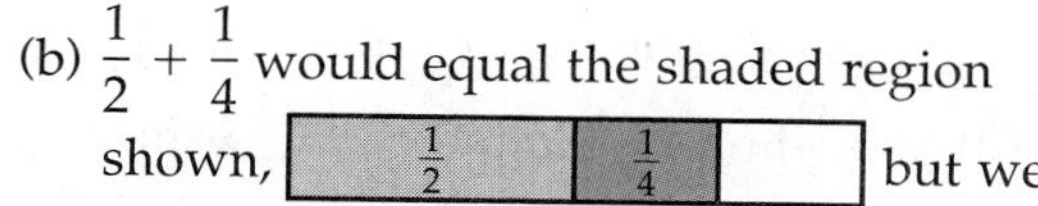

but we cannot name the sum unless we use a common denomination (fourths).

(c)

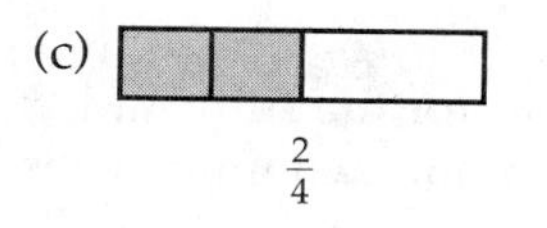

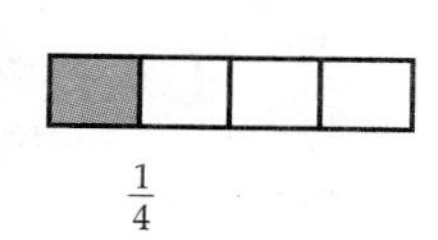

(d) $\frac{1}{2} + \frac{1}{4} = \frac{2}{4} + \frac{1}{4} =$

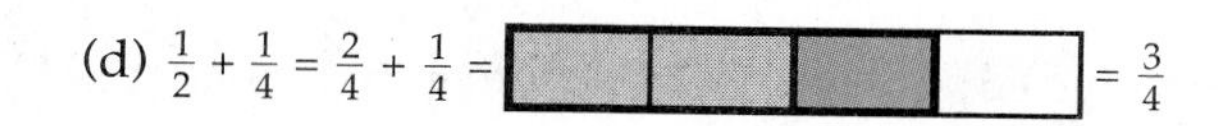

$= \frac{3}{4}$

6.25 (a) 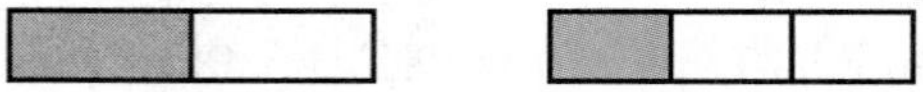

$\frac{1}{2}$ $\frac{1}{3}$

(b) How much is left when we take $\frac{1}{3}$ away from $\frac{1}{2}$? We can't tell. A common denominator is needed.

(c) 

$\frac{3}{6}$ $\frac{2}{6}$

(d) $\frac{1}{2} - \frac{1}{3} = \frac{3}{6} - \frac{2}{6} =$ 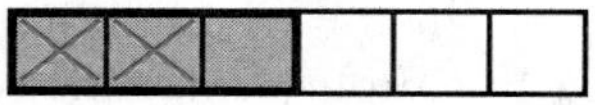 $= \frac{1}{6}$

6.26 (a) $\frac{1}{15}$ (b) $12x$

6.27 (a)

6.28 (a) $6 \times 3 + 2 = 20 \rightarrow \frac{20}{3}$ (b) $\frac{20}{3}$; yes

6.29 Subtraction; compare measures

6.2 HOMEWORK EXERCISES

Basic Exercises

1. Assume that the following fractions are rational.
   (a) $\frac{3}{y} + \frac{1}{y} =$ ________
   (b) $\frac{x}{n} + \frac{2x}{n} =$ ________
   (c) $\frac{4x}{x+1} - \frac{2x}{x+1} =$ ________
2. A child thinks $\frac{1}{2} + \frac{1}{4} = \frac{2}{6}$. Draw a fraction bar for each fraction, and explain why $\frac{2}{6}$ cannot be the answer.
3. A child thinks $\frac{1}{2} + \frac{1}{8} = \frac{2}{10}$. Draw a fraction bar for each fraction, and explain why $\frac{2}{10}$ cannot be the answer.
4. (a) Draw a fraction bar for each fraction in $\frac{1}{2} + \frac{2}{5}$.
   (b) Draw a fraction bar for the sum, and explain why a common denominator is needed.
   (c) Draw a fraction bar for each fraction with the least common denominator.
   (d) Show how to compute $\frac{1}{2} + \frac{2}{5}$ using a picture.
5. (a) Draw a fraction bar for each fraction in $\frac{1}{3} + \frac{1}{4}$.
   (b) Draw a fraction bar for the sum, and explain why a common denominator is needed.
   (c) Draw a fraction bar for each fraction with the least common denominator.
   (d) Show how to compute $\frac{1}{3} + \frac{1}{4}$ using a picture.
6. (a) Draw a fraction bar for each fraction in $\frac{1}{3} - \frac{1}{5}$.
   (b) Explain why a common denominator is needed.
   (c) Draw a fraction bar for each fraction with the least common denominator.
   (d) Show how to compute $\frac{1}{3} - \frac{1}{5}$ using a picture.
7. (a) Draw a fraction bar for each fraction in $\frac{1}{4} - \frac{1}{6}$.
   (b) Explain why a common denominator is needed.
   (c) Draw a fraction bar for each fraction with the least common denominator.
   (d) Show how to compute $\frac{1}{4} - \frac{1}{6}$ using a picture.

*8. (a) Find the least common denominator of $\frac{5}{44}$ and $\frac{3}{28}$.
(b) Give two other common denominators.
(c) Compute $\frac{5}{44} - \frac{3}{28}$.

9. Compute the following.
(a) $\frac{5}{12} + \frac{3}{8}$ (b) $\frac{2}{51} + \frac{1}{21}$
(c) $\frac{2}{9} - \frac{1}{3}$ (d) $\frac{5}{12} - \frac{1}{20}$

10. Compute the following.
(a) $\frac{7}{8} + \left(-\frac{3}{16}\right)$ (b) $\frac{3}{10} - \frac{1}{4}$
(c) $\frac{8}{9} - \frac{5}{6}$ (d) $-\frac{2}{3} - \frac{1}{2}$

11. Solve mentally.
(a) $\frac{1}{8} + \square = \frac{5}{8}$ (b) $3\frac{9}{10} - r = 1\frac{3}{10}$

12. Solve mentally.
(a) $4\frac{1}{8} + x = 10\frac{3}{8}$ (b) $t - \frac{1}{12} = \frac{7}{12}$

13. $\frac{1}{2^3 \cdot 3} - \frac{1}{3^2 \cdot 5^2} =$ ________

14. Assume that the following fractions represent rational numbers.
(a) $\frac{1}{a} + \frac{2}{b} =$ ________
(b) $\frac{3}{2c} - \frac{2}{5c} =$ ________

15. Compute the following.
(a) $397\frac{1}{6} - 286\frac{5}{6}$ (b) $-286\frac{1}{3} + 143\frac{2}{3}$

16. $-43\frac{1}{2} + 32\frac{1}{2} = ?$
(a) $-\left(43\frac{1}{2} + 32\frac{1}{2}\right)$ (b) $\left(43\frac{1}{2} - 32\frac{1}{2}\right)$
(c) $-\left(43\frac{1}{2} - 32\frac{1}{2}\right)$

17. Write $4\frac{1}{5}$ as a sum; show that it equals $\frac{21}{5}$.

18. What operation and category does the following problem illustrate? "You buy $\frac{3}{4}$ lb of Swiss cheese and $\frac{1}{2}$ lb of provolone. How many pounds of cheese have you purchased?"

19. You are working on a project that will take about $4\frac{1}{2}$ hours.
(a) If you have been working on it for $1\frac{3}{4}$ hours, how much more time will it take?
(b) What operation and classification are illustrated here?

20. In an apartment complex, $\frac{7}{8}$ of the people speak English as their native language, and $\frac{1}{16}$ speak Spanish as their native language.
(a) What fraction speak a language other than English or Spanish as their native language?
(b) What operations and classifications are illustrated here?

21. Make up a realistic problem for $5\frac{1}{4} - 2\frac{1}{2}$, and give the answer as a fraction.

22. Use a fraction calculator to
(a) simplify $\frac{20}{48}$. (b) compute $\frac{3}{8} + \frac{3}{4}$.

23. Use a fraction calculator to
(a) simplify $\frac{125}{300}$. (b) compute $\frac{9}{10} - \frac{3}{4}$.

24. Fill in each square with either a + sign or a − sign to complete each equation correctly.

(a) $1\frac{1}{4} \square \frac{1}{4} \square \frac{3}{4} = \frac{3}{4}$

(b) $1\frac{7}{8} \square \frac{1}{4} \square \frac{3}{8} = 1\frac{1}{4}$

25. Pick from the following numbers to complete each equation correctly.

$$2\frac{1}{4} \quad 2\frac{1}{3} \quad 2\frac{1}{2} \quad 1\frac{2}{3} \quad 1\frac{1}{8} \quad 1\frac{3}{4} \quad \frac{7}{8}$$

(a) ____ + ____ $= 3\frac{5}{8}$

(b) ____ − ____ $= \frac{3}{4}$

(c) $2\frac{1}{2} \times$ ____ $= 2\frac{13}{16}$

(d) $2\frac{1}{2} \div$ ____ $= 2\frac{6}{7}$

26. (a) Tell how to find what fraction of the figure is shaded.

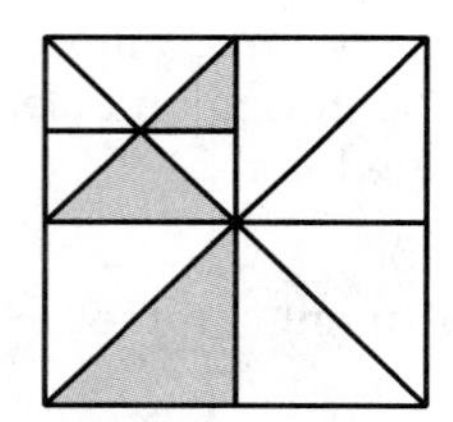

(b) Describe a second way of finding the answer.

27. The sum of two fractions is $\frac{7}{12}$. Their difference is $\frac{1}{4}$. What are the two fractions? (Guess and check.)

Extension Exercises

28. (a) What will the next equation be if the pattern continues? Is the equation true?

$$\frac{1}{2} = \frac{1}{3} + \frac{1}{6}$$

$$\frac{1}{3} = \frac{1}{4} + \frac{1}{12}$$

(b) Complete the general equation showing the pattern in part (a).

$$\frac{1}{N} = \underline{\qquad\qquad}$$

(c) Show that the equation in part (b) is true.

29. The ancient Egyptians represented every elementary fraction other than 0 and $\frac{2}{3}$ as the sum of unequal **unit fractions**, fractions that have a numerator of 1. For example, $\frac{2}{7} = \frac{1}{4} + \frac{1}{28}$. They did this to avoid certain computational difficulties. Write each of the following as the sum of unequal unit fractions.

(a) $\frac{3}{4}$ (b) $\frac{3}{26}$ (c) $\frac{5}{8}$ (d) $\frac{7}{9}$

30. In the December 1991 *Mathematics Teacher*, Arthur Howard describes situations in which the standard algorithm for adding fractions does not yield the correct answer. For example, suppose two third-grade classes go on a field trip. One class has 10 girls out of a class of 25, and the other has 12 girls out of a class of 24. What fraction of the whole group is girls?

$$\frac{10}{25} \oplus \frac{12}{24} = \frac{10 + 12}{25 + 24} = \frac{22}{49}$$

This example illustrates nonstandard fraction addition, denoted by $\oplus$, in which the "reference unit" (denominator) of the sum is the *sum* of the reference units of the addends. In other words, one adds the denominators of the addends!

(a) Make up a problem about fruit that is solved with $\oplus$.

(b) Make up a problem that is solved with an analogous $\ominus$ operation.

31. *Within each part*, the question marks represent the same counting number. Find the missing numbers. All fractions are in simplest terms. (Guess and check.)

(a) $\frac{2}{?} + \frac{1}{2} = \frac{9}{?}$ (b) $\frac{?}{5} - \frac{4}{15} = \frac{1}{?}$

32. A **Farey sequence of order** n lists all rational numbers in simplest-fraction form, in increasing order from 0 through 1, with denominators that do not exceed n. The Farey sequence of order 3 is $\frac{0}{1}, \frac{1}{3}, \frac{1}{2}, \frac{2}{3}, \frac{1}{1}$.

(a) Write the Farey sequence of order 4.

(b) Write the Farey sequence of order 5.

(c) If $\frac{a}{b}$ and $\frac{c}{d}$ are consecutive fractions in a Farey sequence, what is the relationship between ad and bc?

33. Prove, for rational numbers, if $\frac{a}{b} = \frac{c}{d}$, then $\frac{a+b}{b} = \frac{c+d}{d}$.

6.3 MULTIPLICATION AND DIVISION OF RATIONAL NUMBERS

You are planning a chicken dinner. Each person will eat about 1/4 of a chicken. How many chickens will you need for 3 people? Next, you want to dye some shirts. Suppose it takes 3/4 cup of dye to dye a shirt. How many shirts can you dye with 10 cups?

To solve the first problem, you would multiply 1/4 by 3. To solve the second, you would divide 10 by 3/4. Multiplying and dividing rational numbers are the subjects of this lesson.

MULTIPLYING RATIONAL NUMBERS

Now about that chicken dinner. You need 3 servings, each consisting of $\frac{1}{4}$ lb chicken. Using repeated addition, $3 \times \frac{1}{4} = \frac{1}{4} + \frac{1}{4} + \frac{1}{4} = \frac{3}{4}$.

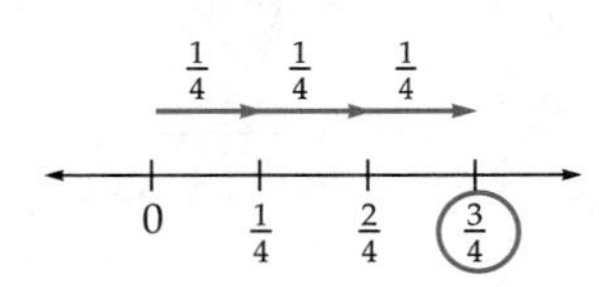

Lesson Exercise 6.32

Show how to compute $5 \times \frac{1}{8}$ using repeated addition.

Children begin studying multiplication of rational numbers with the product of a fraction and a whole number, such as $5 \times \frac{1}{8}$ or $\frac{1}{2} \times 8$. You have already seen how the first of these problems can be approached using repeated addition. The second problem, $\frac{1}{2} \times 8$, can be understood as a fraction of a set.

Just as 2×8 is the same as 2 sets of 8, $\frac{1}{2} \times 8$ is the same as $\frac{1}{2}$ of a set of 8, or, more simply, $\frac{1}{2}$ of 8. Similarly, $\frac{2}{3} \times 6$ is the same as $\frac{2}{3}$ of 6. Even before studying multiplication of fractions, most children know how of find $\frac{1}{2}$ of 8 or $\frac{2}{3}$ of 6.

The product of a fraction and a whole number can be computed using this relationship.

$\frac{1}{2} \times 8 = \frac{1}{2}$ of 8 $\quad$ = 4 $\qquad$ $\frac{2}{3} \times 6 = \frac{2}{3}$ of 6 $\quad$ = 4

1 out of 2 equal groups $\qquad$ 2 out of 3 equal groups

This relationship can also be used in fraction word problems.

Lesson Exercise 6.33

Suppose $\frac{4}{5}$ of a group of 20 people watch television. Write a fraction arithmetic equation that shows how to find the size of this group, and write the result.

After studying how to multiply a whole number by a fraction, children study the product of two fractions, such as $\frac{1}{2} \times \frac{3}{4}$. Example 6.4 shows how to find the result of $\frac{1}{2} \times \frac{3}{4}$ using a diagram.

EXAMPLE 6.4

Suppose $\frac{3}{4}$ of a field is plowed. Then $\frac{1}{2}$ of the plowed field is planted with tomatoes. What fraction of the field is planted with tomatoes? The answer is $\frac{1}{2} \times \frac{3}{4}$. Suppose a child does not know the multiplication rule for fractions. *Explain* how to compute $\frac{1}{2} \times \frac{3}{4}$ using two diagrams.

Solution

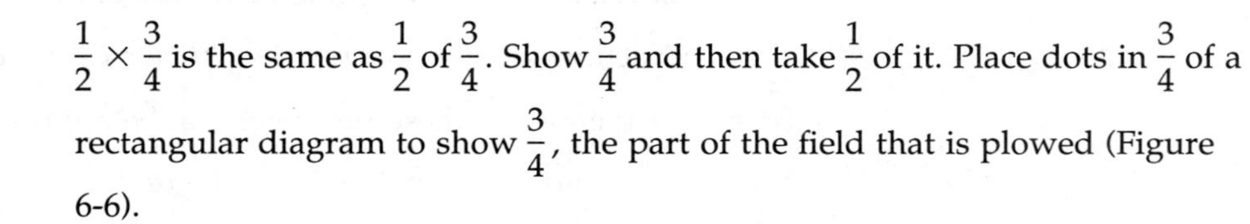

$\frac{1}{2} \times \frac{3}{4}$ is the same as $\frac{1}{2}$ of $\frac{3}{4}$. Show $\frac{3}{4}$ and then take $\frac{1}{2}$ of it. Place dots in $\frac{3}{4}$ of a rectangular diagram to show $\frac{3}{4}$, the part of the field that is plowed (Figure 6-6).

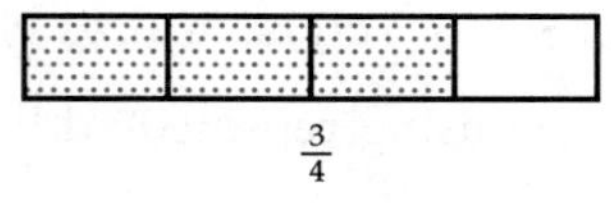

Figure 6–6

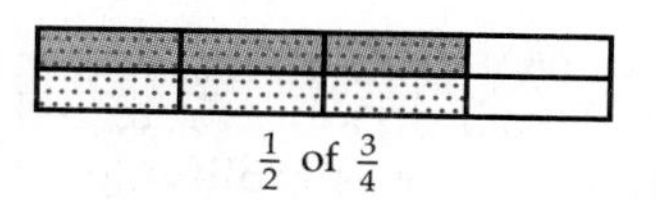

Figure 6–7

To show the part with tomatoes, $\frac{1}{2}$ of $\frac{3}{4}$, darken $\frac{1}{2}$ of the *dotted* part, as in Figure 6-7.

What part of the whole figure is darkened? $\frac{3}{8}$. So $\frac{1}{2}$ of $\frac{3}{4} = \frac{1}{2} \times \frac{3}{4} = \frac{3}{8}$. ■

Lesson Exercise 6.34

Explain how to compute $\frac{1}{2} \times \frac{3}{5}$ using two diagrams, as in Example 6.3.

The results of Example 6.4 and Lesson Exercise 6.34 suggest the multiplication rule for fractions.

Lesson Exercise 6.35

Find the pattern in the preceding results.

(a) Complete the chart, showing results from Example 6.4 and Lesson Exercise 6.34.

Factors	**Product**
$\frac{1}{2} \times \frac{3}{4}$	
$\frac{1}{2} \times \frac{3}{5}$	

(b) The results from part (a) suggest that the general rule for multiplying rational numbers is: $\frac{a}{b} \times \frac{c}{d} =$ ________

Lesson Exercise 6.35 suggests the rule for multiplying rational numbers. This rule is the definition of multiplication of rational numbers.

Definition **Multiplication of Rational Numbers**

$$\frac{a}{b} \times \frac{c}{d} = \frac{ac}{bd}$$

Because the definition is written for fractions, we must *rewrite mixed numbers as improper fractions* before applying this rule.

Another interesting process in multiplying fractions is simplifying the resulting fraction before completing the multiplication. This is no longer referred to as "canceling" in textbooks; it is now called a "shortcut" or "simplifying by dividing by a common factor."

Lesson Exercise 6.36

In multiplying $\frac{3}{7} \times \frac{10}{21}$, why is one allowed to simplify as follows?

$$\frac{\overset{1}{\cancel{3}}}{7} \times \frac{10}{\underset{7}{\cancel{21}}}$$

In school, most children learn to simplify by dividing by a common factor from the numerator of one fraction and the denominator of *another fraction*.

$$\frac{3}{7} \times \frac{10}{21} = \frac{\overset{1}{\cancel{3}}}{7} \times \frac{10}{\underset{7}{\cancel{21}}} = \frac{10}{49}$$

Why does this work?

EXAMPLE 6.5

Explain why, in multiplying $\frac{3}{7} \times \frac{10}{21}$, one can simplify the computation by changing the 3 and 21 to 1 and 7, respectively.

Solution

$$\frac{3}{7} \times \frac{10}{21} = \frac{3 \times 10}{7 \times 21}$$

Now that it's all one fraction, you can divide the numerator and denominator by a common factor, 3. $\left(\text{Then } \frac{3}{21} \text{ becomes } \frac{1}{7}.\right)$

$$= \frac{\overset{1}{\cancel{3}} \times 10}{7 \times \underset{7}{\cancel{21}}}$$

Then we can multiply to obtain the answer $\frac{10}{49}$ in simplest form. ■

When children use the shortcut in school, they simplify before combining the two fractions into one, which makes the process faster but more mysterious. It may not be clear to them why the top of one fraction and the bottom of a *different* fraction can be divided by the same nonzero number.

Lesson Exercise 6.37

Explain why, in multiplying $\frac{4}{9} \times \frac{5}{8}$, one can simplify the computation by changing the 4 and 8 to 1 and 2, respectively.

DIVIDING RATIONAL NUMBERS

"Joe buys 4 lb of Swiss cheese. He and his wife eat a total of 1/2 lb of Swiss cheese each day. For how many days will the cheese last?" Before learning the division rule, children solve problems such as this that model a whole number divided by an elementary fraction. This kind of problem can be solved using pictures and the repeated-measures model of division.

EXAMPLE 6.6

A child has not yet learned the rule for dividing fractions. *Explain* how to compute $4 \div \frac{1}{2}$ using a measurement picture.

Solution

The expression $4 \div \frac{1}{2}$ means "How many $\frac{1}{2}$'s does it take to make 4?" It takes two $\frac{1}{2}$'s to make each whole.

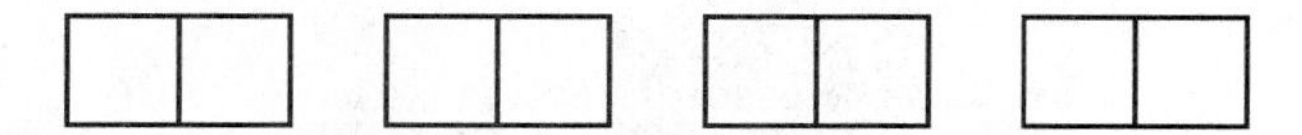

Therefore, it takes eight $\frac{1}{2}$'s to make 4. So $4 \div \frac{1}{2} = 8$. ■

Lesson Exercise 6.38

A child has not yet learned the rule for dividing fractions. *Explain* how to compute $3 \div \frac{1}{4}$ using a measurement picture.

Division involving fractions with whole-number quotients can be done in a similar way.

Lesson Exercise 6.39

Explain how to compute $\frac{3}{4} \div \frac{1}{8}$ using a number line or a fraction-of-a-whole diagram.

Another way to confirm that $\frac{3}{4} \div \frac{1}{8}$ is 6 is to try multiplying $\frac{1}{8} \times 6$. In other words, $\frac{3}{4} \div \frac{1}{8} = 6$ means $\frac{1}{8} \times 6 = \frac{3}{4}$. This method of checking uses the definition of division. Like division of whole numbers and integers, division of rational numbers is defined in terms of multiplication.

Definition **Division of Rational Numbers**

If $\frac{a}{b}$ and $\frac{c}{d}$ are rational numbers and $\frac{c}{d} \neq 0$, then $\frac{a}{b} \div \frac{c}{d} = \frac{e}{f}$ if and only if $\frac{a}{b} = \frac{c}{d} \times \frac{e}{f}$.

Lesson Exercise 6.40

How would you check the result of $3 \div \frac{1}{4}$ by writing it as multiplication?

The results from Example 6.6 and Lesson Exercises 6.38 and 6.39 suggest the shortcut procedure for dividing fractions.

Lesson Exercise 6.41

Can you find the pattern in the preceding results? Refer back to Example 6.6 and Lesson Exercises 6.38 and 6.39 to obtain the answers in the first column. Multiply to find the answers in the second column.

(a) $\frac{4}{1} \div \frac{1}{2} =$ ________ $\frac{4}{1} \times \frac{2}{1} =$ ________

(b) $\frac{3}{1} \div \frac{1}{4} =$ ________ $\frac{3}{1} \times \frac{4}{1} =$ ________

(c) $\frac{3}{4} \div \frac{1}{8} =$ ________ $\frac{3}{4} \times \frac{8}{1} =$ ________

(d) Parts (a)–(c) suggest that for rational numbers $\frac{a}{b}$ and $\frac{c}{d}$ with $\frac{c}{d} \neq 0$,

$\frac{a}{b} \div \frac{c}{d} =$ ________.

Lesson Exercise 6.41 suggests the general invert-and-multiply rule.

Division of Rational Numbers

$$\frac{a}{b} \div \frac{c}{d} = \frac{a}{b} \times \frac{d}{c} = \frac{ad}{bc} \qquad (c \neq 0)$$

Lesson Exercise 6.42

Compute $\frac{5}{6} \div \frac{3}{4}$ using the invert-and-multiply rule. Put the answer in simplest form.

The invert-and-multiply rule is a shortcut for division that uses the reciprocal of a rational number. The **reciprocal** of a rational number $\frac{a}{b}$ in which $a \neq 0$ is $\frac{b}{a}$.

Lesson Exercise 6.43

Any nonzero rational number multiplied by its reciprocal equals ________.

You are familiar with the invert-and-multiply rule for dividing rational numbers and have seen how it can be developed from examples. One can show how the invert-and-multiply rule works in any rational-number division problem. For example, consider $\frac{2}{5} \div \frac{7}{9}$.

EXAMPLE 6.7

A child wants to see why the invert-and-multiply rule works. Use the division model of fractions and the Fundamental Law to show that $\frac{2}{5} \div \frac{7}{9}$ is the same as $\frac{2}{5} \times \frac{9}{7}$.

Solution

$$\frac{2}{5} \div \frac{7}{9} = \frac{\frac{2}{5}}{\frac{7}{9}} = \frac{\frac{2}{5} \times \frac{9}{7}}{\boxed{\frac{7}{9} \times \frac{9}{7}}} = \frac{\frac{2}{5} \times \frac{9}{7}}{1} = \frac{2}{5} \times \frac{9}{7}$$

Division rewritten as a fraction — Fundamental Law of Fractions — Omitted when shortcut is used

So $\frac{2}{5} \div \frac{7}{9} = \frac{2}{5} \times \frac{9}{7}$. It's the invert-and-multiply rule! This rule is simply a shortcut for the longer procedure just shown. ■

Lesson Exercise 6.44

A child wants to see why the invert-and-multiply rule works. Use the division model of fractions and the Fundamental Law to show that $\frac{3}{5} \div \frac{2}{7}$ is the same as $\frac{3}{5} \times \frac{7}{2}$.

The categories of whole-number operations also apply to some situations involving multiplication and division of rational numbers. In Lesson Exercises 6.45 and 6.46, tell what operation and classification are illustrated.

Lesson Exercise 6.45

A computer printer produces a page in 1/2 minute. How many pages would it print in 30 minutes?

Lesson Exercise 6.46

Last year, 20 people signed up for a course called "Mathematics Is Awesome." The word got around, and this year $3\frac{1}{2}$ times as many people signed up. How many people signed up this year?

AN INVESTIGATION: MULTIPLYING ELEMENTARY FRACTIONS

***Cooperative* Lesson Exercise 6.47**

Consider the following problem. "In multiplying an elementary fraction $\frac{a}{b}$ by an elementary fraction $\frac{c}{d}$, tell what conditions would make the product (1) greater than $\frac{a}{b}$, (2) equal to $\frac{a}{b}$, and (3) less than $\frac{a}{b}$."

(a) Devise a plan and solve the problem.

(b) Use the results of this problem to analyze $\frac{a}{b} \div \frac{c}{d}$ (with $c \neq 0$) in a similar way.

Instead of computing the answers, apply your generalizations from the preceding exercise to answer the following.

Lesson Exercise 6.48

$\frac{3}{16} \times \frac{3}{5}$ is (a) greater than $\frac{3}{16}$ (b) equal to $\frac{3}{16}$ (c) less than $\frac{3}{16}$

Lesson Exercise 6.49

$9\frac{1}{2} \div \frac{3}{4}$ is (a) greater than $9\frac{1}{2}$ (b) equal to $9\frac{1}{2}$ (c) less than $9\frac{1}{2}$

ANSWERS TO SELECTED LESSON EXERCISES

6.32 $5 \times \frac{1}{8} = \frac{1}{8} + \frac{1}{8} + \frac{1}{8} + \frac{1}{8} + \frac{1}{8} = \frac{5}{8}$

6.33 $\frac{4}{5} \times 20 = 16$

6.34 $\frac{1}{2} \times \frac{3}{5}$ is $\frac{1}{2}$ of $\frac{3}{5}$. Show $\frac{3}{5}$.

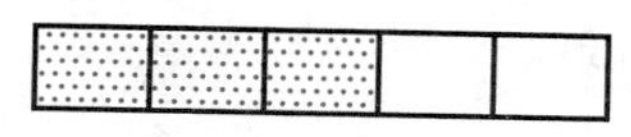

Darken $\frac{1}{2}$ of $\frac{3}{5}$.

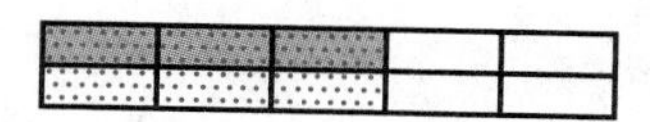

$\frac{3}{10}$ of the figure is darkened. So $\frac{1}{2} \times \frac{3}{5} = \frac{3}{10}$.

6.35 (a) $\frac{3}{8}; \frac{3}{10}$ (b) $\frac{ac}{bd}$

6.37 $\frac{4}{9} \times \frac{5}{8} = \frac{4 \times 5}{9 \times 8}$. Now that it's all one fraction, you can divide the numerator and denominator by a common factor, 4. $\left(\text{Then } \frac{4}{8} \text{ becomes } \frac{1}{2}.\right)$

$$= \frac{\overset{1}{\cancel{4}} \times 5}{9 \times \underset{2}{\cancel{8}}}$$

6.38 $3 \div \frac{1}{4}$ means how many $\frac{1}{4}$'s does it take to make 3? It takes 4 quarters to make 1.

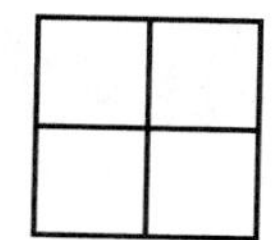
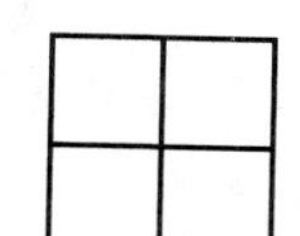
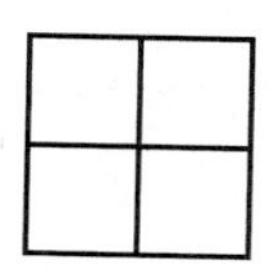

So, it takes 12 quarters to make 3. So $3 \div \frac{1}{4} = 12$.

6.39 $\frac{3}{4} \div \frac{1}{8}$ means "How many $\frac{1}{8}$'s make $\frac{3}{4}$?" The answer is 6. So $\frac{3}{4} \div \frac{1}{8} = 6$.

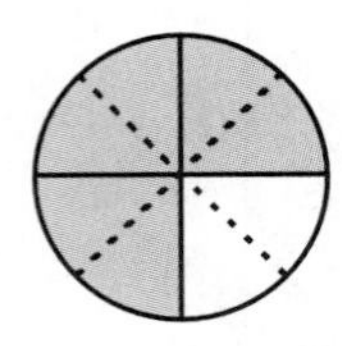

6.40 $3 = \frac{1}{4} \times 12$

6.41 (a) 8; 8 (b) 12; 12 (c) 6; 6 (d) $\frac{ad}{bc}$

6.42 $1\frac{1}{9}$

6.43 1

6.44 $\frac{3}{5} \div \frac{2}{7} = \frac{\frac{3}{5}}{\frac{2}{7}} = \frac{\frac{3}{5} \times \frac{7}{2}}{\frac{2}{7} \times \frac{7}{2}} = \frac{\frac{3}{5} \times \frac{7}{2}}{1} = \frac{3}{5} \times \frac{7}{2}$

Note: The second step in Exercise 6.44 can also be justified as multiplication by 1 in the following form:

$$\frac{\frac{7}{2}}{\frac{7}{2}}$$

6.45 Division; repeated measures

6.46 Multiplication; repeated sets

6.48 (c)

6.49 (a)

6.3 HOMEWORK EXERCISES

Basic Exercises

1. (a) What multiplication is the same as $\frac{1}{2}$ of 8?

   (b) What division is the same as $\frac{1}{2}$ of 8?

2. (a) What multiplication is the same as $\frac{1}{4}$ of 20?

   (b) What division is the same as $\frac{1}{4}$ of 20?

3. You want to budget $\frac{3}{4}$ of your \$1500 monthly paycheck for expenses. Write a fraction arithmetic equation that gives the amount you want to budget.

4. Show how to compute $3 \times \frac{2}{7}$ using repeated addition.

5. Show how to compute $4 \times \frac{1}{5}$ using repeated addition.

6. Write $3\frac{1}{2} + 3\frac{1}{2} + 3\frac{1}{2}$ as a multiplication problem.

7. A child who does not know the multiplication rule for fractions wants to compute $\frac{1}{5} \times \frac{1}{3}$. *Explain* how to compute $\frac{1}{5} \times \frac{1}{3}$ using two diagrams.

8. *Explain* how to compute $\frac{1}{4} \times \frac{2}{3}$ using two diagrams.

9. *Explain* how to compute $\frac{3}{4} \times \frac{4}{5}$ using two diagrams.

10. *Explain* how to compute $\frac{2}{3} \times \frac{5}{6}$ using two diagrams.

11. What fraction product is illustrated by the grid?

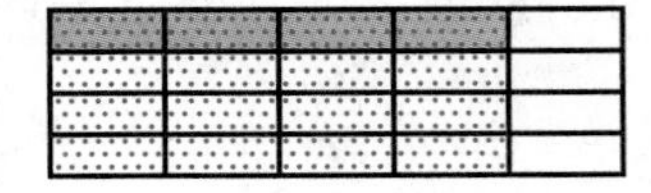

12. *Explain* why, in multiplying $\frac{7}{10} \times \frac{5}{9}$, one can simplify the computation by changing the 10 and 5 to 2 and 1, respectively.

13. *Explain* why, in multiplying $5\frac{1}{2} \times 2\frac{2}{3}$, one can simplify the computation by changing a 2 and 8 to 1 and 4, respectively.

14. A child is confused about when to cross multiply fractions and when to multiply numerator times numerator and denominator times denominator. How would you respond?

15. A child has not yet learned the rule for dividing fractions. *Explain* how to compute $2 \div \frac{1}{4}$ using a measurement picture.

16. *Explain* how to compute $3 \div \frac{1}{2}$ using a measurement picture.

17. Make up a money problem that is modeled by $5 \div \frac{1}{10}$.

***18.** Use the definition of division to rewrite $\frac{10}{21} \div \frac{2}{3} = \frac{n}{m}$ as a multiplication equation and find the answer by inspection.

19. $A \times B = 1$. A is 9 times B. What are A and B? (Guess and check.)

20. Use the division model of fractions and the Fundamental Law to show that $\frac{3}{5} \div \frac{1}{4}$ is the same as $\frac{3}{5} \times \frac{4}{1}$.

21. A child wants to see why the invert-and-multiply rule works. Use the division model of fractions and the Fundamental Law to show that $\frac{1}{2} \div \frac{3}{4}$ is the same as $\frac{1}{2} \times \frac{4}{3}$.

22. Use the division model of fractions and the Fundamental Law to show that $\frac{9}{10} \div \frac{2}{3}$ is the same as $\frac{9}{10} \times \frac{3}{2}$.

23. Compute the following and simplify.

(a) $\frac{2}{3} \times \frac{3}{4}$ (b) $2\frac{1}{3} \times \frac{2}{3}$

(c) $\frac{5}{8} \div \frac{1}{2}$ (d) $-8 \div 3\frac{1}{9}$

24. Compute the following and simplify.

(a) $\frac{7}{8} \times \frac{4}{5}$ (b) $-\frac{1}{4} \times 4\frac{1}{6}$

(c) $\frac{3}{5} \div \frac{7}{10}$ (d) $10 \div 4\frac{2}{3}$

25. Solve mentally.

(a) $\frac{1}{2}x = 10$ (b) $4 \div \frac{1}{2} = y$

26. Solve mentally.

(a) $\frac{1}{4}t = 8$ (b) $9 \div \frac{1}{3} = r$

27. Compute the following for nonzero integers x and y.

(a) $\frac{x}{5} \div \frac{x}{7}$ (b) $\frac{x^2}{y} \div \frac{x}{3}$

28. (a) $1 + \frac{1}{5} =$ ____

(b) $1 + \cfrac{1}{1 + \cfrac{1}{5}} =$ ____

(c) $1 + \cfrac{1}{1 + \cfrac{1}{1 + \cfrac{1}{5}}} =$ ____

(d) What pattern do you see in the results to parts (a)–(c)?

(e) Without computing, guess the answer to the following.

$$1 + \cfrac{1}{1 + \cfrac{1}{1 + \cfrac{1}{1 + \cfrac{1}{5}}}}$$

29. If $A \times \frac{2}{3} = B$, then what number multiplied by B equals A?

30. Suppose $\frac{9}{10}$ of all elementary teachers are women and $\frac{1}{10}$ of these women like mathematics better than any other subject. What fraction of elementary teachers are women whose favorite subject is math?

31. A recipe for 4 people calls for 1/3 cup of olive oil. If you make the same dish for 2 people, how much olive oil should you use?

32. To override a presidential veto, at least 2/3 of the Senate must vote to override. How many votes are needed to ensure passage?

33. An adult pays $231 for plane tickets for himself and a child. If the child goes for half-price, how much is the adult fare?

34. A recipe calls for $\frac{3}{4}$ cup of flour. You have only $\frac{1}{2}$ cup of flour. What fraction of the recipe can you make?

35. A recipe calls for 1/4 cup of flour per person.
(a) If you are cooking for 6 people, how much flour should you use?
(b) What operation and category does part (a) illustrate?

36. Last year a farm produced 1360 oranges. This year they produced $2\frac{1}{2}$ time as many oranges.
(a) How many oranges did they produce?
(b) What operation and category does part (a) illustrate?

37. A wall is $82\frac{1}{2}$ inches high. It is covered with $5\frac{1}{2}$-inch square tiles.
(a) How many tiles are in a vertical row from the floor to the ceiling? (Give an exact answer.)
(b) What operation and category does part (a) illustrate?

38. Norma and Irving want to split a delicious $10\frac{1}{2}$-inch submarine sandwich equally. What division category does this illustrate?

39. Four identical cartons are stacked one on top of the other. The stack is 10 feet high.
(a) If the room is 18 feet high, how many more identical cartons can be stacked on top of the pile?
(b) What operations and categories are illustrated here?

40. A baker takes 1/2 hour to decorate a cake.
(a) How many cakes can she decorate in H hours?
(b) What operation and category does part (a) illustrate?

41. Carol Burbage spent $\frac{1}{2}$ of her money at the movies. Then she spent $\frac{1}{3}$ of what was left at the store. Now she has $4 left. How much did she start with? (*Hint:* Draw a fraction-of-a-whole picture.)

42. Sid spent $6 at the movies. Then he spent 1/3 of what remained for a magazine. Now he has D dollars. What did he start out with?

43. Make up a realistic application problem for $\frac{3}{4} \times 22$.

44. Make up a realistic application problem for $20 \div \frac{3}{4}$.

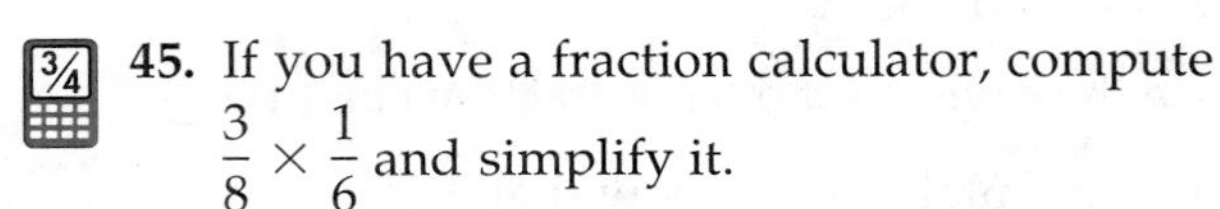

45. If you have a fraction calculator, compute $\frac{3}{8} \times \frac{1}{6}$ and simplify it.

46. If you have a fraction calculator, compute $\frac{3}{4} \div \frac{1}{2}$, simplify it, and convert it to a mixed numeral.

Extension Exercises

47. Crazy King Loopy just died. Loopy's will instructs his attorney, Ward E. Claus, to divide up his prize collection of 17 hogs as follows: $\frac{4}{9}$ of the hogs go to his eldest daughter Wacky, $\frac{1}{3}$

go to his son Harpo, and $\frac{1}{6}$ go to young Loopy II. Ward has no idea how he is going to carry out the will.

Fortunately, the court sage, Wiggy, tells Ward to borrow another hog, and then he will be able to carry out the will. Ward tries it, and it works! Ward returns the extra hog when he is done.

(a) How many hogs does each child receive?
(b) Why does Wiggy's approach seem to work?
(c) In the end, were the conditions of the will fulfilled, or did everyone receive more hog than he or she was supposed to?

48. Suppose that each time a rubber ball is dropped, it rebounds to half the height from which it fell. The ball is dropped from a height of 20 feet.
(a) How high does it bounce on the first bounce?
(b) How high does it bounce on the second bounce?
(c) How high does it bounce on the Nth bounce?

49. Consider the situation described in the preceding exercise. How far has the ball traveled when it hits the ground for the
(a) first time?
(b) second time?
(c) third time?
(d) fourth time?
(e) Estimate the total distance the ball will travel before it stops bouncing.

50. Suppose a pollution control bill is supported by $\frac{3}{5}$ of all voters who are Democrats, $\frac{2}{5}$ of all Republicans, and $\frac{1}{2}$ of all Independents. If $\frac{2}{5}$ of the population is Democratic, $\frac{3}{10}$ is Republican, and $\frac{3}{10}$ is Independent, what fraction of all voters support the bill?

51. Consider the following problem. "A particular ball bounces to $\frac{1}{4}$ of the height it reached on the preceding bounce. After being dropped and bouncing up and down, it hits the ground for the third time, having traveled $37\frac{1}{2}$ feet. What was its original height?" Devise a plan and solve the problem.

52. Consider the following problem. "Ross had a bag of oranges. He gave $\frac{1}{4}$ of them to his mother. Then he gave $\frac{1}{3}$ of what was left to his friend Rita. Next, he gave $\frac{1}{2}$ of what was left to his mathematics professor. Ross has 3 oranges left. How many did he start with?" Devise a plan and solve the problem.

53. In an all-adult apartment building, $\frac{3}{4}$ of the men are married to $\frac{1}{2}$ of the women. What fraction of the residents are married?

54. A student finishes about 2/5 of a paper in one evening. About how many evenings will it take to do the whole paper, working at the same rate?

55. An item is on sale for $\frac{1}{10}$ off the regular price. The regular price is what fraction more than the sale price?

56. Five painters are painting a house. They all work at about the same rate, and they need 7 days to finish the job. However, at the end of 3 days, 2 painters quit. How long will it take the remaining 3 painters, working at their normal rate, to finish the job?

57. Consider the following problem. "Alex has 1/4 as many tapes as Jamie. They have a total of 40 tapes. How many tapes does each have?" Show at least two ways of solving this problem.

58. Units of measurement can be treated like fractions. For example, if a man travels 10 miles per hour for 6 hours, how far does he travel?

$$10\ \frac{\text{miles}}{\text{hour}} \times 6 \text{ hours} = 10 \times 6\ \frac{\text{miles}}{\text{hour}} \times \text{hours}$$
$$= 60 \text{ miles}$$

Carry the units along throughout each of the following exercises.

(a) Swiss cheese costs 4 dollars per pound. How much would $3\frac{1}{2}$ pounds cost?

(b) An ad claims that a car travels 495 miles on a full tank of gas. If the car gets 30 miles to the gallon, how many gallons does the gas tank hold?

59. Consider the following diagram.

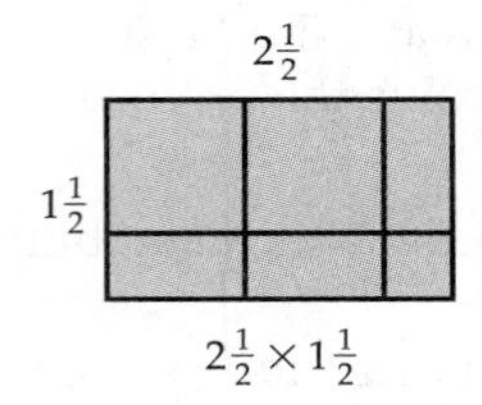

(a) Find $2\frac{1}{2} \times 1\frac{1}{2}$ from the diagram.

(b) Make a similar diagram for $3\frac{1}{2} \times 2\frac{1}{2}$ and give the answer.

(c) Repeat part (b) for $4\frac{1}{2} \times 3\frac{1}{2}$.

(d) Find a shortcut for computing $a\frac{1}{2} \times b\frac{1}{2}$, in which $a = b + 1$ for counting numbers a and b.

(e) Use your shortcut to compute $18\frac{1}{2} \times 19\frac{1}{2}$.

60. Consider the following diagram.

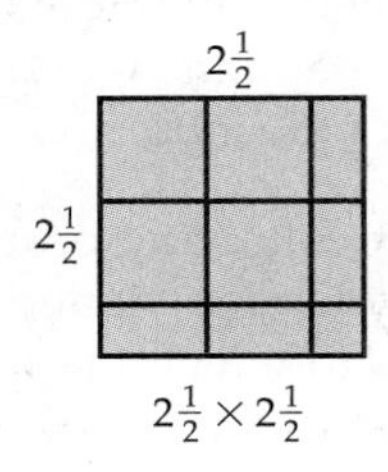

(a) Find $2\frac{1}{2} \times 2\frac{1}{2}$ from the diagram.

(b) Make a similar diagram for $3\frac{1}{2} \times 3\frac{1}{2}$ and give the result.

(c) Repeat part (b) for $4\frac{1}{2} \times 4\frac{1}{2}$.

(d) Find a shortcut for computing $a\frac{1}{2} \times a\frac{1}{2}$ for a counting number a.

(e) Use your shortcut to compute $29\frac{1}{2} \times 29\frac{1}{2}$.

(f) Show algebraically why the shortcut works.

61. If x and y are nonzero rational numbers, with $x > y$, under what conditions is each of the following true?

(a) $\frac{1}{x} > \frac{1}{y}$ (b) $\frac{1}{x} = \frac{1}{y}$ (c) $\frac{1}{x} < \frac{1}{y}$

62. (a) If m and n are counting numbers and $m > n$, then which is larger, $\left(\frac{1}{2}\right)^n$ or $\left(\frac{1}{2}\right)^m$?

(b) If m and n are counting numbers and $m > n$, for what positive rational numbers a is $a^m < a^n$?

63. (a) Write half of 2^3 as a power of 2.

(b) Write half of 2^4 as a power of 2.

(c) Write half of 2^N as a power of 2.

(d) Make up an analogous problem involving 3^N.

64. Sometimes the difference of two fractions equals their product. For example,

$\frac{3}{7} - \frac{3}{10} = \frac{3}{7} \times \frac{3}{10}$ and $\frac{2}{3} - \frac{2}{5} = \frac{2}{3} \times \frac{2}{5}$.

(a) What is the relationship between the two fractions in each example?

(b) Make up two more examples that work.

(c) Show algebraically that the product and difference of two fractions of this type will always be equal. (*Hint:* It takes only two variables to write all the numerators and denominators.)

65. Until the seventeenth century, many people divided fractions *after* finding a common denominator.

(a) Show that this method works for $\frac{1}{3} \div \frac{3}{4}$.

(b) Do you think this method is easier than the standard rule?

66. Assume that b, c, and d are not 0. The statement $\frac{a}{b} \div \frac{c}{d} = \frac{a \div c}{b \div d}$ is

(a) always true
(b) sometimes true
(c) never true

67. A typist completes P pages per day. How many days would it take the typist to complete a paper that is L pages long?

68. Within each part, the question mark represents the same counting number. Find the missing numbers. All fractions are in simplest form. (Guess and check.)

(a) $2\frac{?}{3} \times 4\frac{?}{5} = 11\frac{11}{15}$

(b) $\frac{??}{10} \div 3 = \frac{30}{??}$

Special Exercises

69. How thick is a page in this book?

70. Fill in the chart.

Days	1	$\frac{1}{2}$	$\frac{1}{4}$	$\frac{3}{4}$	$\frac{1}{8}$	$\frac{3}{8}$	$\frac{5}{8}$	$\frac{7}{8}$
Hours	24							

(b) Compute the following arithmetic problems, using hours and then using days. Check one answer against the other.

Days	$\frac{1}{4} + \frac{1}{8} =$ ___	$\frac{7}{8} - \frac{1}{2} =$ ___
Hours	6 + ___ = ___	___ − ___ = ___

6.4 PROPERTIES OF RATIONAL NUMBERS

The properties of whole-number and integer operations discussed in Chapters 3 and 5 hold for rational-number operations. However, the rational-number system possesses some additional properties.

INTEGER PROPERTIES RETAINED?

Do rational-number operations retain the same commutative, associative, identity, inverse, closure, and distributive properties as integer operations?

Lesson Exercise 6.50

Using examples, decide if you think rational-number

(a) addition is commutative.
(b) multiplication is associative.
(c) multiplication is distributive over addition.

The preceding exercise concerns some possible properties of rational numbers. The rational-number operations do retain all the properties of integer operations!

Some Properties of Rational-Number Operations

- Addition, subtraction, and multiplication of rational numbers are closed.
- Addition and multiplication of rational numbers are commutative.
- Addition and multiplication of rational numbers are associative.
- The unique additive identity for rationals is 0, and the unique multiplicative identity for rationals is 1.
- All rational numbers have a unique additive inverse that is rational. For any rational number $\frac{a}{b}$, there is a unique rational number $-\frac{a}{b}$ such that $\frac{a}{b} + -\frac{a}{b} = -\frac{a}{b} + \frac{a}{b} = 0$.
- Multiplication is distributive over addition, and multiplication is distributive over subtraction in the rational number system.

The following exercises make use of these properties.

Lesson Exercise 6.51

(a) What is an easy way to multiply $(9 \times 28) \times \frac{1}{4}$?

(b) What property or properties does this illustrate?

Lesson Exercise 6.52

What is the additive inverse of $\frac{3}{4}$?

The properties also justify some procedures used in algebra.

Lesson Exercise 6.53

If x is rational, what property guarantees that $\left(x + \frac{1}{2}\right) + 2 = x + \left(\frac{1}{2} + 2\right)$?

Lesson Exercise 6.54

Use the distributive property of multiplication over subtraction for a rational number n to simplify $\frac{7}{2}n - \frac{5}{2}n =$ ________ $=$ ________.

What about rational-number subtraction and division? Are they commutative or associative?

Lesson Exercise 6.55

Explain how you know, after studying whole numbers and integers, that rational-number subtraction and division are neither commutative nor associative.

MULTIPLICATIVE INVERSES

Add any rational number to its additive inverse, and the result is the additive identity (0). Try to apply the same idea to a multiplicative inverse in the next lesson exercise.

Lesson Exercise 6.56

Any rational number multiplied by its multiplicative inverse should result in what number?

Do all rational numbers have multiplicative inverses that are rational numbers?

Lesson Exercise 6.57

What is the multiplicative inverse of each of the following?

(a) -4 (b) $\frac{2}{3}$

Lesson Exercise 6.58

(a) Do you think all rational numbers have multiplicative inverses that are rational numbers? Try some examples.

(b) Only one rational number does not have a multiplicative inverse. What is it?

Lesson Exercises 6.57 and 6.58 should convince you that every nonzero rational number has a unique rational multiplicative inverse.

Multiplicative Inverses for Nonzero Rational Numbers

Every rational number except 0 has a unique multiplicative inverse that is rational. That is, for each nonzero rational number $\frac{a}{b}$ there is a unique rational number $\frac{b}{a}$ such that $\frac{a}{b} \times \frac{b}{a} = \frac{b}{a} \times \frac{a}{b} = 1$.

DENSENESS

Consider another property of sets of numbers called denseness. Suppose you pick two numbers from a set of numbers. Can you always find another number (in the set) that is between them?

Lesson Exercise 6.59

(a) If you pick any two whole numbers, is there always a whole number between them?

(b) If you pick any two integers, is there always an integer between them?

Do you think that between any two rational numbers there is another rational number? Example 6.8 shows two ways of finding a rational number between two given rational numbers.

EXAMPLE 6.8

Find a rational number between $\frac{1}{5}$ and $\frac{1}{6}$.

Solution

The Decimal-Conversion Method Although decimals haven't yet been discussed in this book, you already know how to convert fractions to decimals using the division model.

Change both fractions to decimals, using a calculator. Then find a decimal between them and change it into a fraction.

On my calculator, $\frac{1}{5} = \boxed{0.2}$ and $\frac{1}{6} = \boxed{0.1666667}$. The decimal 0.19 is between $\frac{1}{5}$ and $\frac{1}{6}$. The decimal $0.19 = \frac{19}{100}$. So $\frac{19}{100}$ is a rational fraction between $\frac{1}{5}$ and $\frac{1}{6}$.

The Common-Denominator Method Write both fractions with a common denominator: $\frac{1}{5} = \frac{6}{30}$, and $\frac{1}{6} = \frac{5}{30}$. There is no whole number between the numerators 5 and 6, so use a larger common denominator: $\frac{1}{5} = \frac{6}{30} = \frac{12}{60}$ and $\frac{1}{6} = \frac{5}{30} = \frac{10}{60}$. Now it is clear that $\frac{11}{60}$ is between $\frac{10}{60}$ and $\frac{12}{60}$, so it is between $\frac{1}{5}$ and $\frac{1}{6}$.

■

Lesson Exercise 6.60

Find a rational number between $\frac{50}{51}$ and $\frac{51}{52}$.

This "betweenness" property that works for the set of rational numbers but not whole numbers or integers is called the denseness property.

The Denseness Property of Rational Numbers

Between any two rational numbers there is another rational number.

There are infinitely many rational numbers *between* any two rational numbers. It's amazing, isn't it? If you pick any two rational numbers, you can keep finding numbers between them—forever. For example, take the numbers used in Example 6.8. According to this example,

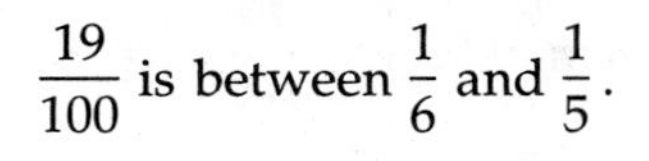

$\frac{19}{100}$ is between $\frac{1}{6}$ and $\frac{1}{5}$.

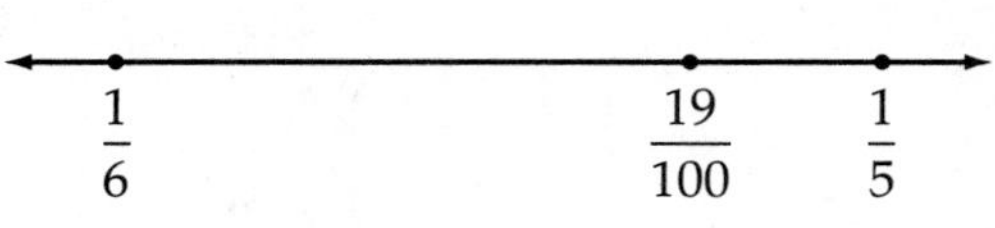

Now $\frac{18}{100}$ is between $\frac{1}{6}$ and $\frac{19}{100}$.

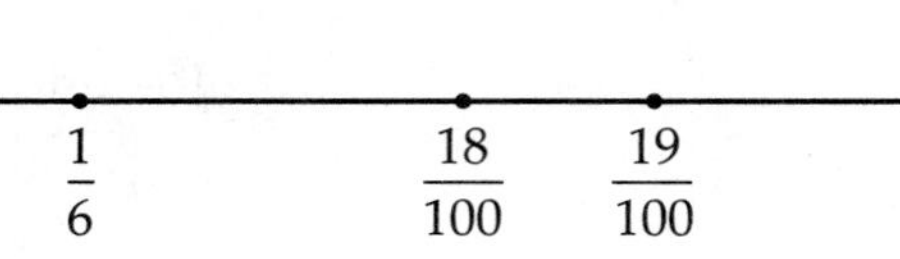

Now $\frac{17}{100}$ is between $\frac{1}{6}$ and $\frac{18}{100}$.

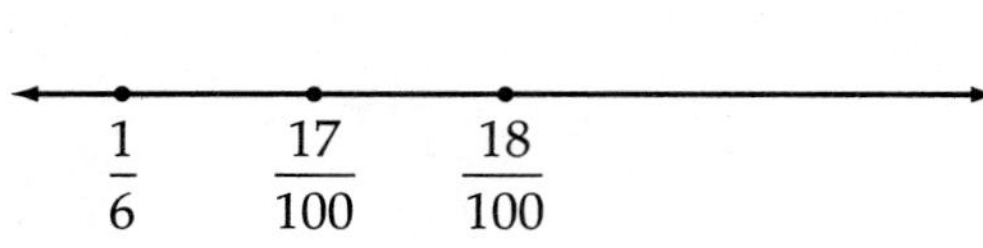

And so on.

ANSWERS TO SELECTED LESSON EXERCISES

6.51 (a) $\left(28 \times \frac{1}{4}\right) \times 9 = 7 \times 9 = 63$

(b) Associative, or commutative and associative for multiplication

6.52 $\frac{-3}{4}$

6.53 Associative property of addition

6.54 $\left(\frac{7}{2} - \frac{5}{2}\right)n$; n

6.55 The counterexamples for whole-number and integer subtraction and division would also be counterexamples for rational numbers. If a rule does not apply to *all* whole numbers, then it cannot apply to *all* rational numbers, since every whole number is also a rational number.

6.56 1

6.57 (a) $-\frac{1}{4}$ (b) $1\frac{1}{2}$

6.58 (b) 0

6.59 (a) No (b) No

6.60 Decimals such as 0.9804 and 0.9805 are between $\frac{50}{51}$ and $\frac{51}{52}$.

6.4 HOMEWORK EXERCISES

Basic Exercises

1. (a) What rational-number operations are commutative?

(b) What rational-number operations are associative?

2. For any rational number x, what property guarantees that

$$\frac{5}{2}x + (2x + 7) = \left(\frac{5}{2}x + 2x\right) + 7?$$

3. According to the associative property of multiplication for a rational number x, $\frac{1}{2} \cdot (4 \cdot x) =$ ________.

4. What is an easy way to multiply $\frac{1}{7} \times 33 \times 14$?

5. What is an easy way to multiply $\frac{1}{3} \times \left(\frac{1}{5} \times 18\right) \times 20$?

6. $\frac{2}{3} + 0 = 0 + \frac{2}{3} = \frac{2}{3}$

$-3\frac{5}{8} + 0 = 0 + -3\frac{5}{8} = -3\frac{5}{8}$

These examples illustrate that _____ is the ____________ for ____________.

7. What is the additive inverse of $-2\frac{1}{2}$?

*8. One reason that we need rational numbers is to provide answers to whole-number and integer division problems.
   (a) Are the rational numbers closed under division? (That is, if x and y are rational numbers, is $x \div y$ a rational number?)
   (b) Would the set of rational numbers, excluding 0, be closed under division?

9. *Explain* how you would know, after studying whole numbers and integers, that rational-number subtraction is not commutative.

10. Give a counterexample showing that rational-number division is not associative.

11. Use the distributive property of multiplication over addition for a rational number n to simplify $-\frac{4}{3}n + \frac{2}{3}n =$ ________ $=$ _____.

12. Use the distributive property to compute $4\frac{1}{2} \times 2\frac{3}{4}$ by rewriting it as $\left(4 + \frac{1}{2}\right)\left(2 + \frac{3}{4}\right)$.

13. What is the multiplicative inverse of $-2\frac{1}{2}$?

14. Any nonzero number multiplied by its multiplicative inverse equals ________.

15. (a) What property do rational numbers have that whole numbers and integers do not have?
    (b) What property do all nonzero rational numbers have that whole numbers and integers do not have?

*16. Find a rational number between $-\frac{1}{9}$ and $-\frac{1}{10}$.

17. True or false? Every rational number is an integer.

18. True or false? Some whole numbers are not rational numbers.

19. True or false? The next largest integer after 6 is 7.

20. $\left(1 + \frac{1}{2}\right)\left(1 + \frac{1}{1}\right) =$ ________

$\left(1 + \frac{1}{3}\right)\left(1 + \frac{1}{2}\right)\left(1 + \frac{1}{1}\right) =$ ________

$\left(1 + \frac{1}{4}\right)\left(1 + \frac{1}{3}\right)\left(1 + \frac{1}{2}\right)\left(1 + \frac{1}{1}\right) =$ ________

   (a) Fill in the blanks.
   (b) What will the next equation be if the pattern continues? Is the equation true?

21. True or false? If a, b, c, and d are nonzero integers, and $a < b$ and $c = d$, then $\frac{a}{c} < \frac{b}{d}$.

Extension Exercises

22. $\frac{1}{3} + \frac{1}{2 \cdot 3} =$ ________

$\frac{1}{4} + \frac{1}{3 \cdot 4} =$ ________

$\frac{1}{5} + \frac{1}{4 \cdot 5} =$ ________

   (a) Fill in the blanks.
   (b) If the pattern continues, what will the next equation be? Is the equation true?
   (c) Write the general formula suggested by parts (a) and (b):

____ + ____ = ____

   (d) Writing a general formula based upon the

examples in parts (a) and (b) requires ____________ reasoning.

(e) Prove that your formula in part (c) is correct. (*Hint:* Find the common denominator on the left side of the equation, and add the two fractions.)

23. Many students are not sure why one can find $\frac{1}{4}$ of 40 by computing $40 \div 4$. Show why $\frac{1}{4}$ of 40 is the same as $40 \div 4$ by changing $\frac{1}{4}$ of 40 first to a multiplication problem and from there to a division problem.

24. Consider the following problem. "You have a piece of construction paper that is 24 in. by 28 in. What is the maximum number of $8\frac{1}{2}$-by-11-in. pieces that you can cut out of it?" Devise a plan and solve the problem.

25. Draw a Venn diagram that includes the following sets: rational numbers, whole numbers, and integers.

26. Consider the sum

$$S = \frac{1}{2} + \frac{1}{2^2} + \frac{1}{2^3} + \frac{1}{2^4} + \cdots + \frac{1}{2^{20}}$$

(a) Write $2S$ as a sum of fractions.

(b) Subtract the given expression for S from your expression for $2S$, and simplify the result. What do you obtain?

(c) How could you write the sum

$$\frac{1}{2} + \frac{1}{2^2} + \frac{1}{2^3} + \frac{1}{2^4} + \cdots + \frac{1}{2^N}$$

in a simpler way?

27. (a) Following steps analogous to those in the last exercise, find a simpler way to write

$$S = \frac{1}{3} + \frac{1}{3^2} + \frac{1}{3^3} + \frac{1}{3^4} + \cdots + \frac{1}{3^{40}}$$

(b) How could you write the sum

$$\frac{1}{3} + \frac{1}{3^2} + \frac{1}{3^3} + \frac{1}{3^4} + \cdots + \frac{1}{3^N}$$

in a simpler way?

28. Show that multiplication of rational numbers $\frac{x}{w}$ and $\frac{y}{z}$ is commutative.

29. Show that multiplication of rational numbers $\frac{u}{v}$, $\frac{w}{x}$, and $\frac{y}{z}$ is associative.

30. Tell whether each of the following is true or false. If the equation is true, prove it. If it is false, give a counterexample. Assume that $\frac{w}{z}$, $\frac{x}{z}$, and $\frac{y}{z}$ are rational numbers.

(a) $\left(\frac{w}{z} - \frac{x}{z}\right) + \frac{y}{z} = \frac{w}{z} - \left(\frac{x}{z} + \frac{y}{z}\right)$

(b) $\left(\frac{w}{z} + \frac{x}{z}\right) \div \frac{y}{z} = \left(\frac{w}{z} \div \frac{y}{z}\right) + \left(\frac{x}{z} \div \frac{y}{z}\right)$ if $y \neq 0$

31. Tell whether each of the following is true or false. If it is false, give a counterexample. Assume x and y are nonzero rational numbers.

(a) $\frac{2}{x} + \frac{2}{y} = \frac{2}{x + y}$ (b) $\frac{x + y}{y} = x$

6.5 ELEMENTARY FRACTIONS: ESTIMATION, MENTAL COMPUTATION, AND ERROR PATTERNS

In the rare instance that you should make an error in arithmetic with elementary fractions, estimation might help you detect your error. For example, if you computed that $3\frac{1}{4} + 2\frac{2}{3} = \frac{11}{12}$ (the error pattern in Lesson Exercise 6.70) or

$8\frac{1}{2} \div 2\frac{2}{3} = \frac{16}{51}$ (the error pattern in Lesson Exercise 6.68), estimating would tell you that the answer could not be right.

THE ROUNDING STRATEGY

The rounding strategy is often the best for estimating in addition, subtraction, and multiplication problems involving elementary fractions. Round the numbers to create a problem that you can compute mentally. In most cases, fractions are rounded to the nearest integer. In addition or subtraction, round fractions between 0 and 1 to 0, $\frac{1}{2}$, or 1.

Use rounding to estimate in the following exercise.

Lesson Exercise 6.61

A factory assembles about $8\frac{1}{4}$ cars per day. How can you estimate the number of cars the factory will assemble in $3\frac{1}{2}$ days?

THE COMPATIBLE-NUMBERS STRATEGY

Are you a flexible person? If so, you'll like using the compatible-numbers strategy with elementary fractions. Rounding to the *nearest* whole number is not always the best approach in fraction multiplication or division.

Rounding may not work in multiplication when at least one factor is close to or less than $\frac{1}{2}$. This factor can be rounded to the closest fraction that has 1 in the numerator and a counting number in the denominator, such as $\frac{1}{2}$, $\frac{1}{3}$, or $\frac{1}{4}$. Then change the other factor to a compatible number.

EXAMPLE 6.9

A new electric eyeglass defogger regularly sells for \$820, but Nutty Mike's is offering it for 2/5 off the regular price. How can you estimate the amount of the discount?

Solution

We need to estimate $\frac{2}{5}$ of \$820. We cannot round $\frac{2}{5}$ to the nearest whole number, 0. Doing so would make the estimate 0 for $\frac{2}{5}$ of any dollar amount! Instead, round $\frac{2}{5}$ to $\frac{1}{2}$ or $\frac{1}{3}$, which is easier to use than $\frac{2}{5}$, and round \$820 to a compatible number.

$$\frac{2}{5} \times \$820 \approx \frac{1}{2} \times \$800 = \$400$$

So $\frac{2}{5}$ of \$820 is about \$400. Or

$$\frac{2}{5} \times \$820 \approx \frac{1}{3} \times \$900 = \$300$$

So $\frac{2}{5}$ of \$820 is about \$300. ■

Lesson Exercise 6.62

Nutty Mike's is offering 3/5 off the price of a \$604 electric bookmark. How can you estimate the discount?

The following example illustrates the use of the compatible-numbers strategy for division.

EXAMPLE 6.10

How can you estimate $83\frac{3}{4} \div 26\frac{1}{4}$?

Solution

Use the compatible-numbers strategy. Rounding to the nearest whole numbers (84 ÷ 26) or to the nearest ten (80 ÷ 30) would not be the best choice. Rounding to compatible numbers such as 80 ÷ 20, 90 ÷ 30, or 75 ÷ 25 is preferable.

$$83\frac{3}{4} \div 26\frac{1}{4} \approx 80 \div 20 = 4 \quad \text{or} \quad 83\frac{3}{4} \div 26\frac{1}{4} \approx 90 \div 30 = 3$$

Both 3 and 4 are good estimates. ■

Lesson Exercise 6.63

You want to lay panels $17\frac{1}{4}$ in. wide across a wall that is $384\frac{1}{2}$ in. wide. How can you estimate how many panels you will need?

MENTALLY MULTIPLYING A MIXED NUMBER BY A WHOLE NUMBER

Do you know a way to do the mental computation in the following exercise?

Lesson Exercise 6.64

Celia bicycles $5\frac{1}{2}$ miles per hour for 8 hours. How can you use a shortcut to compute mentally the exact distance she travels?

Computations like the one in the preceding exercise can be done mentally. Rather than changing $8 \times 5\frac{1}{2}$ to $\frac{8}{1} \times \frac{11}{2}$, you can multiply $8 \times 5 = 40$ and add $8 \times \frac{1}{2} = 4$. So $8 \times 5\frac{1}{2} = 40 + 4 = 44$. This shortcut works because $8 \times 5\frac{1}{2} = 8 \times \left(5 + \frac{1}{2}\right) = (8 \times 5) + \left(8 \times \frac{1}{2}\right) = 40 + 4 = 44$.

Lesson Exercise 6.65

What property guarantees that $8 \times \left(5 + \frac{1}{2}\right) = (8 \times 5) + \left(8 \times \frac{1}{2}\right)$?

Lesson Exercise 6.66

On a car trip, the Durst family averages 50 miles per hour for $6\frac{1}{2}$ hours. How can you mentally compute the total distance traveled?

COMMON ERROR PATTERNS

Your future students will amaze you with their own ingenious procedures for fraction arithmetic (which often do not work). Repetition of incorrect procedures results in error patterns in children's work.

In Lesson Exercises 6.67–6.70, (a) complete the last two examples, repeating the error pattern in the completed examples; (b) write a description of the error pattern; and (c) state how one of the errors might be detected using estimation.

Lesson Exercise 6.67

$\frac{1}{3} + \frac{2}{5} = \underline{\frac{3}{8}}$ $\quad \frac{2}{5} + \frac{1}{5} = \underline{\frac{3}{10}}$ $\quad \frac{2}{3} + \frac{4}{7} = \underline{\frac{6}{10}}$

$\frac{1}{4} + \frac{3}{4} = \_\_\_\_$ $\quad \frac{1}{2} + \frac{2}{3} = \_\_\_\_$

Lesson Exercise 6.68

$\frac{2}{3} \div \frac{1}{4} = \underline{\frac{3}{8}}$ $\quad 8 \div \frac{1}{2} = \underline{\frac{1}{16}}$ $\quad 8\frac{1}{2} \div 2\frac{2}{3} = \underline{\frac{16}{51}}$

$\frac{4}{7} \div \frac{2}{3} = \_\_\_\_$ $\quad \frac{7}{8} \div 2 = \_\_\_\_$

Lesson Exercise 6.69

$\frac{8}{4} = \underline{2}$ $\quad \frac{3}{6} = \underline{2}$ $\quad \frac{3}{9} = \underline{3}$ $\quad \frac{12}{3} = \_\_\_\_$

$\frac{1}{4} = \_\_\_\_$

Lesson Exercise 6.70

$$\begin{array}{r} 3\frac{1}{4} = \frac{3}{12} \\ + 2\frac{2}{3} = \frac{8}{12} \\ \hline \frac{11}{12} \end{array} \qquad \begin{array}{r} 5\frac{1}{4} \\ + 2\frac{1}{2} \\ \hline \end{array} \qquad \begin{array}{r} 6\frac{1}{6} \\ + 4\frac{1}{4} \\ \hline \end{array}$$

ANSWERS TO SELECTED LESSON EXERCISES

6.61 $8\frac{1}{4} \times 3\frac{1}{2} \approx 8 \times 4 = 32$ cars

6.62 $\frac{3}{5}$ of \$604 $\approx \frac{1}{2} \times \$600 = \$300$

6.63 $384\frac{1}{2} \div 17\frac{1}{4} \approx 400 \div 20 = 20$ panels

6.65 Distributive property of multiplication over addition

6.66 $50 \times 6\frac{1}{2} = (50 \times 6) + \left(50 \times \frac{1}{2}\right) =$
$300 + 25 = 325$ miles

6.67 (a) $\frac{4}{8}; \frac{3}{5}$
(b) The child is adding the denominators.
(c) $\frac{2}{5}$ is more than $\frac{3}{10}$. How can $\frac{2}{5} + \frac{1}{5} = \frac{3}{10}$?

6.68 (a) $\frac{14}{12}$ or $1\frac{1}{6}$; $\frac{16}{7}$ or $2\frac{2}{7}$

(b) The child is inverting the first fraction and multiplying it by the second.

(c) $8\frac{1}{2} \div 2\frac{2}{3} \approx 9 \div 3 = 3$. How can the answer be $\frac{16}{51}$?

6.69 (a) 4; 4

(b) The child always divides the smaller number into the larger.

(c) Fractions with denominators that are larger than their numerators cannot be greater than 1.

6.70 (a) $\frac{3}{4}$; $\frac{5}{12}$

(b) The child is adding only the fractional parts of the mixed numbers.

(c) $3\frac{1}{4} + 2\frac{2}{3} \approx 3 + 3 = 6$. How can the answer be $\frac{11}{12}$?

6.5 HOMEWORK EXERCISES

Basic Exercises

1. (a) How can you estimate $\frac{5}{6} + \frac{10}{11}$?

(b) Is your estimate too high or too low?

2. Estimate $\frac{11}{12} + \frac{1}{6} + \frac{7}{12}$ by rounding each number to either 0, $\frac{1}{2}$, or 1.

3. Suppose you nail together two boards that are $\frac{7}{8}$ in. thick and $\frac{1}{16}$ in. thick. Estimate the total thickness by rounding each measurement to either 0, $\frac{1}{2}$, or 1.

4. The Westwood Dribblers have a trip of $836\frac{1}{10}$ miles to their away basketball game. So far, they have traveled $381\frac{7}{10}$ miles. How can you estimate the distance they have left to travel?

5. A recipe calls for $5\frac{3}{4}$ cups of sugar to make 1 lb of belly busters. How can you estimate how much sugar is needed to make $8\frac{1}{4}$ lb of belly busters?

***6.** $26\frac{7}{3600} \times 32\frac{9}{60}$ is about

(a) 7000 (b) 100 (c) 60 (d) 900

7. A, B, C, D, E, and F are counting numbers. You estimate $A\frac{B}{C} - D\frac{E}{F}$ by rounding $A\frac{B}{C}$ up and $D\frac{E}{F}$ down. Is your estimate too high or too low, or is it impossible to tell?

8. A \$327 stereo is selling for 2/5 off during a sale. How can you estimate the sale price?

9. An author takes $14\frac{3}{4}$ hours to write a mathematics lesson. How can you estimate how many lessons she can write in a grueling $62\frac{1}{2}$-hour work week?

10. $286\frac{3}{1900} \div 6\frac{1}{4}$ is about

(a) 280 (b) 1800 (c) 30 (d) 50

11. How can you mentally compute

(a) $8 \cdot \frac{1}{4}$? (b) $\frac{1}{3} \cdot 60$?

12. How can you mentally compute

(a) $\frac{1}{10} \cdot 80$? (b) $45 \cdot \frac{1}{9}$?

13. Suppose the fifth grade has 126 children, and 41 of them bring lunch to school. Give a simple fraction that approximates the fraction of children who bring lunch.

14. Order the following products from smallest to largest, using estimation and number sense.

(a) $\frac{4}{5} \times 17\frac{1}{8}$ (b) $\frac{4}{5} \times 23\frac{2}{5}$

(c) $\frac{1}{2} \times 16\frac{11}{12}$ (d) $\frac{3}{4} \times 16\frac{11}{12}$

15. The product of A, B, and C is approximately

(a) $\frac{1}{10}$ (b) $1\frac{1}{4}$ (c) 3 (d) 4 (e) 5

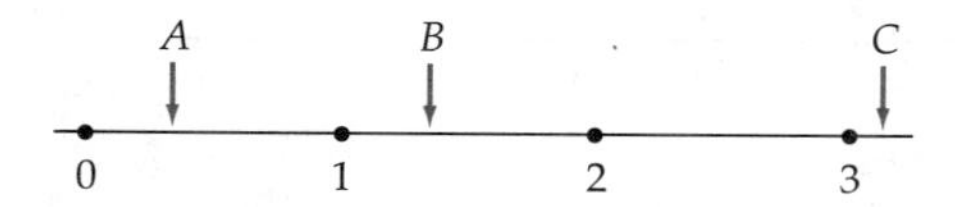

16. What property guarantees that

$$8 \times \left(3 + \frac{1}{4}\right) = (8 \times 3) + \left(8 \times \frac{1}{4}\right)?$$

17. Fatco stock sells for $8\frac{1}{2}$ dollars per share.

Mentally compute the exact cost of 40 shares. Show how you did it.

18. A job pays $40 per hour for $3\frac{3}{4}$ hours.

Mentally compute the exact total pay. Show how you did it.

19. A clothing manufacturer needs 1/2 yd of material to make a T-shirt. How many T-shirts can she make with 50 yd of material?

(a) *Explain* how to compute the answer mentally. (*Hint:* How many $\frac{1}{2}$'s make 50?)

(b) What operation and category does this illustrate?

20. A baker takes 1/4 hour to create a flower decoration for a cake.

(a) How many cakes can he decorate in 32 hours? *Explain* how to compute the answer mentally.

(b) What operation and category does this illustrate?

21. $872\frac{7}{16} \div \frac{3}{8}$ is approximately

(a) $\frac{1}{1{,}000{,}000}$ (b) 40

(c) 320 (d) 2200

22.

$$\frac{\square}{\square} \div \frac{\square}{\square}$$

Place the numbers 2, 3, 7, and 9 in the four boxes to make

(a) the largest possible quotient.

(b) the smallest possible quotient.

23. For each computation, tell which computation method (mental computation, paper and pencil, or calculator) you would use.

(a) $24 \times \frac{1}{8}$ (b) $\frac{2}{7} + \frac{4}{5}$ (c) $\frac{1}{8} \times 3752$

In Exercises 24–29, (a) complete the last two examples, repeating the error pattern in the completed examples, and (b) write a description of the error pattern.

24. $\frac{13}{36} = \underline{\frac{1}{6}}$ $\frac{16}{64} = \underline{\frac{1}{4}}$ $\frac{15}{25} = \underline{\frac{1}{2}}$

$\frac{21}{42} = \_\_\_\_$ $\frac{17}{27} = \_\_\_\_$

25.

$$\begin{array}{r} 4\frac{1}{5} = 3\frac{11}{5} \\ -\,2\frac{2}{5} = 2\frac{2}{5} \\ \hline 1\frac{9}{5} = \end{array} \qquad \begin{array}{r} 3\frac{1}{3} \\ -\,1\frac{2}{3} \\ \hline 2\frac{4}{5} \end{array} \qquad \begin{array}{r} 8\frac{2}{5} \\ -\,3\frac{3}{5} \\ \hline \end{array}$$

26.

$$\begin{array}{r} 14\frac{3}{5} \\ -\,8\frac{1}{2} \\ \hline 6\frac{2}{3} \end{array} \qquad \begin{array}{r} 12\frac{4}{5} \\ -\,6\frac{1}{3} \\ \hline \end{array} \qquad \begin{array}{r} 8\frac{1}{12} \\ -\,5\frac{1}{2} \\ \hline \end{array}$$

27. $\frac{1}{2} + \frac{3}{4} = \frac{2+3}{4+4} = \frac{5}{8}$ $\frac{2}{3} + \frac{1}{6} = \_\_\_\_$

$\frac{1}{5} + \frac{9}{10} = \_\_\_\_$

28. $\dfrac{3 + \overset{2}{\cancel{4}}}{\underset{1}{\cancel{2}}} = \dfrac{5}{1} = 5 \qquad \dfrac{6 + 5}{3} =$ ________

$\dfrac{7 + 8}{4} =$ ________

29. $\dfrac{4 + \cancel{3}}{2 + \cancel{3}} = \dfrac{4}{2} = 2 \qquad \dfrac{5 + 2}{5 + 3} =$ ________

$\dfrac{y + 3}{y + 1} =$ ________.

30. Why might the error pattern in Exercise 26 occur?

31. Why might the error pattern in Exercise 25 occur?

32. Describe two errors children might make in computing $\frac{3}{5} \times \frac{1}{6}$.

33. Fill in the next two fractions that will continue the same pattern.

(a) $\frac{1}{6}, \frac{1}{12}, \frac{1}{24}$, _____, _____

(b) Write a rule for the Nth term.

34. Fill in the blanks, following the rule in the completed examples.

(a)

9	→	6
12	→	8
15	→	___
18	→	___
___	→	20
N	→	___

(b)

4	→	14
8	→	28
10	→	___
5	→	___
___	→	70
N	→	___

Extension Exercises

35. Using mental computation, one can see that $20\frac{1}{2} \times 2\frac{1}{2}$ is

(a) less than 44
(b) between 44 and 50
(c) more than 50

36. People mentally multiply whole numbers by 25 by changing a problem such as 25×44 into $100 \times \frac{44}{4}$ or 100×11.

(a) Show why $25 \times 44 = 100 \times \frac{44}{4}$.

(b) How can you use this same approach to compute 25×84?

37. One can avoid regrouping in some fraction subtraction problems by using equal addition. For example, $3\frac{1}{5} - 1\frac{2}{5} = 3\frac{4}{5} - 2 = 1\frac{4}{5}$. Show how to use this method to compute

(a) $8\frac{1}{3} - 3\frac{2}{3}$. (b) $5\frac{1}{4} - 2\frac{3}{4}$.

38. One can solve subtraction problems that would involve regrouping by changing them to addition. For example, $3\frac{1}{5} - 1\frac{2}{5} = x$ is equivalent to $1\frac{2}{5} + x = 3\frac{1}{5}$. What would you add to $1\frac{2}{5}$ to obtain $3\frac{1}{5}$? $1\frac{2}{5} + \frac{3}{5} + 1 + \frac{1}{5} = 3\frac{1}{5}$. So $x = \frac{3}{5} + 1 + \frac{1}{5} = 1\frac{4}{5}$. Show how to use this method to compute the following.

(a) $8\frac{1}{3} - 3\frac{2}{3}$ (b) $5\frac{1}{4} - 2\frac{3}{4}$

39. Does $1 = 4$? Does $10 = 15$? Find the incorrect step in each of the following.

(a) $1 = \dfrac{2}{1 + 1} = \dfrac{2}{1} + \dfrac{2}{1} = 4$. So $1 = 4$.

(b) $10 = \dfrac{20}{2} = \dfrac{10 + 10}{2} = \dfrac{10}{2} + 10 = 15$.
So $10 = 15$.

40. Consider the following problem. "The Hershey family took a trip. They started out with a full tank of gas. When they were $\frac{2}{3}$ of the way, the gas tank was about $\frac{1}{4}$ full. Can they complete the trip without stopping for gas?"

(a) Select a problem-solving strategy to use.
(b) Solve the problem.

41. $\frac{1}{2} + \frac{1}{3} = \frac{5}{6}$ $\frac{1}{4} + \frac{1}{5} = \frac{9}{20}$
   (a) What is a shortcut for adding the fractions shown?
   (b) Make up two more examples that work the same way.
   (c) Write an algebraic equation, with two variables, that describes the general pattern in each equation.
   (d) Writing a general equation based upon examples requires ________ reasoning.
   (e) Show algebraically that your equation in part (c) is true.

42. A common error pattern in addition of fractions is to add the numerators and the denominators, as in $\frac{1}{2} + \frac{2}{3} = \frac{3}{5}$. In fact, if A, B, C, and D are counting numbers, then $\frac{A}{B} + \frac{C}{D}$ $\underset{(< \text{ or } >)}{\underline{\qquad\qquad}}$ $\frac{(A + C)}{(B + D)}$.

43. Two positive fractions have a sum of 2 and a product of $\frac{7}{16}$. What are they? (Guess and check.)

SUMMARY

"The fraction 2/3 becomes necessary as the only solution to the problem 2 ÷ 3 . . . the need to measure more precisely than to the nearest inch gives rise to numbers like $3\frac{5}{8}$ inches" (NCTM, *Curriculum and Evaluation Standards*, p. 92).

The set of integers is expanded to form the set of rational numbers so that nonzero integer division problems with nonzero divisors will have answers. In elementary school, children study the nonnegative subset of rational numbers that I have called elementary fractions.

Elementary fractions are quite versatile. An elementary fraction can represent a part of a whole, a division problem, a location on a number line, or a part of a set. You can represent any rational number as a division problem or as a location on a number line.

"Area models are especially helpful in visualizing numerical ideas from a geometric point of view. For example, area models can be used to show that 8/12 is equivalent to 2/3" (NCTM, *Curriculum and Evaluation Standards*, p. 88). Algebraically, one can verify the equivalence of fractions by finding a least common denominator or comparing cross products. One can find which of two unequal fractions is greater by these same two methods.

Arithmetic rules for rational numbers are consistent with these models. As with whole-number and integer arithmetic, with rational numbers addition and multiplication are defined, and then subtraction is defined as the inverse of addition and division as the inverse of multiplication. Fraction pictures help children understand the results of fraction arithmetic. The common categories of the four whole-number operations fit many word problems involving rational numbers. Estimation enables students to use their facility with whole numbers to develop their intuition about elementary fraction arithmetic.

Rational numbers include all the numbers most children study in elementary school. Whole numbers, elementary fractions, and integers are all rational numbers. Whole numbers, integers, and rational numbers all possess the commutative and associative properties for addition and multiplication and the distributive properties for multiplication over addition and for multiplication over subtraction. The integers and rational numbers have unique additive inverses as well, and nonzero rational numbers have unique multiplicative inverses. The rational numbers are also dense.

Children now study estimation and mental computation with fractions in school. They learn how to recognize problems that are easy to compute and how to use the rounding and compatible-numbers strategies to estimate.

STUDY GUIDE

To review Chapter 6, see what you know about each of the following ideas or terms that you have studied. You can also use this list to generate your own questions about the chapter.

The NCTM Curriculum Standards and Rational Numbers

Selected NCTM Curriculum Standards

The following standards come from the NCTM document.

- Understand and appreciate the need for numbers beyond the whole numbers.
- Develop concepts of fractions and mixed numbers.
- Develop and use order relations for fractions.
- Recognize relationships among different topics in mathematics.
- Use models to explore operations on fractions.
- Apply fractions to problem situations.
- Develop, analyze, and explain procedures for computation and techniques for estimation.

1. Describe how each standard listed relates to the material you studied in Chapter 6.
2. Select any current elementary-school mathematics textbook series, and describe sample lessons or exercises that illustrate each standard listed.

ELEMENTARY FRACTIONS IN ELEMENTARY SCHOOL

The following chart shows at what grade levels selected elementary fraction topics typically appear in elementary-school mathematics textbooks.

Topic	Typical Grade Level in Current Textbooks
Fraction concepts	1, 2, 3, 4, 5, 6
Fraction addition and subtraction	4, 5, 6
Fraction multiplication	5, 6
Fraction division	5, 6
Fraction estimation	4, 5, 6

REVIEW EXERCISES

1. *Explain* how to compute the exact answer to $24 \times 2\frac{1}{2}$ mentally.

2. A is an integer. If $\frac{A}{20} > \frac{20}{35}$, then A ________.

3. Describe four common meanings of $\frac{5}{6}$. For each meaning, give its name, make a drawing, and describe how it works for $\frac{5}{6}$.

4. Show why $\frac{7}{4} = 1\frac{3}{4}$ using the fraction-of-a-whole model.

5. Name a property that the nonzero rational numbers have that the integers do not have.

6. $-3\frac{1}{2} - \left(-2\frac{1}{4}\right) = ?$
   (a) $3\frac{1}{2} - 2\frac{1}{4}$ (b) $-\left(3\frac{1}{2} - 2\frac{1}{4}\right)$
   (c) $3\frac{1}{2} + 2\frac{1}{4}$

7. $-85\frac{5}{7} + 123\frac{2}{7} =$ ________

8. *Explain* why $\frac{5}{0}$ is undefined.

9. (a) You have 5 ounces of peanuts. Exactly how many $\frac{3}{4}$-ounce servings can you make?
   (b) What operation and category does this problem illustrate?

10. Suppose you give $\frac{1}{10}$ of your earnings to charity and pay $\frac{1}{3}$ of your earnings in taxes.
    (a) What fraction of your earnings is left for other expenses?
    (b) What operations and classifications does part (a) illustrate?

11. Patty bought $1\frac{3}{4}$ pounds of Swiss cheese for 5 lunches during the work week.
    (a) How much cheese will she have for each day? (Give an exact answer.)
    (b) What operation and category does this problem illustrate?

12. (a) Use the least-common-denominator method to show how $\frac{4}{9}$ and $\frac{2}{5}$ compare in size.
    (b) Use the compare-cross-products method to show how $\frac{4}{9}$ and $\frac{2}{5}$ compare in size.

13. (a) Draw a fraction bar for each fraction in $\frac{2}{3} + \frac{1}{6}$.
    (b) Draw a fraction bar for the sum, and explain why a common denominator is needed.
    (c) Draw a fraction bar for each fraction with the least common denominator.
    (d) Show how to compute $\frac{2}{3} + \frac{1}{6}$ using a picture.

14. (a) Draw a fraction bar for each fraction in $\frac{1}{2} - \frac{1}{5}$.
    (b) Explain why a common denominator is needed.
    (c) Draw a fraction bar for each fraction with the least common denominator.
    (d) Show how to compute $\frac{1}{2} - \frac{1}{5}$ using a picture.

15. A child wants to see why the invert-and-multiply rule works. Use the division model of fractions and the Fundamental Law to show that $\frac{2}{5} \div \frac{7}{9} = \frac{2}{5} \times \frac{9}{7}$.

16. (a) What is an easy way to multiply

$$\frac{1}{5} \times 8 \times \left(15 \times \frac{3}{4}\right)?$$

(b) What property or properties did you use?

17. Give an example illustrating the distributive property of multiplication over addition for rational numbers.

18. A child who does not know the multiplication rule for fractions wants to compute $\frac{2}{3} \times \frac{3}{5}$. *Explain* how to compute $\frac{2}{3} \times \frac{3}{5}$ using two diagrams.

19. (a) Complete the last problem, repeating the error pattern from the completed examples.

$$\frac{1}{4} + \frac{2}{3} = \frac{4}{7} + \frac{6}{7} = \frac{10}{7} = 1\frac{3}{7}$$

____

$$\frac{3}{5} + \frac{1}{9} = \frac{12}{14} + \frac{6}{14} = \frac{18}{14} = 1\frac{4}{14} = 1\frac{2}{7}$$

____

$$\frac{2}{5} + \frac{1}{4} =$$

____

(b) Write a description of the error pattern.

(c) Explain how the error pattern in the first example could be detected with estimation. (*Hint:* Why should the answer be less than 1?)

20. $472\frac{1}{4} \div \frac{3}{16}$ is approximately

(a) 2500 (b) 100 (c) 9 (d) $\frac{1}{10{,}000}$

21. Explain why, in multiplying $\frac{1}{4} \times \frac{8}{9}$, one can simplify the computation by changing the 8 and 4 to 2 and 1, respectively.

22. A child thinks $\frac{1}{4} + \frac{1}{3} = \frac{2}{7}$. Draw a fraction bar for each fraction, and explain why $\frac{2}{7}$ cannot be the sum.

23. Write a paragraph that defines whole numbers, integers, and rational numbers and tells how the three sets of numbers are related.

SUGGESTED READINGS

The Arithmetic Teacher. February 1984, Focus Issue on Rational Numbers. Reston, Va.: NCTM, 1984.

Ashlock, R. *Error Patterns in Computation.* 6th ed. Englewood Cliffs, N.J.: Prentice Hall, 1994.

National Council of Teachers of Mathematics. *Mathematics Learning in Early Childhood.* 1975 Yearbook. Reston, Va.: NCTM, 1975.

National Council of Teachers of Mathematics. *Developing Computational Skills.* 1978 Yearbook. Reston, Va.: NCTM, 1978.

National Council of Teachers of Mathematics. *Estimation and Mental Computation.* 1986 Yearbook. Reston, Va.: NCTM, 1986.

National Council of Teachers of Mathematics. *New Directions for Elementary School Mathematics.* 1989 Yearbook. Reston, Va.: NCTM, 1989.

Williams, David. *Fraction Action!* Sunnyvale, Calif.: Stokes Publishing, 1995.

7 Decimals, Percents, and Real Numbers

All rational numbers can be written as fractions or as decimals. Each notation has its advantages.

The ancient Egyptians developed fraction notation over 3500 years ago for measuring and accounting, but fraction notation is often awkward for comparing the sizes of two numbers or for doing computations. Decimal notation is one of the great labor-saving inventions of mathematics. Simon Stevin (1548–1620), a Flemish engineer, was the first to discuss decimal notation and decimal arithmetic in some detail.

Decimal notation uses an extension of our whole-number place-value system to represent numbers. As a result, decimal arithmetic algorithms use whole-number/integer algorithms, with additional rules for shifting and placing decimal points.

Decimals are also significant in mathematics because, as the Pythagoreans discovered over 2000 years ago, rational numbers alone are insufficient for measuring all lengths. The first irrational (that is, "not rational") decimal numbers that the Pythagoreans found were certain square roots.

Although the set of rational numbers is dense, it does not represent every point on the standard number line. The fact that decimal numbers include both rational and irrational numbers makes it possible to label every point on a number line and to measure any length. The union of the set of rational and irrational numbers is called the set of real numbers.

7.1 DECIMALS: PLACE VALUE, ESTIMATION, AND MENTAL COMPUTATION

Sally's Car Clinic **Parts and Labor $362.25**	**GNP Is $3.4 Billion**

Whether it is a Gross National Product (GNP) of $3.4 billion, an atom of diameter 0.00000004 cm, or car repairs costing $362.25, people usually report statistics in decimal notation. In most everyday applications, decimal notation is easier to use than fractions, and decimal computations are easier than fractional computations. Furthermore, calculators and computers generally give output in decimal form, and the metric system employs decimal notation.

PLACE VALUE

What amount does 3426.517 represent? Decimal place value is an extension of whole-number place value, and it has a symmetry as shown in the following diagram. Each place value for whole numbers (extending to the left) corresponds to a place value for decimal places (extending to the right). For example, tens correspond to tenths, and hundreds correspond to hundredths.

3	4	2	6	·	5	1	7
←							→
thousands	hundreds	tens	ones	·	tenths	hundredths	thousandths

The decimal point is read "and." The number 3426.517 is read "three thousand, four hundred, twenty-six and five hundred seventeen thousandths."

Lesson Exercise 7.1

(a) In our numeration system, each place you move to the left multiplies place value by ________.

(b) Each place you move to the right divides place value by ________, which is the same as multiplying by ________.

Pictorial models clarify the relative sizes of decimal numbers. Many elementary-school textbooks use decimal squares to picture 1, 1/10, and 1/100. A decimal square represents 1. Divide the square into 10 equal rectangles, as shown in Figure 7-1, and each rectangle represents 1/10. Divide the square into 100 equal squares, and each small square represents 1/100.

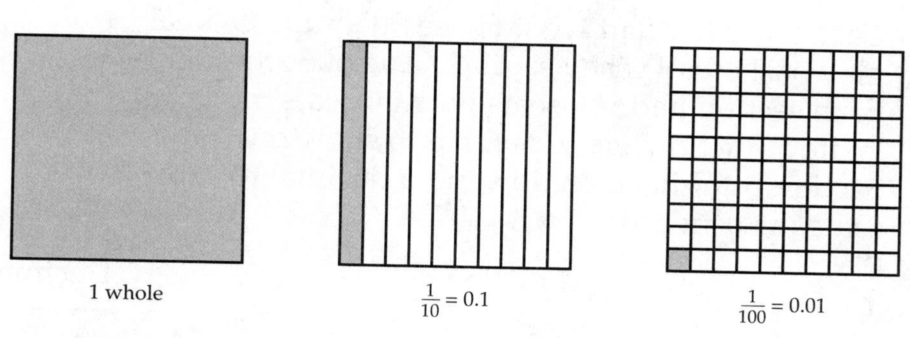

Figure 7–1

Many other decimal numbers can be represented in a similar fashion. For example, 2.3 and 1.34 are shown in an excerpt from *Mathematics Plus*, Grade 4 (Figure 7-3). Some elementary-school textbooks use base-ten blocks to represent decimals. A flat is 1, a long is 0.1, and a cube is 0.01.

Lesson Exercise 7.2

Explain how to use two decimal-square pictures to show that $0.4 > 0.32$.

Another pictorial model that shows the relative sizes of decimals is a number line. Each decimal represents a location on the number line. For example, 0.64 is between 0.6 and 0.7, as shown in Figure 7-2.

As with whole numbers and fractions, if a decimal number u is located to the right of a decimal number v on this number line, $u > v$.

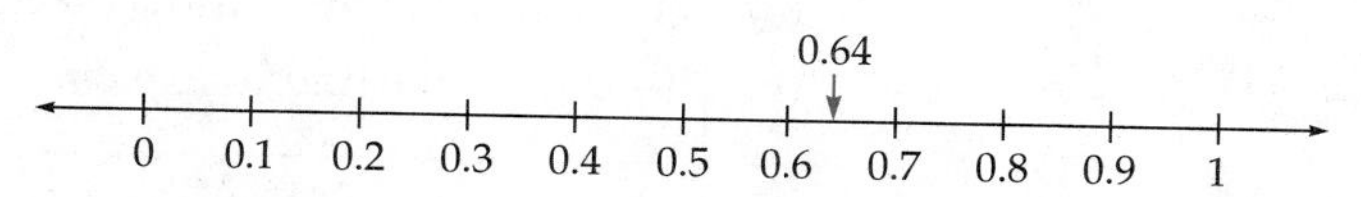

Figure 7–2

Connecting Fractions to Decimals

Carrie and her dad went to see Shaquille O'Neal play basketball. They sat in 2 seats of Section B. Each section had 100 seats with 10 seats in a row. Write a fraction and a decimal to show one row in Section B. Write a fraction and a decimal to show 2 seats in Section B.

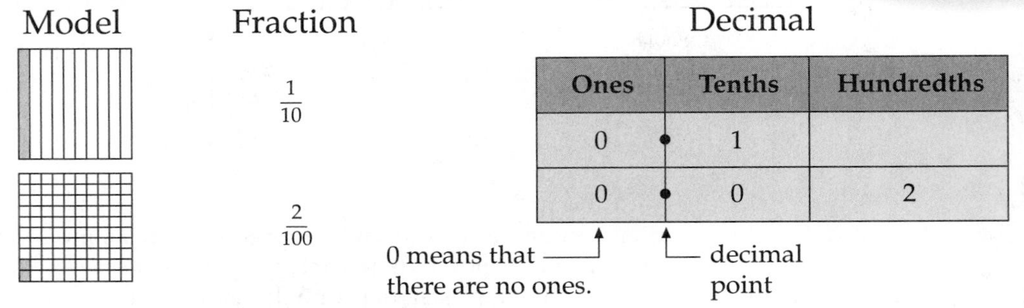

Ones	Tenths	Hundredths
0 •	1	
0 •	0	2

You can write a mixed number as a decimal.

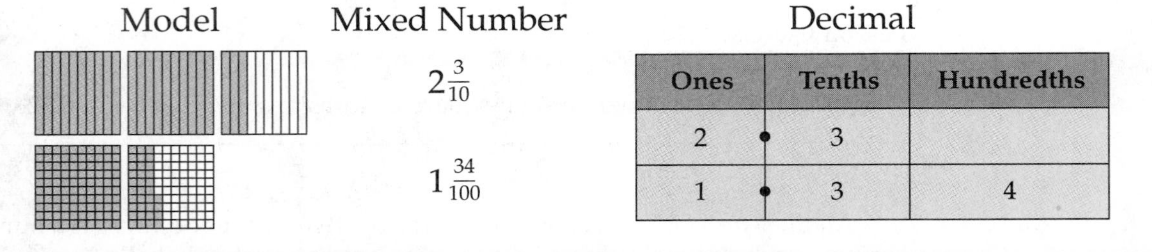

Ones	Tenths	Hundredths
2 •	3	
1 •	3	4

From Mathematics Plus, *Grade 4 (San Diego: Harcourt Brace, 1994), p. 382.*

Figure 7–3

Lesson Exercise 7.3

(a) Why might a child think that $0.52 > 0.8$?

(b) Use a number line to *explain* why $0.8 > 0.52$.

Like whole numbers, other decimal numbers can be written using expanded notation to show place value.

EXAMPLE 7.1

Write 46.28 using expanded notation.

Solution

$$46.28 = (4 \times 10) + 6 + \left(2 \times \frac{1}{10}\right) + \left(8 \times \frac{1}{100}\right)$$

■

Lesson Exercise 7.4

(a) Write 0.317 in expanded notation.

(b) Does 0.317 equal 3 tenths, 1 hundredth, and 7 thousandths, or does it equal 317 thousandths? Answer the question by showing that the sum in part (a) equals 317/1000.

The preceding exercise and example contain examples of terminating decimals. **Terminating decimals** can be written with a finite number of digits to the right of the decimal point. The numbers 8., 0.6, and 0.317 are terminating decimals, but 0.3444 . . . is not (Figure 7-4). The ". . ." in 0.3444 . . . indicates that the 4 repeats infinitely. The number can also be written $0.3\overline{4}$, in which a line is placed over the repeating block of digits.

Figure 7–4

Nonterminating decimals are discussed later in the chapter.

EXPONENTS

Exponents are often used in expressing large and small positive numbers in a briefer format. Decimal place values can be expressed using powers of 10.

Counting-number exponents were used for place value in Chapter 3. Zero and negative-integer exponents are useful in decimal place value. Zero and negative-integer exponents are defined so that certain properties of positive exponents extend to all integer exponents.

Lesson Exercise 7.5

Suppose that a student knows how to compute 10^c, when c is a counting number.

(a) How could you suggest what 10^0, 10^{-1}, and 10^{-2} should equal by extending a pattern? (*Hint:* Start with $10^3 = 1000$.)

(b) Describe the pattern.

Lesson Exercise 7.5 suggests an appropriate definition of zero and negative-integer exponents.

Zero and Negative-Integer Exponents

For all $a > 0$ and integers n:

$$a^0 = 1$$

$$a^{-n} = \frac{1}{a^n}$$

The powers of 10 have a pattern in decimal place value, as follows.

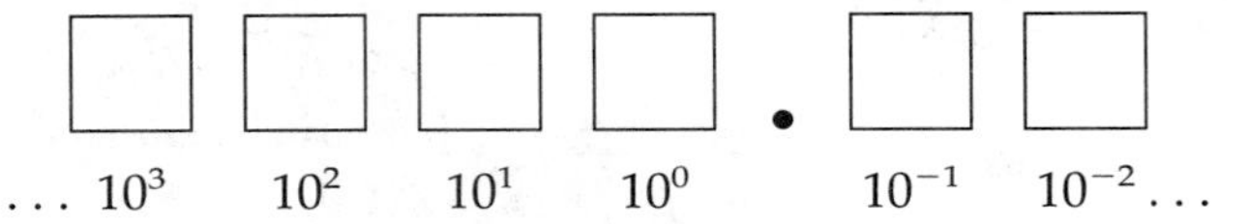

$\ldots\ 10^3 \quad 10^2 \quad 10^1 \quad 10^0 \quad 10^{-1} \quad 10^{-2} \ldots$

These definitions are consistent with exponent properties with which you are familiar.

Lesson Exercise 7.6

Complete the following exercise, to show how the values of 10^0 and 10^{-1} are obtained, by using the rule for adding exponents. Part (a) reviews the rule.

(a) $(10^3) \cdot (10^4) = (10 \cdot 10 \cdot 10) \cdot (\_\_\_\_\_\_\_\_) = 10^?$.

(b) What is the shortcut for multiplying $10^3 \cdot 10^4$?

(c) The rule in part (b) determines what value 10^0 should have. Using the same rule (shortcut), $10^3 \cdot 10^0 = \_\_\_\_\_\_\_\_$.

(d) If $10^3 \cdot 10^0 = 10^3$, then 10^0 must equal $\_\_\_\_\_\_\_\_$.

How nice that letting $10^0 = 1$ and $10^{-1} = 0.1$ makes the pattern in Lesson Exercise 7.5 and the rule in Lesson Exercise 7.6 both work! The general addition rule for exponents states that $a^m a^n = a^{m+n}$, in which a is a nonzero rational number and m and n are integers. The subtraction rule for exponents is reviewed in the homework exercises.

TYPES OF ROUNDING

In preceding chapters, almost all rounding has been done to the nearest appropriate number. Actually, there are three kinds of rounding: rounding up, rounding down, and rounding to "the nearest."

EXAMPLE 7.2

An office-supply store sells pens for \$.1225 each. Round this price

(a) down to the preceding hundredth.

(b) up to the next hundredth.

(c) to the nearest hundredth.

Solution

The hundredths place contains a 2. This means \$.1225 is between \$.12 and \$.13.

(a) \$.12 **(b)** \$.13

(c)

\$.1200 \$.1225 \$.1300

\$.1225 is closer to \$.12 than \$.13. The answer is \$.12. ■

Calculators and computers may round down, or truncate, all long decimal numbers. **Truncate** means to discard digits to the right of a particular decimal place. Note that discarding the 25 in \$.1225 (truncating) is the same as rounding down to the preceding hundredth.

Lesson Exercise 7.7

A mile is 0.621371 kilometer. Round this distance

(a) down to the preceding thousandth.

(b) up to the next thousandth.

(c) to the nearest thousandth.

THE ROUNDING AND FRONT-END STRATEGIES

Decimal estimation, not surprisingly, is very similar to fraction estimation. The rounding strategy is usually best for estimating in decimal addition, subtraction, and most multiplication problems. Decimals are usually rounded to the

ESTIMATING
Decimal Sums and Differences

Sara is saving for a new pair of ice skates. She saved $42.47 during the summer and $11.49 during the fall. About how much has she saved?

Front-end estimation can be used to estimate decimal sums.

Step 1 Add the front-end, or lead, digits.	**Step 2** Adjust the estimate.	**Step 3** Think about the value of the other digits.	**Step 4** Adjust the estimate again.
$4 2 . 4 7 + 1 1 . 4 9 $5 ■.■■	$4 2 . 4 7 + 1 1 . 4 9 $5 3. ■■	$42.47 + 11.49 47 cents + 49 cents ≈ 1.00.	$53.00 + 1.00 $54.00

So, Sara has saved about $54.00.

Another Method

Round to the nearest whole number.

$$\begin{array}{rcr} 42.47 & \rightarrow & 42 \\ +\ 11.49 & \rightarrow & +\ 11 \\ \hline & & 53 \end{array}$$

Round to the nearest ten.

$$\begin{array}{rcr} 42.47 & \rightarrow & 40 \\ +\ 11.49 & \rightarrow & +\ 10 \\ \hline & & 50 \end{array}$$

From Mathematics Plus, *Grade 5 (San Diego: Harcourt Brace, 1994), p. 54.*

Figure 7–5

nearest whole number to create a problem that can be computed mentally. The front-end strategy may also be used for addition, subtraction, and some multiplication problems.

Figure 7-5 shows how the fifth-grade *Mathematics Plus* textbook compares the rounding and front-end strategies.

Lesson Exercise 7.8

(a) You buy 8.6 m of material costing $7.79 per meter. How could you estimate the total cost by rounding each factor to the nearest whole number?

(b) Why would it also be reasonable to estimate the answer using 8 m $\times$ \$8 per meter?

(c) How could you estimate the total cost using the front-end strategy?

THE COMPATIBLE-NUMBERS STRATEGY

The compatible-numbers strategy works for estimating in decimal division just as it does for whole-number and fraction division. Rounding to the *nearest* whole number is not always the best approach.

Lesson Exercise 7.9

Suppose that 6.8 lb of crabmeat costs \$58.20. How could you estimate the cost per pound?

One can use estimation to check results obtained with a calculator.

Lesson Exercise 7.10

I did the following computations on my calculator. Estimate each answer. Then tell whether the given answer must be incorrect.

(a) $86 \div 0.7 = 12.286$ (b) $365 \times 0.05 = 18.25$

MENTALLY MULTIPLYING OR DIVIDING BY POWERS OF 10

The shortcut for multiplying by a counting-number power of 10 is suggested by the pattern in the following lesson exercise.

Lesson Exercise 7.11

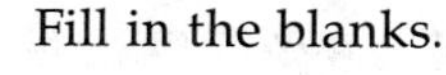

Fill in the blanks.

(a) 9.52×10 or 9.52×10^1 = ________

(b) 9.52×100 or 9.52×10^2 = ________

(c) 9.52×1000 or ____________ = ________

(d) Propose a general rule based upon your answers to parts (a)–(c).

(e) Explain how you used inductive reasoning to answer part (d).

Lesson Exercise 7.11 suggests the following shortcut.

Multiplying a Decimal Number by a Power of 10

Multiplying a decimal number by 10^n ($n = 1, 2, 3, \ldots$) is the same as moving the decimal point n places to the right.

Lesson Exercise 7.12

Explain how to compute 4.6×10^3 mentally.

In 1987, Van Gogh's painting *Irises* sold for a record \$53.9 million. The amount 53.9 million means "53.9 times 1 million," or 53.9×10^6. So \$53.9 million = \$53,900,000, which would buy a lot of real flowers. Figure 7-6 shows how 53.9 million is related to 53 million and 54 million.

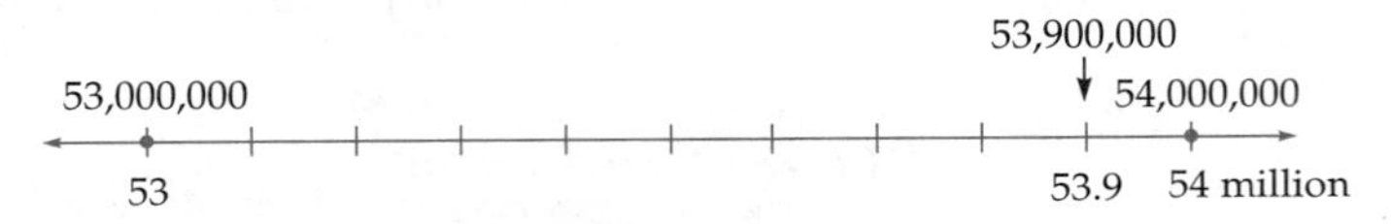

Figure 7–6

Lesson Exercise 7.13

According to the National Safety Council, the cost of motor vehicle accidents in 1994 was \$176.5 billion.

(a) Write this number in decimal notation (without the word "billion").

(b) If there were about 175 million drivers, what was the average cost per driver?

Now consider dividing a number by a counting-number power of 10. What is a shortcut for dividing by a power of 10 such as 100 or 1000?

Lesson Exercise 7.14

Complete the following.

(a) $9.52 \div 10$ or $9.52 \div 10^1 =$ ________

(b) $9.52 \div 100$ or $9.52 \div 10^2 =$ ________

(c) $9.52 \div 1000$ or $9.52 \div 10^3 =$ ________

(d) Based upon parts (a)–(c), dividing a decimal number by 10^n, in which $n = 1, 2, 3, \ldots$, appears to be the same as moving the decimal point ____ places to the ________.

The preceding exercise suggests the following shortcut.

Dividing a Decimal Number by a Power of 10

Dividing a decimal number by 10^n ($n = 1, 2, 3, \ldots$) is the same as moving the decimal point n places to the left.

Lesson Exercise 7.15

A store buys 1000 candy bars for $75. How much did it pay for each candy bar?

(a) *Explain* how to compute the exact answer mentally.

(b) What operation and category does part (a) illustrate?

The shortcut for multiplying or dividing by a power of 10 can be extended to exponents that are negative integers. Consider the following examples.

$$3.47 \times 10^2 = 347$$

$$3.47 \times 10^1 = 34.7$$

$$3.47 \times 10^0 = 3.47$$

$$3.47 \times 10^{-1} = 0.347$$

$$3.47 \times 10^{-2} = 0.0347$$

Lesson Exercise 7.16

Describe a shortcut you can use to multiply by negative-integer powers of 10.

The shortcut for multiplying a decimal number by an integer power of 10 is as follows.

Multiplying a Decimal Number by an Integer Power of 10

To multiply $a \times 10^c$, in which c is an integer, the decimal point in a is moved c places to the right if $c \geq 0$ or $|c|$ places to the left if $c < 0$.

Lesson Exercise 7.17

Write 5.67×10^{-6} using decimal notation.

SCIENTIFIC NOTATION

Archimedes (287–212 B.C.) was one of the first to use very large numbers. He supposedly computed the number of grains of sand needed to fill the entire universe (10^{63}). But why did he do this?

According to my wife, he was at the beach with some of his friends, who taunted him. "If you're so smart Archi, how many grains of sand would fill the universe?" Submitting to peer pressure, Archimedes proceeded to find out.

It is easier to write large numbers such as 10^{63} in shorthand notation. How do calculators deal with such large numbers? Try the following and find out.

Lesson Exercise 7.18

(a) Compute $400{,}000 \times 360{,}000$ by hand.

(b) Compute $400{,}000 \times 360{,}000$ on your calculator. Did you obtain something like [1.44 11]? If you got an error message or "E," your calculator cannot deal with very small or very large numbers. Try this computation on a classmate's calculator that does show the correct answer.

The answer to Lesson Exercise 7.18(b) is in scientific notation. It means 1.44×10^{11}. Some calculators and all computers use scientific notation to abbreviate very large positive numbers and positive numbers that are close to 0. A computer might write 1.44×10^{11} as [1.44 E 11]. Scientific notation shows a number as an integer power of 10 multiplied by a number between 1 and 10 (including 1 but not 10).

Definition **Scientific Notation**

Any positive decimal number can be written in **scientific notation**, $n \times 10^p$, in which $1 \le n < 10$ and p is an integer.

Scientific notation is a useful shorthand for numbers that have many digits. Scientists, calculators, and computers each employ slightly different scientific notation.

People using scientific notation also need to know how to convert numbers in standard form to scientific notation.

EXAMPLE 7.3

Write the 1996 world population, 5,800,000,000, in

(a) scientific notation. **(b)** billions.

Solution

(a) First, move the decimal point to obtain a number between 1 and 10.

5.800000000

How many places was the decimal point shifted? Nine. The 9 gives the *magnitude* of the exponent of 10. Should the exponent be 9 or −9 (5.8×10^9 or 5.8×10^{-9})? Use estimation to decide. Which would equal 5,800,000,000?

$$5{,}800{,}000{,}000 = 5.8 \times 10^9$$

(b) Since 1 billion = 10^9, 5.8×10^9 is 5.8 billion. ■

Lesson Exercise 7.19

A snail moves at a rate of about 0.00036 miles per hour. Write this rate in scientific notation.

ANSWERS TO SELECTED LESSON EXERCISES

7.1 (a) 10 (b) 10; $\frac{1}{10}$

7.2 To show 0.4, shade 4 of the 10 columns. To show 0.32, shade 32 of the 100 squares. The area shaded for 0.4 is larger, so 0.4 > 0.32.

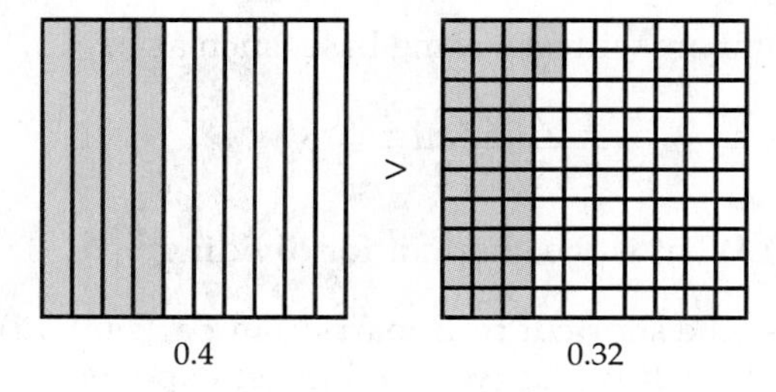

7.3 (a) It reminds the child of 52 > 8.
(b) 0.8 is located to the right of 0.52 on the number line. So 0.8 > 0.52.

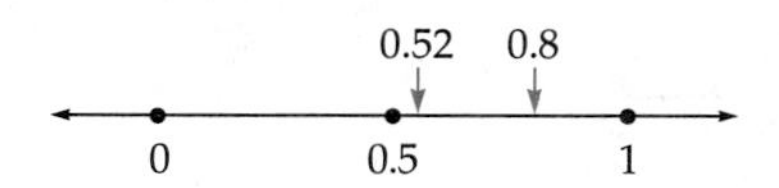

7.4 (a) $\left(3 \times \frac{1}{10}\right) + \left(1 \times \frac{1}{100}\right) + \left(7 \times \frac{1}{1000}\right)$

(b) $\frac{3}{10} + \frac{1}{100} + \frac{7}{1000} = \frac{300}{1000} + \frac{10}{1000} + \frac{7}{1000} = \frac{317}{1000}$

7.5 (a) $10^3 = 1000$, $10^2 = 100$, $10^1 = 10$, $10^0 = 1$, $10^{-1} = 0.1$, $10^{-2} = 0.01$
(b) When the exponent decreases by 1, the result is divided by 10.

7.6 (a) The final answer is 10^7.
(b) Add the exponents.
(c) 10^3
(d) 1

7.7 (a) 0.621 (b) 0.622 (c) 0.621

7.8 (a) $8.6 \times \$7.79 \approx 9 \times \$8 = \$72$
(b) Rounding one factor up and the other down may yield a more accurate estimate.
(c) $8.6 \times \$7.79 \approx 8 \times \$7 = \$56$

7.9 $58.20 ÷ 6.8 ≈ $56 ÷ 7 = $8/lb

7.10 (a) is wrong (b) is reasonable

7.11 (a) 95.2 (b) 952 (c) 9520
(e) A generalization based upon a pattern in some examples

7.12 Move the decimal point in 4.6 three places to the right. The answer is 4600.

7.13 (a) $176,500,000,000 (b) $1009

7.14 (a) 0.952 (b) 0.0952
(c) 0.00952 (d) n; left

7.15 (a) Dividing by 1000 requires moving the decimal point in $75 three places to the left to obtain $.075.
(b) Partition sets/measures

7.17 0.00000567

7.18 (a) 144,000,000,000

7.19 3.6×10^{-4} mph

7.1 HOMEWORK EXERCISES

Basic Exercises

1. If you move the decimal point in a number two places to the left, the value of the number is divided by ________ or multiplied by ________.

2. What is the smallest positive number you can display on your calculator?
3. *Explain* how to use two decimal-square pictures to show that 0.40 = 0.4.
4. *Explain* how to use two decimal-square pictures to show that 0.11 < 0.2.
5. (a) Why might a student think that −2.3 > −2.1?
   (b) Use a number line to *explain* why −2.1 > −2.3.
6. (a) Why might a student think that −0.4 > −0.17?
   (b) Use a number line to *explain* why −0.4 < −0.17.
7. Write each of the following as a decimal number.
   (a) Forty-one and sixteen hundredths
   (b) Seven and five thousandths
8. (a) Write 0.296 using expanded notation.
   (b) The number 0.296 is read "two hundred ninety-six thousandths." Show that 296/1000 equals the sum in part (a).
9. Write the following decimal numbers using expanded notation.
   (a) 0.31 (b) 32.017
10. Write each of the following terminating decimals as a fraction in simplest form.
   (a) 0.27 (b) 3.036 (c) 8.0004
11. Show how to find the value of 10^0 by extending a pattern in positive exponents.

*12. Show how to find the value of 10^{-3} by extending a pattern in positive exponents.

13. Consider the following pattern.

$$5^3 = 125$$
$$5^2 = 25$$
$$5^1 = 5$$

   (a) Each result equals the previous result divided by ________.
   (b) Continuing this pattern,
   $5^0 =$ ____ $5^{-1} =$ ____ $5^{-2} =$ ____

*14. (a) How can you write 10^{-n} (n is a positive integer) with a positive exponent?
   (b) How can you write X^{-n} (X and n are positive integers) with a positive exponent? (See the preceding exercise.)

15. Do you recall the shortcut for dividing numbers with the same base, such as $\frac{2^6}{2^4}$?

   (a) $\frac{2^6}{2^4} = \frac{2 \cdot 2 \cdot 2 \cdot 2 \cdot 2 \cdot 2}{2 \cdot 2 \cdot 2 \cdot 2} = 2^?$

   (b) What is the shortcut for dividing $\frac{2^6}{2^4}$?

   Use the shortcut from part (b) in parts (c), (d), and (e). It works with all integer exponents. Assume that X is a positive number.

   (c) $\frac{10^7}{10^4} =$ ____

   (d) $\frac{5^7}{5^{-3}} =$ ____

   (e) $\frac{X^6}{X^2} =$ ____

16. In everyday life, not all rounding of decimals is done to the nearest whole number. For example, if you mail a 1.1-oz letter, the post office charges you the 2-oz rate. Name another situation in which all fractional amounts are rounded up.

17. Round $.86
(a) up to the next tenth.
(b) down to the preceding tenth.
(c) to the nearest tenth.

18. A job pays $4.35 per hour. How can you estimate how much the job pays for a 32-hour work week?
(a) Estimate the answer using rounding.
(b) Estimate the answer using the front-end strategy.

19. Mount Everest has an altitude of 8847.6 m, and Mount Api has an altitude of 7132.1 m. How much higher is Mount Everest than Mount Api?
(a) Estimate the answer using rounding.
(b) Estimate the answer using the front-end strategy.

20. The answer to 3.74×42.8125 has the digits 16011875. How can you use estimation to determine where the decimal point goes?

21. A 3-line ad in a newspaper costs $1.79 per line per day. The cost to run such an ad for 4 days is about
(a) $5 (b) $8 (c) $12
(d) $22 (e) $54

22. Suppose labels are sold in packs of 100.
(a) If you need 640 labels, how many labels should you order?
(b) Does this application require rounding up, down, or to "the nearest"?

23. A 46-oz can of apple juice costs $1.29. How can you estimate the cost per ounce?

24. Without computing the product, fill in each blank.
(a) $4.6 \times 8 = 2.3 \times$ ____
(b) $8.2 \div 0.3 = 16.4 \div$ ____

25. To estimate the positive decimal problem $a \div b$, a is rounded up and b is rounded down. Will the estimate be too high or too low? Or is it impossible to tell?

26. The length of a table to the nearest tenth of a centimeter is 153.7 cm. The exact length of the table is between ____ cm and ____ cm.

27. I did the following computations on my calculator. Determine by estimating which answers could not be correct.
(a) $2.4 \times 8.6 = 14.4$
(b) $2.13 - 0.625 = 1.505$
(c) $374 \times 1.1 = 41.14$
(d) $43.74 \div 2.2 = 19.88181818$

28. Multiplying a decimal number by 10^n, in which $n = 1, 2, 3, \ldots$, is the same as moving the decimal ____ places to the ______.

29. The 1996 U.S. federal budget included a debt of about $3.9 trillion. Write this number without the word "trillion."

30. Write each of the following population figures in millions (using the word "million").
(a) United States: 266,580,000
(b) Paris: 2,772,000
(c) World: 5,800,000,000

31. The distance from Earth to Saturn is about 800 million miles. About how long would it take to reach Saturn traveling 35,000 mph?

32. Large numbers are difficult to grasp. How long is a million seconds? Is it a week? A month? A year? 5 years?
(a) Guess
(b) Figure out the answer.

33. How long is a billion seconds? Is it a week? A month? A year? 5 years?
(a) Guess.
(b) Figure out the answer.

***34.** Write the following in standard form.
(a) 3.62×10^7 (b) $4268 \div 10^6$

35. Rosa bought 100 board feet of walnut board at $3.29 per board foot. What was the total cost? *Explain* how to compute the exact answer mentally.

36. A store buys 1000 "Honk if you love quiet" bumper stickers for $43. How much did they pay for each bumper sticker? *Explain* how to compute the exact answer mentally.

37. A television signal travels 1 mile in 5.4×10^{-6} second. Write this time interval in standard decimal form.

38. Which is greater, 3.2×10^{-6} or 3.2×10^{-5}?

39. A computer display shows [3.4 *E* 12]. Write this number in scientific notation.

40. (a) If your calculator has an [x^y] key, use it to compute 5^{10} using the following keystrokes: [5] [x^y] [10] [=].

(b) Compute 4^{20} using this approach.

41. Each day, the earth picks up approximately 1.2×10^7 kg of dust from outer space. Does it seem like most of it falls in your house or apartment?

(a) Write this number in standard form.

(b) Name this number.

42. Many scientists believe that the Earth is about 5×10^9 years old.

(a) Write this number in standard form.

(b) Name this number.

43. Why is 48×10^3 not correct scientific notation?

44. The mass of an oxygen atom is 0.000 000 000 000 000 000 000 013 g. Write this number in scientific notation.

45. It would take 3,000,000,000,000,000,000,000,000,000 candles to give off as much light as the sun.

(a) Write this number in scientific notation.

(b) What advantage does scientific notation have over standard decimal form in this case?

(c) Write this number as it would appear on a calculator display.

46. The U.S. government has a large debt. The *interest* on the debt in 1996 was about \$257 billion.

(a) Write this number in standard form.

(b) Write this number in scientific notation.

47. The following chart gives the distance in meters of each planet from the sun.

Planet	Distance from Sun (meters)
Mercury	5.8×10^{10}
Venus	1.1×10^{11}
Earth	1.5×10^{11}
Mars	2.3×10^{11}
Jupiter	7.8×10^{11}
Saturn	1.4×10^{12}
Neptune	2.9×10^{12}
Uranus	4.5×10^{12}
Pluto	5.9×10^{12}

(a) Which planet is about 10 times as far from the sun as the Earth?

(b) Pluto is about ________ times as far from the sun as the Earth is.

(c) Mercury is about ________ times as far from the sun as the Earth is.

48. The average person in the United States discards 6 pounds of trash each day. The population of the United States is about 280 million people. How much trash does the U.S. population discard in a year? Write your answer in scientific notation.

49. The 1996 estimated population of the United States was 267 million. An average of 1800 cigarettes per person were smoked that year.

(a) Compute the total number of cigarettes smoked in the United States (in 1996) in scientific notation.

(b) Write this number in standard form.

(c) Name this number.

50. Fill in the blanks, following the rule from the completed examples.

$$0.6 \rightarrow 1.36$$
$$0.4 \rightarrow 1.16$$
$$0.2 \rightarrow \_\_\_\_$$
$$N \rightarrow \_\_\_\_$$

51. *Explain* how to compute 20×4.25 mentally. (*Hint:* Think of 4.25 as $4\frac{1}{4}$.)

Extension Exercises

52. In some computer languages (e.g., Pascal and FORTRAN), the computer can be commanded to truncate a decimal numeral. The TRUNC command takes the integer part of a real number and discards the decimal part. TRUNC(6.8) = 6, and TRUNC(−236.715) = −236. Truncate the following.

(a) 46.81792 (b) −278.987 (c) 4.325

(d) When does truncating produce a different result from rounding?

53. Use estimation and mental computation to order the following numbers from smallest to largest.

$$x = 0.00211 + 0.00321$$
$$y = 0.00211 - 0.00321$$
$$z = (0.00211)(0.00321)$$
$$w = 0.00211 \div 0.00321$$

54. A job pays \$4.75 an hour. *Explain* how to compute the exact pay for 40 hours mentally.

55. A job pays \$5.75 an hour. *Exlain* how to compute the exact pay for 20 hours mentally.

56. You order 8 books costing \$4.99 each. Show how to compute the exact total cost mentally.

57. You order 12 shirts for \$10.50 each. Show how to compute the exact total cost mentally.

58. (a) Without using a calculator, tell which of the following are less than 42 and which are greater.

(1) $42 \div 2.7$ (2) $42 \div 0.1$

(3) $42 \div 1.01$ (4) $42 \div 0.999$

(b) Check your answers with a calculator.

(c) If $42 \div a$ is greater than 42, then a is ______________.

59. (a) In base five, what would be the place value of the first two digits to the right of the decimal point?

(b) Write 24.21_{five} in expanded notation.

60. Does $(a^b)^c = a^{b^c}$ for all counting numbers a, b, and c?

Special Exercise

61. The following game requires two people and a calculator. Play a game of "100 POINT" with a partner. The rules of 100 POINT are as follows.

1. Player 1 keys any number into the calculator.
2. Then each player in turn multiplies the number on the calculator by another number, trying to obtain 100 or "100 point something" (e.g., 100.54). The first player to succeed wins the round.

7.2 DECIMAL ARITHMETIC AND ERROR PATTERNS

"Where do you want the decimal point?"

Where does the decimal point go in the answer? For someone who understands whole-number arithmetic, decimal points are the major issue in decimal arithmetic. Decimal arithmetic is closely related to both whole-number and fraction arithmetic. Rewriting decimal numbers as fractions helps clarify how decimal points are placed.

ADDING AND SUBTRACTING DECIMALS

You buy two packages of cheese weighing 0.36 lb and 0.41 lb, and you want to compute the total weight. Or you find a CD for $6.42 at one store and for $6.98 at another store, and you want to know the difference in price. These situations call for addition and subtraction of decimal numbers.

One can model simple decimal addition with decimal squares or base-ten blocks.

Lesson Exercise 7.20

A child does not know the rule for adding decimals. Show how to compute $0.3 + 0.4$ in a single decimal square by shading in two different colors.

The decimal-square picture you drew in Lesson Exercise 7.20 indicates how $0.3 + 0.4$ is related to $3 + 4$. The total number of shaded columns in the decimal square is $3 + 4 = 7$. Since each column represents 0.1 (1 tenth), the shaded area represents 0.7 (7 tenths).

The rules for adding and subtracting terminating decimals are similar to those for adding and subtracting whole numbers, since decimal place value is a consistent extension of whole-number place value. The rules for addition and subtraction can be determined by rewriting terminating decimals as fractions.

In computing $0.36 + 0.40$, why is it that we can compute $36 + 40$ and then place the decimal? Try the following.

Lesson Exercise 7.21

Suppose a child knows fraction addition but not decimal addition.

(a) Show that $0.36 + 0.40 = \dfrac{(36 + 40)}{100}$.

(b) How does part (a) suggest the standard rule for adding terminating decimals?

The equation in Lesson Exercise 7.21 indicates that one can compute the whole-number sum 36 + 40 and then place the decimal point. This means that 0.36 + 0.40 can be computed in place-value columns as follows.

36 hundredths		0.36
+40 hundredths	→	+0.40
76 hundredths		0.76

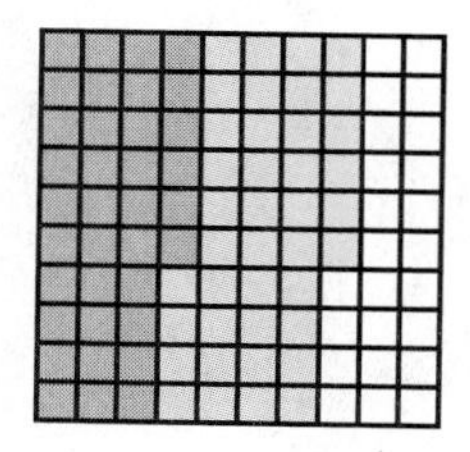

The decimal point in the sum is lined up with the decimal points in the addends. Note that lining up the decimal points corresponds to getting a common denominator in addition of fractions. This is the familiar addition algorithm.

Adding Terminating Decimal Numbers Vertically

Line up the decimal points, add the numbers (ignoring the decimal points), and insert the decimal point in the sum directly below those in the addends.

Like addition, simple decimal subtraction can be modeled with decimal squares or base-ten blocks.

Lesson Exercise 7.22

This decimal-square picture shows 0.3.

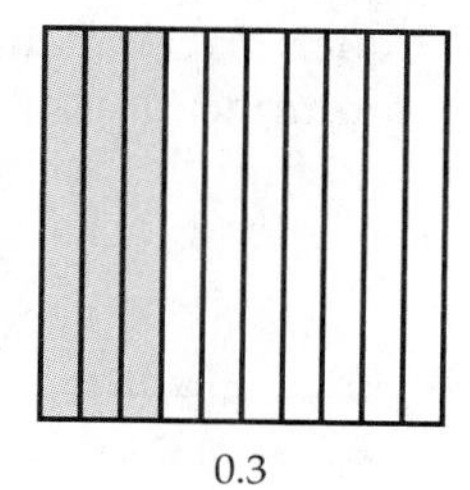
0.3

Use this decimal square to show the result of 0.3 − 0.1, using a take-away approach. Write the result as an equation.

The preceding exercise shows the following connection between decimals and whole-number subtraction.

$$\begin{array}{r} 3 \text{ tenths} \\ -1 \text{ tenth} \\ \hline 2 \text{ tenths} \end{array} \quad \rightarrow \quad \begin{array}{r} 0.3 \\ -0.1 \\ \hline 0.2 \end{array}$$

The procedure for decimal subtraction is just like the procedure for decimal addition. Lining up the decimal corresponds to finding a common denominator in subtraction of fractions.

Subtracting Terminating Decimal Numbers Vertically

Line up the decimal points, subtract the numbers (ignoring the decimal points), and insert the decimal point in the difference directly below those in the other two numbers.

The classifications of whole-number operations also apply to many decimal word problems. Consider one of the problems from the beginning of this section.

Lesson Exercise 7.23

What subtraction category is illustrated in the following problem? "A CD costs $6.42 at one store and $6.98 at another. What is the difference in price?"

MULTIPLYING DECIMALS

What is the cost of 0.6 lb of cole slaw that sells for $.89 per pound? About how much do 5 bananas weigh if 1 banana weighs 0.3 lb? These applications call for multiplication of decimals.

Simple decimal multiplication can be modeled with decimal squares or base-ten blocks.

EXAMPLE 7.4

(a) Show how to compute 3×0.12 with a decimal-square picture.

(b) Work out the same problem using fractions.

(c) Write the same problem using words for place value.

Solution

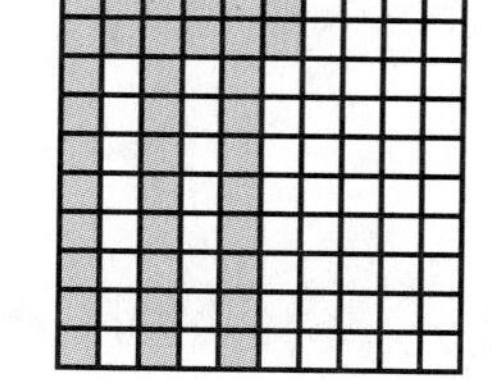

(a) To show 3×0.12, shade 0.12 of a decimal square 3 times. A total of 36 squares are shaded. So $3 \times 0.12 = 0.36$.

(b) $\frac{3}{1} \times \frac{12}{100} = \frac{36}{100}$

(c)

$$\begin{array}{r} 12 \text{ hundredths} \\ \underline{\times\ 3 \text{ ones}} \\ 36 \text{ hundredths} \end{array}$$

■

Use a different method of shading to compute 0.6×0.4 in the next exercise.

Lesson Exercise 7.24

A child knows fraction multiplication but does not know the rule for decimal multiplication.

(a) Explain how to compute 0.6×0.4 using decimal-square pictures. First, dot 0.4 of a decimal square. Then, in a new picture, darken 0.6 (6 tenths) of each column of the 0.4.

(b) Work out the same problem using fractions.

(c) Write out the same problem using words for place value.

The preceding example and exercise indicate how decimal multiplication is just like whole-number multiplication except for having to place the decimal point in the answer. Many elementary-school textbooks now use estimation to place the decimal point in a product. Try it.

Lesson Exercise 7.25

The total cost of 12 shirts that are $7.95 each is

(a) $95.40 (b) $954 (c) $9.54 (d) $.954

How could you use estimation to select the correct choice?

The results of the preceding three decimal multiplication problems suggest the rule for placing the decimal point in the product.

Lesson Exercise 7.26

(a) Complete the following chart.

Numbers			No. of Decimal Places		
Factor 1	**Factor 2**	**Product**	**Factor 1**	**Factor 2**	**Product**
3	0.12	0.36	0	2	2
0.6	0.4	0.24			
12	7.95	95.40			

(b) Use the results to write a rule for determining the number of decimal places in the product.

The preceding exercise suggests the familiar rule for multiplying terminating decimal numbers.

Multiplying Terminating Decimal Numbers

Multiply the two numbers, ignoring the decimal points. The number of decimal places in the product is the sum of the numbers of decimal places in the two factors.

We do not line up the decimal points in multiplication as we do in addition and subtraction. This corresponds to the fact that multiplication of fractions does not require a common denominator.

The categories for whole-number multiplication extend to some decimal applications.

Lesson Exercise 7.27

What operation and category are illustrated in the following problem? "About how much do 5 bananas weigh if 1 banana weighs 0.3 lb?"

DIVIDING DECIMALS

Suppose 4 laps around a track is a distance of 0.8 miles, and you want to know how long each lap is. Children begin decimal division by studying examples like this one that have whole-number divisors and terminating decimals as dividends and quotients.

Decimal squares can be used to compute $0.8 \div 4$.

Lesson Exercise 7.28

In computing $0.8 \div 4$ with a decimal square, first show 8 tenths.

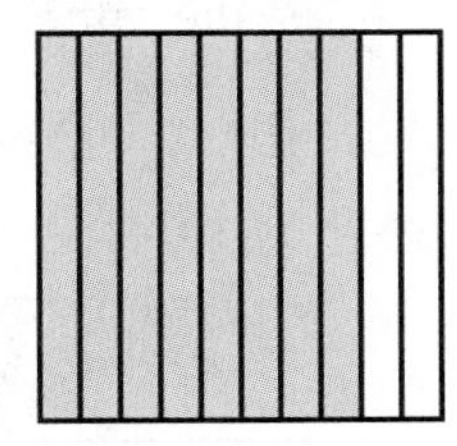

Show how to use this picture to compute $0.8 \div 4$ by partitioning, and write the result as an equation.

In the preceding exercise, each of the 8 columns represents 1 tenth, so the 2-column result represents 2 tenths. This exercise illustrates how $0.8 \div 4$ is computed by dividing 8 by 4 and then placing the decimal point in the quotient above the decimal point in the dividend.

$$4\overline{\smash{)}\,8\text{ tenths}}^{\;2\text{ tenths}} \rightarrow 4\overline{\smash{)}\,0.8}^{\;0.2}$$

This suggests the following procedure.

Dividing a Decimal Number by a Whole Number

Divide, ignoring the decimal point, and place the decimal point in the quotient directly over the decimal point in the dividend.

Now that you have studied problems involving whole-number divisors, what about problems with decimal divisors? These problems are converted to problems with *whole-number* divisors by moving the decimal points in the divisor and the dividend the same number of places to the right. But why does this method work?

$0.4\overline{\smash{)}\,0.08}$	becomes	$4\overline{\smash{)}\,0.8}$
(Unfamiliar, new)		(Hey, we just studied these!)

What allows us to move the decimal points in the dividend and the divisor the same number of places? It had better not affect the answer to the division problem. If it does, we can't go around teaching children to do it!

Lesson Exercise 7.29

Consider $8 \div 2$.

(a) Multiply both original numbers by 10 and divide. Is the quotient still the same?

(b) Multiply both original numbers by 100 and divide. Is the quotient still the same?

(c) Multiply both original numbers by 1000 and divide. Is the quotient still the same?

(d) Part (a) works because $8 \div 2 = \frac{8}{2} = \frac{8}{2} \times \frac{10}{10} = \frac{80}{20} = 80 \div 20$. Show why part (b) works.

(e) For any two decimal numbers a and b where $b \neq 0$, why does $a \div b = (ac) \div (bc)$, in which $c \neq 0$? (*Hint:* Use the same approach as in part (d).)

Lesson Exercise 7.29 verifies the following conclusion.

Equivalent Division

If a, b, and c are decimal numbers, with $b \neq 0$ and $c \neq 0$, then $a \div b = (a \cdot c) \div (b \cdot c)$.

This means that we can multiply both numbers in a division problem by the same nonzero number without affecting the quotient. The equivalent-division property is merely a different form of the Fundamental Law of Fractions! The expression $a \div b = (a \cdot c) \div (b \cdot c)$ is the same as $\frac{a}{b} = \frac{a \cdot c}{b \cdot c}$. This conclusion can also be justified using multiplication by 1 in the form $\frac{c}{c}$. For nonzero b and c,

$$a \div b = \frac{a}{b} = \frac{a}{b} \cdot \boxed{\frac{c}{c}} = \frac{ac}{bc} = ac \div bc$$

Lesson Exercise 7.30

A child wants to know why we can move the decimal point to change $0.4\overline{)0.08}$ to $4\overline{)0.8}$.

(a) By what do you multiply both the divisor and dividend when you change $0.4\overline{)0.08}$ to $4\overline{)0.8}$?

(b) Write $0.08 \div 0.4$ as a fraction, and show how it is converted to $0.8 \div 4$.

(c) Compute $0.08 \div 0.4$.

As Lesson Exercise 7.30 indicates, the equivalent-division property (i.e., the Fundamental Law of Fractions) can be used to convert any decimal divisor problem to a whole-number divisor problem by moving the decimal points in the divisor and the dividend the same number of places to the right.

Decimal division is also used to change fractions to decimals.

Lesson Exercise 7.31

(a) $\frac{3}{11}$ means _____ ÷ _____.

(b) Use paper and pencil to complete the division in part (a), and tell why the decimal representation of $\frac{3}{11}$ is called a repeating decimal.

Decimal division is also useful in certain applications. Can you correctly classify the following division application?

Lesson Exercise 7.32

"A National Motors Finite SST travels 460.8 miles on 16.2 gallons of gas. How many miles does the Finite SST get per gallon?" What division category does this illustrate?

AN INVESTIGATION: MULTIPLYING AND DIVIDING DECIMALS

***Cooperative* Lesson Exercise 7.33**

Consider the following problem. "In multiplying positive decimals x and y, determine what would make the product

(a) greater than x. (b) equal to x. (c) less than x."

Devise a plan and solve the problem.

***Cooperative* Lesson Exercise 7.34**

In dividing positive decimal x by positive decimal y, determine what would make the quotient $x \div y$

(a) greater than x. (b) equal to x. (c) less than x.

Instead of computing the answers, apply your generalizations from Lesson Exercises 7.33 and 7.34 to do Lesson Exercises 7.35 and 7.36.

Lesson Exercise 7.35

Peaches cost \$.69/lb. You buy 0.8 lb. The price you pay is

(a) greater than \$.69 (b) \$.69 (c) less than \$.69

Lesson Exercise 7.36

You need 12 kg of soil, and each bag holds 0.8 kg. How many bags do you need?

(a) More than 12 (b) 12 (c) Fewer than 12

COMMON ERROR PATTERNS

Students regularly make certain errors in decimal arithmetic. The following exercises will help you recognize some common error patterns.

In Lesson Exercises 7.37–7.39, (a) complete the last two examples, repeating the error pattern of the completed examples; (b) write a description of the error pattern; and, (c) if possible, explain how estimation could be used to detect errors.

Lesson Exercise 7.37

$$\begin{array}{r} 0.6 \\ +\ 0.3 \\ \hline \mathit{0.9} \end{array} \qquad \begin{array}{r} 0.8 \\ +\ 0.9 \\ \hline \mathit{0.17} \end{array} \qquad \begin{array}{r} 0.7 \\ +\ 0.6 \\ \hline \mathit{0.13} \end{array} \qquad \begin{array}{r} 0.9 \\ +\ 0.2 \\ \hline \end{array} \qquad \begin{array}{r} 0.5 \\ +\ 0.8 \\ \hline \end{array}$$

Lesson Exercise 7.38

$$\begin{array}{r} 0.6 \\ \times\ 0.9 \\ \hline \mathit{5.4} \end{array} \qquad \begin{array}{r} 0.3 \\ \times\ 0.2 \\ \hline \mathit{0.6} \end{array} \qquad \begin{array}{r} 0.7 \\ \times\ 0.3 \\ \hline \mathit{2.1} \end{array} \qquad \begin{array}{r} 0.8 \\ \times\ 0.7 \\ \hline \end{array} \qquad \begin{array}{r} 0.6 \\ \times\ 0.4 \\ \hline \end{array}$$

Lesson Exercise 7.39

$$\begin{array}{r} \mathit{17} \\ 0.3\overline{)3.21} \end{array} \qquad \begin{array}{r} \mathit{3.5} \\ 5\overline{)15.25} \end{array} \qquad \begin{array}{r} \\ 8\overline{)5.672} \end{array} \qquad \begin{array}{r} \\ 4\overline{)36.16} \end{array}$$

SPREADSHEETS

Many elementary-school classrooms now make use of spreadsheets. A **spreadsheet** is a table with rows and columns. Columns are usually identified by letters, and rows are usually identified by numbers. Figure 7-7 on the next page, shows part of a discussion about spreadsheets in the fourth-grade *Mathematics Plus* textbook.

Each location in the spreadsheet is called a **cell**. The label "Perez," for instance, is located in cell A3. Cells may also contain a value (number) or a formula. The cell that is highlighted by the computer is called the **active cell**. You can move from one cell to another using the four arrow keys.

SPREADSHEET MATH

Working with Formulas

A spreadsheet can be used as an organizer and a calculator. You can type a formula in any of the cells. The formula can tell the computer to add, subtract, multiply, or divide the numbers in any row or column of cells. Both the formula and the answer can live in the same cell, just as you can live at the same address as your family.

In the previous lesson, you calculated how much it cost to print names on T-shirts, and you organized the information on the spreadsheet. In this lesson, you will use spreadsheet formulas to answer these five questions.

- **a.** If the cost per letter is $.65, how much will it cost to print each name?
- **b.** What is the total number of letters?
- **c.** How much will it cost to print everybody's name?
- **d.** What is the average cost of printing a name?
- **e.** What is the average number of letters in a name?

Here is the spreadsheet without the formulas.

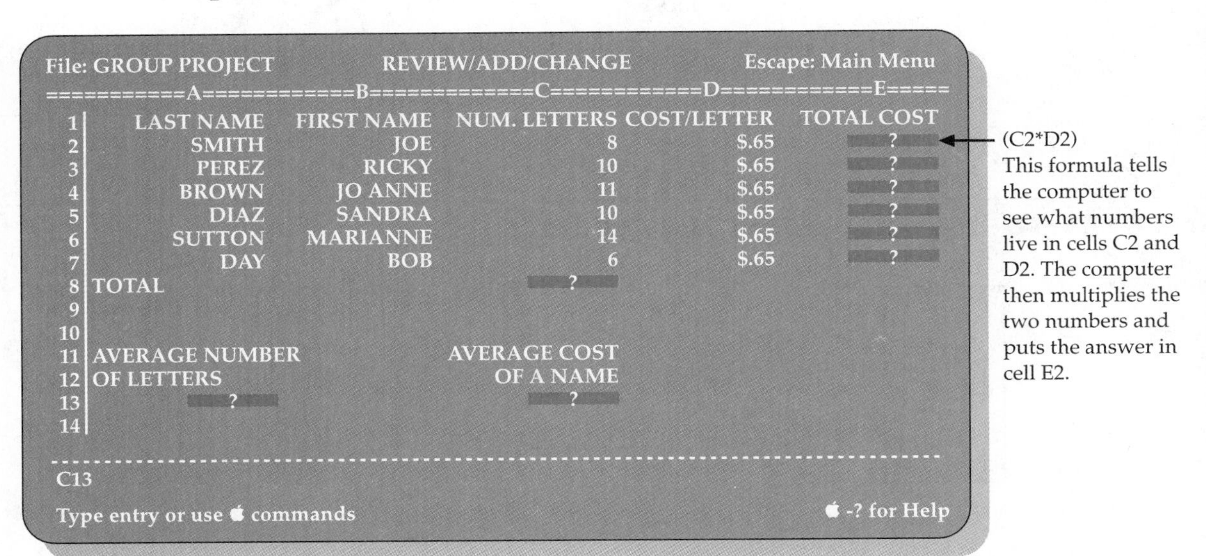

File: GROUP PROJECT REVIEW/ADD/CHANGE Escape: Main Menu

	A	B	C	D	E
1	LAST NAME	FIRST NAME	NUM. LETTERS	COST/LETTER	TOTAL COST
2	SMITH	JOE	8	$.65	?
3	PEREZ	RICKY	10	$.65	?
4	BROWN	JO ANNE	11	$.65	?
5	DIAZ	SANDRA	10	$.65	?
6	SUTTON	MARIANNE	14	$.65	?
7	DAY	BOB	6	$.65	?
8	TOTAL		?		
9					
10					
11	AVERAGE NUMBER		AVERAGE COST		
12	OF LETTERS		OF A NAME		
13	?		?		
14					

C13

Type entry or use commands -? for Help

From Mathematics Plus, *Grade 4 (San Diego: Harcourt Brace, 1994), p. 450.*

Figure 7–7

Lesson Exercise 7.40

Refer to Figure 7-7 on page 347.

(a) Use the column and row headings to give the location of the label "SMITH."

(b) What is the value of the cell C5?

(c) What is a formula for cell E3? (*Note:* The multiplication symbol is *.)

(d) What is a formula for cell C8?

(e) What is a formula for cell A13?

Depending upon your computer spreadsheet, you may have to type in a symbol such as =, +, or @ in front of all formulas. In some programs, the formula in cell C8 can be written SUM(C2. . .C7).

In rows 2–7, column E is column C times column D. After typing the formula C2*D2 in cell E2, you can replicate (copy) this formula into cells E3, E4, E5, E6, and E7.

Lesson Exercise 7.41

(a) Type the labels and values from the first eight rows of Figure 7-7 into your computer spreadsheet, and the formula into cell E2.

(b) Find out how to replicate (copy) the formula from E2 into cells E3, E4, E5, E6, and E7.

(c) Enter the formulas into cells C8 and E8, putting the appropriate symbol in front of each formula.

ANSWERS TO SELECTED LESSON EXERCISES

7.20

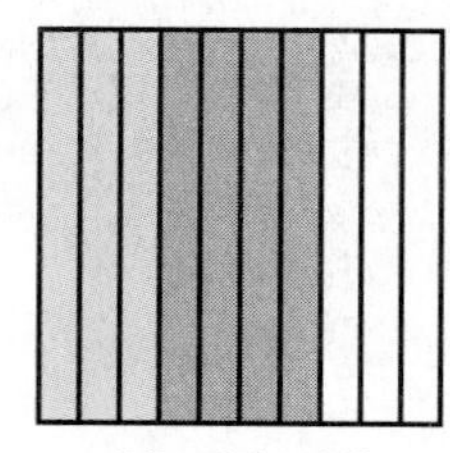

$0.3 + 0.4 = 0.7$

7.22

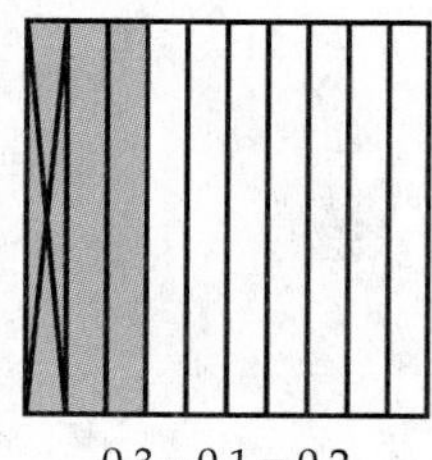

$0.3 - 0.1 = 0.2$

7.21 (a) $0.36 + 0.40 = \frac{36}{100} + \frac{40}{100} = \frac{(36 + 40)}{100}$

7.23 Compare sets/measures

7.24 (a) 0.6×0.4 means 0.6 (6 tenths) of 0.4. Dot 0.4 of a decimal square.

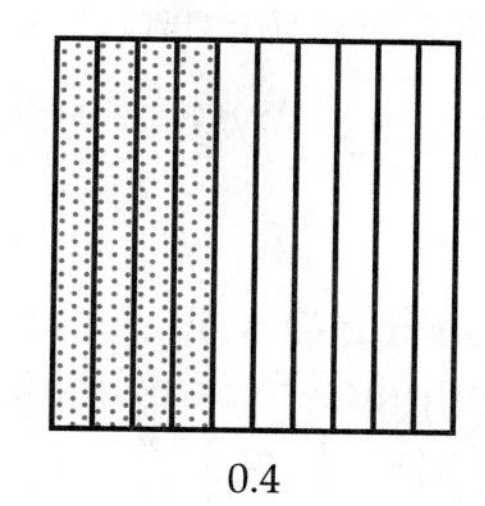
0.4

Now darken 6 tenths of the 0.4.

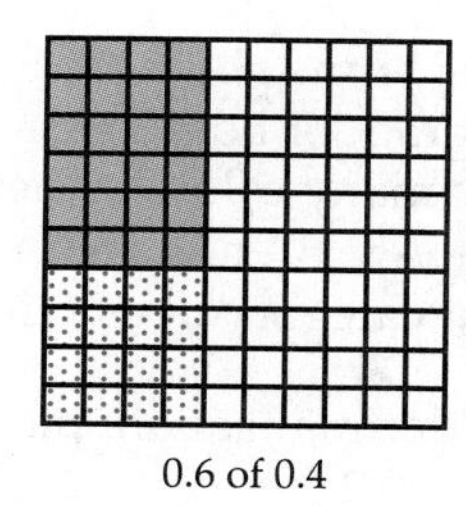
0.6 of 0.4

0.24 of the decimal square is darkened. So $0.6 \times 0.4 = 0.24$.

(b) $0.6 \times 0.4 = \frac{6}{10} \times \frac{4}{10} = \frac{24}{100} = 0.24$

(c) 6 tenths $\times$ 4 tenths = 24 hundredths

7.25 $12 \cdot 7.95 \approx 12 \cdot 8 = 96$. The answer is (a).

7.26 (a)

No. of Decimal Places		
Factor 1	Factor 2	Product
0	2	2
1	1	2
0	2	2

7.27 Multiplication, repeated measures

7.28

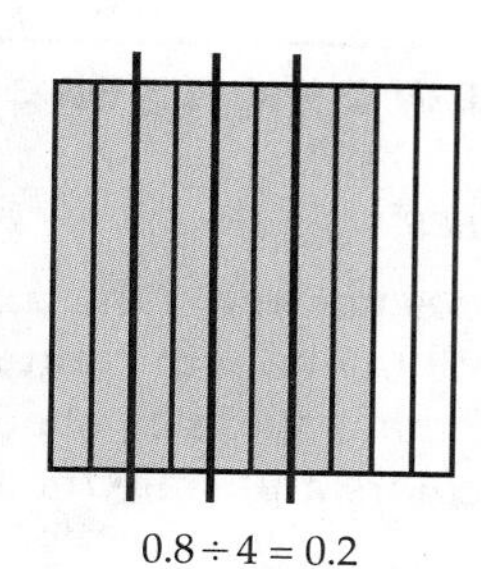
$0.8 \div 4 = 0.2$

7.29 (a) Yes (b) Yes (c) Yes

(d) $8 \div 2 = \frac{8}{2} = \frac{8}{2} \times \frac{100}{100} = \frac{800}{200} = 800 \div 200$

(e) $(a \cdot c) \div (b \cdot c) = \frac{ac}{bc} = \frac{a}{b} = a \div b$

7.30 (a) 10

(b) $0.08 \div 0.4 = \frac{0.08}{0.4} = \frac{0.08 \times 10}{0.4 \times 10} = \frac{0.8}{4} = 0.8 \div 4$

(c) 0.2

7.31 (a) 3; 11

(b) 0.272727 . . . ; the digits 2 and 7 repeat over and over.

7.32 Partition, measures

7.35 (c)

7.36 (a)

7.37 (a) 0.11; 0.13

(b) The sum from the tenths column is all placed to the right of the decimal point in the answer.

(c) $0.8 > 0.17$. How can $0.8 + 0.9 = 0.17$?

7.38 (a) 5.6; 2.4

(b) The product is given the same number of decimal places as each of the factors.

(c) $0.6 \times 0.9 \approx 0.6 \times 1 = 0.6$. How can $0.6 \times 0.9 = 5.4$?

7.39 (a) 0.79; 9.4

(b) Zeroes are omitted from the quotient.

(c) $0.3 \times 17 \approx \frac{1}{3}$ of $17 \neq 3.21$

7.40 (a) A2 (b) 10 (c) C3*D3

(d) SUM(C_2...C_7) (e) C8/6

7.2 HOMEWORK EXERCISES

Basic Exercises

1. A child does not know the rule for adding decimals. Show how to compute 0.2 + 0.3 in a single decimal square by shading in two different colors and state the result in an equation.

2. (a) A child knows fraction addition but not decimal addition. Show that 0.321 + 0.127 $= \dfrac{(321 + 127)}{1000}$.
(b) How does part (a) suggest the standard rule for adding terminating decimals?

3. Consider the following addition problem.

$$\begin{array}{r} {}^{1} \\ 0.36 \\ +\ 0.27 \\ \hline 0.63 \end{array}$$

(a) When you add 6 + 7 and separate the 13 into 3 and 10, this 10 represents 10 __________.
(b) The 1 that is regrouped represents 1 __________.
(c) Are the amounts in parts (a) and (b) equal?

***4.** Draw a decimal-square picture that shows the result of 0.4 − 0.3, using a take-away approach, and show the result in an equation.

5. What operation and category are illustrated in the following problem? "Joe bought a bag of gourmet plant food for $9.89. How much change did he receive back from a $100 bill?"

6. A child knows fraction multiplication but does not know the rule for decimal multiplication. *Explain* how to compute 0.5 × 0.6 using two decimal-square pictures.

7. What decimal multiplication problem does the following decimal-square picture illustrate?

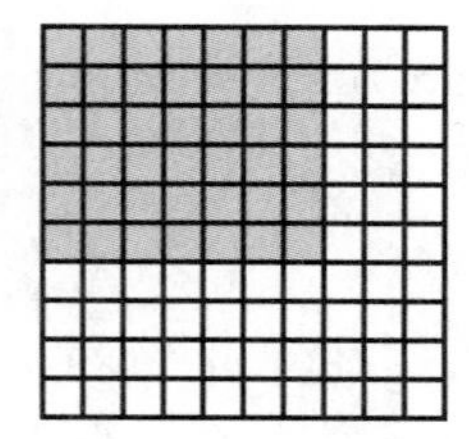

8. In computing (0.2)(0.16), we multiply 2 by 16 and then place the decimal. Change (0.2)(0.16) to $(2 \cdot 16) \cdot \dfrac{1}{1000}$ by writing 0.2 and 0.16 as fractions.

9. In computing (2.64)(0.3), we multiply 264 by 3 and then place the decimal. Change (2.64)(0.3) to $(264 \cdot 3) \cdot \dfrac{1}{1000}$ by writing 2.64 and 0.3 as fractions.

10. Compute the answer to 0.7 × 0.4 by writing both factors as fractions and multiplying.

11. A runner burns about 0.12 calorie per minute per kilogram of body mass.
(a) How many calories does a 60-kg runner burn in a 10-minute run?
(b) What category does this illustrate?

***12.** Draw a decimal-square picture that shows that 0.6 ÷ 3 = 0.2.

13. What do you multiply both numbers by when you change 72 ÷ 3.6 to 720 ÷ 36?

14. Draw a decimal-square picture that shows that 0.8 ÷ 0.2 = 4. (*Hint:* Use repeated measures.)

15. A child wants to know why we can move the decimal point to change 6.4 ÷ 0.32 to 640 ÷ 32. Write $0.32\overline{)6.4}$ as a fraction, and show how the process works.

16. If you were asked to compute 5000 ÷ 12, you could solve a simpler equivalent problem such as 2500 ÷ 6 using the equivalent-division property. Change each problem to a simpler one.

(a) 6000 ÷ 24 (b) $8 \div \dfrac{2}{3}$

17. Which of the following are equal?
(a) 8 ÷ 0.23 (b) 800 ÷ 0.0023
(c) 80 ÷ 2.3 (d) 0.8 ÷ 0.023
(e) 80 ÷ 0.023

18. (a) $\frac{2}{9}$ means _____ ÷ _____.

(b) Complete the division, and explain why the decimal representation of $\frac{2}{9}$ is called a repeating decimal.

19. What operation and category are illustrated in the following problem? "A baker needs 2 kg of flour to make a loaf of bread. How many loaves can he bake from 5 kg of flour?"

20. What operations and categories are illustrated in the following problem? "At the beginning of a 4-day car trip, an odometer read 58,427.7. At the end of the trip, it read 59,271.5. What was the average number of miles driven per day?"

21. What operations and categories are illustrated in the following problem? "Molly bought 4 tires for $37.28 each, an air pump for $11.43, and 2 windshield wipers for $3.10 each. How much did these items cost altogether?" (Assume that tax is already included.)

22. At a Chinese restauant, Szechuan chicken costs $7.50, Hunan shrimp costs $9.25, and moo shi pork costs $6.95. Write two multi-step mathematics problems that could be answered using this information.

23. If $a \times 0.3 = b$, then _____ $\times\ b = a$.

In Exercises 24–26, (a) complete the last two examples, repeating the error pattern in the completed examples, and (b) write a description of the error pattern.

24.

42	8.1	63	4.2
− 3.71	− 3.71	− 5.29	− 3.17
39.71	*4.41*		

25.

6.2	*3.5*		
4⟌26	8⟌29	7⟌36	5⟌93

26.

16.2	14.1	12.3	8.2
− 3.7	− 2.5	− 6.7	− 4.8
13.5	*12.4*		

27. Describe two errors children might make in computing 0.7×0.8.

28. A checkbook can be recorded in a spreadsheet. Enter the following on a computer spreadsheet if you have one. Otherwise, you can do the exercise without a computer.

	A	B	C	D	E	F
1	**Checkbook**					
2	Check #	Date	For	Withd.	Deposit	Balance
3	Starting					$3426.10
4	123	3-12	Rent	$950		
5		3-14			$875.40	
6	124	3-16	Electric	$85.11		
7	125	3-17	Shoes	$62.25		

(a) What is the location of $950?
(b) What is the value of cell C6?
(c) What is the formula for cell F5?

29. A teacher may keep students' grades in a spreadsheet. (Enter the following on a computer spreadsheet if you have one.)

	A	B	C	D	E	F
1	Student	Test1	Test2	Project	Total	Average
2	Nabila	86	90	70		
3	Dennis	73	84	65		

(a) What is the location of the 84?
(b) What does the number in cell B3 represent?
(c) What is the formula for cell E2?
(d) What is the formula for cell F2 if all three scores count equally?

30. Match each fraction with its location on the number line.

$\frac{126}{66}$ $\frac{436}{318}$ $\frac{67}{38}$

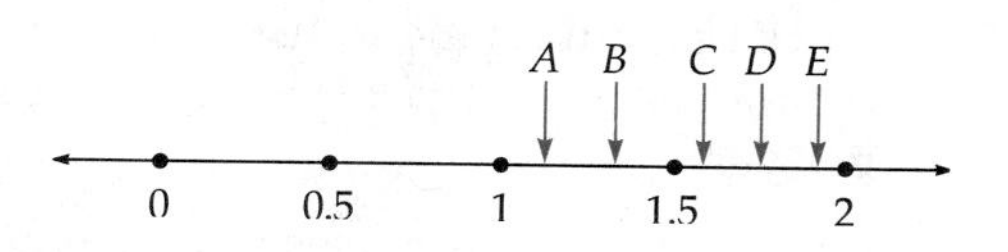

31. Write the next number in the sequence 0.8, 0.16, 0.032, 0.0064, ...

32. Multiply 0.15×0.2 on your calculator. Explain why the answer does not have three decimal places, as the multiplication rule suggests.

33. If a computer performs a computation in 0.0002 second, how many computations can it do in 1 minute?

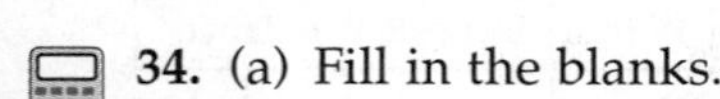

34. (a) Fill in the blanks.

$$0.1089 \times 9 = \_\_\_\_\_\_$$
$$0.10989 \times 9 = \_\_\_\_\_\_$$
$$0.109989 \times 9 = \_\_\_\_\_\_$$

(b) What is the next equation if the pattern continues?
(c) Is your equation in part (b) true?

35. (a) Fill in the parentheses.

$$1 \times 2 + 0.25 = (\ \ )^2$$
$$2 \times 3 + 0.25 = (\ \ )^2$$

(b) What is the next equation if the pattern continues?
(c) Is your equation in part (b) true?

36. The neocortex of a human brain has about 3×10^{10} neurons. A gorilla's neocortex has about 7.5×10^9 neurons.
(a) Which brain has more neurons?
(b) How many times more neurons does the larger brain have than the smaller?

37. In general, a car that has automatic transmission gets about 2 miles per gallon (mpg) less than the same type of car with manual transmission. Assume that a particular car gets 30 mpg with automatic transmission.
(a) If gas costs \$1.10 per gallon and you drive 10,000 miles per year, an automatic transmission increases your gas costs about ________ per year.
(b) Maintenance for an automatic transmission costs about \$160 more per year than for a manual transmission. For this car, the extra cost of an automatic transmission, in maintenance and gas, is about ________ per year.

38. An air conditioner decreases a car's average annual mileage by about 2 mpg. If gas costs \$1.10 per gallon and you drive 10,000 miles per year, an air conditioner increases gas costs about ________ per year. (Make up the car's mileage per gallon.)

39. An intermediate-sized car gets about 5 mpg less than a subcompact car. Driving 10,000 miles per year and paying \$1.10 per gallon of gas would cost about ________ more for gas in the intermediate-sized car than in the subcompact.

Extension Exercises

40. A 22-year-old woman owns a 2-year-old car. Consider the following facts. She drives it about 8000 miles per year. Maintenance on the car costs \$.08 per mile. Gas costs an average of \$1.30 per gallon. The car averages 26 miles per gallon. Insurance costs \$480 per year.
(a) About how much does maintenance cost for a year?
(b) About how much does gas cost for a year?
(c) Excluding car loan payments, about how much does it cost to operate the car for a year?

41. (a) Graph $xy = 16$ for positive decimals x and y.
(b) Graph $xy = 12$ for positive decimals x and y.
(c) Write a generalization about the graph of $xy = k$, in which k is a counting number.

42. Make up two different decimal multiplication problems that have answers of 0.2. Do not use 1 as a factor.

43. Consider the following problem. "In charging a customer, a cashier interchanges dollars and cents in the price of an item and gives the customer an extra \$11.88 in change for a \$20 bill. What is the correct price of the item?"
(a) Devise a plan and solve the problem.
(b) Find all possible solutions.

44. Numbers can be multiplied in scientific notation. Fill in the reason for each step.

$280 \times 260{,}000$
$= (2.8 \times 10^2) \times (2.6 \times 10^5)$ ________________

$= (2.8 \times 2.6) \times (10^2 \times 10^5)$ ________________ (Two properties)

$= 7.28 \times 10^7$ Rules of exponents and multiplication

45. A radar device bounces waves off a plane. The waves take 4×10^{-6} second to travel back and forth. If waves travel 1.8×10^5 miles/s, how far away is the plane?

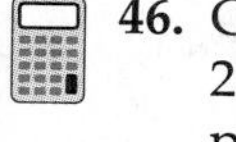

46. Consider the following problem. "The sum of 2 two-digit decimal numerals is 0.42. Their product is 0.0297. What are the numbers?" Devise a plan and solve the problem.

47. If $x > 0$, for what values of x is x^{-3} a negative number?

48. Assume that $x^y = 1$, in which $x > 0$ and $y > 0$. Give all possible solutions.

49. You can use the clustering strategy to estimate total costs. For example, in estimating the total cost of grocery items at $1.29, $.45, $2.45, and $1.09, you can group items that add up to about $1.00 or $2.00. The first two items cost about $2; if you add $2 for the third plus $1 for the fourth, the total is about $5. Mentally estimate the total of each of the following groups of prices.
(a) $1.59, $.30, $3.10, $1.15, $.72, $2.00, $1.59, $.89, $2.29
(b) $3.71, $2.62, $.51, $.30, $26.95, $9.98, $4.25

50. Suppose a long-distance call from Cleveland to Pittsburgh costs $.2696 per minute. The cost of a 12.2-minute call is about
(a) $28 (b) $4000 (c) $.40 (d) $3

51. Ray takes 18 strides to walk across a room. His stride is about 0.68 m long. How can you estimate the distance across the room?

52. Compute $3.12_{\text{five}} + 2.33_{\text{five}}$.

53. (a) Write 1/3 as a decimal.
(b) Write 1/3 in base three.
(c) Write 1/3 in base six.
(d) Name another base in which 1/3 is a terminating decimal, and write it as a decimal in that base.

Special Exercises

54. The following chart gives the energy consumption of some common electric appliances.

Appliance	Usage	Kilowatt-Hours (kwh) Used
Oven	1 hour	1.1
TV (color)	1 hour	0.3
Refrigerator	1 month	120
Frost-free refrigerator	1 month	200
Dishwasher	1 cycle	0.6
Fan	1 hour	0.2
Room air conditioner	1 hour	1.5
Vacuum cleaner	15 minutes	0.2
Clock	1 month	1.8
Iron	1 hour	0.8
Light bulb (100 W)	1 hour	0.1
Baseboard heater	1 hour	2
Clothes washer	1 cycle	0.2
Clothes dryer	1 cycle	3
Water heater	1 month	350

Using a rate of $.08 per kilowatt-hour, find the cost of the following.
(a) Watching color TV for 2 hours
(b) Doing one load of laundry in the washer and dryer
(c) Running a refrigerator for 1 month
(d) Running a water heater for 1 month
(e) Running a room air conditioner for 8 hours

55. Refer to the chart in the preceding exercise, and consider the following. A couple owns an oven, a refrigerator, two clocks, two room air conditioners, a vacuum cleaner, a color TV, a water heater, an iron, and eight 100-watt bulbs. Assume that they run the air conditioner for 100 hours one month.

Estimate how much they will use each other appliance in a month, and then estimate their total electric bill for the month at a rate of $.09 per kilowatt-hour.

7.3 RATIO AND PROPORTION

In selecting a college, you may have been interested in comparing the number of students and the number of professors or the number of men and the number of women.

RATIOS

Suppose Brain Strain College has 800 students and 50 professors. We could say either that there are 750 more students than professors or 16 times as many students as professors. Which would be a more useful comparison?

Lesson Exercise 7.42

A second college, Tom Cruise College, has 775 students and 25 professors. Compare the number of students to the number of professors in two ways.

(a) There are ____ more students than professors.

(b) There are ____ times as many students as professors.

(c) Which would be better for comparing this college to Brain Strain College, part (a) or part (b)?

Both colleges have 750 more students than professors, but Brain Strain has significantly fewer students per professor. The number of students per professor (called the "student-professor ratio") gives the more useful figure.

At Brain Strain College, the student-professor ratio is 16 to 1. At Tom Cruise College, it is 31 to 1. These ratios could also be written using colons (16:1 and 31:1) or as fractions (16/1 and 31/1). The ratios tell us that Tom Cruise College has about twice as many students per teacher as Brain Strain College.

In many cases, the most useful comparison between two sets or two measures is how many times larger one set or measure is than the other. That is what a ratio of two sets shows. Common applications include ratios of students to teachers, miles to gallons, men to women, pounds to cubic feet, and dollars to hours.

A ratio compares two numbers using division. A ratio is another name for a fraction or quotient.

Definition **Ratio**

A **ratio** is a comparison of two numbers a and b, with $b \neq 0$. It can be written a to b, $a : b$, or $\frac{a}{b}$.

In order to have a correspondence between ratios and fractions, the second number in a ratio is not allowed to be 0, as in the ratio of men to women on the New York Yankees, 25 to 0, since $\frac{25}{0}$ is undefined. One can compare the same two groups in accordance with the definition by looking at the ratio of women to men in the New York Yankees $\left(0 \text{ to } 25, \text{ or } \frac{0}{25}\right)$.

People often use ratios to compare the relative sizes of two groups or measurements.

Lesson Exercise 7.43

A car travels 650 miles on 20 gallons. What is the ratio of miles to gallons?

Some middle-school textbooks also use the term "rate." A **rate** is a ratio that compares two quantities with different units, such as 20 miles/gallon. A ratio with no units (that is, in which units cancel out), such as 8 inches/1 inch = 8:1, is not a rate.

PROPORTIONS

Suppose you're driving one of those cars with automatic cruise control. You set it at 50 mph and sit back and relax. ZZZZ. Wake up! You still have to steer the car.

The car travels 50 miles/hour. You will travel 100 miles in 2 hours, 150 miles in 3 hours, and so on. When you double the driving time, the miles traveled also double. When you triple the driving time, the miles traveled also triple.

The ratios of miles to hours—for example, 100 to 2 and 150 to 3—are equal because $\frac{100}{2} = \frac{150}{3}$. The equation $\frac{100}{2} = \frac{150}{3}$ is a proportion. A **proportion** states that two ratios are equal. In this application, one would say that the distance traveled "is proportional to" the driving time.

Lesson Exercise 7.44

The weight of a pile of bricks is proportional to its volume. Suppose that 10 ft^3 of bricks weigh 30 lb.

(a) How much would 40 ft^3 of bricks weigh?

(b) Write a proportion that relates the two piles of bricks.

In Chapter 6, you learned that two rational numbers, $\frac{a}{b}$ and $\frac{c}{d}$, are equal if and only if their cross products are equal, that is, if $ad = bc$. This tells us that $\frac{a}{b} = \frac{c}{d}$ is a proportion if and only if $ad = bc$. The cross-product equation $ad = bc$ is often used to find a missing number in the proportion $\frac{a}{b} = \frac{c}{d}$.

How are proportions used? Proportions with one unknown value occur in a wide variety of applications, including gas mileage, recipes, map scales, currency conversion, and time needed to complete a task.

Sometimes you can find the unknown value in a proportion by looking at the relationships between the numerators and denominators. Then you don't have to find cross products! Try this in Lesson Exercise 7.45(b).

Lesson Exercise 7.45

The ratio of boys to girls in a class is 4:5.

(a) What is the ratio of girls to boys?

(b) There are 20 girls in the class. How many boys are there? (Write a proportion and see whether you can solve it *without* doing the cross multiplication.)

More often, proportion problems are solved using cross products, as in the following example.

EXAMPLE 7.5

Suppose that you want to know how far your car will go on a full 10-gallon tank of gas. Based upon your records, your car recently traveled 109 miles on 4 gallons of gas.

(a) Estimate how far you can go on a full tank.

(b) Compute the exact answer using a proportion.

Solution

The miles traveled are proportional to the number of gallons (Figure 7-8).

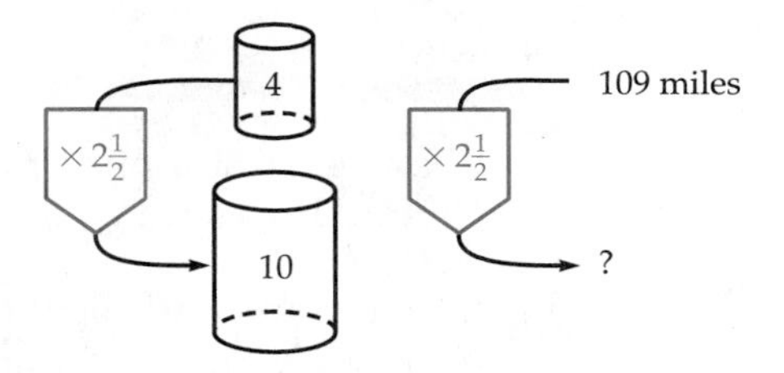

Figure 7–8

(a) Since 10 is $2\frac{1}{2}$ times 4, you will go about $2\frac{1}{2}$ times as far.

$$2\frac{1}{2} \times 109 \text{ miles} \approx 2\frac{1}{2} \times 100 \text{ miles} = 250 \text{ miles}$$

(b) Solve the problem as follows.

(1) Select a variable for the unknown. Let D = the distance you can travel on a full tank.

(2) Write a proportion. Make sure your units match up when you write the proportion. (There is more than one correct way to write it.)

$$\frac{109 \text{ miles}}{4 \text{ gallons}} = \frac{D \text{ miles}}{10 \text{ gallons}}$$

Other correct proportions include

$$\frac{109 \text{ miles}}{D \text{ miles}} = \frac{4 \text{ gallons}}{10 \text{ gallons}}$$

$$\frac{D \text{ miles}}{109 \text{ miles}} = \frac{10 \text{ miles}}{4 \text{ gallons}}$$

and $$\frac{4 \text{ gallons}}{109 \text{ miles}} = \frac{10 \text{ gallons}}{D \text{ miles}}$$

(3) Solve the proportion using cross products.

$$\frac{109}{4} = \frac{D}{10}$$

So $4D = 1090$. (Note that all four proportions have the same two cross products!) Dividing both sides by 4, we obtain $D = \frac{1090}{4} = 272.5$ miles. The estimate in part (a) confirms that this answer is reasonable. ■

Lesson Exercise 7.46

You want to cook a new dish called "seaweed surprise" for 7 people. The recipe for 4 people calls for 18 ounces of seaweed.

(a) Estimate how much seaweed you will need.

(b) Find the exact answer by using a proportion. (Make sure that your units match up.)

(c) Estimate how many friends would accept a subsequent dinner invitation.

When you solve a proportion using cross products, you multiply and then divide. Intuitive readers will be interested in the fact that problems that are solved with proportions can also be worked out without the actual writing of a proportion. You can solve such a problem by using some multiplication and/or division based upon your understanding of the proportional relationships.

EXAMPLE 7.6

Solve the problem in Example 7.5 without writing a proportion.

Solution

Two methods for solving the proportion are as follows.

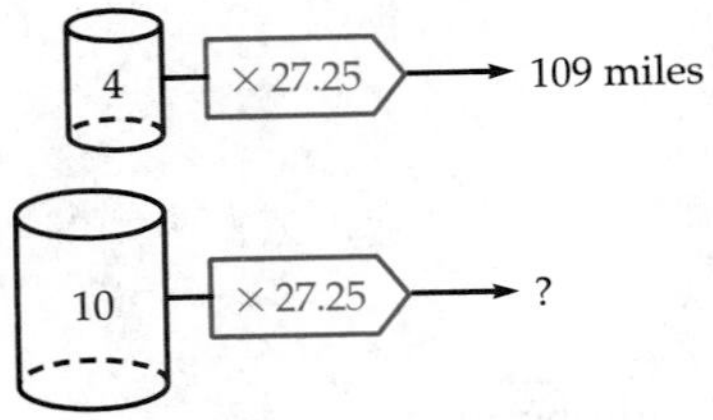

The Unitary Method Figure out how many miles per *unit* (gallon) the car gets by dividing $\frac{109 \text{ miles}}{4 \text{ gallons}} = 27.25$ miles/gallon (Figure 7-9). Then, to figure out the mileage for 10 gallons, compute $27.25 \times 10 = 272.5$ miles.

Figure 7–9

The Multiplier Method This method was used to make the estimate in Example 7.5(a). Figure out *how many times* farther you are going by dividing $\frac{10 \text{ gallons}}{4 \text{ gallons}} = 2.5$. The multiplier is 2.5 (Figure 7-10).

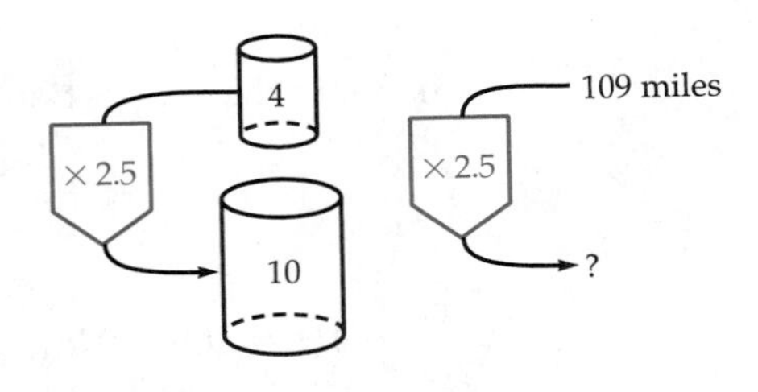

Figure 7–10

To figure out the mileage for 10 gallons, *multiply* $109 \times 2.5 = 272.5$ miles. ■

Lesson Exercise 7.47

Solve Lesson Exercise 7.46 without writing a proportion.

Lesson Exercise 7.48

Luis takes $3\frac{1}{2}$ hours to type 16 pages. How long will it take him to type 44 pages?

(a) Estimate the answer.
(b) Find the exact answer by using a proportion.
(c) Find the exact answer without writing a proportion.
(d) Which was easier for you, part (b) or part (c)?

ANSWERS TO SELECTED LESSON EXERCISES

7.42 (a) 750 (b) 31 (c) Part (b)

7.43 65:2 or 32.5:1

7.44 (a) 120 lb (b) $\frac{10}{30} = \frac{40}{120}$

7.45 (a) 5:4 (b) 16

7.46 (b) $4x = 7 \cdot 18$, so $x = 31.5$ ounces

7.47 Compute $\frac{7}{4} \times 18$ ounces, or compute $\frac{18 \text{ ounces}}{4} \times 7$.

7.48 (b) $16 \cdot x = \left(3\frac{1}{2}\right) \cdot 44$, so $x = 9.625$ hours

(c) *Hint:* Either divide 16 by $3\frac{1}{2}$ or divide 44 by 16.

7.3 HOMEWORK EXERCISES

Basic Exercises

1. Is 4:3 the same as 8:6?
2. Is 7:4 the same as 6:3?
3. Two sets have a ratio of 10 to 3. Determine whether the ratio of the two sets will change if the number of members in each set is
   (a) doubled. (b) increased by 10.
4. An all-female college has 600 students.
   (a) What is the ratio of men to women?
   (b) Why is the ratio of women to men undefined?
5. The *pitch* of a roof is the ratio of its rise (height) to its span (width).

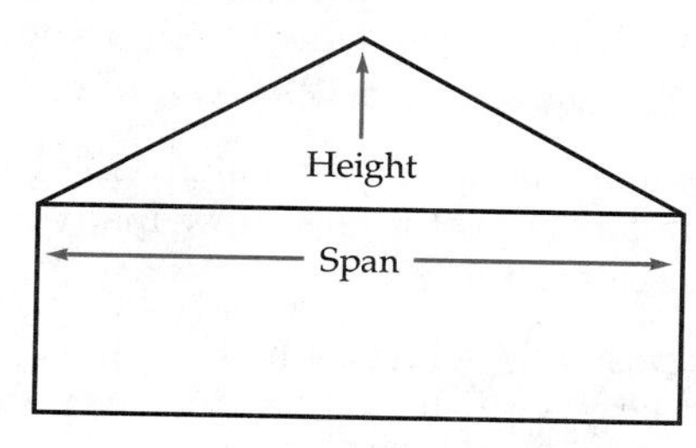

A roof has a rise of 8 ft and a span of 30 ft.
   (a) What is its pitch?
   (b) Write the pitch as a fraction in simplest form.
   (c) What does the pitch tell you about the roof?
6. The following table shows the frequencies of notes in a musical scale. Simple ratios exist between some of the frequencies.

Note	Middle C	D	E	F	G	A	B	High C
Frequency	262	294	330	349	393	440	494	524

   Find each of the following ratios of frequencies, and express it in simplest fractional terms.
   (a) E to A (b) Middle C to G
   (c) Middle C to high C
7. The following table lists the numbers of students and teachers in eight major public school systems, from the 1994 *Digest of Education Statistics*.

Metropolitan Area	Number of Students	Number of Teachers	Student-Teacher Ratio
New York City	983,971	54,376	
Los Angeles	639,781	25,465	
Chicago	411,582	23,342	
Miami	303,346	14,277	
Philadelphia	201,496	10,877	
Dallas	139,711	8,210	
Tampa	132,224	7,660	
San Diego	125,116	5,148	

(a) Complete the last column of the chart, showing the ratio of students to teachers.
(b) Which city has the lowest student-teacher ratio?
(c) Which city has the highest student-teacher ratio?
(d) Would you rather go to a school with a higher or lower student-teacher ratio?

8. Molly is 50, and her grandson Robert is 5. As they get older, the *ratio* of Molly's age to Robert's age will
(a) increase (b) decrease
(c) stay the same (d) do none of these

9. Estimate the answer, and then solve for Q in each proportion.
(a) $\frac{Q}{5} = \frac{7}{3}$ (b) $\frac{Q}{12} = \frac{4}{Q}$

10. Which of the following proportions can be solved easily without computing cross products?
(a) $\frac{3}{7} = \frac{T}{14}$ (b) $\frac{4}{5} = \frac{3}{N}$
(c) $\frac{5}{10} = \frac{13}{J}$ (d) $\frac{9}{7} = \frac{10}{K}$

11. Solve for R in each proportion just by looking at the relationship between each pair of fractions.
(a) $\frac{5}{7} = \frac{R}{14}$ (b) $\frac{3}{R} = \frac{6}{10}$ (c) $\frac{R}{18} = \frac{5}{6}$

12. If a British pound is worth \$1.63, how many pounds can I buy for \$10?
(a) Estimate the answer.
(b) Solve the problem by using a proportion. (Make sure that the units in the proportion match up.)
(c) Solve the problem without writing a proportion.

13. Last year at the Laundromat Users convention, 3216 people ate 1011 chickens at the Saturday afternoon picnic. This year, 3800 people are expected to attend. How many chickens should be ordered?

14. If 15 pounds of fertilizer take care of 2000 ft^2 of lawn, how much fertilizer is needed for a rectangular lawn that is 80 ft by 100 ft?

15. A couple drinks 3 quarts of orange juice and 2 quarts of skim milk each week. At this rate, how much orange juice and skim milk do they drink in 4 days?
(a) Estimate the answer.
(b) Find the exact answer by using a proportion.
(c) Find the exact answer without writing a proportion.

16. A school has 1200 students and a student-teacher ratio of 30 to 1. How many additional teachers must be hired to reduce the student-teacher ratio to 24 to 1?

17. In a large city, 2 million cars are used to commute to work. The average number of people per car is 1.2. If the average number of people per car were increased to 1.5, how many cars would be used? (*Hint:* Should the answer be more or less than 2 million?)

18. Frozen orange juice concentrate is usually mixed with water in a ratio of 3 parts water to 1 part concentrate. How much orange juice can be made from a 12-oz can of concentrate?

19. A college has a male-female ratio of 3 to 2. If there are 1100 students, how many men and women are there?

20. Suppose that a representative survey of adults in a large city shows that 663 support a sales tax increase and 837 are against it. If the adult population of the city is 4,200,000, predict how many adults support the sales tax increase.

21. A 20-ft-long pipe of uniform width and density is cut into two pieces. One piece is 12 ft long and weighs 140 pounds. How much does the other piece weigh?

22. Maria ate 5 times as many cookies as Debbie. They ate a total of 18 cookies. How many did each one eat?

23. Two triangles are **similar** if and only if the lengths of corresponding sides are proportional. The two triangles shown here are similar, and their corresponding sides are in corresponding positions.

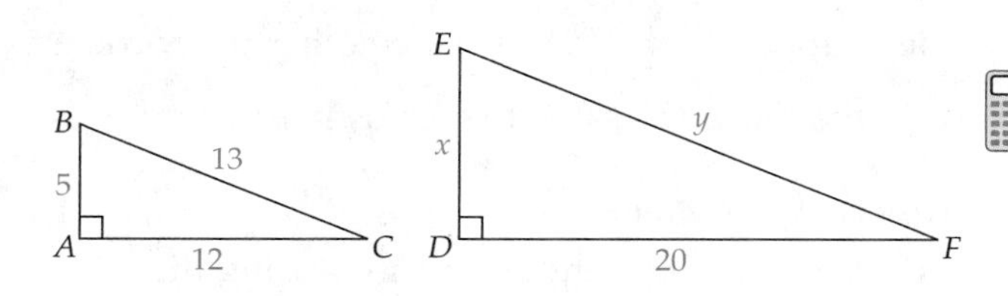

Find x and y by solving proportions.

24. A store sells 3 one-pound bags of pasta for \$1.87. How much would 2 cost? (*Note:* Stores generally round up *all* fractions of a penny.)

25. If 10 oranges cost \$1.89, how much do 12 oranges cost?

***26.** (a) An 18-ounce jar of peanut butter costs \$1.98. What is the cost per ounce (called the **unit price**)?
(b) A 32-ounce jar of peanut butter costs \$3.88. What is the cost per ounce?
(c) Which jar is a better buy (which has the lower unit price)?

27. Consider the following two sizes of cornflakes packages. Which is a better buy?

Item price	Unit price
\$1.25	\$1.67/lb
Soggy's Corn Flakes 12-oz box	

Item price	Unit price
\$1.59	\$1.41/lb
Soggy's Corn Flakes 18-oz box	

28. A 14-oz package of organic mushrooms costs \$2.40. A 20-oz package of organic mushrooms costs \$3.26. Which package is a better buy?

29. Suppose that a 22-oz bag of Giant wheat bread costs \$.69. A 22-oz bag of Wonder wheat bread costs \$1.29, and 1 lb of Pepperidge Farm wheat bread costs \$1.19.
(a) Which product would you buy?
(b) If all three brands had the same unit price, which would you buy?
(c) What is the most you would pay for a 1-lb loaf of Pepperidge Farm bread if the prices of the other loaves were as just listed?

30. Describe how someone might conduct a study to determine approximately how many students at your college are left-handed.

Extension Exercises

31. A 24-oz jar of Lipton iced tea mix costs \$3.29. It contains about $\frac{1}{2}$ oz of tea and $23\frac{1}{2}$ oz of sugar. How much would it cost you to mix your own tea and sugar in the same way, using Lipton tea (8 oz for \$3.79) and sugar (5 lb for \$2.39)?

32.

	Town A	Town B
Population	346,812	182,312
Number of restaurants	641	311

(a) If you had only this information, which town would appear to be a better place to open a restaurant?
(b) What other information should you obtain before deciding?

33. A car averages about 24 miles per gallon, a bus averages about 6 miles per gallon, and an electric train averages about 2 miles per gallon. Under what conditions would each vehicle be the most fuel-efficient?

34. Proportions can be used to estimate populations. A ranger catches and tags 10 fish in Lake Leisure. Then she tosses them back in the lake and lets all the fish mingle. Then the ranger catches 20 fish and finds that 6 of them are tagged. So she guesses that about 6/20 of the fish in the lake are tagged. Estimate the number of fish in Lake Leisure.

35. The ranger in the preceding problem goes to Lake Boggy. She catches and tags 20 fish. Then she releases them back into the lake. Now she catches 50 fish and finds that 4 of them are tagged. Estimate the number of fish in Lake Boggy.

36. It takes 15 minutes to cut a log into 3 pieces. How long would it take to cut a similar log into 4 pieces? (The answer is not 20 minutes.)

37. Consider the following problem. "Three years ago, Francisco and Melissa invested \$3000 and \$5500, respectively, in a business. Today the business is worth \$10,000. What is Francisco's share of the business worth?" Devise a plan and solve the problem.

38. A family has what they estimate to be a 30-day supply of food. However, after 10 days, only an 18-day supply is left. If they continue to eat at the same rate, the food will last a total of how many days?

39. A college class has a male-female ratio of 5:3. Then 3 more women join the class, changing the ratio to 10:7. How many students are now in the class?

40. Two classes have the same numbers of students. Class *A* has a boy-girl ratio of 3 to 5, and class *B* has a boy-girl ratio of 1 to 1. What is the ratio of boys in class *A* to boys in class *B*?

41. A class has 18 students. Which of the following could not be the ratio of girls to boys?
(a) 1:1 (b) 2:1 (c) 3:1 (d) 5:1
(e) 8:1

42. A college has a male-female ratio of M to F. What fraction of all the students are female?

43. A car gets M miles from G gallons. What is the distance it can travel on H gallons?

44. A painter finds that G gallons of paint will cover A square feet. How many gallons are needed to cover B square feet?

45. A family drinks Q quarts of milk in D days. How many quarts will they drink in C days?

46. Suppose $\frac{2}{x} = \frac{3}{y}$. Write three other proportions that have the same two cross products.

Special Exercises

47. Write 16 different proportions using the following numbers. Do not write any fractions that equal 1.

1 2 4 6 8 12

48. Draw a floor plan of your home that is to scale.

49. Go to your local supermarket and determine whether larger sizes of products always have lower unit prices. Write a summary of your findings.

7.4 PERCENTS: NOTATION, MENTAL COMPUTATION, AND ESTIMATION

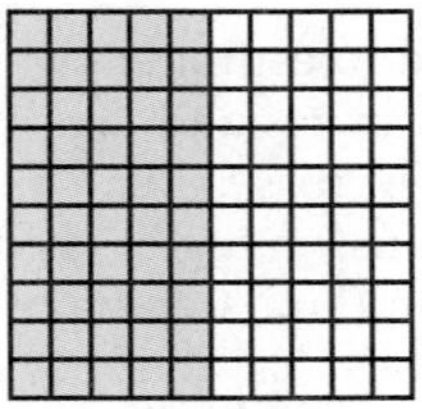

Do you think it is fair that a salesperson who sells twice as much should earn twice the commission, or that someone who owes three times as much money should pay three times as much interest? In situations such as these, many people consider it fair to charge the same rate on amounts of all different sizes. Percents were devised for just such situations!

Could you quickly compute the value of a 50% discount? How about a 10% tax on a \$200 item? The study of percents begins with simpler percents, such as 10%, 50%, and 100%. But first it would be helpful to know what the word "percent" means. "Cent" in "percent" means *100*, just as it does in the words "century" and "centipede." "Percent" means *per 100*. For example, "50 percent" means 50 per 100, or $\frac{50}{100}$, or $\frac{1}{2}$ (Figure 7-11.)

50% (50 per 100)

Figure 7–11

Definition Percent

$n\% = \frac{n}{100}$ where % denotes **percent**.

PERCENTS, FRACTIONS, AND DECIMALS

The concepts of percents, fractions, and decimals are interrelated. The definition of a percent (%), $n\% = \frac{n}{100}$, is used to convert it to a fraction or decimal. Consider the following examples.

$$4\% = \frac{4}{100} = 0.04 \qquad 0.8\% = \frac{0.8}{100} = \frac{8}{1000} = 0.008$$

Lesson Exercise 7.49

Based upon the preceding examples, a shortcut for changing a percent to a decimal involves dropping the percent sign *and* moving the decimal point ____ places to the ________ to compensate.

The process is reversed in changing decimals or elementary fractions to percents. The decimals 0.37 and 0.06 can be converted to percents as follows.

EXAMPLE 7.7

Give the percent equivalents of 0.37 and 0.06.

Solution

Write each number as a fraction with a denominator of 100. Then change it to a percent.

$$0.37 = \frac{37}{100} = 37\% \qquad \text{and} \qquad 0.06 = \frac{6}{100} = 6\%$$

■

Use the results of the preceding example to find a shortcut for converting decimals to percents.

Lesson Exercise 7.50

(a) The results of Example 7.7 suggest that a shortcut for changing a decimal to a percent involves moving the decimal point ____ places to the ________.

(b) Change 1/8 to a decimal, and then use the method of part (a) to change it to a percent.

Why would someone use a percent rather than a fraction or a decimal? A percent gives a fixed rate per hundred. This is especially suitable for financial applications such as tax rates, sales commissions, and interest rates. Percents are also used to describe statistical data in sports, education, and science, because percents make it simple to compare rates in sets of different sizes.

COMPUTING 1%, 10%, 25%, 50%, OR 100% OF A NUMBER

In middle school, the study of percent problems usually begins with easier problems that can be done mentally. Consider the following.

An eight-concert subscription series to the West Dakota Symphony Orchestra sells at a 10% discount off the $160 single-seat price. How much is the savings? 10% of $160. Since 10% of $160 is 1/10 of $160, or 160 ÷ 10, we can mentally obtain the answer: $16.

You should be able to compute 1%, 10%, 25%, 50%, and 100% of many whole numbers mentally, using shortcuts. See if you can describe these shortcuts in Lesson Exercises 7.51 and 7.52.

Lesson Exercise 7.51

(a) Write 50% and 25% as fractions in simplest form.

(b) How do you find 100% of a number?

(c) Computing 50% of a number is the same as dividing the number by _____.

(d) Computing 25% of a number is the same as dividing the number by _____.

Lesson Exercise 7.52

(a) Write 10% and 1% as fractions in simplest form.

(b) Finding 10% of a number is the same as dividing by ______, which means you could move the decimal point in the number ______ place(s) to the ______.

(c) Finding 1% of a number is the same as dividing by ______, which means you could move the decimal point in the number ______ place(s) to the ______.

As Lesson Exercises 7.51 and 7.52 suggest, you can use the following procedures.

Computing 100%, 50%, 25%, 10%, and 1% of a Number

- 100% of a number is that number.
- 50% of a number is half the number, so divide the number by 2.
- 25% of a number is $\frac{1}{4}$ of the number, so divide the number by 4.
- 10% of a number is $\frac{1}{10}$ of the number, so divide the number by 10, moving the decimal point one place to the left.
- 1% of a number is $\frac{1}{100}$ of the number, so divide the number by 100, moving the decimal point two places to the left.

Apply these shortcuts in Lesson Exercises 7.53 and 7.54.

Lesson Exercise 7.53

Mentally compute

(a) 100% of 642. (b) 50% of 286. (c) 25% of 4000.

Lesson Exercise 7.54

A $25 shirt is selling at a 10% discount. Mentally compute the discount and the sale price.

COMPUTING OTHER PERCENTS

Now for some more amazing feats of mental computation! Certain other percents of a number, such as 5%, 30%, 75%, and 200%, can sometimes be computed mentally, using our understanding of 1%, 10%, 25%, and 100%, respectively.

EXAMPLE 7.8

How could one mentally compute

(a) 200% of 14? **(b)** 5% of 300? **(c)** 30% of 620?

Solution

(a) To find 200% of 14, compute 2 times 100% of 14.

$$2 \times 14 = 28$$

(b) To find 5% of 300, compute 5 times 1% of 300.

$$5 \times 3 = 15$$

Note that you could also compute

$$\frac{1}{2} \text{ of } 10\% \text{ of } 300 = \frac{1}{2} \text{ of } 30 = 15$$

(c) To find 30% of 620, compute 3 times 10% of 620.

$$3 \times 62 = 186$$

■

Lesson Exercise 7.55

The number of high-school girls playing basketball in 1996 was 300% of 190,000, the number who played in 1978. How could you mentally compute how many girls were playing high-school basketball in 1996?

Lesson Exercise 7.56

How could you mentally compute the exact value of

(a) 6% of 200? (b) 40% of 300?

ESTIMATING PERCENTS

You can also estimate $a\%$ of b using mental computation. Round the problem and compute mentally or, if you prefer, use the compatible-numbers strategy. The compatible-numbers strategy requires rounding the percent to a nearby fraction. The following table lists some percents and their fractional equivalents.

Percent	5%	10%	20%	25%	$33\frac{1}{3}\%$	50%	$66\frac{2}{3}\%$	75%	100%
Unit Fractions	$\frac{1}{20}$	$\frac{1}{10}$	$\frac{1}{5}$	$\frac{1}{4}$	$\frac{1}{3}$	$\frac{1}{2}$	$\frac{2}{3}$	$\frac{3}{4}$	$\frac{1}{1}$

EXAMPLE 7.9

How could one estimate 32% of 527?

Solution

Method 1: Rounding Round the problem to 30% of 500 and compute mentally. This is 3 times 10% of 500. Since $3 \times 50 = 150$, estimate that 32% of 527 $\approx$ 150.

Method 2: Compatible Numbers Round 32% to a nearby unit fraction $\left(\frac{1}{3}\right)$, so that 32% of 527 is about $\frac{1}{3}$ of 540 = 180. So 32% of 527 $\approx$ 180.

■

Lesson Exercise 7.57

A salesperson earns a 22% commission. Show how to estimate her commission for selling a $648 washing machine, using both the rounding and compatible-numbers methods.

The following example shows how to estimate what percent one number is of another.

EXAMPLE 7.10

John answered 34 questions correctly out of 48. Show how to estimate what percent of the questions he answered correctly.

Solution

34 out of 48 is $\frac{34}{48} \approx \frac{3}{4} = 75\%$. John answered about 75% of the questions correctly. ■

Lesson Exercise 7.58

4 out of 17 children in a class like olive loaf. How could you estimate what percent of the children like olive loaf?

Sometimes people overestimate percents, as the following cartoon shows.

Lesson Exercise 7.59

What is wrong with saying "110%" in the cartoon?

ESTIMATING TIPS

You've just finished a feast. Chef Skippy fixed his special baked zucchini with peanut sauce, buttered apples, and licorice balls. It came to $12.26 plus $.74 tax. What's a reasonable tip?

One common application of percent estimation is estimating the tip in a restaurant. People typically leave about 15% of the bill.

***Cooperative* Lesson Exercise 7.60**

Describe every method you know for computing the tip on a bill of $12.26 plus $.74 tax.

The following example illustrates three methods for estimating a tip.

EXAMPLE 7.11

How could you mentally compute an appropriate tip for the following bill from Skippy's restaurant?

Bill	$12.26
6% tax	.74
Total	$13.00

Solution

Method 1 (10% Base) Estimate 10%, and then add half of the estimate (5%) to obtain 15%.

If the bill is $12.26, 10% is about $1.22, and half of $1.22 is about $.60. A 15% tip is about $1.22 + $.60 ≈ $2.

Method 2 (Unit Fraction) Find a unit fraction close to 15%, such as $\frac{1}{7}$ (14.3%), $\frac{1}{6}$ (16.7%), or $\frac{1}{5}$ (20%). For example, find $\frac{1}{6}$ of the bill after rounding the total to a compatible number.

Estimate $\frac{1}{6}$ of \$12.26. Round it to $\frac{1}{6}$ of \$12, or \$2. So the tip should be about \$2.

Method 3 (Tax) Multiply the tax by a whole number that will make it about 15%.

The tax is 6%. For an 18% tip, triple the tax (\$2.22). Leave \$2 or \$2.25. ■

Lesson Exercise 7.61

Show how to use each of the three methods to estimate the tip on the following bill.

Bill	\$8.79
8% tax	\$ .70
Total	\$9.49

ANSWERS TO SELECTED LESSON EXERCISES

7.49 2; left

7.50 (a) 2; right (b) $\frac{1}{8} = 0.125 = 12.5\%$

7.51 (a) $\frac{1}{2}$ and $\frac{1}{4}$ (b) It is the number.
(c) 2 (d) 4

7.52 (a) $\frac{1}{10}$ and $\frac{1}{100}$ (b) 10; 1; left
(c) 100; 2; left

7.53 (a) 642 (b) 143 (c) 1000

7.54 The discount is \$2.50, and the price is \$22.50.

7.55 $190{,}000 \times 3 = 570{,}000$

7.56 (a) 1% of 200 = 2, and $2 \times 6 = 12$
(b) 10% of 300 = 30, and $30 \times 4 = 120$

7.57 22% of 648 ≈ 20% of \$650 = \$130

7.58 4 out of 17 $= \frac{4}{17} \approx \frac{1}{4} = 25\%$

7.59 A person cannot be more than 100% committed to anything.

7.61 10% of 8.79 ≈ \$.90; \$.90 + $\frac{1}{2}$ (\$.90) = \$1.35

$\frac{1}{6}$ of \$8.79 ≈ $\frac{1}{6}$ of \$9 = \$1.50;

tax is 8% → double it → (\$.70) · 2 = \$1.40

7.4 HOMEWORK EXERCISES

Basic Exercises

1. What does the word "percent" mean?

***2.** 14% means
(a) _____ per 100.
(b) _____ per 50.
(c) Shade a decimal square to represent 14%.

3. Write an equation that generalizes the following pattern.

$$20\% = 20 \cdot 0.01$$
$$37\% = 37 \cdot 0.01$$
$$200\% = 200 \cdot 0.01$$

4. Convert each percent to a fraction and to a decimal.
(a) 17% (b) 0.01% (c) 200%

5. Write each percent as a fraction and as a decimal.
(a) 34% (b) 180% (c) 0.06%

6. Write each decimal as a percent.
(a) 0.79 (b) 5.24 (c) 0.00083

7. Write each fraction as a percent.
(a) $\frac{1}{25}$ (b) $\frac{3}{16}$

8. Write each fraction as a percent.
(a) $\frac{3}{4}$ (b) $\frac{9}{10}$ (c) $\frac{2}{5}$

9. How can you mentally compute each of the following?
(a) 50% of 222 (b) 25% of 6
(c) 10% of 470 (d) 1% of 37

10. A salesperson earns a 10% commission on $88 worth of sales. What is her commission?

11. A $400 television is selling at a 25% discount. Mentally compute its sale price.

12. Many nutritionists recommend getting about 60% of our calories from carbohydrates. How could you mentally compute the exact number of calories that should come from carbohydrates in a 2000-calorie-per-day diet?

13. How could you mentally compute the exact value of each of the following?
(a) 200% of 7 (b) 8% of 400
(c) 75% of 12

14. A Specific Motors Flounder Supersport is discounted by 6% from its $7000 list price. How can you mentally compute the dollar amount of the discount?

15. Compute mentally, and fill in each blank.
(a) 50% of ________ is 22.
(b) 25% of ________ is 80.
(c) 20% of ________ is 110.

16. A typical baby weighs about 200% *more* at age 1 than at birth. Baby Julie weighed 7 lb at birth. Predict her weight at age 1.

17. Compute mentally, and fill in each blank.
(a) 200% of ________ is 36.
(b) 10% of ________ is 180.
(c) 1% of ________ is 74.

18. A computer that is discounted 50% sells for $750. Mentally compute the original price.

19. (a) 50% of ________ is 4.
(b) 20% of ________ is 10.
(c) 25% of ________ is 20.

20. (a) 38 is ________ % of 100.
(b) 11 is ________ % of 50.
(c) 21 is ________ % of 25.
(d) 50 is ________ % of 40.

21. Estimate what percent of the square is shaded.

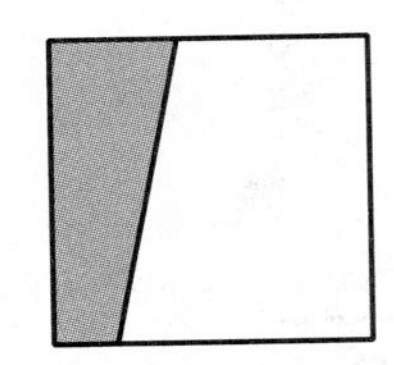

22. The best compatible-numbers estimate for 72% of 87 is
(a) $\frac{1}{4}$ of 88 (b) $\frac{1}{2}$ of 90 (c) $\frac{3}{4}$ of 80

23. $\frac{1}{3}$ $\frac{1}{5}$ $\frac{1}{7}$ $\frac{1}{10}$
(a) Which of these fractions have simple representations as percents?
(b) Which of these fractions do not have simple representations as percents?

24. The Cereal Bowl seats 95,000. The stadium is 64% full for a clash between the Ballerinas and the Fieldmice. Explain how to estimate the attendance mentally.

25. The voting-age population in the United States is about 196 million, of which about 55% vote in presidental elections. Explain how to estimate how many people vote.

26. If you take an extra summer job that pays $4200, you will have 35% taken out of your paycheck in taxes. How can you estimate your take-home pay? (*Hint:* First estimate the amount taken out of your paycheck.)

27. Use estimation to order the following expressions from smallest to largest.

80% of 28	52% of 807
37% of 420	19% of 400

28. Use estimation to determine whether each of the following calculator answers is reasonable. Explain why or why not.
(a) 15% of 724 = [10860]
(b) 22% of 58,000 = [12760]
(c) 86% of 94 = [109.3023256]

In Exercises 29 and 30, estimate the tips on the bills in two different ways.

29.

Bill	\$21.07
7% tax	\$ 1.48
Total	\$22.55

30.

Bill	\$42.78
5% tax	2.14
Total	\$44.92

31. Rick, Mae, Karen, and Pam receive a bill for \$37 plus \$1.85 tax for dinner. They want to divide up the costs evenly, including approximately 15% for a tip. How can they mentally compute how much each person owes?

32. Your luncheon at the Tastee Health Food Restaurant consisted of blanched tofu, holistic corn dogs, and milk-fed guacamole. The meal cost \$9.50 plus tax. If the service was satisfactory, estimate how much you would leave for a tip.

Extension Exercises

33. Start with 100. Decrease it by 50%. Now increase your answer by 50%.
(a) What number did you end up with?
(b) Explain why you did not end up with 100.

34. When you borrow money with simple interest, the interest $I = PRT$, in which P is the principal, R is the rate of interest, and T is the time in years. Show how to estimate the cost of borrowing \$5400, at a rate of 12%, for 2 years and 2 months.

35. Use estimation to order the following expressions from smallest to largest.

31% of 642	11% of 1000
48% of 117	86% of 90

36. Consider the following number trick.
Pick a number.
Multiply by 10.
Add 50.
Take 20% of the result.
Subtract 10.
Divide by 2.
(a) Start with two or three different numbers, and make a conjecture about what the trick does to any number.
(b) Prove your generalization from part (a).

37. Consider the following problem. "The number of boys in a class is 25% of the number of girls. What percent of the whole class is boys?"
(a) Devise a plan and solve the problem.
(b) Make up a similar problem and solve it.

38. If a is 25% of b, then b is ____% of a.

39. If a is 25% more than b, then b is ________ % of a.

7.5 USING PERCENTS

In the last section, you solved some introductory percent problems using mental computation and estimation. In this section, you will compute exact answers to a greater variety of percent problems, such as the following.

"What is 6% sales tax on a \$380 stereo?"

"What percent score is 23 right out of 30?"

"What is the sale price of a \$140 coat that is now 30% off?"

These are all common applications of percents. People have used percents in computing interest and taxes since the late fifteenth century.

BASIC PERCENT PROBLEMS

People who work with tax rates, sales commissions, and interest rates invariably end up solving basic percent problems. Example 7.12 presents two methods for solving the most common type of percent application: finding a percent of a number.

EXAMPLE 7.12

Consider the following problem. "Abby buys a stereo for \$380. The sales tax is 6%. How much sales tax does she have to pay?"

(a) Estimate the answer.

(b) Solve the problem using multiplication.

(c) Solve the problem using a proportion.

Solution

(a) 6% of $380 \approx 6\%$ of $400 = 6 \times (1\%$ of $400) = 6 \times 4 = 24$

(b) ***The Multiplication Method*** As with fractions and decimals, 6% *of* 380 means $6\% \times 380$. In this method, you always have the choice of changing the percent to a fraction or a decimal. Since this problem involves money, using decimals makes more sense: 6% of \$380 = $0.06 \times$ \$380 = \$22.80.

(c) ***The Proportion Method*** A number line helps illustrate the proportional relationship. One can show 6% of \$380 by making a percent scale and a corresponding dollar (\$) scale on opposite sides of the same number line, as in Figure 7-12. The amount \$380 represents 100%, and you want to know what amount 6% would be.

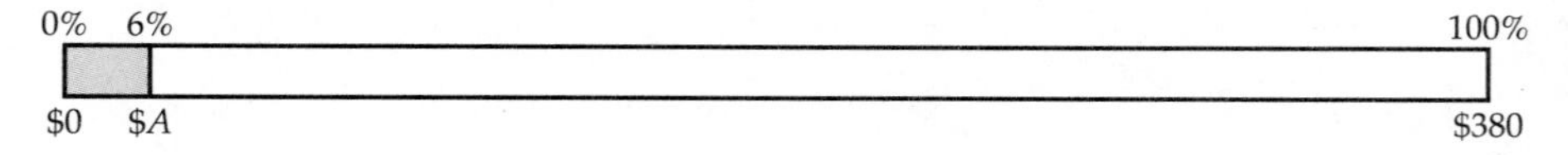

Figure 7–12

This picture suggests some proportions. Traditionally, one ratio is written for the percents and the other for the amounts (dollars, in this case).

$$\frac{6\%}{100\%} \text{ or } \frac{6}{100} = \frac{\$A}{\$380} \text{ or } \frac{A}{380}$$

So $\frac{6}{100} = \frac{A}{380}$. Computing cross products yields

$$6 \cdot 380 = 100 \cdot A$$
$$2280 = 100 \cdot A$$
$$22.80 = A$$

Abby must pay $22.80 in sales tax. ■

Lesson Exercise 7.62

I have a credit-card debt of $300 for one year. How much interest will I be charged if the annual finance rate is 18%?

(a) Solve the problem using multiplication.

(b) Solve the problem using a proportion.

Another fairly common type of percent problem involves finding what percent one number is of another. What percent is represented by each of the following examples?

"16 out of 35 prefer defrosting a freezer."

"His $22,500 salary will be increased by $800."

"The sales tax on $19.99 is $1.20."

In the last section, you estimated the answers to percents like these. Now we will find exact answers. You can represent all of these examples with percents by writing the two numbers that are being compared as a fraction and then converting the fraction to a percent. The denominator of the fraction is either the total amount or the amount before a change, fee, or increase.

EXAMPLE 7.13

May scored 23 out of 30 on a mathematics test. What percent score is this?

Solution

The Division Method 23 out of 30 is $\frac{23}{30}$. To find out what percent this is, change $\frac{23}{30}$ to a percent.

$$\frac{23}{30} = 0.7667 = 0.7667 \times 100\% = 76.67\%$$

Test scores are usually rounded to the nearest percent, so the score is 77%.

The Proportion Method Draw a number-line picture (Figure 7-13). A score of 30 represents 100%, and you want to know what percent 23 represents.

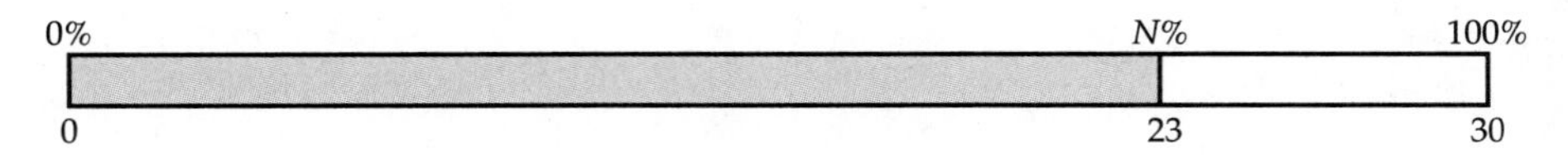

Figure 7–13

$$\frac{23}{30} = \frac{n\%}{100\%} \text{ or } \frac{n}{100}$$

$$23 \cdot 100 = 30 \cdot n$$

$$2300 = 30n$$

$$76.67 = n$$

Test scores are usually rounded to the nearest percent. May's score is 77%. ■

Lesson Exercise 7.63

In a survey asking 35 sixth-grade students whether they would prefer listening to opera, eating squid, or defrosting a freezer, 16 students said that they would prefer defrosting a freezer. What percent does this represent?

(a) Estimate the answer.

(b) Solve the problem using division.

(c) Solve the problem using a proportion.

A less common application is to find a number when a percent of the number is known. The following exercise is just such a situation.

Lesson Exercise 7.64

Consider the following problem. "A group of 105 fifth-grade students attended a school play. This group was 84% of the entire fifth grade. How many students are in the fifth grade?"

(a) Use N to represent the unknown number. Write an equation using N, and solve it.

(b) Draw a number-line diagram showing percent and number scales for this problem. Then write a related proportion and solve it.

PERCENT INCREASES AND DECREASES

Best Sellers at 35% Off
Utility Costs Increase by 14%

Some percent problems involve finding a percent increase or decrease. These can be done as two-step problems. Consider the following example.

EXAMPLE 7.14

A coat at the Outerwear House costs \$140, but today it's 30% off! What is the sale price?

Solution

Method 1: Multiply and Subtract This can be done in two steps. First compute the 30% discount in dollars.

$$30\% \text{ of } \$140 = 30\% \times \$140 = 0.30 \times \$140 = \$42$$

Second, subtract the discount from the regular price: $\$140 - \$42 = \$98$. The sale price is \$98.

Method 2: Subtract and Multiply Make a number-line picture, as in Figure 7-14. As the diagram shows, "30% off the regular price" means 70% of the regular price ($100\% - 30\% = 70\%$). Find 70% of $\$140 = 0.70 \times \$140 = \$98$. The sale price is \$98.

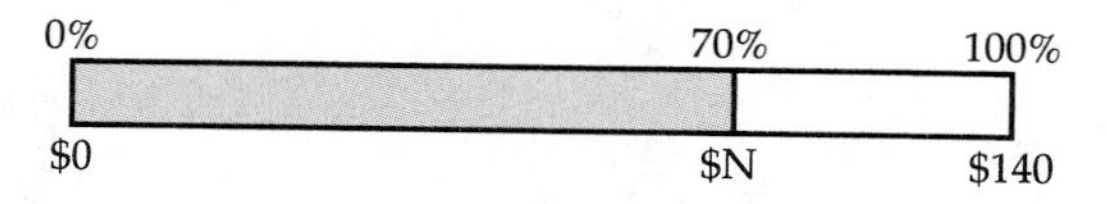

Figure 7–14

Lesson Exercise 7.65

I had \$1000 in the bank earning 12% simple interest for a year. The year is over. How much money do I have in that account now?

(a) Solve this by multiplying and adding.

(b) Draw a number-line picture illustrating the relationship, and solve the problem by adding and multiplying.

Suppose an experienced teacher who earns \$43,870 and a fairly new teacher who earns \$28,800 both received their annual raises. How can the two raises be compared? One way is to compare the percent increase (the change in the number of dollars *per each $100* earned) that each teacher received.

How is the percent change computed? The two-step method is as follows. First, compute the amount of the increase or decrease. Second, compute what percent of the *original* amount this amount is.

EXAMPLE 7.15

This year, Maria Gomez's salary increased from $28,800 to $32,256. What percent increase is this?

Solution

Method 1: Subtract and Divide First, compute the amount of the increase: $32,256 − $28,800 = $3456. Second, compute what percent of the *original* amount this increase is: $3456 is what % of $28,800?

$$\frac{\$3456}{\$28{,}800} = 0.12 = 0.12 \times 100\% = 12\%.$$

Maria received a 12% increase.

Method 2: Divide and Subtract Find the multiplier that produces 32,256 from 28,800 (Figure 7-15).

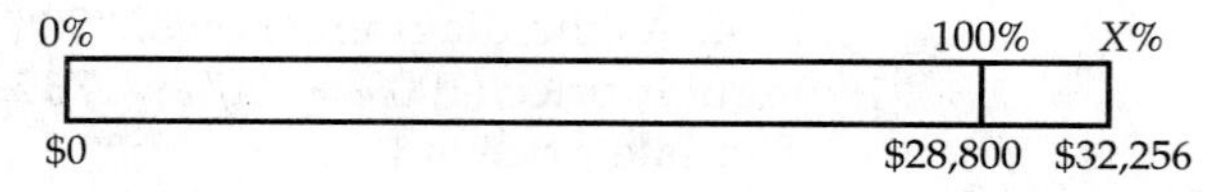

Figure 7–15

$$\frac{32{,}256}{28{,}800} = x\% = 1.12$$

Change 1.12 to a percent: 1.12 × 100% = 112%. So 32,256 is 112% of 28,800. The increase is 112% − 100% = 12%. ■

The following exercise is similar to Example 7.15.

Lesson Exercise 7.66

This year, Sally Martin's salary increased from $43,870 to $48,300. Find what percent increase this is, to the nearest tenth of a percent, by

(a) using the subtract-and-divide method.

(b) using the divide-and-subtract method.

(c) How does this increase compare to Maria Gomez's increase in Example 7.15?

(d) Describe another way of comparing their increases that would result in Sally's increase being larger than Maria's.

(e) Would you say Sally and Maria's raises are comparable, or did one come out better than the other?

You can also find the original amount if you are given the percent increase or decrease and the final result.

Lesson Exercise 7.67

Consider the following problem. This year, a school's enrollment decreased exactly 6%, to 423 students. What was the enrollment before this decrease?

(a) What percent of last year's enrollment is 423?

(b) Write an equation and solve it.

COMPOUND INTEREST

New Rates

Account	Yield	Rate
6-month CD $100 minimum	8.60%	8.60%
1-year CD $100 minimum	8.87%	8.50%

Do you have some money in the bank? In this section, we'll focus on the kind of interest you are earning.

Simple interest pays interest only on the original money (the principal) you deposited. With **compound interest**, you also earn interest on your interest. Most banks pay compound interest on savings, usually more than once a year—either semiannually (twice a year), quarterly (four times a year), monthly, or daily (360 or 365 times a year).

Suppose that, in Lesson Exercise 7.65, I had received 12% interest compounded semiannually. Instead of receiving 12% interest at the end of the year, I received 6% after 6 months and 6% more interest after 12 months. This is semiannual (twice-a-year) compounding. Would this be better than receiving 12% simple annual interest?

EXAMPLE 7.16

Suppose you receive 12% interest on $1000 compounded semiannually. How much will you have after one year?

Solution

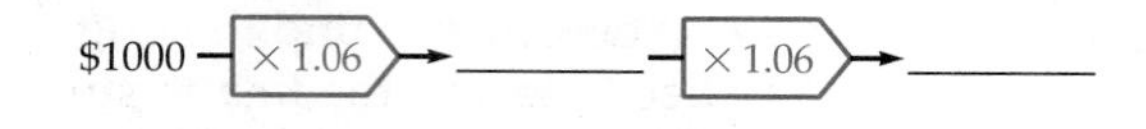

Time	6 months	12 months
Amount	$1000 + 6%($1000)	$1060 + 6%($1060)
	$1000 + $60	$1060 + $63.60
	$1060	$1123.60

So you end up with $1123.60. ■

In Example 7.16, the interest is $123.60, and in Lesson Exercise 7.65, the interest is $120. You receive more money with the semiannual compounding in Example 7.16, because you receive 6% interest on the $60 interest (an extra $3.60) from the first 6 months.

EXAMPLE 7.17

The total annual interest in Example 7.16 is $123.60 on $1000. What percent interest is this?

Solution

$$\frac{\$123.60}{\$1000} = 0.1236 \qquad \text{and} \qquad 0.1236 \times 100\% = 12.36\%$$

So the interest received on 12% interest compounded semiannually is 12.36%. ■

Example 7.17 shows that, when you receive 12% interest compounded semiannually for a year, you actually receive more than 12% of your investment in interest. The actual annual interest received, 12.36% in this case, is called the **effective annual yield**. Bank ads often quote the effective annual yield.

Lesson Exercise 7.68

A bank pays 12% interest compounded quarterly.

(a) What percent interest does the bank pay every 3 months?

(b) Compute the total interest for the year on $1000, assuming you leave all your money and the interest in the bank.
(c) What is the effective annual yield on 12% compounded quarterly?
(d) Is it higher than 12% compounded semiannually?

In computing 12% interest compounded quarterly, you first compute

$$1000\left(1 + \frac{0.12}{4}\right)$$

Then, in the next quarter, you multiply this amount, $1000\left(1 + \frac{0.12}{4}\right)$, by $\left(1 + \frac{0.12}{4}\right)$ again. After 4 quarters, you will have multiplied

$$1000\left(1 + \frac{0.12}{4}\right)\left(1 + \frac{0.12}{4}\right)\left(1 + \frac{0.12}{4}\right)\left(1 + \frac{0.12}{4}\right) = 1000\left(1 + \frac{0.12}{4}\right)^4$$

So the resulting amount A is $1000\left(1 + \frac{0.12}{4}\right)^4$. In general, you accumulate the amount

$$A = P\left(1 + \frac{R}{N}\right)^{T \cdot N}$$

in which 1000 is the principal P, 0.12 is the annual rate R, 4 is the number of payments per year N, and T the number of years in the bank.

Formula for Compound Interest

A formula for computing compound interest is $A = P\left(1 + \frac{R}{N}\right)^{TN}$, in which P is the principal, T is the time in years, A is the amount after time T, R is the rate of interest, and N is the number of times interest is paid each year.

Do not memorize this formula. Look it up when you need it.

EXAMPLE 7.18

A bank pays 12% interest compounded semiannually. Use the compound-interest formula to find out how much you will have at the end of the year if you deposit $1000.

Solution

$$A = P\left(1 + \frac{R}{N}\right)^{TN}$$

$R = 0.12$, $P = 1000$, $N = 2$, $T = 1$, and $A = ?$. Substituting in the formula,

$$A = 1000\left(1 + \frac{0.12}{2}\right)^{(1)(2)} = 1000(1.06)^2 = \$1123.60$$

This is the same answer as in Example 7.16. ■

Lesson Exercise 7.69

Do Lesson Exercise 7.68(b) using the compound-interest formula.

Lesson Exercise 7.70

Many banks compound interest daily, paying interest 360 or 365 times per year. If Ben deposits $1000 for a year, earning 12% interest compounded daily, how much money will he have at the end of the year? (Assume that Ben's bank pays interest 360 times per year.)

ANSWERS TO SELECTED LESSON EXERCISES

7.62 (a) ($300)(.18) = $54

(b) $\frac{x}{300} = \frac{18}{100} \rightarrow x = \$54$

7.63 (a) $\frac{16}{35} \approx \frac{1}{2} \approx 50\%$

(b) $16 \div 35 \approx 0.4571 = 45.71\%$

(c) $\frac{16}{35} = \frac{x}{100} \rightarrow x \approx 0.4571 = 45.71\%$

7.64 (a) $0.84N = 105$, so $N = 125$.

(b) $\frac{84}{105} = \frac{100}{N}$, so $N = 125$.

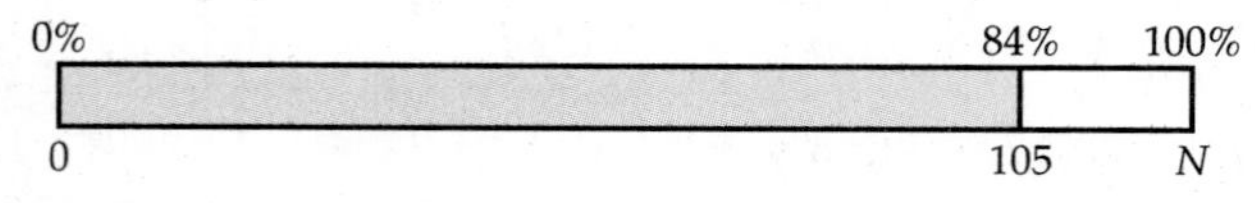

7.65 (a) ($1000)(0.12) + $1000 = $1120

(b)

0% 100% 112%

$0 $1,000 $X

(100% + 12%)($1000) = $1120

7.66 (a) $\$48{,}300 - \$43{,}870 = \$4430$

$\frac{\$4430}{\$43{,}870} \approx 0.101 = 10.1\%$

(b) $\frac{\$48{,}300}{\$43{,}870} \approx 1.101 = 110.1\%$

$110.1\% - 100\% = 10.1\%$

(c) It is lower.

(d) Compare the dollar amount of each increase.

7.67 (a) 94%

(b) $0.94n = 423$, so $n = 450$.

7.68 (a) 3% (b) $125.51

(c) 12.55% (d) Yes

7.69 $A = 1000\left(1 + \frac{0.12}{4}\right)^{(1)(4)} = \$1125.51$

7.70 $1127.47

7.5 HOMEWORK EXERCISES

Basic Exercises

1. A salesperson earns a 12% commission. If she sells a $2400 computer, what is her commission?
   (a) Estimate the answer.
   (b) Solve the problem using multiplication.
   (c) Solve the problem with a proportion, and draw a number-line model.
   (d) If she sells 3 times as many computers, her commission will be _____ times as much.

2. The sales tax on a $9.59 item is $7\frac{1}{2}\%$.
   (a) Estimate the tax.
   (b) Exactly how much is the tax?

3. Each year in the United States, we produce about 110 million tons of air pollution from carbon monoxide. Motor vehicles produce about 82 million tons of this pollutant. What percent of the carbon monoxide do motor vehicles produce?
   (a) Solve the problem using division.
   (b) Solve the problem with a proportion, and draw a number-line model.

4. A department store is considering running a 60-second TV ad. Assume that the average viewer who responds to the ad will bring the store an additional $10 of profit. The marketing department has gathered the following information.

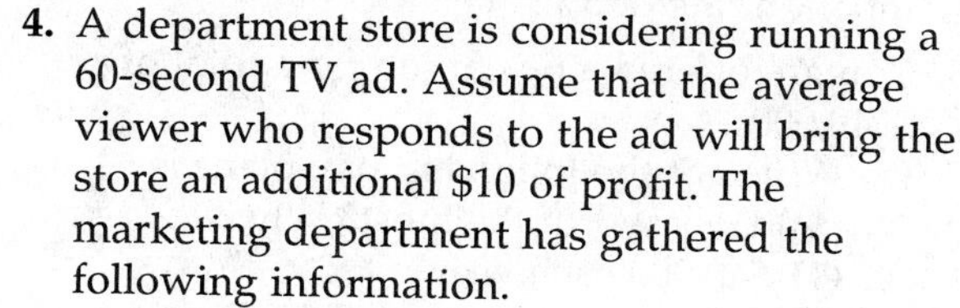

60-Second Television Ad

Time Slot	Cost	Estimated Number of Viewers	Estimated Viewer Response
9:15 A.M.	$1400	13,000	1.0%
1:15 P.M.	900	10,000	1.2%
7:15 P.M.	1600	20,000	0.9%

   (a) Forecast which time slots will be profitable.
   (b) Which time slot should be the most profitable?

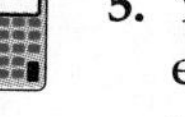

5. You paid $1.20 in tax when you purchased an electric coffee stirrer for $19.99. What percent sales tax is this? (Use either method.)

6. On our last trip to England, my wife and I needed to exchange dollars for pounds. One bank charged a 0.5% commission fee with a minimum charge of 2.50£ (British pounds). Another bank charged a 2% commission with a minimum charge of 2£. If the exchange rate at both banks was $1.76 for 1£ before fees, how many pounds would we have received at each bank for each of the following?
   (a) $100 (b) $1000

7. A fruit juice container shows the following list of ingredients: grape juice, apple juice, and passion fruit juice. Ingredients are listed in order from most abundant to least abundant. At least what percent of the drink is grape juice?

8. Weigh a grape and a raisin (a dried grape), and estimate what percent of the grape is water.

9. Consider the following problem. "A salesperson earns a commission of 8% of total sales. This week she earned $798.40. What were her total sales for the week?"
   (a) Write an equation (that is not a proportion) and solve it.
   (b) Write a proportion and solve it.

10. There are 184 comedy films at Fantasy Island Video. This represents about 18% of their collection. How many films do they have? Give all possible answers. (There are 57 of them!)

11. A credit card company charges 12% simple annual interest on debts.
   (a) What percent interest does the company charge per month?
   (b) What percent interest does the company charge per day?

12. Your new credit card has an annual finance charge of 15% simple annual interest.
   (a) Suppose you borrow $500 for a year. How much interest will you owe?
   (b) Suppose you cannot afford to pay back the $500 for an additional 9 months. How much additional interest will you owe?

13. How much money must I put in an account earning 6% simple annual interest so that I will have $5000 at the end of 1 year?

14. You buy a $5000 savings certificate that pays 8% simple annual interest. How much interest will you earn in 6 months?

15. Sylvia scored 64 out of 70 on a quiz. Write her score as a percent.
(a) Estimate the answer.
(b) Solve the problem.

16. U.S. water use (in billions of gallons) in 1996 was approximately as follows.

Irrigation	Utilities	Domestic	Industrial	Total
153	207	44	46	450

What percent of the water is used for domestic purposes?
(a) Solve the problem using division.
(b) Solve the problem with a proportion.
(c) If domestic water use were cut in half, overall water use would decrease by _____%.

17. A baseball glove sells for $15.98. How much will it cost at 20% off?
(a) Solve the problem by multiplying and subtracting.
(b) Draw a number-line picture illustrating the relationship, and solve the problem by subtracting and multiplying.
(c) If the sales tax is 4%, what is the total price of the glove?

18. In 1988, our annual car insurance for one car rose from $463 to $574. What percent increase is that?
(a) Solve the problem by dividing and subtracting. (Round to the nearest tenth of a percent.)
(b) Solve the problem by subtracting and dividing.
(c) Given that we had no accidents or traffic violations, was this a reasonable increase?

19. My car was worth $16,600 when it was new. Now it is worth only $7000. By what percent has its value depreciated (decreased)?

20. If you want to increase a price by 6%, by what single number can you multiply the old price to obtain the new price?

21. A quart of milk cost $.60 in 1992. By 1996, it had increased 20%. Explain how to compute mentally the 1996 price of a quart of milk.

22. Sam has $20 and Mike has $10. Sam has ________% more money than Mike, and Mike has ________% less money than Sam.

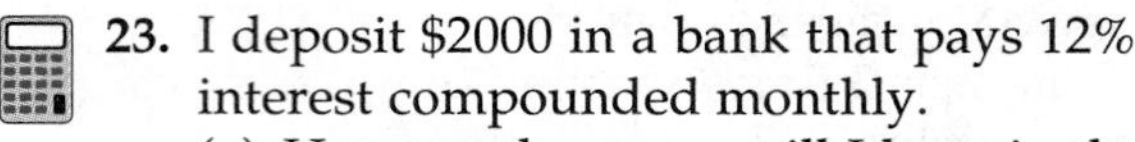

23. I deposit $2000 in a bank that pays 12% interest compounded monthly.
(a) How much money will I have in the account after 1 year?
(b) What is the effective annual yield?

24. I deposit $4000 in a bank that pays 10% interest compounded daily (360 times per year).
(a) How much money will I have in the account after 1 year?
(b) What is the effective annual yield?

25. Five percent annual inflation is the same mathematically as 5% interest compounded annually. If a car cost $10,000 in 1993, what will it cost in 2003, assuming an annual inflation rate of 5%?

26. You are offered a 6-year job for a total of $240,000.
(a) Which of the following pay plans is probably worth the most? Why?
(1) $40,000 each year
(2) Annual payments in the following order: $30,000, $34,000, $38,000, $42,000, $46,000, $50,000
(3) Annual payments in the following order: $50,000, $46,000, $42,000, $38,000, $34,000, $30,000
(b) When given these three choices, people most often selected option 2, which is probably the worst pay plan. Why would someone prefer it?

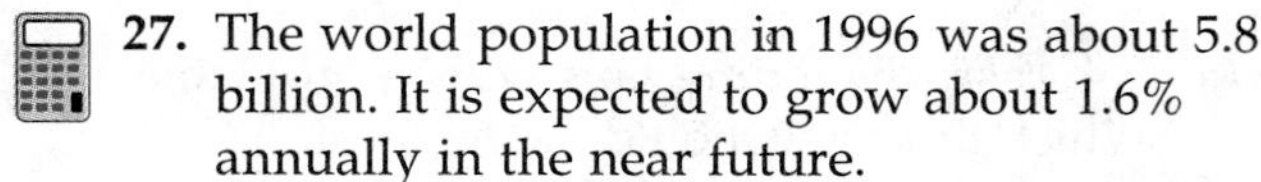

27. The world population in 1996 was about 5.8 billion. It is expected to grow about 1.6% annually in the near future.
(a) Forecast the world population in 1997.
(b) Forecast the world population in 2010. (*Hint:* Use an exponent.)
(c) Forecast the world population in 2050.

28. The U.S. population in 1996 was about 267 million. It is expected to grow about 0.6% annually in the near future.
(a) Forecast the U.S. population in 1997.
(b) Forecast the U.S. population in 2010.
(c) Forecast the U.S. population in 2050.

29. A school has a total of $240,000 to spend on raises for teachers. They are deciding between a raise of $3000 for each teacher and a 7% raise for each teacher. Which teachers would benefit more from each type of raise?

30. An airline ticket costs $218, including 8% tax. What was the base fare?

31. A store owner makes a 25% profit (on the wholesale price) by selling a dress for $80. How much is the profit in dollars?

32. Twenty years ago, Gil had an annual income of $9530. If inflation for the last 20 years has been about 200%, what annual income will he need this year to keep pace with inflation?

33. Driving 40 mph instead of 60 mph on the highway increases gas mileage by 20%. A car that gets 28 mpg at 60 mph will get ________ mpg at 40 mph.

34. Paolo works for a company that manufactures variables. This year, Paolo's salary increased from 8S to 10S. What percent increase is this?

35. Write two multi-step mathematical questions that can be answered using the following data.

	Regular Price	Sale Price
Gazelle Joggers	$78.50	$75.00
Hippo Running Shoes	$29.95	$19.95
Fatiguers Running Shoes	$35.00	$31.99
Bolt Running Shoes	$50.00	$40.00

36. (a) Complete the chart.

Number	0	10	20	30	40
38% of the number	0	3.8			

(b) Plot your points on graph paper, using axes like those shown.

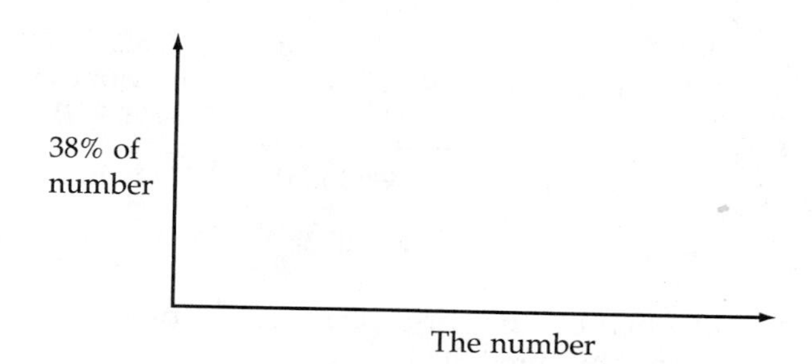

(c) Use your graph to estimate the following: 10 is 38% of ________.

37. A store marks down an item 30%, to a price of $225.60.
(a) Complete the chart.

Original price	$0	$100	$200		
30% off the price	$0				

(b) Plot your points on graph paper, using axes like those shown.

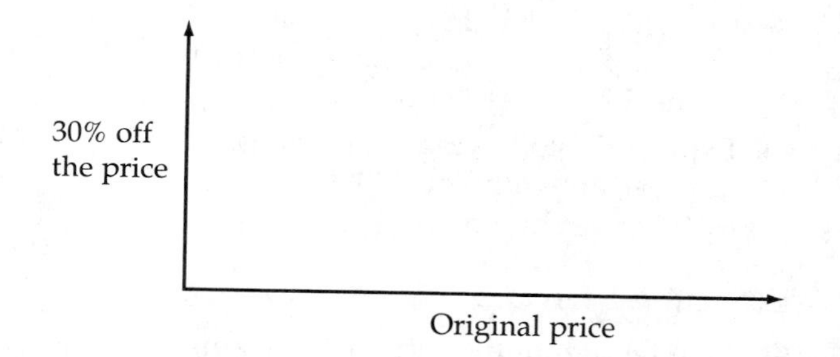

(c) Use your graph to estimate the original price if the sale price is $225.60.

38. You are offered two real estate sales jobs. One pays 9% commission. The other job pays $50 per week and 6% commission. If sales tend to average about $2000 per week, which job would pay more?

39. As of 1996, the United States and Canada have different kinds of health-care systems. In Canada, all citizens are insured; in the United States, about 37 million people (14%) are uninsured. Canada uses a single insurer in each of its ten provinces; the United States has about 1500 different health insurers. Researchers at the Organization for Economic Cooperation and Development compared the two systems.

	Population (1994)	Total Cost (1994)	Total Administrative Costs (1994)
U.S.	261 million	$989 billion	$237 billion
Canada	28 million	$68 billion	$7.5 billion

(a) Compute the per-capita (per-person) costs in the United States and Canada.
(b) What percent of the total cost represents administrative costs in each country?
(c) If the United States reduced its administrative costs from 24% to 11%, how much would be saved?

	Infant Mortality Deaths per 1000 Births (1994)	Persons per Physician (1994)	Life Expectancy (1994)
U.S.	8.1	490	75.9 years
Canada	6.9	465	78.1 years

(d) What do the data suggest about the comparative performance of the two health-care systems?

40. Explain the following expressions.
(a) "We are with you 100%."
(b) "Let's split it 50-50."

Extension Exercises

41. (a) The sign in the cartoon is making a prediction. What sort of prediction is it?

A & M Homewares Inc. Drawing by Joseph Farris; © 1974 the New Yorker Magazine, Inc.

(b) If a is 15% less than b, is b 15% greater than a? (*Hint:* Try substituting numbers for a and b.)

42. Hi-C Orange Drink costs $2.09 for 64 oz. It contains 10% orange juice and about 1.7 oz of sugar, plus water and artificial color. How much would it cost you to make 64 oz of orange drink yourself, using canned orange juice ($2.19 for 46 oz), sugar ($2.39 for 5 lb) and water?

43. Taxes are commonly classified as progressive, intermediate, or regressive. A **progressive tax** requires people with higher income to pay a higher percent of their income as tax. An **intermediate tax** requires people at every income level to pay the same percent income tax, and a **regressive tax** requires people with lower incomes to pay a higher percent of their income as tax. Classify each of the following tax structures as progressive, intermediate, or regressive.

(a)

Total Yearly Income	Percent Tax
$30,000	8%
$25,000	10%
$20,000	12%

(b)

Total Yearly Income	Tax
$30,000	$2000
$25,000	$2000
$20,000	$2000

(c) Maryland has a sales tax rate of 5%. The average person with $30,000 in yearly income buys about $4000 worth of sales-taxed items, and the average person with $10,000 in yearly income buys about $2000 worth of sales-taxed items. Classify this sales tax as progressive, intermediate, or regressive.

44. Fill in tax amounts so that the tax is *regressive* and those with higher incomes pay more money in tax.

Total Yearly Income	Tax Amount
$30,000	________
$20,000	________
$10,000	________

45. Consider the following problem. "A business manager must pay $35,000 per year in fixed expenses, and she pays 40% of her total sales to her workers. How much money must she make to earn a $40,000 profit after paying her expenses and her workers?" Devise a plan and solve the problem.

46. (a) Suppose that Congress is considering raising the highest tax bracket from 40% to 50%. Would this be a 10% increase or a 25% increase? Explain the reasoning behind each interpretation.
(b) Suppose that Congress is considering raising another tax bracket from 30% to 40%. What is the fairest way to compare this increase to that in part (a)?

47. Your salary increases 8%, but prices increase 10%. What percent of your old purchasing power is your new purchasing power?

48. Solve the following. Compute at least one of each pair mentally.
(a) 50% of 36 and 42% of 36
(b) 38% of 28 and 25% of 28
(c) 30% of 200 and 75% of 200

49. (a) Place the digits 4, 5, 6, 8, and 9 in the blanks to obtain an answer as close as possible to 300.

__ __% of __ __ __ ≈ 300

(b) Use a calculator to see how close you came.
(c) Make a second guess and try to get closer to 300.
(d) Check your second guess on a calculator.

50. Use estimation to fill in each blank with a number that will make the statement true. Check your guess with a calculator, and revise it as needed.
(a) 42% of ________ is between 600 and 700.
(b) 64% of ________ is between 1000 and 1100.
(c) 7% of ________ is between 11 and 14.

51. How does the cost of a compact fluorescent bulb compare to that of an incandescent bulb? Use the following information to find the total cost of buying and using each type of bulb for 10,000 hours.

- The fluorescent bulb lasts 10,000 hours and costs $.0014 per hour of electricity.
- The fluorescent bulb lasts 10 times as long as the incandescent bulb.
- The fluorescent bulb uses 75% less energy than the incandescent bulb.
- The fluorescent bulb costs about $16. The incandescent bulb costs about $1.

52. The manager of a music store orders a compact stereo for $200. She wants to price it so she can offer a 10% discount off the posted price and make a profit of 40% of the price she paid. What will the posted price of the stereo be?

53. If inflation is 10% a year for 3 years in a row, the 3-year price increase is
(a) less than 30%
(b) 30%
(c) greater than 30%

54. A $80 dress is marked down 10% one week. The next week, an additional 25% is taken off the sale price. Find a single percent discount equal to these two successive discounts.

55. A car is marked up 30% from its wholesale price. During a sale, the retail price is reduced by 20%. What percent higher is the sale price than the wholesale price?

56. After two annual raises of 20%, Margaret receives an annual salary of $69,840. What was her salary 2 years ago?

57. An investment earns 10% per year. How many years will it take for your investment to double if you leave your interest in the account? (*Hint:* the answer is not 10 years.)

58. How much should you put in a bank that offers 9% interest compounded annually if you want to have $5000 after 5 years? (Guess and check.)

59. If the pattern shown in the diagram continues, what percent of the interior will be shaded in the eighth triangle?

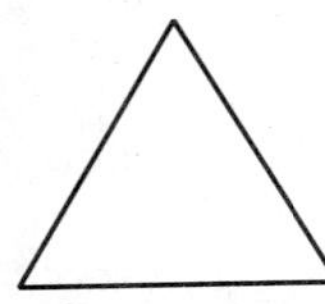
Triangle 1

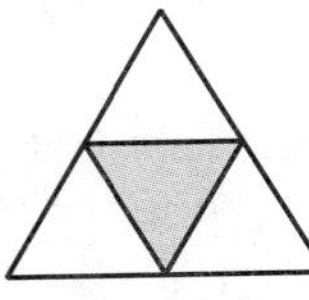
Triangle 2

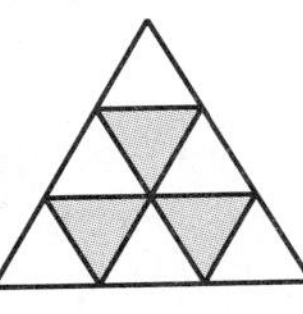
Triangle 3

60. A student answers A out of B questions correctly on a test. If all the questions count the same, write the student's test score as a percent.

61. A car accelerates from a speed of N mph to M mph ($M > N$). What percent increase is this?

Computer Exercises

62. A store is using a spreadsheet to keep track of its discounted prices and the total price. (Use a computer spreadsheet if you have one.)

	A	B	C	D	E	F
1	Item	Retail	Discount	Sale Price	5% Tax	Total Price
2	Shoes	$60	20%			
3	Pants	$48	10%			

(a) What is the value of C2?
(b) What does the value in B3 represent?
(c) What are the formulas for D2, E2, and F2?
(d) If you have a computer, enter the formulas, and give the result for each cell. (Find out how to enter formulas in your spreadsheet program.) If not, use a calculator to find the values in the empty cells.

63. A woman is using a spreadsheet to keep a record of her investments. (Use a computer spreadsheet if you have one.)

	A	B	C	D	E
1	Fund	1-1-96	1-1-97	$ Change	% Change
2	Wings	$8342.61	$8680.10		
3	Rearguard	$12471.82	$11871.17		

(a) What is the value of B2?
(b) What does the value in C3 represent?
(c) What are the formulas for D3 and E3?
(d) If you have a computer, enter the formulas, and give the results for each cell. If not, use a calculator.

Special Exercises

The U.S. Department of Agriculture (USDA) has certain standards for names used to describe various meat products. Some of them follow.

USDA Standards for Meat

Hamburger: no more than 30% fat

Hot dog or **bologna:** no more than 30% fat and 10% added water

Smoke pork sausage: no more than 50% trimmable fat

Egg rolls with poultry: at least 12% poultry meat

Breaded poultry: no more than 30% breading

Poultry chop suey: at least 4% poultry meat

Poultry pies: at least 14% poultry meat

Poultry with gravy: at least 35% poultry meat

Poultry salad: at least 25% poultry meat

Poultry soup (condensed): at least 4% poultry meat

Poultry soup (ready-to-eat): at least 2% poultry meat

64. According to the USDA standards, which is required to have more meat: chicken salad, chicken pot pie, or chicken chop suey?

65. Based upon USDA standards, give the government restrictions on each of the following products.
(a) Chun King chicken egg rolls
(b) Weaver chicken rondelets (breaded chicken patties)
(c) Ball Park franks
(d) Swift Premium turkey roast (boneless meat with gravy)
(e) Campbell's chicken noodle soup (condensed)

7.6 RATIONAL, IRRATIONAL, AND REAL NUMBERS

What is the relationship between rational numbers and decimals? Can all rational numbers be written as decimals? Do all decimals represent rational numbers?

RATIONAL NUMBERS AS DECIMALS

The division model of fractions tells us that a rational number $\frac{p}{q}$ equals $p \div q$. Any rational fraction can be changed to a decimal by dividing the numerator by the denominator.

***Cooperative* Lesson Exercise 7.71**

Change some rational fractions into decimals. Use long division rather than a calculator. What kind of decimals do you obtain?

Decimals come in three forms: terminating decimals, infinite repeating decimals, and infinite decimals that have no infinite repeating block. Examples are 0.3 (terminating), 0.313131 . . . (infinite repeating), and 0.31643847162 . . . (infinite with no repeating block).

Lesson Exercise 7.72

Which of these three forms represent rational fractions?

As you may have surmised, all rational fractions can be represented by terminating or (infinite) repeating decimals. Why is this the case? In converting a rational fraction p/q to a decimal, we divide q into p. If at some point the division works out evenly, we have a **terminating decimal**, as is the case for 3/8.

$$\frac{3}{8} = 8 \overline{)\,3.000\,} \quad \text{quotient } 0.375$$

If each step in the division has a remainder, the remainders repeat at some point. For example, dividing 7 into 2 (to convert 2/7 to a decimal),

$$\frac{2}{7} = 7 \overline{)\,2.0^{6}0^{4}0^{5}0^{1}0^{3}0^{2}0^{6}0^{4}0\ldots} \quad \text{quotient } 0.2\,8\,5\,7\,1\,4\,2\,8\,5\ldots$$

↑ Remainders begin to repeat

Why must the remainders begin to repeat? In this case, with 7 as the divisor, there are only six possible nonzero remainders—1, 2, 3, 4, 5, and 6—so the remainders must start to repeat after no more than six divisions.

A decimal such as 0.285714285714 . . . is a **repeating decimal**, with an infinite number of digits to the right of the decimal point and a repeating block of digits called the **repetend**. A bar indicates that the block of digits beneath it repeats an infinite number of times.

$$0.\overline{285714} = 0.285714285714285714\ldots$$

All rational fractions can be written as terminating or repeating decimals. Is it also true that all terminating or repeating decimals can be written as rational fractions? Do you know how to use place value to convert terminating decimals to fractions? Do so in Lesson Exercise 7.73.

Lesson Exercise 7.73

Write 0.072 as a fraction.

Infinite (repeating) decimals can be converted to fractions using patterns.

***Cooperative* Lesson Exercise 7.74**

(a) Express $\frac{1}{9}$ and $\frac{2}{9}$ as decimals.

(b) Based upon the pattern in part (a), write $\frac{7}{9}$ in decimal form.

(c) Express $\frac{1}{99}$ and $\frac{2}{99}$ as decimals.

(d) Based upon the pattern in part (c), write $\frac{7}{99}$ in decimal form.

(e) Based upon the pattern in part (c), write $\frac{13}{99}$ in decimal form. Use division to check your guess.

(f) Based upon the patterns you have seen, conjecture how to write $\frac{278}{999}$ as a decimal.

The number n of repeating places equals the number of 9's in the denominator (which is $10^n - 1$). The numerator shows what digits are in the repetend. The general formula is

$$0.\overline{a_1a_2a_3\ldots a_n} = \frac{a_1a_2a_3\ldots a_n}{10^n - 1}$$

in which $a_1 \ldots a_n$ are the digits. Apply this formula in the following exercise.

Lesson Exercise 7.75

Write each of the following repeating decimals as a fraction.

(a) $0.\overline{8}$ (b) $0.\overline{37}$ (c) $0.\overline{02714}$

The preceding examples and exercises suggest that any repeating or terminating decimal represents a rational number. Some other types of repeating decimals appear in the homework exercises.

IRRATIONAL NUMBERS

A standard number line includes the location of every rational number. For example, the rational numbers 3/4 and 15/8 are shown in Figure 7-16. Are there points on the number line that are not named by rational numbers?

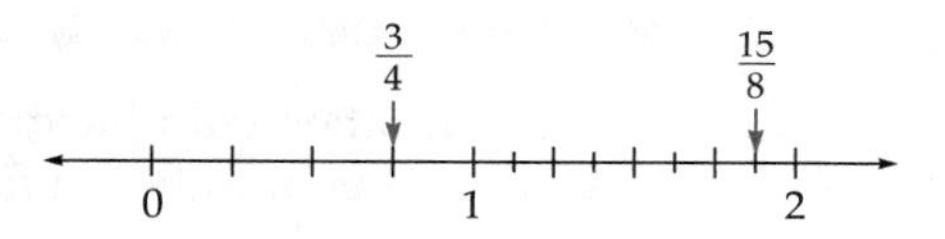

Figure 7–16

Early mathematicians believed that each point on the number line could be named by a rational number. Then, about 2400 years ago, Pythagoras or one of his followers wondered about the answer to a simple-looking geometry problem. The problem was probably how to find the length of the diagonal of a square with sides of length 1. Using the Pythagorean Theorem (see Chapter 10),

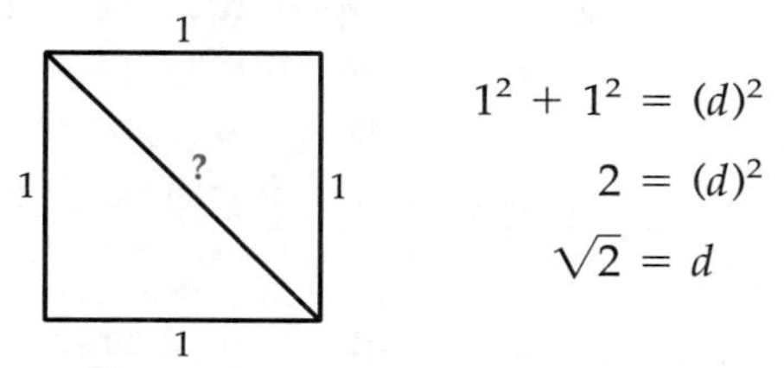

$$1^2 + 1^2 = (d)^2$$
$$2 = (d)^2$$
$$\sqrt{2} = d$$

Where is this number $\sqrt{2}$ on a number line? It is between 1 and 2. A calculator might show $\sqrt{2}$ as 1.414213562. Is this the exact value of $\sqrt{2}$? No, because $1.414213562 \times 1.414213562$ would end in a 4 (from the last digit 2×2). All rational numbers can be written as terminating or repeating decimals. What about $\sqrt{2}$? People have used computers to determine $\sqrt{2}$ to millions of decimal places. And still it has no repeating block of digits!

$\sqrt{2}$ = 1.41421 35624 19339 16628 19759 88713 07959 . . . and so on.

Lesson Exercise 7.76

What other kinds of decimals are there besides terminating and repeating decimals?

A **nonrepeating** (infinite) **decimal** has an infinite number of nonzero digits to the right of the decimal point, but it does not have a repeating block of digits that repeats an infinite number of times. Two more examples begin as follows.

$$3.141592653589783\ldots$$

$$0.010010001\ldots$$

Nonrepeating decimals are called **irrational** (not rational) **numbers** because they cannot be written as rational fractions. How did the ancient Greeks know they had discovered an irrational number? One of them wrote a deductive proof that $\sqrt{2}$ is irrational. Aristotle describes the proof in one of his books.

In order to prove that $\sqrt{2}$ is irrational, we shall use the fact that the square of any counting number greater than 1 has an even number of factors in its prime factorization. For example, $36 = 6 \cdot 6 = 2 \cdot 3 \cdot 2 \cdot 3$, and $400 = 20 \cdot 20 = 2 \cdot 2 \cdot 5 \cdot 2 \cdot 2 \cdot 5$. (The same set of prime factors will repeat twice.)

Lesson Exercise 7.77

Give another example of the square of a counting number, and show that it has an even number of factors in its prime factorization.

Now to show that $\sqrt{2}$ is irrational.

Lesson Exercise 7.78

We'll assume that $\sqrt{2}$ is a *rational* number p/q and show that this is impossible.

(a) Suppose $\sqrt{2} = p/q$, in which p and q are counting numbers. Square both sides and you obtain ____________.

(b) Solve your equation from part (a) for p^2.

(c) So $2q^2 = p^2$. Imagine prime factoring $2q^2$ and p^2. Since p^2 is the square of a counting number, it has an __________ (even, odd) number of prime factors.

Since $2q^2$ is 2 times a perfect square, it has an __________ (even, odd) number of prime factors.

In Lesson Exercise 7.78(c), you found that two equal numbers, $2q^2$ and p^2, appear to have different numbers of prime factors. This is impossible, according to the Fundamental Theorem of Arithmetic. Therefore, our assumption that $\sqrt{2} = p/q$ must be wrong, since all our deductive reasoning from that point on was valid. Thus, $\sqrt{2}$ is irrational!

Lesson Exercise 7.79

(a) Name two more square roots that you think are irrational numbers.

(b) Name a square root that *is* a rational number.

Soon after the Greeks proved that $\sqrt{2}$ is irrational, they proved that the square roots of 3, 5, 6, 7, 8, 10, 11, 12, 13, 14, 15, and 17 are also irrational.

The most famous irrational number of all is π. For thousands of years, people calculated how the distance around a circle (the circumference) compared to the distance across the circle through the center (the diameter).

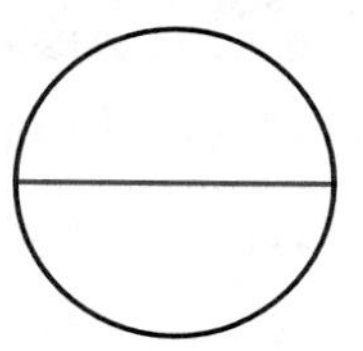

Cooperative Lesson Exercise 7.80

To measure in this problem, you will need a string and a ruler. (A compass would also be useful.)

(a) Draw three fairly large circles of different sizes, or find circular shapes in your classroom.

(b) Measure the circumference C and diameter d of each, and complete the following chart.

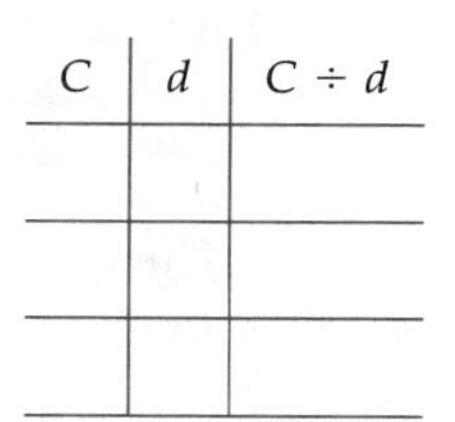

C	d	$C \div d$

(c) What do you notice about the numbers in the last column?

(d) What generalization does part (c) suggest?

Did you find that $\frac{C}{d}$ is always a little more than 3? The exact quotient is represented by the symbol π. By definition, $\pi = \frac{C}{d}$, or $C = \pi d$.

At first, people assumed that π would be a simple number such as 3 or 3.1. The Rhind Papyrus (1650? B.C.) shows that the Egyptians used a value of $\pi \approx 3.16$ to solve problems involving circles.

Archimedes (240 B.C.) found the value of π to two decimal places as 3.14. He drew inscribed and circumscribed polygons around a circle and computed their perimeters (Figure 7-17). He knew that the perimeter of the circle was between these two numbers. But it was not until 1761 that Johann Lambert *proved* that π is irrational. In 1989, Gregory and David Chudnosky of Columbia University used two computers to calculate π to 1,011,196,691 decimal places.

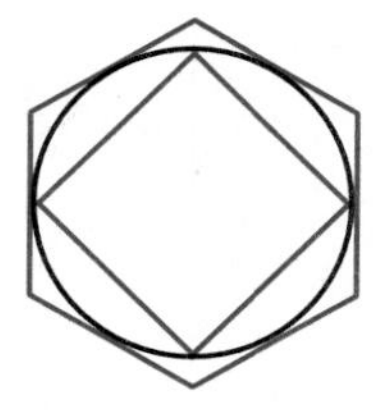

Figure 7–17

$$\pi = 3.14159\ 26535\ 89793\ 23846\ 26434\ \ldots$$

The only way to write the exact value of an irrational number like $\sqrt{2}$ or π is to use the $\sqrt{\ }$ symbol or letter. The exact value cannot be written using a decimal number.

Lesson Exercise 7.81

A wheel has a diameter of 30 in. Approximate its circumference.

REAL NUMBERS

To name every point on a standard number line, we need both rational and irrational numbers. The union of the set of rational numbers and the set of the irrational numbers is the **real numbers**. There is one-to-one correspondence between real numbers and the points of the number line. Real numbers are all numbers that can be represented as decimal numbers.

Lesson Exercise 7.82

Give two examples of real numbers.

The following chart compares the two types of real numbers, rational and irrational.

Real Numbers

Rational numbers	Repeating or terminating decimals
Irrational numbers	Infinite nonrepeating decimals

All of the numbers in this book can be organized into an overall set picture.

Real Numbers

Rational numbers

Integers

Whole numbers

Irrational numbers

You may be wondering if there are any other kinds of numbers besides real numbers. In order to find solutions to $x^2 = a$, in which $a < 0$, we need imaginary numbers such as $\sqrt{-3}$, which are covered in some high-school and college mathematics courses.

EXAMPLE 7.19

Which of the following describe 2.6?

(a) A whole number
(b) An integer
(c) A rational number
(d) An irrational number
(e) A real number

Solution

2.6 is not a whole number or integer because of the .6. But 2.6 is rational because it can be written as $\frac{26}{10}$. Since 2.6 is rational, it cannot be irrational. Since 2.6 is rational, it is also a real number. ■

Lesson Exercise 7.83

Complete the following chart by placing a ✓ in the appropriate columns.

	Whole Number	Integer	Rational Number	Irrational Number	Real Number
2.6			✓		✓
3					
$\sqrt{2}$					
−1					
$-2\frac{1}{2}$					

Because every real number corresponds to a point on the number line, solutions to inequalities for real numbers can be represented on a number line.

Lesson Exercise 7.84

(a) Write the inequality represented by the following number-line graph.

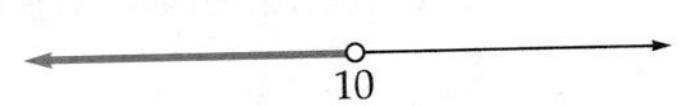

(b) Suppose Christine won't be a teacher for at least 5 years. Graph $n \geq 5$ on a number line, in which n represents the number of years.

MULTIPLE REPRESENTATIONS OF REAL NUMBERS

Why do we need both fractions and decimals to represent quantities? Each notation has its advantages. Sometimes it is easier to represent a number as a fraction than as a decimal, and sometimes it isn't.

Lesson Exercise 7.85

Name a rational number that is easier to represent with a fraction than with a decimal.

It is usually easier to compare the sizes of numbers in decimal form than in fraction form.

Lesson Exercise 7.86

If you have a calculator, what is an easy way to determine whether $\frac{3}{8}$ or $\frac{11}{29}$ is larger?

Some addition problems are easier to do with fractions; some are easier with decimals.

Lesson Exercise 7.87

In parts (a) and (b), state whether you think the addition problem is easier to do with fractions or with decimals.

(a) $\frac{2}{7} + \frac{3}{7}$ or $0.\overline{285714} + 0.\overline{428571}$ (b) $\frac{1}{4} + \frac{7}{100}$ or $0.25 + 0.07$

(c) Describe more generally when it is easier to do addition with fractions and when it is easier to use decimals. (Assume you do not have a calculator.)

In some problems, the preferred notation may depend upon whether you want an exact or an approximate answer.

Lesson Exercise 7.88

Last year, your salary increased from \$21,000 to \$27,000, an increase of $28\frac{4}{7}\%$, or 28.6%.

(a) Classify each percent as exact or approximate.

(b) Describe an advantage of each answer.

Lesson Exercise 7.89

The numbers 1/4, 0.25, and 25% all represent the same amount. Describe a situation in which each representation would be the best choice. For example, 1/4 would be a good notation for a probability.

Why do we use special irrational number symbols (such as $\sqrt{2}$ and π) and decimal representations for the same numbers?

Lesson Exercise 7.90

(a) Why is it easier to write $\sqrt{273}$ than its decimal representation?

(b) A farmer wants to buy some fencing. You determine that the length of fence he needs is $\sqrt{273}$ ft. How much fence would you tell the farmer to buy?

RATIONAL NUMBER PROPERTIES RETAINED!

What properties do real numbers have? In computations with irrational numbers, we find that $(\pi + 2\pi) + 3\pi = \pi + (2\pi + 3\pi)$, $\sqrt{2} \cdot \sqrt{3} = \sqrt{3} \cdot \sqrt{2}$, and the additive inverse of $\sqrt{8}$ is $-\sqrt{8}$. Examples such as these suggest how real-number operations retain all the properties of rational numbers. The following list summarizes these properties.

Some Properties of Real-Number Operations

Addition, subtraction, and multiplication of real numbers are closed.

Addition and multiplication of real numbers are commutative.

Addition and multiplication of real numbers are associative.

The unique additive identity for real numbers is 0, and the unique multiplicative identity for real numbers is 1.

All real numbers have a unique additive inverse that is real, and all nonzero real numbers have a unique multiplicative inverse that is real.

Multiplication is distributive over addition, and multiplication is distributive over subtraction for the set of real numbers.

The following exercises make use of these properties.

Lesson Exercise 7.91

Give an example showing that real-number (decimal) addition is commutative.

Lesson Exercise 7.92

$3.6 + -3.6 = -3.6 + 3.6 = 0$ illustrates what property of real numbers?

Lesson Exercise 7.93

(a) Factor $2x + \sqrt{2}x$.

(b) What property is used to factor in part (a)?

What about properties that did not work for other sets of numbers in this course?

Lesson Exercise 7.94

The real numbers include all the other sets of numbers in Chapters 3, 5, and 6. If a property such as the associative property for subtraction does not work for whole numbers, explain why it could not work for real numbers.

Some of the properties of all the number systems you have studied are summarized in the following chart.

Property	Operation	Whole Numbers	Integers	Rational Numbers	Irrational Numbers	Real Numbers
COMMUTATIVE	+	√	√	√	√	√
	×	√	√	√	√	√
ASSOCIATIVE	+	√	√	√	√	√
	×	√	√	√	√	√
IDENTITY	+	√	√	√		√
	×	√	√	√		√
INVERSES	+		√	√		√
	×			√		√

Lesson Exercise 7.95

Tell why 0 cannot be the additive identity and 1 cannot be the multiplicative identity for irrational numbers.

ANSWERS TO SELECTED LESSON EXERCISES

7.71 Terminating and infinite repeating

7.72 Terminating and repeating

7.73 $\frac{72}{1000} = \frac{9}{125}$

7.74 (a) $0.\overline{1}$ and $0.\overline{2}$ (b) $0.\overline{7}$
(c) $0.\overline{01}$ and $0.\overline{02}$ (d) $0.\overline{07}$
(e) $0.\overline{13}$ (f) $0.\overline{278}$

7.75 (a) $\frac{8}{9}$ (b) $\frac{37}{99}$ (c) $\frac{2714}{99,999}$

7.76 Infinite nonrepeating

7.77 $81 = 3 \cdot 3 \cdot 3 \cdot 3$ (4 factors)

7.78 (a) $2 = \frac{p^2}{q^2}$ (b) $p^2 = 2q^2$
(c) even; odd

7.79 (a) $\sqrt{3}$ and $\sqrt{34}$ (b) $\sqrt{4}$

7.81 $30\pi \approx 94.2$ in.

7.82 3 and 8.765

7.83 3 is whole, integer, rational, and real; $\sqrt{2}$ is irrational and real; -1 is integer, rational, and real; $-2\frac{1}{2}$ is rational and real.

7.84 (a) $n < 10$
(b)
5

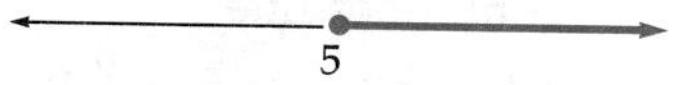

7.85 $\frac{3}{7}$

7.86 For each fraction, divide the numerator by the denominator and compare the decimal representations.

7.87 (a) Fractions (b) Decimals
(c) Fractions are easier when there is a common denominator that has factors other than 2 and 5. Decimals are easier for fractions that are easily converted to decimals.

7.88 (a) The fraction is exact; the decimal is approximate.
(b) The fraction is more precise. The decimal is usually easier to work with.

7.89 The decimal 0.25 could be an amount of money and 25% could be a tax rate.

7.90 (a) The decimal representation is an infinite nonrepeating decimal.
(b) 16.5 ft (a decimal approximation)

7.91 $0.2 + 0.3 = 0.3 + 0.2$

7.92 Additive inverse property

7.93 (a) $(2 + \sqrt{2})x$
(b) Distributive property of multiplication over addition

7.94 All counterexamples for whole numbers are also counterexamples for real numbers.

7.95 They are not part of the set of irrational numbers.

7.6 HOMEWORK EXERCISES

Basic Exercises

1. Write the following fractions as decimals.
(a) $\frac{3}{5}$ (b) $2\frac{2}{9}$

***2.** Write the following decimals as fractions.
(a) 0.731 (b) -13.04

3. Which is greater, $0.\overline{34}$ or $0.3\overline{4}$?

***4.** Write the following decimals as fractions.
(a) $0.\overline{4}$ (b) $0.\overline{32}$ (c) $0.\overline{267}$

5. Write the following decimals as fractions.
(a) $52.\overline{35}$ (b) $0.\overline{3917}$ (c) $0.\overline{9}$

6. One can use the results from the lesson to convert other types of repeating decimals to fractions. For example,

$$0.43\overline{2} = \frac{43\frac{2}{9}}{100} = \frac{\frac{389}{9}}{100} = \frac{389}{900}$$

Use this method to convert the following decimals to fractions.
(a) $0.2\overline{1}$ (b) $0.3\overline{41}$

7. Make up a geometric problem whose answer is $\sqrt{5}$.

8. The origin of π was in computing the ratio of the ______________ to the ______________.

9. A wheel has a circumference of 76 in. What is its diameter?

10. Make up an algebraic equation (other than $x = \sqrt{3}$) with a solution of $x = \sqrt{3}$.

11. (a) Find the value of $\sqrt{5}$ on a calculator.
(b) Why can't this number be the exact value of $\sqrt{5}$?

12. (a) Enter any positive number. Press the square-root key repeatedly. What number do you reach eventually?
(b) Repeat part (a), starting with a different number.
(c) Generalize your results.

13. Classify the following numbers as rational or irrational.
(a) $\sqrt{11}$ (b) $\frac{3}{7}$ (c) π (d) $\sqrt{16}$

14. Classify the following numbers as rational or irrational.
(a) 0.3333333 . . .
(b) 0.817
(c) 0.42638275643 . . .
(d) 0.121121112 . . .

15. Police can estimate the speed s (in miles per hour) of a car by the length L (in feet) of its skid marks. On a dry highway, $s \approx \sqrt{24L}$. On a wet highway, $s = \sqrt{12L}$.
(a) Estimate the speed of a car that leaves 54-ft skid marks on a wet highway.
(b) Estimate the speed of a car that leaves 54-ft skid marks on a dry highway.
(c) Classify your answers to parts (a) and (b) as rational or irrational.

16. On a clear day, the distance D, in kilometers, you can see to the horizon from a height of H meters is given by

$$D = \frac{\sqrt{H}}{3.57}$$

(a) How far can you see from a height of 40 m?
(b) Graph D vs. H for $0 \leq H \leq 100$.

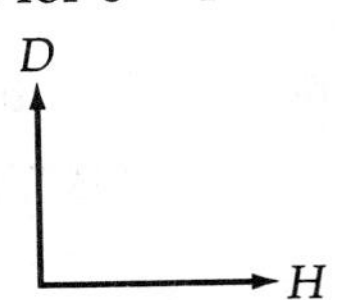

17. A pendulum is constructed by tying a weight on a string. The time to make one complete swing back and forth depends upon the length of the string. Suppose the following procedure can be used to find the time required for a complete swing.

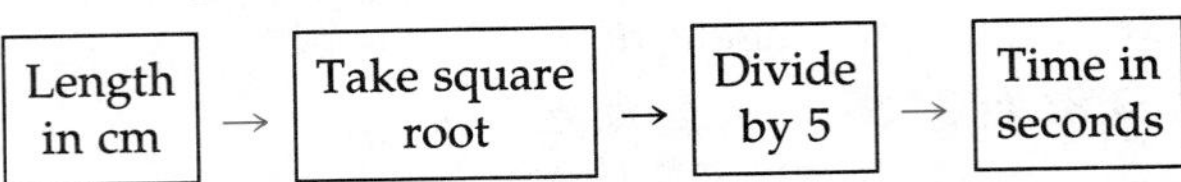

(a) How long would a 40-cm pendulum take to complete one swing?
(b) A pendulum takes 4 seconds to complete a swing. How long is the pendulum?
(c) A formula relating time T to length L is $T =$ __________.
(d) A formula relating length L to time T is $L =$ __________.

18. Give an example of a number on a standard number line that is not a rational number.

19. What set of numbers is used to represent every point on a number line?

20. Match each word in column A to a word in column B.

A	B
Terminating	Rational
Repeating	Irrational
Infinite nonrepeating	

21. Your calculator display shows 0.3333333. Name two different exact values this might represent.

22. Compute $2 \div 3$ on your calculator. Does it round to the nearest or truncate?

23. Complete the chart.

	Whole	Integer	Rational	Irrational	Real
$\frac{1}{3}$			✓		✓
$\sqrt{13}$					
-6					
$\sqrt{9}$					
-0.317					

24. Complete the chart.

	Whole	Integer	Rational	Irrational	Real
$\sqrt{\frac{7}{2}}$				✓	✓
4.21					
$-\sqrt{16}$					
π					
51					

25. Draw a Venn diagram with these sets: real numbers, rational numbers, whole numbers.

26. True or false? All rational numbers are real numbers.

27. (a) How many whole numbers are there between 3 and -3 (not including 3 and -3)?
(b) How many integers are there between 3 and -3?
(c) How many real numbers are there between 3 and -3?

28. The temperature T yesterday was between 54°F and 76°F. Graph this interval on a number line.

29. Graph the following inequalities on a number line.
(a) $x \leq 1$ (b) $x > -3$

30. Tell whether each of the following is more likely to be represented with a fraction or a decimal.
(a) How much of a pizza you ate
(b) The call number for a radio station
(c) The price of a share of stock

31. Name a real number that is not easily represented by either a decimal or a rational fraction.

32. Name a rational number that can be easily represented both as a fraction and as a decimal.

33. Make up an addition problem that is easier to do with fractions than with decimals.

34. Make up a subtraction problem that is easier to do with decimals than with fractions.

35. If you have a calculator, what is an easy way to determine which fraction is larger, $\frac{5}{7}$ or $\frac{12}{17}$?

36. A telephone company plans to raise monthly rates by 3%, which amounts to \$28 million. Which number would probably be more upsetting to consumers?

37. In school, 3.14 and $\frac{22}{7}$ are often used as approximations for $\pi = 3.141592654\ldots$.
(a) Which school approximation is closer to the actual value of π?
(b) If your calculator has a π key, write the decimal value it displays for π.

38. Different mathematicians have used fractional approximations for π. Which of the following is closest to $\pi = 3.141592654\ldots$?
(a) Babylonians (around 1700 B.C.): $\frac{25}{8}$
(b) Archimedes (around 200 B.C.): $\frac{223}{71}$
(c) Tsu Ch'ung-chih (around A.D. 480): $\frac{355}{113}$
(d) Bhaskara (around A.D. 1150): $\frac{3927}{1250}$

39. Why is it easier to use the symbol π than its decimal representation?

40. The distance around a running track is $50\pi + 360$ yd.
(a) What is a more useful way to express this distance?
(b) Is your answer to part (a) exact or approximate?

41. People reporting measurements usually state a margin of error. For example, 16.0 ± 0.2 cm means that the actual measure is somewhere between $16.0 - 0.2 = 15.8$ cm and $16.0 + 0.2 = 16.2$ cm. Give the highest and lowest value of each of the following.
(a) An SAT score of 520 ± 30 points
(b) A candidate preferred by $64\% \pm 2\%$ of the voters
(c) A length of 35 ± 1 ft

42. Which real-number operations are associative?

43. $0.37 + 0.516 = 0.516 + 0.37$ illustrates the ______________ property of ______________.

44. What properties guarantee that $-4x + 3 + 5.5 + 6x = (-4x + 6x) + (3 + 5.5)$ for any decimal x?

45. What is the additive inverse of -0.41798?

46. What is the multiplicative inverse of 0.1?

47. Give a counterexample that shows that real-number subtraction is not commutative.

48. What is the easiest way to multiply $0.6 \times 4.2 \times 5$?

49. How would you mentally compute 6.5×8?

50. According to the distributive property of multiplication over subtraction, $6.4x - 2.1x =$ ________.

51. Give an example showing that real-number addition is not distributive over multiplication.

52. Tell whether each equation is true for no real number, some real number(s), or all real numbers.
(a) $x \cdot 1 = x$ (b) $2x - 4 = 10$
(c) $2 + x = x$ (d) $\sqrt{x + 4} = \sqrt{x} + \sqrt{4}$

53. Tell whether each equation is true for no real numbers, some pairs of real numbers, or all pairs of real numbers.
(a) $x + y = 3$
(b) $xy = yx$
(c) $x + y = x + y + 3$
(d) $\sqrt{x^2 + y^2} = x + y$

54. $\frac{1}{2^1} = 0.5$ (one decimal place)
$\frac{1}{2^2} = 0.25$ (two decimal places)
(a) Write the next two examples that continue the pattern.
(b) Write a generalization about $\frac{1}{2^N}$ based upon your results.

55. $\frac{1}{5^1} = 0.2$ $\frac{1}{5^2} = 0.04$
(a) Write the next two examples that continue the pattern, and see if they are true.
(b) Write a generalization about $\frac{1}{5^N}$ based upon your results.

56. $\frac{1}{2^1 \cdot 5^1} = 0.1$ $\frac{1}{2^2 \cdot 5^1} = 0.05$ $\frac{1}{2^1 \cdot 5^2} = 0.02$
Try some more examples, and hypothesize how many decimal places $2^N \cdot 5^M$ has.

Extension Exercises

57. (a) Which of the following fractions can be rewritten with denominators of 10, 100, 1000, and so on?
(1) $\frac{4}{5}$ (2) $\frac{3}{25}$ (3) $\frac{7}{9}$
(4) $\frac{3}{200}$ (5) $\frac{1}{7}$
(b) Which fractions in part (a) can be represented by terminating decimals?
(c) Prime-factor each denominator in part (a).
(d) Try to generalize your results by filling in the blanks. Fractions (in simplest form) that can be rewritten as terminating decimals have denominators that are divisible by no prime number other than _____ or _____.

58. As the preceding exercise suggests, a rational number a/b in simplest form is a terminating decimal if and only if b is divisible by no prime numbers other than 2 or 5. Factor the denominator of each fraction, and determine whether it is a terminating or repeating decimal.
(a) $\frac{8}{9}$ (b) $\frac{7}{500}$ (c) $\frac{11}{30}$ (d) $\frac{7}{20}$

59. If x/y is a rational number in simplest fraction form, with $y > x > 0$ and $y = 2^a \cdot 5^b$ for whole numbers a and b, how are a and b related to the numbers of digits in the terminating-decimal representation of x/y?

60. According to Einstein's Theory of Relativity, something strange happens to the mass of an object that moves at a velocity close to the speed of light. The mass m at velocity v is

$$m = \frac{m_0}{\sqrt{1 - \frac{v^2}{c^2}}}$$

in which m_0 is the mass at rest and c is the speed of light.
(a) If an object moves at a velocity of $0.8c$, what is its mass?
(b) What do you think happens to the mass as the velocity gets closer to c?
(c) Find the mass at a velocity of $0.99c$.

61. A common algebraic error is to assume that $\sqrt{M + N} = \sqrt{M} + \sqrt{N}$ for all real numbers M and N. For what decimal values of M and N does $\sqrt{M + N} = \sqrt{M} + \sqrt{N}$? Make an educated guess after trying some examples.

62. Consider the following problem. "Name two irrational numbers whose product is 6."
(a) Devise a plan and solve the problem.
(b) Make up a similar problem.

63. Are the real numbers dense? That is, between any two real numbers, is there another real number? Try the following to find out.
(a) Find a real number between 1.41 and 1.42.
(b) Find a real number between 1.41 and $\sqrt{2}$ ($\sqrt{2} = 1.4142135\ldots$) if you can.
(c) Do you think that the real numbers are dense?

64. (a) $0.010110111\ldots + 0.101001000\ldots =$ ______________
(b) Part (a) shows that the sum of two ______________ numbers can be a ______________ number.

65. The sum of a rational number and an irrational number
(a) is rational (b) is irrational
(c) could be either

66. The product of two irrational numbers
(a) is rational (b) is irrational
(c) could be either

67. True or false? If $x > 0$, then $x + \frac{1}{x} \geq 2$. If it is true, prove it. If it is false, find a counterexample.

68. Prove that $\sqrt{3}$ is not rational.

69. Prove that $\sqrt[3]{2}$ is not rational.

Special Exercise

70. Write a short report on the history of π.

SUMMARY

The rules for decimal place value and arithmetic are extensions of the rules for whole-number place value and arithmetic. The primary question in decimal arithmetic is where to place the decimal point in the answer. The rules for placing decimal points can be explained using place value and rational-number properties. The models for whole-number operations apply to many decimal problems.

Decimal estimation relies primarily on the same strategies as fraction estimation—namely, rounding and the compatible-numbers strategy. One can mentally multiply or divide a decimal number by a power of 10 by moving the decimal point of the number. This same technique is used in converting a decimal number to scientific notation, an exponential shorthand used by scientists.

Many everyday applications of mathematics involve proportional relationships. A proportion states that two ratios are equal. Proportions with one missing quantity can be solved using cross products or, more intuitively, with a unitary or multiplier approach. Common ratios that occur in proportion problems include male-female ratios, student-teacher ratios, and miles per gallon. Proportions also occur in situations involving water bills, real estate taxes, map scales, surveys, and foreign currency rates.

Percents developed from the use of hundredths in fractions and decimals. People use percents to describe taxes, inflation, interest rates, finance charges, salary increases, discounts, and test scores. In all these cases, percents give the rate per 100. These applications of percents can usually be solved with algebra or proportions and modeled by number-line pictures.

It is helpful to know the fractional equivalents of some basic percents in order to compute some percent problems mentally using shortcuts. Rounding can be used to estimate solutions to many other percent problems.

The set of all decimal numbers is called the real numbers. Decimals are either terminating, infinite repeating, or infinite nonrepeating decimals. Terminating and repeating decimals represent rational numbers. With rational numbers, one has the option of using either decimal or fraction notation.

Nonrepeating decimals are called irrational numbers. Examples of irrational numbers include $\sqrt{2}$, $\sqrt{5}$, and π. The only way to write the exact value of an irrational square root or π is to use a special symbol.

Real-number operations possess all the properties that whole-number, integer, and rational-number operations possess. Each real number corresponds to a point on the number line.

STUDY GUIDE

To review Chapter 7, see what you know about each of the following ideas or terms that you have studied. You can also use this list to generate your own questions about the chapter.

The NCTM Curriculum Standards and Decimals, Percents, and Real Numbers

Selected NCTM Curriculum Standards

The following standards come from the NCTM document.

- Extend their understanding of whole number operations to decimals.
- Apply decimals to problem situations.
- Develop, analyze, and explain procedures for computation and techniques for estimation.
- Develop number sense for decimals.
- Develop, analyze, and explain methods for solving proportions.
- Understand and apply ratios, proportions, and percents to a wide variety of situations.
- Investigate relationships among fractions, decimals, and percents.
- Understand, represent, and use numbers in a variety of equivalent forms (fraction, decimal, percent, and scientific notation) in real-world and mathematical problem situations.

1. Describe how each standard listed relates to the material you studied in Chapter 7.
2. Select any current elementary-school mathematics textbook series, and describe sample lessons or exercises that illustrate each standard listed.

DECIMALS AND PERCENTS IN ELEMENTARY SCHOOL

The following chart shows at what grade levels selected decimal topics typically appear in elementary-school mathematics textbooks.

Topic	Typical Grade Level in Current Textbooks
Decimal place value	3, 4, 5, 6
Decimal addition and subtraction	3, 4, 5, 6
Decimal multiplication	5, 6
Decimal division	5, 6
Ratio	5, 6
Proportion	6
Percents	5, 6
Decimal estimation	4, 5, 6
Scientific notation	6 (enrichment topic)

REVIEW EXERCISES

1. (a) Write 0.0037 in expanded notation.
(b) Write 0.0037 in scientific notation.

2. Use decimal-square pictures to explain why $0.3 > 0.25$.

3. It took G gallons of paint to paint 3 classrooms. How much paint would it take to paint C similar classrooms?

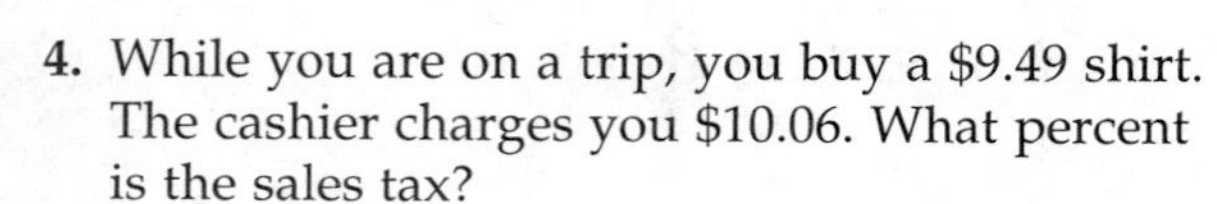

4. While you are on a trip, you buy a \$9.49 shirt. The cashier charges you \$10.06. What percent is the sales tax?

5. If 100 balloons cost \$.76, *explain* how you would mentally compute the charge per balloon.

6. Consider the following problem. "The average 130-pound person needs 2000 calories per day. How many calories does the average 150-pound person need?"
(a) Solve the problem with a proportion.
(b) Show how to solve the problem without writing a proportion.

7. If $\sqrt{m} > m$, what are the possible values of m?

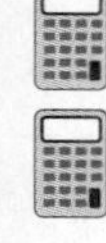

8. In a presidential election, 44 million people voted. This represented 52% of the registered voters. How many registered voters were there?

9. The price of a dress is marked down 30%. Then it is marked down an additional 20%. Find a single percent discount equal to these two successive discounts.

10. Complete the last example, repeating the same error pattern in the completed examples.

$6 \div 0.3 =$ *0.2* $0.8 \div 0.4 =$ *0.02*

$1.2 \div 0.3 =$ ______

11. *Explain* how to compute 1% of \$7250 mentally.

12. A child wants to know why we can move the decimal point to change $42 \div 0.06$ to $4200 \div 6$. Write $42 \div 0.06$ as a fraction, and show how the process works.

13. $42.764764 \div 0.48917563645639$ is about
(a) 9 (b) 40 (c) 1 (d) 90 (e) 20

14. *Explain* how to compute 150% of 46 mentally.

15. What operation and category are illustrated in the following problem? "A man loses 8 lb in 10 days. What is his average weight loss per day?"

16. What operation and category are illustrated in the following problem? "Before driving to work, your odometer reads 26,872.6 miles. When you arrive at work, it reads 26,880.4. How long is the drive?"

17. Draw a decimal-square picture that shows the result of $0.7 - 0.2$, using a take-away approach. Write the result as an equation.

18. Name a property that real numbers have that the whole numbers do not.

19. Give an example of the distributive property of multiplication over subtraction for real numbers.

20. Write 23.7 million in scientific notation.

21. A child knows fraction multiplication but does not know decimal multiplication. Explain how to compute 0.2×0.3 using two decimal-square pictures.

22. Suppose a child knows about positive whole-number exponents. Show how to find the value of 4^0 by either extending a pattern or by using the addition rule for exponents on an example like $4^2 \cdot 4^0$.

23. $\sqrt{17}$ is
(a) a whole number
(b) an integer
(c) a rational number
(d) an irrational number
(e) a real number

24. Classify the following numbers as rational or irrational.

(a) $\frac{\pi}{2}$ (b) $\sqrt{7}$ (c) 3.4 (d) -5

25. Classify the following numbers as rational or irrational.
(a) $0.\overline{23}$ (b) 0.16
(c) 0.143798215 . . . (d) $\sqrt{16}$

26. -5 is
(a) a whole number
(b) an integer
(c) a rational number
(d) an irrational number
(e) a real number

27. Write a paragraph describing how the following are related: real numbers, rational numbers, irrational numbers, terminating decimals, repeating decimals, nonrepeating decimals.

28. Write a sentence or two about π.

SUGGESTED READINGS

Burton, G., and others. *Number Sense and Operations.* Reston, Va.: NCTM, 1993.

Curcio, F., N. Bezuk, and others. *Understanding Rational Numbers and Proportions.* Reston, Va.: NCTM, 1994.

Davis, P. *The Lore of Large Numbers.* New York: Random House, 1961.

National Council of Teachers of Mathematics. *A Sourcebook of Applications of School Mathematics.* Washington, D.C.: MAA, 1980.

National Council of Teachers of Mathematics. *More Topics in Mathematics for Elementary School Teachers.* 1969 Yearbook. Reston, Va.: NCTM, 1969.

National Council of Teachers of Mathematics. 1992 Yearbook. *Calculators in Mathematics Education.* Reston, Va.: NCTM, 1992.

8 Introductory Geometry

The oldest recorded examples of geometry are ancient cave drawings of circles, squares, and triangles. Ancient pottery and weaving bear examples of geometric designs. The root meaning of the word "geometry" is "earth measure," which derives from the use of geometry by the Egyptians and Babylonians in land surveying over 4000 years ago. The Egyptians and Babylonians also used geometry in agriculture, architecture, and astronomy. The Great Pyramid of Gizeh in Egypt, built over 4000 years ago, has a square base with sides that are all 756 ft long, with an error of less than 1 inch!

About 2500 years ago, the Greeks developed an important new approach to geometry, organizing geometric ideas into a logical sequence. They began with the most basic figures and terms and deduced all other concepts of plane geometry.

Today, school geometry follows this historical sequence. Children first study geometry in an informal, concrete manner. Then, in high school, many students study geometry using a more formal, deductive approach.

8.1 BEGINNING GEOMETRY

We all begin exploring geometry when we notice shapes in our surroundings. Children come to elementary school with an awareness of objects and shapes. School geometry builds upon this awareness by beginning with space figures (such as cubes and cylinders) and plane figures (such as rectangles and circles) suggested by objects in the environment. Then a more systematic study of geometry begins with the most basic figures—points, lines, and planes—and proceeds in a deductive sequence.

SHAPES IN OUR WORLD

The informal study of geometry begins with objects in our environment.

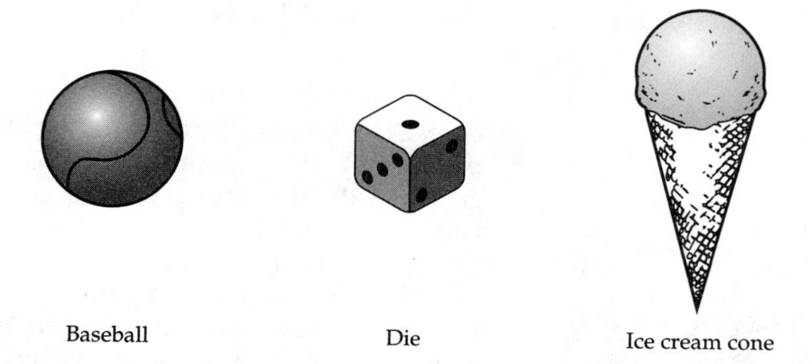

Geometric shapes are idealized versions of objects we see.

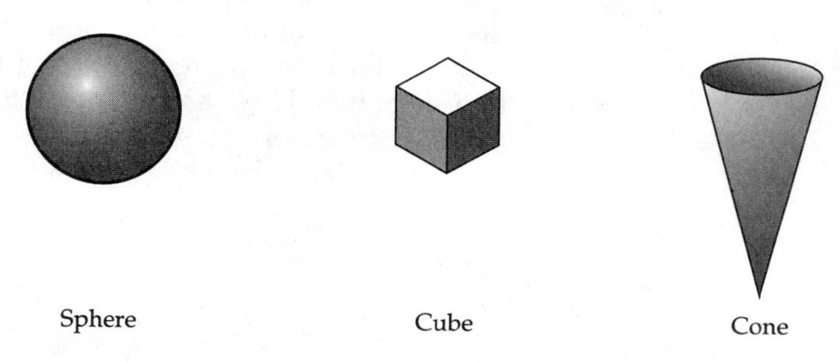

According to Plato, the perfect geometric shapes in our minds are part of the ultimate reality, whereas objects in the world are imperfect.

The surfaces of objects also suggest geometric figures.

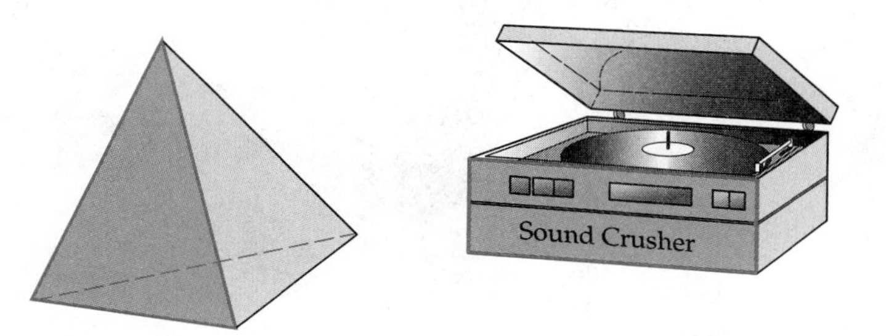

Lesson Exercise 8.1

(a) Find three objects in your classroom that suggest common geometric figures.

(b) Find three objects in your classroom with surfaces that suggest common geometric figures.

(c) Compare your answers to others in your class.

POINTS, LINES, AND PLANES

After an initial introduction to geometry through shapes suggested by objects, how should one begin a more systematic study of geometry? In his famous book *The Elements*, Euclid (c. 300 B.C.) begins **Euclidean geometry** by describing basic geometric terms and stating ten assumptions (postulates). Then he builds geometry deductively from those basic terms and assumptions. About 650 years later, Theon of Alexandria published the version of Euclid's *Elements* that we use in modern translations.

Hypatia

Figure 8–1

Theon's daughter, Hypatia (370–415) (Figure 8-1), is known as the first woman mathematician. In Alexandria, she was a popular lecturer on mathematics and philosophy. Most of Hypatia's mathematical work is lost, but she is known to have written a commentary on the work of Diophantus in number theory. Hypatia was murdered by a Christian mob because of her outspoken defense of scientific methods and Greek paganism.

Today, a formal presentation of Euclidean geometry begins with a description of points, lines, and planes. "Point," "line," and "plane" are undefined terms that we can understand using everyday objects and our intuition. Points, lines, and planes are suggested by our surroundings, as shown in Figure 8-2.

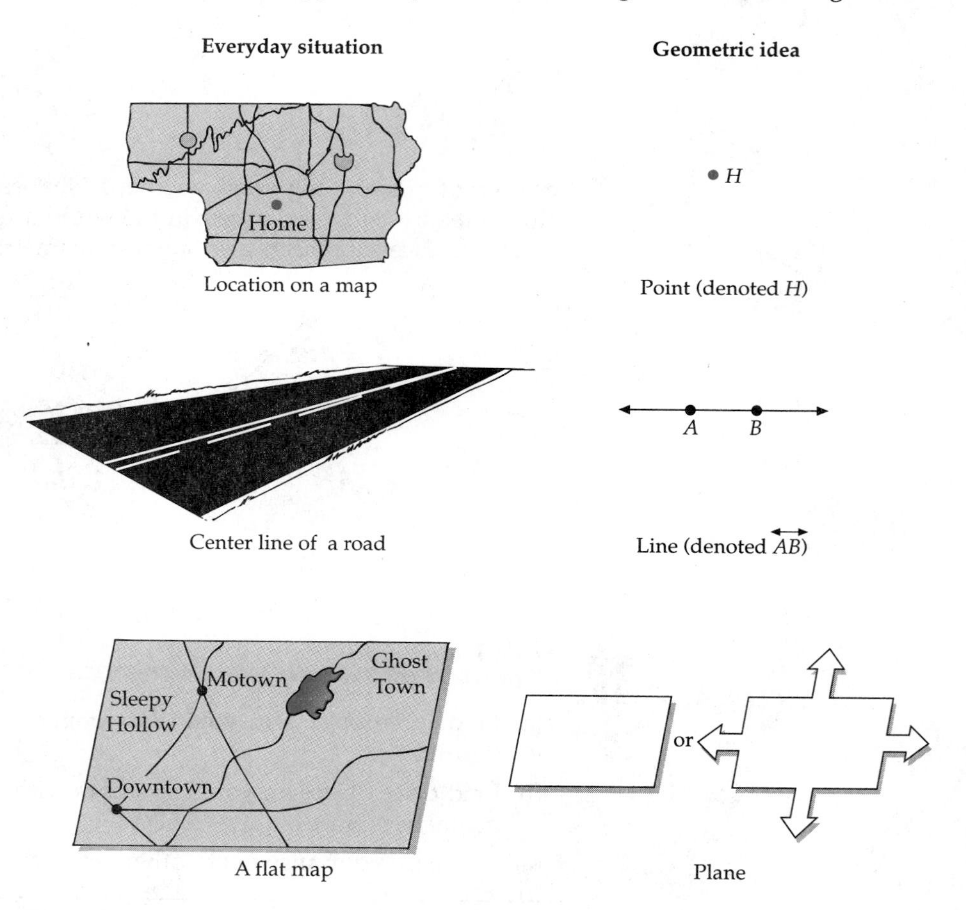

Figure 8–2

A location on a map and a grain of sand each suggest a point. A **point** represents an exact location. A taut wire and the center line on a road suggest a line. A **line** is straight; it extends indefinitely in two directions and has no thickness. The arrows on a line indicate that it extends without limit. The notation $\overleftrightarrow{AB}$ represents the line through points A and B.

A map and a piece of paper each suggest a section of a plane. A **plane** is flat; it has no thickness and extends endlessly in all directions (Figure 8-2).

Mathematicians and philosophers have debated where geometric ideas such as a line come from. Why all the fuss, you ask? Because, in your whole life, you will never see a geometric line!

Lesson Exercise 8.2

How does a geometric line differ from a drawing of a line?

Lesson Exercise 8.3

(a) True or false? If A and B are points in a plane, then $\overleftrightarrow{AB}$ lies entirely in the plane.

(b) Draw a picture that supports your answer.

PLANE FIGURES, SPACE FIGURES, AND DIMENSION

A **plane figure** is a set of points in a plane. Examples of plane figures include lines, line segments, planes, circles, and triangles. Lines and line segments are one-dimensional figures. They have length but no width or height. Planes, circles, and triangles are two-dimensional figures.

Cubes, spheres, and everyday objects are examples of three-dimensional or space figures. A **three-dimensional** or **space figure** does not lie in a single plane, and it has volume. Informally, any shape that can be constructed from modeling clay is three-dimensional.

Lesson Exercise 8.4

What is the dimension (1, 2, or 3) of the geometric figures *suggested* by each of the following?

(a) A very thin string (b) A gorilla (c) The surface of a table

---

LINE SEGMENTS, RAYS, AND ANGLES

The concepts of a "point" and a "line" can now be used to define line segments and rays. Line segments and rays are suggested by our surroundings, as shown in Figure 8-3.

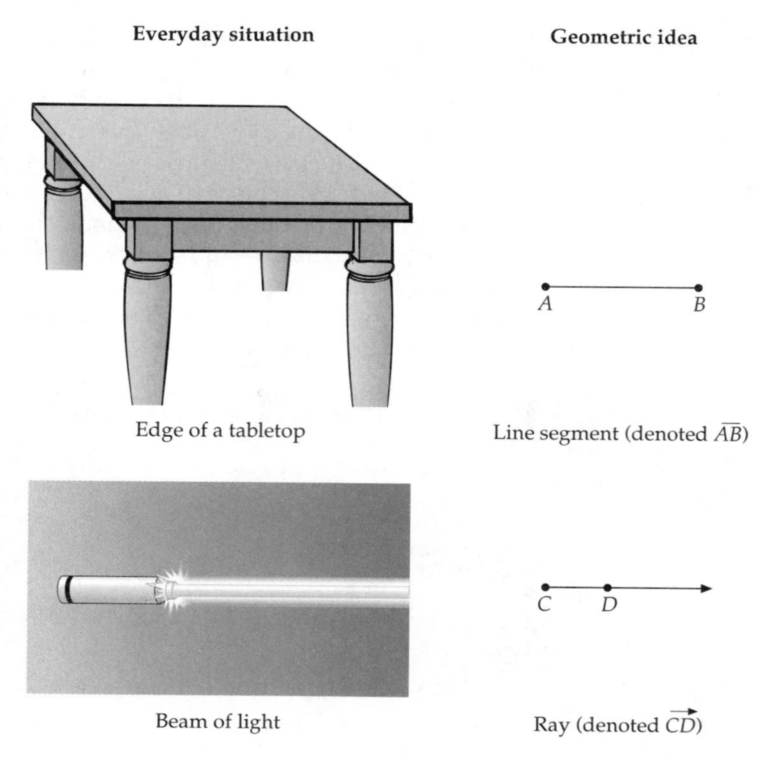

Figure 8–3

Lesson Exercise 8.5

Try to write definitions of "line segment" and "ray" using the terms "point" and "line."

Line segments and rays are subsets of a line. A **line segment** consists of two points on a line and all the points between them. The idea of making a line segment out of "points" is used in dot-matrix printers and stadium scoreboards, as shown in Figure 8-4. The N and Y are each formed by three sets of squares in vertical or diagonal rows. Each row of squares is analogous to a line segment.

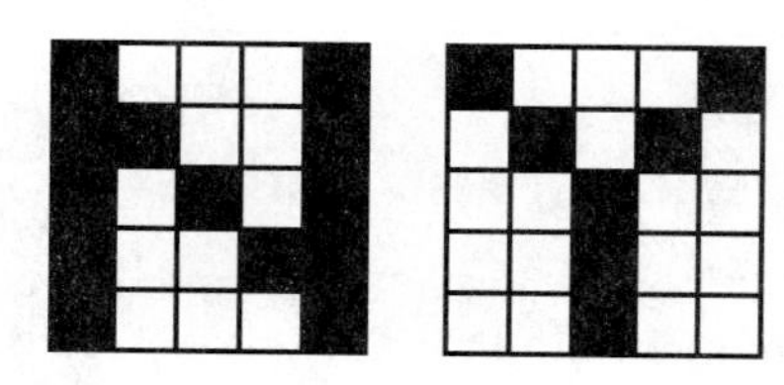

Figure 8–4

A **ray** (for example, $\overrightarrow{CD}$ in Figure 8-3) is a subset of a line that consists of a point (C) together with all points on the line ($\overleftrightarrow{CD}$) on one side of the point (the "D side" of C).

Next, consider the ways in which angles are also suggested by our surroundings (Figure 8-5).

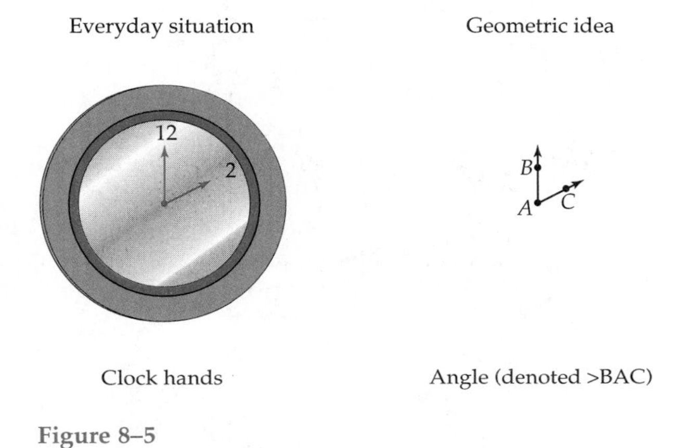

Figure 8–5

Lesson Exercise 8.6

Try to write a definition of "angle" using the term "ray."

An **angle** is formed by two rays that have the same endpoint. The two rays are called **sides**, and the common endpoint is called the **vertex**. The angle $\angle BAC$ in Figure 8-5 has sides $\overrightarrow{AB}$ and $\overrightarrow{AC}$ and vertex A. When three letters are used to name an angle, the middle letter names the vertex.

Lesson Exercise 8.7

Give a simpler name for each of the following.

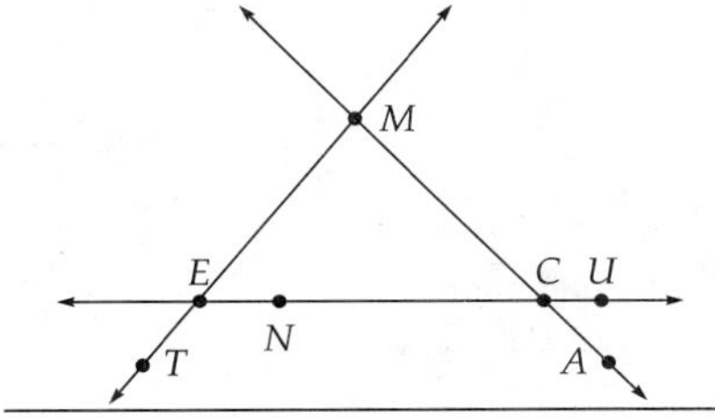

(a) $\overline{ME} \cup \overline{ET}$
(*Hint:* Trace over the two segments in the diagram.)

(b) $\overline{NU} \cap \overrightarrow{CE}$

(c) $\overleftrightarrow{ME} \cap \overline{CN}$

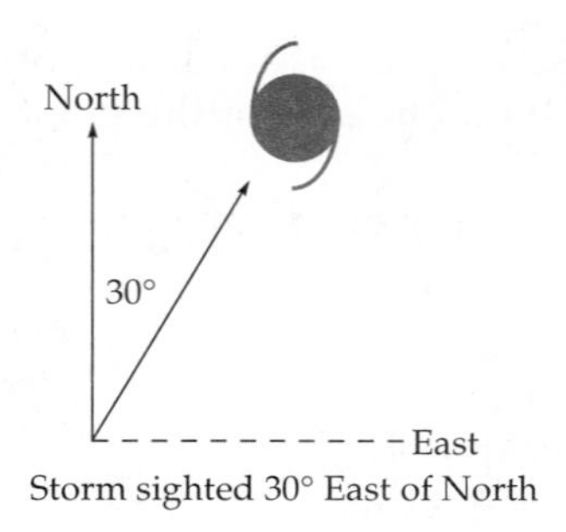

Storm sighted 30° East of North

Figure 8–6

MEASURING ANGLES

To give the position of a storm in relation to a town (see Figure 8-6), a meteorologist might use an angle measure. Surveyors, navigators, and meteorologists all measure angles as a part of their work. Mathematicians use angle measures to classify angles. In Section 8.3, you will study a surprising pattern in the angle measures of polygons. The following background information will prepare you.

Measuring an angle is quite different from measuring a length. The measure of an angle tells the amount of rotation involved in moving from one ray to the other. Over 4000 years ago, the Babylonians chose a degree as the unit for angle measure.

You can measure angles in degrees using a protractor. Place the vertex of the angle (B in Figure 8-7) on the mark at the center of the protractor while lining up one side of the angle ($\overrightarrow{BA}$ in Figure 8-7) with the base of the protractor and the O mark. The other side of the angle ($\overrightarrow{BC}$) will pass through a number on the protractor representing $m\angle ABC$ in degrees.

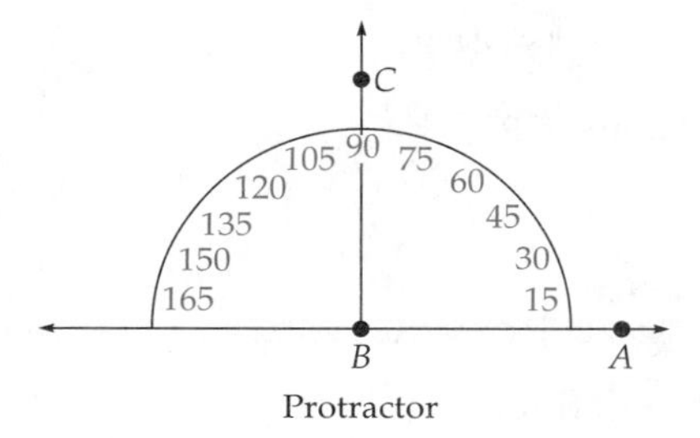

Protractor

Figure 8–7

As you can see, $\angle ABC$ measures 90° on the protractor pictured in Figure 8-7. This is written $m\angle ABC = 90°$.

Lesson Exercise 8.8

A ship S is sighted from point U on the shoreline $\overleftrightarrow{UN}$.

(a) Estimate the measure of $\angle SUN$.

(b) Measure $\angle SUN$ with a protractor.

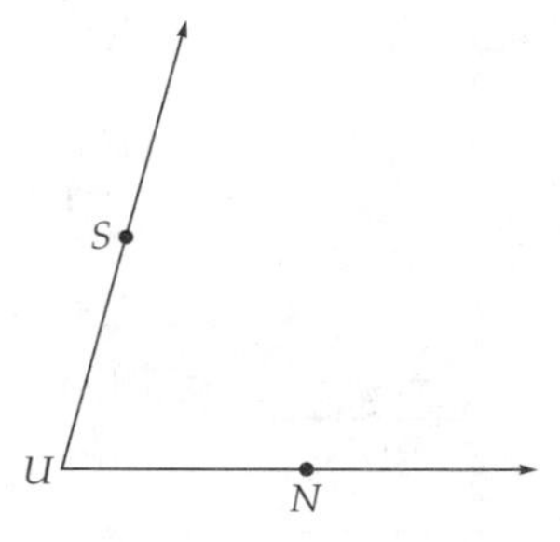

TYPES OF ANGLES

Mathematicians describe and categorize geometric shapes. Then they look for common properties within a category.

Throughout this chapter, geometric figures will be placed into categories. The first example of classifying shapes in this chapter is the categorization of angles according to their measure. Angle categories are suggested by our surroundings.

Lesson Exercise 8.9

What type of angle is most common in the room you are in now?

The most common angles in a classroom are usually **right angles** (90°) and **straight angles** (180°). Right angles appear in everything from books to boxes to corners of rooms (Figure 8-8).

Figure 8–8

Lesson Exercise 8.10

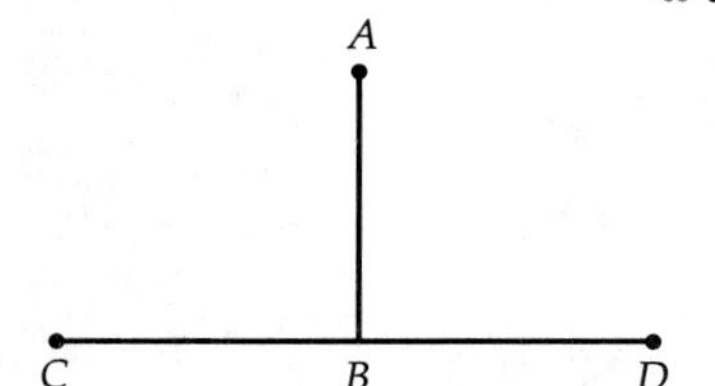

Figure 8–9

Two right angles with a common side, $\angle ABC$ and $\angle ABD$ (Figure 8-9), form a straight angle ($\angle CBD$).

(a) Give an example of an object that illustrates this property.

(b) Does having this property serve any purpose for the object in your example?

Other angle measures between 0° and 180° are not as common. These angles are placed into two groups: **acute angles** (greater than 0° and less than 90°) and **obtuse angles** (greater than 90° and less than 180°). Some examples are shown in Figure 8-10.

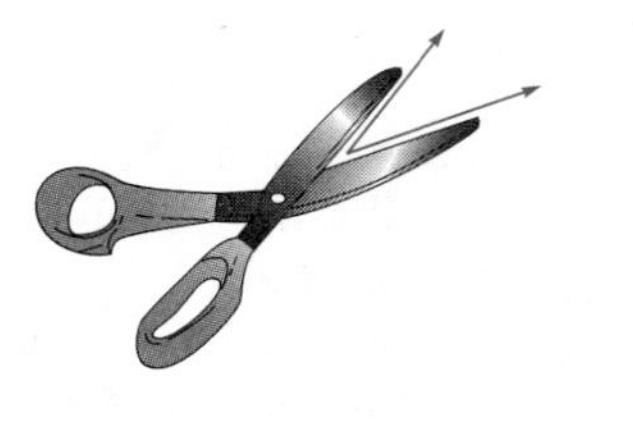

Acute angle

Obtuse angle

Figure 8–10

Lesson Exercise 8.11

Name an object in your classroom that suggests an acute angle.

Certain pairs of angles have special classifications.

Lesson Exercise 8.12

Use the following information to write definitions of "complementary" and "supplementary" pairs of angles.

Angle measures	42°, 48°	60°, 30°	78°, 20°	140°, 40°	82°, 98°
Complementary?	Yes	Yes	No	No	No
Supplementary?	No	No	No	Yes	Yes

Compare your definitions to the following. Two angles whose measures have a sum of 90° are **complementary angles**. Two angles whose measures have a sum of 180° are **supplementary angles**.

Lesson Exercise 8.13

Are $\angle ACE$ and $\angle ACT$ complementary, supplementary, or neither?

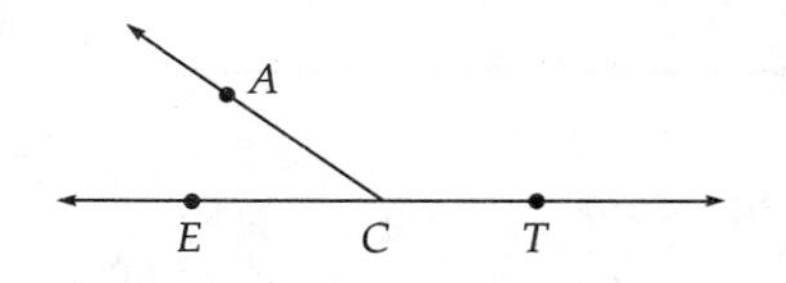

Congruent geometric figures are the same size and shape. How can we tell if two angles are congruent? We can measure them. Two angles with the same measure are **congruent angles**.

Lesson Exercise 8.14

(a) True or false? All acute angles are congruent.
(b) Give an example that supports your answer.

PARALLEL AND INTERSECTING LINES

Our surroundings suggest the common relationships between two distinct lines, as shown in Figure 8-11.

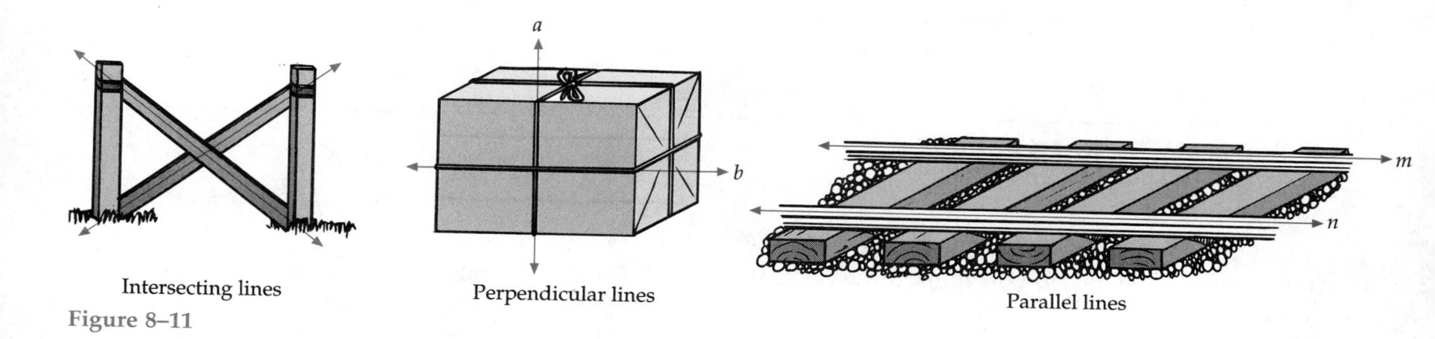

Intersecting lines   Perpendicular lines   Parallel lines

Figure 8–11

Lesson Exercise 8.15

Complete the following definitions.

(a) Two *intersecting lines* ______________________.
(b) Two lines are *perpendicular* if and only if ______________________.
(c) Two *parallel lines* lie in the same plane and ______________________.

Compare your definitions to the following. Two **intersecting lines** have exactly one point in common. Two lines that intersect at right angles are **perpendicular**. Two lines are **parallel** if and only if they lie in the same plane and do not intersect. In Figure 8-11, line a is perpendicular to line b, written $a \perp b$, and line m is parallel to line n, written $m \parallel n$.

Lesson Exercise 8.16

(a) True or false? In a plane, two lines that are both parallel to a third line must be parallel to each other.
(b) In your classroom, identify a model of this situation that supports your answer.

ANSWERS TO SELECTED LESSON EXERCISES

8.2 A geometric line is perfectly straight, extends forever in two directions, and has no width; it has no arrowheads.

8.3 (a) True
(b)

8.4 (a) 1 (b) 3 (c) 2

8.7 (a) $\overline{MT}$
(b) $\overline{NC}$ (what $\overline{NU}$ and $\overrightarrow{CE}$ have in common)
(c) { }

8.8 (b) About 74°

8.10 (a) Doors of a cabinet
(b) The property enables them to fit together and the bottom of the two doors to be level.

8.13 Supplementary

8.14 (a) False
(b) A 30° angle and a 40° angle are acute, but they are not congruent.

8.16 (a) True

8.1 HOMEWORK EXERCISES

Basic Exercises

1. The word "geometry" can be split into "geo-" and "-metry." What do you think these two roots mean?
2. Describe a specific example of each of the following.
   (a) Geometry in nature
   (b) An everyday use of geometry
   (c) Geometry in art
   (d) Geometry in architecture
3. Tell what geometric figure each of the following suggests.
   (a)

Photo courtesy of the Department of the Army.

   (b) A special piece of glass that disperses light into its color components
   (c) The edge of a box
4. Euclid lived about ________ years ago.
   (a) 50 (b) 200 (c) 500
   (d) 2000 (e) 10,000
5. How is a geometric point different from a dot?
6. How does the dictionary define the word "point"?
7. (a) True or false? If a line $\overleftrightarrow{AB}$ intersects a point in plane c, then $\overleftrightarrow{AB}$ lies in plane c.
   (b) Make a drawing that supports your answer.
8. Suppose a point P lies in a plane. Are there any points in the plane that are exactly 100 miles from P? If so, how many?
9. What is the dimension (1, 2, or 3) of the geometric figure *suggested* by each of the following?
   (a) A pillow (b) A piece of very thin wire
   (c) A sheet of paper
10. Give the number of dimensions in each figure.
    (a) Line (b) Cube (c) Rectangle
11. What shapes do you see in each of the structures pictured?
    (a) Water tank

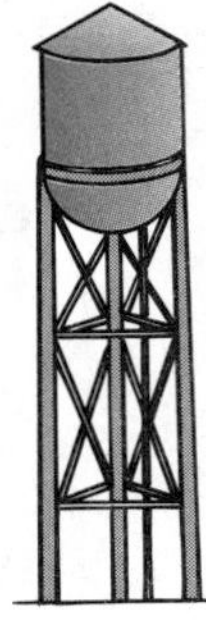

(b) Eskimo igloo (*Hint:* It's half of something.)

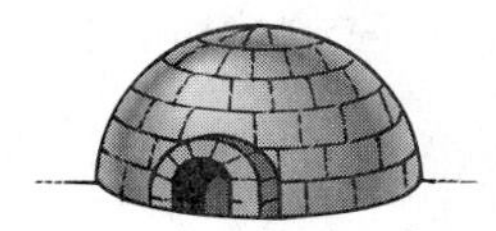

12. (a) Find three objects in your home that suggest common space figures.
(b) Find three objects in your home that suggest common plane figures.

13. How is a line segment different from a line?

14. Give three possible names for the line shown here.

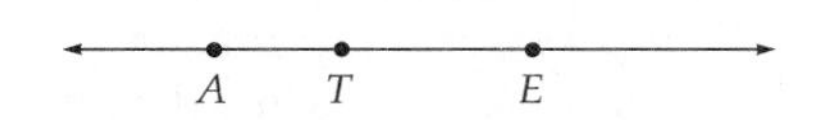

15. (a) What is the vertex of $\angle PAT$?
(b) How is $\overrightarrow{AB}$ different from $\overrightarrow{BA}$?

16. Show all the different patterns you can make by using the following points as the endpoints of one to six line segments. It is not necessary to draw the same pattern in all different positions.

17. Give a simpler name for each of the following.
(a) $\overrightarrow{AB} \cup \overrightarrow{AC}$
(b) $\overline{AB} \cup \overrightarrow{AD}$
(c) $\overleftrightarrow{BA} \cap \overleftrightarrow{AC}$

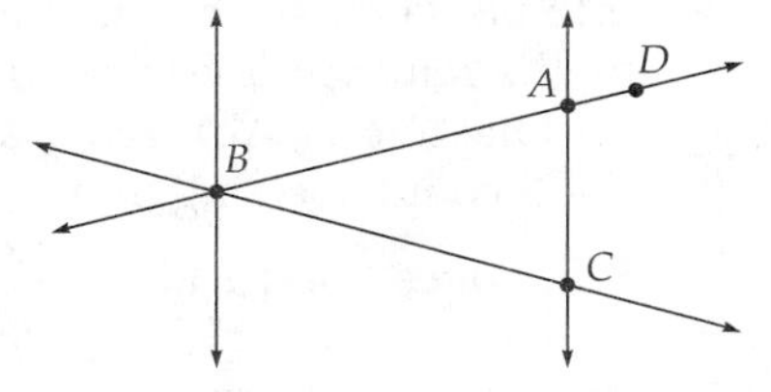

18. Make an enlarged copy of the following drawing. Use a protractor to draw a ray with endpoint C in the given direction.
(a) A storm is observed 30° east of south.
(b) A ship is sighted 25° south of west.

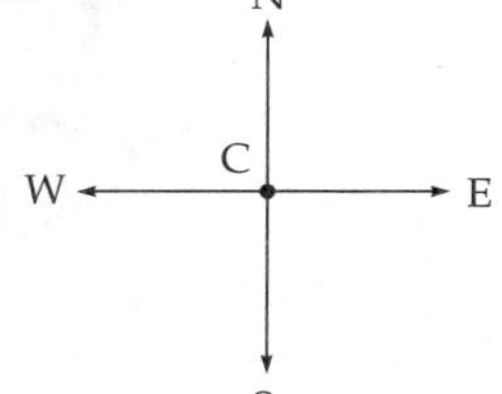

19. Estimate the measure of each marked angle. Then measure the angle with a protractor.
(a) (b)

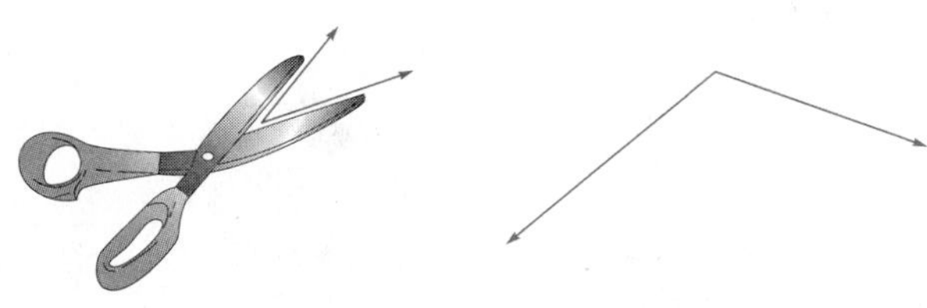

20. Pete and Charissa are on duty at observation towers *A* and *B* in a national forest. If there is a fire at *F*, how can Pete and Charissa measure angles and precisely locate the fire without leaving their towers? Assume that they can communicate by radio and that each has a map of the park and a protractor.

F

A *B*

21.

Station B
Station A
N
W E
S

A ship reports a bearing of 50° north of east from station *A* and 30° west of north from station *B*. Locate the ship on the grid.

22. Which segments are congruent to $\overline{AB}$?

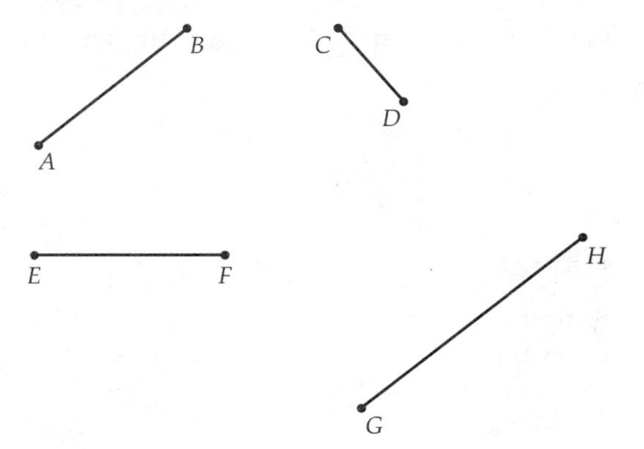

23. Which of the angles are congruent to $\angle A$?

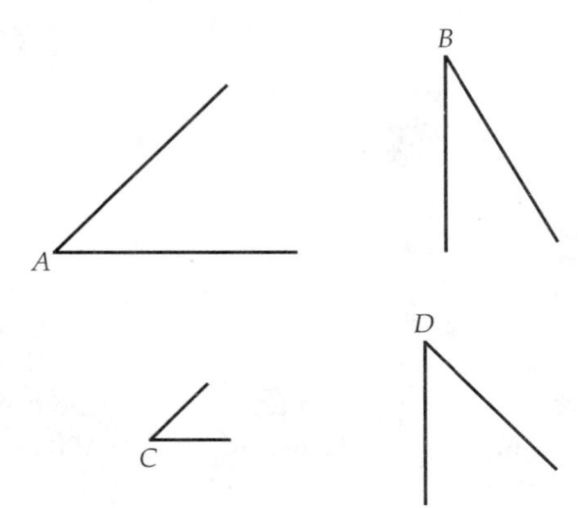

24. Which segments are congruent to $\overline{AB}$?

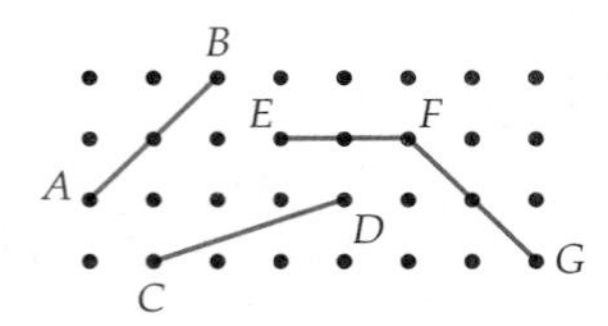

25. Name two capital letters that contain two or more acute angles and no obtuse or right angles.

26. Tell whether each angle is acute, right, or obtuse.
(a) (b) (c)

27. True or false? All right angles are congruent.

28. An angle measures 20°. What is the measure of
(a) its supplement? (b) its complement?

29.

Suppose $m\angle PIN = 90°$.
(a) Name two supplementary angles.
(b) Name two complementary angles.
(c) Name two more supplementary angles.

30. (a) Which of the following road intersection designs makes it easier to see cars coming from other directions?

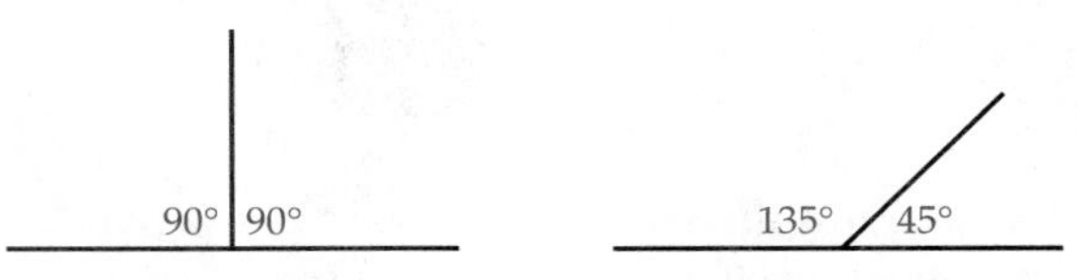

(b) What term describes each pair of angles?

31. True or false? Two complementary angles are congruent.

32. True or false? Two right angles are supplementary.

33. Draw four points A, B, C, and D so that $\overleftrightarrow{AB}$, $\overleftrightarrow{AC}$, $\overleftrightarrow{AD}$, $\overleftrightarrow{BC}$, $\overleftrightarrow{BD}$, and $\overleftrightarrow{CD}$ are
(a) the same line. (b) six different lines.
(c) four different lines.

34. If possible, draw two angles that intersect at exactly
(a) one point. (b) two points.
(c) three points.

35. (a) True or false? In a plane, two lines that are perpendicular to a third line are parallel to each other.
(b) In your home, identify a model of this situation that supports your answer.
(c) How might the property in part (a) be useful to a carpenter?

36. Into how many sections is a plane separated by the removal of each of the following?
(a) One line (b) Two parallel lines
(c) Two intersecting lines

37. What does "perpendicular" mean in the sign?

PERPENDICULAR
PARKING
ONLY

38.

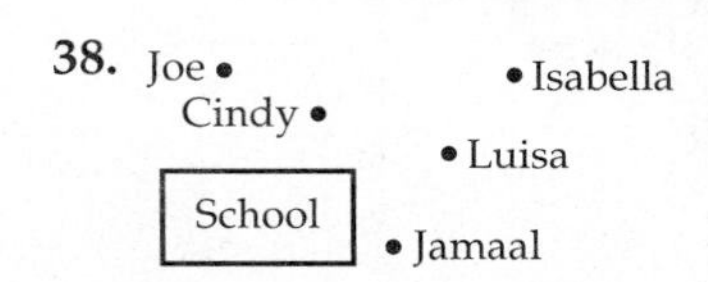

Assume that no one can see anyone else by looking through the school building.
(a) Who can see Joe?
(b) Who can see Cindy?
(c) Who can see Luisa and Jamaal?

39.

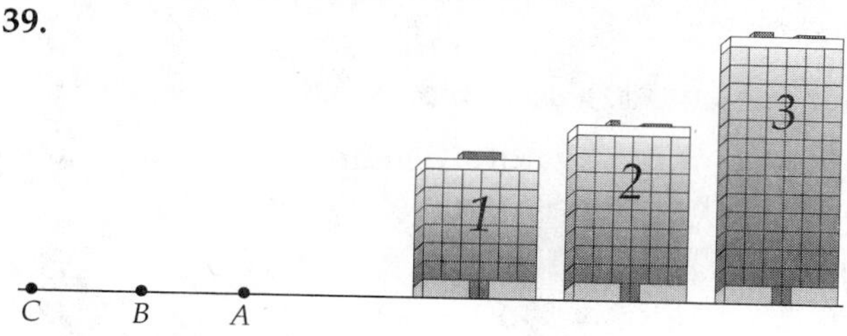

Which buildings can be seen from each of the following?
(a) A (b) B (c) C
(d) Shade the region of points from which only building 1 can be seen.

Extension Exercises

40. Give the angle formed at each of the following times of day by the minute and hour hands of a clock.
(a) 5:00 (b) 3:30 (c) 2:06

41. Name two morning times of day at which the minute and hour hands form a
(a) 150° angle. (b) 105° angle.

42. Place eight dots in the following figure so that there are exactly two dots on each circle and each line.

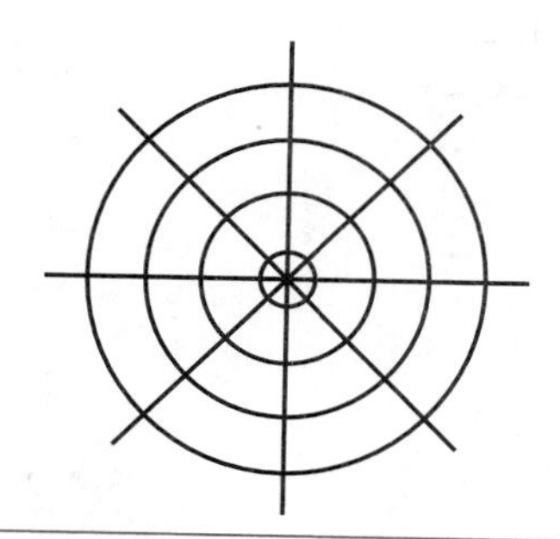

43. If possible, draw four lines in a plane that separate the plane into
(a) 10 sections. (b) 11 sections.

44. When a billiard ball bounces off the side of a billiard table, it forms two congruent angles, as shown.

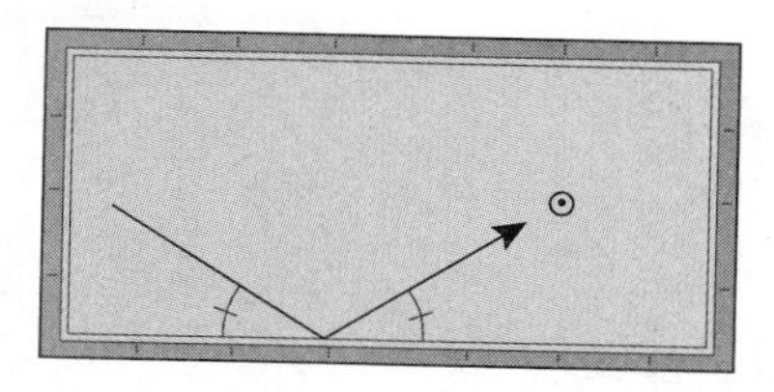

Use a protractor and trial and error to find how billiard ball A should be aimed so that it hits the bottom and right cushions (sides) and then strikes ball B.

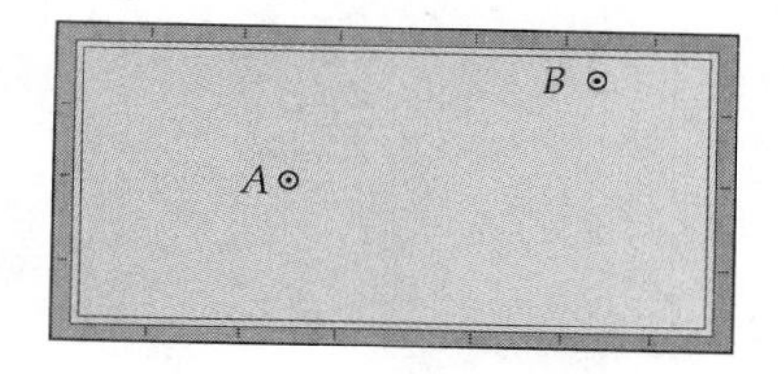

45. Use paper-folding to make an example of each of the following.
(a) Two parallel line segments
(b) Two perpendicular line segments
(c) An angle bisector (for an angle you have drawn)

Special Exercises

46. Read *The Dot and the Line* by Norton Juster, and describe how the author uses geometric shapes as symbols.

47. Measure the angles that different chair legs make with the floor. Write a summary of your results.

8.2 PLANE FIGURES

After seeing how surfaces of objects suggest plane figures and studying basic terms such as "line segment," "angle," "parallel," and "perpendicular," one is prepared for a more detailed study of plane figures.

SIMPLE CLOSED CURVES

In elementary school, children study plane figures that are simple closed curves. See whether you can write a definition of a simple closed curve in the following exercise.

Lesson Exercise 8.17

Based upon the diagrams in Figure 8-12, write a definition of a simple closed curve. (Note that "curves" can be composed of line segments.)

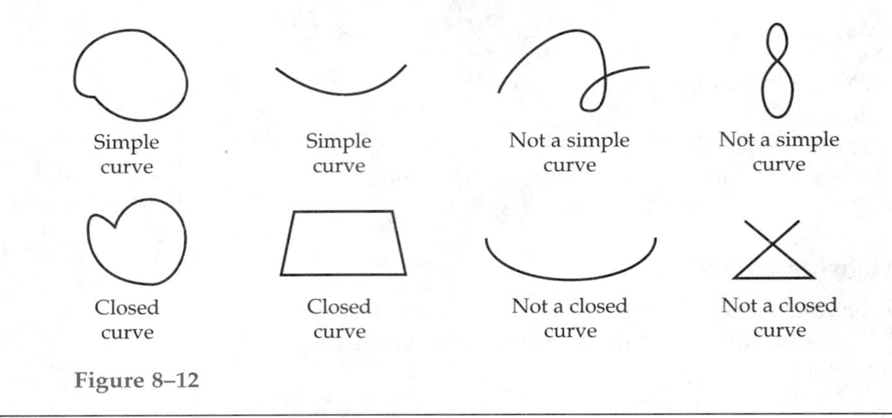

Figure 8–12

A **simple closed curve** does not cross itself and encloses a part of the plane. The **Jordan Curve Theorem** states that a simple closed curve divides the plane into three disjoint sets of points: the interior, the curve itself, and the exterior (Figure 8-13).

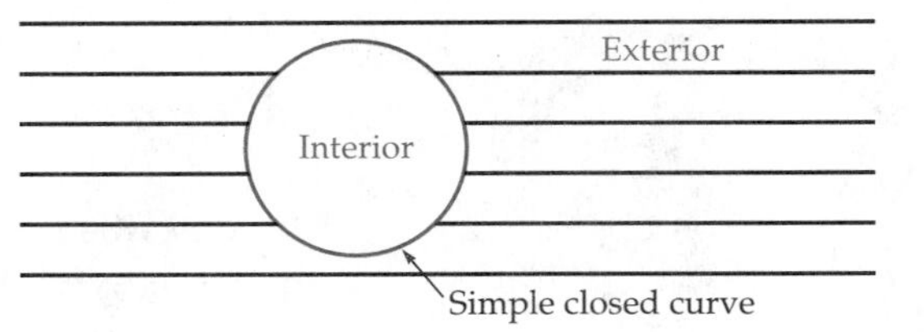

Figure 8–13

In elementary school, children study two main groups of simple closed plane figures: polygons and circles.

POLYGONS

Many surfaces in our environment approximate polygons (Figure 8-14).

See whether you can write a definition of a polygon in the following exercise.

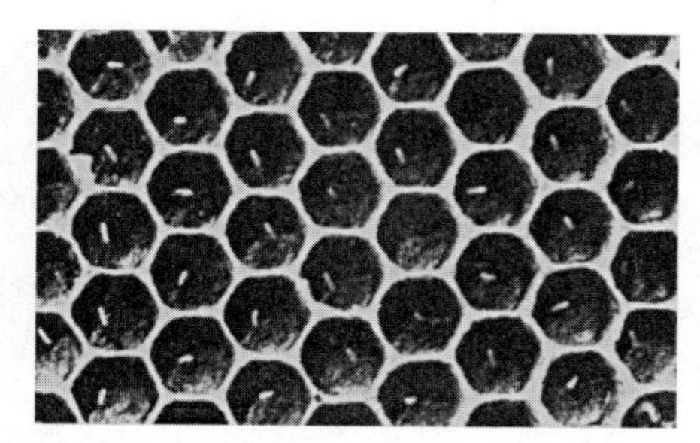

© Photo of school crossing sign by Tom Sonnabend. Beehive photo courtesy of Department of Library Services, American Museum of Natural History.

Figure 8–14

Lesson Exercise 8.18

Based upon the diagrams in Figure 8-15, write a definition of a polygon.

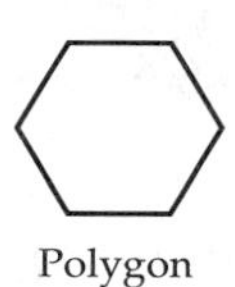

Polygon

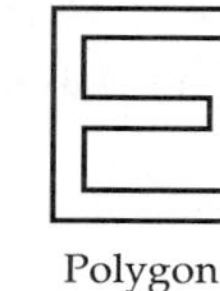

Polygon

Not a polygon

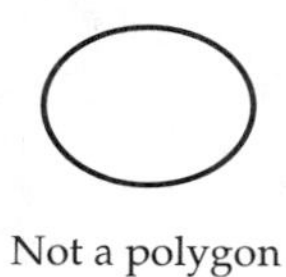

Not a polygon

Not a polygon

Figure 8–15

What is a polygon? After a while, geometry teachers tire of the well-known response "A dead parrot." A **polygon** is a simple closed plane curve formed by three or more line segments.

Lesson Exercise 8.19

Which of the following figures are polygons? If any figure is not a polygon, explain why not.

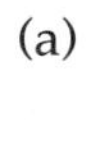

(a)

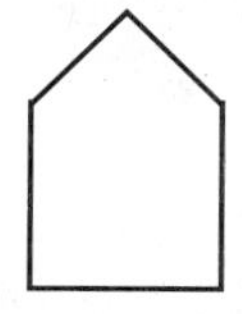

(b)

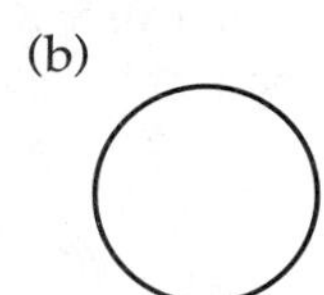

(c)

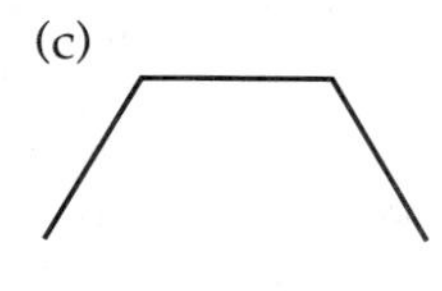

Polygons with three to eight sides are shown in Figure 8-16 on the next page. Elementary-school children usually study all of them except the heptagon.

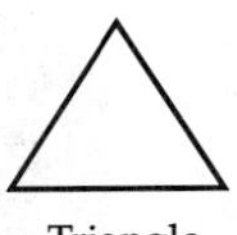
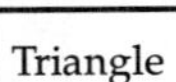

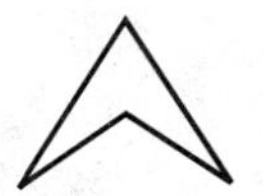

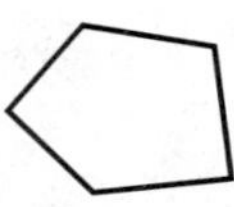

Figure 8–16

These polygons also have three to eight angles, respectively. Each vertex of an angle is also called a **vertex** (plural vertices) of the polygon.

A **geoboard** is a square board that has pegs in grid patterns. Usually there is a square grid pattern on one side, and sometimes there is a circular grid pattern on the other side. By placing rubber bands over the pegs, children can construct various shapes and examine their properties. In Chapters 8–10, you will find numerous exercises that make use of square grid patterns.

Lesson Exercise 8.20

Give the name of each polygon shown on the following geoboard grid patterns.

(a)
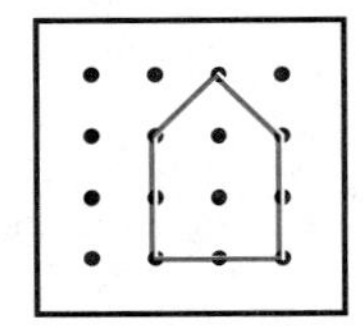

(b)
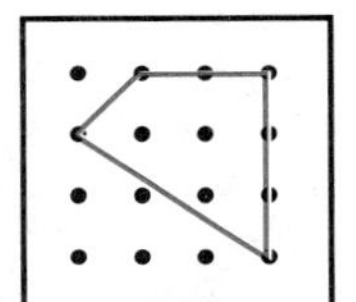

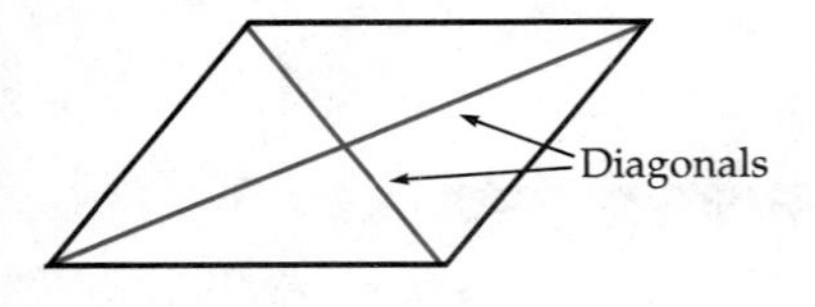

Figure 8–17

A **diagonal** is a line segment other than a side that joins two vertices of a polygon (Figure 8-17).

In buildings and bridges, engineers sometimes use diagonals for reinforcement, as shown in Figure 8-19.

Diagonals are used to define convex polygons, the kind children study in elementary school. No portion of any diagonal of a **convex** polygon lies in its exterior (see Figure 8-18). (All triangles are convex polygons, since they do not have diagonals.)

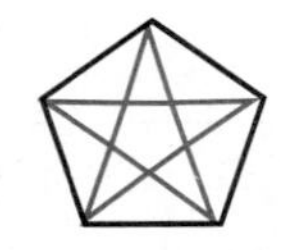

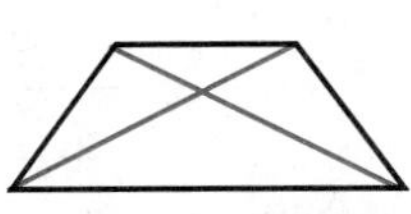

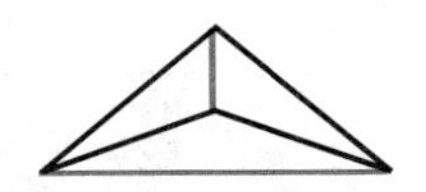

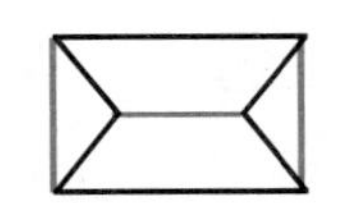

Figure 8–18

© *Photo courtesy of Library of Congress.*

Figure 8–19

Convex lens

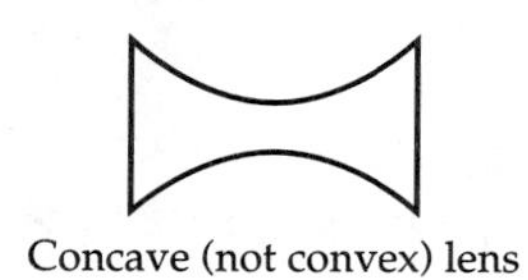

Concave (not convex) lens

Figure 8–20

The term "convex" is also used for certain lenses (Figure 8-20).

AN INVESTIGATION: DIAGONALS OF POLYGONS

The numbers of diagonals in polygons follow a pattern. A triangle has no diagonals. A quadrilateral (Figure 8-21) has a total of two diagonals, one from each vertex.

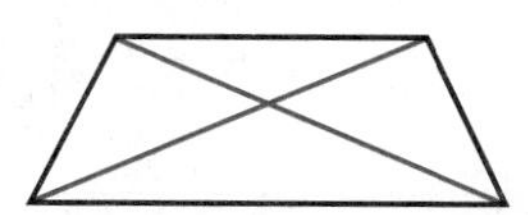
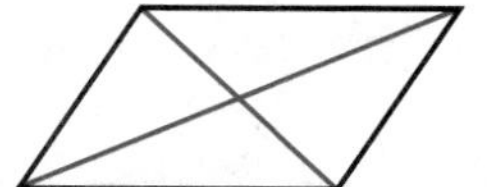
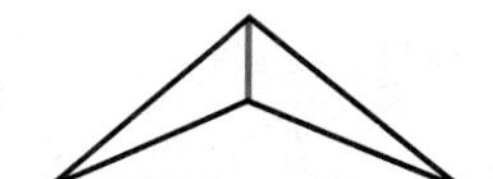

Figure 8–21

These results are included in the following table.

Polygons

Number of vertices	3	4	5	6
Number of diagonals from each vertex	0	1		
Total number of diagonals	0	2		

Lesson Exercise 8.21

(a) Complete the table just presented.

(b) Based upon the pattern in your completed table, predict how many diagonals are in a heptagon (a seven-sided polygon).

(c) Draw a heptagon and its diagonals, and check your prediction.

What is the general formula for the number of diagonals in a polygon that has N sides? The following exercise will help you answer this intriguing question.

Lesson Exercise 8.22

Now consider a polygon that has N sides.

Polygons

Number of vertices	3	4	5	6	7	N
Number of diagonals from each vertex	0	1	2	3	4	
Total number of diagonals	0	2	5	9	14	

(a) How many vertices does it have?

(b) How many diagonals can be drawn *from* a vertex?

(c) How is the total number of diagonals related to the number of vertices and the number of diagonals from each vertex?

(d) How many diagonals does the polygon have?

Were you able to complete Lesson Exercise 8.22? The result is as follows.

Diagonals of a Polygon

A polygon that has N sides has $\dfrac{N(N-3)}{2}$ diagonals.

Lesson Exercise 8.23

(a) Use the formula to find the number of diagonals that a 20-sided polygon has.

(b) The process of assuming that the formula is true and applying it in part (a) involves ______________ reasoning.

The most common polygonal shapes in our environment are triangles and quadrilaterals. These two classes of polygons can be subdivided further.

TRIANGLES

Triangular shapes appear occasionally in nature; they possess a rigid structure that strengthens supports in buildings and furniture (Figure 8-22).

Photo of diatom courtesy of Dr. George W. Andrews.

Photo of bridge courtesy of Library of Congress.

Figure 8–22

One way in which mathematicians classify triangles is by the number of congruent sides. Consider how this could be done in the following exercise.

Lesson Exercise 8.24

(a) How many congruent sides can a triangle have?

(b) If you know them, give the name for each type of triangle you listed in part (a).

The preceding exercise concerned three categories of triangles: **scalene triangles**, which have no congruent sides; **isosceles triangles**, which have at least two congruent sides; and **equilateral triangles**, which have three congruent sides (Figure 8-23).

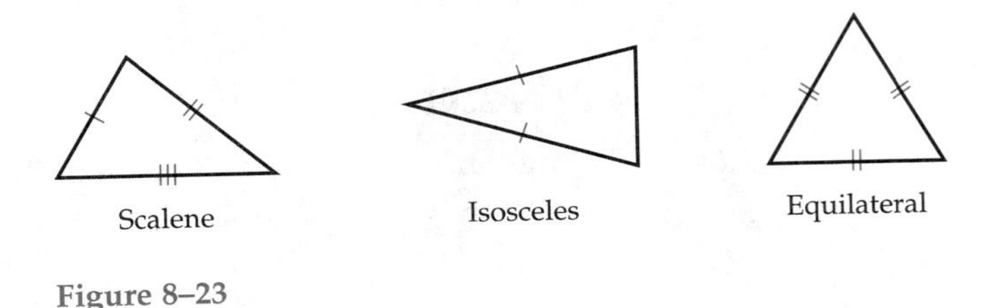

Figure 8–23

You can use these definitions to classify any triangle.

Lesson Exercise 8.25

Which of the following categories include a triangle with three congruent sides?

(a) Scalene (b) Isosceles (c) Equilateral

QUADRILATERALS

How should children learn about geometric shapes? While teaching high-school mathematics in Holland, Dina van Hiele-Geldof and Pierre van Hiele developed a theory about stages of learning in geometry. According to the van Hieles, school geometry should begin with recognition, followed by analysis and then informal deduction.

The first level, **recognition**, concerns recognizing a shape as a whole without analyzing its components (such as sides and angles) and properties. In the second level, **analysis**, students focus on the components and properties of shapes, such as how many sides they have and whether they have some congruent sides or angles. Students use the parts of a figure to describe and define the figure. In the third level, **informal deduction**, students become aware of relationships between different classes of figures (for example, they see that all squares are parallelograms). At this level, students also deduce relationships among the properties of a figure; for example, they determine whether a quadrilateral that has four congruent sides must also have four congruent angles.

The two highest levels are usually developed in high-school and college geometry. At the **formal deduction** level, students employ deduction to develop new ideas in a logical system, such as Euclidean geometry. They study logical systems consisting of axioms, undefined terms, theorems, and definitions and learn to write proofs of new theorems. At the highest level, **rigor**, students work at an abstract level in a variety of axiomatic systems and compare them.

Children in elementary-school mathematics generally work at the first three van Hiele levels: recognition, analysis, and informal deduction. How would a teacher present a unit on quadrilaterals using the first three van Hiele levels?

At the first level, we *recognize* quadrilaterals in our environment (Figure 8-24).

Photos of door and street lamp by Tom Sonnabend.

Photo of halite crystals by A. Singer. Courtesy of Department of Library Services, American Museum of Natural History.

Figure 8–24

At the recognition level, the student compares shapes that are quadrilaterals with shapes that are not quadrilaterals. This same comparison can be made with different types of quadrilaterals. Figure 8-25 on page 428 shows some examples.

At the next level, analysis, the student examines properties of quadrilaterals while still being unaware of any formal definition.

Lesson Exercise 8.26

Which of the five types of quadrilaterals in Figure 8-25 on page 428 have both pairs of opposite sides congruent?

The next phase of the analysis level is to compare and contrast properties of different quadrilaterals.

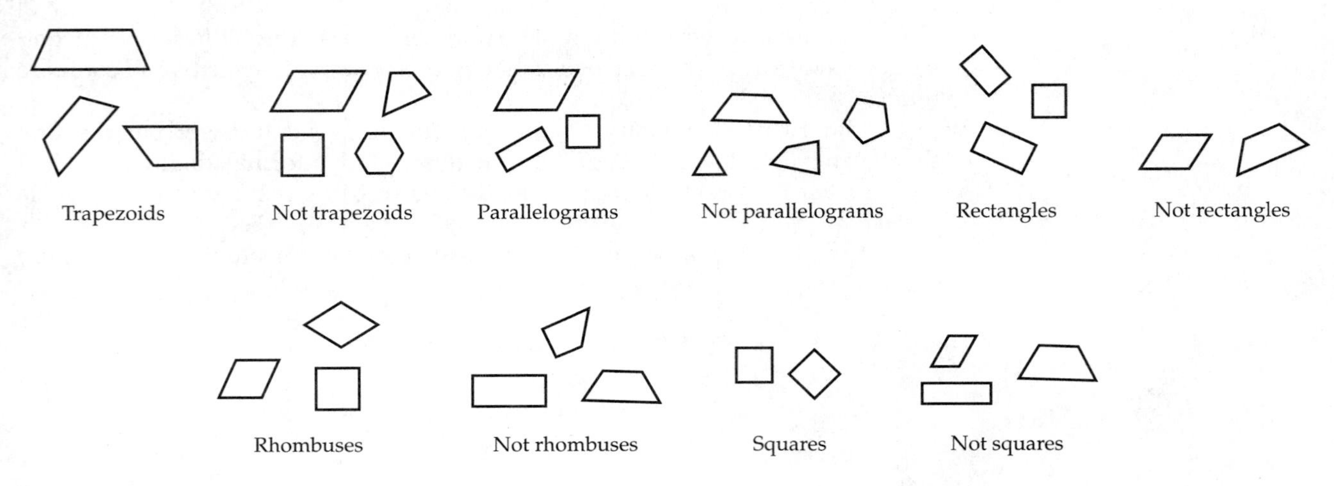

Figure 8–25

Lesson Exercise 8.27

(a) Name two properties of sides or angles that all squares and all rhombuses share.

(b) Name a property of sides or angles that all squares have that some rhombuses do not have.

Lesson Exercise 8.28

(a) Name two properties that all parallelograms and all rectangles share.

(b) Name a property that all rectangles have that some parallelograms do not have.

At the third level, informal deduction, students use properties to define shapes and examine alternative definitions.

Lesson Exercise 8.29

Based upon the preceding exercises and diagrams, write definitions of a trapezoid, a parallelogram, a rectangle, a rhombus, and a square.

One possible set of definitions for common quadrilaterals is as follows.

Definitions: Common Quadrilaterals

- A **trapezoid** is a quadrilateral that has exactly one pair of parallel sides.
- A **parallelogram** is a quadrilateral in which each pair of opposite (non-intersecting) sides is parallel.
- A **rectangle** is a quadrilateral that has four right angles.
- A **rhombus** is a quadrilateral that has four congruent sides.
- A **square** is a quadrilateral that has four congruent sides and four right angles.

Following are illustrations of these shapes. The markings on the angles of the rectangle and square denote right angles.

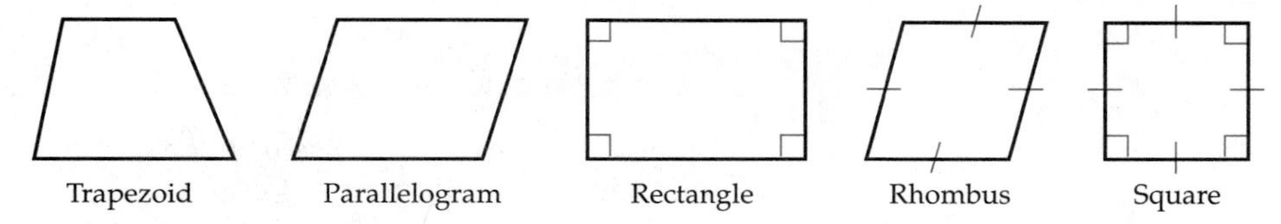

Definitions such as these enable mathematicians to communicate using a shared terminology. A definition *may* give a minimum set of properties that defines a shape; it may also list additional properties that clarify what the shape is like.

For example, the preceding definition of a rectangle does not give a minimum set of properties. The next exercise addresses this issue.

Lesson Exercise 8.30

A rectangle can be defined as a quadrilateral that has right angles. What is the *minimum* number of right angles that makes a quadrilateral a rectangle? (*Hint:* Try to draw quadrilaterals with one, two, and three right angles that are *not* rectangles.)

As Lesson Exercise 8.30 suggests, a rectangle *can* be defined as "a quadrilateral that has three right angles" (the minimum), but elementary and high-school texts usually add other properties that clarify what a rectangle is. Typical definitions are "A rectangle is a quadrilateral that has four right angles" and "A rectangle is a parallelogram that has four right angles."

The five classes of quadrilaterals overlap. At the level of informal deduction, definitions are used to find relationships among classes of figures. See whether you can answer the following questions by referring to the definitions of the five common quadrilaterals.

Lesson Exercise 8.31

(a) According to the preceding definitions, would every square be a type of rectangle?
(b) Is every trapezoid also a parallelogram?
(c) Is every square also a rhombus?

Lesson Exercise 8.32

Suppose P = {parallelograms}, Rh = {rhombuses}, S = {squares}, Re = {rectangles}, T = {trapezoids}, and Q = {quadrilaterals}.

(a) $Rh \cap Re =$ _____ (b) $T \cap P =$ _____

Lesson Exercise 8.33

Using the answers to Lesson Exercises 8.31 and 8.32, decide which of the following Venn diagrams is correct.

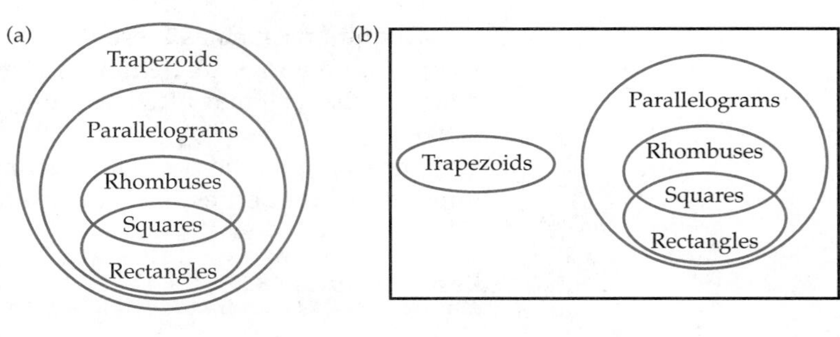

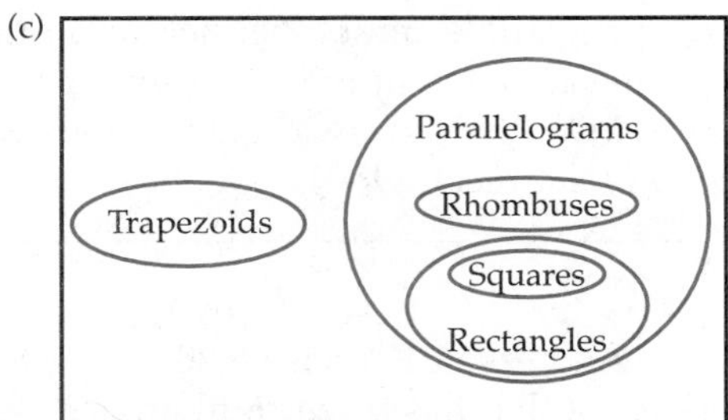

CIRCLES

Another shape sometimes appears in clocks and in nature. It is suggested by the top of a can and the outline of a wheel. You know this plane figure well: it is the circle (Figure 8-26).

Photo of clock courtesy of Library of Congress.

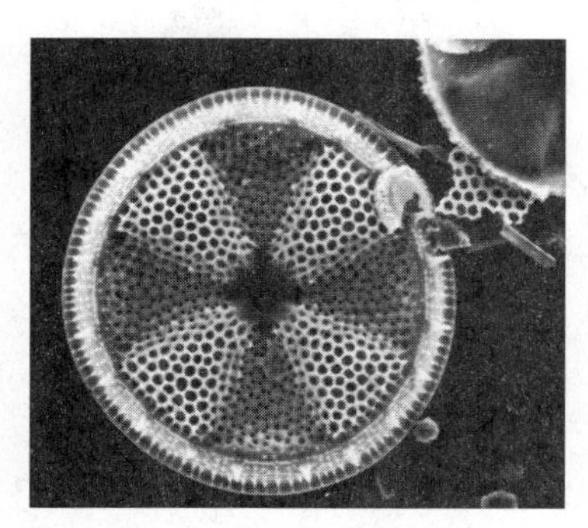

Photo of diatom courtesy of Dr. George Andrews.

Figure 8–26

Lesson Exercise 8.34

(a) Suppose that a radio program can be received anywhere within 20 km of the radio station. If the land is flat and has few obstacles, what shape is the border of the region in which the station can be received?

(b) Based upon part (a), complete the following definition of a circle. A circle is the set of points in a plane that are ________________.

The preceding exercise suggests the following definition of a circle.

Definitions **Circle, Radius, and Center**

A **circle** is the set of points in a plane that are the same distance (the **radius**) from a given point (the **center**).

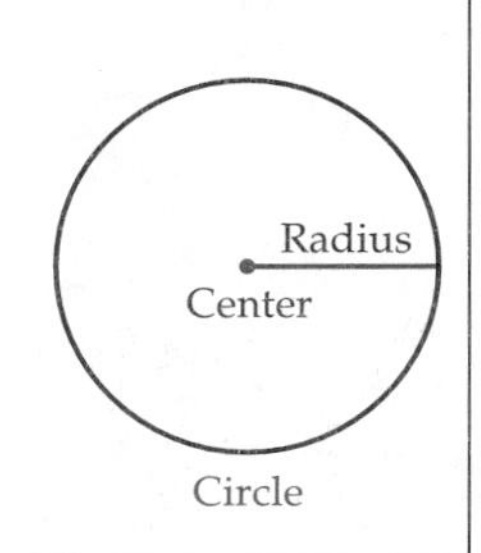

Note that the word "radius" is used to refer to both a *segment* joining the center to the circle and the *length* of that segment.

Lesson Exercise 8.35

What property of a circle makes it work as a wheel?

Can you write a definition of a chord in the following exercise?

Lesson Exercise 8.36

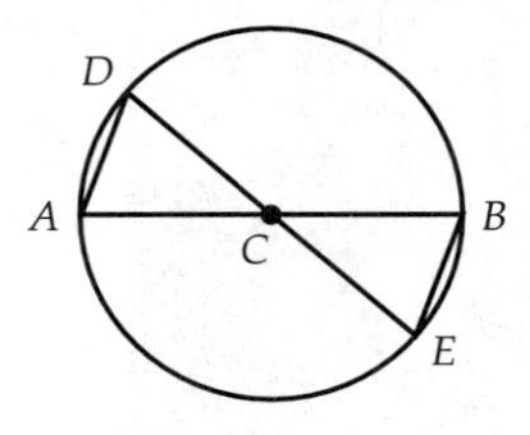

Figure 8–27

In Figure 8-27, $\overline{AB}$, $\overline{AD}$, $\overline{BE}$, and $\overline{AE}$ are chords of a circle. $\overline{AC}$ and $\overline{BC}$ are not chords. Define "chord."

A **chord** is a line segment with endpoints on a circle. What about a "diameter?"

Lesson Exercise 8.37

(a) In Figure 8-27, $\overline{AB}$ and $\overline{DE}$ are diameters of a circle with center C, and $\overline{AD}$ and $\overline{BC}$ are not diameters. Define "diameter."

(b) How could you describe the size of a circular object such as a bicycle wheel?

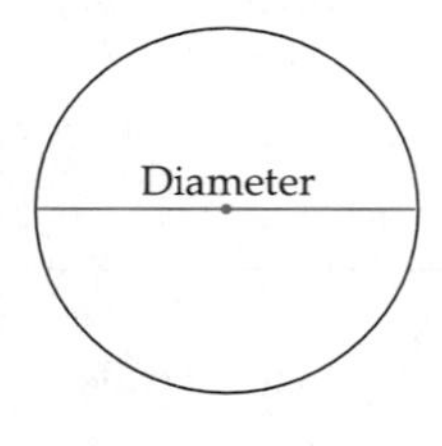

Figure 8–28

A **diameter** is a line segment that (1) has two points of the circle as endpoints and (2) passes through the center of the circle (Figure 8-28). The diameter is a commonly used measurement of circular objects. (Two other common measures, area and circumference, will be examined in Chapter 10.)

Just as with the radius, the term "diameter" can be used to describe the segment or its length. The intersecting diameters of some wheels suggest central angles. So do the hands of a clock (Figure 8-29). A **central angle** is an angle whose vertex is the center of a circle.

Photo courtesy of Library of Congress.

Figure 8–29

Lesson Exercise 8.38

Consider the following problem. "Name a time of day when the minute and hour hands of a clock form a 75° angle." Devise a plan and solve the problem.

DRAWING SOFTWARE

Drawing software enables you to draw and measure angles, segments, polygons, and circles. You can use these drawings to investigate properties of polygons and circles. The computer's drawing area is called the **window**. A **menu** listing the steps and options should appear along one side of the screen. Examples of drawing software are Geometric Supposer (Sunburst), Geometer's Sketchpad (Key Curriculum), Cabri Geometry II (Texas Instruments), Euclid's Toolbox (Addison-Wesley), Geometry Connections (W. K. Bradford), and Geometry Inventor (Wings for Learning).

In the following exercise, you can explore some properties of parallelograms.

Lesson Exercise 8.39

(a) Find out how to draw a parallelogram with your software. Label the parallelogram *ABCD*.

(b) Measure *AB*, *BC*, *CD*, and *DA*, and record the results.

(c) Repeat parts (a) and (b) two more times with different parallelograms.

(d) Make a generalization (conjecture) about the lengths of the sides of a parallelogram.

(e) Draw a parallelogram *ABCD*, and measure its four angles.

(f) Repeat part (e) two more times.

(g) Make a generalization about the angle measures of a parallelogram.

(h) Draw a series of parallelograms with diagonals. Measure the lengths of the diagonals and the lengths of the two parts of each diagonal. Make a generalization based on these measurements.

ANSWERS TO SELECTED LESSON EXERCISES

8.19 (a) is a polygon; (b) is not bounded by line segments; (c) is not closed.

8.20 (a) Pentagon (b) Quadrilateral

8.21 (a)

Number of vertices	3	4	5	6
Number of diagonals from each vertex	0	1	2	3
Number of diagonals	0	2	5	9

(b) 14

8.22 (a) N (b) $N - 3$

(c) Multiply the number of vertices times the number of diagonals from each vertex. But this counts each diagonal twice (at each end). So divide the result by 2.

(d) $\dfrac{N(N-3)}{2}$

8.23 (a) $\dfrac{20(17)}{2} = 170$ (b) deductive

8.24 (a) None, two, or three

8.25 (b), (c)

8.26 Parallelogram, rectangle, rhombus, square

8.27 (a) Both pairs of opposite sides are parallel and all four sides are equal.
(b) Four right angles

8.28 (a) Both pairs of opposite sides are parallel and equal.
(b) Four right angles

8.30 Three

8.31 (a) Yes (every square is a quadrilateral with four right angles)
(b) No (c) Yes

8.32 (a) S (b) { }

8.33 (b)

8.34 (a) Circle

8.35 It has uniform curvature.

8.37 (b) You could measure the diameter, circumference, or area.

8.38 Possible answer: 3:30 (*Hint:* Each minute has a measure of $\frac{1}{60}(360°) = 6°$.)

8.2 HOMEWORK EXERCISES

Basic Exercises

1. Draw a curve that is simple but not closed.

2. Which of the following shapes are polygons? If a shape is not a polygon, explain why not.

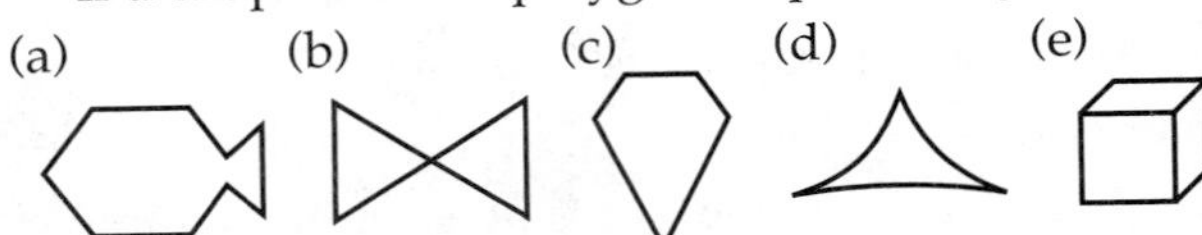

3. What polygon is suggested in each case?

(a)

Photo courtesy of Tom Sonnabend.

(b)

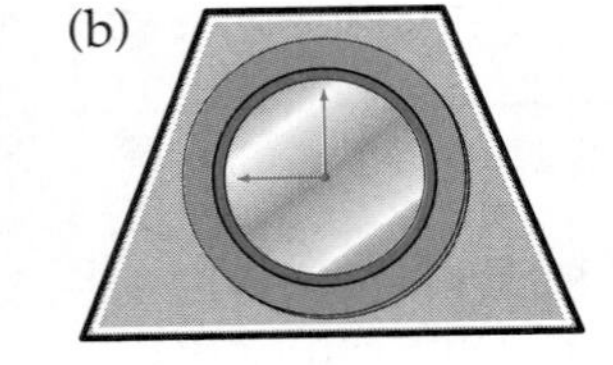

4. Sketch a pentagon that has exactly three right angles.

5. Sketch a hexagon that has exactly two acute angles.

6. Form the following shapes on a geoboard.
   (a) A pentagon that has one pair of parallel sides
   (b) A quadrilateral that has no parallel sides but has two pairs of congruent, adjacent sides

7. Which of the following polygons are convex?

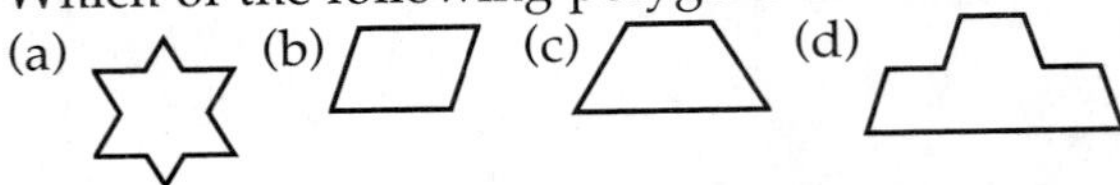

8. Assuming the two figures share no common sides, find the maximum number of intersection points for
   (a) a triangle and a square.
   (b) a triangle and a convex pentagon.
   (c) a square and a convex pentagon.
   (d) a regular polygon that has n sides and a regular polygon that has m sides, $n < m$.

9. Draw all the diagonals of the hexagon.

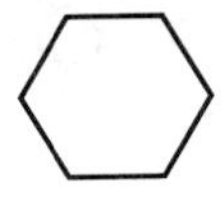

10. A polygon that has N sides has $\frac{N(N-3)}{2}$ diagonals. How many diagonals does an octagon have?

11. Some painters utilize the triangular form. Leonardo DaVinci used it in the *Mona Lisa*, and Hans Memling used it in his *Madonna and Child*. What mood or effect might the triangular form create?

12. True or false? No scalene triangle is isosceles.

13. Tell whether each triangle is scalene, isosceles, or equilateral.

(a)

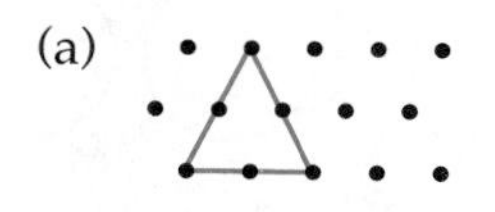

(b)

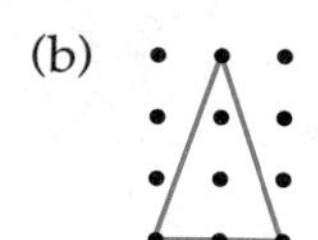

(c)

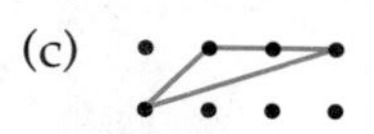

14. Use the dot paper to draw each type of triangle.

(a) (b)

15. Consider the following problem. "How many triangles are contained in the following shape?" Devise a plan and solve the problem.

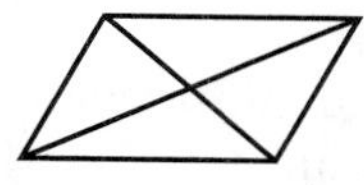

16. How many squares are in the following design?

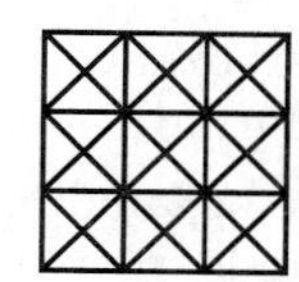

17. Complete a square that has the segment shown as one of its sides.

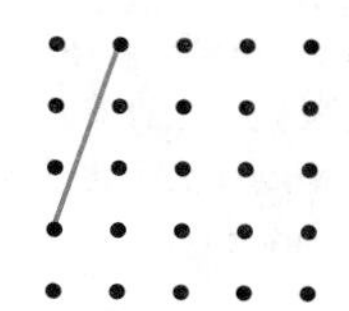

18. If possible, form each of the following shapes on a geoboard. Then make a drawing to record your results.
(a) Parallelogram (b) Trapezoid

19. Which of the following figures are rhombuses?

(a)

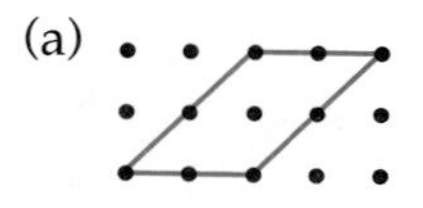

(b)

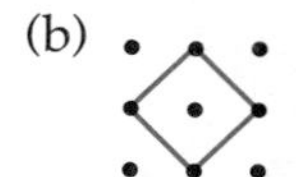

(c)

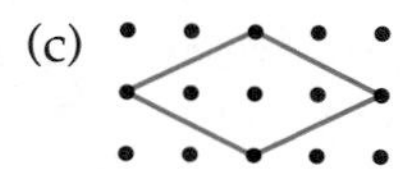

20. What shape is the diamond in a deck of cards?

21. How could you use some of the segments shown to form each of the following?
(a) A parallelogram (b) A trapezoid
(c) A rhombus

Group 1 **Group 2** **Group 3**

22. Give all names that apply to each figure.

(a) (b)

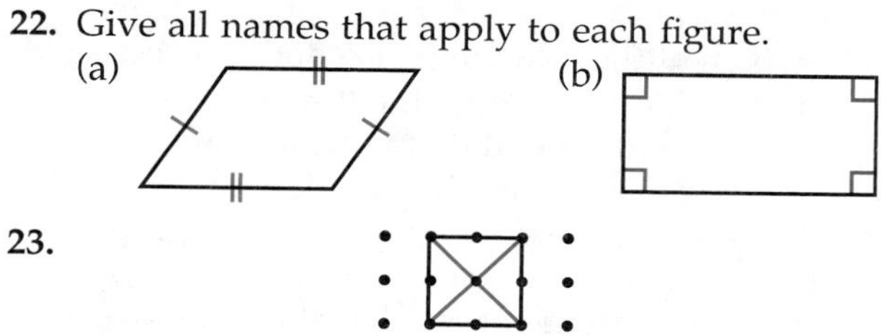

23.

Refer to the drawing. Which of the following appear to be true?
(1) The diagonals have the same length.
(2) The diagonals are perpendicular.
(3) The diagonals bisect each other.

24.

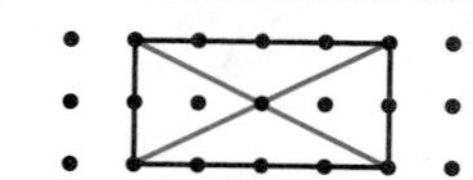

(a) Refer to the drawing. Which of the following appear to be true?
 (1) The diagonals have the same length.
 (2) The diagonals are perpendicular.
 (3) The diagonals bisect each other.
(b) What van Hiele level is represented in part (a)?

25. Tell whether each of the following shapes must, can, or cannot have at least one right angle.
(a) Rhombus (b) Square
(c) Trapezoid (d) Rectangle
(e) Parallelogram

26. In which of the following shapes are both pairs of opposite sides parallel?
(a) Rhombus (b) Square
(c) Trapezoid (d) Rectangle
(e) Parallelogram

27. Cut a rectangular shape along its diagonal to form two congruent pieces. Sketch and name all the shapes you can form by putting the two pieces together in different ways without overlapping them.

28. Fold a sheet of paper to show that the opposite sides of a rectangle are congruent.

29. A carpenter has made a rectangular door. How can he use a tape measure to be sure that the door has right angles?

30. A square is defined in the text as "a quadrilateral that has four congruent sides and four right angles," but this is not a *minimum* set of conditions. Make drawings to determine which of the following also define a square. If a definition is incorrect, tell why.
(a) A quadrilateral that has four congruent sides and three right angles
(b) A quadrilateral that has two congruent sides and four right angles
(c) A quadrilateral that has three congruent sides and four right angles
(d) A quadrilateral that has two *adjacent* congruent sides and four right angles

31. Draw each of the following shapes.
(a) A parallelogram that is not a rhombus (b) A rectangle that is also a square

32. Is every rhombus a parallelogram? If so, give an example. If not, draw a counterexample.

33. A square is also which of the following?
(a) Quadrilateral (b) Parallelogram
(c) Rhombus (d) Rectangle

34. Fill in the blank with "All," "Some," or "No."
(a) ________ rectangles are squares.
(b) ________ parallelograms are trapezoids.
(c) ________ rhombuses are quadrilaterals.

35. "The diagonals of a rectangle are congruent." Why does this statement imply that the diagonals of a square must also be congruent?

***36.** Which set picture best represents the relationship between rhombuses and rectangles? Label the correct circles appropriately.

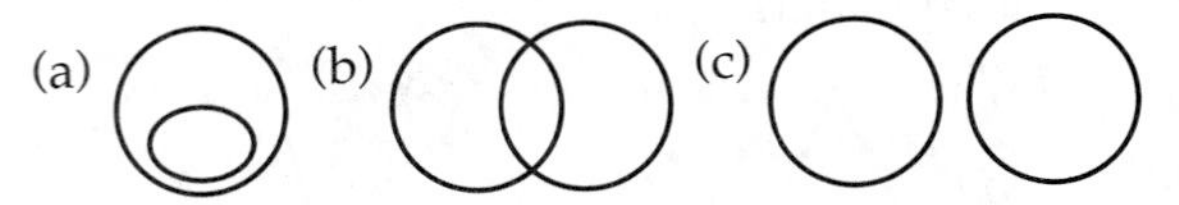

37. Suppose that $P = \{\text{parallelograms}\}$, $S = \{\text{squares}\}$, $T = \{\text{trapezoids}\}$, and $Q = \{\text{quadrilaterals}\}$.
(a) $P \cap S =$ ____
(b) Is $T \subseteq Q$?
(c) $P \cup Q =$ ____

38. Draw a Venn diagram showing the relationship among parallelograms (P), rectangles (RE), and rhombuses (RH).

39. The **midpoint** of a side divides it into two equal lengths.
(a) Draw a large quadrilateral on a sheet of paper.
(b) Use a ruler to locate the midpoint of each side.
(c) Connect the midpoints to form a quadrilateral.
(d) What shape appears to be the result?
(e) Repeat steps (a)–(d) for a second quadrilateral.
(f) Make a generalization based upon your answers to part (d).

40. (a) Draw a large rectangle on a sheet of paper.
(b) Use a ruler to locate the midpoint of each side.
(c) Connect the midpoints to form a quadrilateral.
(d) What shape appears to be the result?
(e) Repeat steps (a)–(d) for a second rectangle.
(f) Make a generalization based upon your answers to part (d).

41. Fill in the blanks to describe the following circle with center N at the top of the next page. Circle N is the set of ________ in a plane that are ____________ from ____________.

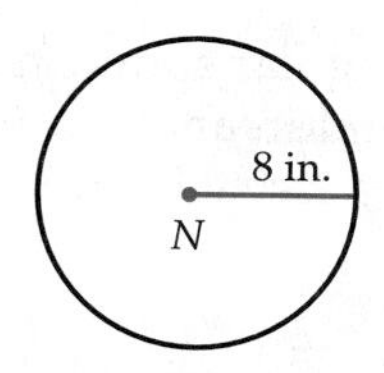

42. The diameter of a circle divides its interior into two congruent regions. How can someone use this property in dividing a circular pizza or pie in half?

43. The diagram of a bicycle wheel has C as its center.

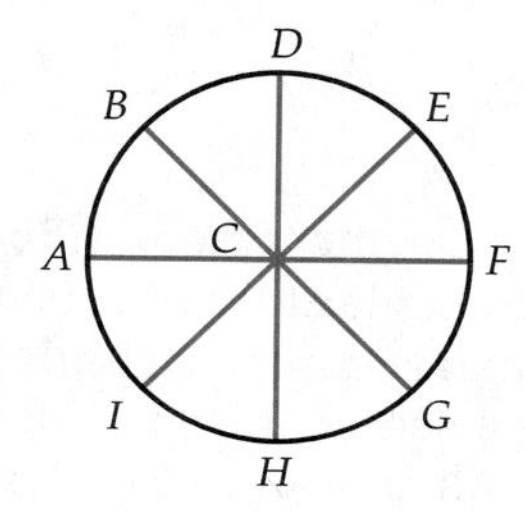

(a) Name a central angle and give its measure.
(b) What kind of triangle is $\triangle CDE$?
(c) Name a chord of the circle.

44. True or false? Every chord is also a diameter of the circle.

Extension Exercises

45. In 1884, Ezra Gilliland designed a phone system that allowed 15 people to speak to one another. How many connections were needed? (*Hint:* How many segments are needed to connect 15 points in pairs?)

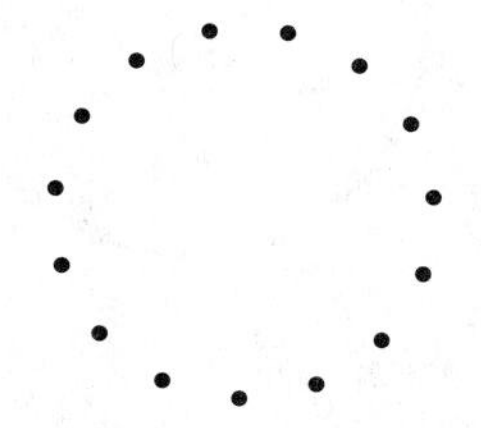

46. At a party, all 15 friendly people want to shake hands with every other person there. How many handshakes would this require?

47. Some mathematics books define a trapezoid as "a quadrilateral that has at least one pair of parallel sides." If this definition is used, which of the following Venn diagrams is correct?

(a)

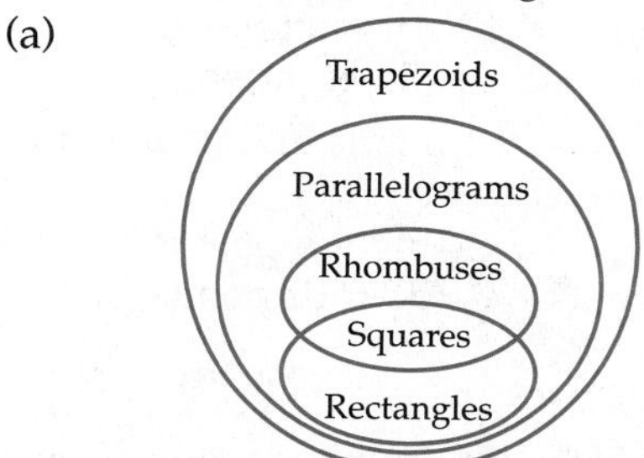

(b)

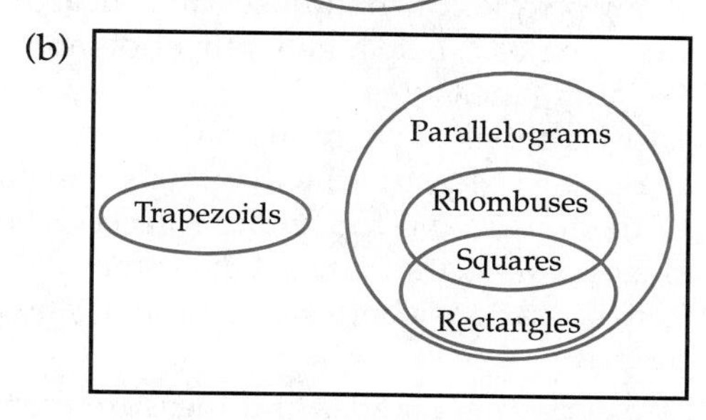

(c)

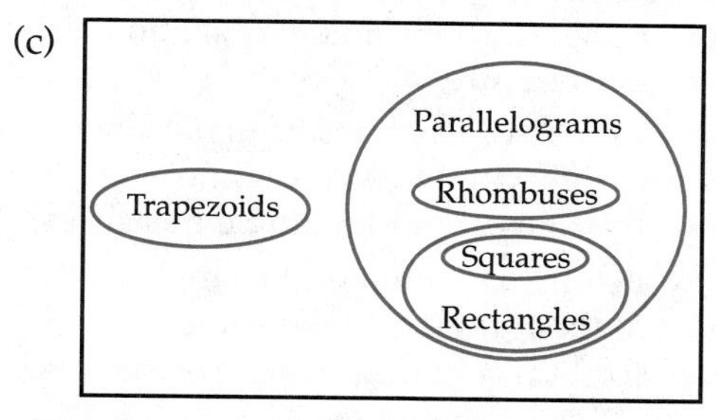

48. Draw on the grid a polygon that meets all of the following conditions.
(a) It is a hexagon.
(b) Not all of its sides are congruent.
(c) It has exactly two acute angles.
(d) It has exactly three dots in its interior.

• • • • •
• • • • •
• • • • •
• • • • •
• • • • •

49. If possible, sketch two parallelograms that intersect at exactly
(a) one point. (b) two points.
(c) three points. (d) four points.

50. If possible, sketch two parallelograms that intersect at exactly
(a) five points. (b) six points.
(c) seven points. (d) eight points.

51. If possible, draw a triangle and a circle that intersect at exactly
(a) one point. (b) two points.
(c) three points. (d) four points.

Computer Exercises

52. (a) Draw a series of rhombuses and measure their angles. Make a generalization about the angle measures.
(b) Draw a series of rhombuses with diagonals. Measure the diagonals, the parts of diagonals, and the angles formed where the diagonals intersect. Make some generalizations about the diagonals and angles.

53. Draw a series of triangles and measure their interior angles. Make a generalization about the angle measures.

54. Draw a series of isosceles triangles and measure their angles. Make some generalizations about the angle measures.

55. Draw a series of equilateral triangles and measure their angles. Make some generalizations about the angle measures.

56. Draw a series of quadrilaterals that are not parallelograms, and measure their angles. Make a generalization about the angle measures.

Special Exercises

57.

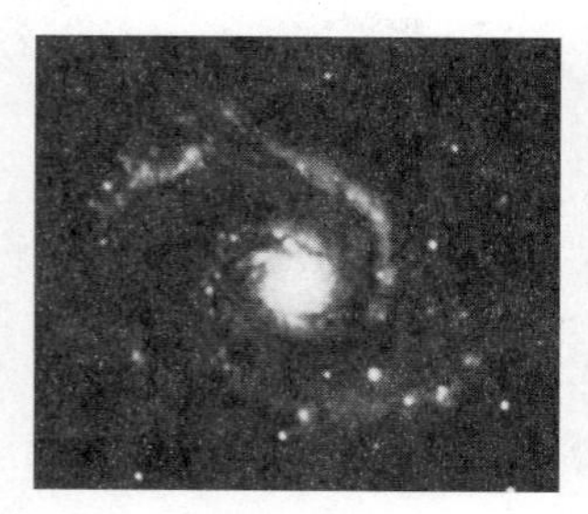

Photo courtesy of NASA.

Photo by A. Singer. Courtesy of Department of Library Services, American Museum of Natural History.

The equiangular spiral shape appears in the stars and in animals. You can graph one using a circular grid.

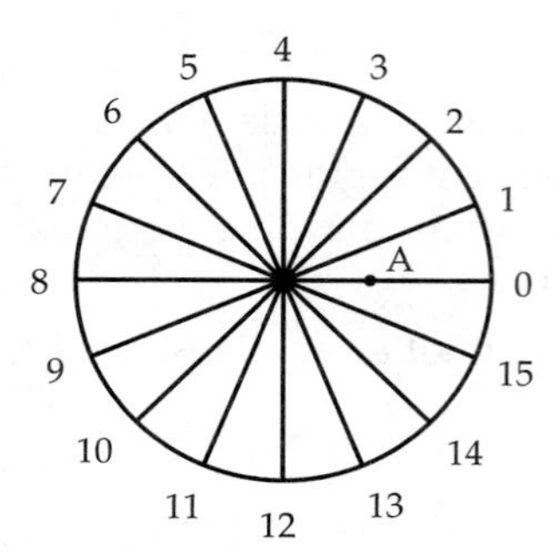

(a) Draw a segment from radius 1 that is perpendicular to radius 0 at point A. Label the intersection point with radius 1 as B.
(b) Draw a segment from radius 2 that is perpendicular to radius 1 at B. Label the intersection point with radius 2 as C.
(c) Continue this pattern for all 15 radii.
(d) Draw a smooth curve through points A, B, C, and so on that resembles an equiangular spiral.

58. A series of straight line segments can be used to define a curved figure. Strange, but true! Trace the following circle on a piece of paper.

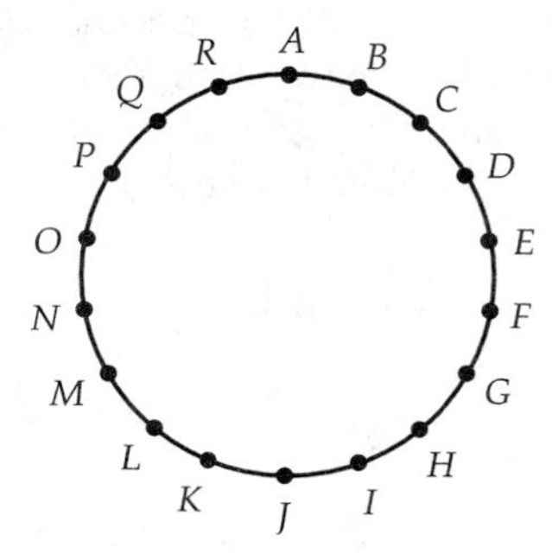

(a) Draw the segments $\overline{AI}$, $\overline{BJ}$, $\overline{CK}$, $\overline{DL}$, . . . , $\overline{RH}$.
(b) Using a different color, draw $\overline{AH}$, $\overline{BI}$, $\overline{CJ}$, $\overline{DK}$, . . . , $\overline{RG}$.
(c) Using another color, draw the next set of segments that continues the pattern.

8.3 ANGLE MEASURES OF POLYGONS

Although most floor tiles are square, many other designs can be used. Figure 8-30 shows some examples. Some quilts also utilize interlocking, repeating patterns. What shapes can be used for floor tiles or quilt patterns?

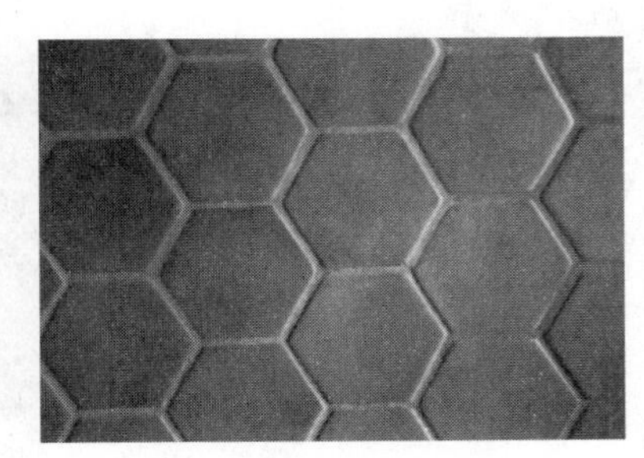

Photos by Tom Sonnabend.

Figure 8–30

We can answer this question by studying the angle measures of polygons, beginning with the simplest polygon, the triangle.

THE ANGLES OF A TRIANGLE

The sum, in degrees, of the angle measures of any triangle is always the same. Do you know what the sum is? Try the following.

Lesson Exercise 8.40

(a) If you have scissors and paper, do the following. If not, skip to part (c). Stack three small sheets of paper, and draw a triangle on the top sheet. Cut out the triangle, creating three copies. Number the angles in each triangle as shown in Figure 8-31.

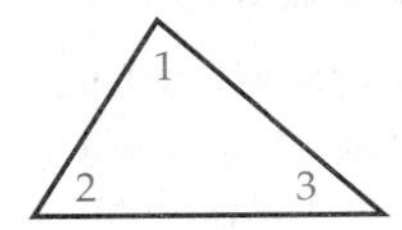

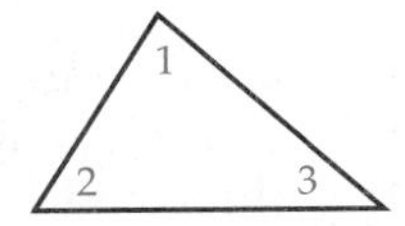

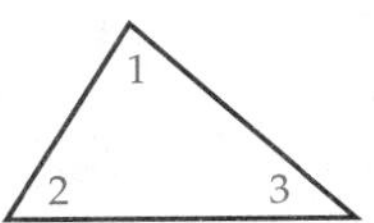

Figure 8–31

(b) Cut out the triangles.

(c) Now place the three triangles together in a tiling pattern, as shown in Figure 8-32.

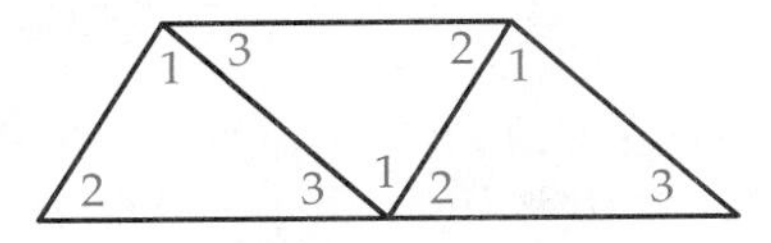

Figure 8–32

(d) What does the sum of the three angle measures ($\angle 1$, $\angle 2$, and $\angle 3$) appear to be?

Lesson Exercise 8.40(d) suggests the following property.

Angle Measures of a Triangle

The sum of the three angle measures of a triangle is 180°.

This property is helpful in finding angle measures of triangles.

Lesson Exercise 8.41

An equilateral triangle has three congruent sides and three congruent angles. What is the measure of each angle?

In Section 8.2, triangles were classified by their sides. A triangle can also be classified by the measure of its largest angle. An **obtuse triangle** has an obtuse angle; a **right triangle** has a right angle; and an **acute triangle** has three acute angles.

Lesson Exercise 8.42

Why is it impossible for a triangle to have two obtuse angles?

THE ANGLES OF A POLYGON

It is possible to find the sum of the angle measures of any polygon without measuring its angles! One can use the sum of the angle measures of a triangle to deduce the sum of the angle measures in any convex polygon. First, consider quadrilaterals.

Lesson Exercise 8.43

What is the sum of the angle measures of any square or rectangle?

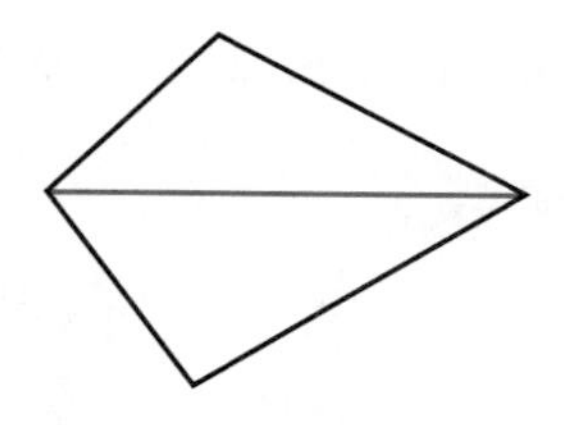

Figure 8–33

The angle measures of all convex quadrilaterals have the same sum! This fact can be proved by drawing a diagonal of any convex quadrilateral to form two triangles, as shown in Figure 8-33.

Lesson Exercise 8.44

(a) What is the sum of the three angle measures of each triangle in the quadrilateral shown in Figure 8-33?

(b) Mark each angle of the triangle with an arc.

(c) What is the sum of the four angle measures of the quadrilateral?

Lesson Exercise 8.44 verifies the following property.

Angle Measures of a Convex Quadrilateral

The sum of the four angle measures of any convex quadrilateral is 360°.

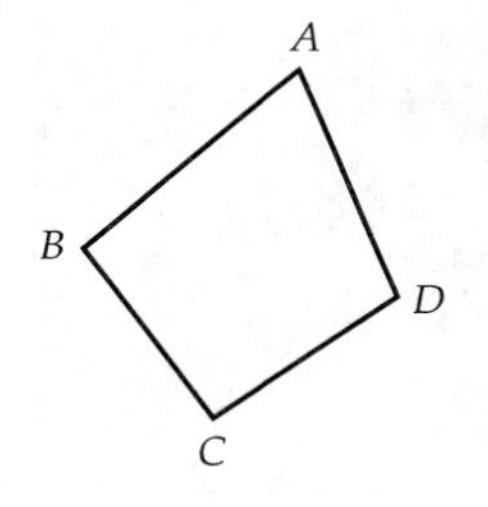

Figure 8–34

Figure 8–35

Lesson Exercise 8.45

(a) Explain why the following is not correct. Consider quadrilateral *ABCD* (Figure 8-34). *ABCD* can be divided into four triangles, as shown in Figure 8-35. The interior angle measures of each triangle add up to 180°, so the interior angle measures of *ABCD* add up to $4 \cdot 180° = 720°$.

(b) How could you compute the sum of the interior angle measures of the quadrilateral using the four triangles?

Lesson Exercise 8.46

Consider the pentagon in Figure 8-36.

(a) What is the sum of its angle measures? Write the answer in the table.

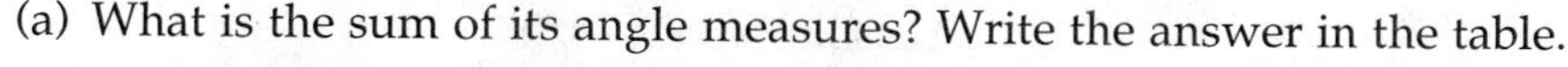

Polygons (convex)

Number of sides	3	4	5	6	7	N
Number of triangles formed	1	2				
Sum of interior angle measures	180°	360°				

(b) Use inductive reasoning or more drawings to complete the table.

(c) Write a generalization of your results.

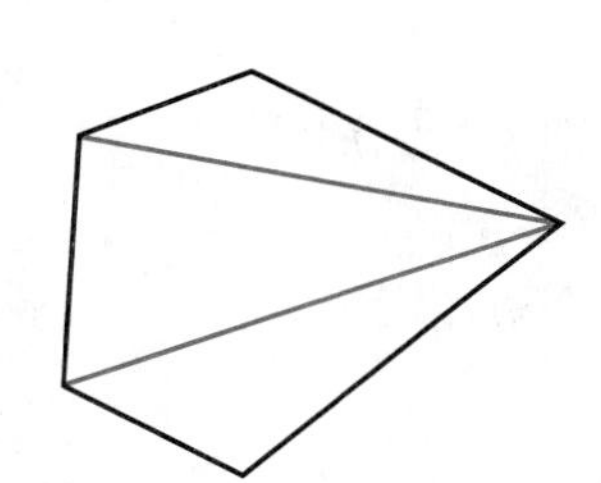

Figure 8–36

As the pattern in Lesson Exercise 8.46 suggests, you can divide any convex polygon that has N sides into $N - 2$ triangles by drawing all the possible diagonals from any single vertex. This result leads to the following conclusion.

Angle Measures of a Convex Polygon

The sum of the angle measures of any N-sided convex polygon is $(N - 2) \cdot 180°$.

This property of angle measures also applies to polygons that are *not* convex; however, such polygons have angles whose measures are greater than 180° (which are not addressed in this textbook).

Lesson Exercise 8.47

What is the sum of the angle measures of a convex octagon?

Many shapes in the world suggest regular polygons (Figure 8-37).

Stop sign photo courtesy of Tom Sonnabend.

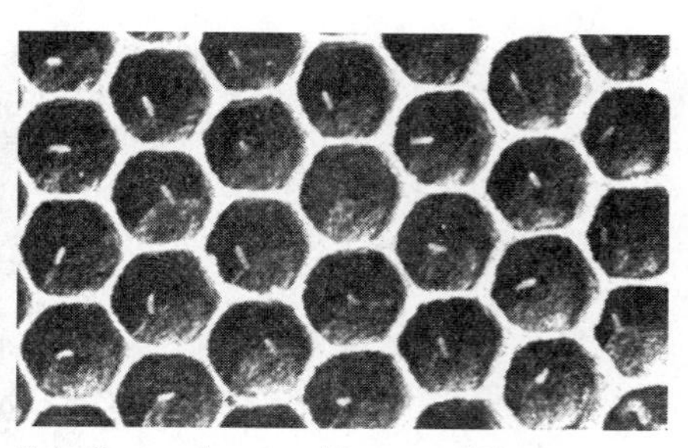

Photo of honeycomb courtesy of Department of Library Services, American Museum of Natural History.

Figure 8–37

A **regular polygon** has sides that are all congruent and angles that are all congruent.

Lesson Exercise 8.48

You want to construct a STOP sign in the shape of a regular octagon. How many degrees are there in each interior angle of a stop sign?

TESSELLATIONS

To return to the opening question, what shapes can be used as tiles?

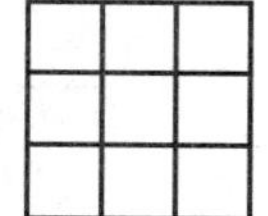

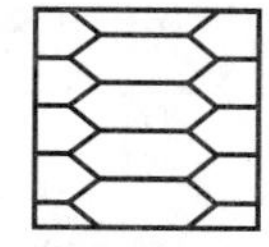

Figure 8–38

All of the patterns in Figure 8-38 are called tesselations. A plane **tessellation** is a complete covering of a plane by shapes in a repeating pattern, without gaps or overlapping.

Figure 8–39

Which *regular* polygons tessellate the plane? First, consider squares. If you've seen a few tile floors, you know that squares can be used. But why?

Pick a vertex of a square (Figure 8-39). Can we cover the plane region around the chosen point with additional squares, with no gaps or overlaps? Yes! Squares tessellate, as shown in Figure 8-40. Since the angle measures of the four squares around the point add up to 360°, the squares fit together with no gaps.

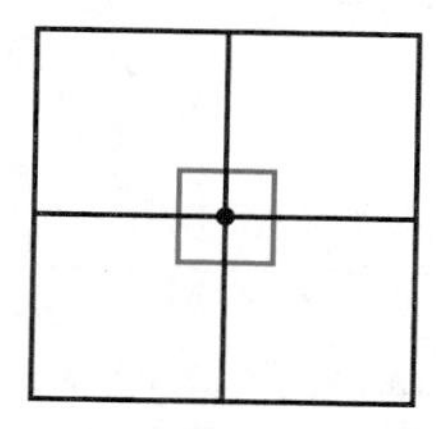

Figure 8–40

Lesson Exercise 8.49

In Figure 8-40, the angle measures around the marked point add up to ________ degrees.

Lesson Exercise 8.50

Do equilateral triangles tessellate? Make a drawing or cut out at least six congruent equilateral triangles. (See if you can find angle measures of triangles around one point that add up to 360°.)

Lesson Exercise 8.51

Do regular pentagons tessellate? (Make sure all five sides are congruent and all five angles are congruent.)

Lesson Exercise 8.52

(a) Complete the table.

Regular Polygons	Total Number of Degrees	Measure of Each Interior Angle
Triangle		
Quadrilateral (square)	360°	90°
Pentagon		
Hexagon		
Heptagon		
Octagon		
Decagon (10 sides)		
Dodecagon (12 sides)		

(b) Without drawing the shapes or making cutouts, tell which figures in the table tessellate the plane.

____

ANSWERS TO SELECTED LESSON EXERCISES

8.41 60°

8.42 Two obtuse angle measures have a sum greater than 180°.

8.43 360°

8.44 (a) 180° (c) 360°

8.45 (a) Together, the angles of the four triangles do not form the four angles of the quadrilateral.
(b) $(4 \cdot 180°) - 360° = 360°$

8.46 (b)

Polygons (convex)

Number of sides	3	4	5	6	7	N
Number of triangles formed	1	2	3	4	5	$N - 2$
Sum of interior angle measures (in degrees)	180°	360°	540°	720°	900°	$(N - 2)180°$

8.47 $(8 - 2) \cdot 180° = 1080°$

8.48 $1080°/8 = 135°$

8.49 360

8.50 Yes

8.51 No

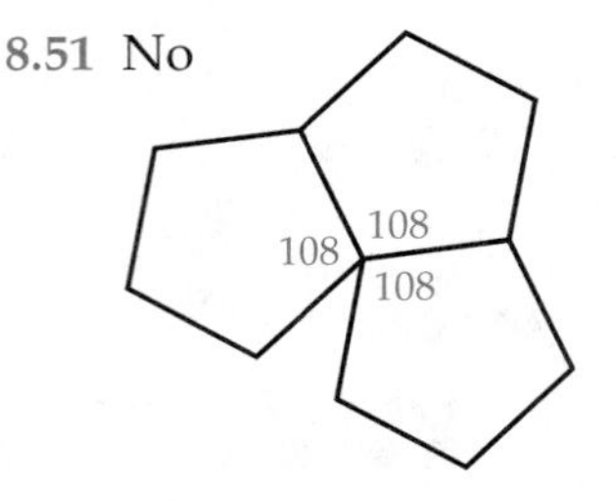

8.52 (a)

Regular Polygons	Total Number of Degrees	Measure of Each Interior Angle
Triangle	180°	60°
Quadrilateral	360°	90°
Pentagon	540°	108°
Hexagon	720°	120°
Heptagon	900°	$128\frac{4}{7}°$
Octagon	1080°	135°
Decagon	1440°	144°
Dodecagon	1800°	150°

(b) Regular triangles, quadrilaterals, and hexagons tessellate the plane.

8.3 HOMEWORK EXERCISES

Basic Exercises

1. (a) Draw a large triangle on a sheet of paper
   (b) Measure each interior angle.
   (c) Do the three angle measures add up to 180°? If not, why not?
2. Explain why a triangle cannot have an obtuse angle and a right angle.
3. $\triangle KAR$ is regular and $m\angle PKR = 108°$. Fill in the missing angle measures.

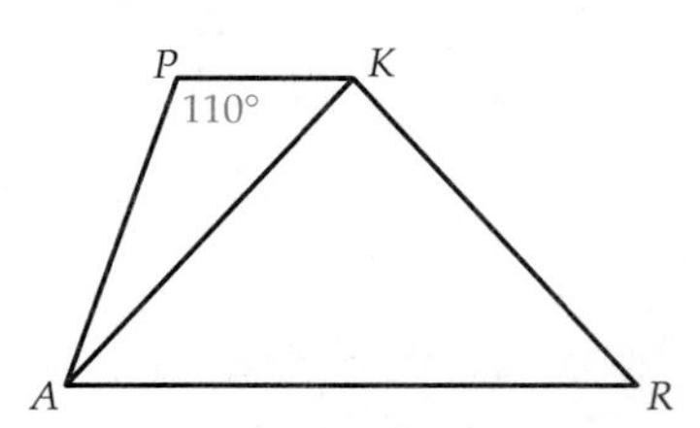

4. Draw each of the following types of triangles.
   (a) Right scalene (b) Acute isosceles
   (c) Obtuse isosceles

5. Is a triangle acute, right, or obtuse if two of its angle measures are
(a) 30° and 50°? (b) 25° and 90°?
(c) 70° and 40°?

6. A quadrilateral has two right angles. What can you deduce about the measures of the other two angles?

7. Use the angle sum property of a triangle and a drawing to explain why the interior angle measures of a convex hexagon add up to 720°.

8. What is wrong with the following explanation?

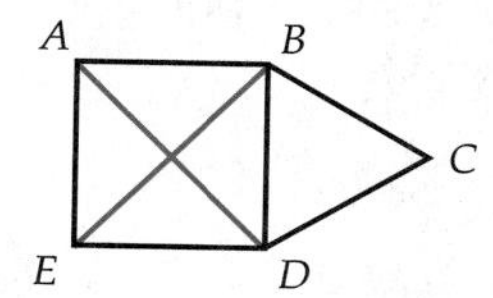

Pentagon *ABCDE* can be divided into five triangles, as shown. Therefore, the sum of the interior angles measures of the pentagon is $5 \cdot 180° = 900°$.

9. What is the sum of the interior angle measures of a 40-sided convex polygon?

10. (a) Is a rhombus a regular polygon? Why or why not?
(b) Is a rectangle a regular polygon? Why or why not?

11. Beehive cells approximate regular hexagons.
(a) When bees make the cells, what size interior angles do they make?
(b) Use a protractor to draw a regular hexagon. Then divide it into three congruent rhombuses.

12. A Canadian nickel has the shape of a regular dodecagon (12 sides). How many degrees are in each angle?

13. In this lesson, you saw that regular triangles, squares, and regular hexagons tessellate the plane. There are also eight possible semiregular tessellations that use two or more different regular polygons to tessellate the plane. The following figure shows a semiregular tessellation that uses two regular triangles and two regular hexagons.

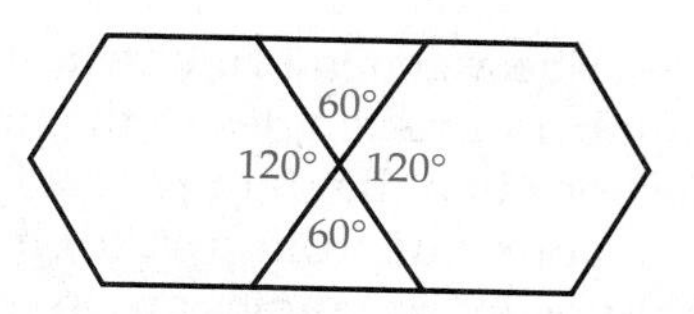

(a) Draw a repeating tile pattern that uses equilateral triangles, squares, and regular hexagons. (*Hint:* Use the angle measures.)
(b) Using the completed table from Lesson Exercise 8.52, try to sketch the other six possible semiregular tessellations.

14. Draw or cut out six congruent copies of each type of figure, and determine if they tessellate.
(a) An isosceles triangle
(b) A scalene triangle

15. Draw or cut out four congruent copies of each type of figure, and determine if they tessellate. (Use pattern blocks if you have them.)
(a) A parallelogram (b) A trapezoid

16. Determine if each figure tessellates.

(a)

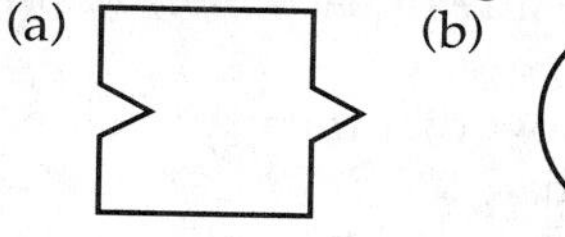

(b)

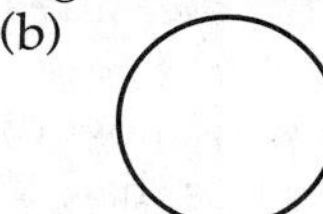

(c)

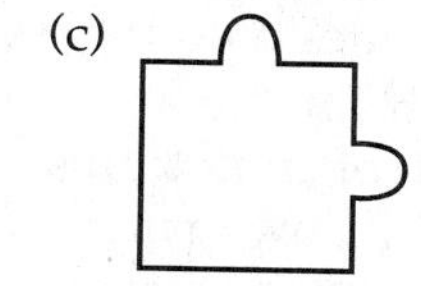

Extension Exercises

17. (a) How many rectangles are in the following diagram?

(b) How many rectangles are in the following diagram?

(c) How many rectangles are in the following diagram?

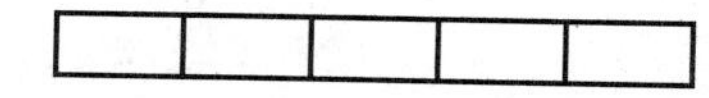

(continues on the next page)

(d) Describe a general rule relating the total number of rectangles to the number of sections in a diagram of this type.

(e) Use your rule from part (d) to determine how many rectangles are in the following diagram.

18. What is the measure of $\angle A$ in the pentagram? (Assume that $FGHIJ$ is regular.)

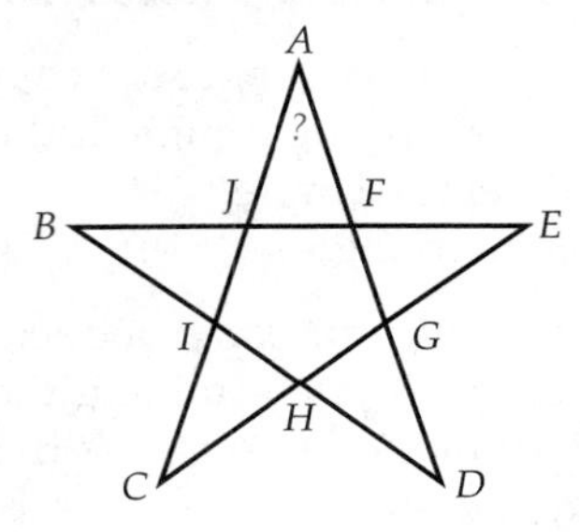

19. A certain regular polygon has n sides.

(a) What is the measure of each interior angle in terms of n?

(b) How do you know that the expression in part (a) is less than 180°?

20. Each angle of a certain regular polygon measures 174°. How many sides does it have?

21. Consider the following problem. "In a basketball tournament, each of eight teams plays every other team once. How many matches are there?" Devise a plan and solve the problem. (*Hint:* Draw a regular octagon.)

22. (a) How many equilateral triangles are shown?

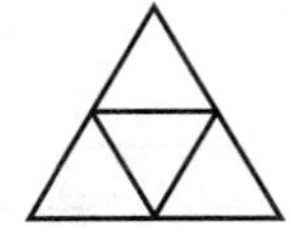

(b) How many equilateral triangles are shown? X

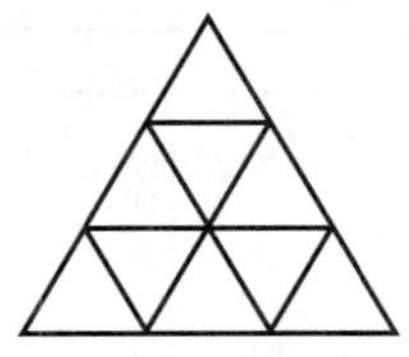

(continues in the next column)

(c) How many equilateral triangles are shown?

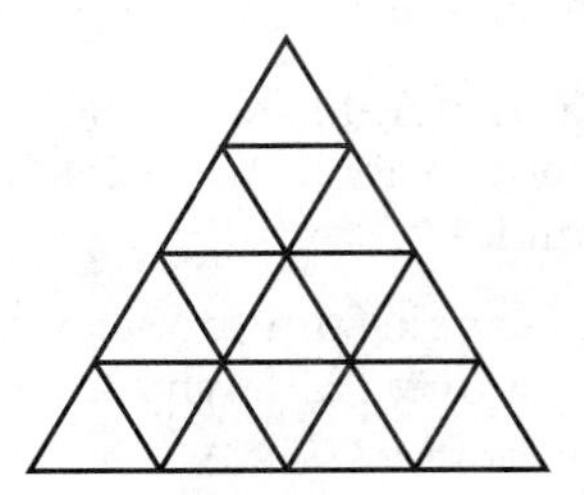

23. On a square grid or a triangular grid, create a tessellation using a polygon or polygons that are not regular.

Special Exercises

24. (a) Trace the square and cut it along the lines into seven pieces.

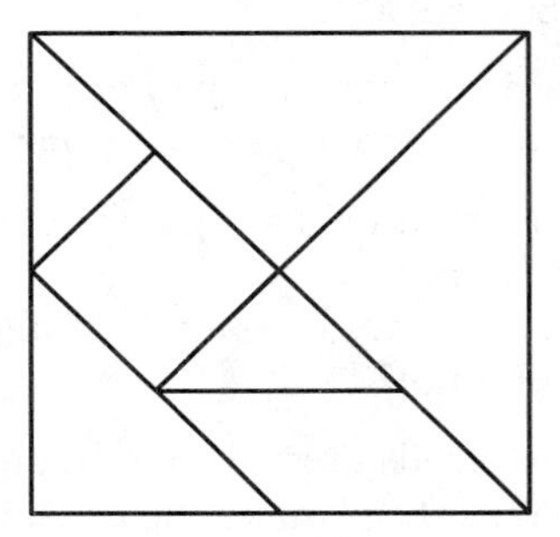

You have just created a Chinese tangram puzzle set. This type of geometric puzzle is at least 4000 years old! How does it work? You can rearrange the seven pieces into a variety of shapes. For example, after taking it apart, you can try to reconstruct a square with all seven pieces.

(b) Try an introductory problem. Cover the following picture with two tangram pieces.

(c) What fraction of the original square is covered by the parallelogram piece?

(d) Cover the following figure with all seven pieces.

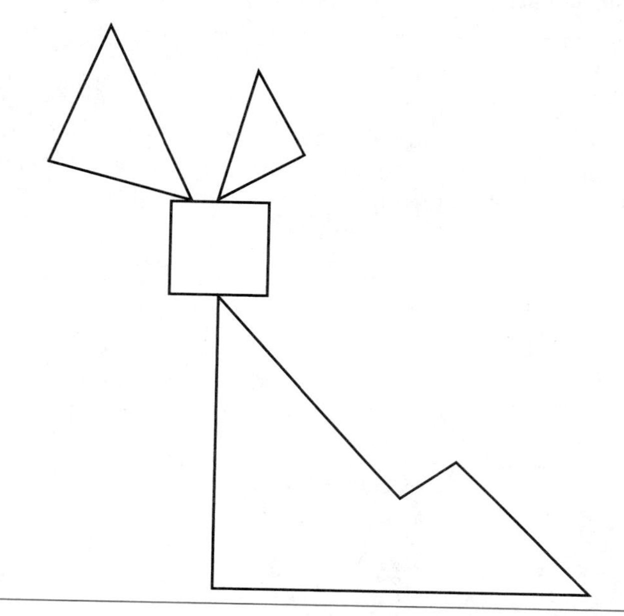

(e) Use all seven pieces to cover a whale.

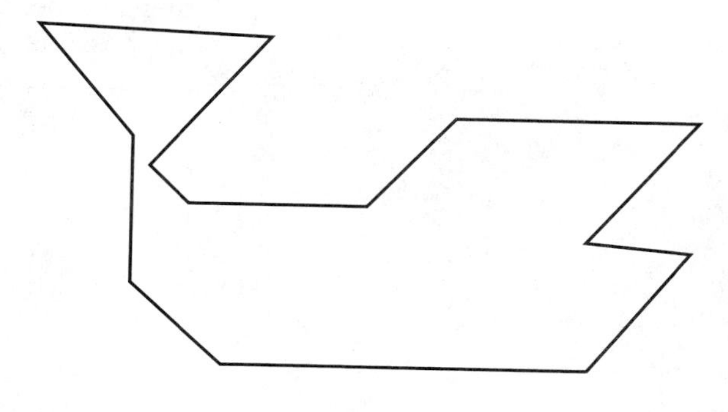

(f) Use all seven pieces to make a triangle.

25. If possible, form each of the following shapes with exactly four tangram pieces.
(a) Triangle
(b) Square
(c) Parallelogram
(d) Trapezoid

8.4 THREE-DIMENSIONAL GEOMETRY

Space figures are the first shapes children perceive in their environment. But children cannot systematically study and categorize three-dimensional shapes in school until they have studied one- and two-dimensional shapes.

In elementary school, children learn about the most basic classes of space figures: prisms, pyramids, cylinders, cones, and spheres. This lesson concerns the definitions of these space figures and the properties of their faces, vertices, and edges.

As a basis for studying space figures, first consider relationships among lines and planes in space. These relationships are helpful in analyzing and defining space figures.

LINES IN SPACE

What are possible relationships between two lines in space? As in two dimensions, two distinct lines can be either parallel or intersecting. But there is another possibility in three dimensions!

Lesson Exercise 8.53

Using two pencils to represent lines in space, see whether you can find another possible relationship in addition to their being parallel or intersecting.

In space, two lines can be parallel, intersecting, or skew. **Skew** lines are two lines that cannot be contained in a plane. Skew lines are not parallel and do not intersect. Figure 8-41 shows two examples of skew lines.

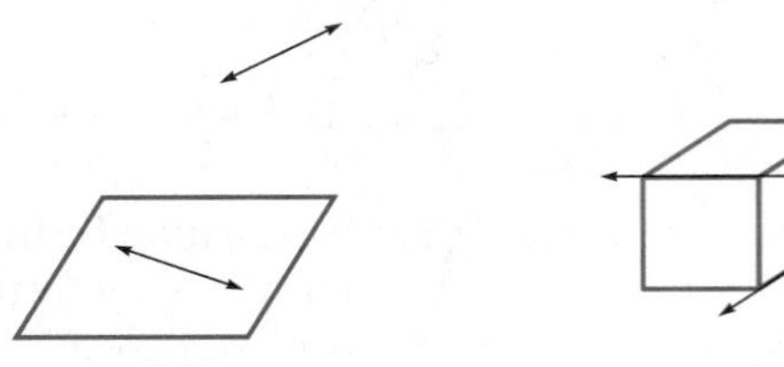

Figure 8–41

Lesson Exercise 8.54

(a) True or false? A line that is parallel to one of two skew lines must intersect the other skew line.

(b) Identify a model of this situation in your classroom that supports your answer.

A LINE AND A PLANE IN SPACE

What are the possible relationships between a line and a plane in space? One possibility is that they are parallel. A line and a plane are **parallel** if and only if they do not intersect.

Lesson Exercise 8.55

Use a piece of paper to represent a plane and a pencil or pen to represent a line. See whether you can find three possible relationships between a line and a plane.

Did you find the following possibilities in Lesson Exercise 8.55?

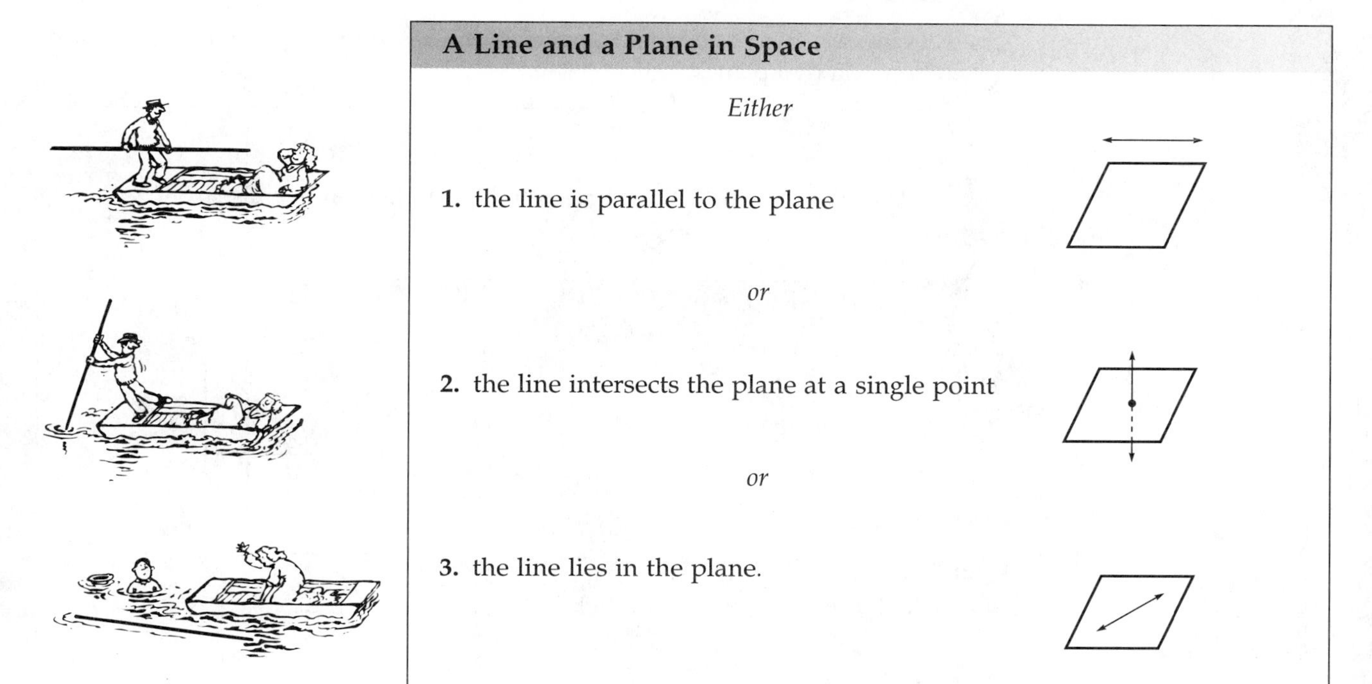

A Line and a Plane in Space

Either

1. the line is parallel to the plane

or

2. the line intersects the plane at a single point

or

3. the line lies in the plane.

Lesson Exercise 8.56

(a) True or false? If a line lies in a plane and a second line intersects the plane, then the two lines must intersect.

(b) In your classroom, identify a model of this situation that supports your answer.

Lesson Exercise 8.57

(a) True or false? If a line is parallel to a plane, and a second line lies in the plane, then the two lines do not intersect.

(b) In your classroom, identify a model of this situation that supports your answer.

PLANES IN SPACE

What are the possible relationships between two planes in three dimensions? One possibility is that the two planes are parallel. Two planes are **parallel** if and only if they do not intersect.

Lesson Exercise 8.58

Using two pieces of paper to represent planes, find the possible relationships between two planes.

Did you find the following possibilities in Lesson Exercise 8.58?

Two Planes in Space

Either

1. the two planes intersect in a line

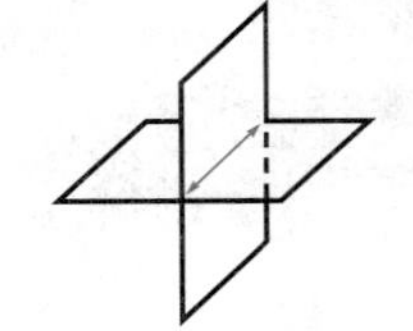

or

2. the two planes are parallel.

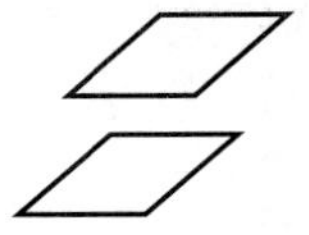

Lesson Exercise 8.59

(a) True or false? Two planes that are parallel to a third plane must be parallel to each other.
(b) In your classroom, identify a model that supports your answer.

Lesson Exercise 8.60

True or false? Two planes may intersect at exactly one point.

Parallel planes are used in defining a prism, the most common type of polyhedron (plural, polyhedra or polyhedrons) in school geometry.

POLYHEDRA

Many everyday objects, such as boxes and crystals (Figure 8-42), resemble polyhedra.

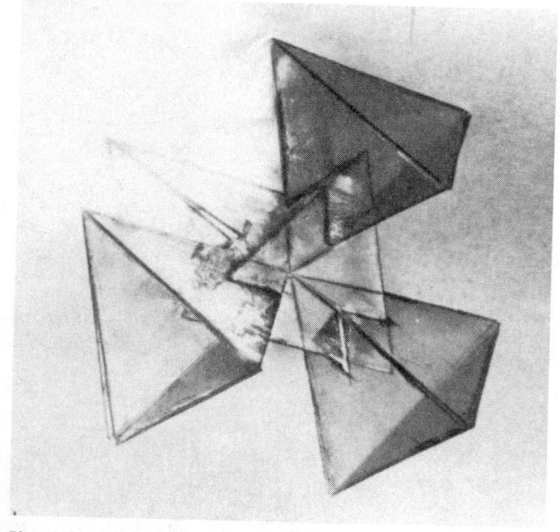

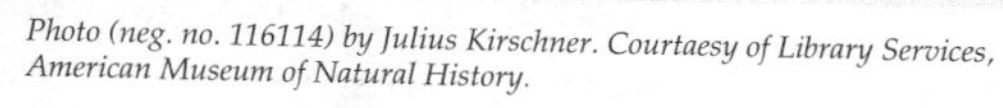
Photo (neg. no. 116114) by Julius Kirschner. Courtaesy of Library Services, American Museum of Natural History.

Figure 8–42

Do you recognize a polyhedron when you see one?

Lesson Exercise 8.61

Which of the following are polyhedra?

(a)

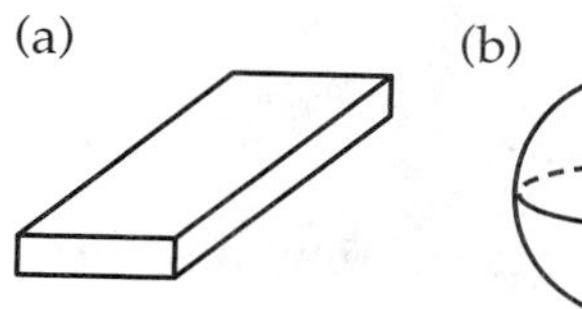

(b)

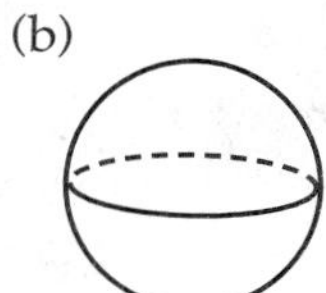

(c)

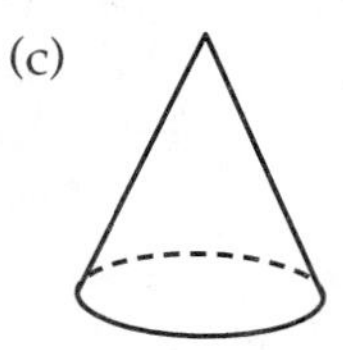

(d)

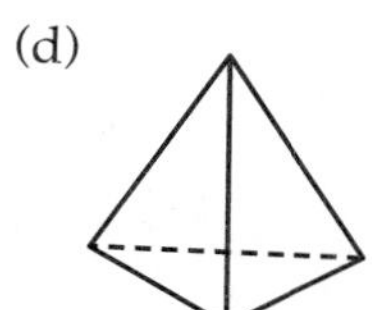

The shapes in Figure 8-43 are polyhedra.

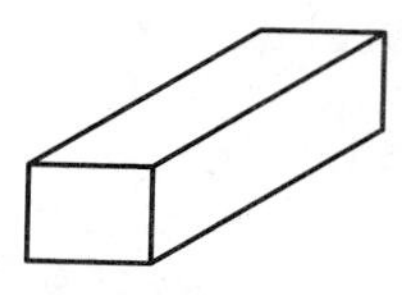

Rectangular prism

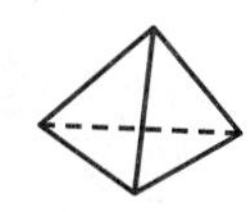

Triangular pyramid

Figure 8–43

The shapes in Figure 8-44 are *not* polyhedra.

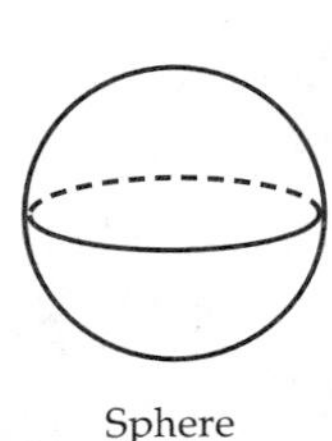

Sphere

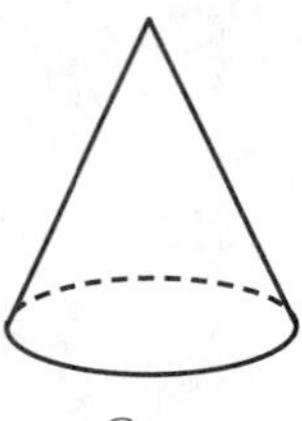

Cone

Figure 8–44

Lesson Exercise 8.62

Name a property of the surfaces of polyhedra.

To define a polyhedron, we will use the terms "simple closed surface" and "polygonal region." A simple closed surface in space is analogous to a simple closed curve in a plane. A **simple closed surface** separates space into three disjoint sets: points inside the surface, points on the surface, and points outside the surface. A **polygonal region** is a polygon together with its interior (Figure 8-45).

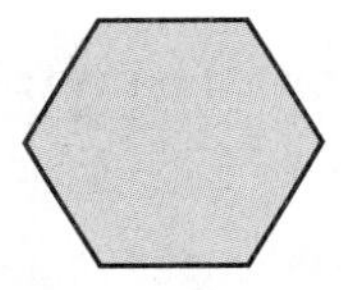

Polygonal region

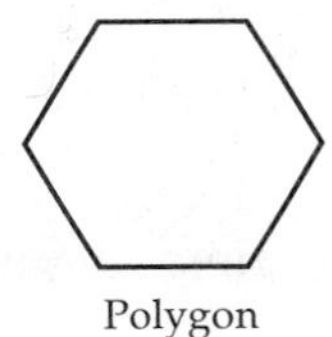

Polygon

Figure 8–45

Lesson Exercise 8.63

Use the preceding information to write a definition of a polyhedron.

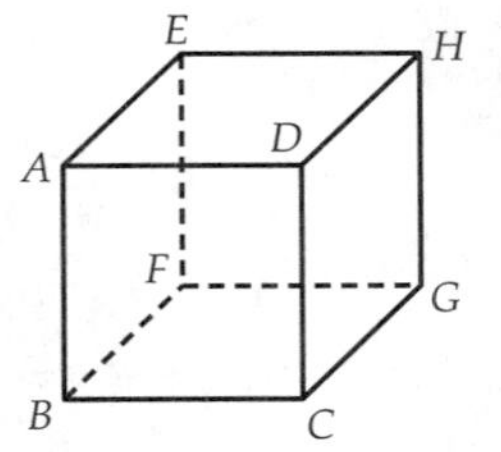

Figure 8–46

A **polyhedron** is a simple, closed space figure bounded by polygonal regions. All of the surfaces of a polyhedron are flat, not curved. The polygonal surfaces of polyhedra are called **faces**. The sides of each face (polygonal region) are called **edges**, and the vertices of the polygons are also **vertices** (corners) of the polyhedron.

For example, in the cube shown in Figure 8-46, square *ADHE* and its interior comprise a *face*, $\overline{AD}$ is an *edge*, and *A* is a *vertex*.

Lesson Exercise 8.64

(a) How many faces does a cube have?

(b) How many vertices (corners) does a cube have?

(c) How many edges does a cube have? Count them in groups, beginning on the top of the cube.

A cube and its interior comprise a solid. A **solid** is the union of a simple closed surface and its interior.

PRISMS

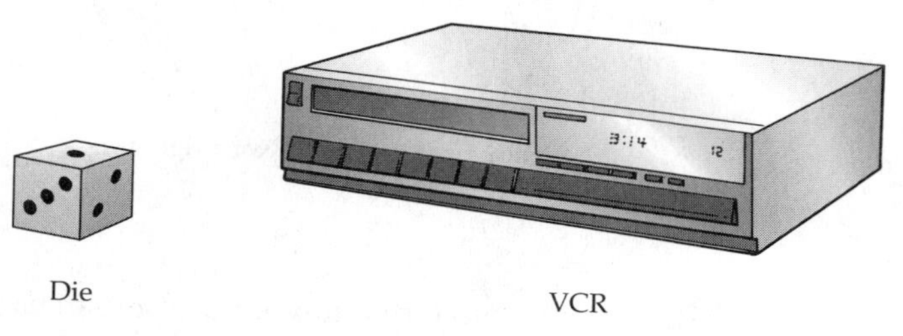

Die VCR

Figure 8–47

A cube and a rectangular solid (Figure 8-47) are both a special type of polyhedron called a prism. A triangular pyramid, a sphere, and a cone, on the other hand, are not prisms.

Lesson Exercise 8.65

Describe a property prisms have that pyramids, spheres, and cones do not have.

A **prism** has two congruent polygonal regions that are opposite faces (usually the top and bottom), and the corresponding vertices are connected by parallel line segments. The opposite faces, called **bases**, lie in parallel planes. The faces that are not bases are called **lateral faces**.

A prism is named by the kind of base it has (Figure 8-48).

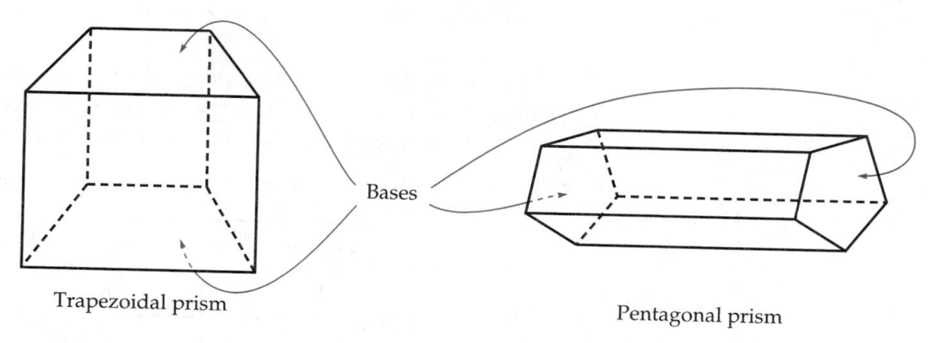

Trapezoidal prism Pentagonal prism

Figure 8–48

In more formal terms, a **prism** is a polyhedron formed by two congruent polygonal bases in parallel planes connected by three or more parallelogram-shaped regions. A prism with lateral faces that are rectangular regions is a **right prism** (Figure 8-49).

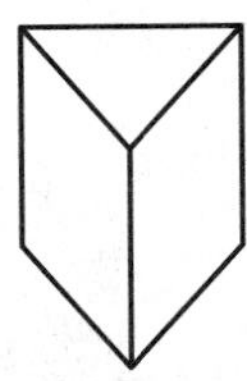

Right prism

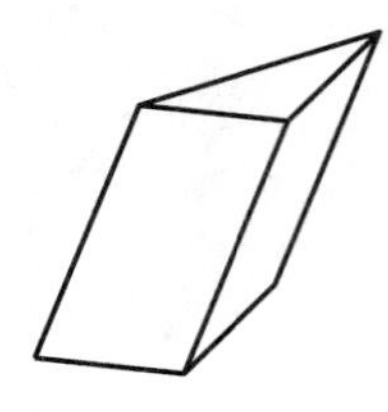

Not a right prism

Figure 8–49

Note that any kind of prism can be split into congruent layers, as shown in Figure 8-50.

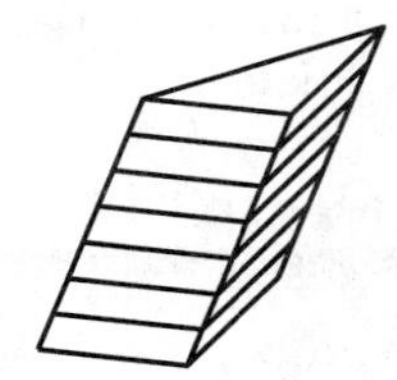

Figure 8–50

Lesson Exercise 8.66

(a) How many faces does a triangular prism have?

(b) How many vertices does it have?

(c) How many edges does it have?

Figure 8-51, on the next page, shows how the fifth-grade textbook of *Mathematics Plus* introduces the topic of faces, vertices, and edges.

Your results from Lesson Exercises 8.64 and 8.66 should have been as follows.

Figure	Faces	Vertices	Edges
Cube	6	8	12
Triangular prism	5	6	9

EXPLORING
Solid Figures

Work Together

Building Understanding

Use patterns to model solid figures.

Use the pattern on page H91 to model a figure that looks like a shoebox.

Your model represents a solid figure called a **rectangular prism.**

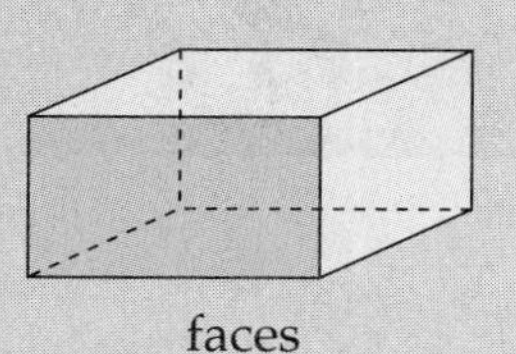

faces

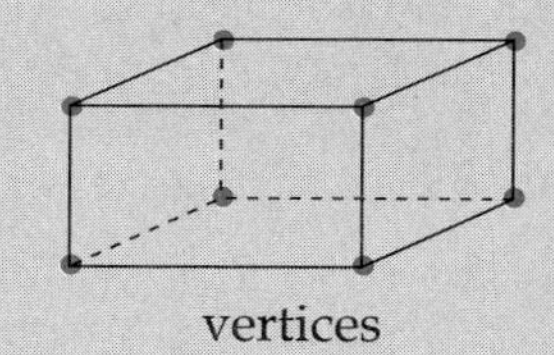

vertices

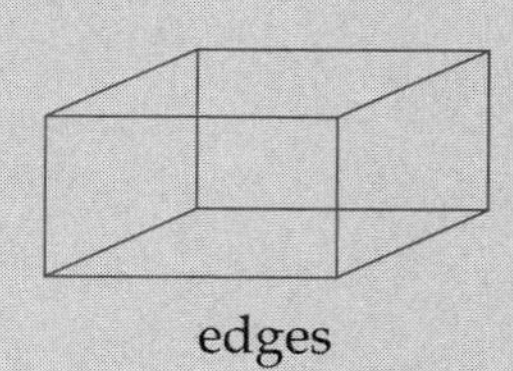

edges

- How many **faces** does the model have?
- What is formed where two faces intersect?
- How many **vertices** does the model have? How many **edges**?

From Mathematics Plus, *Grade 5 (San Diego: Harcourt Brace, 1994), p. 136*

Figure 8–51

Lesson Exercise 8.67

(a) Add two more rows to the preceding table: one for a rectangular prism and one for a pentagonal prism. Then fill in the numbers of faces, vertices, and edges each has.

(b) What pattern do you see in each row of the table?

PYRAMIDS

Photo: *Lowenfish. Courtesy of Department of Library Services, American Museum of Natural History.*

Figure 8–52

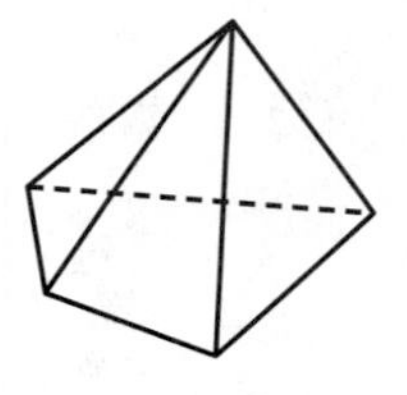

Figure 8–53

The ancient Egyptians constructed some of the largest buildings in the world (Figure 8-52). Some contain over 2 million stones, each weighing at least a ton! These buildings suggest another important group of polyhedra: pyramids. A pyramid can be formed by taking any polygon and a point above the plane of the polygon and connecting all the vertices of the polygon to that point (Figure 8-53). The polygon is the base.

A pyramid is named by the kind of base it has (Figure 8-54). The Great Pyramid of Egypt approximates a square pyramid.

Square pyramid

Triangular pyramid

Figure 8–54

In more formal terms, a **pyramid** is a polyhedron that has a polygonal base. Its lateral faces are triangular regions with a common vertex. The following table summarizes the data we have collected so far about different polyhedra and the numbers of faces, vertices, and edges they have.

Figure	Faces	Vertices	Edges
Cube	6	8	12
Triangular prism	5	6	9
Rectangular prism	6	8	12
Pentagonal prism	7	10	15
Triangular pyramid			
Square pyramid			

Lesson Exercise 8.68

(a) Fill in the last two rows of the preceding table.

(b) Use inductive reasoning to hypothesize how the number of edges of each polyhedron is related to the numbers of faces and vertices it has.

The pattern in Lesson Exercises 8.67 and 8.68 holds for all polyhedra. Euler's formula describes this pattern algebraically.

Euler's Formula

For all polyhedra, $F + V - E = 2$, in which F is the number of faces, V the number of vertices, and E the number of edges.

Photo Courtesy of Library of Congress.

Figure 8–55 Leonard Euler

Leonhard Euler (pronounced "oiler") (1707–1783) was the most prolific mathematician who ever lived (Figure 8-55). Even during the last 17 years of his life, when he was blind, Euler developed many new mathematical ideas.

REGULAR POLYHEDRA AND EULER'S FORMULA

Although there is an infinite number of regular polygons, there are only five regular polyhedra! A **regular polyhedron** is a polyhedron whose faces are congruent regular polygonal regions and in which the number of edges that meet at each vertex is the same. The ancient Greeks proved that the only regular polyhedra are the five shown in Figure 8-56 on the next page.

Natural crystals occur in the shapes of the tetrahedron (for example, chrome alum), the cube (salt), and the octahedron (sodium sulphantimoniate).

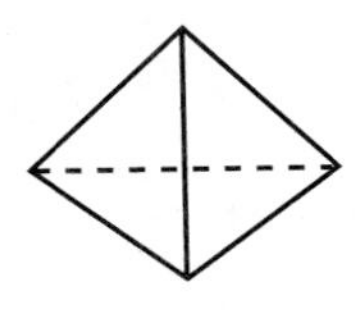

Tetrahedron
(triangular pyramid with
4 equilateral triangles)

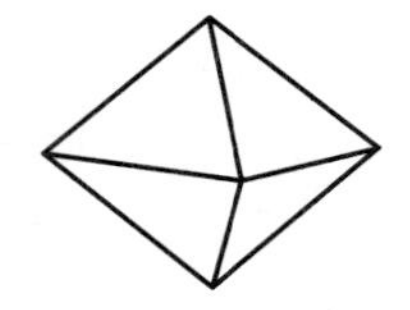

Octahedron
(8 equilateral triangles)

Icosahedron
(20 equilateral triangles)

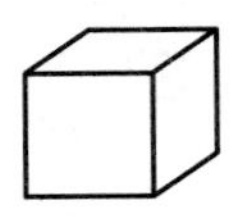

Cube
(6 squares)

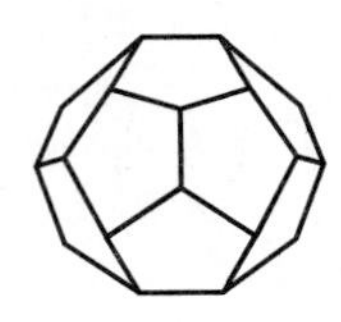

Dodecahedron
(12 regular pentagons)

Figure 8–56

In the following exercise, verify that Euler's formula works for regular polyhedra.

Lesson Exercise 8.69

(a) Complete the table.

Regular Polyhedron	Faces	Vertices	Edges
Tetrahedron			
Octahedron	8		
Icosahedron	20	12	30
Cube			
Dodecahedron	12		30

(b) Does Euler's formula work for all regular polyhedra?

CYLINDERS, CONES AND SPHERES

Some everyday objects suggest space figures that are not polyhedra: cylinders, cones, and spheres (Figure 8-57).

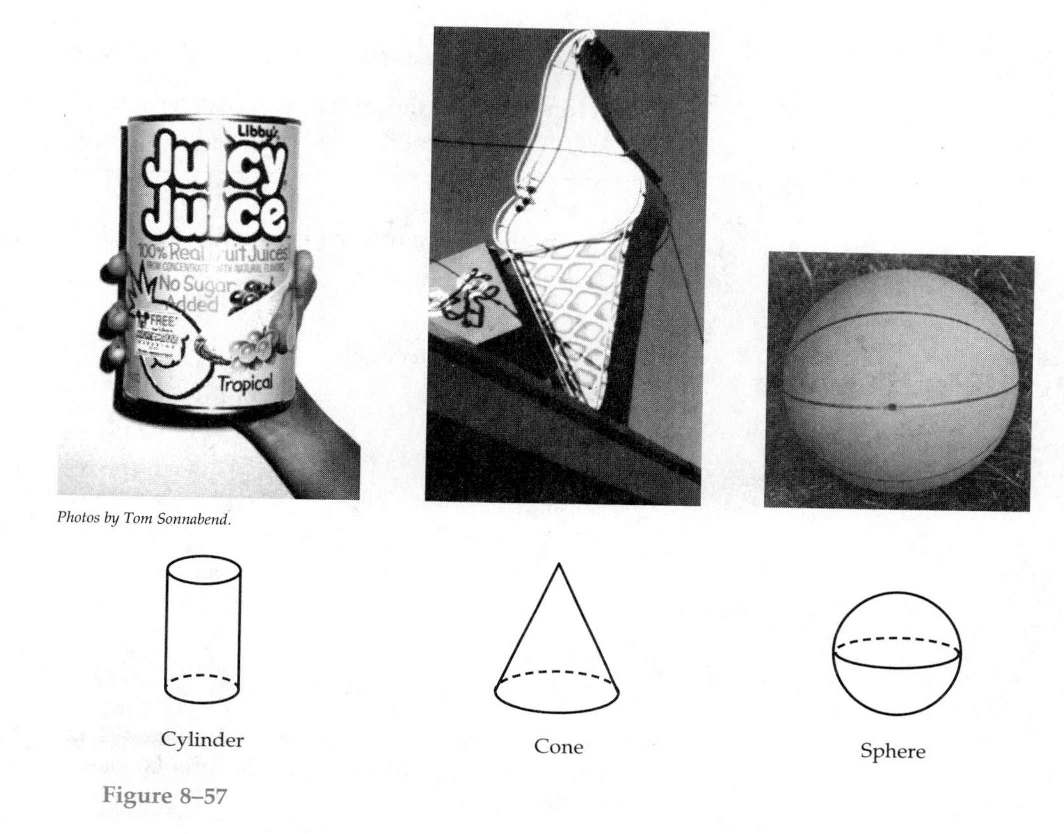

Photos by Tom Sonnabend.

Figure 8–57

Lesson Exercise 8.70

Why aren't cylinders, cones, and spheres polyhedra?

The most common types of cylindrical shapes in our environment are right circular cylinders. Circular cylinders have parallel bases that are circular regions. In a **right circular cylinder** (Figure 8-58), the line segment connecting the centers of the circular bases is perpendicular to the bases.

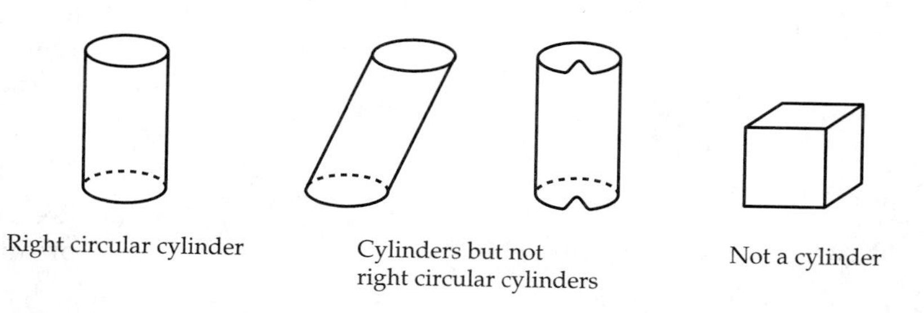

Figure 8–58

Lesson Exercise 8.71

You can construct the lateral surface of a right circular cylinder from a certain polygon. Which polygon is it?

The familiar type of cone is called a right circular cone.

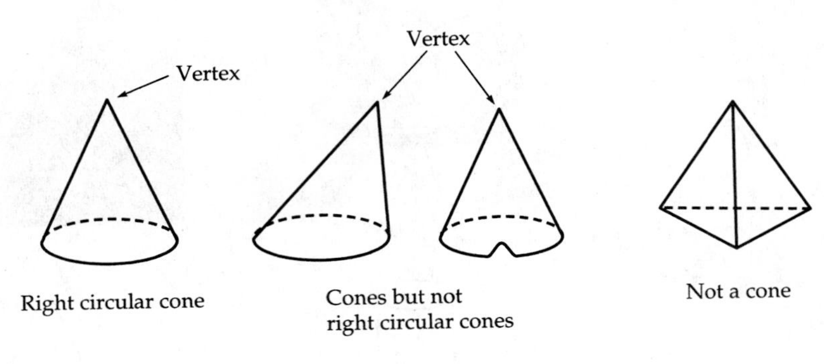

Figure 8–59

A circular cone has a circular region for a base. In a **right circular cone**, the line segment connecting the center of the circular base to the vertex (Figure 8-59) is perpendicular to the base.

Lesson Exercise 8.72

(a) Name a property that all cones and pyramids have in common.
(b) At what van Hiele level is the question in part (a)?

ANSWERS TO SELECTED LESSON EXERCISES

8.54 (a) False

8.56 (a) False

8.57 (a) True

8.59 (a) True

8.60 False (see the preceding diagram)

8.62 They are polygons.

8.64 (a) 6 (b) 8
(c) 12 (4 on top, 4 connecting top to bottom, 4 on the bottom)

8.65 Two congruent bases

8.66 (a) 5 (b) 6 (c) 9

8.67 (a) Rectangular prism: 6, 8, 12; pentagonal prism: 7, 10, 15

8.68 (a) 4, 4, 6 and 5, 5, 8

8.69 (a)

Regular Polyhedron	Faces	Vertices	Edges
Tetrahedron	4	4	6
Octahedron	8	6	12
Icosahedron	20	12	30
Cube	6	8	12
Dodecahedron	12	20	30

(b) Yes

8.70 Their faces are not polygonal regions.

8.71 A rectangle

8.72 (a) They have exactly one base.
(b) Analysis

8.4 HOMEWORK EXERCISES

Basic Exercises

1. (a) True or false? In space, two lines that are perpendicular to a third line are parallel to each other.
   (b) In your surroundings, identify a model of this situation that supports your answer.
2. What are skew lines?
3. In space, point A is not on a line b. How many lines pass through A that
   (a) intersect b?
   (b) are perpendicular to b?
   (c) are parallel to b?
   (d) are skew to b?
4. In space, point A is on line b. How many lines pass through A that
   (a) intersect b?
   (b) are perpendicular to b?
   (c) are parallel to b?
   (d) are skew to b?
5. What are the three possible relationships between a line and a plane in space?
6. (a) True or false? Two lines that are parallel to the same plane are parallel to each other.
   (b) In your surroundings, identify a model of this situation that supports your answer.
7. (a) True or false? A plane contains line m but not line n, and $m \parallel n$. Then the plane is parallel to n.
   (b) In your surroundings, identify a model of this situation that supports your answer.
8. Point A is not on line b. How many different planes contain point A and line b?
9. Draw a sketch of two perpendicular planes.
10. True or false? Two planes, a and b, intersect a third plane in parallel lines. Planes a and b are parallel.
11. True or false? Three planes may intersect at exactly one point.
12. Into how many sections can space be separated by the removal of two planes?
13. (a) The ceiling and floor of a room suggest ________ planes.
    (b) The ceiling and side wall of a room suggest ________ planes.
14. $ABCDEFGH$ is a cube.

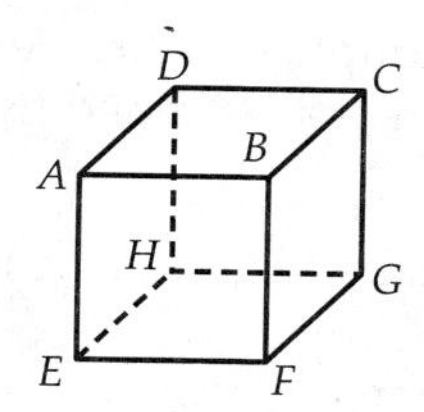

(a) Name two skew lines in the drawing.
(b) The plane that contains $ABEF$ is parallel to the plane that contains ________.
(c) The plane that contains $ABCD$ is ________ to the plane that contains $BCGF$.
(d) What is the intersection of the plane containing E, F, G, and H with the plane containing B, C, F, and G?

15. True or false? If a line contains two points of a triangle, then the line lies entirely in the plane of the triangle.

16. (a) True or false? Every set of four points is contained in one plane.
(b) In your home, identify a model of this situation that supports your answer.

17. (a) True or false? If a line intersects one of two parallel lines, then it intersects the other also.
(b) In your home, identify a model of this situation that supports your answer.

18. True or false? Through a given point not on plane P, there is exactly one line parallel to P.

19. True or false? Through a given point not on plane P, there is exactly one plane parallel to P.

20. True or false? Line m intersects plane P and is not perpendicular to it. Then there is no line in P that is perpendicular to m.

21. Which of the following are polyhedra?

(a)
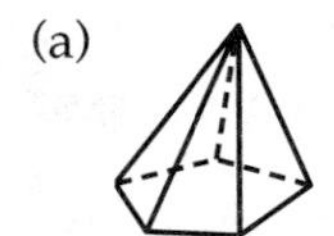

(b)
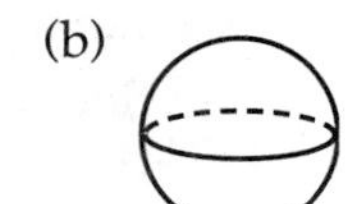

(c) Cylinder (d) Hexagonal prism

22. Give an example of a simple closed space figure that is not bounded by polygonal regions.

23. (a) How many faces does a hexagonal prism have?
(b) How many edges?
(c) How many vertices?

24. (a) Name two properties that prisms and pyramids have in common.
(b) At what van Hiele level is the question in part (a)?

25. The figure shown is a cube.

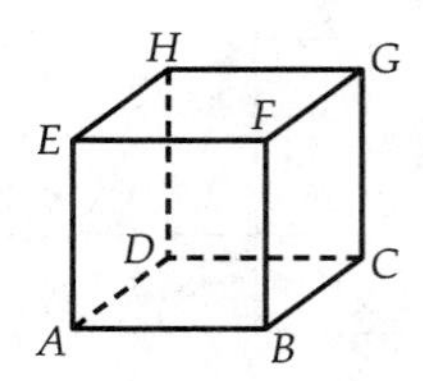

(a) What kind of quadrilateral is $EGCA$?
(b) What kind of triangle is $\triangle EGH$?

26. (a) What simple change in the definition of a circle would make it the definition of a sphere?
(b) What are the possible shapes of the intersection of a plane and a sphere?

27. Name the space figure that would be formed if each shape were folded on the broken lines.

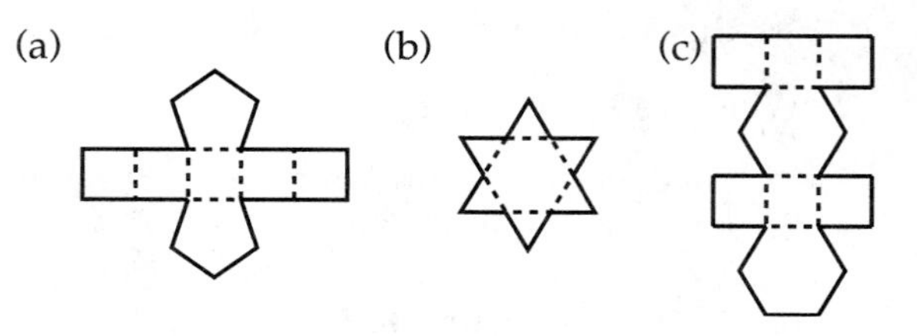

28. Name the space figure that would be formed if each shape were folded on the broken lines.

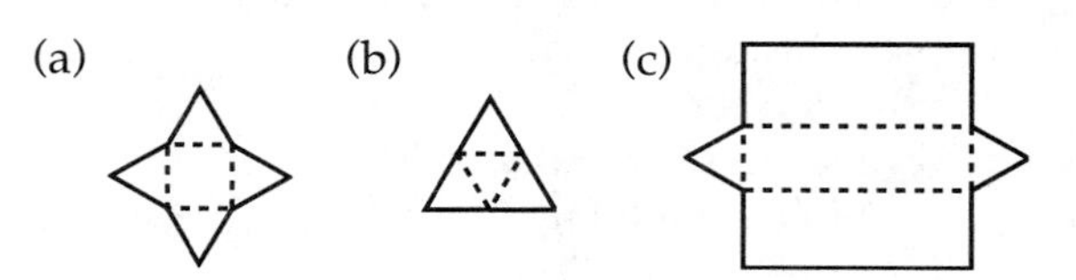

29. Which of the following pictures is the correct layout of the faces of the block? (You could use a model to check.)

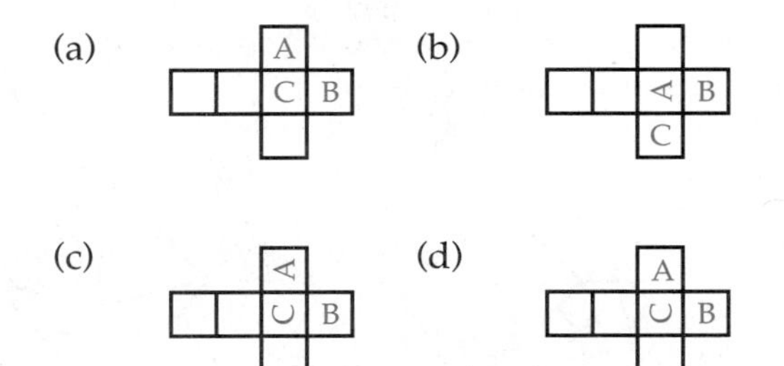

30. Which assembled cube matches the disassembled cube on the left?

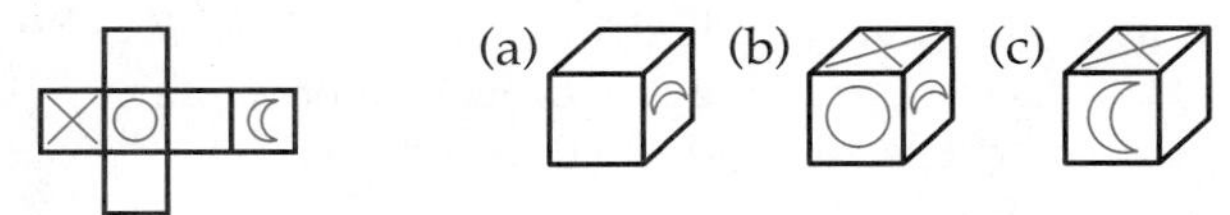

31. Which of the following patterns fold into a cube? (Use paper cutouts as needed.)

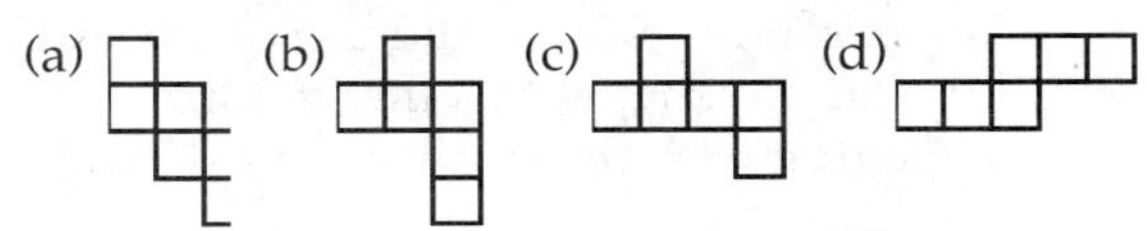

32. Shown here is a layout of the faces of a rectangular prism *ABCDEFGH*. Label the missing vertices.

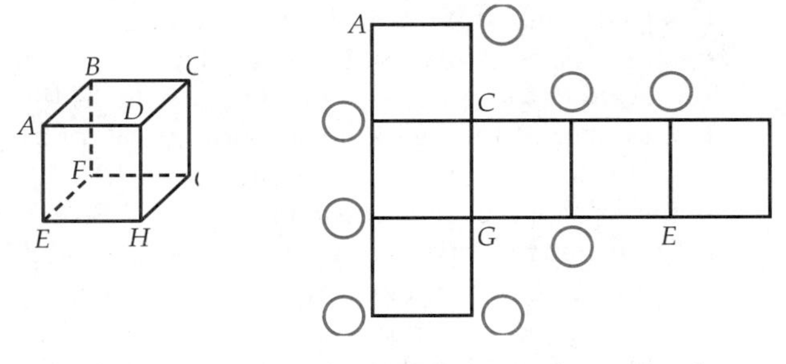

33. (a) Sketch a pyramid that has a hexagonal base.
(b) How many faces does it have?
(c) How many vertices?
(d) How many edges?

34. A pyramid has a heptagonal base. How many faces, vertices, and edges does it have?

35. Name a solid figure that has 1 base, 3 other faces, 6 edges, and 4 vertices.

36. Name a solid figure that has 2 bases, 6 other faces, 18 edges, and 12 vertices.

37. Name a solid figure that has 2 flat faces and can roll.

38. Name a solid figure that has no flat faces and can roll.

39. Determine if Euler's formula is true for the following figures.

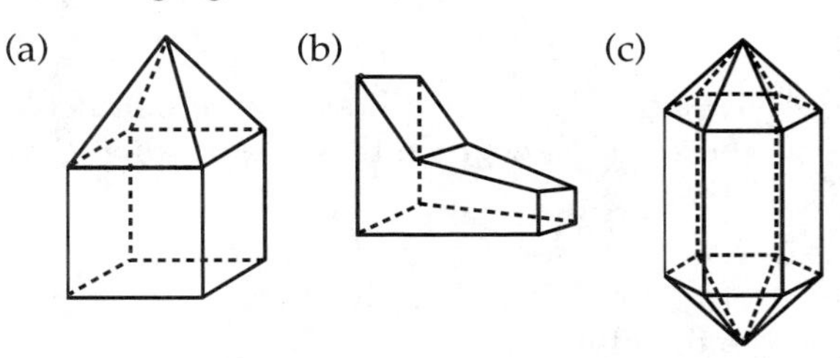

40. Name each figure.

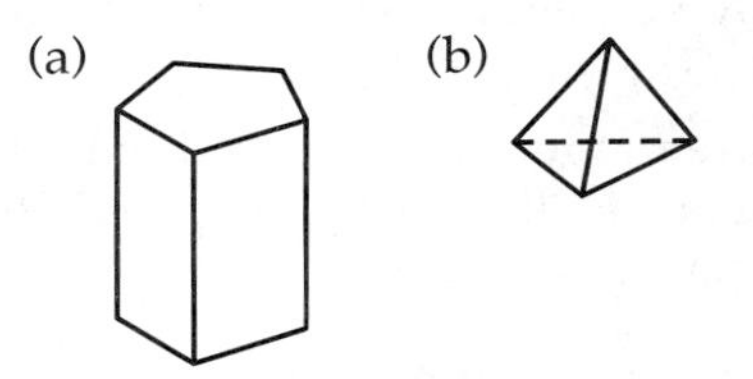

41. Name a space figure that approximates the shape of each of the following.
(a) A new piece of chalk
(b) Your refrigerator at home

42. What property of (right) prisms and cylinders makes them more suitable than pyramids or cones for the shape of a garbage can?

43. What property of (right circular) cones and cylinders makes them more suitable than prisms or pyramids for the shape of an ice cream cone?

44. Write a description of the shape of a milk carton.

Extension Exercises

45. Decide whether each of the following describes a point, line, plane, segment, or ray.
(a) All points that are equidistant from two parallel planes
(b) In space, all points that are equidistant from both endpoints of a line segment
(c) All points that are 3 inches below a line

46. A line and a plane are **perpendicular** if and only if they intersect and the line is perpendicular to every line in the plane that passes through the intersection point.

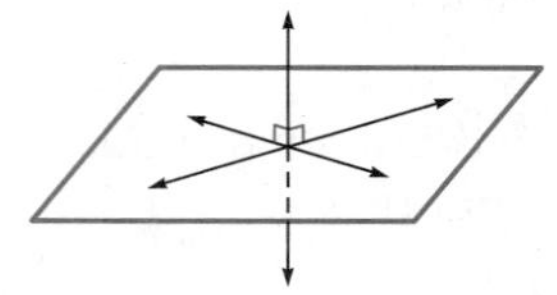

True or false? If two lines are perpendicular to the same plane, then they are parallel to each other.

47. True or false? If a line is perpendicular to one of two parallel planes, then it is perpendicular to the other.

48. Investigate whether chair legs are approximately perpendicular to the floor, and summarize your results.

49. Two planes are **perpendicular** if and only if one plane contains a line that is perpendicular to the other plane.

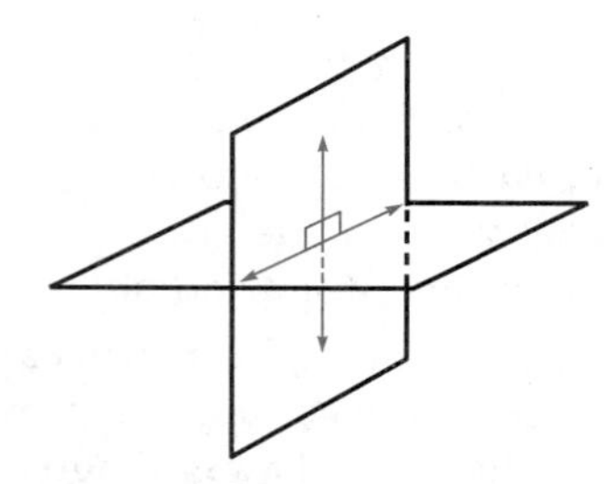

True or false? If two planes are perpendicular to a third plane, then they are parallel to each other.

50. Why are there no such things as skew planes?

51. A prism has a base with n sides.
(a) How many faces does it have?
(b) How many vertices does it have?
(c) How many edges does it have?
(d) Does Euler's formula work for prisms?

52. A pyramid has a base with n sides.
(a) How many faces does it have?
(b) How many vertices does it have?
(c) How many edges does it have?
(d) Does Euler's formula work for pyramids?

53. Take a regular polyhedron. Connect the centers of adjacent faces and you get another polyhedron (called the **dual**)!

What shape is obtained by connecting the centers of adjacent faces of

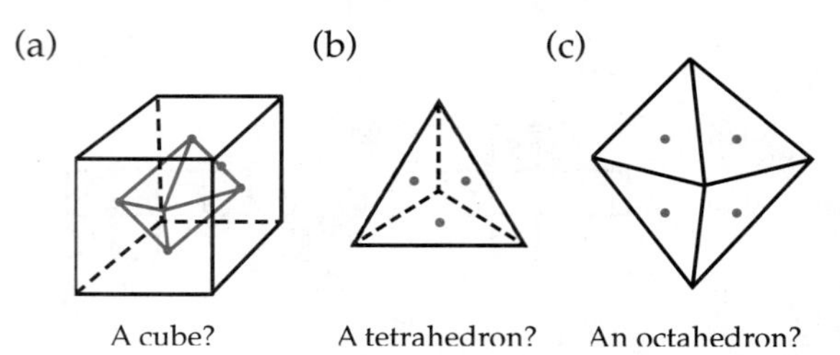

(*Note:* The back of the octahedron is not shown.)
(d) A regular polyhedron has x vertices, y faces, and z edges. Conjecture how many vertices, faces, and edges its dual has.

54. Construct or obtain models of a square pyramid and a triangular pyramid with congruent equilateral triangles for all their triangular faces. (You could use cardboard.) Place the two shapes together to form a polyhedron with the smallest number of faces. How many faces does it have?

Special Exercise

55. A plane surface suggested by a sheet of paper has two sides. In fact, a piece of paper always has two sides, right? Not always. The remarkable Möbius strip has only 1 side! The front cover of this book has a photograph of a Möbius strip. You can make one yourself using scissors, paper, and tape.
(a) Cut out two strips of paper. Tape the ends of one strip to make a regular loop. Before taping together the ends of the second strip, give it a half-twist to create a Möbius strip.

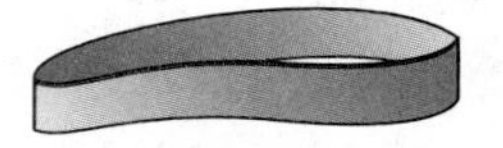
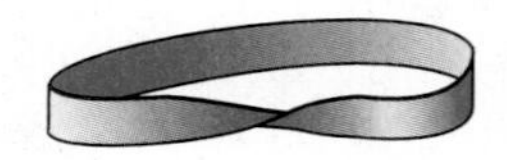

(b) Use a pencil to draw a line down the middle of one side of each paper loop. Continue each line until you reach your starting point.
(c) What is the difference in your results for the two loops in part (b)?

8.5 SPATIAL PERCEPTION

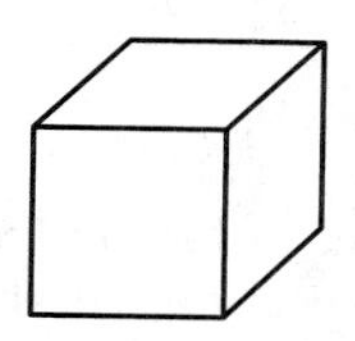

Figure 8–60

It's amazing that a figure sketched on a flat surface can appear to have a third dimension, as in Figure 8-60. Representing three-dimensional figures on a two-dimensional surface posed a serious challenge for artists until the Renaissance. Today, artists have no problem making flat drawings that create the illusion of three dimensions. This knowledge resulted from many years of experimentation.

VISUAL PERCEPTION

Although the M. C. Escher drawing in Figure 8-61 is two-dimensional, it appears to be three-dimensional. Our brains tend to look for depth, the third dimension, even in a flat picture. Some optical illusions are based upon the tendency to perceive depth in a flat drawing.

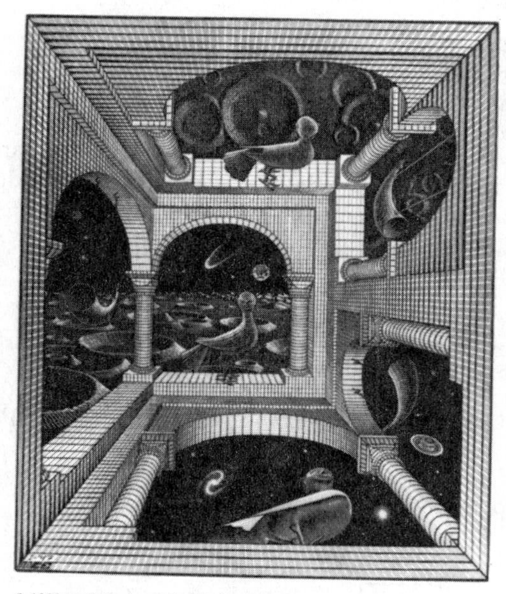

Figure 8–61

In Lesson Exercises 8.73 and 8.74, use your eyes. *No measuring allowed!*

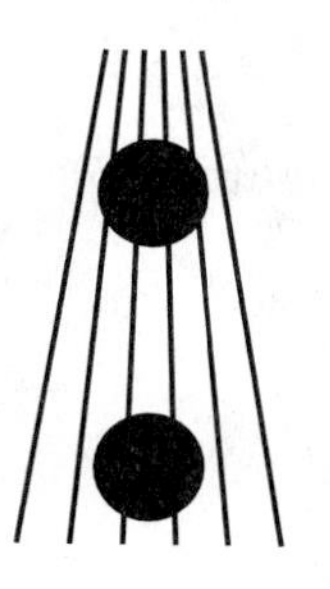

Figure 8–62

Lesson Exercise 8.73

Which one of the following is true?

(a) The circles in Figure 8-62 are the same size.

(b) The top circle is larger.

(c) The bottom circle is larger.

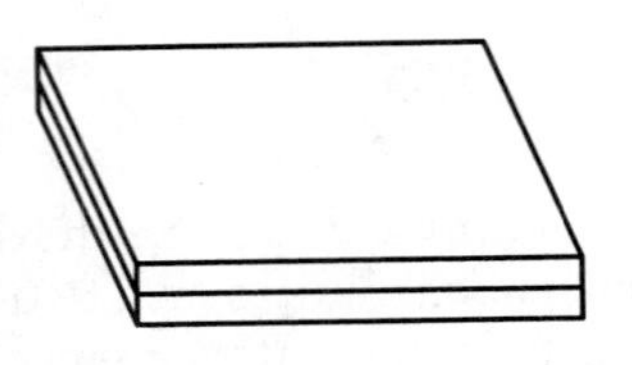

Figure 8–63

Lesson Exercise 8.74

Will a dime fit inside the parallelogram drawing in Figure 8-63?

In Lesson Exercises 8.73 and 8.74 we perceive depth. In Lesson Exercise 8.73, the white ball appears farther back. Our brains expect objects that are farther away to look smaller. Since the ball in back is actually the same size as the "closer" black ball, we see the ball in back as the larger one. In Lesson Exercise 8.74, we assume that the top of the parallelogram is farther back than the bottom, making it seem larger.

Although our eyes help us make sense of our surroundings, these examples illustrate that our visual perception is not always reliable for evaluating spatial relationships. On the other hand, once we understand visual perception, we can make sketches of solids on paper that are helpful in analyzing three-dimensional relationships.

DRAWING SOLIDS

When solids are drawn on paper, some parallel and perpendicular relationships are retained, while others are not. First, consider a cube. What follows is an introductory method for drawing a cube. You will eventually develop your own shortcuts.

Lesson Exercise 8.75

Get a pencil and paper, and follow these steps.

Step 1: Draw a square (Figure 8-64).

Figure 8–64

Step 2: Draw a second identical square "behind" it (Figure 8-65).

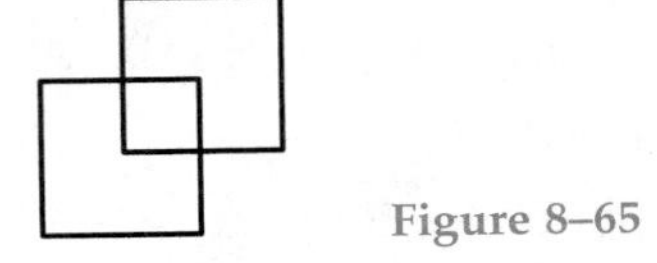

Figure 8–65

Step 3: Connect the corresponding vertices (Figure 8-66).

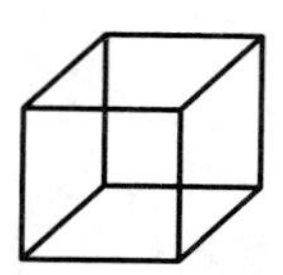

Figure 8–66

Step 4: Erase the three edges that cannot be seen from the front, or draw them as broken lines (Figure 8-67).

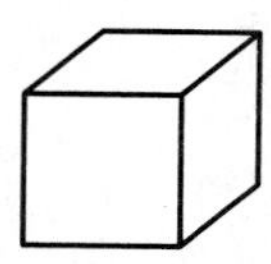

or

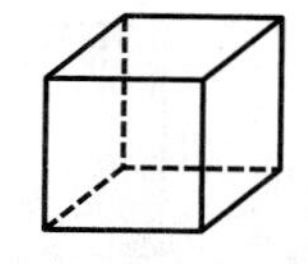

Figure 8–67

Lesson Exercise 8.76

(a) Draw another cube for practice.

(b) Label the vertices of your cube as shown in Figure 8-68.

(c) Name two perpendicular lines in the cube that are drawn perpendicular to each other.

(d) Name two perpendicular lines in the cube that are not drawn perpendicular to each other.

(e) Name two surfaces of the cube that are contained in parallel planes.

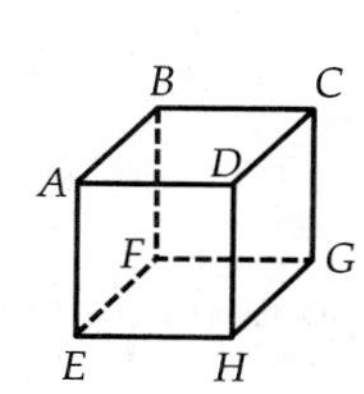

Figure 8–68

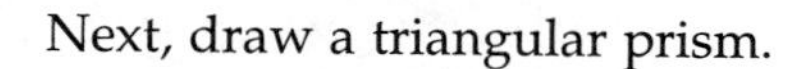

Next, draw a triangular prism.

Lesson Exercise 8.77

Step 1: Draw a triangle (Figure 8-69).

Figure 8–69

Step 2: Draw a second identical triangle above it (Figure 8-70).

Figure 8–70

Step 3: Connect the corresponding vertices (Figure 8-71).

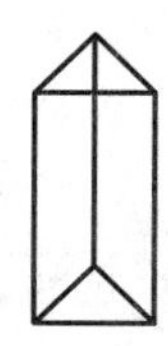

Figure 8–71

Step 4: Redraw as broken lines the part that cannot be seen from the front, as in Figure 8-72.

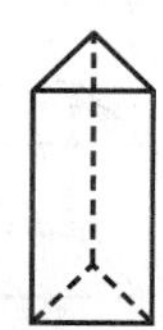

Figure 8–72

Lesson Exercise 8.78

Draw another triangular prism, for practice.

VIEWPOINTS OF SPACE FIGURES

We never see a solid in its entirety; we see it from a viewpoint (Figure 8-73).

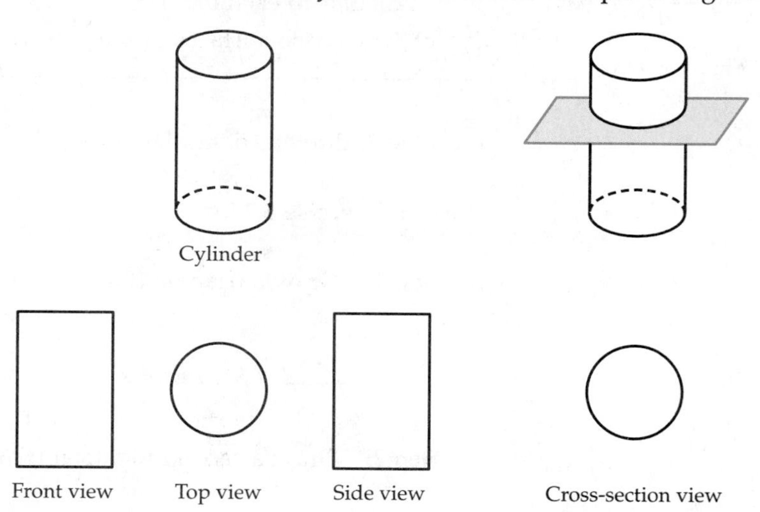

Figure 8–73

Artists, architects, and photographers study how objects appear from different viewpoints. Such knowledge enriches our understanding and our experience in a world replete with geometric shapes and spatial relationships. Architects draw plans using views (called elevations). A view accurately shows the proportions, whereas a perspective drawing does not.

Lesson Exercise 8.79

Name the space figure that has the following front view, top view, and side view.

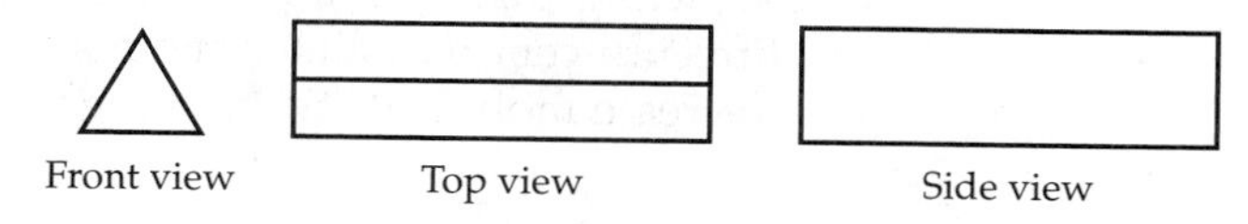

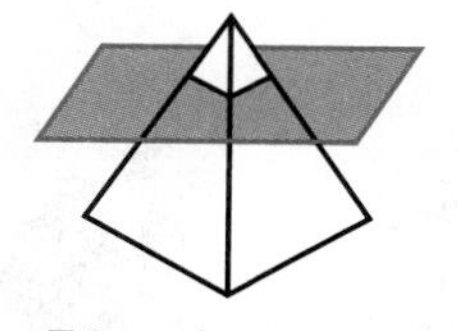

Triangular pyramid

Figure 8–74

Lesson Exercise 8.80

Draw a cross-sectional view of the space figure in Figure 8-74.

PERSPECTIVE DRAWING

In the Middle Ages in Europe, many artists painted scenes that had religious themes. They were not concerned about accurately representing people and objects in these symbolic works. In the late thirteenth, fourteenth, and fifteenth centuries, however, European painters became more interested in accurately drawing their surroundings, but they were not sure exactly how to do it.

LAC 95586, !5th century, Bartholemew the Englishman, Book of Properties, Amiens, Bibliotheque. © Giaraudon/ Art Resource, N.Y.

Figure 8–75

Lesson Exercise 8.81

What is unrealistic about the fifteenth-century painting shown in Figure 8-75?

The painting in Figure 8-75 does not realistically depict how chairs and desks look.

Renaissance painters learned how to portray three dimensions realistically in paintings. They realized that they needed geometry to solve the problem of depicting three dimensions on a flat canvas. Piero della Francesca, the great fifteenth-century painter and mathematician, used the idea of a vanishing point to create more realistic paintings, such as the one in Figure 8-76.

66149. Piero della Francesca, Flagellatum, Urbino. © Alinari/Art Resource, N.Y.

Figure 8–76

When the lines running *from the front to the back* of the picture are extended, they intersect at one point (the vanishing point) just below the middle of the picture.

Lesson Exercise 8.82

Find two parallel lines in the painting in Figure 8-76 that do not pass through the vanishing point. Explain why not.

Figure 8–77

How is the illusion of depth created in a drawing? In a drawing of railroad tracks that go off into the horizon, like those in Figure 8-77, the parallel rails appear to meet at a point on the horizon called the **vanishing point**. Drawings based upon a vanishing point are called **one-point perspective drawings**.

A second method of drawing a cube utilizes one-point perspective. This method is more difficult than the method of Lesson Exercise 8.75, but the result is more realistic.

Lesson Exercise 8.83

To draw a cube with one-point perspective, follow these steps.

Step 1: Draw the front of the cube and choose a vanishing point (Figure 8-78).

Step 2: Use a ruler to connect the vertices to the vanishing point (Figure 8-79).

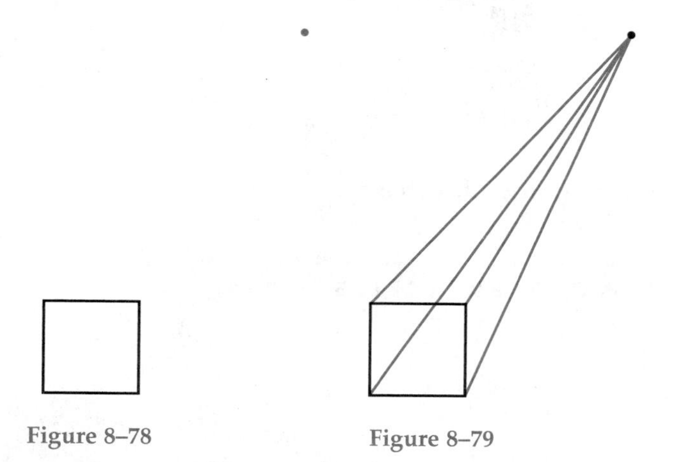

Figure 8–78 Figure 8–79

Step 3: Draw the back of the cube as shown in Figure 8-80.

Step 4: Darken the edges that would be seen from the front (Figure 8-81).

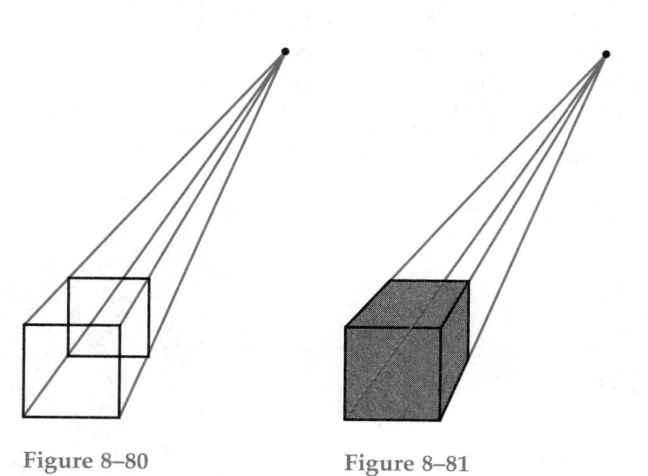

Figure 8–80 Figure 8–81

Figure 8–82

Lesson Exercise 8.84

Where is the vanishing point for the drawing in Figure 8-82?

ANSWERS TO SELECTED LESSON EXERCISES

8.73 (a)

8.74 No

8.76 (c) $\overline{AE}$ and $\overline{EH}$ (d) $\overline{AE}$ and $\overline{AB}$
(e) $ADHE$ and $BCGF$

8.79 Triangular prism

8.80 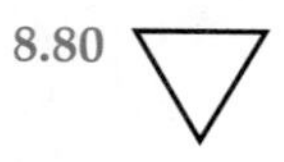

8.81 All of the furniture looks crooked.

8.82 Lines going from left to right across the floor or ceiling are drawn parallel and do not pass through the vanishing point. Only parallel lines from the front to back are drawn so that they meet at the vanishing point.

8.84

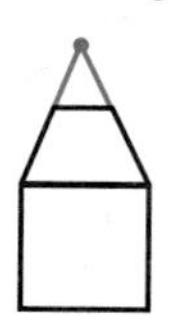

8.5 HOMEWORK EXERCISES

Basic Exercises

1.

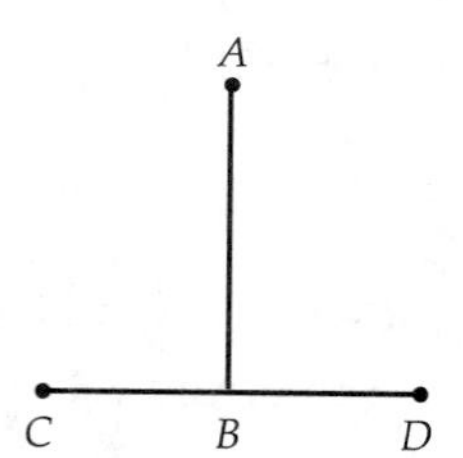

(a) Which appears to be longer, $\overline{AB}$ or $\overline{CD}$?
(b) Measure and find the correct answer.
(c) Why does our visual perception tend to elongate $\overline{AB}$?

2.

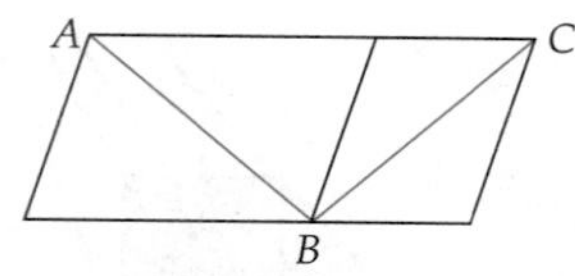

(a) Which one of the following statements appears to be true?
(1) $\overline{AB}$ and $\overline{BC}$ are the same length.
(2) $\overline{AB}$ is longer.
(3) $\overline{BC}$ is longer.
(b) Measure and find the correct answer.

3.

These two photos show the same person on opposite sides of the same room! Explain how this optical illusion is created.

4. Draw a cube.

5. Draw a triangular prism.

6. Here is another way to sketch a rectangular prism or cube.

Draw three line segments like these.

Draw every other edge parallel to one of these segments.

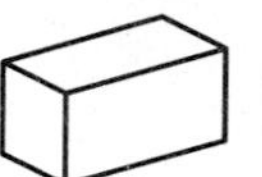

7. Draw a cylinder by following these steps.

Step 1: Draw an oval.

Step 2: Draw a second identical oval above the first one.

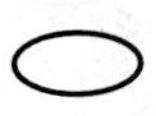

Step 3: Connect the ovals with two line segments.

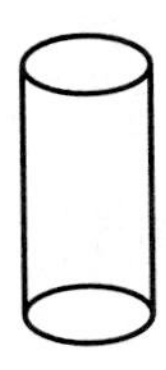

Step 4: Redraw as a broken line the part of the bottom oval that cannot be seen from the front.

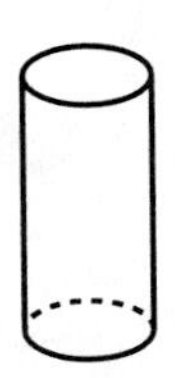

8. Draw a square pyramid by following these steps.

Step 1: Draw a base.

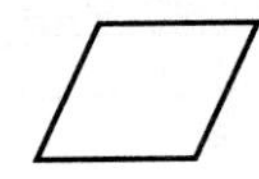

Step 2: Draw a point above it.

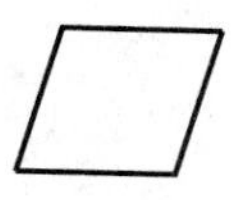

Step 3: Connect the vertices of the base to the point.

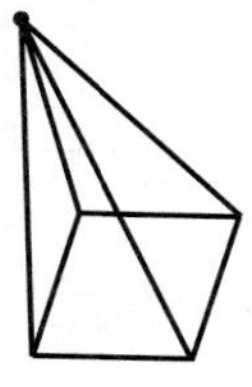

Step 4: Redraw the hidden edges as broken lines (or erase them).

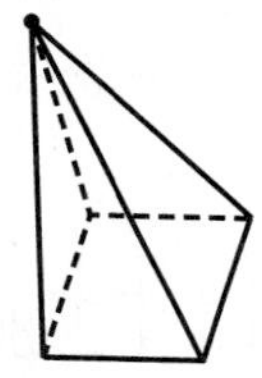

9. Draw a hexagonal prism.

10. Use a grid line like the one shown here to draw a cylinder that has diameter $d = 3$ units and height $h = 4$ units.

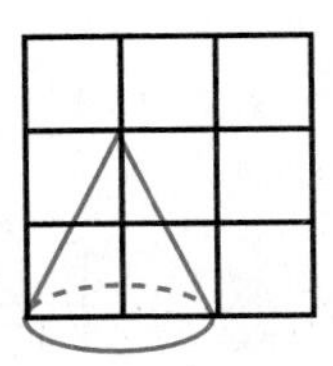

$d = 2 \quad h = 2$

11. (a) Draw a rectangular prism like the following one on a triangular grid.

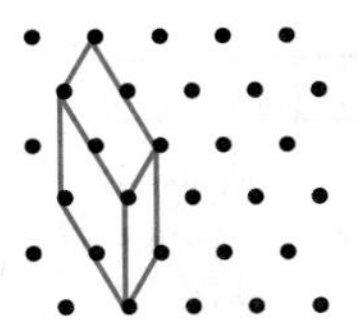

(b) Draw a different rectangular prism on the triangular grid.

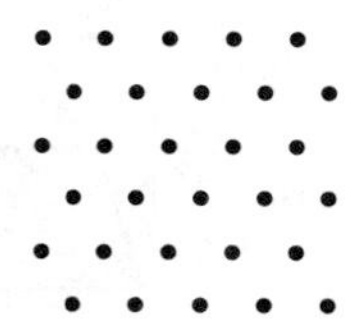

*12. Name the space figure that has the front view, top view, and side view shown.

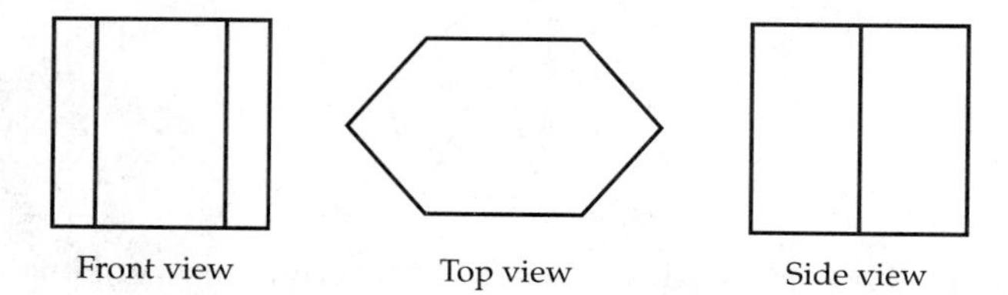

13. Following is a picture of a set of cubes, along with two-dimensional front and bottom views.

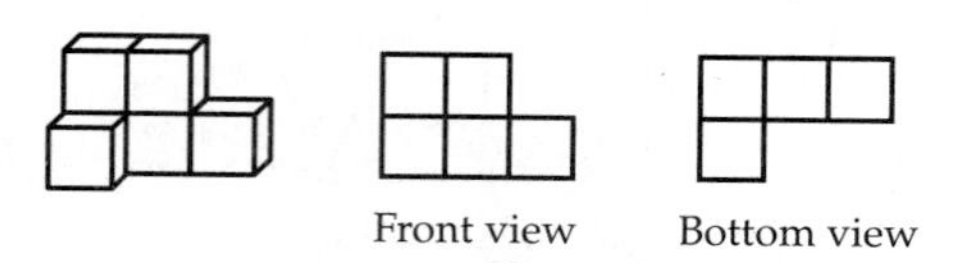

Draw the front and bottom views for each of the following stacks of cubes. (Assume that no cubes are hidden other than those that support visible cubes.)

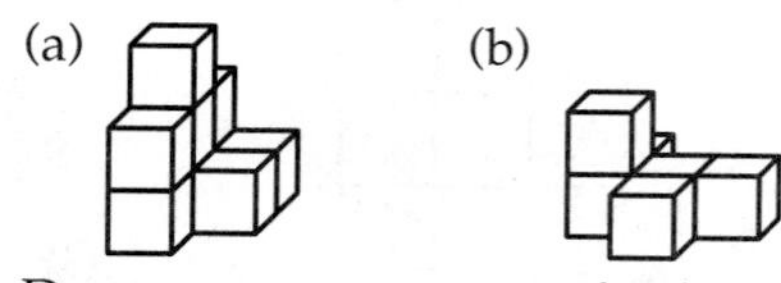

14. Draw an arrangement of cubes that has the following front and bottom views.

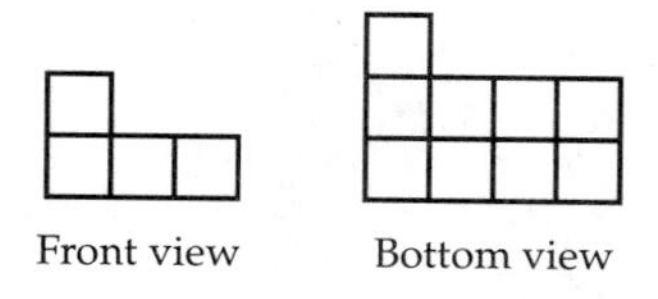

15. The following diagram shows the rear view of a building. Draw the front view.

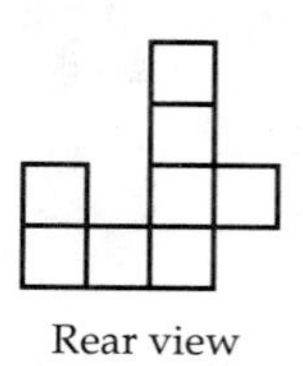

16. Sketch the front and top views of the building in which you live.

17. Draw the front and top views of the buildings.

18. In each case, identify the cross section cut by the plane.

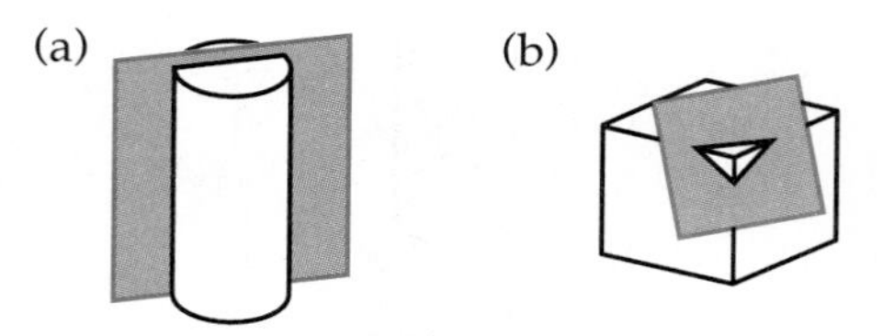

19. A cone is cut by planes as shown. What is the resulting cross section?

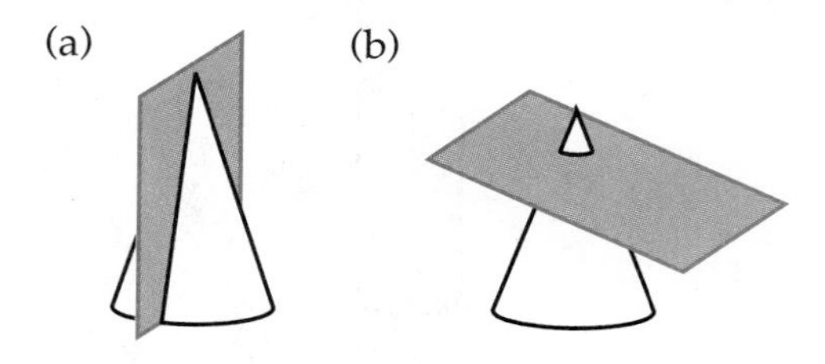

20. An engineering company designs the following piece. Draw a picture of the cross section.

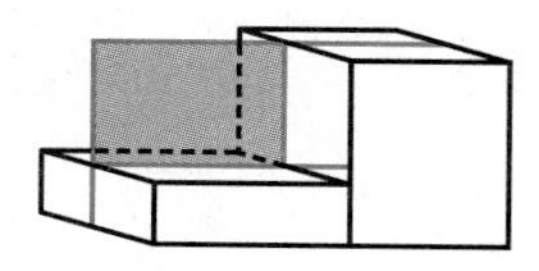

21. If a cube is intersected by a plane passing through the three points shown, what shape is the cross section?

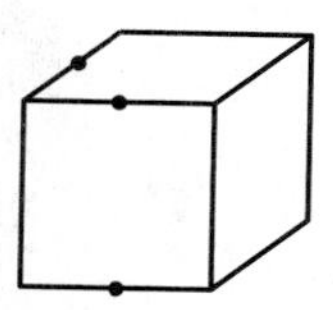

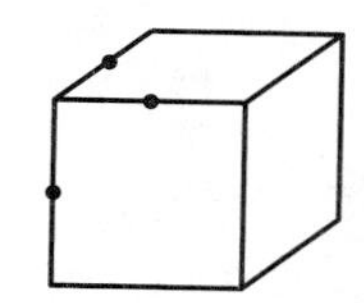

22. Name two figures other than a cube that have a square cross section. Use drawings to support your answers.

23. Why do drinking glasses usually have circular cross sections?

24. On four of its surfaces, a cube has the following four shapes. Look at the four shapes and two views of the cube.

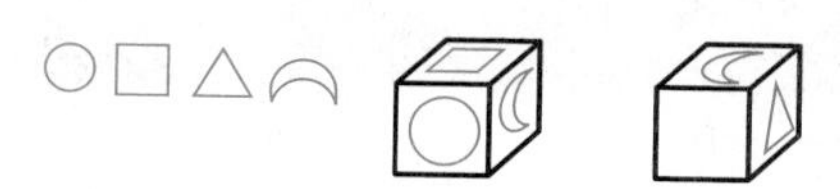

Which of the following also shows the cube correctly?

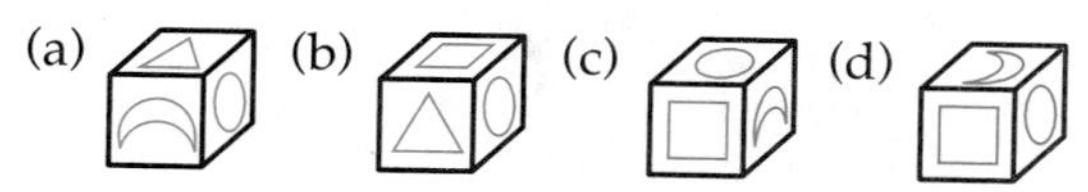

25. Find the vanishing point in the drawing.

26. (a) Locate the vanishing point in Albrecht Dürer's *St. Jerome in His Study.*

Dürer, St. Jerome in His Study, *engraving. © Marburg/Art Resource, N.Y.*

(b) Find two parallel lines that do not pass through the vanishing point. Explain why not.

27. You can draw block letters using a vanishing point, as shown. Draw your first or last name in block letters, using a vanishing point.

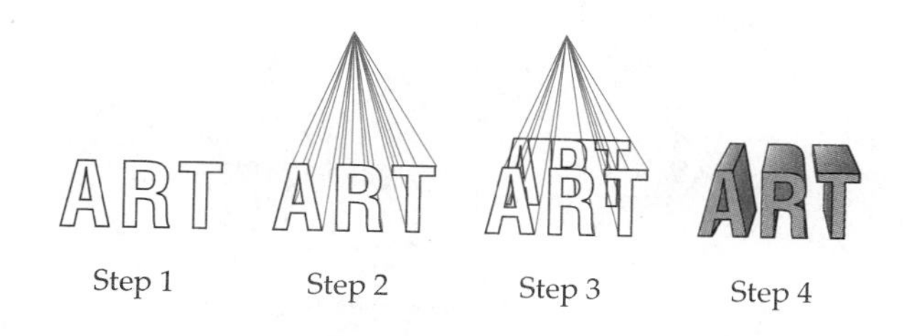

28. Draw a cube with one-point perspective.

29. In the game pick-up sticks, a player picks up the sticks one at a time, always taking the stick on top. In what order should the sticks in the diagram be picked up?

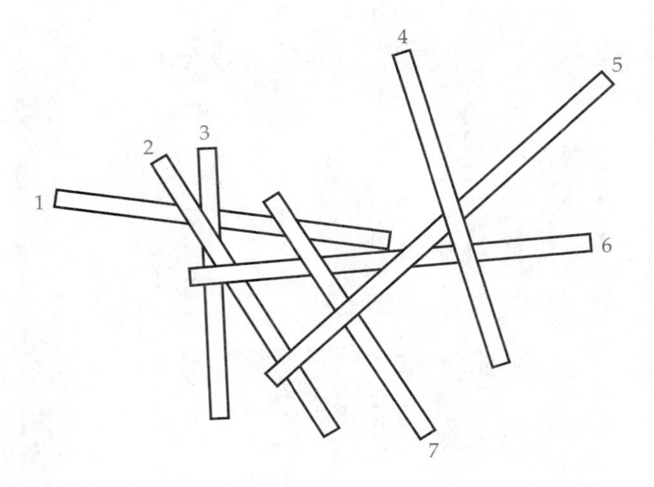

30. Explain why the man reading the book on perspective in the cartoon thinks it's hogwash.

By permission of Johnny Hart and Creators Syndicate, Inc.

31. This drawing by William Hogarth is called *False Perspective*. Make a list of all the perspective errors you can find.

The Bettmann Archives.

32. What is odd about the perspective in the following Escher drawing?

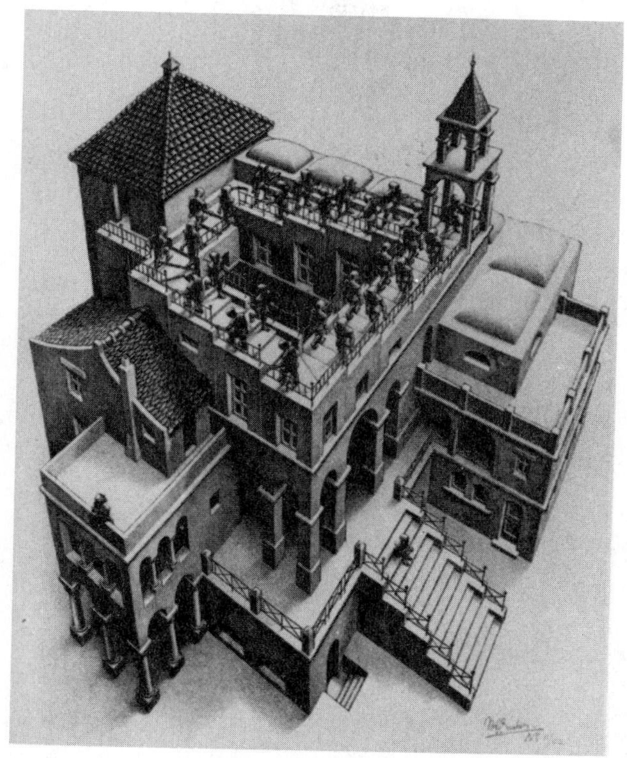

© 1990 M.C. Escher Heirs/Cordon Art—Baarn—Holland.

Extension Exercises

33. (a) Copy the figure.

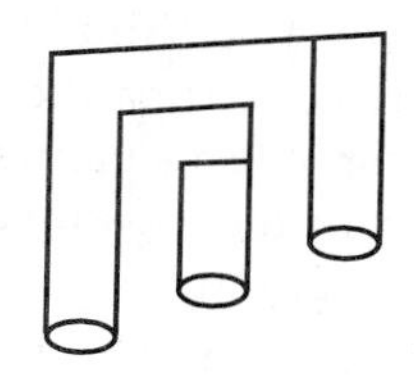

(b) What is odd about the figure?

34. Explain why it is impossible for $\overleftrightarrow{AB}$ to intersect the *congruent* boxes as shown.

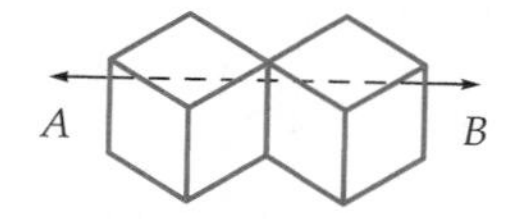

35. Draw a table with one-point perspective.

36. Consider the following problem. "Color six small squares so that no two colored squares are in the same column, the same row, or the same diagonal." Devise a plan and solve the problem.

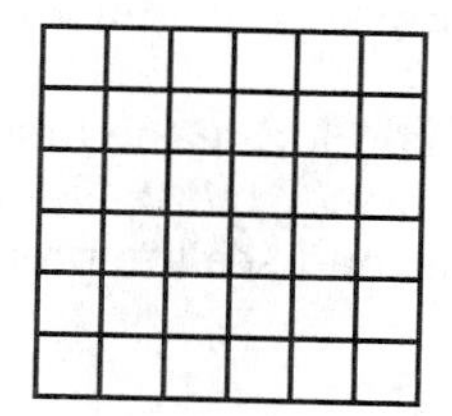

37. In drawing railroad tracks in perspective, how much space should there be between parallel tracks? Look at drawings 1 and 2, and answer the questions that follow.

(a) Which picture of the railroad tracks, 1 or 2, looks more realistic?

(b) Measure the distances between successive horizontal ties in each drawing. Which drawing has evenly spaced ties?

(c) The vanishing point can be used to draw tracks correctly. Following these 4 steps, make your own drawing with a ruler and pencil.

Step 1: Use a vanishing point to draw the rails. Then draw the front and back ties.

Step 2: Label the front $\overline{AB}$ and the back $\overline{CD}$.

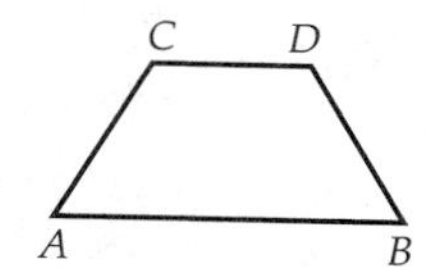

Step 3: To locate the tie between $\overline{AB}$ and $\overline{CD}$, draw $\overline{AD}$ and $\overline{BC}$. Then draw a new tie parallel to $\overline{AB}$, through the point where $\overline{AD}$ and $\overline{BC}$ intersect.

Step 4: Erase your guidelines.

38. How would you locate the first tie in the following drawing? (*Hint:* Use the midpoint of the second tie, and see the preceding exercise.)

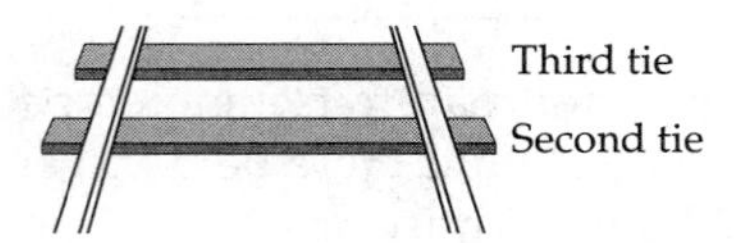

39. Make a perspective drawing with five railroad tracks.

40. Create your own drawing using one-point perspective.

41. A "building" is made by stacking the indicated number of cubes on each square. Draw front, top, and side (right) views of each building.

(a)

3	1
2	2

(b)

2	2	3
2	1	

42. Draw the front, top, and side views of
(a) a cup. (b) a table.

Special Exercises

43. Suppose that a wooden cube is painted blue on the outside, as shown, and then cut into smaller cubes.

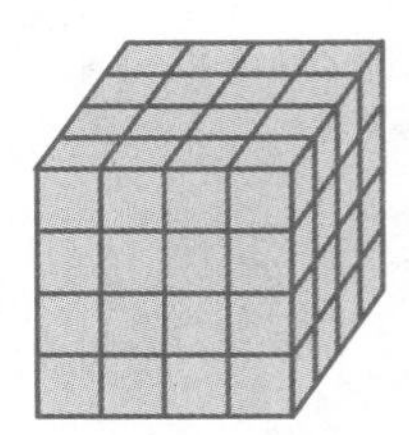

(a) How many small cubes would there be?
(b) How many cubes would have exactly one surface painted blue? (*Hint:* There is a pattern in the locations of these cubes.)
(c) How many cubes would have exactly two surfaces painted blue?
(d) How many cubes would have exactly three surfaces painted blue?
(e) How many cubes would have exactly four surfaces painted blue?
(f) How many cubes would have no surfaces painted blue?

44. Someone builds a 5-by-5-by-5 model of cubes, as shown, and paints the outside of the model.

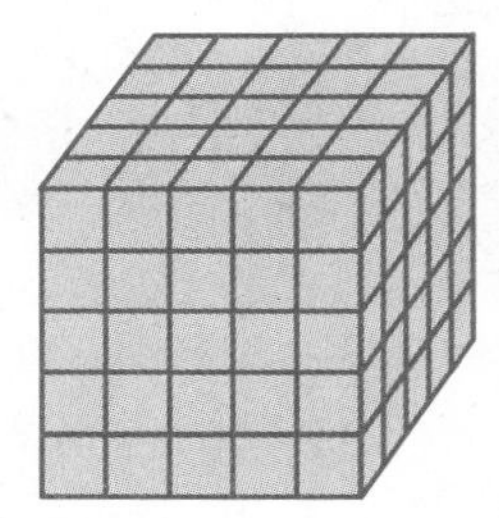

How many of the smallest cubes have each of the following?
(a) Exactly one face painted
(b) Exactly two faces painted
(c) Exactly three faces painted
(d) Exactly four faces painted
(e) Zero faces painted

45. Repeat the preceding question for the following models.
(a) 6-by-6-by-6 model
(b) n-by-n-by-n model

8.6 INTRODUCING LOGO

Logo, an excellent computer programming language for children, was developed by Seymour Papert and his co-workers at MIT in the 1960s. In his book *Mindstorms*, Papert explains why he developed Logo and why you, the reader, should use it.

First, Logo enables children to learn how to solve complex problems and think logically. In Logo, complex tasks can be broken down into simpler components. Second, Logo offers an ideal learning environment. The child has more control over the learning process and receives immediate feedback from the computer.

Logo is easier to begin than most computer languages. After 5 or 10 minutes of instruction, a child can start making drawings in Logo. In elementary schools, children use Logo to draw shapes and examine their properties.

BASIC COMMANDS

Once the computer is set up to do Logo, the following activities will introduce you to the language. Type the following:

```
DRAW
```

and press the RETURN or ENTER key. Pressing this key lets the computer know that you are finished typing a line of information. (*Note:* In Apple Logo, use CLEARSCREEN or CS instead of DRAW.)

Do you now see a triangle in the center of your screen? It is called a "turtle." Experiment with moving the turtle around by typing the following commands. Type:

```
FD 40
```

Be sure to leave a space between the D and the 4. Press RETURN or ENTER. FD stands for "forward." The number tells the turtle how far to move. Type:

```
DRAW
```

and press RETURN or ENTER. (Again, use CS in Apple Logo.) Now try typing an FD command with a different number. Enter:

```
FD 8
```

How far did the turtle move this time? (Throughout this chapter, when you are instructed to *enter* a command, type the command and press the RETURN or ENTER key.)

Lesson Exercise 8.85

Enter DRAW (CS in Apple Logo) and then enter FD 100. What happens this time?

You can also move the turtle by typing BK (back), RT (right turn), or LT (left turn), followed by a blank space and a number. Enter:

BK 60

and press RETURN or ENTER. The RT and LT commands are slightly different. You have to watch the turtle more closely. Enter:

RT 90

What happened? (Did you remember to press RETURN or ENTER afterward?)

Enter:

RT 180

Lesson Exercise 8.86

What does the 90 or 180 after RT stand for?

Lesson Exercise 8.87

Enter DRAW (or CS). Now try to draw a picture using a series of commands.

Lesson Exercise 8.88

Draw a square using Logo.

Figure 8-83 shows how the second-grade *Mathematics Plus* textbook introduces drawing squares and rectangles with Logo.

A square with sides of length 40 can be drawn by repeating the commands FD 40 and RT 90 four times. Logo has a REPEAT command that can save you from having to write the repeated command over and over. Enter:

DRAW
REPEAT 4 [FD 40 RT 90]

The last command repeats the moves inside the brackets four times.

Lesson Exercise 8.89

Draw a 60° angle after filling in the correct number of degrees in the right turn on the second line.

FD 40
RT
FD 40

Try your commands on the computer.

Name ______________________________ Technology Connection
Computer

Use a computer.

Work with a friend.

Use the LOGO turtle to make a line,
a square, and a rectangle.
In the box, draw what you made.

1. Make a line.
Enter RT 90 FD 80

2. Make a square.
Enter FD 70 RT 90 FD 70 RT 90 FD 70 RT 90 FD 70

3. Make a rectangle.
Enter FD 50 LT 90 FD 80 LT 90 FD 50 LT 90 FD 80

4. Make a square or rectangle of your own. Write the commands you use. In the box, draw a shape like the one on the computer. **Answers will vary.**

Enter

FD_____RT_____FD_____RT_____FD_____RT_____FD_____

Figure 8–83

Lesson Exercise 8.90

Consider the following series of commands.

FD 50
RT 120
FD 50
RT 120
FD 50
RT 120

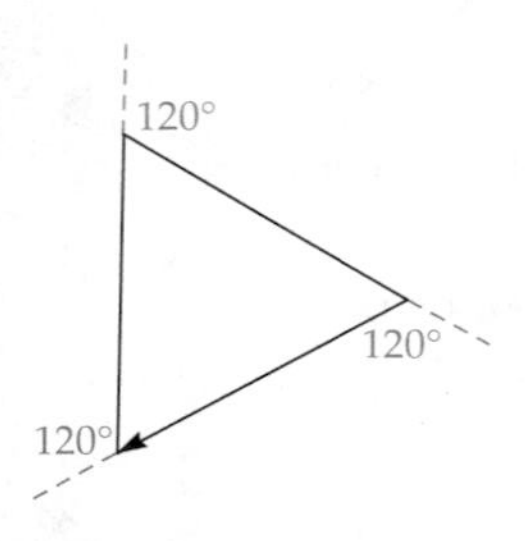

Figure 8–84

(a) Without the aid of the computer, draw the result of these commands.

(b) Using REPEAT, write a single line that is equivalent to the six given lines.

(c) Clear the screen. Enter your REPEAT command and see what results.

DRAWING POLYGONS

So far, you've seen how to draw an equilateral triangle and a square. Were you surprised that the equilateral triangle is drawn with RT 120 instead of RT 60? (If not, *I'm* surprised.) Figure 8-84 shows why RT 120 was used.

The angle measure of the turns is related to the *exterior* angle measure of the regular polygon. In a square, the exterior and interior angles happen to have the same measure (Figure 8-85).

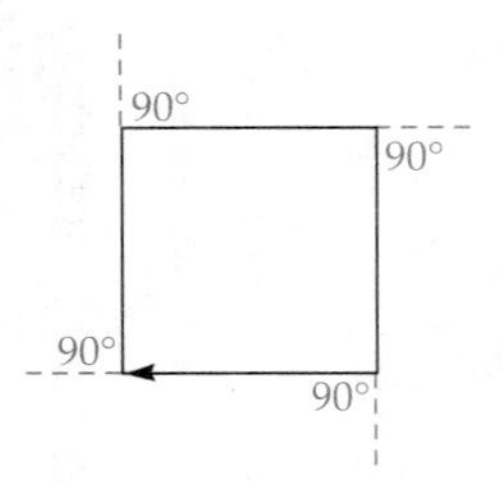

Figure 8–85

Lesson Exercise 8.91

(a) Look at the regular pentagon drawn in Figure 8-86. Write a REPEAT command that will draw it with sides of length 30.

(b) Enter your answer to part (a) into the computer and see if it works.

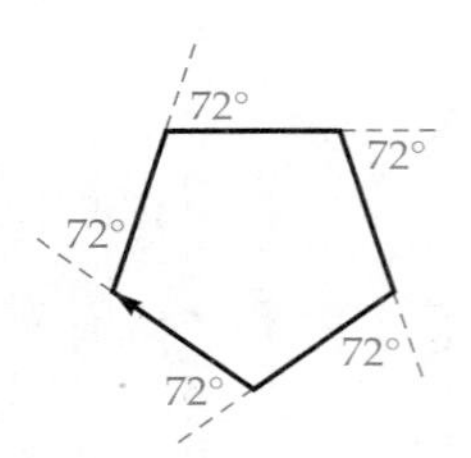

Figure 8–86

How would you draw a regular hexagon?

Cooperative Lesson Exercise 8.92

(a) What is the measure of each interior angle of a regular hexagon (Figure 8-87)? (*Hint:* See Section 8.3.)

(b) What is the measure of each exterior angle in Figure 8-88?

(c) Write a Logo command that draws a regular hexagon.

(d) Enter your answer to part (c) into the computer and see if it works.

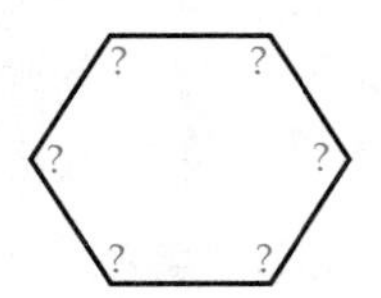

Figure 8–87

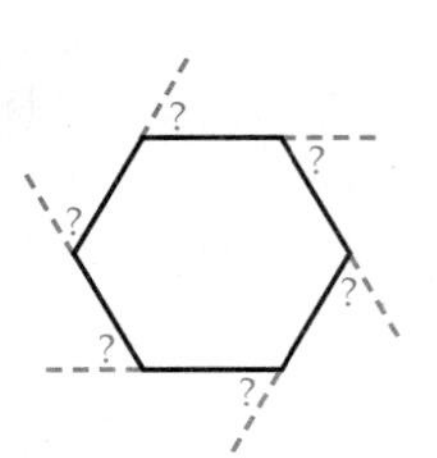

Figure 8–88

Is there a pattern in the angle turns used in drawing regular polygons? You be the judge. Examine the following table.

Drawing Regular Polygons

Sides	Angle of Logo Turn
3	120°
4	90°
5	72°
6	60°
N	?

Lesson Exercise 8.93

What is the pattern in the preceding table? If you understand it, give the angle (in degrees) of Logo turns to draw an N-sided regular polygon.

When you draw a polygon that has a large number of sides, what shape does it resemble? Try drawing a 360-sided polygon!

Lesson Exercise 8.94

(a) Enter REPEAT 360 [FD 1 RT 1].

(b) What shape does the 360-sided polygon approximate?

If you are finished for the day, enter BYE to exit Logo.

```
BYE
```

GLOSSARY OF LOGO COMMANDS

Here is a reference list of Logo commands. When you have some free time at the computer, try out some new commands.

Command	Abbreviation	Function	Example
FORWARD	FD	Moves turtle forward	FD 60
BACK	BK	Moves turtle back	BK 50
RIGHT	RT	Turns the turtle to *its* right	RT 120
LEFT	LT	Turns the turtle to *its* left	LT 90
DRAW (CS in Apple Logo)		Clears the screen, returns turtle to the center	DRAW
PENUP	PU	Turtle moves without leaving a trail	PU
PENDOWN	PD	Turtle leaves a trail when it moves	PD
HOME		Returns turtle to center of screen	HOME
SHOWTURTLE	ST	Shows turtle shape on screen	ST
HIDETURTLE	HT	Conceals turtle shape	HT
ERASE procedure (ERASE "procedure in Apple Logo)		Erases procedure from memory	ERASE SQUARE
BYE		To exit Logo	BYE

The following commands are used in the Logo editor.

Command in Logo Editor	Function
EDIT	Enters edit mode
EDIT procedure (EDIT "procedure in Apple Logo)	Enters edit mode for a procedure
CTRL-C (F2 key in PC Logo)	Computer accepts material from the editor and exits the editor

ANSWERS TO SELECTED LESSON EXERCISES

8.86 The number of degrees in the turn

8.89 120°

8.91 (a) *Hint:* Use RT 72.

8.92 (a) 120° (b) 60°

8.93 $\frac{360°}{N}$

8.94 (b) A circle

8.6 HOMEWORK EXERCISES

Basic Exercises

1. Sketch the figure that each of the following Logo programs would draw. Check your answer on a computer.

(a)
```
RT 45
FD 50
RT 90
FD 50
```

(b)
```
LT 60
FD 20
RT 120
FD 20
RT 120
FD 20
LT 30
FD 30
LT 30
FD 20
RT 120
FD 20
RT 120
FD 20
```

2. Sketch the figure that each of the following Logo programs would draw. Check your answer on a computer.

(a)
```
RT 135
FD 60
LT 45
FD 40
```

(b)
```
FD 40
LT 45
FD 30
RT 90
FD 30
RT 90
FD 30
LT 135
FD 30
```

3. What letter does the following Logo program draw?

```
RT 90
FD 20
BK 40
FD 20
RT 90
FD 50
```

4. What letter does the following Logo program draw?

```
FD 40    RT 90    FD 20    BK 20
RT 90    FD 20    LT 90    FD 15
```

5. (a) What is the measure of each interior angle of a regular octagon?

(b) What is the measure of each exterior angle?

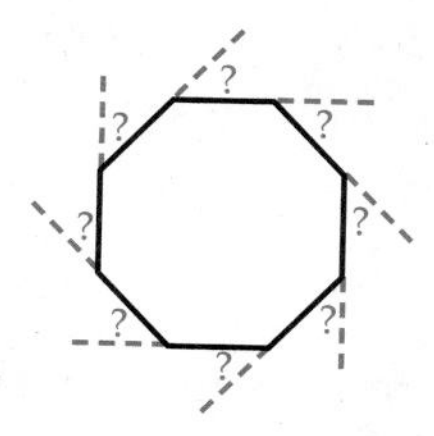

(c) Write a Logo command that draws a regular octagon.

(d) Enter your answer to part (c) into the computer and see if it works.

6. Write a Logo command that draws a 10-sided polygon.

7. Write a Logo program that draws each picture shown.

(a) 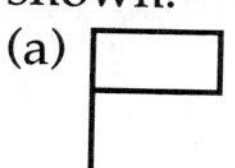

(b) 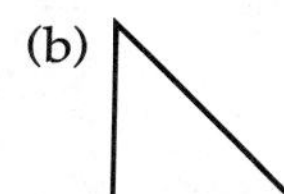

8. Write a Logo command that draws a figure that approximates a circle.

9. Write a Logo program that draws a letter M.

10. Write a Logo program that draws each picture shown.

(a)

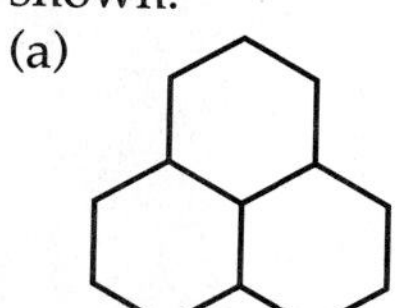

(b) 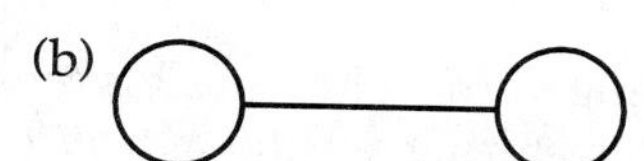

11. Write a single REPEAT command that draws the picture shown.

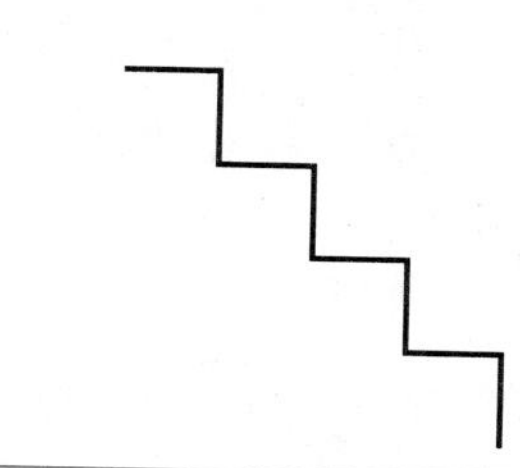

Extension Exercises

12. Experiment with the command REPEAT _____ [FORWARD 1 RIGHT _____] by filling in different numbers in the blanks.

13. Write a single REPEAT command that draws the star.

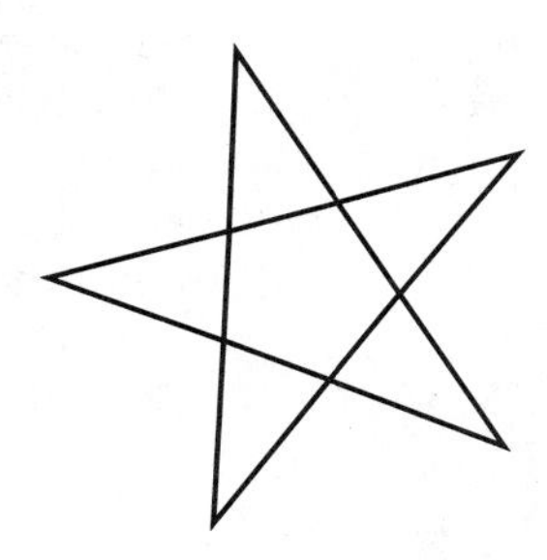

14. Write a Logo program that draws a square and its diagonals.

15. Complete the proof that the sum of the exterior angle measures of a triangle is 360°.

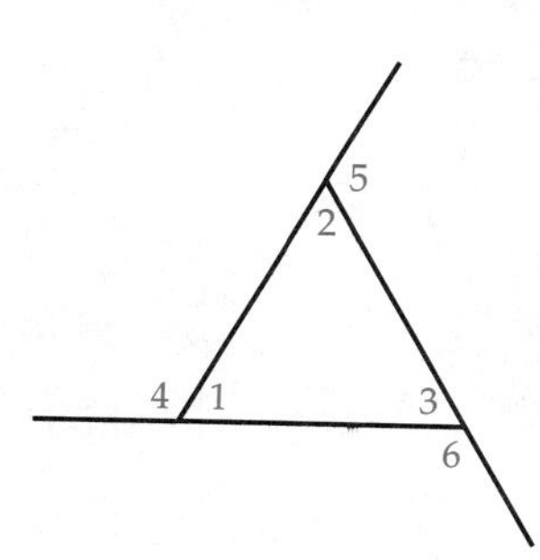

$m\angle 4 + m\angle 1 =$ ________ degrees,
$m\angle 5 + m\angle 2 =$ ________ degrees, and
$m\angle 6 + m\angle 3 =$ ________ degrees.

Combine these equations to obtain $m\angle 4 + m\angle 5 + m\angle 6 + m\angle 1 + m\angle 2 + m\angle 3 =$ _____ degrees. Since $m\angle 1 + m\angle 2 + m\angle 3 =$ _______ degrees, we know that $m\angle 4 + m\angle 5 + m\angle 6 =$ ________ degrees.

16. Prove that the sum of the exterior angle measures of a quadrilateral is 360°. Use the preceding exercise as a model.

8.7 LOGO PROCEDURES

If you want to program the computer to do a more complex task, construct a "procedure" that performs a sequence of simpler tasks.

PROCEDURES

Suppose that you want the computer to draw a square. Instead of telling it to draw each part of the square separately, you can write a Logo program that gives all the commands needed to draw a square. If you store this program, called a **procedure**, in the computer, you can use it whenever you want the computer to draw a square.

The turtle knows the words FORWARD and BACK. Does it know the word SQUARE? Type:

```
SQUARE
```

and press the ENTER or RETURN key.

Logo does not understand this word—yet. You can give the word a meaning in Logo.

Lesson Exercise 8.95

Write a command to draw a square.

To construct a Logo procedure to draw a square, begin by going to the editor in Terrapin Logo. (Skip this step in PC Logo and Apple Logo.) Enter:

```
EDIT
```

Next, in all versions of Logo, start the procedure with a title.

```
TO SQUARE
```

Note how the prompt from the computer has changed from ? to >. This means that you are *inside a program*, and the computer is storing whatever you type into its memory.

Now enter a command to draw a square.

```
REPEAT 4 [FD 40 RT 90]
```

Finally, end the procedure.

```
END
```

If you are in the Terrapin Logo editor, press CTRL and C together to store the program and leave the editor.

(*Note:* If you are ever stuck in a program or want to redo it, the easiest way to exit from a program (and change from > back to ?) is to enter END.)

Now enter SQUARE again and see if the computer understands SQUARE now.

```
SQUARE
```

Aha!

Defining a Logo Procedure

1. EDIT (Terrapin Logo only)
2. Type TO followed by a name.
3. Type all the commands needed to perform the procedure.
4. Type END.
5. Press [CTRL] and [C] (Terrapin Logo only)

You have repeated commands. What happens when you repeat a procedure? Enter:

```
DRAW
REPEAT 6 [SQUARE RT 60]
```

Remember to type CS instead of DRAW if you are using Apple Logo.

Lesson Exercise 8.96

Experiment by combining SQUARE with other commands to make designs.

If you are finished with this procedure, you can erase it from the computer by typing ERASE SQUARE. Enter:

```
ERASE SQUARE
SQUARE
```

The computer has lost its memory!

Lesson Exercise 8.97

Write a procedure TRIANGLE that draws an equilateral triangle with sides of length 40.

VARIABLES

When you want to vary the dimensions of a figure, use variables in your Logo program. Create a new SQUARE procedure that allows you to input the length L of the sides of the square. Enter:

```
DRAW
EDIT (in Terrapin Logo only)
TO SQUARE :L
REPEAT 4 [FD :L RT 90]
END
```

If you are in Terrapin Logo, press CTRL and C together. Now the computer has memorized the SQUARE program. The colon after SQUARE tells the computer that a number will be given to enter in memory location L. Enter:

```
SQUARE 20
SQUARE 40
SQUARE 60
```

To draw a rectangle, you need to input two numbers: a length and a width.

Lesson Exercise 8.98

(a) On a piece of paper, write all the steps needed in Logo to draw the rectangle shown.

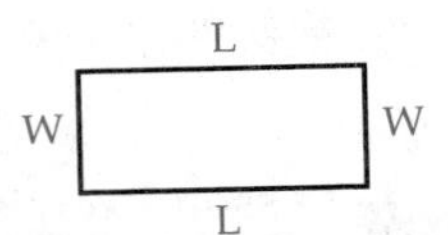

Start with
FD :W

(b) Now type into the computer a procedure RECTANGLE :L :W that will draw a rectangle using *any length and width* that are given as input when the procedure is called. It should begin:

TO RECTANGLE :L :W

(c) Once the procedure is written, run your procedure on the computer using a command such as

RECTANGLE 40 30

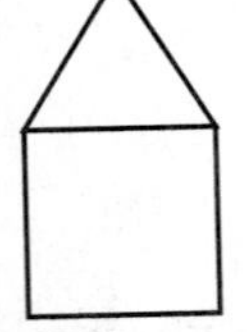

"A house"

Figure 8–89

COMBINING PROCEDURES

Often, a complicated task can be broken down into simpler tasks. You can draw the "house" in Figure 8-89 by drawing a square and a triangle.

Do you still have the TRIANGLE (Lesson Exercise 8.97) and SQUARE :L procedures in the computer's memory? If not, type them in. Then enter:

```
EDIT (in Terrapin Logo)
TO HOUSE
SQUARE 40
FD 40 RT 30      (moves from bottom left corner to top left corner of square)
TRIANGLE
END
HOUSE
```

If you are in Terrapin Logo, press CTRL and C together. Did this program work? Depending upon how your triangle program begins, you may have to make an adjustment. (Adjustments may also be needed with certain computers and monochrome monitors.) The line "TRIANGLE" in this program is a subprocedure. The **subprocedure** is named in the program, but the instructions for it are outside the program.

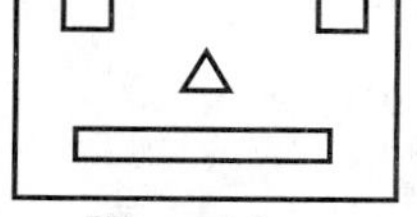

Happy? face

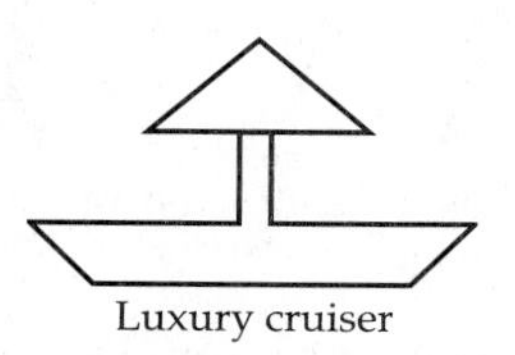

Luxury cruiser

Figure 8–90

A PROJECT: CREATING A DESIGN

Lesson Exercise 8.99

Write a longer procedure that makes a design using triangles, squares, and rectangles. The designs in Figure 8-90 are possibilities. Use them as inspiration to create your own design. (*Hint:* To move the turtle invisibly from one place to another, use PENUP so that nothing is drawn on the screen. Then use PENDOWN when you want to resume drawing on the screen.)

ANSWERS TO SELECTED LESSON EXERCISES

8.95 See the preceding lesson.

8.97 *Hint:* An equilateral triangle is a regular polygon. Find the measure of its exterior angles.

8.7 HOMEWORK EXERCISES

Basic Exercises

1. (a) Type in a procedure SQUARE that draws a square with sides 35 units long.
(b) Try the following.

```
REPEAT 6 [RT 60 SQUARE]
```

2. (a) Type in a procedure PENTA that draws a pentagon with sides 35 units long.
(b) Try the following.

```
RT 90
REPEAT 6 [FD 10 RT 60 PENTA]
```

3. (a) Write a procedure HEXA :S that draws a regular hexagon with sides of any length *S*. The length will be given as input when the procedure is run.
(b) Run your procedure on a computer.

4. (a) Write a procedure OCTA :S that draws a regular octagon with sides of any length *S*. The length will be given as input when the procedure is run.
   (b) Run your procedure on a computer.
5. Guess what each of the following will draw. Then run each one on a computer and see what happens.
   (a) REPEAT 5 [FD 50 RT 144]
   (b) REPEAT 12 [RT 60 REPEAT 3 [FD 30 RT 90]] (Do not press RETURN or ENTER until you finish typing the entire line.)
6. Write a Logo procedure that draws the six-pointed star.

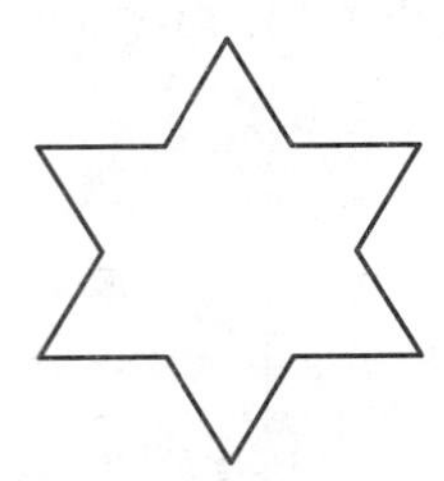

7. Write a procedure that draws an 80° angle and its bisector.

Extension Exercises

8. (a) Write a procedure POLYGON :N that draws a regular polygon with *N* sides. *N* will be input when the procedure is run.
   (b) Run your procedure on a computer.
9. Write a Logo procedure to draw the following.

(a)

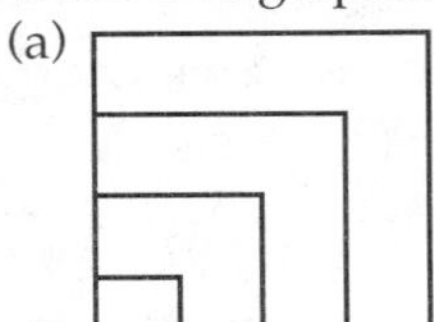

(b)

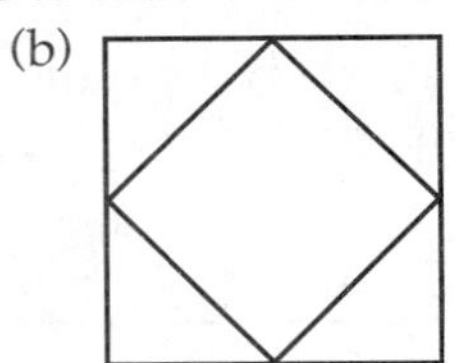

10. On graph paper, draw a cube like the one shown here. Use the coordinates to write a procedure in Logo that draws a cube.

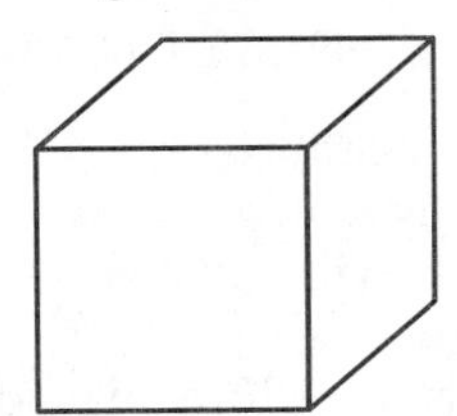

11. Write a Logo procedure to draw the following.

(a) (b)

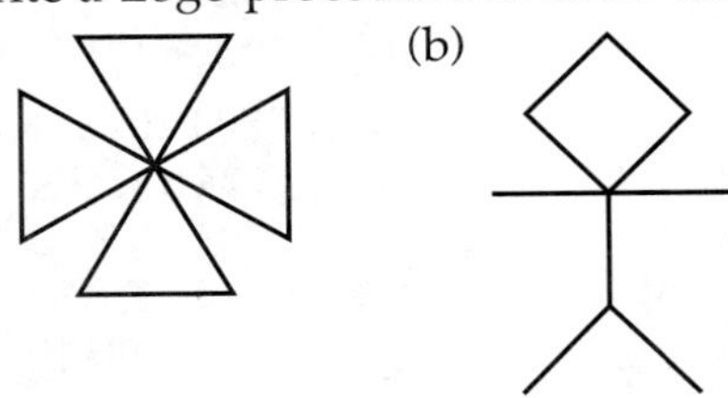

12. Write a program to draw the following.

13. Write a program to draw *one* of the following shapes.

(a) (b)

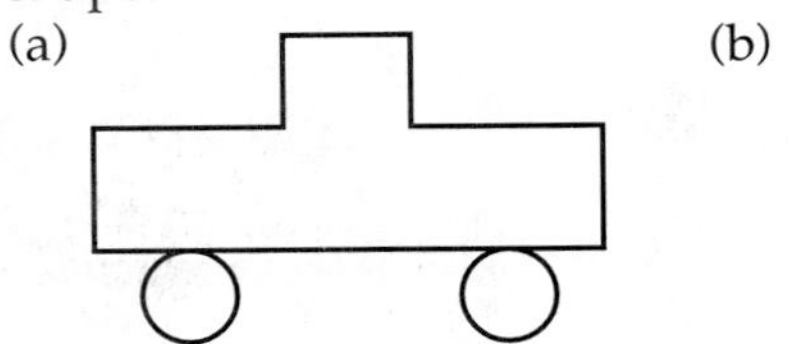

14. Write a procedure that makes a design using triangles, squares, and rectangles.
15. In this exercise, you will create a program that writes the word "Logo."
   (a) Write a procedure to draw an L.
   (b) Write a procedure to draw a letter o; make it a little smaller than the L.
   (c) Write a procedure to draw a letter g; make it about the same size as the o.
   (d) Write a master procedure that writes the word "Logo."
16. (a) Write a program
   PGRAM :SIDE1 :SIDE2 :ANGLE
   that draws a parallelogram with two given lengths and one angle measure.
   (b) Why is only one angle measure needed?

Special Exercise

17. A **recursion** procedure calls the same procedure within itself, causing the program to repeat over and over again. By using a variable, you can change the procedure each time it repeats.

(a) Type the following into a computer.

```
TO TRIANGLE :X
  RT 45
  REPEAT 3 [FD :X RT 120]
  TRIANGLE :X+3
END
```

(b) Type TRIANGLE to perform your program. Press CTRL and G at the same time when you want the program to stop.

(c) Write a similar recursion program that draws a series of squares of different sizes.

18. (a) Without running it, show what the following set of commands would draw.

```
FD 20   RT 15   BK 25   LT 35
```

(b) Type the following program into a computer.

```
TO SUN
FD 20
RT 15
BK 25
LT 35
SUN
END
```

(c) Type SUN to run your program. Press CTRL and G at the same time when you want the program to stop.

(d) Explain the output of the program.

SUMMARY

Geometry began when people first observed three-dimensional shapes in their surroundings. The surfaces of three-dimensional objects suggested two-dimensional shapes. Young children today discover geometry in the same way. "In learning geometry, children [K–4] need to investigate, experiment, and explore with everyday objects and other physical materials" (NCTM, *Standards*, p. 48).

Ideal geometric shapes do not exist in the world. We abstract them from our surroundings. Geometric shapes then take on a life of their own. It was the work of the ancient Greeks, culminating with Euclid's *Elements* around 300 B.C., that gave us the logically organized Euclidean geometry. Euclidean geometry begins with the undefined terms "point," "line," "plane," and "space" and builds geometry from there. Formal geometry proceeds in a certain order because one cannot, for example, define a prism without first being familiar with points, line segments, and polygons.

In geometry, shapes are classified. Within each classification, mathematicians look for common properties. Once these properties are established, they can be applied to all kinds of objects in the world that approximate the shape being studied. This chapter contains classifications of angles, lines, polygons, and space figures.

In learning geometry, the van Hieles suggest first working on recognizing whole figures, then analyzing their properties, and then writing precise definitions and identifying relationships between different classes of figures. The van Hiele levels work especially well in the study of quadrilaterals.

Geometry offers interesting opportunities for investigations. For instance, the total number of diagonals of various polygons is related to the number of sides. The sums of the angle measures in convex polygons follow a pattern. Euler's formula describes another pattern in the numbers of faces, vertices, and edges in any polyhedron.

In studying space figures, we confront the difficulty of visualizing spatial relationships represented by diagrams and the challenge of portraying three-dimensional shapes on two-dimensional surfaces. Renaissance mathematician-artists were the first to develop a system of perspective drawing based upon their understanding of the way we see the world. A teacher needs spatial ability in order to represent solids on the blackboard.

Geometry can be studied on the computer using Logo, probably the best computer language yet developed to teach young children. Children can quickly start using it, and it does not require a large vocabulary. After a short time, a student can grapple with some challenging geometry problems and create some beautiful designs. Logo also teaches the concepts of top-down programming and subroutines (procedures).

STUDY GUIDE

To review Chapter 8, see what you know about each of the following ideas or terms that you have studied. You can also use this list to generate your own questions about the chapter.

The NCTM Curriculum Standards and Geometry

Selected NCTM Curriculum Standards

The following standards come from the NCTM document.

- Identify, describe, compare, and classify geometric figures.
- Recognize and appreciate geometry in their world.
- Understand and apply geometric properties and relationships.
- Visualize and represent geometric figures with special attention to developing spatial sense.

1. Describe how each standard listed relates to the material you studied in Chapter 8.
2. Select any current elementary-school mathematics textbook, and describe sample lessons or exercises that illustrate each standard listed.

GEOMETRY IN ELEMENTARY SCHOOL

The following chart shows at what grade levels selected geometry topics typically appear in elementary-school mathematics textbooks.

Topic	Typical Grade Level in Current Textbooks
Two- and three-dimensional figures	1, 2, 3, 4, 5, 6
Measuring angles	5, 6
Points, lines, and planes	4, 5, 6
Polygons (five or more sides)	3, 4, 5, 6
Radius, diameter, chord	4, 5, 6
Faces, edges, and vertices	4, 5, 6
Drawing space figures	6

REVIEW EXERCISES

1. Why does the *formal* study of geometry begin with points and lines rather than with three-dimensional figures?

2. Name two ways in which a sheet of paper differs from a plane.

3. Give a simpler name for each of the following.
   (a) $\overrightarrow{BA} \cup \overrightarrow{BC}$
   (b) $\overleftrightarrow{AD} \cap \overline{BE}$

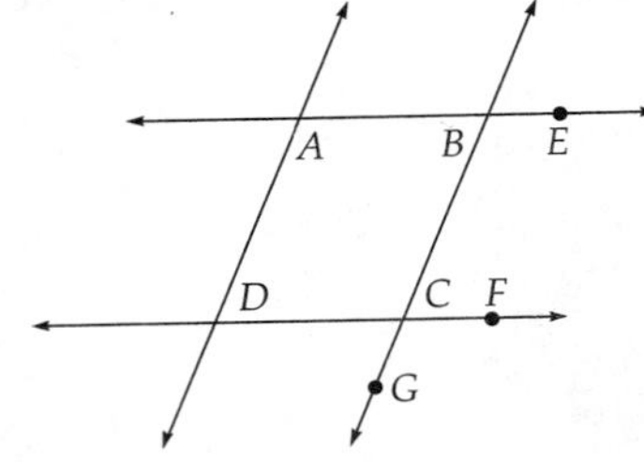

4. Point A is on line b. How many different planes contain point A and line b?

5. True or false? Through a point not on a given line, there is exactly one line that is skew to the given line.

6. (a) True or false? Two intersecting lines cannot both be parallel to a given plane.
   (b) In your home, identify a model of this situation that supports your answer.

7. Make an accurate drawing of a cube using one-point perspective.

8. Locate the next railroad tie in correct perspective.

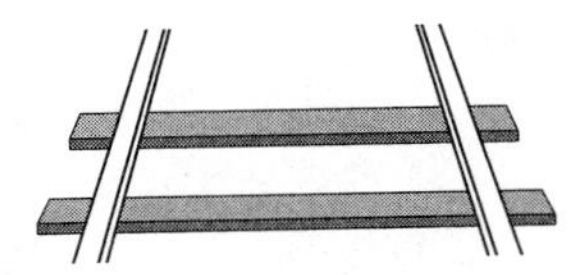

9. Draw a Venn diagram showing the relationship among rectangles, squares, and trapezoids.

10. Is every rhombus a type of square? If not, give a counterexample.

11. Tell whether each of the following shapes must, can, or cannot have four equal angle measures.
    (a) Rhombus (b) Square
    (c) Trapezoid (d) Rectangle
    (e) Parallelogram

12. Fill in each blank with "All," "Some," or "No."
    (a) ______ parallelograms are rhombuses.
    (b) ______ squares are rectangles.

13. Sketch two parallelograms that intersect at exactly five points.

14. Use the angle sum property of a triangle and a drawing to explain why the interior angle measures of a pentagon add up to 540°.

15. Draw a picture and explain why a regular hexagon does or does not tessellate the plane.

16. How many faces, vertices, and edges does a hexagonal pyramid have?

17. A certain prism has 21 edges. How many faces does it have?

18. Assume *ABCDE* is a regular pentagon. Fill in all the missing interior angle measures in the diagram.

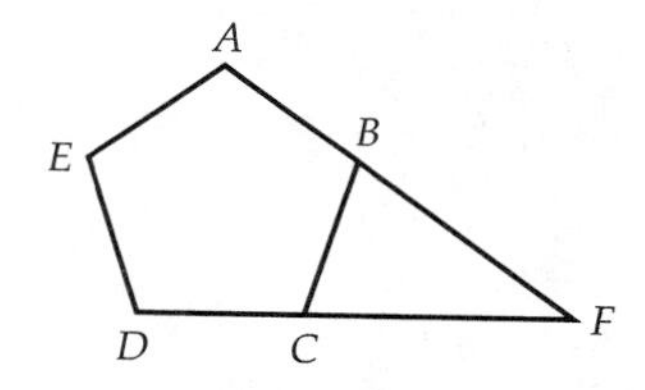

19. How many degrees are in each angle of a regular 9-sided polygon?

20. How many triangles of all sizes are in the figure?

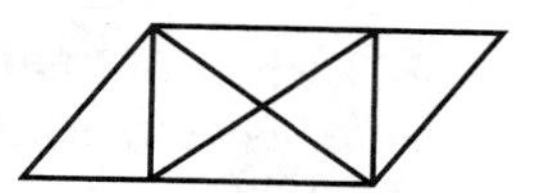

21. Suppose $\angle A$ is supplementary to $\angle B$, and $\angle C$ is supplementary to $\angle B$. What can you deduce about $\angle A$ and $\angle C$?

22. Show what the following sequence of Logo commands would draw.

```
FD 50
LT 45
FD 50
RT 90
FD 50
RT 135
FD 30
```

23. Write a Logo command that draws a regular octagon.

24. Show what the following sequence of Logo commands would draw.

```
REPEAT 4 [FD 50 RT 60]
RT 90
FD 85
```

25. Make a sketch of a repeating tile pattern that uses squares and regular octagons.

SUGGESTED READINGS

Arithmetic Teacher. February 1990 Focus Issue on Spatial Sense. Reston, Va.: NCTM, 1990.

Bergamini, D. (ed.). *Mathematics.* New York: Time, 1963.

Chazan, D., and R. Houde. *How To Use Conjecturing and Microcomputers To Teach Geometry.* Reston, Va.: NCTM, 1989.

Del Grande, J., and others. *Geometry and Spatial Sense.* Reston, Va.: NCTM, 1993.

Equals. *Get It Together.* Berkeley, Calif.: Equals, 1989.

Fuys, D., D. Geddes, and R. Tischler. *The Van Hiele Model of Thinking in Geometry Among Adolescents.* Reston, Va.: NCTM, 1988.

Gardner, M. *Aha!* New York: W. H. Freeman, 1978.

Geddes, D., and others. *Geometry in the Middle Grades.* Reston, Va.: NCTM, 1992.

Hill, J. (ed.). *Geometry for Grades K–6.* Reston, Va.: NCTM, 1987.

Juster, Norton. *The Dot and the Line.* New York: Random House, 1963.

National Council of Teachers of Mathematics. *Geometry in the Mathematics Classroom*. 1973 Yearbook. Reston, Va.: NCTM, 1973.

National Council of Teachers of Mathematics. *Computers in Mathematics Education*. 1984 Yearbook. Reston, Va.: NCTM, 1984.

National Council of Teachers of Mathematics. *Learning and Teaching Geometry, K–12*. 1987 Yearbook. Reston, Va.: NCTM, 1987.

O'Daffer, P., and S. Clemens. *Geometry: An Investigative Approach*. 2nd ed. Reading, Mass.: Addison-Wesley, 1992.

Papert, S. *Mindstorms: Children, Computers, and Powerful Ideas*. 2nd ed. New York: Basic Books, 1993.

9 Congruence, Symmetry, and Similarity

Many artistic and architectural designs utilize patterns of congruent shapes. Sometimes each individual shape is symmetric; that is, all of its parts match up to one another in a certain way.

Sometimes, people use similarity relationships in planning designs. In technical terms, similar figures are the same shape, although they may be of different sizes. For example, a scale model of a building is similar to the planned building.

Congruence, symmetry, and similarity are all related to motion geometry. By moving a geometric figure to different positions or changing its size, one can tell if it is symmetric or if it is congruent or similar to another geometric figure.

Over 2500 years ago, the Greeks used straightedge-and-compass constructions to construct and copy geometric shapes. Many of these constructions can be analyzed using congruence properties.

9.1 CONGRUENCE AND RIGID MOTIONS

In Chapter 8, you learned about congruent line segments and angles. Congruence is one of the most important ideas of geometry. A set of photocopies or a set of mass-produced items would be an example of approximately congruent objects (Figure 9-2, on page 498). The Dutch artist M. C. Escher drew some interesting interlocking congruent shapes (Figure 9-3, on page 498).

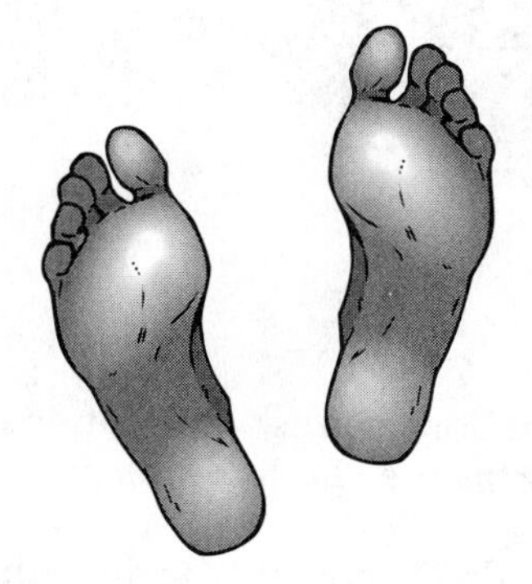

Figure 9–1

RIGID MOTIONS

Ask a child whether two flat shapes are the same (that is, congruent), and the child may place one shape on top of the other to be sure (Figure 9-1). In general, it is true that two plane figures are congruent if and only if one can fit exactly over the other.

The set of **rigid motions** (or **isometries**) consists of the various ways to move a geometric figure around while preserving the distances between points

Photo courtesy of Library of Congress

Figure 9–2

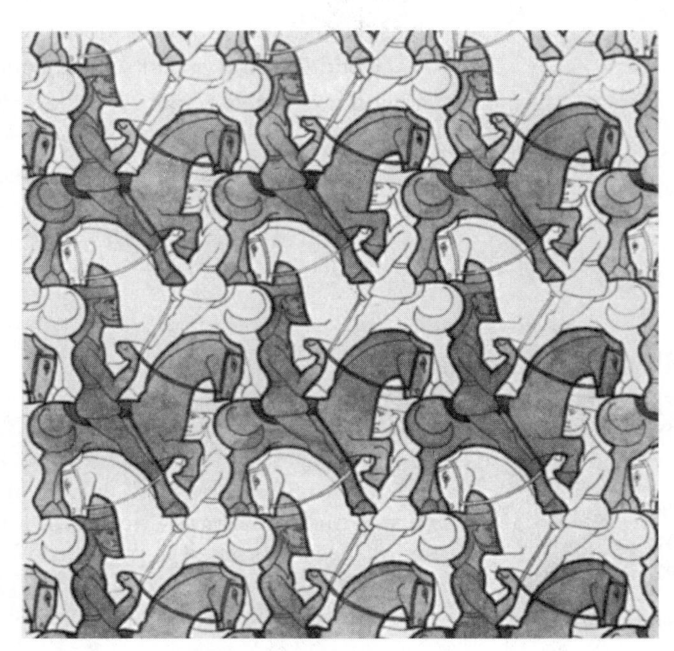

Figure 9–3

in the figure. The three basic rigid motions are rotations (turns), translations (slides), and reflections (flips). The sixth-grade *Mathematics Plus* textbook gives an example of each motion (Figure 9-4).

How do these three basic motions work? In Lesson Exercise 9.1, begin with a turn (rotation) that moves the plane around a fixed point.

Translations, Rotations, and Reflections

This is what a new sign on a window looked like from inside the store. Alex's boss, Mr. Weeks, was not happy. Do you know why?

Mr. Weeks was unhappy because people approaching the store could not read the sign.

You can move a geometric figure three ways.

You can slide the figure along straight lines. This is called a **translation.**

You can turn the figure around a point. This is called a **rotation.**

You can flip the figure over a line. This is called a **reflection.**

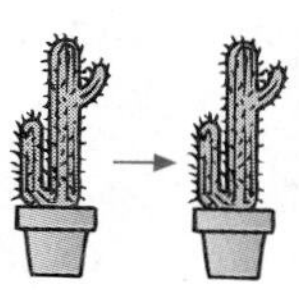

Translation

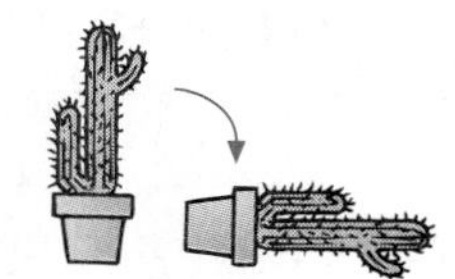

Rotation

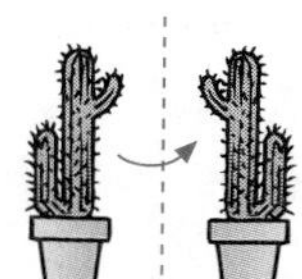

Reflection

Figure 9–4

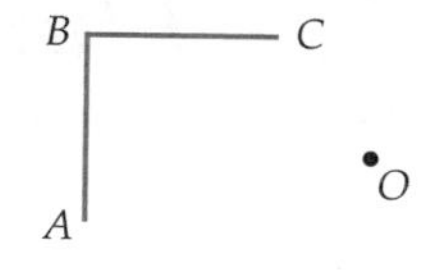

Figure 9–5

Lesson Exercise 9.1

(a) Place a piece of thin paper on this page, and trace the shape ABC and the point O from Figure 9-5.

(b) Now place your pencil point on O and turn the tracing paper a little. You have just done a rotation around the point O!

To describe any clockwise rotation, we specify a fixed center point, the measure of the angle through which we turn the plane around the center point, and the direction of the turn (clockwise).

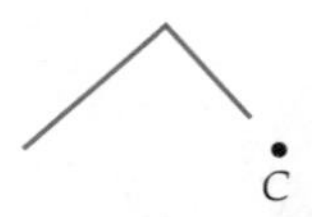

Figure 9–6

Lesson Exercise 9.2

Suppose that you turn ┌ a full turn clockwise around point C (Figure 9-6) so that it ends up back in the same place. How many degrees have you rotated the shape?

Lesson Exercise 9.3

Figure 9–7

Suppose that you rotate the shape clockwise halfway around point C, as shown in Figure 9-7. How many degrees have you rotated the shape?

Lesson Exercises 9.2 and 9.3 illustrate two important rotations: the 360° full turn and the 180° half-turn. Now try another rotation.

Lesson Exercise 9.4

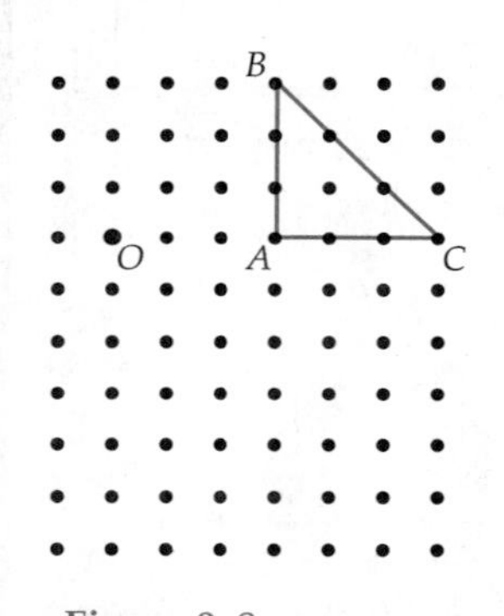

Figure 9–8

(a) Copy the grid and picture from Figure 9-8 onto a piece of paper.

(b) Now use tracing paper to trace O and $\triangle ABC$, and then rotate $\triangle ABC$, 90° clockwise around point O.

(c) After completing the rotation, trace over the triangle in its new position (called the image) so that it shows up on your original paper drawing. Then remove the tracing paper and draw the image in pencil or pen.

Next, try to find more general relationships between a point and its image under a rotation.

Lesson Exercise 9.5

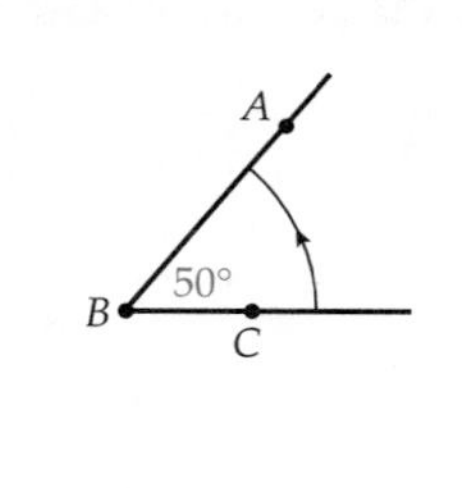

Figure 9–9

Consider $\angle ABC$ and points X and O in Figure 9-9.

(a) Find the image X' of point X under a counterclockwise rotation of $m\angle ABC$ around point O. (*Hint:* first use tracing paper to draw an angle with vertex O that is congruent to $\angle ABC$.)

(b) Describe the relationships you observe among any of the following: XO, $X'O$, BA, BC, $m\angle XOX'$, $m\angle ABC$.

Compare your response to Lesson Exercise 9.5(b) with the end of the following general statement. In the plane, a counterclockwise **rotation** of $m\angle ABC$ around point O maps a point X to X' so that $XO = X'O$ and $m\angle XOX' = m\angle ABC$ (Figure 9-10).

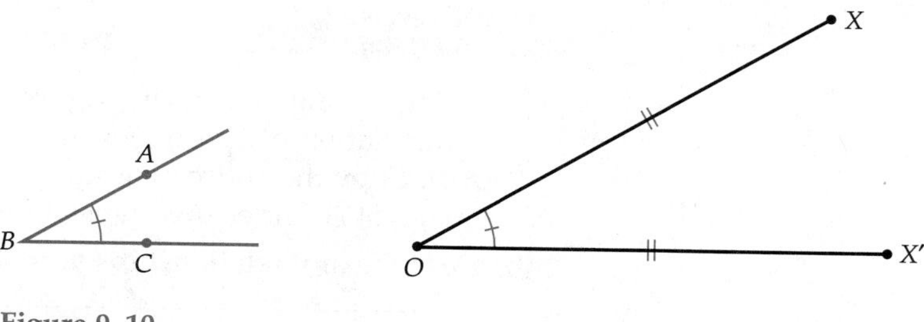

Figure 9–10

There's a new geometric dance called the translation. You stand on one foot and slide 2 inches to the right (Figure 9-11). (It's easier on a recently waxed floor.)

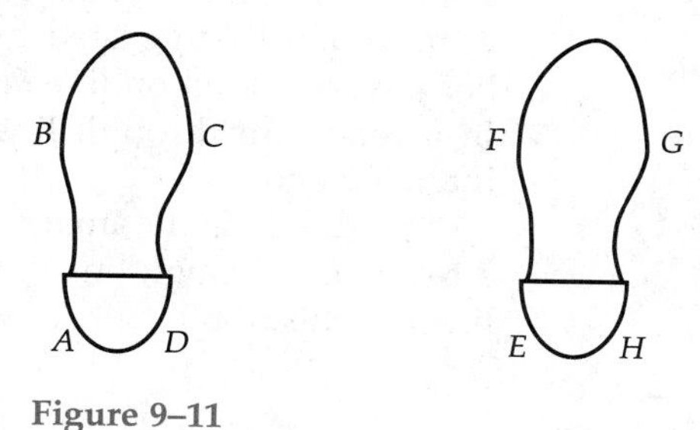

Figure 9–11

How do you do a translation (slide)? In a translation, each point of the plane moves the same distance in the same direction along a line. If footprint (a) in Figure 9-11 slides 4 centimeters to the right, it will coincide with footprint (b). So footprint (b) is the image of footprint (a) under a translation of 4 centimeters to the right. In order to describe a translation, you need a distance and a direction.

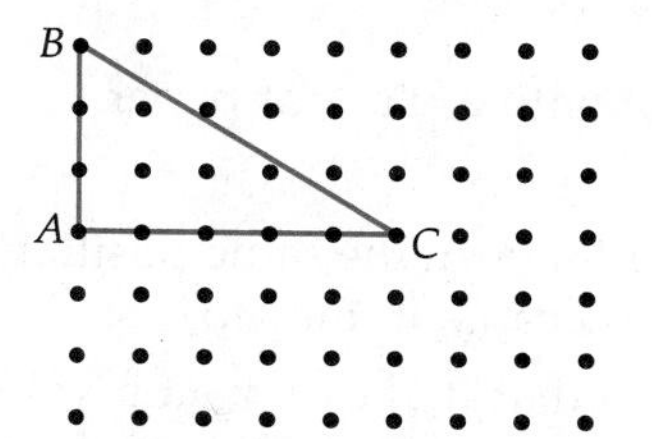

Figure 9–12

Lesson Exercise 9.6

(a) Copy the square grid and picture from Figure 9-12 onto a piece of paper.

(b) Now use tracing paper to trace $\triangle ABC$ and translate (slide) it 2 units down. (Adjacent dots in each column or row are 1 unit apart.)

(c) Trace over the triangle in its new position so it shows up on your original paper. Then remove the tracing paper and draw the image in pencil or pen.

Next, try to find more general relationships between a point and its image under a translation.

Lesson Exercise 9.7

(a) Using tracing paper with Figure 9-13, find the image X' of point X under a translation of length AB in the direction of A to B (**directed segment** $\overrightarrow{AB}$). Copy the entire diagram onto the tracing paper. Then figure out the image of A under this translation.

(b) Describe any relationships you observe between $\overline{XX'}$ and $\overline{AB}$.

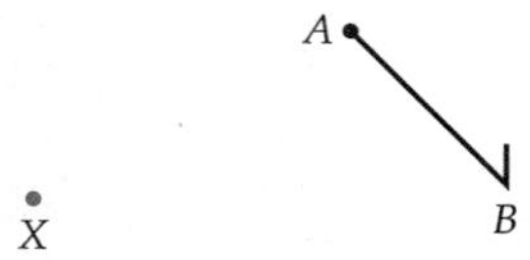

Figure 9–13

Compare your response to Lesson Exercise 9.7(b) with the relationships described in the following general statement. In a plane, a **translation** of length AB in the direction of A to B maps a point X to X' so that $XX' = AB = \overrightarrow{XX'}$ and $\overrightarrow{AB}$ point in the same direction (Figure 9-14).

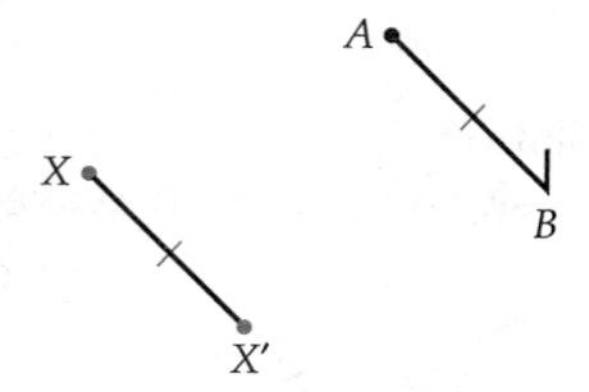

Figure 9–14

Finally, how do you do a reflection (a flip)? Wait a second. Don't start doing body flips over those desks. It's dangerous. Let's use a triangle instead. A reflection is suggested by folding a shape across a line. If this page of the book were folded on line m in Figure 9-15, $\triangle ABC$ would coincide with $\triangle DEF$. In a reflection through line m, the positions of $\triangle ABC$ and $\triangle DEF$ would be interchanged.

So $\triangle DEF$ is the image of $\triangle ABC$, and $\triangle ABC$ is the image of $\triangle DEF$ under a reflection through line m. In order to describe a reflection (a flip), you need a line of reflection.

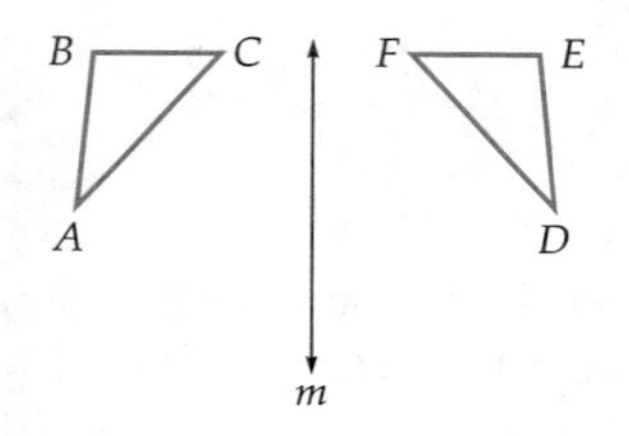

Figure 9–15

Lesson Exercise 9.8

Look at $\triangle ABC$ and its image $\triangle DEF$ in Figure 9-15.

(a) Why is this motion called a reflection?

(b) If you have a mirror available, position it to verify that $\triangle DEF$ is the reflected image of $\triangle ABC$.

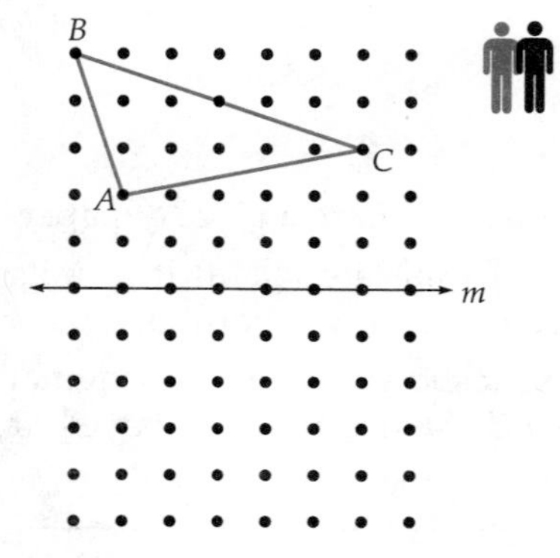

Figure 9–16

Lesson Exercise 9.9

(a) Copy the grid and picture (from Figure 9-16) onto a piece of paper.

(b) Now use tracing paper to trace m and $\triangle ABC$.

(c) Flip the paper over, but make sure that line m stays in the same position. (Do *not* interchange the positions of the arrowheads of line m.)

(d) Trace over the triangle in its new position so that it shows up on your original paper. Then remove the tracing paper and draw the image in pencil or pen.

(e) Explain how to find the image of $\triangle ABC$ without using tracing paper.

Next, try to find more general relationships between a point and its image under a reflection.

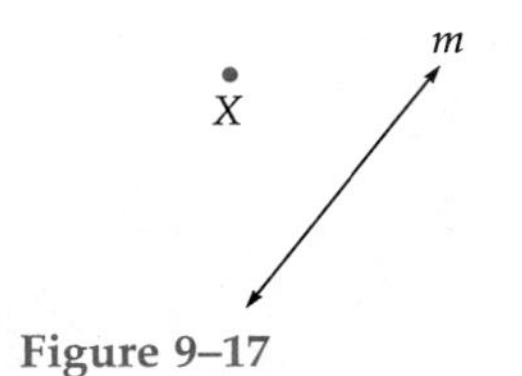

Figure 9–17

Lesson Exercise 9.10

Consider line m and point X in Figure 9-17.

(a) Find the image of X under a reflection through line m.

(b) Describe any relationships you observe between m and $\overline{XX'}$.

Compare your response to Lesson Exercise 9.10(b) with the following definition. In a **reflection** through line m, each point X (not on m) in the plane is paired with X' so that m is the perpendicular bisector of $\overline{XX'}$. If X is on m, then $X' = X$. (*Note:* The **perpendicular bisector** of a line segment is a line that is perpendicular to the line segment and bisects it.)

Lesson Exercise 9.11

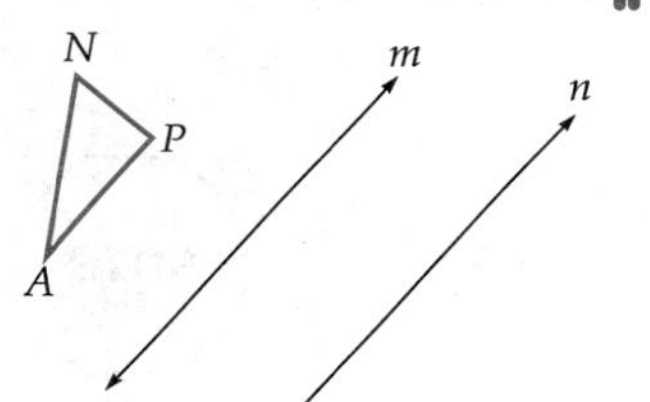

Figure 9–18

In Figure 9-18, assume $m \parallel n$.

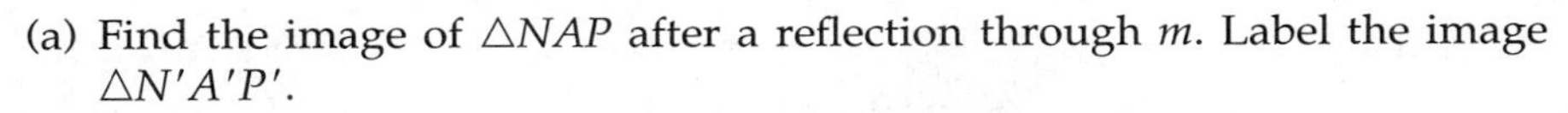

(a) Find the image of $\triangle NAP$ after a reflection through m. Label the image $\triangle N'A'P'$.

(b) Find the image of $\triangle N'A'P'$ after a reflection through n. Label the image $\triangle N''A''P''$.

(c) What single motion maps $\triangle NAP$ to $\triangle N''A''P''$?

(d) How is the distance from N to N'' related to the distance between m and n?

RIGID MOTIONS AND CONGRUENCE

When you perform a rigid motion on a figure, what changes and what stays the same?

Lesson Exercise 9.12

Refer back to Lesson Exercises 9.4, 9.5, 9.7, and 9.8, and compare the image to the original figure.

(a) What is the same about the image and the original figure in each exercise?

(b) How does the image differ from the original figure in each exercise?

Lesson Exercise 9.13

Suppose you have two congruent shapes. Do you think it is always possible to move one shape onto the other using a single rigid motion or some combination of rigid motions?

You can use rigid motions to test for congruence.

> **A Test for Congruence**
>
> Two plane drawings are congruent if one can be moved onto the other using a rotation, a translation, a reflection, or some combination of these motions.

Lesson Exercise 9.14

Trace one of each pair of geometric figures in Figure 9-19, and use a rotation, translation, or reflection to determine if the two shapes are congruent. If the shapes are congruent, state which motion you used to show this.

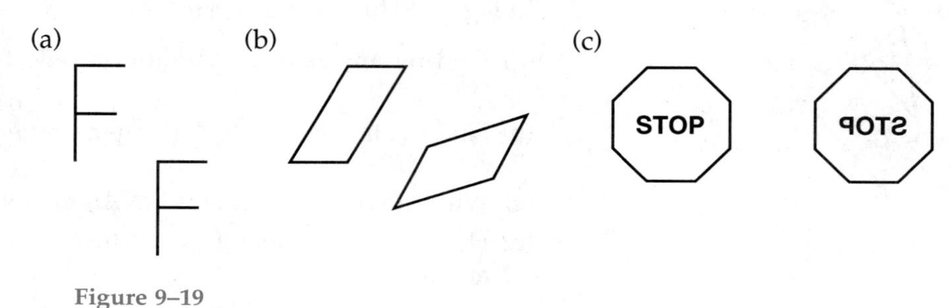

Figure 9–19

Figure 9–20

Lesson Exercise 9.15

What *combination* of motions could be used to map one F onto the other in Figure 9-20?

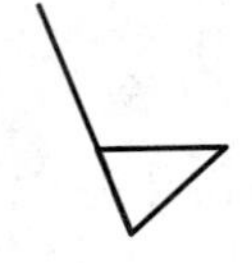

Figure 9–21

GLIDE REFLECTIONS

It is not always possible to move one of two congruent figures onto the other using a *single* rotation, translation, or reflection (Figure 9-21).

This observation led to the introduction of a fourth rigid motion, the glide reflection, which combines a translation and a reflection.

The glide reflection of on $\overrightarrow{AB}$ (direction of A to B) in Figure 9-22 has two steps. First reflect the flag through $\overleftrightarrow{AB}$; then translate the result using $\overrightarrow{AB}$. So a **glide reflection** is a reflection followed by a translation along the line of reflection.

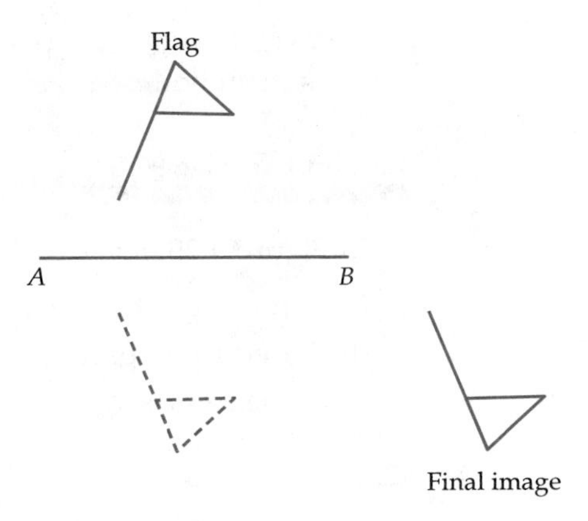

Figure 9–22

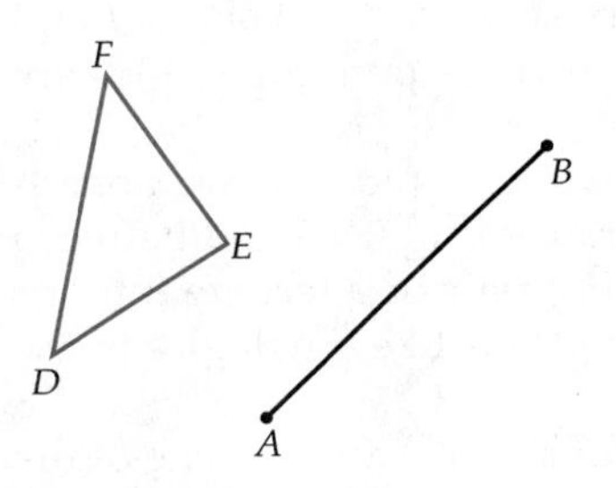

Figure 9–23

Lesson Exercise 9.16

See Figure 9-23, and perform a glide reflection of $\triangle DEF$ on $\overrightarrow{AB}$ (direction of A to B).

Using the set of four rigid motions, we can say: two plane figures are congruent if and only if one figure can be moved onto the other using a *single* rotation, translation, reflection, or glide reflection.

VERTICAL, CORRESPONDING, AND ALTERNATE INTERIOR ANGLES

Rigid motions indicate congruent pairs of angles or line segments. The symbol for congruence is $\cong$.

Lesson Exercise 9.17

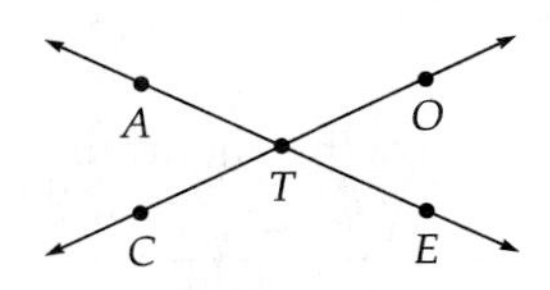

Figure 9–24

(a) Find a specific motion that maps $\angle ATC$ in Figure 9-24 onto another angle.

(b) Based upon part (a), it appears that $\angle ATC \cong$ ________.

(c) Using motion geometry, find another pair of angles in Figure 9-24 that might be congruent.

Angles such as $\angle ATC$ and $\angle ETO$ in Lesson Exercise 9.17 that are formed by two intersecting lines are called **vertical angles**. Vertical angles have a common vertex but no sides in common. $\angle ETC$ and $\angle ATO$ are also vertical angles. Using rotations to *prove* that vertical angles are congruent requires a more rigorous study of rotations.

Next, consider two parallel lines and a line (called a **transversal**) that intersects them both.

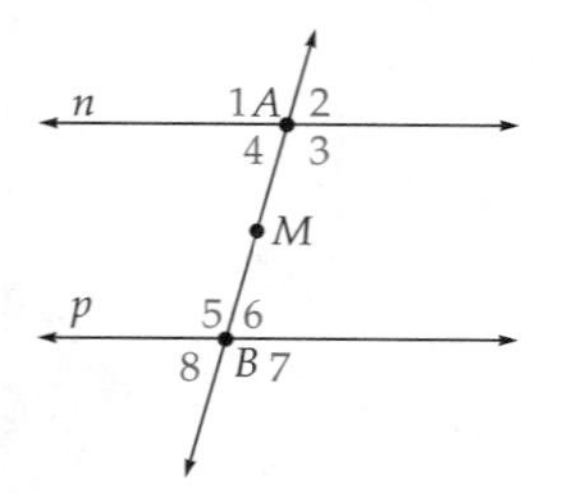

Figure 9–25

Lesson Exercise 9.18

In Figure 9-25, $n \parallel p$, and M is the midpoint of $\overline{AB}$.

(a) Find a specific motion that maps $\angle 1$ to another angle in the diagram.

(b) Based upon part (a), it appears that $\angle 1 \cong$ ________.

(c) Repeat parts (a) and (b) until you have listed all possible sets of congruent angles.

Notice that each of the two parallel lines in Figure 9-25 forms four numbered angles with the transversal. Each angle (such as $\angle 1$) in one group corresponds in position to an angle in the other group (such as $\angle 5$). Thus, $\angle 1$ and $\angle 5$ are called **corresponding angles**. Other pairs of corresponding angles are $\angle 2$ and $\angle 6$, $\angle 3$ and $\angle 7$, and $\angle 4$ and $\angle 8$.

In Figure 9-25, **alternate interior angles** (such as $\angle 3$ and $\angle 5$) are nonadjacent angles on opposite (alternate) sides of the transversal ($\overleftrightarrow{AB}$) and between (interior to) the other two lines (n and p). The other pair of alternate interior angles is $\angle 4$ and $\angle 6$. A pair of alternate interior angles is like the angles in the letter Z.

In Lesson Exercise 9.18, you probably made conjectures about some of the corresponding or alternate interior angles. Propose more general conjectures in the following exercise.

Lesson Exercise 9.19

Look back at the diagram in Lesson Exercise 9.18, and make a conjecture about

(a) alternate interior angles.

(b) corresponding angles.

Lesson Exercises 9.18 and 9.19 should suggest the following theorem.

Parallel Lines and Corresponding and Alternate Interior Angles

If two parallel lines are cut by a transversal, then each pair of corresponding angles is congruent and each pair of alternate interior angles is congruent.

Recall supplementary angles from Section 8.1. Figures 9-24 and 9-25 contain quite a few pairs of supplementary angles.

Lesson Exercise 9.20

Name every pair of supplementary angles in Figure 9-25 that includes $\angle 4$.

CONGRUENT FIGURES

Rigid motions show that one whole figure is congruent to another. We can also show that two figures are congruent by matching up their corresponding parts.

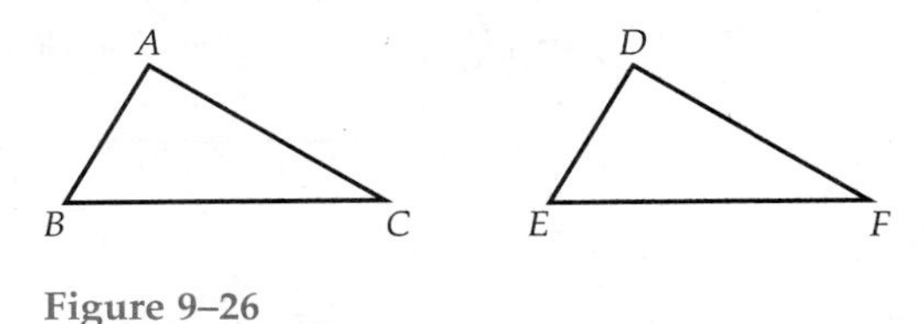

Figure 9–26

Lesson Exercise 9.21

What set of facts about the sides and angles of $\triangle ABC$ and $\triangle DEF$ (Figure 9-26) would guarantee that the two triangles are congruent?

You can determine if two polygons are congruent from information about their corresponding sides and angles. Two **congruent polygons** have all their corresponding sides congruent and all their corresponding angles congruent. Consider $\triangle ABC$ and $\triangle DEF$ in Figure 9-26. If $\triangle ABC$ were translated onto $\triangle DEF$ in Figure 9-26, vertex A would move to D, vertex B would move to E, and vertex C would move to F. To indicate these corresponding vertices, we would write

$$A \leftrightarrow D \qquad B \leftrightarrow E \qquad C \leftrightarrow F$$

We could also say $\triangle ABC \cong \triangle DEF$. In naming congruent triangles such as $\triangle ABC \cong \triangle DEF$, the order of the letters indicates the corresponding vertices. The correspondences can also be used to match up congruent angles or congruent sides. For example, $\angle A \cong \angle D$, and $\overline{BC} \cong \overline{EF}$.

You may be wondering about the difference between the congruence sign ($\cong$) and the equal sign ($=$) in geometry. Congruent figures or parts of figures, such as $\overline{BC}$ and $\overline{EF}$, have the same shape and size, and this congruence is expressed as $\overline{BC} \cong \overline{EF}$. An equal sign is used most commonly to indicate that some *measure* of two geometric figures is the same. The expression "$BC = EF$" means that the length of $\overline{BC}$ equals the length of $\overline{EF}$. The expressions BC and EF represent *numbers* (lengths), whereas the terms $\overline{BC}$ and $\overline{EF}$ represent line segments.

The following chart shows the corresponding notations for equality and congruence of line segments and angles.

	Equality	Congruence
Line segments	$BC = EF$	$\overline{BC} \cong \overline{EF}$
Angles	$m\angle A = m\angle D$	$\angle A \cong \angle D$

Lesson Exercise 9.22

In Figure 9-27, $RANT \cong BCDA$. Complete the following expressions.

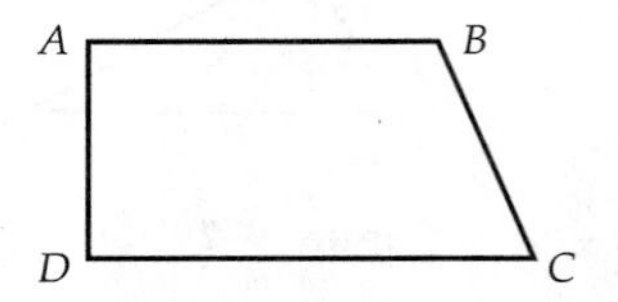

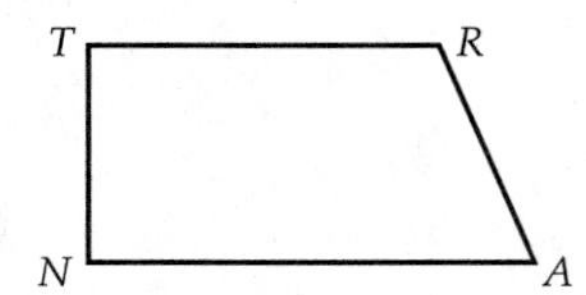

Figure 9–27

(a) $m\angle C =$ ________ (b) $\overline{NT} \cong$ ________

TESSELLATIONS AS ART

Artistic tessellations utilize congruent figures and motions as a basis for creating designs. Recall that all triangles, quadrilaterals, and regular hexagons tessellate the plane.

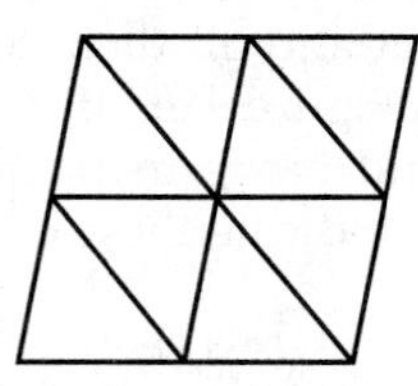
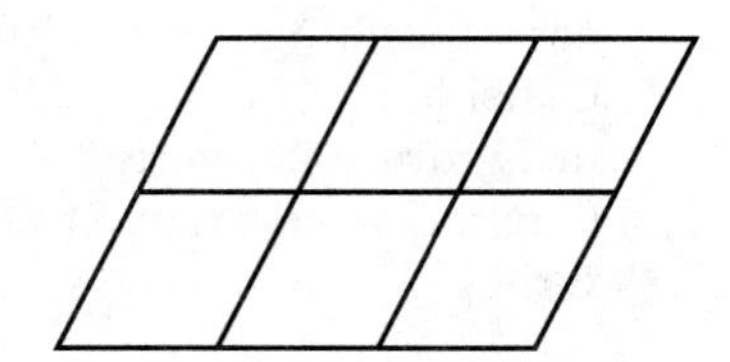
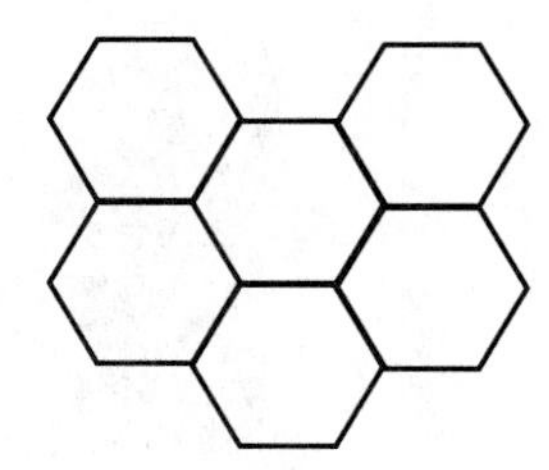

Figure 9–28

The Dutch artist M. C. Escher (1898–1972) used the shapes shown in Figure 9-28 as a basis for many ingenious tessellations (Figure 9-29). Escher was inspired by Moorish art he saw during his travels to Spain in the 1930s.

Lesson Exercise 9.23

If each design in Figure 9-29 covered an entire plane, name a motion that would map the design onto itself.

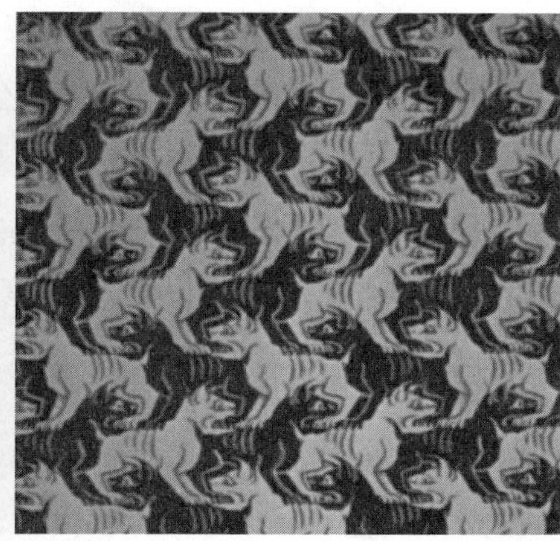

Figure 9–29

How did Escher create his tessellations? By altering regular polygons that tessellate. Consider a square.

Take out a piece from the bottom, and translate or rotate it to create a new figure that tessellates (Figure 9-30 and Figure 9-31).

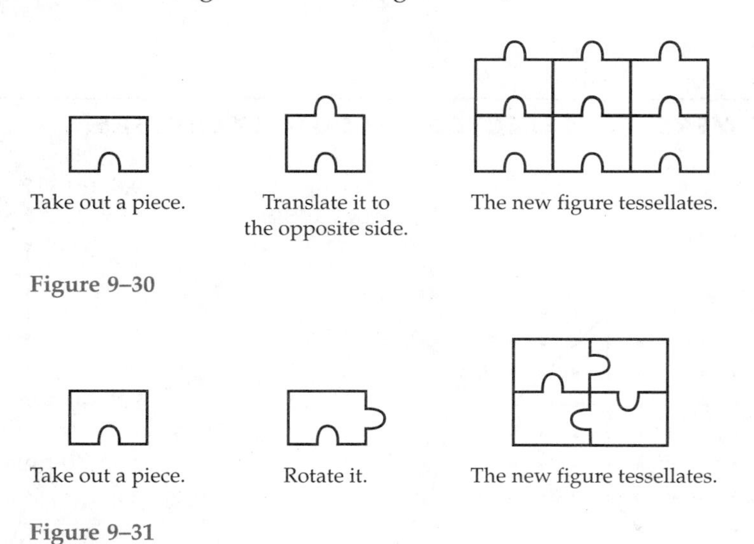

Take out a piece.
Translate it to the opposite side.
The new figure tessellates.
Figure 9–30
Take out a piece.
Rotate it.
The new figure tessellates.

Figure 9–31

Lesson Exercise 9.24

Complete the transformations so that the new shape will tessellate.

(a)

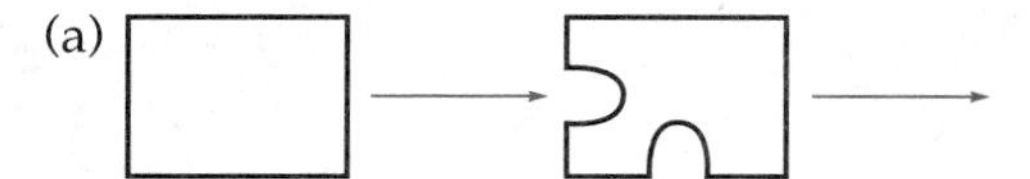

(b)

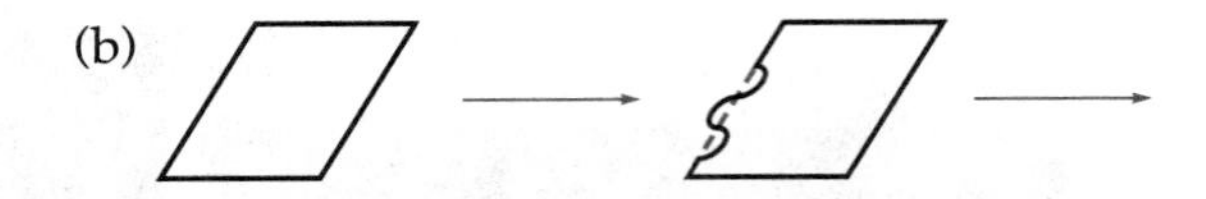

To create a tessellating animal, you can make some alterations in a square (or other tessellating figure) until it starts to look more like an animal. Then add a design inside.

Try one yourself.

Lesson Exercise 9.25

Alter a □ or a ▱ or a △ or a ⬡ to create a new tessellating shape. Then add a design inside it.

ANSWERS TO SELECTED LESSON EXERCISES

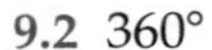

9.2 360°

9.3 180°

9.4 (c)

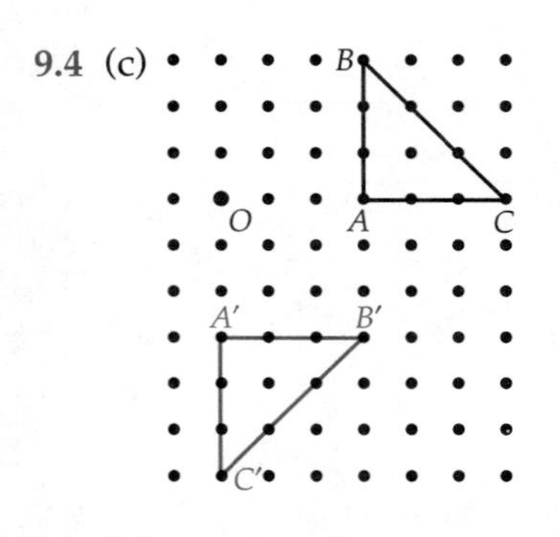

9.5 (a)

9.6 (c)

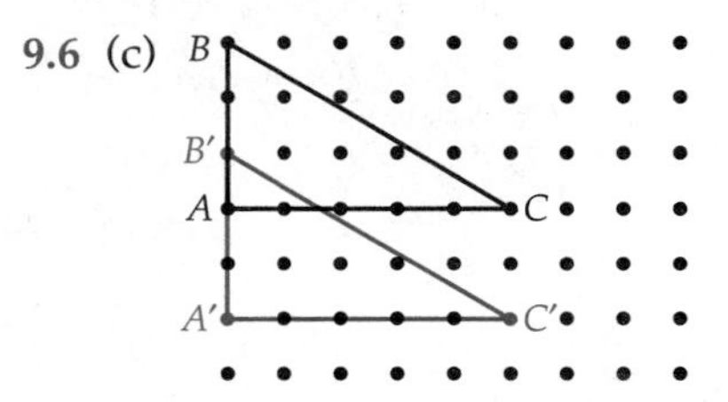

9.8 (a) One triangle looks like a mirror reflection of the other when the line is used as the mirror.

9.9 (d)

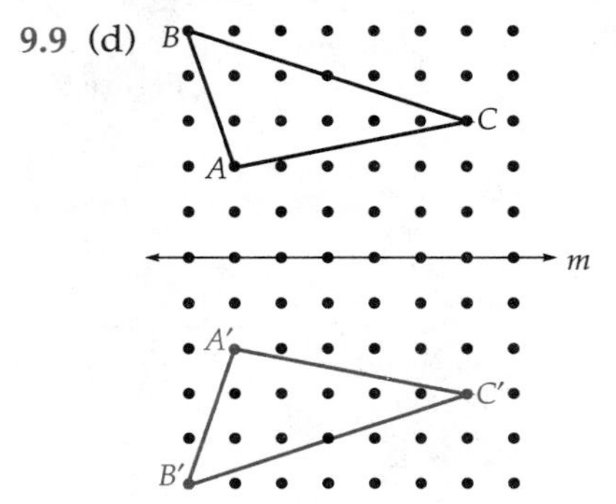

(e) A', B', and C' are on the opposite side of m, the same distances from m, respectively, as A, B, and C.

9.10 (a)

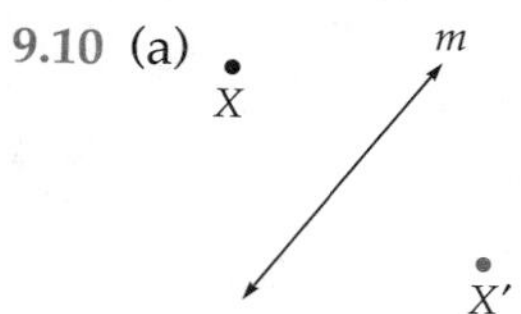

9.11 (b)

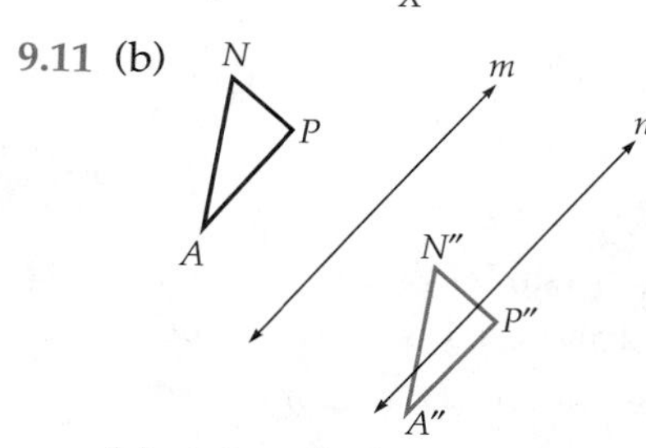

(c) A translation

(d) It is twice as much.

9.12 (a) They have the same size and shape.
(b) Its position is different.

9.13 Yes

9.14 Parts (a) and (c) contain congruent figures by a translation and a reflection, respectively.

9.15 (possible answer) a rotation and a reflection

9.16

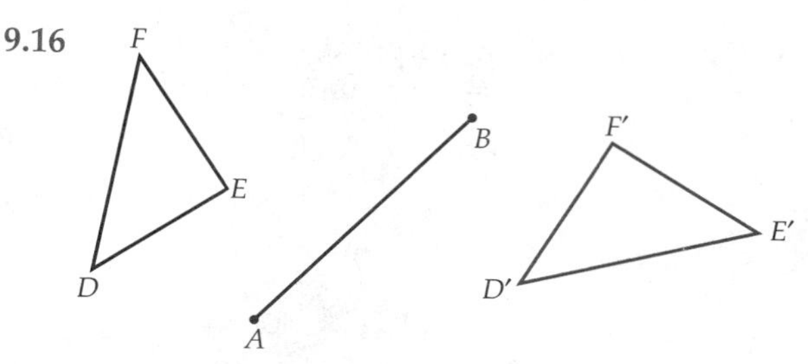

9.17 (a) 180° rotation (b) $\angle ETO$
(c) A half-turn around T appears to map $\angle ATO$ onto $\angle ETC$.

9.18 (c) $\angle 1 \cong \angle 3 \cong \angle 5 \cong \angle 7$ and $\angle 2 \cong \angle 4 \cong \angle 6 \cong \angle 8$

9.20 $\angle 4$ and $\angle 1$, $\angle 4$ and $\angle 3$, $\angle 4$ and $\angle 5$, $\angle 4$ and $\angle 7$

9.21 $\angle A \cong \angle D$, $\angle B \cong \angle E$, $\angle C \cong \angle F$, $\overline{AB} \cong \overline{DE}$, $\overline{AC} \cong \overline{DF}$, $\overline{BC} \cong \overline{EF}$

9.22 (a) $m\angle A$ (b) $\overline{DA}$

9.23 Beetles: a reflection through a line that splits any beetle in half down the middle, or a translation that moves one white beetle's left eye to the left eye of the white beetle above it
Dogs: a translation up or to the right that moves the right ear of one white dog to the right ear of another white dog

9.1 HOMEWORK EXERCISES

Basic Exercises

1. Find the image of the flag after a 270° clockwise rotation around point C.

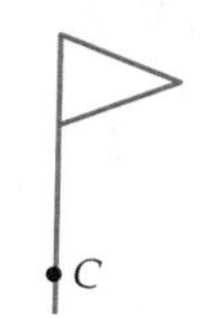

***2.** (a) Copy the grid and picture onto a piece of paper.

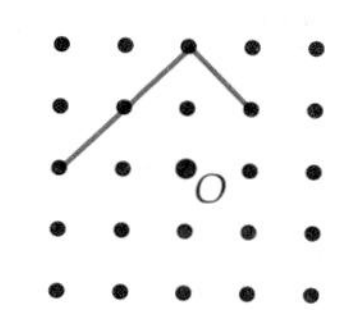

(b) Now use tracing paper to trace point O and the ┌.

(c) Rotate the ┌ 180° clockwise around point O.

(d) Show the image of the ┌ on your original grid.

3. If a point P is rotated clockwise 32° around point O, then $m\angle POP' =$ ________, and $PO =$ ________.

4. Some airport runways are numbered with compass positions in 10-degree units. For example, 36 means 360°. Use a protractor to add a new runway, labeled 11, to the following diagram.

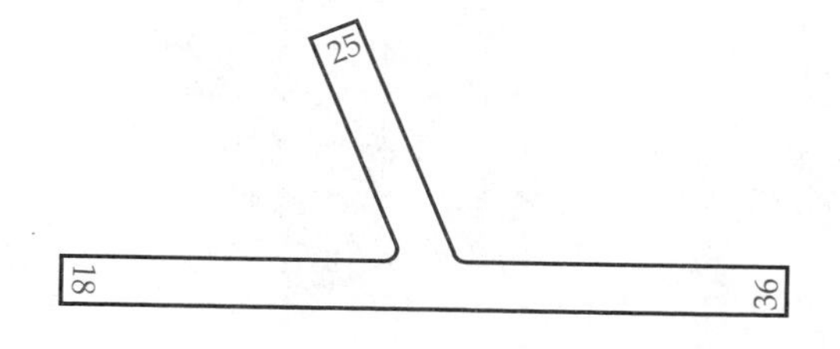

5. Find the image X' of point X after a counterclockwise rotation of $m\angle ABC$ around point O.

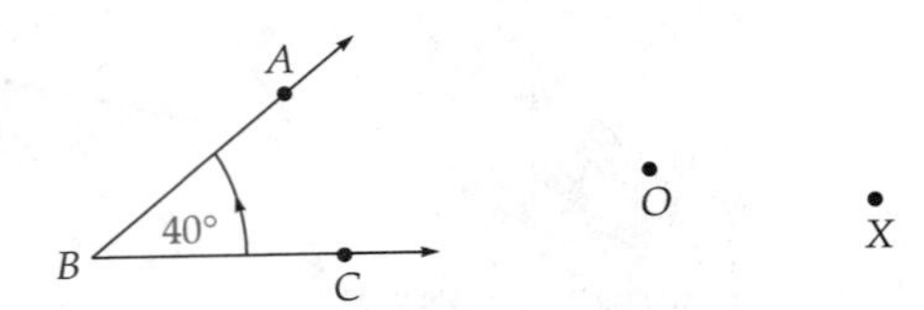

***6.** Using a ruler, find the image of the flag under a translation 4 cm to the left.

7. Show the image of the triangle after a translation 3 units to the right and 2 units down.

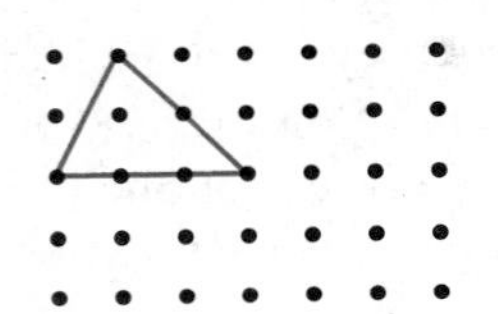

8. Find the image of the triangle after a translation using $\overrightarrow{AB}$.

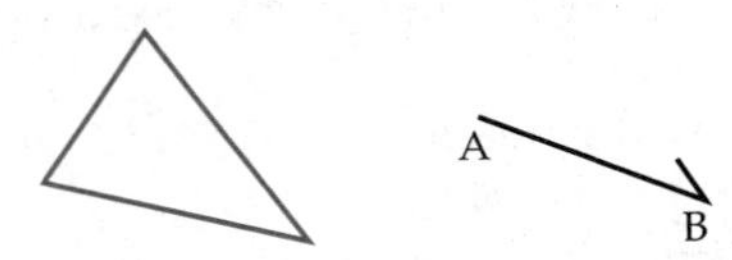

9. A point P is translated using $\overrightarrow{AB}$. If P is not on $\overleftrightarrow{AB}$, then $\overleftrightarrow{PP'} \parallel$ ________, and $\overleftrightarrow{PA} \parallel$ ________.

10. Show the image of the after a reflection through line m.

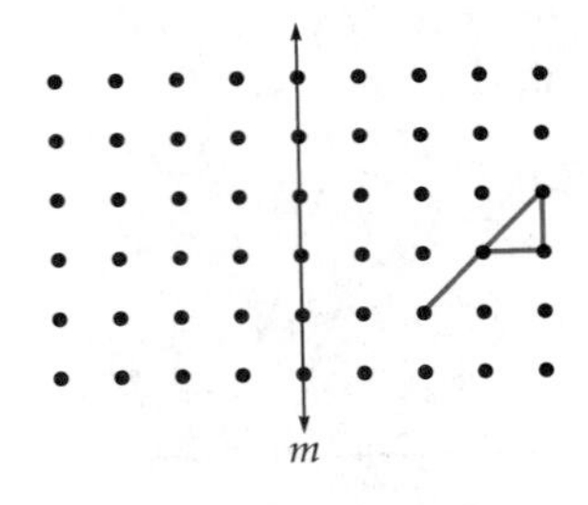

11. (a) Show the image of $\triangle ABC$ after a reflection through line m.

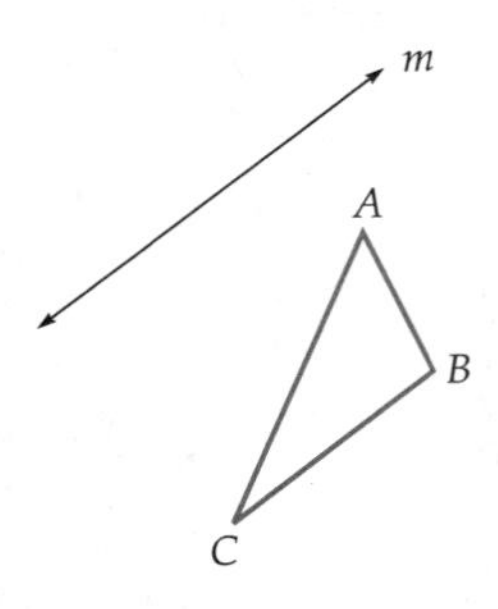

(b) How many image points seem to be needed to draw the image of $\triangle ABC$?

12. Show the image of $ABCD$ after a reflection through line t.

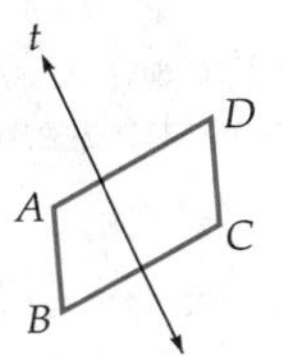

13. If the line shown in the diagram is a mirror, draw the image of the shape on the opposite side of the mirror. Do not use tracing paper.

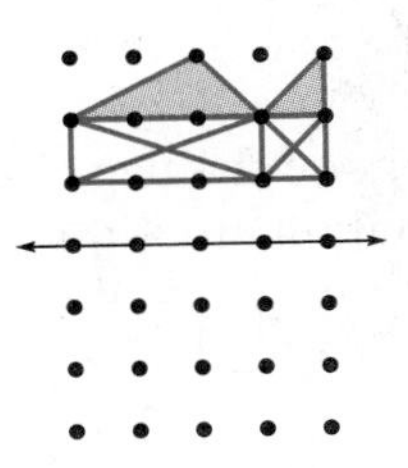

14. What sets of points does a reflection map to itself?

15. Without tracing paper, draw the image of each figure after a 180° rotation around C.

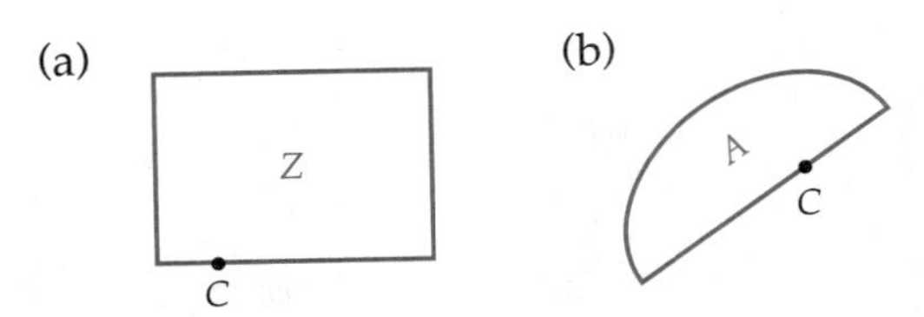

16. $\triangle ABC$ is equilateral, and D is the center point.

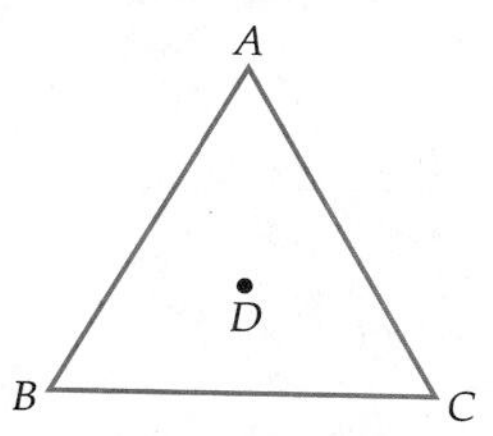

What is the image of B under a
(a) 120° clockwise rotation around D?
(b) 120° counterclockwise rotation around D?
(c) 240° clockwise rotation around D?
(d) reflection through $\overleftrightarrow{CD}$?

17. $ABCDEF$ is a regular hexagon, and G is the center point.

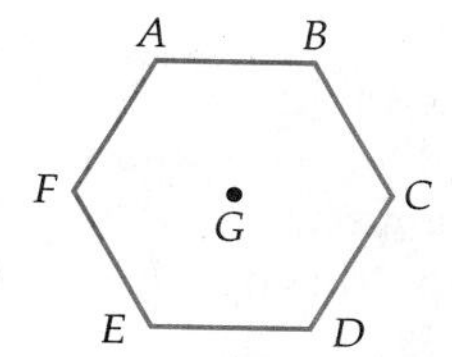

What is the image of A under a
(a) 120° clockwise rotation around G?
(b) 60° counterclockwise rotation around G?
(c) 240° clockwise rotation around G?
(d) reflection through $\overleftrightarrow{CF}$?

18. (a) Give the center point and the degree of rotation that maps $\overline{AB}$ to $\overline{A'B'}$.

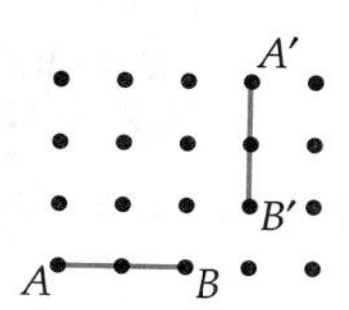

(b) Give the line of reflection that maps $\overline{AB}$ to $\overline{A'B'}$.

19. Find the center of rotation for each object and its image. Use tracing paper and the guess-and-check strategy.

(a)

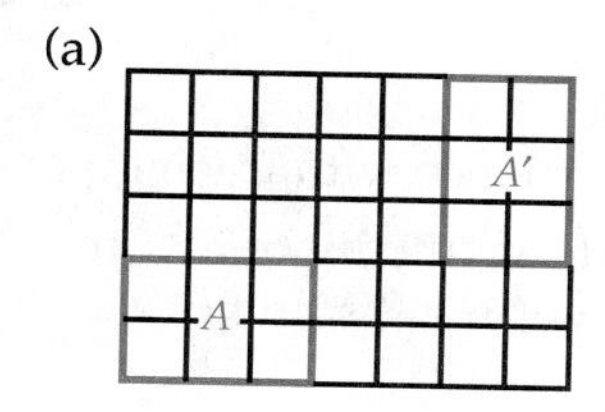

(b)

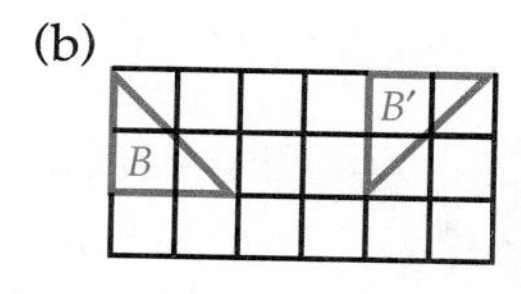

20. Different sets of golf clubs are made for right-handed and left-handed golfers. The two sets of clubs are congruent but have different **orientations**. Name two other kinds of objects that are sometimes made congruent but with different orientations.

21. Some ambulances have ƎƆИA⅃UᙠMA written on the front. Why is the word printed this way?

22. A heavy container must be moved as shown. The easiest way is to move it to an adjacent square by rotating the container around one of its corners.

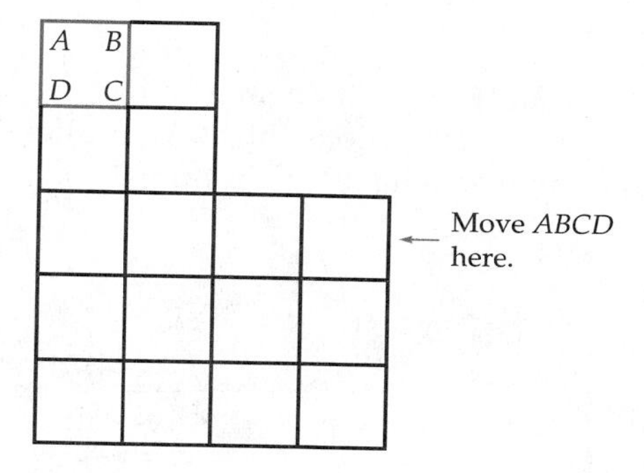

(a) Fill in the squares, including the positions of the corners of the container as it moves from start to finish.
(b) Describe the specific motion used in each step.

23. What rigid motion is suggested by each of the following?
(a) A tree falling to the ground
(b) A pair of shoes
(c) A train traveling straight down a railroad track

24. A dog is tied in front of a house. Draw the limits of the dog's range.

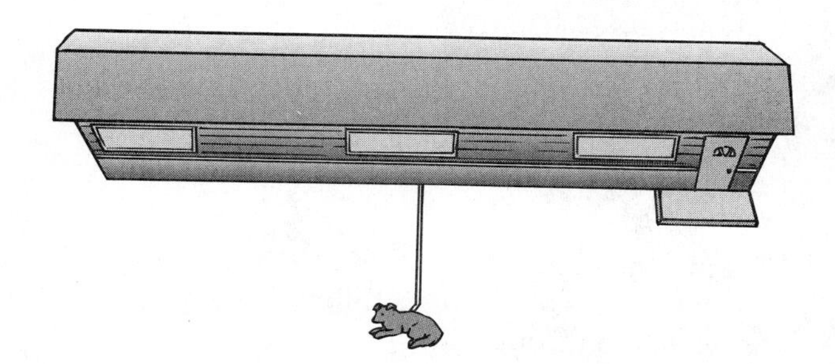

25. You can make a pattern using a rigid motion. Starting at the left, the following pattern was made using a reflection through the vertical line into the next frame.

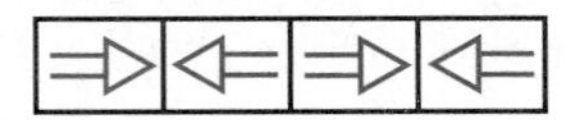

Draw a figure in each left-hand box, and use a rigid motion three times to create a pattern.

(a) (b)

26. The following cartoon shows that two successive reflections through parallel lines are the same as what single motion?

Barbershop mirrors. Drawing by Chas. Addams; © 1957, 1985. The New Yorker Magazine, Inc.

27. Find the image of the flag after a translation 2 centimeters to the right followed by a reflection through line t.

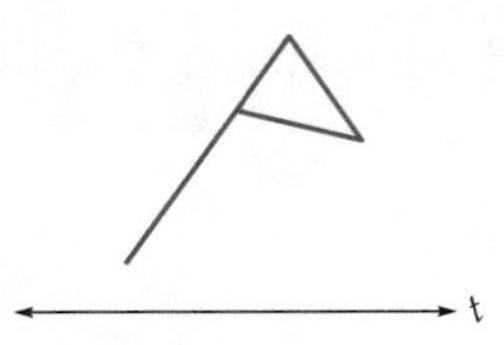

28. Assume that $m \perp n$.

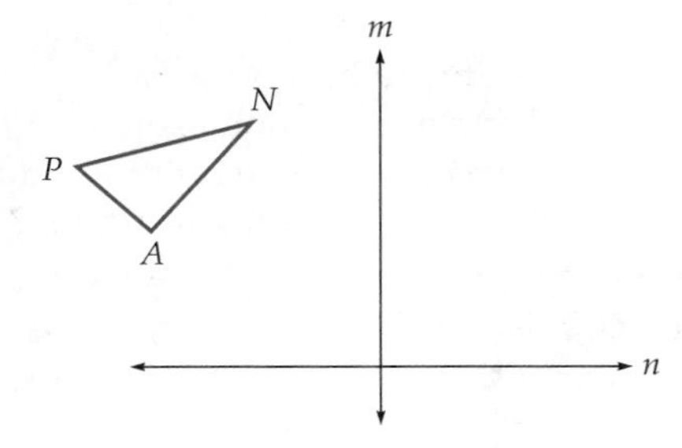

(a) Find the image of $\triangle NAP$ after a reflection through m. Label the image $\triangle N'A'P'$.
(b) Find the image of $\triangle N'A'P'$ after a reflection through n. Label the image $\triangle N''A''P''$.
(c) What single specific motion maps $\triangle NAP$ to $\triangle N''A''P''$?

29. (a) True or false? The image of a vertical line after a reflection is a vertical line.
(b) Give an example that supports your answer.

30. The image of a plane figure under a ________ is congruent to the figure.
(a) rotation (b) translation
(c) reflection (d) all of these

31. (a) Do the segments in each picture appear to be congruent?
(1) $\overline{AB} \cong \overline{BC}$? (2) $\overline{EF} \cong \overline{GH}$?

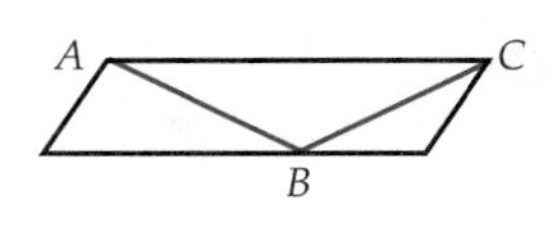

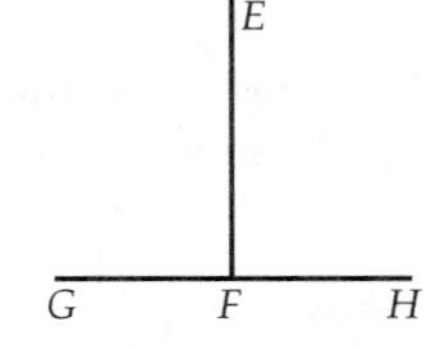

(b) Check their congruence with tracing paper.

***32.** In the following cartoon, the motion of the man and the motion of the Z's do not correspond.

By permission of Johnny Hart and Creators Syndicate, Inc.

(a) Does the man's motion approximate a rotation, a translation, or a reflection?
(b) Which of the three motions is applied to the Z's?

33. Trace one of each pair of figures, and use a rotation, translation, or reflection to determine if the two figures are congruent. If the figures are congruent, state which motion you used to show their congruence. Describe the motion as specifically as possible.

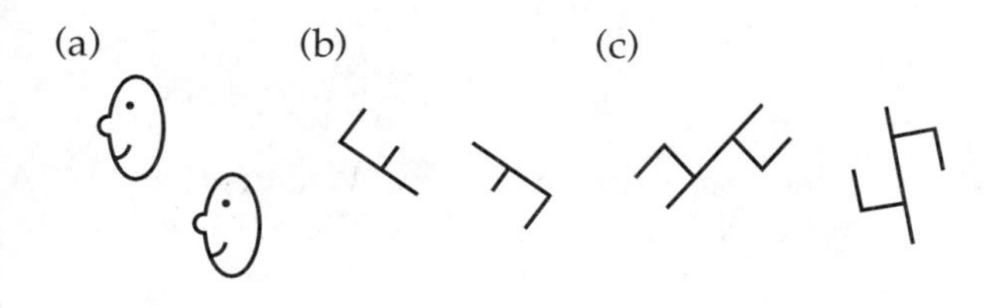

34. The relationship between motions and congruence also applies to space figures. Which of the following pairs of figures are congruent?

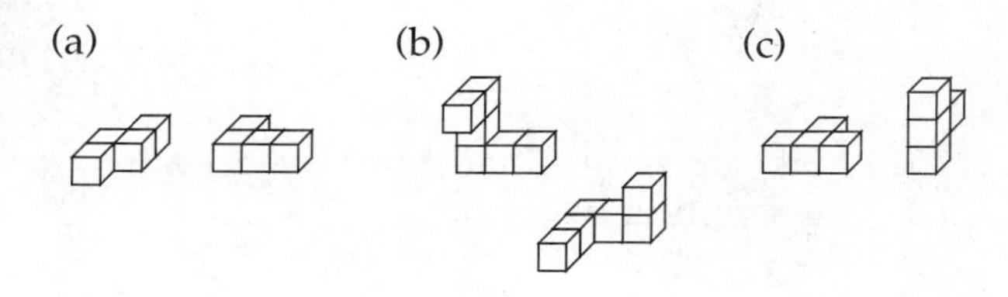

35. Using tracing paper, find the center of rotation that is used to turn $\triangle ABC$ onto $\triangle A'B'C'$. (Guess and check.)

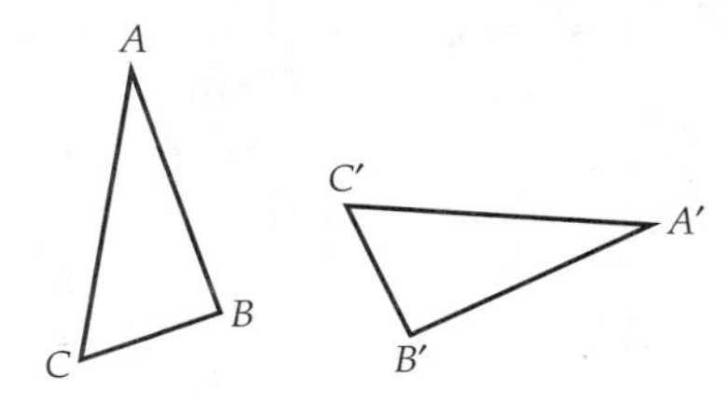

36. Using tracing paper, find the line of reflection that flips one figure onto the other. (Guess and check.)

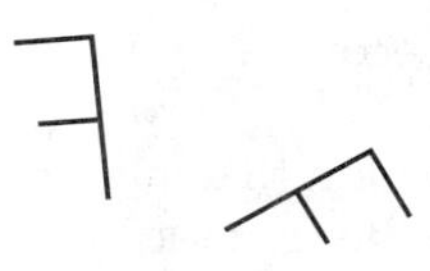

37. Draw two congruent figures such that one figure cannot be mapped onto the other using a *single* rotation, translation, or reflection.

***38.** Find the image of the flag after a glide reflection on $\overrightarrow{AB}$.

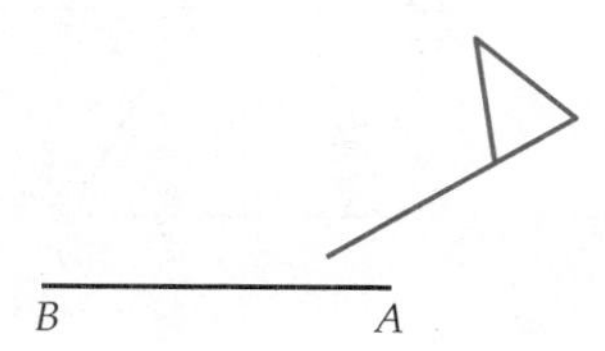

39. *BART* is a parallelogram.

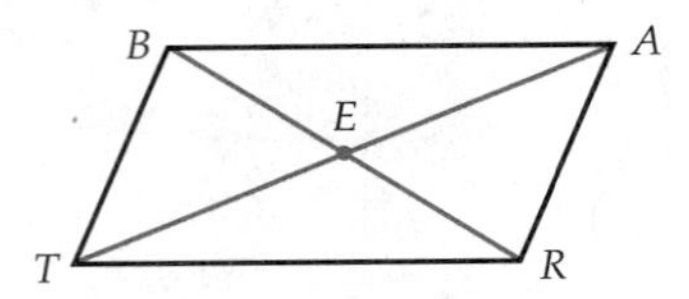

(a) What single motion maps $\overline{BA} \to \overline{RT}$ and $\overline{BT} \to \overline{RA}$?

(b) Based upon part (a), it appears that $\overline{BA}$ ______ $\overline{RT}$ and $\overline{BT}$ ______ $\overline{RA}$.

(c) Find a specific motion that maps $\angle ABT$ to another angle in the diagram.

(d) Based upon part (c), it appears that $\angle ABT \cong$ __________.

(e) Repeat parts (c) and (d) for $\angle BTR$.

***40.** Assume that $\triangle ACT$ is an isosceles triangle, with $\overline{AC} \cong \overline{AT}$ and M the midpoint of $\overline{CT}$.

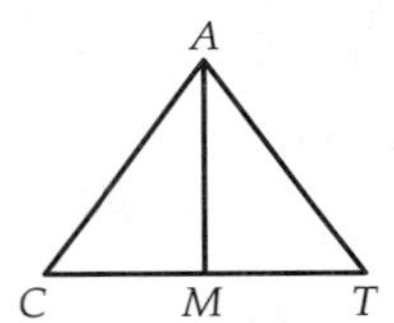

(a) Name a specific motion that maps $\overline{AC} \to \overline{AT}$.

(b) Name a second specific motion that maps $\overline{AC} \to \overline{AT}$.

(c) Find a motion that suggests which two angles in the triangle are congruent, and name them.

41. In the diagram, $\overleftrightarrow{CD} \parallel \overleftrightarrow{GH}$.

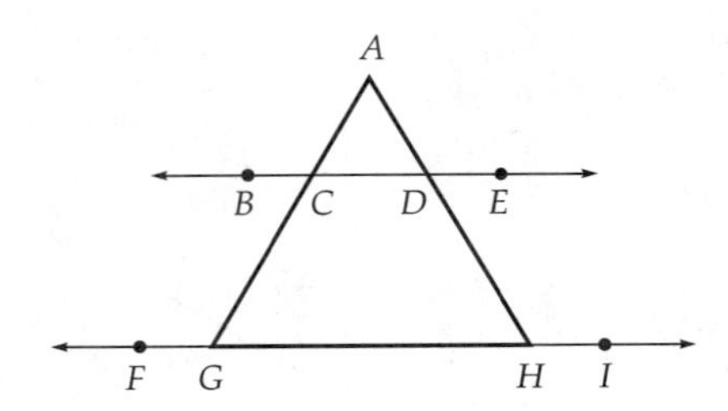

(a) Name four pairs of congruent corresponding angles.

(b) Name four pairs of congruent alternate interior angles.

42. The Parallel Line and Alternate Interior Angles Theorem can be used to prove that the sum of the interior angles of a triangle is 180°.

(a) Draw $\triangle ABC$ and $\overleftrightarrow{AD}$ parallel to $\overleftrightarrow{BC}$, and label the angles as shown.

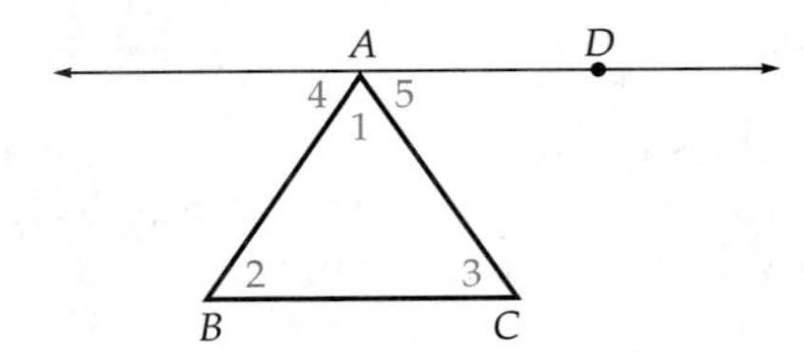

(b) What do we want to prove about $\angle 1$, $\angle 2$, and $\angle 3$?

(c) Name two congruent pairs of alternate interior angles.

(d) Explain why $m\angle 1 + m\angle 2 + m\angle 3 = 180°$. (*Hint:* Start with an equation that includes $m\angle 1$, $m\angle 4$, and $m\angle 5$.)

43. (a) Place a mirror on the line. What is the reflection of the name MATT?

M
A
T
T

(b) Make up another name that matches its own vertical reflection.

44. Rotate the following sign (from *Unexpected Hanging* by Martin Gardner) 180°. What does it say?

NOW NO
SWIMS
ON MON

45. Under what conditions would each of the following be congruent?
(a) Two line segments (b) Two squares
(c) Two rays

46. $\triangle MAY \cong \triangle HUT$.
(a) Which angle of $\triangle HUT$ corresponds to $\angle M$?
(b) $TH =$ ________

47. (a) Select one of the repeated figures in the following drawing by Escher. Find a center point and the number of degrees of a clockwise rotation that maps it onto another figure.

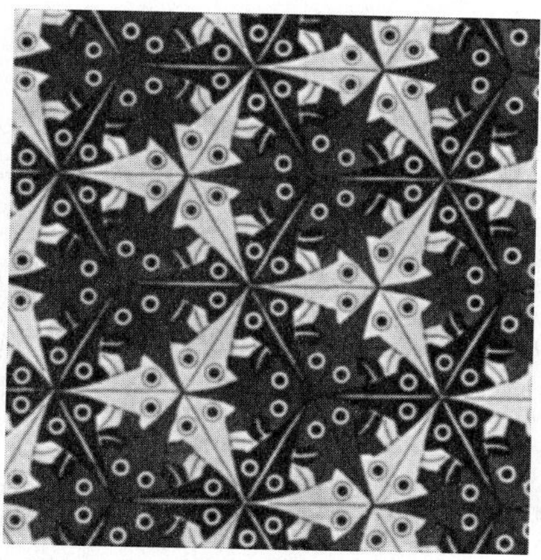

(b) Select one of the repeated figures. Find a line of reflection that maps it onto another figure of the same size and shape.

48.

Select one of the repeated figures. Describe a specific motion that maps it onto another figure of the same size and shape.

49. (a) Begin with a square. Alter one side, and translate the alteration to the opposite side. Does the resulting figure tessellate the plane?

(b) Repeat part (a) with a different tessellating polygon.

50. Complete the following transformations so that the new shape will tessellate.

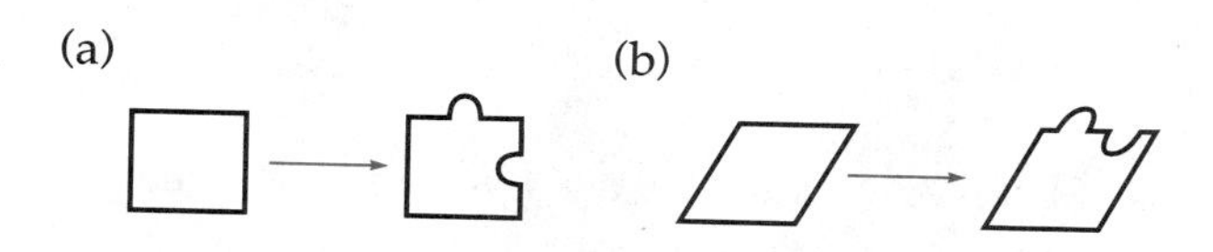

51. Make up your own tessellation.

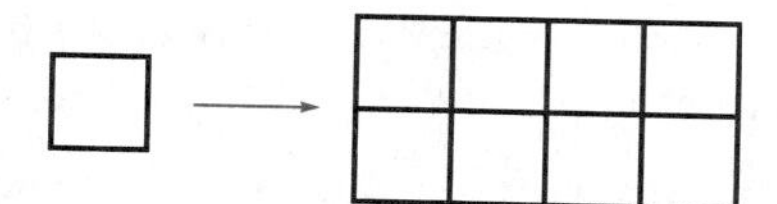

52. Make up your own tessellation.

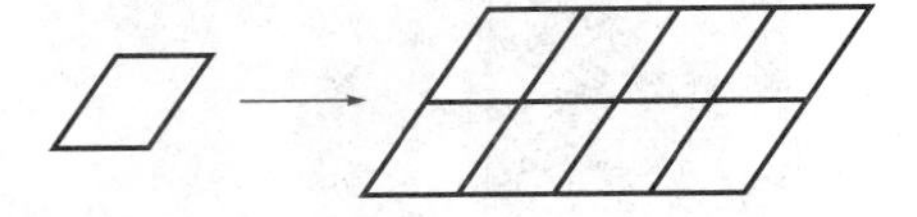

Extension Exercises

53. How was this photograph taken?

Courtesy of the Eames Office and IBM. © The Eames Office 1989.

54. Susan King wants to use two mirrors to see the back of her head. She stands facing one mirror $1\frac{1}{2}$ ft away. She holds a second mirror facing the opposite way, 1 ft behind her head. Her head is about 6 in. thick. How far from her eyes will the back of her head appear to be in the mirror? (Draw a picture.)

55. Suppose that you want to be able to stand in front of a wall mirror and see yourself from head to toe. Find the minimum possible height of the mirror, and tell how it would be positioned.

In Exercises 56–58, fill in the blanks with any of the following words that make the statement true:

rotation translation reflection

56. Under a ________ in a plane, one point is fixed and all other points move.

57. Under a ________ in a plane, no points remain fixed.

58. After a ________ in a plane, two lines that are parallel have images that are parallel to each other.

59. (a) What are the coordinates of A, B, and C?

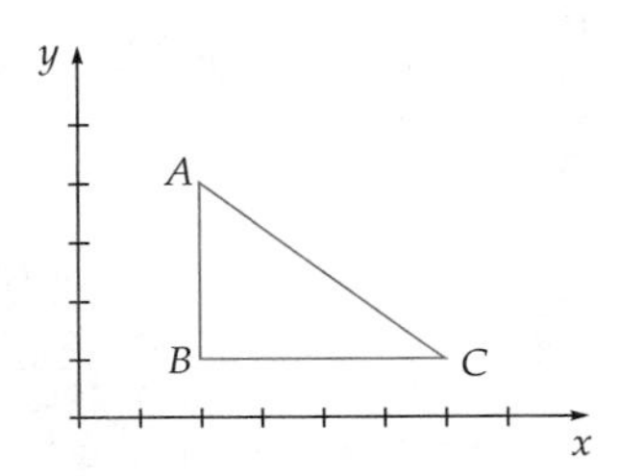

(b) Suppose that a mapping changes the coordinates of each point as follows.

$$(x, y) \rightarrow (x - 4, y + 1)$$

Find image points A', B', and C' using this rule. (Use graph paper or drawing software.)

(c) Plot A', B', and C'.

(d) What motion maps $\triangle ABC$ to $\triangle A'B'C'$?

60. (a) What are the coordinates of A, B, and C?

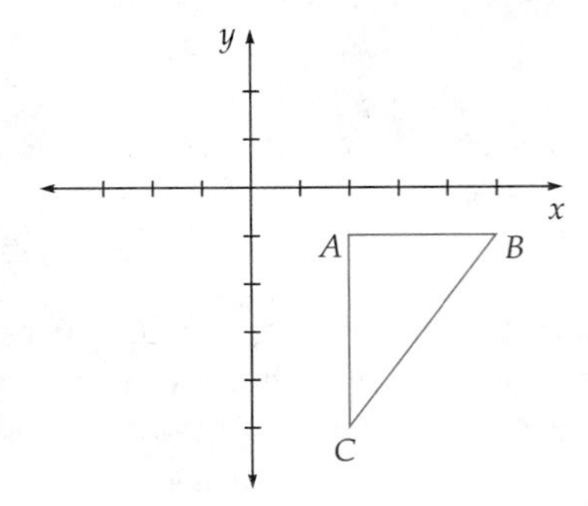

(b) Suppose that a mapping changes the coordinates of each point as follows.

$$(x, y) \rightarrow (y, x)$$

Find image points A', B', and C' using this rule.

(c) Plot A', B', and C'.

(d) What motion maps $\triangle ABC$ to $\triangle A'B'C'$?

61. (a) What are the coordinates of A, B, and C?

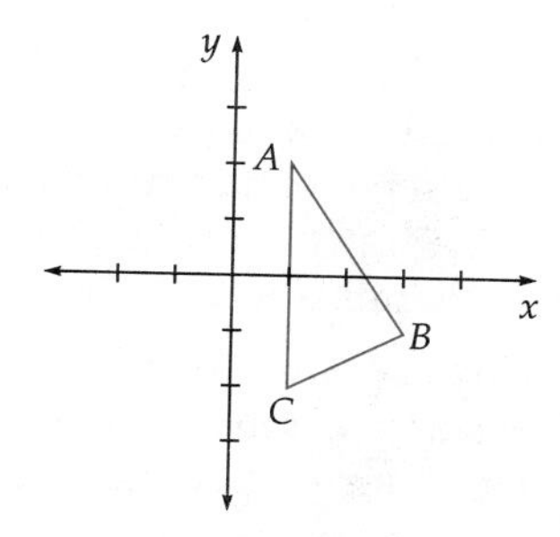

(b) Suppose that a mapping changes the coordinates of each point as follows.

$$(x, y) \rightarrow (-x, y)$$

Find image points A', B', and C' using this rule.

(c) Plot A', B', and C'.

(d) What motion maps $\triangle ABC$ to $\triangle A'B'C'$?

62. Consider the following problem. "What is the effect of the mapping $(x, y) \rightarrow (x, 2y)$ on different figures on a coordinate grid? Check all four quadrants." Devise a plan and solve the problem.

63. A sheet of paper is folded in quarters, and a square is cut out of the center. What will the paper look like when it is unfolded?

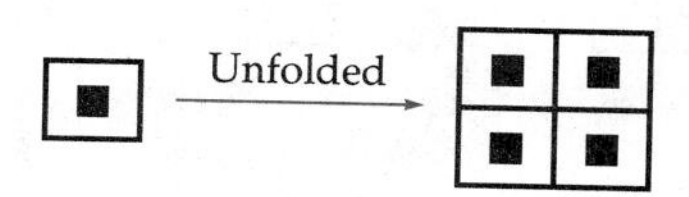

For each of the following designs, the folded and cut paper is shown on the left. Guess what the unfolded design will look like. Then try it, and see if you are right.

(a) Unfolded →

(b) Unfolded →

64. (a) Draw a letter F. Draw and label three different points, A, B, and C, that are not on the letter F.

(b) Rotate the F a measure of 50° clockwise around point A.

(c) Rotate the F a measure of 50° clockwise around point B.

(d) Rotate the F a measure of 50° clockwise around point C.

(e) How are the images in parts (b)–(d) related to one another?

(f) Write a generalization of your results.

65. (a) Draw a letter F. Draw and label two different points, A and B, that are not on the letter F.

(b) Rotate the F a measure of 50° clockwise around point A.

(c) Rotate the image from part (b) 70° clockwise around point B.

(d) What single motion would map the original F to the final image from part (c)? Be as specific as possible.

(e) Repeat parts (b) and (c) a few times, but vary the number of degrees in the rotation in part (c) and see what happens.

(f) Make a generalization of your results.

66. Draw two points, A and A', that are 5 cm apart.

(a) Find parallel lines m and n so that the translation from A to A' is the same as a reflection through line m followed by a reflection through line n.

(b) If your lines m and n are both between A and A', find another solution in which they are not. If they are not both between A and A', find another solution in which they are.

67. (a) How many trapezoids are in the star diagram on the next page?

(b) How many rhombuses are in the star diagram?

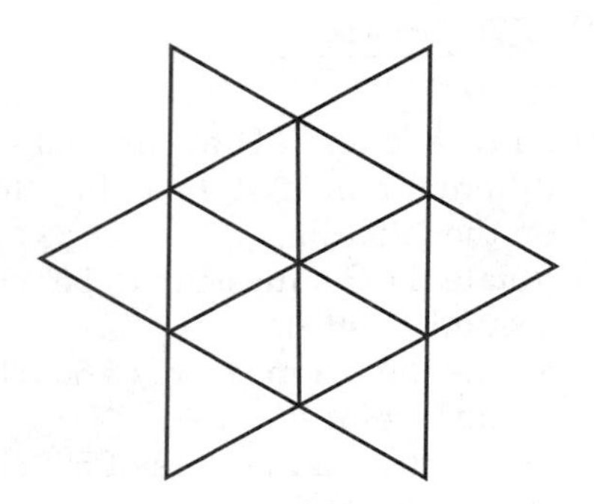

Computer Exercises

68. (a) What motion is illustrated by the following?

```
TO FLAG
FD 60
REPEAT 4 [RT 90 FD 20]
BK 60
END

FLAG
RT 110
FLAG
```

(b) Write a procedure that draws a flag and its image after a 200° counterclockwise rotation.

69. Write a Logo procedure that draws an equilateral triangle and its image after a 140° clockwise rotation around one of its vertices.

70. Use the FLAG procedure (see Exercise 68) in another procedure that draws a flag and then draws its image 30 units to the right. (*Hint:* Use PU between drawing the flag and drawing its image.)

71. Write a Logo procedure that uses another procedure, SQUARE, to draw a series of four squares that tessellate.

72. Write a Logo procedure that uses another procedure, HEXAGON, to draw three regular hexagons that tessellate.

73. Use drawing software to draw two parallel lines and a transversal.

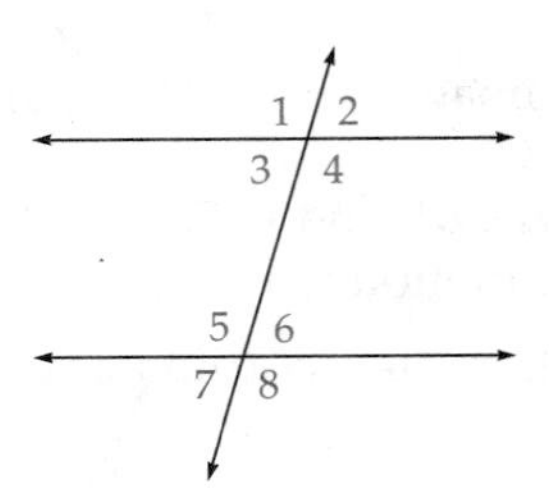

(a) Measure all the angles.
(b) Clear the screen and repeat part (a) after drawing a transversal at a different angle.
(c) Write generalizations of your results.

74. Use drawing software to create a design. Then rotate, translate, or reflect the design to create a repeating pattern (tessellation).

Special Exercise

75. Pentominos are made up of five connected squares. Each square must be connected to some other square on at least one *complete* side.

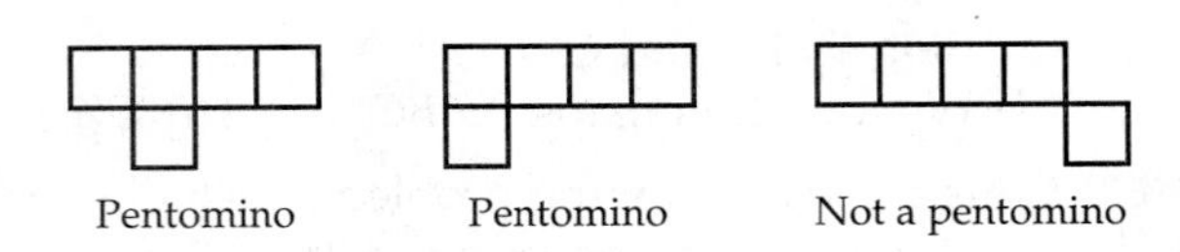

(a) Which two of the following pentominos are congruent? What motion maps one congruent pentomino onto the other?

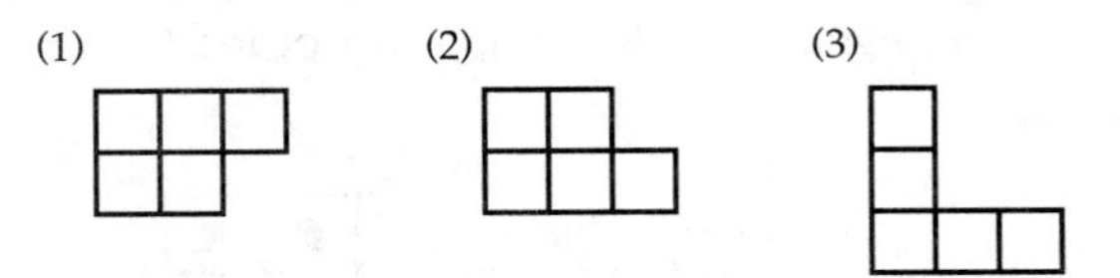

(b) Draw 12 different (noncongruent) pentomino shapes.
(c) Cut out the 12 different pentomino shapes. See whether you can put them together to form a single rectangle.

9.2 CONSTRUCTIONS AND CONGRUENCE

Ancient Greek mathematicians enjoyed the challenge of copying plane figures by drawing a series of line segments (using a straightedge) and circles (using a compass), since they considered lines and circles to be the basic units of geometry. Many plane figures can be copied or divided into two equal parts using only lines and circles.

The ancient Greeks succeeded in constructing nearly everything they attempted, but they were unable to trisect an angle or to construct a square equal in area to a given circle. More recently, it has been proven that these two constructions are impossible.

Today, we do constructions as historical rituals and as thought-provoking puzzles. As you do your constructions, think of the ancient Greeks who puzzled over the same problems, wondering what they could and could not construct with limited tools.

Lesson Exercise 9.26

Before doing anything fancy, draw the basic figures—a circle and a line segment—with your compass and straightedge, respectively.

CONSTRUCTING CONGRUENT FIGURES

What kind of congruent figures can be constructed with a compass and a straightedge? For starters, you should be able to draw a circle and a line segment congruent to those that you drew in Lesson Exercise 9.26.

Lesson Exercise 9.27

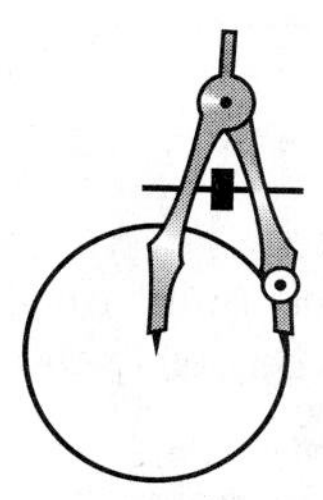

Figure 9–32

Construction: Copying a Circle

(a) Do two circles with radii of equal length have to be congruent?

(b) Part (a) is the basis for copying a circle. Place the point of your compass at the center of the circle (from Lesson Exercise 9.26) and the pencil point on any point of the circle (Figure 9-32).

(c) Hold this distance (the radius), and use it to draw a congruent circle elsewhere on the paper.

Lesson Exercise 9.28

Construction: Copying a Line Segment

(a) What makes two line segments congruent?

(b) Draw a ray with endpoint A that is longer than your line segment from Lesson Exercise 9.26.

A

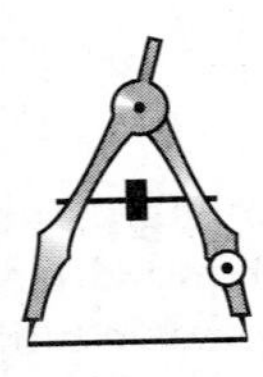

Figure 9–33

(c) Use your compass to span the length of the segment (from Lesson Exercise 9.26) by placing the compass point on one endpoint and the pencil point on the other (Figure 9-33).

(d) Holding the compass width from part (c), move the compass point to A and place the pencil point on the ray, marking its location as B. Segment AB is a copy of your original segment.

The first two constructions of congruent figures are not all that impressive. Let's move on to some more interesting constructions. (*Note:* If you prefer to use drawing software to investigate the conditions that make two triangles congruent, see the last computer exercise in the homework exercises.)

Lesson Exercise 9.29

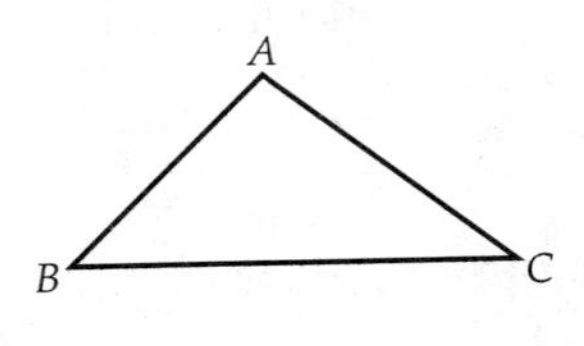

Figure 9–34

Construction: Copying a Triangle (SSS Method)

(a) Can you figure out a way to construct a copy of $\triangle ABC$ (Figure 9-34) by copying each of its sides? If not, go on to part (b).

If your method worked, compare it to the method outlined in parts (b)–(e).

(b) Copy $\overline{BC}$ using the method of Lesson Exercise 9.28. Label the copy $\overline{EF}$ (Figure 9.35).

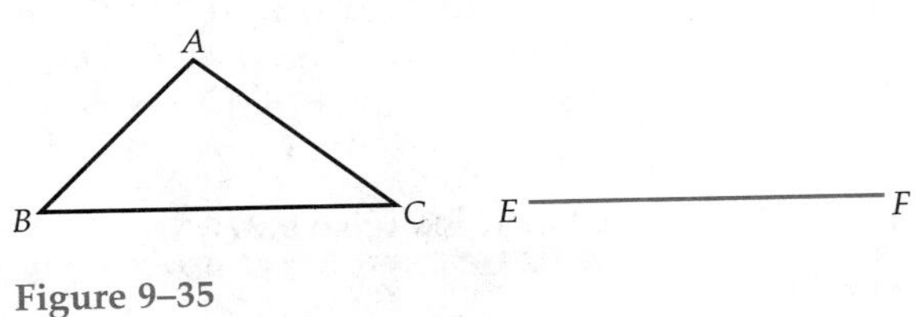

Figure 9–35

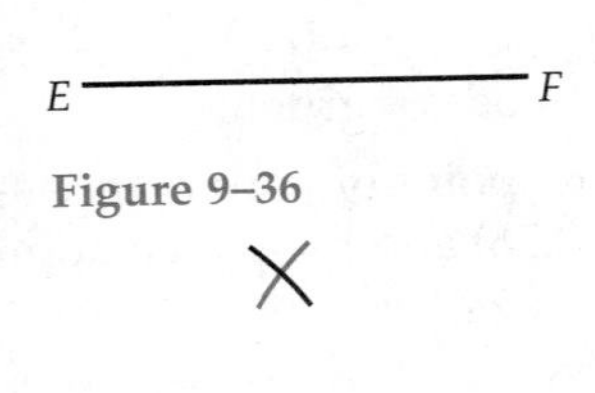

Figure 9–36

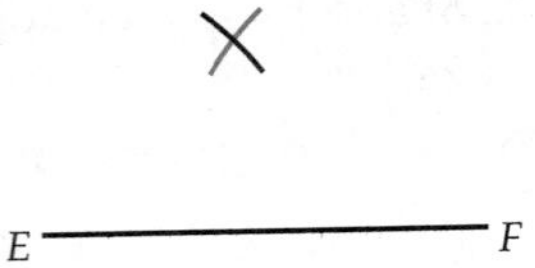

E F

Figure 9–37

(c) Place the point of your compass on B and the pencil point on A. Now use the same compass opening to draw an arc with center E and radius BA (Figure 9-36).

(d) Place the point of your compass on C and the pencil point on A. Now use the same compass opening to draw an arc with center F and radius CA (Figure 9-37) that intersects the last arc you drew.

(e) Label the intersection point D. Draw $\overline{DE}$ and $\overline{DF}$. Voila! You have constructed $\triangle DEF$ (Figure 9-38).

(f) Use tracing paper to check if $\triangle DEF \cong \triangle ABC$.

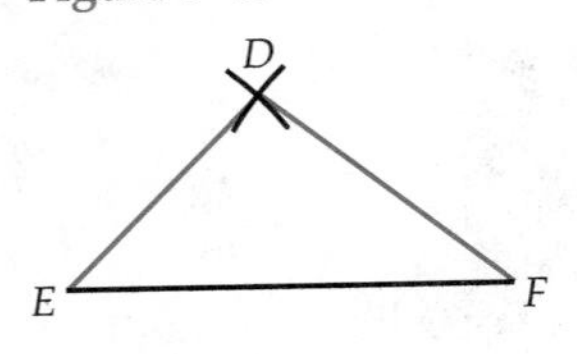

Figure 9–38

The construction in Lesson Exercise 9.29 suggests that making $EF = BC$, $DE = AB$, and $DF = AC$ will make $\triangle DEF \cong \triangle ABC$. This is an example of a property called the SSS (three-sides) property of triangle congruence.

> **SSS Triangle Congruence**
>
> If three sides of one triangle are congruent to the three corresponding sides of another triangle, then the two triangles are congruent.

Lesson Exercise 9.30

Which pairs of triangles must be congruent by SSS?

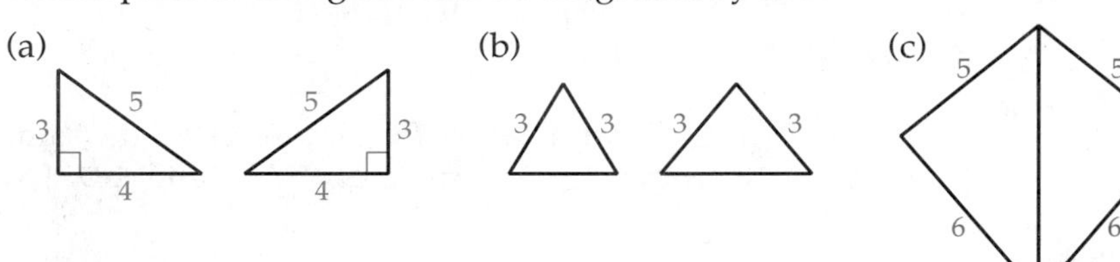

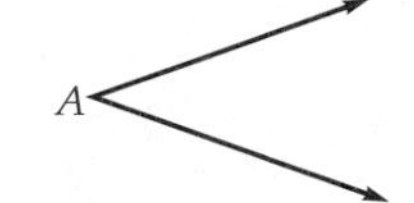

Figure 9–39

You've copied circles, line segments, and triangles. Now try copying an angle.

D

Figure 9–40

Lesson Exercise 9.31

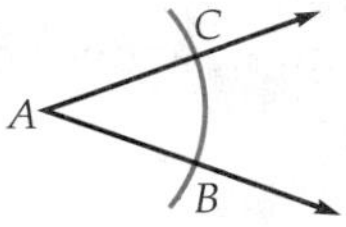

Construction: Copying an Angle (Figure 9-39)

(a) Draw a ray with endpoint D (Figure 9-40).

(b) Draw an arc with center A, and mark points B and C on the sides of $\angle A$ (Figure 9-41).

A C B

Figure 9–41

(c) Using the *same compass opening*, draw an arc with center D that intersects the ray as shown. Label the intersection of the arc and the ray as point E (Figure 9-42).

(d) See whether you can complete the construction on your own. If not, go on to parts (e) and (f).

(e) Now measure BC with the span of your compass, and use the same compass opening as you place the compass point on E and draw an arc intersecting the arc that is already there (Figure 9-43).

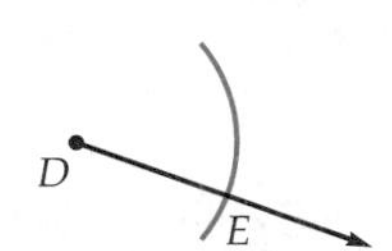

Figure 9–42

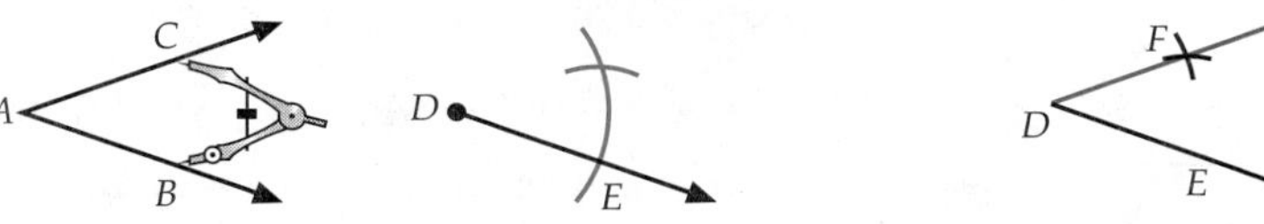

Figure 9–43

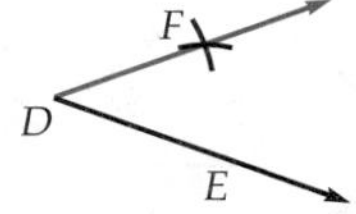

Figure 9–44

(f) Label the intersection of your two arcs F, and draw $\overrightarrow{DF}$ (Figure 9-44). $\angle FDE \cong \angle CAB$!

Why does the procedure for copying an angle work? You begin with $\angle A$ and mark an equal distance on each side (Figure 9-45). That distance is copied on each side of $\angle D$.

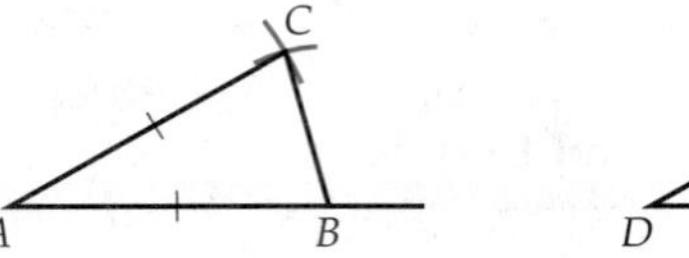

Figure 9–45

You measure $\overline{BC}$ and make $\overline{EF}$ the same length. So you have constructed two congruent triangles, $\triangle ABC$ and $\triangle DEF$ (Figure 9-46).

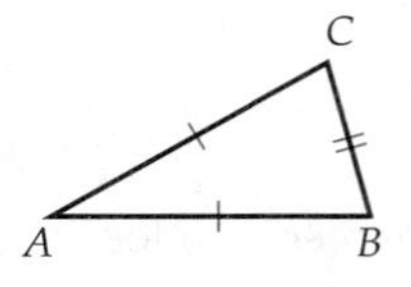

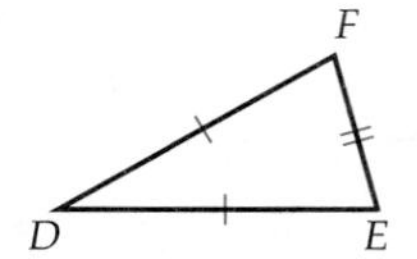

Figure 9–46

Lesson Exercise 9.32

The following questions concern how the preceding construction works.

(a) Why is $\triangle ABC \cong \triangle DEF$?

(b) From part (a), we know that $\triangle ABC \cong \triangle DEF$. We can now deduce that $\angle A \cong$ ________.

(c) Explain why $\angle A \cong \angle D$ in part (b).

So the construction for copying an angle is based upon the SSS triangle property! Now try copying a triangle, given two sides and the angle between them.

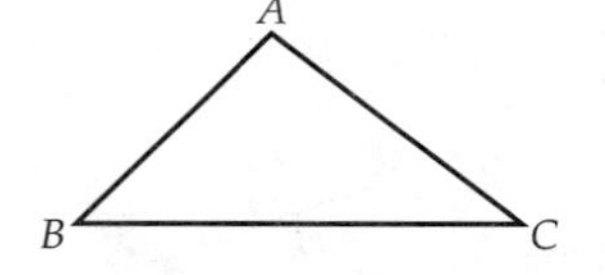

Figure 9–47

Lesson Exercise 9.33

Construction: Copying a Triangle (SAS Method)

(a) Can you figure out a way to construct a copy of $\triangle ABC$ (Figure 9-47) by copying two of its sides and the angle formed by those sides (such as $\overline{AB}$, $\overline{BC}$, and $\angle B$)? If not, go on to part (b).

If your method worked, compare it to the method outlined in parts (b)–(e).

(b) Copy angle B, employing the method of Lesson Exercise 9.31. Label the vertex of the new angle E (Figure 9-48).

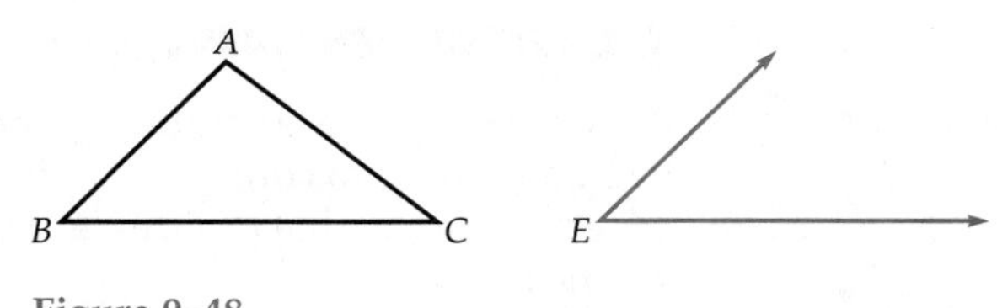

Figure 9–48

(c) See whether you can complete the construction on your own. If not, go on to parts (d) and (e).

(d) Mark the lengths BC and BA from point E on the two rays, and label the points F and D, respectively (Figure 9-49).

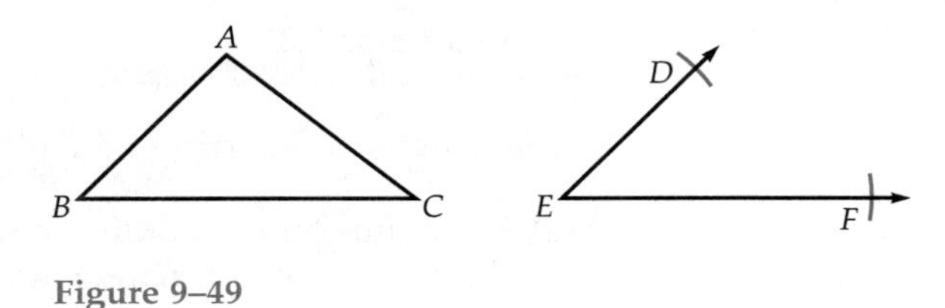

Figure 9–49

(e) Draw $\overline{DF}$ to complete $\triangle DEF$.

(f) Use tracing paper to check if $\triangle DEF \cong \triangle ABC$.

The construction in Lesson Exercise 9.33 suggests that making $m\angle B = m\angle E$, $BC = EF$, and $AB = DE$ will make $\triangle DEF \cong \triangle ABC$. This is an example of the SAS property of triangle congruence.

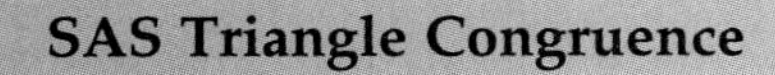

SAS Triangle Congruence

If two sides and the included angle of one triangle are congruent to two corresponding sides and the included angle of another triangle, respectively, then the two triangles are congruent.

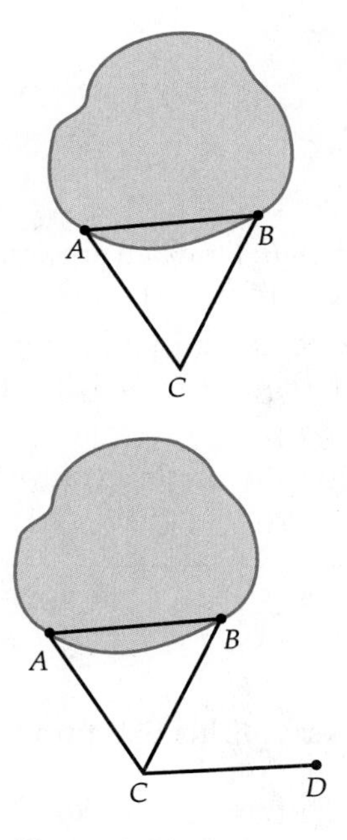

Figure 9–50

Lesson Exercise 9.34

You want to measure indirectly the distance AB across a lake (Figure 9-50). After selecting a point C, measure AC and BC.
Next, construct $\angle BCD$ so that $\angle BCD \cong \angle BCA$ and $CD = CA$ (Figure 9-50).
Explain why BD must be the same as the length across the lake (BA). (*Hint:* First explain why $\triangle ACB \cong \triangle DCB$. Then tell why $BD = BA$.)

CONSTRUCTING BISECTORS

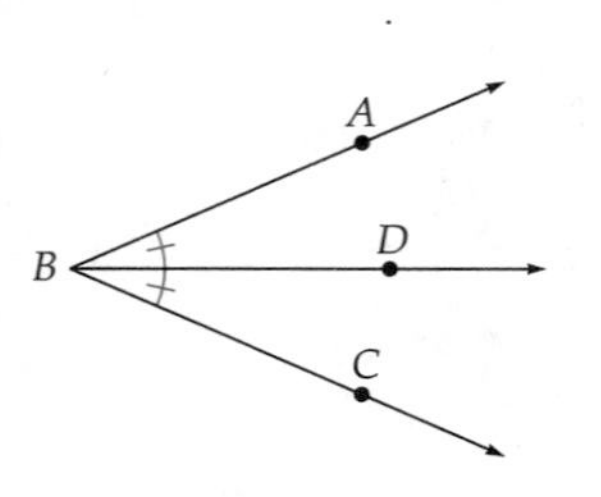

Figure 9–51

A compass and a straightedge can also be used to divide an angle or a line segment into two congruent parts. This procedure is called **bisection**. An **angle bisector** is a ray that divides an angle into two congruent angles. In Figure 9-51, $\overrightarrow{BD}$ bisects $\angle ABC$.

A **midpoint** (bisector) is a point that divides a line segment into two congruent segments. In the following diagram, M is the midpoint of $\overline{AY}$.

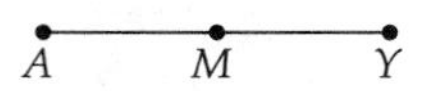

First, try bisecting an angle.

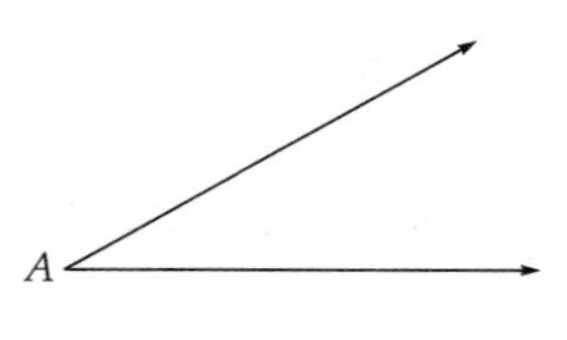

Figure 9–52

Lesson Exercise 9.35

Construction: Bisecting an Angle (Figure 9-52)

(a) Place the compass point on A, and draw an arc intersecting both sides of the angle. Label the intersection points B and C (Figure 9-53).

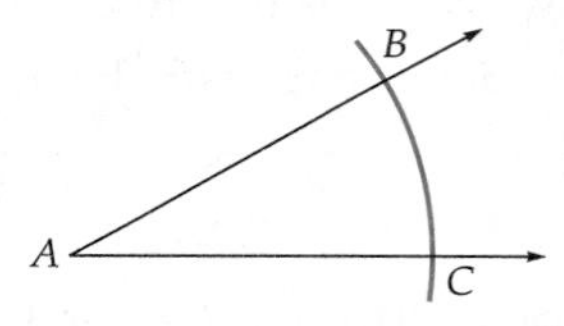

Figure 9–53

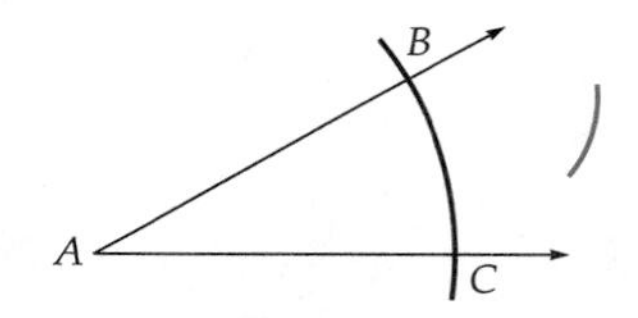

Figure 9–54

(b) Now select a radius and place the compass point on B. Draw an arc across the interior of the angle (on the opposite side of $\overline{BC}$ from A) (Figure 9-54).

(c) Place the compass point on C, and use the same compass opening to draw an arc intersecting the arc that has center B (Figure 9-55). Label the intersection point D.

(d) Connect the intersection point to A. Now $\overrightarrow{AD}$ is the angle bisector of $\angle BAC$ (Figure 9-56).

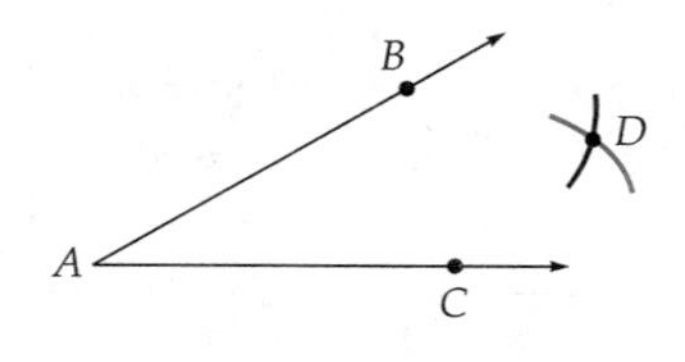

Figure 9–55

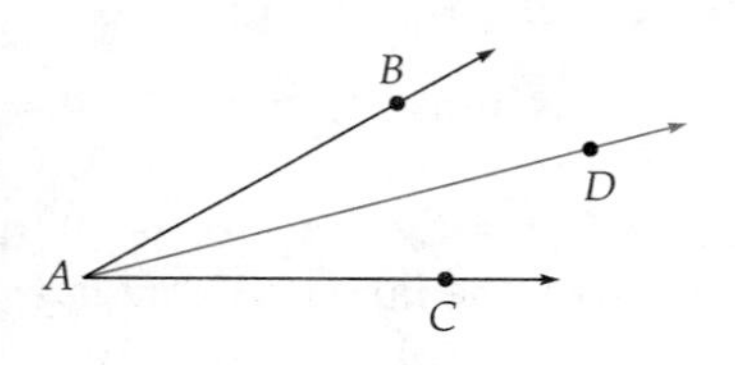

Figure 9–56

This construction also works because of the SSS property. (This SSS property is very handy for explaining constructions!)

In the construction in Figure 9-57, $AB = AC$ and $BD = CD$ because the lengths in each pair were drawn with the same radius.

Lesson Exercise 9.36

(a) In Figure 9-57, why is $\triangle ABD \cong \triangle ACD$ by SSS?

(b) If $\triangle ABD \cong \triangle ACD$, what other corresponding parts of the two triangles are congruent?

(c) Explain why $\overrightarrow{AD}$ must be an angle bisector of $\angle BAC$.

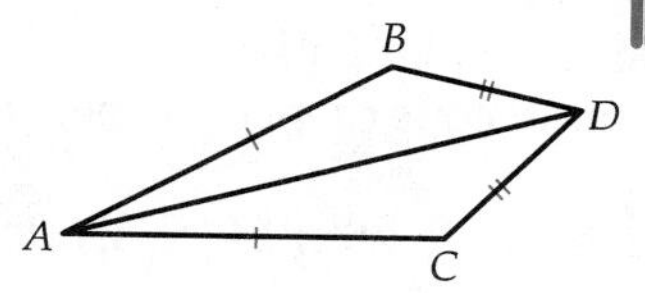

Figure 9–57

In the next construction, you will locate the midpoint of a line segment using a line perpendicular to the line segment. This line is called the **perpendicular bisector** of the line segment. In other words, the construction that bisects the line segment also creates the perpendicular bisector.

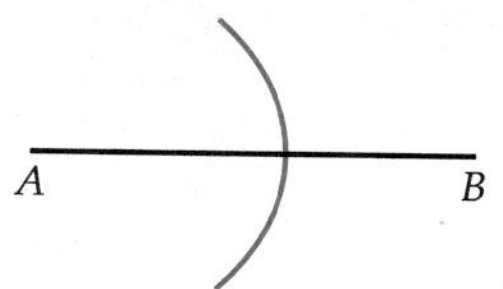
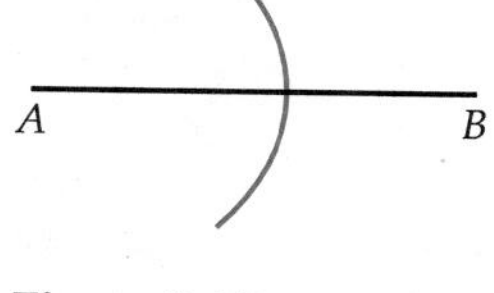

Figure 9–58

Lesson Exercise 9.37

Construction: Finding the Perpendicular Bisector of a Line Segment

(a) Draw a line segment $\overline{AB}$. Select a compass radius slightly less than AB. Place the compass point on A, and draw an arc through $\overline{AB}$ as shown in Figure 9-58.

(b) Use the same radius. Place the compass point on B, and draw an arc above and below $\overline{AB}$ that intersects the other arc at two points (Figure 9-59). Label the intersection points C and D.

(c) Connect those two intersection points with a line segment. Label as M the point of intersection with $\overline{AB}$. Not only is M the midpoint of $\overline{AB}$, but $\overleftrightarrow{CD}$ is the perpendicular bisector of $\overline{AB}$ (Figure 9-60). So this procedure can also be used to construct a 90° angle.

Figure 9–59

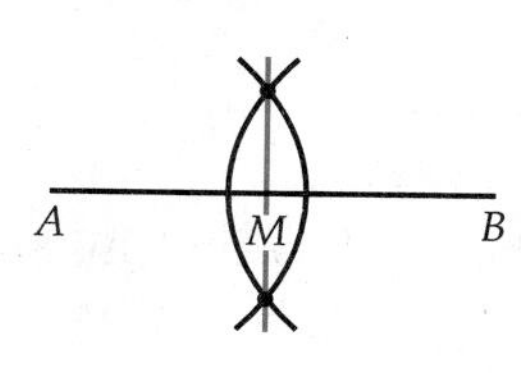

Figure 9–60

You can use the SSS and SAS properties to prove why this construction works.

Lesson Exercise 9.38

In the construction in Figure 9-61, $AC = CB = AD = DB$ because the lengths were all drawn with the same radius.

(a) Why is $\triangle ACD \cong \triangle BCD$?

(b) Why is $\angle ACM \cong \angle BCM$?

(c) Why is $\triangle ACM \cong \triangle BCM$?

(d) Why is M the midpoint of $\overline{AB}$?

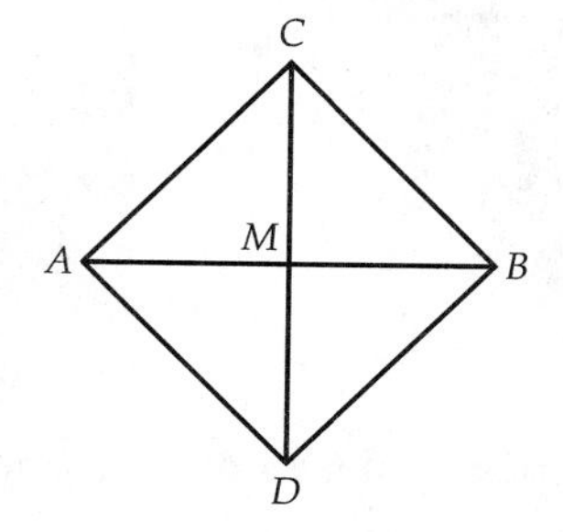

Figure 9–61

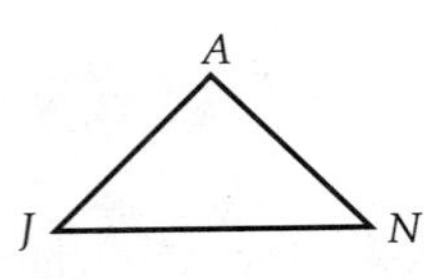

Figure 9–62

THE ISOSCELES TRIANGLE THEOREM

All isosceles triangles have at least two congruent sides. Must an isosceles triangle also have two congruent angles? The SSS triangle congruence property can be used to deduce the answer to this question.

Suppose that $\triangle JAN$ (Figure 9-62) is an isosceles triangle with $JA = AN$. Is $\angle J \cong \angle N$?

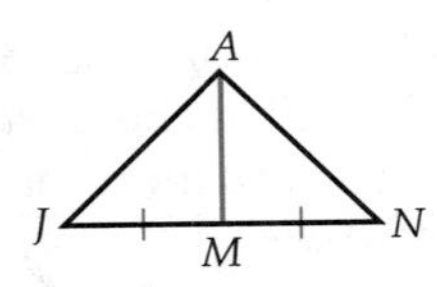

Figure 9–63

Lesson Exercise 9.39

(a) In $\triangle JAN$, draw a line segment from A to the midpoint M of $\overline{JN}$ (Figure 9-63).

(b) Explain why $\triangle JAM \cong \triangle NAM$.

(c) Explain why $\angle J \cong \angle N$.

The angle property you proved in Lesson Exercise 9.39 is called the Isosceles Triangle Theorem.

The Isosceles Triangle Theorem

If two sides of a triangle are congruent, then the angles opposite them are congruent.

The Isosceles Triangle Theorem can sometimes be used to find missing angle measures in polygons.

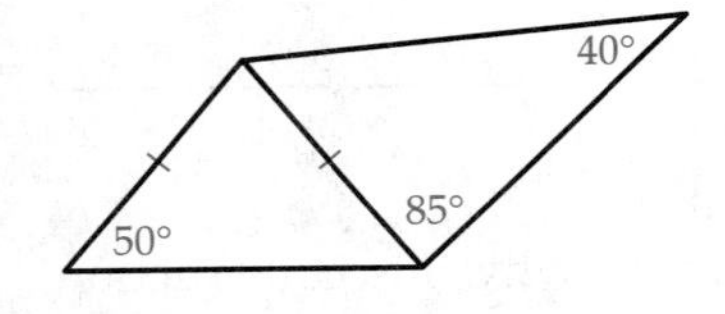

Figure 9–64

Lesson Exercise 9.40

Fill in the missing angle measures in Figure 9-64.

CONSTRUCTING PERPENDICULAR AND PARALLEL LINES

Suppose you have a point B not on a line $\overleftrightarrow{AC}$.

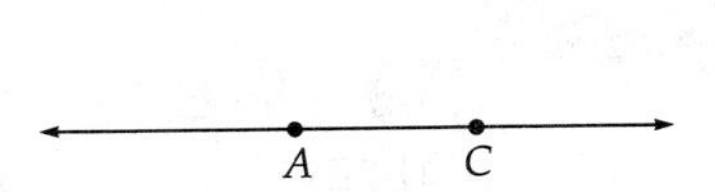

It is possible to construct a line through B that is perpendicular to $\overleftrightarrow{AC}$ or a line through B that is parallel to $\overleftrightarrow{AC}$. Consider the following application.

Lesson Exercise 9.41

You are camping and must return home for an emergency.

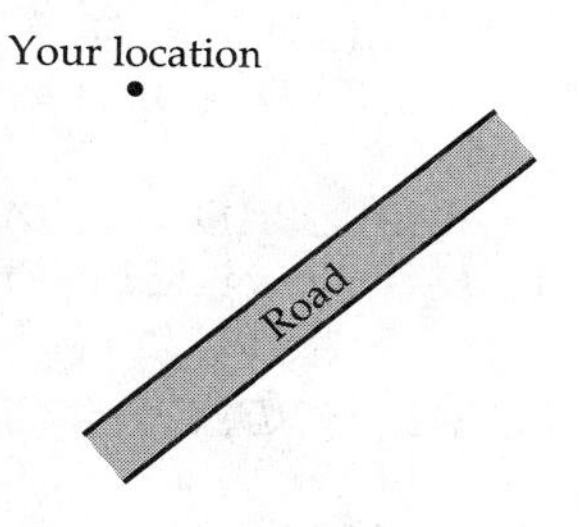

(a) What is the shortest route to the road?

Construction: Finding a Line Perpendicular to a Given Line Through a Point Not on the Line

(b) Construct the shortest route to the road. (*Hint:* Place the point of the compass on "your location," and draw two arcs intersecting the road at points A and B. Then construct the perpendicular bisector of $\overline{AB}$.)

To construct parallel lines, make use of the preceding construction of a perpendicular line.

Lesson Exercise 9.42

Construction: Finding a Line Parallel to a Given Line Through a Point Not on the Line

(a) Suppose B is a point not on $\overleftrightarrow{AC}$. Construct a line through B that is parallel to $\overleftrightarrow{AC}$. (*Hint:* First construct a line through B that is perpendicular to $\overleftrightarrow{AC}$.)

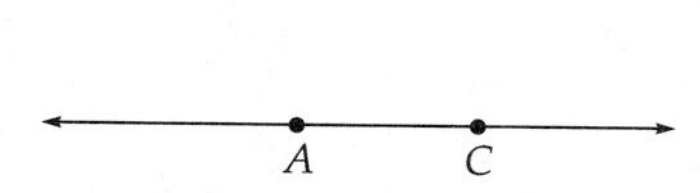

(b) Explain why your construction worked.

ANSWERS TO SELECTED LESSON EXERCISES

9.27 (a) Yes

9.28 (a) They are the same length.

9.30 (a), (c)

9.32 (a) SSS triangle congruence
(b) $\angle D$
(c) The triangles are congruent, so the corresponding parts are congruent.

9.34 Since $AC = CD$, $BC = BC$, and $\angle ACB \cong \angle DCB$, $\triangle ACB \cong \triangle DCB$ by SAS. If the triangles are congruent, then $BD = BA$.

9.36 (a) $AB = AC$, $BD = CD$, and $AD = AD$.
(b) $\angle B \cong \angle C$, $\angle BAD \cong \angle CAD$, $\angle BDA \cong \angle CDA$
(c) Since $\triangle ABD \cong \triangle ACD$, $\angle BAD \cong \angle CAD$. This means that $\overrightarrow{AD}$ is the angle bisector of $\angle BAC$.

9.38 (a) By SSS ($AC = BC$, $AD = BD$, $CD = CD$)
(b) Because $\triangle ACD \cong \triangle BCD$
(c) By SAS ($AC = BC$, $\angle ACM \cong \angle BCM$, $CM = CM$)
(d) Since $\triangle ACM \cong \triangle BCM$, $AM = MB$. So M is the midpoint of $\overline{AB}$.

9.39 (b) By SSS ($JA = NA$, $JM = NM$, $AM = AM$)
(c) Since $\triangle JAM \cong \triangle NAM$, $\angle J \cong \angle N$.

9.40

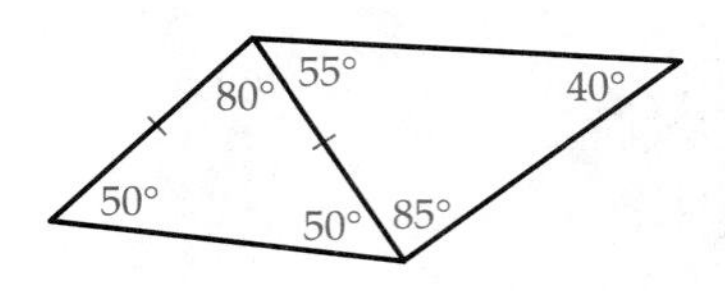

9.42 (a) *Hint 2:* Construct a second line through B perpendicular to the first line that you constructed.
(b) In a plane, two lines that are perpendicular to the same line are parallel to each other.

9.2 HOMEWORK EXERCISES

Basic Exercises

1. The lateral surface of a cone can be built from a region like the following one. Try it.

(a) Draw a circle with a compass, and cut it out.
(b) Draw two radii, and cut out the smaller closed region.

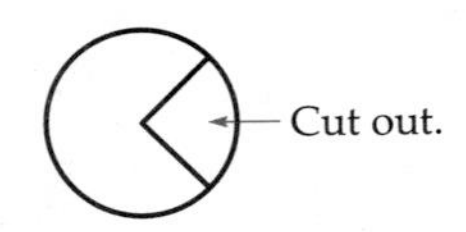

(c) Put the two cut edges together to form the lateral surface of a cone.

***2.** (a) Construct a copy of $\triangle MAN$, called $\triangle BCD$, so that $\triangle MAN \cong \triangle BCD$.

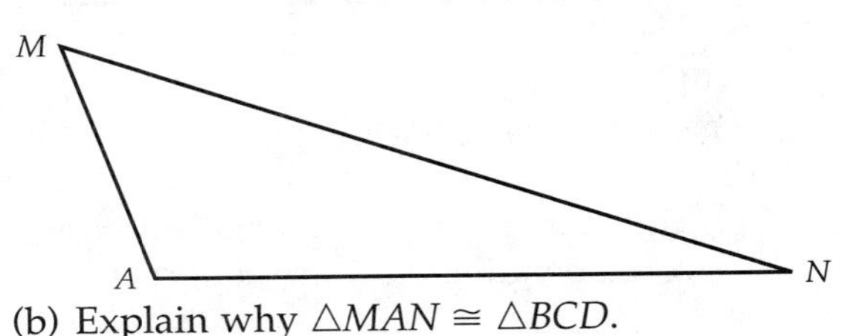

(b) Explain why $\triangle MAN \cong \triangle BCD$.

3. Construct a copy of $\triangle CDE$.

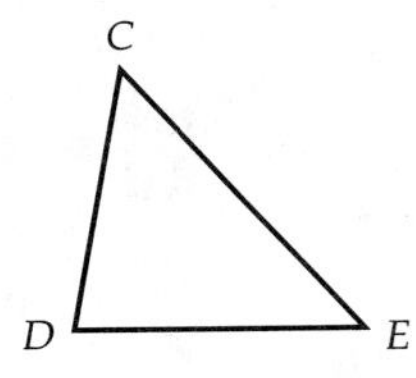

***4.** Which of the following pairs of triangles must be congruent by SSS?

(a) (b) (c)

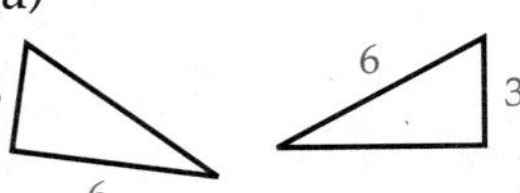

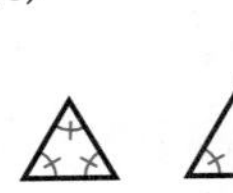

5. An architect constructs triangular roof supports, each with sides of length 0.7 m, 1.8 m, and 2.3 m. Are all the triangular supports congruent? Why or why not?

***6.**

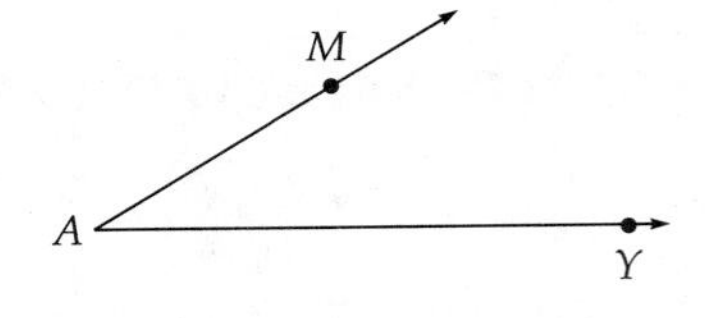

(a) Construct a copy of $\angle MAY$ called $\angle BCD$.
(b) Explain why $\angle MAY \cong \angle BCD$.

7. Construct a copy of $\angle TED$.

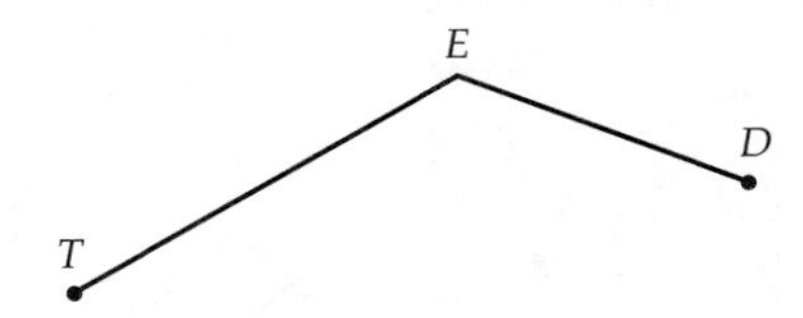

8.

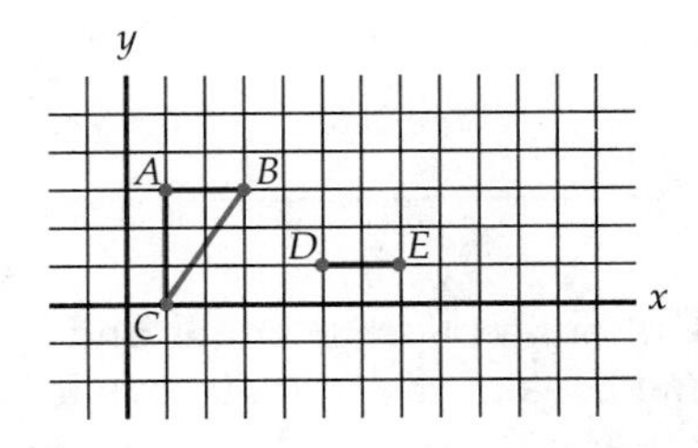

Give all possible coordinates of F that would determine a $\triangle DEF$ that is congruent to the triangle that has vertices A, B, and C.

9. $AB = BC$ and $m\angle ABC = 90°$. Construct the square $ABCD$.

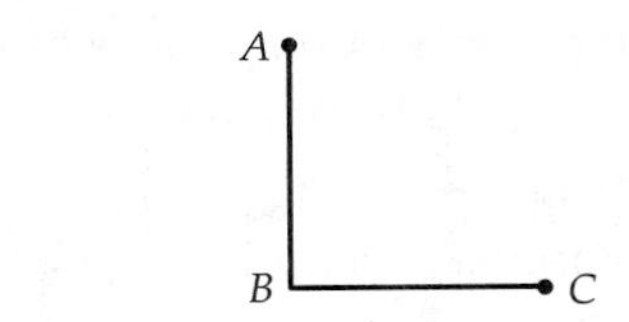

10.

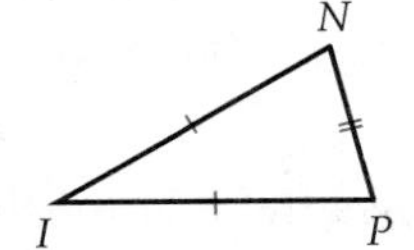

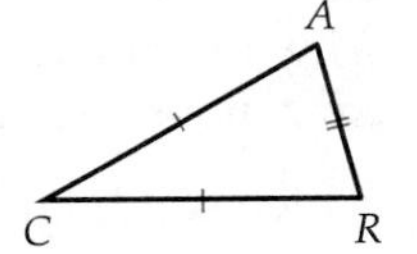

(a) Why is $\triangle NIP \cong \triangle ACR$?
(b) Why is $\angle I \cong \angle C$?

11. Copying a Triangle (ASA Method)

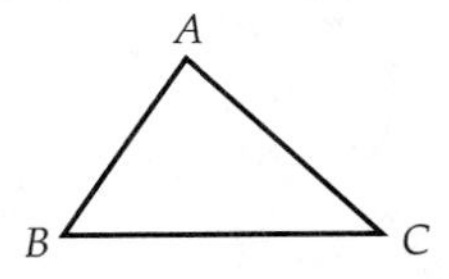

Show how to construct a copy of $\triangle ABC$ by copying two angles and the side between them (such as $\angle B$, $\angle C$, and $\overline{BC}$). This suggests the validity of the ASA triangle congruence property.

12. Would an AAS triangle congruence property work? Use the ASA property to show why. Assume that $\angle A \cong \angle D$, $\angle B \cong \angle E$, and $\overline{BC} \cong \overline{EF}$.

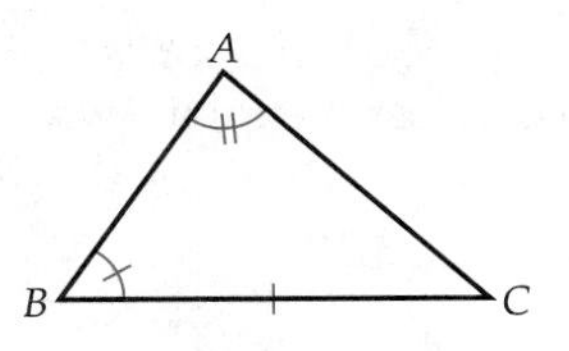

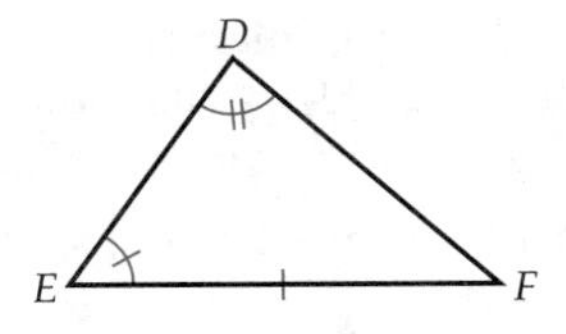

(a) Why is $\angle C \cong \angle F$?
(b) Why is $\triangle ABC \cong \triangle DEF$ (showing that AAS leads to congruent triangles)?

13. Decide whether each pair of triangles must be congruent.

(a)

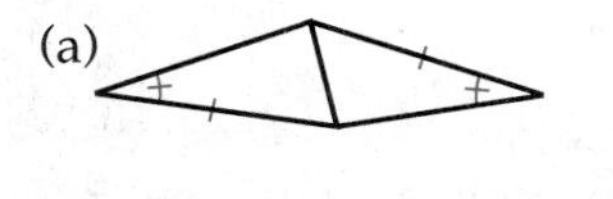

(b)

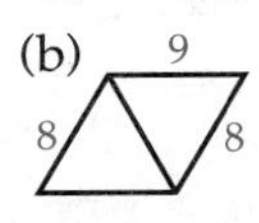

(c)

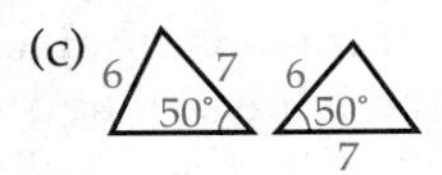

(d)

14. Create a counterexample to the SSA congruence property by completing the following figure. Draw $\triangle DEF$ with $DE = AB$, $DF = AC$, and $m\angle B = m\angle E$ so that $\triangle DEF$ is not congruent to $\triangle ABC$.

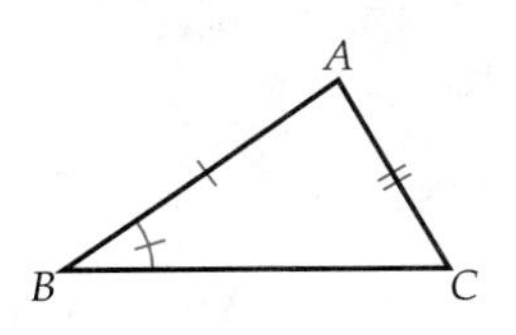

15.

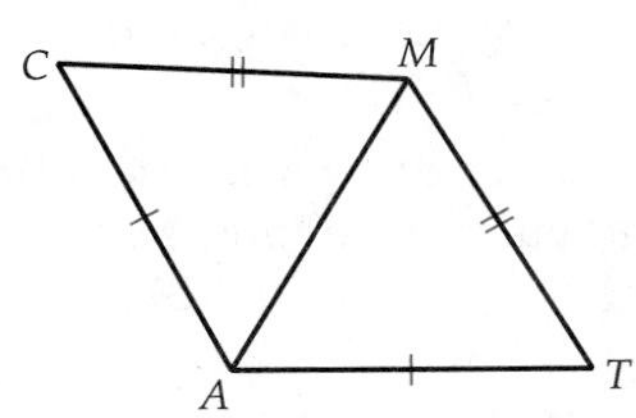

Explain why $\overrightarrow{AM}$ must be the angle bisector of $\angle CAT$.

***16.** (a) Construct the angle bisctor of $\angle CAT$ and call it $\overrightarrow{AB}$.

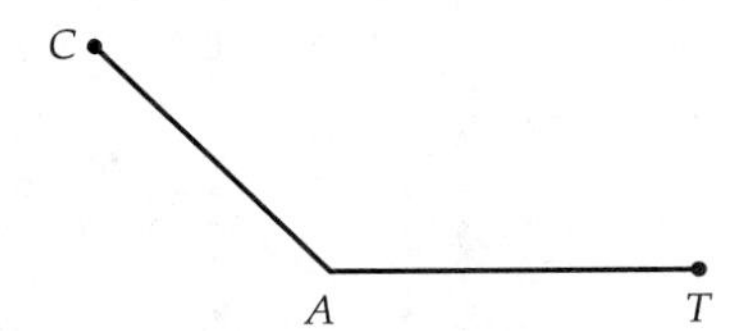

(b) Explain why your procedure works.
(c) Use congruent triangles to explain why $\overrightarrow{AB}$ bisects $\angle CAT$.

17. Assume that $\overrightarrow{EK}$ bisects $\angle MER$.

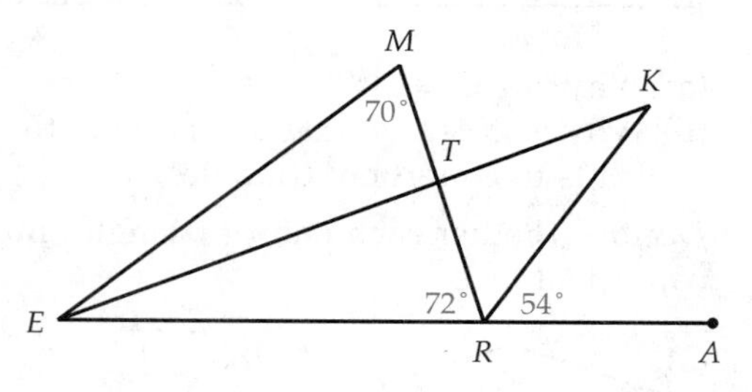

(a) Fill in all the missing angle measures.
(b) Name a pair of supplementary angles.

18. A plane is flying over water from A to B on a clear day with little wind. If there is trouble along the way, construct the point C that can be used to determine whether it would be easier for a plane with engine trouble to go on to B or to return to A.

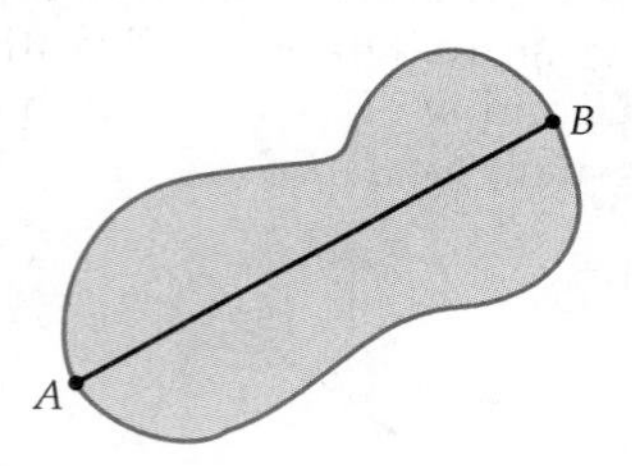

19. Show that a triangle $\triangle ABC$, with $\angle B \cong \angle C$, has two congruent sides. (*Hint:* Draw the angle bisector of $\angle BAC$.)

20. Use a ruler to draw a segment that is perpendicular to $\overline{AB}$ and bisects $\overline{AB}$.
(a) (b)

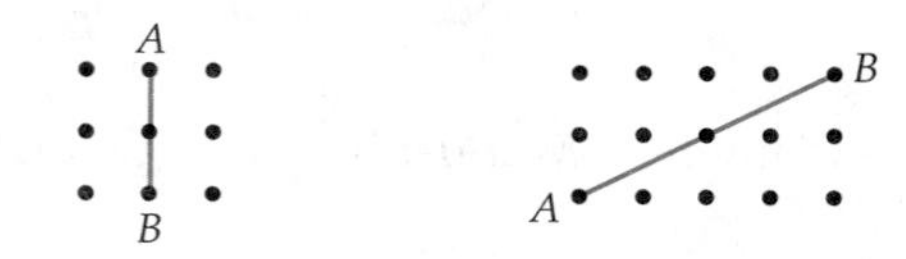

21. A **median** of a triangle connects a vertex to the midpoint of the opposite side.
(a) Use a ruler to draw the three medians of the triangle.

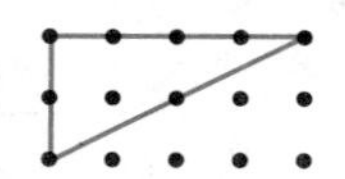

(b) Construct a median from A to the midpoint of $\overline{BC}$.

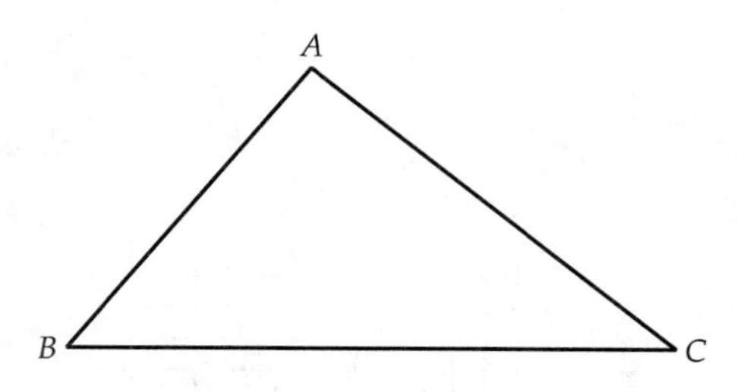

(c) Cut out a paper triangle, and fold the three medians of the triangle. What do you observe?

22. Use a straightedge and a compass to divide a segment into four equal parts.

23. Construct a line segment that is three times the length of $\overline{AB}$.

24. (a) Draw a line segment on a piece of paper.
(b) Locate its midpoint by paper folding.

25. (a) Draw an angle on a piece of paper.
(b) Locate its angle bisector by paper folding.

26. Use the Isosceles Triangle Theorem to explain why an equilateral triangle must have three congruent angles.

27. Which set diagram shows the relationship between equilateral triangles and right triangles?

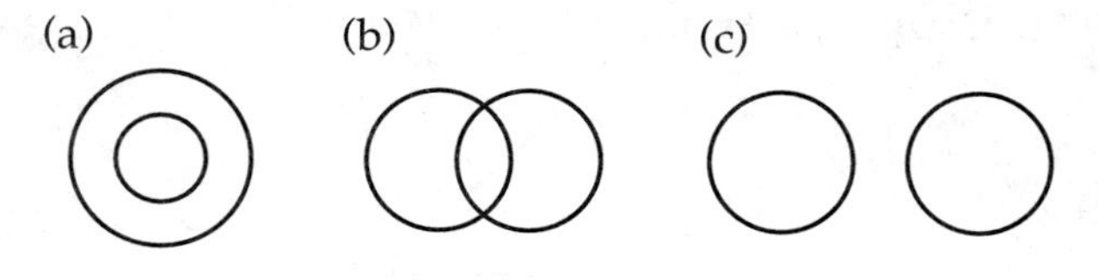

28. Find all missing angle measures in each diagram.

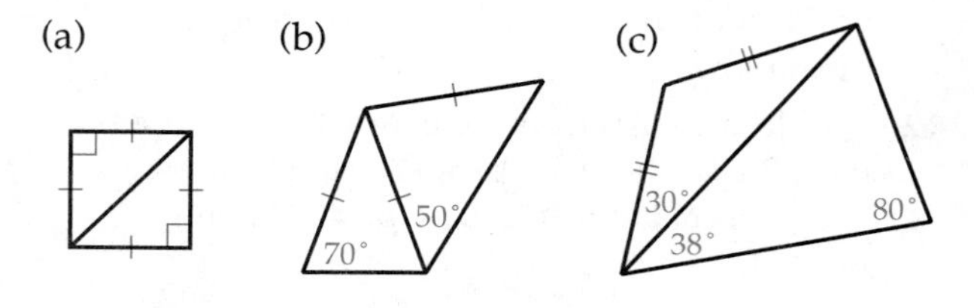

29. Regular hexagon *MARTIN* is inscribed in a circle with center *C*.

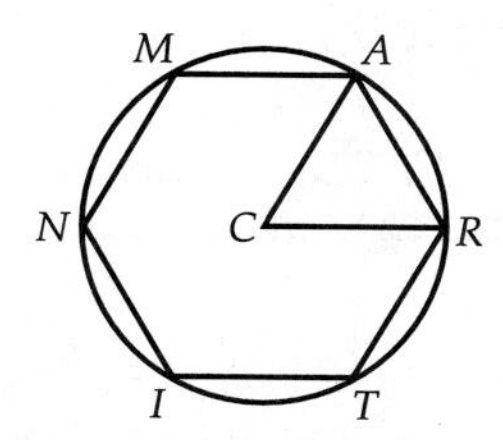

$m\angle ACR =$ __________

30. Find all the angle measures in the diagram.

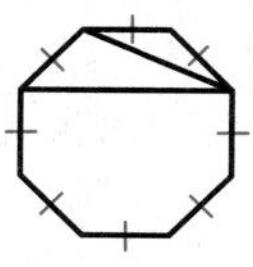

31. Construct a line perpendicular to m through P.

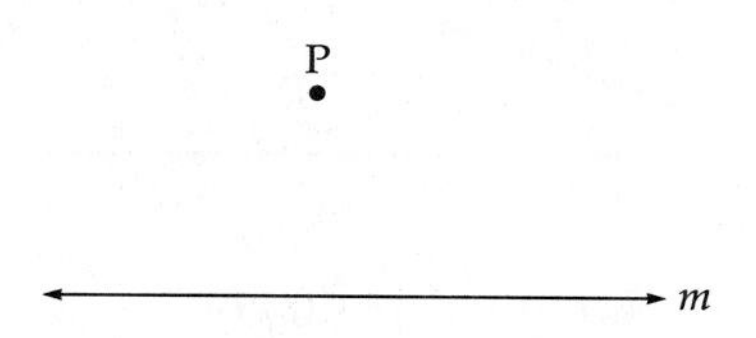

32. Construct a line parallel to m through P.

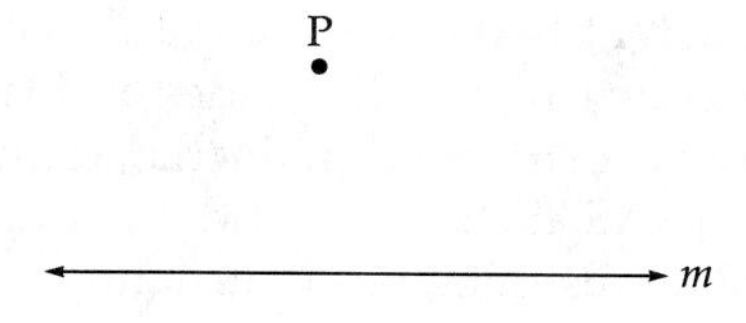

33. The triangular grid is composed of vertices of equilateral triangles. Consider parallelogram *ABCD*.

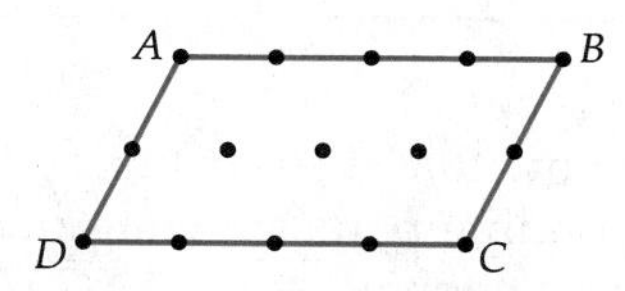

Draw diagonals $\overline{AC}$ and $\overline{BD}$. Check to see if they are congruent or perpendicular or if they bisect each other.

34. The triangular grid is composed of vertices of equilateral triangles. Consider rhombus *ABCD*.

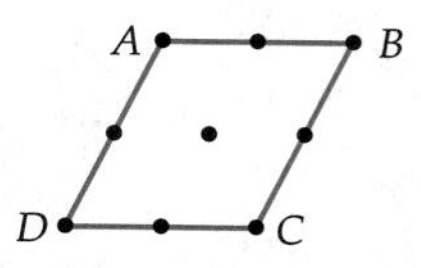

(a) Draw diagonals $\overline{AC}$ and $\overline{BD}$. Check to see if they are congruent or perpendicular or if they bisect each other.

(b) Since a square is a rhombus, part (a) shows that the diagonals of a square are ________.

35. Construct an equilateral $\triangle ABC$ with a side $\overline{AB}$. (*Hint:* C must be a distance AB away from both A and B.)

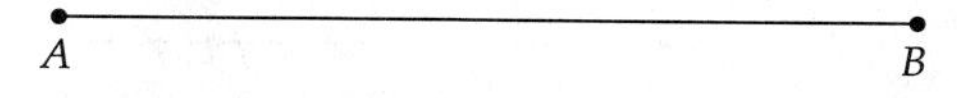

Extension Exercises

36. $m\angle PUN = 60°$. Construct an equilateral triangle $\triangle NUT$ with T on $\overline{UP}$.

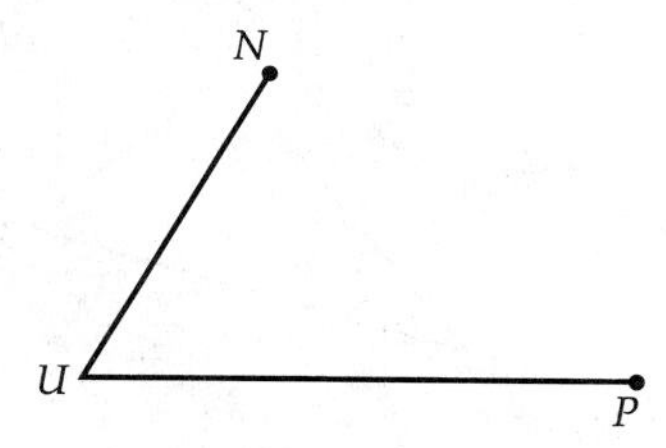

37. (a) Construct a 45° angle.

(b) Construct a 135° angle.

38. Draw a circle. Construct an inscribed regular hexagon as shown.

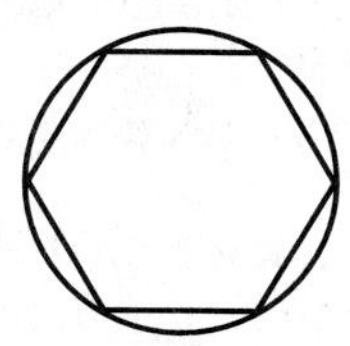

39. A jeweler designs a setting with a turquoise in the center, surrounded by six pearls, as shown. If the turquoise has a radius of 4 mm, what radius should the pearls have?

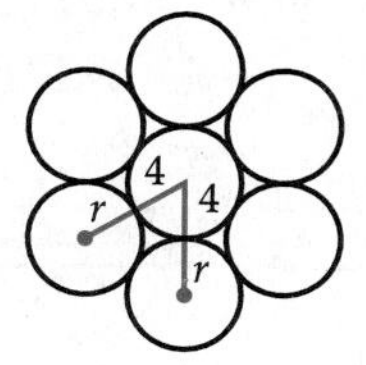

40. Construct $\angle GHI$ so that $m\angle GHI = m\angle ABC - m\angle DEF$.

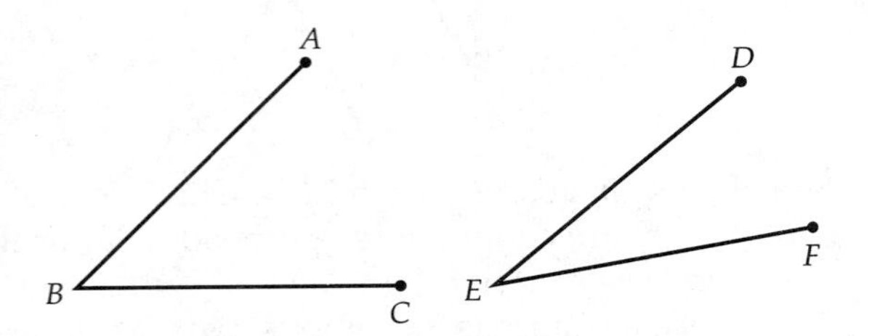

41. Suppose B is on a point $\overleftrightarrow{AC}$. Construct a line through B that is perpendicular to $\overleftrightarrow{AC}$. (*Hint:* Construct an angle bisector for $\angle ABC$.)

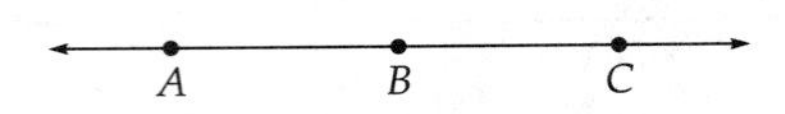

42. Construct a circumscribed circle that passes through the vertices of $\triangle ABC$.

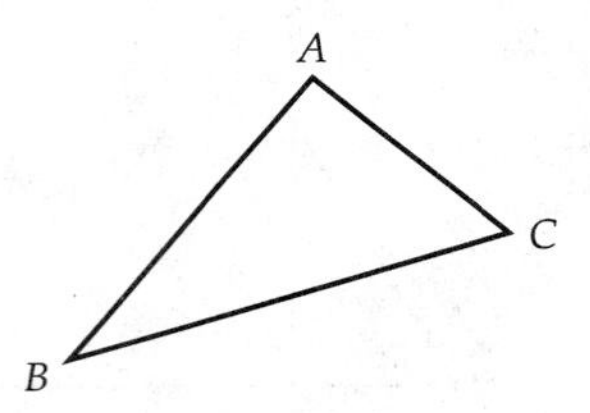

(a) The center of the circle, P, must be the same distance from A, B, and C. To find P, construct the perpendicular bisectors of two sides—say, $\overline{AB}$ and $\overline{BC}$. The intersection of the two perpendicular bisectors is P.

(b) Draw the circumscribed circle with center P.

(c) How do you know that P is the same distance from A and B, from B and C, and therefore from A, B, and C?

43. Imagine that you (A) see a boat at point B and want to know how far it is from the shore. Over 2000 years ago, Thales described the following method for finding the distance. Walk along the shore to a point C and mark it.

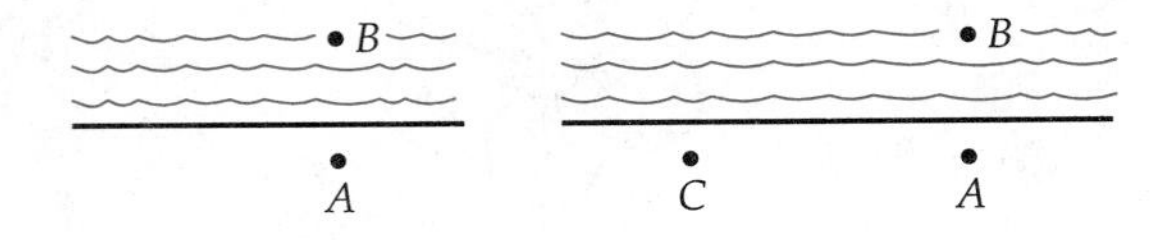

Continue walking an equal distance to point D so that $AC = CD$. Then walk directly away from the water until C is lined up with B, and mark this point E.

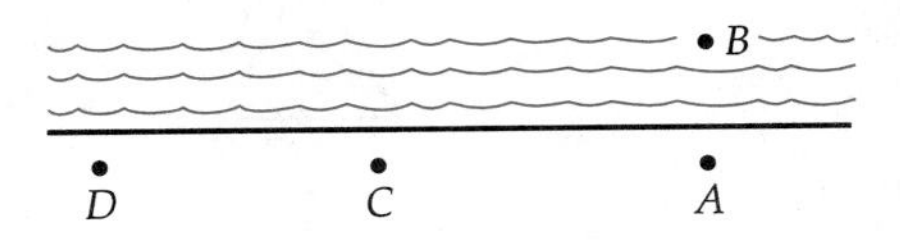

DE equals the distance from the shore to the boat! Explain why $DE = AB$. (*Hint:* Draw $\triangle ABC$ and $\triangle DEC$.)

44. Does a diagonal of a square divide it into two congruent triangles? Give evidence to support your answer.

45. $AB = AC$. Find $m\angle A$.

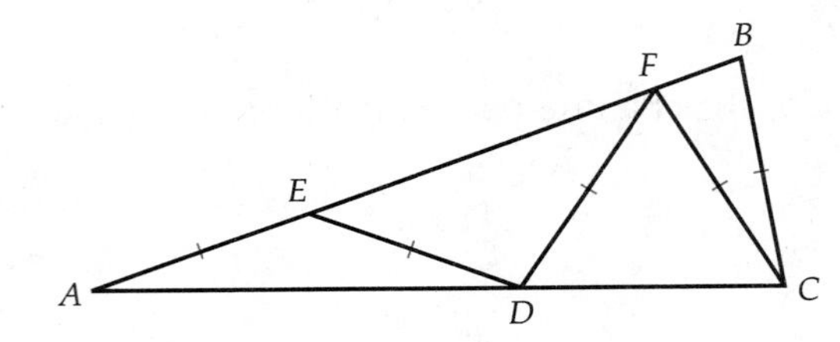

46. $ABCDE$ is a regular pentagon.

(a) How are AC, BD, CE, DA, and EB related? (Draw a picture.)

(b) Prove your answer to part (a).

47. Suppose that two quadrilaterals have three congruent corresponding sides and two congruent corresponding included angles, creating SASAS. Show that the two quadrilaterals must be congruent.

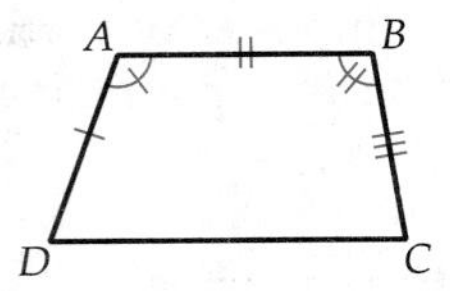

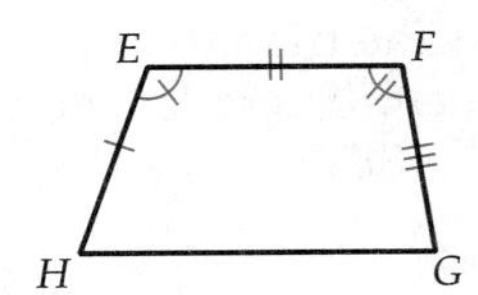

(*Hint:* Show that the remaining three corresponding parts are congruent, after drawing diagonals $\overline{AC}$ and $\overline{EG}$.)

48. Show that an ASASA quadrilateral property works for $ABCD$ and $EFGH$. (*Hint:* Draw $\overline{AC}$ and $\overline{EG}$.)

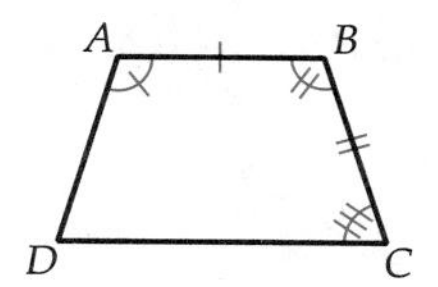

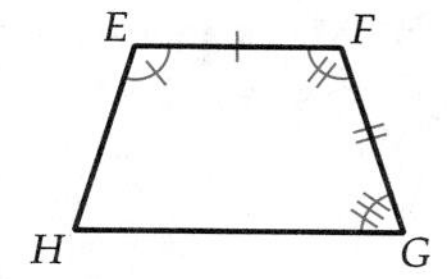

49. (a) Draw two noncongruent quadrilaterals, showing that an SSSS congruence property would not work.
(b) How does part (a) relate to why quadrilaterals (without diagonals) are not used in structures such as bridges?

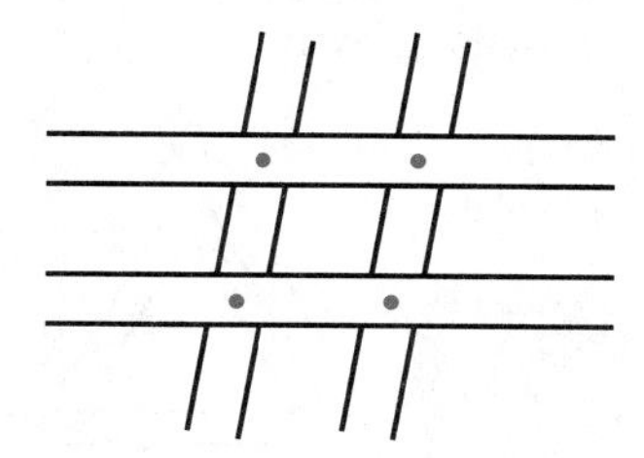

50. A MIRA is a red plexiglass device that acts like a mirror. Use a MIRA to do the following constructions.
(a) Bisect an angle.
(b) Find the midpoint of a line segment.

51. Use a MIRA to do the following constructions.
(a) Copy an angle.
(b) Copy a triangle.

Computer Exercises

52. Write a Logo program that draws a 50° angle and its bisector.

53. Using drawing software, draw $\overline{AB}$ and its perpendicular bisector. Investigate how far a point on the perpendicular bisector is from A and B. Repeat this for different points on the perpendicular bisector.

54. Using drawing software, draw a parallelogram $ABCD$ and its diagonals $\overline{AC}$ and $\overline{BD}$ intersecting at E. Compare the measurements of $\triangle AED$, $\triangle AEB$, $\triangle BEC$, and $\triangle CED$, and make a conjecture about parallelograms.

55. Using drawing software, draw a rectangle $ABCD$ and its diagonals $\overline{AC}$ and $\overline{BD}$ intersecting at E.
(a) Compare AC and BD. Make a conjecture about rectangles.
(b) Compare the measurements of $\triangle AED$, $\triangle AEB$, $\triangle BEC$, and $\triangle CED$, and make a conjecture.

56. Use drawing software to investigate how many side measures or angle measures are needed to determine the shape of a triangle.
(a) Try to draw two noncongruent triangles with sides $AB = 7$ and $BC = 12$ (SS). In other words, draw two segments of lengths 7 and 12. Use the drawing program to form a triangle. See if you can form another triangle with the same two lengths that is not congruent to the first. Confirm your result by measuring the third side and the angles of each triangle.
(b) Try to draw two noncongruent triangles with sides $AB = 7$, $BC = 12$, and $AC = 10$ (SSS).
(c) Try to draw two noncongruent triangles with $m\angle A = 50°$ and $m\angle B = 70°$ (AA).
(d) Try to draw two noncongruent triangles with $m\angle A = 50°$, $m\angle B = 70°$, and $m\angle C = 60°$ (AAA).
(e) Try to draw two noncongruent triangles with $m\angle A = 50°$ and $AB = 7$ (AS).
(f) Try to draw two noncongruent triangles with $AB = 7$, $BC = 12$, and $m\angle B = 50°$ (SAS).
(g) Try to draw two noncongruent triangles with $AB = 7$, $BC = 12$, and $m\angle C = 50°$ (SSA).
(h) Try to draw two noncongruent triangles with $AB = 7$, $m\angle A = 50°$, and $m\angle B = 70°$ (ASA).
(i) Try to draw two noncongruent triangles with $AB = 7$, $m\angle A = 50°$, and $m\angle C = 60°$ (AAS).
(j) List the conclusions of your investigation.

Special Exercises

57. Using only a compass, a straightedge, and a pencil or pen, create one of the following designs or make one of your own.

(a)

(b)
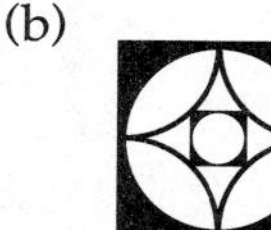

(c)

58. A drawing may appear to represent a three-dimensional object, but that object turns out to be impossible to make! The "Penrose triangle" is such a drawing.

You will need a compass and a ruler to make a Penrose triangle.

(a) Draw an equilateral triangle with sides 12 cm long. Draw the base with a ruler. Locate the top with a compass by measuring 12 cm from each end of the base.

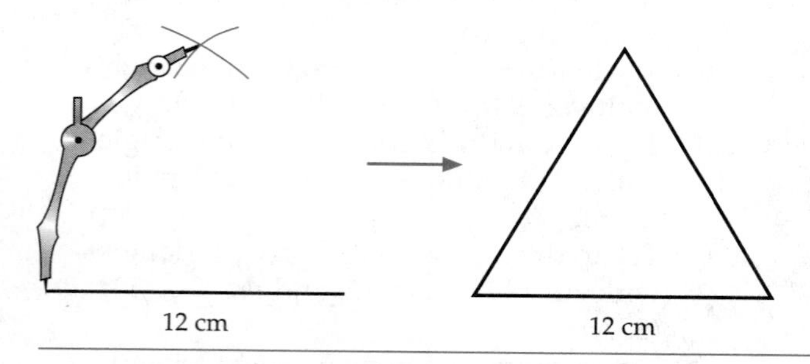

(b) Mark points 1 cm and 2 cm from each corner.

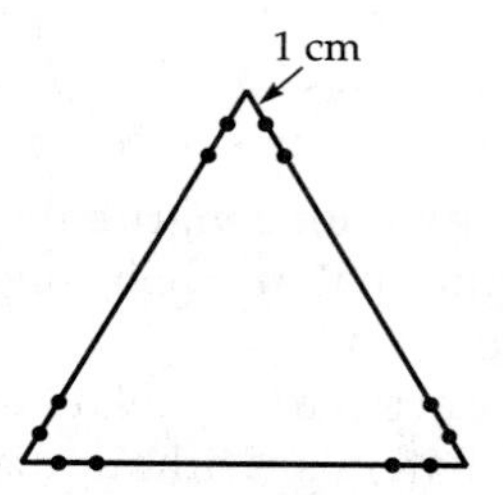

(c) Draw line segments lightly, connecting the points as shown.

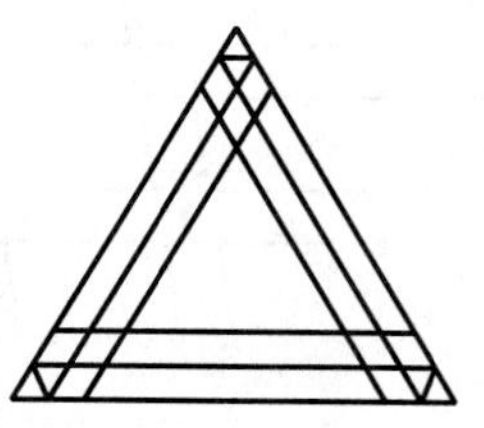

(d) Go over the lines shown. Erase the others.

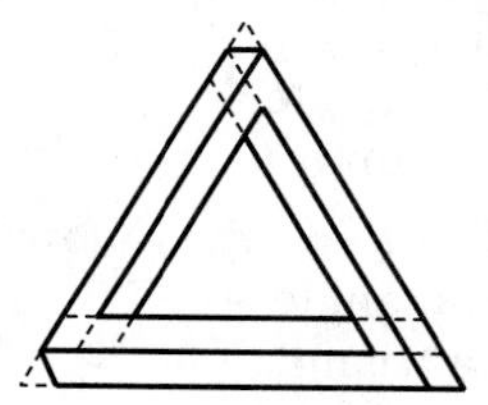

(e) Shade them as shown in the original Penrose triangle.

9.3 SYMMETRY

The term "congruence" describes a relationship between two figures that are the same size and shape. The term "symmetry" describes a way in which a single figure can be divided into parts that are the same size and shape and possess a particular orientation.

SYMMETRY OF PLANE FIGURES

Symmetry adds beauty and balance to natural forms and architectural designs (Figure 9-65).

Photo of butterfly (neg. no. 108780) by J. Kirchner. Courtesy Department of Library Services, American Museum of Natural History. Photo of U.S. Capital courtesy of Library of Congress.

Figure 9–65

Can you recognize plane figures that have symmetry?

Class Lesson Exercise 9.43

Which designs in Figure 9-66 possess some kind of symmetry?

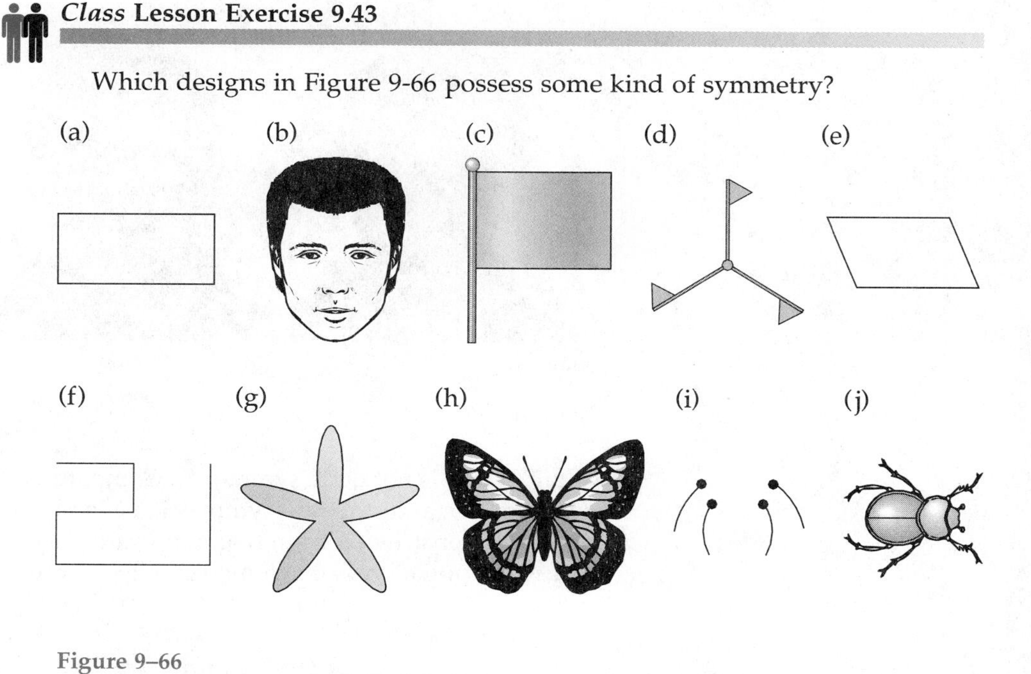

Figure 9–66

In Lesson Exercise 9.43, all but two of the figures are symmetric. You can check your answers with the following test for symmetry.

A Test for Symmetry

A plane figure is symmetric if you can copy it onto tracing paper and find a different position for the tracing paper in which the traced figure coincides with the original figure.

Lesson Exercise 9.44

Use tracing paper to determine which figures in Lesson Exercise 9.43 are symmetric.

For each symmetric figure, you moved the tracing paper to a new position, and the traced figure fit on top of the original figure. The motion you used was either a rotation (other than a full turn) or a reflection.

REFLECTION (LINE) SYMMETRY

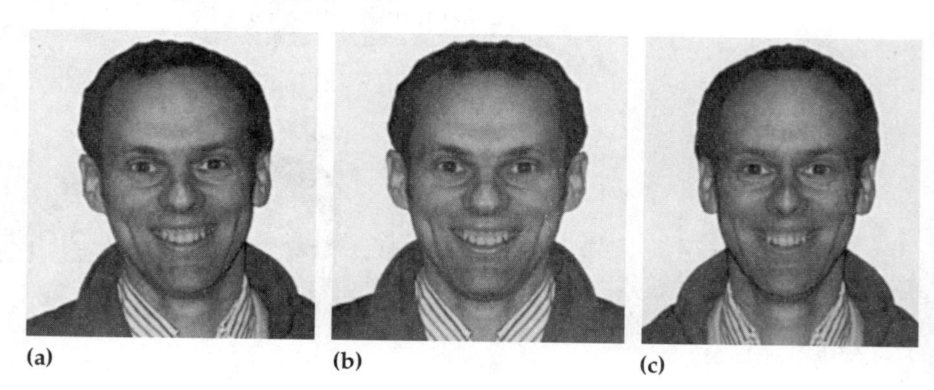

Figure 9–67

A human face is not quite symmetric. Compare the real photograph of my face in Figure 9-67(a) to the two symmetric versions of me. Each of the symmetric versions is constructed from one half of my face. This example illustrates the most well-known type of symmetry: reflection (line) symmetry.

Lesson Exercise 9.45

(a) Which shapes in Lesson Exercise 9.43 can be reflected onto themselves? (These figures have reflection symmetry.)

(b) Select one of the shapes that has reflection symmetry. Draw a line that divides it into two equal parts such that one part is the reflection image of the other through the given line.

(c) Select one of the shapes that has reflection symmetry. How could you use paper folding to show the symmetry?

(d) Select one of the shapes that has reflection symmetry. How could you use a pocket mirror or a Mira (if you have one) to show the symmetry?

A plane figure has **reflection (line) symmetry** if and only if it can be reflected through some line so that its image coincides with its original position. Each point in the figure is the same distance from the line of symmetry as its image point is.

Most elementary-school textbooks refer to reflection symmetry as line symmetry, and they use paper folding to test it. The tracing-paper test in this section is a more general test that also works for other kinds of symmetry.

You will need scissors and paper for the following exercise.

Lesson Exercise 9.46

Fold a piece of paper in half, and cut out the shape along the fold. Predict the shape that will result when you unfold the paper. Then see if your prediction was right.

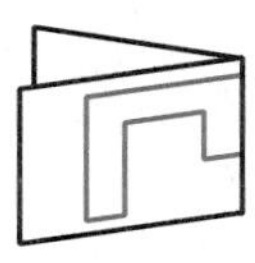

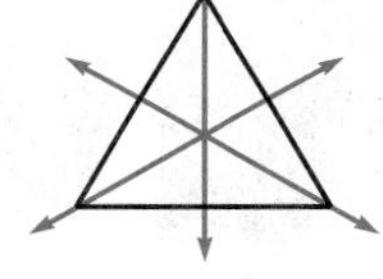

Figure 9–68

Reflection symmetry is also called mirror symmetry, since a line of symmetry acts like a double-sided mirror. Points on each side are reflected to the opposite side. Many figures have several reflection (line) symmetries. The equilateral triangle in Figure 9-68 has three lines of symmetry.

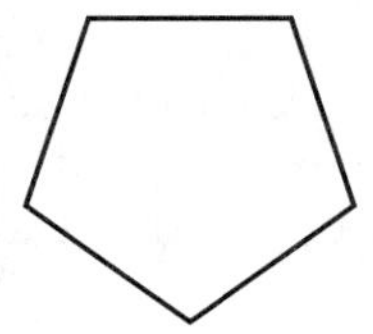

Figure 9–69

Lesson Exercise 9.47

Draw all lines of symmetry for the regular pentagon in Figure 9-69.

Lesson Exercise 9.48

Why isn't the line shown in Figure 9-70 a line of symmetry for the shape in the drawing?

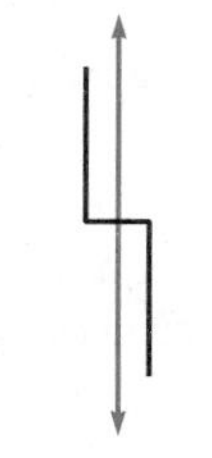

Figure 9–70

Lesson Exercise 9.49

Complete each figure so that the line acts as a line of symmetry.

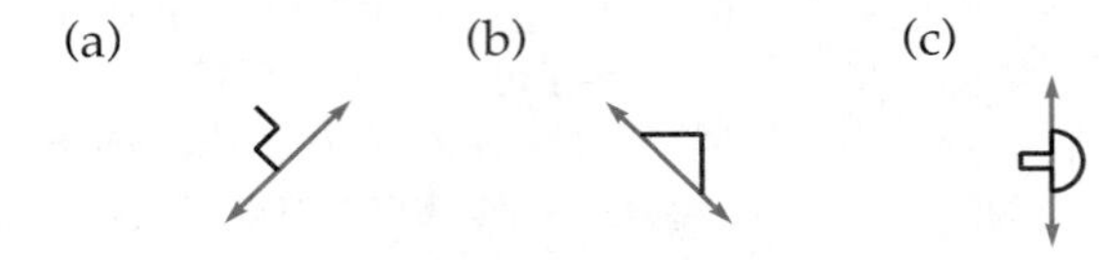

ROTATIONAL SYMMETRY

Courtesy of National Oceanic and Atmospheric Administration.

Figure 9–71

The surfaces of snowflakes (Figure 9-71) suggest a second major type of plane symmetry: rotational symmetry. Natural and manufactured objects that have rotational symmetry possess an order that enhances their beauty and function.

Lesson Exercise 9.50

Which of the shapes in Lesson Exercise 9.43 can be rotated *less than 360°* (a full turn) onto themselves?

The figures you selected in Lesson Exercise 9.50 should have rotational symmetry. A plane figure has **rotational (turn) symmetry** if and only if it can be rotated more than 0° and less than 360° so that its image coincides with its original position.

Many shapes have several rotational symmetries. The equilateral triangle

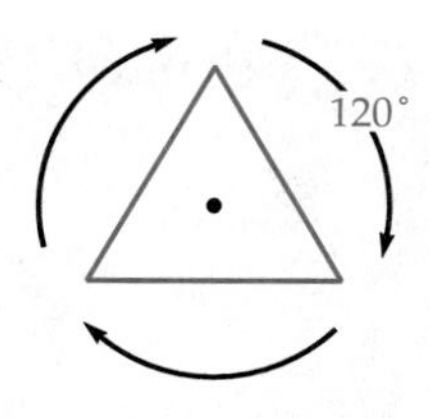

Figure 9–72

in Figure 9-72 can be turned 120° or 240° around its center to coincide with its original position. Therefore, it has 120° and 240° rotational symmetry.

Lesson Exercise 9.51

What rotational (turn) symmetries does a regular pentagon have (greater than 0° and less than 360°)?

Can a plane figure have both rotational and reflection symmetry? Can it have one kind of symmetry without the other? Find out in the following exercises.

Lesson Exercise 9.52

(a) Which shapes in Lesson Exercise 9.43 have rotational *and* reflection (line) symmetry?

(b) Which shapes in Lesson Exercise 9.43 have rotational symmetry but not reflection (line) symmetry?

(c) Which shapes in Lesson Exercise 9.43 have reflection (line) symmetry but not rotational symmetry?

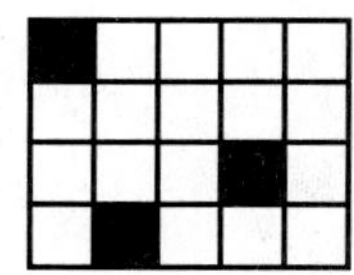

Figure 9–73

***Cooperative* Lesson Exercise 9.53**

Consider the following problem. "Shade the smallest number of squares in Figure 9-73 so that the pattern has reflection *and* rotational symmetry." Devise a plan and solve the problem.

SYMMETRY OF SPACE FIGURES

Some three-dimensional figures have the same kinds of symmetry as two-dimensional shapes. The beetle and the chair in Figure 9-74, on page 542, each have two halves that are virtually identical. These objects have approximate reflection (plane) symmetry.

Reflection (plane) symmetry of space figures exists if a plane divides the figure in half so that one half of the figure is a mirror image of the other. To put it more precisely, each point on one side of the plane must have a corresponding point the same (perpendicular) distance away on the opposite side. This relationship is analogous to the distance relationship for line symmetry.

Lesson Exercise 9.54

The beetle in Figure 9-74, on page 542, has approximate reflection symmetry. Where is the plane of symmetry?

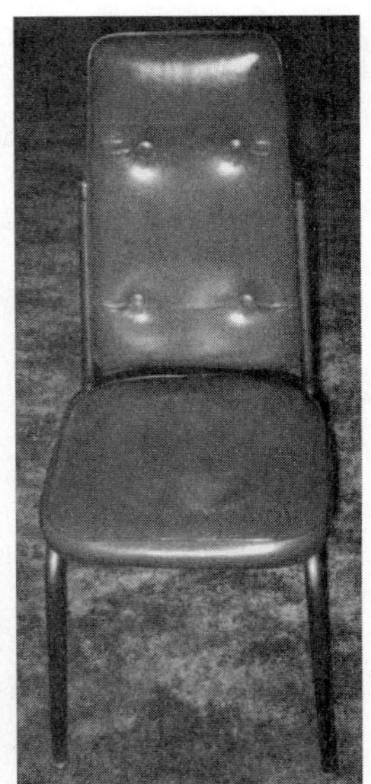

Photo of beetle (neg. no. 122981) by Alex J. Rota. Courtesy Department of Library Services, American Museum of Natural History. Photo of chair by Tom Sonnabend.

Figure 9–74

Reflection symmetry gives animals balance. Many animals and many machines have reflection symmetry.

Lesson Exercise 9.55

Find an object in your classroom that has approximate reflection symmetry that gives it balance.

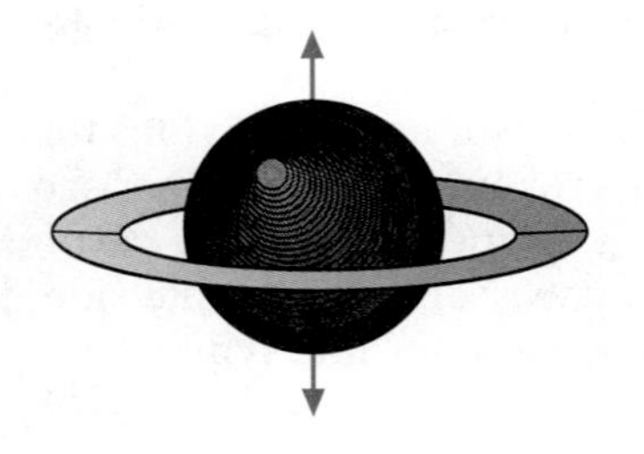

Figure 9–75

Saturn has approximate rotational symmetry. After being rotated any amount around the line shown in Figure 9-75, its image will coincide with its original position.

Rotational symmetry of space figures exists if there is an axis of rotation (a line) around which the figure can be turned (less than a full turn) so that it coincides with itself. The rotational symmetry of an object gives the object "directional flexibility." For example, a square card table can be placed in four different positions that function exactly the same way.

In 1888, the Russian mathematician Sonya Kovalevsky (1850–1891; Figure 9-76) gained prominence with her research paper *On the Rotation of a Solid About a Fixed Point*, which won the French Academy of Sciences prestigious Prix (Prize) Bordin.

Sonya Kovalevsky

Figure 9–76

As a child, Sonya Kovalevsky slept in a room that was wallpapered with her father's calculus notes from college, since the family couldn't afford to buy wallpaper. Perhaps that helped to stimulate her interest in mathematics.

In the nineteenth century, women were not allowed to attend universities in Russia. Kovalevsky married so she could emigrate to Germany, where she studied mathematics. Karl Weierstrass, one of Germany's greatest mathematicians, tutored her privately, since the University of Berlin would not admit her to his lectures. Thanks to Weierstrass's strong recommendation, the University of Göttingen awarded Kovalevsky a doctorate even though she never officially attended school there! Kovalevsky was clearly qualified for the degree on the basis of three outstanding research papers.

In 1883, Kovalevsky was invited to teach at the University of Stockholm. From 1889 until her death in 1891, she worked there as a full professor of mathematics.

Lesson Exercise 9.56

(a) Find an object in your classroom that has approximate rotational symmetry that gives it directional flexibility.

(b) Describe the location of the axis of rotation.

(c) What rotational symmetries between 0° and 360° does the object have?

ANSWERS TO SELECTED LESSON EXERCISES

9.45 (a) a, b, g, h, i, j

9.47

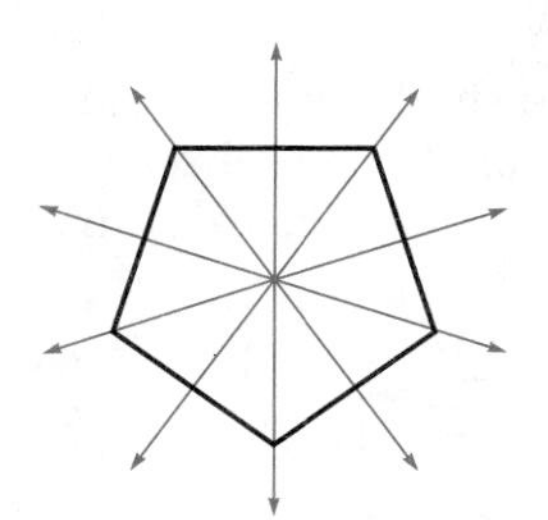

9.48 A reflection through the line does not leave the figure unchanged.

9.49 (a) (b) (c)

9.50 a, d, e, g

9.51 72°, 144°, 216°, 288° clockwise (or counterclockwise)

9.52 (a) a, g (b) d, e (c) b, h, i, j

9.53

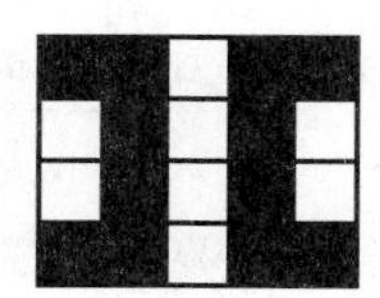

9.54 Perpendicular to the plane of the wings and dividing its body in half

9.55 A garbage can, chair, or table

9.56 (a) A garbage can or table

9.3 HOMEWORK EXERCISES

Basic Exercises

1. Does symmetry enhance the beauty of a shape? Cover up the right half of the butterfly. Is it still as pleasing to the eye?

2. Describe a test for symmetry.

3. Which of the following figures have reflection (line) symmetry?

(a) (b) (c) (d)

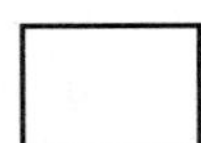

4. Draw a line of symmetry for each figure.

(a) (b)

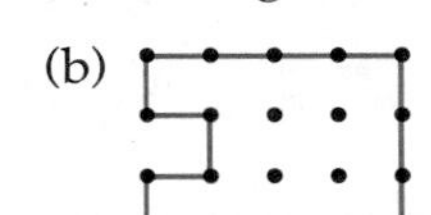

5. How many lines of symmetry does each of the following quadrilaterals have?

(a) Square (b) Rectangle

(c) Rhombus (d) Parallelogram

6. Draw all lines of symmetry for each figure. Verify your results with paper folding.

(a) (b) (c) (d)

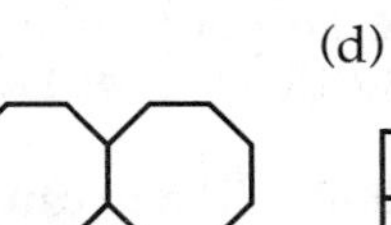
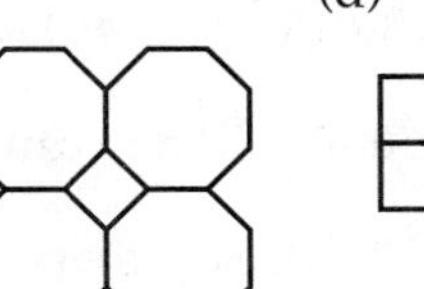

7. Explain why a diagonal is not a line of symmetry for a rectangle.

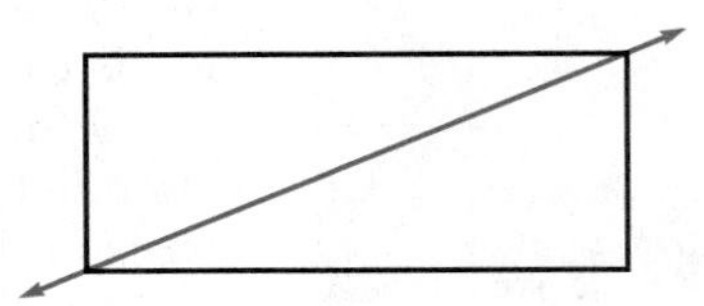

8. Draw an octagon that has exactly two lines of symmetry.

9. (a) Draw two intersecting lines.
(b) Draw the two lines of symmetry for your lines in part (a).
(c) Repeat parts (a) and (b), starting with lines intersecting at a different angle.
(d) Propose a generalization of your results.
(e) Does part (d) involve induction or deduction?

10. Complete the figure so that it is symmetric to the line.

11. For each part, fold a piece of paper in half and cut out the shape shown along the fold. Predict what shape will result when you unfold the paper. Check your prediction.

(a) (b)

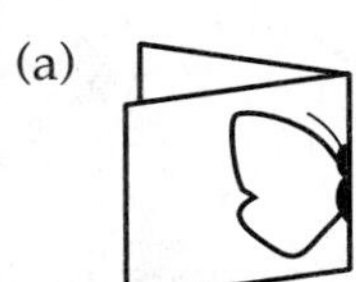
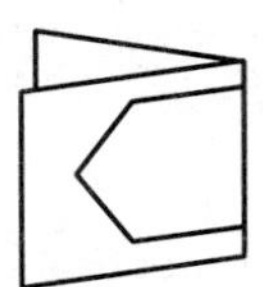

12. (a) Fold a sheet of paper in half. Then make a cut so that the unfolded paper shows a heart.
(b) Repeat part (a), but create a pumpkin.
(c) Repeat part (a) and make your own design.

13. How many rotational (turn) symmetries (greater than 0° and less than 360°) does each of the following quadrilaterals have?

(a) Square (b) Rectangle

(c) Rhombus (d) Parallelogram

14. You are "O" in tic-tac-toe. You have 8 possible moves, but some are equivalent. How many different kinds of moves are there?

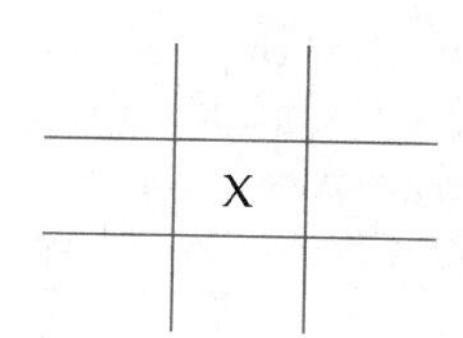

15. What kind of symmetry does each flag have?

(a)

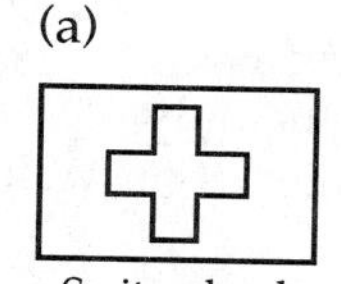

Switzerland

(b)

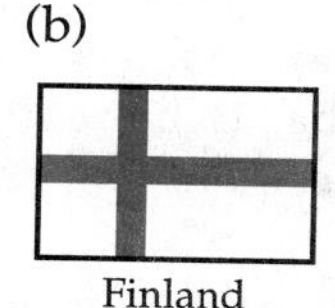

Finland

(c)

Great Britain

16. Three snowflakes follow. Which ones appear to possess (approximate) symmetry?

(a) (b) (c)

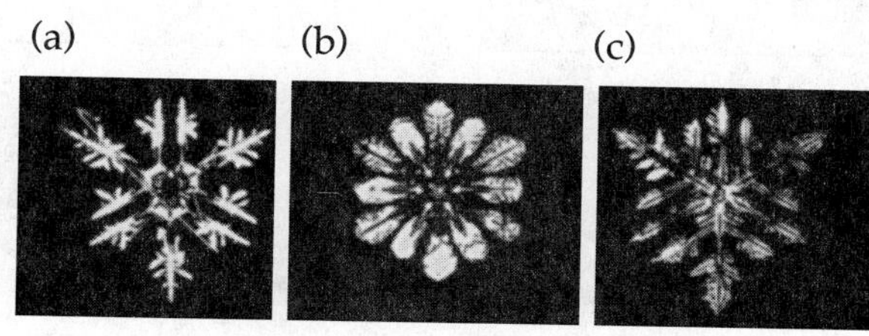

Courtesy of National Oceanic and Atmospheric Administration.

17. (a) Which letters in the word shown have reflection symmetry?
(b) Which letters have rotational symmetry?

S Q U I D

18. The word HIDE has a horizontal line of symmetry. Write three more words that have a horizontal line of symmetry.

19. Examine a deck of playing cards.
(a) Which cards have rotational symmetry?
(b) Which cards have reflection (line) symmetry?

20. Draw a plane figure that has reflection (line) symmetry but no rotational symmetry.

21. For all plane figures having two or more reflection (line) symmetries, there is a relationship between the number of reflection symmetries and the number of rotational symmetries. Devise a plan and find the relationship.

22. Tell the number of rotational and reflection symmetries each hubcap has. (Disregard the logos in the center.)

(a) (b)

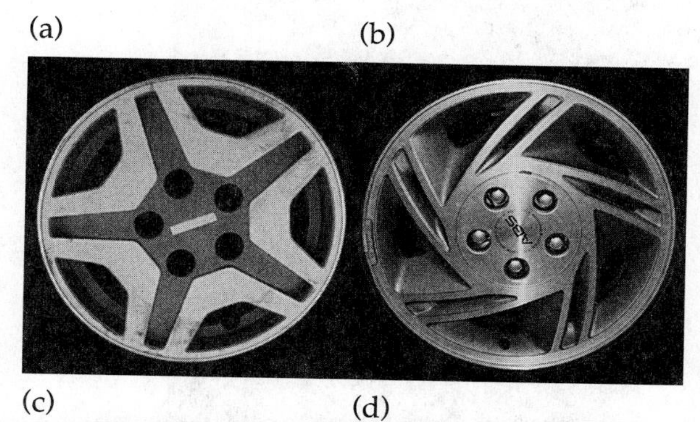

(c) (d)

23. Shade the smallest number of squares so that each of the following diagrams has rotational symmetry.

(a) (b)

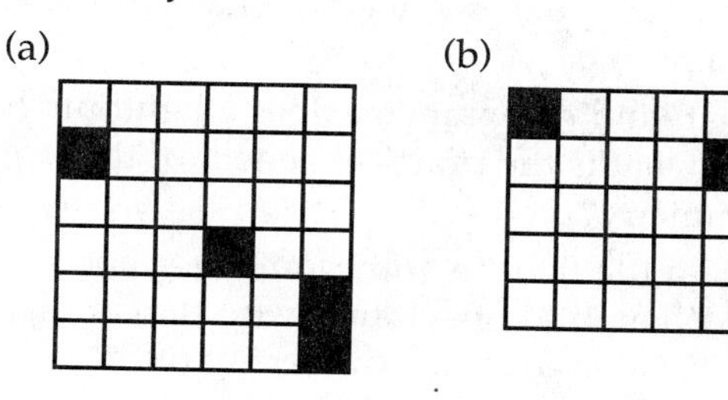

24. Mirror cards (developed by Marion Walters) are used to teach symmetry in elementary school. Using the mirror card at the left, where would you place a mirror on the card to obtain the figures shown in parts (a), (b), and (c)?

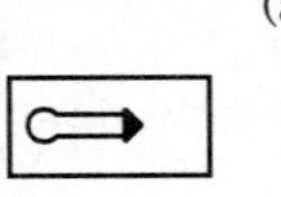

(a)

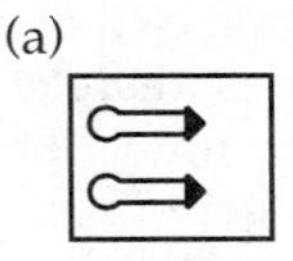

(b)

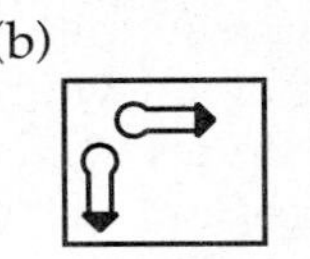

(c)

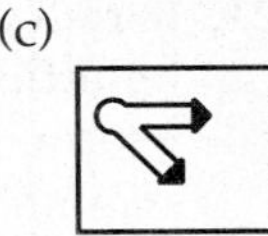

25. A plane figure that looks exactly the same upside down as right side up always has what specific symmetry?

26.

Describe all the symmetries of this pattern.

27. How many planes of symmetry does a right square pyramid have?

28. How many planes of symmetry does a cube have?

29. How many axes (lines) of rotational symmetry does a cube have?

30. A nut ⬡ has approximate rotational symmetry. What is the advantage of this?

31. If you rotate an isosceles triangle in space about its line of symmetry, what solid will you obtain?

32. What kind of symmetry does a lightbulb have, and what is the practical benefit of this symmetry?

33. Find a tile floor in your home or school.
(a) What kind of symmetry do the tiles have?
(b) What is the practical benefit of this symmetry?

34. (a) What kind of symmetry do birds have?
(b) What is the practical benefit of this symmetry?

35. (a) Describe the symmetries of a tennis ball, a tennis racket, and a tennis court.
(b) Explain how the symmetries of each object relate to the function of the object.

36. (a) List five objects in your home that have reflection symmetry.
(b) List five objects in your home that have rotational symmetry.

37. What kind of symmetry does each of the following figures have?

(a) 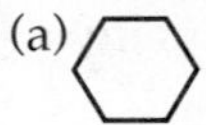
(b) 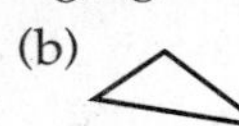
(c) 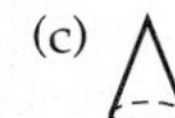
(d) 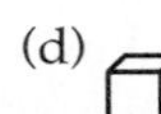
(e) 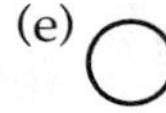

38. Pottery and other objects made on a lathe generally have what kind of symmetry?

39. In how many ways can a rectangular piece of glass fit into a window frame?

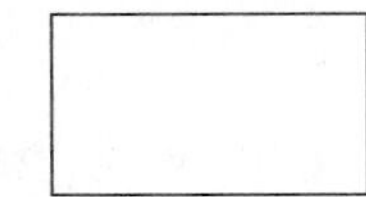

40. The driver's door of a two-door car is damaged and has to be refitted. The only available replacement is a passenger door. Will it fit?

41. In how many ways can each key be inserted into the corresponding hole from one side?

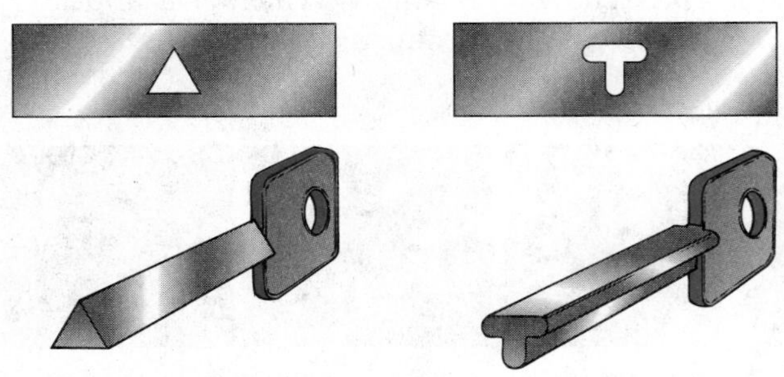

42. A kaleidoscope usually has two mirrors placed at a 60° angle, as shown. One image is drawn. Fill in all the other images. (*Hint:* Include images of images.)

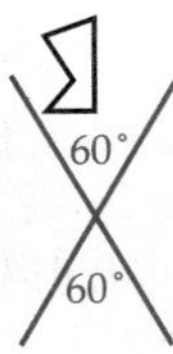

Extension Exercises

43.

+	0	1	2	3	4	5	6	7	8	9
0	0	1	2	3	4	5	6	7	8	9
1	1	2	3	4	5	6	7	8	9	10
2	2	3	4	5	6	7	8	9	10	11
3	3	4	5	6	7	8	9	10	11	12
4	4	5	6	7	8	9	10	11	12	13
5	5	6	7	8	9	10	11	12	13	14
6	6	7	8	9	10	11	12	13	14	15
7	7	8	9	10	11	12	13	14	15	16
8	8	9	10	11	12	13	14	15	16	17
9	9	10	11	12	13	14	15	16	17	18

(a) What kind of symmetry does an addition table have? (Ignore the position of the number within each box.)
(b) What property of addition does this symmetry indicate?
(c) Does a multiplication table for 0 to 9 have the same symmetry?

44. Look at some blank crossword puzzles. Do they have any symmetry?

45. Trace the circle shown here on a sheet of paper. How can you locate its center with paper folding?

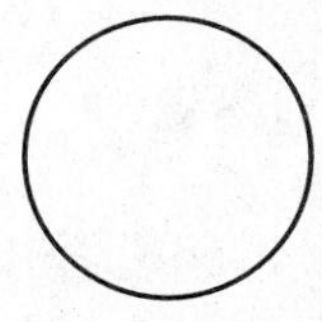

46. (a) Which pentominos have reflection symmetry?
(b) Which pentominos have rotational symmetry?

47. If an isosceles triangle, $\triangle ABC$, has a line of symmetry $\overleftrightarrow{CD}$, what can you say about each of the following?
(a) $\overline{AD}$ and $\overline{DB}$
(b) $\angle A$ and $\angle B$

48. Sketch any of the following seven figures that are possible: hexagons that have exactly zero, one, two, three, four, five, or six lines of symmetry.

49. Add to the drawing so that it has rotational symmetry and exactly one line of symmetry.

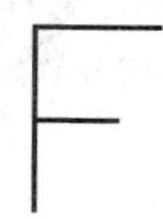

Computer Exercises

50. The following Logo procedures draw fractal snowflakes.
(a) Enter the following lines into a computer and experiment with different inputs. Begin with FLAKE 2 100 and try values of N from 1 to 4. (Values of N greater than 4 may take a long time.) Do not clear the screen between inputs.

```
TO PATTERN :N :L
IF :N = 1 [FD :L STOP]
PATTERN :N = 1 :L/3 LT 60
PATTERN :N = 1 :L/3 RT 120
PATTERN :N = 1 :L/3 LT 60
PATTERN :N = 1 :L/3
END

TO FLAKE :N :L
REPEAT 3 [PATTERN :N :L RT 120]
END
```

(b) What kind of symmetry do your drawings have?
(c) Repeat part (a), but type CS after each run.

51. (a) Enter the following Logo procedure into a computer.

```
TO FOUR
FD 20 RT 80 FD 40 RT 60
FD 20 RT 130
END
```

(b) Now enter:

```
REPEAT 4 [FOUR]
```

(c) How many rotational symmetries does the figure have?
(d) To produce the symmetry, the sum of the turns (270°) must repeat as many times as needed to make the total turn a multiple of 360°. In this case, $4 \cdot 270° = 1080° = 3 \cdot 360°$. Edit the procedure, changing the amounts of turns and forwards, and see if

you can create another figure with rotational symmetry.

(e) Write a procedure that creates four rotational symmetries when you repeat it five times.

Special Exercise

52. Use paper folding and cutting to create your own symmetric design.

9.4 SIMILARITY AND SIZE CHANGES

Scale models of objects are the same shape but not the same size as the object. When you enlarge a photograph, you obtain a new picture in which everything is the same shape but a different size (Figure 9-77).

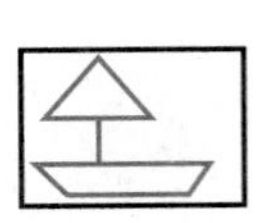

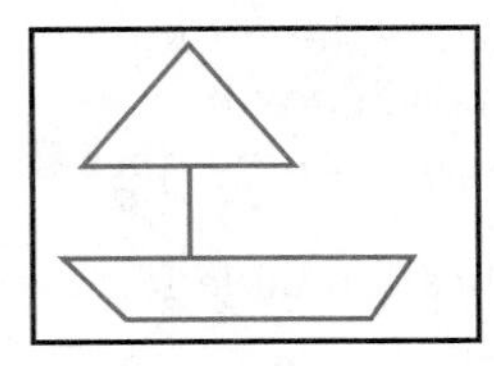

Figure 9–77

Beyond a certain age, dolphins grow larger but retain roughly the same shape (Figure 9-78).

Photo: Rob Mathewson. Courtesy Department of Library Services. American Museum of Natural History.

Figure 9–78

SIZE CHANGES

Suppose that you want to enlarge a drawing without changing its shape. This enlargement would correspond to a geometric motion called a size change. In the next exercise, you will perform a size change in a coordinate plane.

Lesson Exercise 9.57

(a) Plot $A(1, 0)$, $B(4, 0)$, $C(3, 2)$, and $D(1, 2)$ on a graph. Connect A to B, B to C, C to D, and D to A. What shape do you obtain?

(b) To perform a size change, find image points that are twice as far from the origin and in the same direction. For example, A is 1 right and 0 up from $(0, 0)$, so A' is 2 to the right and 0 up from $(0, 0)$. This means that $A' = (2, 0)$. Find B', C', and D', and plot all the image points.

(c) Connect A' to B', B' to C', C' to D', and D' to A'. What shape do you obtain?

(d) How do $ABCD$ and $A'B'C'D'$ compare?

(e) How does $A'B'$ compare to AB? How does $C'D'$ compare to CD?

(f) How do the corresponding angle measures of the two figures compare?

(g) Now find E, F, G, and H that are three times as far from the origin as A, B, C, and D in corresponding directions. Plot E, F, G, and H.

(h) Connect E to F, F to G, G to H, and H to E. How does $EFGH$ compare to the other two figures?

(i) How does EF compare to AB? How does GH compare to CD?

(j) How do the corresponding angles of $EFGH$ compare to $ABCD$?

(k) Propose generalizations of your results.

Lesson Exercises 9.57(b) and 9.57(g) are examples of size changes. A size change has a **center** and a positive **scale factor**. A **size change** in a plane multiplies the distance of each point in the plane from the center by the scale factor. In Lesson Exercise 9.57(b), the size change assigns an image to each point of $ABCD$ that is twice as far from the center point $(0, 0)$. The size change produces an image that has the same shape as the original figure. The size change multiplies the lengths of the original figure by the scale factor but does not change the measures of the angles.

A size change can also be performed on a figure that is not in a coordinate plane.

Lesson Exercise 9.58

Perform the size change $S_{A,1.5}$ with center A and scale factor 1.5, by following these instructions.

C

A B

D

(a) On $\overrightarrow{AB}$, find an image point B' that is 1.5 times as far from A as B is. In the same way, find image points C' on $\overrightarrow{AC}$ and D' on $\overrightarrow{AD}$.

(b) Using a ruler and a protractor, measure the sides and angles of $\triangle BCD$ and $\triangle B'C'D'$.

(c) Describe any relationships that exist between the corresponding sides, angles, and vertices of $\triangle BCD$ and $\triangle B'C'D'$.

Next, try to find more general relationships between a point and its image under a size change.

Lesson Exercise 9.59

Consider a size change $S_{A,k}$ with center A and scale factor k. What is the relationship among a point X, its image X', and point A?

SIZE CHANGES AND SIMILAR FIGURES

A size change produces an image that has the same shape and equal corresponding angle measures. The corresponding lengths are multiplied by the scale factor. These properties describe similar figures. **Similar figures** are the same shape but not necessarily the same size.

The results of the size changes suggest criteria for determining whether two polygons are similar.

Definition **Similar Polygons**

Similar polygons have

1. congruent corresponding angles and
2. proportional corresponding lengths.

The two triangles in Figure 9-79 are similar. They have congruent corresponding angles and proportional corresponding lengths.

The corresponding vertices are $A \leftrightarrow M$, $B \leftrightarrow Y$, and $C \leftrightarrow X$. The relationship can be written symbolically as $\triangle ABC \sim \triangle MYX$. The symbol $\sim$ means

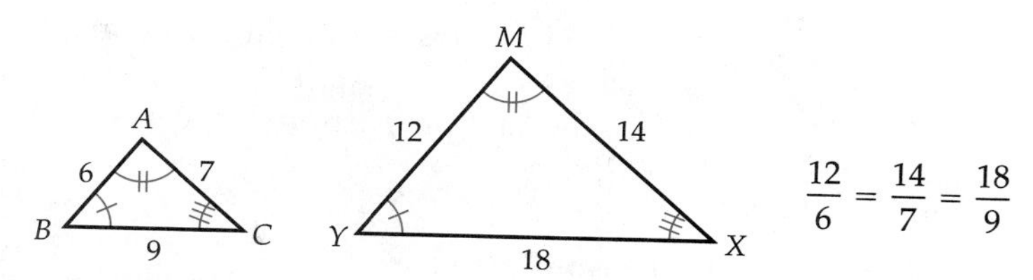

Figure 9–79

"similar to." Note how corresponding vertices are written in corresponding positions in the names of the similar triangles.

You can describe the relationship between corresponding sides with the ratio of corresponding lengths or with the scale factor. The ratio of corresponding lengths is $1/2$ or $1:2$. Going from the left triangle to the right triangle, the scale factor is 2. Going from the right triangle to the left triangle, the scale factor is $1/2$.

Lesson Exercise 9.60

(a) Fill in the blank, putting the vertices for $\triangle DNW$ in the corresponding order (Figure 9-80). $\triangle FAT \sim$ ________.

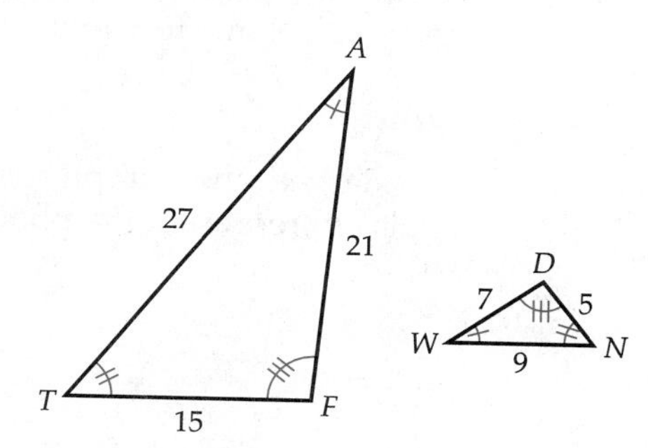

Figure 9–80

(b) What is the ratio of corresponding lengths?

(c) What is the scale factor in going from $\triangle FAT$ to $\triangle DWN$?

Lesson Exercise 9.61

In each case, use the definition of similar polygons to decide if the two figures are similar.

(a)

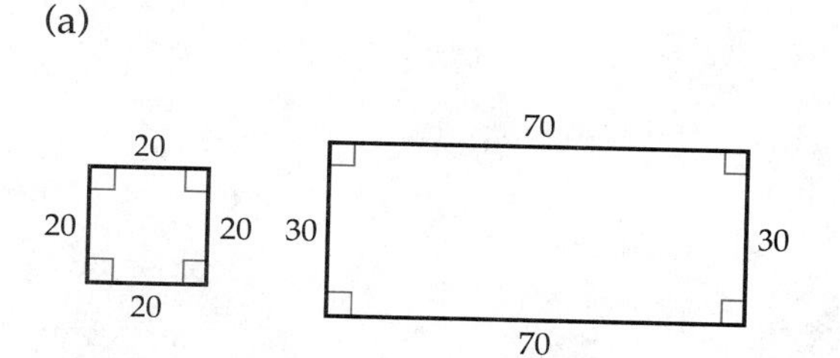

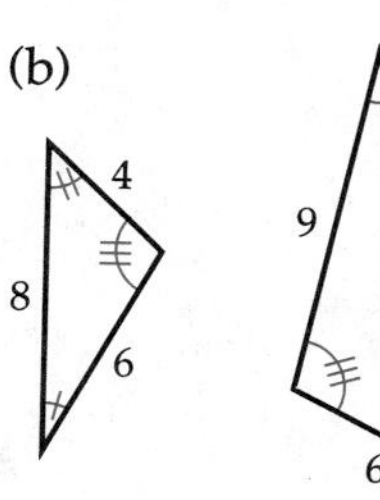

When you are given a statement with the names of two similar figures, such as $ARCX \sim MYTU$, you can tell which angles are congruent and which lengths are proportional because the vertices are written in corresponding order.

Lesson Exercise 9.62

$ARCX \sim MYTU$. Fill in the missing lengths.

$$\frac{RC}{\square} = \frac{RX}{\square} = \frac{\square}{MU}$$

If you know the lengths of some sides of two similar polygons, is it possible to compute the lengths of the remaining sides? In a case like the following, you could use proportions.

EXAMPLE 9.1

You want to enlarge a photograph that is 4 in. by 6 in. so that its longer dimension is 9 in. How long is the shorter dimension?

Solution

An enlarged photograph should retain the same *shape*. Draw two similar rectangles to represent the photograph before and after enlargement (Figure 9-81).

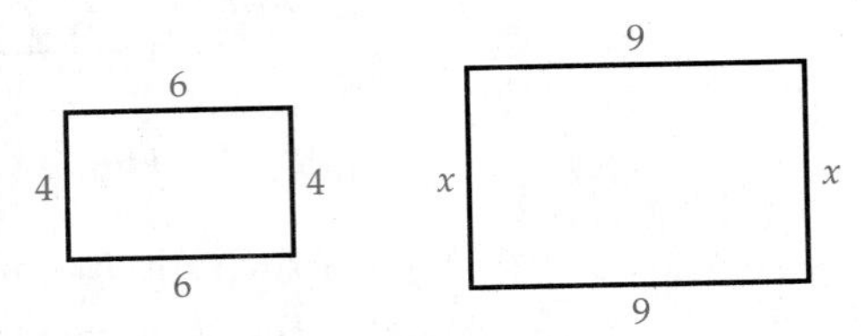

Figure 9–81

Since the rectangles are similar, corresponding lengths are proportional. So

$$\frac{6}{9} = \frac{4}{x}$$

This is the same as

$$6 \cdot x = 4 \cdot 9$$

$$6x = 36$$

$$x = 6 \text{ in.}$$

■

Lesson Exercise 9.63

The two triangles in Figure 9-82 are similar. Find x and y.

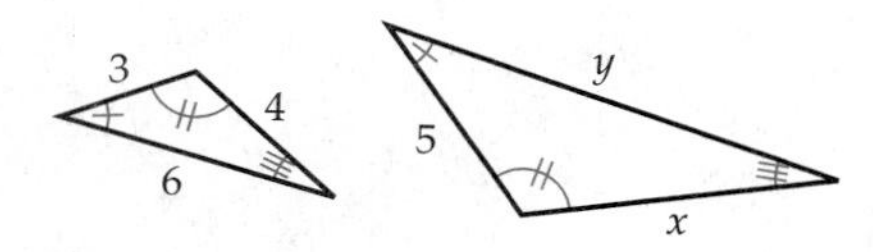

Figure 9–82

SIMILAR SOLIDS

Like similar plane figures, similar solids are the same shape but not necessarily the same size. In similar polyhedra, the lengths of corresponding edges are proportional, and corresponding angles are congruent.

Lesson Exercise 9.64

The two rectangular prisms in Figure 9-83 are similar. AB corresponds to EF, BC to FG, and CD to GH. Find the lengths x and y.

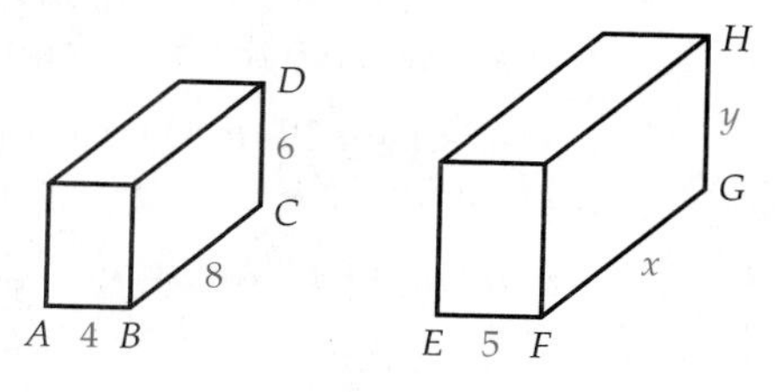

Figure 9–83

APPLICATIONS OF SIMILARITY

On a sunny day, you can use the length of your shadow to compute the height of a building. It's a good way to impress your friends. How does it work?

You and your shadow determine a triangle. This triangle is similar to the triangle determined by nearby objects and their shadows at the same time of day!

In Figure 9-84, on page 554, $\triangle ABC \sim \triangle DEF$. Now suppose that you want to estimate how tall the building is. You can use your height, the length of your shadow, and the length of the building's shadow to compute the height of the building. Example 9.2 illustrates how this is done.

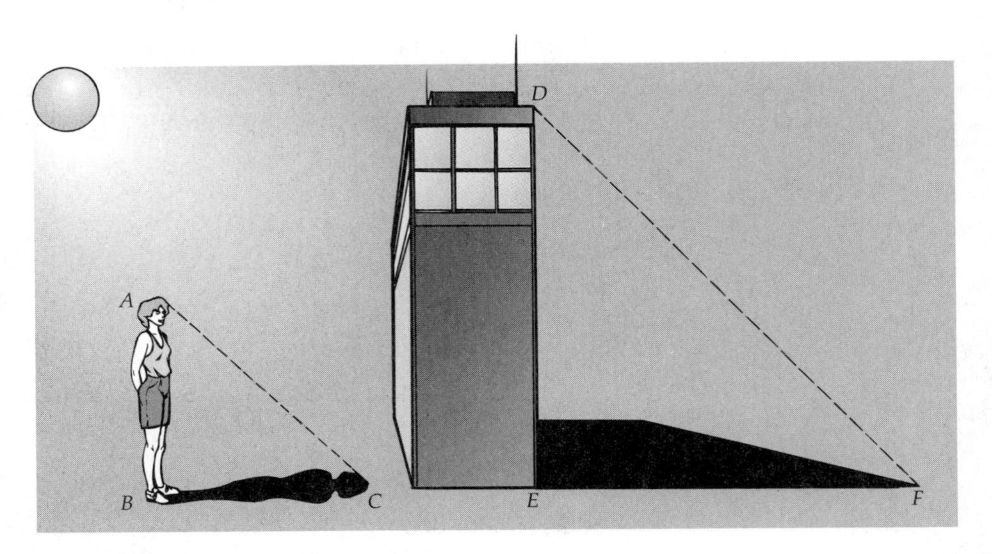

Figure 9–84

EXAMPLE 9.2

Jane is 5 ft 3 in. tall. At 4 P.M., her shadow is 7 ft 6 in. long. She measures the shadow of a nearby building. It is 500 in. long. About how tall is the building?

Solution

Understanding the Problem The length of each object and its corresponding shadow have the same ratio.

Devising a Plan Draw a picture, and set up a proportion with the appropriate units.

Carrying Out the Plan Convert the measurements to inches (Figure 9-85).

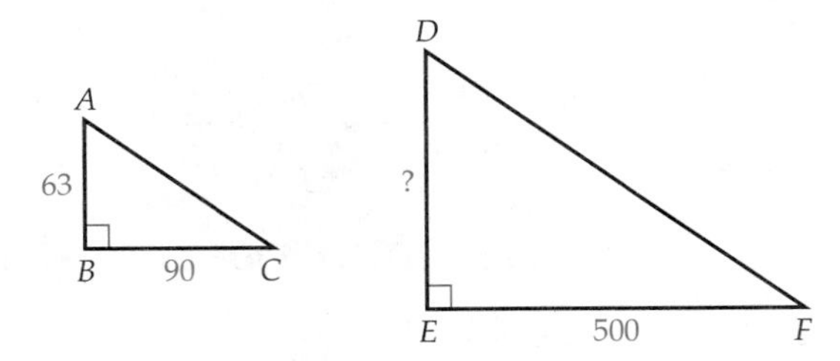

Figure 9–85

$$5 \text{ ft } 3 \text{ in.} = 63 \text{ in.} \quad \text{and} \quad 7 \text{ ft } 6 \text{ in.} = 90 \text{ in.}$$

$$\triangle ABC \sim \triangle DEF \quad \text{so} \quad \frac{AB}{DE} = \frac{BC}{EF} = \frac{AC}{DF}$$

Filling in the given information,

$$\frac{63}{?} = \frac{90}{500} = \frac{AC}{DF}$$

Use the first two ratios and solve.

$$\frac{63}{?} = \frac{90}{500} \quad \text{means} \quad 63 \cdot 500 = 90 \cdot (?)$$

Using a calculator, $63 \times 500 \div 90 = 350$. So the building is about 350 in. or 29 ft 2 in. tall. (On a calculator, $350 \div 12 = 29.166666$ ft or something similar. The .166666 ft = (.16666)(12 in.) = 2 in.)

Looking Back The building's shadow is about 5 times the length of Jane's shadow. So the height of the building should be about 5 times Jane's height. 29 ft 2 in. is about 5 times 5 ft 3 in. ■

Lesson Exercise 9.65

One sunny morning, Arturo, who is 5 ft tall, casts a shadow 7 ft 10 in. long. At the same time, a nearby tree casts a shadow 10 ft 10 in. long. About how tall is the tree?

Computing the distance between two cities on a map involves similar figures. A map shape is approximately similar to the region it represents (although maps are flat and the earth is not). Therefore, map distances are approximately proportional to actual distances.

Lesson Exercise 9.66

A map has a scale ratio of 1 in. ≈ 120 miles. On the map, New York is $1\frac{5}{8}$ in. from Boston. Approximately how far is New York from Boston?

ANSWERS TO SELECTED LESSON EXERCISES

9.57 (a) Trapezoid
(d) They are the same shape.
(e) $A'B' = 2(AB)$; $C'D' = 2(CD)$
(f) They are equal.
(h) It is the same shape but larger.
(i) $EF = 3(AB)$; $GH = 3(CD)$
(j) They are equal.

9.58 (a)

C′
C
A B B′
D
D′

9.58 (c) $\triangle BCD$ and $\triangle B'C'D'$ are the same shape. They have congruent corresponding angles. The corresponding sides of $\triangle B'C'D'$ are 1.5 times as long.

9.59 $AX' = k(AX)$. Points A, X, and X' all lie on a straight line.

9.60 (a) $\triangle DWN$ (b) 3 to 1 (c) $\frac{1}{3}$

9.61 (a) No (b) Yes

9.62 $\frac{RC}{YT} = \frac{RX}{YU} = \frac{AX}{MU}$

9.63 $\frac{3}{5} = \frac{4}{x} = \frac{6}{y}$, so $x = 6\frac{2}{3}$ and $y = 10$.

9.64 $\frac{4}{5} = \frac{8}{x} = \frac{6}{y}$, so $x = 10$ and $y = 7\frac{1}{2}$.

9.65 About 6 ft 11 in.

9.66 About 195 miles

9.4 HOMEWORK EXERCISES

Basic Exercises

1. The two squares shown are similar.

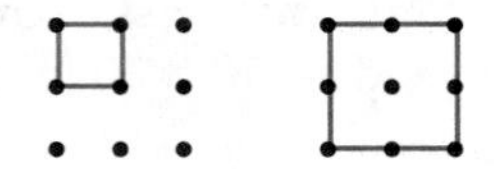

(a) The ratio of corresponding sides is ____ : ____.

(b) The ratio of the areas of the two squares is ________.

(c) The ratio of the perimeters of the two squares is ________.

***2.** Perform the size change $S_{P,2}$ on $\overline{AB}$.

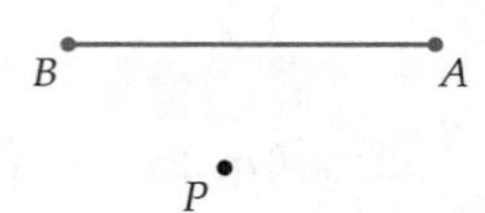

3. Perform the size change $S_{P,1/2}$ on $\triangle ABC$.

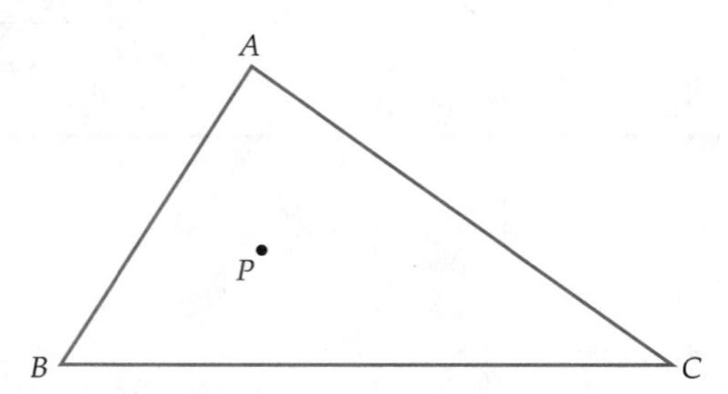

***4.** Look back at the size changes you have done. Under a size change, is the image of a line not containing the center parallel to the original line?

5. (a) Fill in the blank, putting the vertices for $\triangle NMR$ in the corresponding order.

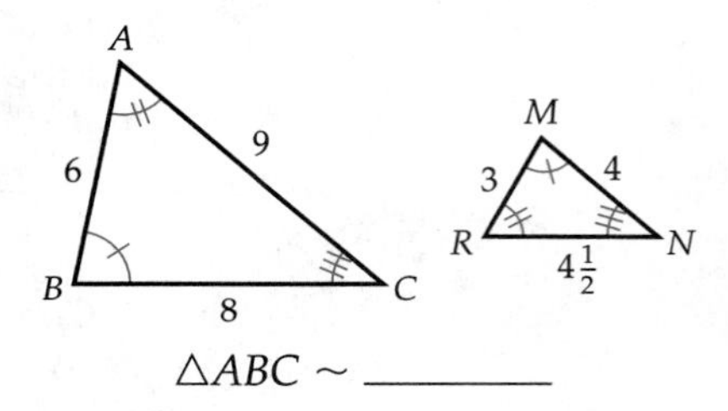

$\triangle ABC \sim$ ________

(b) What is the scale factor in going from $\triangle ABC$ to $\triangle MRN$?

6. In each case, decide if the two figures are similar.

(a)

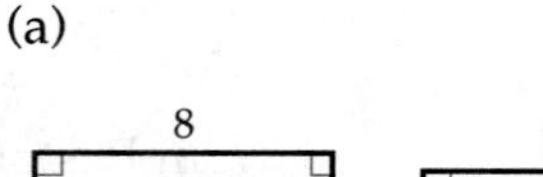

(b)

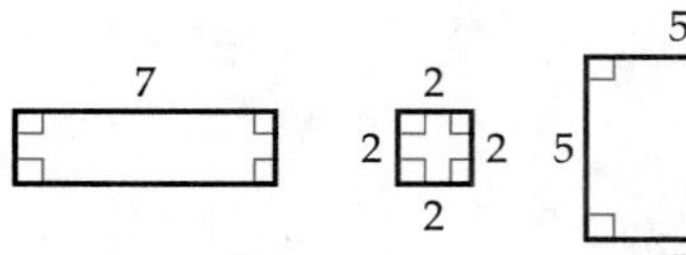

7. Tell why the following two figures are not similar.

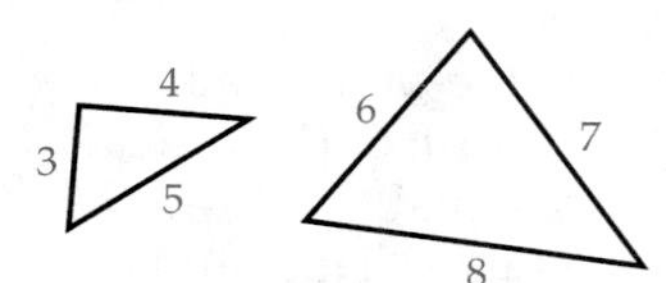

8. Recall that similar figures have the same shape but not necessarily the same size.

(a) Are any two circles similar?
(b) Are any two right circular cylinders similar?

9. (a) Are any two squares similar? Why or why not?
(b) Are any two rhombuses similar? Why or why not?

10. Are two congruent figures also similar?

11. An overhead projector projects the image of a figure from a transparency to a screen. What is the relationship between the figure and its image?

***12.** $\triangle ANT \sim \triangle REP$. Fill in the blanks.

(a) $\angle T \cong$ ________ (b) $\dfrac{ER}{\square} = \dfrac{\square}{TN}$

13. Sketch two similar figures suggested by the equivalent fractions:

$$\frac{8}{10} = \frac{12}{15} = \frac{14.4}{18}$$

14. Suppose you want to enlarge a photograph that is 3 in. by 5 in. so that its longer dimension is 8 in. What will the shorter dimension be?

15. The two triangles shown are similar. Find x and y.

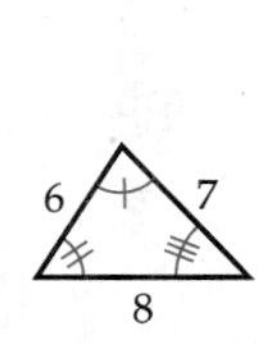

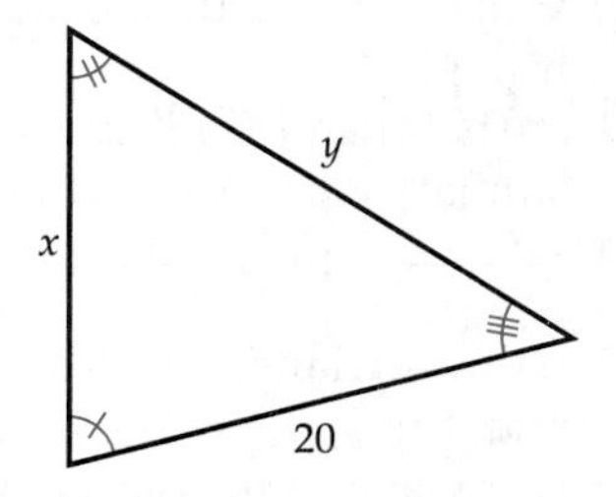

16. Assume that the two triangles in each part are similar, and find x.

(a)

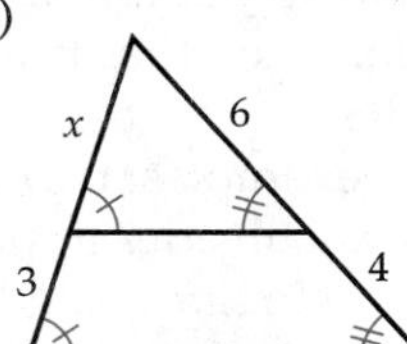

(b)

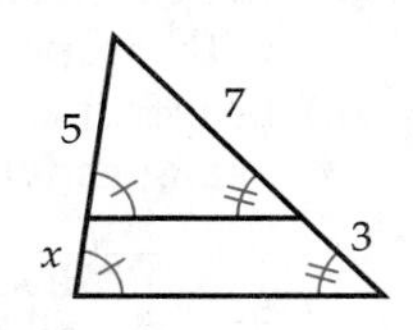

17. Assume that all three triangles are similar. Find x.

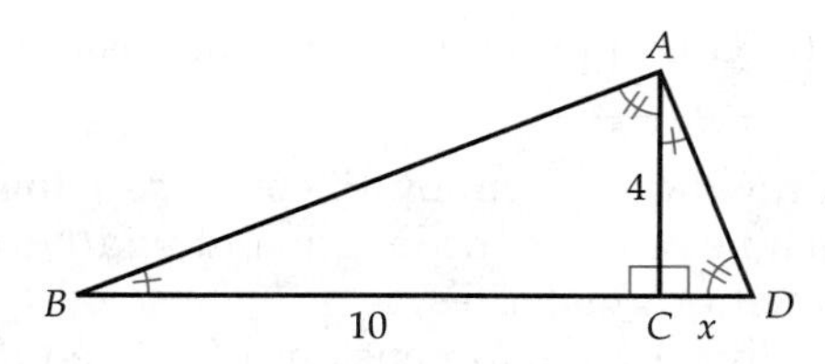

18. Find the missing angles in each pair of similar triangles.

(a) $\triangle ACT \sim \triangle RUG$ (b) $\triangle INK \sim \triangle IPS$

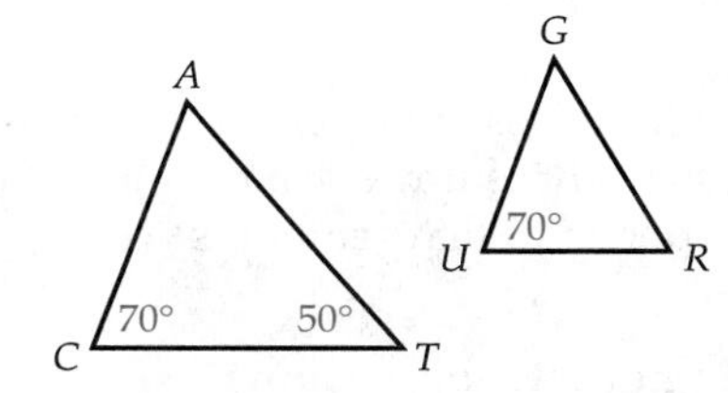

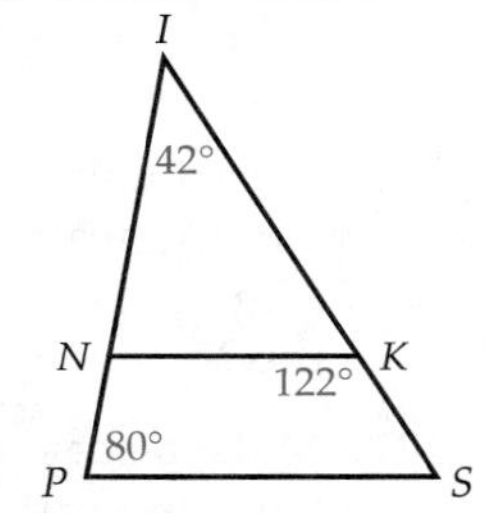

19. The two rectangular prisms shown here are similar; $\overline{AB}$ corresponds to $\overline{EF}$, $\overline{BC}$ corresponds to $\overline{FG}$, and $\overline{CD}$ corresponds to $\overline{GH}$. Find the lengths x and y.

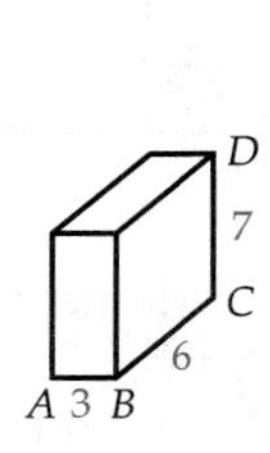

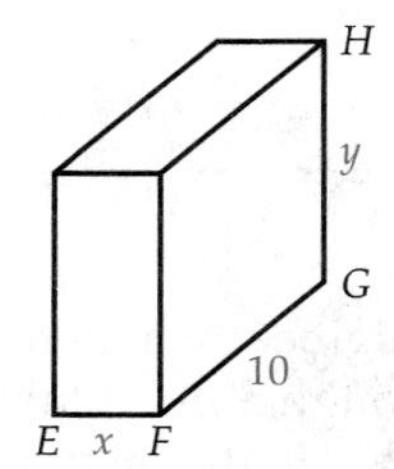

20. Would any two cylinders be similar? Why or why not?

21. Would any two spheres be similar? Why or why not?

22. What is the relationship between a scale model of a bridge and the actual bridge?

23. Bill is 4 ft 4 in. tall. At 3 P.M., his shadow is 5 ft 10 in. long. The shadow of a nearby building is 33 ft 4 in. long. How tall is the building?

24. A map has a scale ratio of 1 in. ≈ 60 miles. Sleepytown is $3\frac{1}{4}$ in. from Swingtown on the map. Based upon this, the actual distance is ____________.

25. A drawing is 10 cm by 15 cm. A copying machine reduces the length and width to 68% of their original sizes.
(a) What are the dimensions of the reduced drawing?
(b) Is the reduced drawing similar to the original?
(c) The area of the reduced image is _______% of the area of the original drawing.

26. A particular magnifying glass enlarges all length measurements to three times their original sizes. What percent enlargement is this?

27. Make a scale drawing of the house shown here. In your drawing, let 1 cm represent 4 m.

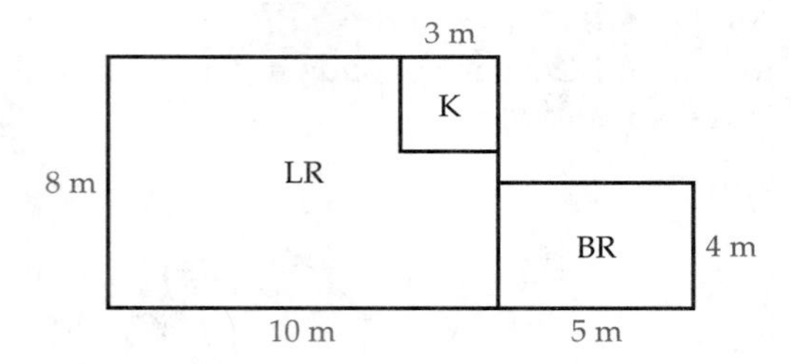

28. An engineer wants to build a tunnel from A to B through a mountain, as shown. To find the length AB, the engineer sights A and B from C. $AC = 60$ ft, $m\angle A = 90°$, $m\angle B = 32°$, and $m\angle C = 58°$. Make a scale drawing and approximate AB.

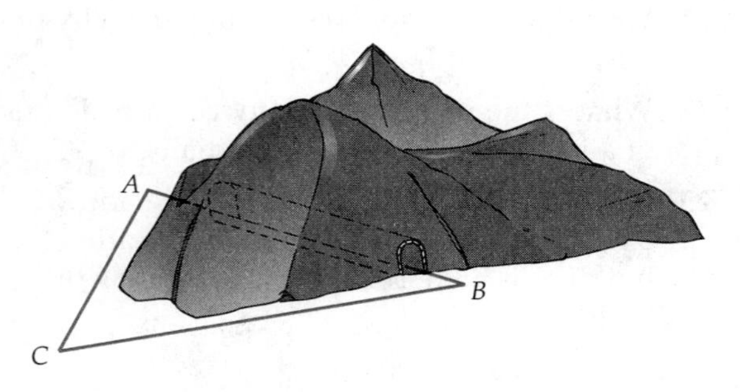

Extension Exercises

29. If two triangles have equal corresponding angle measures, do they have to be similar? Try the following. You will need these materials: one ruler, one protractor, and one live brain.
(a) Draw a triangle and measure its three angles. Now draw a second triangle that has the same three angle measures as the first triangle but has longer or shorter sides.
(b) Now look at your two triangles. What strikes you about their appearance (one compared to the other)?
(c) Measure the sides of each triangle. How do they compare?
(d) Complete the following generalization suggested by this exercise. If three angles of one triangle are congruent to three angles of a second triangle, then the triangles are ________ (called the AAA similarity property).

30. The preceding exercise concerned the AAA similarity property. Now suppose you have two triangles, and two angles in one triangle are congruent to two corresponding angles in the other triangle.
(a) What do you know about the third angle in each triangle?
(b) What can you then conclude about the two triangles?
(c) Parts (a) and (b) illustrate that the AAA similarity property can be simplified as the ________ property.

31. (a) Draw a triangle with sides of 2 in., 3 in., and 4 in.
(b) Draw a second triangle with sides of 1 in., $1\frac{1}{2}$ in., and 2 in.
(c) Do the triangles appear to be similar?
(d) Repeat parts (a)–(c) for two other triangles with proportional sides.
(e) Based upon your two examples, two triangles with proportional sides appear ________________ (to be, not to be) similar (called the SSS similarity property).

32. Find the center and the scale factor for the size change shown. The original figure is black.

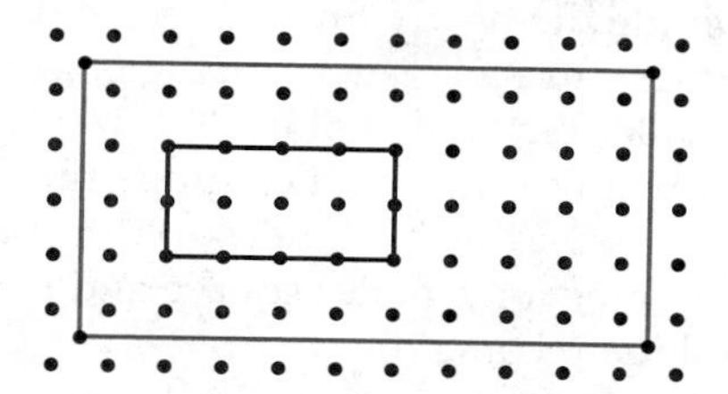

33. Use a size change to demonstrate a property of triangles.
    (a) Draw any $\triangle ABC$, and locate the midpoints of $\overline{AB}$ and $\overline{AC}$. Label them D and E, and draw $\overline{DE}$.
    (b) Under size change $S_{A,2}$, what are the images of D and E?
    (c) Based upon part (b), how is BC related to DE?
    (d) Write a general statement describing the triangle property you found.
34. Go outdoors on a sunny day, and use indirect measurement and shadows to find the height of a tree or pole in your neighborhood.
35. (a) Plot $A(0, 0)$, $B(4, 0)$, $C(2, 2)$, and $D(0, 2)$ on a graph, and connect A to B, B to C, C to D, and D to A.
    (b) What happens if you multiply all the coordinates by $k < 0$, plot the resulting points, and connect them in the same order?
36. A rectangle has dimensions of b and h. A second rectangle has dimensions of $b + 6$ and $h + 6$. Are the rectangles similar?
37. A rectangle has length L and width W. A second rectangle has length KL and width KW, with $K > 0$. Are the rectangles similar?
38. A quadrilateral has sides of length a, b, c, and d. A second quadrilateral has sides of length $2a$, $2b$, $2c$, and $2d$. Must the quadrilaterals be similar?
39. If every teacher's salary increases by 10%, that is like a size change with a scale factor of ________. (*Hint:* The answer is not 0.10.)

Computer Exercise

40. Use drawing software to investigate the minimum conditions for lengths of sides or angle measures that would make two triangles similar.
    (a) Draw two pairs of triangles with all three corresponding lengths of sides proportional. Measure the angles. Are the two triangles similar?
    (b) Repeat part (a) with two different triangles.
    (c) What conclusion do you draw from parts (a) and (b)?
    (d) Draw two pairs of triangles with two equal corresponding angle measures. Are the two triangles similar?
    (e) Repeat part (d) with two different triangles.
    (f) What conclusion do you draw from parts (d) and (e)?
41. (a) Draw any quadrilateral and label it $ABCD$.
    (b) Reduce the quadrilateral, using a scale factor of $\frac{1}{2}$. Label the new quadrilateral $EFGH$.
    (c) Measure the sides and angles of both quadrilaterals. What can you conclude about the quadrilaterals?
    (d) Repeat parts (a)–(c), beginning with a pentagon $ABCDE$.
    (e) Make a generalization of your results.
42. (a) Use a computer drawing program to draw a triangle.
    (b) Make four congruent copies of the triangle, and rotate one of them 180°.
    (c) Move the four triangles together to form a large triangle (with the rotated triangle in the center).

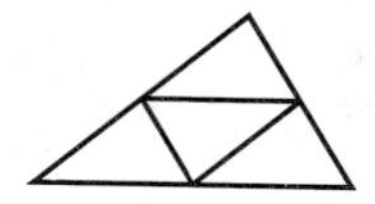

    (d) Measure the angles, sides, and area of the large triangle. How do they compare to the measurements of the small triangles?
    (e) Repeat parts (a)–(d) with another set of triangles.
    (f) Make some generalizations of your results.

Special Exercises

43. Look at the attractive rectangles that follow!

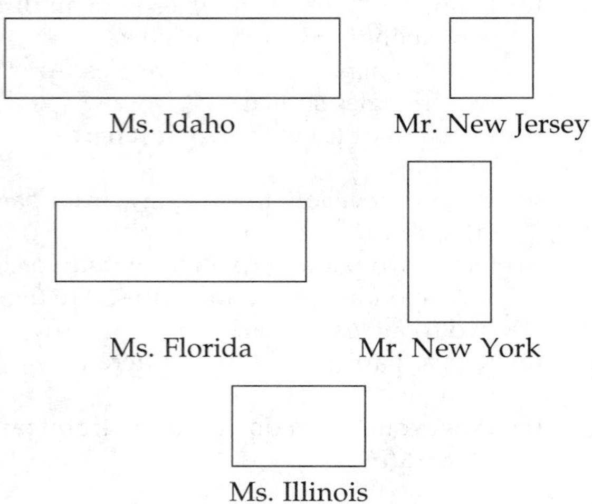

Which rectangle is the most beautiful? To answer this question, consider which one you would like as the front of a house, the frame of a painting, or the shape of a flag.

44. The **golden rectangle** (Ms. Illinois) has been admired since the time of ancient Greece (about 700 B.C.)!

> The **golden rectangle** is any rectangle with a ratio of
>
> $$\frac{\text{Length}}{\text{Width}} = \frac{1 + \sqrt{5}}{2} \approx 1.618$$

(a) If the width of a golden rectangle is 4, give its length to three decimal places.
(b) If the length of a golden rectangle is 10, give its width to three decimal places.

45. Take a typical golden rectangle.

1.618

1

Divide it into a square and a rectangle.

1.618

1 1

1 0.618

(a) Is the new, inside rectangle a golden rectangle?
(b) Divide the newly formed rectangle into a square and a smaller rectangle. Is this smaller rectangle golden?
(c) How long could you continue this process of subdividing each new golden rectangle into an even smaller square and smaller golden rectangle?
(d) Note that $\frac{1}{0.618} = \frac{1 + 0.618}{1}$. The golden ratio comes from the equation $\frac{b}{a} = \frac{b + a}{b} = 1 + \frac{a}{b}$. If r is the golden ratio $\frac{b}{a}$, then $r = 1 + \frac{1}{r}$. Solve this equation for r.

46. The Parthenon was built in Athens, Greece, about 2400 years ago (see photo). Since the shapes in the following drawing are approximately similar to those in the building, we can study the sketch and learn about the building! Use a ruler to measure the length and height of the front, including where the roof would be. What is significant about the ratio of the length to the height?

Photo courtesy of Library of Congress

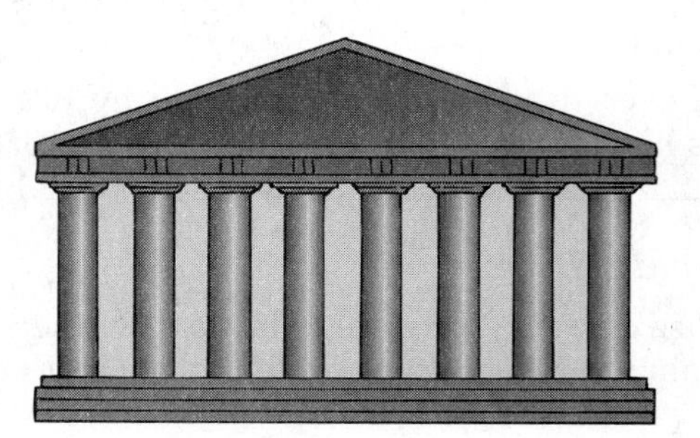

47. Determine how close the following ratios are to the golden ratio.
(a) Your height to the height of your navel
(b) The length of your index finger to the distance from your index fingertip to the middle of the big knuckle
(c) The length of your arm to the distance from your middle fingertip to your elbow
(d) The distance from your left eye to your mouth to the distance from your left eye to the level of your nose
(e) The length of an egg to the width of an egg

48. Find a doll, and measure its height and waist size. Explain whether or not the body proportions are realistic.

SUMMARY

Anyone who spends time studying shapes is likely to notice congruence, symmetry, and similarity. Congruence and similarity are two important relationships between pairs of geometric figures, whereas symmetry is a property of one figure. Congruent figures are identical in shape and size, whereas similar figures are the same shape but not necessarily the same size. Symmetry describes a certain kind of balance possessed by a figure.

Congruence and symmetry of plane figures are related to rotations, translations, and reflections. Ask a young child whether two shapes are identical, and the child will attempt to fit one shape on top of the other. This example illustrates the fact that two plane figures are congruent if and only if one can be mapped onto the other using a rotation, translation, reflection, or some combination of these motions. A plane figure is symmetric if you can find a different position in the plane in which the newly positioned figure coincides with the original figure.

Similarity is related to size changes. A size change uniformly multiplies the dimensions of a figure without changing the angle measures or shape. The resulting image is similar to the original figure.

Symmetry contributes to the beauty of nature. Reflection symmetry gives balance to birds and kites. Rotational symmetry gives directional flexibility to starfish and bolts. Similarity is useful in measuring heights indirectly and in finding distances represented on maps.

Constructions are a geometric ritual that has been passed on for thousands of years. Using only a compass and a straightedge, it is possible to copy or bisect all sorts of plane figures.

STUDY GUIDE

To review Chapter 9, see what you know about each of the following ideas or terms that you have studied. You can also use this list to generate your own questions about the chapter.

The NCTM Curriculum Standards and Congruence, Symmetry, and Similarity

Selected NCTM Curriculum Standards

The following standards come from the NCTM document.

- Explore transformations of geometric figures.
- Develop an appreciation of geometry as a means of describing the physical world.
- Investigate and predict the results of combining, subdividing, and changing shapes.
- Use a mathematical idea to further understanding of other mathematical ideas.

1. Describe how each standard listed relates to the material you studied in Chapter 9.
2. Select any current elementary-school mathematics textbook series, and describe sample lessons or exercises that illustrate each standard listed.

CONGRUENCE, SYMMETRY, AND SIMILARITY IN ELEMENTARY SCHOOL

The following chart shows at what grade levels selected geometry topics typically appear in elementary-school mathematics textbooks.

Topic	Typical Grade Level in Current Textbooks
Slides, flips, turns	4, 5, 6
Congruent figures	1, 2, 3, 4, 5, 6
Line symmetry	2, 3, 4, 5, 6
Rotational symmetry	6
Similar figures	4, 5, 6
Constructions	6

REVIEW EXERCISES

1. What is the relationship between *congruence* and rotations, translations, and reflections?

2. Show the image of the F after a translation using $\overrightarrow{AB}$ followed by a rotation of 90° counterclockwise around point C.

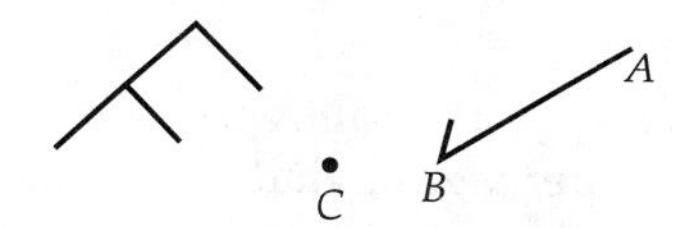

3. (a) Show the image of $\triangle ABC$ after a reflection through line m followed by a reflection through line n.

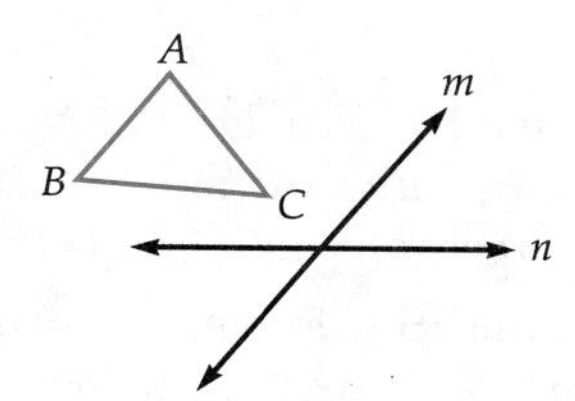

(b) What single motion gives the same image as the two successive reflections?

4. (a) Find the image of A' of point A, using the mapping.

$$(x, y) \rightarrow (x - 3, y - 4)$$

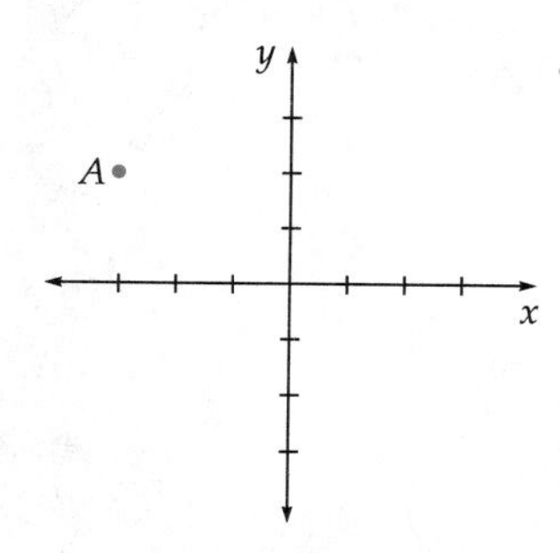

(b) Name the motion that produces the mapping

$$(x, y) \rightarrow (x - 3, y - 4)$$

5. A point P is reflected through line m. If P is not on m, then line m ______ $\overline{PP'}$.

6. $ABCD$ is a rectangle.

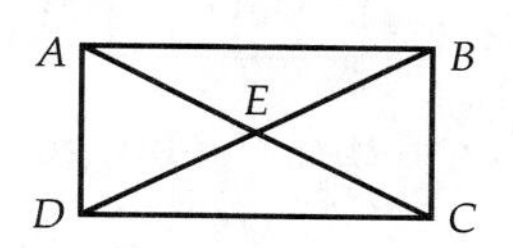

(a) Select a pair of angles that appear to be congruent, and name a motion that maps one angle onto the other.
(b) Repeat part (a) for another pair of angles.

7. (a) Construct the angle bisector of $\angle AIM$ and call it $\overrightarrow{IT}$. (Assume $AI = MI$.)

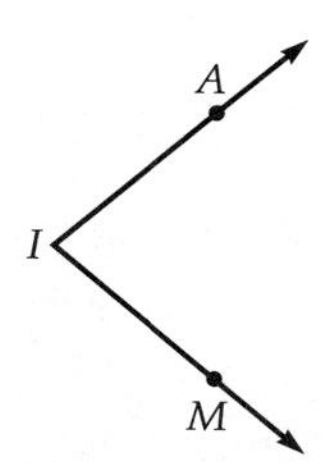

(b) Use congruent triangles to explain why $\overrightarrow{IT}$ bisects $\angle AIM$.

8. Shown here is a regular hexagon. Find all the angle measures in the diagram.

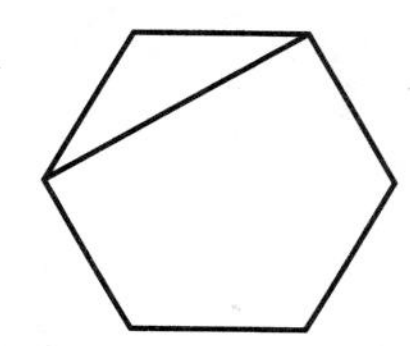

9. Draw all lines of symmetry for the rectangle shown.

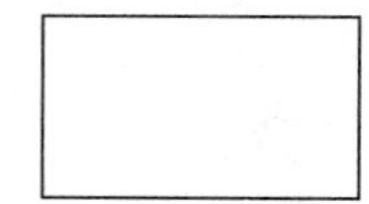

10. Are any two regular pentagons similar? Why or why not?

11. Draw a figure that has four lines of symmetry and three rotational symmetries less than 360°.

12. Shade the smallest number of squares so that the following diagram has rotational symmetry.

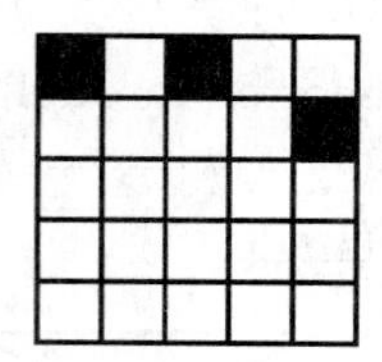

13. Use a ruler to find the image of $\triangle ABC$ after the size change $S_{P,1.5}$.

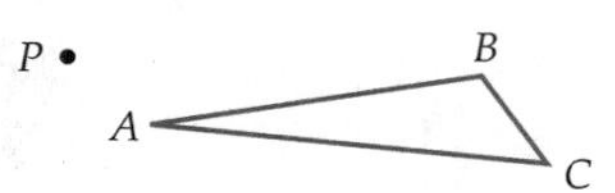

14. If $\triangle RUN \sim \triangle RAC$, find AC.

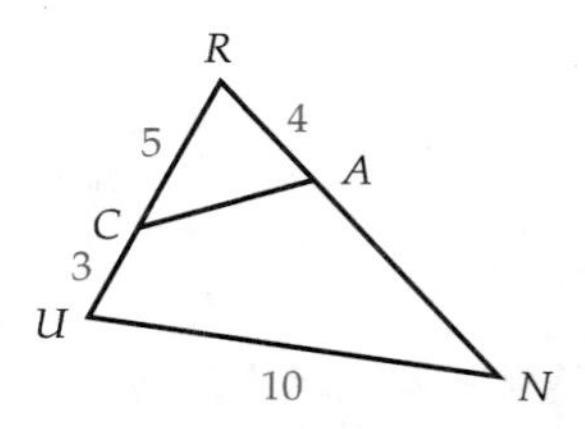

15. Sandy is 4 ft 10 in. tall. At 1 P.M., her shadow is 5 ft 5 in. long. The shadow of a nearby building is 30 ft 2 in. long. How tall is the building?

16. A rectangle has length L and width $L + 6$. A similar rectangle has length 10. What is its width?

17. Sketch any of the following six figures that are possible: pentagons that have exactly zero, one, two, three, four, or five lines of symmetry.

SUGGESTED READINGS

Geddes, D., and others. *Geometry in the Middle Grades.* Reston, Va.: NCTM, 1992.

Hill, J. (ed.). *Geometry for Grades K–6.* Reston, Va.: NCTM, 1987.

Holden, A. *Shapes, Space, and Symmetry.* Mineola, N.Y.: Dover, 1991.

Kroner, L. *Slides, Flips, and Turns.* Palo Alto, Calif.: Creative Publications, 1995.

National Council of Teachers of Mathematics. *Geometry in the Mathematics Classroom.* 1973 Yearbook. Reston, Va.: NCTM, 1973.

National Council of Teachers of Mathematics. *Learning and Teaching Geometry K–12.* 1987 Yearbook. Reston, Va.: NCTM, 1987.

National Council of Teachers of Mathematics. *Connecting Mathematics Across the Curriculum.* 1995 Yearbook. Reston, Va.: NCTM, 1995.

O'Daffer, P., and S. Clemens. *Geometry: An Investigative Approach.* 2nd ed. Reading, Mass.: Addison-Wesley, 1992.

Olson, T. A. *Mathematics Through Paper Folding.* Reston, Va.: NCTM, 1975.

Ranucci, E., and J. Teeters. *Creating Escher-Type Drawings.* Palo Alto, Calif.: Creative Publications, 1977.

Weyl, H. *Symmetry.* Princeton, N.J.: Princeton University Press, 1952.

10 Measurement

People created measurement systems to satisfy practical needs such as finding the distances between places, the heights of horses, and the weights of grains. They developed calendars in order to keep track of days, seasons, and years.

Every measurement requires a unit of measure. Ancient people used body measurements as units of length and seeds and stones as units of weight. The first units of time were the cycles of the sun and moon. In recent times, most countries have developed uniform units of measure. Today, most countries in the world use the metric system. The United States uses both the customary (old English) system and the metric system.

People have developed ingenious formulas and methods for computing the areas and volumes of geometric shapes, and they use these formulas to approximate the measures of many everyday objects.

10.1 METRIC MEASURE

Any object has many attributes that can be measured. When you measure something, you focus on one attribute and ignore all other attributes and qualities. Measurement involves (1) identifying an attribute, (2) selecting a unit of measure, and (3) comparing the attribute of the object to the unit of measure. The first units of measure were nonstandard units.

NONSTANDARD MEASURING UNITS

Historical records indicate that the first units of length were based on people's hands, feet, and arms (Figure 10-1, on page 566).

Lesson Exercise 10.1

(a) How many hands high is this page of your book?
(b) Why might others in the class obtain different answers?

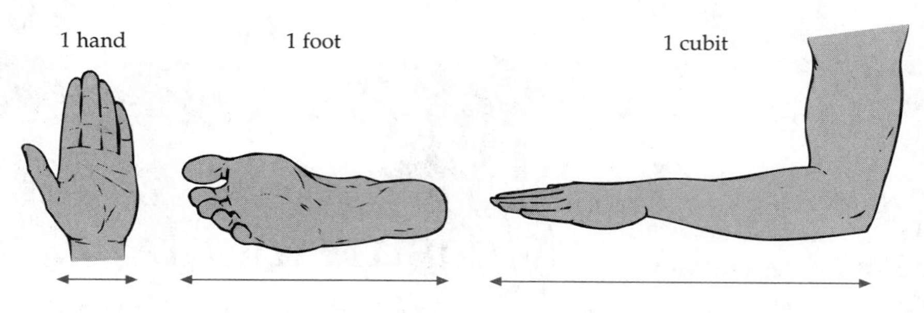

Figure 10–1

Hands, feet, and arms are convenient for measuring, but different people have different-sized hands, feet, and arms. That is why we need standard measuring units such as those of the metric system. By using standard units, scientists and manufacturers around the world communicate easily with one another about measurements.

THE METRIC SYSTEM

The metric system was developed in France in the late 1700s. A meter was originally defined as $\frac{1}{10{,}000{,}000}$ of the distance from the North Pole to the equator. The meter is now defined as the distance light travels in a vacuum in $\frac{1}{299{,}792{,}458}$ second.

Due to the historical connections between the U.S. and England, the U.S. adopted the English system rather than the metric system of England's former enemy, France. Today, nearly every country in the world except the United States and Australia uses the metric system. In the United States, the metric system is widely used in many areas, including science and photography but not carpentry. In what ways is the metric system superior to the customary (U.S.) system?

***Class* Lesson Exercise 10.2**

How many people in your class can correctly answer each of the following questions?

(a) How many yards are in a mile?

(b) How many meters are in a kilometer?

Most of you have been using customary measure all your life and metric measure for only a short time. Despite this fact, more adults in the United States can correctly answer Lesson Exercise 10.2(b) than 10.2(a)!

Two advantages of the metric system are: (1) metric prefixes are the same for length, weight, and liquid volume, and (2) all conversions within the metric system involve powers of 10.

Each pair of students will need a meter stick and a 30-cm ruler to do Lesson Exercises 10.3, 10.4, 10.7, and 10.9.

Cooperative Lesson Exercise 10.3

Examine your meter stick, and complete the following.

(a) Find a centimeter on your meter stick.

(b) How many centimeters are in a meter?

(c) The millimeter is the smallest unit of measure on your meter stick. Find a millimeter on your meter stick.

(d) How many millimeters are in a centimeter?

(e) How many millimeters are in a meter?

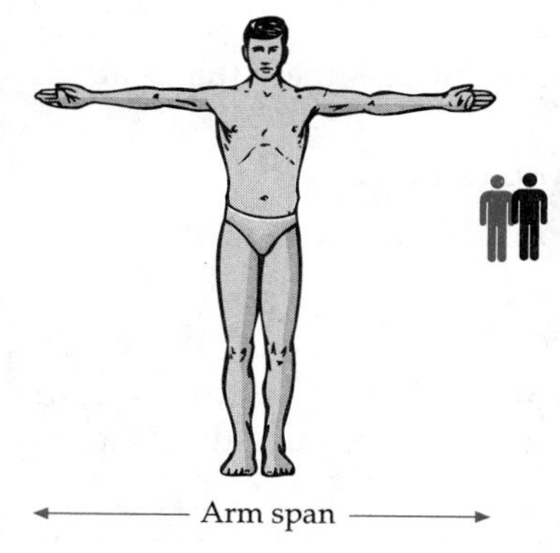

Figure 10–2

How does your arm span compare to your height (Figure 10-2)?

Cooperative Lesson Exercise 10.4

(a) What is your height in centimeters? (Stand against a blackboard and have another student mark your height.)

(b) Give your height, using meters and centimeters (such as 1 m 37 cm).

(c) What is your height in meters?

(d) How long is your arm span?

(e) How does your arm span compare to your height?

In the metric system, people's heights are most commonly recorded in centimeters, as in Lesson Exercise 10.4(a).

REFERENCE MEASURES

A good way to learn the approximate sizes of unfamiliar units of measure is to relate them to measurements of your body or your environment.

Cooperative Lesson Exercise 10.5

(a) Find an object in your classroom that is about 1 m high or 1 m above the floor.

(b) Find an object in your classroom that is about 1 m long.

(c) Find some part of your hand that is about 1 cm long.

(d) Find an object that is about 1 mm long.

Your answers to Lesson Exercise 10.5 are called reference measures. **Reference measures** are common environmental or body measurements that approximate basic measuring units such as a meter. Reference measures help develop an intuitive understanding (number sense) of metric measure.

You can use reference measures to estimate other measurements. In Lesson Exercise 10.6, use a reference measure from Lesson Exercise 10.5 to select the best estimate.

Lesson Exercise 10.6

Using your reference measure, select the best estimate for the length of your professor's shoe.

(a) 25 mm (b) 25 cm (c) 2.5 m (d) 25 m

Lesson Exercise 10.7

(a) Use your reference measure for 1 cm to draw a line segment that is about 8 cm long.

(b) Measure your segment to see how close you came.

Cooperative Lesson Exercise 10.8

Use your reference measure for 1 m to obtain an estimate of the area of your classroom floor, in square meters. (*Note:* Area will be discussed further in the next section.)

How do metric units compare in size to customary units? Examine your rulers to answer the following questions.

Lesson Exercise 10.9

(a) What customary unit of length is about the same length as a meter?

(b) Which is longer, an inch or a centimeter?

Why do we need units of different sizes, such as centimeters, meters, and kilometers? If we measured traveling distances in centimeters, the numbers would be so great that they would be hard to grasp. For example, the distance from Washington, D.C., to Boston is about 66,000,000 cm. Conversely, measuring the length of your foot in kilometers, you might get a result like 0.00021 km, which would not be very meaningful.

CONVERSIONS WITHIN THE METRIC SYSTEM

You have already seen how meters, centimeters, millimeters, and kilometers are related by powers of 10. This same relationship holds for all metric measures, and it makes conversions within the metric system easier than those in the customary system.

The metric prefixes are presented in the following table.

Prefix	kilo-	hecto- †	deka- †	(none)	deci-	centi-	milli-
Symbol	k	h	dk		d	c	m
Meaning	1000	100	10	1	$\frac{1}{10}$	$\frac{1}{100}$	$\frac{1}{1000}$

† Not commonly used.

Every metric length measurement includes one of these prefixes in front of the basic length unit, the meter. For most everyday measures, kilometers, meters, centimeters, and millimeters are sufficient.

Kilometer	**Hectometer†**	**Dekameter†**	**Meter**	**Decimeter**	**Centimeter**	**Millimeter**
km	hm	dkm	m	dm	cm	mm

† Not commonly used in U.S. elementary and secondary schools.

Example 10.1 shows how to convert within the metric system.

EXAMPLE 10.1

350 cm = ________ m

Solution

Compare centimeters to meters. Since 100 centimeters make a meter, divide 350 cm by 100 to convert it to meters: 350 ÷ 100 = 3.50. (Move the decimal point two places to the left.) You can also use a prefix chart and see that, to go from centimeters to meters, one moves two places to the left. This corresponds to the movement of the decimal point in converting from centimeters to meters.

More formally, one can include the units in the computation and multiply 350 cm by 1 in the form $\frac{1 \text{ m}}{100 \text{ cm}}$.

$$350 \text{ cm} \times \frac{1 \text{ m}}{100 \text{ cm}} = \frac{350}{100} \text{ m} = 3.5 \text{ m}$$

This more formal approach with units is part of **dimensional** (unit) **analysis**. ■

As suggested by converting your height from centimeters to meters and by the preceding example (350 cm = 3.5 m), metric conversions require nothing more than moving the decimal point, since they all involve multiplying or dividing by a power of 10.

Try the following conversions.

Lesson Exercise 10.10

(a) 0.5 m = ________ mm (b) 80 mm = ________ cm
(c) A kilometer is ________ meters (about 3/5 of a mile).

MASS AND LIQUID VOLUME

Miracle Diet Program!
Lose 5/6 of Your Weight:
Take a Trip to the Moon

The gram is the basic unit of *mass* in the metric system. People sometimes confuse mass and weight. **Mass** measures the quantity of matter a body contains. **Weight** measures the force exerted by gravity on a body.

If you went to the moon, your mass would remain the same, but your weight would be much less (about 1/6 of your weight on Earth) because the force of gravity is much weaker on the moon. In everyday usage, kilograms (the metric unit of mass) and pounds (the customary unit of weight) are used interchangeably. The conversion 1 kg = 2.2 lb works on the surface of Earth, but on the moon a mass of 1 kg weighs about 0.4 lb.

Lesson Exercise 10.11

(a) If you flew to Jupiter, the largest planet, would your *mass* be more than, the same as, or less than your mass on Earth?

(b) Would your *weight* be more than, the same as, or less than your weight on Earth?

The most commonly used units of metric mass are milligrams (very small), grams, and kilograms. The prefixes *milli-* $\left(\frac{1}{1000}\right)$ and *kilo-* (1000) have the same meaning with grams as with meters. Do you know any reference measures for grams and kilograms?

Lesson Exercise 10.12

Name something that has a mass of about 1 g (gram).

Lesson Exercise 10.13

Name something that has a mass of about 1 kg (kilogram).

A raisin or a solid, well-made paper clip has a mass of 1 g. A textbook you can finish in one semester or two loaves of bread may have a mass of about 1 kg. A milligram is very small, like the mass of a feather or a few grains of salt.

The liter (L) is the metric unit for liquid volume. Do you know any reference measures for a liter?

Lesson Exercise 10.14

Name an everyday object that holds about 1 L (liter).

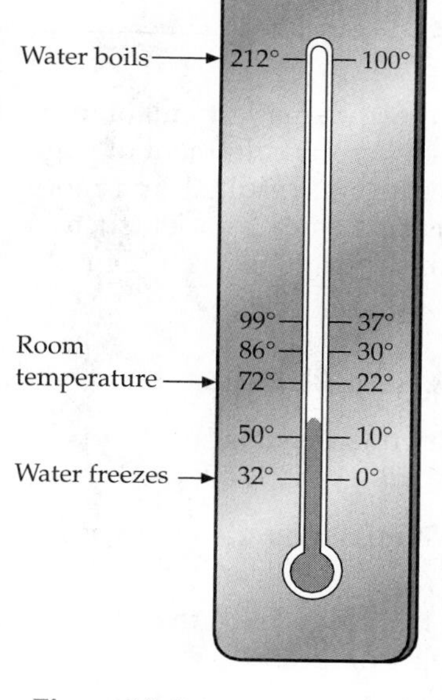

Figure 10–3

A drop of water from an eyedropper is about 1 mL.

Lesson Exercise 10.15

Grams and liters use the same prefixes as meters.

(a) 80 g = ________ kg

(b) N mL = ________ L (N is a positive number.)

TEMPERATURE

Do you know what a 35° Celsius day would feel like? Hot! Metric temperatures are often measured in degrees Celsius (C). In this system, 0°C is the freezing point of water and 100°C is the boiling point of water. These temperatures are easier to remember than those in the Fahrenheit scale, in which water has a freezing point of 32°F and a boiling point of 212°F. The thermometer shown in Figure 10-3 gives some other common temperatures in Fahrenheit and Celsius.

Lesson Exercise 10.16

The temperature inside a working refrigerator is about

(a) 5°C (b) 15°C (c) 25°C (d) 35°C

PRECISION, GREATEST POSSIBLE ERROR, AND RELATIVE ERROR

Whereas exact measurements can be given for perfect geometric shapes, no measurement of an actual object (other than counting a set of objects) is exact.

In theory, perfect geometric figures have exact measures, whereas actual objects have approximate measures. In Chapter 3, this textbook distinguished between sets and measures. In order to be consistent, counting a set of objects will not be referred to as a measurement in this chapter.

Geometric shape → Exact measurement

Everyday object → Approximate measurement

A reported numerical measurement should reflect the precision with which it was made. The precision of an approximate measurement is determined by the unit of measure.

Lesson Exercise 10.17

(a) Measure the height of this page to the nearest centimeter.
(b) Measure the height of this page to the nearest millimeter.
(c) Which measurement is more precise?

The **precision** of a measurement is determined by the smallest unit of measurement used. In the preceding lesson exercise, neither measurement of page height was exact, but using the smaller unit (millimeters) resulted in a more *precise* measure. Similarly, a measurement of height to the nearest foot (such as 5 ft) is less precise than a measurement to the nearest inch (such as 5 ft 4 in.).

Lesson Exercise 10.18

People usually give their weight to the nearest pound or kilogram. Suppose that John's correct mass, rounded to the nearest kilogram, is 53.

(a) What is the smallest decimal number that would be rounded to 53 as the nearest whole number?
(b) Any decimal number greater than or equal to 52.5 and less than ______ would be rounded to 53 as the nearest whole number.
(c) How far from 53 kg could John's exact mass be?

In the preceding exercise, John's exact mass is between 0 and 0.5 kg away from 53 kg. The greatest possible error is 0.5 kg. This could be obtained by taking half of the smallest unit (kg) that was used in measuring 53 kg. The **greatest possible error** (GPE) is half of the smallest unit used in the measurement.

[------Exact measure------)

Rounded measure

↓

52.5 53 53.5

John's actual mass is somewhere between 52.5 and 53.5 kg (including 52.5). In practice, scientists say "between 52.5 and 53.5 kg" without mentioning whether the endpoint measures are included, since no object is likely to weigh *exactly* 52.5 or 53.5 kg.

Lesson Exercise 10.19

Janie's shoe is 27.6 cm long.

(a) What is the smallest unit of measure used?

(b) Her actual shoe length is between ________ cm and ________ cm.

(c) What is the GPE?

Which would be more significant, an error of 0.5 kg in a measurement of John's mass as 53 kg, or an error of 0.05 cm in a measurement of Janie's shoe as 27.6 cm? Relative error answers this by comparing the size of the greatest possible error to the measurement. The **relative error** of a measurement is the GPE divided by the measurement.

Lesson Exercise 10.20

(a) Find the relative error of the 53 kg from Lesson Exercise 10.18.

(b) Find the relative error of the 27.6 cm from Lesson Exercise 10.19.

(c) Which measurement has the smaller relative error?

ANSWERS TO SELECTED LESSON EXERCISES

10.1 (a) About 3 or 4
(b) Their hands are different sizes.

10.3 (b) 100 (d) 10 (e) 1000

10.5 (a) *Hint:* Try a doorknob.

10.6 25 cm

10.9 (a) A yard (b) An inch

10.10 (a) 500 (b) 8 (c) 1000

10.11 (a) The same as (b) More than

10.14 A quart milk carton

10.15 (a) 0.08 (b) $\frac{N}{1000}$

10.16 5°C

10.17 (c) Part (b)

10.19 (a) One tenth of a centimeter
(b) 27.55; 27.65
(c) 0.05 cm

10.20 (a) $\frac{0.5}{53} = 0.0094$ or 0.94%
(b) $\frac{0.05}{27.6} = 0.0018$ or 0.18%
(c) Part (b)

10.1 HOMEWORK EXERCISES

Basic Exercises

1. Make up a measuring unit for length. Use it to measure the height of a page in this book.

2. (a) How long is your pace (one step) in centimeters?
(b) Measure the length of a room in paces.
(c) Convert the length of the room in part (b) to centimeters.

3. (a) How long is your foot in centimeters?
(b) Measure the width of your front door using your foot.
(c) Convert the width of your door in part (b) to centimeters.

4. In what ways is the metric system easier to work with than the customary system?

5. Give a reference measure for
(a) 1 cm. (b) 1 m.

6. The width of an adult woman's hand is about ________ cm.

7. Complete the following conversions.
(a) 0.02 m = ________ mm
(b) 16 cm = ________ mm
(c) 82 m = ________ km

8. A film is 35 mm wide. How many centimeters wide is it?

9. Lynn Matthews wants to swim 1 km in a 50-m-long pool. How many pool lengths must she swim?

10. A town is T km away. How many meters away is it?

11. The length of a new piece of chalk is about
(a) 1 mm (b) 1 cm
(c) 10 cm (d) 1 m

12. (a) Estimate the length of your arm in centimeters.
(b) Measure your arm length in centimeters.

13. The average adult man has a mass of about
(a) 7 kg (b) 700 kg
(c) 700 g (d) 70 kg

14. A pound (on Earth) is about
(a) 5 g (b) 5 kg (c) 50 g (d) 500 g

15. (a) 9.4 L = ________ mL
(b) 37 mg = ________ g
(c) 0.082 kg = ________ g

16. (a) 65 g = ________ mg
(b) 47 g = ________ kg
(c) 346 mL = ________ L

17. A container holds 3.24 liters of milk. How many milliliters is that?

18. A metal strip has a density of 200 g/cm. Express its density in kilograms per meter.

19. A coffee cup holds about
(a) 25 mL (b) 250 mL
(c) 2.5 L (d) 25 L

20. Your heart pumps about 60 mL of blood per heartbeat. About how much blood will it pump in a day?

21. A box contains 4 kg of paper clips. Each paper clip has a mass of about 0.8 g. About how many paper clips are in the box?

22. (a) Complete the third example, repeating the error pattern from the completed examples.
(b) Describe the error pattern.

5 cm = __*500*__ m
20 kg = __*0.02*__ g
3 mm = ________ m

23. True or false?
(a) 1 mm is longer than 1 in.
(b) 1 m is longer than 1 km.
(c) 1 g is heavier than 1 lb.
(d) 1 gallon is more than 1 L.

***24.** For each item, select the most appropriate measuring unit from the following list: mm, cm, m, km, g, kg.
(a) The weight of a penny
(b) Your waist size
(c) The thickness of a page
(d) Your weight
(e) The distance from Miami to Atlanta

25. Someone who tells you that the temperature is 80° outside
(a) is using Fahrenheit (b) is using Celsius
(c) could be using either Fahrenheit or Celsius

26. A temperature of −10°C is about
(a) −20°F (b) 10°F
(c) 40°F (d) 70°F

27. The weather outside is sunny and 15°C. Which of the following would you wear?
(a) A heavy coat (b) A light sweater
(c) A swimsuit

28. A kilometer is about 3/5 of a mile. Mentally estimate conversions of the following speed limits.
(a) 30 miles/h ≈ ________ km/h
(b) 60 km/h ≈ ________ miles/h

29. Can you walk 1 km in 30 minutes?

***30.** List the following measurements in increasing order: 37 m, 46 cm, 871 mm, 3 km, 137 cm.

31. Tell whether each number is probably exact or approximate.
(a) Joe bought 1 kg of apples.
(b) Bill bought 2 ears of corn.
(c) The room is 5 m long.
(d) Sally has 3 brothers.
(e) 8000 people attended the concert.

32. Which measurement in each pair is more precise?
(a) 8 cm or 78 mm
(b) 6 kg or 5820 g

33. I measure the length of my shoe as 28 cm. The actual length of the shoe is between what two measures?

34. Give the GPE of each measurement.
(a) 37 m (b) 5.21 g

35. (a) Measure the following line segment to the nearest millimeter.

____

(b) What is the GPE?
(c) The actual length of the line segment is between ________ and ________.

***36.** Complete the table.

Measurement	Precision	GPE	Actual Length Is Between
32.1 m	Nearest tenth of a meter (or nearest dm)	0.05 m	32.05 m, 32.15 m
(a) 62 g			
(b) 5.12 cm			
(c) 4.6 km			

37. Find the relative error for the measurements in parts (a)–(c) of the preceding exercise.

38. (a) Measure the length and width of a dollar bill to the nearest centimeter.
(b) Give the relative error of each measurement.

Extension Exercises

39. Draw a rectangle with a length of 5 cm and a diagonal 7 cm long.

40. You have a large set of black and white centimeter cubes. You could make two different towers that are 1 cm high, a black tower and a white tower.

(a) How many different towers could you make that are 2 cm high?
(b) How many different towers could you make that are 3 cm high? (Draw pictures.)
(c) How many different towers could you make that are 4 cm high?
(d) How many different towers could you make that are N cm high? (N is a counting number.)

41. A passenger is riding in a train that is going 150 km/h. The passenger sees a second train pass by in 2 seconds. If the second train is moving at a speed of 150 km/h in the opposite direction, how long is the second train?

42. Make a scale drawing of the solar system in which the distance from the sun to Pluto is somewhere between 30 cm and 1 m. The actual distances of planets from the sun, in meters, are Mercury, 5.8×10^{10}; Venus, 1.1×10^{11}; Earth, 1.5×10^{11}; Mars, 2.3×10^{11}; Jupiter, 7.8×10^{11}; Saturn, 1.4×10^{12}; Neptune, 2.9×10^{12}; Uranus, 4.5×10^{12}; Pluto, 5.9×10^{12}.

43. The prefix *micro-* represents 10^{-6}, or $\frac{1}{1{,}000{,}000}$.
(a) 1 m = ________ micrometers
(b) 60 micrometers = ________ m
(c) 821 micrometers = ________ cm
(d) 4800 mm = ________ micrometers

44. The prefix *mega-* means "one million."
(a) 80 megameters = ________ m
(b) 1000 kg = ________ megagrams

45. 54 cm/s = ________ km/h

46. A car that gets 20 km/L of gasoline is getting about ________ miles per gallon.
(a) 10 (b) 20 (c) 30 (d) 45

47. A poster x cm high contains two photographs of the same size, one above the other. There are 3-cm margins at the top, at the bottom, and between the two photographs. What is the height of each photograph?

Special Exercises

48. If you mix measurement units in a problem, you can get some strange equations. For example, when does 1 + 60 = 25? When it's 1 day + 60 minutes = 25 hours! Provide units for the following equations.
(a) 3 + 120 = 5 (b) 8 + 6 = 2
(c) 3 − 12 = 2

49. Make a scale drawing of your classroom on centimeter graph paper.

10.2 PERIMETER AND AREA

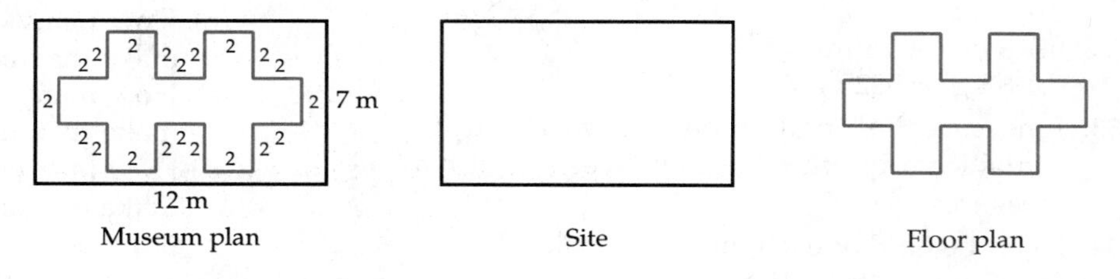

Figure 10–4

Class Lesson Exercise 10.21

In Figure 10-4, which is larger, the site or the floor plan?

To answer the question in Lesson Exercise 10.21, one must first decide whether to compare perimeter or area.

PERIMETER

What length of fence will enclose a field? What length of a decorative border will you need for a classroom bulletin board? These situations both call for measuring the perimeter.

In elementary and secondary school, children study the perimeters of polygons and circles. The **perimeter** of a simple, closed plane figure is the length of its boundary. The perimeter is always measured in units of length, such as feet or centimeters.

Cooperative Lesson Exercise 10.22

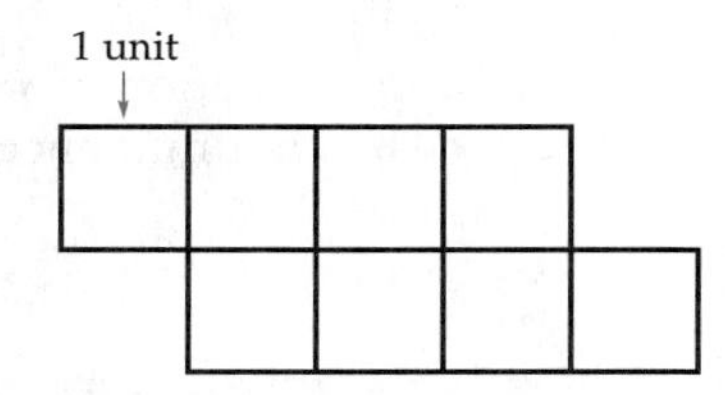

(a) The perimeter is ________.

(b) Arrange the squares in other configurations in which each square shares at least one side with another square. Compute the perimeter of each configuration.

(c) What is the largest perimeter you can obtain?

(d) What is the smallest perimeter you can obtain?

The perimeter of a polygon is the sum of the lengths of its sides. Some polygons have fairly simple perimeter formulas that can be used instead of adding up the lengths of all the sides. Find the appropriate formula for a rectangle in the following exercise.

Lesson Exercise 10.23

A rectangle has a length l and a width w. What is a formula for the perimeter P?

The perimeter formula from Lesson Exercise 10.23 can be used in the following exercise.

Lesson Exercise 10.24

Refer back to the diagram at the beginning of the section. Which has a larger perimeter, the site or the floor plan?

THE CIRCUMFERENCE OF A CIRCLE

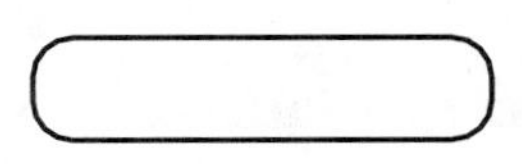

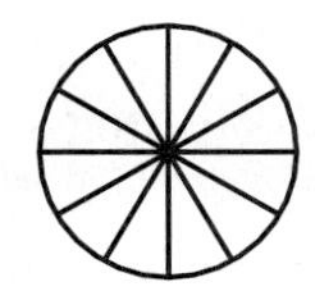

Figure 10–5

How far is one lap around a track with semicircular arcs? What is the perimeter of a bicycle tire with a diameter of 26 in. (Figure 10-5)? To answer these questions, one computes the **circumference** (perimeter) of a circle.

In Section 7.6, you saw how π originated from studies of the relationship between the circumference and the diameter. The circumference C equals π times the diameter d, or $C = \pi d$.

The value of π is an irrational number, an infinite, nonrepeating decimal that begins as 3.141592654 .... In 1987, Hideaki Tomoyori memorized and recited the first 40,000 places of π. Most of us can get by with 3.14 or 22/7 as an approximation of π.

The circumference formula, $C = \pi d$, can be rewritten using $2r$ in place of d as $C = 2\pi r$.

The Circumference Formula

A circle with diameter d and radius r has circumference $C = \pi d$, or $C = 2\pi r$.

You can use this formula to find the circumference when you know the length of the radius or diameter. In computations involving π, use π in situations involving perfect geometric figures, and use an approximate value of π such as 3.14 (or 22/7) in situations involving everyday objects.

Perfect geometric shapes → exact answers → Use π.
Everyday objects → approximate answers → Use a decimal approximation.

Lesson Exercise 10.25

A bicycle wheel has a diameter of 26 in. How far would a rider travel in one full revolution of the tire? (Use 3.14 for π.)

AREA

If you want to know the size of the interior of a field, or which package of gift wrap is a better buy, you measure the area. **Area** is the measure of a closed, two-dimensional region.

Lesson Exercise 10.26

Determine how many of each shape shown are needed to fill in the design in Figure 10-6.

(a) Trapezoids

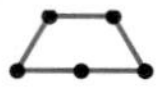

(b) Triangles

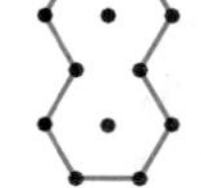

Figure 10–6

Although area can be measured with trapezoids or triangles as units, it is usually measured in square units. An area of "10 square units" means that 10 unit squares are needed to cover a flat surface. A square centimeter (cm^2) is an example of a square unit.

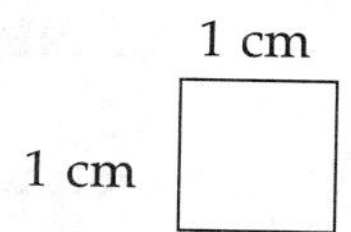

What is the area of your hand? To find out, you will need your hand and some centimeter graph paper.

Lesson Exercise 10.27

(a) Trace an outline of your hand on the centimeter graph paper. (Keep your fingers together.)

(b) Estimate the area of your hand in square centimeters.

(c) Explain how you obtained your answer in part (b).

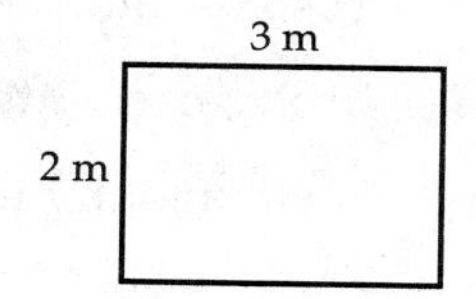

Figure 10–7

Lesson Exercise 10.28

(a) Draw lines to show how many 1-m^2 squares would cover the interior of the rectangle in Figure 10-7.

(b) What is the area of the rectangle?

Lesson Exercise 10.29

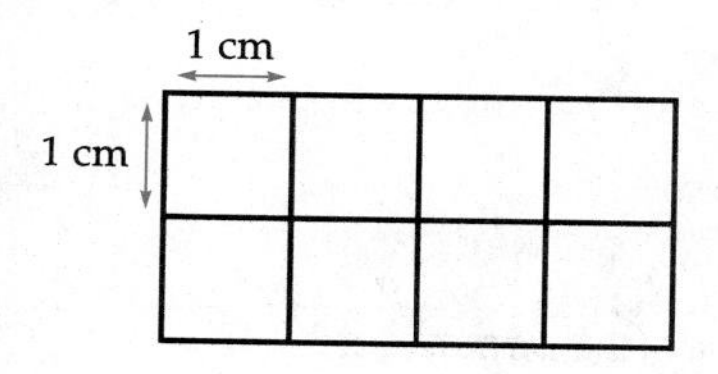

Figure 10–8

Each square inside the rectangle in Figure 10-8 is 1 cm by 1 cm.

(a) How many rows and how many columns of squares are there?

(b) What is the area of the rectangle?

(c) Find the length and width of the rectangle.

(d) How would you compute the area using the length and width?

The rectangle in Lesson Exercise 10.29 has 2 rows and 4 columns of unit squares. The total number of squares is the number of rows times the number of columns (the array model of multiplication). Since the number of columns is determined by the length of the rectangle and the number of rows is determined by the width, the area can be obtained more quickly by multiplying the length (the number of columns) by the width (number of rows).

Area of a Rectangle

The area A of a rectangle that has length l and width w is

$$A = lw$$

Area formulas are especially useful in problems involving larger numbers, when counting squares might take a very long time.

Lesson Exercise 10.30

Refer back to the diagram at the beginning of the section. Which has a larger area, the site or the floor plan?

Consumers use the area formula for a rectangle to decide which package of wrapping is a better buy. Consider the following exercise.

Lesson Exercise 10.31

A store sells two kinds of wrapping paper. Package A costs $4 and has 3 rolls, each $2\frac{1}{2}$ ft by 6 ft. Package B costs $3.25 and has 4 rolls, each 2 ft by 5 ft. Which is a better buy? (*Hint:* Compare the cost per square foot for each package.)

AREA, PERIMETER, AND CONGRUENCE

Do you wonder how area, perimeter, and congruence are related to one another? The following exercises address this question.

Lesson Exercise 10.32

Do two congruent figures have equal area and equal perimeter?

Next, consider the relationship between area and perimeter.

Lesson Exercise 10.33

(a) Draw four different rectangles with perimeters of 8 cm. (Use centimeter graph paper if you have it.)

(b) Do they all have the same area?

(c) Based upon your results, what kind of rectangle tends to have a smaller area?

Lesson Exercise 10.34

(a) Draw three different rectangles with areas of 16 cm^2.

(b) Do they all have the same perimeter?

(c) Based upon your results, what kind of shape tends to have a smaller perimeter?

Lesson Exercise 10.35

Based upon Lesson Exercises 10.32–10.34, which of the following statements about plane figures A and B appear to be true? If you think a statement is false, give a counterexample.

(a) If the area of A equals the area of B, then $A \cong B$.

(b) If $A \cong B$, then the area of A equals the area of B.

(c) If A has a larger perimeter than B, then A has a larger area than B.

(d) If A has a larger area than B, then A has a larger perimeter than B.

As you may have guessed in Lesson Exercise 10.32, if two figures are congruent, then they have equal areas and perimeters. As you saw in Lesson Exercises 10.33 and 10.34, knowing which of two figures has a larger area does not determine which has a larger perimeter.

ANSWERS TO SELECTED LESSON EXERCISES

10.22 (a) 14 (c) 18 (d) 12

10.23 $P = 2l + 2w$

10.24 The floor plan (40 m)

10.25 81.64 in.

10.26 (a) 4 (b) 12

10.28 (a) 3 m, 2 m (b) 6 m^2

10.29 (b) 8 cm^2

10.30 The site

10.31 B (\$3.25 for 40 ft^2 or \$.08125/ft^2)

10.32 Yes

10.33 (b) No (c) A longer, thinner shape

10.34 (b) No (c) A square shape

10.35 Only part (b) is true.

10.2 HOMEWORK EXERCISES

Basic Exercises

1. Measure the perimeter of each figure to the nearest tenth of a centimeter.

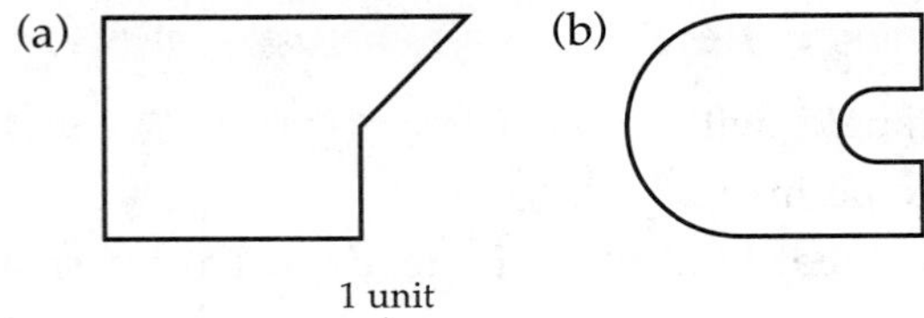

2.

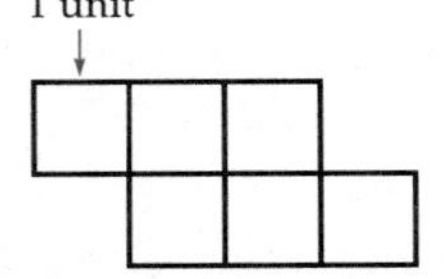

   (a) The perimeter is ________.
   (b) Arrange the squares in other configurations, and compute the perimeter. (Each square must share at least one side with another square.)
   (c) What is the largest possible perimeter?
   (d) What is the smallest possible perimeter?

3. Add a square to the figure that will increase the area of the figure without changing its perimeter.

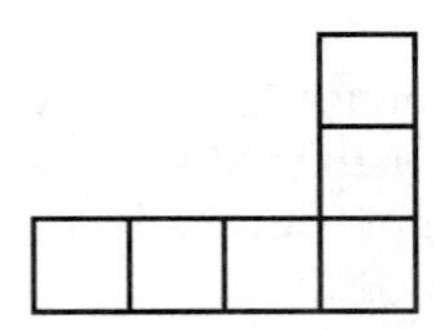

4. A lot is 21 ft by 30 ft. In order to support a fence, an architect wants an upright post at each corner and an upright post every 3 feet in between. How many of these posts are needed?

5. A square field is enclosed by T meters of fence. What are the length and width of the field?

6. A 3-by-5-in. photo print costs $.30. I want an enlargement with twice the length and twice the width. I figure that it should cost about twice as much. Am I right?

7. Draw a square. Draw a second square whose sides are twice as long.
   (a) How do their perimeters compare?
   (b) Start with a new square, and repeat the exercise.
   (c) Make a generalization based upon parts (a) and (b).

8. In a circle, how many times longer is the circumference C than the diameter d?

9. Why is $C = 2\pi r$ the same as $C = \pi d$?

10. The dome of the New Orleans Superdome has a diameter of 680 ft. What is the circumference? (Use 3.14 for π.)

11. The circumference (equator) of the Earth is about 25,000 miles. What is the Earth's approximate diameter?

12. When the radius of a circle increases by 1, the circumference increases by ________.

13. A car has wheels with radii of 40 cm. How many revolutions per minute must a wheel turn so that the car travels 50 km/h?

14. On a circular merry-go-round, one horse is 3.0 m from the center, and the other is 5.0 m from the center.
    (a) How far does each horse travel in one revolution of the merry-go-round? (Use 3.14 for π.)
    (b) If the merry-go-round revolves 3 times per minute, how fast is each horse traveling?

15. A track consists of straightaways and circular arcs.

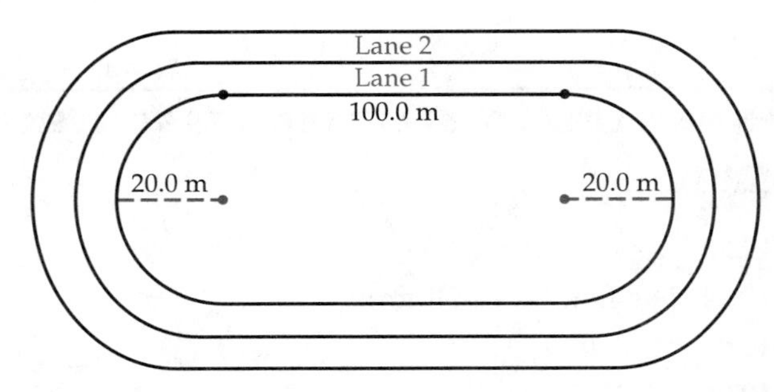

    (a) Find the perimeter of the inside edge of lane 1. Use 3.142 for π.
    b) If lane 1 is 1.0 m wide, how much farther is a lap around the inside edge of lane 2 than a lap around the inside edge of lane 1?

16. A satellite is orbiting 1800 km above Earth's surface. The radius of Earth is about 6400 km. About how fast must the satellite travel to orbit Earth in 10 hours?

17. What aspect of a polygon or circle does area measure?

18.

(a) Estimate the area of the footprint on each grid, in the square units shown.
(b) Which estimate is more precise?

19. Use centimeter graph paper to estimate the area of your foot in square centimeters.

20. Approximate the area of the circle by counting squares.

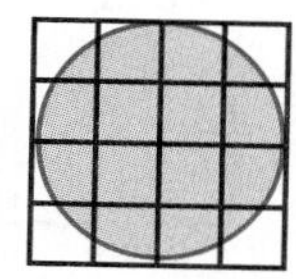

21. What is the area of the following figure in unit hexagons?

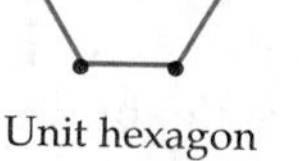

Unit hexagon

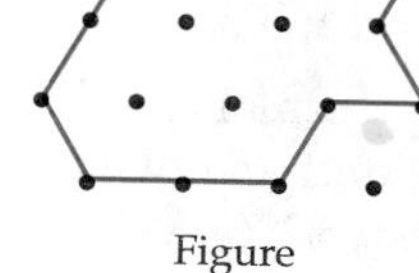

Figure

22. How many of each shape would it take to cover the following hexagon?

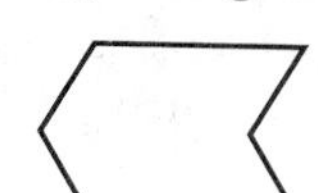

(a)

(b)

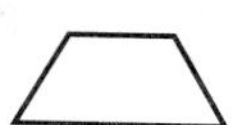

(c)

(d)

23.

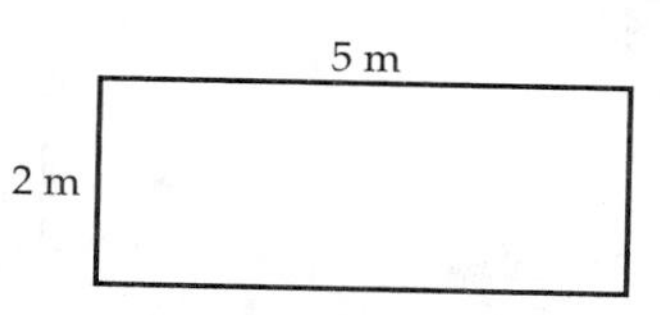

(a) Show how many 1-m squares cover the interior of the rectangle.
(b) What is the area of the rectangle?

24. What fraction multiplication is represented by the picture?

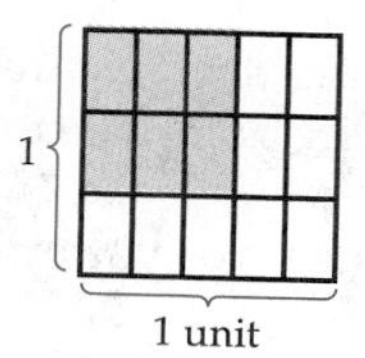

25. What decimal multiplication equation is represented by the picture?

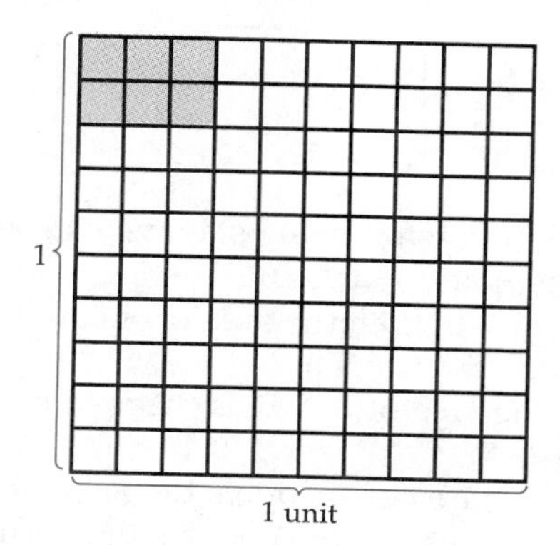

26. Use a geoboard or a geoboard drawing, and show a shape that has
(a) a perimeter of 8 units and an area of 3 square units.
(b) a perimeter of 10 units and an area of 6 square units.

27. The state of Kansas is approximately rectangular. Estimate its area and perimeter.

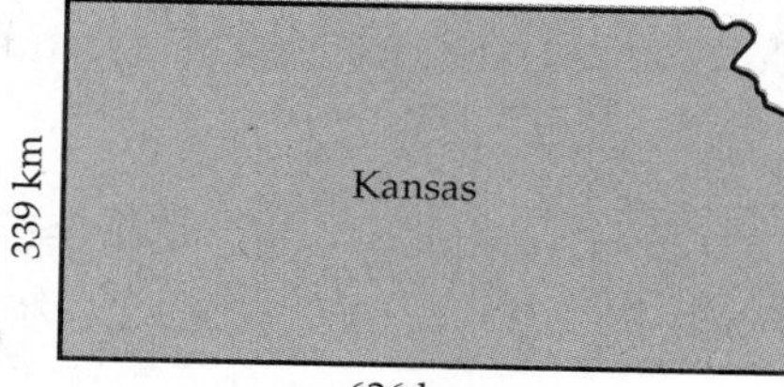

28. Using centimeter graph paper, draw seven different shapes, composed of centimeter squares, that each have a perimeter of 16 cm and an area of 12 cm^2. Each centimeter square should have a common side with at least one other square unit.

29. A rectangular garden is 10 ft by 12 ft. A rectangular sidewalk 1 ft wide is built around the outside. What is the area of the sidewalk? (Draw a picture.)

30. (a) Complete the third example, repeating the error pattern from the completed examples.
(b) Describe the error pattern.

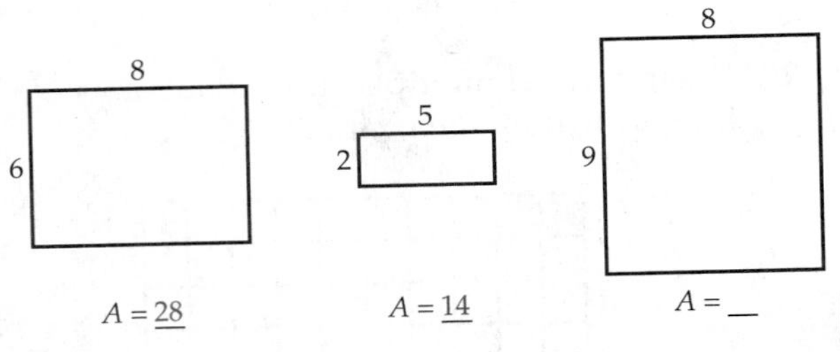

31. Which of the following statements about plane figures A and B are true?
(a) If $A \cong B$, then the area of A equals the area of B.
(b) If the area of A equals the area of B, then $A \cong B$.
(c) If the area of A equals the area of B, then the perimeter of A equals the perimeter of B.

32. Draw rectangles A and B so that they have the same perimeter but A has a larger area than B.

33. A rectangle and a square have equal areas. If the rectangle is 6 ft by 8 ft, how long is a side of the square?

34. The Federal Housing Administration (FHA) requires that the window area of a house be at least 10% of the floor space. How many 15-ft^2 windows are needed to meet FHA requirements for a house with a floor that is 25 ft by 40 ft?

35. Find the area, in square units, of the shaded region.

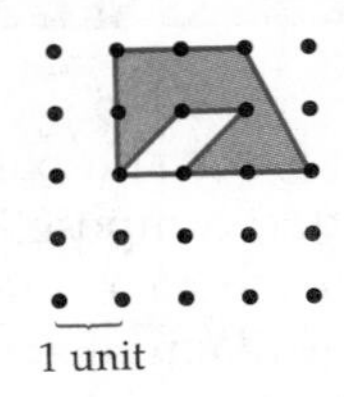

36. Shown here is a scale drawing $\left(\frac{1}{4} \text{ in.} \approx 12 \text{ ft}\right)$ of the floor plan for the bedroom, hall, and living room of your new apartment.

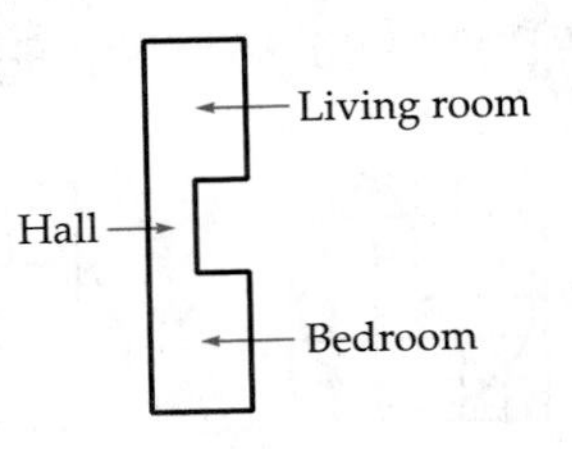

(a) Fill in the missing lengths.
(b) How many square feet of carpet will be needed to cover the floors?
(c) The carpet you want costs \$8 per square *yard*. What will the carpet for your apartment cost?

37. Suppose that the housing authority has valued the house shown here at \$220 per ft^2.

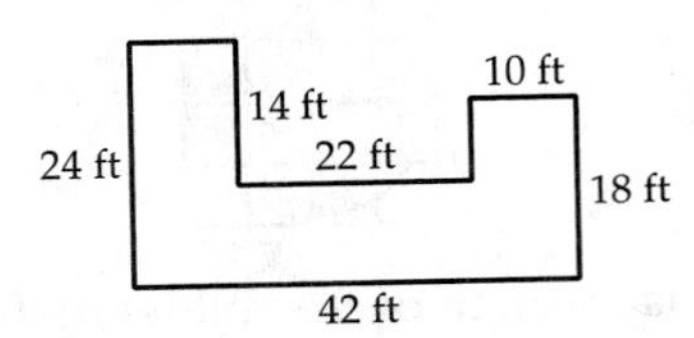

(a) Show two different ways of subdividing the region into rectangles.
(b) Find the assessed value of the house.

38. A photo enlargement that is 6 in. by 9 in. costs \$1.75, and an enlargement that is 8 in. by 10 in. costs \$2.75. Which enlargement costs less per square inch?

39. A store sells two kinds of wrapping paper. Package A costs \$5.25 and has 6 rolls, each 2 ft by 5 ft. Package B costs \$7 and has 4 rolls, each 4 ft by 6 ft. Which package is a better buy?

40. (a) How many square feet are in a square yard? (*Hint:* Draw 1-ft^2 squares in the interior.)

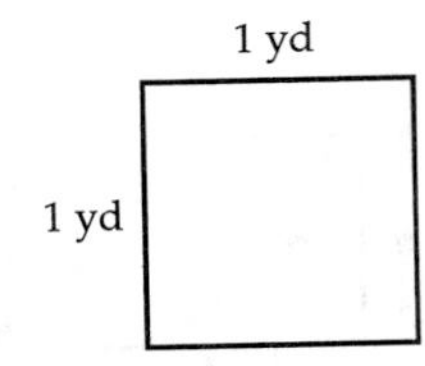

(b) As you know, 3 ft = 1 yd. So (3 ft)(3 ft) = (1 yd)(1 yd). Simplify both sides, and write the resulting equation.

41. A store sells carpet for $8.99 per square yard. How much does a piece of carpet 7 ft by 9 ft cost? (*Hint:* See the preceding exercise.)

***42.** (a) 1 cm^2 = ________ mm^2 (*Hint:* First solve 1 cm = ________ mm and square both sides.)
(b) 1 cm^2 = ________ m^2

43. (a) 8 cm^2 = ________ mm^2
(b) 500 m^2 = ________ cm^2

44. Show that 1 m^2 = 1,000,000 mm^2.

45. (a) Complete the third example, repeating the error pattern from the completed examples.
(b) Describe the error pattern.

7 cm^2 = ___*70*___ mm^2
80 m^2 = ___*8000*___ cm^2
26 mm^2 = ________ cm^2

46. A mapmaker wants to represent an area of roughly 20,000 ft^2 using the scale 1 in. = 20 ft. What area will the map have?

47. How many tiles, each 25 cm by 25 cm, would be needed to cover a floor that is 2 m by 4 m?

48. How many tiles, each 25 cm by 25 cm, would be needed to cover a floor that is 2.2 m by 4.6 m?

49. You plan to make brownies in a pan that is 10 in. by 14 in. If you want to make 24 equal-sized brownies, find possible dimensions of each brownie.

50. Each square of the floor plan is 3 ft^2. How much would it cost to lay this floor at a cost of $15/$ft^2$?

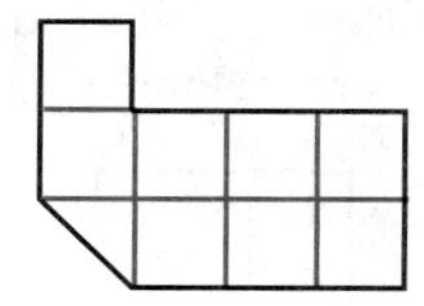

51. A major-league baseball field has an area of about
(a) 6 m^2 (b) 6000 cm^2 (c) 6000 m^2

Extension Exercises

52. The Earth travels around the sun in an approximately circular path with a radius of 9.3×10^7 miles.
(a) How far does the Earth travel in one trip around the sun? (Use 3.14 for π.)
(b) If the Earth takes 365 days to make one revolution, how far does it travel in 1 hour?
(c) If the Earth is moving so fast, why don't we fall off?

53. Recall the set of pentominos (Section 9.1 Special Exercise).
(a) What is the area of any pentomino?
(b) What are all the possible perimeters of pentominos?

54. (a) A rectangle has a whole-number length and a whole-number width, and its area and perimeter are the same number. What are its dimensions?
(b) Find a second solution to part (a).

55. An engineer is designing a track with a semicircular arc on each end, as shown. One lap around the track should be 400 m.

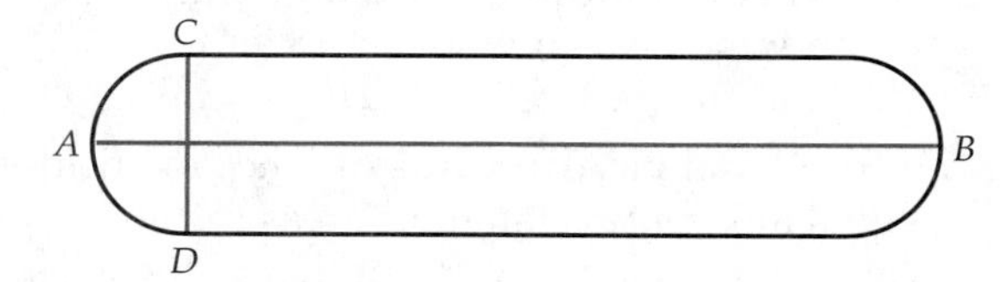

(a) If CD = 40 m, how long is AB?
(b) If AB = 150 m, how long is CD?

56. Try the following if you have a set of 36 square tiles. (Graph paper could also be used.) Suppose a rectangle has an area of 36 square units.
(a) Use the tiles to construct all possible rectangles with a length and a width that are whole numbers. Find the perimeter of each rectangle, and record the results.
(b) What are the length and width of the rectangle with the greatest perimeter?

57. (a) Consider the following problem. "A rectangle has an area of 4 m^2. What is the smallest possible perimeter it could have?" Devise a plan and solve the problem.

(b) Repeat part (a) for a rectangle with an area of 9 m^2.
(c) Generalize your results for a rectangle with an area of N m^2.

58. A rectangle has a length that is 16 m more than its width, and its area is 2961 m^2. What are the length and width? (Guess and check.)

59.

R1 R2 R3

(a) If each side has a length of 1 unit, find the perimeters of R1, R2, and R3.
(b) Draw R4 and find its perimeter.
(c) What is a formula for the perimeter of RN, in which N is a whole number?
(d) Find the perimeter of R10.

60.

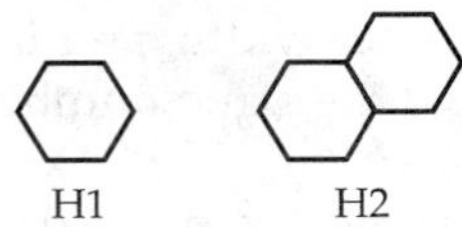

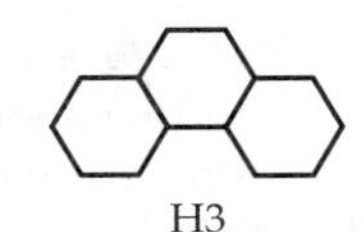

(a) If each side has a length of 1 unit, find the perimeters of H1, H2, and H3.
(b) Draw H4 and find its perimeter.
(c) What is a formula for the perimeter of HN, in which N is a whole number?
(d) Find the perimeter of H10.

61. A cowhand has 350 yards of fence to enclose equal adjacent rectangular areas for horses and cattle. Find an arrangement that will enclose more area than either one large square or two adjacent squares. (Guess and check.)

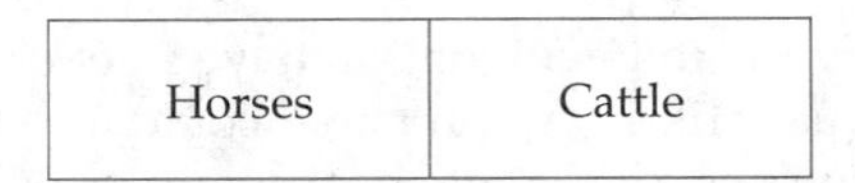

62. Mrs. Cunningham wants to build a pen for her prize-winning hogs next to the barn with a 3-sided, rectangular 24-yard fence. What dimensions will enclose the largest area? If you don't have a computer spreadsheet program, then just guess and check. If you do have a spreadsheet program, make three columns labeled "Base," "Height," and "Area." Fill in all possible whole-number values for the base, and use formulas to generate a table of values in the other two columns.

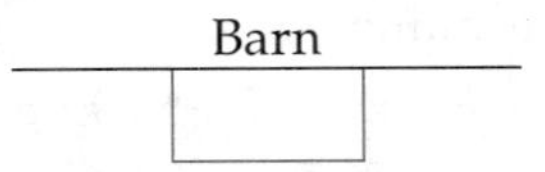

63. (a) Make a graph showing the relationship between length and width for all rectangles that have perimeters of 20.

Width

Length

(b) Make a graph showing the relationship between length and width for all rectangles that have areas of 20.

Width

Length

64. Many algebraic formulas can be represented geometrically. Refer to the figure, and write an equation related to the area of the entire region.

(a)

	a	2
a	a^2	$2a$
2	$2a$	4

(b)

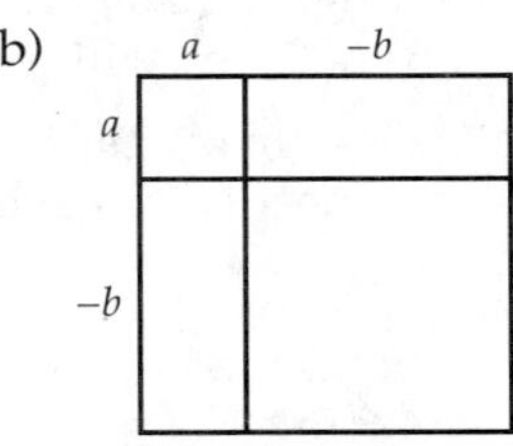

65. What algebraic equation is suggested by each diagram?

(a) a b c

(b)

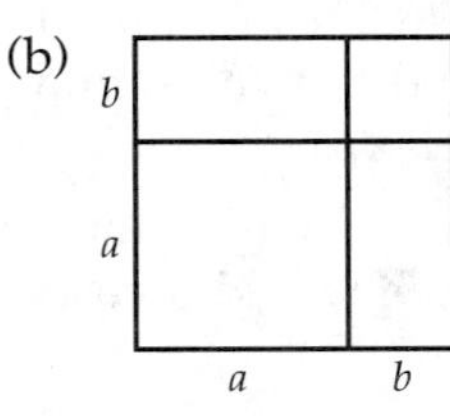

66. A rectangle is divided into four rectangles with the given areas. Find x.

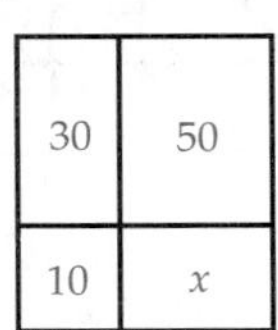

67. Express arc length s in terms of radius r and central angle measure θ.

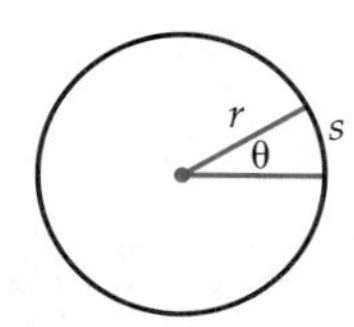

Computer Exercises

68. A store records the price of its rugs on a spreadsheet. Length and width are given in feet. (Use a computer spreadsheet if you have one.)

	A	B	C	D	E
1	**Rugs**				
2	Length	Width	Area	Price/ft^2	Total Price
3	9	12		\$6.99	
4	10	20		\$8.99	

(a) What does the value in cell D3 represent?

(b) What are the formulas for cells C3 and E3?

(c) If you have a computer spreadsheet, enter the correct formulas in the appropriate format. Otherwise, use a calculator. Fill in the values in cells E3 and E4.

69. A student wants to study the areas of various rectangles with perimeters of 16, using a spreadsheet.

	A	B	C
1	**Rectangles**	*P*=16	
2	Length	Width	Area
3	4	4	
4	5	3	
5	6	2	
6	7	1	

(a) What is the formula for cell C3?

(b) Fill in the values in cells C3 through C6.

(c) Describe any patterns you see in the areas.

(d) How is the size of the area related to the length and width?

10.3 AREAS OF QUADRILATERALS, TRIANGLES, AND CIRCLES

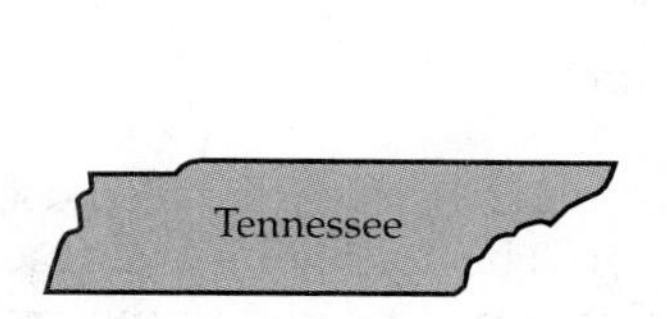

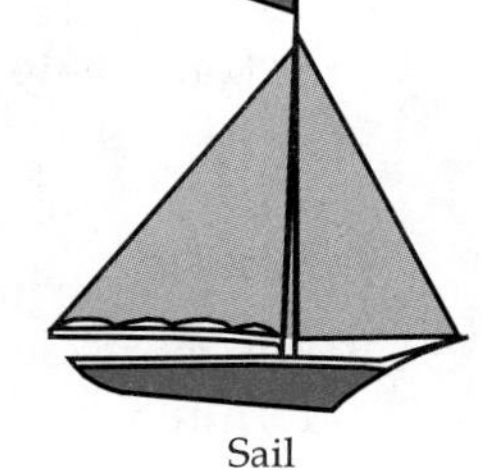

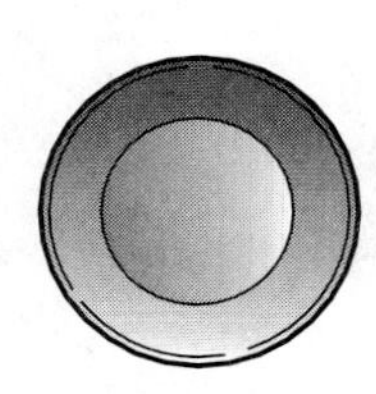

Figure 10–9

To approximate the surface area of each object in Figure 10-9, you could use an area formula. Do you know where such formulas come from? In Section

10.2, you studied the concept of measuring area in square units and how that leads to a more efficient area formula for the rectangle (and square).

We'll pick things up from there. In this section, we will derive in a deductive sequence the area formulas for a triangle, a parallelogram, and a circle.

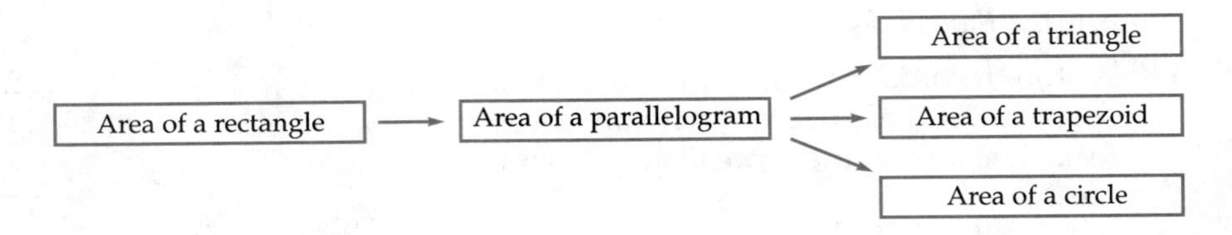

The activities in this section require scissors, a 30-cm ruler, and some paper.

THE AREA OF A PARALLELOGRAM

Before you begin cutting, here is some background information about parallelograms. Any side of a parallelogram can be designated as the **base** (usually it is the bottom side). Then the **height** is the distance from the base to the opposite side. The height is always measured with a segment that is *perpendicular* to the base (or an extension of the base) and the opposite side (Figure 10-10).

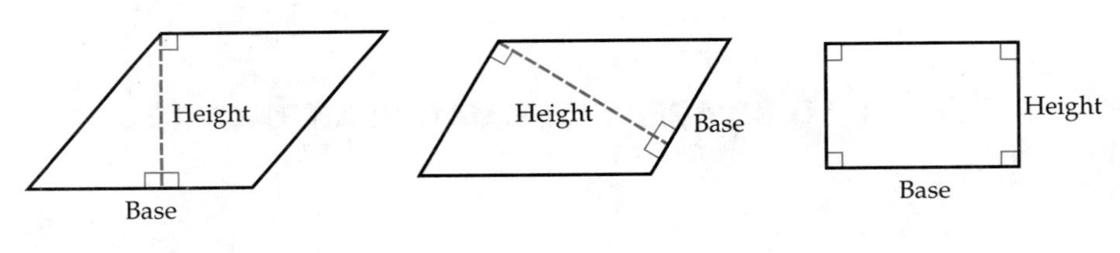

Figure 10–10

The terms "base" and "height" are also used for triangles and trapezoids, but more on that later. The following exercise will show you how the area formula for a parallelogram is related to the area formula for a rectangle.

Cooperative Lesson Exercise 10.36

(a) Cut out a rectangle that is 7 cm by 5 cm. (You may use centimeter graph paper if you have it.)

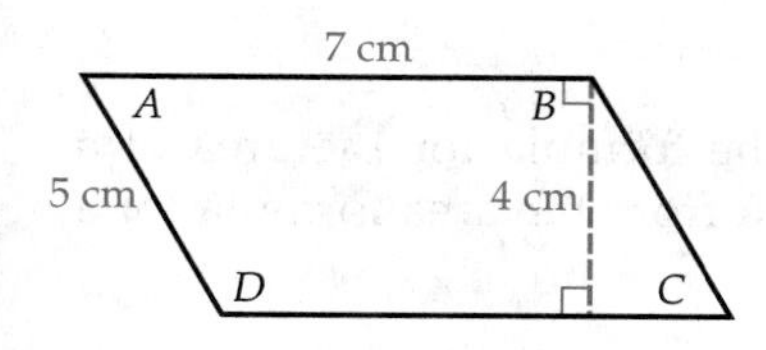

Figure 10–11

(b) Draw a parallelogram with the dimensions shown in Figure 10-11, and label its vertices *A*, *B*, *C*, and *D* inside the parallelogram, as shown. It's tricky to draw a parallelogram with these measurements. *Draw the top first, then the height,* and see whether you can finish it. Make sure all the measurements are accurate. If they are not, ask for help before continuing.

(c) Cut out the parallelogram.

(d) What is the perimeter of the rectangle?

(e) What is the perimeter of the parallelogram?

(f) Compare the areas of the two figures by placing one paper over the other. Is one larger in area, or are they equal in area?

Cooperative Lesson Exercise 10.37

To find the exact area of the parallelogram, do the following.

(a) Cut the parallelogram into pieces by cutting along the perpendicular segment from *B* to $\overline{CD}$. Rearrange these two pieces into a rectangle.

(b) What are the dimensions of the rectangle you formed?

(c) What is the area of the rectangle you formed from the parallelogram?

(d) What is the area of the original parallelogram?

Cooperative Lesson Exercise 10.38

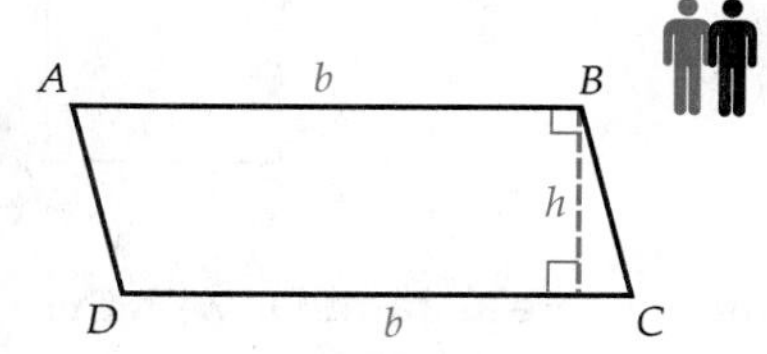

Figure 10–12

(a) Based upon Lesson Exercises 10.36 and 10.37, describe a general method for finding the area of a parallelogram such as the one in Figure 10-12.

(b) What is the area of any parallelogram in terms of its base b and its height h?

The answer to Lesson Exercise 10.38(b) is the parallelogram area formula.

Area of a Parallelogram

The area A of a parallelogram that has a base of length b and height h is

$$A = bh$$

Lesson Exercise 10.39

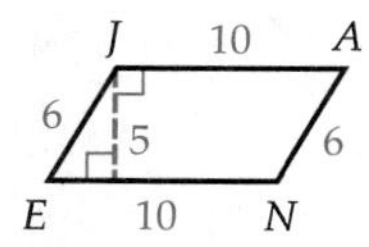

Figure 10–13

Use the area formula for a parallelogram to find the area of *JANE* in Figure 10-13.

THE AREA OF A TRIANGLE

How big are the sails on your new sailboat? The formula for the area of a triangle will tell you. This formula can be derived from the area formula for a parallelogram. Do you have scissors ready?

Cooperative Lesson Exercise 10.40

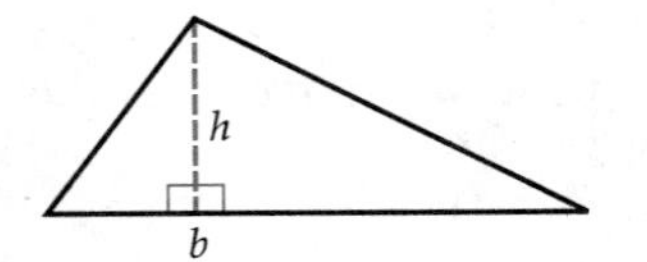

Figure 10–14

(a) Cut out two congruent triangles. Label one side as the base b, and draw the perpendicular height from b to the opposite vertex (Figure 10-14).
(b) Figure out how to put the two triangles together to form a parallelogram.
(c) What is the area of the parallelogram in terms of b and h?
(d) What is the area of each triangle in terms of b and h?

As you saw in Lesson Exercise 10.40, two congruent triangles can be put together to form a parallelogram. Since the two triangles have the same area, each has half the area of the parallelogram.

Area of a Triangle

The area A of a triangle that has a base of length b and height h is

$$A = \frac{1}{2}bh$$

The ancient Egyptians, Babylonians, and Chinese were all familiar with the standard area formulas for rectangles, parallelograms, and triangles.

Lesson Exercise 10.41

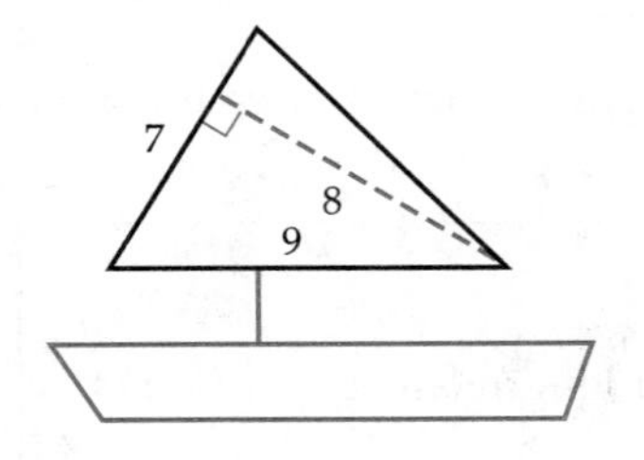

Figure 10–15

Use the area formula for a triangle to find the area of the triangle in Figure 10-15. (*Hint:* The height of a triangle is the perpendicular distance from the base to the opposite vertex.)

The area of any polygon can be found by subdividing it into rectangles or triangles or both! After finding the area of each rectangle and triangle, add up the areas to obtain the area of the polygon.

THE AREA OF A TRAPEZOID

The state of Tennessee is shaped like a trapezoid (Figure 10-16). You could approximate its land area if you had an area formula for a trapezoid.

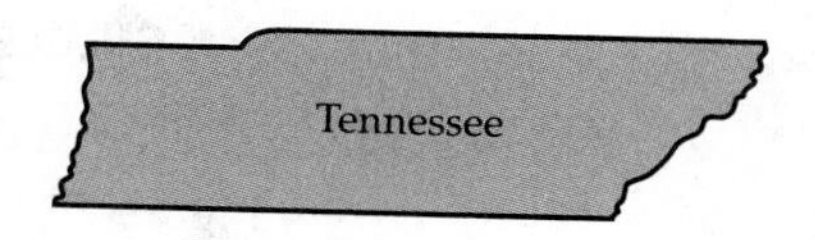

Figure 10–16

The area formula for a parallelogram can be used to deduce the formula for the area of a trapezoid. Two congruent triangles can be put together to form a parallelogram, and so can two congruent trapezoids. You don't believe me? Try Lesson Exercise 10.42.

Cooperative Lesson Exercise 10.42

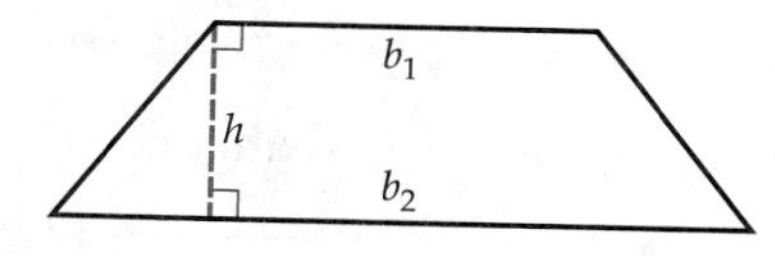

Figure 10–17

(a) Cut out two congruent trapezoids, and label them as shown in Figure 10-17. (The parallel sides of the trapezoids are its bases, b_1 and b_2.)

(b) Put them together to form a parallelogram.

(c) What is the area of the parallelogram in terms of b_1, b_2, and h?

(d) What is the area of the trapezoid?

You have discovered another area formula!

Area of a Trapezoid

The area A of a trapezoid that has parallel sides of lengths b_1 and b_2 and height h is

$$A = \frac{1}{2}(b_1 + b_2)h$$

Lesson Exercise 10.43

Tennessee is approximately a trapezoid (Figure 10-18).

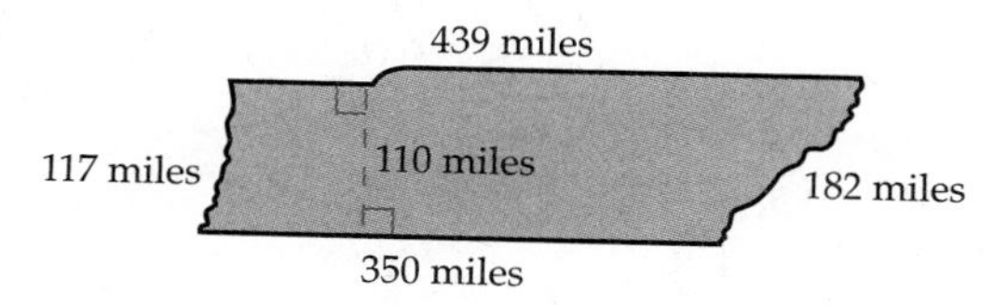

Figure 10–18

(a) What is the approximate surface area of Tennessee?

(b) Land is not perfectly flat. Would this fact make the surface area greater or less than the area of a perfectly flat surface?

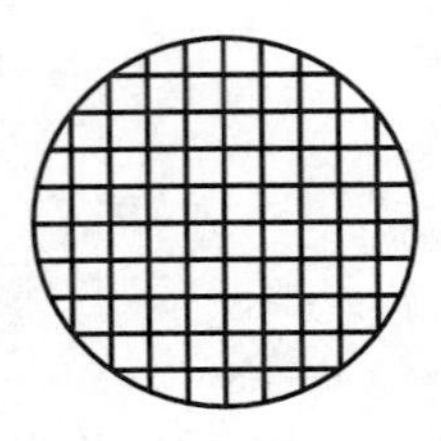
Figure 10–19

THE AREA OF A CIRCLE

What is the surface area of a dinner plate that has a 12-in. diameter? To answer this question, one would compute the area of a circle. It is difficult to count the number of squares inside a circle. Figure 10-19 indicates why circle areas rarely come out to whole numbers of unit squares.

As with the triangle and the trapezoid, the formula for the area of a circle is suggested by its relationship to the area formula for a parallelogram! However, in this case, the parallelogram representation is approximate. Calculus is needed for a precise derivation of the area formula for a circle.

Again, the idea is to make the new figure (a circle) look like a parallelogram. Suppose a circle is cut into six equal parts, as shown in Figure 10-20. The parts can be arranged as shown in Figure 10-21.

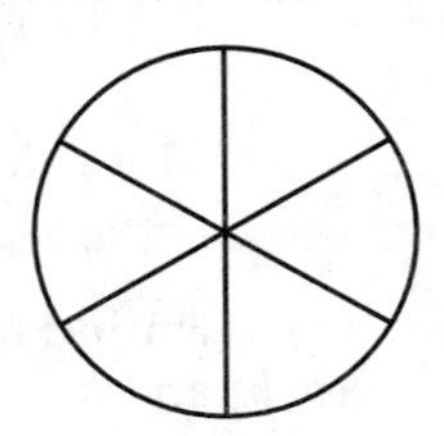
Figure 10–20

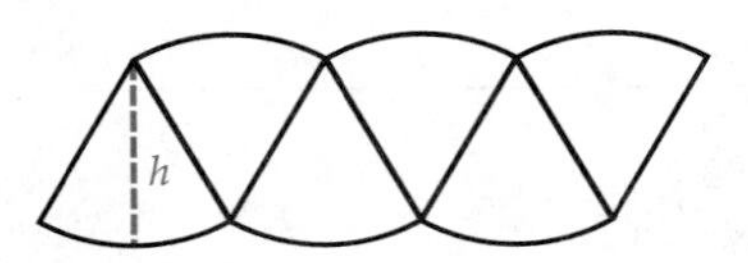

Figure 10–21

Lesson Exercise 10.44

(a) What shape does the rearranged circle in Figure 10-21 approximate?

(b) The "base" of the new figure is about 1/2 the ________ of the original circle, and the "height" of the figure is about the same as the ________ of the original circle.

(c) Using the parallelogram formula $A = bh$, the area of the circle is about ________.

Lesson Exercise 10.45 suggests that the area of the circle is $\frac{1}{2} \cdot C \cdot r$. Using $C = 2\pi r$, we can obtain the familiar area formula from $A = \frac{1}{2} \cdot C \cdot r$. It goes like this (see Figure 10-22).

$$A = \frac{1}{2} \cdot C \cdot r = \frac{1}{2} \cdot 2\pi r \cdot r = \pi r^2$$

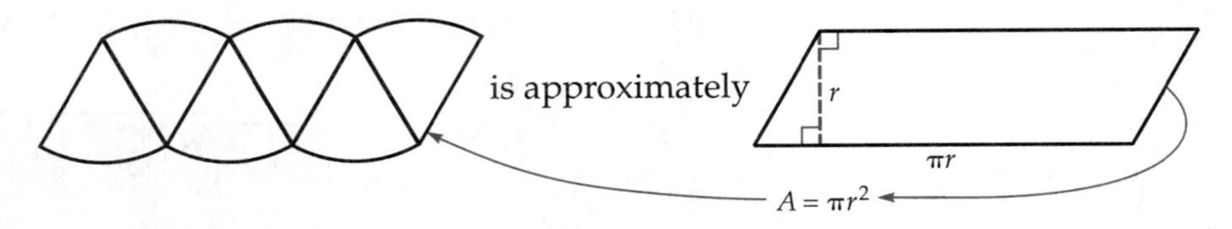

Figure 10–22

Figure 10-23 shows a lesson in the sixth-grade *Mathematics Plus* textbook, which derives the area formula of a circle in a similar way. The more pieces the circle is cut into, the closer the shape is to a parallelogram with base πr and height r. This connection suggests the formula for the area of a circle.

Area of Circles

Now you can develop the formula for the area of a circle. Recall that the area of a parallelogram is $A = bh$.

Think how the base and the height of the approximate parallelogram relate to the parts of the circle.

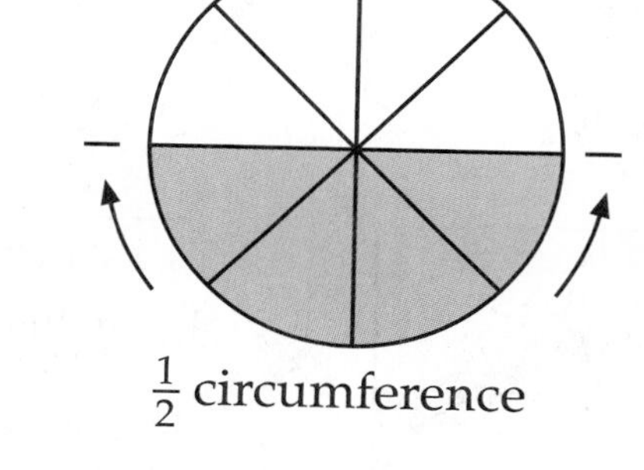

base (b) = $\frac{1}{2}$ the circumference of a circle ($\frac{1}{2} \times C$)
height (h) = radius (r) of a circle

$A = bh$ ← Replace with terms relating to the circle.

$= (\frac{1}{2} \times C) \times r$ ← $C = 2\pi r$

$= \frac{1}{2} \times 2\pi r \times r$ ← $\frac{1}{2} \times 2\pi r = \pi r$

$= \pi r \times r$, or πr^2 Area of a circle → $A = \pi r^2$

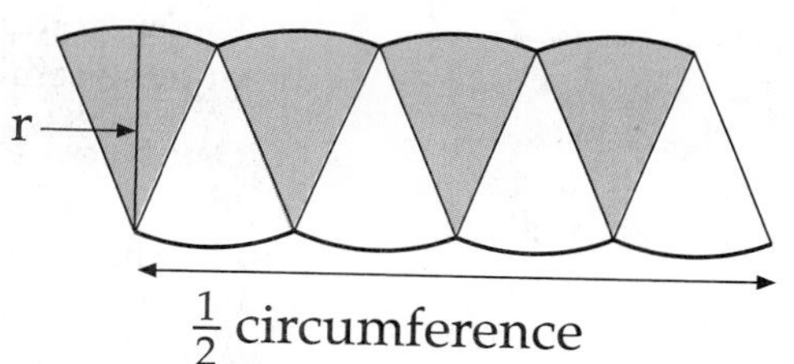

What was the area of Kishi's circle?

$A = \pi r^2$ ← $r = 2$ and $\pi \approx 3.14$

$\approx 3.14 \times 2^2$

$\approx 3.14 \times 4$

≈ 12.56

Figure 10–23

Area of a Circle

The area A of a circle that has radius r is

$$A = \pi r^2$$

You can use this formula to find the area of a circle when you know the radius or diameter.

Lesson Exercise 10.45

How would a pizza with a 16-in. diameter compare in size to two pizzas with 8-in. diameters?

(a) Guess the answer.

(b) Work out the areas, and find out if you're right.

Lesson Exercise 10.46

A dinner plate has a diameter of 12.0 in. The area is ________. (Use 3.1416 for π.)

Some area problems require using more than one area formula.

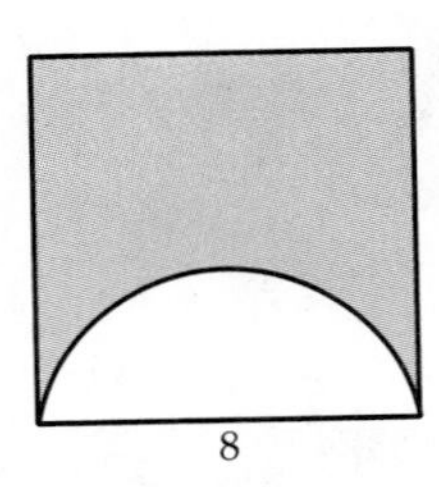

Figure 10–24

EXAMPLE 10.2

Figure 10-24 shows a square and a semicircle. Find the shaded area.

Solution

Understanding the Problem The shaded area is inside the square but outside the semicircle.

Devising a Plan In this type of problem, (1) identify familiar shapes and their area formulas and (2) determine how the shaded area is related to the familiar shapes.

The familiar shapes are a square ($A = s^2$) and half of a circle $\left(A = \frac{1}{2}\pi r^2\right)$. The shaded area is the area of the square minus the area of the half circle.

Carrying Out the Plan

$$A_{\text{shaded}} = A_{\square} - \frac{1}{2}A_{\bigcirc} = (8^2) - \frac{1}{2}\pi \cdot 4^2 = (64 - 8\pi) \text{ square units}$$

Looking Back The same method would work if the square and circle were replaced by other shapes, such as triangles and rectangles. ■

Try one yourself.

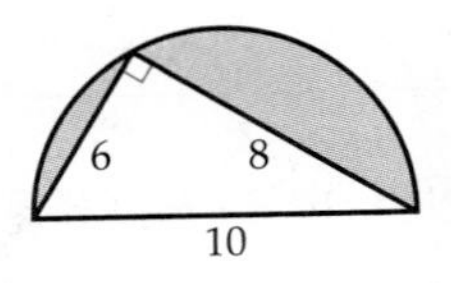

Figure 10–25

Lesson Exercise 10.47

Consider the following problem. "Figure 10-25 shows a semicircle. Find the shaded area." Devise a plan and solve the problem.

ANSWERS TO SELECTED LESSON EXERCISES

10.36 (d) 24 cm (e) 24 cm
(f) The rectangle has a larger area.

10.37 (b) 7 cm by 4 cm (c) 28 cm^2
(d) 28 cm^2

10.38 (a) Translate the right triangle as shown to form a rectangle.

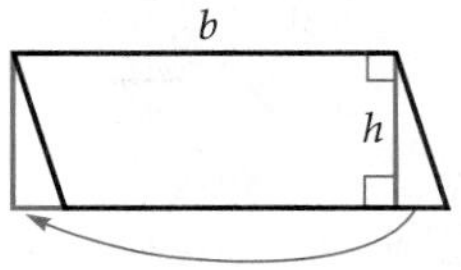

The area of the rectangle is bh.

Parallelogram area = rectangle area = bh

(b) $A = bh$

10.39 50 square units

10.40 (c) bh (d) $\frac{1}{2}bh$

10.41 $\frac{1}{2}(7)(8) = 28$ square units

10.42 (c) $(b_1 + b_2) \cdot h$

10.43 (a) About $\frac{1}{2}(439 + 350) \cdot 110 =$ 43,395 square miles
(b) Greater

10.44 (a) A parallelogram
(b) circumference; radius (c) $\frac{1}{2} \cdot C \cdot r$

10.45 (b) One 16-in. pizza has twice as much pizza as two 8-in. pizzas!

10.46 113.0976 $in.^2$

10.47 $12.5\pi - 24$ square units

10.3 HOMEWORK EXERCISES

Basic Exercises

1. Use the area formula for a rectangle ($A = bh$) to explain why the area of the parallelogram shown is $A = bh$.

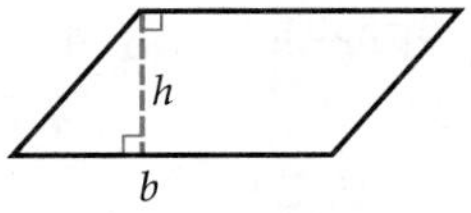

2. Which figure shown has a larger area?

(1)

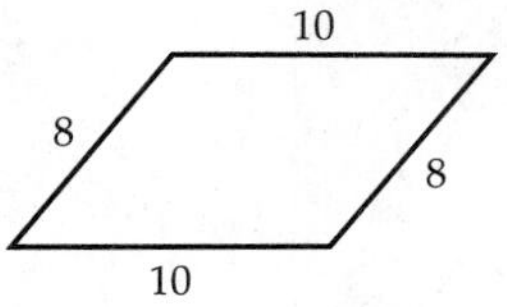

(2)

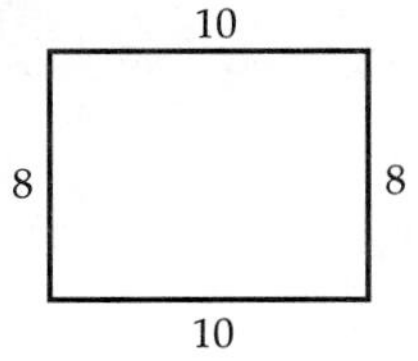

(a) Figure 1 (b) Figure 2
(c) Their areas are equal.

3. Find the area and perimeter of each parallelogram.

(a)

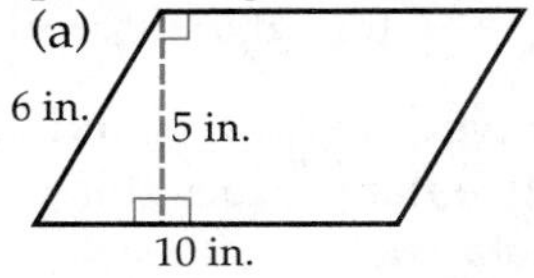

(b)

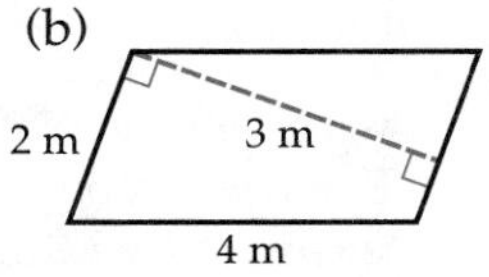

4. A parallelogram has adjacent sides of length 4 cm and 6 cm.
(a) Find the largest possible area the parallelogram could have.
(b) Could the parallelogram have an area of 1 cm^2?

5. Use the area formula for a parallelogram ($A = bh$) to explain why the area of the triangle shown is $A = \frac{1}{2}bh$.

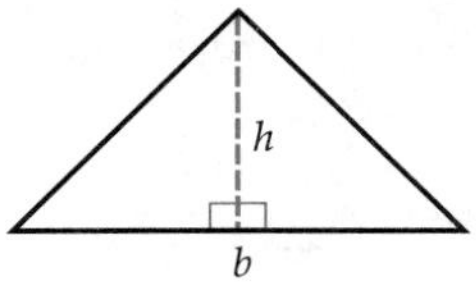

***6.** Find the area of each triangle.

(a)

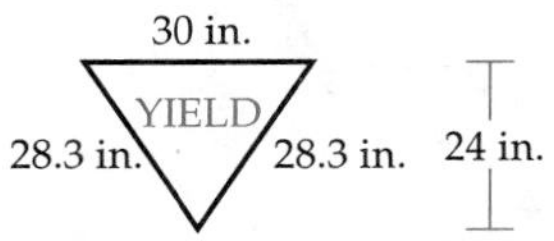

(b)

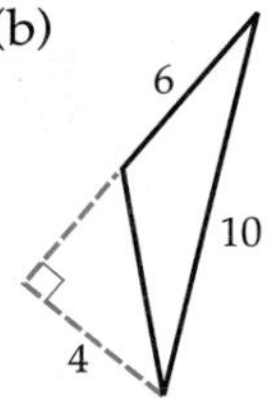

7.

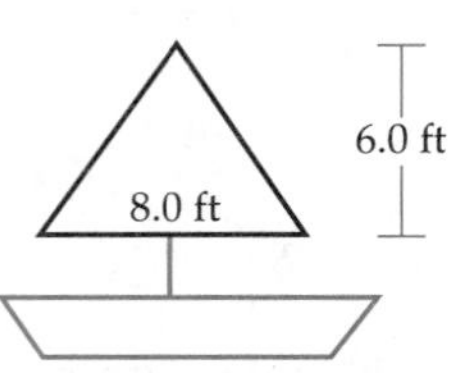

Mentally compute the area of the sail.

8. Use a geoboard or a geoboard drawing, and show a triangle with each of the following areas. (Make a drawing of your answers if you use a geoboard.)
(a) 2 square units (b) 4.5 square units

9. Find the area of the kite.

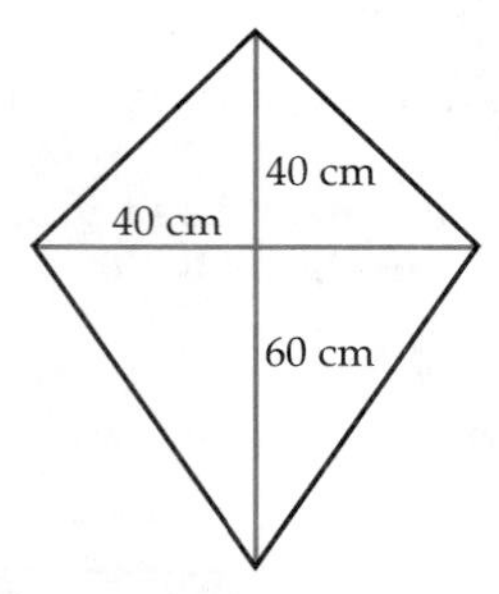

10. Find the area of the triangle.

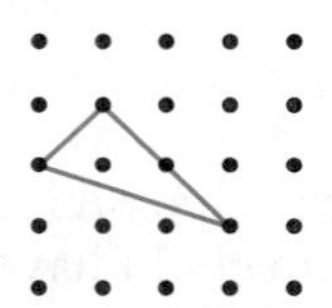

11. Find the area of each polygon. (*Hint:* you could count squares and parts of squares, or you could subdivide each figure.)
(a)

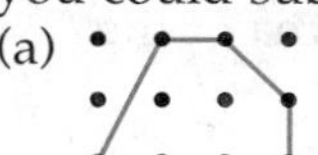

(b)

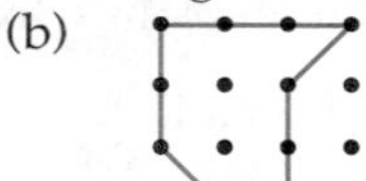

12. Find the area of the trapezoid.

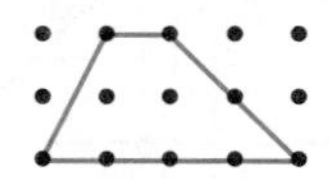

13. Use the area formula for a parallelogram ($A = bh$) to explain why the area of the trapezoid shown is $A = \frac{1}{2}(b_1 + b_2) \cdot h$.

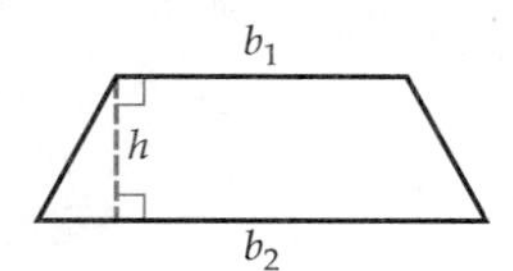

14.

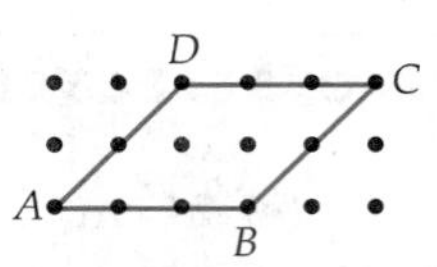

(a) Draw a height of the parallelogram for the base $\overline{AB}$.
(b) Draw a height of the parallelogram for the base $\overline{BC}$.

15. Use a ruler to find the area of the trapezoid in square millimeters.

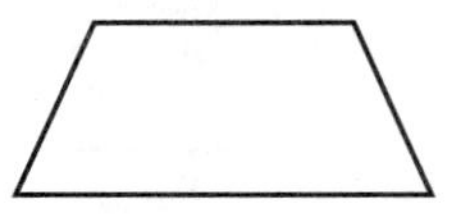

16. Since a square is a type of rectangle, the area formula for a rectangle works for squares. Which of the following have area formulas that would work for any rhombus?
(a) Rectangle (b) Triangle
(c) Parallelogram (d) Square

17. Two adjacent lots are for sale. Lot A costs \$20,000, and lot B costs \$27,000.

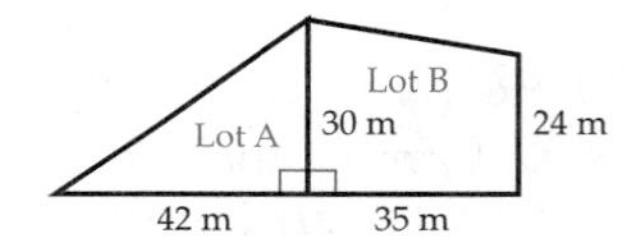

Which lot has a lower cost per square meter?

18. Find the approximate land area of Nevada.

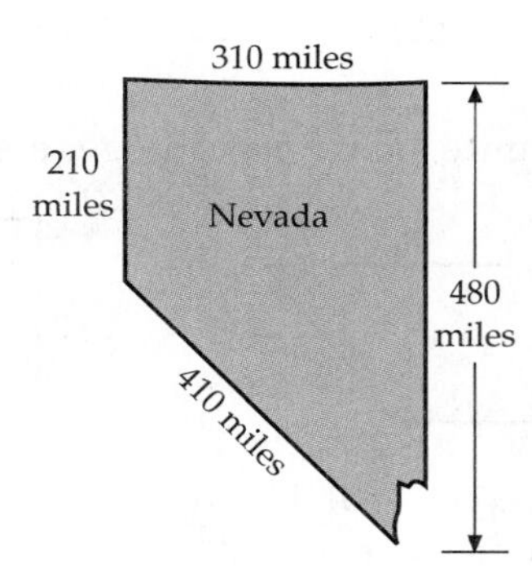

19. About 3600 years ago, Egyptians used the area formula $A = \frac{(a + c)(b + d)}{4}$ for quadrilaterals with lengths of successive sides a, b, c, and d. For what types of quadrilaterals does this formula give the correct area?

20. Following is a drawing of a Boeing 747 airplane wing. What is the area of the top surface of the wing? (Assume that the wing is flat.)

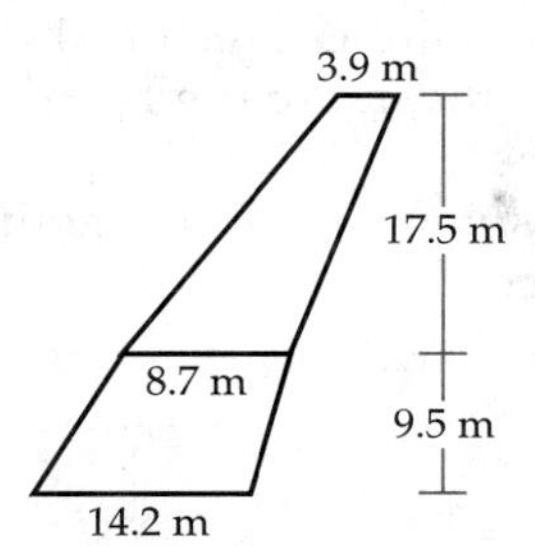

21. An ancient Egyptian document called the Rhind Papyrus (1650 B.C.) shows how to find the area of a circular field with a diameter of 9. It says to take $\frac{1}{9}$ of the diameter, which is 1. Square the remainder of the diameter ($9 - 1 = 8$) to obtain 64, which is the area.
(a) What is the exact answer to the problem?
(b) Approximate the result of part (a), using 3.14 for π.
(c) What area formula did the ancient Egyptians use?
(d) What value of π would make the Egyptians' formula work?

22. A circle is rearranged as shown.

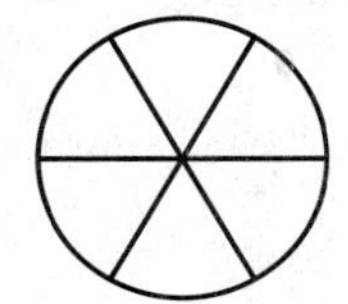

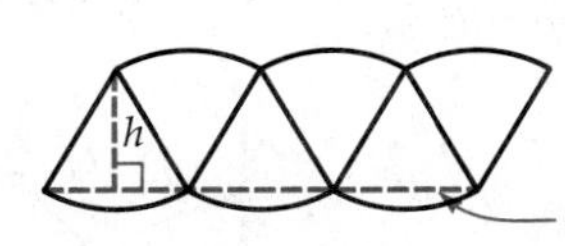

Dashed line has length b

Show how to derive an approximate area formula for a circle using the area formula for a parallelogram.

23. Show why $\frac{1}{2}Cr = \pi r^2$ for any circle.

24. Suppose that you know how to find the area of a square but not the area of a circle. How could you approximate the area of the circle shown?

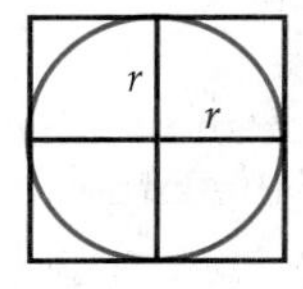

25. (a) When the radius of a circle is doubled, the circumference is multiplied by ________.
(b) When the radius of a circle is doubled, the area is multiplied by ________.

26. (a) Exactly how many times larger is the area of a pizza that has a 14-in. diameter than a pizza that has a 10-in. diameter?
(b) If a 10-in. pizza costs $6, what is a fair price for a 14-in. pizza?

27. A woman wants to water a square plot.

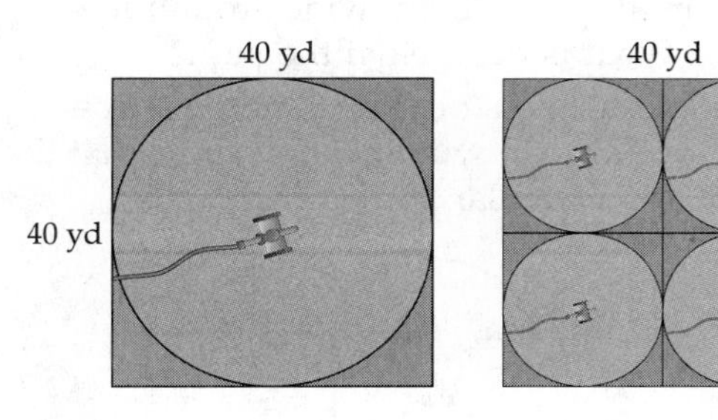

Would 4 small sprinklers reach more area than 1 giant sprinkler? (Assume that the sprinklers spray the circular areas shown.)

28. (a) Complete the third example, repeating the error pattern from the completed examples.

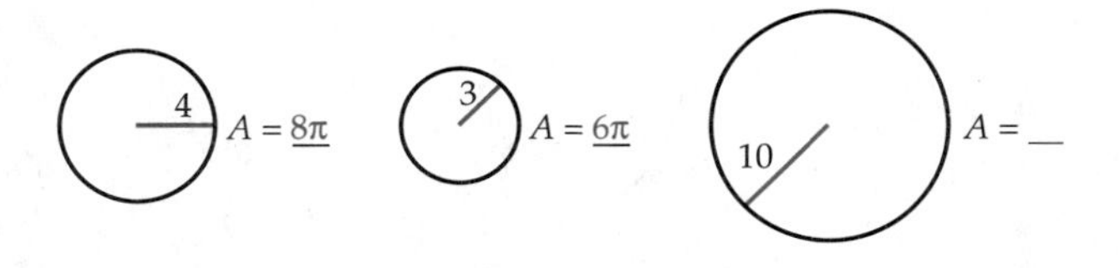

(b) Describe the error pattern.

29. What is the area of the sector of the circle shown? C is the center.

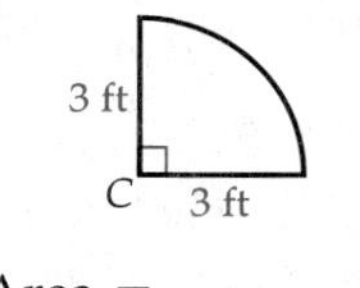

Area = ________

30. The area of a circle is 25π m^2. What is the exact circumference?

31. The circumference of a circle is 20π ft. What is the exact area?

32.

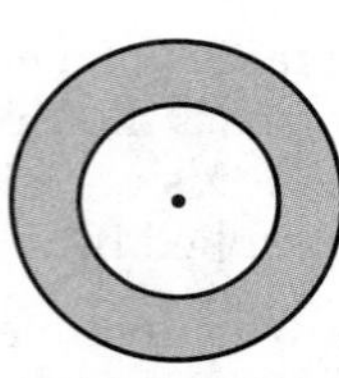

(a) Which do you think is larger, the area of the inner circle or the area of the shaded region?
(b) Using a ruler, compute each area.

33. A circular garden has a diameter of 6.0 m. It has a circular path 1.0 m wide around its border. What is the area of the path?

34. Find the exact area of the geometric shape (used for Norman windows). Assume that the top (BC) is a circular arc.

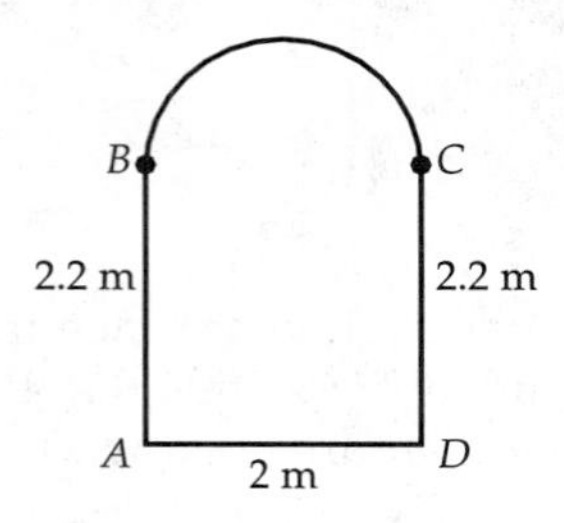

35. C is the center of the circle.

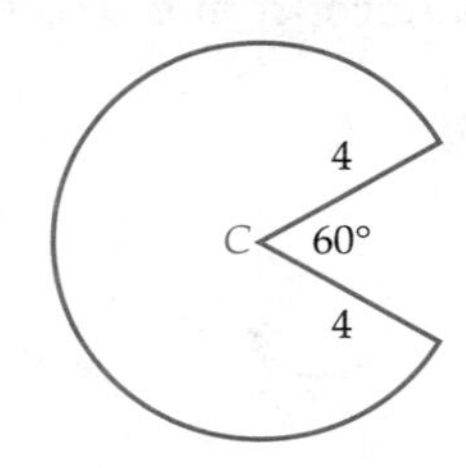

(a) Perimeter = ________
(b) Area = ________

36. If the radius of a circle increases by 30%, then the area of the circle increases by
(a) 9% (b) 30% (c) 69%
(d) 90% (e) 130%

Extension Exercises

37. All of the regions shown have perimeters of 12π. Which has the largest area?

(a)

(b)

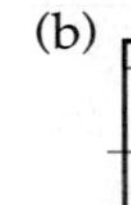

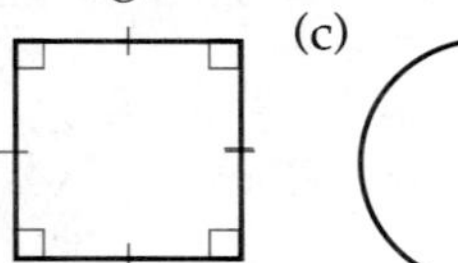

(c)

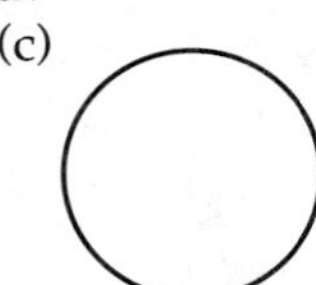

38. You have 40.0 ft of fencing to enclose an area.
(a) What is the maximum rectangular area you can enclose?
(b) What area can you enclose with a circle?

39. A lake has a surface area of 800 km^2.
(a) What is the minimum length the shoreline could be?
(b) What is the maximum length the shoreline could be?

40. The figure shows a square and a semicircle.

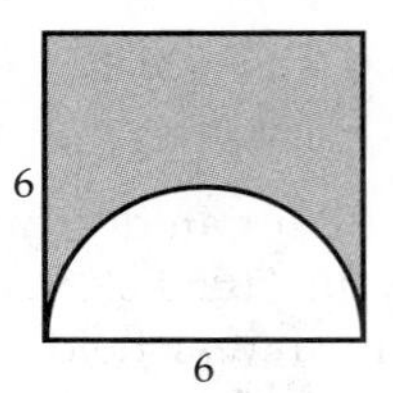

Shaded area = ________

41. C is the center.

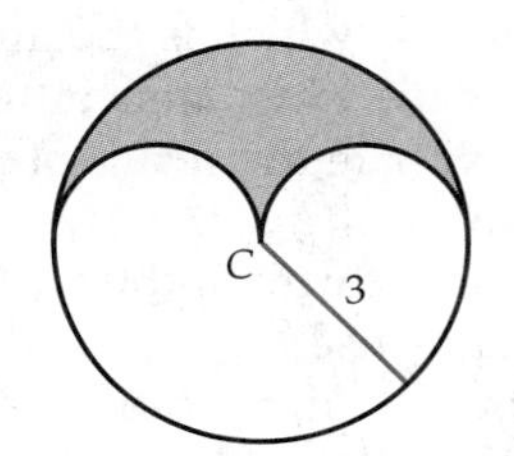

Shaded area = ________

42. A 3.0-m fence is used to enclose a garden with a circular arc.

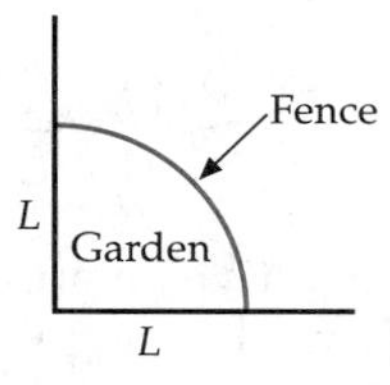

(a) How long is L?? (Use 3.14 for π.)
(b) What is the area of the garden?

43. C is the center. Find the shaded area.

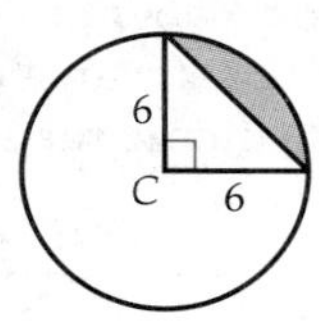

44. The two circles shown have the same center. The radius of the larger circle is 3 times the radius of the smaller circle. What is the ratio of the shaded area to the area of the smaller circle?

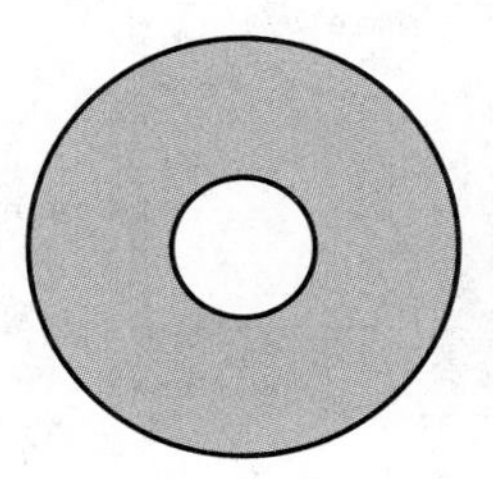

45. Find the area of the region on the square lattice. All curves are circular arcs.

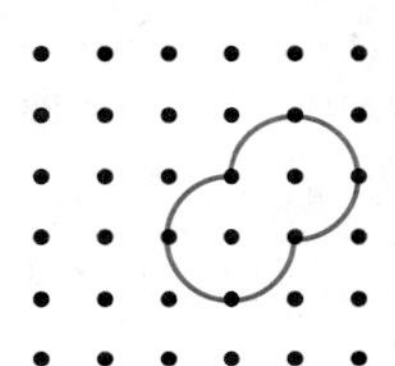

46. Express the area A of a circle in terms of its circumference C.

47. True or false? "The ratio of the areas of two circles equals the ratio of their radii." If the statement is true, prove it. If it is false, give a counterexample.

48. Draw a parallelogram with the measurements shown. (*Hint:* Draw the 8-cm top and the 5-cm height first.)

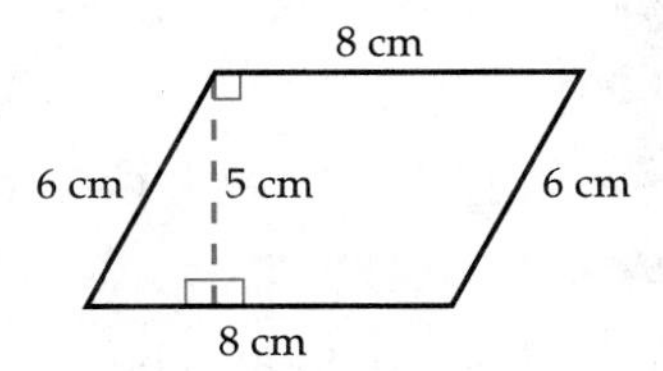

49. Area $ARNO$ = ________

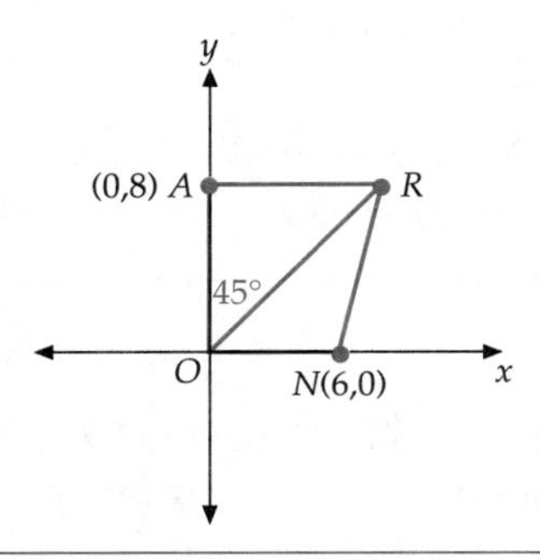

50. Make a graph showing the relationship between the radius and circumference of a circle.

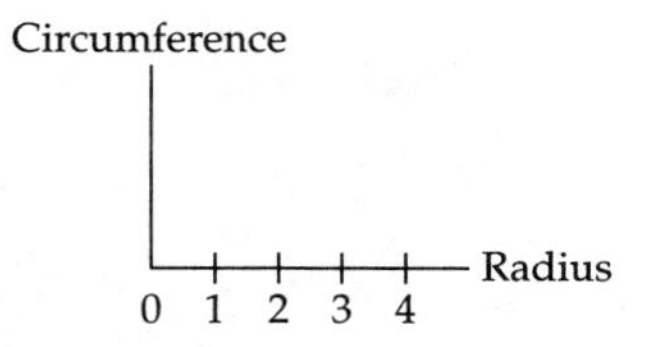

Special Exercises

51. About 2200 years ago, Eratosthenes measured the circumference of the Earth. (That's surprising, considering that many people in more recent civilizations did not even know the Earth was approximately spherical.) Eratosthenes measured the angle of a shadow in Alexandria to be 7.5° at the same time the sun was directly overhead in Aswan, 500 miles away.

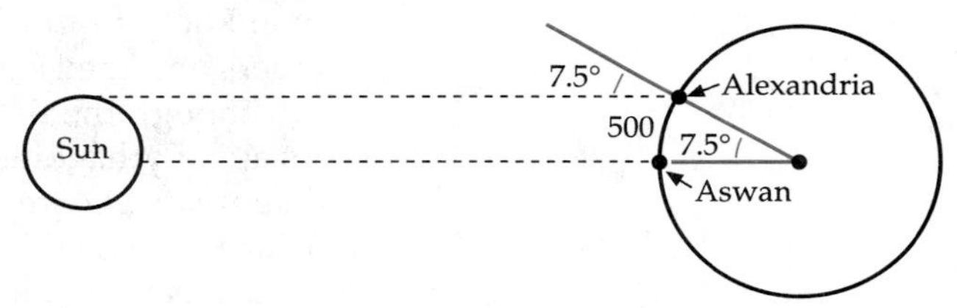

From this, he knew that the central angle shown was 7.5°. Explain how he would compute the circumference of the Earth. (Its actual circumference is about 24,902 miles.)

52. (a) Draw a large quadrilateral $ABCD$ on a sheet of paper. Mark the midpoints of each side E, F, G, and H, and connect E to F to G to H to F.

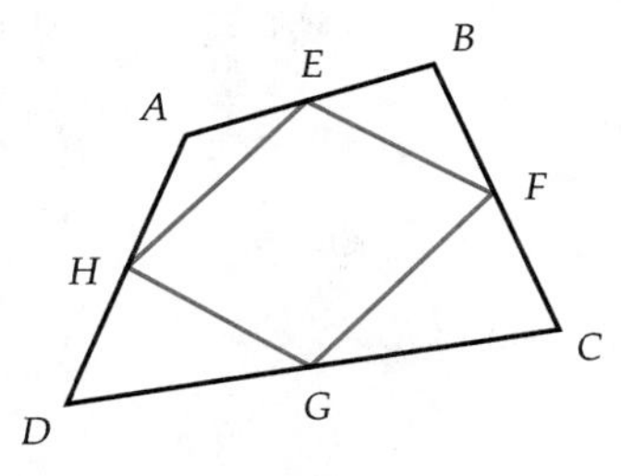

(b) Cut out the four triangles and place them so that they cover $EFGH$.

(c) What relationship does part (b) suggest?

53. Write a report on the history of π.

10.4 THE PYTHAGOREAN THEOREM

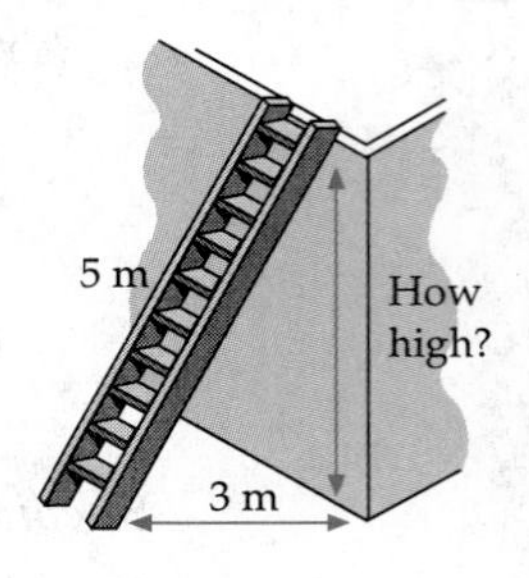

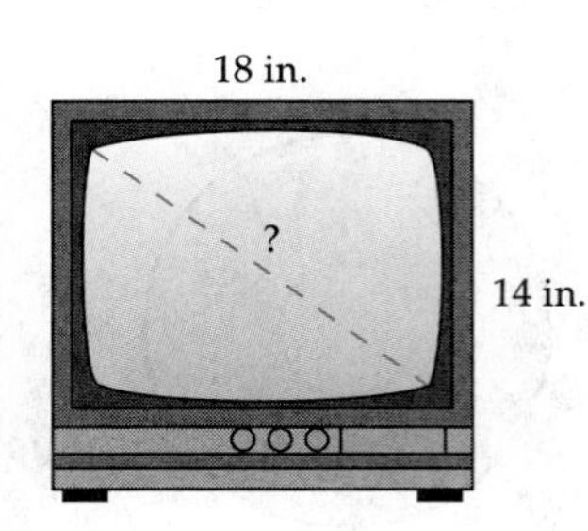

Figure 10–26

The lengths in Figure 10-26 can be found using one of the most famous theorems in mathematics: the Pythagorean Theorem.

PYTHAGOREAN THEOREM

As you know from Chapter 4, Pythagoras and his followers were interested in numbers and their characteristics. They also studied geometry.

Although the Babylonians knew this property of right triangles over 3500 years ago, Pythagoras or one of his followers gave the first known proof of the Pythagorean Theorem about 2500 years ago. The theorem describes the relationships between the lengths of the three sides of a **right triangle** (a triangle that has a right angle).

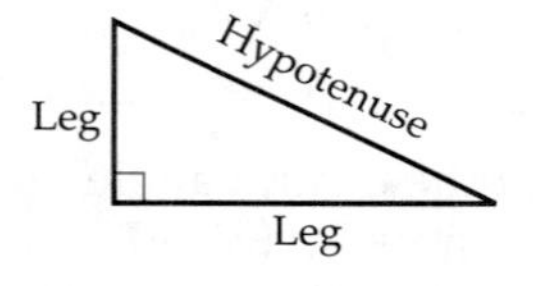

Figure 10–27

The sides of a right triangle have special names. The side opposite the right angle is called the **hypotenuse**, and the other two sides are called **legs** (Figure 10-27).

The lengths of the sides of any right triangle are related by a single formula.

Class **Lesson Exercise 10.48**

(a) Use a ruler to draw a right triangle. (This could also be done on a computer with drawing software.)

(b) Carefully measure the lengths of the three sides.

(c) Square the lengths of the three sides.

(d) How are the squares of the three lengths related?

(e) Compare your results to others in your class.

In spite of measurement error, your results from Lesson Exercise 10.48 may suggest the Pythagorean Theorem. And now, without further ado, here is one of the most famous and useful theorems in all of mathematics.

The Pythagorean Theorem

If a right triangle has legs of lengths a and b and a hypotenuse of length c, then $c^2 = a^2 + b^2$.

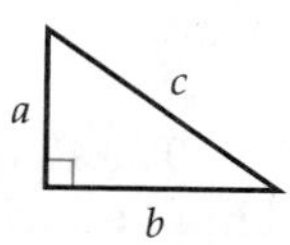

How do we know that this theorem is true? Not only has the theorem been proved; it has been proved about 370 different ways! Even James Garfield (later President Garfield) thought up a new proof while he was in Congress.

One proof uses the area formulas for a triangle and a square and the sum of the angle measures in a triangle. These are all ideas you have studied in this course. Now use them to deduce the Pythagorean Theorem.

Cooperative Lesson Exercise 10.49

The following proof involves putting together four copies of a right triangle to form square $ACEG$ (Figure 10-28).

Figure 10–28

(a) How do we know $ACEG$ is a square?

In parts (b)–(d), show that $BDFH$ is a square.

(b) What is the sum of measures of $\angle 1$ and $\angle 2$?

(c) What is the measure of $\angle 3$?

(d) Why is $BDFH$ a square?

In parts (e) and (f), find the area of the figure in two ways.

(e) Find the area of square $ACEG$ by using the lengths of its sides. Your answer will be in terms of a and b only.

(f) Find the area of square $ACEG$ by adding the area of the four right triangles to the area of square $BDFH$.

Answers to parts (e) and (f) both represent the area of the big square, so they must be equal.

(g) Set your answers to parts (e) and (f) equal to each other, and see whether you can derive the Pythagorean Theorem from your equation.

At first, the ancient Greeks used only whole numbers and rational numbers. They may have discovered a geometric representation of a new type of number by solving the following problem.

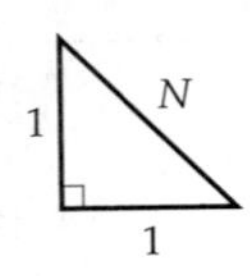

Figure 10–29

Lesson Exercise 10.50

(a) Find N in Figure 10-29.
(b) What new set of numbers is suggested by part (a)?

The Pythagorean Theorem is used in a variety of applications. In any problem in which you know the lengths of two sides of a right triangle, you can use the Pythagorean Theorem to find the length of the remaining side.

Lesson Exercise 10.51

A rectangular screen measures 14.0 in. by 18.0 in. How long is the diagonal, to the nearest tenth of an inch?

The Pythagorean Theorem is sometimes used to find a missing length in an area problem.

EXAMPLE 10.3

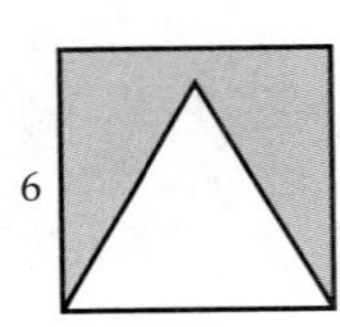

Figure 10–30

Figure 10-30 shows an equilateral triangle inside of a square. Find the shaded area.

Solution

$$\text{Shaded area} = A_{\square} - A_{\triangle}$$

This requires two area formulas.

$$A_{\square} = s^2 \qquad \text{and} \qquad A_{\triangle} = \frac{1}{2}bh$$

$$A_{\square} = 6^2 = 36$$

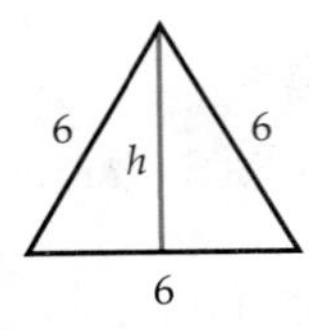

Figure 10–31

To find $A_{\triangle}$, we need b and h. (See Figure 10-31.)

Now $b = 6$, but what is h? Use the Pythagorean Theorem (see Figure 10-32.)

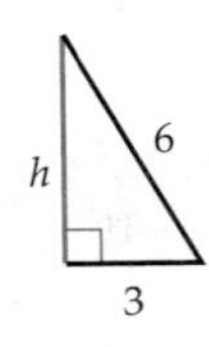

Figure 10–32

$$h^2 + 3^2 = 6^2$$
$$h^2 + 9 = 36$$
$$h^2 = 27$$
$$h = \sqrt{27} \text{ (or } 3\sqrt{3}\text{)}$$
$$A_{\triangle} = \frac{1}{2}(\sqrt{27}) \cdot 6 = 3\sqrt{27}$$

Therefore, $A_{\text{shaded}} = A_{\square} - A_{\triangle} = 36 - 3\sqrt{27}$ square units. ■

Now you try one.

Lesson Exercise 10.52

Find the area of the shaded region; C is the center. (*Hint:* Form a right triangle in the drawing.)

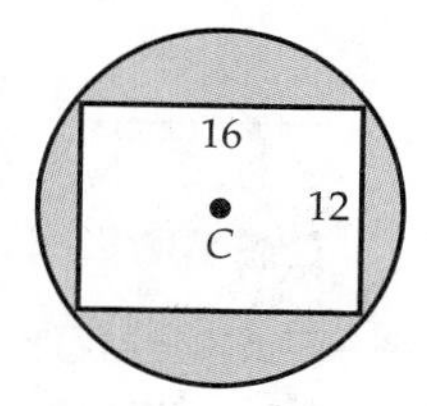

THE RIGHT-TRIANGLE TEST

If you know the lengths of the three sides of a triangle, is there a way to tell whether it is a right triangle? Yes. This is the converse of the Pythagorean Theorem. You can complete the proof in the homework exercises. (No need to thank me for putting it there.)

Lesson Exercise 10.53

Look at the Pythagorean Theorem, and see whether you can write its converse.

According to the converse, any triangle in which the lengths have the relationship $a^2 + b^2 = c^2$ must be a right triangle. The converse, the Right-Triangle Test, can be used to determine whether a triangle with three given lengths is a right triangle.

The Right-Triangle Test
(Converse of the Pythagorean Theorem)

If a, b, and c are the lengths of the sides of a triangle and $a^2 + b^2 = c^2$, then the triangle is a right triangle.

In the Right-Triangle Test, the side of length c would have to be the longest side of the triangle. The Right-Triangle Test is based on the fact that the Pythagorean Theorem works only on right triangles. (The proof is left for the homework exercises.) Use the test in the following exercise.

Lesson Exercise 10.54

Determine whether each of the following is a right triangle.

(a)

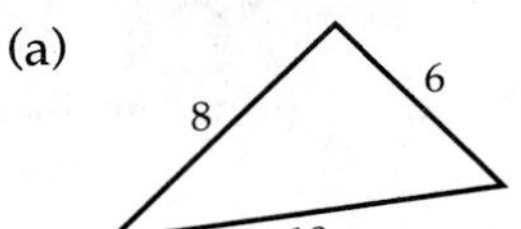

(b)

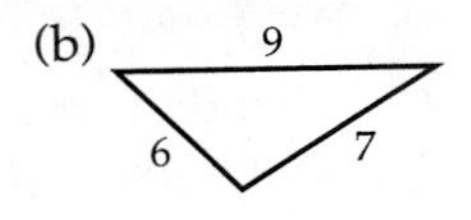

THE SIDES OF A TRIANGLE

The Right-Triangle Test enables us to recognize three lengths that could form a right triangle. Is it possible to tell whether three lengths could form an acute triangle (in which all three angles are acute), an obtuse triangle (in which one angle is obtuse), or no triangle at all?

Lesson Exercise 10.55

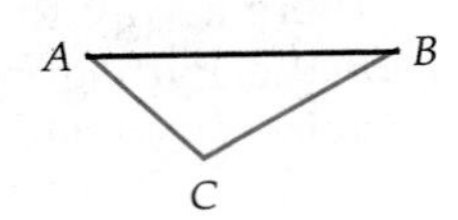

Figure 10–33

Suppose that you are traveling from point A to point B (Figure 10-33).

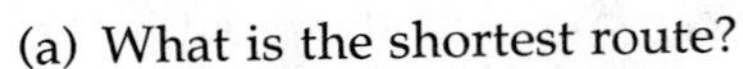

(a) What is the shortest route?

(b) What does part (a) imply about the distance from A to C to B?

(c) Parts (a) and (b) suggest that if C is not on $\overline{AB}$, then $AC + CB$ ________ AB.

The result of Lesson Exercise 10.55(c) is called the Triangle Inequality.

> **The Triangle Inequality**
>
> The sum of the measures of any two sides of a triangle is greater than the measure of the third side.

The following exercise uses the Triangle Inequality and the Right-Triangle Test to produce a more general result.

Lesson Exercise 10.56

(a) Three segments have lengths a, b, and c, with $a < b < c$. Under what conditions could segments of lengths a, b, and c be used to form a triangle? (*Hint:* Find lengths a, b, and c that could not form a triangle.)

(b) A triangle has sides of lengths a, b, and c, with $a \leq b \leq c$. What conditions for a, b, and c would make the triangle a right triangle, an acute triangle, or an obtuse triangle? (*Hint:* Try $a = 3$ and $b = 4$, and vary c.)

You can use the Triangle Inequality to test if three lengths can form a triangle. If they can, you can determine if the triangle is acute, right, or obtuse by comparing $a^2 + b^2$ to c^2.

Lesson Exercise 10.57

Check whether each set of lengths could form a triangle. If it does, determine whether the triangle is right, obtuse, or acute.

(a) 6, 7, 9 (b) 3, 7, 12 (c) 4, 5, 8

ANSWERS TO SELECTED LESSON EXERCISES

10.49 (a) Each side has length $a + b$, and it has four right angles.
(b) 90° (c) 90°
(d) It has four 90° angles and four congruent sides
(e) $(a + b)^2$ (f) $4\left(\frac{1}{2}ab\right) + c^2$
(g) *Hint:* $(a + b)^2 = a^2 + 2ab + b^2$

10.50 (a) $\sqrt{2}$ (b) Irrational numbers

10.51 22.8 in.

10.52 $100\pi - 192$ sq. units

10.54 (a) Yes, since $6^2 + 8^2 = 10^2$
(b) No, since $6^2 + 7^2 \neq 9^2$

10.55 (a) AB
(b) It is longer than AB. (c) >

10.56 (a) $a + b > c$
(b) $a^2 + b^2 = c^2$; $a^2 + b^2 > c^2$; $a^2 + b^2 < c^2$

10.57 (a) Acute triangle, $6^2 + 7^2 > 9^2$
(b) Not a triangle, $3 + 7 < 12$
(c) Obtuse triangle, $4^2 + 5^2 < 8^2$

10.4 HOMEWORK EXERCISES

Basic Exercises

1. Many students know the Pythagorean Theorem as $a^2 + b^2 = c^2$ but do not know anything else about a, b, and c.
(a) To what geometric shape does the Pythagorean Theorem apply?
(b) What do a, b, and c represent?

2. Explain how the following right triangle with a square drawn on each side illustrates the Pythagorean Theorem.

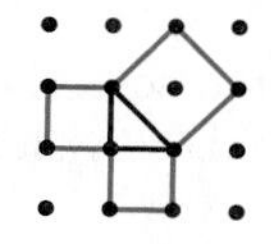

3. $ACEG$ is a square. $\triangle ABH \cong \triangle CDB \cong \triangle EFD \cong \triangle GHF$. Explain why $\angle 3$ must be a right angle.

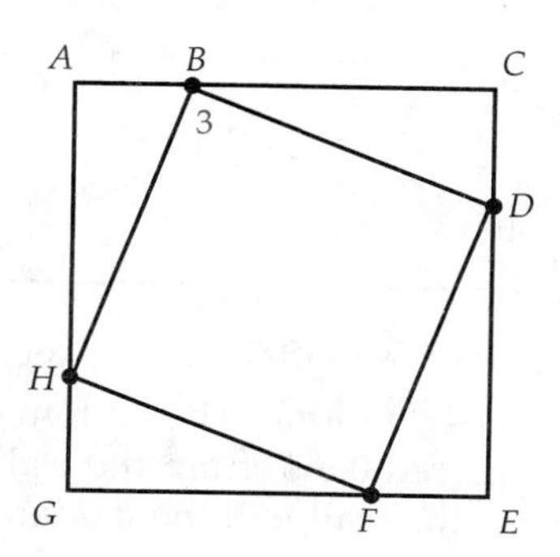

*4. Before becoming president, James Garfield proved the Pythagorean Theorem using the following trapezoid.

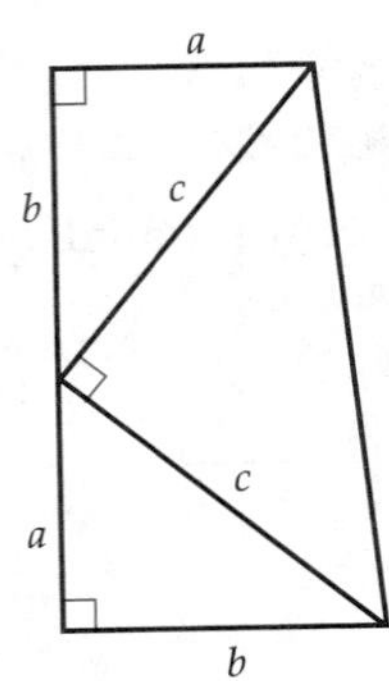

(a) Write formulas for the areas of the three triangles in terms of a, b, and c.
(b) Find the area of the trapezoid.
(c) Set your two areas equal to each other, and see whether you can derive the Pythagorean Theorem.

5. Draw a line segment of length $\sqrt{13}$ units on the square lattice.

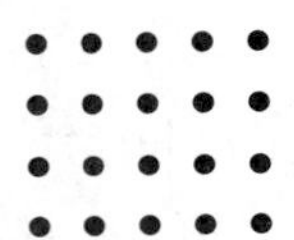

6. Find the area and perimeter of the following right triangle on a square lattice.

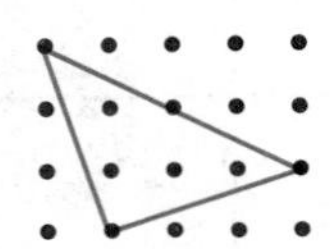

7. Find the perimeter of the following parallelogram on a square lattice.

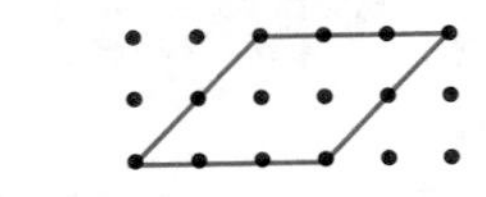

8.

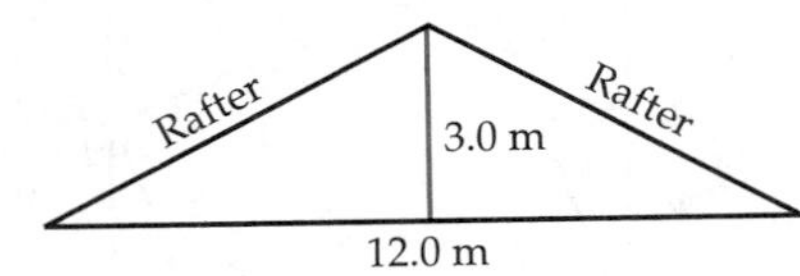

How long are the rafters on the roof?

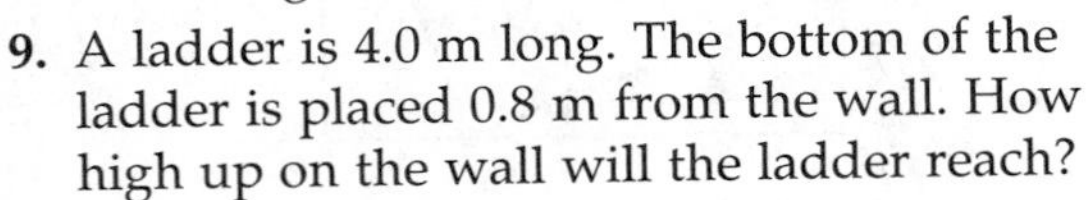

9. A ladder is 4.0 m long. The bottom of the ladder is placed 0.8 m from the wall. How high up on the wall will the ladder reach?

10. You want to hang a microphone 3 m above a stage, using two wires. About how much wire do you need?

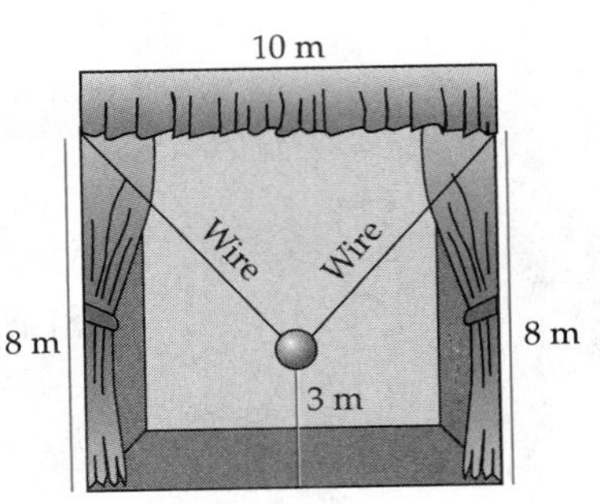

11. In walking from A to B on a square city block, cutting straight across the grass is about ________% the distance of using the sidewalk.

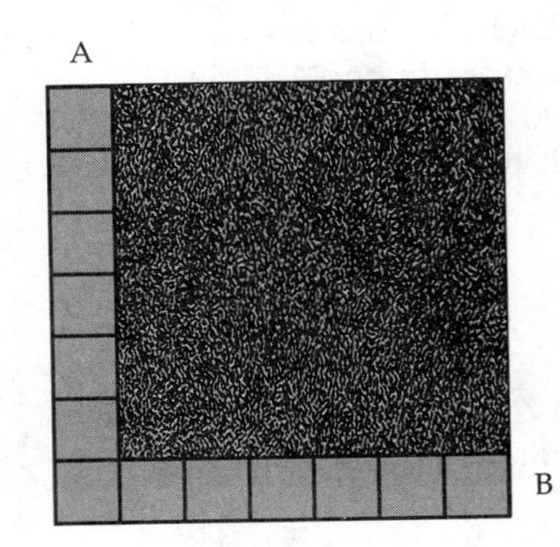

12. A rectangle has sides of length b and h. Compute the length of each diagonal, and show that they are equal in length.

13. Two forces act on an object, one pushing east at 4 mph and one pushing north at 4 mph. Where will the object go, and at what speed?

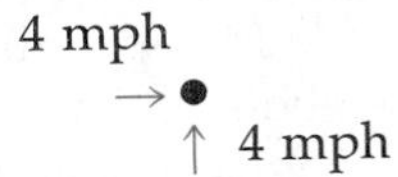

14. Find the perimeter of the square.

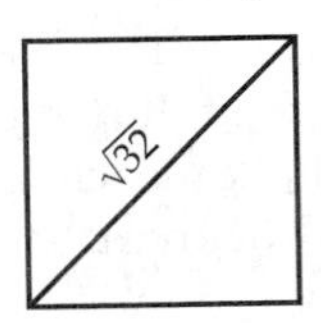

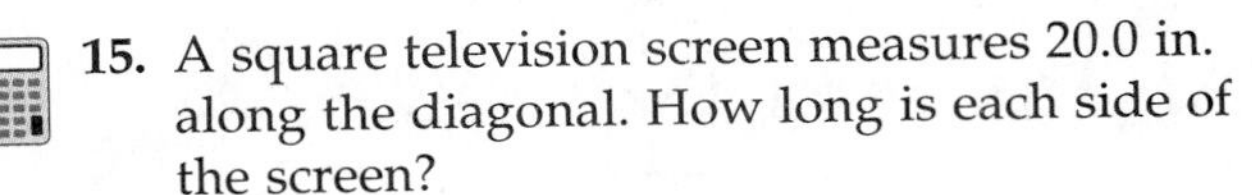

15. A square television screen measures 20.0 in. along the diagonal. How long is each side of the screen?

16. Find the area of the triangle.

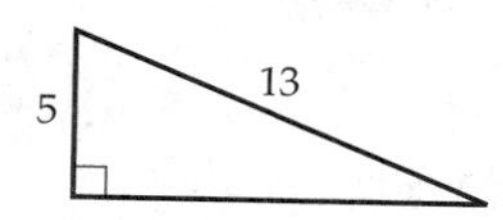

17. Find the area of the rhombus if $AC = 8$ and $AB = 6$.

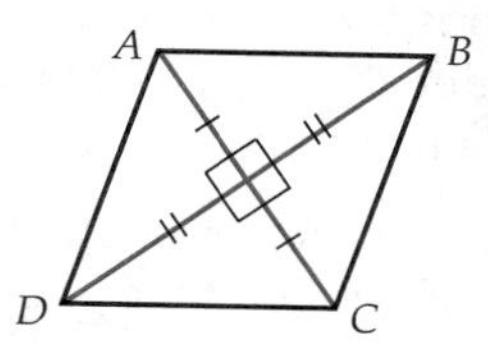

18. Find the area of the shaded region. C is the center of the circle.

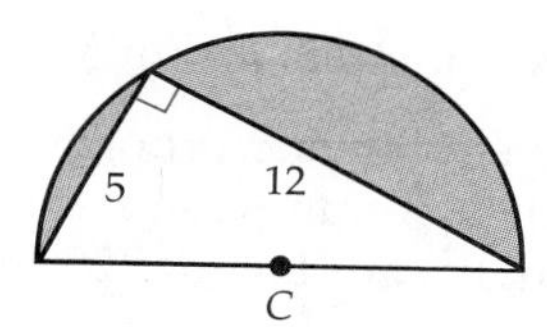

19. After a person in Kansas travels 7 miles north, then 3 miles east, and finally 3 miles south, how far is the person from the starting point?

20. Flo and Ken want to carry a tall piece of glass that is 9 ft square through a rectangular doorway that is 3 ft by 8 ft. Will it fit?

21. (a) Complete the third example, repeating the error pattern from the completed examples.

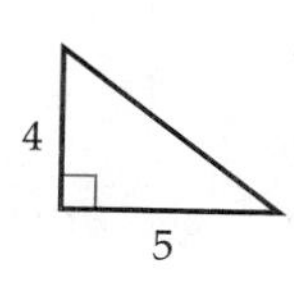

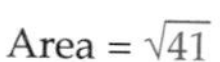

2
3

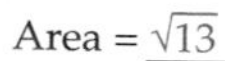

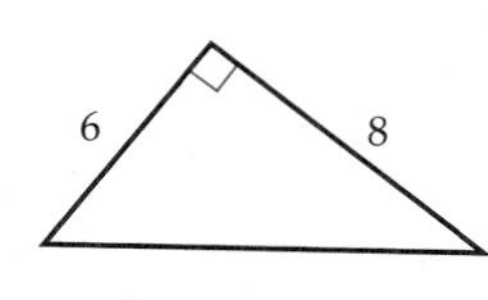

Area = ____

(b) Describe the error pattern.

22. Use the Right-Triangle Test to determine whether each of the following is a right triangle.

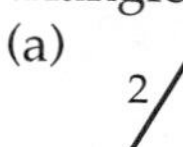

(a)

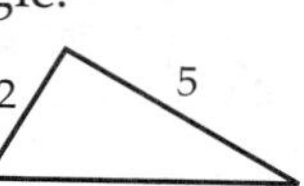

(b)

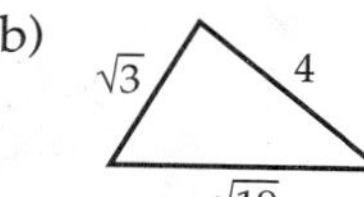

23. Use the Right-Triangle Test to determine whether each of the following is a right triangle.

(a)

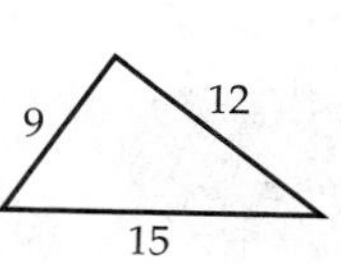

(b)

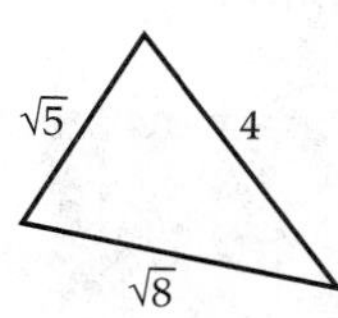

24. (a) Select three lengths that make up a right triangle, and multiply them all by the same positive number. Do the resulting lengths determine a right triangle?

(b) Repeat part (a) using three new lengths.

(c) Propose a generalization of your results.

25. Tell whether each triangle is right, acute, or obtuse.

(a)

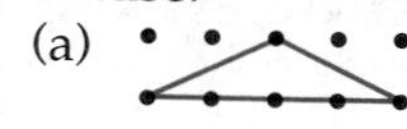

(b)

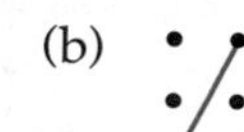

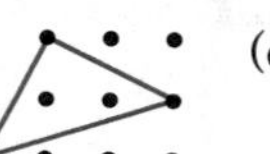

(c)

26. A triangle has sides of lengths 5, 6, x. If the triangle is acute, give all possible values of x. (Use $<$ or $>$ in your answer.)

27. Tell whether each set of lengths could form a right triangle, an acute triangle, an obtuse triangle, or no triangle.

(a) 3, 5, 6 (b) 9, 12, 15

(c) 3, 5, 10 (d) 2, 3, 4

28. Consider the following problem. "A standard stop sign measures 25.0 cm on each side. What is the area of the surface of a stop sign?" Devise a plan and solve the problem.

Extension Exercises

29. **Pythagorean triples** are three counting numbers that satisfy the relationship $a^2 + b^2 = c^2$. Over 2000 years ago, the Chinese knew the smallest triple: 3, 4, 5 (since $3^2 + 4^2 = 5^2$). Another triple is 5, 12, 13 (since $5^2 + 12^2 = 13^2$).

(a) Double 3, 4, and 5. Does this create a new Pythagorean triple?

(b) Name three more Pythagorean triples.

(c) Suppose you know that a, b, c is a Pythagorean triple with $a^2 + b^2 = c^2$. Show that ka, kb, kc, is also a Pythagorean triple for any counting number k.

30. One of the most amazing Babylonian tablets, dating from around 1700 B.C., is called Plimpton 322 (see photo).

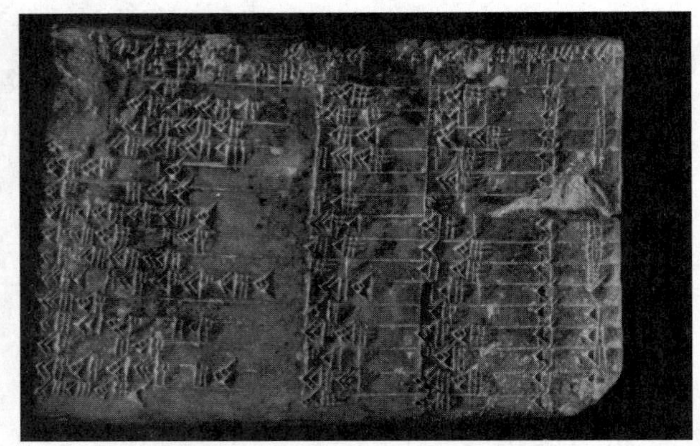

George A. Plimpton Collection, Rare Book and Manuscript Library, Columbia University.

It contains a list of Pythagorean triples! All the triples in the chart are listed in the first three columns (a, b, c). The last two columns list u and v such that $a = 2uv$, $b = u^2 - v^2$, and $c = u^2 + v^2$. Fill in the missing values in the following table, which shows the first five rows of Plimpton 322.

a	b	c	u	v
120	119	169	12	
3,456	3,367	4,825		27
4,800	4,601		75	
13,500		18,541		54
	65	97		

31. One of the most famous problems in mathematics is Fermat's Last Theorem. In 1631, Fermat considered equations of the form $x^2 + y^2 = z^2$, $x^3 + y^3 = z^3$, $x^4 + y^4 = z^4$, and so on. He said that only $x^2 + y^2 = z^2$ has a solution set made up of counting numbers. None of the other equations have any such solutions! After writing this idea down, Fermat wrote in the margin of the paper (in Latin): "I have discovered a truly wonderful proof of this, but the margin is too small to contain it." The theorem remained unproven for over 360 years! In May 1995, Princeton University mathematician Andrew Wiles published a proof of Fermat's Last Theorem.

(a) Give three sets of counting-number solutions to $x^2 + y^2 = z^2$.

(b) Consider $x^3 + y^3 = z^3$. Is $x = 5$, $y = 6$, and $z = 7$ a solution set?

32. People have found formulas for some Pythagorean triples. One is the following. For any odd number $N > 1$, N, $\frac{N^2 - 1}{2}$, and $\frac{N^2 + 1}{2}$ will be a Pythagorean triple. Write four triples using this formula.

33. Consider the following problem. "Find the area A in the diagram. The areas of the two smaller squares are 144 and 25 square units."

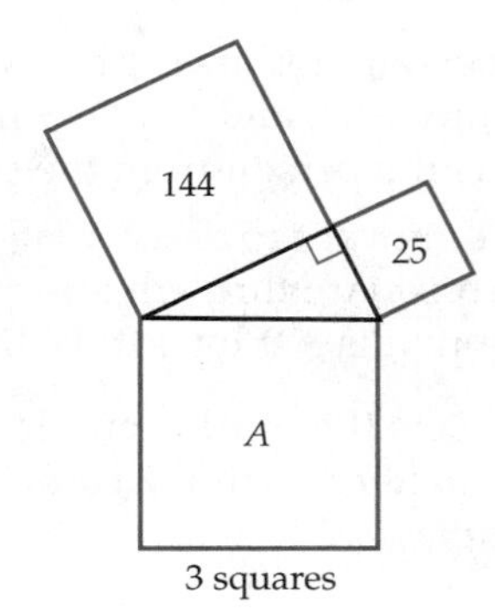

3 squares

Devise a plan and solve the problem.

34. The areas of the two smaller semicircles are 18π and 32π square units. Find A.

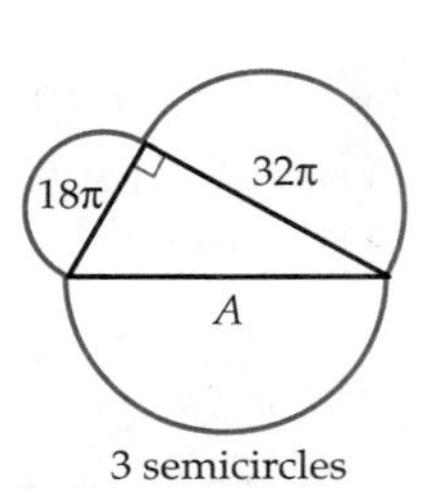

3 semicircles

35. The areas of the two smaller triangles are $\frac{9\sqrt{3}}{4}$ and $4\sqrt{3}$ square units. Find A.

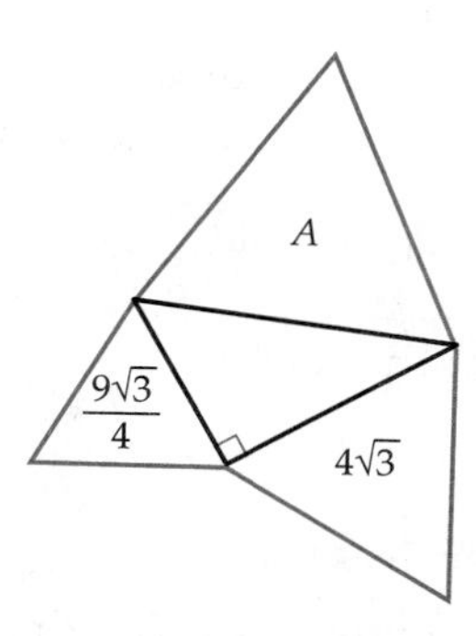

3 equilateral triangles

36. Two congruent 10-by-10 squares overlap so that a vertex of one is the center of the other. One possible drawing follows. What is the largest possible overlapping area?

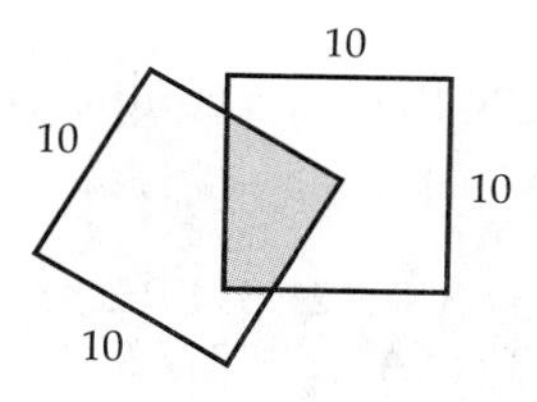

37. A regular hexagon has sides of length 10. Find its area. (*Hint:* See the following diagram.)

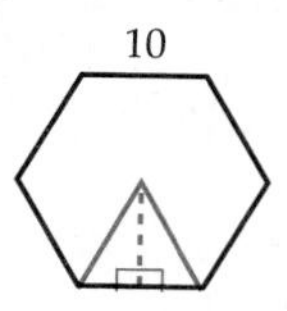

38. The length of each side of the squares, regular hexagons, and equilateral triangles is 1 cm. What is the total area?

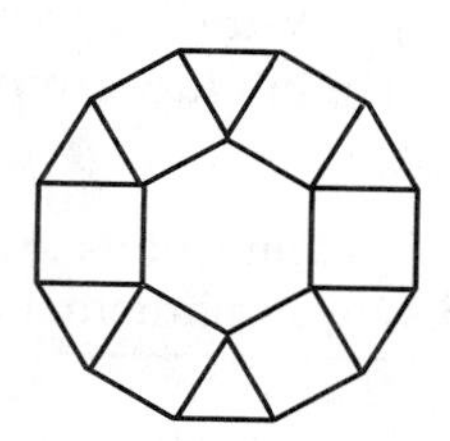

39. Consider the following problem. "Home plate in baseball is cut from a square $ABCD$, as follows.

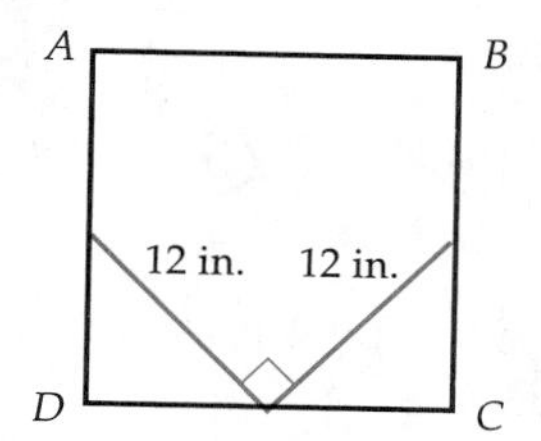

How long is each side of $ABCD$?" Devise a plan and solve the problem.

40. Find the perimeter and the area of the trapezoid.

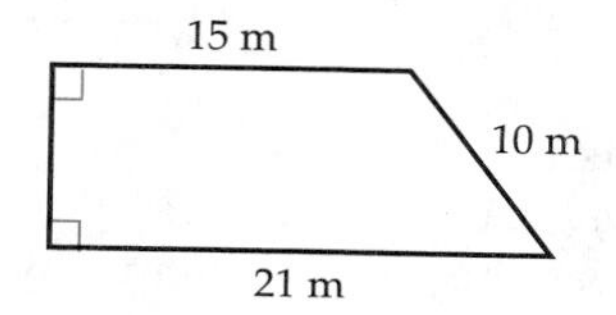

41. C is the center of both circles. $CB = 5$ and $AB = 8$. Find the shaded area.

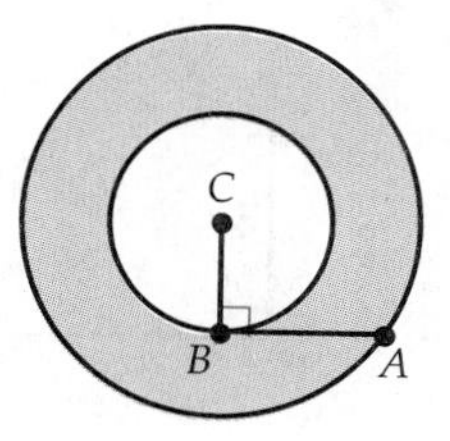

42. How big a circular pan is needed to accommodate four slices of bread, each 4.0 in. by 4.0 in. as shown, with 1.0 in. between them?

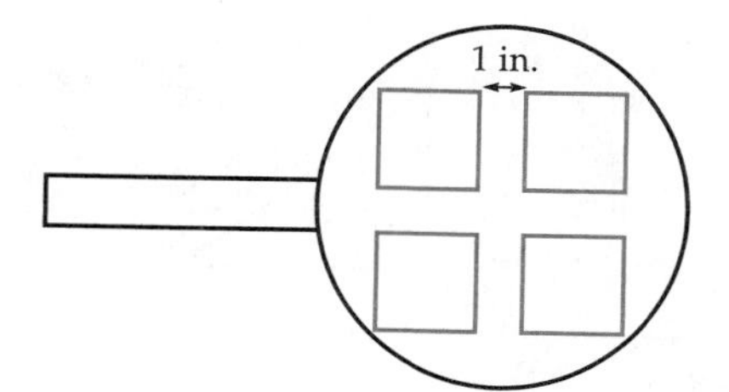

43. Compare the following choices for packing one row of canned drinks in a carton. Assume each can has a height of 1 unit and a diameter of 1 unit. Which package has the least base area *per can*?

(a) 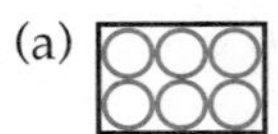 (b) (c) (d)

44. A closet is 8 ft by 4 ft by 6 ft. Will a 10-ft pole fit inside?

45. A cube has edges 2 cm long. How long is a diagonal of the cube? (*Hint:* Draw the diagonal of the base.)

46.

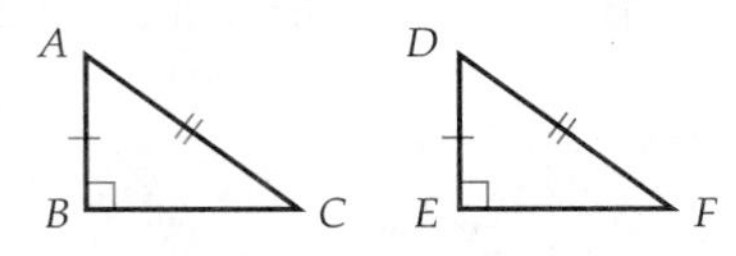

(a) Why is $\overline{BC} \cong \overline{EF}$?
(b) Why is $\triangle ABC \cong \triangle DEF$?
This verifies the *HL* (hypotenuse-leg) congruence property for right triangles.

47. Read the following proof of the Right-Triangle Test, and fill in the missing reasons.

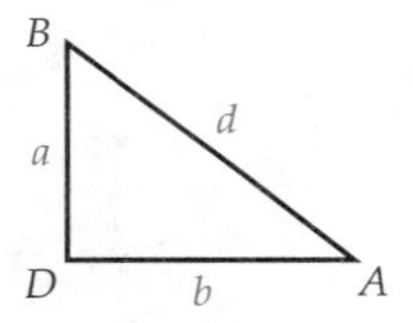

A triangle has lengths *a*, *b*, and *d*, with $a^2 + b^2 = d^2$. How do we show that it must be a right triangle?

Draw another triangle with sides of lengths *a* and *b* and a right angle between them.

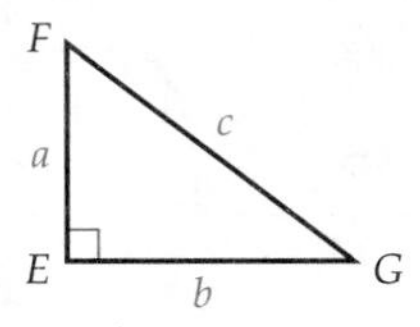

(a) Since $\triangle EFG$ is a right triangle, the Pythagorean Theorem says: ______________________.
(b) Why must $c^2 = d^2$?
(c) Now, if $c^2 = d^2$, then $c = d$. Therefore, $\triangle ABD \cong \triangle$ ________ because of ______________________.
(d) If the two triangles are congruent, then $m\angle E =$ ________ $= 90°$. Therefore, $\triangle ABD$ is a right triangle!

Computer Exercises

48. Write a Logo procedure that draws an isosceles right triangle with legs of length 30.

49. Write a Logo procedure that draws a 30-by-50 rectangle with one of its diagonals.

50. The following BASIC program tests Pythagorean triples of the form $A, \frac{A^2 - 1}{2}, \frac{A^2 + 1}{2}$. RUN this on a computer, and state the results.

```
10 FOR A = 3 TO 21 STEP 2
20    LET B = (A∧2 − 1)/2
30    LET C = (A∧2 + 1)/2
40    PRINT A, B, C
50    IF C∧2 = A∧2 + B∧2 THEN PRINT "IT
      IS A PYTHAGOREAN TRIPLE."
60    IF C∧2 <> A∧2 + B∧2 THEN PRINT
      "NOT A TRIPLE"
70 NEXT A
```

Special Exercises

51. Read "Socrates and the Slave" by Plato (from the *Meno*) in *Fantasia Mathematica*, edited by Clifton Fadiman. Write a report that includes a summary of the main ideas and your reactions to them.

52. Write about the connections the Pythagoreans discovered between mathematics and music.

10.5 SURFACE AREA AND VOLUME

Photo by Tom Sonnabend.

Figure 10–34

How much paper is needed for the label of the can in Figure 10-34? How much juice will the container hold? The first question concerns surface area; the second concerns volume. These are two useful measurements of a space figure.

SURFACE AREA OF RECTANGULAR PRISM

How much will a container cost? How much paint do you need to paint your bedroom? In order to answer these questions, you must compute the surface area (Figure 10-35).

Figure 10–35

The total **surface area** of a closed space figure is the sum of the areas of all its surfaces. A surface area in square units indicates how many squares it would take to cover the outside of a space figure.

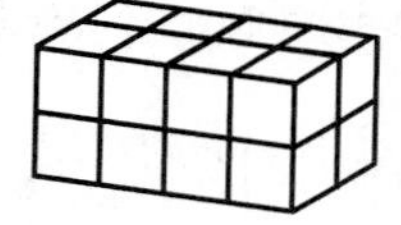

Figure 10–36

Lesson Exercise 10.58

Consider the rectangular prism in Figure 10-36. Compute the total surface area. (In other words, determine how many squares are needed to cover all the faces.)

Computing the surface area of a rectangular prism can help you decide how much paint you need to paint a room.

Lesson Exercise 10.59

A box-shaped room is 18 ft long, 12 ft wide, and 8 ft high. There are two windows that are 3 ft by 5 ft and a doorway that is 4 ft by 7 ft.

(a) Sketch the room.

(b) What is the area of the ceiling?

(c) Suppose you want to paint the walls and ceiling. Excluding the windows and the door, how much area is there to paint?

(d) If 1 gallon of paint covers 400 ft^2, how many gallon cans of paint will be needed to paint the four walls and the ceiling?

SURFACE AREAS OF PRISMS AND CYLINDERS

Computing the total surface areas of other prisms and cylinders also requires finding the area of each surface and adding the areas together.

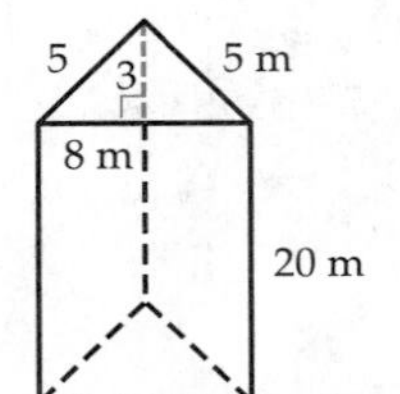

Figure 10–37

EXAMPLE 10.4

Find the total surface area of the triangular right prism in Figure 10-37.

Solution

Understanding the Problem Find the sum of the areas of the five faces of the triangular prism.

Devising a Plan The triangular prism has five faces. There are two isosceles triangles (the bases) and three rectangles (the lateral faces), as shown in the net in Figure 10-38. (A **net** is a pattern that can be used to construct a polyhedron.)

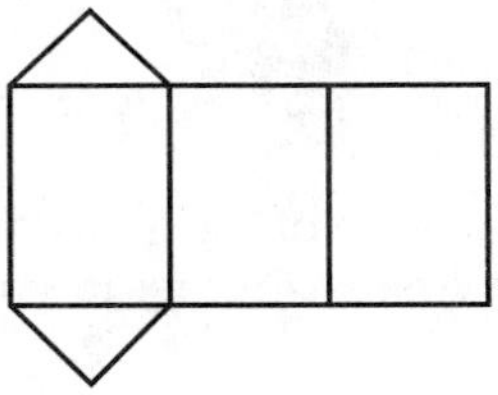

Figure 10–38

$$A_{\text{surface}} = A_{\triangle_1} + A_{\triangle_2} + A_{\square_1} + A_{\square_2} + A_{\square_3}$$

Carrying Out the Plan Each of the triangles has an area of $\frac{1}{2} \cdot 3 \cdot 8 = 12$. The rectangles are 5 by 20, 5 by 20, and 8 by 20. So their areas are 100, 100, and 160.

$$A_{\text{surface}} = 12 + 12 + 100 + 100 + 160 = 384 \text{ m}^2$$

Looking Back The method for finding the total surface area can be used whenever you are able to find the area of each surface. ■

Now it's your turn.

Lesson Exercise 10.60

Consider the following problem. "Find the total surface area of the triangular prism in Figure 10-39." Devise a plan and solve the problem.

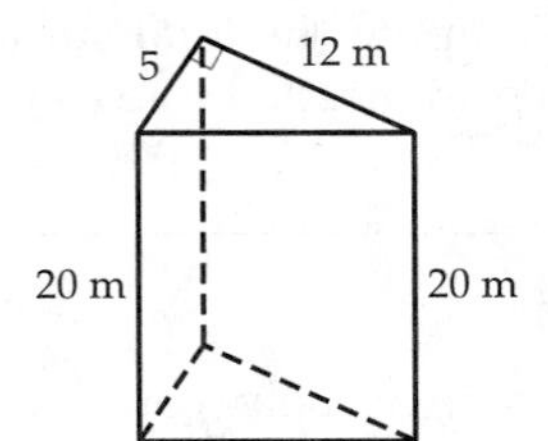

Figure 10–39

How do you find the surface area of a cylinder?

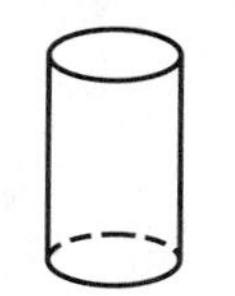

Figure 10–40

Lesson Exercise 10.61

(a) How many surfaces does the cylinder in Figure 10-40 have?

(b) What shape are its bases?

(c) What shape is the lateral surface? (*Hint:* Roll up a regular sheet of paper.)

A cylinder has two circles for bases. Figure 10-41 shows that the lateral surface of a cylinder is a rolled-up rectangle.

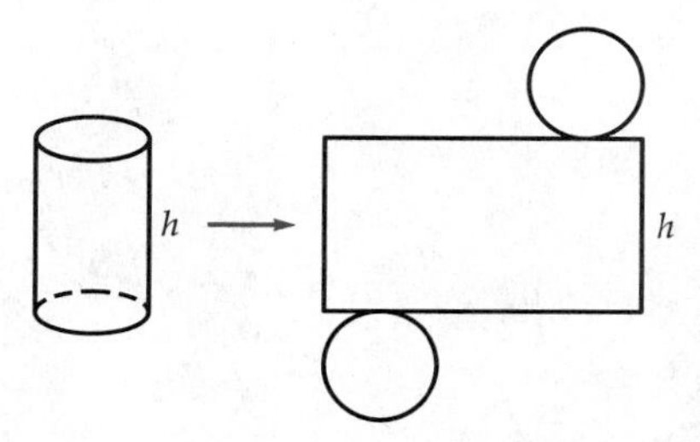

Figure 10–41

Lesson Exercise 10.62

(a) If a cylinder has radius r and height h, what are the dimensions of the rolled-up rectangle (the lateral surface)? It might represent the label of a can.

(b) Compute the area of the two circles and the rectangle, and find a formula for the total surface area of a cylinder.

The surface area of any prism or cylinder can be computed using a more general formula.

Lesson Exercise 10.63

(a) How can the lateral surface area of a prism or cylinder be computed from the perimeter of the base?

(b) If B is the area of the base, p is the perimeter of the base, and h is the height, what is a formula for the surface area of a prism or cylinder?

The general surface-area formula for a prism or cylinder, $A = ph + 2B$, is not commonly used in elementary or middle-school mathematics.

VOLUMES OF RIGHT RECTANGULAR PRISMS

Joe Munroe, Life Magazine. © 1959 Time, Inc.

Figure 10–42

In the late 1950s, college students enjoyed attempting to measure the volume of a phone booth in their own unique way (Figure 10-42). **Volume** is the amount of space occupied by a three-dimensional figure. Volume is usually measured in cubic units. Whereas surface area is the total area of the faces of a solid, volume is the capacity of a solid.

Lesson Exercise 10.64

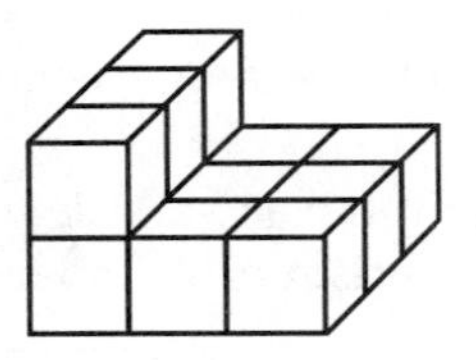

Figure 10–43

(a) What is the volume of the solid in Figure 10-43, in cubic units?

(b) What is the total surface area?

(c) Suppose the 3 cubes on the right-hand side of the first layer were moved up just to the right of the cubes on the second layer. The figure would then have 2 columns and 3 rows in each layer. How would this change the volume and the surface area?

In Lesson Exercise 10.64, one could compute the volume by counting how many cubes fill up the space figure. Children can first study examples like this using materials such as wooden cubes or multilink cubes. With rectangular prisms like the one in the following exercise, there is a shortcut (formula) for counting the cubes.

Lesson Exercise 10.65

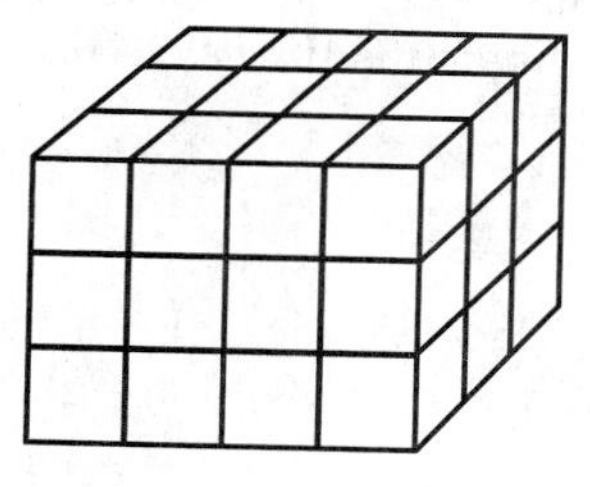

Figure 10–44

Imagine building a solid rectangular prism structure out of wooden cubes as shown in Figure 10-44. Each cube is 1 cm^3.

(a) How many cubes are in each row, going across?
(b) How many cubes are there in a layer?
(c) How many layers are there?
(d) What is the volume of the rectangular prism?
(e) What are the length, width, and height of the rectangular prism, in centimeters?
(f) How can you compute the volume using the length, width, and height?

The results of Lesson Exercise 10.65 suggest a shorter way to compute the volume of a rectangular prism. The formula lw (length times width) gives the number of cubes in a layer of a rectangular prism, and h (height) is the number of layers. So lwh gives the total number of cubes that fill the interior of the rectangular prism in Figure 10-44.

4×3 = number of cubes per layer

3 layers

$4 \times 3 \times 3 = 36$ cubes

Volume of a Right Rectangular Prism

The volume V of a right rectangular prism that has dimensions l, w, and h is

$$V = lwh$$

Computing the volume of a right rectangular prism can help you decide which freezer is the best buy.

Lesson Exercise 10.66

A store sells two types of freezers. Freezer A costs \$310 and measures 1.5 ft by 1.5 ft by 5.0 ft. Freezer B costs \$400 and measures 2.0 ft by 2.0 ft by 3.5 ft. Which freezer has a lower unit cost?

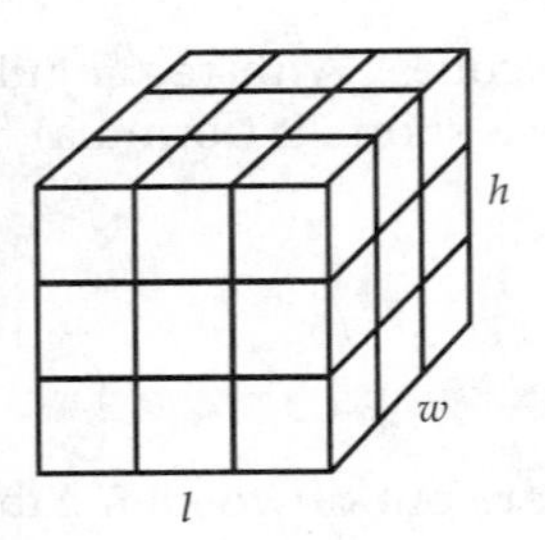

Figure 10–45

VOLUMES OF PRISMS AND CYLINDERS

Are you ready for some good news? And no bad news? All right prisms and right circular cylinders have the same general volume formula!

As you have seen, you can compute the volume of a rectangular prism (Figure 10-45) by counting the number of cubes in each layer, which is the *area of the base (lw)*, and multiply by the number of layers, which is the *height (h)*.

$$V = (l \cdot w) \cdot h$$

$$V = (\text{base area}) \cdot \text{height}$$

We can apply the same idea to all right prisms and cylinders. Each has two congruent bases. Imagine that the three solids in Figure 10-46 are glass containers.

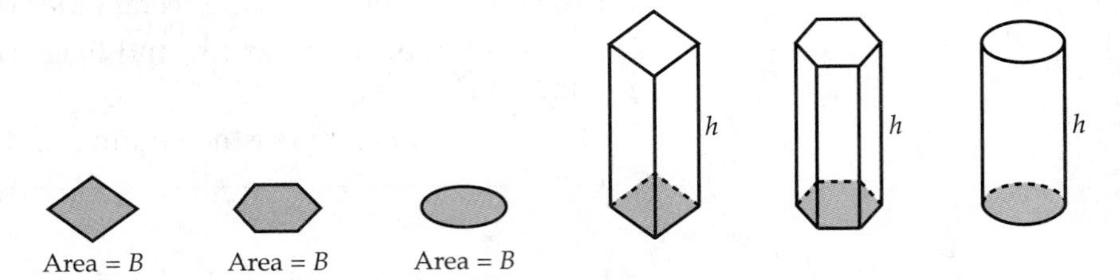

Figure 10–46

Visualize filling them up with some water (or orange juice, if you prefer). The total amount of liquid can be measured by moving the base area from the bottom to the top of the container. In other words, the volume is the area of the base multiplied by the height.

Volume of Any Right Prism or Right Cylinder

The volume V of a right prism or a right cylinder that has a base of area B and height h is

$$V = Bh$$

Most geometry books use b to denote the *length* of a base (side) of a polygon and B to denote the *area* of a base that is the face of a polyhedron, cone, or cylinder. The general formula for prisms and cylinders can be used to derive specific volume formulas.

Lesson Exercise 10.67

Consider a right circular cylinder with radius r and height h.

(a) What shape is the base?

(b) What is the area of the base?

(c) What is a formula for the volume obtained by substituting in $V = Bh$?

The following example illustrates how to find the volume of a triangular prism by using the general volume formula.

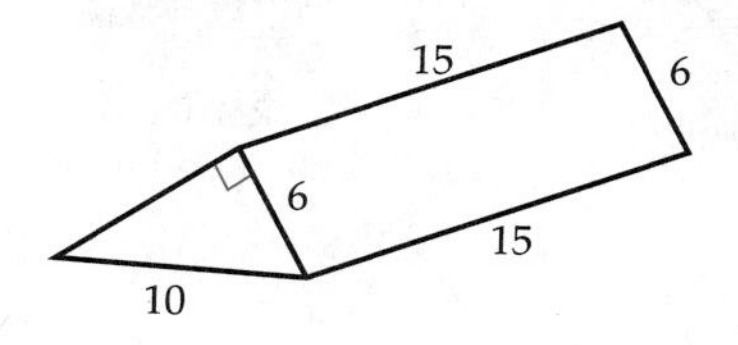

Figure 10–47

EXAMPLE 10.5

Find the volume of the right triangular prism shown in Figure 10-47.

Solution

$$V \text{ (volume)} = B \text{ (base area)} \times h \text{ (height of prism)}$$

The bases are right triangles. First, use the Pythagorean Theorem to find the length of the other leg.

$$6^2 + x^2 = 10^2$$

$$36 + x^2 = 100$$

$$x^2 = 64$$

$$x = 8$$

The base has area $B = \frac{1}{2} \cdot 6 \cdot 8 = 24$. The height of the prism is 15.

$$V_{\text{prism}} = 24 \cdot 15 = 360 \text{ cubic units}$$

■

Lesson Exercise 10.68

Refer to Figure 10-48.

(a) Find the volume of the cylinder (it could be the design for a pipe).

(b) Find the volume of the prism (it could be the shape of a wedge of cheese).

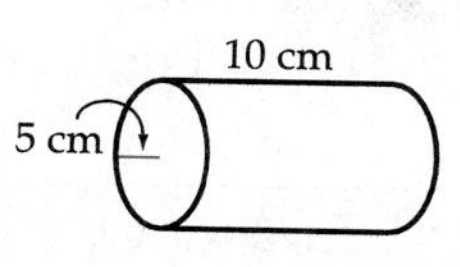

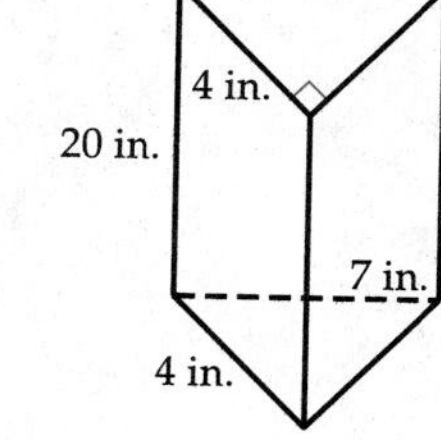

Figure 10–48

Think of the right prism in Figure 10-49 as a stack of extremely thin sheets of paper. The related oblique prism would be obtained by shifting the stack. The right cylinder and the oblique cylinder in Figure 10-49 are related in the same way.

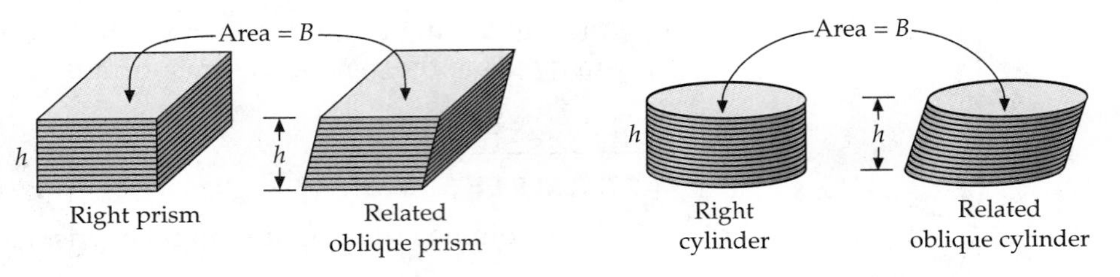

Figure 10–49

Lesson Exercise 10.69

(a) How do the volumes of the right prism and the oblique prism in Figure 10-49 compare?

(b) How do the volumes of the right cylinder and the oblique cylinder in Figure 10-49 compare?

Zu Chongzhi (A.D. 429–500) and his son Zu Geng of China were probably the first mathematicians to use the general relationship between a pair of figures like those in Figure 10-49. Bonaventuri Cavalieri (1598–1647), an Italian mathematician, later stated the general principle, and it is named after him.

Cavalieri's Principle

Two solids with bases in the same plane have equal volumes if every plane parallel to the two bases intersects the two solids in cross sections of equal area.

Lesson Exercise 10.69 and Cavalieri's Principle suggest why the volume formula for right prisms and right cylinders applies to all prisms and cylinders.

Volume of Any Prism or Cylinder

The volume V of a prism or cylinder that has a base of area B and height h is

$$V = Bh$$

VOLUMES OF PYRAMIDS AND CONES

Pyramids and cones have the same general volume formula. Their volume formula is closely related to the volume formula for cylinders and prisms.

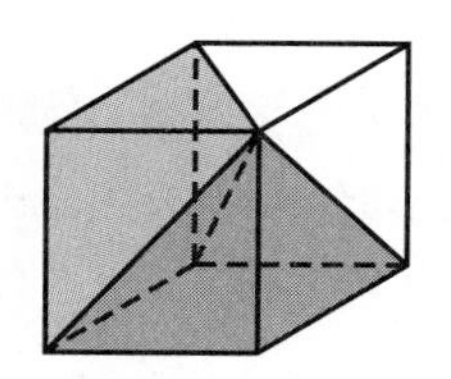

Figure 10–50

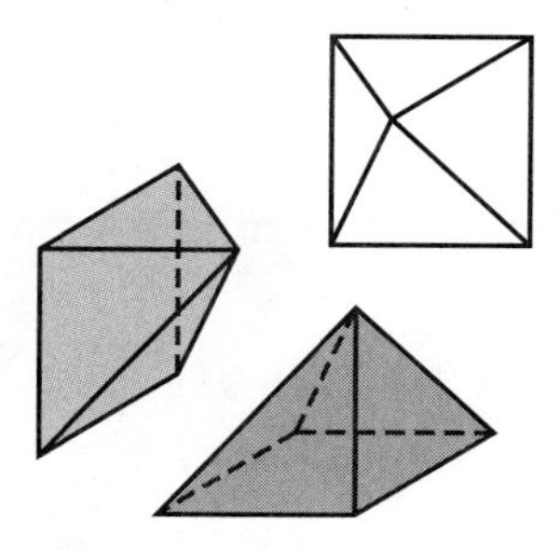

Figure 10–51

The following example shows how the volume of a pyramid is related to the volume of a prism with the same base and height. Take a cube, and draw the three diagonals from one vertex (Figure 10-50). The diagonals divide the cube into three congruent pyramids (Figure 10-51). Since the three pyramids are congruent, the volume of each pyramid is $\frac{1}{3}$ the volume of the cube. So

$$V_{\text{pyramid}} = \frac{1}{3}V_{\text{cube}}$$

This suggests the general relationship between the volumes of pyramids and prisms that have the same base and height:

$$V_{\text{pyramid}} = \frac{1}{3}V_{\text{prism}} = \frac{1}{3}Bh$$

If you have models available, you can see that a prism holds about three time as much water or rice as a pyramid with the same base and height. The volumes of cylinders and cones are related in the same way (Figure 10-52).

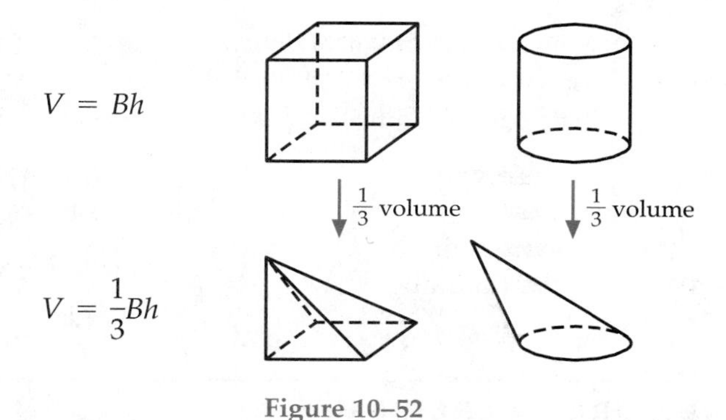

Figure 10–52

Volume of a Pyramid or Cone

The volume V of a pyramid or cone that has a base of area B and height h is

$$V = \frac{1}{3}Bh$$

The general volume formula $V = \frac{1}{3}Bh$ can be used to obtain the volumes of cones and pyramids.

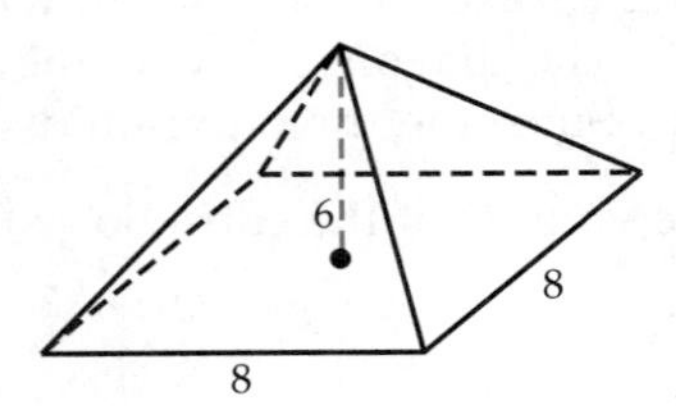

Figure 10–53

Lesson Exercise 10.70

What is the formula for the volume of a cone that has radius r and height h?

Lesson Exercise 10.71

Find the volume of the square pyramid in Figure 10-53. (It could be a building design.)

ANSWERS TO SELECTED LESSON EXERCISES

10.58 40 square units

10.59 (b) $18 \cdot 12 = 216$ ft^2
(c) $18 \cdot 12 + 2 \cdot 12 \cdot 8 + 2 \cdot 18 \cdot 8 - 2 \cdot 3 \cdot 5 - 4 \cdot 7 = 638$ ft^2
(d) 2

10.60 660 m^2

10.61 (a) 3 (b) Circular (c) Rectangular

10.62 (a) $2\pi r$ by h
(b) $A = 2\pi rh + 2\pi r^2$ square units

10.63 (a) ph
(b) $A_{\text{surface}} = ph + 2B$

10.64 (a) 12 (b) 38 square units
(c) V and A would be the same.

10.65 (a) 4 (b) 12 (c) 3 (d) 36 cm^3
(e) 4 cm, 3 cm, 3 cm
(f) 4 cm $\times$ 3 cm $\times$ 3 cm = 36 cm^3

10.66 Freezer A (\$27.56/ft^3)

10.67 (a) Circular (b) πr^2 (c) $V = \pi r^2 h$

10.68 (a) 250π cm^3
(b) $20\left(\frac{1}{2} \cdot 4\sqrt{33}\right) = 40\sqrt{33}$ in.^{3}

10.69 (a) They are equal. (b) They are equal.

10.70 $V = \frac{1}{3}\pi r^2 h$

10.71 $\frac{1}{3}(8 \cdot 8 \cdot 6) = 128$ cubic units

10.5 HOMEWORK EXERCISES

Basic Exercises

1. What is the surface area of the following rectangular prism?

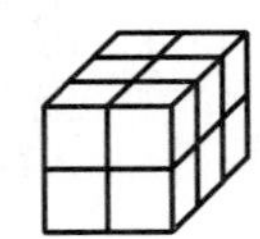

2. What is the surface area of the following rectangular prism?

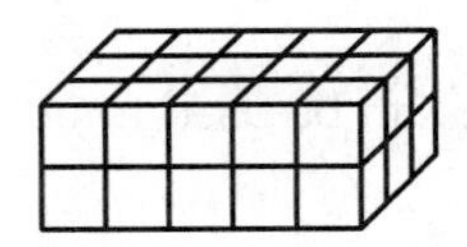

3. Show how you would arrange 20 congruent cubes to obtain a rectangular prism with
(a) the largest possible surface area.
(b) the smallest possible surface area.
4. Explain in terms of measurement why a potato will bake more quickly if you cut it into smaller pieces and cook them separately.
5. How much paper is needed to cover the box shown?

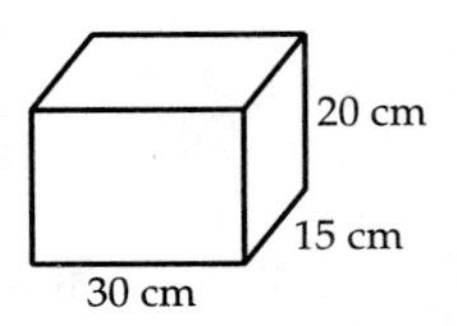

6. A rectangular prism has dimensions l ft, w ft, and h ft. What is its total surface area?

7. You want to carpet the steps as shown by the colored areas.

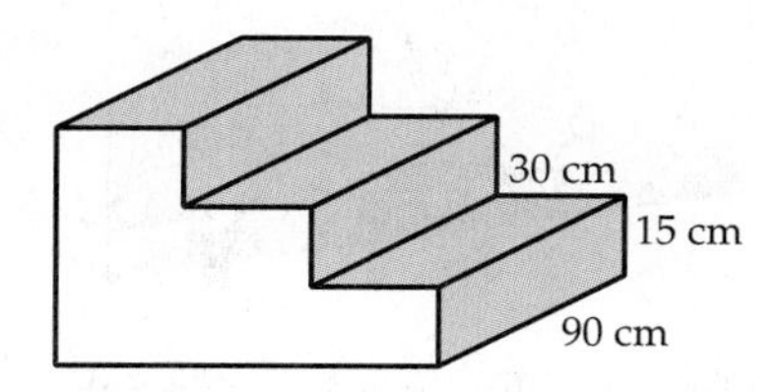

(a) How many square centimeters of carpet will you need?
(b) If the carpet costs $20 per square *meter*, how much will your carpet cost?

8. A box-shaped room is 16 ft long, 10 ft wide, and 8 ft high. There is one window that is 3 ft by 6 ft and a doorway that is 4 ft by 7 ft. If 1 gallon of paint covers 450 ft^2, how many gallon cans would you buy to paint the four walls and the ceiling?

9. Carpet A sells for $6.99 per square yard. Carpet B is a remnant 9 ft by 12 ft, selling for $99. If you want a carpet that is 9 ft by 12 feet, will it cost less to buy the remnant or to buy carpet A by the square yard?

10. How much nylon (in square feet) would be needed to make the tent shown, including the bottom? (You may approximate square roots to one decimal place.)

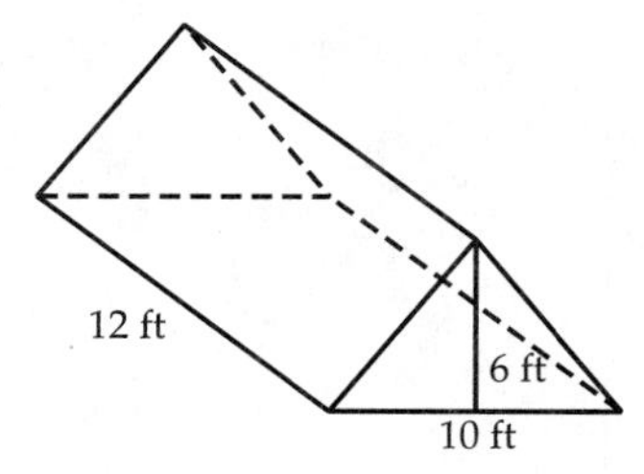

11. Find the total surface area of the triangular prism.

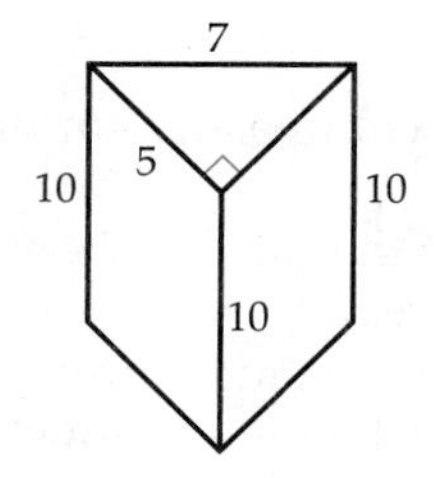

*12.

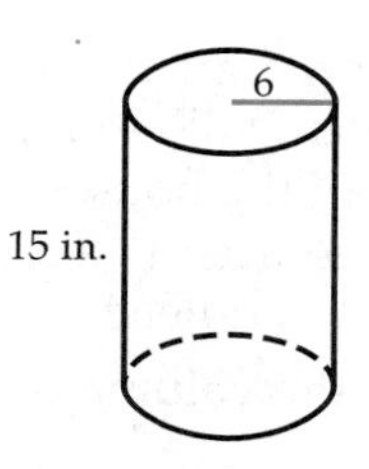

(a) Find the total area of the bases.
(b) The lateral surface can be unwrapped to form a rectangle. What is its area?
(c) What is the surface area of the cylinder?

13. Give two possible radii and heights for a cylinder with a surface area of 40π.

14. Four basic roof designs used in the U.S. are flat, gable, hip, and pyramid hip. Find the total area of each roof shown. Exclude the shaded parts.

(a)

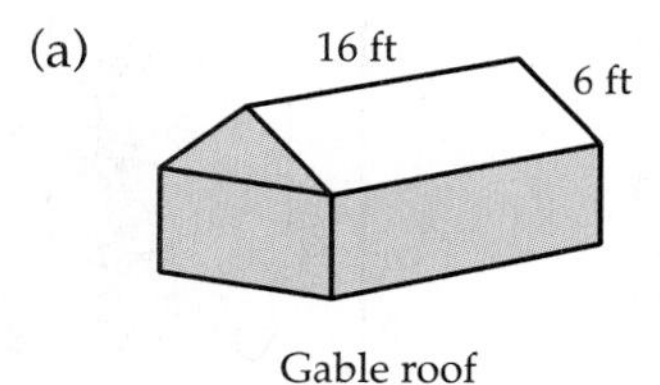

Gable roof

(b)

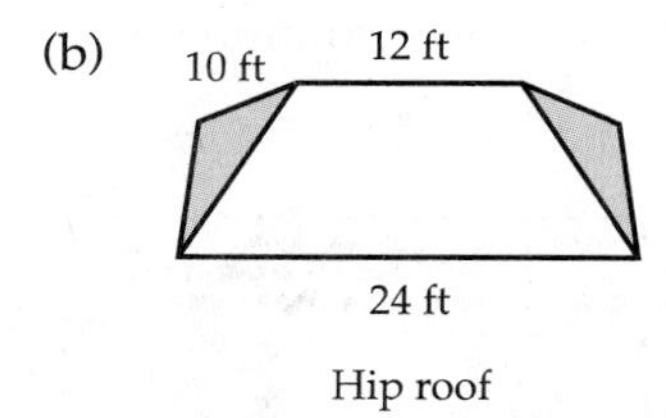

Hip roof

15. (a) Find the surface area of the sides, front, and back of the house. (Include the two triangles but not the roof.)

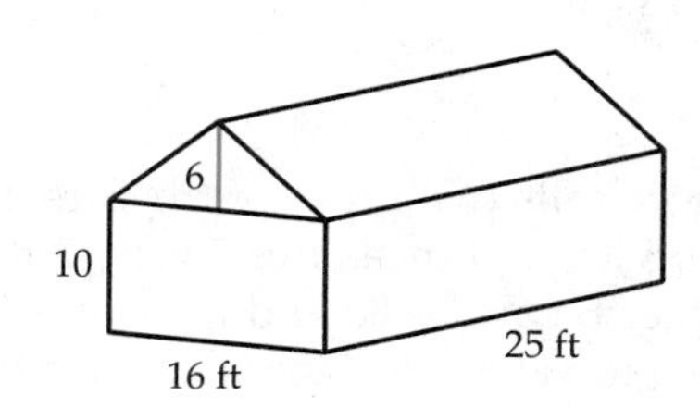

(b) If a gallon of paint covers 400 ft^2, how much paint will you need to paint the outside of the house?
(c) If paint comes in gallon cans, how many cans will you need?

*16. (a) What is the volume of the rectangular prism?

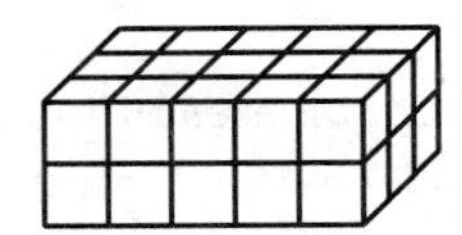

(b) The most common incorrect answer to part (a) is 31. How would a child obtain this answer?

17. A sketch of an apartment building follows. Find its volume.

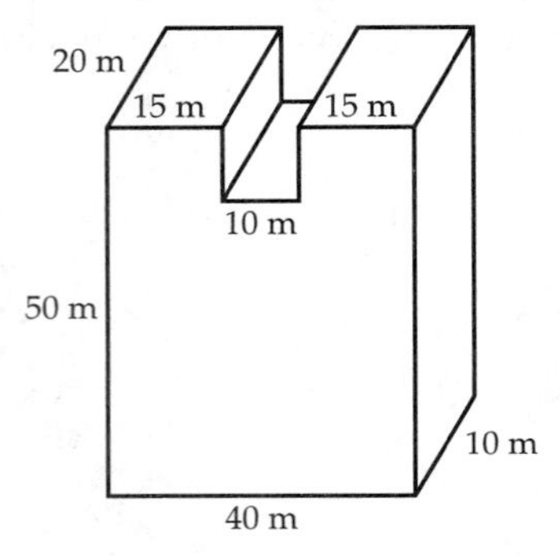

18. What is the volume of the swimming pool?

19. A store sells two types of freezers. Freezer A costs $350 and measures 2 ft by 2 ft by 4.5 ft. Freezer B costs $480 and measures 3 ft by 3 ft by 3.5 ft. Which freezer is a better buy?

20.

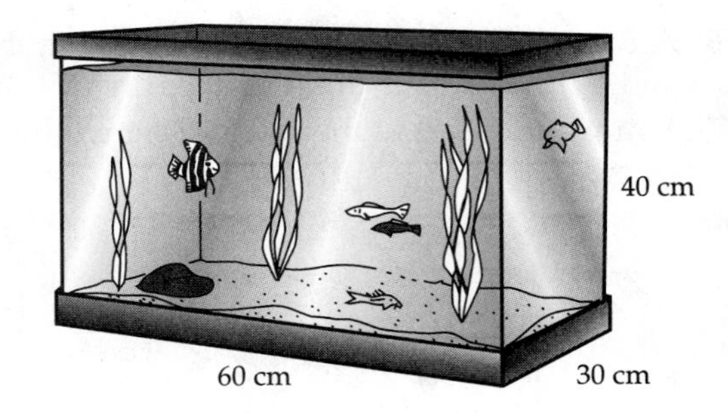

(a) How much water can the fish tank hold?
(b) 1 L = 1000 cm^3. How many liters of water does the fish tank hold?
(c) Small tropical fish need about 1500 cm^3 of living space. How many fish can live in this tank?

21. Is each quantity related more to volume or to surface area?
(a) The amount of paper needed to make a bag
(b) The amount a bag will hold
(c) Your weight

22. Estimate the volume of your refrigerator in cubic meters.

23. (a) Find an open container at home, and compute its surface area.
(b) What volume does the container hold?
(c) If you put a top on the container, would this change the answer to part (a) or (b)?

24. (a) Find the volume and surface area of a box of tissues.
(b) Compute the ratio of the volume to the surface area.
(c) Would the manufacturer want the ratio in part (b) to be high or low?

25. The formula $V = Bh$ applies to what kind of figures?

26. Find the volume of a nickel, to the nearest cubic millimeter.

27. A standard 46.00-oz can of juice has a radius of 5.3 cm and a height of 17.5 cm. What is its volume?

28. One can of juice is twice as tall as a second can but only half as wide. How do their volumes compare?

29. (a) Roll an $8\frac{1}{2}$-by-11-in. sheet of paper into a cylindrical tube. What is the diameter?
(b) Roll the sheet of paper into a cylindrical tube of a different size. What is the diameter?
(c) Which cylinder has the greater volume?

30. Which would appear to hold more, a cube-shaped container or a tall, thin, box-shaped container with the same volume?

31. Find the volume of the oblique prism shown. Its base area is 50 m^2.

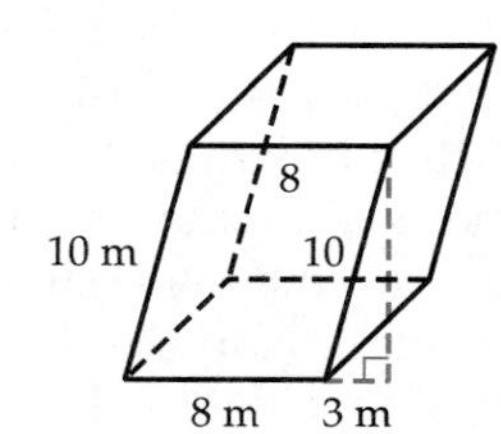

32. What is the volume of the piece of cheese shown?

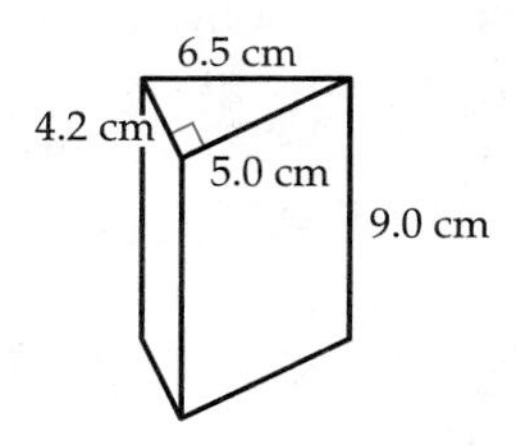

33. Find the volume of the watering trough.

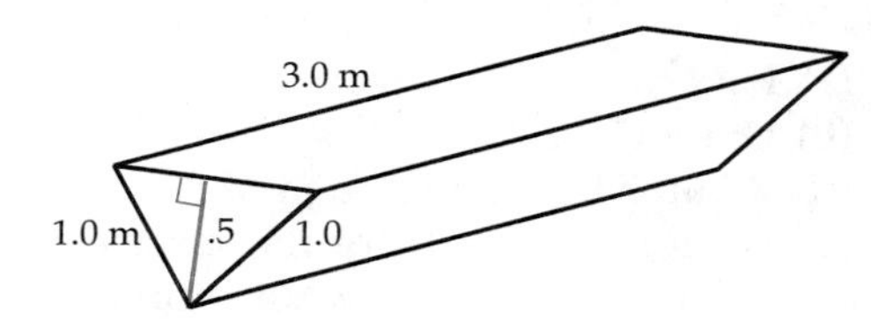

34. A cylindrical water tank has a radius of 6.0 m. About how high must it be filled to hold 400.0 m^3?

35. The formula $V = \frac{1}{3}Bh$ applies to what figures?

36. A cylinder-shaped drinking cup holds ______ times more water than a cone-shaped cup that has the same radius and height.

37. A 15-step staircase is made out of concrete. Three of the steps are shown. What is the volume of the staircase?

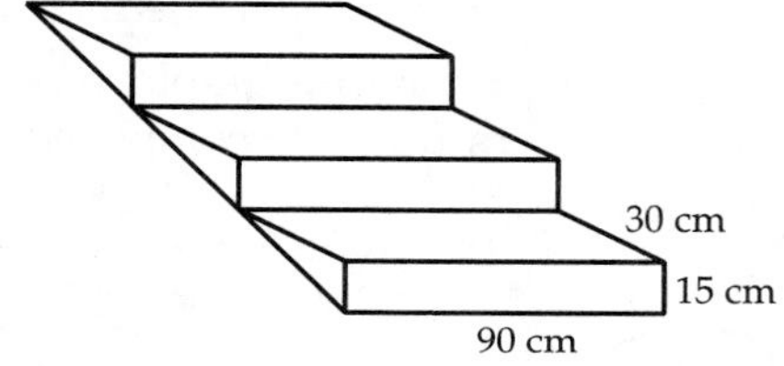

38. (a) How much land is needed for the house itself?
(b) What is the volume of the house?

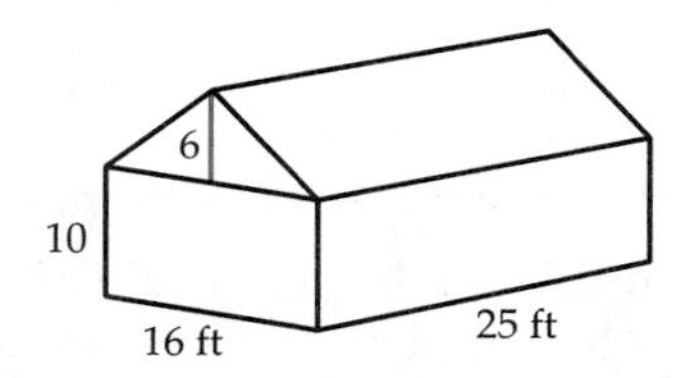

39. If 200 in.^{3} of icing is spread evenly onto a cake to a thickness of 0.5 in., about how much area will it cover?

40. A cylindrical hot water heater has a radius of 35 cm and a height of 170 cm. How many liters of water can it hold? (1 cm^3 = 1 mL.)

41. Find the volume of a 12-oz (355-mL) soda can by
(a) measuring the radius and height and using a formula.
(b) using the conversion 1 mL = 1 cm^3.

42. Find the approximate volume of a standard soup can in cubic centimeters.

43. The largest pyramid in the world is in Cholula, Mexico. Its base area is 1,960,000 ft^2, and its height is 177 ft.
(a) What is its volume?

(b) About how many apartments, each 30 ft by 25 ft by 10 ft, would it take to make the same volume?

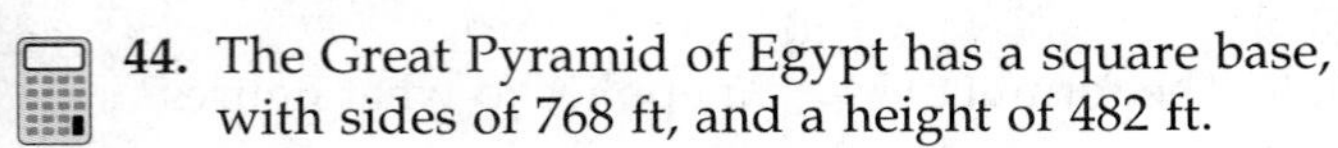

44. The Great Pyramid of Egypt has a square base, with sides of 768 ft, and a height of 482 ft.
(a) What is its volume?
(b) About how many apartments, each 30 ft by 25 ft by 10 ft, would it take to make the same volume?

45. An ice cream cone has a diameter of 2.0 in. and a height of 5.5 in. What is its volume?

46. A cone-shaped cup is filled to half its height. What fraction of the cup is filled?

47. (a) Complete the third example, repeating the error pattern from the completed examples.

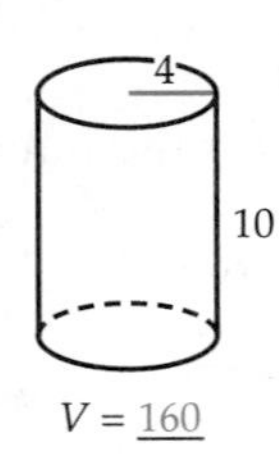

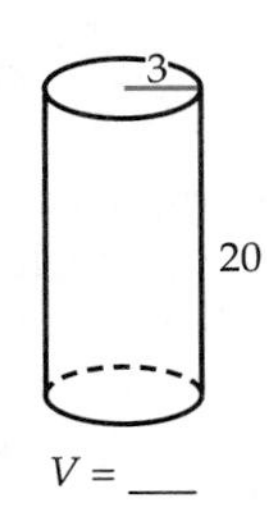

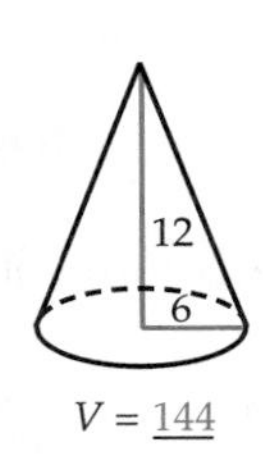

(b) Describe the error pattern.

48. A conical tank has an inside diameter of 20 ft and a height of 12 ft. The tank is filled with liquid to a height of 9 ft. How much liquid is in the tank?

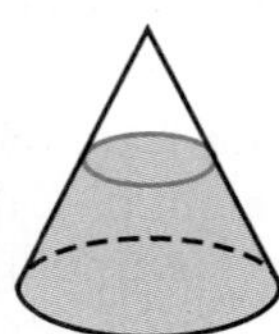

49. A can of 3 tennis balls is 19.3 cm high with a radius of 3.8 cm, and each tennis ball has a radius of 3.2 cm. What percent of the can is occupied by the tennis balls? For a sphere, volume $V = \frac{4}{3}\pi r^3$. Use $\pi = 3.14$.

50. A balloon takes 3 seconds to inflate to a radius of 4 in. Assuming a constant flow of air, after 6 seconds, it has a radius of ________ in. (The answer is not 8.)

Extension Exercises

51. Consider the following problem. "How heavy is a firefighter's hose? A large hose is 100 ft long and 5 in. in diameter when full. Empty, the hose weighs 200 lb. If 1 gallon (231 in.^{3}) of water weighs 8 lb, how much does a full hose weigh?" Devise a plan and solve the problem.

52. Given the following floor/wall plan for an apartment, find the cost of redecorating it if you cover the ceiling with 1-ft^2 tiles costing \$.70 each, carpet the floor with carpet costing \$8 per square yard, and paint the walls with paint costing \$9 per gallon. Each gallon of paint covers 250 ft^2.

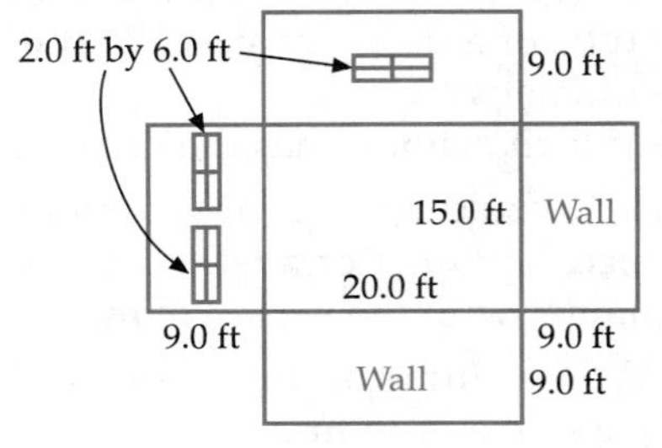

53. Suppose that you want to manufacture boxes with square bases as cheaply as possible. The volume of each box must be 2000 cm^3. The stronger material on the top and bottom costs \$.0002/cm^2, and the material for the sides costs \$.0001/cm^2. Find the dimensions of the box that minimize the total cost. (Guess and check.)

54. A plumber needs to deliver 150 steel pipes to an office building. Her truck can carry 800 kg. Each pipe is 120 cm long, with an outer diameter of 7 cm and an inner diameter of 5 cm. The steel weighs 7.7 g/cm^3. How many trips will the plumber need to make?

55.

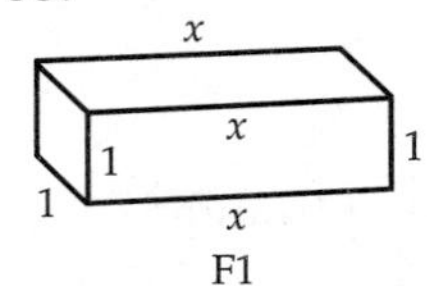

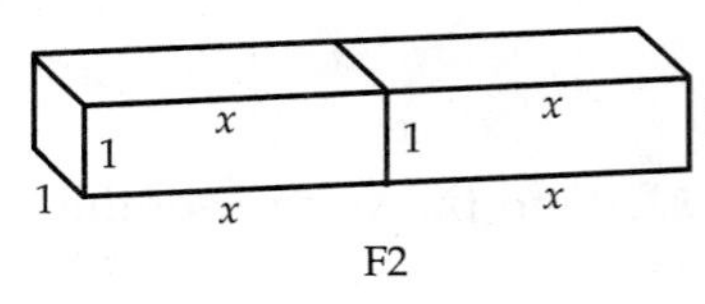

(a) Find the surface areas of F1 and F2.
(b) Draw F3 and find its surface area.
(c) What is the surface area of F10?
(d) What is a formula for the surface area of FN, in which N is a whole number?

56.

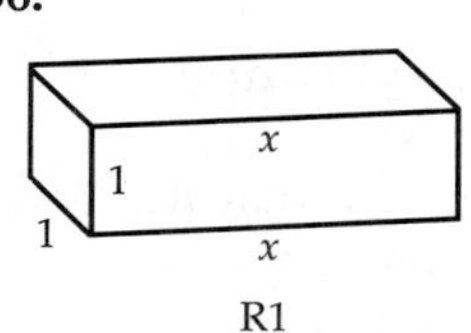

R1

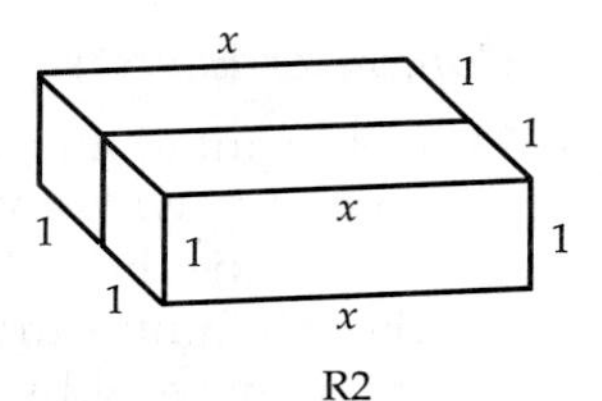

R2

(a) Find the surface areas of R1 and R2.
(b) Draw R3 and find its surface area.
(c) What is the surface area of R10?
(d) What is a formula for the surface area of RN, in which N is a whole number?

57.

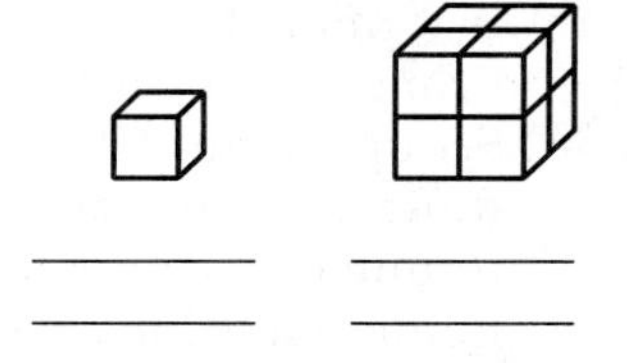

Length of edge ________ ________
Volume ________ ________
Surface area ________ ________

(a) Fill in the blanks.
(b) As the cube grows in size, which increases faster, surface area or volume?

58. The lengths of the edges of a cube are increased by 20%.
(a) The surface area increases by ________ %.
(b) The volume increases by ________ %.

59. A box with height h contains 6 solid metal cylinders, each with radius r. What percentage of the box is empty?

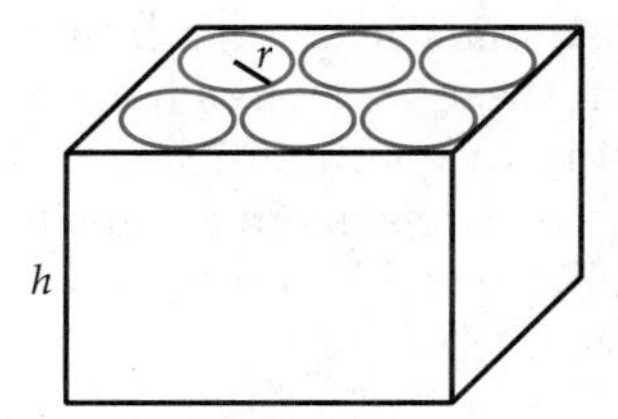

60. A log has a diameter of 36 cm and a height of 60 cm. Give the dimensions of the largest rectangular solid with a square base that can be cut from it.

61. A cylindrical glass jar contains juice. Without measuring, how can you tell when the jar is half full?

62. Consider the following problem. "What size open boxes can you construct from a piece of cardboard 25 cm by 25 cm if you cut off 4 squares from the corners and fold?"

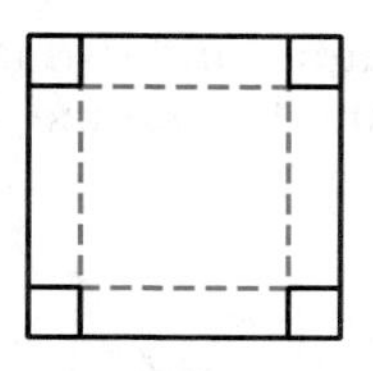

(a) What size squares can be cut from the corners?
(b) Devise a plan and solve the problem.

63. A cylindrical pipe is hollow inside. What is the volume of the pipe material?

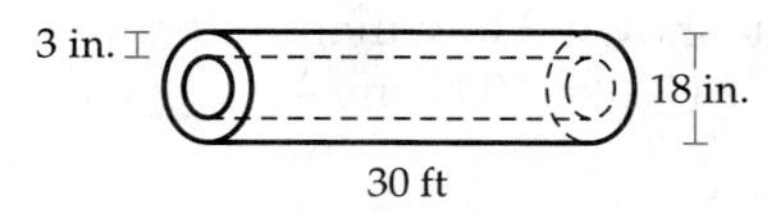

64. A cylindrical pipe has length L, inner radius I, and overall radius R. What is the volume of the pipe material?

65. A page in this book is a rectangular prism.
(a) What are its dimensions? (*Hint:* To find the thickness of a page, measure more than one page.)
(b) What is the volume of a page?

66. A city water inspector who is 6.0 ft tall inspected a spherical water tank. His head touched the tank when his feet were 18.0 ft away from the lowest point. How did he use this information to find the volume of the tank? (*Hint:* Use the Pythagorean Theorem.)

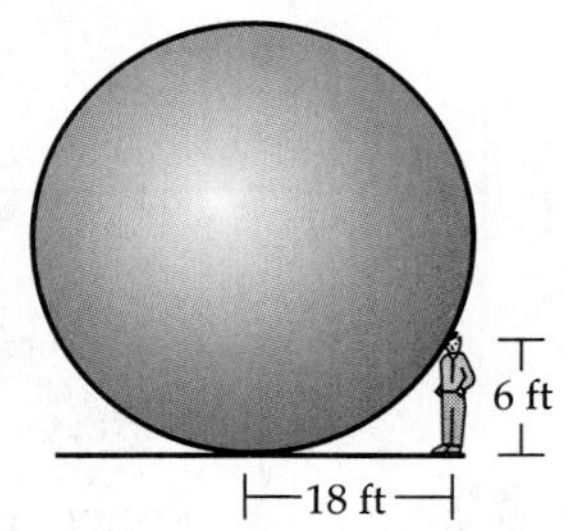

67. A garbage can has the dimensions shown. Approximate its volume.

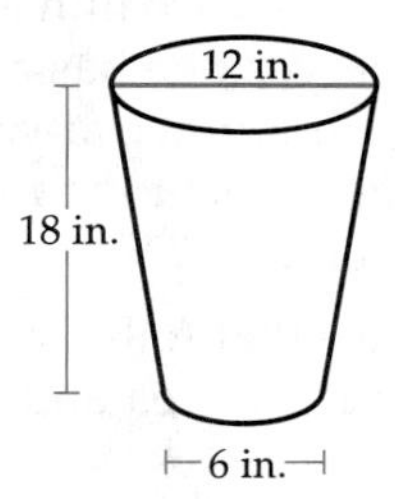

68. Find the volume and surface area of the square pyramid. (Assume that it has rotational symmetry.)

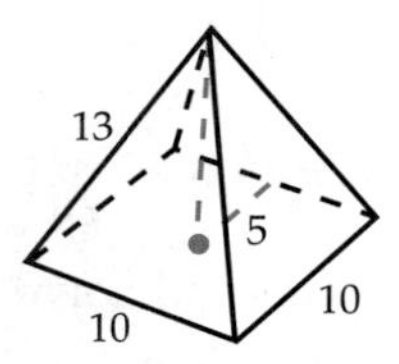

69. A swimming pool 20 m wide has the cross section shown. How many liters of water does it hold? ($1 \text{ L} = 1000 \text{ cm}^3$.)

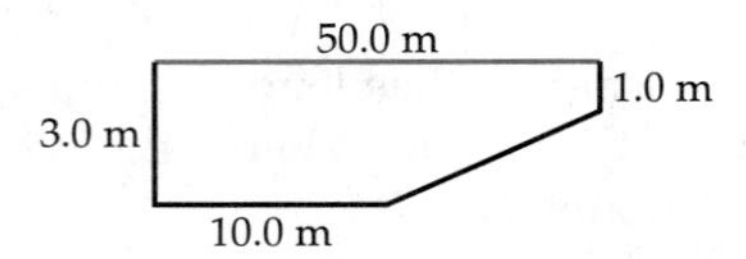

70. In each graph, the region is revolved around the x-axis. Find the volume of the resulting figure.

(a)

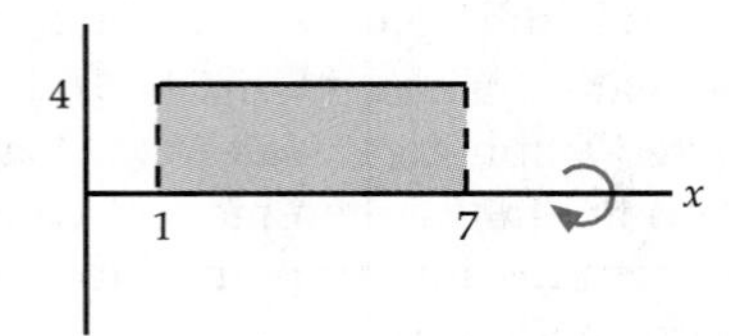

(b)

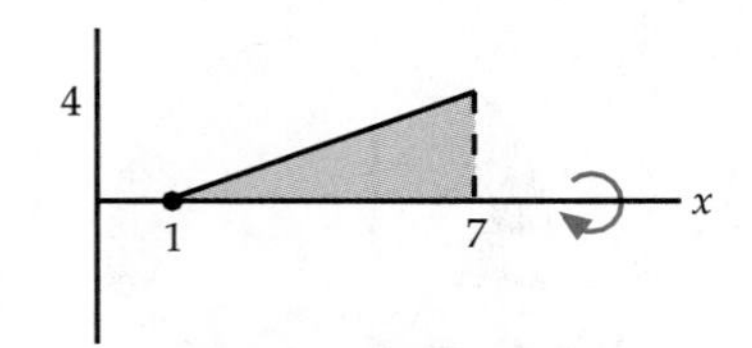

71. The surface area A of a sphere of radius r is $A = 4\pi r^2$. Suppose you are manufacturing two different-sized balls of the same material and thickness. The smaller ball has a radius of 2 in., and the larger one has a radius of 4 in.

(a) How do their surface areas compare?

(b) How do the costs of the material for the two balls compare?

72. Write a sentence that tells the difference between the surface area and volume of a prism.

Computer Exercise

73. You want to manufacture a cylindrical container with a volume of 84 cm^3. To determine the radius and height that will give the minimum surface area, use a spreadsheet.

(a) Why would you want to find the minimum possible surface area of a container?

(b) A formula for the height h of a cylinder in terms of the volume V and radius r is $h =$ ________.

(c) A formula for the surface area A of a cylinder is $A =$ ________.

(d) Set up a spreadsheet as follows. Try different values of r, and use your formulas to compute h and A. Find the minimum value of A.

	A	B	C	D
1	**Volume**	r	h	A
2	84			
3	84			
4	84			

(e) Give the values of r and h that result in the minimum surface area.

Special Exercises

74. In the customary system, 1 pint = 1 lb of water, but these measures have no simple relationship to volume measures (such as 1 ft^3). The metric system relates volume (cm^3), liquid volume (mL), and mass (g) of water in a simple way.

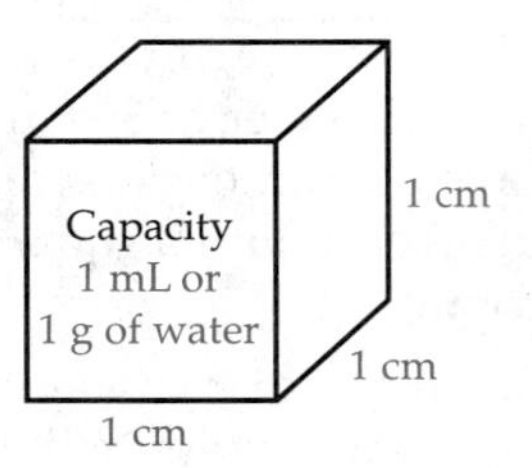

(a) What is the volume of the container, in cubic centimeters?

So 1 mL of water = 1 g of water = 1 cm^3 of water! Use this information to answer the following questions.

(b) How much would 4 L of water weigh?
(c) A fish tank is 70 cm by 50 cm by 40 cm. About how many liters of water are needed to fill it?
(d) How much would 1 m^3 of water weigh?
(e) What is the liquid volume of 1 m^3 of water?

75. If 1 cm of rain falls uniformly on a field 100 m by 100 m, how much does this rainwater weigh? (1 cm^3 weighs 1 g.)

76. In order to calibrate a cylindrical beaker with an inside diameter of 2.2 cm to show cubic centimeters, how far apart should the calibration marks be?

77. (a) Design a floor plan for a one-story, two-bedroom apartment. Use a scale in which 1 cm represents 1 m. The apartment must fit on a 12-by-12-m square.
(b) Decide how high the walls will be and how many windows and doors the apartment will have. All the walls should be the same height.
(c) Estimate the cost of painting the interior of the apartment.

78. (a) Take an $8\frac{1}{2}$-by-11-in. piece of paper and cut out off a 1-inch square from each corner.
(b) Fold and tape to make an open box (no top).
(c) Find the volume of the box.
(d) Repeat parts (a)–(c), but cut off a 2-inch square from each corner.
(e) Can you make a different open box with a larger volume?

79. Go to the supermarket and collect data on the radii and heights of cylindrical cans. Compute the ratio of h to r for each can.

10.6 LENGTHS, AREAS, AND VOLUMES OF SIMILAR FIGURES

Do you want to invest in some imaginary real estate? It's safer than investing in real real estate. Consider the following situation.

Lesson Exercise 10.72

You are offered two right-triangular pieces of land that are the same shape. One has sides measuring 30 ft, 40 ft, and 50 ft. The other has sides measuring 60 ft, 80 ft, and 100 ft. The second piece of land costs 3 times as much as the first. Guess which piece of land is a better buy.

SIMILAR PLANE FIGURES

The preceding exercise involves two similar figures. The two triangles are the same shape, but all the sides of the second triangle are twice as long as the corresponding sides of the first. How are the areas of the triangles related?

Lesson Exercise 10.73

(a) If you have studied the Right-Triangle Test, show that both triangles in Figure 10-54, on page 628, are right triangles.
(b) Find the area of each right triangle.

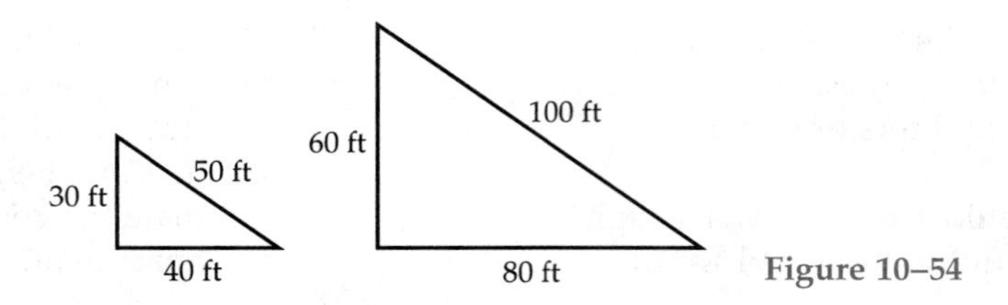

Figure 10–54

(c) The area of the larger triangle is ________ times the area of the smaller triangle.

(d) Find the perimeter of each triangle.

(e) The perimeter of the larger triangle is ________ times the perimeter of the smaller triangle.

(f) The second triangular piece of land costs 3 times as much as the first. Which is a better buy?

The triangle with twice the lengths had four times the area. What is the general relationship between all corresponding lengths and areas of similar figures? If the suspense is too much for you, do not hesitate to try the next exercise.

Lesson Exercise 10.74

The drawings in Figure 10-55 show two sizes of a photocopy. Enlarging a photocopy creates a *similar* rectangle.

(a) The ratio of corresponding sides is ____:____.

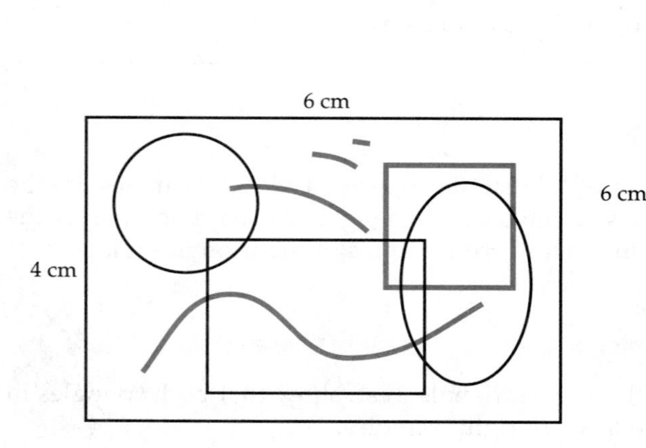

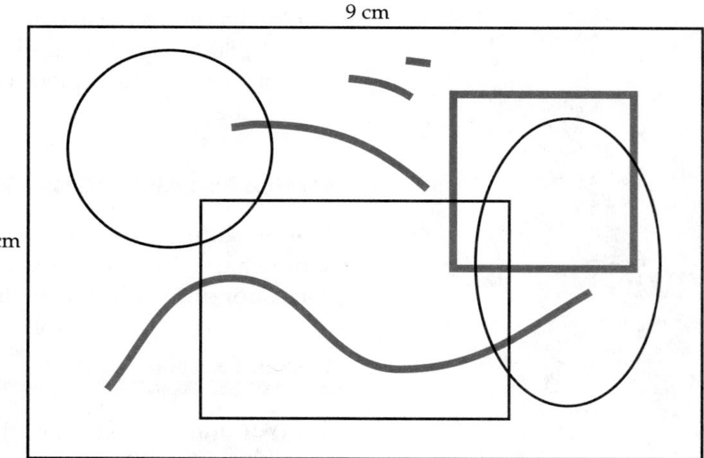

Figure 10–55

(b) The areas of corresponding rectangles are ______ and ______.

(c) The ratio of their areas is ________.

(d) The perimeters of the two rectangles are ______ and ______.

(e) The ratio of their perimeters is ________.

Try to generalize the results from Lesson Exercises 10.73 and 10.74 in the following exercise.

Lesson Exercise 10.75

Two similar figures have a ratio of corresponding sides $m:n$.

(a) What is the ratio of their areas?

(b) What is the ratio of their perimeters?

The results of the preceding exercises suggest the following rule.

Lengths and Areas of Similar Plane Figures

Figures A and B are similar plane figures with a ratio of corresponding length measurements $m:n$. The ratio of corresponding area measurements is $m^2:n^2$, and the ratio of their perimeters is $m:n$.

Use these properties in the following exercise.

Lesson Exercise 10.76

Two similar triangles have a ratio of corresponding sides of 4:1. What are the ratios of their areas and their perimeters?

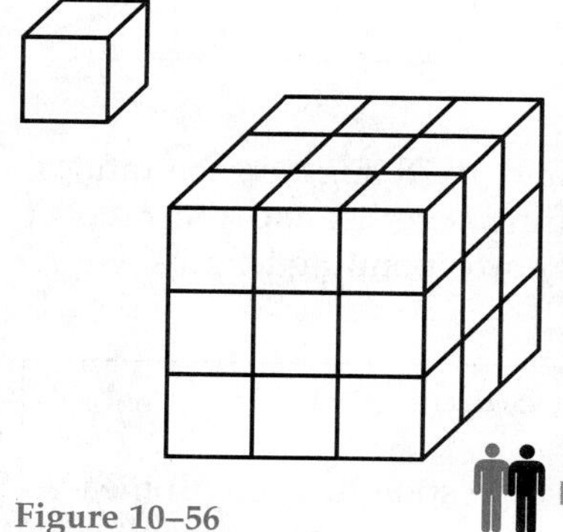

Figure 10–56

SIMILAR SPACE FIGURES

Relationships between corresponding areas and volumes in similar solids indicate the relative strengths, heat loss, and weights of animals with similar shapes. After examining relationships in similar geometric solids, you will use this knowledge to study similar shapes in nature.

Lesson Exercise 10.77

Suppose that you have a 1-cm cube and a 3-cm cube. (Use wooden blocks if they are available. See Figure 10-56.)

(a) What is the ratio of corresponding edge lengths?

(b) The surface areas of the two figures are ________ and ________.

(c) What is the ratio of their surface areas?

(d) Compute the volume of each figure.

(e) What is the ratio of their volumes?

Next, consider two similar rectangular prisms and see whether you can generalize the results.

Lesson Exercise 10.78

Consider the two rectangular prisms in Figure 10-57.

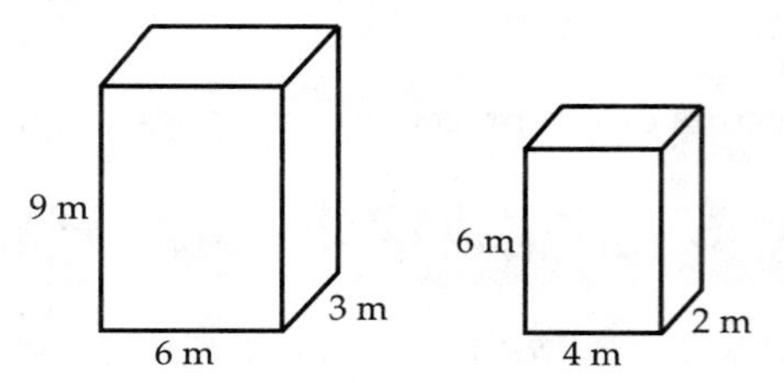

Figure 10–57

(a) What is the ratio of corresponding edge lengths?

(b) What is the ratio of their surface areas?

(c) Guess how the ratio of surface areas is related to the ratio of corresponding edges in similar solids.

(d) What is the ratio of their volumes?

(e) Guess how the ratio of volumes is related to the ratio of edges in similar solids.

As your results in Lesson Exercises 10.77 and 10.78 suggest, the ratio of volumes is the cube of the ratio of corresponding lengths, and the ratio of corresponding areas is the square of the ratio of corresponding lengths.

Lengths, Areas, and Volumes of Similar Solids

A and *B* are similar solids with a ratio of all corresponding length measurements of $m:n$. The ratio of all corresponding area measurements is $m^2:n^2$, and the ratio of volumes is $m^3:n^3$.

Use these properties to answer the following questions.

Lesson Exercise 10.79

Two similar solids have corresponding heights of 3 m and 12 m.

(a) What is the ratio of their total surface areas?

(b) What is the ratio of their volumes?

SIMILARITY IN NATURE

Photo courtesy of Film Stills Archives.

Figure 10–58

Do you ever worry that a giant ape might visit your neighborhood (Figure 10-58)? Is this possible? Similar solids provide the answer! But first we'll consider some friendly dolphins (Figure 10-59, on page 632).

The properties of similar solids can be applied to pairs of animals that are approximately the same shape. Young dolphins and adult dolphins are approximately the same shape.

Photo: Rob Mathewson. Courtesy of Department of Library Services, American Museum of Natural History.

Figure 10–59

Lesson Exercise 10.80

What measurements of dolphins are length measurements?

In Lesson Exercises 10.81–10.88, suppose that an adult dolphin is 2 times as long as a young dolphin of the same shape.

Lesson Exercise 10.81

Guess how many times stronger the adult dolphin is.

According to biologist D'Arcy Thompson, "The strength of a muscle depends upon the size of a cross section (slice)" (Figure 10-60), just as the strength of a steel beam depends upon the cross section of the beam.

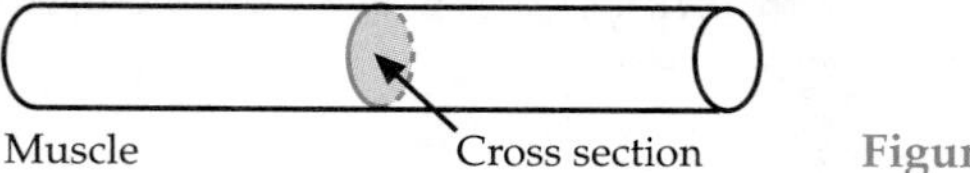

Figure 10–60

The following exercise concerns the relationship between the strength of a cylindrical muscle and the area of its circular cross section.

Lesson Exercise 10.82

(a) What is the formula for the area of a circle?

(b) If the cross section of the young dolphin's muscle is circular, with a radius of 1 in., the area of the cross section is ________.

(c) The radius of the adult's corresponding muscle is 2 in. Its area is ________.

(d) The adult has a muscle cross section that is ________ times larger in area than that of the young dolphin's, so the adult is ________ times stronger.

Lesson Exercise 10.83

The ratio of corresponding lengths in the young dolphin and the adult dolphin is 1:2.

(a) The ratio of strengths is ________.

(b) The ratio of strengths is the same as the ratio of
(1) corresponding lengths (2) surface areas (3) volumes

Another characteristic of interest is weight.

Lesson Exercise 10.84

(a) Guess how many times heavier the adult dolphin in the preceding exercises is than the young dolphin.

(b) Would you guess that weight relationships are the same as length relationships, area relationships, or volume relationships?

(c) Would you like to revise your guess in part (a)? (Last chance.)

In fact, weight corresponds—is proportional—to volume. The following chart shows how a variety of characteristics relate to length, area, or volume.

Classification of Measurements

Lengths—sides, edges, perimeters, heights

Areas—surface area, amount of skin, strength

Volumes—weight

Now it is possible to deduce other facts about the two dolphins. The next two lesson exercises and the next example refer again to the young and adult dolphins.

Lesson Exercise 10.85

How many times more material would it take to make a rubber sweater (same thickness) for the adult dolphin than for the young dolphin?

EXAMPLE 10.6

The adult's waist size is 27 guppy fins (dolphins don't use inches). What is the young dolphin's waist size?

Solution

First determine whether waist size corresponds to length, area, or volume. It corresponds to length. The length ratio for these two dolphins is 2:1. So the young dolphin's waist size is $\frac{1}{2}$ of 27, or 13.5, guppy fins. ■

Lesson Exercise 10.86

The young dolphin can pull a weight of 8 shells with its dorsal fin. The adult can pull a weight of ________ of the same shells with its dorsal fin.

That's enough of the dolphins. Try the following more general question, and then we'll get back to the giant animals that may visit your neighborhood.

Lesson Exercise 10.87

In general, when an animal triples all of its length measurements, its weight becomes about ________ times as much.

STRENGTH VS. WEIGHT

Could a giant ape or giant insect from a horror film exist in real life? The following exercise will help you find out.

Lesson Exercise 10.88

In a horror film, a cockroach eats radioactive brussels sprouts and grows 100 times longer in terms of all of its length measurements!

(a) How many times stronger are the cockroach's legs now?

(b) How many times heavier is the cockroach?

(c) Explain why such a giant cockroach could not even stand on its legs, let alone walk around your kitchen.

As animals get larger, supporting their weight with their own legs becomes more difficult. Consequently, heavy animals such as hippos tend to have legs that are thick compared to the rest of their bodies, and giant fictional apes such as King Kong would be unable to stand.

The tallest person on record was Robert Wadlow. He grew to a height of 8 ft 11 in. Unfortunately, his legs could not support his weight. He wore a leg brace to help support his weight, but his joints became diseased, and he died at the age of 22.

ANSWERS TO SELECTED LESSON EXERCISES

10.73 (a) $30^2 + 40^2 = 50^2$ and $60^2 + 80^2 = 100^2$
(b) 600 ft^2 and 2400 ft^2 (c) 4
(d) 120 ft and 240 ft (e) 2
(f) The second

10.74 (a) 2:3 (c) 4:9 (e) 2:3

10.75 (a) $m^2:n^2$ (b) $m:n$

10.76 Areas, 16:1; perimeters, 4:1

10.77 (a) 1:3 (b) 6 cm^2; 54 cm^2
(c) 1:9 (d) 1 cm^3, 27 cm^3 (e) 1:27

10.78 (a) 3:2 (b) 9:4 (d) 27:8

10.79 (a) 1:16 (b) 1:64

10.82 (b) π in.^{2} (c) 4π in.^{2} (d) 4; 4

10.83 (a) 1:4 (b) 2

10.85 4

10.86 32

10.87 27

10.88 (a) 10,000 (b) 1,000,000
(c) The cockroach's weight has increased 100 times more than its strength.

10.6 HOMEWORK EXERCISES

Basic Exercises

1. The two rectangles shown are similar.

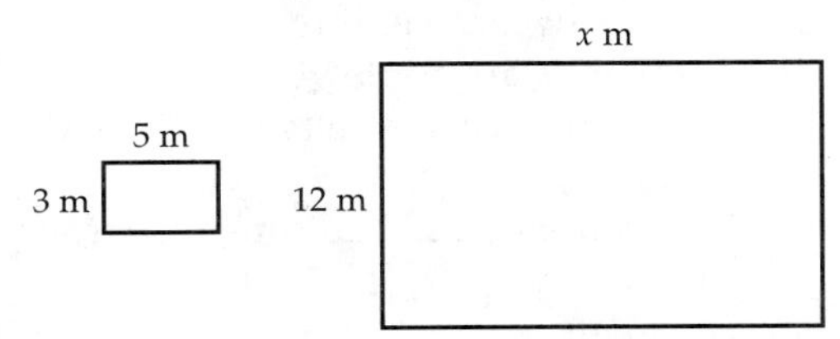

(a) Find x.
(b) The ratio of corresponding sides is ____:____.
(c) The ratio of the areas of the two rectangles is ________.
(d) The ratio of the perimeters of the two rectangles is ________.

2. The two triangles shown are similar.

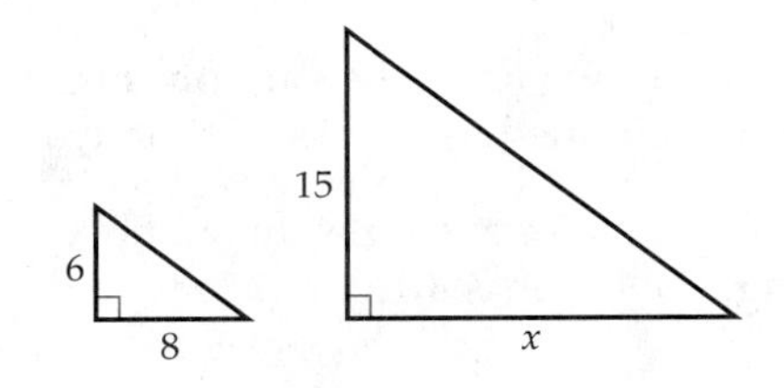

(a) Find x.
(b) The ratio of corresponding sides is ________.
(c) The ratio of the areas of the two triangles is ________.
(d) The ratio of the perimeters of the two triangles is ________.

3. A scale drawing of a stop sign and the actual stop sign have a ratio of corresponding sides $m:n$.
(a) What is the ratio of their perimeters?
(b) What is the ratio of their areas?

***4.** At the Pizza Chalet, a small pizza has an 8-in. diameter, and a medium pizza has a 12-in. diameter.
(a) What is the ratio of their areas?
(b) What is the ratio of their perimeters?
(c) The medium pizza should cost about ________ times more than the small pizza.

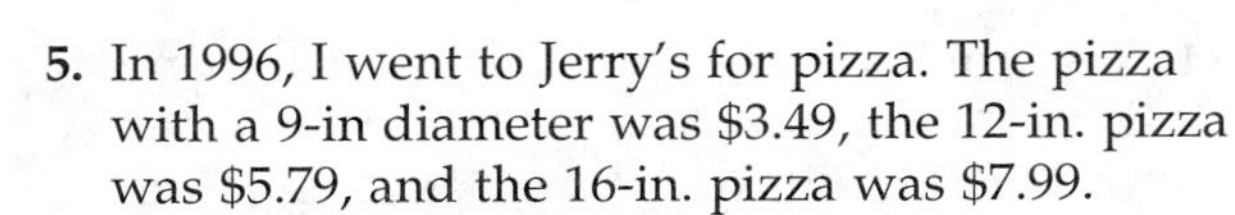

5. In 1996, I went to Jerry's for pizza. The pizza with a 9-in diameter was $3.49, the 12-in. pizza was $5.79, and the 16-in. pizza was $7.99.
(a) Which was the best buy?
(b) Which was the worst buy?

6. A television screen with a 20-in. diagonal is a rectangle about 12 in. by 16 in. A television screen with a 12-in. diagonal is a rectangle about 7.2 in. by 9.6 in. Which would be a better buy, a 20-in. color television for $380 or a 12-in. color television for $125?

7. Consider the two rectangular prisms shown.

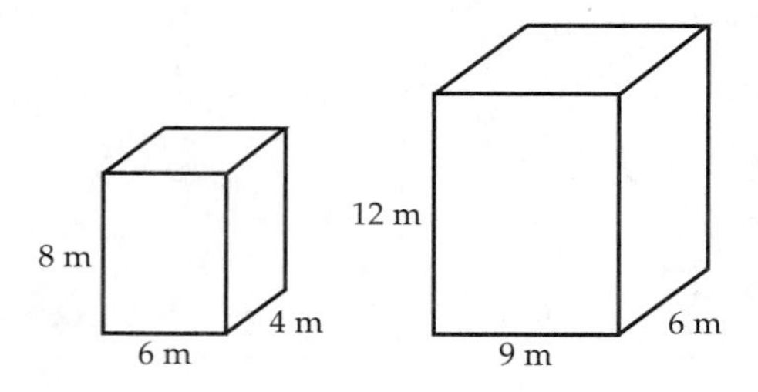

   (a) What is the ratio of their corresponding edge lengths?
   (b) The surface areas of the two figures are ________ and ________.
   (c) What is the ratio of their surface areas?
   (d) How is the ratio of their surface areas related to the ratio of their corresponding edges?
   (e) Compute the volume of each figure.
   (f) What is the ratio of their volumes?
   (g) How is the ratio of volumes related to the ratio of edges?

8. Two similar solids have a ratio of corresponding heights $x:y$.
   (a) What is the ratio of their surface areas?
   (b) What is the ratio of their volumes?

9. The Earth has about 16 times the surface area of the moon.
   (a) What is the ratio of their circumferences?
   (b) What is the ratio of their volumes?

10. Two similar hexagons have corresponding heights of 3 m and 4 m. If the area of the smaller hexagon is 15 cm^2, what is the area of the larger hexagon?

11. Two similar cones have heights of 4 m and 7 m.
   (a) If the radius of the smaller cone is 2 m, what is the radius of the larger cone?
   (b) The surface area of the larger cone is ________ times larger than the surface area of the smaller one.
   (c) What is the ratio of the volume of the smaller cone to the volume of the larger cone?

12. Consider the following problem. "Two similar prisms have total surface areas of 100 cm^2 and 225 cm^2. If the smaller prism has a height of 10 cm, what is the height of the larger prism?" Devise a plan and solve the problem.

13. The ratio of strengths in similar solids is the same as the ratio of corresponding
   (a) lengths (b) areas (c) volumes

14. If a child grows so that all of her length measurements double, her new clothes will require ________ times as much of the same material as her old ones.

15. A scale model of a car is built in the same shape and of the same materials as the actual car, but it is only 1/8 as long. How many times heavier is the actual car?

16. Suppose a giant exists who is the same shape as you, except that all of the giant's length measurements are triple yours.
   (a) The giant's waist size is ________ times yours.
   (b) The giant is ________ times as strong as you are.
   (c) The giant weighs ________ times as much as you do.

17. Two animals are similar in shape. One weighs 125 lb; the other weighs 64 lb.
   (a) How many times longer is the larger animal?
   (b) How many times stronger is the larger animal?
   (c) How many times heavier is the larger animal?
   (d) If the two animals walk on their legs, which one has an easier time walking?

18. Two animals are similar in shape. One is 6 ft long; the other is 4 ft long.

(a) How many times heavier is the larger animal?
(b) How many times more skin surface does the larger animal have?

19. In a scale model of a building, 1 ft represents 40 ft. The volume of the scale model is 18 ft^3. What is the volume of the building?

20. Explain why a giant cockroach could not exist.

21. In *Gulliver's Travels*, Gulliver visits the tiny Lilliputians. The Lilliputian emperor finds out that all of Gulliver's length measurements are about 12 times the corresponding measurements of an average Lilliputian. The emperor says that Gulliver will need to eat as much food as 1728 Lilliputians! Explain the emperor's reasoning.

***22.** In another scene in *Gulliver's Travels* (see the preceding exercise), the emperor wants to make a suit for Gulliver using the same fabric that the tiny Lilliputians wear. How many times more material will he need for Gulliver's suit than for an average Lilliputian's suit?

23. Lilliputians (see the preceding exercises) are 6 inches tall. About how much does each weigh?

24. A regular can of Blando ravioli is 12 cm tall and sells for $1.25. A large can is the same shape and 16 cm tall, and it sells for $2.
(a) How many times more ravioli does the large can hold?
(b) Which can is a better buy?

25. A child's bicycle might have a 20-in. wheel diameter, while an adult's bicycle might have a 26-in. wheel diameter.
(a) Is the cost of a bicycle proportional to length, area, or volume?
(b) If the bicycles are similar and the child's bicycle costs $60, what is a fair price for the adult's bicycle?

26. Similar triangles can be used to measure objects indirectly. Lyle wants to measure the height of a tree. He stands 20 yards from the tree and sights the top of the tree through a tube at an angle of 35°.

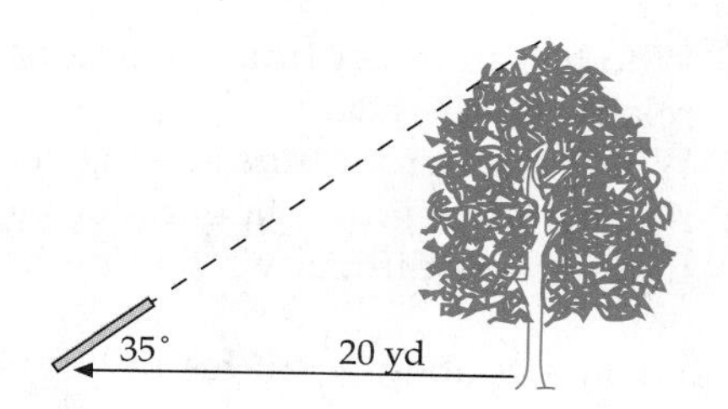

(a) Using a ruler and protractor, make a scale drawing, with 1 in. representing 5 yd.
(b) Explain why the triangle in your scale drawing is similar to the triangle it represents.
(c) About how tall is the tree?

Extension Exercises

27. Warm-blooded animals such as humans must maintain a reasonably constant body temperature. Animals must balance the heat gained from eating (calories) with the heat lost through their skin.
(a) Heat gained from eating is proportional to an animal's
(1) length (2) area (3) volume
(b) Heat lost through skin surfaces is proportional to an animal's
(1) length (2) area (3) volume

28.

(a) Baby and Daddy are similar in shape. Complete the chart.

	Length	Surface Area (where heat is lost)	Weight (how heat is gained)
Baby	1 m	1 square unit	1 cubic unit
Daddy	4 m		

(b) Who has an easier time gaining heat relative to the heat lost?
(c) Who has a harder time staying cool?
(d) As an animal grows larger, does it have more or less difficulty retaining heat?

29. A person eats about $\frac{1}{50}$ of his or her weight each day. Would a smaller animal such as a mouse need to eat a larger or smaller fraction of its body weight each day in order to stay warm enough?

30. A large animal such as a hippopotamus or an elephant has more weight relative to surface area than a person. What problem does this create for these large animals?

31. The following table gives men's results from the 1989 U.S. Weightlifting Championships.

Maximum weight of weightlifter (kg)	60.0	67.5	75.0	82.5	90.0
Snatch event weight lifted (kg)	102.5	120.0	132.5	145.0	150.0

Explain how this table supports the ideas presented in this section about weight and strength.

Computer Exercises

32. Use drawing software to investigate the relationship between the lengths of sides and the perimeters of two similar triangles. Write a generalization of your results.

33. Use drawing software to investigate the relationship between the lengths of sides and the areas of two similar triangles. Write a generalization of your results.

SUMMARY

Nearly every industrialized nation except the United States uses the metric system. In the United States, the metric system is used in most businesses other than companies dealing in building trades or consumer goods. Why? Because it is easier to work with. All metric conversions involve powers of 10, and the same set of prefixes is used for most metric measurements. In learning a new measurement system such as the metric system, it helps to begin by learning some reference measures for the units.

Some carpentry, building, and consumer problems require finding length, area, or volume. Area is the measure of the surface of a solid or the space inside a polygon or circle. Volume is the measure of the capacity of a solid. Area and volume can be measured by counting squares and cubes, respectively, but formulas exist that simplify these computations for many common figures.

"Students should develop multiplicative procedures and formulas for determining measures. The curriculum should focus on the development of understanding, not on the rote memorization of formulas" (NCTM, *Standards*, p. 116). A number of area formulas are logically related. The parallelogram formula follows from the rectangle formula. The triangle and trapezoid formulas can be deduced from the parallelogram formula.

If you know the lengths of two sides of a right triangle, you can find the length of the third side using the Pythagorean Theorem. The Pythagorean Theorem is frequently applied to right triangles that appear in other figures.

There is one general volume formula and one general surface-area formula for all prisms and cylinders. There is also one general volume formula for all pyramids and cones.

All corresponding area measurements of two similar figures are in the same ratio; all corresponding volume measurements of two similar figures are also in a single ratio. These ratios are helpful in explaining the relationship between weight and muscle strength and in explaining how well warm-blooded animals of various sizes retain heat.

STUDY GUIDE

To review Chapter 10, see what you know about each of the following ideas or terms that you have studied. You can also use this list to generate your own questions about the chapter.

The NCTM Curriculum Standards and Measurement

Selected NCTM Curriculum Standards

The following standards come from the NCTM document.

- Extend their understanding of the concepts of perimeter, area, and volume.
- Make and use estimates of measures.
- Make and use measurements in problem and everyday situations.
- Recognize and apply deductive and inductive reasoning.

1. Describe how each standard listed relates to the material you studied in Chapter 10.
2. Select any current elementary-school mathematics textbook series, and describe sample lessons or exercises that illustrate each standard listed.

MEASUREMENT IN ELEMENTARY SCHOOL

The following chart shows at what grade levels selected measurement topics typically appear in elementary-school mathematics textbooks.

Topic	Typical Grade Level in Current Textbooks
Metric measure	1, 2, 3, 4, 5, 6
Perimeter	2, 3, 4, 5, 6
Rectangle area	4, 5, 6
Volume	3, 4, 5, 6
Surface area	6
Parallelogram area	6
Triangle area	5, 6
Circumference	5, 6
Circle area	6

REVIEW EXERCISES

1. On a hot summer day in Washington, D.C., the temperature is about
   (a) 5°C (b) 35°C (c) 65°C
   (d) 95°C (e) 125°C

2. A rectangular prism is 8 m by 6 m by 7 m. How long is the diagonal shown?

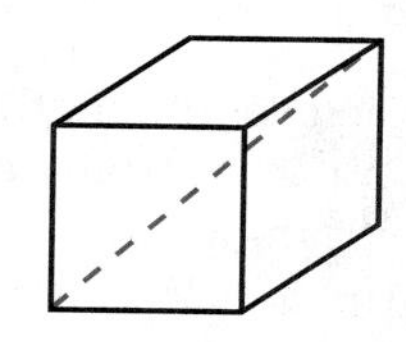

3. 8 cm^2 = ________ m^2

4. What is the surface area of the figure?

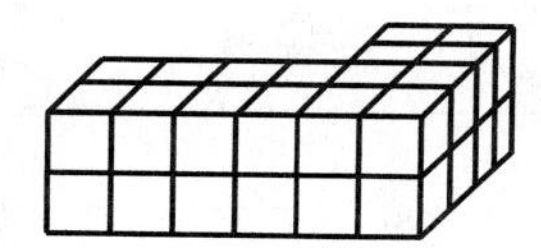

5. The room shown has an area of 58 m^2. What is the length across the back of the room?

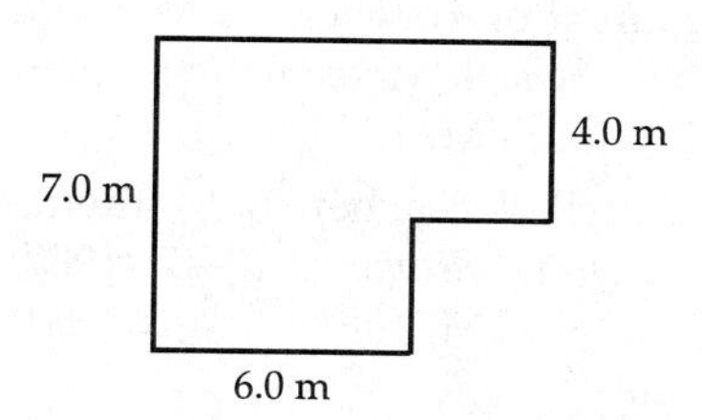

6. A store sells two kinds of wrapping paper. Package A costs $5 and has 4 rolls, each 2 ft by 7 ft. Package B costs $4 and has 3 rolls, each $3\frac{1}{2}$ ft by 4 ft. Which is a better buy?

7. A rectangular solid has dimensions of 4 m, 5 m, and W m. Its total surface area is 166 m^2. What is W?

8. If the length, width, and height of a rectangular prism are each multiplied by 4, how does the surface area change?

9. Use the area formula for a parallelogram ($A = bh$) to explain why the area of the triangle shown is $A = \frac{1}{2}bh$.

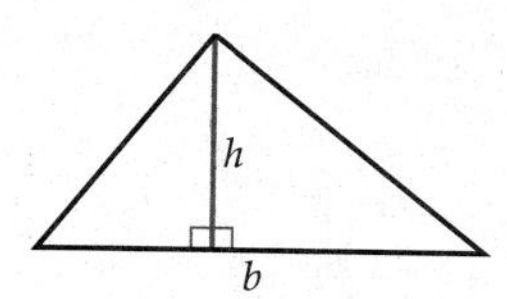

10. A triangular sail is measured as shown. What is the best approximation of its area?

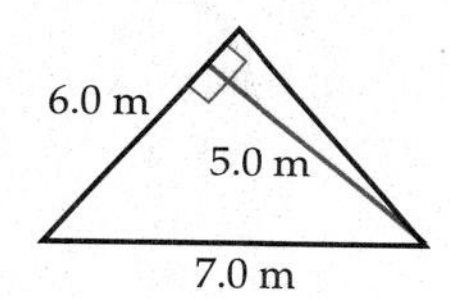

11. Find the volume and surface area of the triangular prism shown.

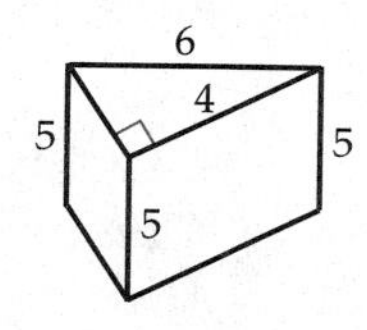

12. Find the shaded area.

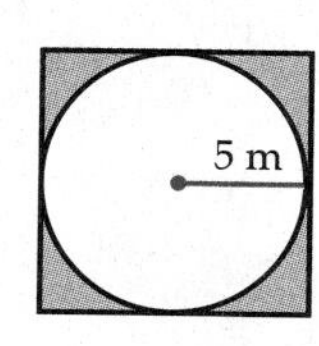

13. The quarter circle has an area of 25π m^2. What is its perimeter?

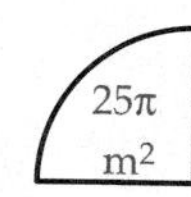

14. Find the shaded area. $\overline{AB}$ is a diameter.

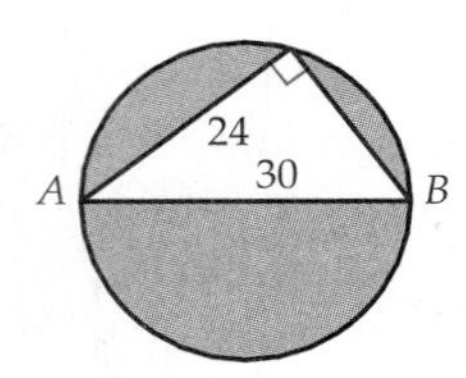

15. The circumference of a circle is 50 cm. What is its exact area?

16. A 9-in. pizza costs \$5, and a 14-in. pizza costs \$11. Which is a better buy?

17. A cylinder has a radius of 6 ft and a height of 12 ft. Find its volume and surface area.

18. Two similar spheres have radii of 2 ft and 3 ft. What is the ratio of their surface areas?

19. Two similar solids have total surface areas of 400 m^2 and 900 m^2. The smaller solid has a volume of 200 m^3. What is the volume of the larger solid?

20. Two animals of the same type are the same shape. The larger animal has a waist size of 20 cm; the smaller animal has a waist size of 8 cm.
(a) How many times heavier is the larger animal?
(b) If the larger animal can lift 40 lb, how many pounds can the smaller animal lift?

21. Tell whether each set of lengths could form a right triangle, an acute triangle, an obtuse triangle, or no triangle.
(a) 2, 3, 7 (b) 5, 6, 7 (c) 4, 5, $\sqrt{41}$

22. Write a sentence that tells the difference between the perimeter and area of a convex polygon.

SUGGESTED READINGS

Andrini, B. *Cooperative Learning and Mathematics.* San Juan Capistrano, Calif.: Kagan, 1992.

Beckmann, P. *A History of Pi.* 5th ed. Boulder, Colo.: Golem Press, 1977.

Equals. *Get It Together.* Berkeley, Calif.: Equals, 1989.

Geddes, D., and others. *Measurement in the Middle Grades.* Reston, Va.: NCTM, 1994.

Jacobs, H. *Geometry.* 2nd ed. New York: W. H. Freeman, 1987.

Loomis, E. *The Pythagorean Propositions.* Washington, D.C.: NCTM, 1972.

Mathematics Teaching in the Middle School. March–April, 1996 Focus Issue on Mathematical Connections. Reston, Va.: NCTM, 1996.

National Council of Teachers of Mathematics. *Measurement in School Mathematics.* 1976 Yearbook. Reston, Va.: NCTM, 1976.

National Council of Teachers of Mathematics. *Learning and Teaching Geometry K–12.* 1987 Yearbook. Reston, Va.: NCTM, 1987.

11 Algebra and Coordinate Geometry

Two-dimensional graphs are widely used in mathematics and everyday life. These graphs show relations between two sets of variables.

Graphing combines ideas from geometry and algebra. Although the ancient Greeks had developed the major ideas of geometry by 300 B.C., algebra developed far more slowly after originating in Babylonia about 4000 years ago. Algebra was done first with words and later with a combination of words and symbols. Around A.D. 250, Diophantus was the first to use letters for variables and exponents. It was not until the late 1500s that mathematicians such as François Viète succeeded in solving fairly complicated algebraic equations.

The stage was then set for Descartes and Fermat to connect algebra and geometry using coordinate graphs. In coordinate geometry, algebraic equations are classified according to the shapes of their graphs, and geometric shapes can be represented by equations.

Amalie Emmy Noether

Figure 11–1

More recently, Emmy Noether (1882–1935), the greatest woman mathematician of her time (Figure 11-1), developed more advanced algebra that is now a fundamental part of graduate-school mathematics. Noether battled discrimination throughout her life. One of only a few female students in the entire University of Erlangen (Germany), she completed a doctorate in mathematics at the age of 23. Noether then did research and lectured at the university for no pay. In 1922, Noether was appointed an "extraordinary" professor at the University of Göttingen and received a modest salary. Eleven years later, in 1933, the Nazis came to power and dismissed her because she was Jewish. Noether emigrated to the United States and was appointed as a visiting professor at Bryn Mawr College in Pennsylvania in the fall of 1933. She died suddenly in April 1935 after routine surgery.

In this chapter, you will use formulas, graphs, and tables to represent relations between variables and to solve problems. You will also see how graphs offer another way to study some geometric concepts.

11.1 RELATIONS AND FUNCTIONS

How far would a 1998 Honda Civic travel on 10 gallons of gas? How much does it cost to mail a 1.2-oz first-class letter? Each question involves two related quantities. The distance a car travels can be related to the number of gallons of gas used, and the cost of mailing a letter can be related to its weight. As the word "related" indicates, these are examples of two-set relations and functions, the subjects of this lesson.

REPRESENTING RELATIONS BETWEEN TWO SETS

Relations between two quantities can be described in words.

Lesson Exercise 11.1

For each pair of variables, tell what y generally does when x increases.

(a) x = gallons, y = distance traveled in a car

(b) x = supply of apples, y = price of apples

Numerical data can reveal a more precise relation between two sets. Consider a National Motors Garish that gets 36 miles per gallon. We can represent this mileage relation with a table, an equation, or a graph (Figure 11-2). A **relation** pairs the members of two sets (or of a set and itself), usually according to some criterion.

G (gallons of gas)	D (distance traveled–miles)
0	0
1	36
2	72
3	108
4	144

Table

$D = 36G$

Equation

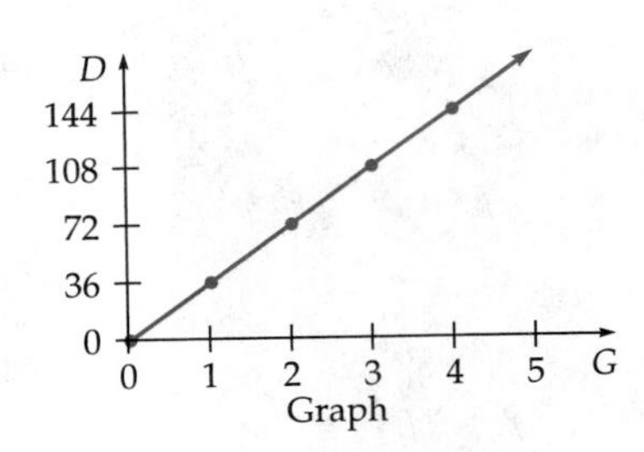

Graph

Figure 11–2

The graph is a ray because 0 is the lowest value for D and G, and G can represent any positive decimal number of gallons. When a relation involves countable sets such as the supply of apples, the graph is a **discrete** set of dots, as in Figure 11-3(a). When a relation involves a measurement such as distance or weight, the graph is a **continuous** segment or set of segments, as in Figure 11-3(b).

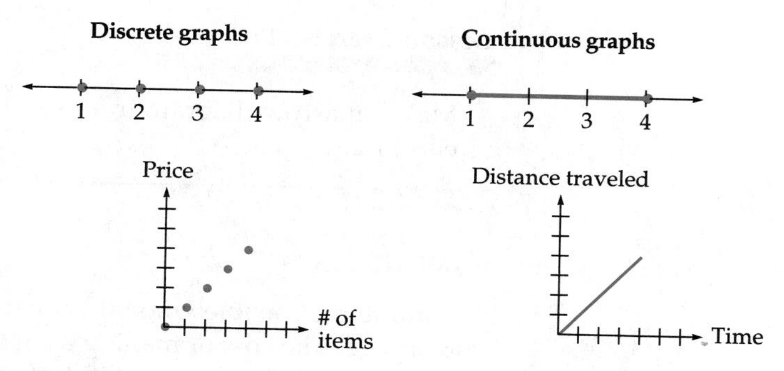

Figure 11–3

Three of the most important mathematical relations are "is less than," "is equal to," and "is greater than." The following exercise concerns "is less than."

Lesson Exercise 11.2

Let the expression "is less than" relate $A = \{1, 2, 3\}$ to $B = \{1, 2, 3, 4\}$.

(a) Complete the table by listing all possible ordered pairs under the relation "is less than."

A Is Less Than *B*	
1	2
1	3
1	4

(b) Graph the set of points from your table in part (a).

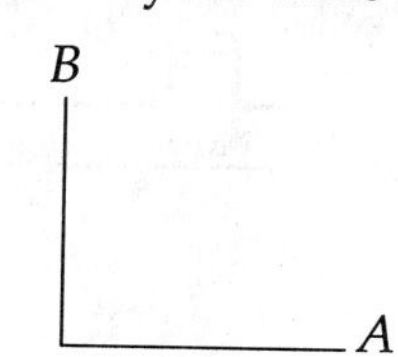

(c) Is your graph discrete or continuous?

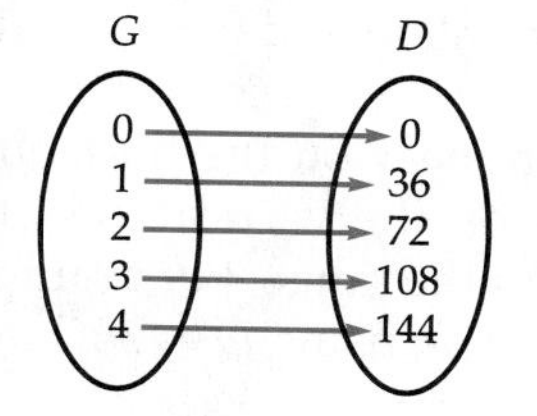

Figure 11–4

A relation can also be represented with an arrow diagram. An **arrow diagram** matches the corresponding elements of two sets of arrows. The arrow diagram for the selected values from the introductory gas mileage problem is shown in Figure 11-4.

The arrow between each pair of elements indicates the order of the two elements in an ordered pair. For example, 1 gallon → 36 miles corresponds to the ordered pair (1, 36).

Lesson Exercise 11.3

Make an arrow diagram from set A to set B for the relation in Lesson Exercise 11.2.

FUNCTIONS

The number of people who buy bananas at your store is a function of the price you charge. The cost of mailing a first-class letter is a function of its weight.

A function is a special kind of relation for which each value of the first variable (called the *input*) is related to *exactly one* value of the second variable (called the *output*). Functions result from our observation and analysis of patterns.

Lejeune Dirichlet, a German mathematics professor, defined a function in 1837.

> ***Definition*** **A Function**
>
> A **function** from set A to set B is a relation in which each member of A is paired with exactly one member of B.

Whereas a relation may assign any number of outputs to an input, a function gives a single, definite output for each input. The example illustrated in Figure 11-4 is a function because each number of gallons G is assigned exactly one distance in D. On the other hand, the relation in Lesson Exercise 11.3 is not a function, because numbers such as 1 in set A are assigned to more than one number in set B. A function is often illustrated with a "function machine" (Figure 11-5).

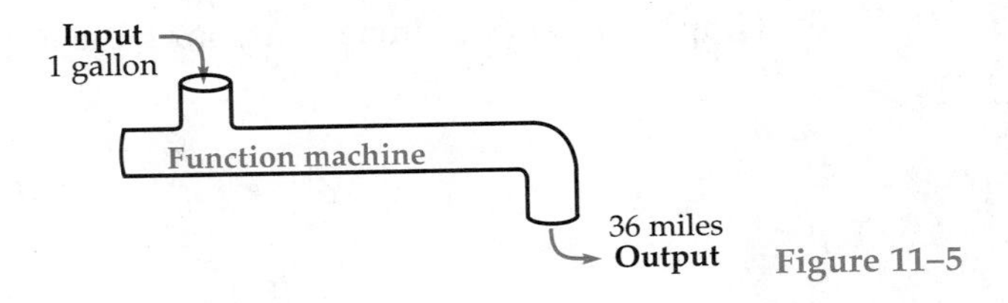

Figure 11–5

The set of input values is the **domain**, and the resulting set of output values is the **range**. In Figure 11-4, 0, 1, 2, 3, and 4 are in the domain and 0, 36, 72, 108, and 144 are in the range.

In a standard graph, the values in the domain appear on the horizontal axis, the x-axis, and the values in the range appear on the vertical axis, the y-axis. Usually, the equation is solved for the variable of the range. For example, given the value of G, it is easier to compute the value of D, using $D = 36G$ than it would be using $G = \frac{D}{36}$.

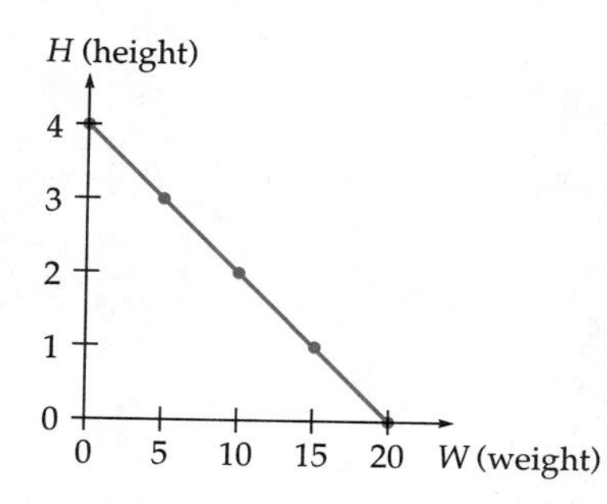

Figure 11–6

Lesson Exercise 11.4

A weight of W (pounds) pulls a spring to a position H (inches) above the ground according to $H = 4 - \frac{W}{5}$ (Figure 11-6).

(a) What is the domain (input set)? (b) What is the range (output set)?
(c) Explain why the graph exhibits a function from W to H.

Not all relations between two sets are functions.

EXAMPLE 11.1

Tell whether each of the diagrams illustrates a function from set A to set B.

(a)

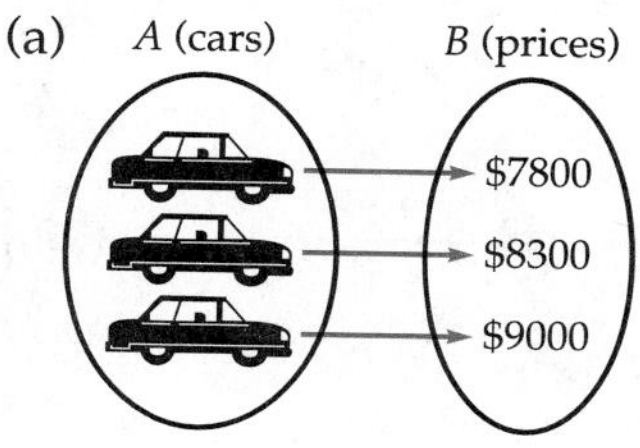

(b)

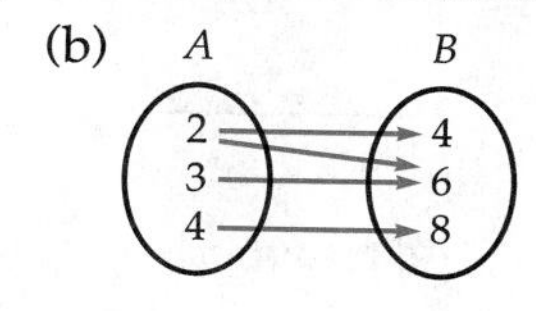

(c)

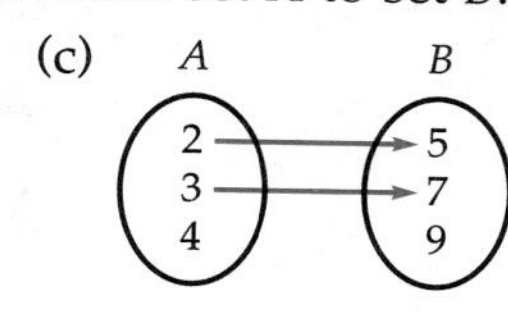

Solution

A function must assign a single, definite output to each input. Note that an input or output set can contain objects (or people) as well as numbers.

(a) Is each element in set A, "Cars," assigned exactly one element in set B, "Prices?" Yes. So part (a) represents a function.

(b) No. The value 2 in set A is assigned to two elements, 4 and 6, in set B. So part (b) does not represent a function.

(c) No. The value 4 in set A is not assigned to any number in set B. So part (c) does not represent a function. ■

Now it's your turn.

Lesson Exercise 11.5

Which of the diagrams illustrates a function from set A to set B?

(a)

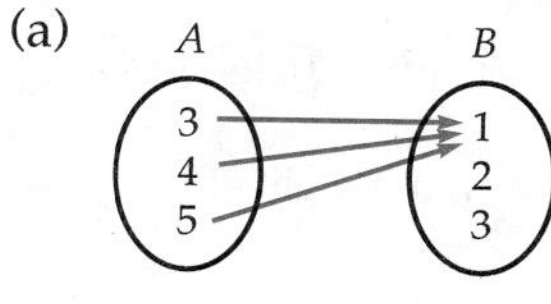

(b)

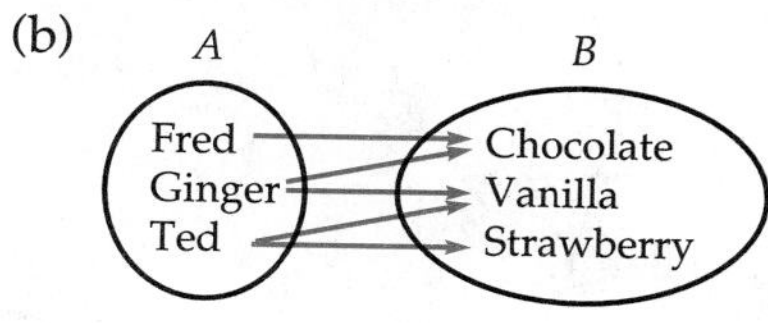

(c)

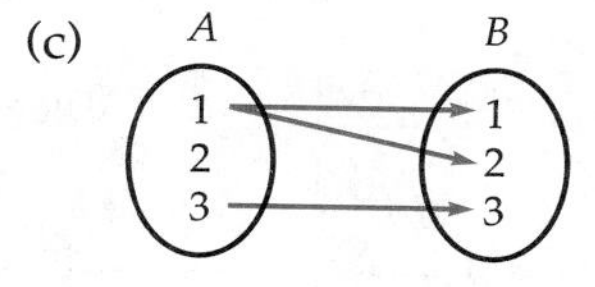

Lesson Exercise 11.6

Explain why the relation "is less than" in Lesson Exercise 11.2 is not a function from set A to set B.

GRAPHS OF RELATIONS AND FUNCTIONS

Some graphs represent functions; others do not. In the graph of a function, there is exactly one corresponding value of y for each possible value of x.

EXAMPLE 11.2

Tell whether each graph represents a function.

(a)

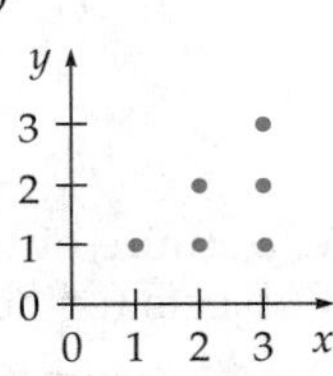

(b)

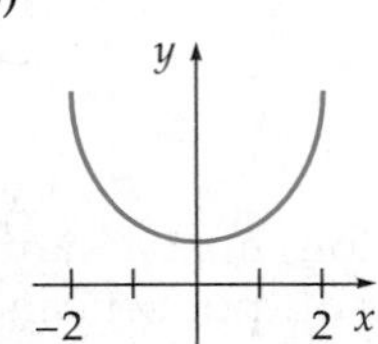

(c) Speedometer reading (mph)

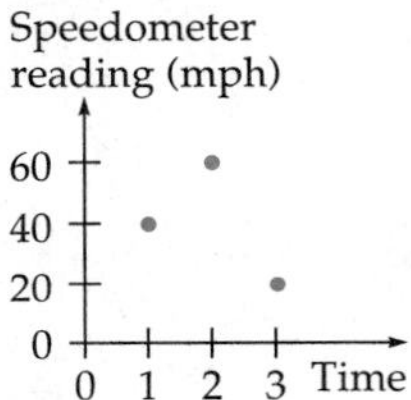

Solution

(a) No, because $x = 2$ is assigned to two y values and $x = 3$ is assigned to three y values.

(b) Yes. Each x value on $-2 \leq x \leq 2$ is assigned exactly one y value.

(c) Yes. Each time value is assigned exactly one speedometer reading. ■

Lesson Exercise 11.7

State whether each graph represents a function.

(a)

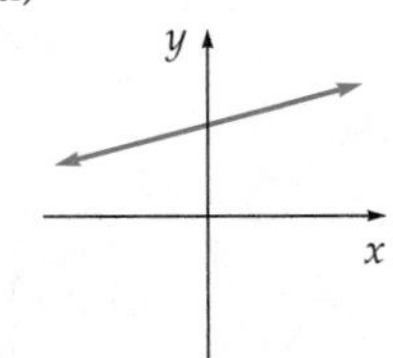

(b) Height of baseball

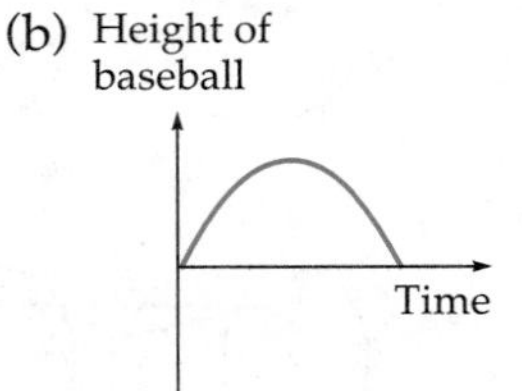

(c)

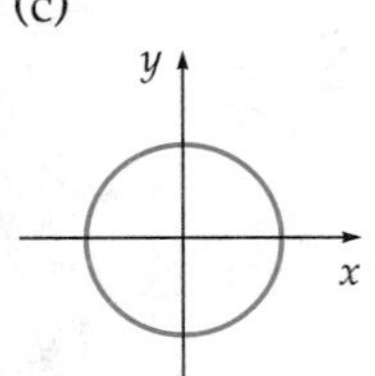

ANSWERS TO SELECTED LESSON EXERCISES

11.1 (a) The distance y generally increases when x increases.
(b) The price y generally decreases when x increases.

11.2 (a)

A Is Less Than *B*	
1	2
1	3
1	4
2	3
2	4
3	4

(b)

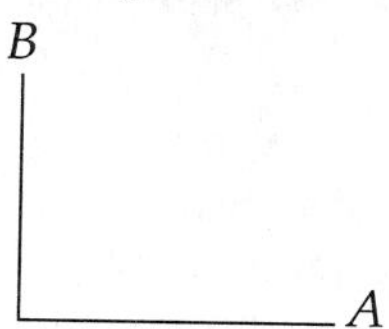

(c) Discrete

11.3

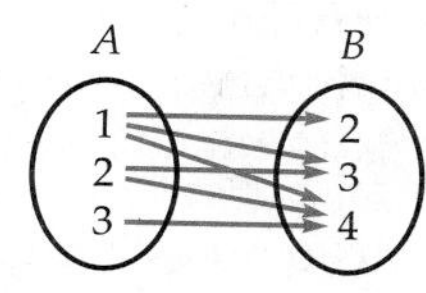

11.4 (a) $0 \leq W \leq 20$ (b) $0 \leq H \leq 4$
(c) Each value for W is assigned exactly one value for H.

11.5 (a)

11.6 Some values in set A, such as 1, are assigned to more than one value in set B.

11.7 (a) Yes (b) Yes (c) No

11.1 HOMEWORK EXERCISES

Basic Exercises

1. Let "is greater than or equal to" relate $A = \{0, 1, 2, 3, 4\}$ to $B = \{0, 2, 4, 6\}$.
   (a) Complete a table of values or a set of ordered pairs.
   (b) Graph the set of points from part (a).

   B

   A

2. Let "is a factor of" relate $A = \{1, 2, 3, 4\}$ to $B = \{10, 15, 20\}$.
   (a) Complete a table of values or a set of ordered pairs.
   (b) Graph the set of points from part (a)

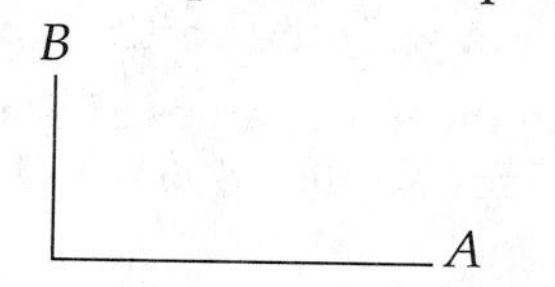

3. Make an arrow diagram of the data in Exercise 1.

4. (a) What set of ordered pairs corresponds to the following arrow diagram?

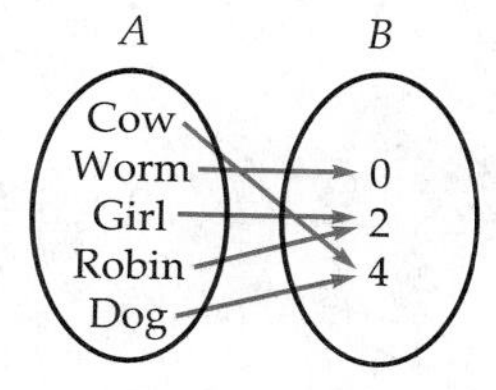

   (b) Describe a rule that relates the two sets.

5. A construction company charges \$60,000 plus \$65 per square foot to build a house. The total cost $C = 60{,}000 + 65F$, in which F equals the floor space in square feet.

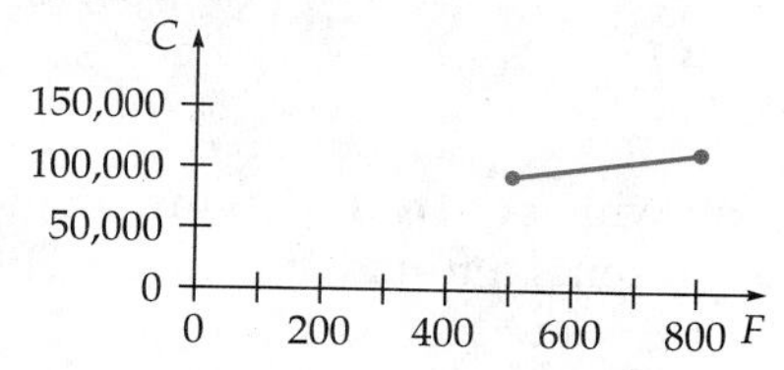

(a) Based on the graph, what is the domain (input set)?
(b) What is the range (output set)?
(c) Explain why the graph exhibits a function from set F to set C.

6. Your car gets 36 miles per gallon. You fill up the tank with 15 gallons and drive. The number of gallons, g, left after you drive m miles is given by $g = 15 - \frac{m}{36}$.

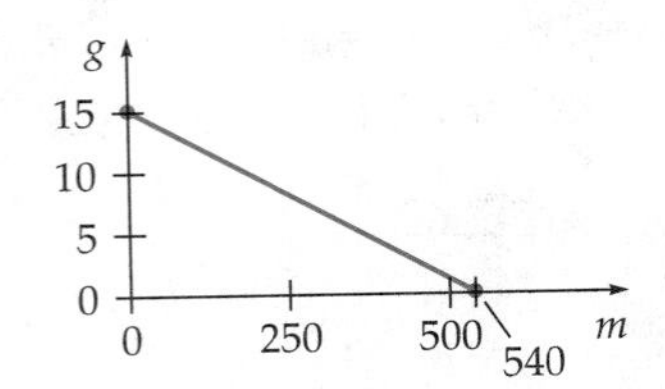

(a) Based on the graph, what is the domain (input set)?
(b) What is the range (output set)?
(c) Explain why the graph exhibits a function from set m to set g.

7. (a) Tell which of the following diagrams represents a function from set A to set B.
(b) Give a possible rule relating each set of ordered pairs.

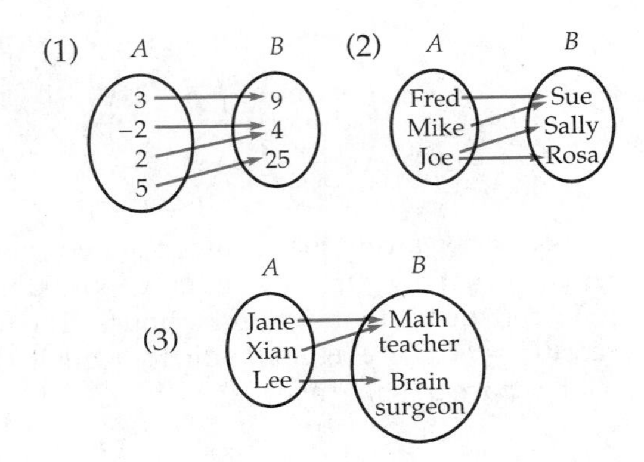

8. (a) Tell which of the following diagrams represents a function from set A to set B.
(b) Give a possible rule relating the ordered pairs.

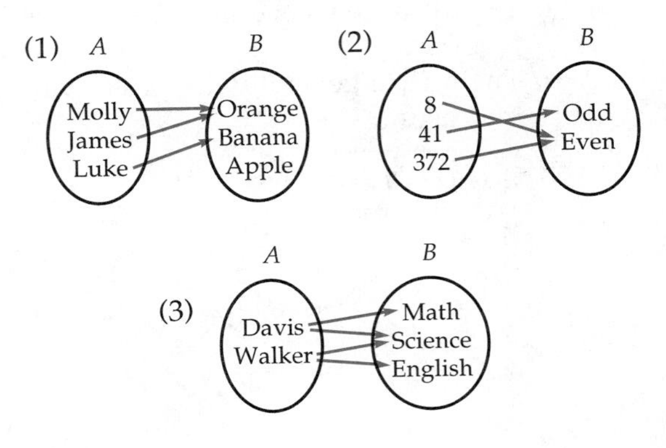

9. Explain why the relation "is greater than" from $A = \{10, 20, 30\}$ to $B = \{10, 20, 30\}$ is *not* a function from set A to set B.

10. Is the relation y "is the brother of" x a function from set x to set y if the input set x is all of the students in your mathematics class?

11. Which of the following are functions from x to y? (Assume that the entire domain is given.)
(a) $\{(4, 2)\ \ (4, 3)\ \ (4, 4)\ \ (4, 5)\}$
(b) $\{(1, 7)\ \ (2, 8)\ \ (3, 9)\ \ (4, 10)\}$
(c) $\{(2, 4)\ \ (3, 4)\ \ (4, 4)\ \ (5, 4)\}$

12. Tell whether each graph represents a function.

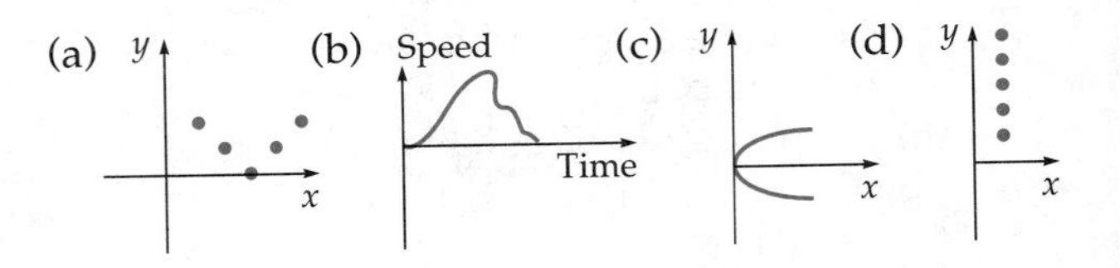

13. The vertical-line test is a method for determining whether a graph represents a function. A graph does not represent a function if a vertical line can be drawn through two or more points of the graph.
(a) Draw a vertical line in each of the graphs of the preceding exercise that is not a function.
(b) Explain why the vertical-line test works.

14. A movie theater charges \$2 for children ages 5–17, \$6 for adults 18–64, and \$4 for seniors

(65 and up). There is no charge for children under 5. Which of the following sets of ordered pairs creates a function?
(a) (age, price) (b) (price, age)

15. Which of the following assignments creates a function?
(a) Each student in a school is assigned a teacher for each course.
(b) Each dinner in a restaurant is assigned a price.
(c) Each person is assigned a birthdate.

16. Propose a possible rule for each function shown.

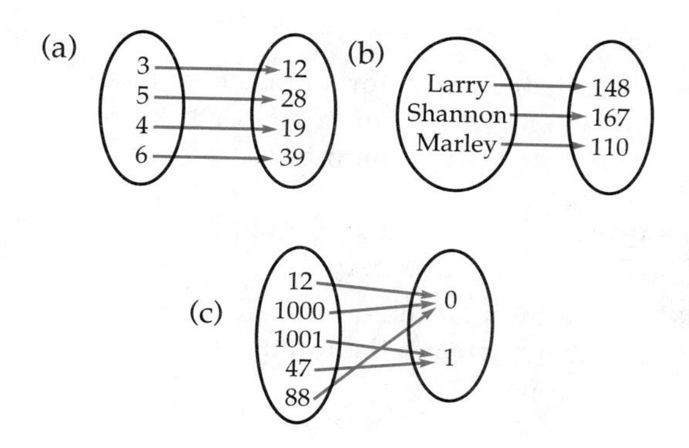

17. Propose a possible rule for each function shown.

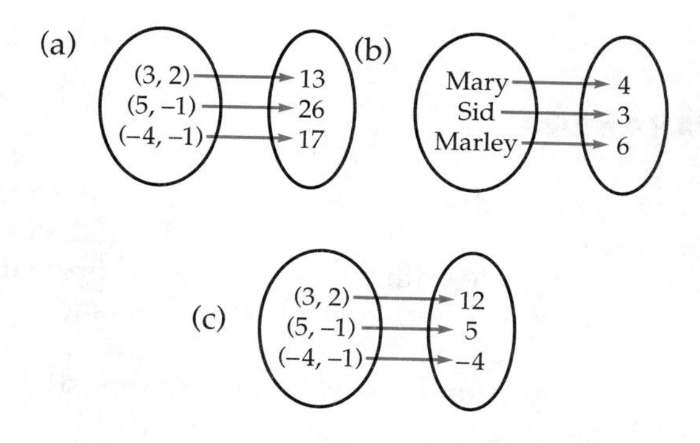

18. Give an equation relating y to x in each set of ordered pairs.
(a) $\{(5, 24)\quad (6, 35)\quad (7, 48)\}$
(b) $\{(7, 18)\quad (9, 22)\quad (11, 26)\}$

19.

x	0	1	2	3
y	7	9	11	

(a) Fill in the last value for y, continuing the pattern in the table.
(b) A formula relating y to x is $y =$ ________.

20. Fill in the blanks, finding a rule that works for each number pair.

Regular Price		Sale Price
\$10	→	\$ 8
\$20	→	\$16
\$30	→	\$24
\$40	→	____
\$*N*	→	____

21. Consider the following function for the speed and force of crash impact for a particular car.

Speed (mph)		Force
20	⟶	100
30	⟶	225
40	⟶	400
50	⟶	625
60	⟶	____

(a) When the speed doubles, the force is multiplied by ________.
(b) Fill in the force for 60 mph that continues the pattern.
(c) Find a formula relating force to speed.
(d) What factors in addition to speed would affect the impact force of a car?
(e) What is a reasonable domain for this function?

22. A function has the rule $P = 8N - 50$. The range (output set) for P is $\{46, 62, 78\}$. What is the domain (input set)?

23. Measuring creates a relation between two sets. Suppose a 30-cm ruler is used to measure the lengths of people's feet. This would create a measuring function.
(a) What could be the domain of the function?
(b) What numbers might be in the range?

24. Does the following represent a function for the domain $\{0, 1, 2, \ldots, 90\}$?

x = Number of Students in a Section of Math 130	y = Number of Sections of Math 130
0–7	0
8–32	1
33–62	2
63–90	3

Extension Exercises

25. Some relations are **transitive**. The relations "is equal to" and "is less than" are transitive, as illustrated by the following. For real numbers x, y, and z, if $x = y$ and $y = z$, then $x = z$; and if $x < y$ and $y < z$, then $x < z$.
    The relation "is a sibling of" is also transitive. If person x is a sibling of person y and person y is a sibling of person z, then person x is a sibling of person z.
    Name two other relations that are transitive.

26. $y = x + 2$ and $z = 3y$.
    (a) Describe (in words) the rule that relates x to y.
    (b) Describe (in words) the rule that relates y to z.
    (c) Find a function that relates z directly to x.
    (d) Describe (in words) the rule that relates z to x.

27. Which of the following represents a function if the domain is the set of all real numbers?
    (a) $y = x^2 + 1$ (b) $y^2 = x^2 + 1$

28. What real numbers *cannot* be used for x in each of the following functions, if y must be a real number?
    (a) $y = \frac{1}{x}$
    (b) $y = \sqrt{x}$
    (c) $y = \sqrt{25 - x}$

29. What is the range for y in each equation?
    (a) $y = 8 - x^2$ (domain for x: all real numbers)
    (b) $y = \frac{1}{x}$ (domain for x: all real numbers except 0)

30. Assuming that the domain for x is all real numbers, what is the range for y in each equation?
    (a) $y = x^2$ (b) $y = x^3$

31. $A = \{5, 10, 15, 20\}$, and $B = \{9, 18\}$. List the elements (x, y) of $A \times B$ such that $x < y$.

11.2 GRAPHS OF LINES AND PARABOLAS

The idea of using graphs to represent functions is about 350 years old. In the seventeenth century, two Frenchmen, René Descartes (1596–1650) and Pierre de Fermat (1601–1665), (Figure 11-7), had the brilliant idea of using coordinates to relate algebraic equations (such as $y = 2x$) to geometric shapes (such as a line).

Descartes was a brilliant but sickly youngster who entered a prestigious private school at age 8. Because of his ill health, his teachers allowed him to lie in bed in the morning for rest and reflection. Descartes continued this practice throughout his life and came up with many of his philosophical and mathematical ideas during his morning meditations.

Although he was one of the great mathematicians of his time, Fermat was a lawyer who did mathematics as a hobby! He was not well known during his lifetime because he did not try to publish his work. Fermat's achievements

René Descartes

Pierre de Fermat

Photo of Descartes (left) courtesy of Library of Congress. Photo of Fermat (right) from the David Eugene Smith Collection, Rare Book and Manuscript Library, Columbia University.

Figure 11–7

include major discoveries in number theory, the invention (along with Descartes) of coordinate geometry, and helping to lay the groundwork for both probability and calculus.

Today, people use coordinates to locate places and to portray relationships between quantities (Figure 11-8).

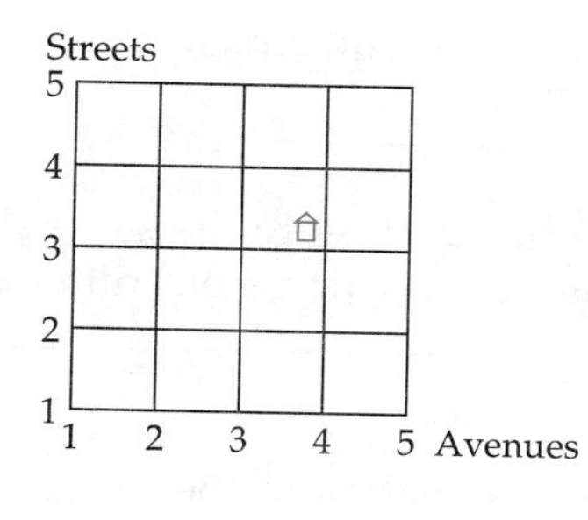

The house is on 4th Avenue and 3rd Street.

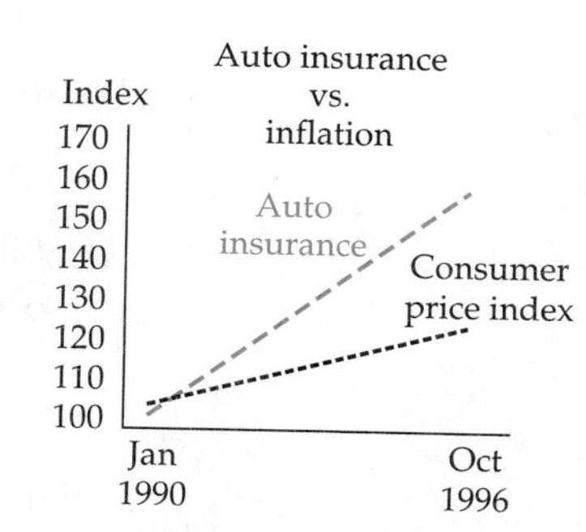

Figure 11–8

$y = mx$

Why are equations of the form $y = mx$ of interest? Many applications have formulas in which one variable (y) equals another variable (x) multiplied by a number ($y = mx$). Try the following exercise.

Lesson Exercise 11.8

Whole-wheat flakes have 1.5 calories per gram.

(a) Complete the table.

x (number of grams)	0	20	40	60	80
c (number of calories)					

(b) Graph your points, with c on the vertical axis and g on the horizontal axis.

(c) Would it be appropriate to connect your points to make a line segment, a ray, or a line?

(d) An equation relating c and x is $\frac{c}{x}$ = ________ or c = ________.

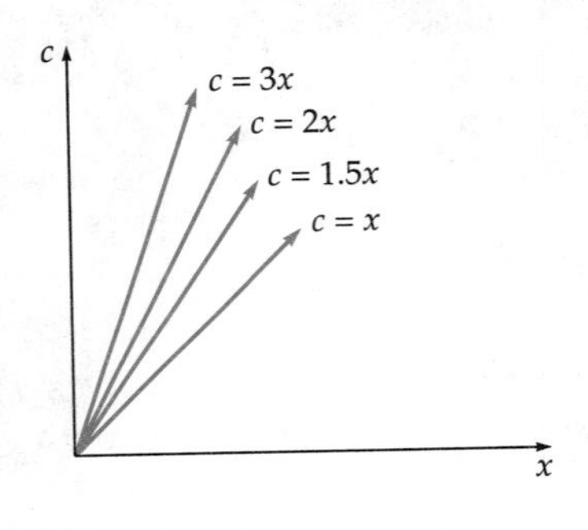

Figure 11–9

The preceding exercise illustrates how points that satisfy $c = 1.5x$ all lie on a straight line. The graphs of $c = x$ (tomato catsup), $c = 2x$ (chicken breast), and $c = 3x$ (whole-wheat toast) are similar. Their graphs are shown in Figure 11-9. (You could plot these graphs using a graphing calculator.)

How are the graphs and equations alike, and how do they differ? That's the subject of the following exercise.

Lesson Exercise 11.9

(a) Are the graphs of $c = x$, $c = 1.5x$, $c = 2x$, and $c = 3x$ all the same shape?

(b) The graphs $c = x$, $c = 1.5x$, $c = 2x$, and $c = 3x$ all pass through the point ________.

(c) In all four equations, $c = mx$ for a constant m. If $c = mx$, then $\frac{c}{x}$ = ________.

(d) All four graphs are the same shape, and they pass through the origin. How do the four graphs differ from one another?

For each equation in the calorie example, $\frac{c}{x}$ is always the same number. This means that the number of calories is proportional to the number of grams. For example, $\frac{c}{x} = \frac{3}{2}$ is a proportion for whole-wheat flakes.

The graphs of all four equations are rays passing through the origin. The calorie examples are rays rather than lines because g and c cannot be negative. If the negative numbers were included in the input set, all of these equations would have graphs that are lines.

The graphs of all equations of the form $y = mx$ (in which m is any real constant) have the same shape. This is an example of the relationship between the shape of a coordinate graph and the form of the corresponding algebraic equation.

Equations of the Form $y = mx$

The graph of $y = mx$ for all real numbers x and constant m is a line through the origin (0, 0).

A line (or ray) through the origin has an equation in two variables in which one variable is proportional to the other. Examples include the total pay and the hours worked when you earn the same pay for each hour, the number of dollars and the equivalent number of British pounds, and the length of a particular type of wire and its weight.

You have seen graphs relating calories and grams of a given food. Consider the example of total pay and hours worked.

Lesson Exercise 11.10

Suppose you earn \$6 per hour on a job. If you work a fraction of an hour, you earn that fraction of \$6. Make a graph of total pay (p) vs. time (hours) worked (t), with time on the x-axis.

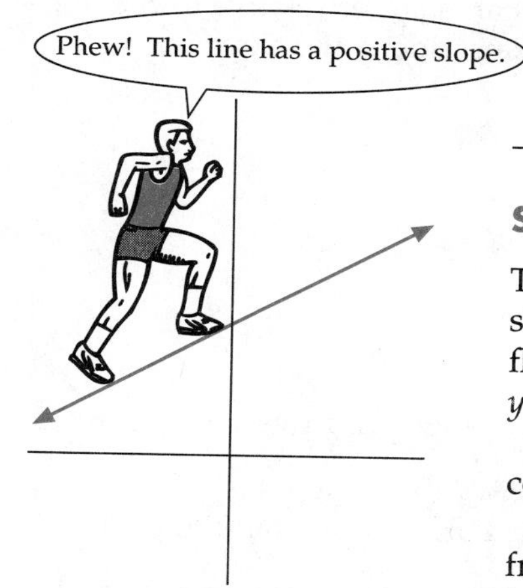

Figure 11–10

SLOPE

The four graphs in the calorie example are the same shape, but they differ in steepness. The graph of $c = 3x$ is the steepest, and the graph of $c = x$ is the flattest. The steepness of a graph is called its **slope**. When the value of m in $y = mx$ changes, the slope of the graph changes.

Mathematicians use the slope to describe lines precisely. Engineers must compute the slope of a road, a ramp, or a roof when they design it.

Imagine walking along a line from left to right (Figure 11-10). Lines rising from left to right (uphill) have a positive slope. Lines going downward from left to right (downhill) have a negative slope. A horizontal line has a slope of 0. Figure 11-11 shows some examples.

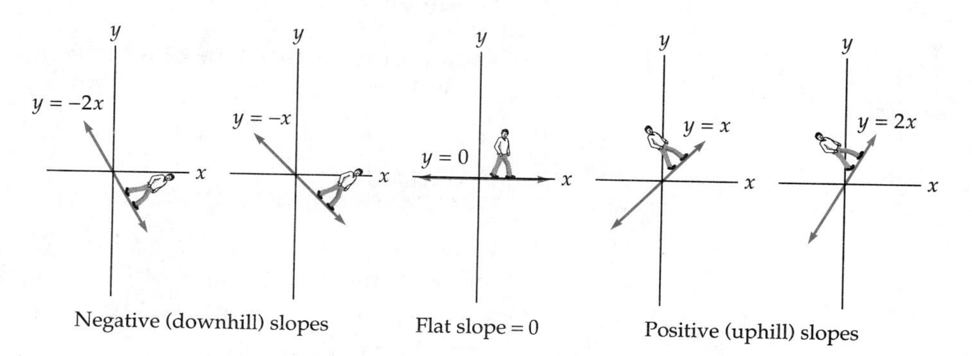

Figure 11–11

Lesson Exercise 11.11

Classify the slope of each line as slope < 0, slope $= 0$, or slope > 0.

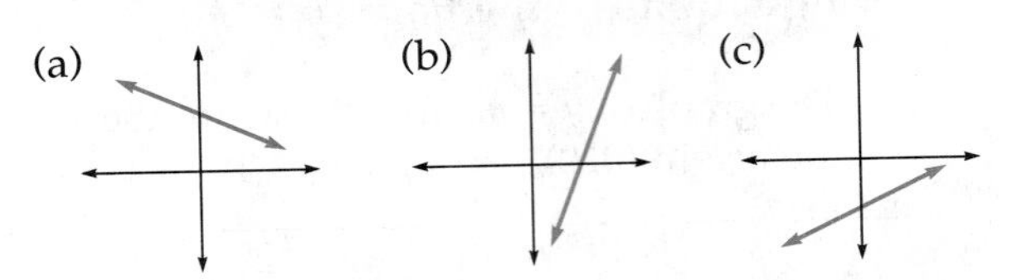

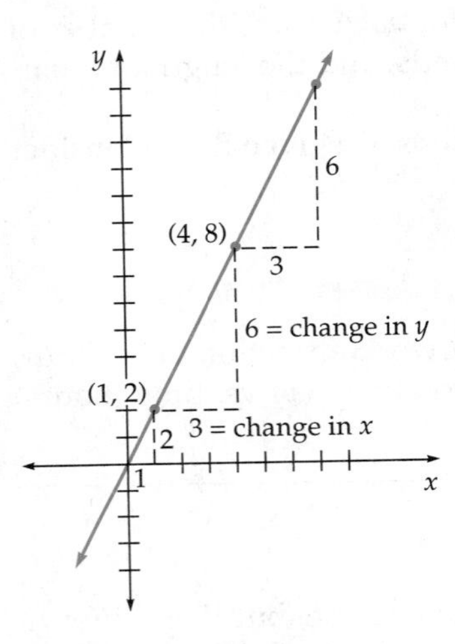

Figure 11–12

You can recognize visually whether the slope is positive or negative, but can you compute the numerical value of the slope of a line? Since the slope of a line is the same everywhere, you can compute it from any two points on the line! For example, (1, 2) and (4, 8) are on the line $y = 2x$, as shown in Figure 11-12.

Traveling from (1, 2) to (4, 8) involves moving vertically (up 6 in this case) and horizontally (right 3 in this case) a certain distance. For lines with a positive slope, a steeper line has a vertical change (6) that is larger relative to the horizontal change (3). For this reason, mathematicians define the slope as

$$\frac{\text{Change in } y}{\text{Change in } x}$$

or

$$\frac{\text{Rise}}{\text{Run}}$$

Dividing the vertical change by the horizontal change produces a greater result for steeper lines (with positive slope).

In this example, the slope is $\frac{6}{3} = 2$. Note that 2 is the value of m in the equation in $y = 2x$! In the case of $c = 2x$, the slope 2 corresponds to the $\frac{\text{change in calories}}{\text{change in grams}}$, so it indicates that each change of 1 gram results in a change of 2 calories.

You can compute the slope of a line joining two points by using the coordinates of the points.

Lesson Exercise 11.12

(a) How can the slope of the line joining (1, 2) and (4, 8) be computed from the coordinates 1, 2, 4, and 8 (without making a drawing)?

(b) How can the slope of the line joining (x_1, y_1) and (x_2, y_2) be computed from the coordinates?

Lesson Exercise 11.12 concerns the following two-point slope formula.

The Two-Point Slope Formula

The slope of the line joining two points (x_1, y_1) and (x_2, y_2), in which $x_1 \neq x_2$, is

$$m = \frac{\text{change in } y}{\text{change in } x} = \frac{y_2 - y_1}{x_2 - x_1}$$

Use the slope formula in the following exercise.

Lesson Exercise 11.13

(a) Points $(-1, -3)$ and $(1, 3)$ lie on the line $y = 3x$. Compute the slope of $y = 3x$ from these two points.
(b) Does the slope equal the value of m in $y = 3x$?
(c) Find two points on $y = 4x$, and see if the slope equals m.
(d) Generalize the results of parts (b) and (c).

As you may have guessed from the preceding exercise, the slope of $y = mx$ is m. What does the slope tell in an application?

Lesson Exercise 11.14

(a) In the formula $c = 1.5x$ (for whole-wheat flakes), what is the slope?
(b) What does the slope tell about whole-wheat flakes?

Not all lines have a defined slope. Consider the following exercise.

Lesson Exercise 11.15

(a) Why can't $x_1 = x_2$ in the two-point slope formula?
(b) Graph (2, 3) and (2, 4). What kind of line passes through these points?

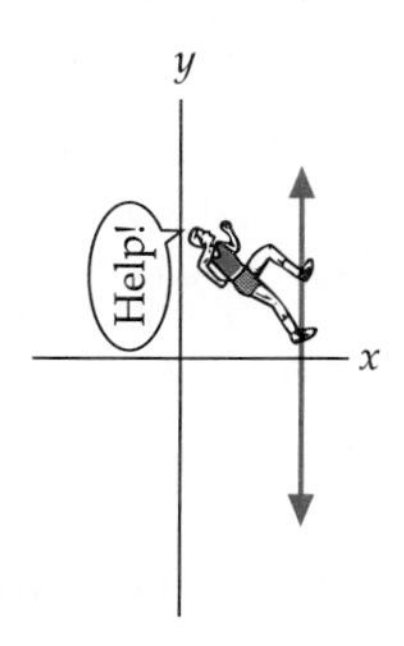

Figure 11–13

When $x_1 = x_2$, the line passing through the two points is vertical. Imagine trying to walk up a vertical line (Figure 11-13)!

Vertical lines have an undefined (infinite) slope, corresponding to the fact that $x_1 = x_2$ makes the two-point slope formula come out undefined.

$y = mx + b$

A rental car costs $89 per week plus $.26 per mile. The cost C for a week's rental is $C = 0.26D + 89$, in which D is the distance driven (in miles). This equation has the form $y = mx + b$.

You have seen that an equation of the form $y = mx$ represents a line (or part of a line), with slope m passing through the origin, and that y is proportional to x. What happens when you add a constant b to the equation so that it has the form $y = mx + b$?

Lesson Exercise 11.16

(a) Graph $y = 2x$. (This exercise can be done with a graphing calculator. See the end of the section for more information.)

(b) On the same graph, plot three points that are solutions of $y = 2x + 2$, and connect them.

(c) On the same graph, plot three points that are solutions of $y = 2x - 2$, and connect them.

(d) What is the same about all three of the graphs?

(e) What is the relationship between the three equations you graphed?

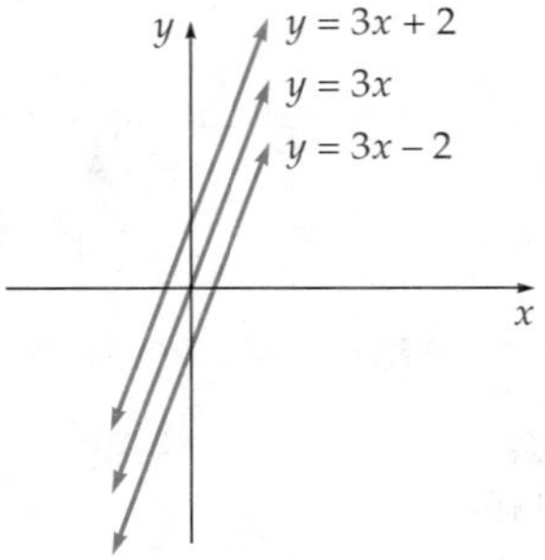

y x $y = 3x + 2$ $y = 3x$ $y = 3x - 2$

Figure 11–14

As the preceding exercise suggests, lines $y = mx + b$ with the same value of m have the same slope and are parallel. Consider the graphs of $y = 3x$, $y = 3x + 2$, and $y = 3x - 2$ shown in Figure 11-14. All three lines have the same slope, and they are parallel.

The graph of $y = 3x + 2$ is the same as the graph of $y = 3x$ translated up 2 units. The graph of $y = 3x - 2$ is the same as the graph of $y = 3x$ translated down 2 units. Lesson Exercise 11.16 and Figure 11-14 suggest the following two properties.

Slopes of Parallel Lines

Lines that have the same slope are parallel.

Relationship Between $y = mx$ and $y = mx + b$

The graph of $y = mx + b$ is the same as the graph of $y = mx$ translated b units vertically.

Use the relationship between $y = mx$ and $y = mx + b$ to do the following exercise.

Lesson Exercise 11.17

(a) The graph of $y = 5x - 3$ is the same as the graph of $y = 5x$ translated ________.

(b) If you have a graphing calculator, check your answer by graphing both equations.

How can you tell where a line will cross the y-axis? The following exercise addresses this question.

Lesson Exercise 11.18

(a) A point on the y-axis has an x-coordinate of ________.

(b) Give the coordinates of the point where $y = mx + b$ crosses the y-axis.

The preceding exercise shows that b in $y = mx + b$ tells the y-coordinate of the point $(0, b)$ where the line intersects the y-axis. This point is called the ***y*-intercept**.

The Slope-Intercept Form of a Line

The graph of $y = mx + b$ is a line that has slope m and y-intercept $(0, b)$.

Some applications have a relationship of the form $y = mx + b$.

Lesson Exercise 11.19

A car has a ground clearance of 40 cm when it carries no weight. Every 10 kg of weight decreases the ground clearance by 1 cm.

(a) Complete the table.

Added weight, w (kg)	0	10	20	30
Ground clearance, g (cm)	40			

(b) A formula for g in terms of w is $g = 40 -$ ________. (*Hint:* Use the first two sentences in the exercise or use the two-point slope formula.)

(c) Find the slope and g-intercept of your formula.

(d) What does the slope tell you about the relationship between ground clearance and added weight?

(e) Graph your equation from part (b).

To find a linear formula using two or more data points, you might first see if y is (directly) proportional to x. Then $\frac{y}{x} = m$, or $y = mx$. Otherwise, find the slope using the two-point slope formula. Then b is the y value when $x = 0$. You can also find b by substituting any point on the line into $y = mx + b$ and solving for b.

TABLES AND EQUATIONS OF LINES

The application in the preceding lesson exercise is a linear function. What kind of pattern in the table of points shows that it is linear? Try the following exercise to find out.

Lesson Exercise 11.20

(a) Tell whether each set of data lies on a line.

(1)

x	0	1	2	3
y	10	8	6	4

(2)

x	0	2	4	6
y	20	40	80	160

(3)

x	10	20	30	40
y	6	12	18	24

(b) Write a general rule for recognizing a table of data points that lie on a line.

You can recognize a linear pattern in data if y changes by a fixed amount when x changes by a fixed amount. How can you recognize an equation whose graph is a line? The graph of a function $y = mx + b$ is a line. What about other functions, such as $y = \sqrt{x}$ and $y = x^2$?

Lesson Exercise 11.21

(a) Graph $y = x^2$ for $x = -2, -1, 0, 1$, and 2. Is it a linear equation?

(b) Graph $y = \sqrt{x}$ for $x = 0, 1, 4$, and 9. Is it a linear equation?

Lesson Exercise 11.22

Consider a line through (x_1, y_1) with slope m.

(a) Using the two-point slope formula, any other point (x, y) on the line must satisfy $m =$ ________.

(b) Solve the equation in part (a) for y.

In the preceding exercise, $y = mx - mx_1 + y_1$ has the form $y = mx + b$, with $b = -mx_1 + y_1$. Equations such as $y = x^2$ and $y = \sqrt{x}$ that cannot be put into the form $y = mx + b$ do not have straight-line graphs. (*Note:* $x = a$ is the equation of a vertical line.)

A Test for the Equation of a Line

A function has a straight-line graph if it can be put in the form $y = mx + b$.

For example, $4x + 2y = 12$ is the equation of a line, since it can be written as $y = -2x + 6$ (after solving the equation for y).

Lesson Exercise 11.23

Which of the following have straight-line graphs?

(a) $y = -2x + 7$ (b) $h = -16t^2$ (c) $2R + 4S = 20$

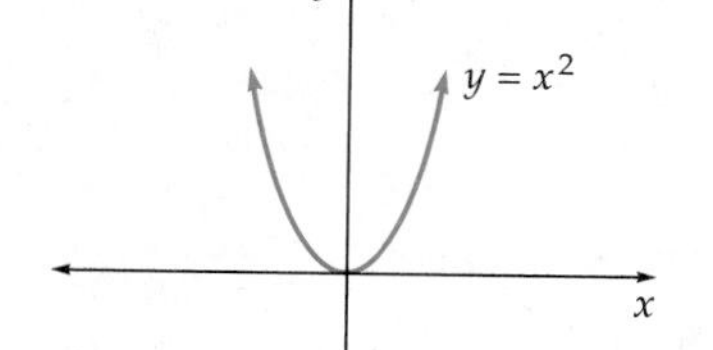

Figure 11–15

PARABOLAS

All equations of the form $y = mx + b$ have straight-line graphs. What happens if x is squared, as in the equation $y = x^2$? Your graph in Lesson Exercise 11.21 suggests the general shape shown in Figure 11-15.

Lesson Exercise 11.24

(a) Graph $y = x^2 - 2$ after completing the following table, or use a graphing calculator.

x	−2	−1	0	1	2
y					

(b) How is this graph related to the graph of $y = x^2$?

(c) Figure 11-16 shows a graph of $y = x^2 + 4$. Use the graphs in this problem to write a generalization about the relationship between the graph of $y = x^2$ and the graph of $y = x^2 + c$.

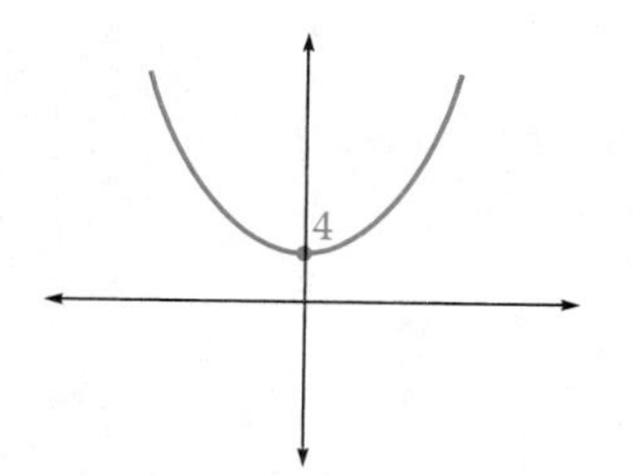

Figure 11–16

The preceding exercise illustrates how $y = x^2 + c$ is the curve $y = x^2$ translated c units vertically. The preceding graphs $y = x^2$, $y = x^2 - 2$, and $y = x^2 + 4$ are all the same shape: a parabola.

A **parabola** is a special type of U-shaped curve. The equation of a parabola that opens up or down has the form $y = ax^2 + bx + c$, where $a \neq 0$.

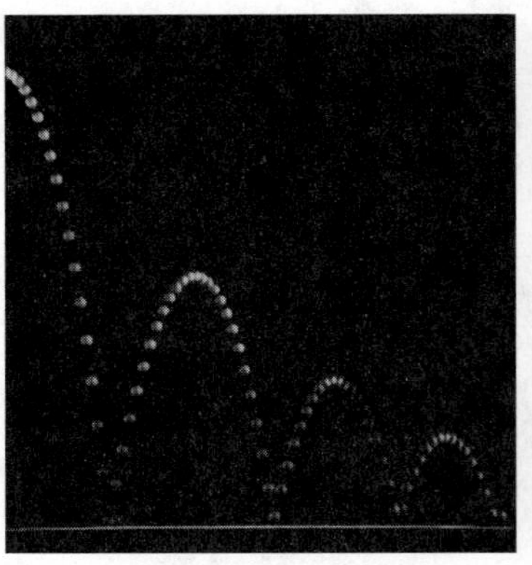

Photo of bouncing ball from PSCC, Physics, 2d ed., 1965; © Education Development Center, Inc., and D.C. Heath & Co.
Photo of Golden Gate Bridge Courtesy of Library of Congress.

Figure 11–17

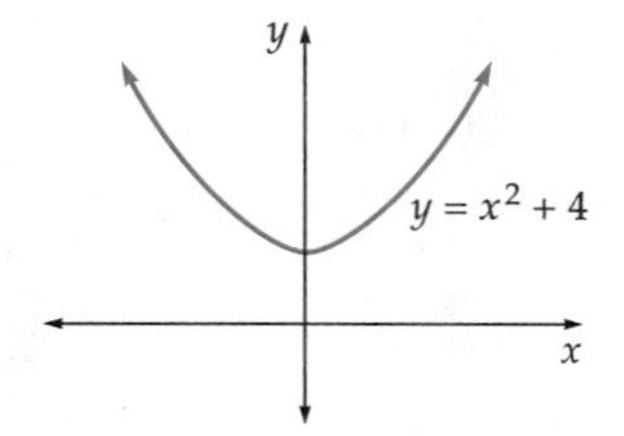

Figure 11–18

Parabolas describe the path of a ball when it is thrown and the shape of the cable on some suspension bridges (Figure 11-17).

Parabolas with the form $y = x^2 + c$ (such as $y = x^2$, $y = x^2 - 4$, and $y = x^2 + 2$) all open up (Figure 11-18).

What happens with parabolas of the form $y = -x^2 + c$?

Lesson Exercise 11.25

(a) Graph $y = -x^2$ for $x = -3, -2, -1, 0, 1, 2, 3$. (You may use a graphing calculator.)

(b) How is $y = -x^2$ related to $y = x^2$?

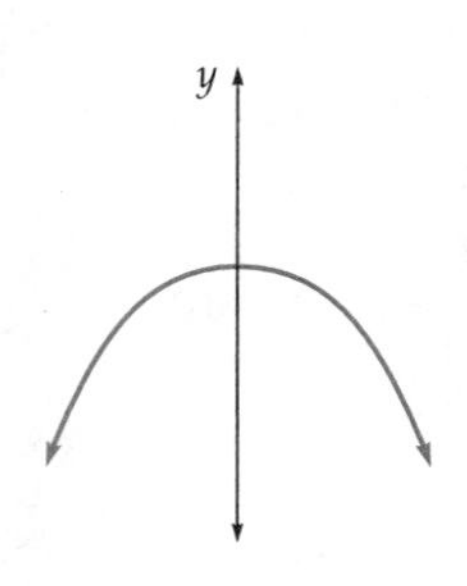

Figure 11–19

Parabolas of the form $y = -x^2 + c$ open down (Figure 11-19).

What kinds of applications are modeled by parabolas?

Lesson Exercise 11.26

(a) A formula for the area A of a square in terms of the length L of a side is $A =$ ________.

(b) Which graph illustrates this relationship with length on the horizontal axis?

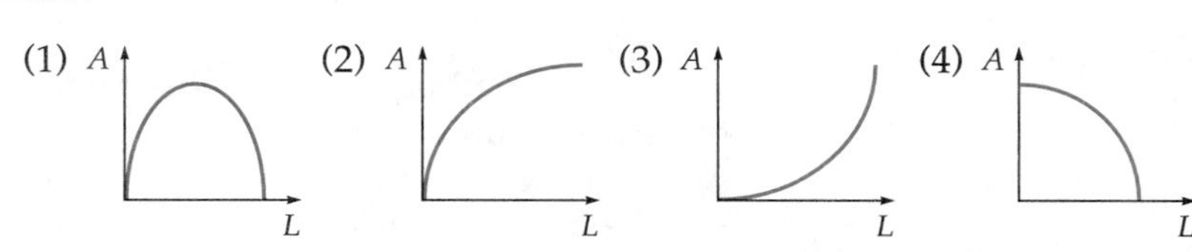

One of the amazing discoveries of coordinate geometry was that equations of the same form generally have graphs of the same shape. This discovery reveals a fundamental connection between algebra and geometry.

GRAPHING CALCULATORS AND COMPUTER GRAPHERS

If you have a graphing calculator or computer software that graphs, you can enter an equation such as $y = 2x$ or $y_1 = 2x$. Usually, the equation must be a formula for y in terms of x.

Next, tell the machine what part of the coordinate graph to display, the **window**. For example, you might want $-2 \leq x \leq 2$ and $-4 \leq y \leq 4$. The calculator or computer requires lowest and highest x values (sometimes called XMIN and XMAX) and lowest and highest y values. If you don't set a window, the machine sets one for you, but it may not be suitable.

Then tell the calculator or computer to make a graph. If you don't like the domain (x values) or range (y values) you chose, go back, change the window, and graph again. If you want to compare two graphs, such as $y = 2x$ and $y = 2x + 2$, you can graph them together, using notation such as $y_1 = 2x$ and $y_2 = 2x + 2$.

Lesson Exercise 11.27

(a) Graph $y = 2x$, using the (default) window chosen by your calculator or computer.

(b) Graph $y = 2x$ again, using a window of $-2 \leq x \leq 2$ and $-4 \leq y \leq 4$.

(c) Now graph $y = 2x + 2$ as y_2 (second equation) so that $y = 2x$ and $y = 2x + 2$ both appear on your graph. Change the window as needed.

(d) What is the same about the graphs of the two equations?

Another feature you might find useful is the TRACE or a similar key that moves the cursor (a blinking symbol) along your graph (sometimes using arrow keys). As you move the cursor, it displays the coordinates.

Lesson Exercise 11.28

Use the trace feature to approximate the coordinates of the point where $y = 2x + 2$ intersects the x-axis.

You may also want to substitute numbers into an algebraic expression or formula. This requires typing in the expression and then assigning a number to each variable. The calculator or computer can then determine the result. Refer to your manual or ask your professor or classmates for further directions.

ANSWERS TO SELECTED LESSON EXERCISES

11.8 (a)

g	0	20	40	60	80
c	0	30	60	90	120

(b)

(c) Ray (d) 1.5; $1.5x$

11.9 (a) Yes (b) (0, 0) (c) m

11.10

11.11 (a) Slope < 0 (b) Slope > 0
(c) Slope > 0

11.13 (a) 3 (b) Yes

11.14 (a) 1.5
(b) They are 1.5 calories per gram.

11.15 (a) It would make the fraction undefined.
(b) Vertical

11.16 (d) Their slopes
(e) Their graphs are parallel lines.

11.17 (a) 3 units down

11.18 (a) 0 (b) $(0, b)$

11.19 (a)

w	0	10	20	30
g	40	39	38	37

(b) $\frac{w}{10}$

(c) Slope $= \frac{-1}{10}$; g-intercept is (0, 40)

(d) The ground clearance decreases by 1 cm each time the weight increases by 10 kg.

(e)

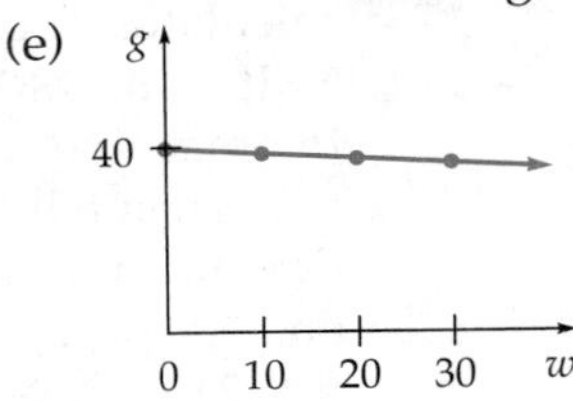

11.20 (a) (1) Yes, (2) no, (3) yes

11.21 (a) No (b) No

11.22 (a) $\frac{y - y_1}{x - x_1}$
(b) $y = mx - mx_1 + y_1$

11.23 (a) and (c)

11.24 (a)

x	−2	−1	0	1	2
y	2	−1	−2	−1	2

(b) It is the same graph translated down 2 units.

11.25 (b) The graph of $y = -x^2$ is the reflection of $y = x^2$ through the x-axis.

11.26 (a) L^2 (b) Graph 3

11.27 (d) Their slopes

11.28 $x = -1, \quad y = 0$

11.2 HOMEWORK EXERCISES

Basic Exercises

1. A marathon runner runs 12 km/hr. So $d = 12t$ for distance d (in kilometers) and time t (in hours).
(a) Why must $t \geq 0$?
(b) Why must $d \geq 0$?
(c) Graph $d = 12t$. Put t values on the horizontal axis and d values on the vertical axis.
(d) What shape is the graph?
(e) Is this formula exact or approximate?

2. A pump fills up an empty swimming pool at the rate of 4 m^3/min.
  (a) A formula for the volume of water, V (in cubic meters), in the swimming pool, in terms of the elapsed time T (in minutes) is $V =$ ________.
  (b) Graph V vs. T.

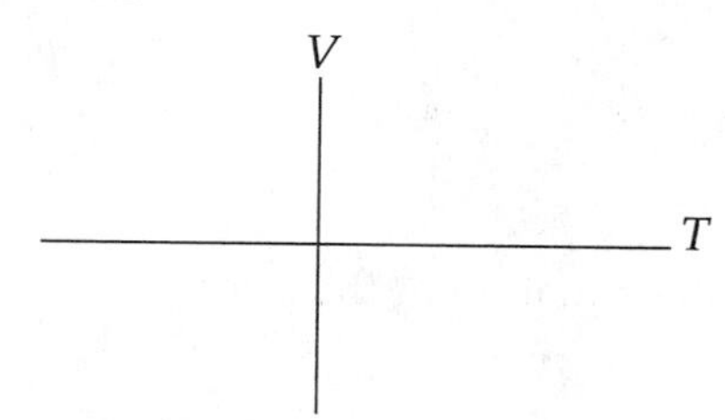

  (c) What shape is your graph?

3. A store estimates that the daily revenue (in dollars) $R = 4C$, in which C is the number of customers that day.
  (a) Describe in words what the formula says about customers and revenue.
  (b) Show why R is proportional to C.

4. What do the graphs of equations of the form $y = mx$ have in common?

5. Classify the slope of each line as slope < 0, slope $= 0$, or slope > 0.

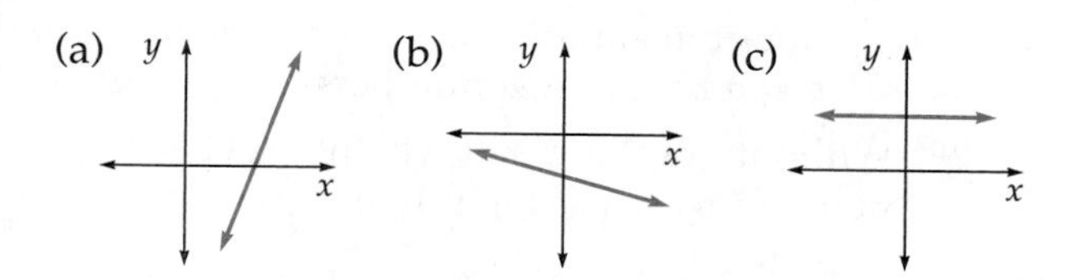

6. Which is steeper, a line in which a change in x of -3 results in a change in y of 6, or a line in which a change in x of -3 results in a change in y of 3?

7. Give the slope of each line.

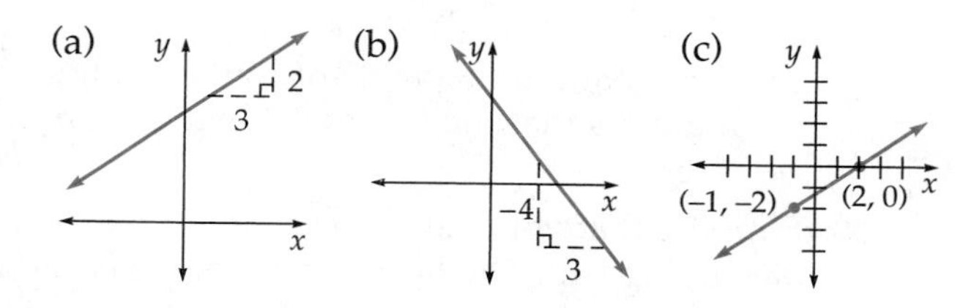

8. What is the slope of each segment?

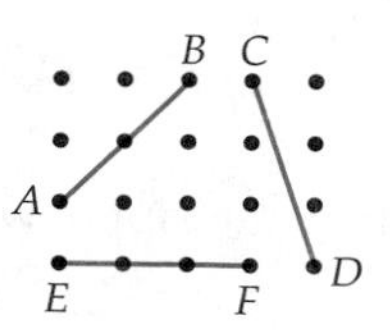

9. What is the slope of each segment?

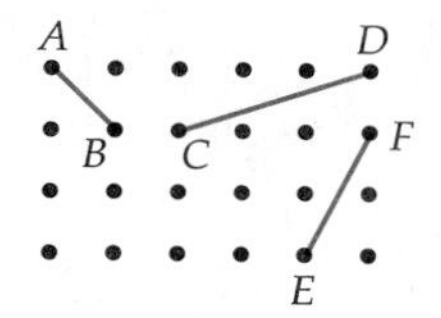

10. If a line goes down from left to right, what do you know about the slope?

11. A building entrance is 3 ft above ground level. A wheelchair ramp with slope 1/12 would have to be how long to go from the ground to the entrance?

12. A highway grade of 3% means that the highway rises 0.03 mile for every 1 mile of horizontal distance. How much does a highway with a 4% grade rise in 2 miles?

13. Water is being pumped into a tank. Readings are taken every 3 minutes.

Time (min)	0	3	6	9	12
Quarts of water	0	90	180	270	360

  (a) Plot these 5 points.
  (b) What is the slope of the line joining the 5 points?
  (c) Estimate how much water is in the tank after 1 minute.
  (d) At what rate is the water being pumped in?

14. The table gives Annabel's height at different ages. During which time period was she growing the fastest?

Age (yr)	3	4	5	6
Height (in.)	30	38	44	48

15. (a) Plot $A(2, 3)$, $B(4, 3)$, and $C(5, 6)$.
(b) Find the coordinates of D so that $ABCD$ is a parallelogram.
(c) Find the slope of $\overline{BC}$ and $\overline{AD}$.

16. (a) In computing the slope, would $\frac{y_2 - y_1}{x_2 - x_1}$ and $\frac{y_1 - y_2}{x_1 - x_2}$ come out the same?
(b) Explain the significance of your answer to part (a).

17. The graph of $y = 4x + 5$ is the same as the graph of $y = 4x$ translated ______________.

18. (a) Graph $y = 2x$.
(b) Draw the image of $y = 2x$ after a translation 3 units to the right.
(c) What is the equation of the image line?

19. What is the relationship between the graphs of $y = 5x + 2$ and $y = 5x + 6$?

20. Tell whether or not the graph of each equation passes through the origin. Explain how you know.
(a) $y = x + 4$ (b) $y = 3x$
(c) $y - 2x = 0$ (d) $y - x = 2$

21. How could you show that the two line segments on the square lattice are parallel?

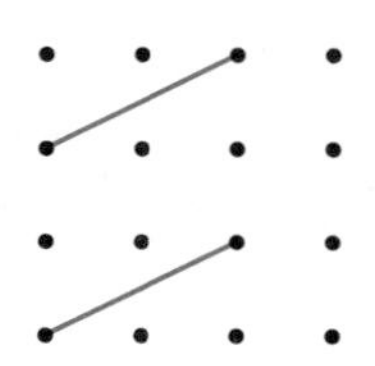

22. Without graphing, tell whether $y = 3x - 2$ and $y = x - 2$ represent parallel lines.

23. (a) Draw the graph of $y = x + 1$. (You may use a graphing calculator for this exercise.)
(b) Graph $y = 2x + 1$ on the same graph.
(c) Graph $y = 3x + 1$ on the same graph.
(d) What is the same about the three graphs?
(e) Which graph is the steepest?

***24.** Find the slope and y-intercept of
(a) $y = 2x - 5$. (b) $5x + 2y = 10$.

25. What is the equation of the line?

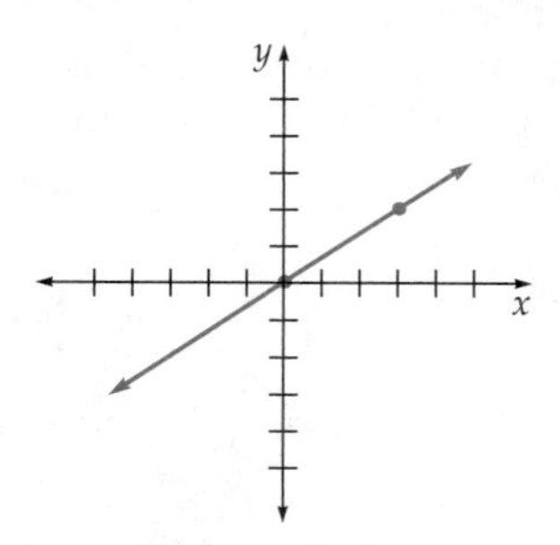

26. What is the equation of the line?

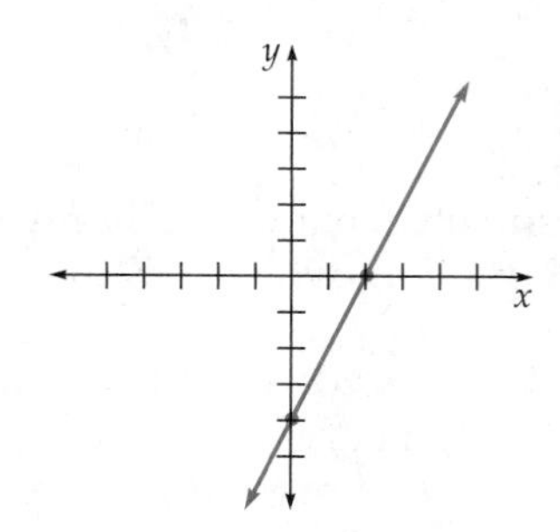

27. A line passes through (0, 5) and (4, −3). What is its equation?

28. (a) Name a point on the graph of $2x - y = 8$.
(b) Is (2, −4) on the graph of $2x - y = 8$?
(c) How many different pairs of solutions does the equation $2x - y = 8$ have if x and y can be any real numbers?

29. Without graphing, tell whether each set of points lies on a straight line.

(a)

x	1	2	3	4
y	5	7	9	11

(b)

x	1	2	3	4
y	1	2	4	7

(c)

x	1	2	3	4
y	11	8	5	2

(d) What pattern in a table of x and y values indicates that the points lie on a straight line?

30. Suppose that you want to cook a hot meal on a camping trip. The higher you go, the easier it is to boil water. At sea level, the boiling point of water is 212°F. For every additional 500 ft above sea level, the boiling point decreases by 1°.

(a) Complete the table.

Altitude, A (ft)	0	500	1000	1500
Boiling point, B (°F)	212	211		

(b) A formula for B in terms of A is $B = 212 -$ ________.

(c) Graph your equation from part (b) with A on the x-axis.

31. You can time the interval between the lightning flash and the accompanying thunder to estimate how far it is from you. Every 5 seconds in the interval indicates an additional distance of about 1 mile.

(a) A formula for distance D (in miles) in terms of time T (in seconds) is $D =$ ________.

(b) Graph your formula.

(c) What does the slope tell you about the relationship between distance and time?

(d) The formula is based upon the difference between the speed of light (lightning) and the speed of sound (thunder) in reaching us. The speed of light is 186,000 miles per second, and the speed of sound is 0.2 mile per second. Explain how the formula $D = \frac{T}{5}$ was obtained.

32. In scuba diving, the water pressure increases as a diver descends. For each additional 20 ft of depth, the pressure increases another 9 pounds per square inch (psi).

(a) Assuming that water pressure P equals 0 at the water surface, a formula for P in terms of depth D is $P =$ ________.

(b) Graph your formula.

(c) What does the slope tell you about the relationship between water pressure and depth?

(d) Suppose that you do not want to exceed 25 psi added pressure. Use your graph to estimate how deep you could go.

33. (a) Plot $W(1, 4)$, $E(3, 6)$, and $T(6, 9)$.

(b) Find two more points on the same line as W, E, and T.

(c) What pattern do you see in the coordinates of the five points?

(d) Write an equation for y in terms of x.

34. Which of the following have straight-line graphs?

(a) $x^2 + y^2 = 9$ (b) $y = 4x^2 + 2$

(c) $5x - 2y = 13$

35. (a) Graph $y = x^2 + 1$. Use $x = -3, -2, -1, 0, 1, 2,$ and 3.

(b) Write three ordered pairs that are solutions to $y = x^2 + 1$.

(c) For what values of x, $-3 \le x \le 3$, is it true that as x increases, y increases?

(d) Use your graph to estimate x when $y = 7$.

***36.** How is the graph of $y = x^2 + 3$ related to the graph of $y = x^2$?

37. How is the graph of $y = x^2 + 5$ related to the graph of $y = x^2 + 1$?

38. How is the graph of $y = -x^2 - 1$ related to the graph of $y = x^2 + 1$?

39. When you drop a heavy object, its speed increases at every instant until it hits the ground. Which of the following graphs shows the correct relationship between h, the height above the ground, and t, the time elapsed since you dropped the object?

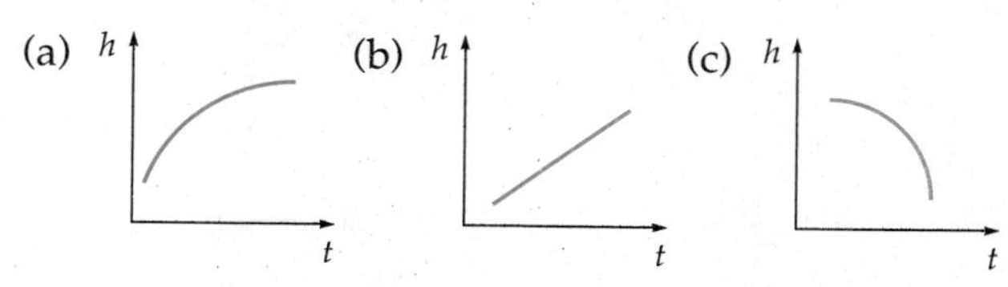

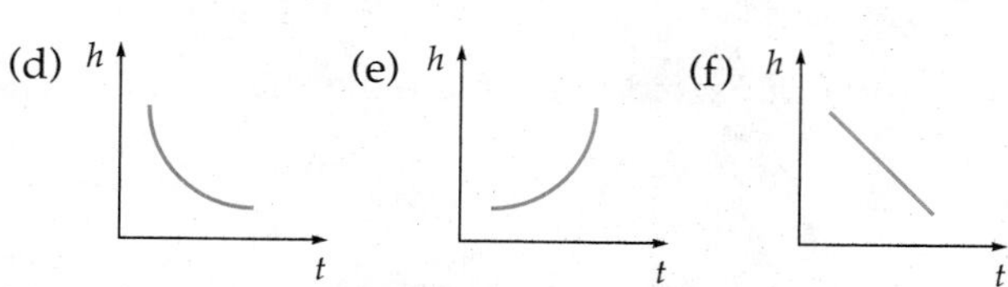

40. The graph of the price p of a television vs. the length of its screen diagonal d is approximately a parabola. Which graph is correct?

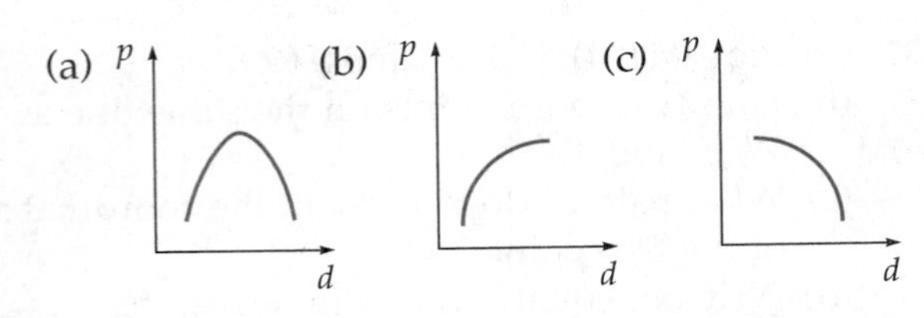

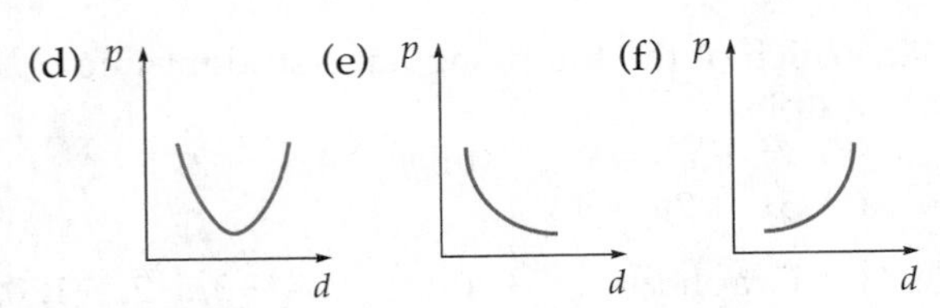

41. A ball is dropped from a height of 64 ft. Its height H (in feet) at time T (in seconds) is $H = 64 - 16T^2$.
(a) Graph this equation for $T = 0, 0.5, 1, 1.5,$ and 2.
(b) What shape does the graph appear to be?
(c) Why would it be wrong to use $T = 3$ in this equation?

42. Revenue R from ticket sales for a concert depends upon the price P that you charge, according to $R = -60P^2 + 840P$.
(a) What values of P should be excluded?
(b) Graph this equation.

43. (a) Plot the points $M(4, 8)$, $A(-2, 8)$, $R(4, 2)$, and $K(-2, 2)$. Connect M to A, A to R, R to K, and K to M.
(b) What shape results?
(c) Plot $MARK$ on axes like the one shown here. How is it different?

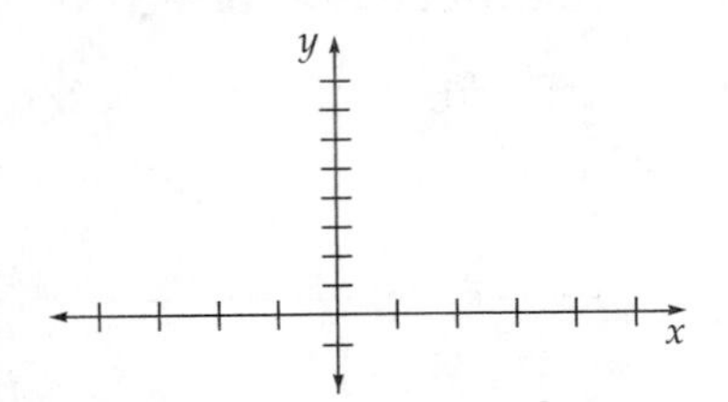

44. Use a trapezoid to estimate the shaded area.

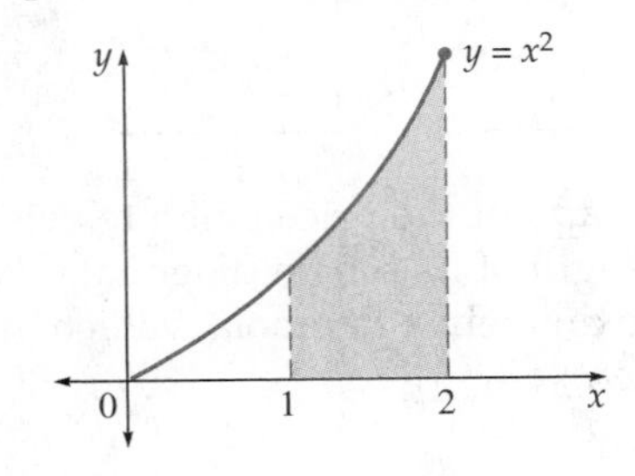

45. (a) Graph $y = x^3$. (You may use a graphing calculator for this exercise.)
(b) Guess what the graph of $y = x^3 + 2$ will look like.
(c) Graph $y = x^3 + 2$, and check your guess.

46. (a) Plot $M(0, 0)$, $A(3, 0)$, and $N(3, 2)$, and connect M to A, A to N, and N to M.
(b) What shape do you obtain?
(c) Add 3 to the y-coordinates of points M, A, and N to obtain coordinates for S, K, and Y, respectively.

$S = (\ \ ,\ \ )$ $K = (\ \ ,\ \ )$

$Y = (\ \ ,\ \ )$

(d) Plot S, K, and Y, and connect S to K, K to Y, and Y to S. What shape do you obtain?
(e) What motion would map $\triangle MAN$ to $\triangle SKY$?
(f) How does $\triangle MAN$ compare to $\triangle SKY$?

Extension Exercises

47.

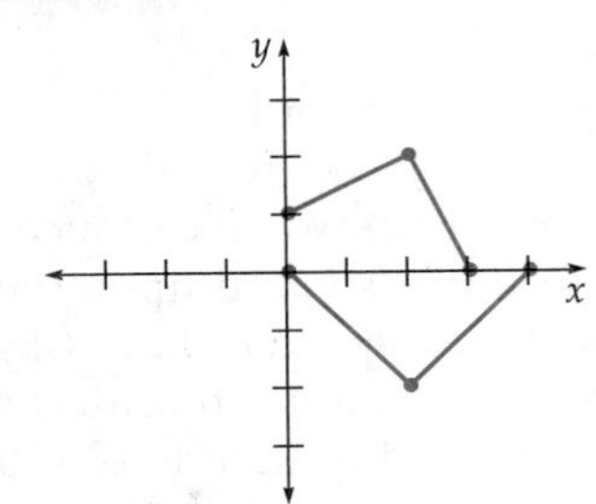

(a) Find the slopes of each pair of perpendicular lines.
(b) Make a conjecture about the relationship between the slopes of perpendicular lines.

48. Use the properties of similar triangles to show that the slopes m_1 and m_2 of perpendicular lines L_1 and L_2, respectively, satisfy $m_1 m_2 = -1$.

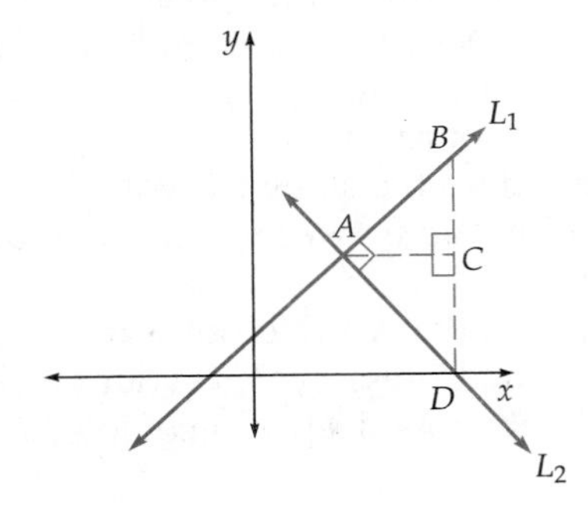

49. Motion geometry can be used to show why two perpendicular lines (which are not horizontal and vertical) have slopes with a product of -1. In the diagram that follows, the legs of $\triangle T$ show the slope of p. Answer parts (a) and (b) to see why the slopes of lines p and q have a product of -1.

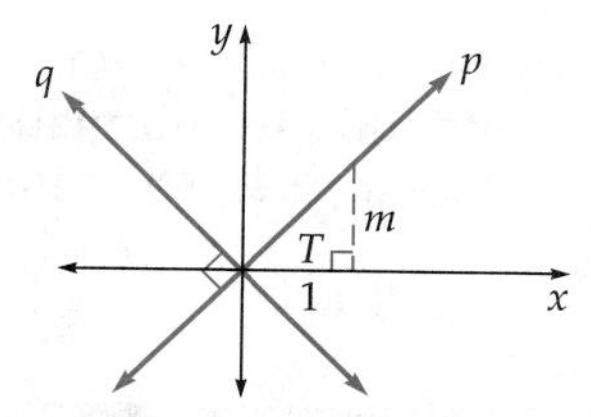

(a) What is the slope of line p?
(b) Rotate line x and $\triangle T$ counterclockwise 90°. Label the image of $\triangle T$ as $\triangle T'$. What is the slope of line q?

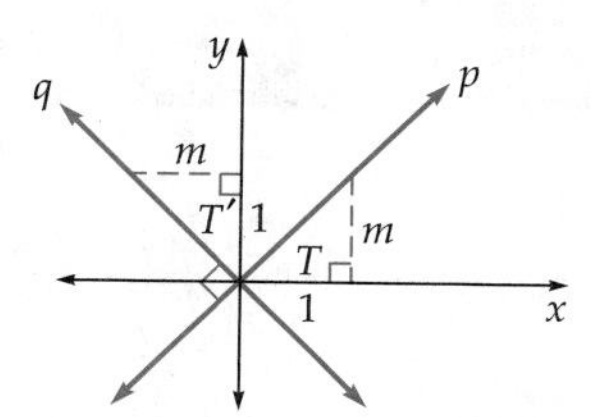

50. Use the result of the preceding exercise to explain why the two segments on the following square lattice form a right angle.

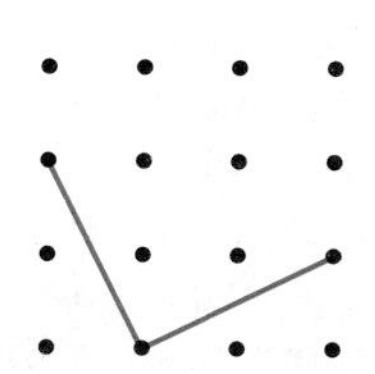

51. Use slope properties to determine if $ABCD$ is a rectangle for $A(0, 0)$, $B(3, 6)$, $C(8, 3)$, and $D(5, -3)$.

52. (a) Graph $y = 2x$.
(b) Draw the image of $y = 2x$ after a reflection through the y-axis.
(c) What is the equation of the image line?

53. (a) Graph $y = 3x - 2$.
(b) Draw the image of $y = 3x - 2$ after a reflection through the y-axis.
(c) What is the equation of the image line?

54. (a) Graph $y = 2x$.
(b) Draw the image of $y = 2x$ after a 90° clockwise rotation around (0, 0).
(c) What is the equation of the image line?

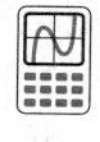

55. Use a graphing calculator (or graphing software) to investigate the relationship between the graphs of $y = mx + b$ and $y = -mx - b$ by graphing pairs of equations such as $y = 2x + 3$ and $y = -2x - 3$.

56. (a) Use a graphing calculator (or graphing software) to graph $y = \frac{1}{x}$.
(b) Investigate the significance of a real number a in the graph of $y = \frac{1}{x - a}$.

Computer Exercises

57. The Logo command SETX 50 moves the turtle from its current position to horizontal coordinate 50. The y-coordinate does not change. There is a similar SETY command.
(a) Start with the turtle at HOME. Enter the following.

```
SETX 30
SETY 20
SETX 0
SETY 0
```

What is the result?
(b) Write a procedure using SETX and SETY to draw a rectangle with vertices (−20, 50), (−20, 10), (10, 50), and (10, 10).

58. Investigate the SETXY (Terrapin or PC Logo) or SETPOS (Apple Logo) command that accepts x- and y-coordinates as in SETXY 10 20 (Terrapin Logo), SET XY [10 20] (PC Logo), or SETPOS [10, 20] (Apple Logo). Try to draw a rectangle or parallelogram with four given coordinates. Use PENUP before the first coordinate and PENDOWN after the first coordinate. Then continue entering coordinates.

11.3 WORDS, ALGEBRA, TABLES, AND GRAPHS

Mathematicians communicate with words, equations, tables, and graphs. Each format offers a different way of describing a relation. This lesson makes use of these formats and shows how to translate from one another.

TRANSLATING WORDS INTO ALGEBRA

In applying mathematics to everyday situations, one may translate words into algebra. Algebra offers a precise way to communicate information about quantities. Some people enjoy the precision and logic of algebra. But not everyone has this reaction (Figure 11-20).

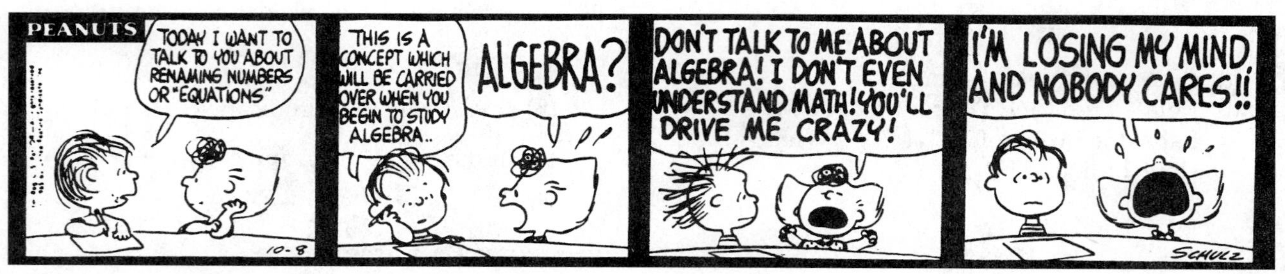

Figure 11–20

Mary Fairfax Sommerville (1780–1872), a self-educated Scotswoman, had a more positive reaction to algebra. Reading a fashion magazine at a tea party, she wondered about a puzzle filled with x's and y's. After learning that it was algebra, Sommerville decided to learn more about the subject. It was considered inappropriate for a young woman to study mathematics so Sommerville had a man secretly obtain a copy of Euclid's *Elements* for her.

Later in life, Sommerville published some articles on mathematics. Her finest work was her 1831 translation of and commentary on Pierre LaPlace's *Méchanique Céleste*, a difficult text on the mathematics of astronomy. Sommerville continued doing mathematics until the day she died at age 92.

Algebra serves as a mathematical language. It is more concise, more abstract, and less ambiguous than English. Algebra is also limited; it can only describe relationships between quantities. I couldn't say $H = M + 5$, in which H = Horace and M = Mitzie. You'd say, "Now hold on, pal! Horace and Mitzie are more than just numbers." And you'd be right.

So how and when do mathematicians translate words into algebra? First, consider English phrases. Some English phrases can be translated into algebraic expressions in two steps.

Step 1: Select variables for unknown quantities.

Step 2: Write an expression or equation using the variable(s).

EXAMPLE 11.3

Translate "five years older than Mitzie" into algebra.

Solution

1. Mitzie's age is the unknown quantity. So let M = Mitzie's age in years.
2. Then "five years older than Mitzie" is $M + 5$.

So the answer is: Let M = Mitzie's age

$$M + 5$$

■

EXAMPLE 11.4

Translate "funnier than a lead balloon" into algebra.

Solution

This phrase does not involve relationships between *quantities,* so it cannot be translated into algebra. ■

See whether you can translate each of the following English phrases into an algebraic expression.

Lesson Exercise 11.29

Translate "25% of the recommended daily allowance of iron" into algebra.

Lesson Exercise 11.30

Translate "a shower of commanded tears" into algebra.

You have seen how some English *phrases* translate into algebraic **expressions** (an arithmetic operation or operations involving variables and numbers). What does an English *sentence* look like after being translated into algebra?

EXAMPLE 11.5

Translate the following into an algebraic sentence. "The total cost of the repair is \$80 for parts plus \$30 per hour for labor.

Solution

The unknowns are total cost and hours of labor. Let C = total cost and H = hours of labor. Then translate the sentence. The phrase "the total cost of the repair is" translates to "C =", and this equals "\$80 plus \$30 per hour of labor."

If you are not sure how to compute the cost from the number of hours, try it with a number such as 10 hours and see how you would find the cost. To compute the cost of 10 hours, multiply 30 by 10 and add 80. To compute the total cost, multiply 30 by the number of labor hours (H) and add 80.

$$C = 30H + 80$$ ■

EXAMPLE 11.6

Translate the following into an algebraic sentence. "Sally's income is at least twice as much as Bill's income."

Solution

Let S = Sally's income, and B = Bill's income. The phrase "at least" translates to "$\geq$."

$$S \geq 2B$$ ■

The preceding examples illustrate the ways in which English sentences that describe quantities correspond to equations or inequalities. If possible, translate the English sentences in Lesson Exercises 11.31 and 11.32 into equations or inequalities.

Lesson Exercise 11.31

Translate the following statement into an equation or inequality. "Monthly local phone service costs $5 plus $.08 for each call."

Lesson Exercise 11.32

Translate the following statement into an equation or inequality. "The sum of the money spent by the U.S. government on transportation and environmental protection in fiscal year 1996 was less than 1/6 the amount spent on national defense."

One important reason for translating English into algebra is to obtain concise descriptions of rules and properties. For example, instead of saying, "We can add two real numbers in either order and obtain the same sum," we could write "$x + y = y + x$ for real numbers x and y."

Lesson Exercise 11.33

Translate the following into an algebraic equation. "When two rational numbers are multiplied, the result is the product of the numerators divided by the product of the denominators."

A second important reason for translating English into Algebra is to obtain a formula. A formula can be obtained by translating an English-language description into an algebraic formula or by determining how the quantities in a problem are related mathematically, as in the following example.

EXAMPLE 11.7

A gas station sells regular unleaded gas for $1.20 per gallon and premium unleaded gas for $1.30 per gallon. Write a formula for total sales in terms of the amount of each type of gas sold.

Solution

Let S = total sales in dollars, R = gallons of regular unleaded gas sold, and P = gallons of premium sold. How does one compute S in this problem? By multiplying R by 1.20 and P by 1.30 and then adding the results together. In symbols,

$$S = 1.20R + 1.30P$$

Lesson Exercise 11.34

A job pays $100 per week plus 5 percent of sales. Write a formula for total weekly salary in terms of sales.

An algebraic equation or inequality focuses upon some quantifiable aspect of a situation. In Example 11.6, the inequality does not tell us what Bill and Sally do, whether they enjoy their jobs, or whether their work is beneficial to society.

Lesson Exercise 11.35

What are some questions that the equation in Lesson Exercise 11.34 does not answer about the quantities involved?

TRANSLATING ALGEBRA INTO WORDS

You have practiced translating words into algebra. Now try translating algebra into words (Figure 11-21, on page 674).

EXAMPLE 11.8

Translate the following expression into English: "$H = 0.2R$, in which H = cost of heat in dollars and R = rent in dollars."

Figure 11–21

Solution

An equal sign can be translated to "is" or "is equal to." So the heat is 0.2 times the rent. Or, in more common English, the heat costs $\frac{2}{10}$ or 20% of the rent. ■

Try the following exercise.

Lesson Exercise 11.36

Translate into an English sentence the equation $T = 0.15B$, in which T = the tip in dollars and B = the bill in dollars.

TRANSLATING GRAPHS INTO WORDS

Like equations and inequalities, a graph displays a relationship that can also be written in words.

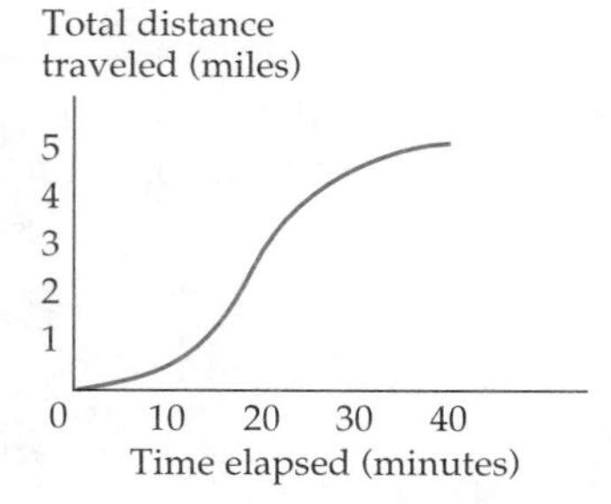

Figure 11–22

Lesson Exercise 11.37

Figure 11-22 shows a graph of the progress of a runner. Which of the following describes the run?

(a) She started slowly and went faster and faster.

(b) She started fast, slowed down, and sped up again.

(c) She started slowly, sped up, and slowed down again.
(d) She started slowly and sped up at the end.

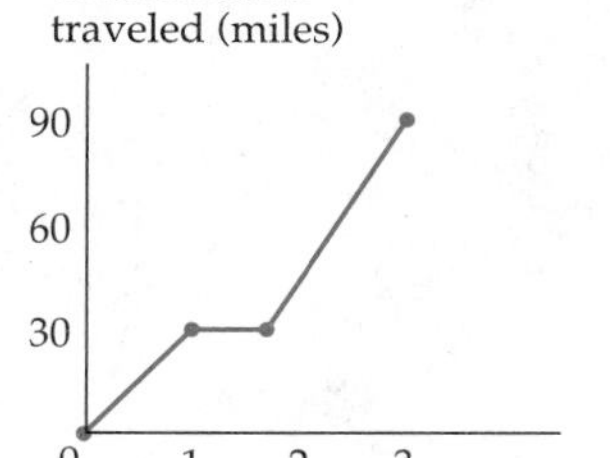

Figure 11–23

In the next exercise, see whether you can interpret the graph in words.

Lesson Exercise 11.38

Figure 11-23 shows a graph of a journey by car.

(a) At what point might the driver have stopped to rest?
(b) When was the car traveling the fastest?
(c) Describe the progress of the car during the journey.

TRANSLATING TABLES AND GRAPHS INTO EQUATIONS

You have probably solved a lot of algebraic equations in your time but have less experience finding equations related to everyday applications. Where do such equations come from? Sometimes a formula (equation) can be derived from a table or graph.

How much would you weigh on the moon (Figure 11-24)? In order to answer this question, you need a formula. The following data relate people's weights on the Earth and the moon.

Photo courtesy of NASA

Figure 11–24

Weight on the Earth	100 lb	150	200
Weight on the moon	16 lb	24	32

Lesson Exercise 11.39

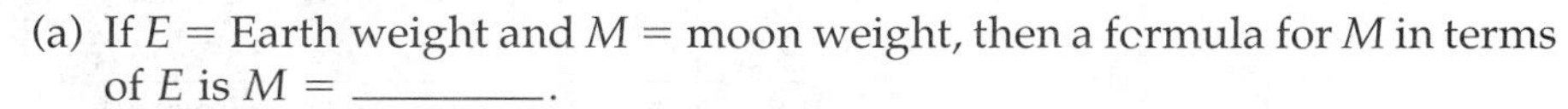

(a) If E = Earth weight and M = moon weight, then a formula for M in terms of E is $M =$ ________.
(b) Did you use induction or deduction to find the formula?
(c) The author of this book weighs 138 pounds. How much would he weigh on the moon? Would you like to send him there?

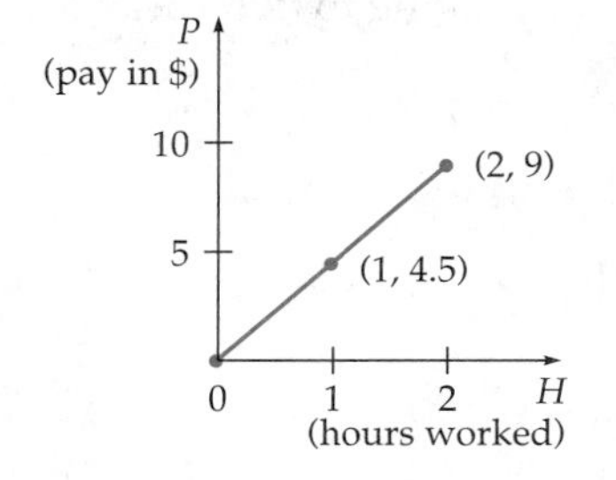

Figure 11–25

You can sometimes translate a graph into an equation. The shape of the graph (for example, a line or parabola) may suggest the appropriate form of the equation. If the graph is a linear function, the equation has the form $y = mx + b$, and if the graph is a parabolic function, the form may be $y = ax^2 + c$.

Lesson Exercise 11.40

Write an equation for the graph in Figure 11-25.

Lesson Exercise 11.41

Write an equation for the following graph.

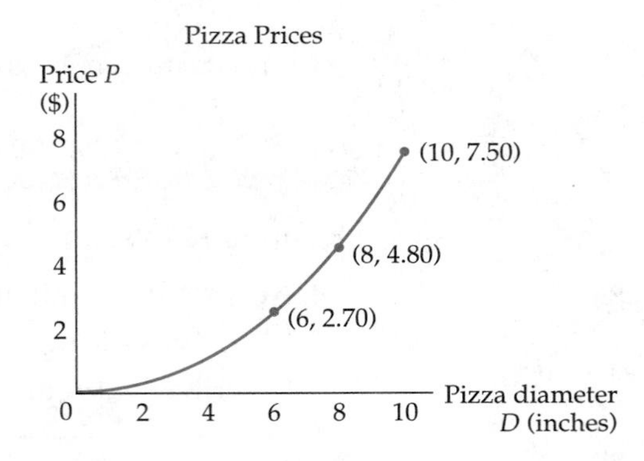

A scatterplot is a useful display of the relationship between two variables, as shown in Figure 11-26. As height increases, weight also tends to increase (Figure 11-26a). This relationship shows a **positive correlation**. A graph of the price of apples versus supply (Figure 11-26b) shows a **negative correlation**, since price tends to decrease as supply increases.

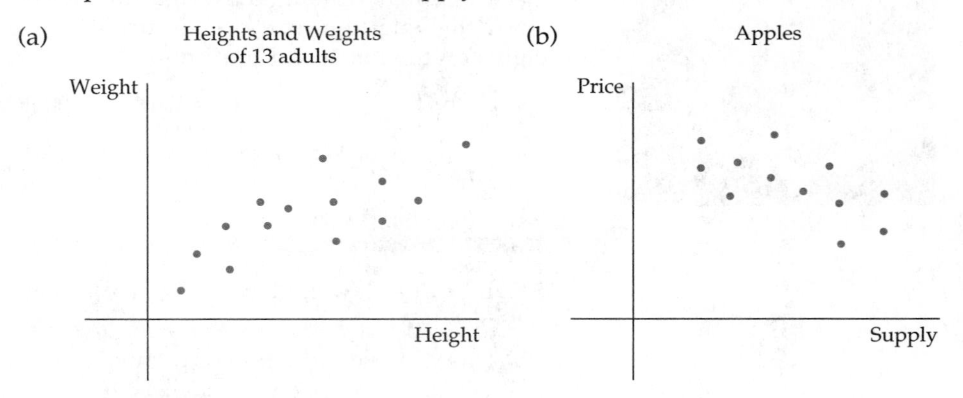

Figure 11–26

If a scatterplot approximates a shape such as a line segment or a parabola, the corresponding equation may be tried as a formula relating to two variables.

Lesson Exercise 11.42

A sample of 20 married couples yields the following data in the form (husband's age, wife's age).

(35, 30)	(31, 31)	(42, 37)	(29, 26)	(20, 21)	(25, 23)	(36, 34)
(40, 39)	(34, 32)	(28, 31)	(47, 47)	(29, 27)	(40, 31)	(37, 36)
(37, 36)	(38, 34)	(38, 31)	(24, 20)	(37, 35)	(29, 25)	

(a) Plot the 20 points on a two-dimensional graph.
(b) Does this graph show a positive correlation or a negative correlation?
(c) Draw a straight line (a "regression line") that seems to fit the data best. Draw it so that about half of the points are below the line and half are above it.
(d) Select two points on your line, and use them to find the equation of your line.
(e) If it's available, use a graphing calculator or computer software to find the best-fitting line, and compare the results to part (c).

TRANSLATING WORDS INTO TABLES

Suppose you are planning a concert at your school and want to see how much profit you might make in relation to ticket sales. The following exercise illustrates how a computer spreadsheet can be used to generate a table of possible outcomes after you translate from words into equations.

Lesson Exercise 11.43

The concert will cost \$2200 to set up and \$2 in cleanup costs per person. Tickets will sell for \$18 each.

(a) Write a formula for the total cost C of the concert in relation to the number of tickets sold, x.
(b) Write a formula for the total revenue R from the sale of x tickets.
(c) How is profit related to revenue and cost?
(d) Write a formula for the profit P in terms of x.
(e) If you have a computer spreadsheet, set it up as follows. Type the numbers 50, 100, 150, . . . 500 in column A. The top part of the spreadsheet follows.

	A	B	C	D
1	**Concert**			
2	Tickets sold	Cost	Revenue	Profit
3	50	\$2300	\$900	–\$1400
4	100			

(f) Enter appropriate formulas in cells B3, C3, and D3. Then copy the formulas through the rest of your columns, B, C, and D. Compare your results for B3, C3, and D3 with those given in the spreadsheet.
(g) Comment on the results in each column.

ANSWERS TO SELECTED LESSON EXERCISES

11.29 Let I = recommended daily allowance of iron, $0.25I$ or $\frac{1}{4}I$.

11.30 Cannot be translated

11.31 $M = 5 + 0.08N$, in which M = the monthly charge in dollars and N = the number of local calls made.

11.32 Let T = transportation budget in dollars.
Let E = environmental protection budget in dollars.
Let D = defense budget in dollars.

$$T + E < \frac{1}{6}D$$

11.33 $\frac{a}{b} \times \frac{c}{d} = \frac{a \times c}{b \times d}$, in which $\frac{a}{b}$ and $\frac{c}{d}$ are rational numbers.

11.34 $P = 100 + 0.05S$, in which P = weekly pay and S = sales in dollars.

11.35 What is the job? What do others get paid? Is the job enjoyable or interesting? Is the job beneficial to society?

11.36 The tip is 0.15 times the bill; or, more commonly, the tip is 15% of the bill.

11.37 (c)

11.38 (a) After 1 hour
(b) From 2 to 3 hours into the trip
(c) The car traveled for an hour at 30 mph. Then it stopped for a little less than an hour. Then the car traveled for a little over an hour at about 50 mph.

11.39 (a) $0.16E$ (b) Induction
(c) About 22 lb

11.40 $P = 4.5H$

11.41 $P = 0.075D^2$

11.42 (a)

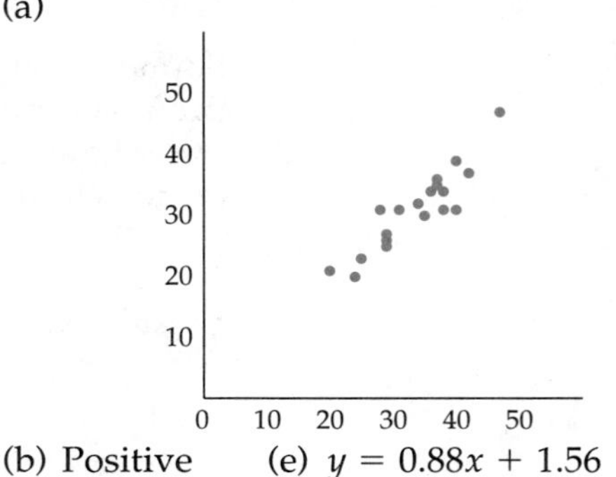

(b) Positive (e) $y = 0.88x + 1.56$

11.43 (a) $C = 2200 + 2x$
(b) $R = 18x$
(c) Profit = $R - C$
(d) $P = 16x - 2200$
(g) The operation shows a profit for 150, 200, 250, . . . , 500 tickets sold.

11.3 HOMEWORK EXERCISES

Basic Exercises

1. Translate "twelve degrees warmer than the average for this day" into an algebraic expression.

2. Translate "40% off the regular price" into an algebraic expression.

3. (a) Translate "Twenty dollars is 85% of the list price" into an equation.
   (b) What doesn't this equation tell us about the quantities involved?

4. Which of the following could be represented by an algebraic variable?
   (a) Mary's age
   (b) Transportation
   (c) Mike

5. In the cartoon, why can't X = Shobert Zangox?

© 1972 Walt Kelly. Reprinted with permission of Los Angeles Times Syndicate.

If possible, translate the English sentences in Exercises 6–12 into algebraic equations or inequalities.

6. "The fool thinks he is wise, but a wise man knows he is a fool." (Shakespeare, *As You Like It*)

7. The perimeter of a special rectangle is three times its length.

8. There are, at most, 30 more girls than boys.

9. Katie is at least twice as tall as she was last year.

10. The ratio of students to teachers is 12 to 1.

11. The total cost of movie tickets is \$5 per adult ticket plus \$3 per child's ticket.

12. Sixty percent of the students at Stellar University are women.

13. Translate into algebra.
(a) The former price P increased by \$6 and the price is now \$20.
(b) A car has N gallons in the tank. After 8 gallons are added, it has G gallons in the tank.

14. When an English sentence is translated into algebra, it becomes an ________ or ________.

15. Translate the following sentence into algebra. "The sum of any real number and zero is that real number."

16. In each case, decide whether the quantity measured is (1) a significant or (2) a trivial feature of the person or object.
(a) What your score S on the last mathematics test tells about how much you know about the material covered
(b) What your score S on the last mathematics test tells about you
(c) What the time T it takes you to finish a mathematics test tells about your general mathematical ability

17. Translate the following sentence into algebra. "The product of a real number and the sum of two other real numbers is the sum of the products of the real number and each of the other two real numbers."

18. Fine Line phone company charges a monthly rate R of \$7 plus \$.08 per call.
(a) What is the total monthly rate R for 10 calls? $R =$ ________.
(b) What is the total monthly rate R for N calls? $R =$ ________. (This is a general formula for R.)
(c) For what values of N can you use the formula?
(d) Is your general formula exact, or is it an approximation?
(e) Graph R (vertical axis) vs. N (horizontal axis).

19. The annual cost C of owning some cars can be estimated as \$1100 plus \$.10 per mile.
(a) What is the annual cost C for M miles? $C =$ ________.
(b) Graph C (vertical axis) vs. M (horizontal axis).

20. A rental car from Nightmares costs \$11 per day plus \$.07 per mile. What is the total cost C for renting a car for D days and driving M miles? (*Hint:* If you're not sure, make up numbers for D and M and see how you compute C.)

21. After 5 years, a car is worth \$3600. After 6 years, the same car is worth \$2700. Assume that its decline in value (depreciation) is the same amount each year. What is a formula relating its worth W to its age in years, Y?

22. A salesperson earns \$120 per week plus a 10% commission on her total sales.
(a) What is the formula relating total weekly earnings E to total sales S?
(b) Graph the formula for E vs. S, with E on the vertical axis and S on the horizontal axis.

23. A man who is 5 ft 8 in. tall weighs 200 pounds.
(a) If he loses 5 pounds per week, what does he weigh after W weeks?
(b) Estimate a maximum reasonable value for W.

24. An airplane at an altitude of 33,000 ft begins descending at a rate of 1200 ft per minute.
(a) Write a formula relating altitude A (in ft) to M, the number of minutes after the descent begins.
(b) What is the domain for M?

Translate the algebraic information in Exercises 25–28 into English.

25. $R = 22D + 10$, in which R = rental rate in dollars and D = number of days

26. $0.1P$, in which P = Iowa's population

27. $C = 0.025S$, in which C = commission in dollars and S = sales in dollars (use the word "percent" in your translation)

28. $C > 2P$, in which C = current U.S. population and P = U.S. population in 1940

29. Here is a graph of a journey by car.

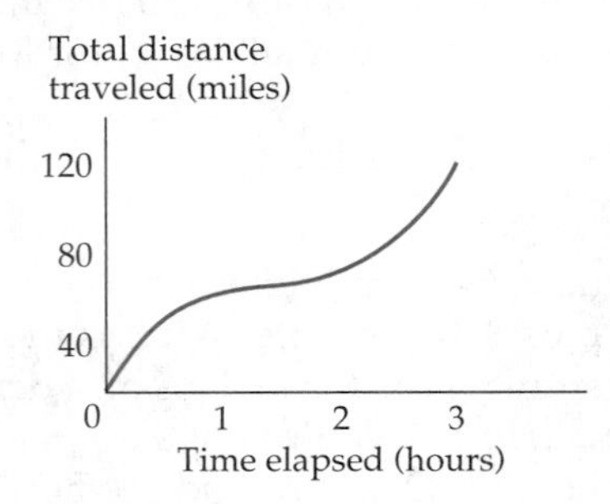

Describe the progress of the car during the journey.

30.

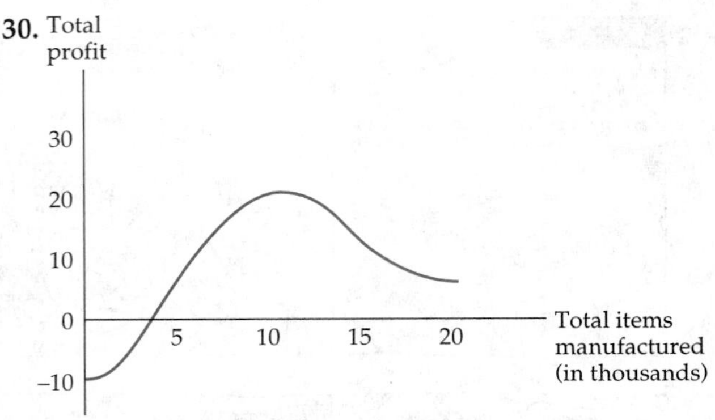

Describe the general relationship between the total number of items sold and the total profit.

31. Match each graph with its description.
(a) Speed of a pitched baseball
(b) Falling object

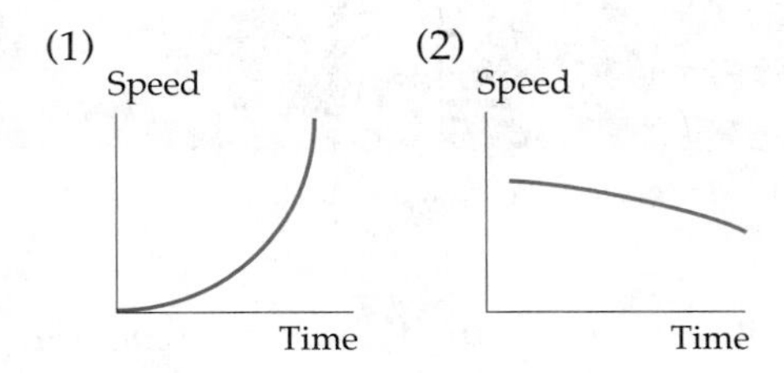

32. What does the graph suggest about the relationship between motivation and the difficulty of a lesson?

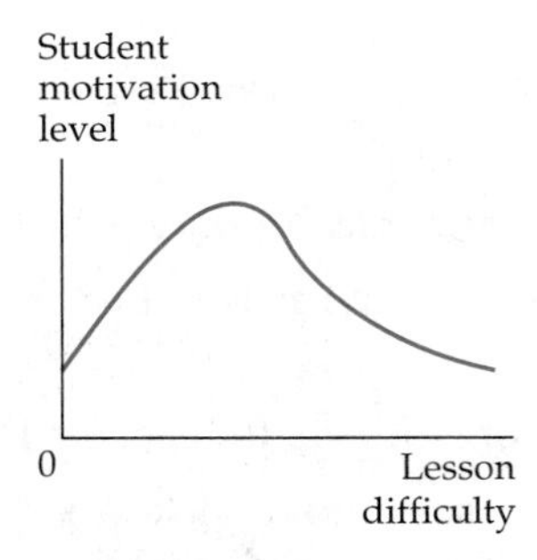

33. Match each graph with its description.
(a) Outside temperature from 7:00 A.M. to 11:00 P.M.
(b) U.S. population from 1950 to 1990
(c) Speed of an airplane during a 2-hour flight

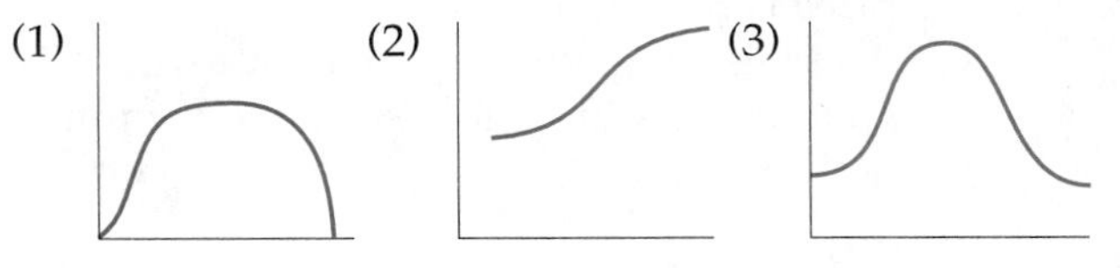

***34.** Make a reasonable sketch of each graph, showing a typical relationship between the two variables.

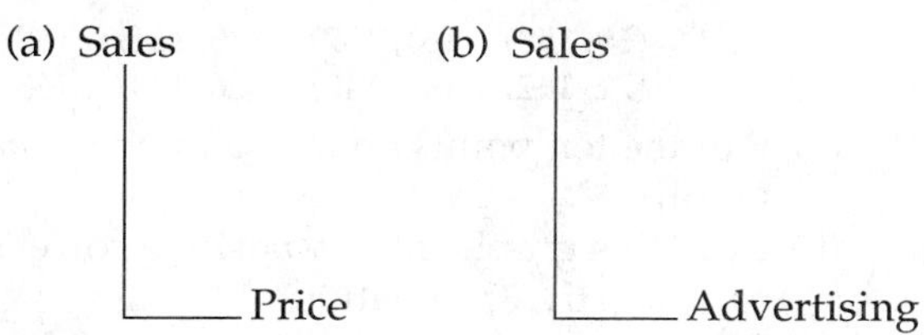

(c) Value of a car

Age of car

35. A roller-coaster car moves from left to right over the course shown in the following drawing. Make a sketch of speed versus time, with time on the horizontal axis.

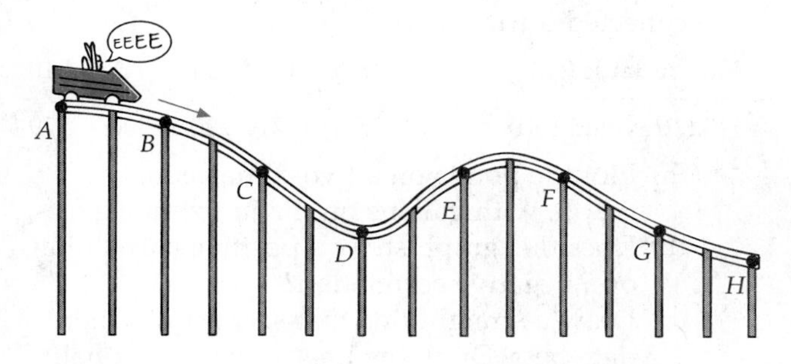

36.

D (km)	0	5000	10,000	15,000	20,000
C ($)	1800	2100	2400	2700	3000

A formula for the annual cost (C) of driving a National Motors Cheetah SST, in terms of the distance driven (D), is $C =$ ________.

37.

Sales, x	0	100	200	300	400
Salary, y	100	110	120	130	140

A formula for salary in terms of sales is $y =$ ________.

38. Julio works in the bakery at the Scrumptious Food Mart. Data on his weekly pay follow.

Hours worked, H	20	36	30
Pay, P	$84	$151.20	$126

(a) A formula for P is $P =$ ________.

(b) Did you use induction or deduction to figure out part (a)?

39.

Hours awake, x	14	15	16	17	18	19
Hours asleep, y	10	9	8	7	6	5

A formula for hours asleep is $y =$ ________.

40. An object is dropped from a height of 200 ft. Its height H, in feet, after T seconds is given in the table. A formula for H is $H =$ ________. (*Hint:* It involves T^2.)

T	0	1	2	3
H	200	184	136	56

41. The average weights of 6-year-old boys with selected heights are given in the table.

Height, H (in.)	42	44	46	48
Weight, W (lb)	43	49	55	61

A formula for W is $W =$ ________. (*Hint:* Find the slope.)

42. The average weights of 21-year-old women with selected heights are given in the table.

Height, H (in.)	60	62	64	66
Weight, W (lb)	112	120	128	136

A formula for W is $W =$ ________.

43. Write an equation for the graph.

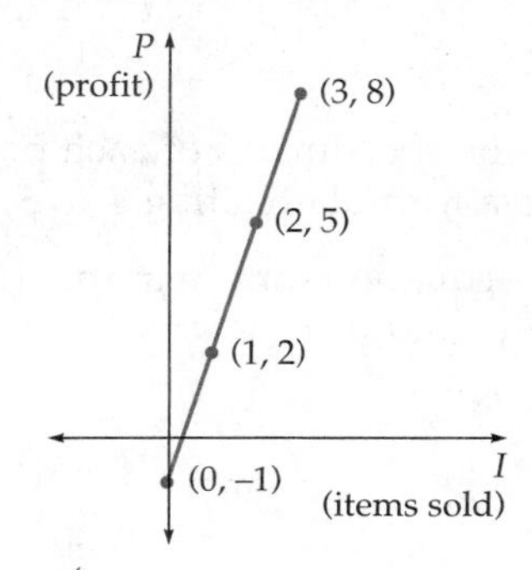

44. Write an equation for the graph.

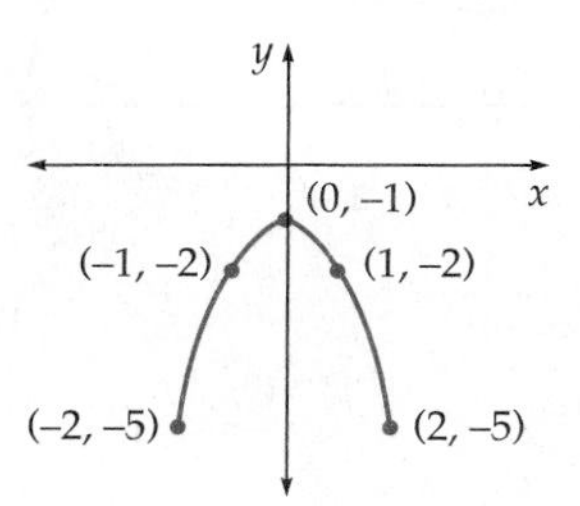

45. The rates for a cab company are shown on the graph. A formula for computing the total cost C in terms of the distance D is $C =$ ________.

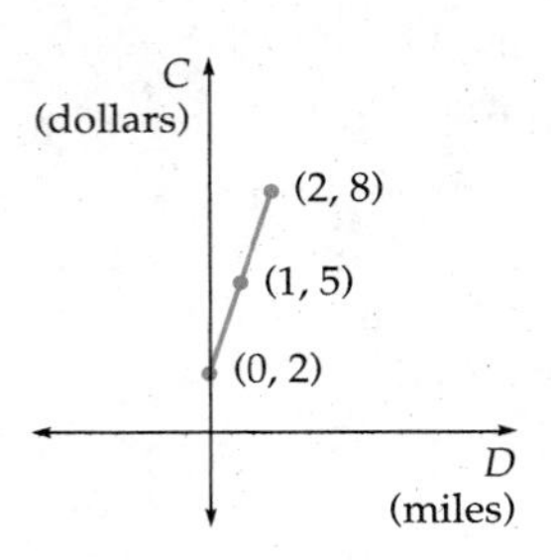

46.

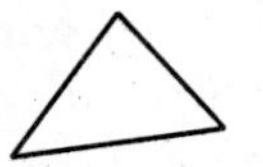
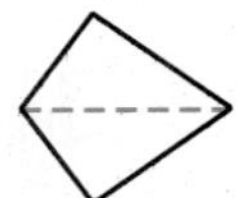
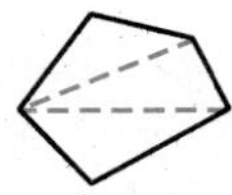
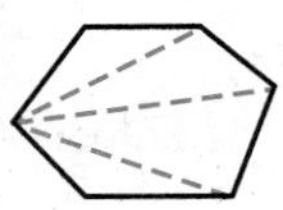

Any polygon can be divided into a *minimum* number of triangular regions by its diagonals. The following graph shows the relationship between S (the number of sides of a polygon) and T (the minimum number of triangular regions formed by diagonals).

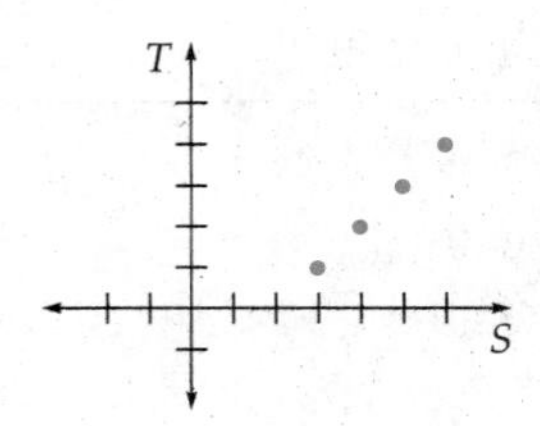

(a) Give the coordinates of each point.
(b) Write a formula relating T to S.

47. Write the equation for the graph. (*Hint:* It involves x^2 and y^2.)

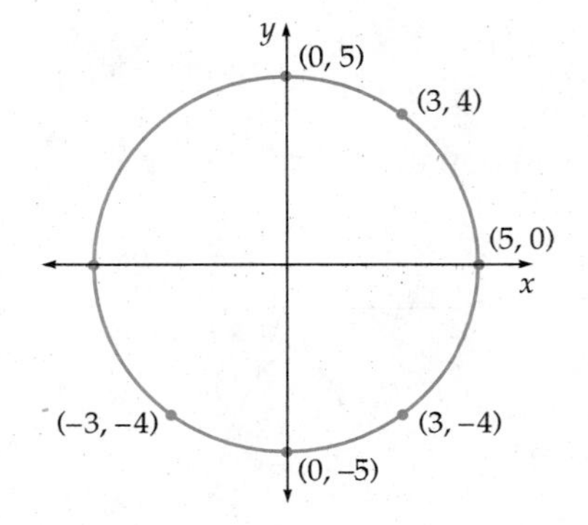

48. The heights (in inches) and weights (in pounds) of ten people in a sample are as follows.

(63, 142) (69, 152) (61, 137) (70, 161) (68, 148)
(73, 166) (72, 162) (64, 140) (66, 144) (65, 160)

(a) Plot the ten points on a two-dimensional graph.
(b) Does this graph show a positive correlation or a negative correlation?
(c) Draw a straight line that seems to fit the data best. Draw the line so that about half of the points are below it and half are above it.
(d) What is the equation of your line?
(e) If you have a graphing calculator or statistical software, use it to find the regression line, and compare it to your answer to part (d).

49. The gas tank of a National Motors Titan holds 20 gallons of gas. The following data are collected during a week.

Fuel in tank (gal)	20	18	16	14	12	10
Dist. traveled (mi)	0	75	157	229	306	379

(a) Plot the points on a two-dimensional graph, with gallons on the horizontal axis.
(b) Does this graph show a positive correlation or a negative correlation?
(c) Draw a straight line that seems to fit the data best. Draw the line so that about half of the points are below it and half are above it.
(d) What is the equation of your line?
(e) If you have a graphing calculator or statistical software, use it to find the regression line, and compare it to your answer to part (d).

50. A community tries different speed limits on a stretch of road and records the following numbers of accidents per month.

Speed limit, S (mph)	40	45	50	55	60	65
Accidents per month, A	3	7	13	18	21	25

(a) Plot the six points on a two-dimensional graph.
(b) Do these data show a positive correlation or a negative correlation?

(continued on the next page)

(c) Draw a straight line that seems to fit the data best. Draw the line so that about half of the points are below it and half are above it.
(d) What is the equation of your line?
(e) If you have a graphing calculator or statistical software, use it to find the regression line, and compare it to your answer to part (d).

51. Tell whether each of the following pairs of variables has a positive correlation or a negative correlation.
(a) The number of sleeping pills taken and the degree of drowsiness you feel
(b) The length of the side of a square and the area of the square
(c) The size of a car's gas tank and the number of stops needed on a long car trip

52. Tell whether each of the following pairs of variables has a positive correlation or a negative correlation.
(a) The winter temperature setting of a home thermostat in Detroit and the amount of heating oil used
(b) The number of people in a car pool and the cost per person of driving to work
(c) The speed of a car and the time it takes to travel 1 mile

Extension Exercises

53. Suppose that you want to fence in a rectangular area of 30 m^2. Complete the following to investigate the length and width such a rectangular area could have.
(a) Fill in the table.

L	3	5	7		
W	10				

(b) Write a formula for W.
(c) Graph your formula.
(d) Describe the shape of the graph. (It is called a **hyperbola**.)
(e) What are the possible values of L and W?

54. You plan to go on a 200-mile car trip. The time it takes is related to the rate at which you drive.
(a) Fill in the table.

Rate, R (mph)	20	30	40	50	60
Time, T (h)	10				

(b) Graph R (vertical axis) vs. T.

55. The time needed to cook a piece of chicken in a microwave oven at a setting of 500 watts is about 6 minutes. Additional information is provided in the table.

W (watts)	200	300	400	500	600
Time, T (min)	15	10	7.5	6	

(a) Write a formula for T.
(b) How long would it take to cook the chicken at 600 watts?
(c) Graph the formula for T vs. W.

56. (a) In a vacuum, the product of the pressure P and the volume V is constant. For example, PV could always equal 4. Graph $PV = 4$, using positive numbers for P and V.
(b) The graph shows that, as P increases, V ________.

57. Compare $x^2 - 9$ and $(x + 3)(x - 3)$ by doing the following.
(a) Plot $y = x^2 - 9$ and $y = (x + 3)(x - 3)$ with a graphing calculator.
(b) Compare the values of $x^2 - 9$ and $(x + 3)(x - 3)$ for $x = -2, -1, 0, 1,$ and 2.
(c) Multiply out $(x + 3)(x - 3)$.
(d) Which part or parts—(a), (b), or (c)—*prove* that $x^2 - 9 = (x + 3)(x - 3)$?

58. Use the information in the table to write an algebraic equation in three variables.

Total cost (T)	\$11	14	41
Pounds of peanuts (P)	1	2	3
Pounds of cashews (C)	1	1	4

59. Bill made 50% more money this year than he did last year. Translate this sentence into algebra. (It cannot be translated word for word.)

60. (a) The length of a rectangle is 4 cm more than its width. Express this relationship with an equation.
(b) Write the area of this rectangle in terms of its width.

61. The graph shows the speed of a roller coaster throughout the ride. Draw a sketch of the roller-coaster track.

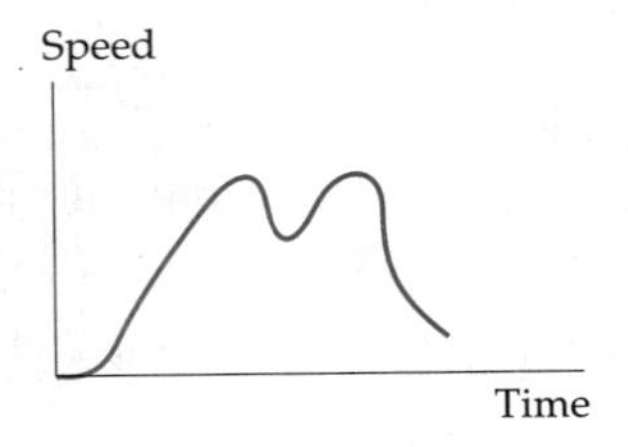

62. The graph shows the speed of a car vs. time. Sketch a graph of distance vs. time for the same trip.

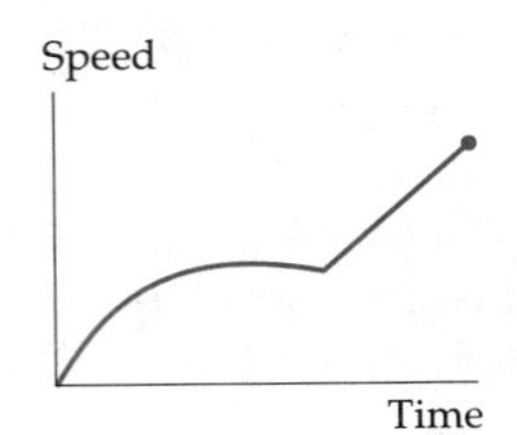

63. In its February 1989 issue, *Consumer Reports* rated 26- and 27-inch color television sets. The following table gives the price and ratings of each set.

Set	Score	Price	Set	Score	Price
Hitachi	90	$609	Zenith	79	$639
Panasonic	86	$680	GE	76	$519
Magnavox	84	$729	Sears	76	$740
Sylvania	84	$572	Sony	74	$747
JC Penney	83	$800	Toshiba	73	$539
Mitsubishi	83	$799	Emerson	71	$525
RCA	83	$579	Fisher	70	$659
NEC	81	$640	JVC	70	$659

(a) Plot price (vertical axis) vs. score.
(b) How strong is the relationship between price and score?
(c) Suppose that it were always true that the higher the price, the higher the score. What would the graph look like?

64. One of the most famous formulas is $E = mc^2$.
(a) What do E, m, and c represent?
(b) What is the formula used for?

Computer Exercises

65. Statistical software such as MINITAB can be used to display data, analyze data, perform simulations, or find a best-fitting line. Further exercises on MINITAB occur in Chapter 12. (This exercise can also be done with a graphing calculator.)

If you have MINITAB, here is how to find a best-fitting line for the data in the last exercise from the lesson (husband's and wife's ages). First, type in the data as follows.

MTB > READ HUSBAND IN C1, WIFE IN C2
DATA > 35 30
DATA > 31 31
⋮

Then type:

mtb > REGRESS 'WIFE' ON 1, 'HUSBAND'

and summarize the output. (Note: Data also can be entered in an array in later versions of MINITAB.)

66. In basketball, a player scores 1 point for a free throw, 2 points for a regular field goal, and 3 points for a long field goal. Find a formula for the total points T in terms of shots made, and use it to generate column E in the following spreadsheet.

	A	B	C	D	E
1	Name	Free Throws	2-Pt Goals	3-Pt Goals	Total
2	Nancy	5	7	2	
3	Tiwana	3	8	3	
4	Hakeem	6	11	0	
5	Shaq	9	10	0	
6	Penny	4	6	4	

67. You plan to go on a 300-mile trip and stop for a 1-hour lunch. You want to find how the total time T for the trip is related to your average driving speed S.
(a) Find a formula for T in terms of S.
(b) Set up a computer spreadsheet as follows. Type the numbers 30, 35, 40, 45, 50, and 55 in column A. The top part of the spreadsheet follows.

	A	B
1	**Trip**	
2	Speed	Total time
3	30	

(c) Enter an appropriate formula in cell B3, and copy until the end of column B.
(d) Summarize the results.

11.4 SOLVING PROBLEMS WITH FORMULAS, TABLES, AND GRAPHS

Consider the following word problems.

- Jane's age is 1 less than one-third her grandmother's age. Six years ago, Jane's age was one-fourth her grandmother's age. How old are they both today?
- I have 74 coins in all, consisting of dimes and quarters that total $14.75. How many dimes and quarters do I have?

Please don't try to solve these problems. Age and coin problems make nice puzzles, but they are not important enough to be done by the pageful, as many algebra classes do.

In this lesson, you will see more realistic applications of algebra that people encounter in everyday life. Such applications should be the focus of algebra classes.

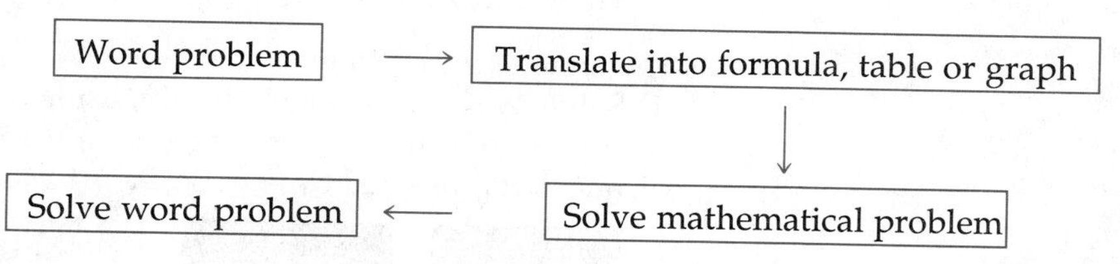

SOLVING PROBLEMS WITH FORMULAS

Are you ready to solve some word problems? Wait—come back here. You can do it. You've studied everything you need to know: (1) translating words into algebra, (2) writing an equation relating the quantities in a problem, and (3) solving an equation for an unknown.

You can use the following five-step procedure.

Understanding the Problem

1. Read the problem and assign a variable to the quantity you are looking for.

Devising a Plan

2. Write an English sentence describing how the quantities in the problem are related.
3. Write an equation (or equations) relating the quantities in the problem.

Carrying Out the Plan

4. Solve for the unknown variable (or variables) in the equation (or equations), and answer the original question.

Looking Back

5. Check to see whether your answer makes sense.

The next example is simple enough that you don't need to use algebra, but it usefully illustrates how manipulatives and pictorial models are used to introduce equation-solving in middle-school mathematics.

EXAMPLE 11.9

Nestor, the dog, just had 2 puppies of equal weight. The 3 animals weigh a total of 12 kg. Nestor weighs 8 kg. How much does each puppy weigh?

Solution

Although you could solve this without algebra, use algebra for practice. Follow the five steps.

Understanding the Problem

1. We are asked to find the weight of each puppy. Let x = weight of each puppy.

Devising a Plan

2. The total weight is 2 times the weight of each puppy plus 8.
3. Therefore, $2x + 8 = 12$.

Carrying Out the Plan

4. Solve $2x + 8 = 12$ algebraically, using a pictorial model of a balance scale in which x is represented by a bar (or bag or carton) and each one is represented by . So $2x + 8$ is represented by

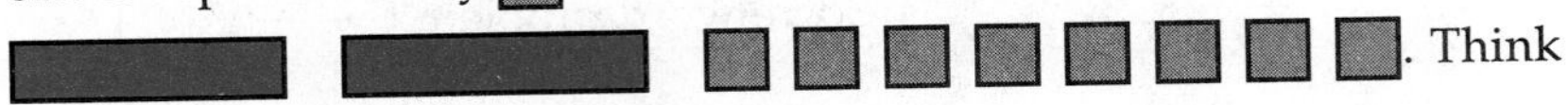

. Think of an equation as a balanced scale.

Pictorial Model

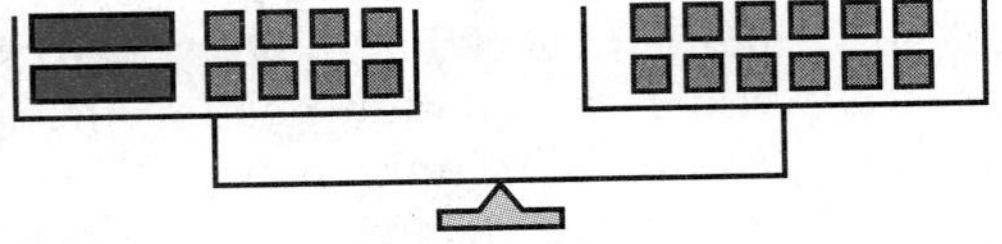

Abstract Model

$2x + 8 = 12$

We want to find the weight of each bar.

So remove 8 squares from each side. This leaves

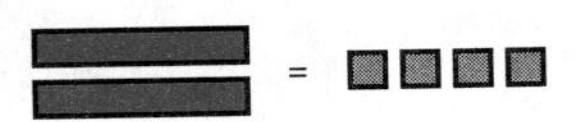

Subtract 8 from each side of the equation.

$2x = 4$

(Note that the model can be shown without a scale.) Since the two bars balance with 4, one bar balances with 2.	Divide each side by 2.
▬ = ■■	$x = 2$

Each puppy weighs 2 kg.

Looking Back

5. Check the answer: 2(2 kg) + 8 kg = 12 kg. ■

Now you try one.

Lesson Exercise 11.44

Consider the following problem. "A radio show has 14 minutes left. The producer plans to play 2 minutes of commercials and 3 songs of about the same length. How long should each song be?" Show how to solve the problem both algebraically and with a pictorial model for solving equations.

The following example illustrates three methods for solving an algebraic word problem: the guess-and-check strategy, writing an equation, and using a proportion.

EXAMPLE 11.10

A saleswoman earns a commission of 6% of her sales. How much must she sell to earn a commission of $1000?

(a) Solve it using the guess-and-check strategy.

(b) Write an equation.

(c) Use a proportion.

Solution

(a) Guess an answer for sales, and compute the commission on it. See how close you are to $1000. Then adjust your answer accordingly.

For example, I might guess $20,000 in sales. The commission would be ($20,000)(0.06) = $1200. Since $1200 > $1000, $20,000 is a little too high. I would guess a little lower than $20,000 and continue in this fashion. I should end up with an answer around $16,700. This method avoids the use of algebra.

(b) Follow the five steps.

Understanding the Problem

1. We are asked to find total sales. Let S = total sales.

Devising a Plan

2. To find the saleswoman's earnings, we multiply 0.06 by her total sales.

3. Therefore, $1000 = 0.06S$.

Carrying Out the Plan

4. Solve $1000 = S \cdot 0.06$ for S.

$$\frac{1000}{0.06} = S$$

Since S represents money, round it to two decimal places.

$$\$16{,}666.67 = S$$

She must sell \$16,666.67 of goods to earn \$1000.

Looking Back

5. Is the answer reasonable? Yes, her earnings are a small fraction of the amount she sells. Six percent is about $\frac{1}{20}$.

$$6\% \text{ of } \$16666.67 \approx \frac{1}{20} \text{ of } \$20{,}000 = \$1000$$

(c) Follow the five steps.

Understanding the Problem

1. We are asked to find the total sales. Let S = total sales.

Devising a Plan

2. She earns \$6 for each \$100 in sales.

Sales	\$100	S
Commission	6	\$1000

This suggests a proportion (four ways to write it) such as $\frac{100}{S} = \frac{6}{1000}$.

Then $6S = 100{,}000$ and $S = \$16{,}666.67$.

Carrying Out the Plan

3. Solve $\frac{100}{S} = \frac{6}{1000}$ for S.

$$100{,}000 = 6S$$

$$\frac{100{,}000}{6} = S$$

Since S represents money, round it to two decimal places.

$$\$16{,}666.67 = S$$

Looking Back

4. Refer to the same part of the solution to part (b). ■

Now try some exercises using the guess-and-check strategy or the five-step procedure.

Lesson Exercise 11.45

A college has 300 places in the entering class. In the past, about 46% of the students who were accepted attended the school. How many students should the college accept?

(a) Solve the problem with the guess-and-check strategy.

(b) Solve it with an equation or a proportion.

(c) Explain how your formula is similar to the one in Example 11.10.

Lesson Exercise 11.46

Joe sells squirrel-powered vacuum cleaners. He earns 230 a week plus 14% of his sales. How much must he sell in a week in order to earn $400?

Some problems can be solved algebraically with two equations.

EXAMPLE 11.11

Some state inspection centers check automobile exhaust for carbon monoxide (CO) and carbon dioxide (CO_2). CO is 43% carbon, and CO_2 is 27% carbon.

A sample of 1200 mg of exhaust has 31% carbon. How many milligrams of the sample are CO, and how many milligrams are CO_2?

Solution

Understanding the Problem

1. We are asked to find milligrams of CO and CO_2 in the sample. Let m = milligrams of CO and d = milligrams of CO_2.

Devising a Plan

2. A problem with two unknowns generally requires two sentences or equations. The milligrams of CO in the sample plus the milligrams of CO_2 add

up to 1200 mg; also, 43% of the milligrams of CO plus 27% of the milligrams of CO_2 make 31% of 1200.

3.

$$m + d = 1200$$

$$0.43m + 0.27d = (0.31)(1200)$$

Carrying Out the Plan

4. Using substitution, find values of m and d that satisfy both equations,

$$m + d = 1200 \rightarrow d = 1200 - m$$

and substitute for d in the other equation.

$$0.43m + 0.27(1200 - m) = (0.31)(1200)$$

$$0.43m + 324 - 0.27m = 372$$

$$0.16m + 324 = 372$$

$$0.16m = 48$$

$$m = 300 \text{ mg}$$

$$d = 1200 - 300 = 900 \text{ mg}$$

The sample contains 300 mg of CO and 900 mg of CO_2.

Looking Back

5. Check by substituting in the second equation.

$$(0.43)(300) + (0.27)(900) \stackrel{?}{=} (0.31)(1200)$$

$$129 + 243 \stackrel{?}{=} 372$$

$$372 = 372$$

■

Lesson Exercise 11.47

You are planning to buy a dishwasher. Choice A sells for $280 and costs $12 per month to operate. Choice B sells for $220 and costs $20 per month to operate.

(a) Find the number of months for which the total cost of purchasing and operating each machine is the same. (*Hint:* Write a formula for the total cost of purchasing and operating each machine.)

(b) Solve the problem without using the equations. (*Hint:* Choice A costs $60 more to start. How long does it take to make up this cost?)

SOLVING PROBLEMS WITH TABLES

Suppose that you invest $1000 earning 8% interest compounded annually. How long will it be before your account is worth $1500? You can solve this problem by making a table. (Solving the problem by using the equation $1000(1.08)^T = 1500$ would require logarithms.) Example 11.12 illustrates how to solve the problem using a table.

EXAMPLE 11.12

You invest $1000 earning 8% interest compounded annually. About how long will it be before your account is worth $1500?

Solution

Understanding the Problem An account starts with $1000 and earns 8% per year.

Devising a Plan Make a table of values. Experiment until you reach values above and below $1500.

Carrying Out the Plan Each year, the account will have 108% of the previous year's amount. Multiply each amount by 1.08 to obtain the amount for the next year.

Time (years)	0	1	2	3	4	5	6
Amount	$1000	1080	1166.40	1259.71	1360.49	1469.33	1586.88

It will take a little more than 5 years.

Looking Back 1500 represents a 50% increase. At 8% annual interest, it should take about 6 years to gain 50%, so the answer is reasonable. ■

Lesson Exercise 11.48

You invest $1000 earning 10% interest compounded annually. About how long will it be before your account is worth $2000?

(a) Guess the answer.

(b) Solve the problem using a table.

A spreadsheet can be used to generate a table that compare the costs of two or more alternatives.

Lesson Exercise 11.49

You decide to rent a copier from either Crisp Copy or Repro-Man. Crisp Copy charges $60 per month plus $.02 per copy. Repro-Man charges $40 per month plus $.025 per copy.

(a) Set up a computer spreadsheet (or use a calculator and make a table) as follows. Enter 1000, 2000, 3000, . . . , 10,000 in column A. The top part of the spreadsheet follows.

	A	B	C
1	Copies	Crisp Copy	Repro-Man
2	1000		
3	2000		
4	3000		
5	4000		

(b) Enter appropriate formulas in cells B2 and C2. Copy these formulas in rows 3 through 11.

(c) Tell how you would decide where to rent a copier.

SOLVING PROBLEMS WITH GRAPHS

You can solve a problem with information from a graph.

Lesson Exercise 11.50

Here is a graph of a journey by car.

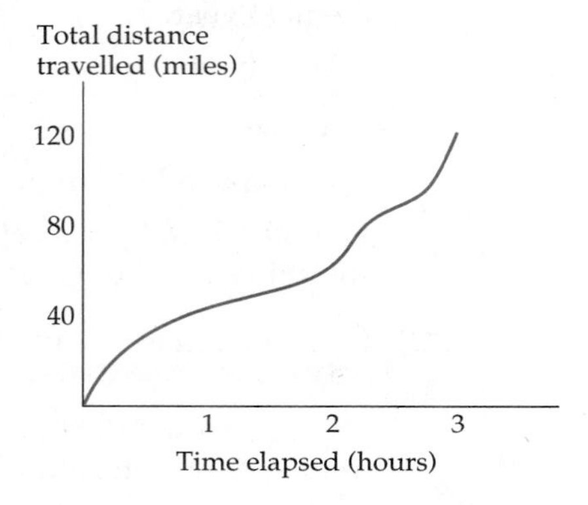

(a) What was the average speed of the car?

(b) Around what time was the car going slowest?

In Section 11.3, you found the equation of a line that seemed to fit a set of data. You can make predictions using such equations.

Lesson Exercise 11.51

Fifteen people are surveyed. The following data give the average miles driven per month and the average cost per month of maintaining a 3-year-old car.

(800, $70)	(1200, $110)	(1000, $120)	(900, $85)	(900, $105)
(800, $100)	(900, $95)	(800, $80)	(400, $40)	(1400, $130)
(800, $75)	(1100, $105)	(700, $50)	(700, $65)	(600, $60)

(a) Plot the 15 points on a two-dimensional graph. (Use graph paper.)

(b) Draw a straight line that seems to fit the data best.

(c) Estimate where the graph passes through the y-axis.

(d) What is the equation of your line?

(e) What monthly expense does your equation predict for someone who drives a 3-year-old car 1300 miles per month?

You can also use graphs to compare the alternatives represented by two equations. Graphing the two equations together may clarify the relative merits of the alternatives.

Lesson Exercise 11.52

You are currently paying an average of $52 per month for heat. After installing solar panels at a cost of $610, you can reduce your monthly heating costs to about $20 per month (this would vary depending upon how much sun your town gets).

(a) What is a formula for the total heating cost in terms of the number of months, with the current system?

(b) Graph your equation from part (a) on graph paper [or you could use a graphing calculator in parts (b), (d), and (e)].

(c) What is a formula for the total heating cost using solar panels (including the installation cost)?

(d) Graph your equation from part (c) on the same axes you used in part (b).

(e) Estimate the intersection point of the two graphs.

(f) Use the intersection point and state the conditions that make one system more economical than the other.

Earlier in the lesson, you studied how to solve algebraically problems that involve two equations. These problems can also be solved by graphing the two equations.

Lesson Exercise 11.53

(a) Solve Lesson Exercise 11.47 by graphing the two equations on the same graph. (Use graph paper or a graphing calculator.)

(b) Which method do you think is easier, using algebra or using a graph?

AN INVESTIGATION: LENGTH AND PERIOD OF A PENDULUM

Scientists and statisticians search for patterns in experimental data in order to answer questions such as "How is the length of a pendulum related to the time it takes to complete a full swing?"

To do this experiment, you need a string about 4 feet long and some weights of uniform size (for example, a set of four washers).

Lesson Exercise 11.54

(a) Attach a weight to the end of a string, and hang the string over the edge of a table. Time ten swings, and use this information to compute how long it takes for the pendulum to complete one full swing. Vary the length of the string, and collect data for the following table.

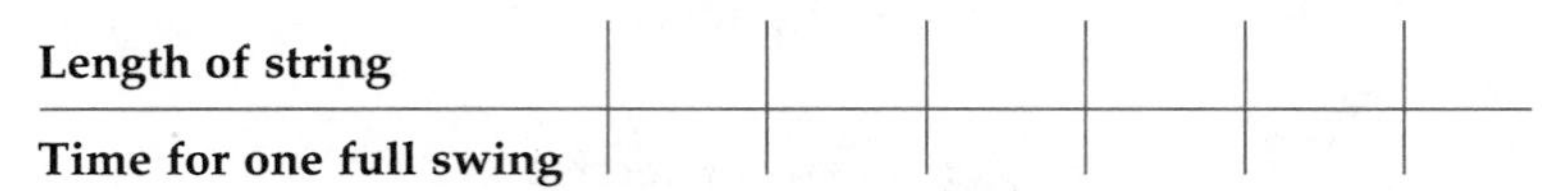

Length of string					
Time for one full swing					

(b) Graph the data on a two-dimensional graph.

(c) Describe any pattern you see in the data or the graph.

(d) Make a conjecture in words about the relationship between the length of a pendulum and the time required for one full swing.

Lesson Exercise 11.55

Conduct a similar investigation of the relationship between the weight and the time required for ten swings (keeping the length the same).

ANSWERS TO SELECTED LESSON EXERCISES

11.44

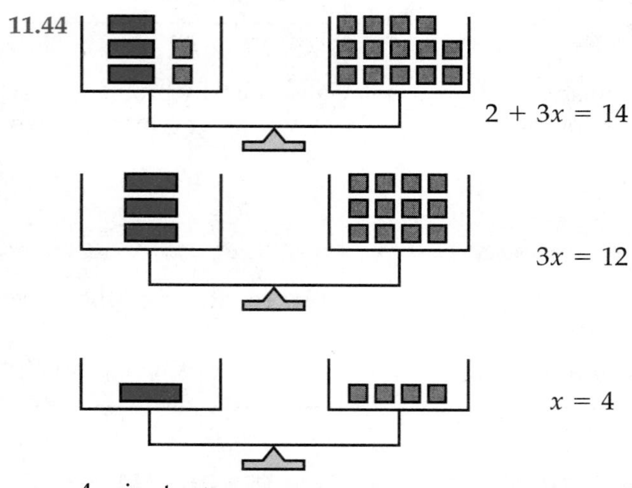

4 minutes

11.45 (b) $300 = 0.46x$

$x = 652$ students

(c) It has the form $ax = b$.

11.46 $1214.29

11.47 (a) 7.5 months

(b) Choice A costs $60 more to start. You save $8/month. In $\frac{60}{8} = 7.5$ months, they cost the same.

11.48 (b) A little over 7 years

11.49 (b)

Copies	Crisp Copy	Repro-Man
4000	140	140

(c) Repro-Man for less than 4000 copies in a month; Crisp Copy for more than 4000 copies in a month

11.50 (a) $\frac{120}{3} = 40$ mph

(b) After about 1 hour

11.51 (a)

Cost

$150

$100

$50

0 500 1000 1500 Miles

11.52 (a) $H = 52M$, in which M is the number of months and H is the total heating cost

(c) $H = 610 + 20M$

(e) Around $M = 19$

(f) The current scheme is cheaper for a period of less than 19 months. Installing solar panels would be cheaper for time periods longer than 19 months.

11.4 HOMEWORK EXERCISES

Basic Exercises

1. A person's intelligence quotient (I.Q.) is 100 times his or her mental age divided by actual age.
   (a) Write a formula for I.Q.'s.
   (b) How do psychologists attempt to measure people's mental ages?
   (c) Find the I.Q. of an 8-year-old child with a mental age of 10.

2. Light travels at a speed of 186,000 miles per second.
   (a) Write a formula for distance traveled, D (miles), in terms of time, T (seconds).
   (b) The world's fastest human runs 1 mile in about 4 minutes. How long does it take light to travel 1 mile?
   (c) The distance from the sun to the Earth is about 93,000,000 miles. How long does it take for light from the sun to reach the Earth?

3. The cost of printing photos at Blurry's is $.27 per picture plus $1.80 for development. So $C = .27P + 1.80$, in which C is cost and P is the number of pictures.
   (a) Suppose that you got a bill for $3.69. How would you figure out, *without* using a formula, how many pictures were printed?
   (b) Solve for P in the equation.
   (c) Suppose that you bring in 12 negatives featuring your newborn nephew Drew Lalotte. Your total bill is $3.69 plus tax. Using one of the two formulas, compute how many pictures, P, of the baby they were able to print.

4. Tickets for the Sonic Boom's rock concert sell for $12. The expenses for setting up the gig are $30,000. So the profit $P = 12 \cdot T - 30{,}000$, in which T is the number of tickets sold.
   (a) Solve for T.
   (b) Select one of the two formulas to solve the following problem. "The profit for the concert was $3000. How many tickets were sold?"

5. Consider the following problem. "A family has 16 oz of cheese. The parents each have 5 oz of cheese. The 3 children each have the same amount. How much cheese does each child have?" Show how to solve the problem both algebraically and with a pictorial model for solving equations.

6. Show how to solve $2x + 3 = 11$ algebraically and with a pictorial model.

7. Solve $4x - 3 = 9$ using the cover-up method.
   (a) Cover the term that contains the x. You have $\square - 3 = 9$. What number goes in the box?
   (b) If $\square = 4x$, what is x?
   (c) Try solving $-3x + 2 = 8$ using the cover-up method.

8. Solve the following equations using the cover-up method. (See the preceding exercise.)
   (a) $-5x - 3 = 12$ (b) $8x + 4 = 9$

9. Did you know that you can estimate the Fahrenheit temperature by counting cricket chirps

By permission of Johnny Hart and Creators Syndicate, Inc.

   (a) Let N equal the number of chirps per minute, and T equal the temperature in Fahrenheit. Derive a formula for T from the first frame of the cartoon.
   (b) Solve for N.
   (c) Use one of the formulas to figure out what the temperature is when crickets chirp 100 times per minute.
   (d) It's 95°F outside. About how many times per minute are the crickets chirping?

10. Density $D = \frac{M}{V}$ (mass ÷ volume).
(a) Solve for V.
(b) Use one of the formulas to find the volume if the mass is 8 g and the density is 10 g/cm^3.

11. A rental van costs $120 per day plus $.32 per mile. What is the range of distances you could drive the van for less than $280?
(a) Solve it with the guess-and-check strategy.
(b) Solve it with an equation.

12. The total bill in a restaurant, including 5% tax, was $16.17. How much was the meal without the tax? (*Hint:* The 5% tax was computed on what amount?)
(a) Solve it with the guess-and-check strategy.
(b) Solve it with an equation.

13. An item is marked up 30% to a retail price of $442. What is the wholesale price?

14. A dress is on sale for $26. This is 35% off the regular price. What is the regular price?

15. (a) Women make up 36% of a college student body. There are 261 women. How many students are at the college?
(b) This exercise has the same form as which lesson exercise?

16. An ice cream store sells a cone with one scoop for $1.25 and a cone with two scoops for $2.10.
(a) Assume that each scoop costs the same amount. How much is the cone?
(b) Write a formula for the price P of a complete ice cream cone in terms of the number of scoops, S.

17. Tickets for a college basketball game were $10 for adults and $4 for children. The paid attendance was 2460, and total ticket sales were $21,060. How many adult tickets and how many children's tickets were sold?

18. A sample of 1400 mg of automobile exhaust contains 36% carbon (from CO and CO_2). If CO is 43% carbon and CO_2 is 27% carbon, how many milligrams of exhaust are CO, and how many milligrams are CO_2?

19. A chicken recipe includes a coating that contains 3 oz of bread crumbs for every 2 oz of parmesan cheese. If you want 16 oz of coating, what amounts of bread crumbs and parmesan cheese do you need?

20. Consider the following problem. "Suppose that you invest $1000 earning 6% interest compounded annually. About how long will it be before your account is worth $1500?" Devise a plan and solve the problem.

21. Suppose that you invest $1000 earning 12% interest compounded annually. About how long will it be before your account is worth $1500?

22. Suppose that you invest $5000 earning 9% interest compounded annually. About how long will it be before your account is worth $8000?

23. Suppose that a certain new car costs about $16,000 in 1998. If inflation is about 7% per year, in about how many years will a comparable new car cost about $32,000?

24. A store rents an average of 200 videos at $3 per day. For each increase of $.10 in the price (up to $4), the store owners estimate that they will lose an additional 10 rentals per day. Each $.10 decrease in price (down to $2) is expected to result in a gain of 10 rentals per day. What rental price will maximize their daily revenue (total money collected from sales)?

25. You must choose between two different rental car companies for a 3-day trip. Drive-U-Crazy charges $39.95/day with unlimited mileage. Used-a-Bit charges $24.95/day and gives 100 free miles/day; mileage beyond that costs $.10/mile. The sales tax rate at both places is 8%.
(a) Set up a computer spreadsheet (or use a calculator and make a table) as follows. Enter 500, 700, 900, 1100, 1300, and 1500 in column A. The top part of the spreadsheet follows.

	A	B	C
1	Miles	Drive-U-Crazy	Used-a-Bit
2	500		
3	700		
4	900		

(b) Enter appropriate formulas in cells B2 and C2. Copy these formulas in rows 3 through 7.
(c) Tell how you would decide which car to rent.

26. You are offered two sales jobs. Fast Talkers pays \$180/week plus 5% commission on sales. Wheeler Dealers pays \$225/week plus 3% commission on sales.
(a) Set up a computer spreadsheet (or use a calculator and make a table) as follows. Enter 1000, 2000, 3000, 4000, and 5000 in column A. The top part of the spreadsheet follows.

	A	B	C
1		Total	Pay
2	Weekly sales	Fast Talkers	Wheeler Dealers
3	1000		
4	2000		

(b) Enter appropriate formulas in cells B3 and C3. Copy these formulas in rows 4 through 7.
(c) Tell how you would decide which job to take.

27. A woman is driving home from work. The graph shows her speeds throughout the trip.

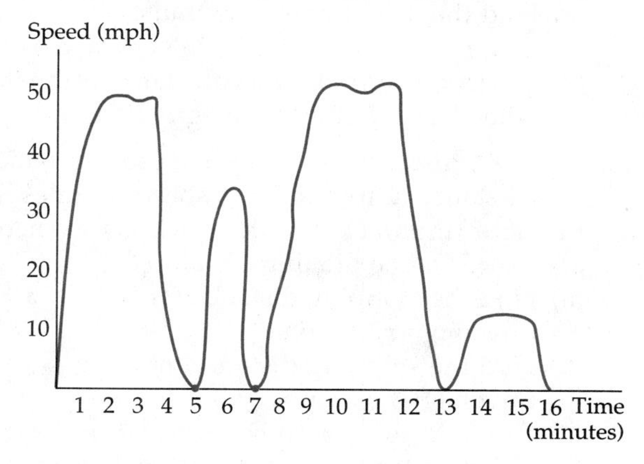

(a) Describe her trip.
(b) About how many miles long is the drive?

28. The graph shows the cost of producing a certain type of sneakers and the revenue from selling them.

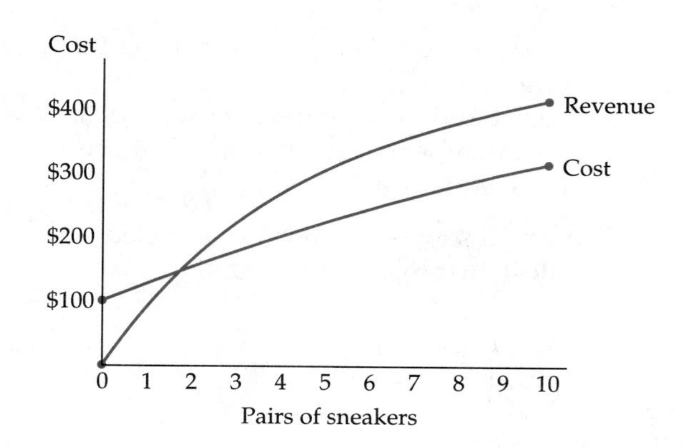

(a) What is the cost of producing 5 pairs of sneakers?
(b) What is the profit or loss on 5 pairs of sneakers?
(c) What is the profit or loss on 8 pairs of sneakers?
(d) At least how many pairs must be sold in order to make a profit?

29. The graph shows the speed of a car on a highway.

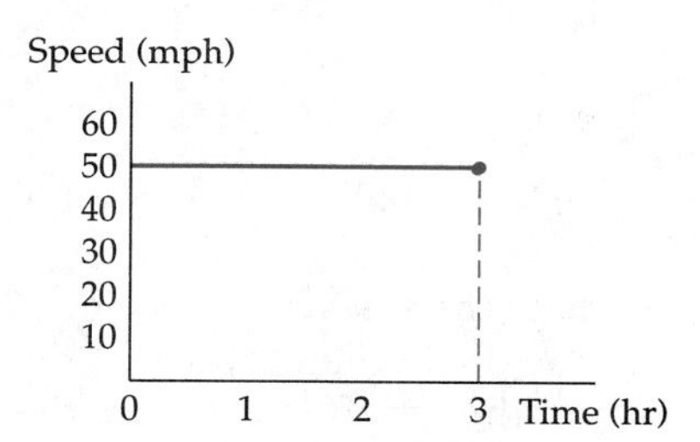

(a) How far did the car travel?
(b) What is the area of the rectangle on the graph?

30. The graph shows the speed of a car on a highway.

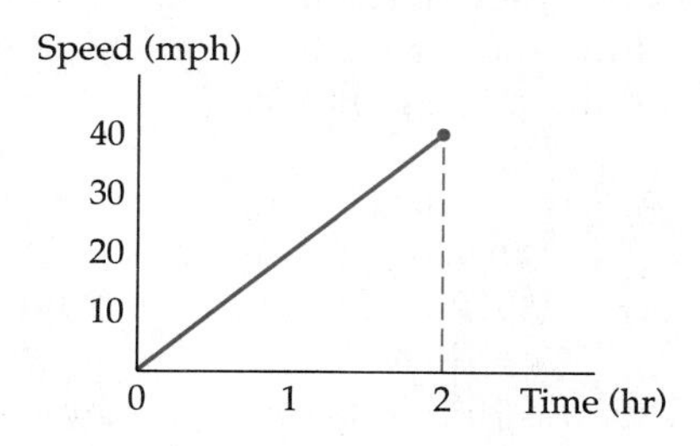

(a) How far did the car travel?

(b) What is the area of the triangle on the graph?
(c) Based upon the preceding exercise and this one, it appears that the total distance traveled is equal to ________.

31. The graph shows the distance traveled, y (miles), by a plane in x hours.

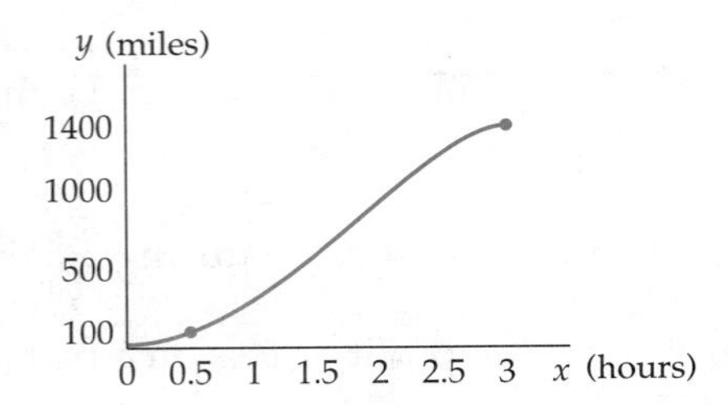

(a) About how far did the plane travel in the first 1/2 hour?
(b) About how far did it travel altogether?
(c) What was the average speed of the plane?

32. A marketing research department makes the following forecast.

Price	\$4	\$5	\$6	\$7	\$8
Sales	18	15	13	11	9

(a) Estimate the sales at a price of \$6.50.
(b) Plot the data on a graph.
(c) Extend the graph to estimate the sales at a price of \$10.
(d) Find an approximate formula relating price to sales.

33. You sample 15 businesses in an industry and ask them how many thousands of dollars they spent on advertising and how many millions of dollars they made in sales. The results are as follows (advertising dollars in thousands, sales dollars in millions):

(35, 10)	(48, 11)	(26, 8)	(28, 7)	(0, 3)
(8, 5)	(32, 9)	(15, 5)	(40, 12)	(38, 12)
(37, 11)	(30, 10)	(56, 15)	(52, 14)	(50, 13)

(a) Plot the 15 points on a two-dimensional graph.
(b) Draw a straight line that seems to fit the data best.
(c) What is the equation of your line?
(d) What amount of sales does your equation predict for a company that spends \$20,000 in advertising?

34. Suppose that you sample 20 people who began full-time work last year, and ask them about their years of education and their starting salaries (in thousands of dollars). The results are as follows (years of education, salary in thousands of dollars).

(14, 18)	(16, 22)	(10, 13)	(20, 27)
(12, 10)	(16, 21)	(16, 20)	(11, 12)
(17, 22)	(12, 16)	(12, 18)	(18, 24)
(12, 15)	(16, 21)	(12, 14)	(16, 23)
(19, 23)	(12, 16)	(14, 20)	(20, 25)

(a) Plot the 20 points on a two-dimensional graph.
(b) Draw a straight line that seems to fit the data best.
(c) What is the equation of your line?
(d) What starting salary does your equation predict for someone who has 17 years of school?

35. An object is tossed up in the air with a velocity of 80 ft/s. Its height h, in feet, after t seconds is $h = -16t^2 + 80t$.
(a) Find the height after 3 seconds.
(b) Graph h vs. t, with h on the vertical axis.
(c) Using your graph or a calculator, estimate when the object is at a height of 50 ft.

36. Stopping distance $D = 1.1S + 0.05S^2$, in which D is distance in feet and S is speed in miles per hour (mph). ($1.1S$ is the reaction distance, and $0.05S^2$ is the braking distance.)
(a) Find the stopping distance for a car traveling at 30 mph.
(b) Find the stopping distance for a car traveling at 50 mph.
(c) Graph D vs. S, with D on the vertical axis.
(d) Using your graph or a calculator, estimate the speed of a car that took 136 ft to stop.
(e) This formula is only an approximation. What factors in addition to speed should affect stopping distance?

37. A rectangle has a perimeter of 10 ft.
(a) Write a formula for the width W in terms of the length L.
(b) Is this formula exact or approximate?
(c) Use the formula to find the width when the length is 3 ft.
(d) What values can the length and width have?
(e) Graph W vs. L, with W on the vertical axis.

38. Scientists can estimate a person's height from the length L, in centimeters, of the femur (thigh bone). For a female, height (in centimeters) $H \approx 2.3L + 61.4$.
(a) Why would anyone want to estimate a women's height from the length of her femur?
(b) Estimate the height of a female with a femur of length 42.1 cm.
(c) Graph the equation.
(d) What does the slope tell you about the relationship between the length and height?

39. Revenue R from ticket sales for a concert depends upon the price charged, according to $R = -50P^2 + 600P$.
(a) Graph this equation.
(b) Approximately what price gives the maximum revenue?

40. A ball is dropped from a height of 144 ft. Its height H (in feet) at time T (in seconds) is $H = 144 - 16T^2$.
(a) Graph this equation.
(b) What shape is the graph?
(c) When does the ball hit the ground?
(d) How high does the ball go?

41. You light a 25-cm candle. Its height H after T hours is shown.

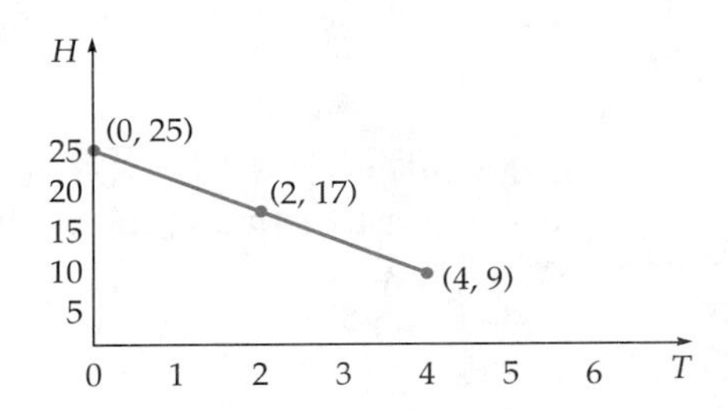

(a) Extend the graph to estimate when the candle will burn out.
(b) Write an equation relating H to T.

42. You are choosing between two health insurance plans that cost the same amount. One pays 80% of all expenses *beyond* the first \$500 (a \$500 deductible). The second pays 60% of all expenses, with no deductible.
(a) Write a formula for the amount A that each policy pays on a claim of C dollars.
(b) Graph both equations on the same graph. (You may use a graphing calculator for parts (b) and (c).)
(c) Find the intersection point.
(d) Use the intersection point and state the conditions that make one plan more economical than the other.
(e) What factors besides price might you consider in choosing a plan?

43. You are planning to buy a new air conditioner. Choice A sells for \$400 and costs \$30 per month to run. Choice B sells for \$550 and is more energy-efficient, so it costs \$20 per month to run.
(a) Write a formula for the cost C of purchasing and operating each air conditioner for M months.
(b) Graph both equations on the same graph. (You may use a graphing calculator for parts (b) and (c).)
(c) Find the intersection point on the graph.
(d) Find the intersection point algebraically.
(e) Use the intersection point and state the conditions that make one air conditioner a better buy than the other.
(f) What factors besides price might you consider in deciding which air conditioner to buy?

44. The Screeching Tires have been offered two different recording contracts. Contract A includes \$20,000 for signing and \$1 royalty per tape or CD. Contract B includes \$5000 for signing and \$2 royalty per tape or CD.
(a) Write a formula for the total pay (y) for each contract in terms of the number of tapes and CDs sold (x).

(b) Graph both equations on the same graph. (You may use a graphing calculator in parts (b) and (c).)
(c) Find the intersection point.
(d) Use the intersection point and state the conditions that make each contract better than the other.

45. You are offered two sales jobs. Job A pays \$800 per week plus 5% commission. Job B pays \$300 per week plus 15% commission.
(a) Write a formula for the weekly pay for each job.
(b) Graph both equations on the same graph. (You may use a graphing calculator in parts (b) and (c).)
(c) Find the intersection point.
(d) Use the intersection point and state the conditions under which each job pays more than the other.
(e) What factors besides pay might you consider in choosing a sales job?

46. A job pays \$7.25 an hour. If 34% is deducted for taxes, union dues, and benefits, how many hours must you work at this job in order to take home \$500?

Extension Exercises

47. Consider the following problem. "A job pays \$6.10 an hour. If 23% is deducted for taxes, union dues, and benefits, how many hours must you work at this job in order to take home \$100?" Devise a plan and solve the problem.

48. Your employer deducts 37% of your salary for taxes and benefits. You take home \$312 for 40 hours of work. What does your job pay per hour?

49. In 1996, first-class mail costs \$.32 for the first ounce and \$.23 for each additional ounce or fraction thereof. (Fractional amounts are rounded up to the next whole ounce.)
(a) Find the cost of a 2.6-ounce first-class letter in 1996.
(b) A postal clerk in 1996 weighs your first-class package and says the charge is \$1.70. How much might the package weigh?
(c) Let C be the cost of a W-ounce letter. Graph C (on the vertical axis) vs. W (on the horizontal axis).
(d) What is the input set in part (c)?
(e) The formula $C = 0.23W + 0.09$ works for what values of W?
(f) What shape is the graph?

50. A school group already has \$600 in a bank account. They need a total of \$900 to finance a concert. If they sell concert tickets for D dollars, how many tickets must they sell to break even?

51. In deciding between a National Motors Integer Maxima with a gas engine and one with a diesel engine, you collect the following information. First, the gas car costs \$7200 and the diesel car costs \$8500. Second, the gas car gets 38 miles per gallon (mpg) and the diesel car gets 45 mpg. Third, gas costs \$1.10 per gallon and diesel fuel costs \$1.18 per gallon. Under what conditions would each car be a better buy?

52. A manufacturer of television sets has weekly expenses $E = 1500 + 100T$ dollars for manufacturing T televisions. The total revenue (money received) from selling T televisions is $180T$.
(a) What is the *profit or loss* on manufacturing and selling 100 sets in a week?
(b) What is the *profit or loss* on manufacturing and selling 20 sets in a week?
(c) What number of sets is the break-even point?

53. The graph shows $y = x^2 - 9x + 4$. Use it or a graphing calculator to estimate the solutions to
(a) $x^2 - 9x + 4 = 0$
(b) $x^2 - 9x + 4 = -8$

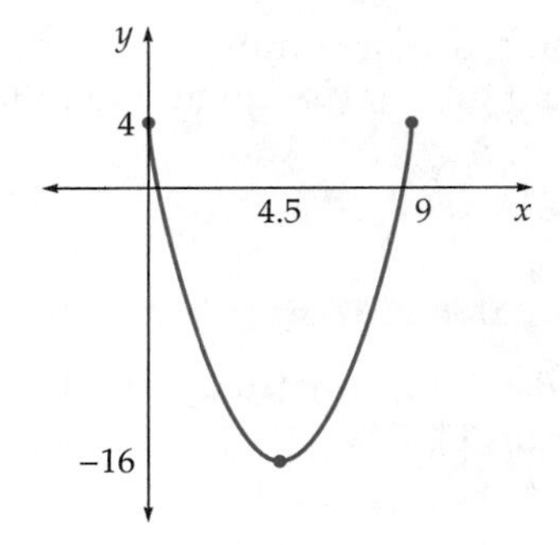

54. Do you know the day of the week on which you were born? The following formula gives the day of the week for any given date. D is the number of days in the latest year up to and including the date; Y is the year.

$$W = D + Y + \left[\frac{Y-1}{4}\right] - \left[\frac{Y-1}{100}\right] + \left[\frac{Y-1}{400}\right]$$

The notation [ ] means take the quotient without any remainder. For example,

$$\left[\frac{11}{4}\right] = 2$$

To obtain the day of the week, take the remainder when W is divided by 7 (1 = Sunday, 2 = Monday, . . . , 0 = Saturday).

On what day of the week was the author born (January 11, 1952)?

$$W = 11 + 1952 + \left[\frac{1951}{4}\right] - \left[\frac{1951}{100}\right] + \left[\frac{1951}{400}\right]$$

$$= 11 + 1952 + 487 - 19 + 4 = 2435$$

$$7\overline{)2435} \quad 347 \text{ R } 6 \rightarrow \text{sixth day} = \text{Friday}$$

So the author was born on a Friday.

(a) Use the formula to find out the day of the week on which you were born.

(b) On what day of the week will January 1, 2020, fall?

55. Write a word (story) problem that could be solved with the equation $2x + 4 = 10$.

56. Write a word (story) problem that could be solved with the equation $20 - 2x = 10$.

Special Exercise

57. Read Chapters 1–7 of *The Education of T. C. Mits* by Lillian Lieber, and write a report that includes a summary of the main ideas and your reactions to them.

11.5 GEOMETRY WITH COORDINATES

Coordinate geometry establishes a connection between algebra and geometry, enabling mathematicians to describe geometry problems algebraically.

THE DISTANCE FORMULA

How can we find the length of a line segment using coordinates?

Lesson Exercise 11.56

(a) Plot $D(2, 1)$, $E(5, 3,)$, and $F(5, 1)$ on a graph, and draw right triangle $\triangle DEF$.

(b) The legs of $\triangle DEF$ have lengths ________ and ________.

(c) Find DE using the Pythagorean Theorem.

The Pythagorean Theorem is the basis for computing the distance between any two points on a two-dimensional graph. Complete the following exercise to derive the distance formula.

Lesson Exercise 11.57

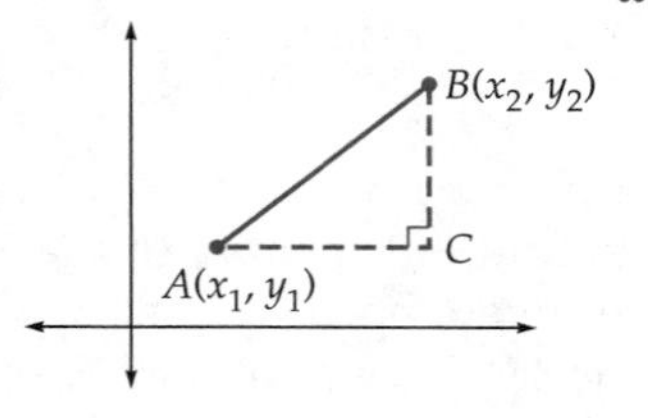

Figure 11–27

(a) For two points $A(x_1, y_1)$ and $B(x_2, y_2)$, construct a right triangle with hypotenuse $\overline{AB}$ as shown in Figure 11-27.

(b) What are the coordinates of C?

(c) Give the lengths AC and BC in terms of the coordinates.

(d) Using the results of part (c), $(AB)^2 = (AC)^2 + (BC)^2 =$ ________.

(e) Solve for AB, and write it in terms of the coordinates.

If the slope of $\overline{AB}$ is negative, the derivation is similar and the resulting formula is the same. The **distance formula** tells us how to find the length of a line segment using the coordinates of its endpoints.

The Distance Formula

The distance between two points $A(x_1, y_1)$ and $B(x_2, y_2)$ is

$$AB = \sqrt{(x_2 - x_1)^2 + (y_2 - y_1)^2}$$

Lesson Exercise 11.58

A quadrilateral $ABCD$ has coordinates $A(0, 0)$, $B(5, 0)$, $C(8, 4)$, and $D(3, 4)$. Use the distance formula to show that $ABCD$ is a rhombus.

The distance formula can also be used to find the algebraic equation of a circle.

Lesson Exercise 11.59

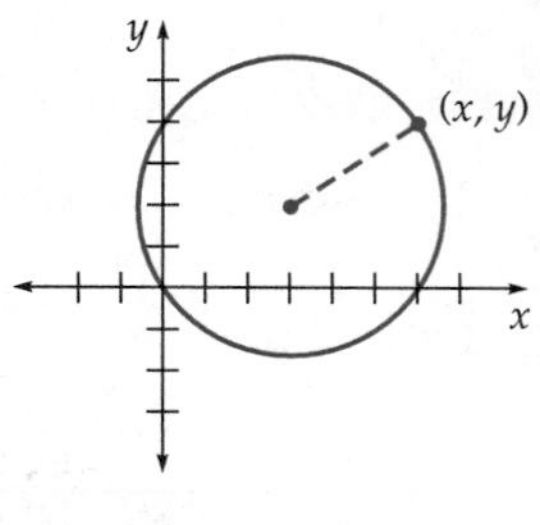

A circle has center (3, 2) and a radius of 4.

(a) This circle is the set of all points that are ________ from ________.

(b) Use the distance formula to write an algebraic equation for all points (x, y) that are 4 units from (3, 2). [*Hint:* Represent the distance between the center and a point on the circle (x, y) by applying the distance formula to the circle in Figure 11-28. Then square both sides of the equation.]

Figure 11–28

Use the same procedure to find a more general equation for a circle in the following exercise.

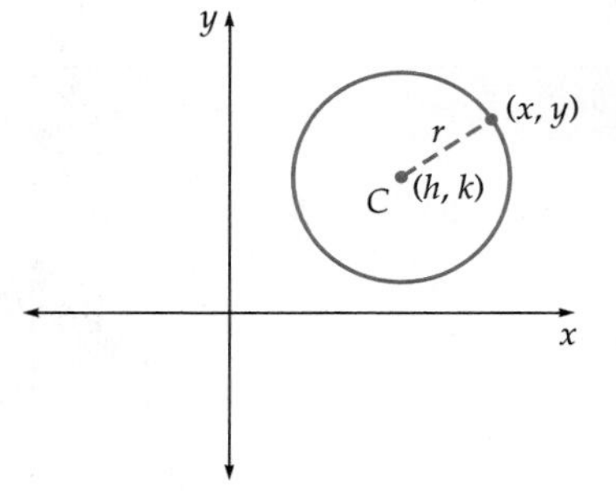

Figure 11–29

Lesson Exercise 11.60

A circle has center (h, k) and radius r (Figure 11-29). Find an algebraic equation for the circle.

The result of Lesson Exercise 11.60 is the general equation of a circle.

The Equation of a Circle

The equation of a circle with center (h, k) and radius r is

$$(x - h)^2 + (y - k)^2 = r^2$$

Use this general formula in the following exercise.

Lesson Exercise 11.61

Find the equation of a circle with center $(2, -3)$ and a radius of 4.

THE MIDPOINT FORMULA

How are the coordinates of the midpoint of a line segment related to the coordinates of the endpoints?

Lesson Exercise 11.62

(a) Plot the points $A(3, 2)$, $B(-1, 4)$, and $C(5, 6)$, and draw $\triangle ABC$ on graph paper.
(b) Find the coordinates of the midpoint of $\overline{AC}$.
(c) Find the coordinates of the midpoint of $\overline{AB}$.
(d) Find the coordinates of the midpoint of $\overline{BC}$.
(e) What is the relationship between the coordinates of the endpoints of a line segment and the coordinates of the midpoint?
(f) Given two points $A(x_1, y_1)$ and $B(x_2, y_2)$, what are the coordinates of the midpoint of $\overline{AB}$?
(g) Using parts (b)–(d) to answer parts (e) and (f) is an example of what kind of reasoning?

The answer to Lesson Exercise 11.62(f) is the midpoint formula.

The Midpoint Formula

Given $A(x_1, y_1)$ and $B(x_2, y_2)$, then the midpoint M of $\overline{AB}$ is

$$\left(\frac{x_1 + x_2}{2}, \frac{y_1 + y_2}{2}\right)$$

The midpoint and distance formulas can be used to find properties of triangles and quadrilaterals.

Lesson Exercise 11.63

$\triangle ABC$ is a right triangle with $A(0, 0)$, $B(8, 0)$, and $C(0, 6)$.

(a) Plot $\triangle ABC$ on a coordinate graph.

(b) Find the coordinates of M, the midpoint of the hypotenuse.

(c) Compare the lengths AM, BM, and CM.

(d) Repeat parts (a)–(c) with a new right triangle.

(e) Make a generalization based upon your results.

In Chapter 8, you studied properties of parallelograms. The following exercise shows how to use coordinate geometry to prove one such property: that the diagonals of a parallelogram bisect each other.

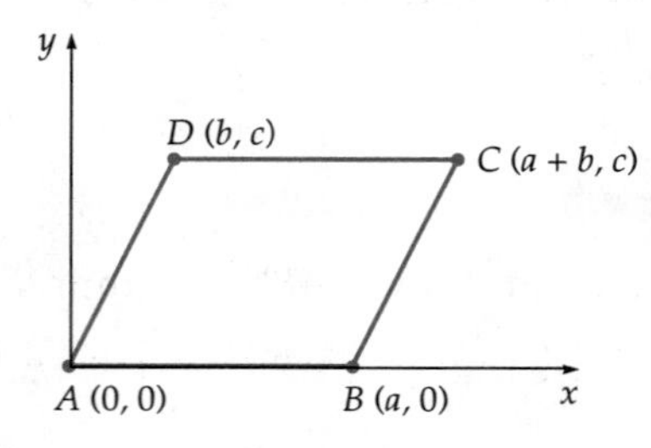

Figure 11–30

Lesson Exercise 11.64

(a) What is the definition of a parallelogram?

(b) Prove that $ABCD$ in Figure 11-30 is a parallelogram.

(c) Prove that the midpoint of $\overline{AC}$ is the midpoint of $\overline{BD}$.

TAXICAB GEOMETRY

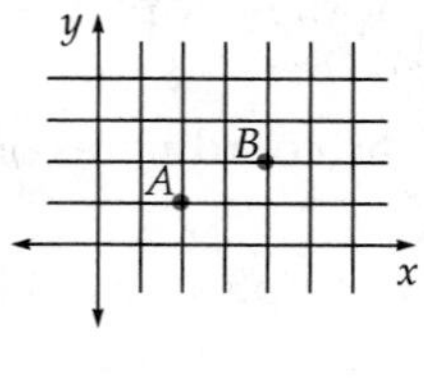

Figure 11–31

A town has its streets arranged in a beautiful grid pattern, as shown in Figure 11-31. How far is it from A to B by car? The answer, 3 blocks, is called the **taxicab distance** from A to B and is written $d_T(AB) = 3$. The distance formula at the beginning of this section would not give the correct taxicab distance from $A(2, 1)$ to $B(4, 2)$ along "streets."

Lesson Exercise 11.65

How many different routes from A to B of taxicab distance 3 are there?

Lesson Exercise 11.66

(a) Find the taxicab distance from $(-2, 3)$ to $(4, 1)$.

(b) Which is longer, the taxicab distance or the standard (Euclidean) distance from $(-2, 3)$ to $(4, 1)$?

Are taxicab and standard distance ever the same? Try the following exercise.

Lesson Exercise 11.67

If possible, draw points A and B on a graph such that

(a) $d_T(AB) = AB$. (b) $d_T(AB) < AB$. (c) $d_T(AB) > AB$.

As you saw in the preceding exercise, $d_T(AB) \geq AB$ for all points A and B. The following exercise presents the taxicab equivalent of a geometric shape you studied earlier in this lesson.

Lesson Exercise 11.68

Point E has coordinates $(2, 4)$.

(a) Plot all points that are a taxicab distance of 2 from point E.

(b) What shape do these points form?

(c) If you plotted the corresponding set of points in standard two-dimensional coordinate geometry, what shape would you obtain?

ANSWERS TO SELECTED LESSON EXERCISES

11.56 (b) 2; 3 (c) $\sqrt{13}$

11.57 (b) (x_2, y_1) (c) $x_2 - x_1$ and $y_2 - y_1$
(d) $(x_2 - x_1)^2 + (y_2 - y_1)^2$
(e) $\sqrt{(x_2 - x_1)^2 + (y_2 - y_1)^2}$

11.58 $AB = BC = CD = DA = 5$, so $ABCD$ is a rhombus.

11.59 (a) 4 units; $(3, 2)$
(b) $(x - 3)^2 + (y - 2)^2 = 16$

11.61 $(x - 2)^2 + (y + 3)^2 = 16$

11.62 (b) (4, 4) (c) (1, 3) (d) (2, 5)
(e) The coordinates of the midpoint are the averages of the coordinates of the two endpoints.
(f) $\left(\frac{x_1 + x_2}{2}, \frac{y_1 + y_2}{2}\right)$
(g) Inductive

11.63 (b) $M(4, 3)$ (c) $AM = BM = CM = 5$
(e) The midpoint of the hypotenuse is equidistant from the three vertices.

11.64 (b) *Hint:* Show that the slopes of the opposite sides are parallel.
(c) The midpoints of both diagonals are $\left(\frac{a + b}{2}, \frac{c}{2}\right)$.

11.65 3

11.66 (a) 8 (b) Taxicab distance

11.67 (a) A and B must have the same x- or y-coordinate.
(b) Impossible
(c) All pairs of points besides those in part (a).

11.68 (b) A square (c) A circle

11.5 HOMEWORK EXERCISES

Basic Exercises

1. What is the distance between (4, 6) and (5, 10)?

*2. Explain why the distance between $A(x_1, y_1)$ and $B(x_2, y_2)$ is

$$AB = \sqrt{(x_2 - x_1)^2 + (y_2 - y_1)^2}$$

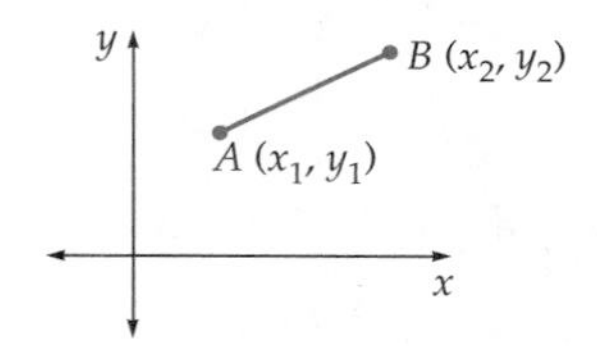

3. $ABCD$ is a rectangle.

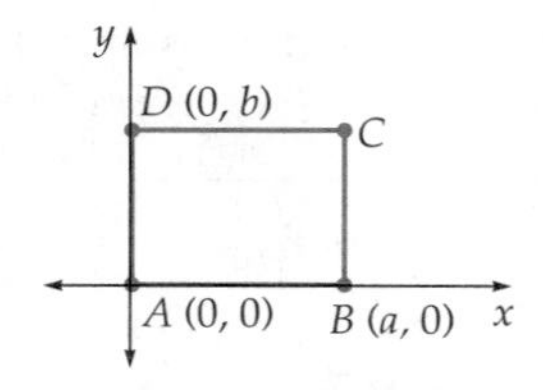

(a) What are the coordinates of C?
(b) Show that $AC = BD$.

4. Three vertices of a parallelogram are (1, −2), (0, 2), and (2, 0). Find three possible coordinates for the fourth vertex.

5. Find the perimeter of $\triangle ABC$ with coordinates $A(3, 6)$, $B(3, 9)$, and $C(5, 7)$.

6. Determine whether or not (0, 4), (4, 5), and (6, −3) are the vertices of a right triangle.

7. A circle has center (2, −1) and a radius of 3. Find the equation of the circle.

8. A circle has the equation $(x + 4)^2 + (y - 1)^2 = 16$. What are its center and radius?

9. A circle has center (2, 3), and the point (−2, 7) is on the circle. Find the equation of the circle.

10. A modern-day pirate wants to locate a treasure in a park. The water fountain is at (3, −2), and the swings are at (1, −4). The treasure is somewhere 4 units east of the water fountain and 2 units north of the swings. Where is the treasure?

11. Point A has coordinates (4, −1), and point B has coordinates (2, 7). What are the coordinates of the midpoint of $\overline{AB}$?

12. Point M is the midpoint of $\overline{AB}$. Point A has coordinates (3, 7), and M has coordinates (−2, 1). Find the coordinates of B.

13.

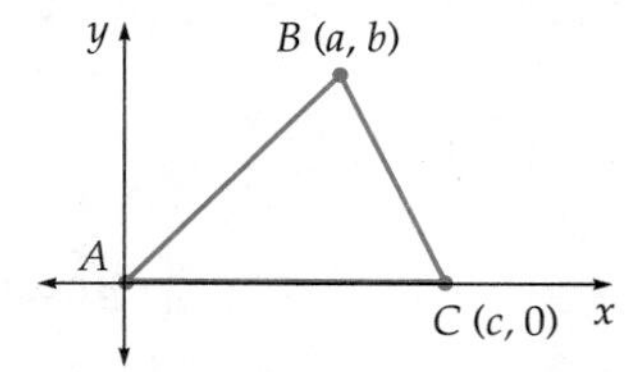

(a) Find the coordinates of the midpoints D of $\overline{AB}$ and E of $\overline{BC}$.

(b) Show that $\overline{DE} \parallel \overline{AC}$.

(c) Show that $DE = \frac{1}{2}AC$.

14.

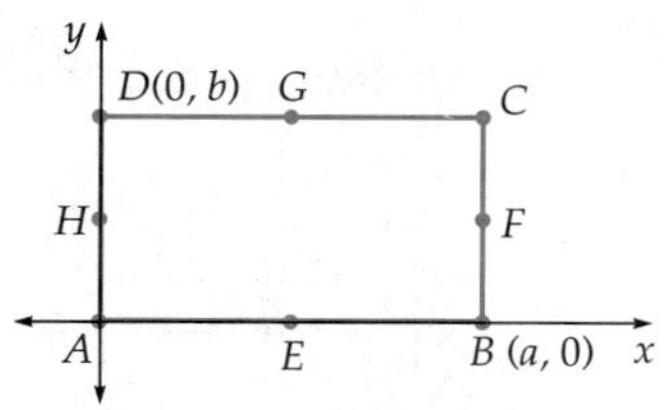

(a) Find the coordinates of the midpoints of the four sides, E, F, G, and H.

(b) Show that EFGH is a rhombus.

15. A circle has center (4, −2), and (3, 1) is a point on the circle. Find another point on the circle. (*Hint:* Find a diameter of the circle.)

16. Points (3, −1) and (5, 3) are endpoints of the diameter of a circle. Find the equation of the circle.

17. Find the taxicab distance from (2, 4) to (1, −2).

18. If $d_T(AB) = AB$, then describe the relationship between the coordinates of A and B.

19. Point E has coordinates (1, −1). Plot all points that are a taxicab distance of 4 from E.

20. Find points A, B, C, and D on a graph such that $d_T(AB) = d_T(CD)$ and $AB > CD$.

21. You are driving from your home to school. The numbers on the grid give the time, in minutes, needed to cover that segment of the route. Which route from home to school is the quickest?

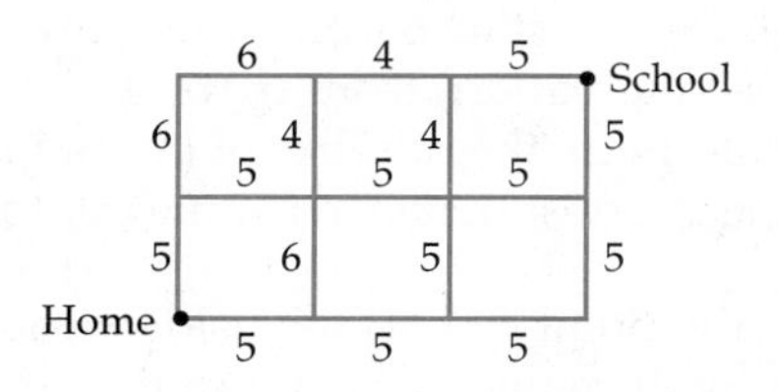

22. Roy (R) and Paula (P) each live three blocks from school. Where could the school be?

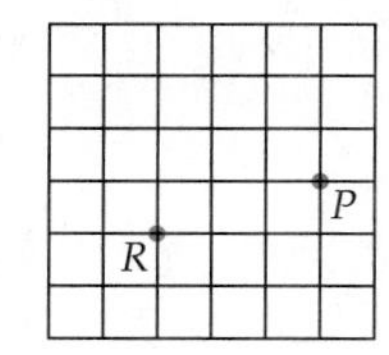

Extension Exercises

23.

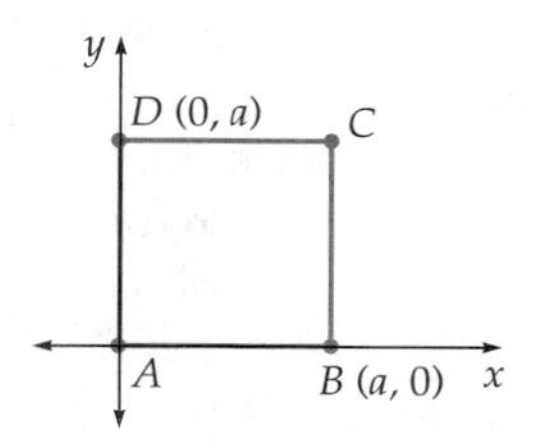

ABCD is a square.

(a) What must the coordinates of C be?

(b) Show that $AC = BD$.

(c) Use slopes to show that $\overline{AC} \perp \overline{BD}$.

24. The **center of gravity** of a polygon on a two-dimensional graph is $(\bar{x}, \bar{y})$, in which $\bar{x}$ is the average (mean) of the x-coordinates of the vertices and $\bar{y}$ is the average (mean) of the y-coordinates of the vertices.

(a) Find the center of gravity of a quadrilateral with vertices at coordinates A(−1, −3), B(2, 4), C(5, 4), and D(6, −3).

(b) Cut out a piece of cardboard that has the shape of ABCD, and balance it on the tip of a pin. Is the center of gravity very close to the location you obtained in part (a)?

25. A city has four retirement homes at A, B, C, and D. Where should the city build a medical center M so that the sum of the four distances from M to the four homes is as small as possible?

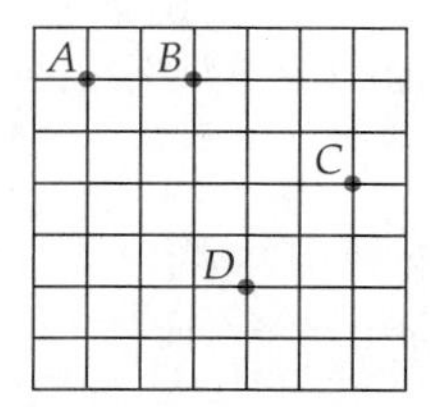

26. A **parabola** is the set of points equidistant from a line and a point not on that line. Follow steps (a)–(b) to find the equation of the parabola that is equidistant from $y = -2$ and (0, 4).

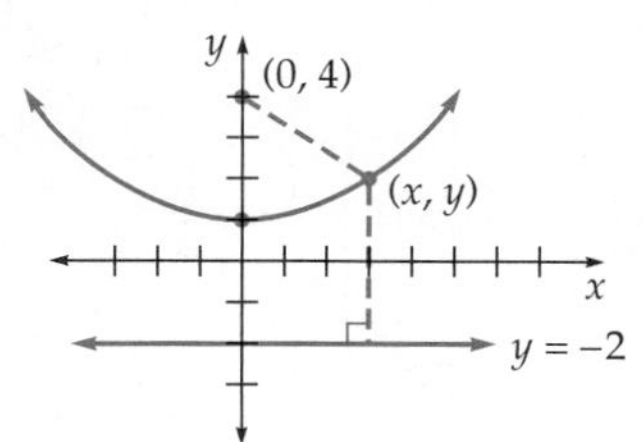

(a) Any point (x, y) on the parabola is equidistant from $(0, 4)$ and $y = -2$. What are the coordinates of the point where $y = -2$ intersects the perpendicular?

(b) Complete the following equation.

$$(0, 4) \text{ to } (x, y) \text{ dist.} = (x, -2) \text{ to } (x, y) \text{ dist.}$$

$$\sqrt{(\quad)^2 + (\quad)^2} = \sqrt{(\quad)^2 + (\quad)^2}$$

(c) Square both sides and combine like terms to obtain the equation of the parabola.

27. Follow the instructions to create a curve by folding waxed paper.

(a) Crease a line segment $\overline{AB}$ and mark a point C on the waxed paper as shown.

.C

A——————B

(b) Fold the point onto the segment and crease. Repeat this about 20 times, folding the point onto different places on the segment.

(c) What shape is suggested by the creases?

(d) How does this work? (*Hint:* See the preceding exercise.)

28. If $(3, k)$ is equidistant from $A(5, -3)$ and $B(-1, -5)$, find the value of k.

29. Find a point that is 3/4 of the way from $(2, 5)$ to $(4, 1)$.

30. Find the area of a triangle with vertices $(-1, 4)$, $(3, 1)$, and $(7, 2)$. (*Hint:* Enclose the triangle in a rectangle.)

31. A **taxicab midpoint** M of A and B is any point such that if $d_T(AB) = c$, then $d_T(AM) = d_T(MB) = \frac{1}{2}c$. Find all taxicab midpoints of A and B.

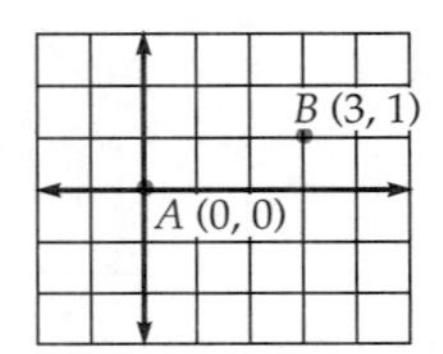

32. Devise a formula for computing the taxicab distance from the coordinates of a point. If A has coordinates (x_1, y_1) and B has coordinates (x_2, y_2), then $d_T(AB) =$ ________. (*Hint:* Look at the change in x and the change in y.)

33.

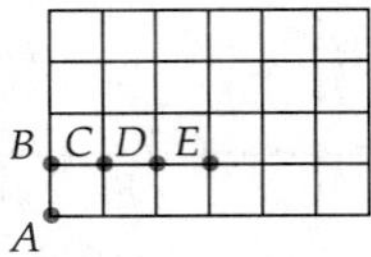

(a) How many different taxicab routes of the shortest distance from A to B are there?

(b) How many different taxicab routes of the shortest distance from A to C are there?

(c) How many different taxicab routes of the shortest distance from A to D are there?

(d) How many different taxicab routes of the shortest distance from A to E are there?

(e) Use the pattern from parts (a)–(d) to determine the number of taxicab routes of the shortest distance from $(0, 0)$ to $(10, 1)$.

34.

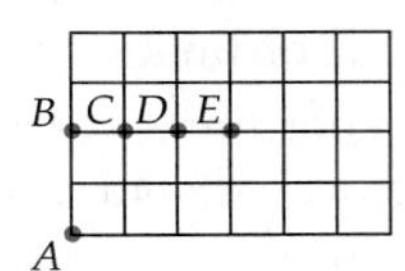

(a) How many different taxicab routes of the shortest distance from A to B are there?

(b) How many different taxicab routes of the shortest distance from A to C are there?

(c) How many different taxicab routes of the shortest distance from A to D are there?

(d) How many different taxicab routes of the shortest distance from A to E are there?

(e) Use the pattern from parts (a)–(d) to determine the number of taxicab routes of the shortest distance from $(0, 0)$ to $(6, 2)$.

35. Does the point $(-1, 4)$ lie inside or outside the circle with center $(0, 1)$ and radius 3?

SUMMARY

In many mathematics problems, one examines a relation between two sets of numbers. A relation can be represented with a graph, a table, an equation, an arrow diagram, or words.

Algebra and calculus focus upon a special group of relations called functions, in which each x value has exactly one corresponding y value. Linear functions are a particularly useful class of functions with many applications. In linear functions of the form $y = mx$, y is proportional to x, and the graph passes through the origin and has a slope of m.

Equations can be grouped by their forms so that they correspond with graphs of a particular shape. All equations of the form $y = mx + b$ represent lines, and all equations of the form $y = ax^2 + bx + c$ (in which $a \neq 0$) represent parabolas. Some other forms of equations also correspond to specific geometric shapes.

Algebra is a mathematical language used to express relationships between quantities. Algebra cannot be used to analyze nonquantifiable relationships or the merit of relationships that exist between quantities.

The most common applications of algebra involve formulas. Formulas can be derived from words, tables, or graphs. After they are derived, such formulas may be used to solve problems in which all but one of the quantities are known. Tables and graphs offer alternative ways to solve problems.

Two thousand years after Euclid died, Fermat and Descartes devised coordinate geometry. Coordinate geometry unifies algebra and geometry. Coordinate geometry gives an algebraic perspective to many basic geometry concepts, including length, slope, and parallelism, and a new way of examining midpoints, circles, right triangles, and quadrilaterals.

STUDY GUIDE

To review Chapter 11, see what you know about each of the following ideas or terms that you have studied. You can also use this list to generate your own questions about the chapter.

The NCTM Curriculum Standards and Algebra

Selected NCTM Curriculum Standards

The following standards come from the NCTM document.

- Represent situations and number patterns with tables, graphs, verbal rules, and equations, and explore the interrelationships of these representations.
- Analyze functional relationships to explain how a change in one quantity results in a change in another.
- Relate their everyday language to mathematical language and symbols.
- Understand the concepts of variable, expression, and equations.
- Use functions to represent and solve problems.

1. Describe how each standard listed relates to the material you studied in Chapter 11.
2. Select any current elementary-school mathematics textbook series, and describe sample lessons or exercises that illustrate each standard listed.

ALGEBRA AND COORDINATE GEOMETRY IN THE ELEMENTARY SCHOOL

The following chart shows at what grade levels selected algebra and graphing topics typically appear in elementary-school mathematics textbooks.

Topic	Typical Grade Level in Current Textbooks
Variables	5, <u>6</u>
Formulas and equations	5, <u>6</u>
Graphing in the coordinate plane	5, 6

REVIEW EXERCISES

1. Where possible, translate the following phrases or sentences into algebraic expressions, equations, or inequalities.
   (a) Joe weighs more than twice as much as his sister.
   (b) Sara Lee is the sweetest woman in Texas.
2. Tell whether each graph represents a function.

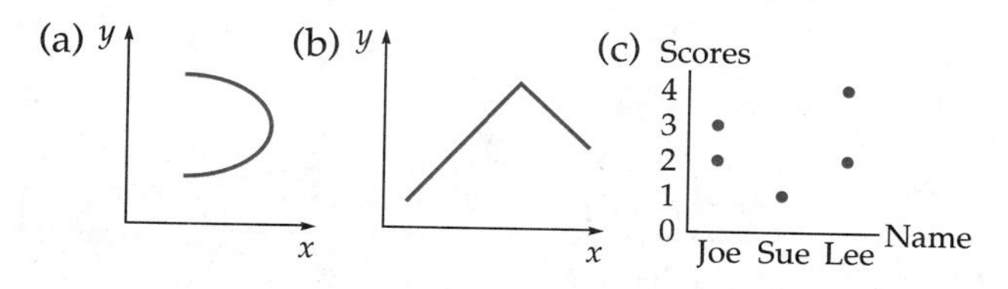

3. What real numbers cannot be used for x if $y = \sqrt{3 - x}$?
4. (a) Graph $y = 2x + 1$.
   (b) What is the equation of the image of $y = 2x + 1$ after a reflection through the y-axis?
5. What is the slope of $5x - 2y = 6$?
6. The height of an automobile jack is a function of the number of times you crank it. Before cranking, a jack might be 10 in. high. Each crank raises the jack (and car) 0.5 in.
   (a) Write a formula for the height of the jack, h, in terms of the number of cranks, n.
   (b) Graph your formula.
   (c) What shape is your graph?
7. A waiter earns \$1 per hour plus tips. Tips average \$2 per table.
   (a) Write a formula for the total earnings E in terms of the number of hours H and number of tables T.
   (b) How many tables must he serve in an 8-hour shift to earn \$40?

8. A salesperson earns a commission of 8% of her sales. Her boss then deducts 26% of the commission for taxes. How much must the salesperson sell in order to take home 500?
9. How is the graph of $y = x^2 + 2$ related to the graph of $y = x^2$?
10. The typical weight W (in pounds) of a man of height H (in inches) can be estimated by $W = 5.5H - 220$.
    (a) What does the slope of the equation tell about the height and weight?
    (b) Solve for H.
    (c) Which formula would be easier to use to find the typical weight of a man 5 feet 8 inches tall?
11. Suppose that you collect the following data on a group of 50 boys.

Age	3	4	5	6	7	8
Avg. height (cm)	100.0	106.6	112.8	119.5	126.0	133.0

    (a) Plot the points on a two-dimensional graph.
    (b) Draw a straight line that seems to fit the data best.
    (c) What is the equation of your line?
    (d) What average height does your equation predict at age 15?
12. The graph shows distance traveled vs. time for a car trip. Draw a graph of speed vs. time for the car trip.

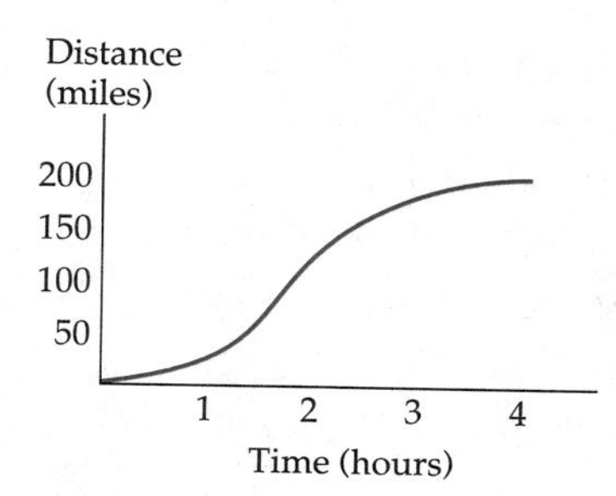

13. Explain why the distance between $A(x_1, y_1)$ and $B(x_2, y_2)$ is

$$AB = \sqrt{(x_2 - x_1)^2 + (y_2 - y_1)^2}$$

$B\ (x_2, y_2)$

$A\ (x_1, y_1)$

14.

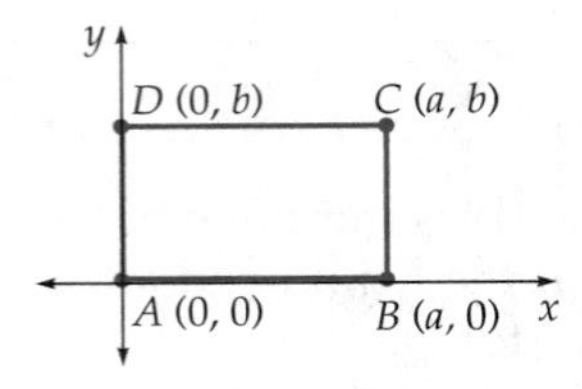

Show that the midpoint of $\overline{AC}$ is the midpoint of $\overline{BD}$.

15. How many different taxicab routes of the shortest distance from A to B are there?

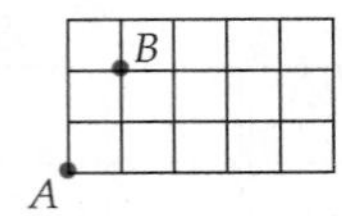

16. Suppose the daily profit P (in \$) from selling x neckties is $P = 10x - 25$.

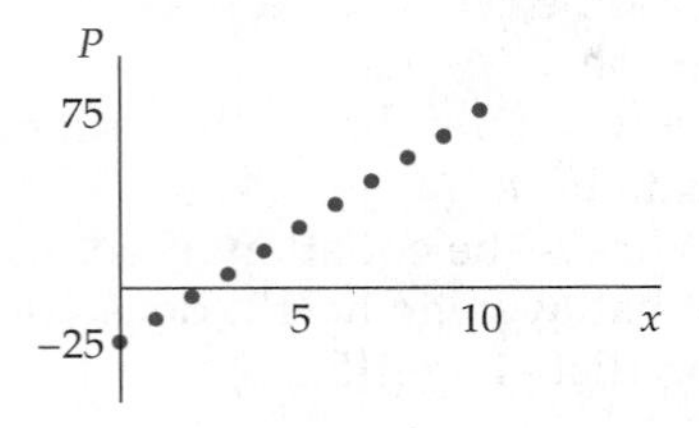

(a) According to the graph, what is the domain (input set)?
(b) What is the range (output set)?
(c) Explain why the graph exhibits a function from x to P.

17. (a) Tell which of the following diagrams represents a function from set A to set B.
(b) Give a possible rule relating each ordered pair.

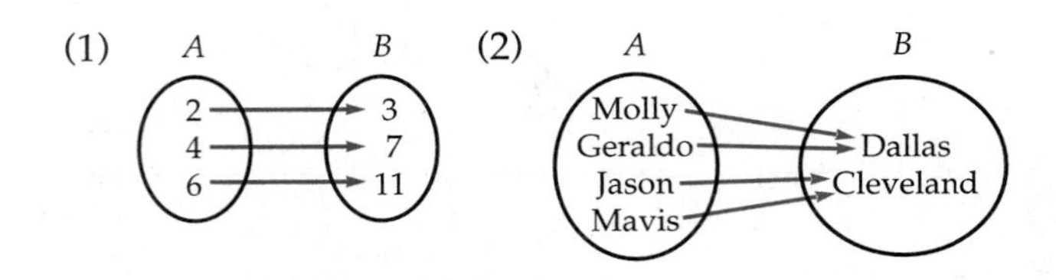

18. A taxicab company charges \$2.10 plus \$.80/mile.
(a) A formula relating the total cost C (in \$) to the distance d (in miles) is $C =$ ________.
(b) Graph your equation from part (a), with d on the x-axis.

SUGGESTED READINGS

Davis, R. *Discovery in Mathematics*. New Rochelle, N.Y.: Cuisenaire, 1980.

Edwards, R. *Algebra Magic Tricks* (2 vols.). Pacific Grove, Calif.: Critical Thinking Press, 1992.

Gardner, M. *Aha!* New York: W. H. Freeman, 1978.

Heid, M., J. Choate, C. Sheets, and R. Zbiek. *Algebra in a Technological World*. Reston, Va.: NCTM, 1995.

Jacobs, H. *Algebra*. New York: W. H. Freeman, 1979.

National Council of Teachers of Mathematics. *A Sourcebook of Applications of School Mathematics*. Washington, D.C.: MAA, 1980.

National Council of Teachers of Mathematics. *The Ideas of Algebra, K–12*. 1988 Yearbook. Reston, Va.: NCTM, 1988.

National Council of Teachers of Mathematics. *Calculators in Mathematics Education*. 1992 Yearbook. Reston, Va.: NCTM, 1992.

National Council of Teachers of Mathematics. *Communication in Mathematics, K-12 and Beyond*. 1996 Yearbook. Reston, Va.: NCTM, 1996.

Willicutt, B. *Manipulative Activities to Build Algebraic Thinking* (3 vols.). Pacific Grove, Calif.: Critical Thinking Press, 1995.

12 Statistics

Pick up a newspaper and you'll read statistics about what the typical citizen eats, earns, and believes. Statistics help us predict election results and tomorrow's weather. People use statistics to evaluate TV shows, schools, and government programs. An educated citizen should be able to tell whether statistics are being used appropriately in all of these situations.

Data collection is as old as written history, but the first comprehensive application of statistics was completed in 1662, when John Graunt tabulated and published information about births and deaths in England. Insurance companies used those data to set their rates.

The term "statistics" refers to both a set of data and the methods used to analyze the data. Statistical methods enable us to organize and simplify a set of data to reveal essential characteristics and relationships.

First we select some quantifiable aspect to measure. Then data are collected. From there, statistical methods enable us to organize and interpret the data.

Throughout this process, certain information is discarded. For example, a graph or table removes the original order of the numbers. An average reduces a whole set of data to just one number. These methods can reveal underlying patterns, or they can obscure reality. That is the difference between good and bad statistics.

12.1 STATISTICAL GRAPHS AND TABLES

Statisticians often communicate information about data by using a graph or a table (Figure 12-1, on page 714). Reading and interpreting graphs and tables is the most important part of statistics for most people. Graphs and tables can organize numerical data in a simple, clear way. Because graphs and tables simplify information, they also have the potential to mislead.

Graphs and tables are part of descriptive statistics. **Descriptive statistics** involves all of the techniques that are used to describe and summarize the characteristics of numerical data.

HERMAN

"That's the last time I go on vacation."

Figure 12–1

BAR GRAPHS, LINE GRAPHS, AND CIRCLE GRAPHS

Three basic types of statistical graphs are bar graphs, line graphs, and circle graphs. Each type of graph is more suitable for presenting certain types of information.

Lesson Exercise 12.1

Which graph, (a) the line graph or (b) the bar graph, is more appropriate for displaying the data?

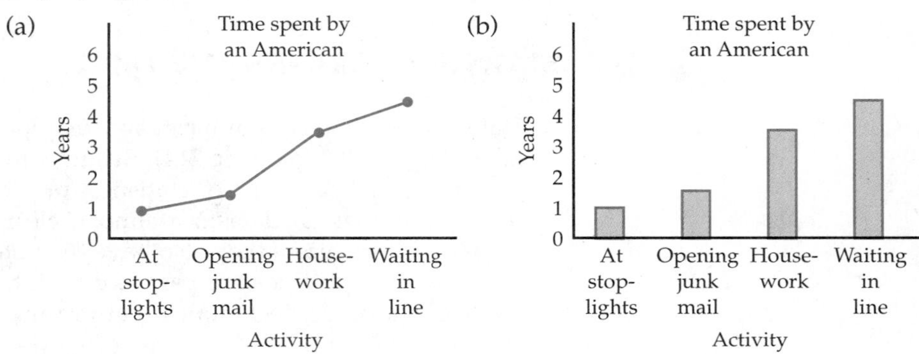

Statisticians generally use **bar graphs** to compare the values of several variables. Each bar shows the frequency of one of the variables. In most vertical bar graphs, a simple comparison of the heights of the bars reveals the relationships among the variables.

Three variations on bar graphs are line plots, pictographs, and stem-and-leaf plots. A **line plot** is a number line with X's or dots placed above specific numbers to show their frequency. The data set {2, 3, 3, 3, 4} is shown as a line plot in Figure 12-2. If the bars are replaced with pictures, the result is a pictograph. A **pictograph** employs pictures or symbols for quantities (Figure 12-2). The third variation on bar graphs, the stem-and-leaf plot, is introduced later in the section.

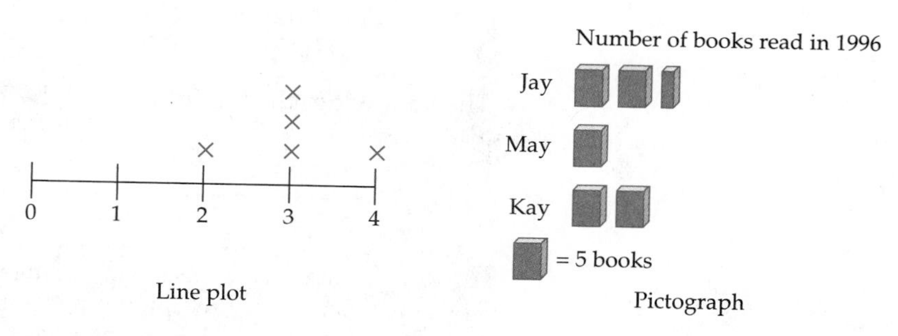

Figure 12–2

Statisticians most often use **line graphs** to show changes over time (Figure 12-3).

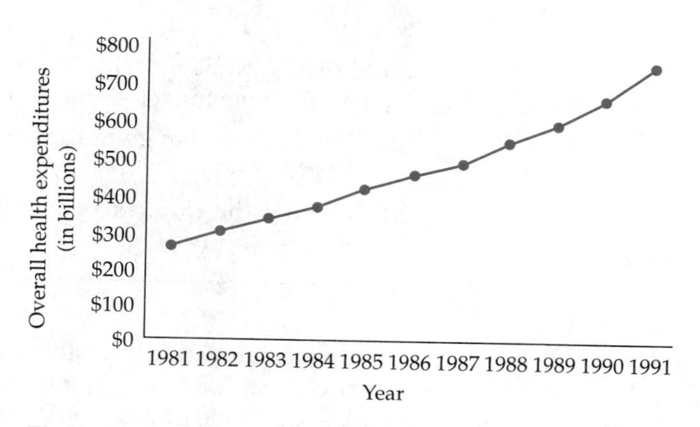

Figure 12–3

Circle graphs are useful for showing what part of the whole falls into each of several categories. Governments often use circle graphs to show how they distribute their funds (Figure 12-4). If there are more than six or seven parts, a circle graph becomes difficult to subdivide and label.

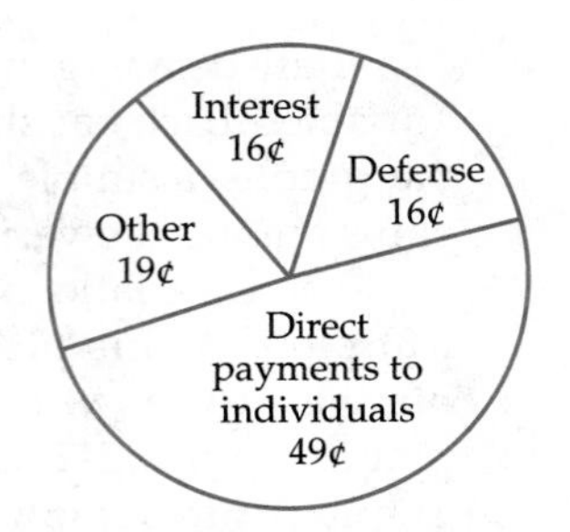

1997 federal budget dollar (estimated)

Figure 12–4

The following chart summarizes the types of graphs and their common uses.

Graph	Best for
Line	Showing changes over time
Bar	Comparing values of several categories
Circle	Showing parts of a whole

Lesson Exercise 12.2

For each of the following, would the best choice be a line graph, a bar graph, or a circle graph?

(a) Showing what percent of a family budget is devoted to each of the following: housing, clothing, food, taxes, and other

(b) Showing the change in the Consumer Price Index during the 12 months of 1997

(c) Showing the number of people who picked each of six sports as their favorite

A statistician would probably display the graph in Lesson Exercise 12.2(a) with a circle graph. However, if the budget were subdivided into eight or more categories, the circle graph would be difficult to read, so a bar graph would be better.

READING AND INTERPRETING GRAPHS

Do you feel confident reading and interpreting a graph? In reading and interpreting a graph, you should be able to:

1. Read specific data from the graph.
2. Summarize the overall pattern of the graph and any significant deviations from that pattern.
3. Assess the reliability of the data.
4. Note any important information that is not included in the graph.
5. Make decisions or predictions using the graph.

Try the following exercises.

Lesson Exercise 12.3

Estimate the following from the double bar graphs in Figure 12-5.

(a) What percent of people in the U.S. were under 15 years of age in 1995?
(b) What percent change is expected in the Japanese population 65 years and older between 1995 and 2025?

Percent of population in selected age groups: 1995 and 2025

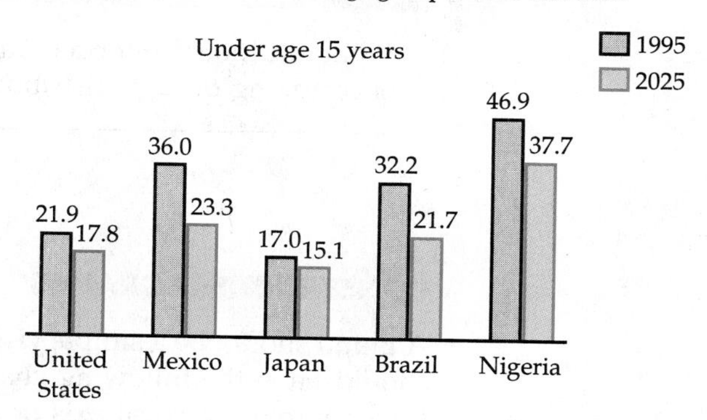

Age 65 years and over

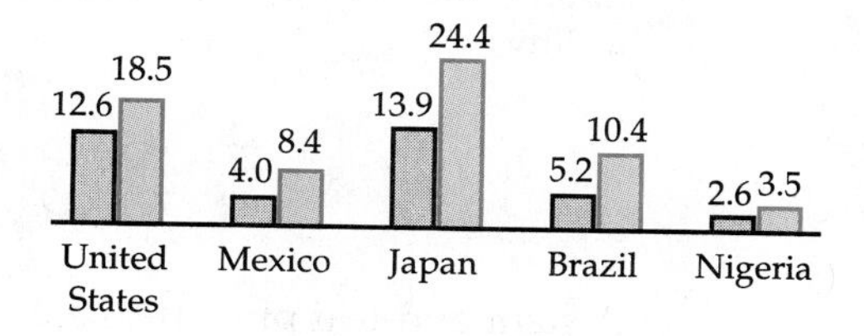

Source: *International Marketing Data and Statistics 1995*

Figure 12–5

In describing the overall pattern in each graph, note the highest points and the trends of increase and decrease.

Lesson Exercise 12.4

(a) Compare the age distributions of younger and older people in each country in 1995 (Figure 12-5).

(b) Speculate about why the age distributions differ as they do.

(c) Compare the predicted changes in age distribution in each country.

Lesson Exercise 12.5

Do you think the data in these graphs are accurate? Why or why not?

Lesson Exercise 12.6

What additional information would be useful for understanding the difference among the age distributions in the various countries?

CONSTRUCTING GRAPHS

A graph should be a simple visual summary of a set of data. A statistical graph should have the following characteristics: (1) a title and labels on both axes, (2) a complete vertical axis or a break showing that the axis is not complete, and (3) vertical bars (for bar graphs) unless there is a good reason not to use them.

A researcher selects 20 U.S. families at random and records their annual family incomes (in thousands of dollars). Suppose that the results are as follows.

27	52	40	98	10	50	37	72	112	58
62	134	19	46	26	81	12	48	23	32

A stem-and-leaf plot offers a quick way to display a small set of positive numbers such as this one. A **stem-and-leaf plot** uses the actual data numbers to make a type of bar graph.

EXAMPLE 12.1

Make a stem-and-leaf plot of the family incomes.

Solution

Plot each score with a stem and a leaf. For example, 27, with a stem of 2 and a leaf of 7, is

2|7

Now plot the trunk and each stem (in this case, the hundreds and tens digits).

```
13|
12|
11|
10|
 9|
 8|
 7|
 6|
 5|
 4|
 3|
 2|
 1|
```

Then plot the leaves for all the scores

```
Stem → 13|4 ← Leaf
       12|      ← No data
       11|2
       10|      ← No data
        9|8
        8|1
        7|2
        6|3
        5|208 ← For 52, 50, and 58
        4|068
        3|72
        2|763
        1|092
```

■

A stem-and-leaf plot looks like a sideways bar graph. Note that each individual score can be read from the stem-and-leaf plot. Some statisticians arrange the leaves in increasing numerical order for each stem.

Now try one yourself.

Lesson Exercise 12.7

A second researcher selects 20 U.S. families at random and records their annual family incomes (in thousands of dollars) as follows.

41	74	22	151	42	29	60	115	13	56
45	48	99	17	48	25	68	72	7	38

(a) Make a stem-and-leaf plot of the family incomes.

(b) How do the distributions of the two samples of 20 family incomes compare?

Stem-and-leaf plots are more cumbersome for large data sets. In addition, stem-and-leaf data must be grouped according to place value, although a statistician might prefer different groupings.

To avoid these limitations, statisticians usually select their own intervals to group data and display the frequency distribution in a histogram. To illustrate this process, let's use the data from the last lesson exercise. The first step is to decide how to put the data into 6 to 15 intervals of equal width.

Lesson Exercise 12.8

Suppose you want to put the data into 6 subintervals of equal size.

(a) The lowest subinterval starts around the number ________, and the highest subinterval ends around the number ________.

(b) About how big should each subinterval be?

(c) How many whole numbers are there from 1 to 3 (inclusive)?

(d) How many whole numbers are there from 7 to 151 (inclusive)?

In Lesson Exercise 12.8, you may have gotten a good idea of how to create 6 subintervals. Statisticians have a procedure for doing it. First they find how far apart the scores are. The **range** is the difference between the highest score and the lowest score. In this case, the range is $151 - 7 = 144$. Are there 144 whole numbers from 7 to 151 (inclusive)? No. The number of whole numbers from 7 to 151 (inclusive) is the range plus 1. There are $151 - 7 + 1 = 145$ scores.

To divide a width of 145 into 6 classes (subintervals) of equal width, compute $\frac{145}{6} = 24.2$. Round up so that each subinterval has a width of 25.

Lesson Exercise 12.9

Complete the grouped frequency distribution table for the income data.

Annual Income (thousands of dollars)	Frequency
7–31	
32–56	
57–81	
82–106	
107–131	
132–156	

Why is it reasonable to use 6 intervals? Ideally, the data values within each interval represent similar results and differ from the results in other intervals. A much larger or much smaller number of intervals would not give a meaningful display of the overall distribution. It would have many short bars or a few very big bars, respectively.

Statisticians usually select classes of equal width, making it possible to compare frequencies in different intervals by looking at the heights of the bars.

After tabulating the data in a table, statisticians usually display the frequency distribution of a *single* variable using a histogram. A **histogram** is a bar graph that shows the numbers of times data occur within a certain interval. It has the following properties: (1) the bars are always vertical, (2) the width of each bar is based upon the size of the interval it represents, and (3) there are no gaps between adjacent bars (if bars of height 0 are included).

The histogram in Figure 12-6 shows the distribution of frequencies of family income. The histogram shows a pattern in which most families have incomes at the lower or middle level ($7000 to $81,000) and a few are at the upper end.

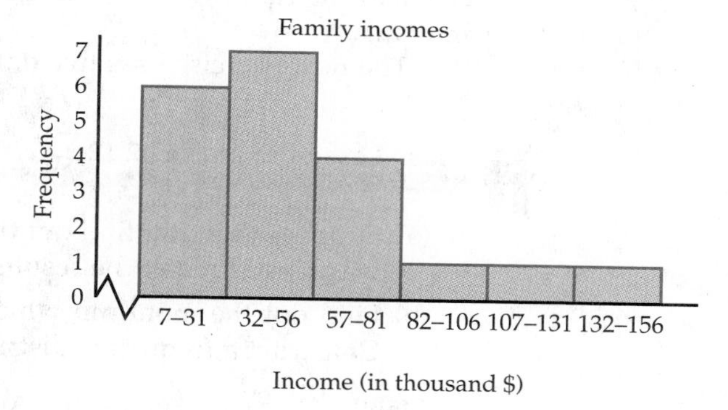

Figure 12–6

How would the graph change if the number of classes were changed?

Lesson Exercise 12.10

(a) Put the same income data into 9 classes of equal width, and construct a histogram.

(b) How does your histogram compare to the one with the 6 intervals?

Histograms have no gaps because their bases cover a continuous range of possible values of a variable, such as the days of the year. If all days of the year are included, there are no gaps between them. In a bar graph, the gaps between bars indicate that each bar describes a different item.

Lesson Exercise 12.11

Which of the following graphs are histograms?

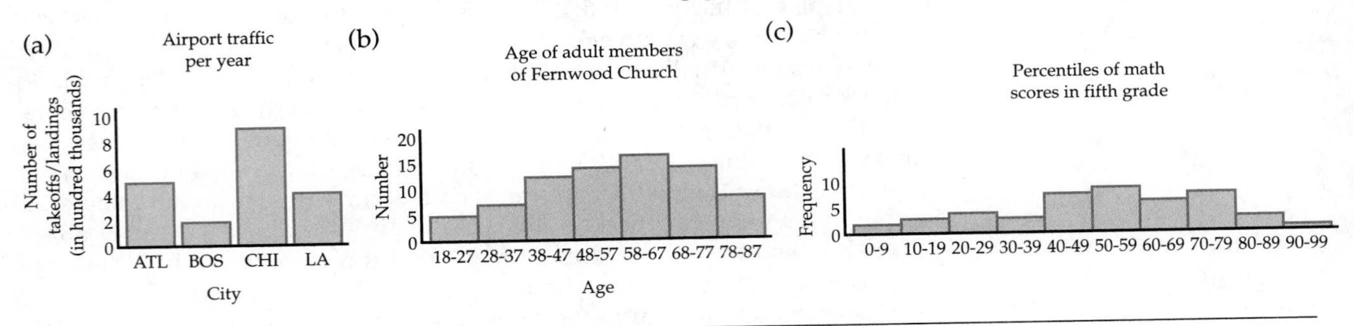

The *area* of each bar is proportional to the number of items or the length of the corresponding interval. Usually intervals are equal in size, so the reader can look at the *heights* of the bars to compare the frequencies of different intervals.

The next exercise uses live data from your class!

Class Lesson Exercise 12.12

(a) Suppose that you find out the month in which each person in your class was born. Predict the results.

(b) Find out the month in which each person in your class has a birthday. Construct a frequency distribution table.

Month	Jan	Feb	Mar	Apr	May	Jun	Jul	Aug	Sep	Oct	Nov	Dec
Number of birthdays												

(c) What would be the most appropriate type of graph—a line graph, a histogram, or a circle graph?

(d) Construct a bar graph of the frequency distribution. Assume that all intervals are the same size (although they differ slightly).

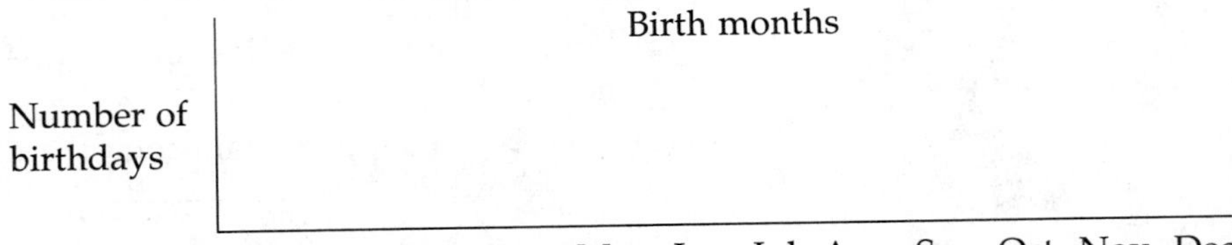

(e) Do you see any pattern in the results? If so, why do you think the pattern occurs?

The preceding exercise illustrates a common process in statistics. Begin with a statistical question of interest and a hypothesis about the answer. Then collect data. Next, organize and display the data. Finally, analyze the data in relation to your hypothesis.

DISTORTED GRAPHS

Graphs convey visual information in a simple, dramatic way, but sometimes the display is distorted. People most commonly distort graphs by stretching and shrinking scales on the vertical axis without indicating what they have done (Figure 12-7).

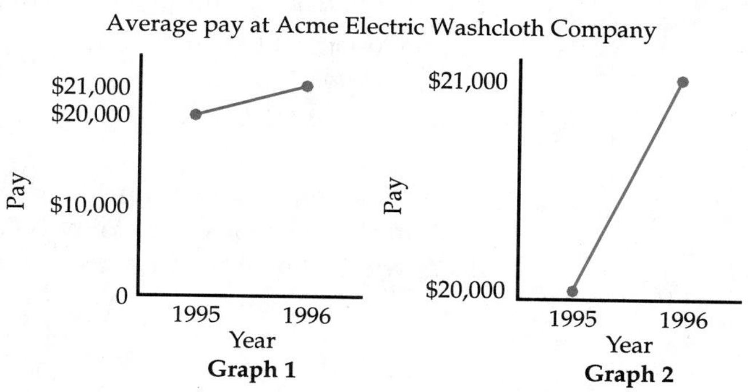

Figure 12–7

Lesson Exercise 12.13

(a) Suppose that you manage the Acme Electric Washcloth Company and want to show that wages have increased sharply. Which graph in Figure 12-7 would you prefer?

(b) Suppose you want to present the data in an undistorted way. Which graph would you use?

(c) What additional information would be helpful for assessing the wage situation at Acme?

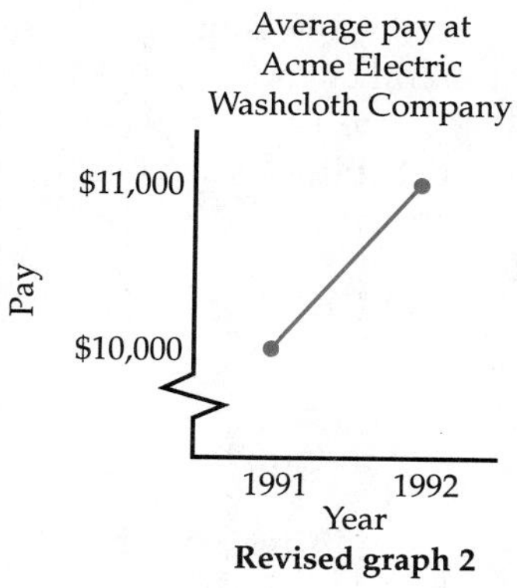

Figure 12–8

The most common method of distorting graphs is to start from a number other than 0 on the vertical axis, as in Graph 2 in Figure 12-7. Why might someone use Graph 2 instead of Graph 1 in Figure 12-7? The person may want to mislead you. Many times, however, a distorted graph is used because it looks better and is easier to read. Notice the large empty space in Graph 1. If you use a distorted scale on a graph, alert the reader by distorting the y-axis with the ⋦ symbol. The ⋦ indicates a distortion in the numbering scale between 0 and the next number (Figure 12-8).

In deciding on the overall width and height of a graph, many statisticians use the "3/4 high" rule, making the height about three fourths of the width, as is done in Graph 1 and the revised Graph 2, but not in the original Graph 2.

Lesson Exercise 12.14

Average Teachers' Salaries (thousands of dollars)

Year	1990	1992	1994	1996
Amount	30.9	33.1	35.4	38.2

(a) Construct an undistorted line graph of these data.

(b) Construct a distorted line graph that makes the salary increase from 1990 to 1996 look greater.

If someone glances quickly at a distorted graph without reading the numbers on the axes, he or she may be deceived by the shape of the graph. If you don't want to be fooled by distorted graphs, read the numbers on the axes.

ANSWERS TO SELECTED LESSON EXERCISES

12.1 The bar graph

12.2 (a) A circle graph (b) A line graph (c) A bar graph

12.3 (a) 21.9% (b) An increase of 75.5%

12.4 (a) The U.S. and Japan have relatively low percentages of people in the younger age group and relatively high percentages of people in the older age group. Mexico, Brazil, and Nigeria have relatively high percentages in the younger age group and relatively low percentages in the older age group.
(b) People in Mexico, Brazil, and Nigeria live shorter-than-average lives. People in the United States and Japan live longer-than-average lives.
(c) All age distributions are predicted to change in the same way, with all the populations becoming somewhat older.

12.5 The data for 1995 should be fairly accurate. The data for 2025 are highly speculative.

12.6 The life expectancy in each region and the age distributions for the groups between 15 and 65 years old

12.7 (a)

```
15 | 1
14 |
13 |
12 |
11 | 5
10 |
 9 |
 8 |
 7 | 24
 6 | 08
 5 | 6
 4 | 125889
 3 | 8
 2 | 259
 1 | 37
 0 | 7
```

(b) The group in the exercise has a greater range and more families in the $40,000 group. The two distributions are fairly similar.

12.8 (a) 7; 151 (b) 24 or 25
(c) $3 - 1 + 1 = 3$
(d) $151 - 7 + 1 = 145$

12.9

Annual Income (thousands of dollars)	Frequency
7–31	6
32–56	7
57–81	4
82–106	1
107–131	1
132–156	1

12.11 (b) and (c)

12.13 (a) Graph 2 (b) Graph 1
(c) The rate of inflation, the benefits, the kind of training required, and what other similar companies pay

12.14 (b) *Hint:* Do not start the vertical axis at 0.

12.1 HOMEWORK EXERCISES

Basic Exercises

1. For each of the following, determine whether the best choice would be a line graph, a bar graph, or a circle graph.
(a) Showing the percent of people who sleep an average of 5 or less 6, 7, 8, or 9 or more hours per night
(b) Showing the annual U.S. government budget for environmental protection from 1978 to 1997
(c) Showing the total sales of raffle tickets by each of grades 3, 4, and 5

2. For each of the following, determine whether the best choice would be a line graph, a bar graph, or a circle graph.
(a) Showing the percent of people who favor each of three candidates in an election
(b) Showing the favorite days of the week of students in your class
(c) Showing the change in the number of students taking high-school algebra over the last 10 years

3.

Major and starting salary
for bachelor's degree in 1993

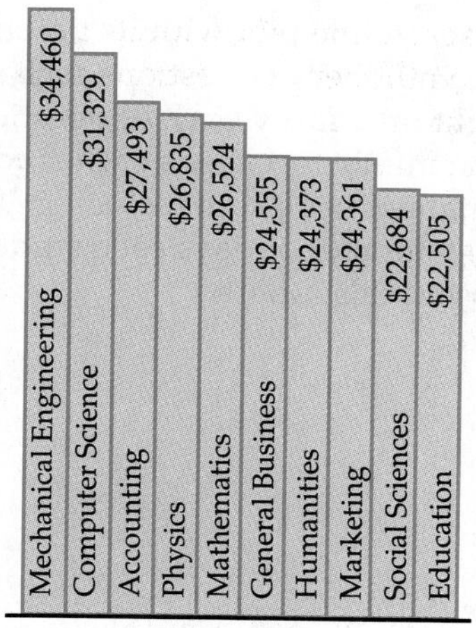

Mechanical Engineering $34,460
Computer Science $31,329
Accounting $27,493
Physics $26,835
Mathematics $26,524
General Business $24,555
Humanities $24,373
Marketing $24,361
Social Sciences $22,684
Education $22,505

(a) Which degree earned the lowest average starting salary?
(b) Which degree earned the highest average starting salary?
(c) Summarize the overall trend in starting salaries and the degrees shown.

(continued on the next page)

(d) Is there any degree whose average salary surprises you?
(e) Do you think the data are accurate? Why or why not?
(f) What additional information would be useful?

4. The following pictograph shows how many books each child read in March.

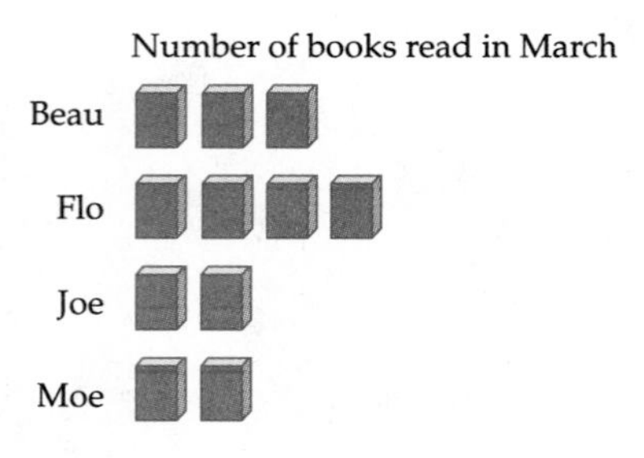

(a) How does a pictograph differ from a bar graph?
(b) What additional information would be useful?

5. Match each line plot with its description.
(a) The number of questions each student got right on a fairly easy 10-question quiz
(b) The number of heads each student obtained in 10 coin tosses
(c) The number of days each student has been absent this month

```
(1)                    (2)                   (3)
X
X X                          X                  X
X X       X                X X                X X
X X X     X              X X X              X X X
X X X X X     X     X    X X X X X        X X X X   X
0 1 2 3 4 5 6 7 8   4 5 6 7 8 9 10        3 4 5 6 7 8
```

6. In *State of the World* (1984), Lester Brown lists data on causes of premature deaths in the United States. (Data for 1996 would be similar.)

Cause of Death	Number of Premature Deaths
Tobacco use	375,000
Alcohol use	100,000
Motor vehicle accident	50,000
Use of hard drugs	30,000
Suicide	27,500
Murder	19,000

(a) Make a bar graph of the data.
(b) What conclusions could be drawn from the data?
(c) What additional information would be useful?

7. Complete the double bar graph of fall-semester enrollments for the two years given in the table.

Student Enrollment	Math	English	Business	Psych.
Fall 1996	2421	2260	1572	874
Fall 1997	2580	2501	1610	710

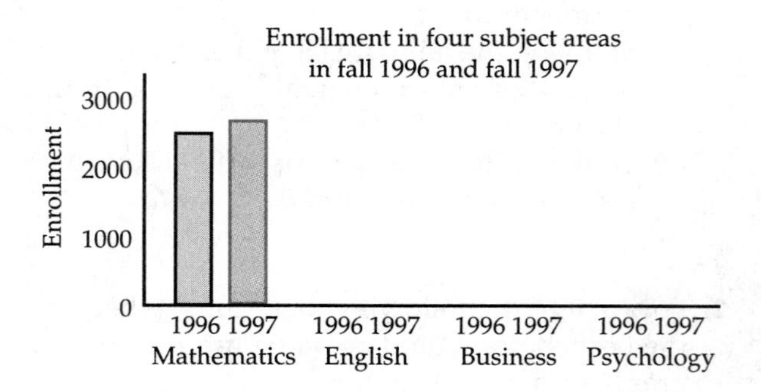

8. The following double line graph shows the average daily prices of the Dow Jones stock index and the Standard and Poor index from 1994 through January 1996. How do the two indices compare?

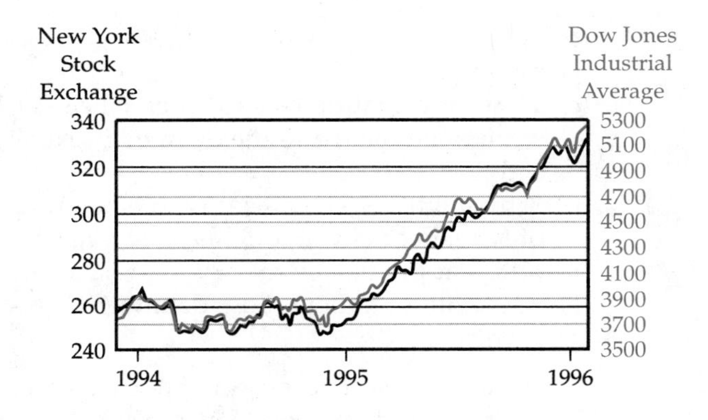

9. The following line graph shows the percent of persons living below the poverty level from 1977 to 1993.

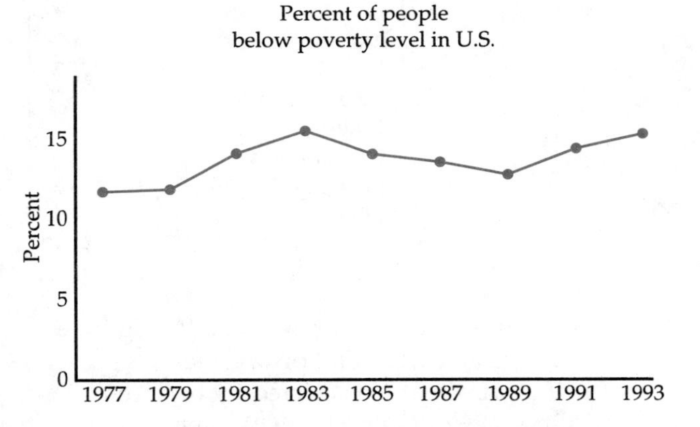

Source: U.S. Bureau of the Census, *1995 Stat. Abstract*

(a) Describe the overall trends.
(b) Can you explain why these trends occurred?
(c) What additional information would be useful?

10. The data in the following table are from *How You Rate* by T. Bicaree.
(a) Why is an older person's life expectancy higher?
(b) A woman's life expectancy is about ________ years longer than a man's.
(c) Speculate why women tend to live longer than men.

Age	Man's Life Expectancy	Woman's Life Expectancy
20	71	78
30	72	79
40	73	79
50	74	80
60	77	82
70	81	84
80	87	89

11. The Perky Academy annual budget for 1996–1997 is as follows.

Revenues	
Tuition and fees	$4,374,000
Alumni gifts	194,000
Endowment interest	109,000
Bookstore and vending machines	204,000
Summer day camp	90,000
Other	93,000
Total	$5,064,000

Expenditures	
Teachers' salaries and benefits	$2,269,000
Administration salaries and benefits	769,000
Maintenance and renovation	654,000
Scholarships	504,000
Bookstore and vending machines	194,000
Summer day camp	63,000
Student activities	21,000
Utilities, phone, and postage	501,000
Insurance	114,000
Other	130,000
Total	$5,219,000

(a) What percent of the money spent goes for teachers' salaries and benefits?
(b) How much profit does the school make running the summer camp?
(c) What is the school's total deficit (loss) for the year?

12. The following circle graph shows the marital statuses of U.S. women ages 20–24 in 1993. The data are from the *Universal Almanac 1995* (John Wright, ed.).

Marital status of U.S. women ages 20–24 in 1993

Married 31.0%

Single 66.8%

Divorced or widowed 2.2%

(a) According to the graph, what percent of women ages 20–24 have ever been married?
(b) Guess what the data for men ages 20–24 would look like.

13.

Pregnancies per 1000 15- to 19-Year-Olds (Alan Guttmacher Institute)	
United States	96/yr
England/Wales	45
Canada	44
France	43
Sweden	35
Netherlands	14

Percent of Teenagers Who Are Sexually Active

	Age		
	16	*17*	*19*
United States	31%	44%	72%
Canada	14%	19%	41%
England/Wales	22%	37%	66%
France	16%	28%	74%
Netherlands	No data	35%	64%
Sweden	34%	58%	89%

(a) Which countries have the most sexually active teenagers?
(b) Which country has the most teenage pregnancies?
(c) To what extent do teenage pregnancies seem to be related to the degree of sexual activity?
(d) Why might two countries with equal rates of sexual activity among teenagers have different pregnancy rates?
(e) Speculate as to how these data were collected.

14. The data in the second column are based on a Gallup poll conducted on November 9–12, 1984. Voter participation in 1992 was virtually identical (61%).

Eligible Voters in the 1984 Presidential Election

Voted	60%		
Did not vote	40%	Did not like candidates	7%
		Not interested in politics	6%
		Working	5%
		Inconvenient	5%
		Illness	5%
		Out of town	4%
		New resident	4%
		Other	4%

(a) About ________ out of every 10 eligible voters actually vote for president.
(b) If a president was elected by between 50% and 60% of *those who voted*, then about ________ out of every 10 eligible voters voted for him.
(c) About ________ out of every 10 eligible voters do not vote for president.
(d) For the 4% who gave other reasons for not voting, name a reason they might have given.

Use your judgment to estimate answers to parts (e) and (f).

(e) ________% of the people did not vote because they had more important commitments.
(f) ________% of the people did not vote because of a lack of interest in the election.

15. Test results for two mathematics classes are as follows.

Class A: 58, 62, 62, 70, 72, 75, 80, 81, 85, 92, 98

Class B: 42, 65, 65, 68, 75, 80, 82, 90, 91, 99

(a) Make a stem-and-leaf plot for class A.
(b) Make a stem-and-leaf plot for class B.
(c) Look at the stem-and-leaf plots from classes A and B, and describe the difference in the distribution of scores in the two classes.

16. Following are the ages at inauguration of all U.S. presidents from Washington to Clinton.

57	61	57	57	58	57	61	54	68
51	49	64	50	48	65	52	56	46
54	49	50	47	55	55	54	42	51
56	55	51	54	51	60	62	43	55
56	61	52	69	64	46			

(a) Make a table of these data, showing the frequency distribution.
(b) Make a stem-and-leaf plot of these data.
(c) Make a bar graph of these data.

17. The following list shows the retirement ages of the last 15 teachers who retired at Brain Power High.

64	52	68	72	65	47	59	43
56	48	49	51	58	66	58	

(a) Make a stem-and-leaf plot of the data.
(b) Summarize in words what the stem-and-leaf plot indicates about the retirement ages.
(c) Make a bar graph of the same data.
(d) State one advantage of using the stem-and-leaf plot and one advantage of using the bar graph.

18. (a) Time the next 20 commercials you see on television.
(b) Select intervals for a frequency distribution of the commercial lengths.
(c) Make a graph showing the frequency distribution of commercial lengths.

Frequency

Length of commercial, in seconds

19. Following are the 1994 average annual salaries (in thousands of dollars) of U.S. elementary-school teachers, by state, from the *Statistical Abstract*.

AL	28.7	IL	37.5	MT	28.0	RI	39.2
AK	47.5	IN	35.6	NE	29.6	SC	29.2
AZ	31.8	IA	29.7	NV	33.4	SD	24.9
AR	27.3	KS	33.9	NH	34.1	TN	30.0
CA	39.8	KY	31.2	NJ	43.7	TX	29.9
CO	33.3	LA	26.3	NM	27.4	UT	27.8
CT	49.1	ME	30.5	NY	44.0	VT	33.9
DE	36.9	MD	38.4	NC	29.7	VA	32.0
DC	41.9	MA	40.9	ND	25.5	WA	35.5
FL	31.9	MI	45.2	OH	35.2	WV	30.2
GA	30.7	MN	35.6	OK	26.3	WI	35.6
HI	36.6	MS	24.7	OR	37.0	WY	30.8
ID	27.6	MO	29.5	PA	41.7		

(a) Select intervals, and make a table showing the frequency distribution.
(b) Make a histogram of the data.

20. Suppose you wanted to collect data to rate how livable the 100 largest cities in the United States are.
(a) Name five categories of data you would collect.
(b) If it is available, look at *The Places Rated Almanac* by R. Boyer and D. Savageau. What categories do they use?

21.

USA SNAPSHOTS®

A look at statistics that shape your finances

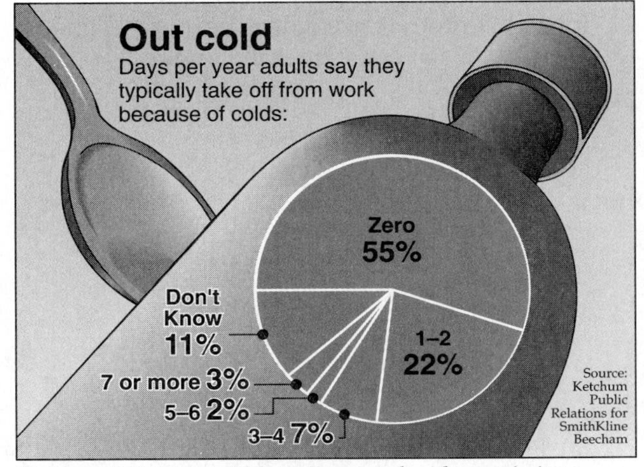

The central angles for each region of the circle graph are obtained by computing the corresponding percent of 360°. For example, 55% of the respondents said that they take a total of zero days per year because of colds, so the central angle of the region representing families is 55% of 360° = 198°. How big is the central angle for the region representing people who take off 1 or 2 days per year because of colds?

22. In 1994, each dollar in the federal budget was divided up as follows: 37¢ for income/social security, 18¢ for defense, 14¢ for interest, 19¢ for health/Medicare, and 12¢ for other programs.
(a) What percent of the budget was spent on health/Medicare?
(b) How big should the central angle for health/Medicare be?
(c) Use a protractor to draw a circle graph of the 1994 federal budget.
(d) What are some governmental programs included in the part labeled "other"?
(e) What additional information would be useful?

23. Juan Gomez's average monthly budget is broken down as follows:

Food	$150	Entertainment	$40
Clothing	50	Other	60
Rent	300		

(a) Find out what percent of the total monthly budget each category is.
(b) Use a protractor to construct a circle graph of Juan's average monthly budget.

24.

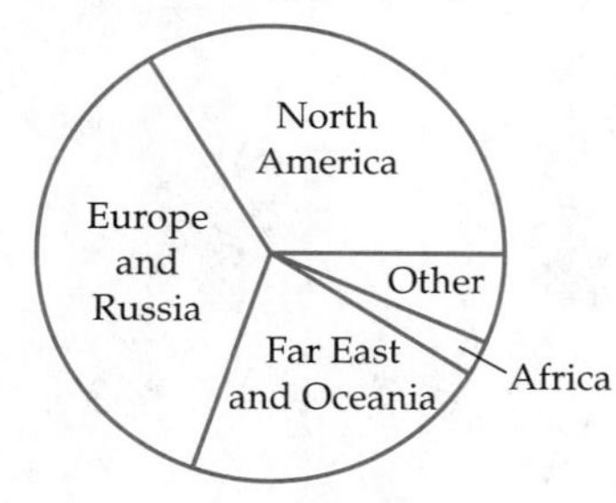

(a) Estimate the percent of energy consumption for each place.
(b) Measure each angle in the circle graph, and compute what percent of the 360° circle each place has.

25. Construct a circle graph showing how you spend time during an average school week. Include the following categories: sleeping, eating, working, going to school, watching television, recreation, and other.

26. Which of the following graphs are histograms?

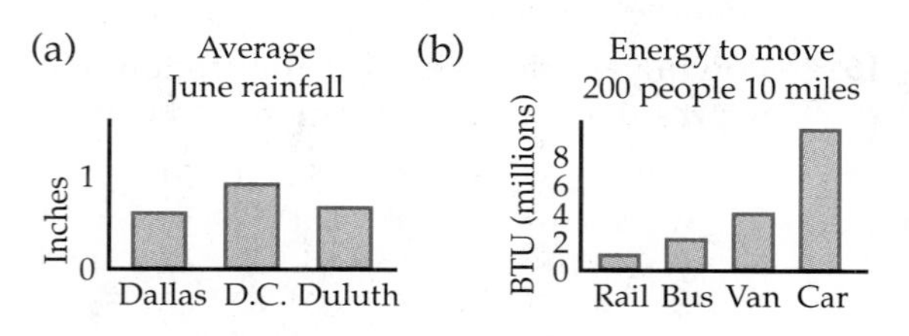

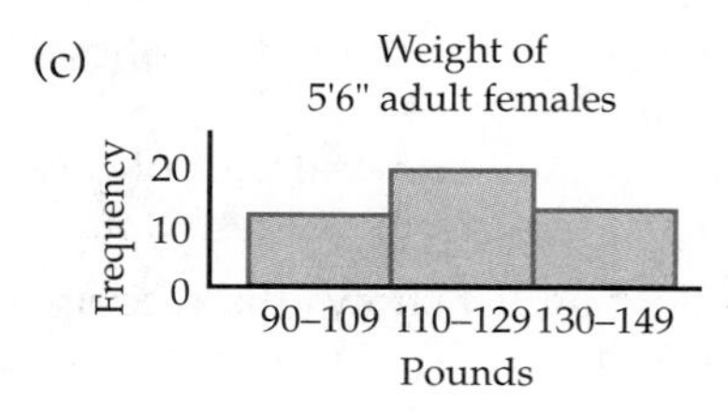

27. The following graph appeared in the May–June 1989 issue of *Learning 89*. It compares the lengths of the school years in six countries.

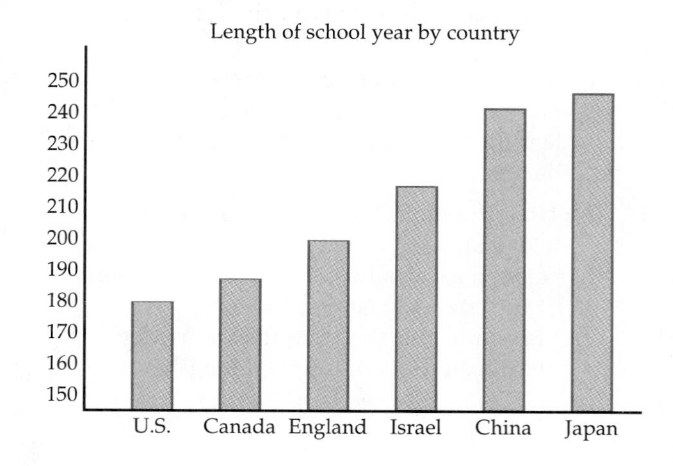

(a) Which country *appears* to have a school year twice as long as the U.S.? (Use a ruler.)

(b) Graph the same data on an undistorted graph.
(c) Which country is made to look the worst by the distortion in the original graph?

28. The following graph of average hourly earnings of factory workers appeared in the February 14, 1996, issue of *The Wall Street Journal*. The graph appears to show a major increase in hourly earnings from 1993 to 1996.

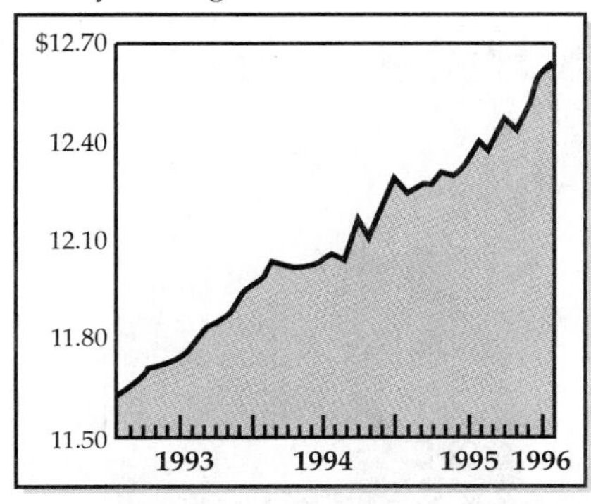

(a) How did the statistician distort this graph?
(b) How many times higher do the June 1995 earnings *appear* to be than the June 1993 earnings? (Use a ruler.)
(c) Make an undistorted graph of the same data.

29. What is misleading about the sizes of the regions in the following graph?

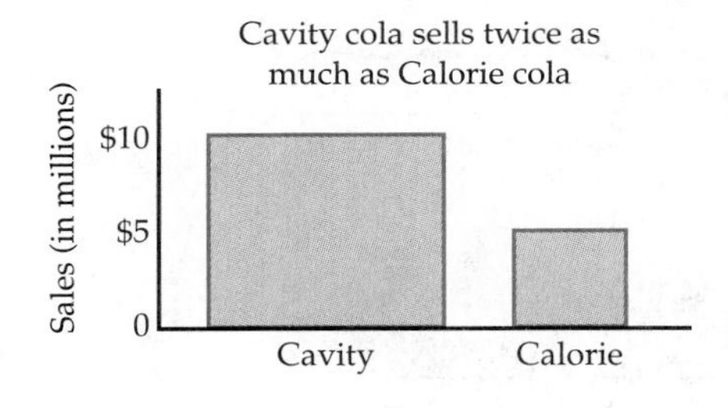

30. Find a graph in a magazine or newspaper. Write a paragraph telling what the graph is supposed to show and whether it is accurate and useful or misleading.

Extension Exercises

31. A women's group claims that a university discriminates against women in its graduate-student admissions. The university admits 40% of the men who apply to graduate school and only 30% of the women who apply.

	Applied	Admitted	Percent Admitted
Men	150	60	40%
Women	150	45	30%

The university has two graduate departments. The university administration claims that departmental breakdowns of admissions data show that there is no discrimination.

		Applied	Admitted	Percent Admitted
Scuba Dept.	Men	50	10	20%
	Women	100	20	20%
Astrology Dept.	Men	100	50	50%
	Women	50	25	50%

In this situation, which argument is more valid, that of the women's group or that of the university?

32. A company hires men and women for jobs in three categories: executive, bureaucrat, and laborer.

	Men		Women	
Category	*Number Applied*	*Number Hired*	*Number Applied*	*Number Hired*
Executive	600	60	100	5
Bureaucrat	100	50	400	100
Laborer	100	20	200	30
Total	800	130	800	135

The company claims that it is not discriminating against female applicants, since it hires about the same overall proportions of men and women (130/800 compared to 135/800).

(a) A women's group claims otherwise. Fill in the table to obtain a different perspective.

(b) In this situation, which argument is more valid, that of the women's group or that of the company?

Category	Percentage of Applicants Hired	
	Men	*Women*
Executive		
Bureaucrat		
Laborer		

(for exercise 35)

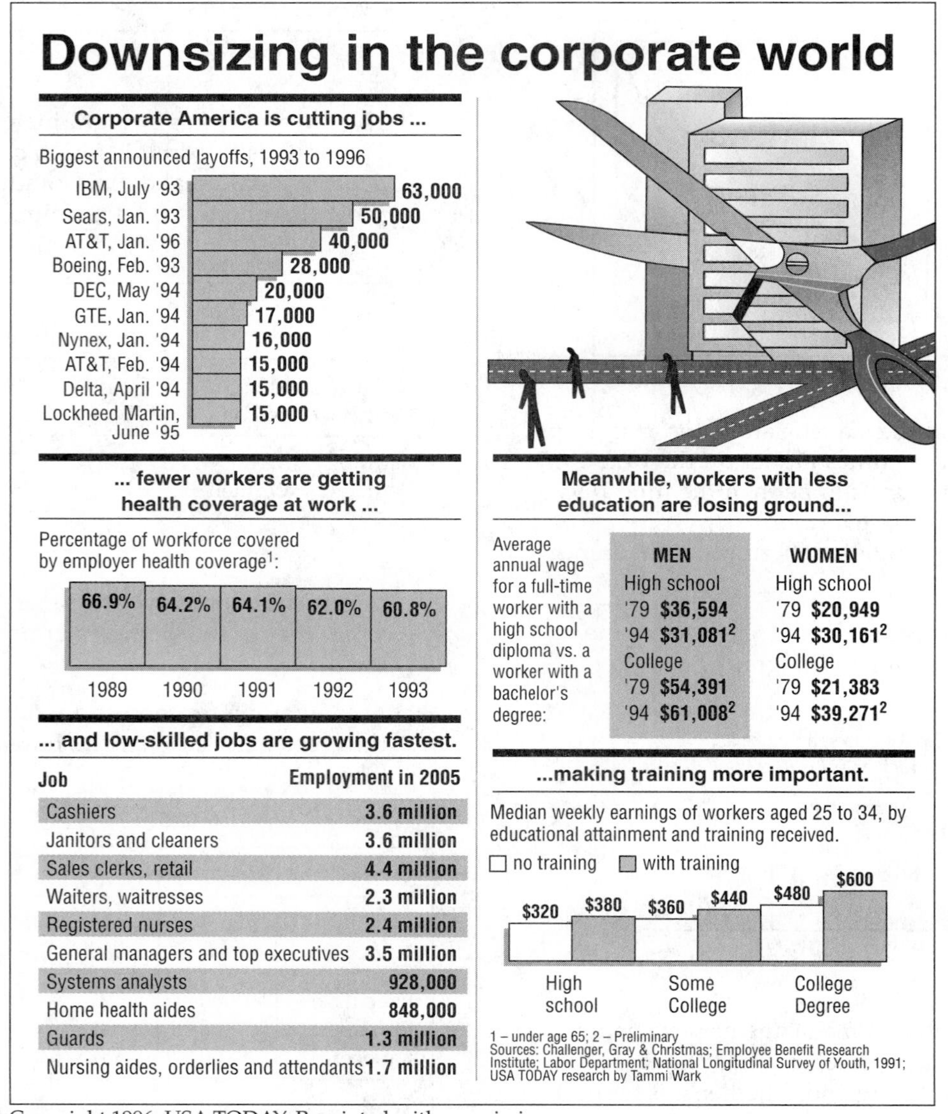

33. A class received the following test scores: 61, 63, 65, 68, 69, 75, 81, 82, 91, 92, 94, 97.
(a) Construct a table with score intervals that show that most of the scores were either high or low.
(b) Construct a shorter, deceptive table, using intervals that make the scores appear to be evenly distributed.

34. A poll is taken to see how people in different age groups feel about candidate Hope. One hundred people in each age group are asked whether they would vote for Hope rather than his opponent. The results are as follows.

Age	<21	21–30	31–40	41–50	51–60	>60
Votes for Hope	6	66	40	36	40	20

Make a table with different groupings that makes it appear that Hope receives similar support from young, middle-aged, and older voters.

35. Write a paragraph or two giving an overall summary that combines the information from all the graphs and tables from *USA Today* (2-21-96) on the following page.

Computer Exercises

36. (a) The following BASIC program will print 100 random numbers. Run it (or a similar program in Logo or Minitab).

```
10 FOR I = 1 TO 100
20     LET N = RND(1)
30     PRINT N
40 NEXT I
```

(b) Select class intervals, and construct a frequency distribution table of your results.
(c) Construct a histogram of the frequency distribution.
(d) Describe any pattern you see in your results.

37. Use MINITAB to construct a histogram and a stem-and-leaf plot for the family income data from the lesson. (This could also be done with certain graphing calculators.)

```
MTB> SET THE FOLLOWING DATA INTO
     C1
DATA> 34 62 18 126 35 24 50 96 11 46
DATA> 37 40 83 14 40 21 57 60 6 32
MTB> HISTOGRAM OF C1
MTB> STEM C1
MTB> STOP
```

38. Most computer spreadsheet programs can create bar charts (graphs) and pie charts (circle graphs) from spreadsheet data. A simple bar chart plots the values from one row or column of data in a spreadsheet. A pie chart tells what percent of the total in a row or column each entry represents. Consider the following spreadsheet.

	A	B	C
1	**Chandler Family Expenses in 1996**		
2	Rent/utilities		$13200
3	Food/clothes		$ 8950
4	Automobiles		$ 7280
5	Taxes		$11850
6	Other		$12710
7	Total		$53990

(a) Enter the data in a computer spreadsheet.
(b) Display a bar chart of the family expenses.
(c) Display a pie chart of the family expenses.

Special Exercises

39. The following table gives the percent frequencies of each letter in written English.

Letter	A	B	C	D	E	F	G
Percent frequency	8%	1%	3%	4%	13%	3%	2%

Letter	H	I	J	K	L	M	N
Percent frequency	5%	7%	0.2%	0.5%	4%	0.3%	4%

Letter	O	P	Q	R	S	T	U
Percent frequency	8%	2%	0.2%	6%	7%	9%	3%

Letter	V	W	X	Y	Z
Percent frequency	1%	2%	0.3%	2%	0.1%

(a) Name the four most frequently occurring letters.

(b) Plot the data for the 11 most frequently occurring letters as a bar graph.

Frequency

Letters

(c) Codes can sometimes be deciphered through analysis of the frequencies of the letters. Tabulate the frequency of each letter in the following coded message, and make a bar graph.

UAC QCCUWPT IWNN
XC GU UAC ZWCS

(d) The most frequent letter in the coded message in part (c) represents E. Which letter is this?

(e) Try to decode the whole message. (*Hint:* W stands for I in the coded message.)

40. The frequencies of letters in Scrabble are as follows.

A—9	B—2	C—2	D—4	E—12	F—2
G—3	H—2	I—9	J—1	K—1	L—4
M—2	N—6	O—8	P—2	Q—1	R—6
S—4	T—6	U—4	V—2	W—2	X—1
Y—2	Z—1				

(a) Construct a bar graph for the frequencies of the 11 most frequently occurring Scrabble letters.

(b) Compare this graph to the one in the preceding exercise. Which letters are significantly more frequent in Scrabble than in written English? Which letters are significantly less frequent in Scrabble?

12.2 STATISTICAL DECEPTIONS

Miracle Yellow Food Diet!
Eat only yellow foods . . . as much as you like.
LEMONS POTATO CHIPS SQUASH
GRAPEFRUITS BANANAS EGG YOLKS
Lose as much as 8 pounds in a week!

If you don't learn to recognize common statistical deceptions, some people will deceive you. By learning about the deceptions presented in this chapter, you will become a more shrewd individual who can cut through the b.s. (butchered statistics).

In Section 12.1, you learned about some common deceptions involving graphs and tables. Some other kinds of deceptions are described here.

DECEPTIONS INVOLVING PERCENTS

Some statistical deceptions involve percents of numbers of different sizes. In Lesson Exercises 12.15–12.17, try to find the mistake in the italicized conclusion and correct it. If you can't find the deception, go on to the next problem.

(*Hint:* All of the deceptions in Lesson Exercises 12.15–12.17 are based on the assumption that taking the same percent of different numbers will give you the same answer.)

Lesson Exercise 12.15

Suppose the United States and an African nation both have GNP (gross national product) growth of 5% in 1997. *The two nations have increased their wealth (GNPs) by the same amount.*

Lesson Exercise 12.16

Suppose food prices went up 10% last year and 10% again this year. Over the 2 years, food prices went up *20%* (*italicized*). (*Hint:* Assume that food prices started at 100, and see where they end up.)

Lesson Exercise 12.17

Last year, profits in a business went down 20%. This year, profits rose 20%. Now profits are *the same amount* they were 2 years ago.

Lesson Exercise 12.18

Your landlord's heating bill went up 25% this year. Does this justify his raising your rent 25%? Why or why not?

Lesson Exercise 12.19

"SUDS 'N CRUD IS AMERICA'S FASTEST GROWING BEER!"

(a) Fill in the last column of the table.

Product	1995 Sales	1996 Sales	Percent Increase in Sales
Blubber Beer	$2,000,000	$2,200,000	
Lo-Cal Beer	1,500,000	2,100,000	
Slob's beer	800,000	1,000,000	
Suds 'n Crud	300	600	

(b) What is deceptive about the headline at the beginning of this problem?

Lesson Exercises 12.15–12.19 concern deceptions involving percents. Remember that taking the same percent of a larger number will give a larger amount, so "all 50%'s are not equal." Beware of a percent standing alone. Every percent is a percent of *something*.

Percents of Numbers of Different Sizes

A, B, and C are positive numbers. If $A > B$, then $C\%$ of A is greater than $C\%$ of B.

DECEPTIONS INVOLVING MATHEMATICAL LANGUAGE

The following ad copy appeared in an issue of *Seventeen* magazine.

BEAUTIFUL THICK HAIR

Your hair will be up to 136% thicker in 10 days. We guarantee it, or your money back.

Certain phrases, such as "up to" and "as much as," enable advertisers to include highly unlikely results in their slogans.

Lesson Exercise 12.20

"LOSE AS MUCH AS 8 POUNDS IN A WEEK!"

(a) If the claim is true, what is the maximum amount of weight you may lose in a week?

(b) If the claim is true, what is the minimum amount of weight you may lose in a week?

"Up To" and "As Much As"

The phrase "up to a number A" can mean any number $\leq A$. The phrase "as much as a number A" also can mean any number $\leq A$.

The statement "Nothing is better" makes it sound as if a product is the best.

Lesson Exercise 12.21

"NOTHING CLEANS SHOES BETTER THAN WASHOUT."

(a) Suppose that there are five different shoe-cleaning products on the market, all equally effective. Is the advertising headline true?

(b) Assume that there are five equally effective products. Would it also be true to say that "NO OTHER SHOE CLEANER IS WORSE AT CLEANING SHOES THAN WASHOUT"?

"Nothing Is More Than"

The statement "Nothing is more than a number A" can mean all numbers $\leq A$.

Lesson Exercises 12.20 and 12.21 concern deceptions involving mathematical language. As you have seen, the phrases "as much as a number" and "up to a number" mean that any amount is possible from 0 "up to" the number mentioned. "Nothing is better" means that other products may be the same as this one.

DECEPTIONS INVOLVING COMPARISONS AND TIME INTERVALS

In making a comparison based upon statistics, consider unmentioned factors that are also important.

Lesson Exercise 12.22

"THE NATIONAL MOTORS PARAKEET GETS THE BEST MILEAGE OF ANY COMPACT CAR!"

Oh, Come on, I move faster than that!

What else should a prospective buyer find out about the Parakeet when comparing it to other compact cars?

People presenting historical statistics often select time intervals that are not representative. In examining data from a specific time period, consider whether there is anything special about that period that would affect the data.

Lesson Exercise 12.23

"SALARIES INCREASE MORE SLOWLY IN THESE HARD TIMES." A politician notes that U.S. salaries have increased an average of 25% from 1980 to 1984, compared to an increase of 34% from 1976 to 1980. What additional information is needed?

Lesson Exercises 12.22 and 12.23 illustrate deceptions involving comparisons and time intervals. Advertisers often choose a particular factor that makes their products compare favorably by ignoring other important factors. In discussing trends over time, people sometimes select a time period that supports their viewpoint but is not truly representative.

ANSWERS TO LESSON EXERCISES

12.15 *Therefore, the United States has increased its wealth by a greater amount.*

12.16 *21%* (100 → 110 → 121)

12.17 *4% lower than* (100 → 80 → 96)

12.18 No. The heating bill is only a small part of the total expenses.

12.19 (a) Last column has 10%, 40%, 25%, 100%
(b) Suds 'n Crud has the highest percent increase but a much lower dollar increase in sales than the other three companies.

12.20 (a) 8 lb (b) 0 lb

12.21 (a) Yes (b) Yes

12.22 Price, repair record, safety record, trunk space

12.23 The rate of inflation during each time period; whether anything unusual happened between 1976 and 1984

12.2 HOMEWORK EXERCISES

Basic Exercises

1. Oil prices went up 20% one year and 30% the next. Over the two years, oil prices rose ________%.

2. My rent went down 10% last year and then rose 20% this year. Over the two years, my rent went up ________%.

3. New computers enable a company to increase productivity by 20% (assuming that profits increase by the same amount). Labor demands a 20% increase in wages. Management claims that if wages increase at the same rate as productivity, no additional profits will be made. Is management right? (*Hint:* Make up a situation with numbers.)

4. The numbers c, x, and y are positive. If $x < y$, then $c\%$ of x ________ $c\%$ of y.
(<, >, =)

5. Last year, a company made a 10% profit. If its profit margin decreased by 60%, what percent profit did the company make this year?

6. Sidney's expenses for food, rent, and miscellaneous all increased 20% this year. Sidney figures that his total expenses went up 60%. Explain what is wrong with his reasoning.

7. What important information is missing from the following statement? "After 100,000 miles of test driving, less than 1% of National Motors Cars needed any repairs."

8. Last year, the Social Security budget rose 12%. This year, the rate of increase is down 25%. This year's Social Security budget is ________% ________ than last year's budget.
(higher, lower)

9. What important information is missing from the following statement? "Eighty percent of all dentists surveyed recommend sugarless gum for their patients who chew gum."

10. "FALL SALE: SAVINGS UP TO 50%!"
(a) If the claim is true, what is the most you will save on an item?
(b) If the claim is true, what is the least you will save on an item?

11. In *USA Today*, March 17, 1989, Dodge took out a full-page ad with the following announcement in large type.

> **HERE'S TO YOU**
> **AMERICA.**
> UP TO
> **$3000**
> **SAVINGS**

Then, in very small type at the bottom of the page was the following:

*Total savings = pkg. savings + cash back. $50–$1,200 pkg. savings (depending on pkg.), based on list prices of pkg. items if sold separately, avail. on all model lines, but not on each individual model. $300–2,000 cash back (depending on model) or 4.9% annual percentage rate financing on all new vehicles in stock except Caravans & diesel pickups.

What does the phrase "up to $3000" really mean in this ad?

12. Find any misleading statements in the following advertisement that once appeared in *Seventeen* magazine.

> LONDON UNIVERSITY CRASH-BURN
> WONDER DIET
> Of all medically safe and sound reducing programs
> **FASTEST WEIGHT-LOSS METHOD**
> **KNOWN TO MEDICAL SCIENCE!**
> (EXCEPT FOR TOTAL STARVATION)
> Compared to the Scarsdale diet, weight-watchers, Duke Univ., even Atkins or Pritikin, it burns away as much as:
> 4 TIMES FASTER THAN HIGH SPEED DIETS
> 11 TIMES FASTER THAN EXERCISE!"

13. The statement "A, B, and C are not more than X" means

(a) $A = B = C = X$
(b) A, B, and C are less than X
(c) A, B, and C are greater than or equal to X
(d) A, B, and C are less than or equal to X

14. "NO DETERGENT CLEANS CLOTHES BETTER THAN SLUDGE." According to the preceding headline, which of the following statements is true?

(a) Sludge cleans clothes better than other detergents.
(b) Sludge removes chocolate stains.
(c) Sludge cleans clothes at least as well as other detergents.

15. "MORE PEOPLE DRINK JITTERS INSTANT COFFEE THAN ANY OTHER BRAND." If there are ten brands of instant coffee, the percent of all instant coffee drinkers drinking Jitters is between ________% and ________%.

16. "TANDY COMPUTERS ARE CLEARLY SUPERIOR." According to this slogan, which of the following is (are) true?

(a) Tandy computers are the best computers.
(b) Tandy computers are superior to Apple computers.
(c) Tandy computers work well.

17. Explain the flaw in each claim or suggestion made about a product.

(a) A store sells tapes with a list price of $10.98 for $7.98. You save 27%.
(b) Alpo's beef-flavored dinner has "as much meat protein in a serving as 7 ounces of the finest steak."

18. An ad by the Beef Industry Council states that a 4-oz serving of beef contains the following percents of the recommended daily allowances: 76% of the protein, 35% of the niacin, 51% of the zinc, 19% of the iron, 100% of the vitamin B-12, and 13% of the calories. What important nutritional information about beef is omitted?

19. The following report describes an actual study involving Aim toothpaste.

> **Lever Brothers Company**
>
> Lever Brothers clinically tested Aim with fluoride. . . . Thousands of children were enrolled in two studies. . . . The dentists reported their findings as the reduction of dental decay of Aim with fluoride compared to Aim without fluoride in all of the teeth and their surfaces. After two years the dentists concluded that a significant reduction of dental decay resulted from the use of Aim with fluoride.

Summary of Clinical Trials of Aim with Fluoride

	Study I	Study II
Duration	2 years	2 years
Number of children	1107	1154

Reduction of Cavities of Aim with Fluoride vs. Aim Without Fluoride

Whole mouth	25%	29%
Between teeth	38%	37%
Newly erupted teeth	21%	25%

(a) What two products are being compared in the study?
(b) Does the study show that Aim is better than any other fluoride toothpaste? If not, what does it show?

20. In a speech on November 23, 1982, former president Reagan used the following graph to illustrate the declining share of the federal budget spent on defense. Note that the defense budget was especially high from 1962 to 1967.

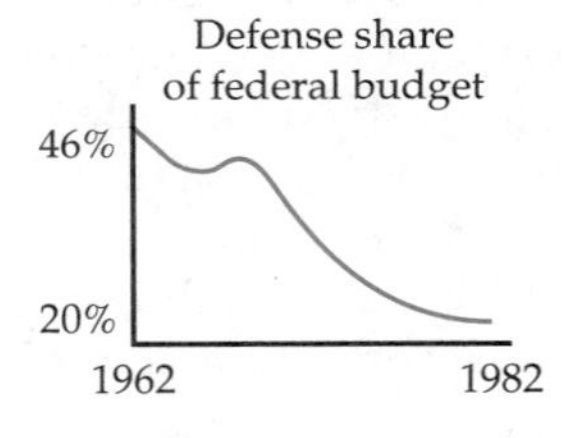

(a) Is there anything special about this time period that accounts for the higher defense spending?
(b) How is the graph distorted?
(c) Does the distortion favor or undermine Reagan's viewpoint?

21. In 1970, roughly 331,000 people died from cancer in the United States. In 1995, roughly 560,000 people died from cancer in the United States. Explain why it is possible that deaths from cancer increased even though medical treatments for cancer were improving.

22. More students are taking mathematics courses than ever before. One could conclude that mathematics has become a well-liked subject. Why might the conclusion be incorrect?

23. "More people die in hospitals than at home." From this statement you might conclude that you are better off staying at home when you are sick. Explain why this conclusion is unjustified.

Extension Exercises

24. In the United States, women's average income is 60% of men's average income. Speculate about what this statistic shows.

25. On the same day, two New York newspapers published conflicting headlines. One said "ELECTRIC POWER USAGE INCREASES." The other said "ELECTRIC POWER USAGE DECLINES." Both were right! How is this possible?

26. Consider the following problem. "A price P rises 20% and then rises 30%. Prove that the overall change is an increase of 56%. (*Hint:* Show that the final price is $1.56P$.)" Devise a plan and solve the problem.

27. A price P rises 10% and then declines 20%. Prove that the overall change is a decline of 12%.

12.3 MEASURES OF THE CENTER

It is possible to use a single number to describe a whole set of numbers. Sounds amazing, doesn't it? Whereas graphs and tables are useful for presenting statistical data, a set of numbers can be described more simply with an "average." However, information is also lost in this simple description of a data set.

THE MEAN, MEDIAN, AND MODE

What is the "average" amount of time it takes people in your class to go from home to class?

Class Lesson Exercise 12.24

(a) Find out how long (in minutes) it takes each person in your class to go from home or from a dorm room to class. (If you have a large class, select 10 students.)

(b) Describe ways to find the "average" travel time for your data.

See which of the following measures of the center you found.

The most commonly used measure of the center is called the mean or average.

> ***Definition*** **Mean or Average**
>
> The **mean** or **average** of a set of numbers is their sum divided by how many numbers there are. The mean $\bar{x}$ of the numbers $x_1, x_2, x_3, \ldots x_N$ is
>
> $$\bar{x} = \frac{x_1 + x_2 + x_3 + \cdots + x_N}{N}$$

The mean (short for *arithmetic mean*) gives equal weight to all measurements. The word "average" usually refers to the mean. In my class, the commuting times, in minutes, were 30, 15, 12, 36, 12, 15, 15, 20, 45, 30, 15, and 20. The mean is the sum of the travel times, 265, divided by 12—or $\frac{265}{12} = 22\frac{1}{12}$ minutes.

Another useful measure of the center is the median.

> ***Definition*** **Median**
>
> The **median** is the middle value of a set of numbers when the numbers are arranged in order. If there is an even number of values, the median is the mean of the two middle values.

To find the median in my class, arrange the numbers in order.

$$\{12, 12, 15, 15, 15, 15, 20, 20, 30, 30, 36, 45\}$$
$$\uparrow \uparrow$$

The two middle numbers are 15 and 20, so the median is $\dfrac{15 + 20}{2} = 17\dfrac{1}{2}$ minutes.

Besides the mean, people may use "average" to refer to the median or to most typical score, the mode. The **mode** is the score that occurs most frequently in a set. For my class, the mode is 15 minutes. The mode is used less often than the mean or the median and is most useful for nonnumerical data such as peoples' favorite colors. Use the mode only for larger data sets, since slight changes in smaller data sets sometimes cause significant changes in the mode.

Lesson Exercise 12.25

(a) Find the mean, median, and mode of your class data if you haven't already.

(b) Which of the mean, median, and mode do you feel is the best measure of your class's average travel time?

Lesson Exercise 12.26

The Nielsen ratings estimate the percent of all U.S. television sets that are tuned in to each show. The hourly prime-time Nielsen ratings for two of the major networks for the week of 1/8/96 were as follows.

NBC: 10.3, 12.4, 12.4, 13.7, 14.4, 11.0, 8.1, 12.2, 11.9, 16.7, 19.8, 17.6, 10.3, 11.4, 9.7, 7.4, 5.1, 6.2, 10.6, 11.3, 11.3

ABC: 10.8, 10.8, 9.1, 14.6, 16.1, 15.3, 11.4, 13.3, 13.2, 10.1, 8.8, 8.8, 12.2, 11.1, 14.2, 6.2, 6.2, 6.2, 10.9, 9.6, 9.6

How would you determine which network did the best in that week?

FINDING THE MEAN, MEDIAN, AND MODE OF A FREQUENCY DISTRIBUTION

Statisticians usually tabulate larger data sets as frequency distributions. The following example shows how to compute the mean and median of a frequency distribution: my students' travel times.

EXAMPLE 12.2

The following table shows the frequency distribution of travel times for my class. Find the mean, median, and mode. The frequency represents the number of students with a given travel time.

Time	Frequency
12	2
15	4
20	2
30	2
36	1
45	1

12 12
15 15 15 15
20 20
30 30
36
45

Solution

The *mode* is the easiest to find. The most frequent time, 15, has the highest number in the frequency column.

The *mean* is the total number of minutes divided by the total number of students. There are

$$2(12 \text{ minutes}) + 4(15 \text{ minutes}) + 2(20 \text{ minutes}) + 2(30 \text{ minutes}) + 1(36 \text{ minutes}) + 1(45 \text{ minutes}) = 265 \text{ minutes}$$

There are $2 + 4 + 2 + 2 + 1 + 1 = 12$ students. So the mean is $\frac{265}{12} = 22\frac{1}{12}$ minutes. Note how the mean of a frequency distribution can be computed more quickly by multiplying each data number by its frequency, adding them together, and dividing by the total number of scores.

The *median* is the middle number. Since there are 12 numbers, the middle numbers are the sixth and seventh numbers, which are 15 and 20. Writing out the students' travel times in increasing order confirms this.

$$\{12, 12, 15, 15, 15, \underline{15}, \underline{20}, 20, 30, 30, 36, 45\}$$

So the median is $\frac{15 + 20}{2} = 17\frac{1}{2}$ minutes. ■

The formula for finding the mean of a frequency distribution is as follows.

The Mean of a Frequency Distribution

The mean $\bar{x}$ of the frequency distribution

Score	Frequency
x_1	f_1
x_2	f_2
$\vdots$	$\vdots$
x_N	f_N

is $\bar{x} = \dfrac{x_1 f_1 + x_2 f_2 + \cdots + x_N f_N}{f_1 + f_2 + \cdots + f_N}$.

Lesson Exercise 12.27

A class made the following scores on a test.

Score	Frequency
100	1
90	3
80	4
70	3
60	0
50	2

(a) How many students received a 70?

(b) How many students are in the class?

(c) Find the mode.

(d) Find the mean.

(e) Find the median. (*Hint:* You could write out all 13 scores from lowest to highest.)

MEAN OR MEDIAN?

How does a statistician decide whether to represent the center with the mean or the median? In Example 12.2, the median travel time $\left(17\frac{1}{2}\right)$ was slightly lower than the mean $\left(22\frac{1}{12}\right)$. Which is a better indicator of the "average" travel time? How do you know which one to use with a particular set of data?

Sometimes a set of data has a small subset of scores, called **outliers**, that are quite different from the rest of the scores. Would this group of scores be better represented by the mean or by the median? It depends upon whether you want to include or minimize the effect of the outliers. In the travel-time example, the 36- and 45-minute travel times are particularly high. To give these scores equal weight, one would use the mean rather than the median.

Lesson Exercise 12.28

Pat played in six basketball games, scoring the following numbers of points: 2, 22, 26, 28, 30, 30. She scored 2 points during a game in which she had the flu and played for only 3 minutes. Which would be a better indicator of her typical scoring game, the mean or the median score?

As Lesson Exercise 12.28 illustrates, the median is a better choice when you want to minimize the effect of an extreme value. Since the median involves only the middle value or values, it is not affected by a change in size of an extreme value. Economists usually compute "average" family income as a median so that a small number of very wealthy families do not have a large impact on the average.

The mean, on the other hand, gives equal weight to all scores. The mean is a better choice if you want to include the effect of extreme values. When scores are scattered or there are very few scores, the mean may be a better indicator than the median. For example, the mean (72) of a set of test scores such as {30, 40, 90, 100, 100} is more representative than the median (90). When there are no extreme scores, the mean and median are usually about the same.

Lesson Exercise 12.29

For each of the following sets, (1) without computing, tell whether the mean or median would be higher; and (2) tell which would be a better indicator of the "average."

(a) {23, 25, 28, 68, 71, 75, 78}

(b) {10, 12, 14, 16, 16, 23, 95} (Assume that you want to minimize the impact of the 95.)

People sometimes choose the "average" that promotes their viewpoint. They may even neglect to say whether they are using the mean or the median. Lesson Exercise 12.30 illustrates such a situation.

Lesson Exercise 12.30

The Washington Black-and-Blueskins have salaries (in thousands of dollars) as follows:

100 100 150 200 200 400 400 600 900

(a) The owner wants to demonstrate how high the players' "average" salary is. Would he prefer the mean or the median?

(b) The players' association wants to demonstrate how low the players' "average" salary is. Would they prefer the mean or the median?

ANSWERS TO SELECTED LESSON EXERCISES

12.26 Compute the means or medians for the two networks, and compare them.

12.27 (a) 3 (b) 13 (c) 80 (d) $76\frac{12}{13}$

(e) 80 (the 7th highest score)

12.28 The median score

12.29 (a) (1) Median; (2) mean

(b) (1) Mean; (2) median

12.30 (a) The mean (b) The median

12.3 HOMEWORK EXERCISES

1. Tom bowls 136, 158, and 129. Jean bowls 160, 120, 118, and 137. Find a way to determine who bowled better.

2. A set contains five numbers. Each number is either 1, 2, or 3. What must the five numbers be if the mean is 1?

3. (a) A teacher wants to find the mean of test scores V, W, X, Y, and Z. What is the mean?
   (b) The mean of a set of five test scores is M. What is the *sum* of the five test scores?

4. If $\bar{x}$ is the mean of $x_1, x_2, \ldots, x_n$, and $\bar{y}$ is the mean of $y_1, y_2, \ldots, y_m$, what is the formula for the mean $\bar{z}$ of $x_1, x_2, \ldots, x_n, y_1, y_2, \ldots, y_m$ in terms of $\bar{x}$ and $\bar{y}$?

5. How much lower is the mean cost of the three generic drugs shown than the mean cost of the three brand-name drugs?

Brand Name	Cost of 100 Units	Generic	Cost of 100 Units
Librium	$13.98	Chlordiazepoxide	$3.98
Sumycin	$ 7.85	Tetracycline	$3.75
Polycillin	$25.50	Ampicillin	$9.98

6.

Courtesy of Columbia Classical Club, CBS Records.

If postage and handling average 50 cents per tape and the regular club price for a tape is $8.98, what is the mean cost per tape for the minimum number of tapes you must buy?

7. Find the mean and median of $\{4, 7, 10, 7, 5, 2, 7\}$.

8. Match the expressions in columns A and B.

A	B
Uses order of scores	Mean
Most frequent	Median
Equal weight to each score	Mode
Middle	

9. You have 5 stacks of books with 3, 5, 7, 2, and 3 books, respectively.
   (a) If you rearrange the books so that the stacks are all the same size, how many books will be in each stack?
   (b) What type of "average" does part (a) illustrate?
   (c) How would you rearrange the stacks to show the median number of books in a stack?

10. Meat consumption per capita (in kg), 1993

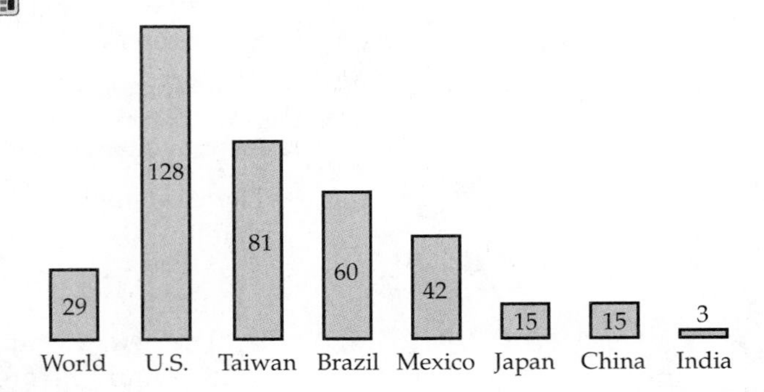

The world population was about 5.5 billion in 1993.
   (a) How much meat did all the countries in the world consume in 1993?
   (b) What percent of all the meat was consumed in the United States if the U.S. population was about 260 million?

11. Where have you heard the word "median" used in everyday life? How does it relate to a median in mathematics?

12. Lisa has a scoring average (mean) of 20 points per game for 5 games. She scored more than 22 points in each of the first 4 games. What

can you tell about how many points she scored in the fifth game?

13. Which four numbers in the following group have a mean of 8 and a median of 8.5?

3 5 5 7 8 9 10 10

14. (a) Use a shortcut to find the mean of {9, 10, 11}.
(b) Use a shortcut to find the mean of {20, 21, 22, 23, 24, 25}.
(c) Use a shortcut to find the mean of {36, 37, 38, . . . , 100}.

15. Find the mean for each set *without* adding up all the numbers and dividing.
(a) {7, 7, 7, 7, 7} (b) {56, 57, 58}
(c) {5, 7, 9, 11, 13} (d) {42, 42, 46, 46}

***16.** (a) Find the mode of the following set.

{6, 8, 8, 9, 10, 10, 12, 12, 15, 15, 15, 15, 18, 19, 20, 22}

(b) How would you significantly lower the mode by changing just two scores?

17. A newspaper editorial says, "About half of our fifth-graders scored below average on a standardized mathematics test. A major remediation effort must begin." Explain why these test results are not very alarming. (*Hint:* Look at the cartoon.)

© David Pascal, 1978.

***18.** A class obtained the following test scores.

Score	Frequency
90	2
80	4
70	9
60	5
50	3
40	1

(a) How many students are in the class?
(b) Find the mean, median, and mode for the class.
(c) Using the mean, how many students scored above average?
(d) Using the median, how many students scored above the center?

19. In 1988, Senate incumbents running for reelection received the following campaign contributions over a two-year period.

Total Two-Year Donations for U.S. Senate Campaign (to nearest million dollars)	Frequency
1	5
2	8
3	4
4	5
5	2
6	2
7	1
8	1
9	1
10	0
11	1

Find the mean, median, and mode campaign contributions to an individual senator.

20. The following chart gives the heights of students in a particular class.

Height (cm)	Frequency
177	1
175	2
173	2
171	4
170	3
169	2
168	3
166	2
163	1
160	1
158	1
156	1

Find the mean, median, and mode heights for the class.

21. Following are Karen Svenson's grades for the fall semester. Compute her grade-point average (A = 4, B = 3, C = 2, D = 1, F = 0).

Course	Credits	Grade
Calculus	4	B
Biology	4	C
Literature	3	B
French II	3	A
Badminton	1	D

22. A class of 23 students had a mean of 78 on a mathematics test. The 10 boys in the class had a mean of 76.2. What was the mean of the girls' test scores?

23. Ross wants to prepare a 20-quart orange-juice mixture that will sell for $3/qt. He will use Valencia OJ that sells for $3.75/qt and Parson OJ that sells for $2.50/qt. How many quarts of each should he use? (Make a table, and guess and check.)

24. Blair's bowling average after 3 games is 150.
(a) Is a bowling average a mean or a median?
(b) Blair then bowls 170. How much does his average go up?
(c) If he bowls 170 again, how much does his average go up?
(d) Why does the second 170 raise his average less than the first 170?
(e) Blair has an average of A after N games. If he bowls 170, what is his new average?

25. Without computing, tell which would be higher, the mean age or the median age in your mathematics class.

26. The salaries of the executives of a certain company follow. Find the mean and the median. Which statistic seems to describe the "average" salary best?

Position	Salary
President	$90,000
First vice-president	35,000
Second vice-president	35,000
Supervisor	30,000
Accountant	24,000
Personnel manager	24,000

27. A class has 10 students with brown hair, 6 with black hair, and 3 with blonde hair. Would you use the mean, median, or mode to find the "average" hair color?

28. A survey shows that green is a class's favorite color. Is this a mean, median, or mode?

29. In 1983 Chicago civil cases, the median award was about $8000 and the mean award was about $69,000. Were most awards closer to $8000 or to $69,000?

30. An article says that the "average" number of persons in each car on the road is 1.3 persons. Is this figure a mean, median, or mode?

31. Each set shows the ages of a group of college students in a seminar. Which would be a better indicator of the "average" age, the mean or the median?
(a) {19, 20, 20, 20, 42, 47, 48}
(b) {18, 19, 19, 20, 20, 21, 62} (Assume that you want to minimize the effect of the 62.)

32. A class has 8 students with heights, in inches, as follows:

{60, 62, 64, 66, 68, 68, 70, 72}

(a) Find the mean and the median.
(b) Suppose that a basketball recruit who is 86 inches tall joins the class. Find the new mean and median.
(c) Which "average" changed the most because of the new student?

33. Which would be higher, the mean height or the median height of the three people in the following photograph?

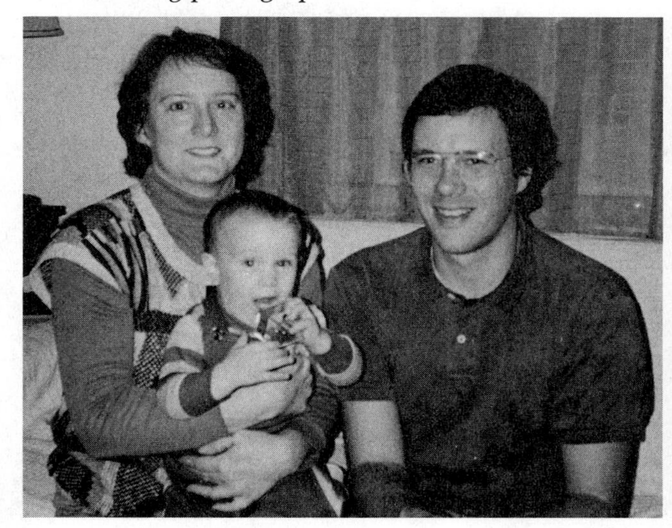

34. A business has five employees. The owners say that the "average salary" is \$30,000. The workers say that the "average salary" is \$20,000. Make up a set of data that makes both groups right.

35. The teachers' salaries (in thousands of dollars) at a school are

20 22 22 24 26 28 30 35 40

(a) The teachers' union wants to show how low the "average salary" is. Would they prefer the mean or the median?

(b) The school board wants to show that the "average salary" is high enough. Would they prefer the mean or the median?

36. Make up a set of data for which *neither* the mean nor the median is representative.

37. A man bought desserts for 10 people, including himself. Five of the desserts cost \$1 apiece, three desserts cost \$2 apiece, and two desserts cost \$3 apiece. He told his friends, "The desserts cost \$1, \$2, and \$3. So the average cost was $\$(1 + 2 + 3)/3 = \$2$. If each of you gives me \$2, we'll be even."

(a) What is the actual mean price of the desserts?

(b) How much extra money will the man have after charging everyone, including himself, \$2?

38. A business pays \$100 to Cindy at \$20 per hour, \$100 to Max at \$10 per hour, \$100 to Bart at \$5 per hour, and \$100 to Wanda at \$4 per hour. Is the mean wage paid by the business

$$\frac{20 + 10 + 5 + 4}{4} = \$9.75/\text{h}$$

or is it some other amount?

39. In the 1988 campaign, candidate George Bush said, "We've created 17 million new jobs the past 5 years . . . [that] paid an average of more than \$22,000 per year." Bush's claim was based on the fact that at least half the new jobs were in fields in which the average pay was \$22,000 or more. Why was his claim inaccurate?

40. In the math class you teach, the student's average quiz score counts for 10% of the grade, each unit test counts for 20%, and the final exam counts for 30%. Compute the mean for each student. (*Hint:* This is similar to computing the mean of a frequency distribution.)

Student	Quiz Average	Unit Test	Unit Test	Unit Test	Final Exam
Jose Cruz	80	70	60	90	90
Kathy Heid	75	80	92	93	70

41. You have made test scores of 68, 79, 88, 74, and 82. The final test counts as 1/4 of your grade. What is the minimum score you need on the final for an 80 (B) average?

Extension Exercises

42. (a) During a 20-mile commute, Jill averages 50 mph for the first 10 miles on the expressway and 30 mph for the last 10 miles in town. What is her mean speed for the whole trip? (*Hint:* It's not 40 mph! Figure out the total time for the trip.)

(b) During a 20-mile commute, Jill averages M mph for the first 10 miles and N mph for the next 10 miles. What is her mean speed for the 20 miles?

43. In the first half (distance) of a trip, a truck travels at a speed of 60 km/hr. How fast must it go during the second half of the trip in order to average 80 km/hr for the entire trip?

44. A professor gives five exams. Two students have the same mean score, although one student scored higher on four of the five exams. Make up a set of possible test scores for each student.

45. (a) A newspaper reports that "average family income in the United States in 1992 is \$37,480." What important detail has been omitted from this report?

(b) Which would be higher, the mean family income or the median family income in the United States?

46. In the United States, the median age at first marriage is 22 for women and 25 for men. Would the mean ages be higher or lower?

47. Construct a set of numbers for which the mean and median are 5 and the mode is 4.

48.

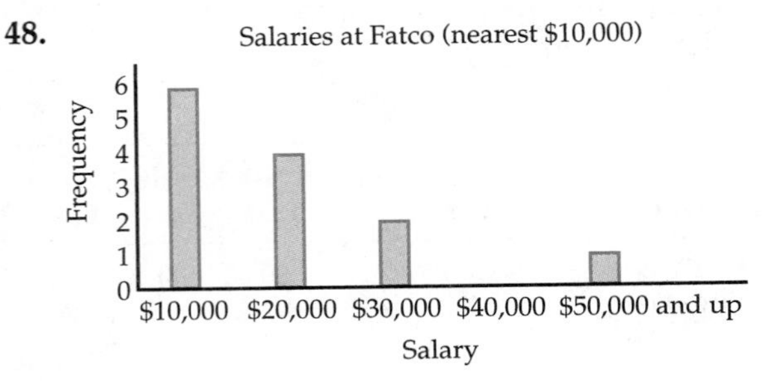

(a) Which would be highest, the mean, median, or mode?
(b) Which would be lowest, the mean, median, or mode?

49. Two math classes take a test. The mean is 70 in class A and 80 in class B. What circumstance would guarantee that the mean for the total group is 75?

50. Another way to compute the mean is to use the *assumed-mean method*. Suppose that you want to find the mean of {72, 86, 89, 90}. You might guess (assume) that the mean is 82. Then see how far each score is from the assumed mean.

Score	Assumed Mean	Difference
72	82	−10
86	82	4
89	82	7
90	82	8
	Total	+9

The sum of the differences for the four scores is 9 above the assumed mean. If each score contributed equally to the +9, each score would be $\frac{9}{4}$ above the assumed mean. Thus, the actual mean is $82 + \frac{9}{4} = 84\frac{1}{4}$.

(a) Find the mean of {65, 71, 75, 84} using the assumed-mean method.
(b) Find the mean of {25, 40, 42, 51, 58} using the assumed-mean method.
(c) Show why $\frac{72 + 86 + 89 + 90}{4}$ is the same as $82 + \frac{-10 + 4 + 7 + 8}{4}$. (*Hint:* The 82 results from $\frac{82 + 82 + 82 + 82}{4}$.)

51. One day in the *New York Times*, NBC and CBS advertised their success in the Nielsen ratings. Both said they had "the Biggest Average Nighttime Audience!" How is this possible?

Computer Exercise

52. A company keeps daily sales records for five employees in a spreadsheet.

	A	B	C	D	E	F	G
1	Name	MON	TUES	WED	THURS	FRI	TOTAL
2	Ranjit	8	2	4	5	6	
3	Miguel	7	5	3	2	3	
4	Mavis	2	4	6	1	2	
5	Wang	3	5	6	8	9	
6	Janice	3	1	5	3	4	
7	MEAN						
8	MEDIAN						

(a) Enter these data into your computer along with appropriate formulas for cells G2 and B7. Use parentheses.
(b) Copy your formulas down column G and across row 7.
(c) Find out how to sort (arrange) data in a column. Arrange the data in column B in decreasing order. Then enter the median in B8. Repeat this process in columns C–G.

Special Exercises

53. (Adapted from Walter Penney's problem in the 12/79 *Scientific American*)

A New Tax Plan

A society has four economic classes, each containing one quarter of the people. Class 1 is the richest, class 2 is the second richest, class 3 is the second poorest, and class 4 is the poorest.

Senator Trickledown proposes that each pair of consecutive classes, beginning with the richest (1 and 2, then 2 and 3, then 3 and 4), have their wealth averaged and redistributed evenly to everyone in those two classes. Senator Dragdown proposes the same thing but wants to begin the averaging with the poorest classes (3 and 4, then 2 and 3, then 1 and 2).

Suppose that the money is distributed as follows.

Class	Money, in Trillions of Quaggles
1	20
2	12
3	6
4	4

(a) How would the money be distributed after the completion of Senator Trickledown's proposed redistribution?
(b) How would the money be distributed after the completion of Senator Dragdown's proposed redistribution?
(c) Which classes will prefer which plan?
(d) Which class would lose under both plans?
(e) Which classes would gain under both plans?

54. Use comparative data for ready-to-eat cereals, and develop a system for rating the nutritional value of these cereals.

55. Read "Milo and the Mathemagician" by Norton Juster, in *The Mathematical Magpie*, edited by Clifton Fadiman, or in *The Phantom Tollbooth* by Norton Juster. Write a report that includes both a summary of the story and your reaction to it.

12.4 MEASURING SPREAD

If two people have the same test average, can there be any significant differences in their test scores? Milly and Billy have each taken six mathematics quizzes. Their scores are as follows.

Milly	2, 4, 6, 7, 7, 10
Billy	4, 5, 6, 7, 7, 7

Lesson Exercise 12.31

Find the mean and median for Milly and Billy.

According to the statistics in Lesson Exercise 12.31, Milly's and Billy's scores are the same.

Lesson Exercise 12.32

How do Milly's and Billy's scores differ?

As this example suggests, a mean or a median does not always sufficiently distinguish between two sets of data. Did you use the range in Lesson Exercise 12.32? The range is a simple way to measure the difference in spread between

Milly's and Billy's scores. Milly's range is $10 - 2 = 8$, and Billy's range is $7 - 4 = 3$.

The range is not a very useful measure of spread, because it uses only the maximum and minimum values. Fortunately, statisticians have developed better ways to measure the spread.

PERCENTILES AND BOX-AND-WHISKER PLOTS

Statisticians usually describe a set of data using a measure of the center (mean or median) and a measure of the spread. In the case of Milly and Billy, their average scores are the same, but the spreads of their scores are different.

Statisticians sometimes describe the spread with percentiles. A score is in the ***p*th percentile** if about $p\%$ of the scores in the distribution are below that score. When we figure the percentile of someone's test score, there is usually more than one person with the same score. In such a case, half of the others with that score are assumed to have better scores and half are assumed to have worse scores than the score being ranked.

In describing the spread with percentiles, statisticians use a five-number summary that includes the median, the maximum and minimum scores, and the first and third quartiles. The median indicates the 50th percentile. The **first (lower) quartile** indicates the 25th percentile, and the **third (upper) quartile** indicates the 75th percentile.

The Five-Number Summary

1. Find the minimum, median, and maximum scores.
2. The first quartile is the median of the scores below the location of the median.
3. The third quartile is the median of scores above the location of the median.

Statisticians use these five numbers to construct a picture called a **box-and-whisker plot**.

EXAMPLE 12.3

Graph Milly's distribution in a box-and-whisker plot.

Solution

Milly's minimum, median, and maximum are 2, 6.5, and 10, respectively. The first quartile is the median of {2, 4, 6}, which is 4. The third quartile is the median of {7, 7, 10}, which is 7. The five-number summary is 2, 4, 6.5, 7, and 10. The box-and-whisker plot uses these five numbers (Figure 12-9). The ends of the box are at the quartiles. The median determines the line segment inside the box. Line segments (called *whiskers*) extend outside the box to the minimum and maximum data values. ■

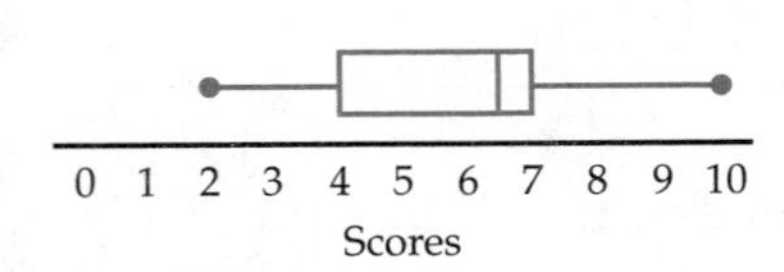

Figure 12–9

The preceding example used a very small data set to simplify computations. Statisticians only draw box-and-whisker plots for much larger data sets. Visually, the box-and-whisker plot shows the spread of the data. The thicker box section shows the center part of the distribution, and the skinny line segments extend to the high and low values.

Box-and-whisker plots provide an excellent way to compare two distributions.

Lesson Exercise 12.33

(a) Draw box-and-whisker plots for Milly's and Billy's scores (one above the other) on the same graph.

(b) Describe how the box-and-whisker plots for Milly and Billy differ.

Although box-and-whisker plots are useful for comparing distributions, a bar graph displays a single distribution with greater clarity and detail.

THE STANDARD DEVIATION

The standard deviation is the most common measure statisticians use to describe the spread. The standard deviation complements the mean, and like the mean, it takes all scores into account.

The standard deviation measures how far scores tend to be from the mean. Consider two sets that both have means of 24: {22, 23, 24, 24, 24, 25, 25, 25} and {2, 13, 17, 24, 31, 33, 48}. The standard deviation is 1 for the first set; and about 14 for the second set.

Visually, a frequency distribution that has a lower standard deviation has more scores closer to the mean, and a distribution that has a higher standard deviation has more scores farther from the mean. In Figure 12.10, compare a set of scores or a graph with a lower standard deviation to one with a higher standard deviation.

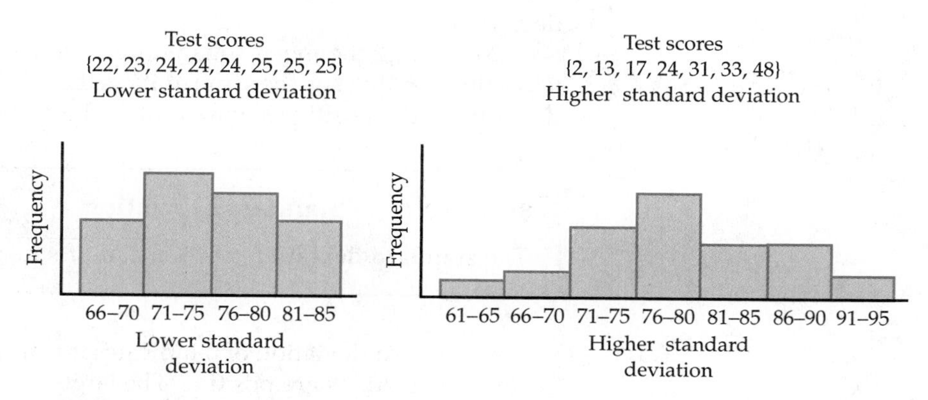

Figure 12–10

Determining the standard deviation usually requires some lengthy calculations with a calculator or computer. You can gain some understanding of the standard deviation by computing it for small sets of numbers.

The standard deviation is based upon the deviation of each score from the mean.

Definition **Deviation**

The **deviation** of a score x from the mean $\bar{x}$ is $x - \bar{x}$.

For example, Milly's mean is 6. Her score of 2 has a deviation of $2 - 6 = -4$.

Lesson Exercise 12.34

Compute the deviation for each of Milly's scores.

Once you compute the deviations, is it possible simply to add the deviations and take the average (mean) of them?

Lesson Exercise 12.35

Compute the sum of the deviations for Milly's scores.

The sum of the deviations is always 0. The balance property of the mean suggests why positive and negative deviations cancel each other out.

How can we change the deviations so they won't cancel each other out? We can make them all positive by using absolute values or by squaring. Statisticians square the deviations. (Squaring the deviations gives greater weight to scores that have larger deviations.) Next, add the squared deviations and divide the sum by the total number of scores to obtain a mean of the squared deviations.

Computing the mean squared deviation squares the units of data. To obtain a number that has the same units as the original data, compute the **standard deviation**—the positive square root of the mean squared deviation.

Definition **Standard Deviation**

The **standard deviation** $s = \sqrt{\text{mean of squared deviations}}$

A standard deviation of 0 indicates no variations from the mean. All other standard deviations are positive. The larger the standard deviation, the greater the spread of the data. Like the mean, the standard deviation includes all scores and is significantly affected by extreme values.

The following example illustrates how to compute the standard deviation.

EXAMPLE 12.4

Find the standard deviation of Milly's scores.

Solution

Step 1: The mean $\overline{x} = 6$.

Step 2: Find the deviation of each score from the mean.

Score	Deviation from the Mean
2	−4
4	−2
6	0
7	1
7	1
10	4

Step 3: Square each of the deviations.

Score	Deviation from the Mean	Squared Deviation
2	−4	16
4	−2	4
6	0	0
7	1	1
7	1	1
10	4	16

Step 4: The sum of the squared deviations is 38. The mean squared deviation is $\frac{38}{6} = 6.3$.

Step 5: The standard deviation $s = \sqrt{6.3} \approx 2.5$. ■

The five steps needed to compute the standard deviation are as follows.

Computing the Standard Deviation for *N* Scores

1. Find the mean.
2. Find the deviation of each score from the mean.
3. Square each of the deviations.
4. Add up the squared deviations and divide by N.
5. Take the square root to obtain the *standard deviation*.

The standard deviation measures how far scores are from the mean. It is something like an average deviation. (*Note:* In some situations, statisticians divide by $N - 1$ instead of N in step 4. We shall not concern ourselves with this distinction in this course.) Most scores will be somewhere around 1 standard deviation from the mean. Few scores will be more than 2 standard deviations from the mean.

Lesson Exercise 12.36

(a) Would you expect Billy's standard deviation to be higher than, the same as, or lower than Milly's?

(b) Follow the five-step procedure and compute Billy's standard deviation.

Lesson Exercise 12.37

The Teenytown Hamsters (basketball players) have heights of 60, 60, 64, 68, and 68 inches. The starting five for the Leantown Lizards have heights of 62, 62, 64, 65, and 67 inches.

(a) What is the mean height of each team?

(b) Without computing, tell which team's heights have a larger standard deviation.

Together, the mean and standard deviation tell a lot about a set of measurements that have an approximately normal distribution.

NORMAL DISTRIBUTION

Large random samples of measurements from homogeneous populations often have the same pattern! Consider the following exercise.

Lesson Exercise 12.38

Suppose that you took a random sample of 500 women at your college and measured their heights.

(a) Describe the pattern you would expect to find in the results.

(b) Sketch a histogram of your guess.

In the 1830s, the Belgian statistician Quetelet collected measurements, including heights, weights, and limb lengths, from a large sample of people. For each measurement, Quetelet found that most of the measurements clustered around the mean and that there were fewer scores farther from the mean in either direction.

The data on 500 women's heights, in centimeters, might be as shown in the table and graphed in Figure 12-11.

Interval (cm)	Frequency	Relative Frequency
140–149	4	0.8%
150–159	90	18%
160–169	285	57%
170–179	112	22.4%
180–189	9	1.8%

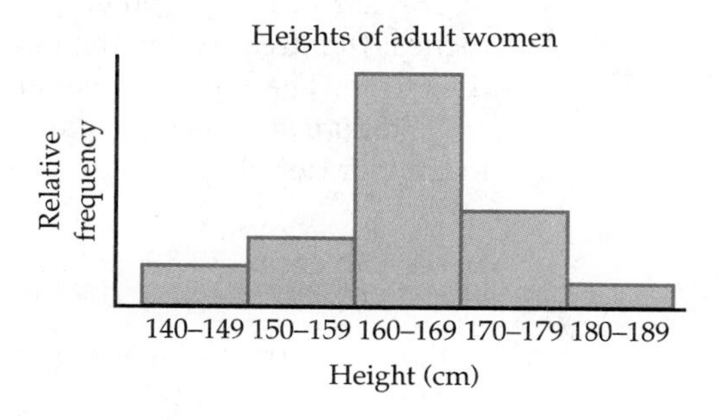

Figure 12–11

If progressively larger samples were taken and the interval sizes decreased, the relative frequency histogram of women's heights would be expected to approach a certain shape, as shown in Figure 12-12.

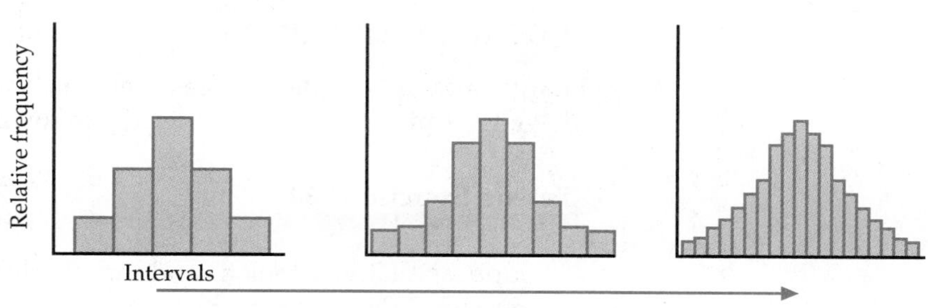

Figure 12–12

The histograms approach a theoretical graph, called a **normal distribution** or bell curve, that is continuous and perfectly symmetric on each side of the mean. Three examples of normal distributions are shown in Figure 12-13.

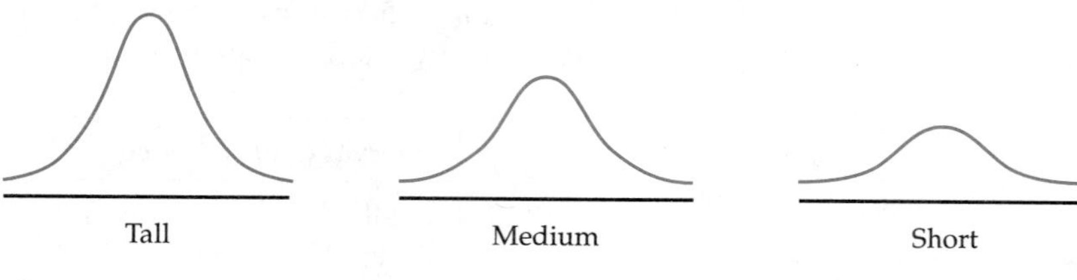

Figure 12–13

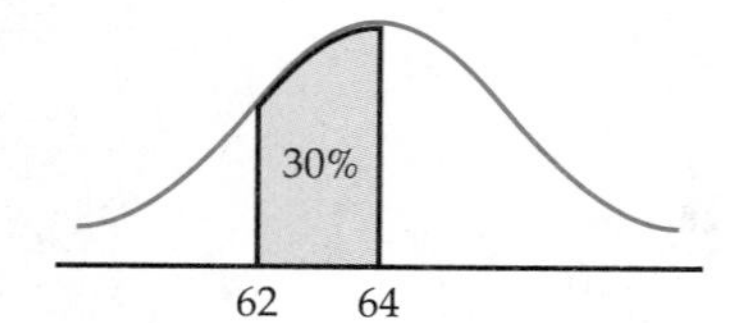

Figure 12–14

As with a histogram, the area under the curve for an interval is proportional to its frequency. Figure 12-14 shows the distribution of heights (in inches) of 16-year-old-boys. Since 30%, or 0.30, of the heights fall between 62 and 64 inches, the area under the curve from 62 to 64 is 30% of the total area under the curve. The region looks like a bar with a curved top.

Statisticians use the standard deviation to describe the normal distribution in greater detail.

Lesson Exercise 12.39

College women have a mean height of about 5 ft 5 in. and a standard deviation of about 2.5 in.

(a) Estimate the percent of college women who are within 2.5 in. (one standard deviation) of 5 ft 5 in. That is, estimate the percent of women with heights between 5 ft 2.5 in. and 5 ft 7.5 in. (You could survey your class.)

(b) Estimate the percent of college women who are within 5 in. (two standard deviations) of 5 ft 5 in.

The theoretical normal distribution provides approximate answers to the preceding two questions: 68% and 95%. The normal distribution has the following properties (Figure 12-15).

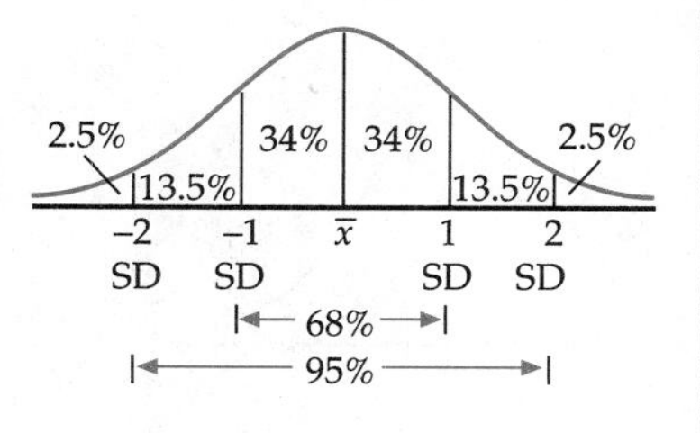

Figure 12–15

Properties of the Normal Distribution

1. It is symmetric around the mean.
2. About 68% of all measurements fall within one standard deviation of the mean (34% on each side).
3. About 95% of all measurements fall within two standard deviations of the mean (47.5% on each side).

Although the normal distribution was developed for continuous data (for example, all lengths between 0 and 10 cm), it can also be applied to large sets of discrete data (such as quiz scores that are all whole numbers between 0 and 10) (Figure 12-16).

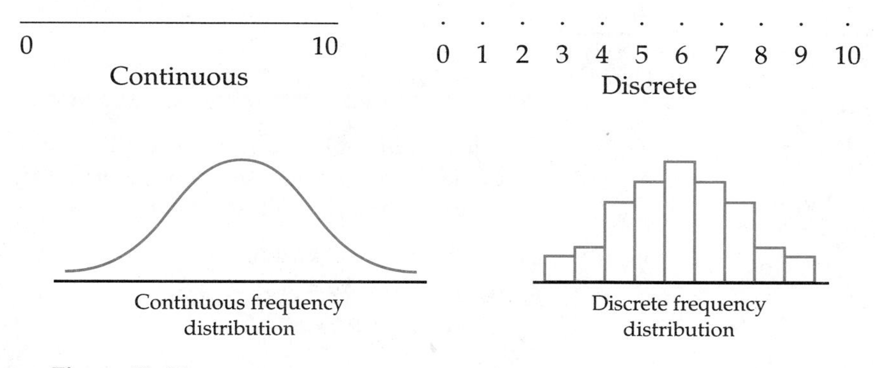

Figure 12–16

Measurements, including heights, weights, IQs, and shoe sizes, have approximately normal distributions for large homogeneous populations. Just knowing the mean and standard deviation of a normal population enables one to estimate the percent of the population data that falls between two measurements.

EXAMPLE 12.5

The mean height of U.S. adult men is 68.0 in., with a standard deviation of 2.5 in. Assume a normal distribution. What percent of adult men have the following heights?

(a) Between 65.5 and 70.5 in.

(b) Between 63 and 70.5 in.

(c) Greater than 73 in.

Solution

Sketch a graph of the distribution. It is a normal distribution, so it is bell-shaped, with the mean, 68, in the center (Figure 12-17). Since $s = 2.5$, count 2.5 units in each direction from the mean (Figure 12-18).

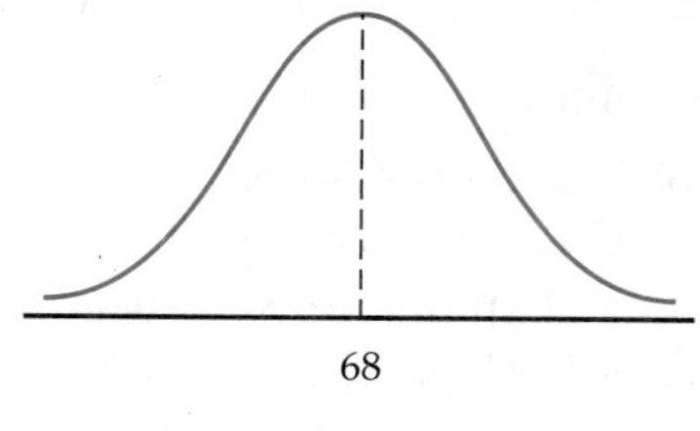

Figure 12–17

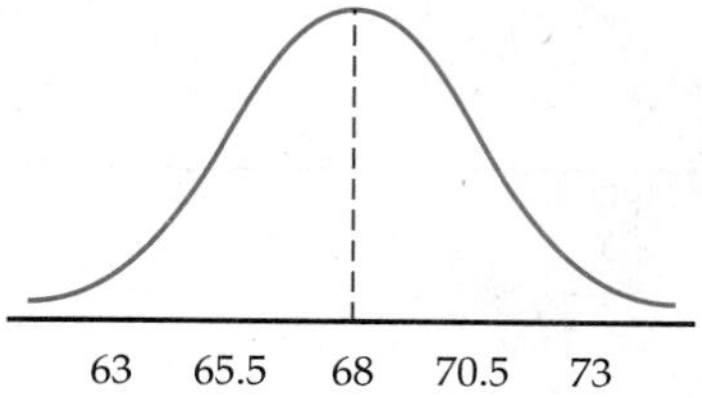

63 65.5 68 70.5 73

Figure 12–18

Fill in the percents for the normal distribution (Figure 12-19).

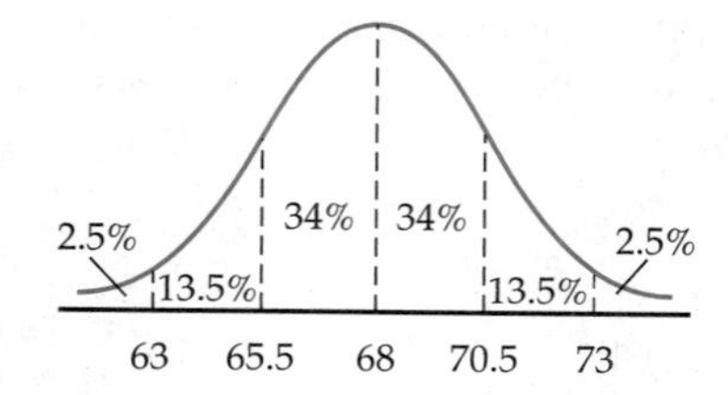

Figure 12–19

(a) 68% of the heights are between 65.5 and 70.5 in.

(b) 81.5% of the heights are between 63 and 70.5 in.

(c) 2.5% of the heights are greater than 73 in. ■

Lesson Exercise 12.40

The heights of U.S. adult women have a mean of 64.5 inches, with a standard deviation of 2.5 in. Assume a normal distribution. What percent of adult women have the following heights?

(a) Between 62 and 67 in.

(b) Between 59.5 and 64.5 in.

(c) Above 64.5 in.

You can also report measurements from normal distributions in percentiles.

Lesson Exercise 12.41

Using the data from Example 12.5, give the percentile for each man.

(a) N. B. Tween, 68 in. tall

(b) Hy Mann, 73 in. tall

(c) Stan Dup, 63 in. tall

A data distribution can be classified as approximately **symmetric** (for example, the normal distribution) or **skewed** in one direction. When a nonsymmetric curve has a longer left-hand "tail," as shown in Figure 12-20(a), it is

skewed to the left. A curve with a longer "tail" on the right, as shown in Figure 12-20(b), is **skewed to the right**.

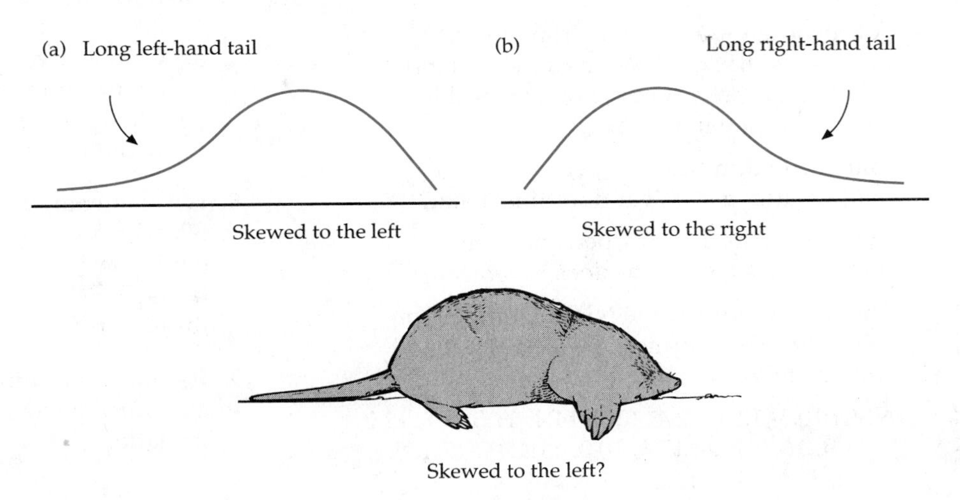

Figure 12–20

Lesson Exercise 12.42

Decide whether each set of discrete data is most likely to be approximately symmetric, skewed to the left, or skewed to the right.

(a) Second-grade students' test scores on a fourth-grade test

(b) Fourth-grade students' test scores on the same test

(c) Sixth-grade students' test scores on the same test

ANSWERS TO SELECTED LESSON EXERCISES

12.31 Median = 6.5, mean = 6

12.32 Milly's scores are more spread out than Billy's.

12.33 (a) The five-number summary for Billy is 4, 5, 6.5, 7, 7.
(b) Milly's scores have a considerably larger spread than Billy's.

12.34 −4, −2, 0, 1, 1, 4

12.35 0

12.36 (a) Lower (b) $\sqrt{\frac{8}{6}} \approx 1.2$

12.37 (a) T. H. mean = 64, L. L. mean = 64
(b) The Teenytown Hamsters

12.40 (a) 68% (b) 47.5% (c) 50%

12.41 (a) 50th (b) 97.5th (c) 2.5th

12.42 (a) Skewed to the right
(b) Symmetric
(c) Skewed to the left

12.4 HOMEWORK EXERCISES

Basic Exercises

1. A brochure for a new lakefront resort, Muddy Lake, says that the lake has a mean depth of 6 ft. Explain why it may not be possible to swim or canoe in the lake.

2. Mary scored in the 80th percentile on a mathematics test. What does this mean?

3. Lyle scored in the 43rd percentile on a mathematics test. What does this mean?

4. The hourly prime-time Nielsen ratings for NBC and ABC during the week of 1/8/96 were as follows.

   NBC: 10.3, 12.4, 12.4, 13.7, 14.4, 11.0, 8.1, 12.2, 11.9, 16.7, 19.8, 17.6, 10.3, 11.4, 9.7, 7.4, 5.1, 6.2, 10.6, 11.3, 11.3

   ABC: 10.8, 10.8, 9.1, 14.6, 16.1, 15.3, 11.4, 13.3, 13.2, 10.1, 8.8, 8.8, 12.2, 11.1, 14.2, 6.2, 6.2, 6.2, 10.9, 9.6, 9.6

   (a) Make box-and-whisker plots for the two networks on the same graph.
   (b) Describe how their ratings compare, based upon the box-and-whisker plots.

5. Two mathematics classes have the following test scores.

   $$A = \{56, 67, 73, 78, 81, 84, 90, 92, 97\}$$
   $$B = \{68, 73, 75, 79, 80, 82, 84, 89, 91\}$$

   (a) Make box-and-whisker plots for the two classes on the same graph.
   (b) Describe how their scores compare, based upon the box-and-whisker plots.

6. (a) Give the five-number summary for the box-and-whisker plot shown.

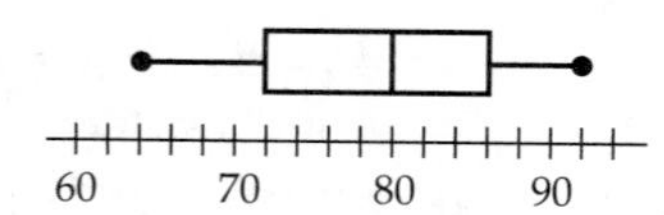

   (b) Draw a second box-and-whisker plot showing a 10% increase in all values in the set.
   (c) Compare the two box-and-whisker plots, and tell how they are related.

7. (a) Make up a set of numbers with a mean of 20.
   (b) Change two of the numbers without changing the mean.
   (c) Change three of the original numbers without changing the mean.

8. Without computing the standard deviation, tell which set of scores has a higher standard deviation.
   (a) $\{4, 5, 6, 7, 8\}$
   (b) $\{0, 3, 6, 9, 12\}$

9. Without computing the standard deviation, tell which histogram has a higher standard deviation.

(a)

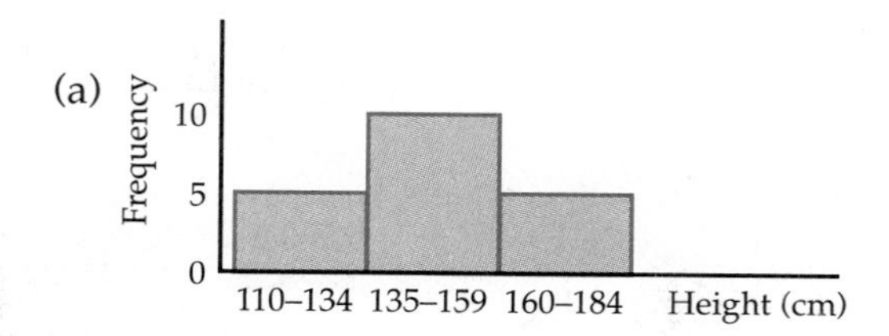

(b)

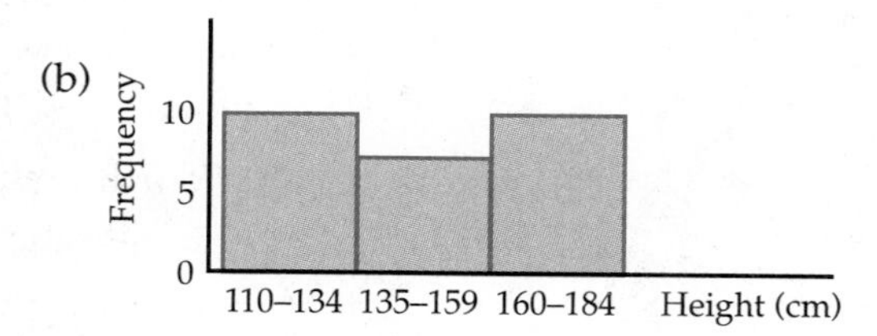

10. My wife has two routes to work. She can drive on back roads that never have much traffic. Or she can drive on an expressway that is faster when there is not much traffic but can take a lot longer when there is a traffic jam. Which would tend to have a higher standard deviation, a set of daily travel times on the back roads or a set of daily travel times on the expressway?

11. (a) Find the mean and standard deviation of $\{80, 90, 90, 90, 100\}$.
    (b) The range, from Section 12.1 (highest score minus lowest score), is the simplest measure of spread. What is the range of this set?

12. (a) Find the mean and standard deviation of $\{3, 7, 8, 10\}$.
(b) If your calculator has a σ_n (standard deviation) key, compute the standard deviation of $\{3, 7, 8, 10\}$ and compare it to your answer.

13. The following set of heights has a mean of 60 and a standard deviation of 7.7.

$$\{50, 50, 53, 56, 60, 64, 67, 70, 70\}$$

(a) Which heights are within one standard deviation of the mean?
(b) Which heights are more than two standard deviations from the mean?

14. Each number in a set of five numbers is either 0, 1, or 2.
(a) If the mean is 1 and the standard deviation is 0, what numbers are in the set?
(b) List the members of a set with the largest possible standard deviation.

15. A population has a mean weight of 140 lb with a standard deviation of 21 lb.
(a) Explain what this means so that someone with no statistics background could understand it.
(b) What weights would be more than two standard deviations from the mean?

16. What are the characteristics of the normal distribution?

17. Take six coins.

Frequency

0H 1H 2H 3H 4H 5H 6H

(a) Toss them 40 times and record the number of heads each time.
(b) Graph a frequency distribution of the results.
(c) Do your results suggest a normal curve?

18. Surveys indicate that U.S. adults watch television for a mean of 25 hours per week, with a standard deviation of 3 hours. Assume a normal distribution. What percent of adults watch television for the following lengths of time?
(a) Between 22 and 28 hours per week
(b) Less than 19 hours per week
(c) More than 28 hours per week

19. A survey shows that the mean income in Boomtown is \$21,000, with $s = \$2600$. Assume a normal distribution. What percent of people have the following incomes?
(a) Over \$21,000
(b) Between \$18,400 and \$26,200
(c) Under \$18,400

20. The length of a human pregnancy has a mean of 267 days and a standard deviation of 16 days. Assume a normal distribution. What percent of pregnancies
(a) last over 283 days?
(b) last over 235 days?

21. Brand A lightbulbs have a mean life of 1000 hours, with a standard deviation of 20 hours. Brand B lightbulbs have a mean life of 1050 hours, with a standard deviation of 150 hours. Describe the advantages of each lightbulb.

22. A standardized test has normally distributed scores with $\bar{x} = 50$ and $s = 5$. Give the percentiles of the following scores.
(a) 55 (b) 40 (c) 54 (Estimate it.)

23. A group of gifted third-grade students takes a regular third-grade test. Would you expect their score distribution to be approximately symmetric, skewed to the right, or skewed to the left?

24. Describe the distribution of each of the following as approximately symmetric, skewed to the right, or skewed to the left. (*Hint:* Sketch a graph of each distribution.)
(a) Typical scores on a classroom unit test graded 0 to 100%
(b) Heights of adult males in the United States
(c) Family incomes in the United States
(d) Amounts of time students in this course will study mathematics in a given week

25. Suppose a health maintenance organization (HMO) keeps track of the numbers of visits made in one year by all of its members.

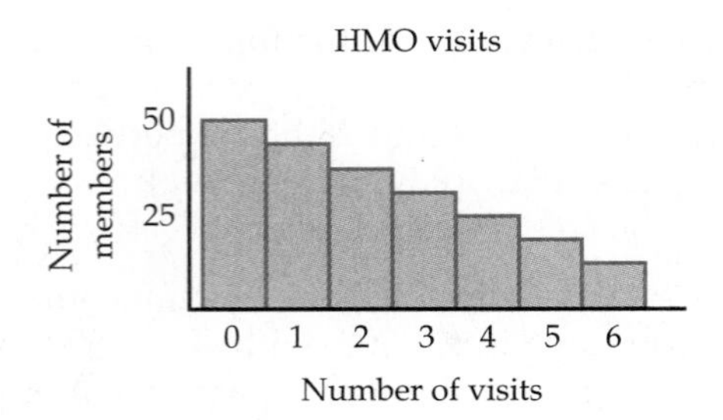

Is this distribution approximately symmetric, skewed to the right, or skewed to the left?

26. Select a page in any book.
(a) Tabulate the length of each word on the page, and display the results in a bar graph.
(b) Is your graph approximately symmetric, skewed to the right, or skewed to the left?

Extension Exercises

27. An ERB standardized test report might include the following information for a sixth-grade class.

MEAN	S.D.	Q1	MED	Q3
426.7	6.2	422	428	431

Tell what all the numbers represent.

28. Suppose that you read an article about a disease you have. The article says that a person with the disease lives an average (mean) of 4 years beyond diagnosis. You have already lived 3 years since being diagnosed.
(a) Explain why this article may be bad news.
(b) Explain why this article may be good news.

29. Match each class in column A to all appropriate descriptions in column B.

A: Class	B: Typical Test Scores
Class with wide ability range	High mean
Honors class	Low standard deviation
Students with very similar abilities	High standard deviation
Remedial class	Low median

30. (a) If you add 2 to each number in a set, how does this affect each of the following?
(1) The mean
(2) The standard deviation
(b) If you add 10 to each number in a set, how does this affect each of the following?
(1) The mean
(2) The standard deviation
(c) Based upon parts (a) and (b), if you increase each score in a data set by a positive number k, how does this affect the mean and the standard deviation?

31. (a) If the salary of each employee in a company is increased by \$2000, how does this change the mean and the standard deviation of the company's salaries?
(b) If the salary of each employee in a company is increased by 10%, how does this change the mean and the standard deviation of the company's salaries?

32. Each of the following sets of numbers has a mean of 10. Determine whether the standard deviation is closer to 1, 5, or 10 without actually computing it.
(a) {9, 9, 9, 9, 11, 11, 11, 11,}
(b) {0, 0, 0, 0, 20, 20, 20, 20}
(c) {3, 3, 6, 9, 11, 14, 17, 17}
(d) {8, 9, 9, 10, 10, 11, 11, 12}

33. Prove that the sum of the deviations for a set of five numbers, A, B, C, D, and E, is 0.

34. Suppose that the mean 100-watt Brite lightbulb burns for 2200 hours, with a standard deviation of 200 hours. Assume a normal distribution. What is the chance that a Brite lightbulb will burn for each of the following periods of time?
(a) More than 2200 hours
(b) Fewer than 1800 hours
(c) More than 2400 hours

35.

Sugar Fix
Cereal
MILK
Wheat-flavored
candies with lots
of added vitamins

Sugar Fix cereal comes in a 16-oz package. The actual weights of the filled boxes are normally distributed, with a mean of 16.3 oz and $s =$ 0.3 oz. What is the chance that a randomly selected box has each of the following weights?
(a) Under 16.3 oz (b) Under 16.0 oz
(c) Above 16.9 oz

36. Show that the standard deviation of $x_1 + k, x_2 + k, \ldots, x_n + k$ is the same as the standard deviation of $x_1, x_2, \ldots, x_n$.

37. There is another way to compute the standard deviation of a set. For four numbers, A, B, C, and D,

$$s = \sqrt{\frac{A^2 + B^2 + C^2 + D^2}{4} - (\bar{x})^2}$$

(a) Write a formula for $\bar{x}$ in terms of A, B, C, and D.
(b) Write s as the square root of a mean sum of squared deviations, using A, B, C, D, and $\bar{x}$.
(c) Square the binomials and combine terms to obtain the formula given at the beginning of the problem. (*Hint:* $A + B + C + D = 4\bar{x}$.)

Computer Exercises

38. The following BASIC program uses a shortcut for computing the standard deviation from the preceding exercise. In the program, the computer adds the squares of the scores and divides by N, the number of scores. Then it subtracts the square of the mean and takes the square root of the result.
(a) Enter the following program into the computer.

```
10 PRINT "HOW MANY NUMBERS DO
   YOU HAVE?"
20 INPUT N
30 LET T = 0: LET S = 0
40 FOR I = 1 TO N
50    PRINT "TYPE A DATA NUMBER
      AND PRESS ENTER/RETURN."
60    LET T = T + D
70    LET S = S + D * D
80 NEXT I
90 PRINT "THE MEAN IS "; T / N
100 LET V = S / N − (T / N) * (T / N)
110 PRINT "THE STANDARD
    DEVIATION IS "; SQR(V)
```

(b) Run the program for {65, 74, 80, 82, 86, 95}.

39. Use MINITAB to produce data analysis for Milly's scores from the lesson.

```
MTB> SET THE FOLLOWING DATA INTO
     C1
DATA> 2 4 6 7 7 10
MTB> DESCRIBE DATA IN C1
MTB> STOP
```

12.5 SAMPLING

Whom do people want as the next president? How do they feel about government spending on education? In order to answer questions such as these, would you need to ask every single adult what he or she thinks?

Fortunately, the answer is no. Taking a representative sample of about 1500 will usually give a good estimate of how 190 million U.S. adults stand on an issue! Sampling is a part of inferential statistics. **Inferential statistics** involves systematically drawing inferences (conclusions) about a larger group (population) based upon information about a subgroup (sample). The preceding lessons in this chapter covered descriptive statistics—techniques used to organize and summarize data. Descriptive statistics and inferential statistics are the two main areas of statistics.

Education, for 1st time, is top voter concern

And 'until the picture becomes clearer, a low-grade anxiety persists'

By Richard Benedetto
USA TODAY

The highly volatile mood of the electorate appears to have shifted from angry to anxious as the nation enters this pivotal 1996 election year.

Voters vented years of pent-up anger by kicking George Bush out of the White House in 1992, and then stung Democrats by giving Congress to the Republicans in 1994.

But the budget, health care, welfare and other top issues are still at an impasse. And voters approach the upcoming campaign not at all sure if the sweeping changes wrought in the last two elections are working the way they envisioned.

Now, a new USA TODAY/CNN/Gallup Poll finds voters rallying around issues that touch closest to home: the quality of their children's education, crime and the economy. All are perennial top concerns, but this is the first time education, with 67% calling it a high priority, has topped the list.

Figure 12–21

SURVEYS

Newspapers frequently report survey results that give us information about our attitudes (Figure 12-21). A survey is one of the best ways to determine the political preferences, moral beliefs, and favorite television programs of a society.

The following experiment will help you understand how surveys work.

Class Lesson Exercise 12.43

The class will need a paper bag with an identical small slip of paper for each class member.

(a) Select a question of interest to the class that has two possible responses, about which people in the class are likely to disagree. (For example, "Do you prefer plastic or paper bags at the supermarket?" "Would you prefer ________ or ________ in the upcoming election?")

(b) Have everyone write a response on a slip of paper and put it in the paper bag.

In a survey, the **population** is the total set of responses the researcher wants to study. In this example, the entire collection of responses from your class is the population.

Class **Lesson Exercise 12.44**

Appoint three poll takers. Each in turn will take five slips of paper from the bag, record the results, and return the slips to the bag.

A **sample** is a subset of the population, chosen to represent it. In this case, each poll taker collects a sample of five responses. Samples are used because they can provide simple, fairly accurate information about a much larger population. Samples are sometimes the only way to obtain information about a population. For example, would you expect a car company to crash-test every car it manufactures?

Class **Lesson Exercise 12.45**

(a) Ask each poll taker to give his or her sample results and to make a prediction about the population results based upon *his or her sample.*

(b) Have the class predict the response totals for the class, based upon all three samples.

(c) Take all the papers out of the bag and tabulate the actual results.

Discuss the following questions about the simulated survey.

Class **Lesson Exercise 12.46**

Was each poll taker's sample chosen randomly?

Class **Lesson Exercise 12.47**

(a) Was the sample large enough?

(b) How large a sample would be needed to predict with certainty the class's *majority* response?

(c) How large a sample would be needed to predict with certainty the *exact results* for the class?

Although a mathematical discussion of sample size is beyond the scope of this course, you probably have an intuitive feeling for it. If reliable sampling techniques are used, a larger sample tends to be more representative of a population than a smaller one. In nationwide surveys, pollsters usually select about 1000 to 2000 people to obtain results with an error of about 2% to 4%. If the sampling procedure is a biased one, such as voluntary responses, a larger sample will not improve it.

Class Lesson Exercise 12.48

How could the wording of the survey question or the response choices be modified to yield different results?

Figure 12-22 shows a sampling activity from the fourth-grade *Mathematics Plus* textbook.

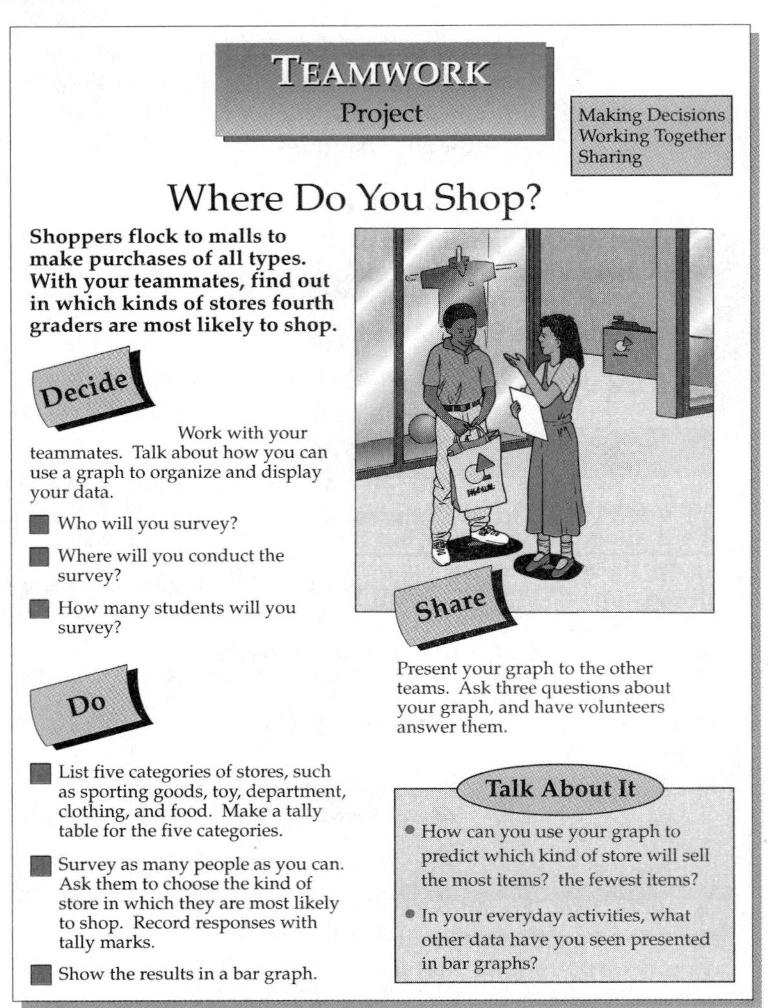

TEAMWORK
Project

Making Decisions
Working Together
Sharing

Where Do You Shop?

Shoppers flock to malls to make purchases of all types. With your teammates, find out in which kinds of stores fourth graders are most likely to shop.

Decide

Work with your teammates. Talk about how you can use a graph to organize and display your data.

- Who will you survey?
- Where will you conduct the survey?
- How many students will you survey?

Do

- List five categories of stores, such as sporting goods, toy, department, clothing, and food. Make a tally table for the five categories.
- Survey as many people as you can. Ask them to choose the kind of store in which they are most likely to shop. Record responses with tally marks.
- Show the results in a bar graph.

Share

Present your graph to the other teams. Ask three questions about your graph, and have volunteers answer them.

Talk About It

- How can you use your graph to predict which kind of store will sell the most items? the fewest items?
- In your everyday activities, what other data have you seen presented in bar graphs?

Figure 12–22

SAMPLING TECHNIQUES

To be useful, a sample must be representative of the population. A large random sample has a good chance of being representative. In a **simple random sample of size *n*** from a population, each subgroup of size n has an equal chance of being selected. Also, each individual in a random sample has an equal chance of being selected. A random sample is the ideal. In practice, it is virtually impossible to obtain a perfectly random sample.

Lesson Exercise 12.49

In a survey of adults in Washington, D.C., telephone numbers are randomly selected from the city phone directory and are dialed. Why isn't the sample obtained in this way a perfectly random sample of adults in Washington, D.C.?

It is usually easier to reach some people in the population than others. It is also difficult to persuade everyone you do reach to respond.

The preceding exercise describes an attempt at a simple random sample of size n. In national samples, pollsters sometimes use another type of random sampling called stratified random sampling. In **stratified random sampling**, the pollster groups the population by a classification (strata) such as region, community size, or gender. Then the pollster takes an appropriately sized random sample from each group.

Design a stratified random sample in the following exercise.

Lesson Exercise 12.50

A pollster wishes to select a stratified random sample totaling 200 children from three schools, in numbers proportional to the sizes of the schools. If the schools have 641, 410, and 549 students, respectively, how many children should be selected from each school?

Surveys sometimes make use of **cluster sampling**, in which the population is divided into subgroups, each of which should be representative of the population. Then a random sample of these subgroups is selected. The sample includes everyone in the subgroups that are selected.

Lesson Exercise 12.51

A researcher wants to interview students from 10 sections of statistics at a college. Using cluster sampling, the researcher randomly selects 2 classes from the 10. Why isn't each class perfectly representative of the population?

You have seen three types of random sampling that can work well. Two types of sampling that are much more likely to be biased are voluntary-response sampling and convenience sampling. A **voluntary-response sample** is composed of people who have volunteered to be part of the survey. In a **convenience sample**, the pollster selects a sample of people who are readily available.

Lesson Exercise 12.52

(a) In a voluntary-response sample, a TV station tells viewers to call one of two toll numbers to register their opinions about raising the sales tax. Why wouldn't we expect the results to be representative of what all adults think about raising the sales tax?

(b) In a convenience sample, a researcher questions some of the students coming out of the cafeteria one afternoon to determine their opinions about school parking facilities. Why wouldn't we expect the results to be representative of all the students in the college?

ANSWERS TO SELECTED LESSON EXERCISES

12.46 Yes

12.47 (c) The entire class

12.49 Not all adults have telephones, and there is not exactly one adult per telephone number. Also, some people are not at home to answer the phone and some have unlisted numbers.

12.50 80, 51, and 69

12.51 Classes vary by chance, by time of day, and because of the teacher.

12.52 (a) People who call tend to have stronger opinions about the sales tax or more positive feelings about this type of telephone survey.
(b) It is likely that the time of day and the location outside the cafeteria would eliminate certain groups of students from the sample.

12.5 HOMEWORK EXERCISES

Basic Exercises

1. Patti Englander is running for the U.S. Senate in Colorado. In a poll of 1200 voters, 322 support her.
(a) What is the population for this poll?
(b) If P people vote in the election, what is a good prediction of the number that will vote for Patti?

2. The following data were collected from a variety of surveys.
- About 45% of Americans do not read books at all.
- The average number of books read by the reading population is 16 per year.
- About 53% of Americans read fewer than 12 books a year.

- About 25% of Americans read at least 20 books a year.

Assume that all of these results are accurate.
(a) What percent of Americans read at least one book a year?
(b) What percent of Americans read 12 to 19 books (inclusive) a year?

3. An aerial photograph of a forest is divided into 40 sections of equal area. Five sections are randomly selected, and they have 46, 38, 39, 52, and 47 trees, respectively. Estimate the total number of trees in the forest.

4. After President Bush's 1992 State of the Union Address, CBS asked viewers with touch-tone phones to call in and rate the speech. Why didn't the 314,786 responses constitute a perfectly random sample?

5. A telephone survey asking whether people favor or oppose the legalization of marijuana finds that 15% are in favor and 78% are opposed. Why are these results unreliable?

6. A soft drink company says, "In our taste test, the majority of adults preferred our cola to theirs." You find out that they commissioned five different taste tests. Why might their claim be misleading?

7. In a door-to-door daytime survey, an interviewer asks any adult who is at home whether or not he or she uses Bubbly laundry detergent. In a similar survey, an interviewer asks the adult at home to display the box of laundry detergent he or she is using. Would the results of the two surveys tend to be the same? Why or why not?

* **8.** A college has 250 freshmen, 220 sophomores, 200 juniors, and 180 seniors. If a stratified sample of 100 students is chosen, how many should be chosen from each class?

9. A survey question reads as follows: "Would you favor or oppose building a peace shield that might protect us from nuclear attack?" Change the wording of the question so that more people will oppose it.

10. Consider the following survey question, with no choices given for the response. "What should be the three main priorities of the federal government?" How would the results change if the following choices were given?
(a) Gun-control laws
(b) Balancing the budget
(c) Developing solar power
(d) Reducing taxes
(e) Aid to the poor and elderly
(f) Other

11. People are asked to respond to the following survey item. "Schoolteachers are more highly respected than stockbrokers.
(a) Agree (b) Disagree"
How would the results be affected if we added the following third choice?
(c) Don't know

12. A random sample of 2000 U.S. adults shows that 80% like grapes. Which of the following statements is most accurate?
(a) Eighty percent of all U.S. adults like grapes.
(b) Close to 80% of all U.S. adults like grapes.
(c) One can be pretty sure that about 80% of all U.S. adults like grapes.

13. A Gallup poll (April 22, 1993) reported the following survey results. The question was, "Do you think abortions should be legal under any circumstances, legal only in certain circumstances, or illegal in all circumstances?" The results were that 32% said legal under all circumstances, 51% said only certain circumstances, and 13% said illegal in all circumstances.

A reporter who supports legalized abortions might say the following about the results. "83% of all Americans favor legal abortions under some or all circumstances." Write a similarly distorted headline that an anti-abortion reporter might write.

14. Ann Landers writes an advice column. She asked her readers, "If you had it to do over again, would you have children?" About 70% of roughly 10,000 respondents said no. Then a national survey was taken, using a random sample of 1400. About 90% of the respondents said yes. Which results should be more reliable? Why?

15. In a random sample of 50 people who are asked whether they favor stricter gun-control legislation, 58% favor it and 36% oppose it. Why are these results unreliable for showing that most people favor stricter gun control?

16. We all make judgments based upon sampling. For example, if you wanted to decide whether new Squeezie grapefruit juice is fresh and tasty, you might buy two cartons, try them, and then make your decision.

Give an example of a sample you took and a conclusion you reached about
(a) a restaurant.
(b) a person.

17. On May 25–June 15, 1995, a Phi Delta Kappan/Gallup poll asked 1311 adults the following question. "Students are often given the grades A, B, C, D, and FAIL to denote the quality of their work. Suppose the public schools themselves in this community were graded the same way. What grade would you give the public schools here?" The results were as follows.

A	B	C	D	FAIL	Don't Know/ No Answer
8%	33%	37%	12%	5%	5%

Why is this survey question a bad one?

18. Classify each of the following as stratified sampling, cluster sampling, voluntary-response sampling, or convenience sampling.
(a) A dean interviews all the students from each of 12 randomly selected classes.
(b) A dean selects 5 men and 5 women from each of 4 English classes.
(c) A researcher interviews all AIDS patients in the hospital closest to the researcher's home.

19. Classify each of the following as stratified sampling, cluster sampling, voluntary-response sampling, or convenience sampling.
(a) A student newspaper asks readers to send in their opinions on an issue.
(b) A student newspaper interviews proportional numbers of freshmen, sophomores, juniors, and seniors.
(c) A researcher interviews all AIDS patients in each of 10 randomly selected counties.

20. Describe how to obtain a sample of 30 fifth-graders from a large elementary school, using
(a) stratified sampling.
(b) cluster sampling.

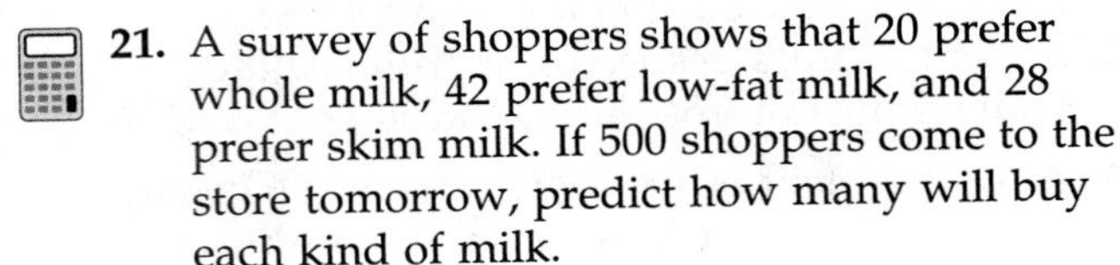

21. A survey of shoppers shows that 20 prefer whole milk, 42 prefer low-fat milk, and 28 prefer skim milk. If 500 shoppers come to the store tomorrow, predict how many will buy each kind of milk.

***22.** "IS OUR FACE RED!"
So read the headline of the November 14, 1936, issue of *The Literary Digest*. The magazine had conducted a poll just before the 1936 presidential election between Alfred Landon and Franklin Roosevelt. The sample came from a list of 10 million names, primarily from telephone directories, automobile registrations, and club membership lists. Each of these people was mailed a questionnaire. Of the 2.4 million people who sent in the questionnaires, 43% preferred Roosevelt, and 57% preferred Landon. This was one of the worst errors ever made in a major survey. Roosevelt won the election 62% to 38%!
(a) Was the sample large enough?
(b) Was the sample representative and random?
(c) Why do you think the prediction was so far off?

23. You want to conduct a survey at your school to determine student preferences in the next presidential election. Describe in detail how you would obtain a representative sample.

Extension Exercises

24. The Dow Jones Average is the most widely quoted stock market average. It attempts to represent the trends in the stock market by averaging the prices of the following 30 stocks (as of 1996).

Allied Signal	Alcoa	American Express
AT&T	Beth. Steel	Boeing
Caterpillar	Chevron	Coca Cola
Disney	Dupont	Eastman Kodak
Exxon	General Electric	General Motors
Goodyear	IBM	Int. Paper
McDonald's	Merck	Minn. Mining
J.P. Morgan	Phillip Morris	Procter & Gamble
Sears	Texaco	Union Carbide
United Tech.	Westinghouse	Woolworth's

Why might the Dow Jones average not be representative of the New York Stock Exchange as a whole?

25. You own a clothing store and want to decide how many of each color pants of a certain style to order. How could you decide?

26. You own an ice cream store and serve 10 flavors at any one time. How could you decide what flavors to serve?

27. **DOONESBURY** **by Garry Trudeau**

(a) In the cartoon, what does Doonesbury mean when he says polls "often become self-fulfilling prophecies"?

(b) Do you agree with Doonesbury?

28. A police department checked the colors of clothing worn by pedestrians who were killed in traffic accidents at night. They found that 4/5 were wearing dark-colored clothes and 1/5 were wearing light-colored clothes. The police department concluded that pedestrians wearing white are less likely to be killed in accidents at night. Explain why this conclusion may not be correct.

29. Following are the Nielsen ratings for the top ten regular network prime-time television shows during the week 1/8–1/14/96.

	Rating	Share
1. *Seinfeld*	21.0	31
2. *Caroline in the City*	18.6	29
3. *Friends*	17.9	28
4. *Home Improvement*	17.9	26
5. *ER*	17.6	29
6. *Frasier*	15.7	23
7. *Single Guy*	15.5	24
8. *NYPD Blue*	15.3	25
9. *3rd Rock from the Sun*	15.2	23
10. *Chicago Hope*	14.9	23

The Nielsen ratings are based on the television-viewing habits of a sample of 4000 U.S. homes (out of 95 million). A box is attached to each television set in every home in order to record what show is watched each minute. The **rating** is the percentage of the 4000 homes in which a show is watched. The **share** considers only the homes with a television set on; it is the percent of these homes that are tuned to a particular show.

(a) Explain why the share is always higher than the rating.

(b) The Nielsen company attempts to obtain a random sample of 4000, but only about 65% of those they ask are willing to participate. How could they make a good case that the 4000 homes they end up with are representative?

(c) How many different shows could be counted as being watched in one home in a 30-minute time slot?

30. Suppose the 4000 homes in the Nielsen sample have 8000 television sets. One Tuesday night, 2700 households have sets in use, and 900 of these households tune in to *Friends*.
    (a) What share does *Friends* receive?
    (b) What percent rating does *Friends* receive?

31. *ER* had a 29 share on 1/11/96. Which of the following must be true?
    (a) 29% of all Americans watched *ER*.
    (b) *ER* is the fifth-best show on television.
    (c) Of all televisions in use, 29% were tuned to *ER*.
    (d) Of all the Nielsen households, 29% had sets tuned to *ER*.
    (e) Of the people living in Nielsen households, 29% were watching *ER*.

32. During the Monday-night 8:00 P.M. slot, assume that 4000 television sets are counted as being watched. Try to answer the following questions.
    (a) How many of the 4000 homes have televisions on?
    (b) How many people are watching television?
    (c) How many people watched the whole program they were watching?
    (d) How many television sets were on?

33. In a 6-year study in Seattle, all firearm deaths were investigated. There were 743 firearm deaths, and investigators collected the following information.
    I. There were 469 suicides, 256 homicides, 11 accidents, and 7 deaths of unknown cause.
    II. Of the deaths, 473 occurred inside a house (343 involved a handgun).
    III. Of those 473 deaths, 398 happened in a house involving a firearm that was kept in the house.
    IV. The 398 deaths were divided as follows.

    333 suicides (68% from handguns)
    50 homicides (42 during a fight in the house)
    11 accidents
    4 self-inflicted deaths (not known whether they were suicide or accident)

    V. Of the 65 nonsuicides, 63 involved families or friends and 2 involved strangers (burglars).

    What conclusions can be drawn from the study?

Special Exercises

34. (a) Keep a log of your television watching, including hours spent watching, shows watched, and hours of commercials seen.
    (b) Based upon your data from part (a), estimate the number of hours of television you have watched in your lifetime.
    (c) How does your answer in part (b) compare to the number of hours you have slept, eaten, or attended school?
    (d) Estimate how many hours of commercials you have seen in your lifetime.
35. (a) Compile the class data for all three sets of data in the preceding exercise.
    (b) Present these data in appropriate graphs or tables.
36. (a) Find an advertisement that makes a claim based upon a survey or research study.
    (b) Write to the company for further details about the study.
    (c) If the company responds, analyze the study and decide whether the company has proved its claim and whether the claim justifies buying the product.
37. After the class has composed a list of items that are representative of what the class eats, have each person find the costs of these items at a neighborhood food store. Bring the results to class for a comparison.
38. Conduct a college or community survey on a topic of your choice. Consider the following suggestions.
    What is your favorite radio station?
    Do you prefer paper bags or plastic bags at the grocery store?
    What can be done to improve public schools in your county?
39. Find an example of a survey in a newspaper. Evaluate the survey questions, the sampling technique, the presentation of the data, and the interpretation of the data.

12.6 STANDARDIZED TEST SCORES

As a teacher, you will receive standardized test reports about your students similar to the one in the following table (which is based on the Stanford Achievement Test). Most standardized test scores are based upon the concepts of the median, percentiles, and the normal distribution.

Test Score Report	Score Type	Math Comp.	Reading	Vocabulary	Math Appl.	Spell.	Lang.
Gr 5 Norms Gr 5.2	RS/NO POSS	35/44	42/60	26/36	38/40	33/40	32/53
Fran Tikley Age 10-3	NATL'L PR-S LOCAL PR-S	68-6 60-6	59-5 48-5	64-6 65-6	97-9 92-8	73-6 63-6	48-5 36-4
Test date 10/25/96	GRADE EQUIV	5.4	5.1	5.6	8.4	6.4	4.1

Cooperative Lesson Exercise 12.53

Try to guess the meanings of the numbers and symbols in the test score report just presented.

Fran Tikley's score report includes raw scores, percentiles, stanines, and grade-level equivalents. The **raw score** is the student's actual score on the exam. Fran received a raw score of 33/40 on spelling, which means he got 33 questions right out of 40.

Lesson Exercise 12.54

(a) What is Fran's raw score in math computation?

(b) What does this score mean?

The test report also gives percentiles. Fran's national percentile for spelling is 73. This means that he scored better in spelling than about 73% of the fifth-grade second-month students in the United States on this exam.

Lesson Exercise 12.55

(a) What is Fran's local percentile score in math computation?

(b) What does this score mean?

Lesson Exercise 12.56

On the same test as Fran, Hy Skor had a national vocabulary percentile of 66. Does this show that Hy knows more of the vocabulary tested than Fran? Why or why not?

How accurate are these raw scores and percentiles? As you know from personal experience, a student's test score varies somewhat according to how the student feels on a particular day and which questions are chosen for the test. So a percent or percentile on an exam has some error in it, just as a percent in a survey does. Repeating a test or a survey on another day usually results in a slightly different outcome.

Percentile scores can be misleading because they appear more accurate (measuring to the nearest percent) than they really are. This problem led to the development of stanine scores. The term "stanine" is a contraction of "standard nine." **Stanines** divide normally distributed student scores into nine groups based on their percentiles, as shown in Figure 12-23.

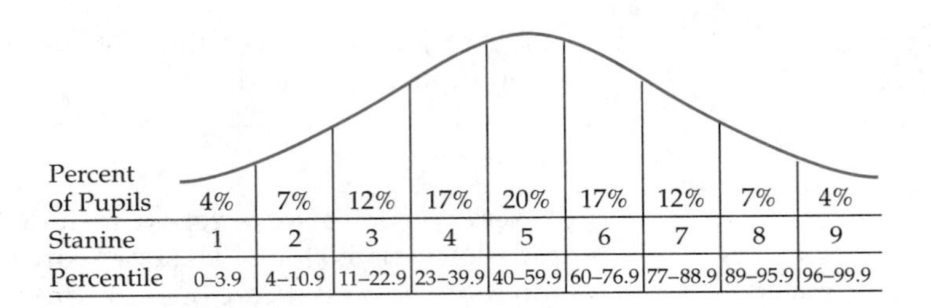

Percent of Pupils	4%	7%	12%	17%	20%	17%	12%	7%	4%
Stanine	1	2	3	4	5	6	7	8	9
Percentile	0–3.9	4–10.9	11–22.9	23–39.9	40–59.9	60–76.9	77–88.9	89–95.9	96–99.9

Figure 12–23

For example, a stanine of 6 means that the student is in the 60th to 76.9th percentile range. [*Note:* A 77th-percentile score on a test could represent a percentile such as 76.8 (6th stanine) or 77.3 (7th stanine) rounded to 77.]

In Fran Tikley's score report, the stanines are given to the right of the percentiles.

Lesson Exercise 12.57

What is Fran's national stanine score in math computation?

Stanine scores do not attempt to make "the fine distinctions which are frequently but improperly made" using percentile scores (NACOME report, 1975).

Lesson Exercise 12.58

One student has a stanine of 6, and another has a stanine of 7.

(a) What is the smallest possible difference between their percentile scores?
(b) What is the largest possible difference between their percentile scores?

Another type of score reports the student's standing in relation to grade levels. The **grade-level score** indicates the grade level, in years and months, at which the student's score would be average. Fran has a grade level of 6.4 in spelling. This indicates that he scored the same as the average student in the fourth month of sixth grade.

Lesson Exercise 12.59

(a) What is Fran's grade-level score in math computation?
(b) What does it mean?

Many educators question the value of grade-level scores because they are commonly misinterpreted in two ways. First, people expect most children to score at or above grade level even though about half the students should score below grade level.

Second, very high or low grade-level scores have little meaning. Fran Tikley scored so well on math applications that his grade-level score was 8.4. Just because he did extremely well on fifth-grade word problems, can we tell whether he can do eighth-grade word problems? A fifth-grade test would have so few eighth-grade-level questions that it could not measure Fran's ability at that level.

These drawbacks of grade-level scoring have led some educators to recommend that it be abandoned.

Lesson Exercise 12.60

A sixth-grade student receives a grade-level score of 3.2. Why is this score misleading?

A test score estimates a student's level. To account for possible error, some standardized test reports give a score such as 80 and a *confidence interval* such as 77–83. This means that the student scored an 80 and it is very likely that the student's true score is between 77 and 83 (inclusive).

Lesson Exercise 12.61

A student has a test score of 420 and a confidence interval of 390–450. Tell what the confidence interval means.

CLASSROOM USE OF STANDARDIZED TESTS

Using standardized test scores in conjunction with other information, you can make a general evaluation of a new student's ability when he or she first comes to your school. The standardized test gives a general score that assesses student progress in subjects that are taught in most schools.

Lesson Exercise 12.62

The following student is transferring to your classroom. His score report (based on the Metropolitan Achievement Test) is given.

Name Robert Mark　　Grade 5

School Fargo Elementary, N.D.　　Date of Test 10/2/96

Score Summary Box

Test	Number Possible	Number Right	Grade Equiv.	Percentile	Stanine
Reading	60	56	6.1	85	
Mathematics	50	25	4.3	30	
Language	60	47	5.7	74	
Science	45	0	4.1	32	
Social Studies	45	8	5.3	62	

(a) Use the percentile information to fill in the stanine scores for each test.

(b) Give your general assessment of how this student will do at your school.

Some schools routinely use standardized tests to evaluate individual students. Since these tests do not measure student achievement in *specific* areas, and since they are not designed to test the particular curriculum the student has been studying, the test results have little value for measuring an individual student's progress in a school's curriculum.

IQ SCORES

Psychologists disagree about what intelligence is and how to measure it. In spite of these disputes, some standardized tests attempt to measure intelligence.

The results of intelligence tests usually are reported as IQ scores. The IQ score compare a child's results with other children the same age. The average (mean) raw score is assigned a standardized IQ score of 100. A raw score one standard deviation above the mean is assigned a standardized IQ score of 115. **Standardized IQ** scores are normally distributed with a mean of 100 and a standard deviation of 15 (Figure 12-24).

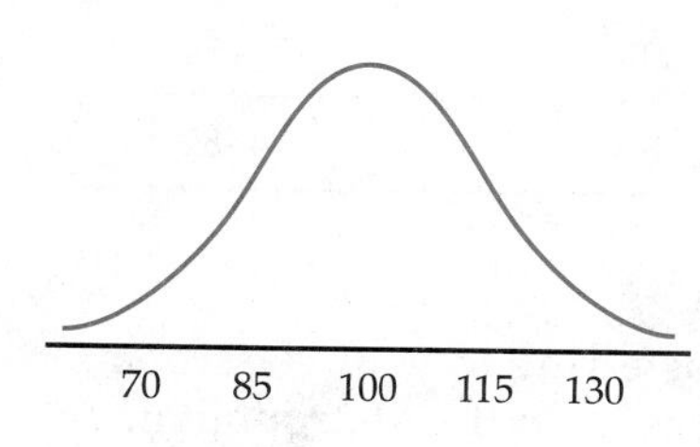

Figure 12–24

Lesson Exercise 12.63

People who score more than two standard deviations below the mean are classified as mentally retarded. In the distribution in Figure 12-24, what IQ score is two standard deviations below the mean?

Lesson Exercise 12.64

What percent of children in an age group have IQs between 85 and 130?

ANSWERS TO SELECTED LESSON EXERCISES

12.54 (a) 35/44 (b) 35 right out of 44

12.55 (a) 60
(b) Fran scored higher in math computation than 60% of the students in his age group.

12.56 No. Their percentile scores are too close to be considered significantly different. Since the tests are multiple choice, guessing accounts for part of each child's score.

12.57 6

12.58 (a) 0.1 (b) 28.9

12.59 (a) 5.4
(b) He scored as well as the average student in the fourth month of fifth grade.

12.60 The student is being compared to third-grade students based upon a test with very few third-grade-level questions.

12.61 It is very likely that the student's true score is between 390 and 450 (inclusive).

12.62 (a) The stanines are 7, 4, 6, 4, and 6.
(b) If your school is average, this student will be fairly average. He may be a little stronger than average in English and social studies and a little below average in science and math.

12.63 $100 - 2 \cdot 15 = 70$

12.64 81.5%

12.6 HOMEWORK EXERCISES

Basic Exercises

Exercises 1–5 refer to the following test report.

Test Score Report	Score Type	Math Comp.	Reading
Gr 6 Norms Gr 6.2	RS/NO POSS	39/44	36/60
Susan Porter	NAT'L PR-S	75-6	41-5
Age 11–5	LOCAL PR-S	82-7	48-5
Test Date 10/22/96	GRADE EQUIV	7.4	5.8

1. (a) What is Susan's raw score in reading?
   (b) What does this score mean?

*2. (a) What is Susan's local percentile score in reading?
   (b) What does this score mean?

3. The local percentiles are higher than the national percentiles. What does this indicate?

4. (a) What is Susan's national stanine score in math computation?
   (b) What does this score mean?

5. (a) What is Susan's grade-level score in math computation?
   (b) What does this score mean?

6. What are two common ways in which grade-level scores are misinterpreted?

7. If Alice scores in the 70th percentile in science, does this mean that she got 70% of the questions right?

8. Why do percentiles tend to be more misleading than stanines?

9. Complete the table. Refer to Figure 12-23 on page 776.

	Math	English	Science
Percentile	92	28	65
Stanine			

10. A student has an IQ score of 102 with a confidence interval of 94–110. Tell what the confidence interval means.

11. Nancy's and Patty's mathematics scores follow. Is one score clearly better than the other? Explain why or why not.

	Math Score	Confidence Band
Nancy B.	80	76–84
Patty G.	78	74–82

12. Izzy Wright is transferring to your classroom. His score report (based on the Metropolitan Achievement Test) follows.
    (a) Use the percentile information to fill in the stanine score for each test.
    (b) Give your general assessment of how this student will do at your school.

Name Izzy Wright   Grade 4

Score Summary Box

Test	Grade Equiv.	Percentile	Stanine
Reading	3.2	42	
Mathematics	3.4	43	
Language	2.9	31	
Science	2.7	24	
Social Studies	5.1	70	

13. People who score more than two standard deviations above the mean on intelligence tests are sometimes labeled "gifted." What is the minimum standardized IQ score for a "gifted" student?

14. About what percent of children in an age group score over 115 on an IQ test?

15. What percentile is an IQ score of 85?
16. Joe scored in the 75th percentile in math, and Moe scored in the 73rd percentile. Does this show that Joe knows more of the math that was tested than Moe? Why or why not?

Extension Exercises

17. Is it possible to miss 3 out of 60 questions on a standardized test and still score in the 99th percentile?
18. Why isn't there a 100th percentile?
19. The percentile of an IQ score of 110 is about
    (a) 30th (b) 50th (c) 60th (d) 75th
20. On a standardized test, the first quartile was 35, the median was 42, and the third quartile was 47.
    (a) Sally scored in the 39th percentile. Estimate her score.
    (b) Mike scored in the 9th percentile. Estimate his score.
21. A standardized test has a mean of 65 and a standard deviation of 5. You want to place students in the upper 2.5% and the lower 2.5% into special programs. Which of the following scores qualify for special programs?
    (a) 60 (b) 77 (c) 50 (d) 72

SUMMARY

Our society uses statistics to analyze trends and make decisions. Economists analyze financial trends, psychologists measure intelligence, pollsters ascertain our opinions, and educators assess achievement. Politicians and advertisers try to persuade us with statistics.

These situations make it vital for citizens to understand different ways of collecting, organizing, and interpreting data. People regularly collect data from samples and organize the results into graphs and tables. Data can be further analyzed by computing averages or measures of spread. People who understand these processes will not be fooled by those who collect, organize, or interpret data in a careless or deceptive manner.

"Students need to be actively involved in each of the steps that comprise statistics, from gathering information to communicating results. . . . Statistics should focus on the active involvement of students in the entire process: formulating key questions; collecting and organizing data; representing the data using graphs, tables, frequency distributions, and summary statistics; analyzing the data; making conjectures; and communicating information in a convincing way" (NCTM, *Standards*, p. 106).

Even when they are reasonably accurate, statistics tell only part of the story. One also should know what other kinds of important information must be included to give a more complete picture.

Statistical methods are used to design and analyze surveys. After taking a sample that is as representative and random as possible, a well-made survey will still have an error of around ±2%. In many large random surveys, the results approximate a special distribution called the normal distribution. In this case, the mean and the standard deviation can be used to approximate the percent of scores that fall within a particular interval.

Teachers regularly receive a collection of statistics about each student—a standardized test report. Since any test score has a certain amount of error, less precise stanine scores tend to be less misleading than percentiles. Grade-level scores that are much above or below the student's grade level are misleading, since most test items are related to the student's actual grade level.

Standardized test scores are useful for making general comparisons among schools or students. Standardized tests do not give any specific information about how well a student understands a particular topic in the student's school curriculum.

STUDY GUIDE

To review Chapter 12, see what you know about each of the following ideas or terms that you have studied. You can also use this list to generate your own questions about the chapter.

The NCTM Curriculum Standards and Statistics

Selected NCTM Curriculum Standards

The following standards come from the NCTM document.

- Construct, read, and interpret displays of data.
- Evaluate arguments that are based on data analysis.
- Collect, organize, and describe data.
- Make convincing arguments that are based on data analysis.
- Apply mathematical thinking and modeling to solve problems that arise in other disciplines.

1. Describe how each standard listed relates to the material you studied in Chapter 12.
2. Select any current elementary-school mathematics textbook series, and describe sample lessons or exercises that illustrate each standard listed.

STATISTICS IN THE ELEMENTARY SCHOOL

The following chart shows at what grade levels selected statistics topics typically appear in elementary-school mathematics textbooks.

Topic	Typical Grade Level in Current Textbooks
Reading graphs	1, 2, 3, 4, 5, 6
Constructing graphs	2, 3, 4, 5, 6
Mean and median	4, 5, 6
Mode	5, 6
Range	5, 6
Surveys	2, 3, 4, 5, 6

REVIEW EXERCISES

1. Use the table to answer the questions that follow.

Price per Pound, Whole Chicken	**Chicken Parts Are an Equally Good Buy If the Price per Pound Is**			
	Breast Half with Rib	*Thigh*	*Drumstick*	*Wing*
$ .49	$ .65	$ .55	$ .50	$ .39
.59	.78	.66	.61	.48
.69	.91	.77	.72	.57
.79	1.04	.88	.83	.66
.89	1.17	.99	.94	.75

(a) What is the next row in the table?

In parts (b)–(d), tell which is a better buy.

(b) A whole chicken for \$.59/lb or drumsticks for \$.79/lb

(c) A whole chicken for \$.72/lb or breasts with ribs for \$1.29/lb

(d) Drumsticks for \$.89/lb or thighs for \$.89/lb

2. The third-biggest item in the federal budget is interest expense. The graph shows what percent of the U.S. government budget was spent on interest expense from 1974 to 1996. In 1996, the amount was about \$250 billion.

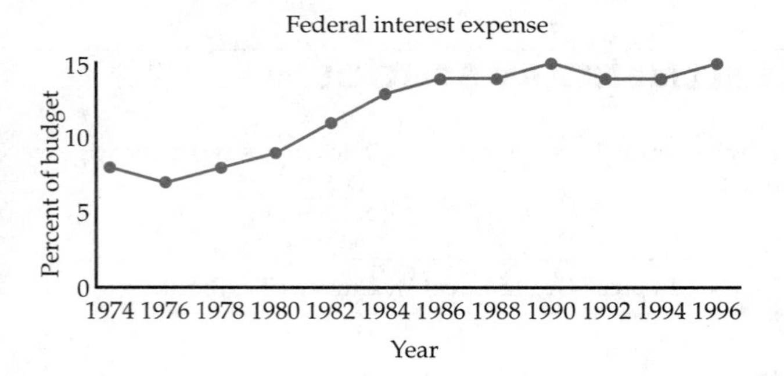

(a) Describe the overall trend and any significant exceptions to it.

(b) Why does the government have to pay interest?

3. Sugar prices rose 40% one year and declined 20% the next year.

(a) Over the 2 years, prices rose ________%.

(b) Prove your answer, using P to represent the original price.

4. A Listerine ad says "98% of the people reading this ad have plaque. They can reduce it as much as 50% with Listerine." Explain what is deceptive about the ad.

5. The following table gives the median salary for employees at a restaurant in 1985, 1990, and 1995.

Median salary	19,800	21,500	24,000
Year	1985	1990	1995

(a) Draw a graph that makes the salary increases look large.

(b) Draw a graph that makes the salary increases look small.

6. The following table gives police response times to emergency calls.

Response time (minutes)	5	10	15	20	25	30
Frequency	3	8	7	2	1	2

Find the mean and median response times.

7. A college has 2000 undergraduates who are ages 18 to 22, 500 undergraduates who are ages 23 to 40, and 100 undergraduates who are ages 41 to 60. Without computing them, explain which would be greater, the mean student age or the median student age.

8. In a nutrition experiment, two groups want to lower their blood cholesterol level. Group C tries to do it by lowering the cholesterol in their diet. Group S tries to do it by lowering the saturated fat in their diet. The change in blood cholesterol level is shown for the ten people in each group.

$$C = \{-10, 8, 0, 2, 5, -2, -5, 4, 1, -2\}$$
$$S = \{-14, -7, 2, -7, -3, 0, -2, 1, -5, -4\}$$

(a) Draw box-and-whisker plots for the two groups on the same graph.
(b) Briefly describe how the two groups compare.

9. A school's monthly budget (in thousands of dollars) is broken down as follows.

Salaries	$420	Maintenance	$120
Insurance	$40	Utilities	$100
Supplies	$40	Other	$80

(a) Find what percent of the total monthly budget each category is.
(b) Use a protractor to construct a circle graph of the school's budget.

10. For each of the following, determine whether the best choice would be a line graph, a bar graph, or a circle graph.
(a) Showing the number of cases of tuberculosis in the U.S. in 1940, 1950, 1960, 1970, 1980, and 1990.
(b) Showing the average life spans of ten different animals

11. The diameter of a telephone cable is normally distributed with a mean of 1.20 cm and a standard deviation of 0.02 cm. Determine the percent of the cables that have diameters
(a) between 1.16 and 1.24 cm.
(b) less than 1.18 cm.

12. The mean weight of an adult woman is normally distributed with a mean of 146 lb and a standard deviation of 22 lb. What percentile is each of the following weights?
(a) 168 lb (b) 190 lb

13. According to an intelligence test, Mike Goldman has an IQ of 115. What is the percentile rank of his score?
(a) 34th (b) 65th (c) 84th (d) 16th

14. Describe a situation in which grade-level scores are misleading, and explain why they are misleading.

15. Consider the following test score report.

Test Score Report	Score Type	Spell.	Lang.
Gr 4 Norms Gr 4.2	RS/NO POSS	29/40	34/53
Kay O'Brien	NATL'L PR-S	39-4	53-5
Age 11–5	LOCAL PR-S	41-5	54-5
Test Date 10/22/96	GRADE EQUIV	3.7	4.5

(a) What is Kay's raw score in spelling?
(b) What does this score mean?
(c) What is Kay's local percentile score in language?
(d) What does this score mean?

16. A county reports two "averages" for home prices in 1997—$77,920 and $93,422. Which is the mean and which is the median? Tell how you know.

17. Without computing the standard deviation, tell which set of scores has the highest standard deviation.
(a) $\{8, 9, 9, 10, 10, 11, 11, 12\}$
(b) $\{9, 9, 9, 10, 10, 11, 11, 11\}$
(c) $\{5, 8, 8, 10, 10, 12, 12, 15\}$

18. Show how to compute the mean and standard deviation of $\{5, 6, 10\}$.

19. In a random sample of voters, a pollster selected proportional numbers of people from the following groups: men, ages 18–40; women, ages 18–40, men, ages 41 and up; and women, ages 41 and up. Is this an example of stratified sampling, cluster sampling, convenience sampling, or voluntary-response sampling?

20. Write a paragraph telling what the mean and median are and how you would decide which one to use.

SELECTED READINGS

Freedman, D., R. Pisani, R. Purves, and A. Adhikari. *Statistics*. 2nd ed. New York: Norton, 1991.

Gnandeskian, M., and others. *Quantitative Literacy Series*. 5 vols. Palo Alto, Calif.: Dale Seymour, 1994.

Huff, D., and I. Geis. *How To Lie with Statistics*. New York: Norton, 1954.

Lindquist, M., and others. *Making Sense of Data*. Reston, Va.: NCTM, 1992.

Mathematics Teacher. February 1990 Minifocus Issue on Data Analysis. Reston, Va.: NCTM, 1990.

Moore, D. *The Basic Practice of Statistics*. New York: W. H. Freeman, 1995.

Mosteller, F., and others. *Statistics: A Guide to the Unknown*. 3rd ed. New York: Holden Day, 1989.

National Council of Teachers of Mathematics. *Teaching Statistics and Probability*. 1981 Yearbook. Reston, Va.: NCTM, 1981.

Teaching Children Mathematics. February, 1996 Focus Issue on Data Exploration. Reston, Va.: NCTM, 1996.

13

Probability

Our futures are uncertain. Will it rain today? Will you enjoy teaching for many years? Will the Cubs win the World Series next season? These questions can be answered only with words such as "unlikely" or "probably" or with numerical probabilities.

People estimate probabilities using intuition and personal experience, but in some cases it is possible to give a more precise numerical probability. If a future event is a repetition of (or very similar to) past events, a numerical estimate of its likelihood can be found. Probability is the study of random phenomena that are individually unpredictable but that have a pattern in the long run.

Although archaeologists have found dice that are 5000 years old, the formal study of probability did not begin until the sixteenth and seventeenth centuries in Italy and France. In 1654, a French gambler asked the mathematician Blaise Pascal (1623–1662) a question about a gambling game. Pascal wrote to another mathematician, Pierre de Fermat (1601?–1665), initiating a correspondence in which they solved gambling problems and developed probability theory (Figure 13-1).

Blaise Pascal

Photos courtesy of Library of Congress.

Pierre de Fermat

Figure 13–1

You already know Fermat as one of the inventors of coordinate geometry, but who was Blaise Pascal? Pascal's mathematical ability was evident at a young age. He excelled in geometry and wrote a major paper on the subject at the age of 25. But at 27, Pascal, already in poor health, decided to give up mathematics and devote his life to religion. During the rest of his life, he occasionally returned to mathematics. However, because he lived to be only 39, Pascal is known as the greatest "might-have-been" in the history of mathematics, and his best-known works concern religion.

Pascal and Fermat studied probability in order to understand gambling. People today use probability to estimate the chances of all kinds of repeated events that have patterns in the long run. Actuaries in insurance companies use probabilities of catastrophes to determine how much to charge for insurance. Testing services design and score standardized tests according to the probabilities of guessing correctly. Meteorologists utilize past frequencies of rain to predict the chance of rain tomorrow. Gambling casinos and some state governments use probabilities to design gambling games and lotteries.

13.1 EXPERIMENTAL AND THEORETICAL PROBABILITY

When someone flips a coin, the result is unpredictable. How likely is it that it will rain tomorrow?

We use probability terms such as "unpredictable" and "likely" in an informal way. Numerical probabilities are more precise measures of the likelihood of events. How do we find numerical probabilities of events?

Lesson Exercise 13.1

(a) You are going to flip a coin 10 times. Predict how many heads you will get.
(b) Flip a coin 10 times and record the total number of heads.

Your results from Lesson Exercise 13.1(b) yield an experimental probability for getting heads on a coin toss. For example, if I tossed a coin 20 times and obtained heads 7 times, my experimental probability for heads would be 7 out of 20, or 7/20. The **experimental probability** of A, written $P(A)$, in an experiment of N trials is $\dfrac{\text{Number of times } A \text{ occurs}}{N}$.

Lesson Exercise 13.2

What is your experimental probability for getting heads on a coin toss, based on your results from Lesson Exercise 13.1(b)?

If someone asked you for the probability of heads on a coin flip, you wouldn't take out a coin and start flipping it, would you? You know another method. This method requires developing some basic ideas of probability.

When you flip a coin, there are two possible results: heads (H) and tails (T). Each possible result, H or T in this case, is an **outcome**. The set of all possible outcomes, $\{H, T\}$, is called the **sample space** of an experiment.

Lesson Exercise 13.3

Complete the sample space for flipping 2 coins. HH means a head on the first coin and a head on the second coin.

$$\{HH, \_\_\_\_, \_\_\_\_, \_\_\_\_\}$$

Any subset of the sample space, such as $\{HT, TH\}$, is called an **event**. Probability questions deal with the probability of an event.

EQUALLY LIKELY OUTCOMES

Probabilities for experiments that have equally likely outcomes (such as a coin flip) are easier to compute. For this reason, it is important to determine whether or not outcomes are equally likely.

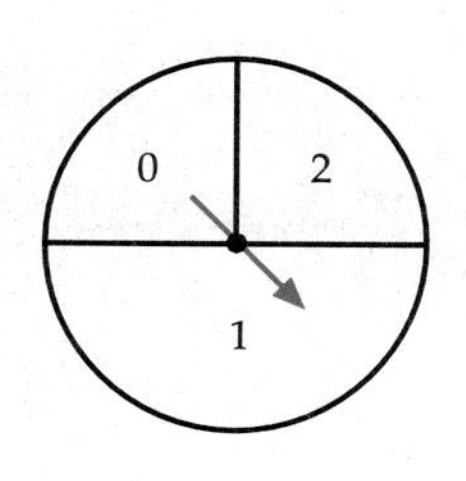

Figure 13–2

Lesson Exercise 13.4

(a) Are the outcomes 0, 1, and 2 equally likely to occur on the spinner in Figure 13-2?

(b) Sketch a spinner with regions numbered 0, 1, and 2 so that each outcome is equally likely.

It is often possible to find the probability of spinning a number such as 2 on a spinner without actually spinning. This is a **theoretical probability** that assumes ideal conditions and is determined from the sample space.

Lesson Exercise 13.5

What is the theoretical probability of spinning a 2 on each of the spinners in the preceding lesson exercise?

The second spinner has 3 equally likely outcomes. The probability of each outcome is 1/3.

Theoretical probabilities for an experiment involving equally likely outcomes can be computed as follows.

Definition **Probability with Equally Likely Outcomes**

If all outcomes in a sample space S of an experiment are equally likely, the **probability** of an event A is

$$P(A) = \frac{\text{number of outcomes in } A}{\text{number of outcomes in } S}.$$

To use this definition, one must have a sample space that lists equally likely outcomes.

Lesson Exercise 13.6

(a) You want to determine the experimental probability of getting 2 heads when you flip 2 coins. Flip 2 coins 20 times and obtain an experimental probability for 2 heads.

(b) Which of the following two explanations for determining the theoretical probability of 2 heads is correct?

(1) The possible outcomes are $\{HH, HT, TH, TT\}$, so the probability of 2 heads is 1/4.

(2) The possible outcomes are {0 heads, 1 head, 2 heads}, so the probability of 2 heads is 1/3.

(c) Which of the two spinners in Lesson Exercise 13.4 is a model for this coin-flipping experiment?

Probability originated with gambling games. The next exercise concerns probabilities in roulette, a casino gambling game (Figure 13-3).

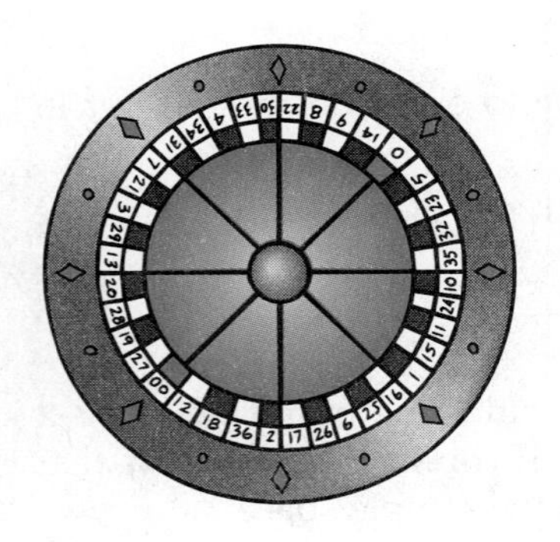

Figure 13–3

Lesson Exercise 13.7

A roulette wheel has 38 numbers. Two numbers, 0 and 00, are green (blue in Figure 13-3). The other slots are numbered 1 through 36. Exactly 18 of the numbers are red (white in Figure 13-3); 18 are black. Each number is equally likely to occur after a spin of the wheel. What is the probability of each of the following?

(a) A 10

(b) A 10 if a 10 came up on the previous 2 spins

(c) A black or green number

As with coins and spinners, the probabilities of outcomes with regular dice can be determined experimentally or theoretically. Consider, for example, the probability of rolling a sum of 7 on 2 dice (that is, hexahedral random digit generators).

Cooperative and Class Lesson Exercise 13.8

(a) Estimate the probability of rolling a sum of 7 on 2 dice.

(b) Roll the dice 30 times and obtain an experimental probability for rolling a sum of 7.

(c) If possible, collect results from the whole class and determine the class's experimental probability for a sum of 7.

To find the theoretical probability of rolling a sum of 7 on 2 dice, complete the following lesson exercise.

Lesson Exercise 13.9

(a) Complete the following list of all possible results for rolling 2 dice.

Sum of 2 Dice

First Die

Second Die	1	2	3	4	5	6
1	2	3	4	5	6	7
2						
3						
4						
5						
6						

(b) How many (different) possible outcomes are there?

(c) Are they equally likely?

(d) The theoretical probability of rolling a sum of 7 is ________.

(e) How does your answer to part (d) compare to the experimental probability obtained by the class?

Lesson Exercise 13.10

(a) Use your table of possible dice outcomes to compute the theoretical probability of rolling a sum of 11.

(b) Predict how many sums of 11 you would get if you rolled 2 dice 50 times.

To predict how many times an event will occur, multiply the probability of the event times the total number of trials. In the preceding exercise, $\frac{1}{18} \cdot 50 = 2.\overline{7} \approx 3$ times.

You have been finding both experimental and theoretical probabilities in a variety of situations. What is the relationship between these two types of probabilities?

Lesson Exercise 13.11

Suppose you select 1 card at random from a regular deck of 52 cards. The theoretical probability of selecting the ace of spades is 1/52. Which of the following conclusions does this probability lead you to make?

(a) If I repeat this experiment 52 times, I will pick an ace of spades 1 time.

(b) If I repeat this experiment 520 times, I will pick an ace of spades 10 times.

(c) If I repeat this experiment a large number of times, I *will* pick an ace of spades 1/52 of that number of times.

(d) If I repeat this experiment a large number of times, I *expect* to pick an ace of spades 1/52 of the time.

(e) If I repeat this experiment a large number of times. I *expect* to pick the ace of spades *about* 1/52 of the time.

The correct statement in the preceding exercise contains three subtle but important parts: "repeat . . . a large number of times," "expect," and "about." Each is essential. First, the theoretical probability relates to a large number of trials. Second, no matter how many trials we perform, we can expect only certain kinds of results, since very unusual results are also possible. Third, we don't expect the experimental results to match the theoretical probability exactly. We must say "about" or "close to."

The following statement describes the relationship between the theoretical and experimental probabilities of an event.

Theoretical and Experimental Probability

If the theoretical probability of an event is $\frac{a}{b}$, then $\frac{a}{b}$ approximates the fraction of the time this event is expected to occur when the same experiment is repeated many times under uniform conditions.

It's amazing that the overall results of a *large number* of coin flips are fairly predictable, but an individual coin toss is completely unpredictable! That is why it is both fair and unpredictable to toss a coin at the beginning of a football game to decide who gets the ball.

In some situations, such as weather forecasting, there is no theoretical probability. A probability of rain tomorrow, such as 80%, is an experimental probability. The meteorologist looks at past records of similar weather conditions and sees that it rained the following day about 80% of the time.

Lesson Exercise 13.12

The principal of a school is expecting a visitor 2 weeks from today. She wants to estimate the probability that all her teachers will be in school that day. How can she do it?

A survey is another common way to obtain experimental probabilities.

Lesson Exercise 13.13

A survey of 1000 elementary-school teachers regarding their preferred subject areas yielded the following results.

	Math/Science	Language/History
Male	14	22
Female	212	752

Based upon this survey, estimate the probability of each of the following.

(a) An elementary-school teacher is female.
(b) An elementary-school teacher prefers the math/science area.
(c) A female elementary-school teacher prefers the math/science area.

ANSWERS TO SELECTED LESSON EXERCISES

13.2 The number of heads you got divided by 10.

13.3 *HT*; *TH*; *TT*

13.4 (a) No, 1 is more likely than 0 or 2.
(b)

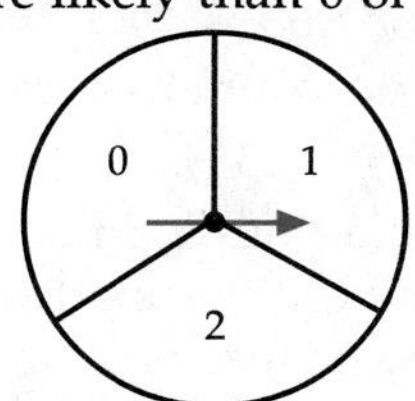

13.5 $\frac{1}{2}$ and $\frac{1}{3}$

13.6 (b) (1)
(c) The spinner from part (a)

13.7 (a) $\frac{1}{38}$ (b) $\frac{1}{38}$ (c) $\frac{20}{38} = \frac{10}{19}$

13.9 (b) 36 (c) Yes (d) $\frac{6}{36} = \frac{1}{6}$

13.10 (a) $\frac{1}{18}$ (b) About 3

13.11 (e)

13.12 She can look at attendance records for the past month of school days and see what percent of the days all her teachers were in school.

13.13 (a) $\frac{964}{1000} = 0.964$
(b) $\frac{226}{1000} = 0.226$
(c) $\frac{212}{964} = 0.22$

13.1 HOMEWORK EXERCISES

Basic Exercises

1. You roll a fair die and read the result.
   (a) What is the sample space?
   (b) Is each outcome equally likely?

2. In an experiment, people rate 3 brands of orange juice, A, B, and C, from best to worst.
   (a) What is the sample space?
   (b) Is each outcome equally likely?

3. In In-Between, you are dealt 2 cards from a standard deck of 52 cards. Then you pick a third card from the deck. In order for you to win, its value must be between the values of the other 2 cards. What is the probability of winning if you are dealt each of the following?
   (a) A 6 and a king (b) A 3 and a 7

4. In a poker game, you have 4 hearts and a spade and you do not know what anyone else has. You discard the spade and pick 1 card. What is the probability of picking a heart (making a "flush")?

5. A certain experiment consists of spinning a spinner like the one shown.

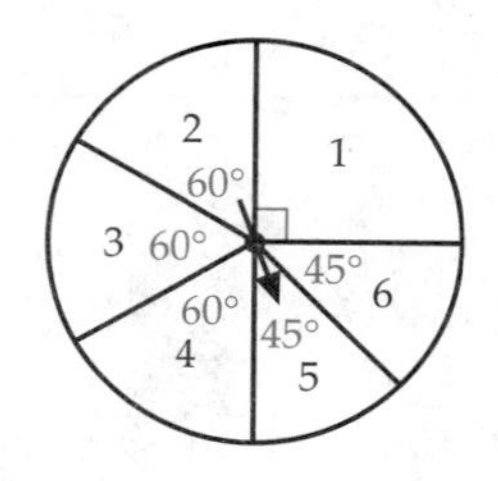

   What is the probability of spinning each of the following?
   (a) 1 (b) 2 (c) 5
   (d) An even number

6. A scientist was asked to explain the greenhouse effect, which is causing our weather to get warmer. He said that in the 1980s, determining whether you would have a hot, average, or cool summer was like rolling a die with 2 sides labeled "hot," 2 sides labeled "average," and 2 sides labeled "cool." In the 1990s, the die would have 4 sides labeled "hot," 1 side labeled "average," and 1 side labeled "cool."
   (a) Explain what this means in terms of probabilities.
   (b) Explain in plain English what this means.

7. The following exercise requires flipping 3 coins.
   (a) Flip 3 coins 20 times and record the results.

Result	Number of Times
3 heads 0 tails	
2 heads 1 tail	
1 head 2 tails	
0 heads 3 tails	

   (b) Based upon part (a), what is your *experimental* probability for getting exactly 2 heads?
   (c) List the different possible equally likely outcomes when flipping 3 coins all at once.

First Coin	Second Coin	Third Coin

   (d) What is the *theoretical* probability of getting exactly 2 heads?
   (e) How do your answers to parts (b) and (d) compare?

8. How would you find the experimental probability of your mathematics professor continuing to teach beyond the end of the period?

9. Two dice are rolled.
   (a) List the theoretical probability of each sum.

Sum	2	3	4	5	6	7	8	9	10	11	12
Probability											

(b) What is the pattern in your answers to part (a)?
(c) If you rolled 2 dice 50 times, predict how many times you would roll a sum of 4.

10. When you roll 2 dice, the sum can be 2, 3, 4, 5, 6, 7, 8, 9, 10, 11, or 12. There are 11 possibilities. So $P(\text{sum is } 7) = 1/11$. What is wrong with this reasoning?

11. You have vulnerable backgammon men 4 and 6 spaces away from your opponent's pieces. What are your opponent's chances of knocking you off on the next roll of the dice? (Your opponent needs a sum of 4 or 6 on the two dice or a 4 or a 6 on either of the two dice.)

12. In Chapter 8, you were introduced to regular polyhedra. Usually dice are shaped like cubes, but any of the regular polyhedra could be used.

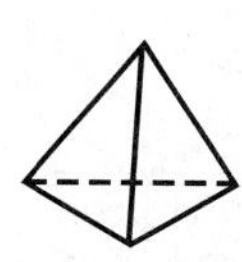
Tetrahedron

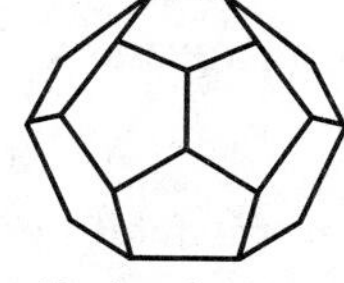
Dodecahedron

(a) Each of two tetrahedral dice has an equal chance of rolling a 1, 2, 3, or 4. What is the probability of rolling a sum of 6 on the 2 tetrahedral dice?
(b) The faces of 2 dodecahedra are numbered 1 through 12. What is the probability of rolling a sum of 6 on 2 dodecahedral dice?

13. Two dice are numbered 1, 2, 3, 4, 5, and 6 and 1, 1, 2, 2, 3, and 3, respectively.
(a) Find the probability of rolling each sum from 2 through 9.
(b) If you rolled these dice 100 times, predict how many times you would roll a sum of 3.

14. (a) Guess how many cards you would have to select from a regular deck of cards (without replacement) to obtain 2 cards of the same denomination (for example, two 7's).
(b) Try it 20 times with a deck of cards and obtain an experimental probability.

15. The IRS audited 2 million out of 82 million taxpayers in 1983. Of those audited, 1.5 million had to make additional payments and 247 were convicted of fraud.
(a) What is the estimated probability of having to make additional payments if you are audited?
(b) What is the estimated probability of being convicted of fraud if you are audited?
(c) The average IRS auditor collects $10 in back taxes for every $1 he or she is paid. Based upon this fact, do you think it would pay to hire more IRS auditors?

16. A meteorologist reports that the chance of rain tomorrow is 60%. What does this mean?

17. When we say that the theoretical probability of rolling a 5 on a die is 1/6, what does this mean?

18. The probability for getting a head when you flip a coin is 1/2. What does this mean?

19. (a) Guess the probability that a thumbtack lands point up.
(b) Devise an experiment and obtain an experimental probability.

20. Open this book to any page and find the first complete paragraph. Use that paragraph to estimate the probability that a word begins with a consonant.

21. You toss a fair coin 1000 times. Which of the following probably happens?
(a) You get 500 heads and 500 tails.
(b) You get between 450 and 550 heads.
(c) You get 500 to 600 heads.
(d) None of the above

22. A fair coin is tossed 5 times. Two possible results are *HHHHH* and *HTHTH*. Which is more likely to occur?
(a) *HHHHH* (b) *HTHTH*
(c) Both are equally likely.

23. A college has 1020 male and 1180 female students. What is the probability that a student selected at random will be female?

24. A school has a raffle for the 130 students in the fifth and sixth grades. There are 62 fifth-graders. What is the probability that the grand-prize winner selected at random from the 130 students will be a sixth-grader?

25. Sam and Mike want a fair way to decide who will take out the garbage. How could they roll a fair die and decide?

26. A survey of 400 community college students yielded the following results.

	Democrat	Republican	Other
Freshmen	95	70	40
Sophomores	86	84	25

Based upon this survey, what is the probability of each of the following?
(a) A college student is a Republican.
(b) A freshman is a Democrat.

27. The following table shows the results of a survey of 1000 buyers of new or used cars of a certain model.

	Satisfied	Not Satisfied
New	300	200
Used	220	280

Based upon this survey, what is the probability of each of the following?
(a) A new car buyer is satisfied.
(b) A used car buyer is satisfied.
(c) Someone who is not satisfied bought a used car.

28. Write a sample space for flipping a coin 3 times, and find the probability of getting *at least* 2 heads ("at least 2 heads" means 2 or more heads).

29. Assume that the probability of a pregnant woman's having a baby boy is 50%. If she has 3 children, what are the chances of having 2 boys and 1 girl in any order? (*Hint:* See the preceding exercise.)

30. If a couple plans to have 4 children, estimate the probability of their having 4 girls.

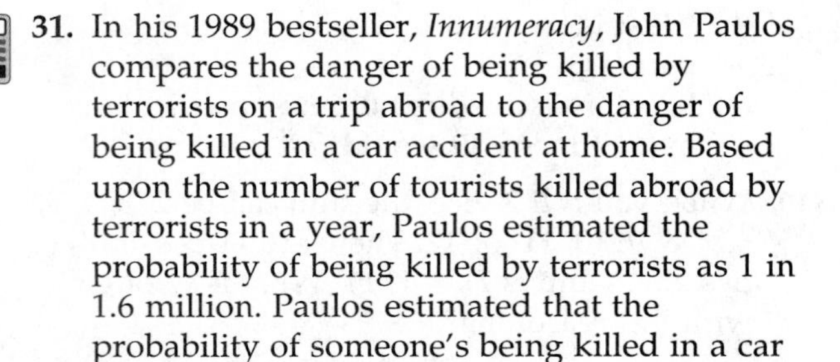

31. In his 1989 bestseller, *Innumeracy*, John Paulos compares the danger of being killed by terrorists on a trip abroad to the danger of being killed in a car accident at home. Based upon the number of tourists killed abroad by terrorists in a year, Paulos estimated the probability of being killed by terrorists as 1 in 1.6 million. Paulos estimated that the probability of someone's being killed in a car crash in a year is 1 in 5300.
(a) How many times more likely is death in a car crash than death on a trip abroad?
(b) How could you make a fairer comparison by considering the time intervals involved?

32. At the Rockville train station, trains run hourly in each direction. If you arrive at the train station at a random time, the next train is 3 times more likely to be northbound than southbound. Explain how this can be the case.

Extension Exercises

33.

	1980	1993
Total traffic fatalities	51,091	40,115
Number involving alcohol	22,480	14,040

Source: U.S. Department of Transportation

(a) Estimate the probability that a traffic death can be attributed to a drunken driver.
(b) If about 2% of drivers are legally drunk, a drunken driver is about how many times more likely to have an accident than a sober driver?

34. WIN, designed by Bradley Efron, is a game that uses unusual cubical dice. The faces are shown here.

A: 3; 3 3 3; 3; 3
B: 0; 4 0 4; 4; 4
C: 5; 1 1 1; 5; 5
D: 2; 2 2 2; 6; 6

You pick 1 die, and your opponent picks 1 die. Whoever rolls the higher number wins.
(a) Suppose you pick *A*. Which die would a smart opponent then pick?

(b) Suppose your opponent picks B first. Which die would you then pick?
(c) Suppose your opponent picks C first. Which die would you then pick?

35. Sicherman described the only nonstandard pair of cubical dice with counting numbers on each face that have the same probabilities for the sums 2 to 12 as regular dice. Can you figure out what numbers are on the faces? (*Hint:* The two dice have different numbers. One of them has only counting numbers from 1 to 4, inclusive.)

36. Design 2 cubical dice so that the probability of each whole-number sum from 1 to 12 is $\frac{1}{12}$. (*Hint:* Use fractions.)

37. Pascal's triangle relates to a variety of mathematics problems.

```
      1   1
    1   2   1
  1   3   3   1
1   4   6   4   1
```

(a) Add a fifth row, continuing the same pattern.
(b) In flipping 1 coin, there is 1 way to get 0 heads and 1 way to get 1 head. In flipping 2 coins, how many ways are there to get 0 heads? 1 head? 2 heads?
(c) How does part (b) relate to Pascal's triangle?
(d) Explain how the third row of the triangle relates to flipping 3 coins.

38. (a) $(x + y)^1 =$ ________
(b) $(x + y)^2 =$ ________
(c) How are the results of parts (a) and (b) related to Pascal's triangle?
(d) Compute $(x + y)^4$ using Pascal's triangle.

39. Consider the following table.

Probabilities of Birth Control Methods Being Effective

	Theoretical	Actual
Abstinence	100%	?
Vasectomy	99.8	99.8
Pill	99	97
Condom	97	83
IUD	97	94
Diaphragm	97	78
Calendar rhythm	85	65
Chance	20	20

Source: John Lobell, *Little Green Book*, p. 32.

(a) How are the actual probabilities obtained?
(b) Why are the actual probabilities usually lower than the theoretical probabilities?

40. Consider the following problem. "One player rolls 2 dice and multiplies the numbers. The other player rolls 1 die and squares the number. The higher number wins. Who wins more often?" Devise a plan and solve the problem.

41. What is the probability of rolling a sum of 6 on three regular dice?

42. Ten people are randomly selected from planet Earth. Use a reference book, such as *World Statistics in Brief* from the United Nations, to estimate the following.
(a) How many speak English as their native language?
(b) How many of their homes have flush toilets?
(c) How many have sufficient food to eat?
(d) How many out of 10 people over 15 years old can read and write?
(e) How many live under each of the following types of government: free, partly free, and not free?

Special Exercise

43. Read "Inflexible Logic" by Russell Maloney, in *Fantasia Mathematica*, edited by Clifton Fadiman. Write a report that includes a summary of the story and your reaction to it.

13.2 SOME BASIC PROBABILITY RULES AND SIMULATIONS

Probability has its own set of mathematical rules. For example, most numbers cannot represent probabilities. How large or small can a probability be? Next, consider arithmetic of probabilities. Under what circumstances can we add or subtract probabilities of events? (Multiplication of probabilities is covered later in the chapter.)

This section also presents another method for obtaining experimental probabilities, called simulation. Simulations are used because they are easier to perform than the actual experiments.

PROBABILITY VALUES

What numbers can represent probabilities? Read the cartoon in Figure 13-4, and then try Lesson Exercise 13.14.

Figure 13–4

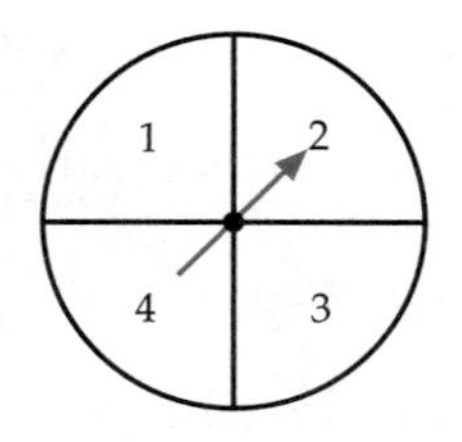

Figure 13–5

Lesson Exercise 13.14

(a) Suppose that an event has no chance of happening (for example, spinning a 5 on the spinner in Figure 13-5). What is its probability?

(b) Suppose that an event will definitely occur (for example, spinning a number less than 5). What is its probability?

(c) Parts (a) and (b) suggest that all probabilities are between ________ and ________ (inclusive).

Lesson Exercise 13.14 illustrates why probabilities can be as low as 0 or as high as 1. In the equally-likely-outcomes formula

$$P(A) = \frac{\text{number of outcomes in } A}{\text{number of outcomes in } S}$$

the number of outcomes in A can be as low as 0, making $P(A) = 0$, or as high as the number of outcomes in S, making $P(A) = 1$. Put differently, a probability tells what percent of the time an event is likely to happen, so it could be any number between 0% (or 0) and 100% (or 1).

Probability Values of an Event

If A is any event, then $0 \leq P(A) \leq 1$.

Probabilities are normally expressed as fractions, decimals, or percents. (Odds and ratios are covered in Section 13.4.) For example, if I flip a coin, the probability of heads is given by

$$P(\text{heads}) = \frac{1}{2} = 0.50 = 50\%$$

Lesson Exercise 13.15

Which of the following could not be the probability of an event?

(a) $\frac{2}{3}$ (b) -3 (c) $\frac{7}{5}$ (d) 15% (e) 0.7

We often discuss probabilities of events using words such as "impossible," "likely," and "certain." How are these words related to numerical probabilities? Figure 13-6 shows possible probabilities of events and some corresponding verbal descriptions.

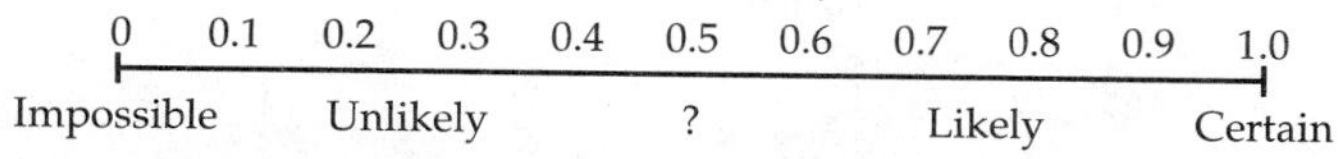

0	0.1	0.2	0.3	0.4	0.5	0.6	0.7	0.8	0.9	1.0
Impossible		Unlikely			?		Likely			Certain

Figure 13–6

MUTUALLY EXCLUSIVE EVENTS

How are probabilities that involve more than one event computed? One of the simplest cases involves two events that do not intersect.

Lesson Exercise 13.16

Consider a college with 25% freshman, 25% sophomores, and 25% women.

(a) What is the probability that a student is a freshman *and* a sophomore?

(b) Name two nonintersecting groups at the college besides freshmen and sophomores.

(c) What is the probability of selecting a single student who has both of the characteristics you mentioned in part (b)?

Lesson Exercise 13.16, parts (a) and (c), suggests the definition of non-overlapping—that is, mutually exclusive—events. Mutually exclusive events cannot occur at the same time. Events A and B are **mutually exclusive** if and only if $P(A \text{ and } B) = 0$ or $A \cap B = \phi$. The expression "$P(A \text{ and } B)$" means the probability that both events A and B occur.

$P(A \text{ or } B)$ is easier to compute for mutually exclusive events than for other events.

Lesson Exercise 13.17

A college has 25% freshman, 25% sophomores, and 25% women. A student is picked at random.

(a) Which of the following are mutually exclusive groups?
 (1) Freshmen and sophomores (2) Freshmen and women

(b) What is the probability that a randomly chosen student is a freshman or a sophomore?

(c) What is the probability that a randomly chosen student is a freshman or a woman?

In part (b), you could add the probabilities (percentages) because the two groups, freshmen and sophomores, do not overlap. When two events A and B do overlap, as in part (c), you cannot simply add their probabilities to compute $P(A \text{ or } B)$.

The general rule for $P(A \text{ or } B)$ when A and B are mutually exclusive is as follows.

The Addition Rule for Mutually Exclusive Events

If A and B are mutually exclusive events, $P(A \text{ or } B) = P(A) + P(B)$.

The expression "A or B" combines all the outcomes from event A and event B. For this reason, $P(A \text{ or } B)$ can also be written as $P(A \cup B)$. However, current middle-school textbooks do not use set symbols in probability.

Probabilities have already been defined for equally likely outcomes. For a sample space with outcomes that are not equally likely, the addition rule can also be used to compute probabilities.

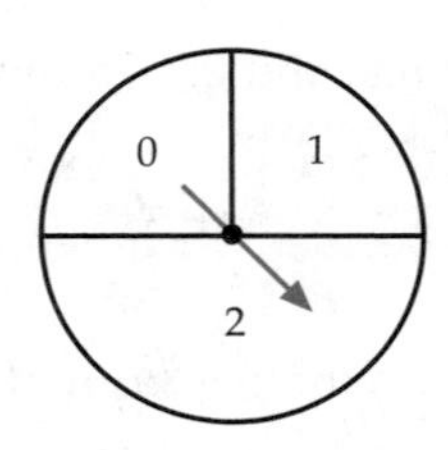

Figure 13–7

Lesson Exercise 13.18

How would you compute the probability of spinning a 1 or a 2 on the spinner in Figure 13-7?

The preceding exercise illustrates the following.

Definition The Probability of an Event

If A is any event, $P(A)$ is the sum of the probabilities of all outcomes in set A.

A set of outcomes is mutually exclusive. This is why we can add the probabilities of the outcomes. The addition rule for mutually exclusive events can also be used to derive a formula for events that are complements.

COMPLEMENTARY EVENTS

The probability that an event will occur and the probability that it will not occur have a simple relationship.

Lesson Exercise 13.19

Suppose that the probability of rain tomorrow is 1/4. What is the probability that it will *not* rain tomorrow?

The event "not rain" in the preceding exercise is the complement of the event "rain." The **complement** of any event A is the event that A does *not* occur, written "not A" or $\overline{A}$. (Some textbooks use the notation A'.) Complementary events have no outcomes in common (are mutually exclusive), and together they encompass all possible outcomes. In set notation, $A \cap \overline{A} = \phi$ and $A \cup \overline{A} = S$, the sample space.

Lesson Exercise 13.20

If the probability of an event is n, what is the probability of the complement of the event?

Lesson Exercise 13.20 generalizes the rain example. The probability of the complement of an event is 1 (or 100%) minus the probability of the event.

Probabilities of Complementary Events

$$P(\overline{A}) = 1 - P(A)$$

The preceding equation follows from the addition rule for mutually exclusive events.

Lesson Exercise 13.21

(a) A and $\overline{A}$ are mutually exclusive. By the addition rule, $P(A \text{ or } \overline{A}) =$ ________.

(b) What is the numerical value of $P(A \text{ or } \overline{A})$?

(c) Combine the equations in parts (a) and (b).

(d) How would you derive the probability rule for complementary events from your equation in part (c)?

SIMULATIONS

When the theoretical probability is difficult or impossible to compute, and it is impractical to find an experimental probability, a simulation can be designed. A **simulation** is a probability experiment that has the same kind of probabilities as the real-life event. Simulations are based upon connections between mathematical models and everyday life.

A series of coin flips can be used to simulate whether a series of newborns will be boys or girls. The result of a coin flip has the same kind of probabilities as the sex of a baby. The probability of heads is 1/2 and the probability of tails is 1/2, just as the probability of a boy is about 1/2 and the probability of a girl is about 1/2. Furthermore, each coin flip has no effect on subsequent coin-flip probabilities. The same is usually true of the sexes of babies.

In the following simulation, each coin flip represents the birth of a new child. Furthermore, suppose that each head represents having a girl and each tail represents a boy.

***Cooperative and Class* Lesson Exercise 13.22**

A couple wants to have a baby girl. They are willing to have up to 3 children in an effort to have a girl. You will use coin flipping to simulate this in pairs.

(a) Using H = girl and T = boy, simulate the event of the couple's having a family. (For example, if you flip heads on the first try, they have 1 girl and stop having children.) What family do you end up with?

(b) Repeat this experiment 19 more times so that you have created 20 families.

(c) Out of the 20 families, how many have a girl?

(d) Out of the 20 families, how many have 3 children?

(e) Based upon your results, estimate the probability of the family's having a girl.

(f) Based upon your results, estimate the probability of the family's having 3 children.

(g) If you have data from others in the class, use these data to revise your answers to parts (e) and (f).

A simulation is usually much easier to perform than the event it represents. The results of a simulation provide an experimental probability that is used to estimate the theoretical probability of the simulated event. A greater number of trials is likely to provide a better estimate.

AN INVESTIGATION: A TRAFFIC SIMULATION

A simulation can also be used to analyze a traffic intersection, as in Lesson Exercises 13.23–13.25.

Imagine that you want to decide how long the traffic signal should stay green in each direction at a particular intersection. First, consider the cars going from east to west (on a one-way street). A counter can be used to determine how many cars pass through the intersection during a given period of time. Suppose you find out that 200 cars pass by during a 10-minute period.

Lesson Exercise 13.23

One car passes east to west every ________ seconds.

Using a timer, you also compute that cars leave the intersection at a rate of 1 per second when the light is green. What if you make the light alternately red and green every 15 seconds (disregard yellow)? The east-west traffic pattern can be simulated with a die. This will give you an idea of how many cars will get backed up at the traffic light, assuming the traffic and the die both have the same random pattern.

Lesson Exercise 13.24

One car passes every 3 seconds.

(a) If each roll of the die represents 1 second, and the die has ________ possible results, how many of these results should count as a second in which a car passes?

(b) Decide which die results represent "car" and which represent "no car," and write this information in the table.

	Result on Die
Car	
No car	

Cooperative Lesson Exercise 13.25

(a) Now for the simulation. Start with a red light, and assume that 1 car can leave the intersection each second. Roll 1 die 120 times (for 120 seconds), and fill in the results in a table like the one shown here. Compute how many cars are lined up each second. (*Note:* This simulation could also be done with a computer.)

Second	**1**	**2**	**3**	**4**	**5**	**6**	**7**	**8**	**9**	**10**	**...**
Result on die											
New car arrives?											
Car leaves? (when light is green)											
Total number of cars lined up											

(b) What was the largest number of cars you had backed up? (Get results from other members of the class.) Is this result satisfactory? If not, suggest an alternative timing for the signal.

AN INVESTIGATION: A COMPUTER SIMULATION

A company is buying 10 automobiles from a manufacturer. From past experience, they know that about 1 out of every 6 automobiles needs some kind of adjustment. You can simulate determining whether or not automobiles need adjustment by rolling a die or by using a calculator or computer program.

Cooperative Lesson Exercise 13.26

Suppose you buy 10 automobiles, and you expect that about 1 out of every 6 will need an adjustment. What is the probability that 3 or more automobiles will need adjustments?

Use a die to simulate 20 purchases of 10 automobiles like the one just described.

(a) What roll or rolls on the die will represent an automobile needing adjustment?

(b) If you have a computer available, use the following BASIC program for the simulation. (A similar program can be written with Logo, with MINITAB, or on a graphing calculator.) RND(1) randomly selects a decimal number between 0 and 1, including 0 but excluding 1. (In Logo, RANDOM 6 would pick a number at random from {0, 1, 2, 3, 4, 5}.)

```
10 FOR I = 1 TO 10
20    LET ROLL = INT (RND(1) * 6) + 1
30    PRINT ROLL
40    IF ROLL = 1 THEN PRINT "DEFECTIVE"
50 NEXT I
```

(c) If you have studied INT in BASIC, explain why INT (RND(1) * 6) + 1 simulates the roll of a die.

(d) Roll the die or use a computer to simulate 20 orders.

(e) Make a frequency graph of the number of automobiles needing adjustment in each of your 20 shipments.

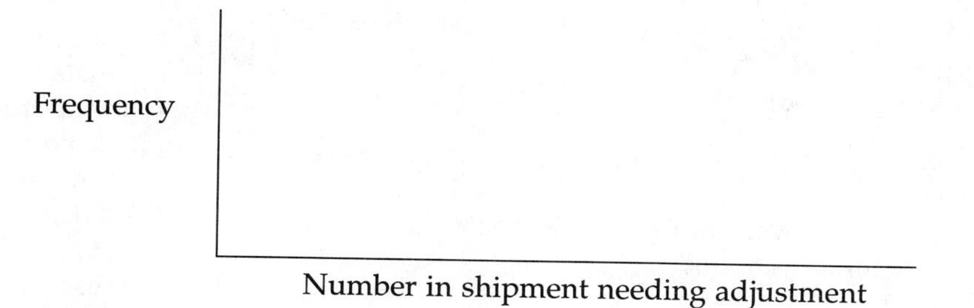

(f) What is your experimental probability that 3 or more cars in a shipment will need adjustments?

Statisticians often use simulations to approximate probabilities that are difficult to compute using theory or for which there is no theory. Computers offer the possibility of simulating a large number of trials in a short amount of time.

ANSWERS TO SELECTED LESSON EXERCISES

13.14 (a) 0 (b) 1 (c) 0; 1

13.15 (b) and (c)

13.16 (a) 0 (b) Men and women (c) 0

13.17 (a) (1) (b) 50%
(c) It cannot be determined, since some of the freshmen are also women.

13.18 $P(1 \text{ or } 2) = P(1) + P(2) = \frac{1}{4} + \frac{1}{2} = \frac{3}{4}$

13.19 $\frac{3}{4}$

13.20 $1 - n$

13.21 (a) $P(A) + P(\overline{A})$ (b) 1
(c) $P(A) + P(\overline{A}) = 1$
(d) Subtract $P(A)$ from both sides of the equation.

13.23 3

13.24 (a) 6; 2

13.26 (c)

```
0 ≤ RND(1) < 1
So 0 ≤ 6*RND(1) < 6
and 1 ≤ 6*RND(1)+1 < 7
Int (6 * RND(1)+1) produces
1, 2, 3, 4, 5, and 6 with
equal probability
```

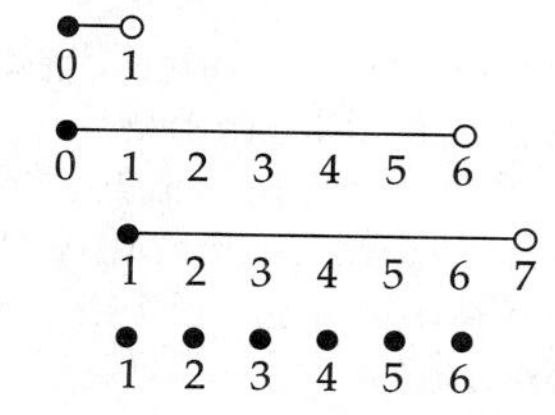

13.2 HOMEWORK EXERCISES

Basic Exercises

1. What is the probability that a resident of Iowa lives in the United States?

2. Which of the following numbers cannot be a probability?
(a) 0.6 (b) -1 (c) 3 (d) 0%
(e) $\frac{1}{1000}$

3. An event is very unlikely to happen. Its probability is about
(a) $\frac{1}{10}$ (b) $\frac{3}{10}$ (c) $\frac{1}{2}$
(d) $\frac{7}{10}$ (e) $\frac{9}{10}$

4. Estimate the probability that you will eat chicken in the next week.
(a) 0 (b) 0.3 (c) 0.5

5. Determine whether or not each pair of events is mutually exclusive.
(a) Getting an A in geometry and a B in English this semester
(b) Getting an A in geometry and a B in geometry this semester

6. What is wrong with each of the following statements?
(a) The probability that a person has black hair is 22%. The probability that a person has brown eyes is 72%. So the probability that a person has either black hair or brown eyes is 94%.
(b) Since there are 7 continents, the probability of being born in North America is 1/7.

7. Suppose that you flip a coin once. Name two events, A and B, that are mutually exclusive.

8. If $P(N) = 0.03$, then $P(\text{not } N) =$ ________.

9. At a college, the probability that a randomly selected student is a freshman is 0.35, and the probability that the student is a sophomore is 0.2. What is the probability that a randomly selected student is neither a freshman nor a sophomore?

10. Suppose Gordon Gregg and Debra Poese are the only two candidates for president. The probability that Gordon wins is 0.38. What is the probability that Debra wins?

11. Use a coin flip or a computer program to simulate the birth of a boy or girl. Investigate four-children families by doing the following.
(a) If there are 20 four-children families, guess how many will have exactly 1 girl.
(b) Create 20 four-children families, using H = girl and T = boy. Write down the result for each family.
(c) Use your results to complete the table.

Four-Children Families

Number of girls	0	1	2	3	4
Estimated probabilities					

(d) Of the 20 families, how many have exactly 1 girl?

12. Consider the following problem. "On a game show, a contestant gets to pick box 1, 2, or 3. One box contains $10,000. The other 2 boxes are empty. Use a die to simulate 20 groups of 5 shows (weekly), and estimate the probability that at least 4 people will win during a particular 5-show week." Devise a plan and solve the problem.

13. New Raisin Crumbles come with 1 of 6 different famous-mathematics-teacher cards in each box.
(a) Guess the approximate number of boxes you would have to buy to get all 6 different teacher cards.
(b) Using a die, simulate collecting all of the cards 8 times to obtain an experimental estimate.

14. A particular forward on the Phoenix Suns makes 50% of his shots. Using a coin, someone simulates the player taking 6 shots. H (heads) means that he made the shot and T (tails) means that he missed the shot. Based upon the following results, what is the experimental probability that the player makes exactly 3 out of 6 shots?

H H T T H T
H T T T T H
H T H T T T
H H T T H H
H H H H H T
H T T T H H
H T T H T T

15. Two contestants on a game show must choose door 1, 2, 3, or 4. Three of the locations have prizes, and the other does not. A simulation is carried out with a spinner to find the probability that both contestants will win a prize. It is assumed that doors 1, 2, and 4 lead to prizes and door 3 is a dud. The results follow.

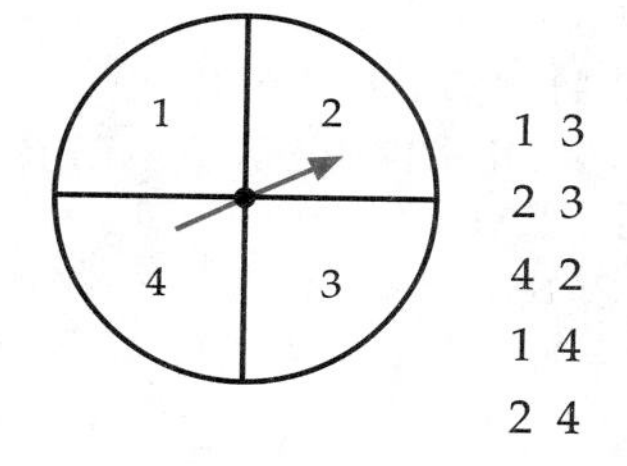

1 3
2 3
4 2
1 4
2 4

(a) What does the first row of numbers represent?
(b) Based upon these results, what is the probability that both contestants will win a prize?

16. Create a dice game that simulates baseball. Use probabilities something like the following.

STRIKE OUT 17% WALK 9%
SINGLE 16.5%
DOUBLE 4% TRIPLE 0.5%
HOME RUN 3%
FLY OUT 22%
OUT OR DOUBLE PLAY
(IF RUNNER IS ON FIRST) 6%
GROUNDOUT (RUNNERS ADVANCE
1 BASE) 22%

Extension Exercises

17. Draw a spinner, with sectors labeled 2, 3, 4, . . . , 12, that could be used to simulate rolling 2 dice and computing the sum.

18. Draw a spinner that has all the following characteristics.

(1) The probability of spinning a 2 is $\frac{1}{2}$.
(2) The probability of spinning an odd number is $\frac{1}{4}$.
(3) The probability of spinning a sum of 5 on 2 spins is $\frac{1}{8}$.

19. Recall the traffic-intersection problem from this section. Suppose that cars traveling in the north-south direction (one way) arrive at the intersection at the rate of 1 every 6 seconds. Do a traffic simulation, using the light timing of your choice, for cars traveling in the north-south and east-west directions.

20. The following supermarket-checkout simulation is adapted from NCTM's "Student Math Notes" of March 1986.

You decide to open a supermarket. How many checkout lanes should you have? A die will be used to simulate the arrival of shoppers at the checkout counter. Assume that a new customer arrives at the checkout counter during 1 out of every 3 minutes. Assume that it takes the cashier 3 minutes to process each customer.

(a) What die results would represent a customer arriving during a given minute?
(b) What die results would represent a customer not arriving during a given minute?
(c) Roll the die 30 times and simulate 30 minutes at a checkout counter (or use a computer program). See how long the line gets. You could use a chart like the following one to keep a record.

END OF MINUTE	1	2	3	4	5	6	. . .
Customer arrives	*A*	—	—	*B*	*C*		
Checking out	—	*A*	*A*	*A*	*B*	*B*	
Waiting	—	—	—	—	—	*C*	

(continued on p. 808)

(d) Complete the following record of your results.

Customer	A	B	C	D	E	F	. . .
Minute of checkout arrival							
Minute checkout completed							
Minutes wait before checkout began							

(e) Give the following estimates based upon your 30 minutes of data.
Number of customers arriving ________
Total customer waiting time ________
Average waiting time ________
Total clerk idle time ________
Total time to process everyone ________

(f) Repeat parts (c)–(e) using two checkout counters.

(g) Which number of checkout counters works out better? Consider the waiting time and the cost to the store.

Computer Exercises

21. Recall that the RND(1) statement randomly selects a number between 0 and 1, including 0 but excluding 1.

(a) What are the possible values of INT(2 * RND(1))?

(b) Enter the following program into a computer.

```
10 PRINT "HOW MANY COIN FLIPS?"
20 INPUT N
30 LET H = 0
40 FOR I = 1 TO N
50    LET F = INT(2 * RND(1))
60    IF F = 0 THEN PRINT "HEADS"
70    IF F = 0 THEN H = H + 1
80    IF F = 1 THEN PRINT "TAILS"
90 NEXT I
100 PRINT "YOU GOT "; H;" HEADS
      AND "; N - H;" TAILS."
```

(c) RUN the program 20 times for 5 coin flips, and use your results to estimate the probability of getting 2 heads and 3 tails.

22. You own two National Motors cars. Suppose that, on a given morning, the probability that the Squealer starts is 0.4 and the probability that the Alley Cat starts is 0.2. Use the following computer simulation to estimate the probability that at least one of your cars will start.

(a) Complete the following program.

```
10 PRINT "HOW MANY TIMES DO YOU
     WANT TO RUN THIS PROGRAM?"
20 INPUT N
30 FOR I = 1 TO N
40    LET CARS = 0
50    IF RND(1) < 0.4 THEN 80
60    PRINT "OOPS THE SQUEALER
      WON'T START."
70       GOTO 100
80       CARS = 1
90       PRINT "THE SQUEALER IS
         HUMMING TODAY."
100    IF RND(1) < 0.2 THEN 130
110
120
130
140
150    IF CARS > 0 THEN PRINT
       "YOU'VE GOT A CAR!"
160 NEXT N
```

(b) In line 50, what is the probability that RND(1) < 0.4?

(c) RUN the program 10 times to obtain an experimental probability that at least one car will start.

(d) Repeat part (c).

(e) Repeat part (c) again.

(f) Estimate the probability that at least one car will start.

23. (a) Enter the following program.

```
10 PRINT "HOW MANY ROLLS OF THE
   DICE?"
20 INPUT N
30 LET S = 0
40 FOR I = 1 TO N
50    LET F = INT(6 * RND(1)) + 1
60    LET G = INT(6 * RND(1)) + 1
70    LET R = F + G
80    IF R = 7 THEN S = S + 1
90    PRINT "YOU ROLLED A "; R
100 NEXT I
110 PRINT "YOU ROLLED "; S;
    "SEVENS."
```

(b) RUN the program for 40 rolls, and use it to estimate the probability of rolling a sum of 7.

(c) Edit the program so that you can use it to compute the number of 10's in 100 rolls.

(d) Estimate the probability of rolling a sum of 10.

24. Twenty people are at a party.

(a) Guess the probability that at least 2 people will have the same birthday.

(b) Use MINITAB (or BASIC, Logo, or a calculator) to generate 20 random integers between 1 and 365 (inclusive), and see if at least 2 match. Repeat this 30 times to obtain an experimental probability.

Special Exercise

25. Design a simulation game that uses dice or a spinner.

13.3 COUNTING

To compute theoretical probabilities, one often counts the number of equally likely results in a sample space. Sometimes the sample space is so large that shortcuts are needed to count all the possibilities.

Figure 13–8

How do state officials know how many different license plates they can make using three letters followed by three digits. (Figure 13-8)? How does the telephone company know how many people can have the same area code? To answer these questions, they count all possible arrangements of a sequence.

ORGANIZED LISTS AND TREE DIAGRAMS

In order to understand the shortcut for counting, we'll begin by analyzing some simpler counting problems.

EXAMPLE 13.1

You have 3 shirts and 2 pairs of pants. How many different choices of outfits do you have (assuming everything matches)?

Solution

If the shirts are s_1, s_2, and s_3, and the pants are p_1 and p_2, you can count the outfits in an organized list or a tree diagram. A tree diagram can be used when an event has two or more steps. Each step is shown with a set of branches.

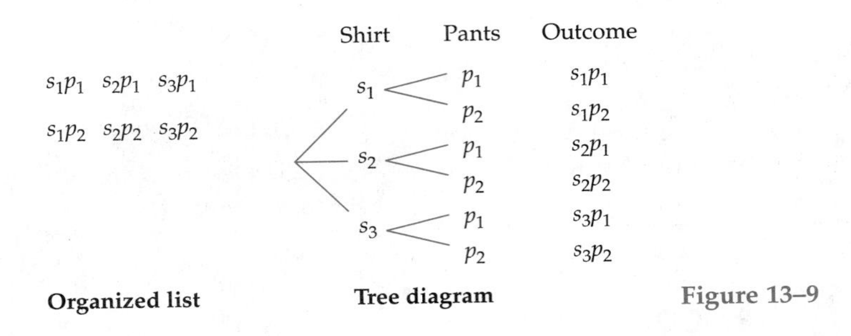

Figure 13–9

There are 6 different outfits (Figure 13-9). ■

Lesson Exercise 13.27

A cafe offers the following menu. Each customer chooses one of two appetizers and one main dish.

LUNCHEON MENU $4

Canned fruit du jour
Cream of bamboo soup

Blowfish thermidor
Woodchuck pilaf
Semi-boneless falcon
Hippo in a blanket

(a) How many different selections can a customer make?

(b) What is a shorter way to compute the answer to part (a), without using a list or diagram?

Although the two preceding examples can be solved by making an organized list or a tree diagram, there is a shorter way.

THE FUNDAMENTAL COUNTING PRINCIPLE

Did you recognize the Cartesian product model of multiplication from Chapter 3? In both preceding problems, you could multiply the number of ways to make the first choice by the number of ways to make the second choice to obtain the total number of arrangements.

In the outfit problem, there were 3 ways to select a shirt and 2 ways to select a pair of pants. This resulted in $3 \times 2 = 6$ outfits.

(3)	×	(2)	=	6
Shirt choice		Pants choice		Outfit choices

Lesson Exercise 13.28

If you didn't already do this in the preceding exercise, show how you would use multiplication to count choices in the menu problem.

Lesson Exercise 13.29

Suppose I want to order a dessert at a restaurant. The menu offers 3 kinds of pie and 2 kinds of cake. How many dessert choices do I have?

In Lesson Exercise 13.29, you do not multiply 3 by 2 because you are not picking a cake and then a pie *in sequence*. You are picking only one dessert. The outfit and menu situations are both examples of the Fundamental Counting (Multiplication) Principle, whereas the dessert choice is not.

The Fundamental Counting Principle

If an event E can occur in e ways and, after it has occurred, an event F can occur in f ways, then event E followed by event F can occur in $e \cdot f$ ways.

The Fundamental Counting Principle works not only for two events in sequence but also for any number of events in sequence. The Fundamental Counting Principle can be applied to a complex event if the event can be thought of as a *series of steps* with a specified order.

EXAMPLE 13.2

The state of Maryland has automobile license plates consisting of 3 letters followed by 3 digits. How many possible license plates are there?

Solution

Understanding the Problem How many ways are there to choose 3 letters followed by 3 digits? A license plate can be created in 6 steps: picking each of the 3 letters and then selecting each of the 3 digits. There is a certain number of choices for each step.

Devising a Plan Using the Fundamental Counting Principle, one can compute the total number of choices for each step and then multiply these numbers together.

Carrying Out the Plan There are 26 choices for each of the 3 letters and 10 choices for each of the 3 digits. The total number of possible license plates is

(26)	(26)	(26)	(10)	(10)	(10) = 17,576,000
Letter	Letter	Letter	Digit	Digit	Digit

Looking Back The same technique would work for many other license plate designs. ■

Lesson Exercise 13.30

A state has license plates with 2 letters followed by 4 digits. How many possible license plates can be made?

Lesson Exercise 13.31

How many possible telephone numbers can you form (in one area code) if you may choose any 7 digits, except that you may not select 0 or 1 as the first or second digit? (*Hint:* Write an example of a phone number, and see how many choices you have for each step.)

One special type of counting problem involves a series of steps in which each step uses up one of the choices.

EXAMPLE 13.3

A club has 10 members. How many ways can they choose a president and vice-president if everyone is eligible?

Solution

Understanding the Problem Think of the elections as 2 steps. The first step is to elect a president. Then the club elects a vice-president from the remaining members.

Devising a Plan Use the Fundamental Counting Principle.

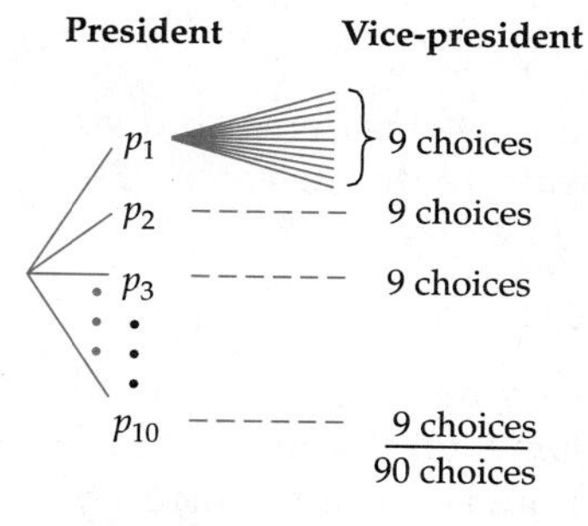

Figure 13–10

Carrying Out the Plan How many choices are there for president? 10. Once the president is selected, how many choices are there for vice-president? 9. Using the Fundamental Counting Principle, there are $10 \cdot 9 = 90$ ways to select a president followed by a vice-president.

Looking Back This procedure works for any election in which each position is different. ■

An ordered arrangement of people or objects (such as a choice of president and vice-president) is called a **permutation**. In the preceding example, the number of permutations is 90, as shown by the tree diagram in Figure 13-10. There are 10 branches for the first step, and each of these 10 choices has 9 branches for the second step ($10 \cdot 9 = 90$).

Lesson Exercise 13.32

A principal must select a head teacher and an assistant head teacher from her faculty of 30 people. In how many ways can she do this?

FINDING PROBABILITIES USING THE FUNDAMENTAL COUNTING PRINCIPLE

People who design multiple-choice tests must figure out the probability that an examinee will guess a correct answer. A test is not very useful if someone who knows nothing about the material does well on it.

Suppose you give a 5-question multiple-choice quiz with 4 choices for each question. What is the probability that someone who randomly selects an answer from each set of choices will do well (that is, will get at least 4 out of 5 questions right)?

Example 13.4 shows how to solve part of this problem. You can complete the solution in the exercise that follows Example 13.4. You will need to use the Fundamental Counting Principle on various sets of test responses and then *add* together the number of ways of doing each set of responses.

EXAMPLE 13.4

A quiz has 5 multiple-choice questions. Each question has 4 answer choices, of which 1 is the correct answer and the other 3 are incorrect. Suppose that you guess all the answers.

(a) How many ways are there to answer the 5 questions?

(b) What is the probability of getting all 5 questions right?

(c) What is the probability of getting exactly 4 questions right and 1 wrong?

(d) What is the probability of doing well (getting at least 4 right)?

Solution

(a) In how many ways can you answer the 5 questions? Consider each question as a step in completing a test. There are 4 ways to do each step. Using the Fundamental Counting Principle,

$$(4)(4)(4)(4)(4) = 4^5 = 1024$$

(b) How many ways are there to get all 5 questions right? There is 1 way to answer each question correctly. Using the Fundamental Counting Principle, $(1)(1)(1)(1)(1) = 1$. There is 1 way to answer all 5 questions correctly out of 1024 possibilities. So $P(\text{all 5 right}) = \frac{1}{1024}$.

(c) There is more than one way to get 4 right and 1 wrong, depending upon *which question you get wrong*. In the following table, which lists all possible responses that involve at least 4 right answers, R stands for a right answer and W stands for a wrong answer.

Each type of response (for example, *WRRRR*) can occur in 3 ways based upon the Fundamental Counting Principle. Since each of the 5 sets of responses is a complete test result, we add the five 3's together.

Five Responses	**Number of Ways to Fill Out the Test**
WRRRR	(3)(1)(1)(1)(1) = 3
RWRRR	(1)(3)(1)(1)(1) = 3
RRWRR	(1)(1)(3)(1)(1) = 3
RRRWR	(1)(1)(1)(3)(1) = 3
RRRRW	(1)(1)(1)(1)(3) = 3
	15 ways

So there are 15 ways out of the 1024 possible ways that result in 4 right answers and 1 wrong answer.

$$P(\text{4 right, 1 wrong}) = \frac{15}{1024} \approx 1.5\%$$

The chances of getting exactly 4 right by guessing are not very good. It makes more sense to study.

(d) "At least 4 right" means you can get either 4 right and 1 wrong or all 5 right. We add the probabilities (of mutually exclusive events).

$$P(\text{at least 4 right}) = P(\text{4 right, 1 wrong}) + P(\text{5 right})$$
$$= \frac{15}{1024} + \frac{1}{1024} = \frac{16}{1024} \approx 0.016$$

■

People who construct multiple-choice tests know the probabilities that someone will get various numbers of questions right by guessing randomly. In order to decrease the probability of someone's doing well just by guessing, test constructors use a lot more than 5 questions.

Lesson Exercise 13.33

A true-false test has 6 questions.

(a) How many ways are there to answer the 6-question test?

(b) What is the probability of getting at least 5 right by guessing the answers at random?

The preceding example and lesson exercise both required multiplication *and* addition. Multiply numbers that represent the number of ways to do *each step,* such as answering each question on an exam. Add numbers that represent *complete results,* such as how many ways to answer 5 right on the whole test and how many ways to answer 6 right on the whole test.

PERMUTATIONS AND COMBINATIONS

In how many different ways can 8 horses finish in a race (assuming there are no ties)? Using the Fundamental Counting Principle, the number is

$$(8)(7)(6)(5)(4)(3)(2)(1) = 40{,}320$$

This is another example of an ordered arrangement called a permutation. Products such as (8)(7)(6)(5)(4)(3)(2)(1) can be written in a shorthand notation called factorial. In this case, $8 \cdot 7 \cdot 6 \cdot 5 \cdot 4 \cdot 3 \cdot 2 \cdot 1 = 8!$ (read "8 factorial").

Factorial Notation

$n!$ is called ***n* factorial**, and $n! = n(n-1)(n-2)\ldots 3 \cdot 2 \cdot 1$, in which n is a positive integer. By definition, $0! = 1$.

The horse-race example illustrates that the number of permutations of n objects is $n!$.

Lesson Exercise 13.34

(a) Compute 5! by hand.

(b) If your calculator has a factorial key, compute 5! on it.

Example 13.3 showed that the number of ways to pick ordered arrangements of 2 from a group of 10 is $10 \cdot 9 = 90$. By rewriting $10 \cdot 9 = \dfrac{10 \cdot 9 \cdot 8 \cdot 7 \cdot 6 \cdot 5 \cdot 4 \cdot 3 \cdot 2 \cdot 1}{8 \cdot 7 \cdot 6 \cdot 5 \cdot 4 \cdot 3 \cdot 2 \cdot 1} = \dfrac{10!}{8!} = \dfrac{10!}{(10-2)!}$, we obtain a formula for permutations. Notice that the formula can be written using only the 2 (the number of selections) and the 10 (the total number) from the problem.

Permutation Formula

If n objects are chosen r at a time, the number of permutations (ordered arrangements) is ${}_nP_r = \frac{n!}{(n-r)!}$

Try this formula out on the permutation problem you solved earlier in the lesson.

Lesson Exercise 13.35

(a) Solve Lesson Exercise 13.32 using the permutation formula.

(b) Find out the easiest way to do permutation computations on your calculator.

Consider a similar but different question.

EXAMPLE 13.5

Suppose a principal wants to select 2 people for a committee from her faculty of 30 people. In how many ways can she do this?

Solution

There won't be as many possibilities as in Lesson Exercise 13.32, because choosing teachers named Mr. Alonso and Ms. Brunett is the same as choosing Ms. Brunett and Mr. Alonso. These are *not ordered arrangements*. Now $30 \cdot 29 = 870$ counts each pair in both orders. Each two-person choice can be arranged in 2! or 2 ways. To obtain the correct result, compute

$$\frac{{}_{30}P_2}{{}_2P_2} = \frac{{}_{30}P_2}{2!} = \frac{30 \cdot 29}{2} = 435.$$

■

Example 13.5 illustrates a combination. A **combination** is a group of items in which the order does not make a difference. In the example,

$${}_{30}C_2 = \frac{{}_{30}P_2}{2!} = \frac{30!}{2!(30-2)!}$$

Notice how easily you can cancel the factorials if you are not using a calculator.

$$\frac{30!}{2!28!} = \frac{30 \cdot 29 \cdot 28 \cdot 27 \cdot \ldots \cdot 2 \cdot 1}{2!28 \cdot 27 \cdot \ldots \cdot 2 \cdot 1} = \frac{30 \cdot 29}{2!} = \frac{30 \cdot 29}{2} = 15 \cdot 29 = 435$$

Lesson Exercise 13.36

Suppose you obtained the result $\frac{12!}{9!3!}$ in a combination problem. Simplify it without a calculator. (*Hint:* Cancel the factors in the largest factorials first.)

The formula for combinations is as follows.

Combination

The number of ways of selecting a subset of r objects (order does not matter) from a set of n objects is

$$_nC_r = \frac{n!}{r!(n - r)!}$$

Apply this formula to the following exercise, noting that the order of the arrangements does not matter.

Lesson Exercise 13.37

(a) A class has 10 members. In how many ways can they choose 2 people to attend a national meeting?

(b) Find out the easiest way to solve this problem with your calculator.

As you can tell, permutations and combinations are quite similar. Both involve choosing r objects from n objects. See if you can distinguish between them. In order to do so, make up an arrangement you might choose and see if the order matters. If the order does matter, it's a permutation. If not, it's a combination.

Lesson Exercise 13.38

In each part, tell whether the problem concerns a permutation or a combination. Then find the answer using the appropriate formula.

(a) A college wants to hire a mathematics tutor, a calculus teacher, and a statistics teacher from a group of 10 applicants, each of whom is qualified for all of the jobs. In how many ways can the college fill the 3 positions?

(b) How many different 4-member committees can be formed from 100 U.S. senators?

(c) A meeting is to be addressed by 5 speakers: Al, Baraka, Clyde, Dana, and Elvira. In how many ways can the speakers be ordered? (*Hint:* $0! = 1$.)

ANSWERS TO SELECTED LESSON EXERCISES

13.27 (a) 8 (b) Continue reading.

13.28 Multiply the number of possible main dishes by the number of desserts.

13.29 5

13.30 $26^2 \cdot 10^4 = 6{,}760{,}000$

13.31 $8^2 \cdot 10^5 = 6{,}400{,}000$

13.32 $30 \cdot 29 = 870$

13.33 (a) $2^6 = 64$ (b) $\frac{6}{64} + \frac{1}{64} = \frac{7}{64}$

13.34 (a) 120

13.35 (a) $\frac{30!}{28!} = 30 \cdot 29 = 870$

13.36 $\frac{12!}{9!3!} = \frac{12 \cdot 11 \cdot 10}{3!} = 2 \cdot 11 \cdot 10 = 220$

13.37 (a) ${}_{10}C_2 = 45$

13.38 (a) Permutation; ${}_{10}P_3 = 720$
(b) Combination; ${}_{100}C_4 = 3{,}921{,}225$
(c) Permutation; ${}_5P_5 = 120$

13.3 HOMEWORK EXERCISES

Basic Exercises

1. Jane has 4 shirts and 2 skirts that match.
   (a) How many different outfits can she make?
   (b) Draw a tree diagram to support your answer.
2. A lottery allows you to select a two-digit number. Each digit may be either 1, 2, 3, 4, or 5. How many different numbers can be selected?
3. The Grain Barn is known for its healthy 3-course dinner consisting of appetizer, entree, and vegetable. How many different dinners are possible if you choose 1 item for each course?

DINNER AT THE GRAIN BARN

Sponge bread
Yogurt gumbo
Buckwheat balls

Twice-baked kelp
Oat bran surprise
Flax hash
Marinated bulgur

Scalloped kale
Lima bean puree

4. A car dealer offers the National Motors Nag in 5 colors with a choice of 2 doors or 4 doors and a choice of 2 different engines. How many different possible models are there?

5.

PICK 4 NUMBERS GAME

"STRAIGHT" bet is when your number matches the winning number selected in *exact order.* "BOX" bet is when your number matches the winning number selected in *any order.*

DESCRIPTIONS OF POSSIBLE BETS.	EXAMPLE if you bet:	YOU ARE A WINNER if any of these combinations are selected (Example)
STRAIGHT. *To win, the four digit number you select must match in* exact order *the number drawn on TV. 1 way to win.* (ODDS 1 IN 10,000)	1234	1234 ONLY EXACT MATCH WINS.
4 WAY BOX. *Of the four digits you select, three are the same. To win, all four digits of your number must match the winning number in* any order. *4 ways to win.* (ODDS 1 IN 2,500)	1112	1112 1121 1211 2111

Explain how the probabilities (incorrectly called "odds") for the 2 types of payoffs are obtained.

6. In theory, a monkey randomly selecting keys on a typewriter would eventually type some English words. If a monkey is at a keyboard with 48 keys, what is the probability that 5 random keystrokes will produce the word "lucky"?

7. A club has 15 members. In how many different ways can they select a president and a vice-president?

8. If 6 horses are entered in a race and there can be no ties, how many different orders of finish are there?

9. My wife's new bicycle lock is a combination lock with 5 dials, each numbered 1 to 9. If we forget the combination, how many possible combinations are there to try?

10. Until 1995, area codes could not have a first digit of 0 or 1, and the second digit had to be a 0 or a 1. The third digit could be any number.
    (a) How many possible area codes were there? (In fact, some codes, such as 411 and 911, are used for other purposes.)
    (b) In 1996, there were about 140 different area codes in the United States and Canada. Were the countries close to running out of possible area codes under the old system?
    (c) Starting in 1995, the second digit was no longer restricted. How many possible area codes are there now?

11. A state license plate consists of 1 letter followed by 5 digits.
    (a) How many different license plates can be made?
    (b) If no digit could be repeated, how many possible license plates would there be?

12. A true-false test has 4 questions. What is the probability of getting at least 3 right by guessing the answers randomly?

13. A quiz has 4 multiple-choice questions. Each question has 5 choices, 1 being correct and the other 4 being incorrect. Suppose you guess on all 4 questions.
    (a) How many ways are there to answer the 4 questions?
    (b) If someone got "at least 3 right," exactly how many questions could he or she have gotten right?
    (c) What is P(at least 3 right)?

14. (a) In the preceding exercise, how many ways are there to get no questions right?
    (b) What is P(none right)?

15. Compute the following by hand, and check the answers on a calculator.
    (a) $6!$ (b) $\frac{9!}{7!}$ (c) $\frac{11!}{4!7!}$

16. Compute the following by hand, and check the answers on a calculator.
    (a) $7!$ (b) $\frac{40!}{38!}$ (c) $\frac{50!}{48!2!}$

17. A fifth-grade class of 30 students wants to elect a president, vice-president, treasurer, and secretary. How many different ways are there to fill the 4 positions?

18. A photographer wants to arrange 3 children in a row for a family photograph. In how many ways can this be done?

19. How many possible 5-card poker hands are there? (*Note:* A regular deck has 52 cards.)

20. Suppose a test allows a student to pick which questions to answer. In how many different ways can a student choose a set of 8 questions to answer from a group of 10 questions?

21. When do we use permutations rather than combinations in counting?

22. Which is usually greater, the number of combinations of a set of objects or the number of permutations?

23. A consumer group plans to select 2 televisions from a shipment of 8 to check the picture quality. In how many ways can they choose 2 televisions?

24. A baseball manager wants to arrange 9 starting players into a batting order. How many different batting orders are possible?

25. How many ways are there to rank (as first and second) the 2 greatest U.S. presidents in history from a list of 12 choices?

26. In a state lottery game, you choose 6 different numbers from the set of counting numbers 1 to 50 (inclusive). The order does not matter. How many different choices can you make?

27. Substitute $r = n$ in the formula ${}_nP_r$ and simplify it to obtain the formula for ${}_nP_n$.

Extension Exercises

28. At Risk High, the football coach also teaches probability. He knows that when the opposition calls the coin flip, they call heads 80% of the time. Since heads comes up only 50% of the time, he always lets the visiting team call the coin flip. Should he expect to come out ahead this way?

29. A slot machine has the following symbols on its 3 dials. Each dial is spun independently at random and lands on 1 of 20 possibilities.

Dial 1	Dial 2	Dial 3
1 bar 1 bell 2 cherries 7 lemons 8 oranges 1 plum	1 bar 2 bells 2 cherries 1 lemon 6 oranges 8 plums	1 bar 2 bells 1 cherry 7 lemons 4 oranges 5 plums

The payoff for each nickel is as follows.

Payoffs

1 cherry	$.10	3 bells	$5
2 cherries	$.50	3 bars	$10
3 cherries	$1		

(a) How many possible outcomes are there for the 3 dials taken together?
(b) How many ways are there to get 3 cherries (1 on each dial)?
(c) How many ways are there to get 3 bars?
(d) How many ways are there to get 3 bells?
(e) Give the probabilities of the events in parts (b), (c), and (d).
(f) How many ways are there to spin exactly 1 cherry?
(g) How many ways are there to spin exactly 2 cherries?
(h) Give the probabilities of the events in parts (f) and (g).

30. Photocopy the 3 characters shown here, and cut each one into 3 separate sections (head, midsection, legs).
(a) Construct some new characters using the pieces.
(b) How many possible characters could you construct?

31. You plan to visit London, Paris, Avignon, and Nice on a trip. You can travel from London to Paris by airplane or by the so-called "boat train." You can travel from Paris to Avignon and Avignon to Nice by car, bus, or train. However, if you rent a car, you will use it for both journeys. Otherwise, you will ride buses or trains between Paris, Avignon, and Nice. How many possible travel arrangements are there for the trip from London to Paris to Avignon to Nice?

32. A quiz has 5 multiple-choice questions. Each question has 4 choices; 1 is the correct answer and the other 3 are incorrect. Suppose that you guess all the answers.
(a) P(at least 1 right) $= 1 - P($________$)$.
(b) Use the equation in part (a) to find the probability of getting at least 1 right.

33. Three couples are sitting together at a show. In how many ways can they sit without any of the couples being split up? Find out by answering the following.
(a) How many choices are there for the aisle seat?
(b) Once the aisle seat is filled, how many choices are there for the seat next to it?
(c) Now how many choices are there for the next seat in?
(d) Now, how many choices for the fourth seat?

(e) Now, how many choices for the fifth seat?
(f) Now, how many for the sixth seat?
(g) How many ways are there, then, to seat all 6 people?

34. Follow the method of the preceding question to find how many ways there are to seat two couples (male-female) at a concert if a woman sits on the aisle and the remaining people are allowed to sit in any possible arrangement.

35. Three couples go to a concert. In how many ways can they arrange themselves if all 3 men sit together and all 3 women sit together?

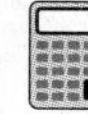

36. A survey asks you to list your first 3 choices for the Democratic presidential nomination from a list of 6 contenders and your first 3 choices for the Republican presidental nomination from a list of 5 contenders. In how many ways can the survey be completed? (Assume that you choose all of your responses from the lists.)

37. Consider a circular table.
(a) In how many ways can 3 people be arranged at a circular table?
(b) In how many ways can 4 people be arranged at a circular table?
(c) In how many ways can 5 people be arranged at a circular table?
(d) In how many ways can N people be arranged at a circular table?

13.4 EXPECTED VALUE AND ODDS

Have you ever played a lottery or gambled at a casino? Do you know how to compute how much you are going to *lose* in the long run? Do you know what the term "odds" means? This lesson introduces the mathematics of the expected value and odds.

In a casino, people risk their money and usually end up losing some of it. Although individual outcomes vary quite a bit, casinos consistently come out ahead at their games.

Many people are not interested in gambling, but the same mathematics applies to another area that does affect nearly everyone—insurance. In buying insurance, we also risk our money and usually end up losing some.

FAIR VS. UNFAIR GAMES

Would you play a gambling game in which you roll 2 dice and you win the amount you bet whenever the sum is greater than 10? This is an example of an unfair game.

In a **fair game** involving money, a player can expect to come out even in the long run. In a fair game between two players, neither player has an advantage due to the rules.

Lesson Exercise 13.39

In an odd-even dice game, you roll 2 dice. If the product is odd, you win. If it is even, your opponent wins. Is this a fair game? If not, whom does it favor?

Lesson Exercise 13.40

Make up a fair game for 2 players that involves rolling 2 dice.

EXPECTED VALUE

Casino games are not fair games. If they were, casino owners would not make any money. Consider the simple model of a casino game in the following example.

EXAMPLE 13.6

Suppose that you pick 1 card at random from the box shown, and you win the amount of money printed on the card. The cards are replaced and reshuffled for each new game.

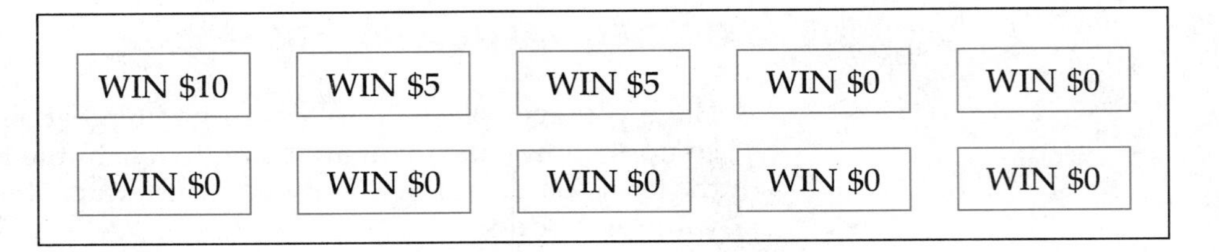

If you played this game often, how much would you expect to be paid on the average?

Solution

In the long run, you would expect to pick each of the 10 cards with about the same frequency. You would win about

$$\$10 + \$5 + \$5 + \$0 + \$0 + \$0 + \$0 + \$0 + \$0 + \$0 = \$20$$

every 10 draws. The expected winnings per draw would be $\frac{\$20}{10} = \$2.$ ■

In Example 13.6, the expected average result in the long run was computed. This result is called the **expected value**.

Lesson Exercise 13.41

Consider a lottery game in which 7 out of 10 people lose, 1 out of 10 wins $25, 1 out of 10 wins $10, and 1 out of 10 wins $5. This can be modeled by selecting 1 card at random from the group shown.

(a) What is the expected (average) value of the payoff?

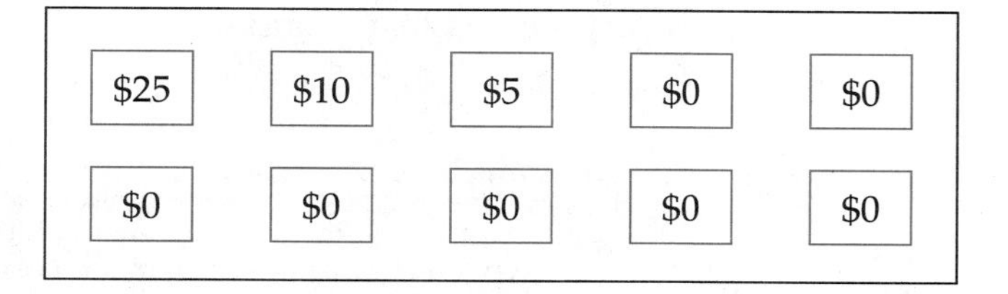

(b) Complete the following table, showing the outcomes.

Payoff	Frequency
$25	1
10	
5	
0	

(c) Compute the average from the frequency table.

$$\text{Expected average} = \frac{25 \cdot 1 + 10 \cdot 1 + \underline{\qquad\qquad} + \underline{\qquad\qquad}}{10} = \underline{\qquad\qquad}$$

The first state lottery, held in New Hampshire in 1964, created a lot of controversy. Now about 36 states use lotteries to raise money for state programs, as an alternative to raising taxes. Is a gambling game a good way to raise money? Many people in the states without lotteries don't think so.

State officials in lottery states need to know how much money they can expect to raise in the long run—the expected value. People playing the lottery are also interested in the expected value of their lottery tickets (Figure 13-11).

Figure 13–11

The next exercise develops a more widely applicable method for computing expected value in gambling games. This method uses the probability of each possible financial outcome.

Lesson Exercise 13.42

Expected values are usually computed from probability tables.

(a) Complete the probability table for the lottery game described in the preceding exercise.

Payoff	Probability
\$25	$\frac{1}{10}$
10	
5	
0	

(b) Compute the expected payoff from the probability table.

$$\text{Expected payoff} = 25 \cdot \frac{1}{10} + 10 \cdot \underline{\hspace{2em}} + \underline{\hspace{2em}} + \underline{\hspace{2em}} = \underline{\hspace{2em}}$$

(c) If the game costs \$5 to play, how much should you expect to lose, on the average, per play?

The formula used in part (b) of the preceding exercise is as follows.

Definition **Expected Value**

If an experiment has the possible numerical outcomes $n_1, n_2, n_3, \ldots, n_r$ with corresponding probabilities $p_1, p_2, p_3, \ldots, p_r$, then the **expected value** E of the experiment is

$$E = n_1 p_1 + n_2 p_2 + n_3 p_3 + \cdots + n_r p_r$$

A simpler example makes it clear why we multiply each payoff times its probability to compute the expected value. Suppose a coin-flipping game pays

\$10 for a head and nothing for a tail. What is the average payoff? You would (expect to) be paid about $\frac{1}{2}$ the time.

$$\text{Average payoff} = (\$10)\left(\frac{1}{2}\right) = \$5$$

This is the payoff times its probability.

Use the expected-value formula in the following exercise.

Lesson Exercise 13.43

Suppose a lottery game allows you to select a 2-digit number. Each digit may be either 1, 2, 3, 4, or 5. If you pick the winning number, you win \$10. Otherwise, you win nothing.

(a) What is the probability that you will pick the winning number?

(b) What is $E(\text{winning})$?

(c) If the lottery ticket costs \$1, how much should you expect to lose on the average (per play)?

Insurance companies also use expected values. In order to determine their rates and payoffs, they estimate the probabilities of particular catastrophes.

EXAMPLE 13.7

An insurance company will insure your dorm room against theft for a semester. Suppose the value of your possessions is \$800. The probability of your being robbed of \$400 worth of goods during a semester is $\frac{1}{100}$, and the probability of your being robbed of \$800 worth of goods is $\frac{1}{400}$. Assume that these are the only possible kinds of robberies. How much should the insurance company charge people like you to cover the money they pay out and to make an additional \$20 profit per person on the average?

Solution

Understanding the Problem The insurance company wants \$20 per person to be left over after paying out all the claims.

Devising a Plan Compute the expected payout for the insurance. Then, to make \$20 profit, charge the expected payout plus an additional \$20.

Carrying Out the Plan

$$E(\text{payout}) = (\$400) \cdot \frac{1}{100} + (\$800) \cdot \frac{1}{400} = \$4 + \$2 = \$6$$

They should charge $6 + $20 = $26 for the policy in order to make an average gain of $20 per policy.

Looking Back The answer seems reasonable. ■

Lesson Exercise 13.44

Consider the following problem. "In a carnival game, you toss a single die. If you roll a 3, you win $3. If you roll a 6, you win $6. Otherwise you win nothing. How much should the carnival operators charge you in order to have an expected gain of $.50 per game?" Devise a plan and solve the problem.

ODDS

In gambling games, probabilities are often stated using odds. Odds compare your chances of losing and winning on each play. For example, on the race card in the following table, the odds against Turf Tortoise winning are 12–1 (12 to 1). This expression means that Turf Tortoise is expected to lose about 12 times for every 1 time he wins. Whereas a probability compares your chance of losing or winning to the *total number of possibilities*, odds compare your possibilities of losing and winning to *each other*.

Actually, the real chance of Turf Tortoise winning is slightly worse than 12 to 1. The track pays off bettors based upon the odds *after* taking out about 20% of the money for expenses. For this reason, the track adjusts the odds by a factor of about 20%.

Post Time 1 P.M.

Race 1: 6 Furlongs 3YO; CLM $5000

Hot Air	6–5
Dog Lover	3–1
Wet Blanket	4–1
Lost Marbles	5–1
Lead Balloon	7–1
Turf Tortoise	12–1
Obstacle	20–1

Odds (against) can be defined as follows.

> ***Definition*** **Odds of Experiments with Equally Likely Outcomes**
>
> Odds against = number of unfavorable outcomes to number of favorable outcomes

Lesson Exercise 13.45

In the table just presented, the odds against Wet Blanket's winning are 4–1 (4 to 1). What does this statement mean?

Probabilities describe the frequency of a (favorable) result in relation to all possible outcomes. "Odds against" compare unfavorable and favorable results, such as losing and winning. Any probability can be converted to odds, and any odds can be converted to a probability.

EXAMPLE 13.8

What is the probability that the director whose door is shown in Figure 13-12 is in his office at 1 P.M.?

Solution

Odds of "25 to 1 against" mean that the director will not be there 25 times for each 1 time the director is there. The probability of the director's being there is 1 out of 26, or $\frac{1}{26}$.

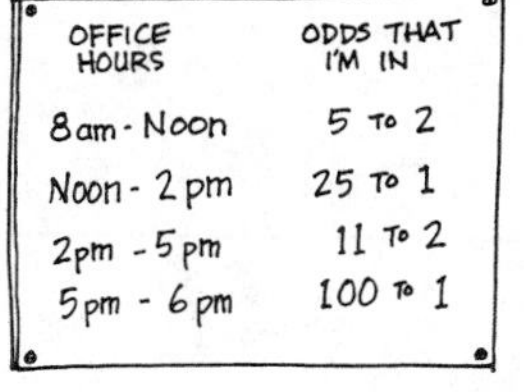

DIRECTOR OF PROBABILITY RESEARCH
OFFICE HOURS
ODDS THAT I'M IN
8am - Noon
5 TO 2
Noon - 2pm
25 TO 1
2pm - 5pm
11 TO 2
5pm - 6pm
100 TO 1

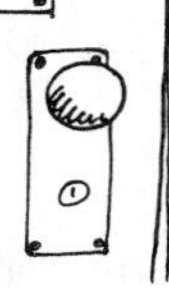

Figure 13–12

The preceding example illustrates the relationship between the odds against E and the probability of E. You can compute the odds against E using the formula $\frac{P(\overline{E})}{P(E)}$.

Lesson Exercise 13.46

The odds against Dead Weight winning the Crabgrass Derby today are 7 to 1. Express the probability of winning using a fraction, a decimal, and a percent.

Lesson Exercise 13.47

A roulette wheel has 38 numbers. If I bet on 1 number, what are the odds against my winning?

The odds are the basis for payoffs on bets. If you were paid off on the basis of the *true* odds, the game would be fair. You would tend to come out even in the long run. For example, in the preceding lesson exercise, suppose you won \$37 for each \$1 bet (from odds of 37 to 1) when you picked the winning number. The expected value could be computed from the following chart.

Gain	Probability
37	$\frac{1}{38}$
−\$1	$\frac{37}{38}$

$$E(\text{Gain}) = \$37\left(\frac{1}{38}\right) + (-\$1)\left(\frac{37}{38}\right) = \$0$$

When the expected value is \$0, it is a fair bet, since you would not expect to win or lose money in the long run.

Since casinos want to make money, they pay out the money at a slightly lower rate. If you made this bet in roulette, you would actually gain \$35, rather than \$37, on a \$1 bet.

ANSWERS TO SELECTED LESSON EXERCISES

13.39 Not fair; favors your opponent

13.41 (a) \$4
(b) Other frequencies are 1, 1, and 7
(c) $5 \cdot 1$; $0 \cdot 7$; \$4

13.42 (a) Other probabilities are $\frac{1}{10}$, $\frac{1}{10}$, and $\frac{7}{10}$.
(b) $\frac{1}{10}$; $5 \cdot \frac{1}{10}$; $0 \cdot \frac{7}{10}$; \$4
(c) \$1

13.43 (a) $\frac{1}{25}$ (b) \$.40 (c) \$.60

13.44 $E(\text{Payoff}) = \frac{1}{6}(\$3) + \frac{1}{6}(\$6) = \$1.50$
\$.50 + \$1.50 = \$2

13.45 If the odds are accurate, Wet Blanket can be expected to lose 4 times for every 1 time he wins under these conditions.

13.46 $\frac{1}{8}$, 0.125, 12.5%

13.47 37 to 1

13.4 HOMEWORK EXERCISES

Basic Exercises

1. In a dice game, you roll 2 dice. If the sum is divisible by 3 or 5, you win. Otherwise, your opponent wins. Whom does this game favor?

2. Make up a fair game for 2 players that involves flipping 2 coins.

3. A box contains the cards shown. You randomly select 1 card and win the amount written on that card. What is the expected (average) value of the payoff for a single draw?

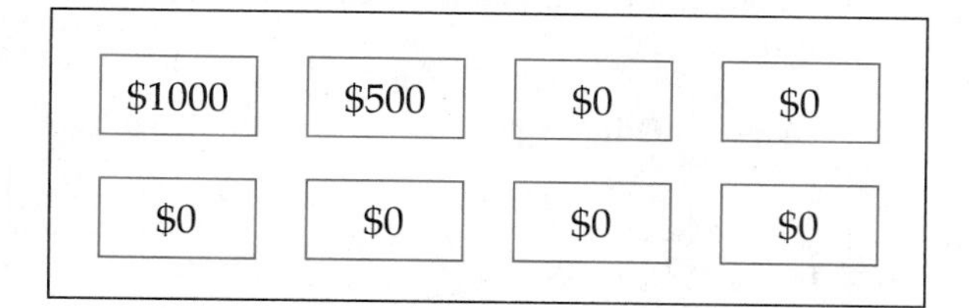

4. A lottery game allows you to select a 3-digit number using the digits 0, 1, 2, and 3. You may use the same digit more than once. If you pick the winning number, you win $25 on a $1 bet.
   (a) What is the average payoff?
   (b) How much would you expect to lose, on the average, per $1 bet?

5. Recall that a roulette wheel has 38 slots. Eighteen are red, 18 are black, and 2 are green.
   (a) What is the probability of black?
   (b) If you bet $1 on black, you receive a $2 payoff if the result is black and $0 if the result is red or green. What is the average loss on a $1 bet on black?
   (c) You can also bet on 6 different numbers. If any of them comes up, you receive $6 back for each $1 bet. What is the expected gain or loss on a $1 bet?

6. A particular oil well costs $40,000 to drill. The probability of striking oil is 1/20.
   (a) If an oil strike is worth $500,000, what is the expected gain from drilling?
   (b) Would you go ahead and drill?

7. (a) The Unshakeable Insurance Company will insure your dorm room against theft. The insurance costs $25 per year. If the probability of getting robbed of an average of $500 worth of possessions is 1/100, how much profit does the insurance company expect to make from your $25 insurance premium?
   (b) The company sells $100,000 worth of automobile liability insurance that costs $50 per year. In the past, about 1 out of every 1000 people have collected on a claim. The average (mean) claim is $30,000. What is the expected payoff on your $50 investment?
   (c) Why might the insurance in part (b) be worth the investment, whereas the insurance in part (a) is probably not worth it?

8. Madame Wamma Jamma will predict the sex of a future baby for $5. If she is wrong, she even has a money-back guarantee! If Madame Wamma Jamma has no psychic power (Perish the thought!), what is her expected gain per customer? Assume that she cheerfully refunds the $5 whenever she is wrong.

9. An apartment complex has 20 air conditioners. Each summer, a certain number of them have to be replaced.

Number of Air Conditioners Replaced	Probability
0	0.21
1	0.32
2	0.18
3	0.11
4	0.11
5	0.07

   What is the expected number of air conditioners that will be replaced in the summer?

10. The probability that a particular operation is successful is 0.32.
    (a) If the operation is performed 600 times a year at a hospital, about how many successful operations can they expect?

(b) If you decide to have this operation, give some reasons why the probability of its success for you might be somewhat higher or lower than 0.32.

11. A set of 3 coins is tossed 32 times. How many times would you expect the result to be 3 heads?

12. You run a check-cashing service. About 1 out of every 500 checks is bad. The average (mean) bad check is for $100. How much of a charge per person would offset the cost of the bad checks themselves? (This does not include other costs related to bad checks.)

13. State Penn Insurance Company will insure your apartment against theft for 1 year. The value of your insured possessions is $4000. In the past, about 1 out of every 100 people has collected on a claim. The average (mean) claim is $1200. How much should the insurance company charge you in order to cover payouts and make an average gain of $20 per policy?

14. A carnival game costs $2 to play. You roll 2 dice. If the sum is 5, you receive a $5 payoff. If the sum is 10, you receive a $10 payoff. What is the expected payoff, *E*($2 bet)?

15. In a gambling game, you pick 1 card from a standard deck. If you pick an ace, you win $10. If you pick a picture card (J, Q, or K), you win $5. Otherwise, you win nothing. How much should a carnival booth charge you to play this game if they want an average profit of $.40 per game?

16. *T* tickets are sold for $1 each for a lottery drawing with a prize of $100. What is the expected payoff of each ticket in terms of *T*?

17. In a game, the number you spin is the number of dollars you win. Which of the following spinners would be the best one to spin?

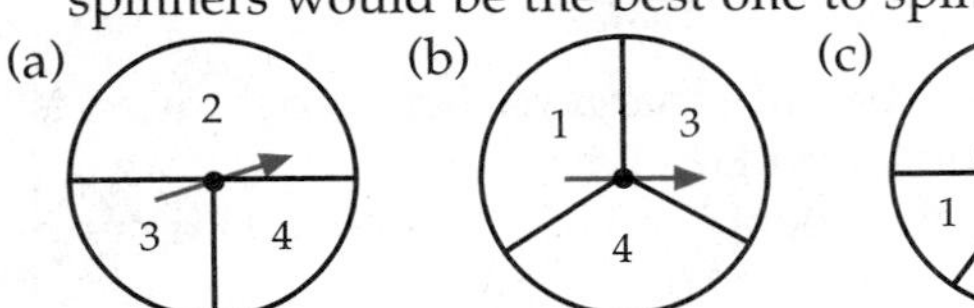

18. A roulette wheel has 38 slots, of which 18 are red, 18 black, and 2 green. If I bet on red, what are the odds against my winning?

19. Data indicate that about 3 out of 4 marriages last at least 20 years. The odds against a marriage lasting at least 20 years are ________.

20. Of the people who purchase handguns, 11 times as many use them on themselves or their families as use them on intruders. This means that about ________ of every ________ gun owners who use handguns will use it on an intruder.

21. Are you ready for a cheery problem? The following table gives the odds against dying in the next year.

Odds Against Dying in the Next Year

Age	*Male*	*Female*
15–24	576–1	1814–1
25–34	560–1	1489–1
35–44	384–1	743–1
45–54	140–1	269–1
55–64	57–1	111–1
65–74	25–1	49–1
75–84	11–1	17–1
85–up	6–1	7–1

What is the probability that you will live another year?

Extension Exercises

22. Refer to the slot-machine homework exercise from the last section. Use the probabilities you found to compute the expected payoff on a $.05 bet.

23. A company wants to test its employees for a drug. The test is 98% accurate. If someone is using the drug, that person tests positive 98% of the time. If someone is not using the drug, that person tests negative 98% of the time.

(a) Suppose the company randomly tests all 10,000 of its employees. Also, assume that 50 people are actually using the drug. How many people who are not using the drug will test positive (false positive)?

(b) In part (a), what is the probability that someone who tests positive is actually using the drug?

(c) Suppose that, instead, the company tests only people who exhibit suspicious behavior. Suppose 100 people act suspiciously, and 30 of them are using the drug. How many people who are not using the drug will test positive (false positive)?

(d) In part (c), what is the probability that someone who tests positive is actually using the drug?

24. A company wants to test its employees for a drug. The test is 95% accurate.

(a) Suppose that the company randomly tests all 5000 of its employees. Furthermore, assume that 100 people are actually using the drug. How many people who are not using the drug will test positive (false positive)?

(b) In part (a), what is the probability that someone who tests positive is actually using the drug?

(c) Suppose that, instead, the company tests only people who exhibit suspicious behavior. Suppose 150 people act suspiciously and 80 of them are using the drug. How many people who are not using the drug will test positive (false positive)?

(d) In part (c), what is the probability that someone who tests positive is actually using the drug?

25. A patient has a stroke and must choose between brain surgery and drug therapy. Of 100 people who have the surgery, 20 die during surgery, 15 more die after 1 year, and 5 more die after 3 years. Of 100 people who receive drug therapy, 5 die almost immediately, 20 more die after 1 year, and 30 more die after 3 years. Which treatment would you advise someone to take? Why?

26. A lottery game allows you to select any 3-digit number. The first digit cannot be a 0. If you pick just the first 2 digits in the correct order, you win $50 for each $1 bet. If you pick the exact winning number, you win $250 for each $1 bet. What is the expected payoff of a $1 bet?

27. You are designing a spinner game for a carnival. You want to charge people $1 and estimate that 500 people will play. You would like to make about $100. Sketch a spinner, and give the rules and payoffs.

28. Read the Beetle Bailey cartoon, and answer the following questions.

(a) If Beetle lives another 50 years, how much money will he win (after taxes)?

(b) Assume that Beetle has a 1-in-5,000,000 chance of winning. What is the expected payoff for his mailing in the entry form?

(c) Is this payoff worth the price of a stamp?

Reprinted with special permission of King Feature Syndicate, Inc. © Features Syndicate, Inc., 1971. World rights reserved.

Special Exercise

29. If your state has a lottery, describe how it works. Include probabilities and expected values in your description.

13.5 INDEPENDENT AND DEPENDENT EVENTS

If you feel sick in the morning, it affects the probability that you will go to school or work. In the study of probability, it is important to know whether the outcome of one event affects the outcome of another.

Lesson Exercise 13.48

Consider the following events.

R = rain tomorrow

U = you carry an umbrella tomorrow

H = coin flipped tomorrow lands on heads

(a) Does the probability of R affect the probability of U?

(b) Does the outcome of the coin flip affect whether or not it will rain tomorrow?

In Lesson Exercise 13.48, R and H are independent events, whereas R and U are dependent events. Two events are **independent** if the probability of one remains the same regardless of how the other turns out. Events that are not independent are **dependent**.

Lesson Exercise 13.49

You roll a regular red die and a regular green die. Consider the following events.

A = a 4 on the red die

B = a 3 on the green die

C = a sum of 9 on the two dice

Tell whether each pair of events is independent or dependent.

(a) A and B (b) B and C

In the preceding exercise, the outcome of the red die is not affected by the outcome of the green die. The events A and B are totally separate (independent). The probability of B is the same no matter how A turns out. (Similarly, the probability of A is the same no matter how B turns out.)

INDEPENDENT EVENTS

What is the probability of two independent events occurring in succession? Try the following exercise.

Lesson Exercise 13.50

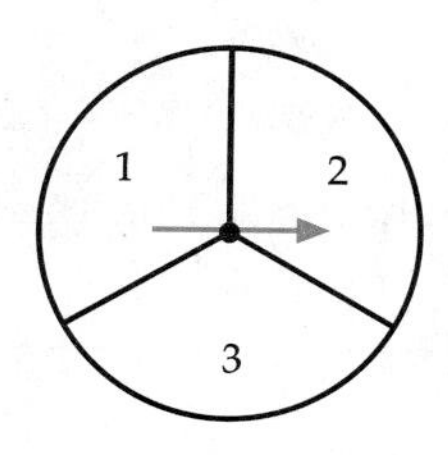

Figure 13–13

Suppose a spinner contains equal-sized regions numbered 1, 2, and 3 (Figure 13-13).

(a) In an experiment, you spin twice. Are the results of the two spins independent or dependent?

(b) Write a sample space of equally likely outcomes for the experiment. (Use an organized list or tree diagram.)

(c) Suppose A is the event that the first spin is a 1 and B is the event that the second spin is a 2. Compute $P(A)$, $P(B)$, and $P(A \text{ and } B)$.

(d) Suppose you get a 1 on the first spin $\frac{1}{3}$ of the time. Next, suppose that $\frac{1}{3}$ of those times that you get a 1 on the first spin, you also get a 2 on the second spin. The fraction of the time that you get a 1 on the first spin *and* a 2 on the second spin is $\frac{1}{3}$ of $\frac{1}{3}$ = ________ × ________ = ________.

(e) What formula relating $P(A)$, $P(B)$, and $P(A \text{ and } B)$ is suggested by parts (c) and (d)?

Another way to see why you can multiply probabilities to find $P(A \text{ and } B)$ is to look at the sample space:

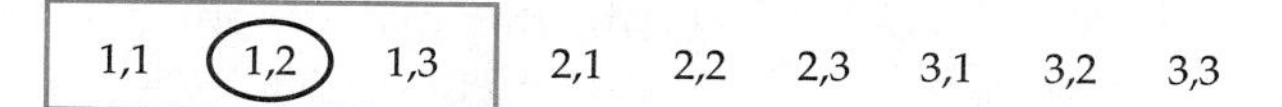

The boxed-in outcomes show that a 1 on the first spin occurs in $\frac{1}{3}$ of the equally likely outcomes. The circled outcome shows that $\frac{1}{3}$ of the boxed-in outcomes would result in a 1 followed by a 2. The circled outcome is $\frac{1}{3}$ of $\frac{1}{3} = \frac{1}{3} \times \frac{1}{3} = \frac{1}{9}$ of the total sample space. This spinner exercise is an example of the independent-events formula.

The Independent-Events Formula

If A and B are independent events, $P(A \text{ and } B) = P(A) \cdot P(B)$.

Apply this formula in the following exercise.

Lesson Exercise 13.51

A fair die is tossed twice. What is the probability of getting a 3 on the first toss followed by an odd number on the second toss?

You can find the probability of 3 or more independent events occurring in succession in the same way, by multiplying the probabilities of the individual events.

DEPENDENT EVENTS

Now investigate the probability of two dependent events occurring in succession.

Lesson Exercise 13.52

A box contains cards numbered 1, 2, and 3. You pick 2 cards in succession *without replacement*. ("Without replacement" means you do not put back a card after you pick it.)

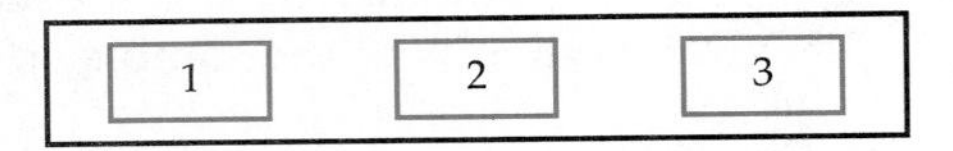

(a) In this experiment, are the first and second draws independent or dependent events?

(b) Write a sample space of equally likely outcomes for the experiment. (Use an organized list or a tree diagram.)

(c) Suppose A is the event that the first card is a 2 and B is the event that the second card is a 3. The $P(B \text{ given } A)$ means the probability of B, given that A has happened. Compute $P(A)$, $P(B \text{ given } A)$, and $P(A \text{ and } B)$.

(d) Suppose A occurs $\frac{1}{3}$ of the time. Next, suppose that $\frac{1}{2}$ of those times A occurs, B also occurs. The fraction of the time that A and B both occur is $\frac{1}{2}$ of $\frac{1}{3}$ = ________ × ________ = ________.

(e) What formula relating $P(A)$, $P(B \text{ given } A)$, and $P(A \text{ and } B)$ is suggested by parts (c) and (d)?

In Lesson Exercise 13.52(e), you may have guessed the formula for the Multiplication Rule for Probabilities.

Multiplication Rule for Probabilities

$P(A \text{ and } B) = P(A) \cdot P(B \text{ given } A)$

Use this formula in the next exercise.

Lesson Exercise 13.53

You have a drawer with 10 black socks and 6 white socks. If you select 2 socks at random, what is the probability of picking 2 white socks? (*Hint:* What is the probability that the first sock will be white?)

What is the connection between the independent-events formula and the Multiplication Rule for Probabilities?

Lesson Exercise 13.54

The independent-events formula is $P(A \text{ and } B) = P(A) \cdot P(B)$. The Multiplication Rule for Probabilities is $P(A \text{ and } B) = P(A) \cdot P(B \text{ given } A)$.

(a) What is the difference between the two formulas?

(b) If A and B are independent events, why would $P(B \text{ given } A) = P(B)$?

The preceding lesson exercise shows that the independent-events formula is a special case of the Multiplication Rule for Probabilities in which $P(B \text{ given } A)$ is simplified to $P(B)$.

Now see whether you can find an experimental and a theoretical probability for the following situation.

Lesson Exercise 13.55

You have 6 black socks and 4 white socks in a drawer. You pick 2 socks at random.

(a) Without computing probabilities, guess the probability of picking a matching pair.

(b) Simulate the experiment 30 times, using 2 colors of chips and a bag. What is your experimental probability for matching a pair of socks?

(c) Find the theoretical probability of picking a matching pair.

ANSWERS TO SELECTED LESSON EXERCISES

13.48 (a) Yes (b) No

13.49 (a) Independent
(b) Dependent, since a 3 on the green die makes it more likely $\left(\frac{1}{6}\right)$ that you will get a sum of 9.

13.50 (a) Independent
(b) First draw Second draw Outcome

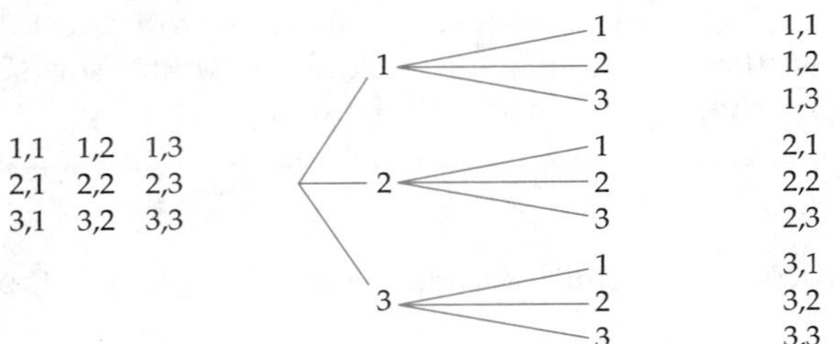

Organized list **Tree diagram**

(c) $P(A) = \frac{1}{3}$, $P(B) = \frac{1}{3}$, $P(A \text{ and } B) = \frac{1}{9}$
(d) $\frac{1}{3} \cdot \frac{1}{3} = \frac{1}{9}$

13.51 $\frac{1}{6} \cdot \frac{3}{6} = \frac{1}{12}$

13.52 (a) Dependent
(b)

1,2 1,3
2,1 2,3
3,1 3,2

First draw	Second draw	Outcome
1	2	1,2
	3	1,3
2	1	2,1
	3	2,3
3	1	3,1
	2	3,2

Organized list **Tree diagram**

(c) $P(A) = \frac{1}{3}$, $P(B \text{ given } A) = \frac{1}{2}$, $P(A \text{ and } B) = \frac{1}{6}$
(d) $\frac{1}{2}; \frac{1}{3}; \frac{1}{6}$

13.53 $\frac{6}{16} \cdot \frac{5}{15} = \frac{1}{8}$

13.55 (c) $\frac{4}{10} \cdot \frac{3}{9} + \frac{6}{10} \cdot \frac{5}{9} = \frac{7}{15}$

13.5 HOMEWORK EXERCISES

Basic Exercises

1. Consider the following events.

C = you eat cereal tomorrow
B = you wear blue shoes tomorrow
M = you drink milk tomorrow

(a) Are C and M independent or dependent events?
(b) Are M and B independent or dependent events?

2. A fair coin is tossed twice.

A = heads on the first toss
B = heads on the second toss
C = heads on both tosses

Tell whether each pair of events is independent or dependent.
(a) A and B (b) A and C (c) B and C

3. Suppose A = rolling a sum of 7 with two regular dice. Make up an event B so that
(a) A and B are independent.
(b) A and B are dependent.

4. Consider the following events.

A = college soccer team wins

B = college soccer team wins

Suppose the college baseball team wins $1/2$ of its games, and the college football team wins $1/3$ of its games. What fraction of the time would you expect both teams to win when they play on the same day?

5. A box contains cards numbered 1 and 2. You pick 2 cards in succession with replacement. ("With replacement" means that the number selected first is replaced in the box, and the cards are shuffled before the second draw is made.)

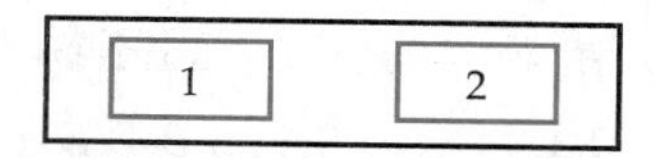

(a) Are the first and second draws independent or dependent events?

(b) Write a sample space of equally likely outcomes.

(c) Complete the tree diagram, showing all the possible outcomes.

First draw **Second draw** **Outcome**

(d) Suppose $A = 1$ on the first pick and $B = 2$ on the second pick. $P(A \text{ and } B) =$ ________.

(e) Show how to find the result of part (d) using a formula.

6. A game requires that you spin two different spinners.

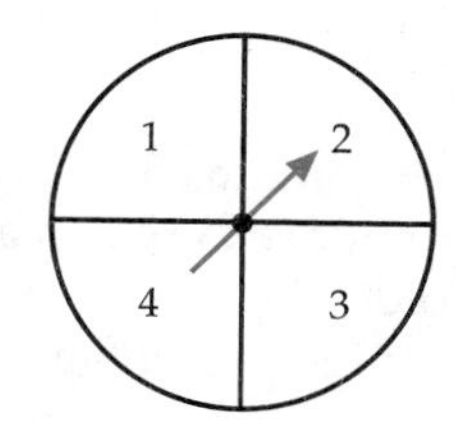

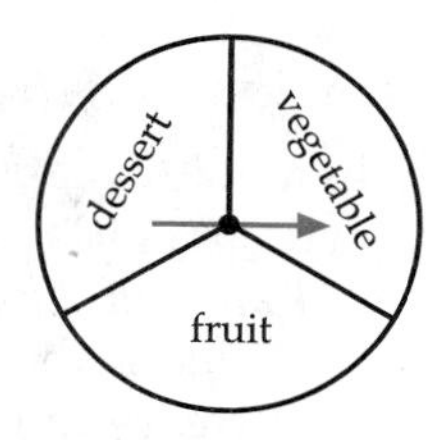

You win the number of food items that corresponds to what you spin.

(a) Write a sample space of equally likely outcomes for the experiment. (Use an organized list or a tree diagram.)

(b) What is the probability you will win 4 desserts?

(c) Show how to find the result of part (b) using a formula.

7. Two dice are rolled. What is the probability that both dice show a 5 or a 6?

8. Some drug tests are about 98% reliable. This means that there is a probability of 0.98 that the test will correctly identify a drug user or a nonuser. To be safe, each person is tested twice.

(a) What is the probability that a drug user will pass both tests?

(b) What is the probability that a nonuser will fail at least 1 of the tests?

9. A man has 2 cars, a Recall and a Sea Bass Brougham. The probability that the Recall starts is 0.1. The probability that the Sea Bass Brougham starts is 0.7. (Assume that the cars operate independently of each other.)

(a) What is the probability that both cars start?

(b) What is the probability that neither car starts?

(c) What is the probability that exactly 1 car starts?

***10.** A box contains cards numbered 1, 2, 3, and 4. You pick 2 cards in succession without replacement.

1	2	3	4

(a) In this experiment, are the first and second draws independent or dependent?

(b) Write a sample space of equally likely outcomes. (Use an organized list or a tree diagram.)

(c) Suppose $A = 3$ on the first pick and $B = 2$ on the second pick. Find $P(A)$, $P(B \text{ given } A)$, and $P(A \text{ and } B)$.

(d) Show how to find $P(A \text{ and } B)$ using a formula.

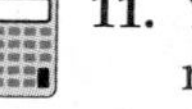

11. You pick 2 cards from a regular deck *without* replacement. What is the probability of picking 2 aces?

12. A drawer contains a mixture of 10 black socks, 8 white socks, and 4 red socks. You randomly select 2 socks to wear. Determine the probability that
(a) both are black.
(b) both are white.
(c) you pick a matching pair.

13. (a) When drawing at random with replacement, are the draws independent or dependent?
(b) Without replacement, are the draws independent or dependent?

14. Two draws are made from a standard deck of cards without replacement. Are the draws independent or dependent?

15. Assume that, for a nuclear power plant to have an accident, 4 systems must fail. The probabilities that the 4 individual systems fail are 0.01, 0.006, 0.002, and 0.002, respectively. What is the probability that the plant will have an accident?

16. Suppose that an international phone cable has 300 underwater amplifiers placed at regular intervals. Underwater amplifiers are very difficult to replace. If the probability that an amplifier lasts 20 years is 0.999998, what is the probability that all 300 amplifiers last 20 years? (Assume that all 300 amplifiers operate independently of one another.)

17. A slot machine has 3 independent wheels set up as follows.

Dial 1	
1 bar	2 cherries
1 bell	7 lemons
1 plum	8 oranges

Dial 2	
1 bar	2 bells
1 lemon	6 oranges
2 cherries	8 plums

Dial 3	
1 bar	4 oranges
1 cherry	5 plums
2 bells	7 lemons

What is the probability of getting each of the following?
(a) 3 lemons (b) 3 limes (c) 0 cherries

18. In order to repair his knee, a football player must successfully go through 2 operations. The probability that the first operation will be successful is 0.8. If the first operation is successful, the probability that the second operation will be successful is 0.7. What is the probability that both operations will be successful?

19. A group of mothers and their grown daughters were surveyed.

$$A = \text{mother attended college}$$

$$B = \text{daughter attended college}$$

$P(A) = 0.3$, $P(B) = 0.6$, and $P(B \text{ given } A) = 0.7$. What is the probability that a mother and her daughter both attended college?

20. On the average, it rains or snows about 114 days a year in Spokane, Washington.
(a) If you visit for a day, what is the probability that there will be precipitation (rain or snow)?
(b) If there is precipitation one day, the probability of precipitation the next day is 0.5. If you visit for 2 days, what is the probability that it will rain or snow both days?

21. A shipment of radios contains 97 that work and 3 that are defective. An inspector selects 5 at random. What is the probability that they all work?

Extension Exercises

22. In a mathematics class that contains 50% boys and 50% girls, 60% of the children are 8 years old and 40% are 9 years old.
(a) What is the largest possible percentage of 9-year-old girls in the class?
(b) What is the smallest possible percentage of 9-year-old girls in the class?
(c) If age and sex are independent, then ________% of the class is 9-year-old girls.

23.

Pair 1
H = a fair coin flip lands on heads today
T = your TV works today

Pair 2
B = you eat a breakfast today
L = you eat a big lunch today

(a) Tell whether or not each pair of events is independent. Explain your answers.
(b) Tell whether or not each pair of events is mutually exclusive. Explain your answers.

24. Consider the following problem. "A football betting service picks 1 game each week. They charge $100 per pick, but only if they pick the game correctly. The service starts with 200 customers. They tell 100 customers to bet on one team and 100 customers to bet on the other team. Assume that this continues for 10 weeks, with the following results. Each week, only the customers who win continue the next week. Furthermore, the service attracts 40 new customers each week. How much can the service expect to make if none of the games are ties?" Devise a plan and solve the problem.

25. Suppose tulip bulbs have a probability of 0.6 of flowering. How many bulbs should you buy to have a probability of 0.9 of obtaining at least 1 flower?

26.

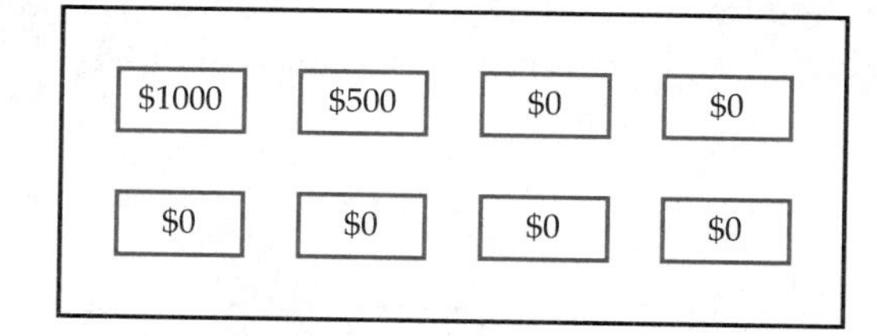

Two cards are drawn at random without replacement. What is the expected value of the total amount drawn?

27. Two basketball teams play a best-2-out-of-3 series, meaning that the first team to win two games is the champion. Suppose that team A has a probability of 0.6 of winning each game. Find the probability that team A wins
(a) in 2 games. (b) in 3 games.
(*Hint:* Draw a tree diagram.)

28. A World Series is a best-4-out-of-7 series. If the American League team has a probability of 0.5 of winning each game, determine the probability that they win the World Series in
(a) 4 games. (b) 5 games.

29. A basketball team has a probability of 0.7 of making each free throw. If they shoot 5 free throws at the end of a game, determine the probability that they make
(a) all 5 shots. (b) at least 4 shots.

30. Write a sentence that tells the difference between two independent events and two dependent events.

31. Consider how $P(A \text{ and } B)$ is related to $P(A)$. Let A = Mike will take math next semester. If possible, make up an event B so that
(a) $P(A \text{ and } B) < P(A)$
(b) $P(A \text{ and } B) = P(A)$
(c) $P(A \text{ and } B) > P(A)$

Computer Exercise

32. An airline has booked 50 people for a flight that has 45 seats. If the probability that a passenger shows up is 0.86, RUN the following BASIC program at least 10 times to estimate the probability that the flight will be overbooked. N represents the number of people who show up for the flight.

```
10 LET N = 0
20 FOR P = 1 TO 50
30    IF RND(1) < 0.86 THEN N = N + 1
40 NEXT P
50 IF N < 46 THEN PRINT "OKAY"
60 IF N > 45 THEN PRINT "OVERBOOKED"
```

SUMMARY

Uncertainty is part of our everyday lives. It has been said that nothing in life is certain but death and taxes. For this reason, probability theory is helpful for studying the likelihood of everyday events.

Probabilities tell us approximately what we can expect to happen when the same event is repeated many times under the same conditions. If a sample space of equally likely events can be written for an experiment, it may be possible to compute a theoretical probability.

Experimental probabilities are based upon experimental results under identical or similar conditions. If it is impractical or costly to find an experimental probability of a random event, one can sometimes study a simulation of the event, using coins, dice, or a computer.

"Students should actively explore situations by experimenting and simulating probability models. Such investigations should embody a variety of realistic situations" (NCTM, *Standards*, p. 109).

In order to decide how many letters and digits to put in license plates, phone numbers, and codes, people use counting techniques. These same techniques indicate how likely one is to guess correctly on a multiple-choice test. In probability, counting techniques are used to determine the sizes of sample spaces.

How do casinos design gambling games, states design lotteries, and insurance companies set fees? They all use probabilities to estimate the expected average payoffs per person. They then use those payoffs to determine a fee that will cover the payoffs and other expenses.

In computing the probability that events A and B will occur, one multiplies probabilities. If A and B are dependent, one must calculate how much one event affects the probability of the other:

$$P(A \text{ and } B) = P(B \text{ given } A) \cdot P(A)$$

If events A and B are independent, meaning they do not influence each other's probabilities, one can compute $P(A \text{ and } B)$ simply by multiplying $P(A) \cdot P(B)$.

STUDY GUIDE

To review Chapter 13, see what you know about each of the following ideas or terms that you have studied. You can also use this list to generate your own questions about the chapter.

The NCTM Curriculum Standards and Probability

Selected NCTM Curriculum Standards

The following standards come from the NCTM document.

- Model situations by devising and carrying out experiments or simulations to determine probabilities.
- Model situations by constructing a sample space to determine probabilities.
- Appreciate the power of using a probability model by comparing experimental results with mathematical expectations.
- Apply mathematical thinking and modeling to solve problems that arise in other disciplines.

1. Describe how each standard listed relates to the material you studied in Chapter 13.
2. Select any current elementary-school mathematics textbook series, and describe sample lessons or exercises that illustrate each standard listed.

PROBABILITY IN ELEMENTARY SCHOOL

The following chart shows at what grade levels selected probability topics typically appear in elementary-school mathematics textbooks.

Topic	Typical Grade Level in Current Texbooks
Basic probability	3, 4, 5, 6
Experimental and theoretical probability	5, 6
Simulations	5, 6
Predictions (expected value)	3, 4, 5, 6
Counting Principle	6
Independent events	5, 6

REVIEW EXERCISES

1. (a) What is the probability of rolling a product less than 10 on 2 dice?
   (b) Describe how you could determine the same probability experimentally.
2. What is the probability of getting 3 heads and 1 tail when you flip 4 coins?
3. A true-false quiz has 3 questions. If I guess at random, what is the probability that I will get at least 2 right? (Assume that I answer each question either *true* or *false*.)
4. If you toss 5 coins, the theoretical probability of getting 3 heads and 2 tails is 5/16. Explain what this probability tells us about what to expect when we do such a coin-tossing experiment.
5. A judge rates the best and second-best orange juices out of 4 brands, *A*, *B*, *C*, and *D*. What is the sample space for his pair of choices?
6. A state has 2 kinds of license plates: plates with 2 letters followed by 2 digits, and plates with 2 letters followed by 3 digits. How many different plates can the state make?
7. You have a spinner like the one shown. You want to simulate each *second* at a one-way traffic intersection. In preliminary work, you found that 150 cars passed in 10 minutes. How would you use the spinner in the simulation?

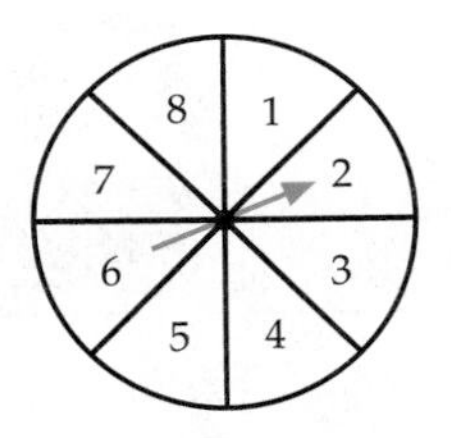

8.

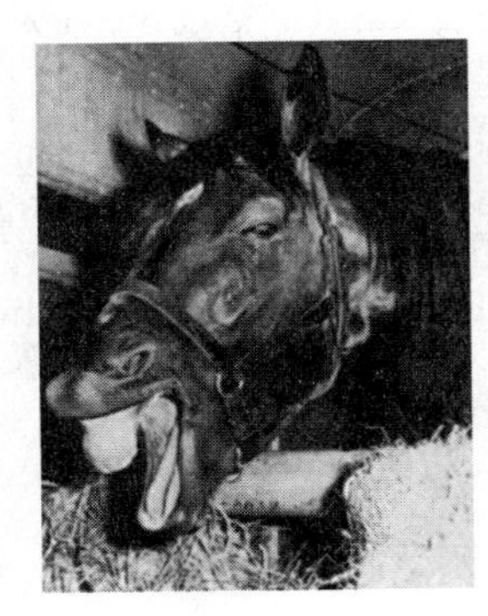

Sure Thing	1–2
Hopeless	4–1
Bad Breath	6–1
Unsettled	7–1
Lethargic	10–1
Why Bother	50–1

(a) According to the odds, Unsettled (see photo) wins about 1 out of ____ races.
(b) The probability of Lethargic winning is about ____.

9. A dice game pays \$3 back on a \$1 bet if you roll a sum of 7 on two dice and \$10 back on a \$1 bet if you roll a sum of 11. Otherwise, you lose the \$1.
(a) What is *E*(payoff on \$1 bet)?
(b) What is *E*(gain on \$1 bet)?

10. A defective undergarment is inspected by inspector 12 and inspector 14. Each of them has a 0.9 chance of finding the defect. What is the probability that neither one of them will find the defect?

11. An insurance company will insure your home against theft. The value of your possessions that are insurable is \$1000. Suppose the probability of your being burglarized of \$500 worth of goods is $\frac{1}{200}$, and the probability of your being burglarized of \$1000 worth of goods is $\frac{1}{1000}$. Assume that these are the only kinds of burglaries possible. How much should the insurance company charge people like you in order to make an average profit of \$10 per policy?

12. New Hay Checks come with one of 4 different collectible famous-thoroughbred cards in each box. Using a spinner with equal regions numbered 1 to 4, I simulated buying boxes until I had collected all 4 cards. The results follow.

1 3 1 2 4
2 4 1 1 2 2 3
4 3 1 2
4 3 4 1 2
1 3 3 1 4 1 3 2

Based upon these results, how many boxes would you expect to buy to collect all 4 cards?

13. A fortune teller claims that the card she selects is related to your future. She would say the card and your future are ________ events. In fact, they are ________ events.

14. A motel chain has found that the probability that a customer pays by check is 0.2. The probability that the check is no good, given that a customer pays by check, is 0.1. What is the probability that
(a) a customer does not pay by check?
(b) a customer pays by check and it is no good?

15. A jury of 12 is to be selected from 20 eligible jurors. How many different juries are possible?

16. A teacher wants to award a first prize, a second prize, and a third prize to a class of 25 children. How many ways are there to do this?

17. A pollster asked 300 students which of the following pizza toppings they prefer: mushrooms, pepperoni, or spinach.

	Mushroom	**Pepperoni**	**Spinach**
High school	50	80	25
College	50	60	35

Based upon these results, what is the probability of each of the following?
(a) A college student prefers pepperoni.
(b) A student prefers spinach.

18. Write a paragraph that defines experimental and theoretical probability and tells how these two kinds of probabilities are related.

SELECTED READINGS

Bergamini, D., ed. *Mathematics.* New York: Time, 1963.

Jacobs, H. *Mathematics: A Human Endeavor.* 3rd ed. New York: W. H. Freeman, 1994.

Logo Probability. Portland, Me: Terrapin Software, 1990.

National Council of Teachers of Mathematics. *The Teaching of Statistics and Probability.* 1981 Yearbook. Reston, Va.: NCTM, 1981.

Packel, E. *The Mathematics of Games and Gambling.* Wasington, D.C.: MAA, 1981.

Paulos, J. *Innumeracy.* New York: Hill and Wang, 1988.

Appendix A: Programming in BASIC

A.1 INTRODUCTION TO BASIC

BASIC is particularly well suited to number theory and algebra. Many microcomputers come with BASIC installed.

If you have a computer available, type

```
PRINT 21*42
```

and press THE RETURN or ENTER key.

PRINT is a BASIC word, and "*" is the symbol for multiplication.

Now type each of the following, and press RETURN or ENTER.

```
PRINT 20/3
PRINT 3 ∧ 4
```

(The last command does not work in all versions of BASIC. Some versions use 3 ↑ 4.)

```
PRINT SQR(10)
PRINT ABS(−2)
```

The following chart summarizes the arithmetic symbols in BASIC.

Arithmetic Symbols in BASIC

Mathematical Notation	BASIC Notation
$3 + 4$	3 + 4
$3 - 4$	3 − 4
3×4 or $3 \cdot 4$	3 * 4
$3 \div 4$	3/4
3^4	________ (Fill in for your computer.)
$3\frac{1}{2}$	3.5 or 7 / 2
$\sqrt{3}$	SQR(3)

The PRINT command can also be used to type messages. Enter the following commands.

```
PRINT "HELLO GORGEOUS"
PRINT "21*42"
```

In each case, the computer prints whatever is written between the quotation marks (without acting on it in any way).

RUNNING AND EDITING PROGRAMS

A computer program enables you to give a computer a set of instructions. A BASIC computer program has numbered lines. Line numbers are used in program editing.

Type the following program into the computer. Press RETURN or ENTER after each numbered line. If you make a typing error, you can press RETURN or ENTER and simply type that line over again. Your instructor may show you other ways to correct errors.

```
10 PRINT "HELLO."
20 PRINT "COMPUTERS ARE LIKE PARROTS."
```

Notice that when the lines are numbered, the computer does not perform them. It has not printed anything yet, but the computer has recorded the lines in its memory. To see what lines are in memory, type

```
LIST
```

and press RETURN or ENTER. The computer should have listed your two-line program. Now enter

```
RUN
```

The computer should have performed your program. As you have seen, numbered statements are stored in memory. To ascertain what is in memory, enter LIST. To perform statements in memory, enter RUN.

You can correct or rewrite a line by typing in a new version of it.

Enter

```
20 PRINT "PEOPLE ARE ILLOGICAL."
LIST
```

See how the program in memory changed? Now enter

```
RUN
```

When you are finished with a program, enter NEW to erase the program from memory. Try it.

```
NEW
LIST
```

Nothing is left in memory to list. If you don't enter NEW before starting a new program, some old program lines may get mixed into the new program.

Enter the following:

```
10 PRINT "NOT"
20 PRINT "LUCKY"
RUN
```

Now you decide to embellish the phrase.

```
15 PRINT "VERY"
```

Lesson Exercise A.1

What will happen if you LIST and RUN the program now?

```
LIST
RUN
```

As you saw, the computer puts numbered statements in order in its memory. Now suppose that you want to erase line 10 because your luck has changed.

```
10
LIST
RUN
NEW
```

To erase any line in a program, type the line number with nothing after it, and press RETURN or ENTER.

To see how your computer handles large numbers, type the following program. The number in line 10 has a 3 followed by 16 zeroes. Do *not* insert commas in the number.

```
10 PRINT 30000000000000000
RUN
```

Your computer probably wrote its version of scientific notation for 3×10^{16}.

ALGEBRA ON THE COMPUTER

Chapter 11 covers applications that use algebraic formulas. With their large memories and extraordinary computational capabilities, computers are well suited to work with formulas. In businesses, computers sometimes deal with hundreds of equations containing hundreds of variables.

Enter the following program into the computer.

```
NEW
10 LET A = 3
15 LET C = 4
20 LET P = 7 * A + 4 * C
25 PRINT P
```

Lesson Exercise A.2

What do you think the computer will print when you RUN this program?

LIST it and RUN it.

```
LIST
RUN
```

LET is another BASIC word. LET is always followed by an equation. The equal sign in BASIC has a different meaning from that in algebra. The computer performs two steps with a LET statement:

Step 1: It finds the value on the right side of the equal sign.

Step 2: It assigns this value to the variable address (actually a location inside the computer) on the left side.

So, for example, 10 LET A = 3 tells the computer to take 3 and put it in a memory mailbox labeled A. You can model this result with a drawing.

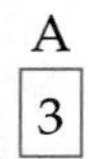

Edit the program as follows:

```
15 LET C = 2
LIST
```

Lesson Exercise A.3

Now what will the output be?

RUN it and see if you are right.

```
RUN
```

Try the following program.

```
NEW
100 LET R = 2
200 LET PI = 3.14159
300 LET A = PI * (R ^ 2)
400 PRINT "AREA = "; "SQUARE UNITS"
```

Lesson Exercise A.4

What do you think the output will be?

LIST it and RUN it.

```
LIST
RUN
```

Did your program print 12.56636SQUARE UNITS without a space? If so, you did not put a space *between the quotation mark and the first S* in SQUARE UNITS.

The preceding program always uses $R = 2$. There is another BASIC statement that would enable the person running the program to INPUT any value of R he or she wanted! Try the following.

```
80 PRINT "AREA OF A CIRCLE"
90 PRINT "ENTER THE RADIUS."
100 INPUT R
```

LIST the edited program and RUN it.

```
LIST
RUN
```

You should see a question mark.

```
?
```

This tells you that the computer has read line 100 and is waiting for you to input a value for R.

Lesson Exercise A.5

(a) Pick any decimal number for R and enter it. What happens?

(b) Run the program again and use a different value for R.

The INPUT statement allows you to pick the values for variables in a program after the program is running. Usually these variables are part of the right side of a formula.

Enter the following.

```
NEW
10 PRINT "MAY I SERVE YOU?"
20 PRINT "PROGRAM FINDS SLOPE OF LINE JOINING 2 POINTS"
30 PRINT "ENTER FIRST POINT-WRITE TWO COORDINATES SEPARATED
   BY A COMMA"
40 INPUT X1, Y1
50 PRINT "ENTER SECOND POINT"
60 INPUT X2, Y2
70 LET S = (Y2 - Y1) / (X2 - X1)
80 PRINT "THE SLOPE IS"; S
```

This time, each input is *two numbers separated by a comma.* RUN the program and select input.

RUN

Lesson Exercise A.6

RUN the program again and select different input values.

Many programs using formulas have the following parts.

1. A PRINT statement describing the purpose of the program
2. A PRINT statement asking for input
3. An INPUT statement
4. A LET statement to process input
5. A PRINT statement to print the result

Summary of BASIC Language

Command	**Function**	
LIST	Lists all statements currently in computer memory	
RUN	Performs the program currently in computer memory	
NEW	Erases all statements currently in computer memory	
Statement	**Function**	**Example**
PRINT	Types out values and messages	10 PRINT ''I LIKE YOU.''
LET	Computes and assigns values	35 LET X = A * B
INPUT	Allows program user to select numerical values for variables	12 INPUT C, T

A.1 HOMEWORK EXERCISES

1. Suppose you type the following into the computer.

```
NEW
10 PRINT ''MAD''
20 PRINT ''MAN''
20 PRINT ''DOG''
```

(a) What will the computer type as output when you enter LIST?

(b) What will the computer type when you enter RUN?

2. What would you type into the computer to insert PRINT ''NOT'' between the two following statements?

10 PRINT "IT IS"
20 PRINT "RIGHT."

3. What will be in the memory of the computer after you enter the following lines?

10 PRINT "MIKE"
20 PRINT "SALLY"
30 PRINT "ROSA"
20 PRINT "LIKES"
10 PRINT "TONY"

4. What would you type into the computer to erase line 20?

10 PRINT "PLEASE"
20 PRINT "STOP"
30 PRINT "HER"

5. What is the output of each of the following lines?
(a) 10 PRINT 5 * 3 (b) 10 PRINT "5*3"
(c) 10 PRINT "5*3="; 5 * 3 (*Note:* the semicolon is used to separate the two expressions, and it will not be printed.)

6. Write each of the following in BASIC notation.
(a) $4x^2$ (b) $\frac{2}{3}$ (c) $\frac{1}{2}(x \div 3)$
(d) $D = RT$

7. Write a BASIC expression to evaluate the 30th power of the sum of 28.9 and 37.6 and 0.8.

8. (a) Predict what the computer will give as the answer to 7 + 3 * 6 − 1.
(b) Enter a statement on a computer so it will calculate and write the result.

9. Tell whether it is most appropriate to do each of the following mentally, with a calculator, or with a computer, assuming that you want an exact answer.
(a) $8.617 \div 2.3$ (b) 30×700 (c) 5^{153}

10. What is the output of the following?

PRINT 3000000 * 4000000

11. Write each decimal number in computer *E*-notation.
(a) 468,000 (b) 0.0013

12. Write each of the following in decimal notation.
(a) 4.31 *E* −05 (Apple)
4.31 *D* −05 (IBM)
(b) 3.2 *E* 03 (Apple)
3.2 *D* 03 (IBM)

13. Write a one-line BASIC program that calculates and prints $(4 + 0.1)^{60}$, and RUN it on a computer.

14. Write a one-step BASIC program that has the computer write the message

"I'M A PRETTY COMPUTER."

15. Write a single PRINT statement that finds the balance on $5000 earning 8% interest compounded annually for 10 years.

16. In the seventeenth century, Marin Mersenne, a French monk, examined numbers of the form $2^p - 1$, now known as Mersenne numbers. Mersenne substituted different values for p and obtained the following results.

$2^2 - 1 = 3$ $2^3 - 1 = 7$
$2^4 - 1 = 15$ is not prime. $2^5 - 1 = 31$
$2^6 - 1 = 63$ is not prime. $2^7 - 1 = 127$, a prime
$2^8 - 1 = 255$ is not prime. $2^9 - 1 = 511$ is not prime.
$2^{10} - 1 = 1023$ is not prime. $2^{11} - 1 = 2047$, a prime
$2^{13} - 1$ and $2^{17} - 1$ are prime.

(a) Based upon these results, it appears that if p is ________, then $2^p - 1$ is prime.
(b) Mersenne proposed that $2^p - 1$ is prime when p is prime. In 1903, F. N. Cole said that Mersenne was wrong for $2^{67} - 1$. He said that $2^{67} - 1$ is divisible by 193,707,721. Check this on a computer.

17. Which set of pictures correctly illustrates what is contained in the computer's memory after the following program is run?

10 LET M = 3
20 LET T = 2 * M + 5
30 PRINT T

(a) M: [3] T: [2M + 5] (b) M: [3] T: [T]
(c) M: [3] T: [11]

18. Draw mailboxes that show what is contained in the computer's *memory* after the following program is run.

```
10 INPUT C
20 LET T = 3.10 + 0.05 * C
30 PRINT "YOUR MONTHLY RATE IS $"; T
RUN
? 10
```

19. Correct the following BASIC programs.

(a)
```
10 LET A = 4
20 LET 3 = B
30 PRINT A * B
40 RUN
```

(b)
```
10 SET A = 3
14 LET T = 3A
18 PRINT TOTAL PRICE IS T
```

20. What is the output of the following program?

```
10 LET A = 5
15 PRINT 4 * A
10 LET A = 8
RUN
```

21. What is the output of the following programs?

(a)
```
20 INPUT N
30 LET T = 4 + N / 40
40 PRINT T ; " DEGREES"
RUN
? 100
```

(b)
```
5 PRINT "BASE, HEIGHT"
10 INPUT B, H
15 LET A = 0.5 * B * H
20 PRINT "THE AREA IS "; A
RUN
BASE, HEIGHT
? 3, 6
```

22. What is the output of the following programs?

(a)
```
10 LET A = 3
20 LET B = 4
30 LET C = 4 * A - 5 * B
40 PRINT C
```

(b)
```
100 LET A = 5
200 LET B = 2 + 3 * A
300 PRINT B
```

23. (a) Run the following BASIC program, using numbers of your choice.

```
10 PRINT "TYPE IN FRACTIONS A/B
   AND C/D BY TYPING A, B, C, D
   WITH B AND D ≠ 0."
20 INPUT A, B, C, D
30 PRINT A * C; "/"; B * D
```

(b) What is the purpose of the program?

(c) Revise line 30 so the program prints the result of $\frac{A}{B} \div \frac{C}{D}$.

24. Many banks pay interest "daily" (usually 360 days a year or exactly 30 times each month). The following program computes interest compounded daily.

(a) Enter the following program into a computer.

```
10 PRINT "HOW MUCH MONEY WILL
   YOU DEPOSIT?"
20 INPUT P
30 PRINT "HOW MANY YEARS IS IT IN
   FOR?"
40 INPUT T
50 PRINT "TYPE IN THE ANNUAL %
   INTEREST RATE AS A DECIMAL
   (LIKE 0.065)."
60 INPUT I
70 LET A = P * (1 + I / 360) ∧ (360 * T)
80 PRINT "YOU WILL END UP WITH $";
   A
```

(b) RUN the program using the input of your choosing.

(c) Find the output if $P = 1000$, $T = 3$, and $I = 0.06$.

(d) Have the program compute interest compounded annually instead of daily, by changing line 70 to

```
70 LET A = P * (1 + I) ∧ T
```

Use the data from part (c) in your new program. How does daily compounding compare to annual compounding?

(e) Use the program in part (d) to find out how many years it would take your investment to double if you earn 8% interest compounded annually.

25. Write a program that asks for the total number of test questions and the number right and prints out the percent of questions that are right.

26. Write a program that asks for an old price and a new price and prints out the percent increase or decrease.
27. (a) What is the surface area A of a rectangular prism that has length L, width W, and height H?
    (b) Write a program that asks a person for the length L, width W, and height H of a rectangular prism and computes the total surface area A.
28. (a) Write a program that accepts input for the radius and height of a cylinder and prints out the volume.
    (b) Modify your program so that it also prints out the surface area.
29. Write a program that asks for the lengths of the two legs of a right triangle and prints out the length of the hypotenuse. (*Hint:* Use SQR.)
30. Write a program that will print out the rental charge for a car when someone types in the number of days and the number of miles driven. Assume that the rental charges are $26.50 per day and $.22 per mile driven.
31. The following BASIC program computes phone bills based upon a monthly rate R (in $) plus a charge per call C (in $).

```
10 PRINT "TYPE THE MONTHLY RATE
   AND CHARGE SEPARATED BY A
   COMMA (LIKE 6.50, .09)."
20 INPUT R, C
30 PRINT "HOW MANY CALLS WERE
   MADE?"
40 INPUT N
50 PRINT "THE MONTHLY BILL IS $";
```

    (a) Complete line 50 with an appropriate formula.
    (b) RUN the program and find the cost of making 63 calls when the monthly rate is $8.40 and the cost per call is $.08.
    (c) Modify the program so that it also adds 6% tax to the bill.

A.2 BRANCHES AND LOOPS

In all computer languages, programs rely on two basic structures: branches and loops.

BRANCHES

A **branch** tests a statement to see if it is true or false. If it is true, the program branches to one group of statements. Otherwise, the program branches to a second group of statements.

To write a branch, you need two new BASIC statements: IF-THEN and END.

IF-THEN AND END INSTRUCTIONS

The IF-THEN instruction checks a condition, and if the condition is true, the computer jumps to another part of the program or performs the instructions after THEN. The END instruction stops a program.

The following program utilizes these two new BASIC instructions in a branch.

Lesson Exercise A.7

Consider the following program.

```
10 PRINT "HOW MANY SHIRTS DO YOU WANT?"
20 INPUT N
30 IF N > 4 THEN PRINT "THE COST IS $";N*8
40 IF N <= 4 THEN PRINT "THE COST IS $";N*10
50 END
```

Without using a computer, do parts (a) and (b).

(a) What is the output of the program if $N = 2$?
(b) What is the output of the program if $N = 6$?
(c) RUN the program on a computer, and check your answers to parts (a) and (b).

When $N = 2$, the program skips over line 30 and goes to line 40. It prints 20. When $N = 6$, the program performs the computation in line 30. It prints 48.

LOOPS

A FOR-NEXT loop is a simple way to repeat an action in a program. RUN the following program.

```
10 FOR N = 1 TO 20
20    PRINT 5 * N
30 NEXT N
```

How does this program work? The FOR statement tells the computer to start with $N = 1$ and consider in turn the numbers 1, 2, 3, . . . , 20 for N. For each of these numbers, the computer proceeds line by line until it encounters the NEXT statement. So, when $N = 1$, the computer performs line 20 and prints 5. Then it reads line 30, which picks the next N, 2.

Since 2 is in the range of numbers given in the FOR statement, the computer begins the loop again at line 10 with $N = 2$. It performs line 20 and prints 10. Then it reads line 30, which assigns 3 to N.

The FOR-NEXT loop is then repeated in this fashion for $N = 1, 2, 3, \ldots, 20$, and line 20 prints out multiples of 5.

THE INT INSTRUCTION

BASIC has a special rounding instruction (INT) that is useful in number theory programs. The INT instruction rounds all numbers *down* to the nearest integer. What use is such a statement? After an introduction to INT statements, you'll see how they are used to find divisors and prime numbers.

EXAMPLE A.1

Find the value of each computer expression.

(a) INT(3.6) **(b)** INT(7) **(c)** INT(−4)
(d) INT(−4.2) **(e)** INT(5 / 3) **(f)** INT(−5 / 3)

Solution

INT rounds the value in the parentheses *down* to the nearest integer.

(a) INT(3.6) = 3 **(b)** INT(7) = 7 **(c)** INT(−4) = −4

(d) INT(−4.2) = −5

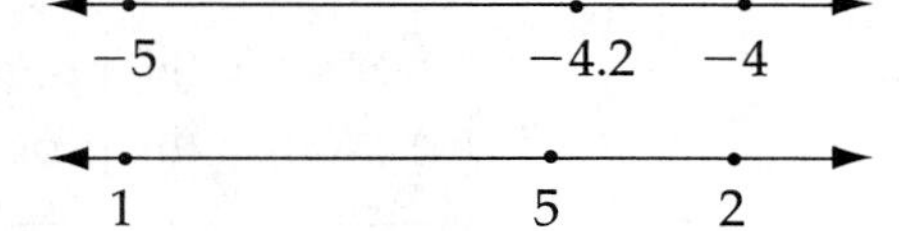

(e) INT(5 / 3) = 1

(f) INT(−5 / 3) = −2

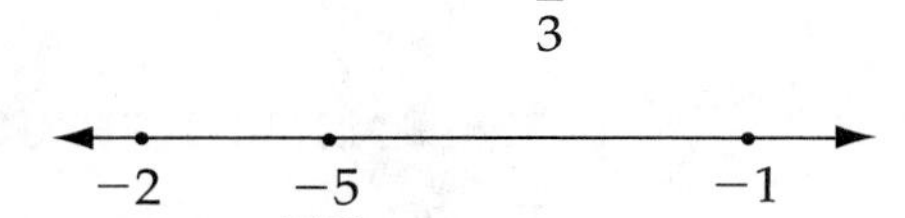

■

Lesson Exercise A.8

Find the value of each computer expression.

(a) INT(20 / 3) (b) INT(3) (c) INT(−3.2)

Lesson Exercise A.9

If $A = 27$ and $B = 5$, then INT(A / B) = ________.

Lesson Exercise A.10

(a) If X = INT(X), then X is ________.
(b) If $X \neq$ INT(X), then X is ________.

In programs involving factors, multiples, and primes, it is often helpful to compare INT(A / J) to A / J.

Lesson Exercise A.11

(a) Suppose $A = 8$. For what counting numbers J in the set {1, 2, 3, 4, 5, 6, 7, 8} does A / J = INT(A / J)?
(b) In part (a), what mathematical term describes how the numbers in the set you listed are each related to 8?
(c) Based upon part (b), it appears that if A / J = INT(A / J), then ________.

The following program contains a branch within a loop.

Lesson Exercise A.12

(a) RUN the following program on a computer.

```
10 FOR J = 1 TO 8
20    IF 8 / J = INT(8 / J) THEN PRINT J
30 NEXT J
```

(b) What is the purpose of this program?

(c) Modify the program to print all factors of 50.

(d) Modify the program so that the user can input any number.

THE RND(1) INSTRUCTION

RND(1) randomly selects a decimal number between 0 and 1, including 0 but excluding 1. The number of decimal places in the number depends upon the particular computer.

Lesson Exercise A.13

(a) What values could 6*RND(1) have?

(b) What values could INT(6*RND(1)) have?

(c) What values could INT(6*RND(1)) + 1 have?

INT(6*RND(1)) + 1 will have the value 1, 2, 3, 4, 5, or 6 with equal likelihood, just like a regular die!

Lesson Exercise A.14

(a) Guess what the following program will do.

```
10 FOR I = 1 TO 10
20    LET ROLL = INT(RND(1)*6) + 1
30    PRINT ROLL
40 NEXT I
```

(b) RUN the program and check your guess.

The preceding program simulates 10 rolls of a die and prints the results.

ANSWERS TO SELECTED LESSON EXERCISES

A.8 (a) 6 (b) 3 (c) −4

A.9 5

A.10 (a) an integer (b) not an integer

A.11 (c) *J* is a factor of *A*

A.13 (b) 0, 1, 2, 3, 4, and 5

A.2 HOMEWORK EXERCISES

1. Consider the following program.

```
10 PRINT ''WHAT IS YOUR GPA?''
20 INPUT G
30 IF G >= 3.5 THEN PRINT ''HONORS''
40 IF G < 3.5 THEN PRINT ''NOT QUITE''
50 END
```

Without using a computer, answer the following questions.

(a) What is the output of the program if G = 2?

(b) What is the output of the program if G = 3.5?

(c) In answering parts (a) and (b), did you use inductive or deductive reasoning?

2. The following BASIC program performs a branch for police officers' heights. Type the program into a computer and RUN it, trying some different heights.

```
10 PRINT ''PLEASE ENTER YOUR HEIGHT
   IN INCHES AND PRESS RETURN.''
20 INPUT H
30 IF H >= 66 THEN PRINT ''YOU
   QUALIFY.''
40 IF H < 66 THEN PRINT ''I AM SORRY
   BUT YOU ARE TOO SHORT.''
50 END
```

3. Dress shirts are $18 each if fewer than 5 are purchased and $16 each when at least 5 are purchased. Fill in line 40 to complete the program so that it computes the total cost of any number of shirts.

```
10 PRINT ''TYPE IN THE NUMBER OF
   SHIRTS YOU WANT.''
20 INPUT N
30 IF N < 5 THEN C = 18*N
40 ________________
50 PRINT ''THE TOTAL COST IS $''; C
60 END
```

4. Write a program that asks a person's age and tells the price of admission. If the person is under 18, the cost is $3. If the person is 18 or over, the cost is $5.

5. (a) RUN the following BASIC program on a computer. What is the output?

```
10 FOR N = 1 TO 7 STEP 2
20 PRINT N
30 NEXT N
```

In parts (b) through (d), type in a new line 10 to create a program that prints the sequence in the given order.

(b) 2, 4, 6, 8

(c) 7, 11, 15, 19, 23

(d) 11, 8, 5, 2, −1 (*Hint:* Use a negative number after STEP.)

6. (a) RUN the following BASIC program on a computer, and find the output.

```
10 FOR T = 1 TO 15 STEP 2
20     PRINT T
30 NEXT T
```

(b) Write a FOR statement that counts the multiples of 5 from 5 to 100.

(c) Predict the output of the following. Then check it on a computer.

```
10 FOR T = 10 TO 0 STEP -2
20     PRINT T
30 NEXT T
```

7. A job pays $5.76 per hour for the first 40 hours during a week and time-and-a-half for each hour worked over 40.

(a) Write a formula for T, total pay, for someone who worked over 40 hours, in terms of H, the number of hours worked.

(b) Write a program that prints out the total weekly pay for any given number of hours worked.

8. In Washington, D.C., one type of monthly phone service costs $10.25 for up to 60 local calls and $.09 for each additional call.

(a) Write a formula for the total rate in terms of the number of calls, for a person who makes over 60 calls.

(b) Write a program that prints out the monthly rate for any given number of local phone calls.

9. The following program computes the slope using the two-point formula.

```
10 PRINT "MAY I SERVE YOU?"
20 PRINT "PROGRAM FINDS SLOPE OF
   LINE JOINING 2 POINTS"
30 PRINT "ENTER FIRST POINT-WRITE
   TWO COORDINATES SEPARATED BY A
   COMMA."
40 INPUT X1, Y1
50 PRINT "ENTER SECOND POINT"
60 INPUT X2, Y2
70 LET S = (Y2 − Y1) / (X2 − X1)
80 PRINT "THE SLOPE IS "; S
```

Modify the program so that it checks for the special case of undefined slope and prints an appropriate message.

10. Write a program that asks a person for the radius of a circle. Then ask the person to enter a 1 for the circumference or a 2 for the area. The program should then type out the appropriate measurement. (Use 3.14159 as an approximation of π.)

11. An object is tossed in the air from a height of 192 ft. The height H of the object after T seconds is given by $H = -16T^2 + 64T + 192$.

(a) RUN the following program.

```
10 FOR T = 0 TO 10
20    PRINT T; " SECONDS", −16*T*T +
      64*T + 192; " FEET"
30 NEXT T
```

(b) Use the program to estimate when the object is at its high point and when it hits the ground.

(c) What do negative heights mean?

12. (a) RUN the following program on a computer.

```
10 PRINT " L", " W", " P"
20 FOR L = 1 TO 16
30 LET W = 16 / L
40 LET P = (2 * L) + (2 * W)
50 PRINT L, W, P
60 NEXT L
```

(b) What shape is this program about?

(c) What is true about the area of this shape?

(d) Explain what the program is about and what it shows.

13. Write a program that prints all multiples of 9, beginning with 9 and ending with 180.

14. RUN the following program on a computer.

```
10 PRINT "X                    Y"
20 FOR X = −3 TO 3
30    LET Y = X ∧ 2 + 3 * X − 2
40    PRINT X; "                "; Y
50 NEXT X
```

(a) What is the output of the program? (If necessary, edit line 10 or 40 to line up the table.)

(b) Use the table to graph $y = x^2 + 3x - 2$.

15. You are choosing between two health insurance plans. One pays 60% of all expenses beyond the first $100. The second pays 80% of all expenses beyond the first $500.

(a) Write a formula for A, the amount the first policy pays on a claim of C dollars.

(b) Write a formula for A, the amount the second policy pays on a claim of C dollars.

(c) Write a computer program that prints the total amount paid under each plan for claims of $250, $500, $750, . . . , $2000.

(d) Describe the conditions under which each plan pays more.

16. Find the value of each computer expression.
(a) INT(10.2) (b) INT(−8.1)
(c) INT(21.7)

17. If INT(T) = T, then T is ______________.

18. Suppose $R = 20$. For what counting numbers does INT(R / S) = R / S?

19. (a) Without using a computer, find the output of the following program.

```
10 INPUT A
20 FOR J = 2 TO A
30    IF INT(A / J) = A / J THEN PRINT J
40 NEXT J
RUN
? 186
```

(b) Check your answer on a computer.

20. (a) Enter the following BASIC program into a computer.

```
10 PRINT "TYPE TWO WHOLE
   NUMBERS SEPARATED BY A
   COMMA AND PRESS RETURN. TYPE
   THE SMALLER NUMBER FIRST."
20 INPUT M, N
30 FOR D = M TO 1 STEP −1
40    IF M / D = INT(M / D) AND N / D
      = INT(N / D) THEN 60
50 NEXT D
60 PRINT "THE GREATEST COMMON
   FACTOR OF "; M; " AND ";N;" IS "; D;
   "."
```

(b) RUN the program and type in 20, 36. What is the output?
(c) Use the program to find the greatest common factor of 1457 and 1581.

21. (a) RUN the following BASIC program a few times, inputting a different number each time.

```
10 PRINT "TYPE A WHOLE NUMBER
   AND PRESS RETURN/ENTER."
20 INPUT N
30 IF N > 9 AND N < 100 THEN PRINT
   N
```

(b) What numbers will this program print?
(c) Change AND to OR in line 30 to create a new program. Predict which numbers it will print. Then RUN the program a few times to check your hypothesis.

22. The LCM of two numbers can be computed by multiplying the two numbers and dividing by their greatest common factor.
(a) Revise the program in the preceding exercise so that it prints the GCF and the LCM.
(b) RUN the program for 1581 and 1457.

23. Write a program that uses INT to round a decimal number to the nearest integer, using the rounding rules from Chapter 3. (*Hint:* Add 0.5.)

24. (a) RUN the following nested loops program on a computer, and find the output.

```
10 FOR I = 1 TO 3
20    FOR J = 2 TO 3
30    PRINT J
40    NEXT J
50 NEXT I
```

(b) Explain the resulting output, using mailboxes for I and J.

Appendix B: Databases

A **database** is a set of information organized by category. For example, a database might contain the name, phone number, social security number, and math grade for each student in a class. Each category, such as a phone number, is called a **field**. The set of information for each person is called a **record**.

NAME	PHONE NUMBER	SS NUMBER	MATH GRADE
Simpson, Lisa	842-9061	361-84-1299	A
Batman	555-8723	482-66-8408	D
Crocker, Betty	214-1982	217-32-7989	C

Lesson Exercise A.15

Refer to the preceding database.

(a) Which fields contain numbers?

(b) How many records are there?

Material in a database can be sorted by field. Text or letters can be alphabetized, and numbers can be ordered from greatest to least or least to greatest.

Lesson Exercise A.16

Name two useful ways to sort information in the preceding database example.

Lesson Exercise A.17

Consider the following database.

NAME	GRADE LEVEL	MATH SCORE	READING SCORE	TEACHER
Sward, Gil	3	80	70	Lopez
Guyton, Gar	4	72	80	Sahu
Wee, Leben	3	85	88	Lopez
McGee, Bruce	2	70	81	Lopez
Dalton, Pat	3	90	72	Usiskin

(a) How many records are there?
(b) Which fields contain letters?
(c) Name five useful ways to sort the information in this database.

If you have database software, try the following.

Lesson Exercise A.18

(a) Create your own database with at least 4 fields and 5 records. It could be a phone directory or an imaginary set of student records.
(b) Have the computer sort the material in two different ways.
(c) Have the computer select some of the records based on some characteristic.

ANSWERS TO SELECTED LESSON EXERCISES

A.15 (a) Phone number, social security number
(b) 3

A.16 Alphabetize the names. Put the social security numbers in increasing order. Alphabetize the grades.

A.17 (a) 5 (b) NAME, TEACHER
(c) Alphabetize the names or teachers. Put the grade levels, math scores, or reading scores in decreasing order.

Answers to Selected Exercises

Chapter 1

1.1 *Homework Exercises*

1. (b) You end up with 6 more than the original number.

3. Yes

5. With a counterexample

6. False;

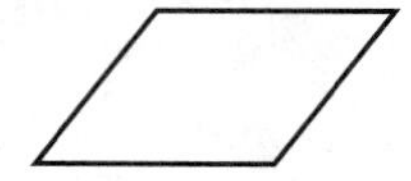

7. False; Mike does not drink beer.

8. Reasonable

9. Reasonable

11. (a) 88,831
(b) The answer has the same digits in reverse order.
(c) 68,952; 25,986
(d) Reversing the order of the digits in two three-digit factors will reverse the order of the digits of their product.
(e) False

13. (a) 16 units
(b) The length of the pendulum is the square of the period of the swing.
(c) $L = P^2$ (d) 2.2 sec

15. (a) For the last 3 weeks, my teacher has given one pop quiz per week. I generalize that my teacher will always do this.
(b) A mathematician sees three examples in which the sum of the two odd numbers is an even number. The mathematician generalizes that this will always occur.

17. (a) Generalization 1 is based upon the observation that all through recorded history there have been wars. Generalization 2 could be based upon teaching a few classes in which the boys did better than the girls. Generalization 3 could be based upon seeing a few movie sequels and

19. (a) True (b) True (c) True
(d) The product of any N consecutive whole numbers is divisible by N.

21. No

1.2 *Homework Exercises*

1. All rectangles are quadrilaterals.

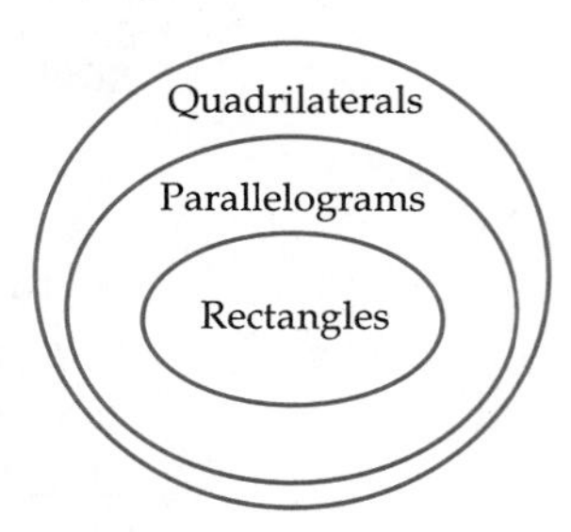

3. (a) $\overline{AB}$ is parallel to $\overline{CD}$ and $\overline{AD}$ is parallel to $\overline{BC}$.
(b) *Hypotheses:* $ABCD$ is a rectangle. The opposite sides of a rectangle are parallel.
Conclusion: $\overline{AB}$ is parallel to $\overline{CD}$ and $\overline{AD}$ is parallel to $\overline{BC}$.

5. (e)

7. Yes

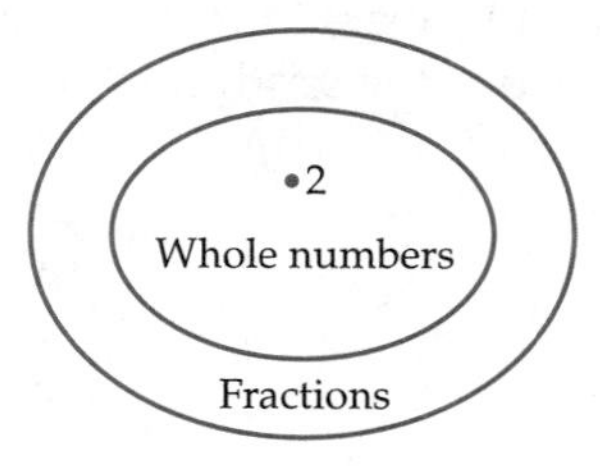

9. No

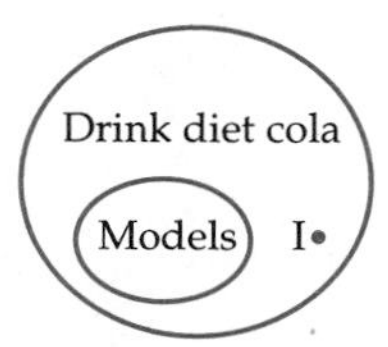

10. Yes

11. No

12. 5, since one correct digit was lost by changing 5 to 4.

13. (a) 0, 1, 2, 6, 7, 9
(b) 8. The only digit that is dropped from the third guess to make the fourth guess is the 8.
(c) 3, 5 (e) 358

15. If you have your teeth cleaned twice a year, then you will lose fewer teeth.

17. Some adults don't have time to read.

19. I will have more dates. (false)

21. (a) All mathematics teachers love mathematics.
(b) Time flies when you are having fun.

23. Sandy is a female dachshund that is not white.

25. Nancy, Ma, Igor, Migraine, Lurch

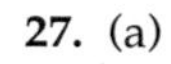

27. (a)

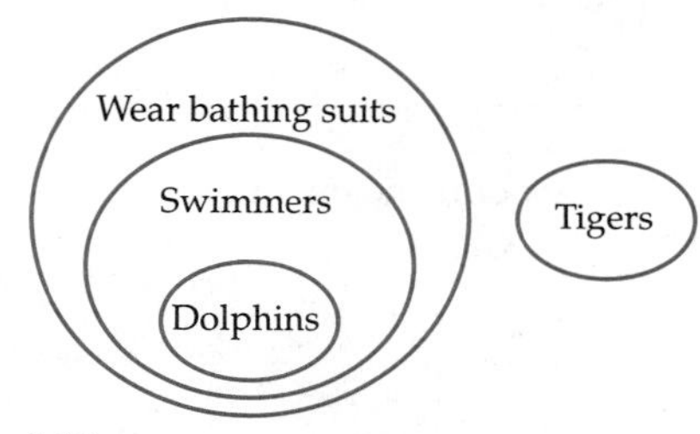

(b) (1) and (4)

29. 1 honest, 99 crooked

31. Select one ball from "W and Y" and you know if it's really "Y" or "W" and then you can work out the rest.

33.

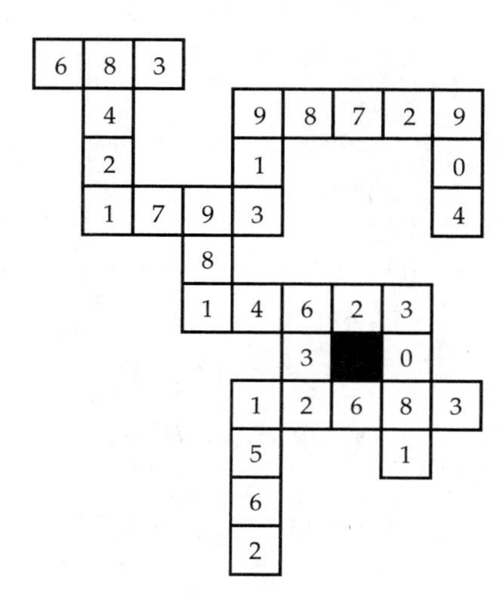

1.3 *Homework Exercises*

1. Inductive

3. *Hypotheses:* All boys play baseball. All baseball players chew gun.
Conclusion: All boys chew gum.

5. Induction

7. Induction

9. Inductive

11. (a) It will increase by 4.
(b) Because it may not work for some number that you did not try
(c) $N \to 3N \to 2N \to 2N + 8 \to N + 4$
(d) If you have $2N$ and then you add 8, you can deduce that the result is $2N + 8$.

13.

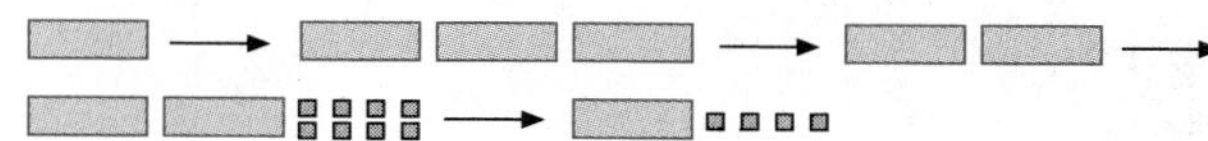

15. Interchange the hypothesis and the conclusion.

17. (a) Yes (in general)
(b) If you are sick, then you have a fever.
(c) No

19. (a) If you are smart, then you may be successful.
(b) If a triangle is equilateral, then it is isosceles.

21. If a triangle has two equal sides, then it has two equal angles. If a triangle has two equal angles, then it has two equal sides.

23. $2x = 4$ if and only if $x = 2$.

27. (a) Canaries like to sing. Blue jays like to sing. Cardinals like to sing. Therefore, all birds like to sing.
(b) Tweety is a canary. Canaries like to sing. Therefore, Tweety likes to sing.

29. (a) If $x + 3 \neq 5$ then $x \neq 2$.
(b) If I do not confess, then I am not guilty.
(c) Possible

31. (a) If $x \neq 2$ then $x + 3 \neq 5$.
(b) If I am not guilty, then I do not confess.
(c) Impossible

33. (a) If I drove to work, then it was raining.
(b) If I did not drive to work, then it was not raining.
(c) If it was not raining, then I did not drive to work.
(d) Part (b)

35. (c)

37. No

39.

A 5	B 16	C 3	D 10
E 4	F 9	G 6	H 15
I 14	J 7	K 12	L 1
M 11	N 2	O 13	P 8

1.4 *Homework Exercises*

1. (a) 625
(b) Induction to find the general rule of raising to the fourth power and deduction to apply the general rule to 5 and compute 5^4.

3. 12

5. 132

7. (a) 8 or 7
(b) Multiplying by 2 or adding 1 more each time

9. (a) 1, 1, 2, 3, 5, 8, 13, 21, 34, 55
(b) The sum is 1 less than the term.
(c) The sum of the first n terms is 1 less than the $(n + 2)$nd term.

11. (a)

Time (years)	0	5600	11,200	16,800	22,400
Fraction of C-14 left	1	1/2	1/4	1/8	1/16

(b) About 16,000 years (since 1/7 is between 1/4 and 1/8)

13. (a) 2 (b) 4 (c) 8 (d) $2^{12} = 4096$

15. 9, 8, 7, 6, 5

17. (a) 2, 4, 6, 8, . . . (b) $2n$
(c) Each term is 2 more than the preceding.

18. (a) 1, 3, 5, 7, . . . (b) $2n - 1$
(c) Each term is 2 more than the preceding.

19. (a) $2(50) - 1 = 99$ (b) $50^2 = 2500$

21. (a) 19 (b) $2n + 9$ (c) 89
(d) Deduction

23. (a) 38 (b) $n^2 + 2$ (c) 902

25. (a) $7n - 5$ (b) 695

27. 8 since $3^9 > 10{,}000$

28. (a) $P = 2N + 1$ (b) 9 (c) Inductive

(d)

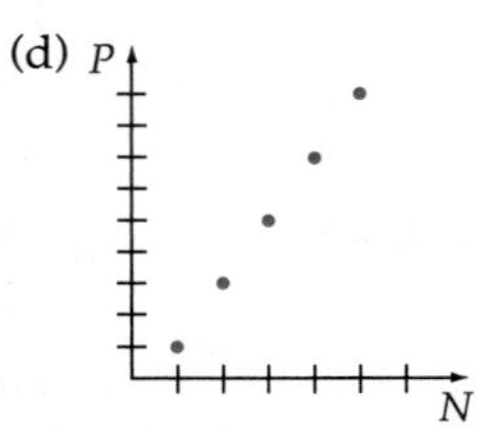

(e) The points all lie on a line.
(f) $2n - 1$

29. (a) $V = L^3$ (b) 125 (c) n^3

31. (a) $y = 8$ (b) □□□□; $y = 10$
(c)

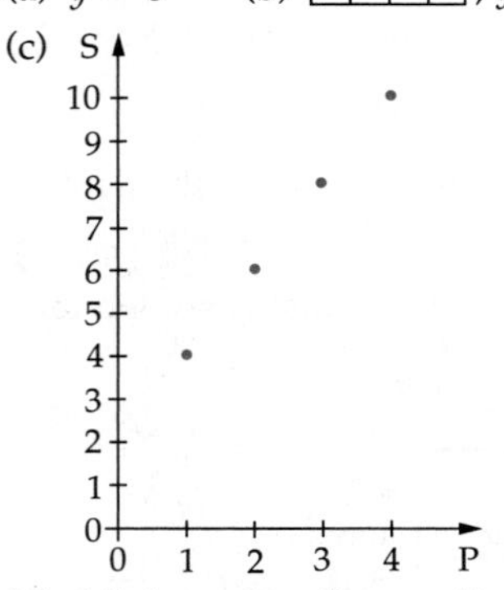

(d) All the points lie on a line.
(e) $y = 2N + 2$

33. 5; $8 - N$

35. (b)

37. (a) 5 and 6
(b) $3^4 + 4^4 + 5^4 + 6^4 = (7)^4$. It is false.

39. (a) $1 + 8 \cdot 10 = 9^2$; yes
(b) $(2n + 1)^2$
(c) $1 + 8\left(\frac{n(n + 1)}{2}\right) \stackrel{?}{=} (2n + 1)^2$

$$1 + 8\left(\frac{n^2 + n}{2}\right) \stackrel{?}{=} 4n^2 + 4n + 1$$

$$1 + 4n^2 + 4n = 1 + 4n^2 + 4n$$

41. (a) 3, 7, and 13 (b) $4^2 + 5^2 + 20^2 = (21)^2$, yes
(c) $N^2 + (N + 1)^2 + [N(N + 1)]^2 = [N(N + 1) + 1]^2$
(d) $N^2 + (N + 1)^2 + [N(N + 1)]^2 \stackrel{?}{=} [N(N + 1) + 1]^2$

$$N^2 + N^2 + 2N + 1 + [N^2 + N]^2 \stackrel{?}{=} [N^2 + N + 1]^2$$

$$2N^2+2N+1+N^4+2N^3+N^2 \stackrel{?}{=} N^4+2N^3+3N^2+2N+1$$

$$N^4 + 2N^3 + 3N^2 + 2N + 1 \stackrel{?}{=} N^4 + 2N^3 + 3N^2 + 2N + 1$$

43. (a) 12th row (b) Down
(c) 22nd row (d) Down

45. (a) 0 (b) 1 (c) $1 - 1 + 1 - 1 + \ldots$
(d) $1 - 1 + 1 - 1 + \ldots$ (e) Yes

47. $(161 \times 4) + 7 = 651$

49. $74^2 - 40 = 5436$

53. (a) 1 (b) 3 (c) 12

55. (a) 128 (b) 2

57. No

59. (a) The height of each bounce decreases geometrically.

61. (a) They add up to 101. (b) 50 (c) 5050

63. (a) 21 (b) $N(N + 1)/2$

65. $(15 + 29) \cdot 15/2 = 330$

1.5 *Homework Exercises*

1. Multi-step translation

3. Puzzle

5. The doctor first tries to understand what is wrong with you. Next, the doctor devises a treatment. Then, the doctor implements the treatment. Finally, the doctor checks to see whether the treatment worked.

7. No, it's too high. $(2 \times \$13) + 9 = \$35$.

11. *Understand:* How are the cuts made? Vertically if the log is lying on the ground. *Plan:* Start making cuts and see what happens. *Solve:* 4 cuts. *Look Back:* The first cut creates two pieces, and each additional cut adds another piece.

13. $15 + 8 + 3 = 26$

15. (a) 5 (b) $T + N$

17. (a) 3 (b) 6

19. (a) Bea Young is 12 years old. Her father is 2 less than 3 times her age. How old is he?
(b) A room has 3 rows of 12 chairs each. During a film, only 2 of the chairs are empty. How many people are attending the film?

21. Move the two dots at the ends of the top row to the ends of the third row. Then move the bottom dot to the top.

23. Divide the coins into 3 groups: 6, 5, and 5. Put 5 coins on each side of the balance and find which of the three groups has the lighter coin. Take the group of 5 or 6 with the lighter coin and divide it into 3 groups of 2, 2, 2 or 2, 2, 1. Place two coins on each balance and find which of the three remaining groups has the lighter coin. If this group has more than 1 coin, place 1 coin on each balance to find the lighter coin.

25. (a) 27 (b) 9

27. (a) 1, 2, 1; total = 4
(b) 1, 3, 3, 1; total = 8 (c) 2^{N-1}

29. Briefly put, Polya's four steps for problem solving are: understand, plan, solve, and check. First, read the problem and make sure you understand all the information. Next, devise a plan to solve the problem. Then, carry out your plan. Finally, see if your result makes sense and review what you have learned from working on the problem.

1.6 *Homework Exercises*

1. Devising a plan

3. 32 and 37

5. 4.6 or −9.6

7. 60

9. 33rd

11. A is cheaper for 44 or fewer calls; B is cheaper for 45 or more calls.

13. (possible answer) The assistant manager is off weeks 1 and 2; one secretary is off weeks 3 and 4; the other is off weeks 5 and 6. One agent is off weeks 1 and 2; the other is off weeks 3 and 4.

15. 66 ft (draw a picture)

17. 17 (make a table)

19. (a) 40 yd (a square) (b) $4\sqrt{N}$ yd

21. First, he takes the goose across. Then he takes the fox (or corn) across and picks up the goose when he drops off the fox (or corn). He brings the goose back to the starting side and picks up the corn (or fox). Finally, he goes back and brings the goose across again.

23. 2, 3, 4, 5, 6, 7, 8, 9, or 10

25. Minimum of $E - (C - G)$ and a maximum of G

27. $C = 4T + 2$

29. $C = 2T + 4$

31. (a) 2 (b) 6 (c) 24
(d) $N \cdot (N - 1) \cdot (N - 2) \cdots 2 \cdot 1$
(e) inductive

33. In the guess-and-check strategy, you start by making a guess. Then check to see how accurate it is. Make a new guess using what you learned from your check. Continue guessing and checking until you find the answer.

Chapter 1 *Review Exercises*

1. The process of reaching a necessary conclusion from given hypotheses

2. Induction

3. Deduction

4. Deduction

5. First, the mechanic would find out what was wrong with the car. Next, the mechanic would devise a plan to fix the problem. Then, the mechanic would try out the repair plan. Finally, the mechanic would check to see if the problem no longer occurred.

6. No

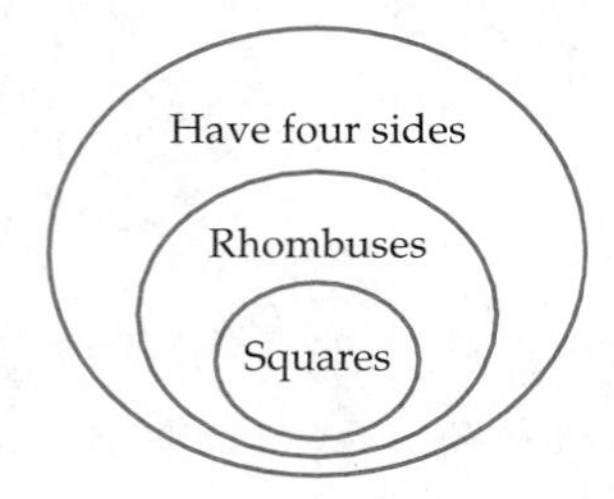

7. If a teacher started class by checking attendance 10 days in a row, a student would generalize that the teacher always starts class this way.

8. (a) All dogs are animals.
All poodles are dogs.
Therefore, all poodles are animals.
(b) All dogs are cats.
All poodles are dogs.
Therefore, all poodles are cats.

9. Yes

10. (a) If the ground gets wet, then it is raining.
(b) No

11. (a) If it has four congruent sides, then a rectangle is a square.
(b) If I call, then there is a problem.

12. $N \to N + 5 \to 3N + 15 \to 3N + 6 \to 2N + 6 \to N + 3$

13. (a) $21 - n$ (b) -19

(c)

x	1	2	3	4
y	20	19	18	17

(d) Inductive

14. (a) 1.5 (b) $T = \dfrac{90}{S}$

15. (a) 1, 2, 3 (b) $13 + 15 + 17 + 19 = 4^3$; yes

16. (a) $7^2 - 4^2 = 3 \cdot 11$
(b) $N^2 - (N - 3)^2 = 3 \cdot (2N - 3)$
(c)
$$N^2 - (N - 3)^2 \stackrel{?}{=} 3 \cdot (2N - 3)$$
$$N^2 - (N^2 - 6N + 9) \stackrel{?}{=} 6N - 9$$
$$N^2 - N^2 + 6N - 9 \stackrel{?}{=} 6N - 9$$
$$6N - 9 = 6N - 9$$

17. I had \$10. I spent \$3 and then found a \$5 bill. How much do I have now?

18. 18

19. (a) Guess and check
(b) 3 tuna and 8 buckwheat

20. 6

21. 144

22. 9 days

23. (a) (possible answer) A well is 40 ft deep. A snail goes up 7 ft and slides back 5 ft each day. On what day does it reach the top?
(b) (possible answer) A well is 40 ft deep. A snail goes up 5 ft and slides back 3 ft each day. On what day does it reach the top?

24. (a) Mack owns the sea turtle, Rosie the gorilla, Mona the crocodile, and Ahmad the ant.
(b) Deduction

25. (a) $142{,}857 \times 4 = 571{,}428$
(b) The answer always contains the digits 1, 2, 4, 5, 7, and 8.
(c) The pattern works for 1 to 6 as second factor.

26. Inductive reasoning involves making a reasonable generalization from specific examples. Deductive reasoning is the process of drawing a necessary conclusion from given assumptions. To tell the difference, see if the conclusion requires a leap from specific to general (induction) or if the conclusion follows automatically from what is given to be true (deduction).

27. First see if it's a simple nth term formula based on $n \pm k$ or kn. If not, see if there is a common difference between successive terms. If so, then the nth term is kn plus or minus a constant. If this doesn't work, try squaring or cubing n combined with addition, subtraction, or multiplication.

Chapter 2

2.1 *Homework Exercises*

1. Group, batch, bunch, pile, flock

3. $\{0, 2, 4, 6, 8\}$

4. (a) T (b) F (c) T (d) F

5. (a) and (b)

7. A, D, F, and H; E and N

9. (a) F (b) $\{1, 2\}$ and $\{3, 4\}$

11. (a) Infinite (b) Finite (c) Infinite

13. $\overline{A} = \{2, 4, 6, 8, 10\}$
$\overline{B} = \{6, 7, 8, 9, 10\}$

15. (a), (c), and (d)

16. (a) $A \subseteq B$ (b) $2 \notin T$

17. (a) T (b) F (c) T (d) T

19. (a) $\subseteq$ (b) $\in$

21. {B, R, J}, {B, R, M}, {B, S, J}, {B, S, M}, {B, J, M}, {R, S, J}, {R, S, M}, {R, J, M}, {S, J, M}, {B, R, S, M}, {B, R, J, M}, {B, S, J, M}, {R, S, J, M}, {B, R, S, J, M}

23. (a) 2 (b) 6 (c) 24
(d) $N \cdot (N - 1) \cdot (N - 2) \cdots 2 \cdot 1$

25. (a) 3 (b) 5 (c) 7 (d) $2N - 1$

27. $KN - K + 1$

29. (b) Match each number in W with twice itself in E.

31. (a) Positive multiples of 4
(b) All whole numbers that end in a 3

2.2 *Homework Exercises*

1. (a) $\{3, 9\}$ (b) $\{1, 3, 5, 6, 7, 9, 11, 12, 15, 18\}$
(c) Yes (d) No

3. The two streets pass through the intersection, which is the region that is common to both streets.

5. (a) Yes
(b)

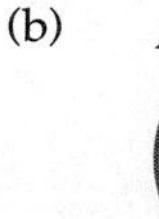

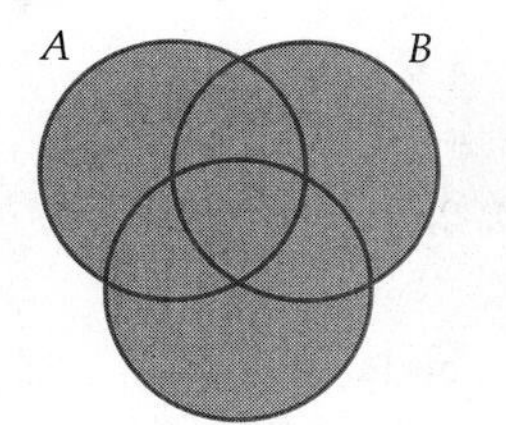

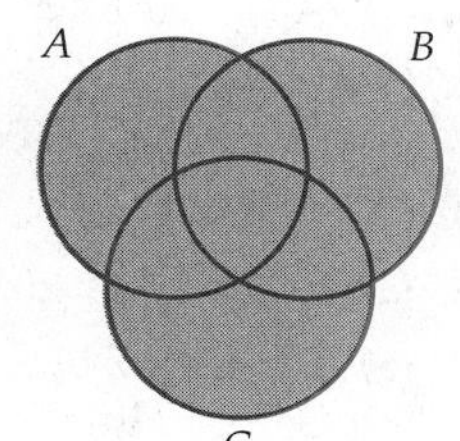

6. (a) 5 (b) 3

7. (a) $a + b$ (b) a

9. (a)

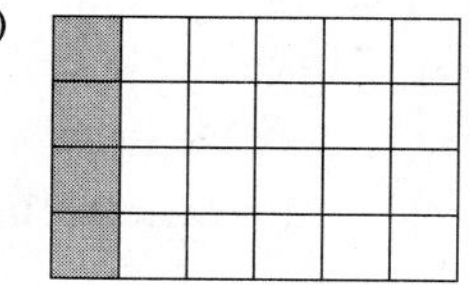

(b)

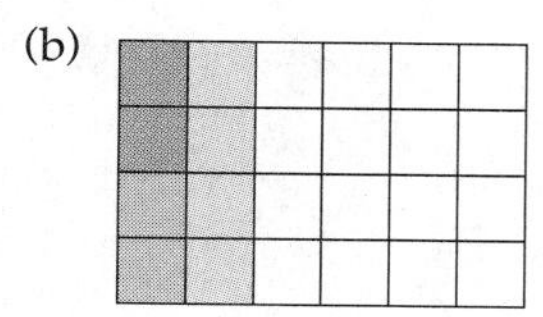

(c) 16

11. (a) $E = F \cap G$ (b) $A \cup E = E$

13. (a), (b), (c)

15. (a) $\in$ (b) $\cap$ (c) $\subseteq$

17. (a) Impossible (b) $A = \{1\}, B = \{2\}$
(c) $A = \{1\}, B = \{1\}$

19. (a)

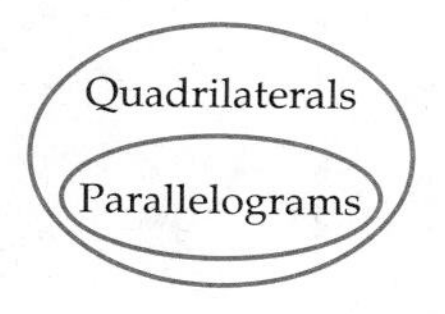

(b)

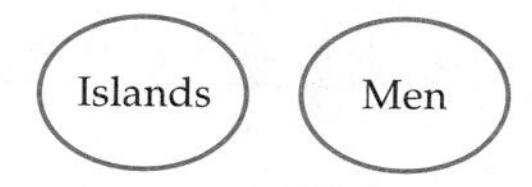

21. (a) 125 (b) 350 (c) 10

23. (a)

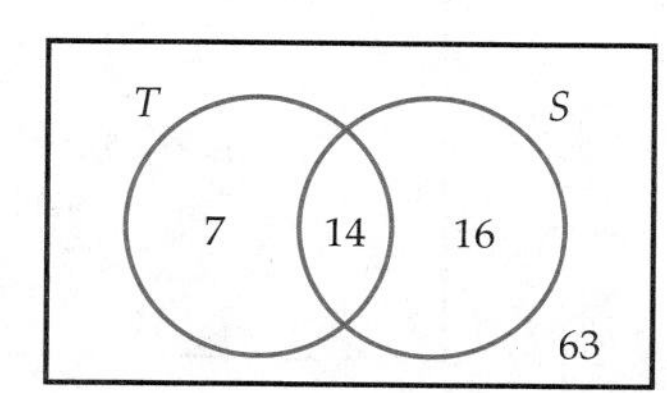

(b) 63

25. (a) Gray blocks and triangle blocks
(b) Gray triangles

27. I is small rectangles that are not blue; II is small blue shapes that are not rectangles; III is small, blue rectangles; and IV is blue rectangles that are not small.

29. (a) The left inner circle is BLUE TRIANGLES, the left outer circle is TRIANGLES, and the right-hand circle is RECTANGLES.
(b) The left circle is GRAY, the middle circle is TRIANGLES, and the right circle is BLUE.

31. (a) 2 (b) 7 (c) 8 (d) 3

33. (c); P contains polygons with the interior shaded

35. (a) $\{(1, 0), (1, 2), (1, 4), (1, 6), (2, 0), (2, 2), (2, 4), (2, 6)\}$
(b) 8

37. 12

39. 90

41. (a) 136%
(b) Some people put down more than one response.

43. (a) Female education majors
(b) College students 22 years or older who are not education majors
(c) Set U (d) { }

45. (a)

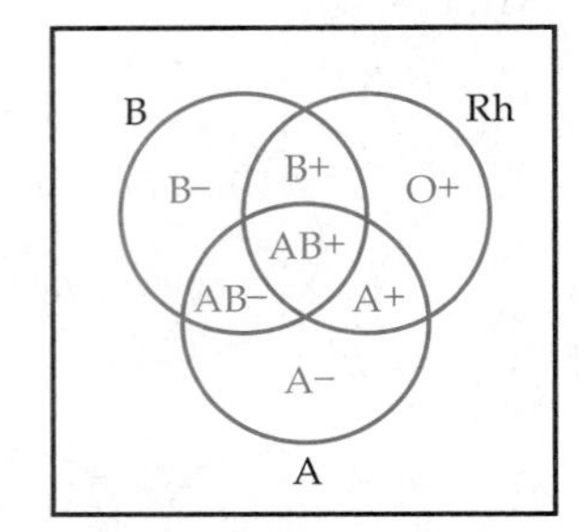

(b)

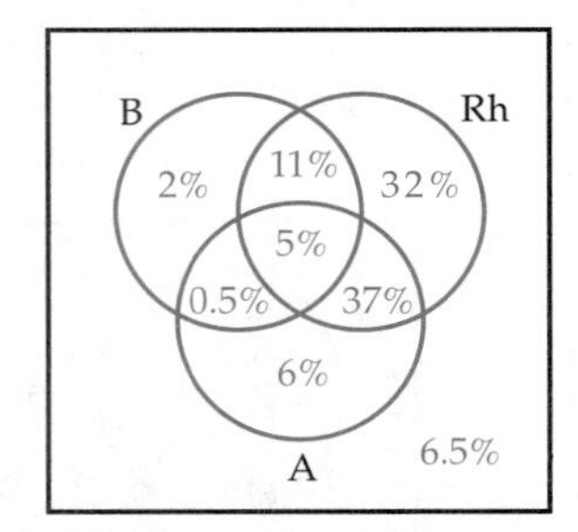

(c) 0

47. (b)

49. Each line perpendicular to $\overline{AC}$ will match up a point from $\overline{AC}$ with a point from $\overline{AB} \cup \overline{BC}$.

51. Could be

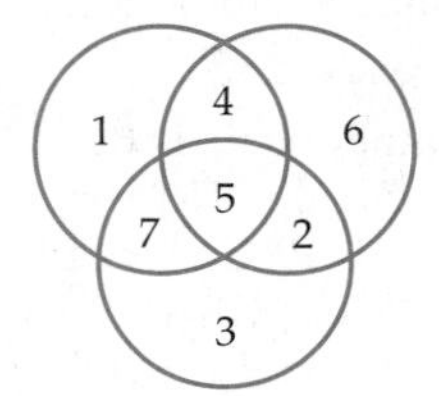

53. (b)

Chapter 2 *Review Exercises*

1.

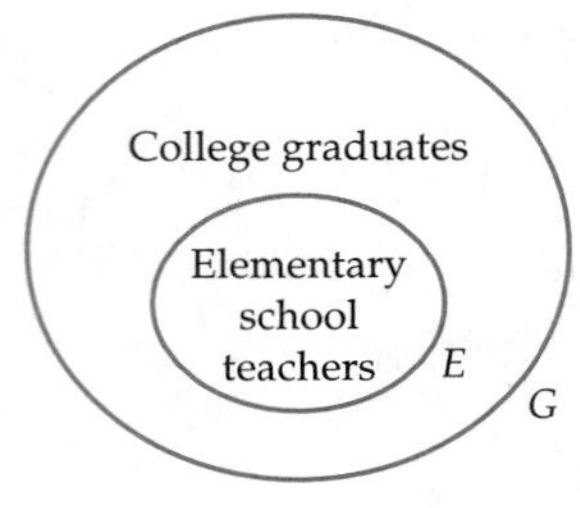

$E \subseteq G$

2. Not thick, gray, and not triangles

3. All are true.

4. (a) $\cap$ (b) $\in$

5. (a) 8 (b) A set with 3 elements
(d) A set with the two elements from U that are not in Set A.

6. (a) 700 (b) 1000 (c) 300

7. (a) 8 (b) 6

8. 5

9. (a) 4 (b) 14

Chapter 3

3.1 *Homework Exercises*

1. (a) Numeral (b) Number

3. 𓍢𓍢𓍢∩∩IIIIIIII

5. A symbol for zero, ten digits to represent all numbers, place value

7. (a) 14 (b) 40 (c) 1613 (d) 1964

9. (a) 642 (b) 12 (c) 6

11. (a) $4 \cdot 100 + 7$
(b) $3 \cdot 1000 + 1 \cdot 100 + 2 \cdot 10 + 5$

13. $(A \times 1000) + (B \times 100) + (C \times 10) + D$

15. (a) 1324 (b) 207

17. (a) 3 greens, 7 blues, and 2 yellows (b) Green
(c) Chip trading is more abstract because chips for 1, 10, 100, and 1000 are all the same size, whereas the value of a base-ten block is proportional to its size.

19. 1 hundred, 2 tens, 7 ones; 1 hundred, 1 tens, 17 ones; 1 hundred, 0 tens, 27 ones

21. (a) Measure (b) Count (c) Measure

22. It's in between. Dollars are like a measure, but cents are like a count.

23. (a) \$600 (b) \$36,400

25. (a) 500 (b) 400 (c) 500

27. (a) Ten (b) Hundred (c) Ten thousand

29. (a) \$120 (b) Down

31. (a) 400 (b) 4,000,000
(c) 4 (d) 4000 (e) 400,000,000

33. 6,210,001,000

35. Sixty

3.2 *Homework Exercises*

1. (a)

3. (a) Combine measures
(b) The sale price, the current value, the size and condition of the house, the location of the house

4. (a) $2 + 3 = 5$
(b)

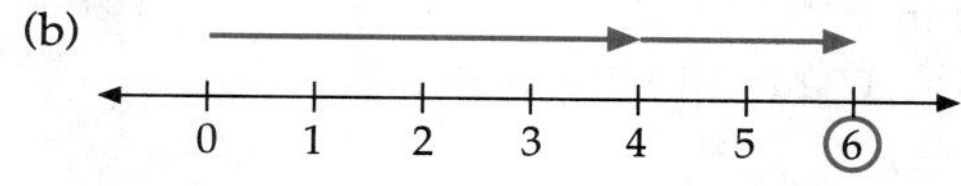

5. (a) $3 + 5 = 8$ (b) Combine measures

7. Compare measures

8. Take away sets

9. 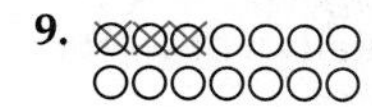

$14 - 3 = 11$.

10. (a) $7 - 5 = 2$
(b)

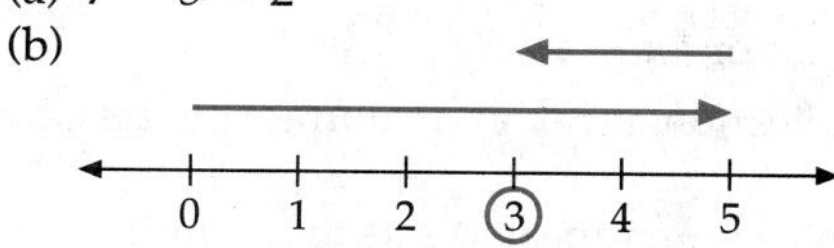

11. (a) $5 - 3 = 2$ (b) Compare measures

13. (a) ⊗⊗OOOOOOO

(b) OOOOOOOOO OO → OOOOOOOO OOOOOOOO

(c) □□□□□□□□□ □□ → □□□□□□□□□ □□□□□□□□□

14. No

15. Problem (a) is about sharing and emphasizes cooperation. Problem (b) has a materialistic and competitive theme.

17. Addition, combine sets

18. Addition, combine measures

19. I have 10 oranges. I give 3 to Bill and 5 to Susan. How many do I have left?

21. $A - F - C + P$

23. (a) $10 - x$ (b) $x - 2$ (c) $x + 6$

25. $c - 1$

27. (a) $8 = 6 + 2$
(b) For whole numbers a and b, $a < b$ when $a = b - k$ for some counting number k.

29. (a) $3 = 1 + 2$, $5 = 2 + 3$, $6 = 1 + 2 + 3$, $7 = 3 + 4$, $9 = 2 + 3 + 4$, $10 = 1 + 2 + 3 + 4$
(b) Any counting number that is not a power of 2 can be written as the sum of two or more consecutive counting numbers.

31. The 2 should be subtracted from the 27, not added. The women first paid \$30 and then got \$3 back. So they ended up paying \$27, \$2 to the bellperson and \$25 for the room.

33. If $(a - b) \geq c$

35. (a) $A - B + X - Y$
(b) Last year, your gross pay was \$$A$. You spent \$$T$ on taxes and \$$E$ on other expenses. Write an expression showing how much money you had left.

37. (a)

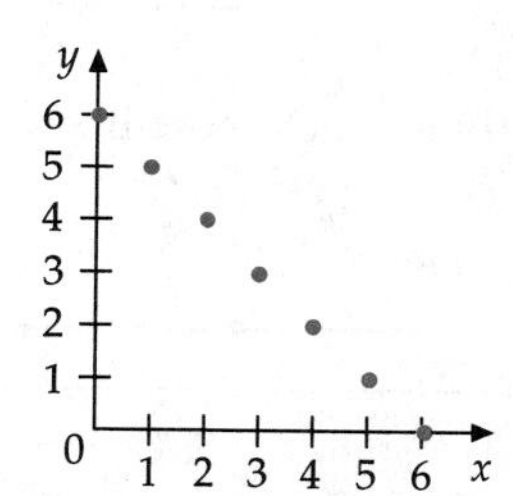

(b) They all lie on a straight line.
(c)

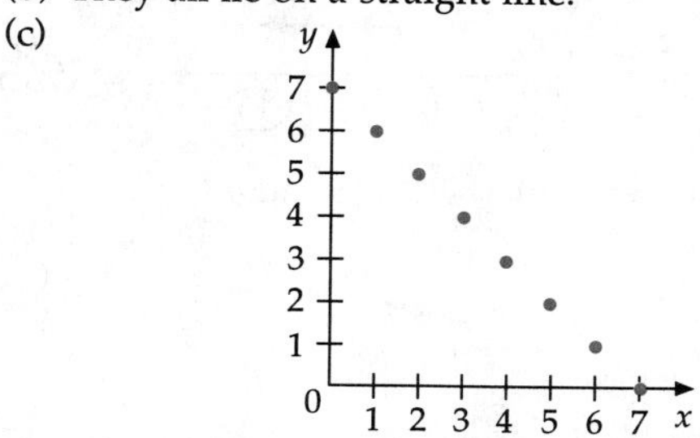

(d) They all lie on a straight line.
(e) If the ordered pairs of any fact family are plotted on a graph, the points will lie on a straight line.
(f) Induction

39. (a)

4	3	8
9	5	1
2	7	6

(b) Same problem but change "1 to 9" to "5 to 13" and change "15" to "27"

41. $A = 9, B = 2$

43. Yes

3.3 *Homework Exercises*

1. $3 \cdot 7$

3. (a) No, $2 \cdot 2 = 4$ (b) Yes (c) Yes

5. Repeated measures

7. Repeated sets

8. Divide the interior into 1-ft squares as shown. It takes 6 squares to cover the interior, so the area is 6 ft^2.

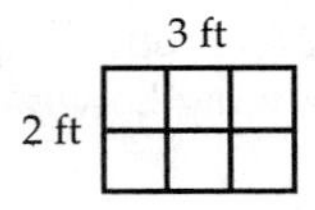

9. Problem (b) is much more likely to occur in everyday life.

10. (a) $3 \times 2 = 6$
(b)

11. (a) (b) (c)

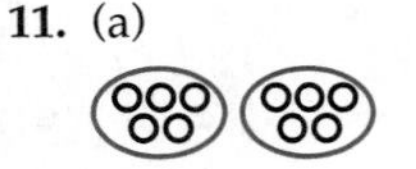

13. (a) 74 and $X^2 + 10$ (b) 38 and $4X - 2$
(c) Inductive

15. (a) 3^2 (b) 4^3

17. Partition measures

19. Repeated measures

21. (a) $8 \div 4 = 2$
(b)

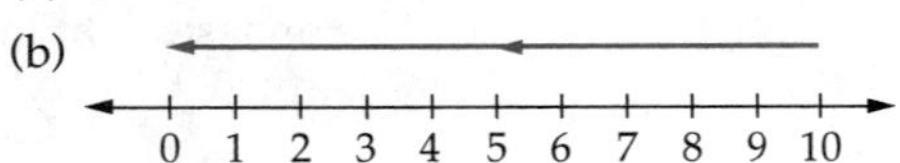

22. (a) $8 \div 2 = 4$ (b)

23. (a) (b) (c)

25. (a) If he puts a maximum of 4 on a page, how many pages will he need?
(b) If he puts 4 on each page, how many extra photographs will be left over?
(c) If 4 photographs fill a page, how many pages can he fill?

27. (a) 1 (b) 2 (c) Quotient (d) Remainder

29. (a) Undefined
(b) The expression $7 \div 0 = ?$ is the same as $? \times 0 = 7$, which has no answer. So we make $7 \div 0$ undefined.

31. Some sort of error message; the problem is undefined.

32. (a) 5 and 20 (b) Deductive

33. $N/3 - 2$

35. 36

37. (a) 10 (b) 67 (c) 9

39. (a) $2t$ (b) $t - 5$ (c) $5t$

41. (a) 100 (b) 3 (c) 7

43. I have 40 pistachio nuts to share equally with you. How many will we each get?

45. 608

46. No

47. $n^2 - (n - 2)^2$ or $4n - 4$

49. (a) Multiplication, array and subtraction, take away sets
(b) Multiplication, repeated sets and addition, combine sets

51. (a) Addition, combine sets and subtraction, take away sets $(10 - (2 + 3))$, or both steps are subtraction, take away sets $(10 - 2$ and $8 - 3)$
(b) Multiplication, array and subtraction, take away sets

53. A school is charging \$3 for lunch and \$6 for a concert. How much will it cost to buy 7 tickets for both events?

54. $AB + CD$ dollars

55. $c/2$

57. (a) 5 (b) 1, 5, or 6

59. Pour 3 cupfuls into the glass to get 6 oz. Then fill one cup. Pour from one cup into the other until both cups are at the same level. Then each has 1 oz. Pour one of them into the glass.

61. They are both odd.

63. (a) Correct (b) Incorrect

65. (e) The first and last columns each contain 5, 4, 6, and 9. The second column contains the same four numbers each multiplied by 2. The third column contains the same 4 numbers each multiplied by 4. The product will always be $5 \cdot 4 \cdot 6 \cdot 9 \cdot 2 \cdot 4 = 8640$.

(f)

1	3	2
3	9	6
5	15	10

67. Take out 1 chip. On successive turns, reduce the pot to 7 chips, 4 chips, and finally 1 chip.

69. (a) Multiplication, repeated sets
(b) Division, partition sets

71. (a) 36
(b) Find a two-digit number that is 3 times the product of its digits.

3.4 *Homework Exercises*

1. Mutiply any two of the numbers first. Then multiply the answer times the remaining number.

3. (a) Commutative, multiplication

(b)

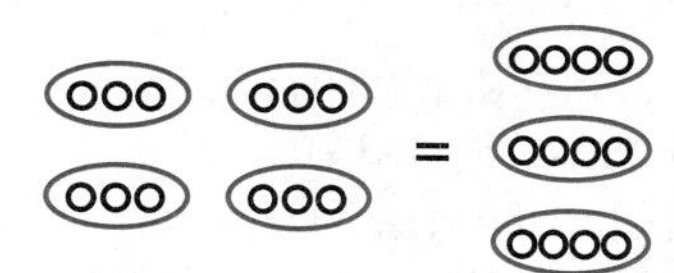

5.

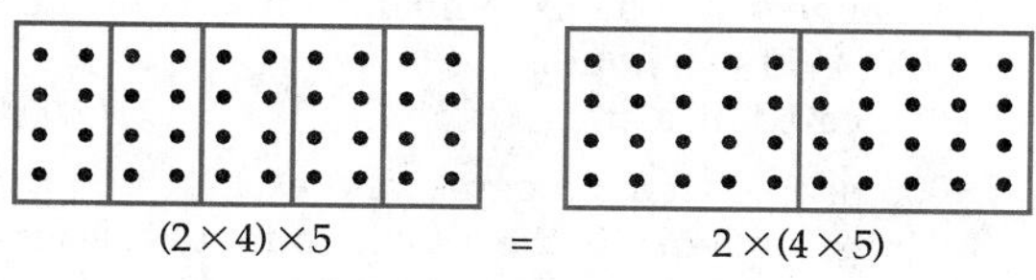

6. (a) Commutative property of division
(b) In $8 \div 2$, 2 is the divisor, and the result is 4. In $2 \div 8$, 8 is the divisor, and the result is 1/4.

7. $(8 - 4) - 2 \neq 8 - (4 - 2)$

9. Multiply $6 \times \$5$ first and then multiply $\$30 \times 9 = \$270$.

11. Each fact that you learn in one order has the same answer when the order of the factors is reversed.

13. (b)

15. (a) *Hint:* $37 \times 12 = (37 \times 3) \times 4$

17. 0, identify, addition

18. Any fact involving 1 as a factor.

19. Order property = Commutative;
Zero property = Identity for addition;
Grouping property = Associative;
Property of one = Identity for multiplication

21. (a) (b) $10(8 + 4) = 120$ and $10 \cdot 8 + 10 \cdot 4 = 120$
(c) $10(8 + 4) = 10 \cdot 8 + 10 \cdot 4$ by the distributive property of multiplication over addition.

23. (a) Associative property of multiplication
(b) Distributive property of multiplication over addition
(c) Commutative property of multiplication
(d) Associative property of addition

25. "Multiplication over addition is distributive" goes in all three blanks.

27. Start with 1 hundred, 8 tens, and 2 ones, and 3 hundreds, 3 tens, and 6 ones. Combine 2 ones and 6 ones to obtain 8 ones. Combine 8 tens and 3 tens to obtain 11 tens. Trade 10 tens for 1 hundred, leaving 1 ten. Combine 1 hundred, 1 hundred, and 3 hundreds to obtain 5 hundreds. The answer is 5 hundreds, 1 ten, 8 ones = 518.

29. Start with 3 hundreds, 3 tens, and 6 ones. Take away 2 ones, leaving 4 ones. You can't take away 8 tens from 3 tens. Regroup 3 hundreds and 3 tens as 2 hundreds and 13 tens. Take away 8 tens from 13 tens, leaving 5 tens. Take away 1 hundred from 2 hundreds, leaving 1 hundred. The difference is 1 hundred, 5 tens, and 4 ones = 154.

31. (a) 25 (b) 445

33. 24×4 means 4 sets of 24. Show this.

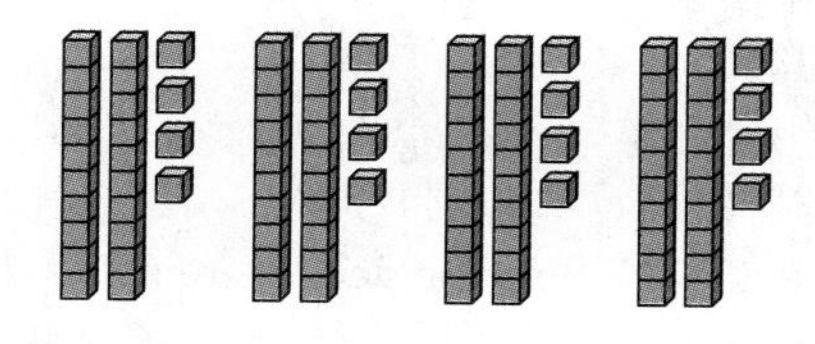

Combine the ones together. You have 16 ones.
Trade in 10 ones for 1 ten.
Combine the tens together. You end up with 9 tens and 6 ones = 96. So $24 \times 4 = 96$.

35. Distributive property of multiplication over addition

37.
$$\begin{array}{r l} 86 & \\ \times\ 42 & \\ \hline 172 & \leftarrow\ 2 \times 86 \\ 3440 & \leftarrow\ 40 \times 86 \\ \hline 3612 & \end{array}$$

39. (a) 62 R 3 (b) 241 R 4

41. $246 \div 2$ means to divide 246 into 2 equal groups. Start with 2 hundreds, 4 tens, and 6 ones.

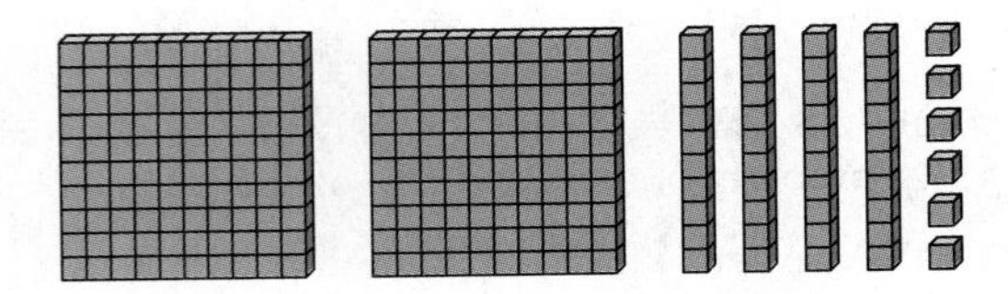

First, put the 2 hundreds into 2 equal groups. Each group gets 1 hundred. Next, put the 4 tens into 2 equal groups. Each group gets 2 tens. Finally, divide the 6 ones between the 2 equal groups. Each group ends up with 1 hundred, 2 tens, and 3 ones = 123. So $246 \div 2 = 123$.

43. (a) 175
(b) Adding the number of ones in the second addend to the number of tens in the first addend

45. (a) 75
(b) Borrowing 1 ten without subtracting it from the tens column

47. (a) 638
(b) Failing to carry the tens from the right-hand multiplication to the tens column

49. (a) 46
(b) Placing the digits in the quotient from right to left

51. A child might forget to put the 0 in the product, or a child might make an error on a multiplication fact.

53. $300 \times 7 = 2100 \rightarrow 80 \times 7 = 560 \rightarrow 2 \times 7 = 14 \rightarrow 2100 + 560 + 14 = 2674$

55. (a) 29 (b) 167

57. (a) 1456 (b) 2286

59. (a) 27, 28, 29 (b) 136, 137, 138
(c) $N/3 - 1, N/3, N/3 + 1$

61. 372 R 13 (Multiply the decimal remainder by 23.)

63. (a) True for $A = B$; false for $A \neq B$
(b) True if $B = 1$ or $C = 0$; false if $B \neq 1$ and $C \neq 0$
(c) True if $A = 1$ or $A = 0$ or $B = 0$ or $C = 0$; false if $A \neq 1$ and $A \neq 0$ and $B \neq 0$ and $C \neq 0$
(d) True if $A \geq B$; false if $A < B$
(e) True if A is divisible by B; false if A is not divisible by B
(f) True if $A = 1$ or $C = 0$; false if $A \neq 1$ and $C \neq 0$
(g) (possible answer) True if $B = 1$, $A = 2$, and C is any number; false for $A = B = C$ and many other cases

65. (a) It doubles. (b) $2a + 2b = 2(a + b)$
(c) Inductive (d) Deductive

67. Multiplication over addition is distributive; multiplication over addition is distributive; addition is commutative; addition is commutative.

69. (e) The tens digit is 9 and the sum of the ones and hundreds digits is 9.
(f) Inductive

71. The dividend could be 1230 or 1236, and the quotient would be 205 or 206, respectively.

73. (a) Yes (b) No (c) No (d) No

3.5 *Homework Exercises*

1. (a) $39 + 90 = 129$ and $129 + 7 = 136$ miles or $97 + 3 + 36 = 136$ miles
(b) Commutative property of addition

3. $5^3 = 125$ and triple the number of zeroes to obtain 6 zeroes in the answer, 125,000,000.

5. $4 \times 3 = 12$ and add the number of zeroes in the two factors to obtain the number of zeroes (5) in the product, 1,200,000 people.

6. (a) You must estimate when you predict the future.
(b) You cannot tell exactly how much each person will weigh each time that people board the elevator. You also want to incorporate a safety factor.
(c) You don't need to know the exact time. An estimate is good enough.

7. Round it to $6000 + 8000 + 4000$. Add $6 + 8 + 4 = 18$ and attach three zeroes. 18,000.

9. (a) Change $437 \div 55$ to compatible numbers like $400 \div 50 = 8$ hours.
(b) Repeated measures

11. If 9000 people attend a game, and each person paid \$8 admission, what is the total amount of sales in dollars?

12. $59 + 42$ is about 100, plus 97 is about 200, plus 32 makes about 230.

13. Too high

15. (a) 8 (b) 28 (c) 36 (d) 26 (e) 17

17. Add $3 + 1 + 21 \approx \$25$ million (rounding)

19. (a) $32 \times 48 \approx 30 \times 40 = 1200$ lightbulbs
(b) $32 \times 48 \approx 30 \times 50 = 1500$ lightbulbs
(c) Rounding

21. (a) Estimate (b) Exact
23. Multiply 18 × 24 on the calculator and place two zeroes on the end.
27. (a) 225, 625, 1225
(b) The product ends in 5. Then multiply the tens digit t times $(t + 1)$ to obtain the left-hand digits of the product.
29. (a) $n = 14$ (b) $x = 177$ (c) $y = 7$
31. (a) 631 × 542 (guess and check)
(b) Same problem, but use 1, 2, 3, 7, 8, and 9 as the digits
33. (a) Wrong; the answer should be about $3 \cdot (600) = 1800$
(b) Reasonable, since $30 \times 50 = 1500$
(c) Wrong; the answer should be about $4000 \div 20 = 200$
35. (a) Calculator; too time-consuming to do another way
(b) Mental computation (add 750 + 250 first)
(c) Paper-and-pencil or calculator, depending upon which is more convenient
37. (a) $389 + 895 = 384 + 900 = 1284$
(b) $4990 + 3627 = 5000 + 3617 = 8617$
(c) $762 - 89 = 763 - 90 = 673$
39. (a) $8 \times 1000 - 8 = 7992$ (b) 5994
(c) $a \times 999 = a \times (1000 - 1) = a \times 1000 - a$
41. (a) 24 plus half of 24, and add a 0 on the right. $24 + 12 = 36 \rightarrow 360$
(b) 630
43. 10 million × 70 × 30. Now 7 × 3 = 21, and there are three zeroes in the factors. The answer is 21,000 million or 21 billion gallons.

3.6 *Homework Exercises*

1. (a) 1 eight and 4 ones (b) 2 fives and 2 ones
3. 1_{three}, 2_{three}, 10_{three}, 11_{three}, 12_{three}, 20_{three}, 21_{three}, 22_{three}, 100_{three}, 101_{three}, 102_{three}, 110_{three}
5. (a) 61 (b) 113 (c) 139
6. (a) 56_{eight} (b) 101_{five} (c) 2244_{five}
7. (a) 3 quarters, 2 nickels, and 1 penny
(b) $86 = 321_{five}$
8. (a) The last digit of an odd number is odd; the last digit of an even number is even.
(b) An odd number has an odd number of odd digits; an even number has an even number of odd digits.
9. The last digit is 0.
11. 1011_{five}
13. 1314_{five}
14. (a) 112_{five} (b) 1011_{five} (c) 1220_{five}
15. (a) 13_{five} (b) 124_{five} (c) 234_{five}
16. Combine 3 longs and 1 unit with 2 longs and 2 units. First combine the units to obtain 3 units. Then combine the longs to obtain 5 longs. Now 5 longs can be traded in for 1 flat. The sum is 1 flat and 3 units. So $31_{five} + 22_{five} = 103_{five}$.
17. Start with 4 longs (1 by 5) and 3 ones. Now take away 24_{five} starting with the ones. You cannot take 4 ones from 3 ones, so trade 1 of the 4 longs for 5 ones. Now you can take away 24_{five} from 3 longs and 8 ones. 8 ones − 4 ones = 4 ones and 3 longs − 2 longs = 1 long. There is 1 long and 4 ones left, so $43_{five} - 24_{five} = 14_{five}$.
18. (a) 1442_{five} (b) $243{,}302_{five}$
19. (a) 24_{five} R 2 (b) 1204_{five} R 1
21.

+	0	1	2	3	4	5	6	7
0	0	1	2	3	4	5	6	7
1	1	2	3	4	5	6	7	10
2	2	3	4	5	6	7	10	11
3	3	4	5	6	7	10	11	12
4	4	5	6	7	10	11	12	13
5	5	6	7	10	11	12	13	14
6	6	7	10	11	12	13	14	15
7	7	10	11	12	13	14	15	16

×	0	1	2	3	4	5	6	7
0	0	0	0	0	0	0	0	0
1	0	1	2	3	4	5	6	7
2	0	2	4	6	10	12	14	16
3	0	3	6	11	14	17	22	25
4	0	4	10	14	20	24	30	34
5	0	5	12	17	24	31	36	43
6	0	6	14	22	30	36	44	52
7	0	7	16	25	34	43	52	61

22. (a) 120_{eight} (b) 44_{eight} (c) 1104_{eight}
23. (a) 25_{eight} (b) 244_{eight} R 2
25. (a) 1, 2, 3, 4, . . . 15 pounds
(b) 1, 2, 4, and 8 pound weights
(c) Same problem with 1 to 100 pounds
27. (a)

+	0	1
0	0	1
1	1	10

×	0	1
0	0	0
1	0	1

(b) $10{,}010_{two}$ (c) $1{,}000{,}001_{two}$

29. (a) 182 (b) $10{,}110{,}110_{two}$
(c) The ones digit occupies the first 4 columns in base two and six $\rightarrow$ 0110. The sixteens digit occupies the next 4 columns to the left in base two and $B \rightarrow$ 1011. So $B6_{sixteen} = 10{,}110{,}110_{two}$.

31. (a) 10 is on the second and fourth cards, so you would add $2 + 8 = 10$.
(b) The numbers in the upper left-hand corners of the four cards correspond to the first four place values of base two: ones, twos, fours, eights. The numbers from 1 to 15 are converted to base two and written on the corresponding place-value cards for which they have a ones digit.

33. (a) 464 (b) $328_{twelve} = 464$

35. (a) 2434_{six} (b) $121{,}110_{four}$

37. 1024_{five}

39. $30{,}424_{six}$

41. (a) 4 (b) $26_{five} = 121_b$. What base is b?

43. b is any whole number greater than 7.

Chapter 3 *Review Exercises*

1. (a)

(b) Commutative property of multiplication

2. $362 - 187 = 375 - 200 = 175$

3. True for $A = 0$ and $B = C = 1$ and false for $A = B = C = 1$

4. (a) $8 \times 40 = 320$
(b) Distributive property of multiplication over addition

5. Nearest hundred

6. In a place-value system, the position of the digit determines its value. In a system that does not have place value, the value of a digit is the same regardless of its position.

7. (a) Undefined
(b) $3 \div 0 = ?$ is the same as $? \times 0 = 3$, which has no solution. So we make $3 \div 0$ undefined.

8. (a) 37 (b) 2 (c) 6

9.

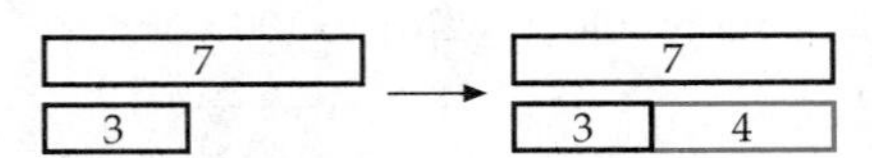

10. (a) Change to compatible $2400 \div 40 = 60$ gallons.
(b) Repeated measures

11. Subtraction, compare sets

12. Division, repeated sets

13. (a) No (b) $8 \div (4 \div 2) \neq (8 \div 4) \div 2$.

14. Multiplication, repeated sets and subtraction, take away sets

15. Subtraction, take away measures and division, repeated measures

16. (a) 1513 (b) 240 (c) 2836

17. Represent the numbers with base-ten blocks. The first is 3 flats, 2 longs, and 6 units; the second is 2 flats, 9 longs, and 3 units. Add the ones, then the tens, and then the hundreds. 6 units plus 3 units makes 9 *units*. 2 longs plus 9 longs makes 11 longs. Trade in 10 longs for 1 flat and you have *1 long left over*. Finally, combine this 1 flat with 3 flats plus 2 flats for a total of 6 *flats*. The answer is 6 flats, 1 long, and 9 units or 619.

18. (a) $77 + 666 = 743$

19. $HD - S$ dollars

20. \$46

21. 244_{six}

22. 150_{seven} R 4_{seven}

23. b is whole number ≥ 3; $a = 2b + 1$

24. A repeated sets problem asks how many times you must repeat the amount of the divisor to obtain the total amount. A partition sets problem asks you to divide up the total into the number of equal-sized groups specified by the divisor.

25. In the rounding strategy, each number in the problem is rounded to the nearest number by place value. In the compatible numbers strategy, each number is rounded to a number close to it in size, and the resulting numbers must create a simple problem (usually division).

Chapter 4

4.1 *Homework Exercises*

1. (a) 1, 2, 4, 7, 14, 28
(b) There is no whole number c such that $6 \cdot c = 28$.

3. True

4. (a) No (b) No (c) Yes

5. 1210

7. True; $A = 4$, $B = 8$, $K = 3$ is an example.

9. True; $A = 2$, $B = 10$, $C = 20$ is an example.

11. True; $A = 2$, $B = 4$, $C = 6$ is an example.

13. 3, B, A, B

15. Divisibility-of-a-Product Theorem

17. $8 \mid (C + 16)$

19. $A \mid 8C$

21. (a) 1 by 20, 2 by 10, and 4 by 5
(b) Factoring 20

23. False; $B = 5$

25. True, by the Divisibility-of-a-Product Theorem, if $A \mid B$ then $A \mid (B \cdot B)$ or $A \mid B^2$.

27. True
1. $A \mid B$, $B \mid C$, and $A \neq 0$
2. $A \cdot K = B$ and $B \cdot L = C$, in which K and L are whole numbers
3. $A \cdot K \cdot L = C$ and KL is a whole number
4. $A \mid C$

29. True
1. C and D are even numbers.
2. $C = 2W$ and $D = 2X$ where W and X are whole numbers.
3. $CD = 2W \cdot 2X$
4. $CD = 2(2WX)$ where $2WX$ is a whole number.
5. CD is an even number.

31. $x \mid x \rightarrow x \mid xy^2 \rightarrow x \mid (xy^2 \cdot x)$ or $x \mid x^2y^2$

33. $N^3 - N = N(N + 1)(N - 1)$. Either $N - 1$, N, or $N + 1$ must be divisible by 3. By the Divisibility-of-a-Product Theorem, $N^3 - N$ must be divisible by 3. So $3 \mid (N^3 - N)$.

35. 4 in. by 36 in.; 3 in. by 48 in.; 18 in. by 8 in.; 6 in. by 24 in.; 12 in. by 12 in.; 9 in. by 16 in.

4.2 *Homework Exercises*

1. 0. 7, 14, 21, . . .

3. There is no whole number C such that $6 \cdot C = 3$.

5. (a) Factor (b) Factor (c) Multiple
(d) Factor

7. No

9. (a) 1, 4, or 7 (b) The last digit is 0.
(c) 1, 4, or 7 followed by 0

11. (a) No
(b) Any amount made with $.25 and $.15 stamps must be divisible by 5 cents.

13. (a) False (b) 15 (c)

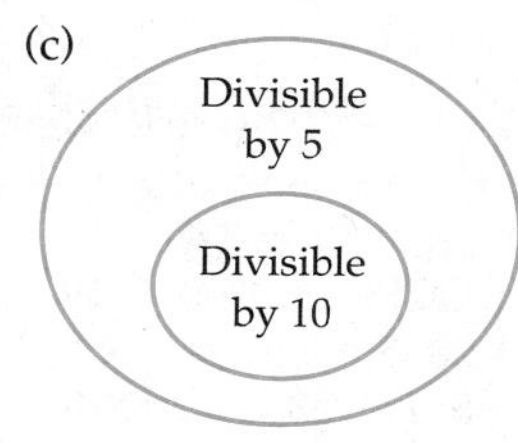

15. (a) False (b) 24 (c)

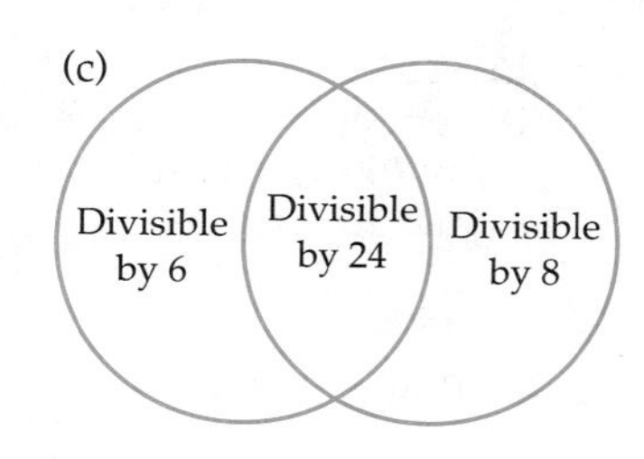

17. 1, 3, and 5

19. 100A + 10B + C; 10B; C; 100A + 10B + C

21. $\underline{A}\,\underline{B}\,\underline{C}\,\underline{D}$ = 1000A + 100B + 10C + D
= 999A + 99B + 9C + A + B + C + D
999A, 99B, 9C, and (A + B + C + D) are each divisible by 9. So 999A + 99B + 9C + A + B + C + D or $\underline{A}\,\underline{B}\,\underline{C}\,\underline{D}$ is divisible by 9.

23. If the last two digits of a number form a number that is divisible by 4, then the number is divisible by 4.

25. (b) If a number is divisible by 2 and 3, then it is divisible by 6.

27. (a) Last 2 digits are 0 (b) (1), (3), (4)

29. (a) y is divisible by a and b. (b) True

31. 222,456,564

33. False; 15 with $A = 2$ and $B = 3$

35. True
1. A is a multiple of B.
2. $A = KB$ for some whole number K.
3. $2A = 2KB$ where $2K$ is a whole number.
4. $2A$ is a multiple of B.

37. (a) or (b)

39. (a) Yes (b) Yes
(c) Let the larger number be $\underline{A}\,\underline{B}$. Then $\underline{A}\,\underline{B} - \underline{B}\,\underline{A}$ = (10A + B) − (10B + A) = 9A − 9B = 9(A − B), which must be divisible by 9.

41. (a) Yes
(c) All 8-digit numbers of the form $\underline{A}\,\underline{B}\,\underline{C}\,\underline{D}\,\underline{A}\,\underline{B}\,\underline{C}\,\underline{D}$ are divisible by 73.
(d) $\underline{A}\,\underline{B}\,\underline{C}\,\underline{D}\,\underline{A}\,\underline{B}\,\underline{C}\,\underline{D}$ = A · 10,000,000 + B · 1,000,000 + C · 100,000 + D · 10,000 + A · 1000 + B · 100 + C · 10 + D = A · 10,001,000 + B · 1,000,100 + C · 100,010 + D · 10,001
The numbers 10,001,000; 1,000,100; 100,010; and 10,001 are all divisible by 73, so A · 10,001,000 + B · 1,000,100 + C · 100,010 + D · 10,001, which equals $\underline{A}\,\underline{B}\,\underline{C}\,\underline{D}\,\underline{A}\,\underline{B}\,\underline{C}\,\underline{D}$, is divisible by 73.

43. (a) The last digit must be 0 or 4.
(b) For 2, the last digit must be 0, 2, 4, or 6; for 7, the sum of the digits must be divisible by 7.

45. (c) Yes (d) Yes (e) Yes (g) 1001

(h) 382,382
(i) Any number of the form A B C A B C is divisible by 1001, so it must also be divisible by 7, 11, and 13.

4.3 *Homework Exercises*

1. (a) Prime

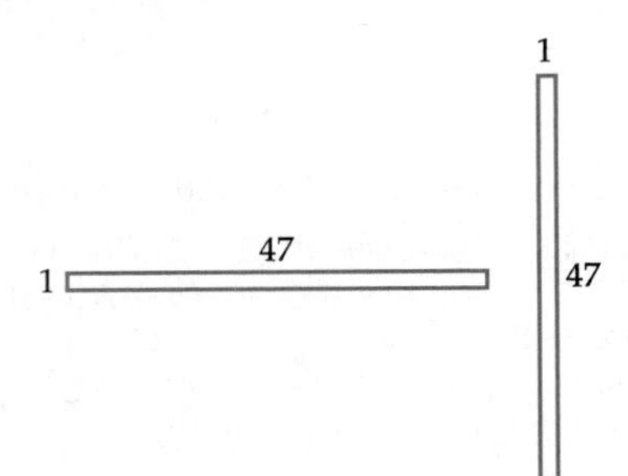

(b) Composite

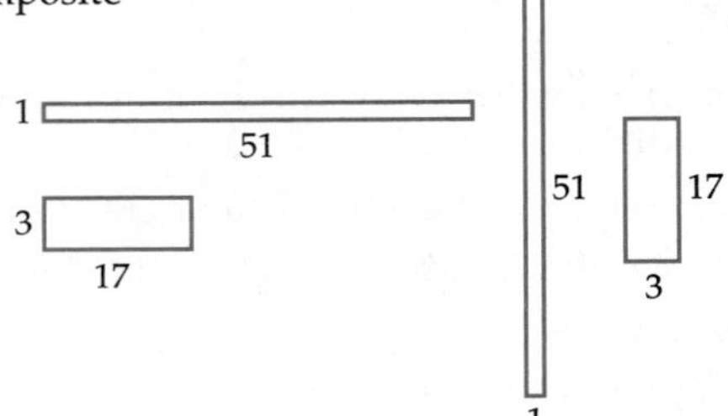

3. All the results are prime.

5. (a) 257
(b) 4,294,967,297 and 4,294,967,297 ÷ 641 = 6,700,417

7. Neither

9. 2, 3, 5, 7, 11, 13, 17, 19 (since $\sqrt{431} \approx 21$)

11. (a) Prime (b) Composite ($17 \cdot 41$)

13. exactly one prime factorization

15. (a) $76 = 2^2 \cdot 19$ (b) $320 = 2^6 \cdot 5$

17. (a) $6 \cdot 3 \cdot 2 = 36$ divisors
(b) $148 = 2^2 \cdot 37$. It has $3 \cdot 2 = 6$ divisors.

19. 101 and 103, 107 and 109, 137 and 139

21. (a) 5 + 7 (b) 13 + 17 or 11 + 19 or 23 + 7
(c) 5 + 103 is one answer.

23. $17 = 1 + 4^2$, $37 = 1 + 6^2$, $101 = 1 + 10^2$

25. (a) 6 (b) 6
(c) Either at the beginning or second from the end of a row
(d) $(6n + 1)^2 + 17 = 36n^2 + 12n + 18 = 12(3n^2 + n + 1) + 6$, which will have remainder 6 when divided by 12. $(6n - 1)^2 + 17 = 36n^2 - 12n + 18 = 12(3n^2 - n + 1) + 6$, which will have remainder 6 when divided by 12.

27. Start at 27,722

29. (a) $2 \times 3 \times 5 \times 7 + 1 = 211$,
$2 \times 3 \times 5 \times 7 \times 11 + 1 = 2311$
(b) Yes
(c) $2 \times 3 \times 5 \times 7 \times 11 \times 13 + 1 = 30{,}031$ and $30{,}031 \div 59 = 509$

31. (a) No (b) Yes

33. (a) It is divisible by 3, 5, 7, 11, 13, etc.
(b) It is divisible by 2. (c) It is divisible by 5.

35. The lockers with perfect square numbers

4.4 *Homework Exercises*

1. (a)

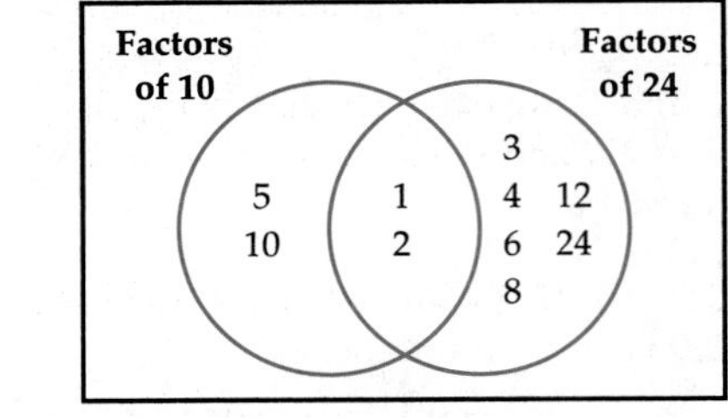

(b) 1 and 2 (c) 2
(d)

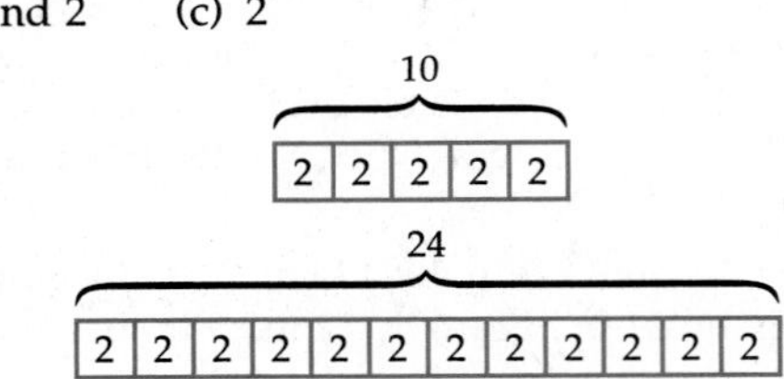

2. (a) Factors of 42: 1, 2, 3, 6, 7, 14, 21, 42
Factors of 120: 1, 2, 3, 4, 5, 6, 8, 10, 12, 15, etc.
GCF(42, 120) = 6
(b) $42 = 2 \cdot 3 \cdot 7$ and $120 = 2^3 \cdot 3 \cdot 5$. GCF $= 2 \cdot 3 = 6$.

3. 18

5. (a) 1, 2, 1, 4, 1, 2, 1, 4
(b) The sequence of GCF's will have the repeating pattern 1, 2, 1, 4, 1, 2, 1, 4, . . .

6. The GCF must also include 3 and 5. It is $3 \cdot 5 \cdot 11 = 165$.

7. $2^2 \cdot 3 \cdot 5^6$

9. (a) and (c)

11. (a) 665 ÷ 627 = 1 R 38
627 ÷ 38 = 16 R 19
38 ÷ 19 = 2
GCF(627, 665) = 19
(b) 2035 ÷ 851 = 2 R 333
851 ÷ 333 = 2 R 185

$333 \div 185 = 1 \text{ R } 148$
$185 \div 148 = 1 \text{ R } 37$
$148 \div 37 = 4$
$\text{GCF}(851, 2035) = 37$

(c) $609 \div 551 = 1 \text{ R } 58$
$551 \div 58 = 9 \text{ R } 29$
$58 \div 29 = 2$
$\text{GCF}(551, 609) = 29$

13. (a) 60, 120, 180 (b) An infinite number
(c) 60

15.

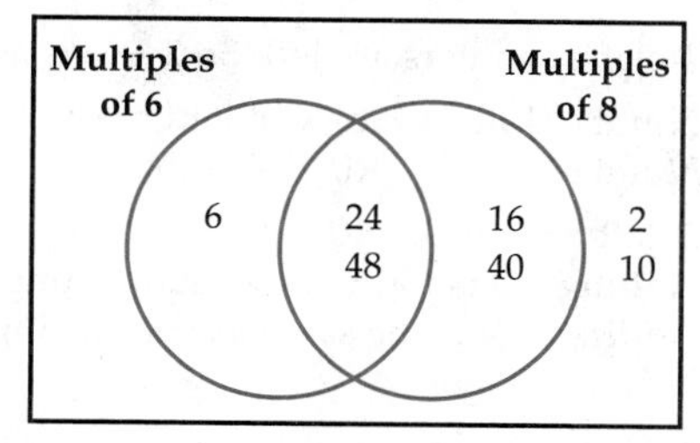

17. (a) Multiples of 20: 20, 40, 60, 80, 100, 120, 140, 160, . . .
Multiples of 32: 32, 64, 96, 128, 160, . . .
LCM(20, 32) = 160

(b) $20 = 2^2 \cdot 5$ and $32 = 2^5$.
So $\text{LCM} = 2^5 \cdot 5 = 160$.

(c)

20	20	20	20	20	20	20	20
32	32	32	32	32			

19. 4320

21. (a) $2^3 \cdot 3^4 \cdot 5^7 \cdot 7 \cdot 11$ (b) $2 \cdot 3^2 \cdot 5 \cdot 11$

23. $2 \cdot 3^2 \cdot 5^4 \cdot 7$

25. None

27. (b)

29. (a) 2^{10} (b) 10^6

31. True; $A = 2$ and $B = 8$ is an example

33. $3y$

35. (b) and (c)

37. 6 in. by 6 in. (GCF(30, 48) = 6)

39. (a) Sometimes true; true for $x = 10$, $y = 20$; false for $x = 20$, $y = 10$
(b) Never true; $x = 10$, $y = 20$
(c) Sometimes true; true for $x = 10$, $y = 20$, $z = 20$; false for $x = 10$, $y = 20$, $z = 30$

41. (b) $ab = \text{GCF}(a, b) \cdot \text{LCM}(a, b)$

Chapter 4 *Review Exercises*

1. 840

2. There is a whole number C such that $8 \cdot C = N$.

3. $312 = 2^3 \cdot 3 \cdot 13$

4. 3 and 7

5. 4, GCF

6. $x - 2$ is a factor of any whole number multiplied by $x - 2$, including $(x + 3)(x - 2)$.

7. True; $A = 2$, $B = 4$, $C = 5$ is an example

8. True; GCF(3, 7) = 1 is an example

9. True; $A = 3$ and $B = 6$ is an example

10. True, $XY \mid Z$ and $XY \mid XY$. So $XY \mid (XY + Z)$.

11. False; 2 and 4

12. 0, 3, 6, or 9 followed by 0

13. (a) $2 \cdot 5^2 \cdot 11^2$ (b) $2^3 \cdot 3 \cdot 5^4 \cdot 11^3$

14. 2, 3, 5, 7, 11, 13, 17, 19, 23 (since $\sqrt{577} \approx 24$)

15. $\underline{A}\,\underline{B}\,\underline{5} = A \cdot 100 + B \cdot 10 + 5$ and $\underline{A}\,\underline{B}\,\underline{0} = A \cdot 100 + B \cdot 10$. Now $A \cdot 100$, $B \cdot 10$, and 5 are all divisible by 5. Therefore, $A \cdot 100 + B \cdot 10 + 5$ and $A \cdot 100 + B \cdot 10$ are divisible by 5.

16. 120 cm, 240 cm, 360 cm, . . .

17. Write the hypothesis as the first step and the conclusion as the last step. Then convert the hypothesis and conclusion into equations and write these as the second and next-to-last steps, respectively. Finally, use properties of equations to work from the second step to the next-to-last step.

Chapter 5

5.1 *Homework Exercises*

−1. (a), (b), and (d)

1. −1

3. (a) Loss of yardage
(b) More expenditures than receipts
(c) Below sea level

5.

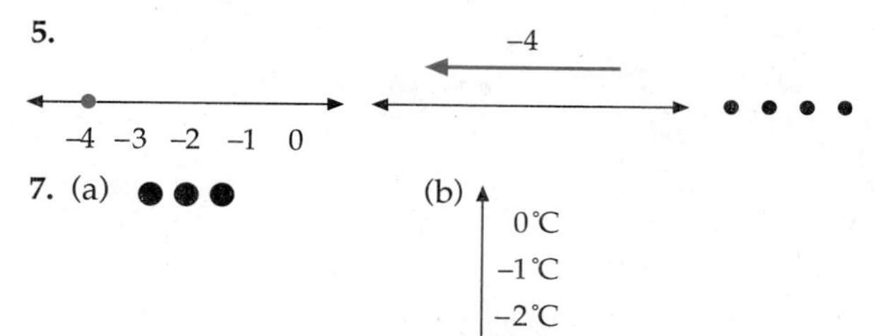

9. (a) 20 seconds before liftoff; −20
(b) A gain of \$3; 3 (c) 3 floors lower; −3

11. (a) 3 (b) 7 (c) $-x$

13. $(-4) + (-2) = -6$

15. (a) Go to −5. To add 3, move 3 to the right. You end up at −2. So $-5 + 3 = -2$.

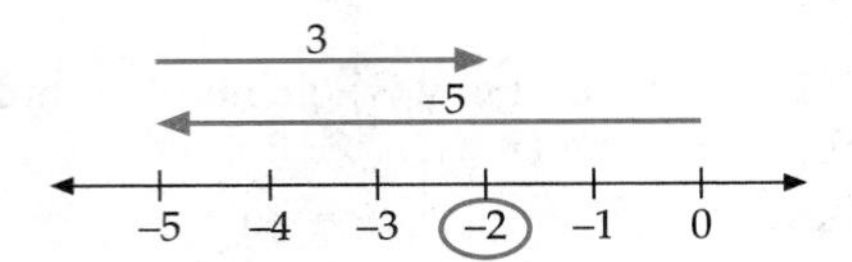

(b) Combine measures
(c) Show −5 as 5 black counters and 3 as 3 blue counters. Each blue counter cancels out a black one. You are left with 2 black counters.
So $-5 + 3 = -2$.

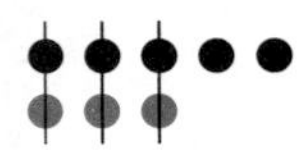

17. (c)

19. (a) $86 + (-30) + (-20) = \$36$
(b) Combine sets/measures

21. $2 + (-1) = 1$, $2 + (-2) = 0$

23. (b)

25. It is −6° and the temperature drops 4°. What is the new temperature? −10°. So $-6 - 4 = -10$.

27. Go to 5. Subtracting −2 is the opposite of adding −2, so go 2 to the right. You end up at 7.
So $5 - (-2) = 7$.

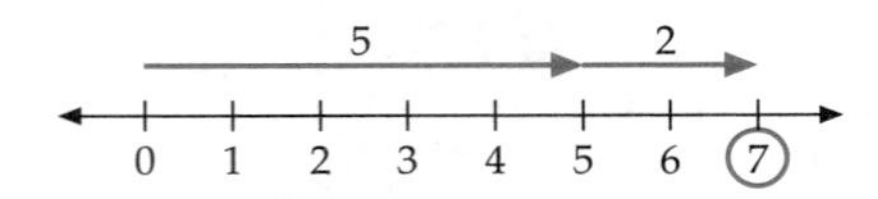

29. (a) or

(b) or

30. (a) $-5 - (-3) = -2$ (b) $2 - 5 = -3$
(c) $-3 - 2 = -5$

31. (a) Show −6 as 6 black counters.

Now take away −2.

4 black counters are left. So $-6 - (-2) = -4$.

(b) Show 2 as 2 blue counters. To be able to take away 6, we must change 2 to 6 blue counters and 4 black counters.

Now take away 6 blue counters.

4 black counters are left. So $2 - 6 = -4$.

33. (a) Number line (b) Colored counters
(c) Number line (d) Debatable

35. (a) 3, left (b) 3, left
(c) Adding −3 is the same as subtracting 3.
(d) Adding $-N$ is the same as subtracting N, where N is an integer.

37. (a) −73 (b) 38

39. (a) −14 (b) 8 (c) 10

41. (c)

43. (a)

45. (a) I (b) { }

47. (a) $-10 - 30 = -40$ (b) −40 ft
(c) Take away measures

49. (a) Loss of 7 yards
(b) Addition, combine measures or subtraction, take away measures

51. Wang: change = −3, Texas Oil and Gas: Tues. price = \$83, K-Mart: Mon. price = \$55

53. $9 - 7 = 2$
$8 - 7 = 1$
$7 - 7 = 0$
$6 - 7 = -1$
$5 - 7 = -2$
$4 - 7 = -3$

55. (a) $6 = -2 + n$; $n = 8$
(b) $-3 = 2 + n$; $n = -5$

57. (a) All odd integers
(b) All positive and negative multiples of 3

59. 7 and −3

61. (a) True, $x = 3$, $y = 2$ (b) False if $x < y$

63. (a) (possible answer) $a = 1$, $b = -1$, $c = 2$
(b) (possible answer) $a = 1$, $b = 2$, $c = 3$
(c) Impossible

65. (possible answer) First row: −2, −4, 6; second row: 8, 0, −8; third row: −6, 4, 2

67. $0 \leftrightarrow 0, 1 \leftrightarrow -1, 2 \leftrightarrow 1, 3 \leftrightarrow -2, 4 \leftrightarrow 2$, and so on.

5.2 *Homework Exercises*

1. $-5 \cdot (-1) = 5, -5 \cdot (-2) = 10$

3. (a) -24 pounds (b) $-6 \times 4 = -24$
(c) Repeated measures

5. Clem loses 5 pounds a week for 4 weeks in a row. What is the total change in his weight?

7. It is now 0°C. The temperature has been dropping 6°C per hour. What was the temperature 4 hours ago?

9. (a) At least one of a and b is 0.
(b) One of them is negative and the other one is positive.

11. $x \cdot x \cdot x \cdot x$, which is always positive.

12. (a) $-6 \times$ ______ $= -54$; 9
(b) $-4 \times$ ______ $= 32$; -8

13. (a) $-3 \times$ ______ $= 0$, 0
(b) $0 \times$ ______ $= -3$ has no solution so $-3 \div 0$ is undefined.

15. (a) $-\$40{,}000$ (b) $-480{,}000 \div 12 = -40{,}000$
(c) Partition sets/measures

17. You hope to lose 40 pounds at the rate of 5 pounds per week. How long will it take?

19. (a) -24 (b) 240 (c) -3 (d) 1

21. (a) -7 (b) -37

23. (a) 0 (b) 2 or -2 (c) -5

25. (a) 5, 7
(b) The sum of two negatives is a positive.

27. Mixing up the addition and multiplication rules

29. 2

31. (a)

33. (a) $-8 + (-8) + (-8) + (-8) = -32$
(b) $-5 + (-5) = -10$

35. For all x and y

37. Positive, negative, positive

39. $(-2 \div -2) + (-2 \div -2) = 2$;
$(-2 \times -2) - (-2 \div -2) = 3$;
$(-2 \times -2) + -2 - (-2) = 4$;
$(-2 \times -2) + (-2 \div -2) = 5$;
$-2 \times [-2 - (-2 \div -2)] = 6$;
$(-2 \times -2) + (-2 \times -2) = 8$

5.3 *Homework Exercises*

1. (a) Addition and multiplication
(b) Addition and multiplication

3. Compute $-5 \times (-8)$ first and then multiply by 7.

5. Commutative and associative properties of addition

6. (a) -3 (b) 11
(c) Commutative and associative properties of addition

7. Compute $(-20 \times 6) + (-7 \times 6) = -162$.

9. $(2 - 5)n$

11. (a) $(-12 \times 100) + (-12 \times (-1)) = -1188$
(b) Distributive property of multiplication over subtraction
(c) -3366

13. Any counterexample showing that whole-number division is not associative would also show that integer division is not associative.

15. 1; identity element; multiplication

17. Closure property of subtraction

19. (b)

21. 0.1

23. (a) $0 = 2 \cdot 4 + (-2 \cdot 4)$
(b) $0 = 2 \cdot 4 + (-2 \cdot 4) = 8 + (-2 \cdot 4)$.
So $-2 \cdot 4 = -8$ because each integer has a unique additive inverse.

25. $(2m)(2n) = 4mn = 2(2mn)$, which is even.

Chapter 5 *Review Exercises*

1. $8 - (4 - 2) \neq (8 - 4) - 2$

2. Addition and multiplication

3. The temperature is now -3°C. If the temperature drops 6°C, what will the new temperature be? -9°C. So $-3 - 6 = -9$.

4. I now have \$0. I lost \$5 per hour. How much did I have 3 hours ago?

5. (a) Show -4 as 4 black counters. Now take away -2.

2 black counters are left. So $-4 - (-2) = -2$.
(b) Go to -4. Subtracting -2 is the opposite of adding -2, so move 2 to the right. You end up at -2. So $-4 - (-2) = -2$.

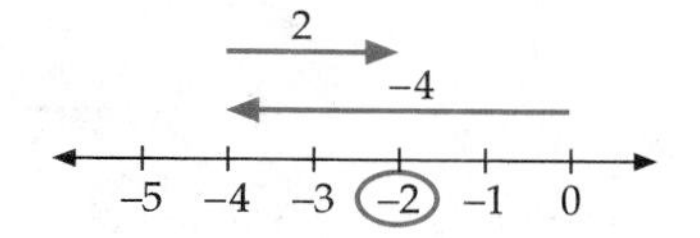

6. $4 - 1 = 3, 4 - 0 = 4, 4 - (-1) = 5, 4 - (-2) = 6$

7. (a) 500 soldiers better
   (b) $-300 - (-800) = 500$ (c) Compare sets

8. (a) $-18 \div 9 = -2$ lb/month
   (b) Partition measures

9. (a) $-30 + 20 = -10$ ft
   (b) Addition, combine measures

10. (a) Go to 2. To subtract 4, move 4 to the left.

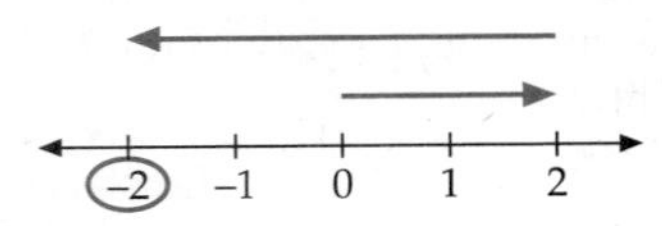

You end up at -2. So $2 - 4 = -2$.

(b) Show 2 as 2 blue counters. To be able to take away 4, we must change 2 to 4 blue counters and 2 black counters.

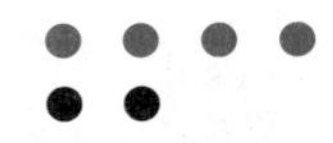

Now take away 4.

2 black counters are left. So $2 - 4 = -2$.

11. (c)

12. (b) and (e)

13. Additive inverses

14. The set of whole numbers is {0, 1, 2, 3, . . .}. The set of whole numbers is a subset of the set of integers. The set of integers is the union of the set of whole numbers and their opposites: {. . . , −3, −2, −1, 0, 1, 2, 3, . . .}.

Chapter 6

6.1 *Homework Exercises*

1. (a) Shade 3 out of every 5.

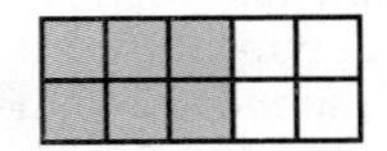

(b) Go $\frac{3}{5}$ of the way from 0 to 1.

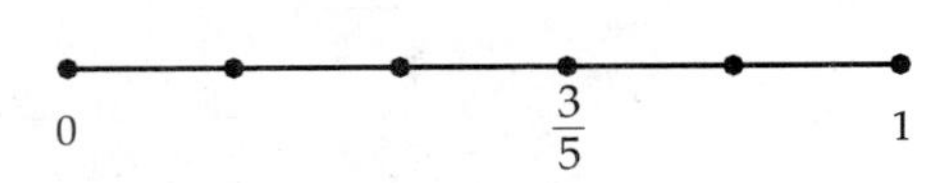

3. The five parts are unequal.

5. 1/3 represents:

Part of a whole

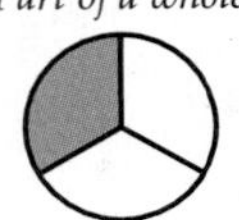

1 of 3 equal parts

Part of a set

1 out of every 3

Division
$1 \div 3$

Numerator÷
denominator

Number line location

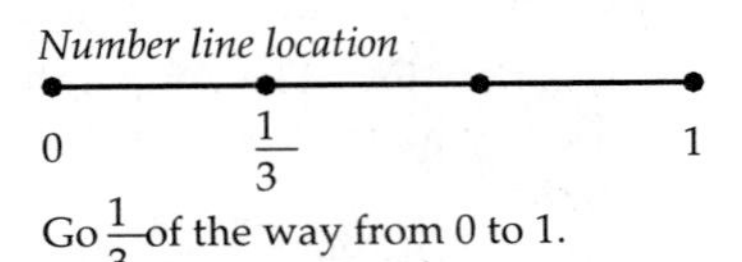

Go $\frac{1}{3}$ of the way from 0 to 1.

7. (a) (2) (b) (4) (c) (1) (d) (3)

9. (a) 3/1 (b) −3/1 (c) 9/2
   (d) −56/10 (e) 25/100

10. (a), (b), and (d)

11. (a) Yes (b) W

13. (a) $13 \div 3 = 4\frac{1}{3}$
   (b)

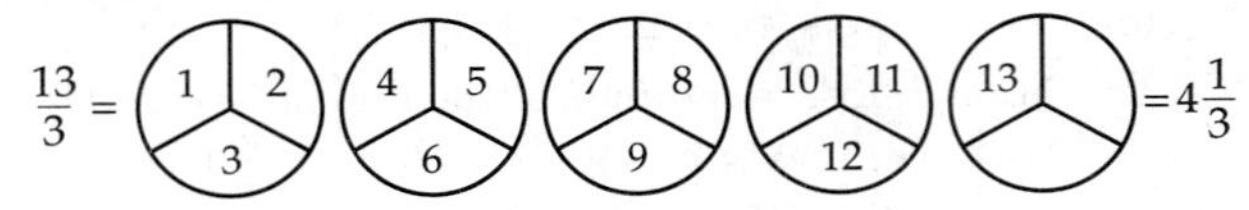

15. $\frac{0}{5}$ means $0 \div 5$, which is 0.

18. $\frac{-4}{7} = -4 \div 7 = 4 \div (-7) = \frac{4}{-7}$

19. (a) and (c)

20. (a) = (b) $\frac{1}{2} = \frac{1 \cdot 2}{2 \cdot 2} = \frac{2}{4}$

21. Eating 4 pieces of a 4-slice pizza is 4/4, and eating 8 slices of an 8-slice pizza is 8/8. 4/4 = 8/8.

23. $\frac{x}{y} = \frac{x \cdot 2x}{y \cdot 2x} = \frac{2x^2}{2xy}$

25. (a) Simplifying a fraction at the end of a problem
   (b) In addition or subtraction

27. (a) 21/58 (b) $1/yz$ (c) 2/11 (d) 2

29. Both fractions represent the same amount.

31. (a) $\frac{15}{20} \neq \frac{14}{20}$ so $\frac{3}{4} \neq \frac{7}{10}$
   (b) $3 \cdot 10 \neq 4 \cdot 7$ so $\frac{3}{4} \neq \frac{7}{10}$

33. (a) $2\frac{1}{2}$ (b) 6 or -6 (c) $\frac{5}{6}M \neq 0$

35. 1600 say yes in the second group.

37. $\frac{60}{105} > \frac{56}{105}$; $4 \cdot 15 > 8 \cdot 7$ so $\frac{4}{7} > \frac{8}{15}$

39. $\frac{14}{23} > \frac{17}{30}$ since $14 \cdot 30 > 17 \cdot 23$

41. (a) $4/52 = 1/13$ (b) $12/52 = 3/13$

43. (a) 6 (b) 1/6 (c) 1/2 (d) 2/3

45. 1/2

47. (a) Yes (b) Reasonable

49. $a > b \rightarrow \frac{a}{ab} > \frac{b}{ab} \rightarrow \frac{1}{b} > \frac{1}{a}$

51.

$1\frac{1}{3}$, $\frac{1}{4}$	$\frac{2}{8}$, $\frac{1}{3}$, $\frac{2}{6}$	$\frac{1}{3}$, $\frac{4}{10}$
$\frac{4}{3}$, $\frac{3}{4}$, $\frac{2}{16}$	$\frac{6}{8}$, $\frac{1}{3}$, $\frac{10}{12}$, $\frac{3}{6}$	$\frac{2}{5}$, $\frac{1}{2}$, $\frac{7}{7}$
$\frac{1}{8}$, $\frac{1}{2}$	$\frac{2}{4}$, $\frac{5}{6}$, $\frac{0}{3}$	1, 0

6.2 *Homework Exercises*

1. (a) $4/y$ (b) $3x/n$ (c) $2x/(x+1)$

3.

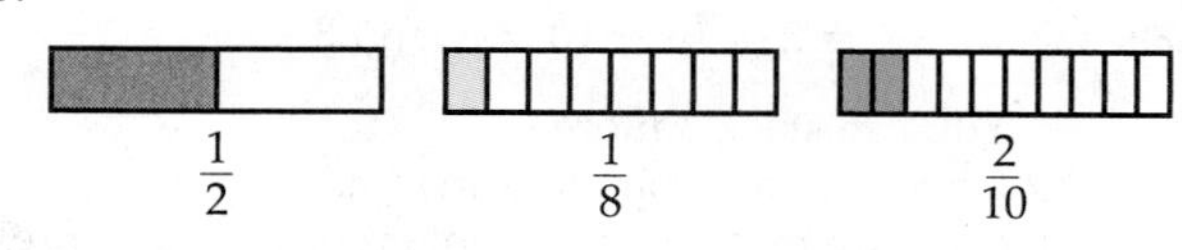

$\frac{1}{2}$ $\frac{1}{8}$ $\frac{2}{10}$

$\frac{1}{2} > \frac{2}{10}$ so $\frac{1}{2} + \frac{1}{8}$ does not equal $\frac{2}{10}$.

5. (a)

$\frac{1}{3}$ $\frac{1}{4}$

(b) $\frac{1}{3} + \frac{1}{4}$ would equal the shaded region shown, but we cannot name it unless we use a common denominator (twelfths).

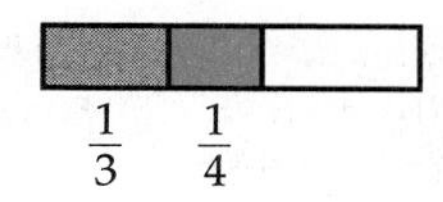

$\frac{1}{3}$ $\frac{1}{4}$

(c)

$\frac{4}{12}$ $\frac{3}{12}$

(d) $\frac{1}{3} + \frac{1}{4} = \frac{4}{12} + \frac{3}{12} = $ $= \frac{7}{12}$

7. (a)

$\frac{1}{4}$ $\frac{1}{6}$

(b) How much is left when we take 1/6 away from 1/4? We cannot tell. A common denominator is needed.

(c)

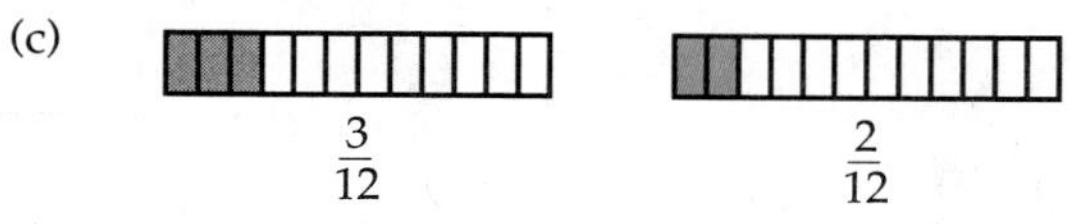

$\frac{3}{12}$ $\frac{2}{12}$

(d) $\frac{1}{4} - \frac{1}{6} = \frac{3}{12} - \frac{2}{12} = $ $= \frac{1}{12}$

8. (a) 308 (b) 616; 924 (c) $\frac{2}{308} = \frac{1}{154}$

9. (a) 19/24 (b) 31/357 (c) 1/9 (d) 11/30

11. (a) $\frac{1}{2}$ (b) $2\frac{6}{10} = 2\frac{3}{5}$

13. 67/1800

15. (a) $110\frac{1}{3}$ (b) $-142\frac{2}{3}$

17. $4\frac{1}{5} = 4 + \frac{1}{5} = \frac{20}{5} + \frac{1}{5} = \frac{21}{5}$

19. (a) $2\frac{3}{4}$ hours

(b) Subtraction, take away measures

21. You have a $5\frac{1}{4}$-hour job to do. You have been working for $2\frac{1}{2}$ hours. How much more time is needed? $2\frac{3}{4}$ hours.

23. (a) 5/12 (b) 3/20

25. (a) $2\frac{1}{2}$; $1\frac{1}{8}$ (b) $2\frac{1}{2}$; $1\frac{3}{4}$ (c) $1\frac{1}{8}$ (d) $\frac{7}{8}$

27. 5/12 and 1/6

29. (a) $\frac{1}{2} + \frac{1}{4}$ (b) $\frac{1}{13} + \frac{1}{26}$ (c) $\frac{1}{2} + \frac{1}{8}$

(d) $\frac{1}{9} + \frac{1}{2} + \frac{1}{6}$

31. (a) 14 (b) 3

33. $\frac{a}{b} = \frac{c}{d} \rightarrow \frac{a}{b} + \frac{b}{b} = \frac{c}{d} + \frac{d}{d} \rightarrow \frac{a+b}{b} = \frac{c+d}{d}$

6.3 *Homework Exercises*

1. (a) $\frac{1}{2} \times 8$ (b) $8 \div 2$

3. $\frac{3}{4} \times \$1500 = \$1125$

5. $4 \times \frac{1}{5} = \frac{1}{5} + \frac{1}{5} + \frac{1}{5} + \frac{1}{5} = \frac{4}{5}$

7. $\frac{1}{5} \times \frac{1}{3}$ means $\frac{1}{5}$ of $\frac{1}{3}$. First show $\frac{1}{3}$.

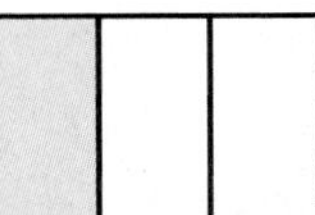

Now darken $\frac{1}{5}$ of $\frac{1}{3}$.

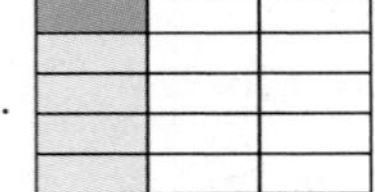

$\frac{1}{15}$ of the figure is darkened. So $\frac{1}{5} \times \frac{1}{3} = \frac{1}{15}$.

9. $\frac{3}{4} \times \frac{4}{5}$ means $\frac{3}{4}$ of $\frac{4}{5}$. First show $\frac{4}{5}$.

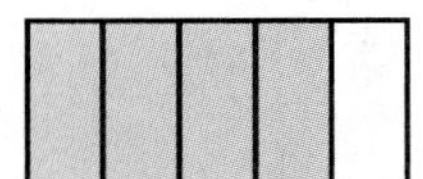

Now darken $\frac{3}{4}$ of $\frac{4}{5}$.

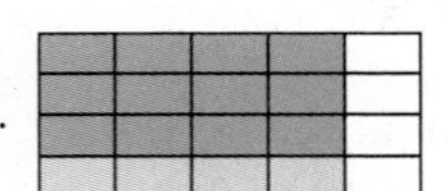

$\frac{12}{20}$ of the figure is darkened. So $\frac{3}{4} \times \frac{4}{5} = \frac{12}{20} = \frac{3}{5}$.

11. $\frac{1}{4} \times \frac{4}{5} = \frac{1}{5}$

13. $5\frac{1}{2} \times 2\frac{2}{3} = \frac{11}{2} \times \frac{8}{3} = \frac{11 \times 8}{2 \times 3}$. Now that it's all one fraction, you can divide the numerator and the denominator by a common factor, 2. $\left(\text{Then } \frac{8}{2} \text{ becomes } \frac{4}{1}.\right)$

$$= \frac{11 \times \overset{4}{\cancel{8}}}{\underset{1}{\cancel{2}} \times 3}$$

15. $2 \div \frac{1}{4}$ means how many $\frac{1}{4}$'s does it take to make 2?

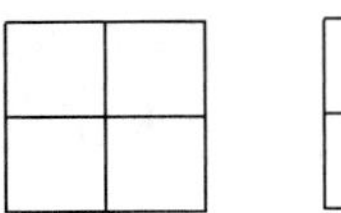

It takes 8 quarters to make 2. So $2 \div \frac{1}{4} = 8$.

17. How many dimes would make \$5?

18. $\frac{10}{21} = \frac{2}{3} \times \frac{n}{m}$; $\frac{n}{m} = \frac{5}{7}$

19. 3 and 1/3

21. $\frac{1}{2} \div \frac{3}{4} = \frac{\frac{1}{2}}{\frac{3}{4}} = \frac{\frac{1}{2} \times \frac{4}{3}}{\frac{3}{4} \times \frac{4}{3}} = \frac{\frac{1}{2} \times \frac{4}{3}}{1} = \frac{1}{2} \times \frac{4}{3}$

23. (a) $\frac{1}{2}$ (b) $1\frac{5}{9}$ (c) $1\frac{1}{4}$ (d) $-2\frac{4}{7}$

25. (a) 20 (b) 8

27. (a) $1\frac{2}{5}$ (b) $\frac{3x}{y}$

29. 3/2

31. 1/6 cup

33. \$154

35. (a) $1\frac{1}{2}$ cups
(b) Multiplication, repeated measures

37. (a) 15 (b) Division, repeated measures

39. (a) $10 \div 4 = 2\frac{1}{2}$, $18 - 10 = 8$, and $8 \div 2\frac{1}{2} = 3\frac{1}{5}$.
You can stack 3 more.
(b) Division, partition measures; subtraction, compare measures; division, repeated measures.

41. \$12

43. If it takes 3/4 of an hour to write a page, about how long will it take to write 22 pages?

45. 1/16

47. (a) 8 for Wacky, 6 for Harpo, and 3 for Young Loopy II
(b) Because the fractions $\frac{4}{9} + \frac{1}{3} + \frac{1}{6} = \frac{17}{18}$ and $\frac{17}{18}$ of 18 is 17
(c) Debatable

49. (a) 20 ft (b) 40 ft (c) 50 ft (d) 55 ft
(e) a little less than 60 ft

51. $23\frac{1}{13}$ ft

53. 3/5

55. 1/9

57. (Algebra) $\frac{1}{4}x + x = 40 \rightarrow \frac{5}{4}x = 40 \rightarrow x = 32$
Alex has 8 tapes, Jamie has 32 tapes.
(Guess and check) Two numbers add up to 40 and one is 4 times the other. Alex has 8 tapes. Jamie has 32 tapes.

59. (a) $3\frac{3}{4}$ (b) $8\frac{3}{4}$

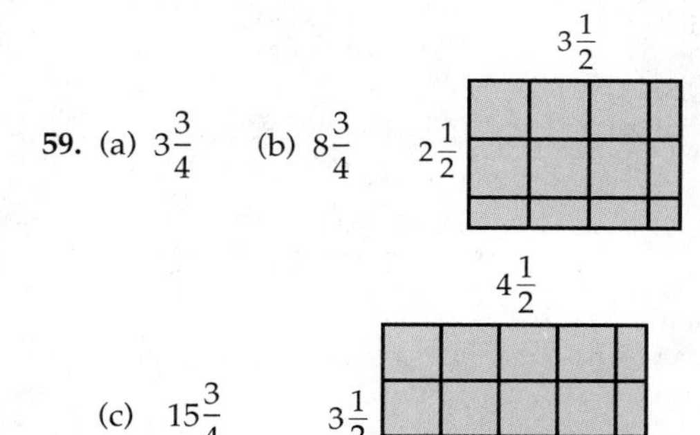

(c) $15\frac{3}{4}$ $3\frac{1}{2}$

(d) The whole-number part is $b(b + 2)$, and the fraction part is 3/4.

(e) $(18 \times 20) = 360 \rightarrow 360\frac{3}{4}$

61. (a) $x > 0$ and $y < 0$ (b) Never
(c) x and y both positive or both negative

63. (a) 2^2 (b) 2^3 (c) 2^{N-1}
(d) Write $\frac{1}{3}$ of 3^5 as a power of 3. Next, do the same for 3^6. Finally, generalize your results for 3^N.

65. (a) $\frac{1}{3} \div \frac{3}{4} = \frac{1}{3} \times \frac{4}{3} = \frac{4}{9}$ or $\frac{1}{3} \div \frac{3}{4} = \frac{4}{12} \div \frac{9}{12} = \frac{4}{12} \times \frac{12}{9} = \frac{4}{9}$.

67. L/P days

69. Measure the thickness and divide by the number of pages. It should be less than 1/100 of a centimeter.

6.4 *Homework Exercises*

1. (a) Addition and multiplication
(b) Addition and multiplication

3. $\left(\frac{1}{2} \cdot 4\right) \cdot x$

5. Multiply $\frac{1}{3} \times 18$ and $\frac{1}{5} \times 20$ first. Then $6 \times 4 = 24$.

7. $2\frac{1}{2}$

8. (a) No (b) Yes

9. A counterexample for whole numbers (like $2 - 3 \neq 3 - 2$) would also be a counterexample for whole numbers, integers, and rational numbers.

11. $\left(-\frac{4}{3} + \frac{2}{3}\right)n = -\frac{2}{3}n$

13. $-2/5$

15. (a) Denseness (b) Multiplicative inverses

16. (possible answer) $-11/100$

17. False

19. True

21. False for $a = -4$ and $b = c = d = -2$

23. $\frac{1}{4}$ of 40 is the same as $\frac{1}{4} \times 40 = 40 \times \frac{1}{4} = 40 \div 4$ or $\frac{1}{4} \times 40 = \frac{40}{4} = 40 \div 4$

25.

Rational numbers
Integers
Whole numbers

27. (a) $S = \dfrac{1 - \frac{1}{3^{40}}}{2}$ (b) $\dfrac{1 - \frac{1}{3^N}}{2}$ or $\dfrac{3^N - 1}{2 \cdot 3^N}$

29. $\left(\frac{u}{v} \cdot \frac{w}{x}\right) \cdot \frac{y}{z} = \frac{uw}{vx} \cdot \frac{y}{z} = \frac{uwy}{vxz}$ and
$\frac{u}{v} \cdot \left(\frac{w}{x} \cdot \frac{y}{z}\right) = \frac{u}{v} \cdot \frac{wy}{xz} = \frac{uwy}{vxz}$

31. (a) False for $x = 1$ and $y = 1$
(b) False for $x = 1$ and $y = 1$

6.5 *Homework Exercises*

1. (a) $\frac{5}{6} + \frac{10}{11} \approx 1 + 1 = 2$ (b) Too high

3. $\frac{7}{8} + \frac{1}{16} \approx 1 + 0 = 1$ in.

5. $5\frac{3}{4} \times 8\frac{1}{4} \approx 6 \times 8 = 48$ cups

6. (d)

7. Too high

9. $62\frac{1}{2} \div 14\frac{3}{4} \approx 60 \div 15 = 4$ lessons

11. (a) $8 \div 4 = 2$ (b) $60 \div 3 = 20$

13. $\frac{41}{126} \approx \frac{1}{3}$

15. (b)

17. $(8 \times 40) + \left(\frac{1}{2} \times 40\right) = 320 + 20 = \$340$

19. (a) 100, 2 halves make 1 yard, so 50×2 or 100 halves make 50 yards.
(b) Division, repeated measures

21. (d)

23. (a) Mental computation (b) Paper-and-pencil
(c) Calculator

25. (a) $1\frac{9}{3} = 4$; $4\frac{9}{5} = 5\frac{4}{5}$
(b) In regrouping, the child adds 10 to the numerator of the fraction.

27. (a) $\frac{5}{12}$; $\frac{11}{20}$
(b) After obtaining the common denominator, the child adds the numerators and denominators.

29. (a) $\frac{2}{3}$ and 3
(b) The child cancels common terms rather than common factors in the numerator and denominator.

31. The child is confusing fraction regrouping with whole-number regrouping.

33. (a) $\frac{1}{48}, \frac{1}{96}$ (b) $\frac{1}{(3 \cdot 2^N)}$

35. (c)

37. (a) $8\frac{1}{3} - 3\frac{2}{3} = 8\frac{2}{3} - 4 = 4\frac{2}{3}$
(b) $5\frac{1}{4} - 2\frac{3}{4} = 5\frac{1}{2} - 3 = 2\frac{1}{2}$

39. (a) $\frac{2}{(1 + 1)} \neq \frac{2}{1} + \frac{2}{1}$ (b) $\frac{(10 + 10)}{2} \neq \frac{10}{2} + 10$

41. (a) Add and multiply the denominators to get the new numerator and denominator, respectively.
(b) $\frac{1}{3} + \frac{1}{5} = \frac{8}{15}$ and $\frac{1}{2} + \frac{1}{6} = \frac{8}{12} = \frac{2}{3}$
(c) $\frac{1}{N} + \frac{1}{M} = \frac{M + N}{NM}$ (d) inductive
(e) $\frac{1}{N} + \frac{1}{M} = \frac{M}{NM} + \frac{N}{NM} = \frac{M + N}{NM}$

43. $1\frac{3}{4}$ and $\frac{1}{4}$

Chapter 6 *Review Questions*

1. Compute $24 \times 2 = 48$ and $24 \times \frac{1}{2} = 12$.
$48 + 12 = 60$.

2. $A > 11\frac{3}{7}$

3. 5/6 represents:

Part of a whole — 5 of 6 equal parts

Part of a set — 5 out of every 6

Division — $5 \div 6$ — Numerator ÷ denominator

Number line location — 0, $\frac{5}{6}$, 1 — Go $\frac{5}{6}$ of the way from 0 to 1.

4.

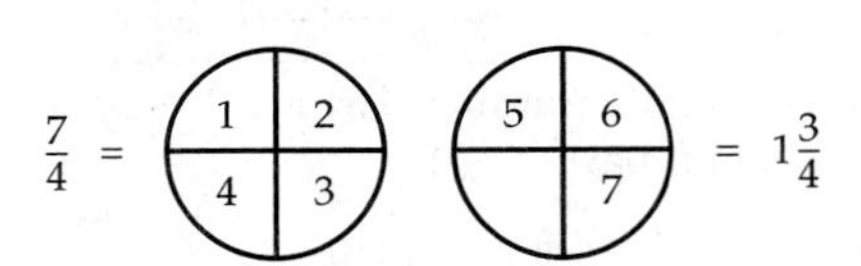

5. Denseness or multiplicative inverses

6. (b)

7. $37\frac{4}{7}$

8. $\frac{5}{0}$ means $5 \div 0$. No number multiplied by 0 equals 5, so $5 \div 0$ and $\frac{5}{0}$ are both undefined.

9. (a) $6\frac{2}{3}$ (b) Division, repeated measures

10. (a) 17/30
(b) Addition, combine sets/measures and subtraction, take away measures/sets

11. (a) 7/20 lb (b) Division, partition measures

12. (a) $\frac{4}{9} = \frac{20}{45}$ and $\frac{2}{5} = \frac{18}{45}$ so $\frac{4}{9} > \frac{2}{5}$
(b) $4 \cdot 5 > 2 \cdot 9$ so $\frac{4}{9} > \frac{2}{5}$

13. (a)

$\frac{2}{3}$ $\frac{1}{6}$

(b) $\frac{2}{3} + \frac{1}{6}$ would equal the shaded region shown, but we cannot name it unless we use a common denominator (sixths).

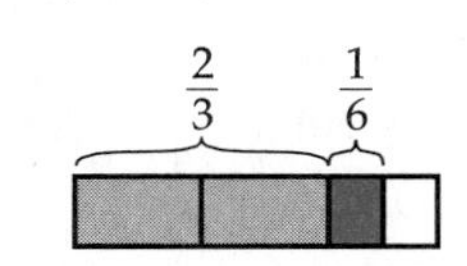

(c)

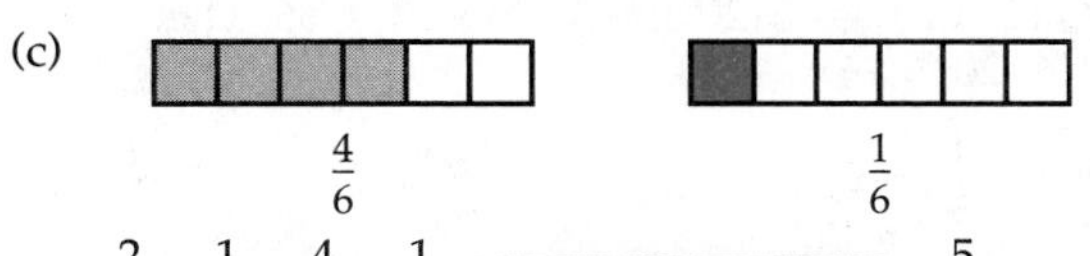

(d) $\frac{2}{3} + \frac{1}{6} = \frac{4}{6} + \frac{1}{6} =$ [bar] $= \frac{5}{6}$

14. (a)

$\frac{1}{2}$ $\frac{1}{5}$

(b) How much is left when we take 1/5 away from 1/2? We cannot tell. A common denominator is needed.

(c)

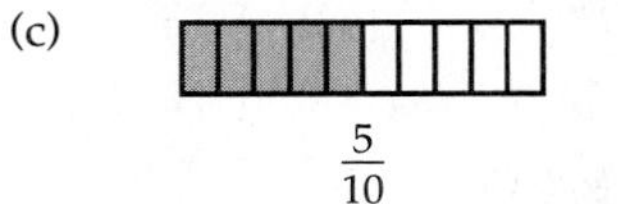

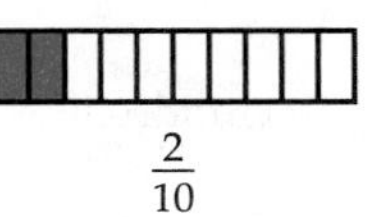

(d) $\frac{1}{2} - \frac{1}{5} = \frac{5}{10} - \frac{2}{10} =$ [bar] $= \frac{3}{10}$

15. $\frac{2}{5} \div \frac{7}{9} = \dfrac{\frac{2}{5}}{\frac{7}{9}} = \dfrac{\frac{2}{5} \times \frac{9}{7}}{\frac{7}{9} \times \frac{9}{7}} = \dfrac{\frac{2}{5} \times \frac{9}{7}}{1} = \frac{2}{5} \times \frac{9}{7}$

16. (a) Multiply $\frac{1}{5} \times 15$ and $\frac{3}{4} \times 8$. Then $3 \times 6 = 18$.

(b) Commutative and associative properties of multiplication

17. $5 \times (3 + 2) = (5 \times 3) + (5 \times 2)$

18. $\frac{2}{3} \times \frac{3}{5}$ means $\frac{2}{3}$ of $\frac{3}{5}$. Show $\frac{3}{5}$.

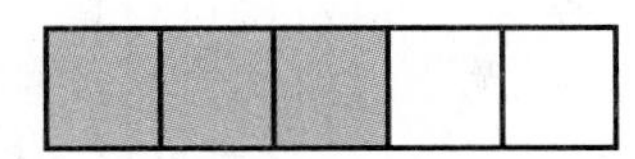

Now darken $\frac{2}{3}$ of $\frac{3}{5}$.

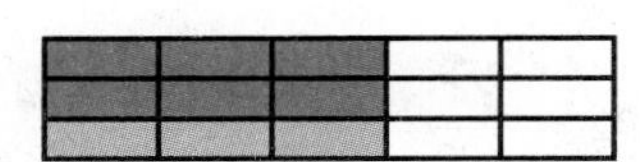

$\frac{6}{15}$ or $\frac{2}{5}$ is darkened. So $\frac{2}{3} \times \frac{3}{5} = \frac{2}{5}$.

19. (a) $1\frac{1}{3}$

(b) The child adds the denominators to obtain the common denominator. Then the child uses addition to obtain the new numerators.

(c) $\frac{1}{4} + \frac{2}{3} < \frac{1}{3} + \frac{2}{3}$ so the answer should be less than 1.

20. (a)

21. $\frac{1}{4} \times \frac{8}{9} = \frac{1 \times 8}{4 \times 9}$. Now that it's all one fraction, you can divide the numerator and the denominator by a common factor, 4. $\left(\text{Then } \frac{8}{4} \text{ becomes } \frac{2}{1}.\right)$

$$= \frac{1 \times \overset{2}{\cancel{8}}}{\underset{1}{\cancel{4}} \times 9}$$

22.

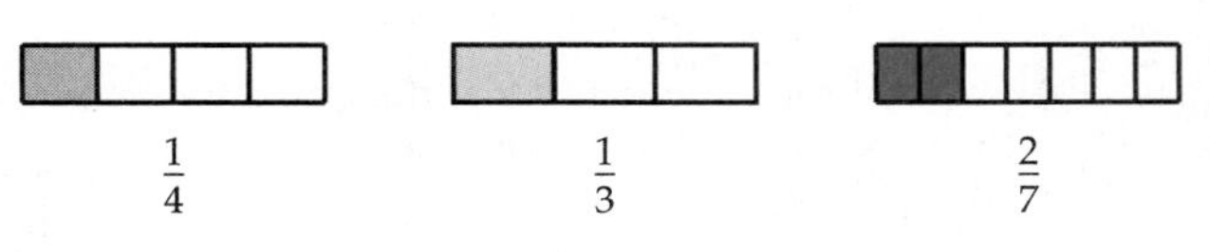

$\frac{1}{3} > \frac{2}{7}$ so $\frac{1}{3} + \frac{1}{4}$ does not equal $\frac{2}{7}$.

23. The set of whole numbers is {0, 1, 2, 3, . . .}. The set of whole numbers is a subset of the set of integers. The set of integers is the union of the set of whole numbers and their opposites: {. . . , −3, −2, −1, 0, 1, 2, 3, . . .}. The set of integers is a subset of the set of rational numbers. A rational number is a number that can be written in the form p/q where p and q are integers and $q \neq 0$.

Chapter 7

7.1 *Homework Exercises*

1. 100; $\frac{1}{100}$

3. To show 0.40, shade 40 out of 100 squares. To show 0.4, shade 4 out of 10 columns. The area shaded for 0.40 is equal to the area for 0.4 so 0.40 = 0.4.

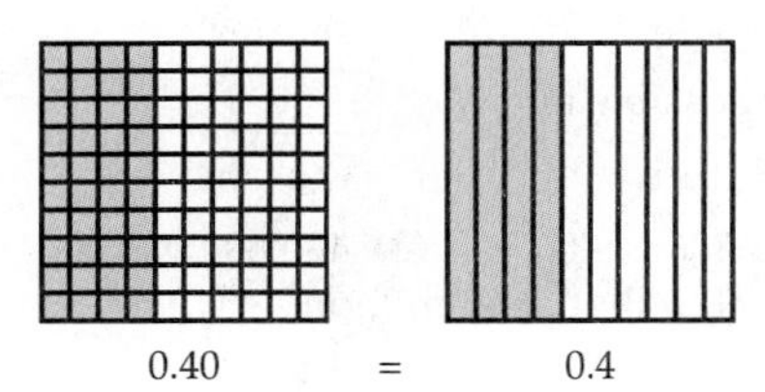

5. (a) Since $2.3 > 2.1$
(b) -2.1 is to the right of -2.3 so $-2.1 > -2.3$.

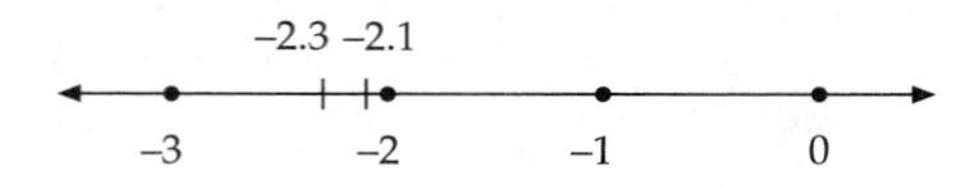

7. (a) 41.16 (b) 7.005

9. (a) $3 \times (1/10) + 1 \times (1/100)$
(b) $(3 \times 10) + 2 + (1 \times 1/100) + (7 \times 1/1000)$

11. $10^3 = 1000$, $10^2 = 100$, $10^1 = 10$, $10^0 = 1$

12. $10^2 = 100$
$10^1 = 10$
$10^0 = 1$
$10^{-1} = 0.1$
$10^{-2} = 0.01$
$10^{-3} = 0.001$

13. (a) 5 (b) $1; \frac{1}{5}; \frac{1}{125}$

14. (a) $1/10^n$ (b) $1/X^n$

15. (a) 2^2
(b) Subtract the exponent of the denominator from the exponent of the numerator.
(c) 10^3 (d) 5^{10} (e) x^4

17. (a) \$.90 (b) \$.80 (c) \$.90

19. (a) $8847.6 - 7132.1 \approx 8800 - 7100 = 1700$ m
(b) $8847.6 - 7132.1 \approx 8000 - 7000 = 1000$ m

21. $(\$1.79)(3)(4) \approx \$24$; (d)

23. $129 \div 46 \approx 120 \div 40 = 3¢$

25. Too high

27. (a) and (c)

29. \$3,900,000,000,000

31. 20,000 hours

33. (b) 31.7 years

34. (a) 36,200,000 (b) 0.004268

35. Move the decimal two places to the right. Answer: \$329.

37. 0.0000054

39. 3.4×10^{12}

41. (a) 12,000,000 (b) 12 million

43. 48 is greater than 10.

45. (a) 3×10^{27} (b) It's shorter!
(c) (possible answer) | 3 27 |

47. (a) Saturn (b) 40 (c) 0.4

49. (a) 4.806×10^{11} (b) 480,600,000,000
(c) 480 billion, 600 million

51. 20×4.25 is $(20 \times 4) + \left(20 \times \frac{1}{4}\right)$, which is $80 + 5 = 85$.

53. $y < z < x < w$

55. $5.75 \times 20 = (5 \times 20) + \left(\frac{3}{4} \times 20\right) = 100 + 15 =$ \$115

57. $12 \times \$10 = \$120$ and $12 \times \$.50 = \$6$. Then $\$120 + \$6 = \$126$.

59. (a) $\frac{1}{5}$ and $\frac{1}{25}$ (b) $2 \cdot 5 + 4 + 2 \cdot \frac{1}{5} + 1 \cdot \frac{1}{25}$

7.2 *Homework Exercises*

1.

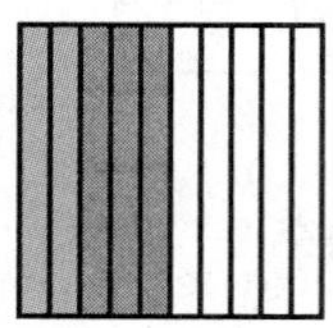

0.2 + 0.3 = 0.5

3. (a) Hundredths (b) Tenths (c) Yes

4.

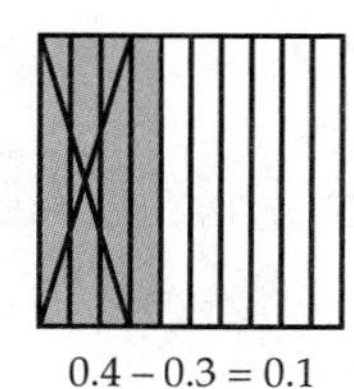

0.4 – 0.3 = 0.1

5. Subtraction, take away measures/sets

7. $0.6 \times 0.7 = 0.42$

9. $(2.64)(0.3) = \left(2\frac{64}{100}\right)\left(\frac{3}{10}\right) = \left(\frac{264}{100}\right)\left(\frac{3}{10}\right) = \frac{264 \cdot 3}{1000} = (264 \cdot 3) \cdot \frac{1}{1000}$

11. (a) 72 calories (b) Repeated measures

12.

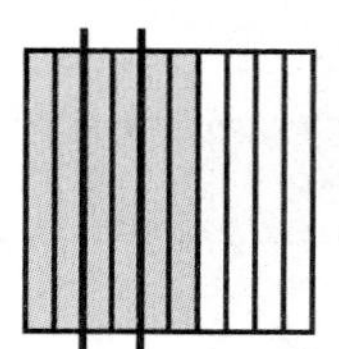

13. 10

15. $6.4 \div 0.32 = \frac{6.4}{0.32} = \frac{6.4}{0.32} \times \frac{100}{100} = \frac{640}{32} = 640 \div 32$

17. (a), (c), and (d)

19. Division, repeated measures

21. Multiplication, repeated sets/measures and addition, combine sets/measures

23. 10/3

25. (a) 5.1, 18.3
(b) The remainder is written as the first place to the right of the decimal point.

27. A child might line up the decimal in the answer with the decimals in the factors, or a child might not use a decimal point in the answer.

29. (a) C3 (b) Dennis's grade on test 1
(c) B2 + C2 + D2 (d) E2/3

31. 0.00128

33. 300,000

35. (a) 1.5, 2.5 (b) $3 \times 4 + 0.25 = (3.5)^2$
(c) Yes

37. (a) \$23 (b) \$183

39. About \$40 to \$75 (depends upon gas mileage)

41. (a)

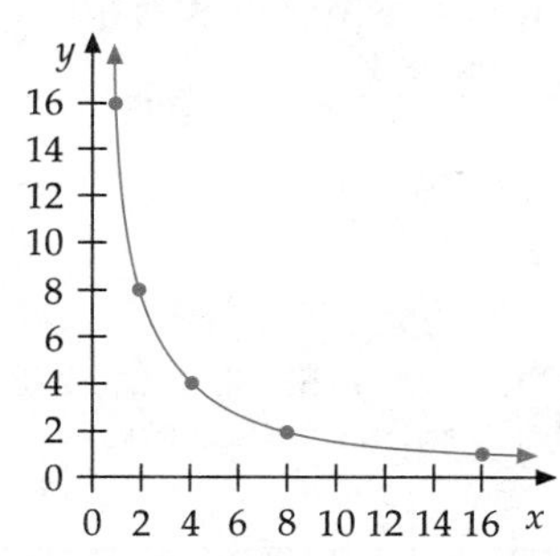

(b)

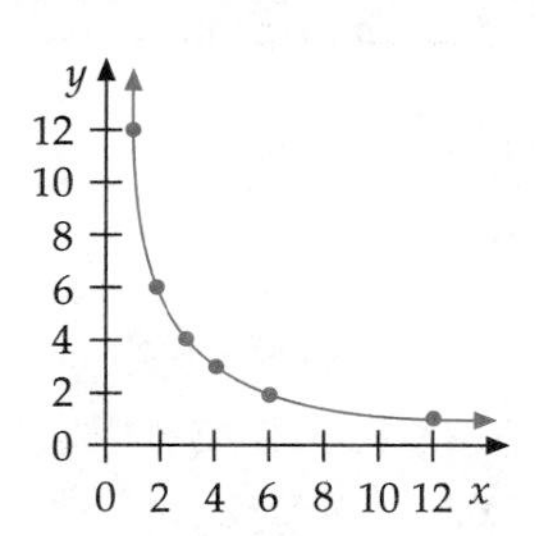

(c) The graph always has the same general shape. It approaches the positive y-axis at one end and the positive x-axis at the other end.

43. (b) \$12.00, \$13.01, \$14.02, \$15.03, \$16.04, \$17.05, \$18.06, or \$19.07

45. 0.36 mile

47. None

49. (a) \$13 (b) \$48

51. $0.68 \times 18 \approx \frac{2}{3} \times 18 = 12$ m

53. (a) 0.33333 . . . (b) 0.1_{three} (c) 0.2_{six}
(d) Nine; 0.3_{nine}

55. About \$90

7.3 *Homework Exercises*

1. Yes

3. (a) No (b) Yes

5. (a) 8 to 30 (b) 4:15
(c) The greater the fraction, the steeper the roof

7. (a) 18.1:1, 25.1:1, 17.6:1, 21.2:1, 18.5:1, 17.0:1, 17.3:1, 24.3:1
(b) Dallas (c) Los Angeles (d) Lower

9. (a) $11\frac{2}{3}$ (b) $\pm\sqrt{48}$

11. (a) 10 (b) 5 (c) 15

13. 1195

15. (b) $1\frac{5}{7}$ quarts of orange juice and $1\frac{1}{7}$ quarts of skim milk

17. 1.6 million cars

19. 660 men, 440 women

21. $93\frac{1}{3}$ lb

23. $x = 8\frac{1}{3}, y = 21\frac{2}{3}$

25. \$2.27

26. (a) \$.11/oz (b) \$.12/oz (c) 18-oz jar

27. 18-ounce box

31. \$.94

33. The bus is more efficient than the car when it carries more than 4 times as many people. The train is more efficient than the bus when it carries more than 3 times as many people as the bus.

35. 250 fish

37. \$3529.41

39. 51

41. (c)

43. MH/G

45. QC/D

47. $1/2 = 4/8$, $1/2 = 6/12$, $4/8 = 6/12$, $1/2 = 2/4$, $2/4 = 4/8$, $2/4 = 6/12$, $1/6 = 2/12$, $1/4 = 2/8$, and the fractions on both sides of each proportion can be inverted to create a new proportion.

7.4 *Homework Exercises*

1. Per 100

2. (a) 14 (b) 7 (c)

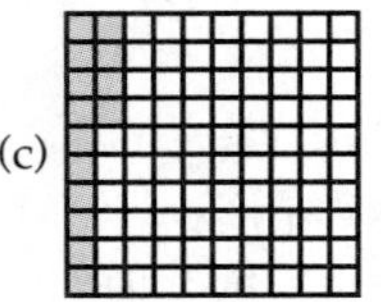

3. $x\% = x(0.01)$

5. (a) $\frac{34}{100}$, 0.34 (b) $1\frac{4}{5}$, 1.8 (c) $\frac{6}{10000}$, 0.0006

7. (a) 4% (b) 18.75%

9. (a) $1/2$ of $222 = 111$ (b) $1/4$ of $6 = 1.5$
(c) $1/10$ of $470 = 47$ (d) $1/100$ or $37 = 0.37$

11. $\$400 - \$100 = \$300$

13. (a) $2 \cdot 7 = 14$ (b) 1% of $400 = 4$ and $8 \cdot 4 = 32$
(c) $3/4$ of $12 = 9$

15. (a) 44 (b) 320 (c) 550

17. (a) 18 (b) 1800 (c) 7400

19. (a) 8 (b) 50 (c) 80

21. About 20%

23. (a) $\frac{1}{5}, \frac{1}{10}$ (b) $\frac{1}{3}, \frac{1}{7}$

25. 55% of 196 million ≈ 50% of 200 million = 100 million

27. 80% of 28, 19% of 400, 37% of 420, 52% of 807

29. Double tax to \$3 and add a little to get \$3.25;
10% of bill is \$2.10 plus half (\$1.05) makes \$3.15;
$1/6$ of \$21.00 ≈ \$3.50

31. The tip is 15% of \$37 ≈ \$3.70 + \$2 ≈ \$6.
\$38.85 + \$6 ≈ \$45. Each person leaves
\$45 ÷ 4 = \$11.25.

33. (a) 75
(b) The second 50% is of a smaller number (50) than the first 50% (of 100).

35. 48% of 117, 86% of 90, 11% of 1000, 31% of 642

37. (a) 20%
(b) The number of boys in the class is 80% of the number of girls. What percent of the whole class is boys? 44%

39. 80%

7.5 *Homework Exercises*

1. (b) $(0.12)(\$2400) = \$288$
(c) $\frac{12}{100} = \frac{x}{\$2400}$; $x = \$288$

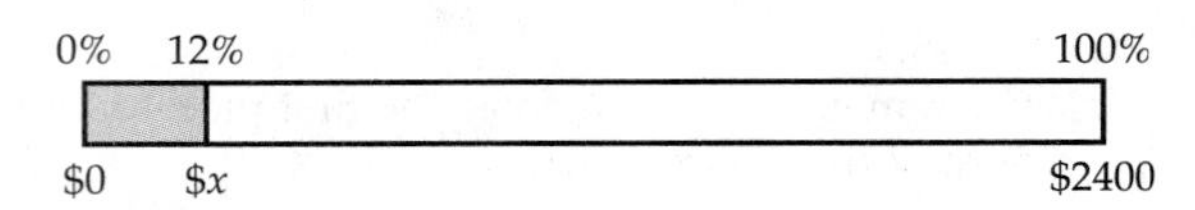

(d) 3

3. (a) $82 \div 110 = 0.745 = 0.745 \times 100\% = 74.5\%$
(b) $\frac{82}{110} = \frac{x}{100\%}$; $x = 74.5\%$

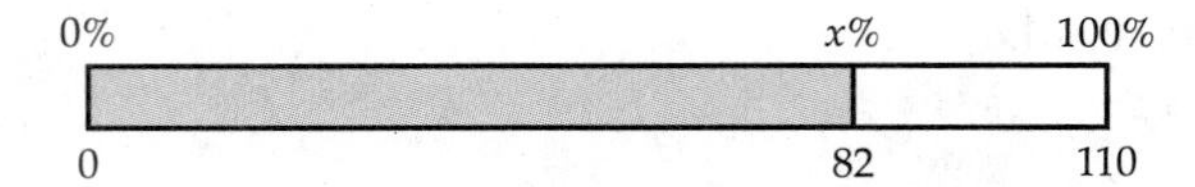

5. 6%

7. $33\frac{1}{3}\%$

9. (a) $0.08x = \$798.40$; $x = \$9980$
(b) $\frac{8}{100} = \frac{\$798.40}{x}$; $x = \$9980$

11. (a) 1% (b) 0.033%

13. \$4716.98

15. (b) 91%

17. (a) $(.20)(\$15.98) = \$3.20$ and $\$15.98 - \$3.20 = \$12.78$
(b) $1 - .20 = .80$ and $(.80)(\$15.98) = \$12.78$

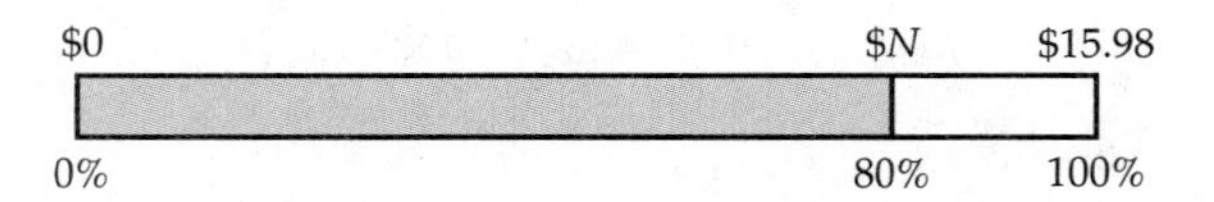

(c) \$13.29

19. About 58%

21. Find 20% or $\frac{1}{5}$ of \$.60 which is \$.12.
Then \$.60 + \$.12 = \$.72

23. (a) \$2253.65 (b) 12.68%

25. \$16,289

27. (a) 5.9 billion (b) 7.2 billion (c) 13.7 billion

29. Teachers with salaries of \$42,857.14 and under would benefit more from the \$3000 raise. Other teachers would benefit more from a 7% raise.

31. \$16

33. 33.6 mpg

35. *Example:* Which has a higher percent markdown, Hippo or Bolt shoes?

37. (a) First row: \$0, \$100, \$200, \$300, \$400; second row: \$0, \$70, \$140, \$210, \$280

(b)

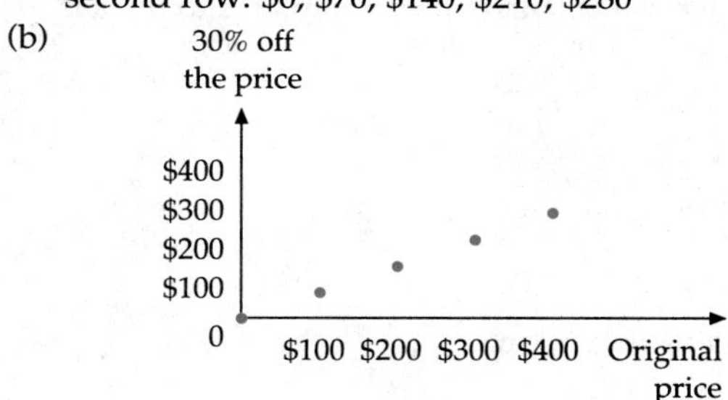

(c) About \$320

39. (a) \$3789 in the U.S.; \$2429 in Canada
(b) 24% in the U.S.; 11% in Canada
(c) \$144 billion
(d) Canadians tend to live a little longer than people in the U.S., and Canadians have fewer people per physician.

41. (a) Prices will go up in the next year.
(b) No, it is 17.6% higher.

43. (a) Regressive (b) Regressive (c) Regressive

45. \$125,000

47. About 98.2%

49. 45% of 689 is close.

51. Fluorescent bulb costs \$30; incandescent bulbs cost \$66

53. (c)

55. 4%

57. About 7 years

59. $\frac{28}{64} = \frac{7}{16}$

61. $100(M - N)/N$

63. (a) \$8342.61
(b) The value of the REARGUARD fund on 1-1-97.
(c) D3 = C3 − B3
E3 = (C3 − B3) * 100/B3
(d)

D	E
\$337.49	4.0
−\$600.65	−4.8

65. (a) At least 12% chicken
(b) No more than 30% breading
(c) No more than 30% fat and 10% added water
(d) At least 35% turkey
(e) At least 4% chicken

7.6 *Homework Exercises*

1. (a) 0.6 (b) $2.\overline{2}$

2. (a) $\frac{731}{1000}$ (b) $-13\frac{1}{25}$

3. $0.3\overline{4}$

4. (a) $\frac{4}{9}$ (b) $\frac{32}{99}$ (c) $\frac{267}{999} = \frac{89}{333}$

5. (a) $52\frac{35}{99}$ (b) $\frac{3917}{9999}$ (c) 1

7. A square has an area of 5. How long are its sides?

9. 24.2 in.

11. (b) $\sqrt{5}$ is an infinite decimal.

13. (a) and (c) are irrational; (b) and (d) are rational.

15. (a) 25 mph (b) 36 mph
(c) Irrational and rational

17. (a) About 1.3 s (b) 400 cm
(c) $\sqrt{L}/5$ (d) $25T^2$

19. Real numbers

21. 1/3 or 3,333,333/10,000,000

23. $\sqrt{13}$ is irrational and real; −6 is an integer, rational, real; $\sqrt{9}$ is whole, integer, rational, real; −0.317 is rational, real.

25.

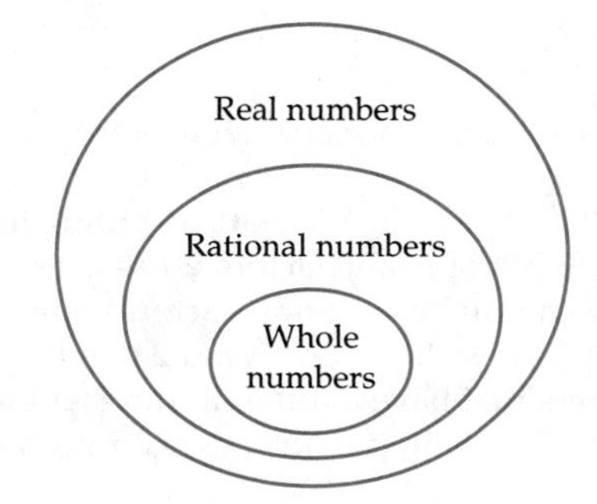

27. (a) 3 (b) 5 (c) An infinite number

29. (a)

−1 0 1 2

(b)

−4 −3 −2 −1 0 1

31. π or $\sqrt{2}$

33. 1/3 + 1/6

35. Convert them both to decimals by dividing the numerator by the denominator. Then compare the size of the decimal values.

37. (a) 22/7

39. π is an infinite nonrepeating decimal.

41. (a) 550, 490 (b) 66%, 62% (c) 36, 34

43. Commutative, addition

45. 0.41798

47. $3 - 2 \neq 2 - 3$

49. $6 \times 8 = 48$ plus $0.5 \times 8 = 4$ and $48 + 4 = 52$

51. $8 + (3 \cdot 4) \neq (8 + 3) \cdot (8 + 4)$

53. (a) Some pairs (b) All pairs
(c) No real numbers (d) Some pairs

55. (a) $\frac{1}{5^3} = 0.008$; $\frac{1}{5^4} = 0.0016$; both true

(b) $\frac{1}{5^N} = 2^N \times 10^{-N}$

57. (a) (1), (2), and (4) (b) (1), (2), and (4)
(c) 5, 5^2, 3^2, $2^3 \cdot 5^2$, 7 (d) 2, 5

59. The number of digits is equal to the larger of a and b.

61. Only if $M = 0$ or $N = 0$

63. (a) 1.414 (b) 1.412 (c) Yes

65. Irrational

67. True, $x + \frac{1}{x} \geq 2 \rightarrow x^2 + 1 \geq 2x \rightarrow$
$x^2 - 2x + 1 \geq 0 \rightarrow (x - 1)^2 \geq 0$, which is true because all real numbers squared are greater than or equal to 0.

69. Assume $\sqrt[3]{2}$ is rational. Then $p/q = \sqrt[3]{2}$, in which p and q are counting numbers. Then $p = \sqrt[3]{2}q$ or $p^3 = 2q^3$. If p^3 is the cube of a counting number, then the number of prime factors it has is a multiple of 3. $2q^3$ has a number of prime factors equal to a multiple of 3 plus 1. So p^3 and $2q^3$ are equal numbers that have a different number of prime factors. This is impossible, so $\sqrt[3]{2}$ must be irrational.

Chapter 7 *Review Exercises*

1. (a) $(3 \times 1/1000) + (7 \times 1/10{,}000)$
(b) 3.7×10^{-3}

2. Represent 0.3 as 3 columns. Represent 0.25 as 25 small squares.

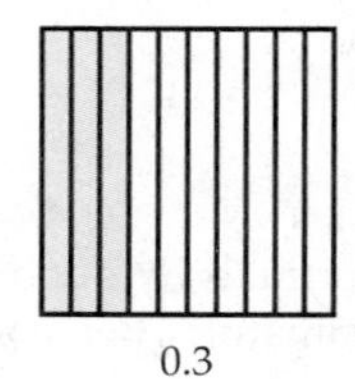
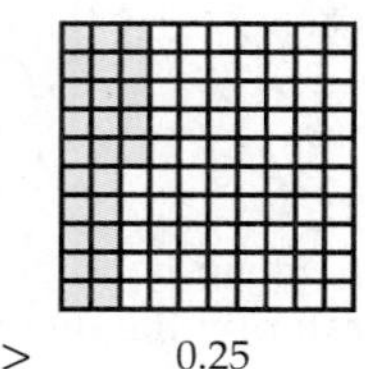

0.3 > 0.25

3. $GC/3$ gallons

4. 6%

5. Move the decimal point two places to the left.
Answer: $.0076.

6. (a) $\frac{2000}{130} = \frac{x}{150}$ and $x = 2308$ calories

(b) $2000 \cdot \frac{150}{130} = 2308$ calories

7. $0 < m < 1$

8. About 85 million

9. 44%

10. 0.04

11. Move the decimal two places to the left.
Answer: $72.50

12. $42 \div 0.06 = \frac{42}{0.06} = \frac{42}{0.06} \times \frac{100}{100} = \frac{4200}{6} = 4200 \div 6$

13. (d)

14. 46 + half of 46 = 69.

15. Division, partition measures

16. Subtraction, compare measures

17.

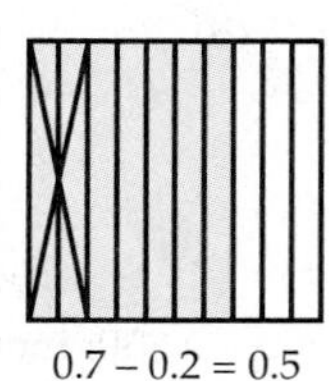

0.7 – 0.2 = 0.5

18. Additive inverses

19. $3(4 - 2) = (3 \cdot 4) - (3 \cdot 2)$

20. 2.37×10^7

21. 0.2×0.3 means 0.2 of 0.3.
Shade 0.3 of a decimal square.

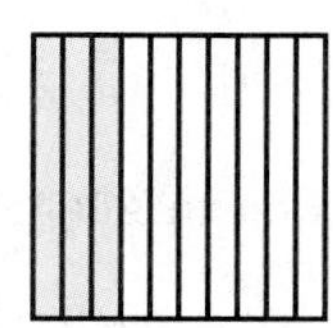

Now, darken 0.2 of the 0.3.

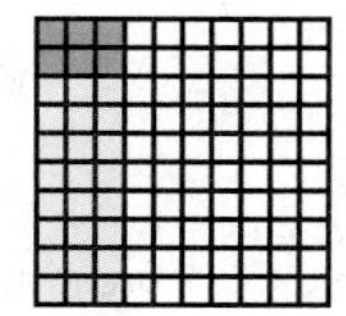

0.06 of the decimal square is darkened.
So $0.2 \times 0.3 = 0.06$.

22. By the addition rule, $4^2 \cdot 4^0 = 4^2$. Therefore $4^0 = 1$. $4^3 = 64$, $4^2 = 16$, $4^1 = 4$. Each answer is 1/4 of the previous one. So $4^0 = 1$.

23. (d) and (e)

24. (a) Irrational (b) Irrational
(c) Rational (d) Rational

25. (a) Rational (b) Rational
(c) Irrational (d) Rational

26. (b), (c), (e)

27. The set of real numbers is the union of the set of rational numbers and the set of irrational numbers. The set of rational numbers contains all terminating and repeating decimals, while the set of irrational numbers contains all nonrepeating decimals.

28. The irrational number π is the ratio of the circumference of a circle to its diameter.

Chapter 8

8.1 *Homework Exercises*

1. Earth measure

3. (a) Pentagon (b) Triangular prism
(c) Line segment

5. A point has no length or width.

7. (a) False (b)

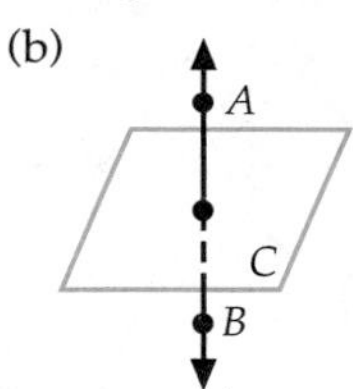

9. (a) 3 (b) 1 (c) 2

11. (a) Cone, cylinder, triangle, square, line segment, angle, rectangle, pyramid, prism, hemisphere
(b) Semicircle, hemisphere, line segment

13. A line segment has two endpoints; a line has none.

15. (a) *A*
(b) Different endpoints, opposite directions

17. (a) $\angle BAC$ (b) $\overrightarrow{BD}$ (c) *A*

19. (a) 32° (b) 121°

21.

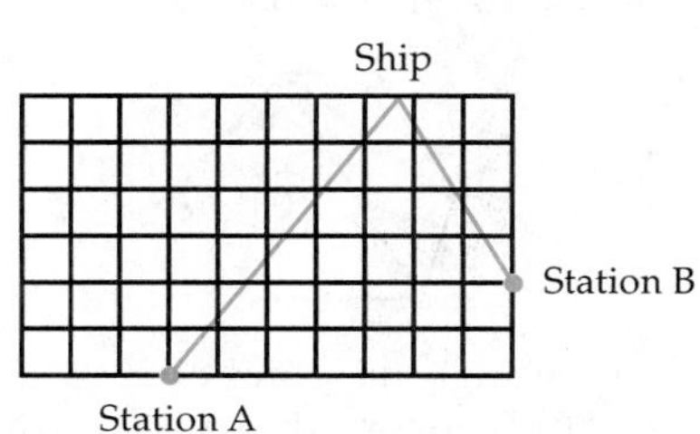

23. $\angle C$ and $\angle D$

25. *W*, *M*, *Z*, *N*

27. True

29. (a) $\angle PIN$ and $\angle PIG$ or $\angle NIR$ and $\angle RIG$
(b) $\angle PIR$ and $\angle RIG$
(c) See the answer to part (a).

31. False

33. (a)

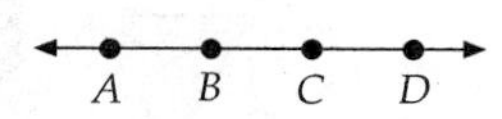

(b)

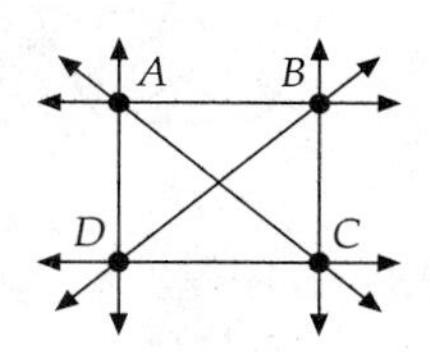

(c)

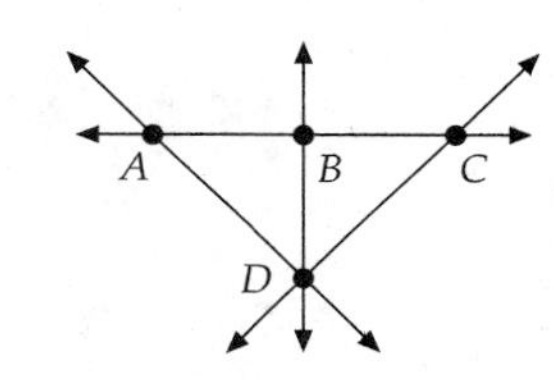

35. (a) True
(b) Two parallel lines on the floor of a room that are perpendicular to a third line
(c) A carpenter could tell that two lines are parallel if they both form a right angle with a straight-edged tool.

37. A line running straight back from the front to the back of the parked car is perpendicular to the line of the curb.

39. (a) 1 (b) 1 (c) 1 and 3

41. (a) 5 AM and 7AM (b) 2:30 AM and 9:30 AM

43. (a)

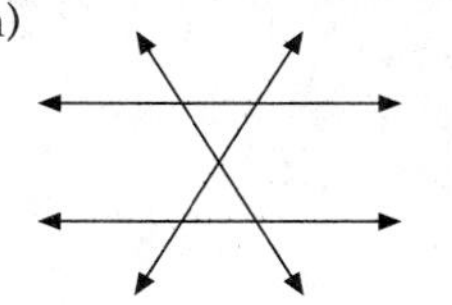

(b)

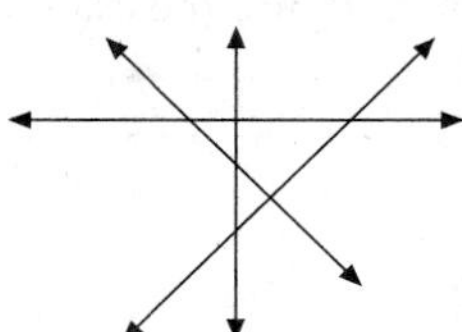

45. (c) Fold one side of the angle into the other.

8.2 *Homework Exercises*

1.

3. (a) Octagon (b) Trapezoid

5.

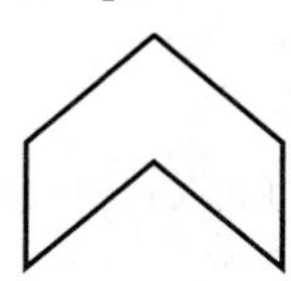

7. (b) and (c)

9.

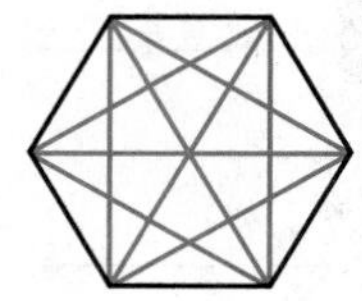

11. Stability

13. (a) Isosceles and equilateral
(b) Isosceles (c) Scalene

15. 8

17.

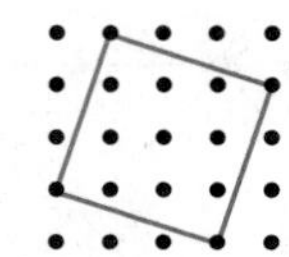

19. (b) and (c)

21. (a) 4 from Group 1, or 2 from Group 1 and 2 from Group 2
(b) 1 from Group 3 with 1 or 2 from Group 1 and 2 or 1 from Group 2
(c) 4 from Group 1

23. All of them

25. (a) Can (b) Must (c) Can
(d) Must (e) Can

27. Rectangle, parallelogram, pentagon, hexagon, triangle, kite, heptagon

29. Measure the diagonals and see if they are equal.

31. (a) (b)

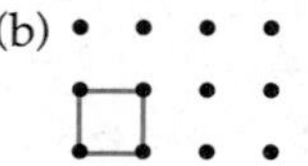

33. All of them

35. All squares are rectangles.

36. (b); Rectangles Rhombuses

37. (a) S (b) Yes (c) Q

39. (d) A parallelogram

41. Points; 8 in.; N

43. (a) $m\angle DCF = 90°$ (b) Isosceles (c) $\overline{BG}$

45. $15 + [15(12)/2] = 105$

47. (a)

49. (a)

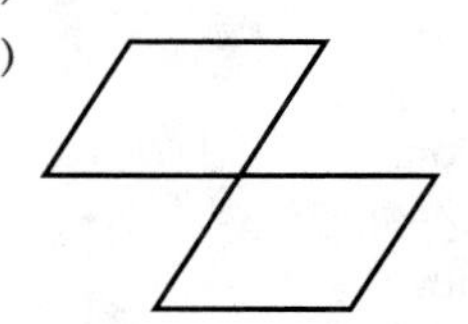

(b)

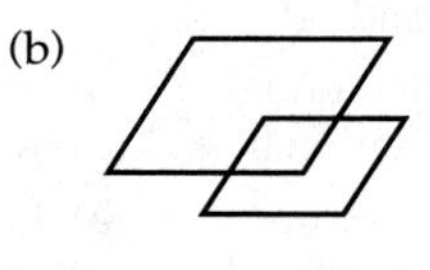

(c) (d)

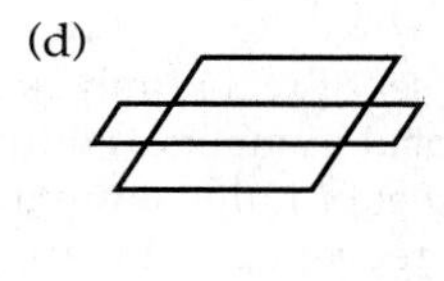

51. (a) (b) (c) (d)

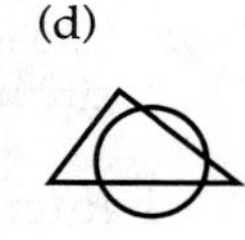

53. The sum of the three angle measures is 180°.

55. All three angles measure 60°.

57. (c)

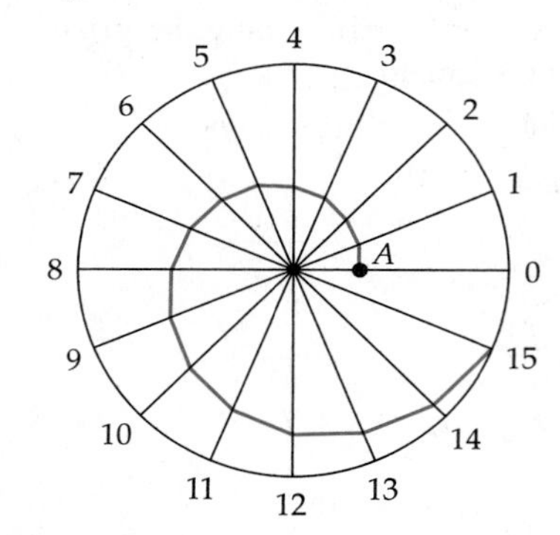

8.3 *Homework Exercises*

1. (c) The sum of the angles should be close to 180°, but it may not be exactly 180° because of drawing and measuring error.

3. $m\angle PKA = 48°$, $m\angle PAK = 22°$, $m\angle AKR = m\angle R = m\angle KAR = 60°$

5. (a) Obtuse (b) Right (c) Acute

7. Divide the hexagon into 4 triangles as shown.

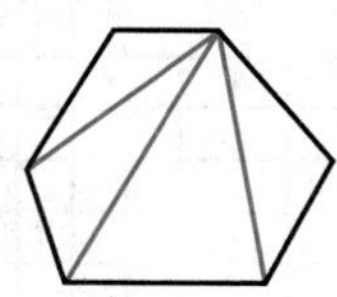

The sum of the angle measures of the 4 triangles is $4 \cdot 180° = 720°$. Therefore, the sum of the angle measures of the hexagon is also 720°.

9. $38(180°) = 6840°$

11. (a) 120° (b)

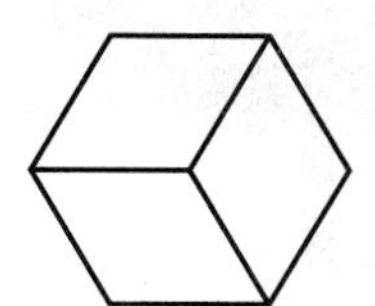

13. (a)

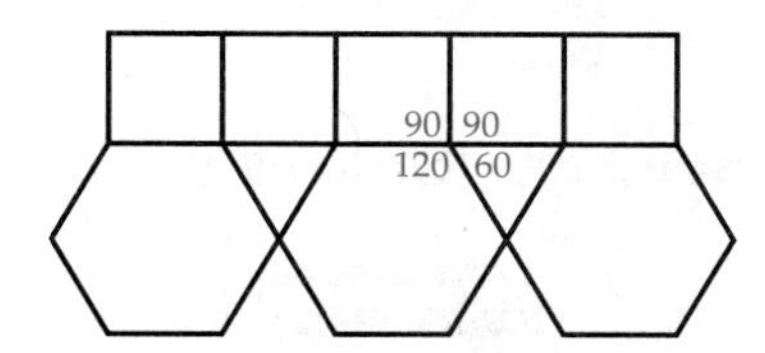

(b) The other six are: (1) 2 octagons, 1 square, (2) 1 square, 1 hexagon, 1 dodecagon, (3) 2 squares, 3 triangles, (4), 2 squares, 3 triangles (a second way), (5) 1 hexagon, 4 triangles, (6) 2 dodecagons, 1 triangle.

15. (a) Yes (b) Yes

17. (a) 6 (b) 10 (c) 15
(d) $\frac{n\ (n + 1)}{2}$ for n small rectangles
(e) 36

19. (a) $\frac{(n - 2)180°}{n}$ (b) Because $\frac{(n - 2)}{n}$ is less than 1

21. 28

25. (a)

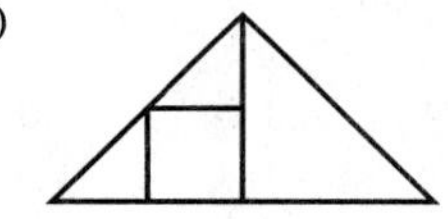

(b)

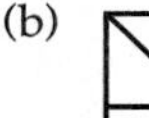

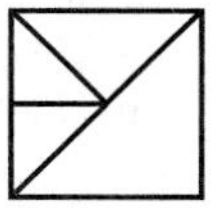

(c)

(d)

8.4 *Homework Exercises*

1. (a) False
(b) The three line segments that meet at the corner of a room

3. (a) An infinite number (b) One (c) One
(d) An infinite number

5. Parallel, intersecting in one point, or the line lies in the plane

7. (a) True
(b) A line on the floor and a line on the ceiling

9.

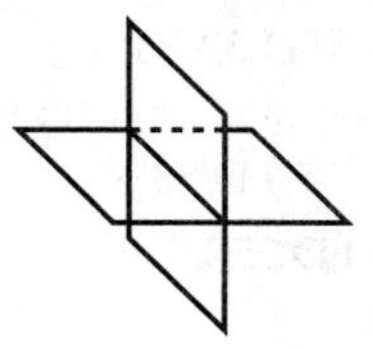

11. True

13. (a) Parallel (b) Perpendicular

15. True

17. (a) False
(b) Two parallel lines on the floor and a line from the ceiling to the floor that intersects one of them

19. True

21. (a) and (d)

23. (a) 8 (b) 18 (c) 12

25. (a) Rectangle (b) Isosceles

27. (a) Pentagonal prism (b) Hexagonal pyramid
(c) Hexagonal prism

29. (c)

31. (a), (c), (d)

33. (a) (b) 7 (c) 7 (d) 12

35. Triangular pyramid

37. (Right circular) cylinder

39. (a) $F = 9, V = 9, E = 16$
(b) $F = 7, V = 10, E = 15$
(c) $F = 18, V = 14, E = 30$

41. (a) Right circular cylinder (b) Rectangular prism

43. Uniform curvature

45. (a) Plane (b) Plane (c) Line

47. True

49. False

51. (a) $n + 2$ (b) $2n$ (c) $3n$ (d) Yes

53. (a) Octahedron (b) Tetrahedron (c) Cube

55. (c) The line on the regular loop is on one side. The line on the Möbius strip goes all around it on both "sides." It has only one side!

8.5 *Homework Exercises*

1. (b) CD is longer than AB.
(c) We tend to look for depth, a third dimension.

3. The room is not rectangular. The back of the room on the right side is closer to the camera.

9.

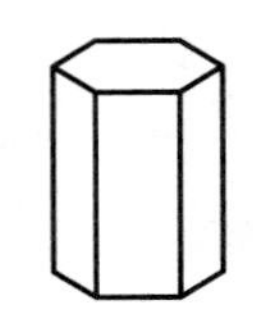

12. Hexagonal prism

13. (a)

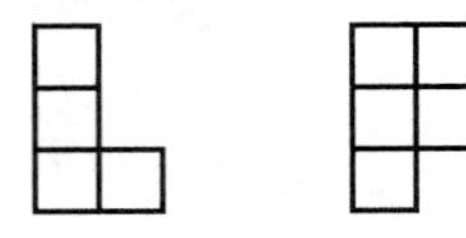

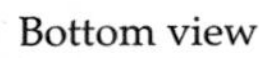

Front view   Bottom view

(b)

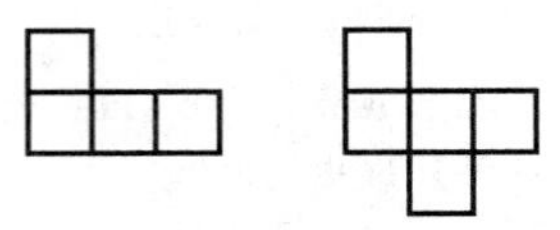

Front view   Bottom view

15.

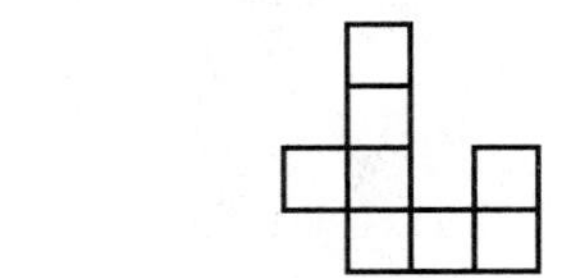

17.

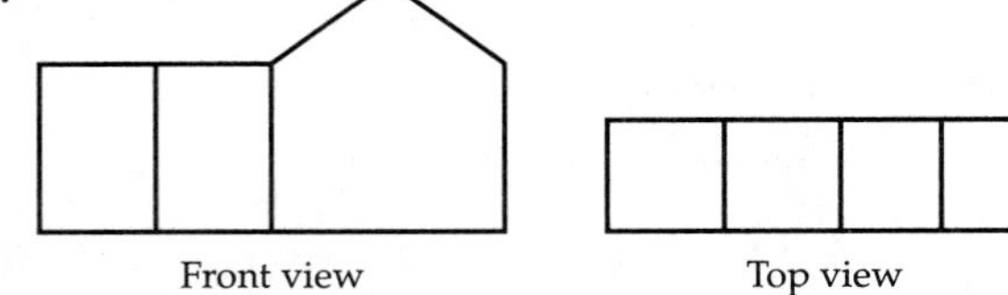

Front view   Top view

19. (a) Triangle (b) Ellipse

21. (a) Rectangle (b) Triangle

23. So we don't injure ourselves drinking from them

25.

29. 4, 5, 7, 6, 2, 3, 1

31. The fisherman in the bottom right reaches the water with his pole. The man in the upper right on the hill lights the woman's lantern. The man on one side of the bridge shoots his rifle on the other side. The banner from the building hangs out among the trees.

33. (b) It could not exist as the three-dimensional figure it appears to be.

35.

37. (a) Drawing 2 (b) Drawing 1

39.

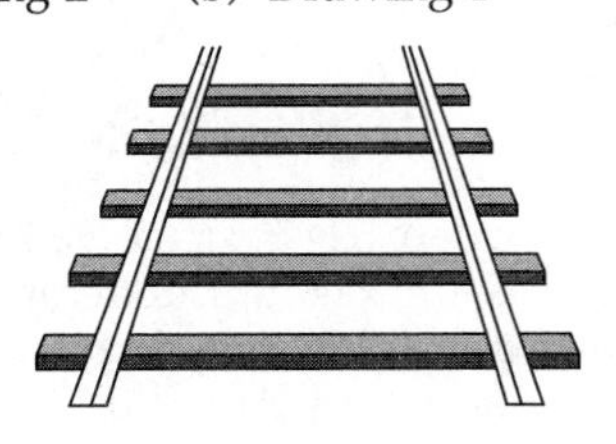

41. (a)

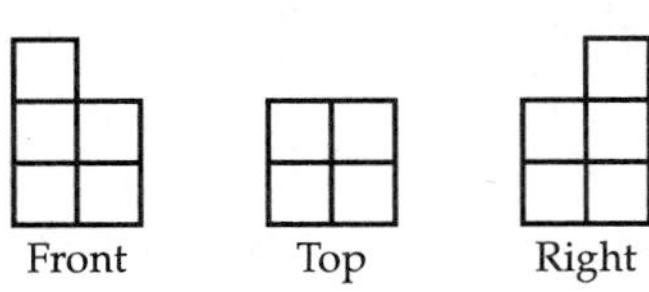

Front   Top   Right

(b)

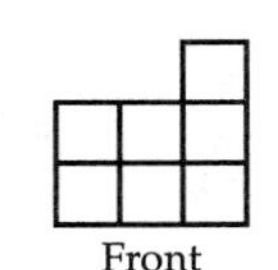

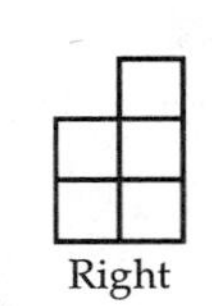

Front   Top   Right

43. (a) 64 (b) 24 (c) 24 (d) 8 (e) 0
(f) 8

45. (a) 96, 48, 8, 0, 64
(b) $6(N-2)^2$, $12(N-2)$, 8, 0, $(N-2)^3$

8.6 *Homework Exercises*

1. (a) (b)

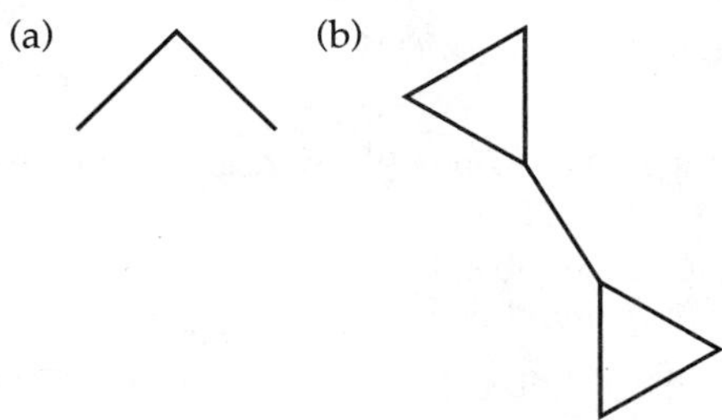

3. T

5. (a) 135° (b) 45°
(c) REPEAT 8 [FD 30 RT 45]

7. (a) FD 60 RT 90 FD 50 RT 90 FD 25 RT 90 FD 50
(b) FD 50 RT 135 FD 70 RT 135 FD 50

9. FD 50 RT 135 FD 25 LT 90 FD 25 RT 135 FD 50

11. REPEAT 4 [FD 30 LT 90 FD 30 RT 90]

13. REPEAT 5 [FD 60 RT 144]

15. 180° in first 3 blanks; 540°, 180°, 360°

8.7 *Homework Exercises*

1. (a)
```
TO SQUARE
  REPEAT 4 [FD 35 RT 90]
  END
```
(b) A rotational design with 6 squares

3. (a)
```
TO HEXA :S
  REPEAT 6 [FD :S RT 60]
  END
```

5. (a) A pentagram (b)

7.
```
TO ANGBI
  FD 50 RT 100 FD 50 BK 50 RT 40 FD 50
  END
```

9. (a)
```
TO SQUARE :N
  REPEAT 4 [FD :N RT 90]
  END
TO GROW
  SQUARE 10
  SQUARE 20
  SQUARE 30
  SQUARE 40
  END
```
(b)
```
TO BOX
  REPEAT 4 [FD 30 RT 90]
  FD 15 RT 45
  REPEAT 4 [FD 21 RT 90]
  END
```

11. (a)
```
TO TRI
  REPEAT 3 [FD 30 RT 120]
  END
TO PW
  LT 30
  REPEAT 4 [TRI RT 90]
  END
```
(b)
```
TO HUMAN
  LT 45
  REPEAT 4 [FD 30 RT 90]
  LT 45 FD 30 BK 60 FD 30
  LT 90 FD 45 RT 45 FD 30
  BK 30 LT 90 FD 30 BK 30
  END
```

13. (a)
```
TO WHEEL
  REPEAT 50 [FD 2 RT 9]
TO CAR
  PU LT 90 FD 60 RT 90 PD
  WHEEL
  LT 180 FD 30 BK 120 RT 90 FD 30
  LT 90 FD 40 RT 90 FD 20
  LT 90 FD 40 LT 90 FD 20 RT 90 FD 40
  LT 90 FD 30 LT 90 FD 90
  WHEEL
  END
```
(b)
```
TO CONE
  FD 50 RT 135 FD 50
  END
TO SCOOP
  REPEAT 70 [FD 3 RT 9]
  END
TO DESSERT
  PU FD 50 PD
  SCOOP
  LT 180
  SCOOP
  REPEAT 22 [FD 3 RT 9]
  CONE
  END
```

15. (a)
```
TO L
  RT 180 FD 60 LT 90 FD 35
  END
```
(b)
```
TO O
  REPEAT 20 [FD 4 RT 18]
  END
```
(c)
```
TO G
  REPEAT 30 [FD 4 RT 18]
  FD 30
  REPEAT 10 [FD 4 RT 18]
  END
```
(d)
```
TO MASTER
  PU FD 30 LT 90 FD 80
  RT 90 PD
  L
  PU FD 25 LT 90 PD
  O
  PU RT 90 FD 50 LT 90 PD
  G
  PU RT 90 FD 50
  LT 90 FD 30 PD
  O
  END
```

17. (c)
```
TO SQUARE :X
  RT 45
  REPEAT 4 [FD :X RT 90]
  SQUARE :X + 3
  END
```

Chapter 8 *Review Exercises*

1. In order to define space figures properly, one must first study the components of these figures.
2. It is finite and it has thickness.
3. (a) $\angle ABC$ (b) { }
4. An infinite number
5. False
6. (a) False
   (b) Two intersecting lines on the ceiling are both parallel to the floor.
7. 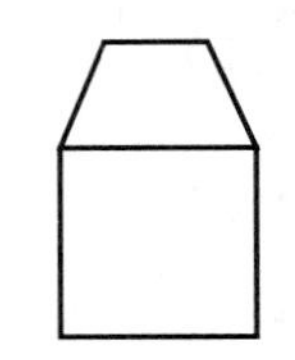
8. 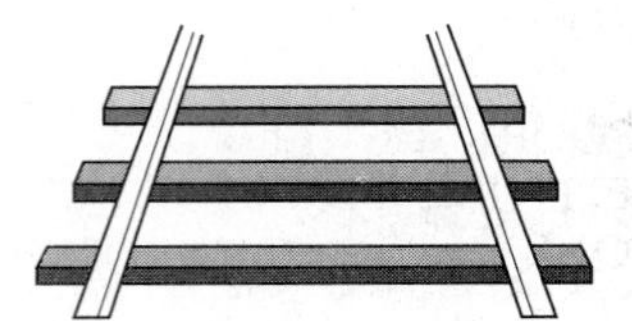
9. 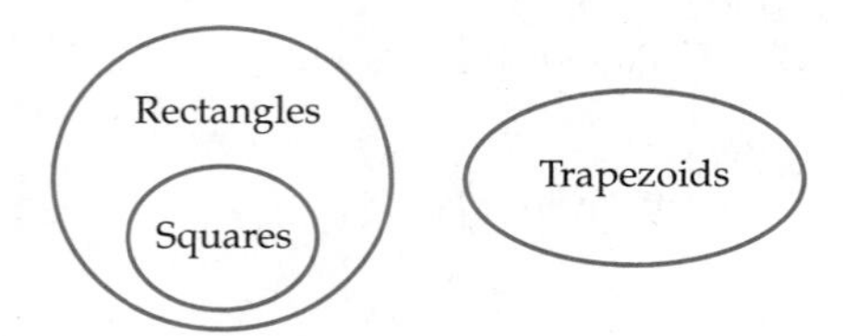

10. No
11. (a) Can (b) Must (c) Cannot (d) Must (e) Can
12. (a) Some (b) All
13. 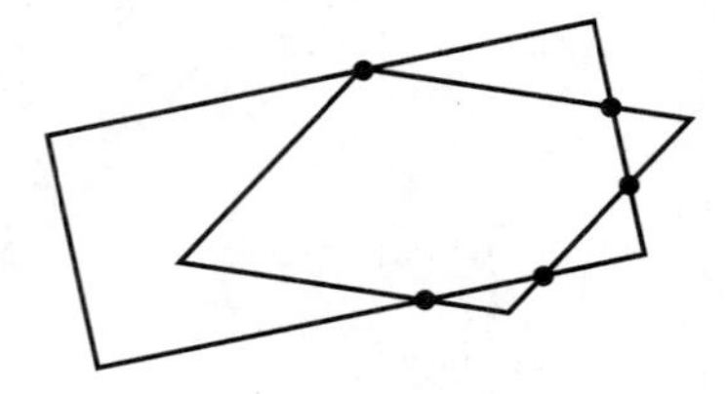
14. A pentagon can be divided into three triangles as shown. The angle sum of each triangle is 180°, so the angle sum of the pentagon is $3 \cdot (180°) = 540°$.

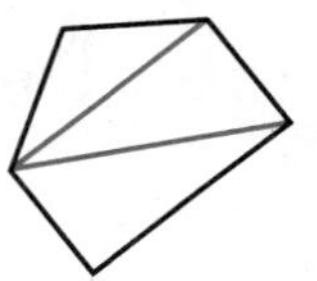

15. Each angle of a regular hexagon is 120°, so 3 regular hexagons will fit together around a point as shown.

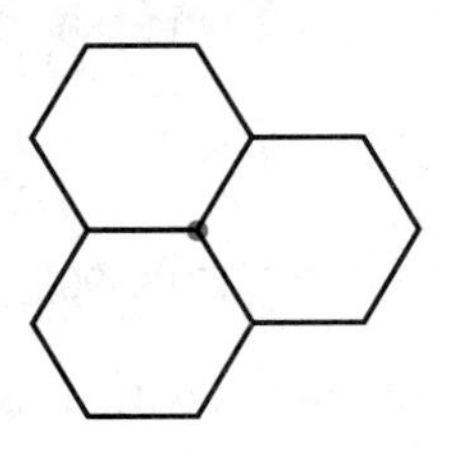

16. 7 faces, 7 vertices, and 12 edges
17. 9
18. 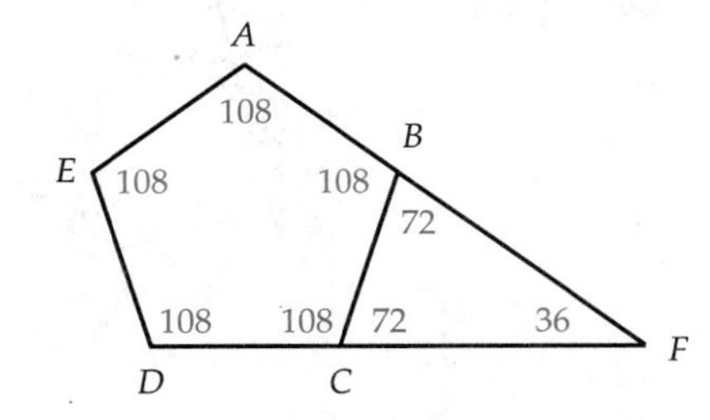

19. $\frac{7(180°)}{9} = 140°$
20. 12
21. $m\angle A = m\angle C$ or $\angle A \cong \angle C$
22. 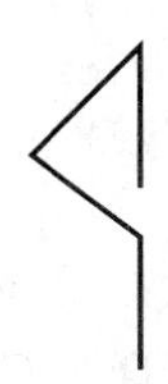
23. REPEAT 8 [FD 50 RT 45]
24. 
25. 2 hexagons $(2 \cdot 135°) + 1$ square $(90°)$ around each point

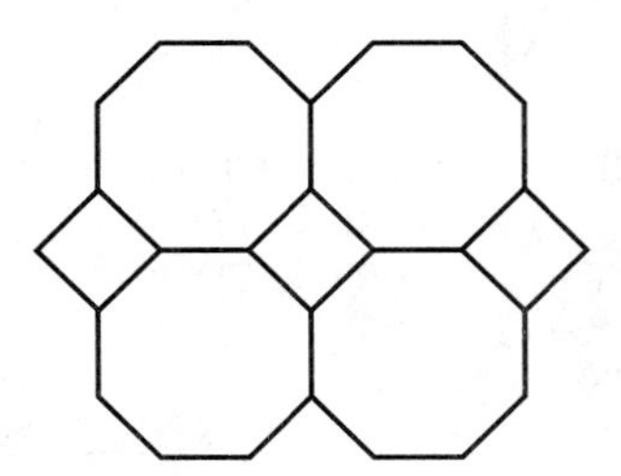

Chapter 9

9.1 *Homework Exercises*

1.

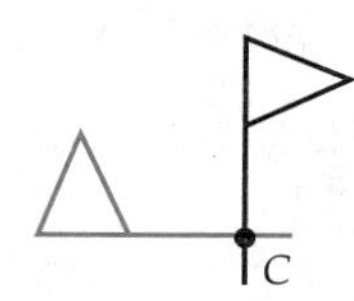

2. (d)

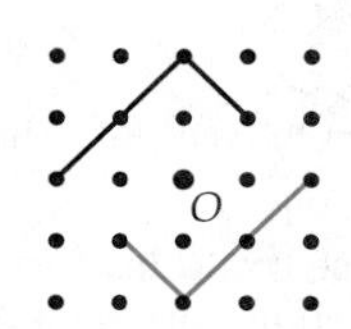

3. 32°, $P'O$

5.

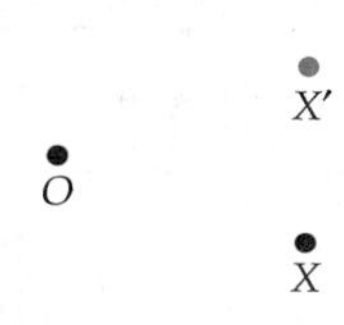

6. *Hint:* Move the flag 4 cm to the left.

7.

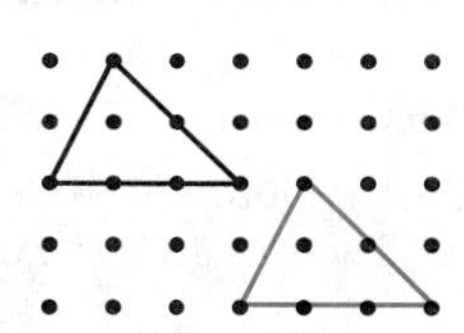

9. $\overleftrightarrow{AB}$; $\overleftrightarrow{P'B}$

11. (a) (b) 3

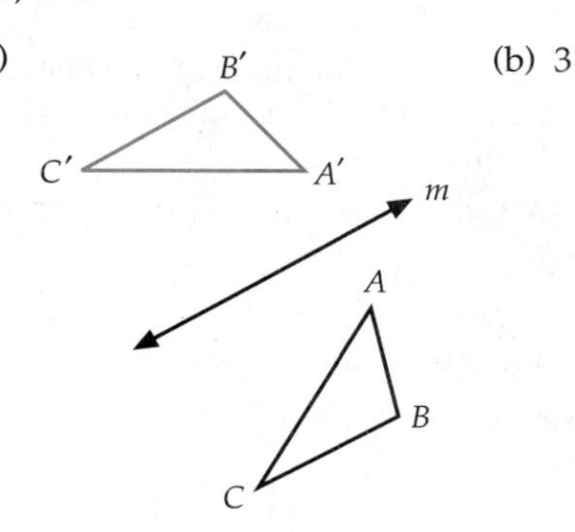

13.

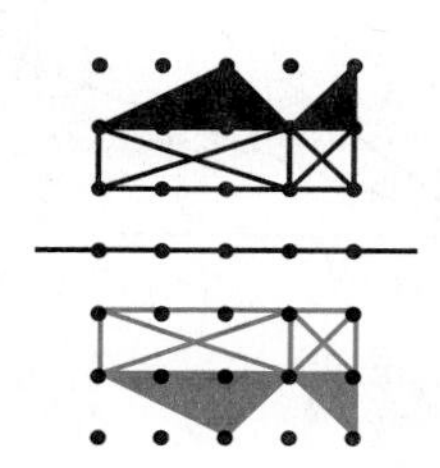

15. (a) (b)

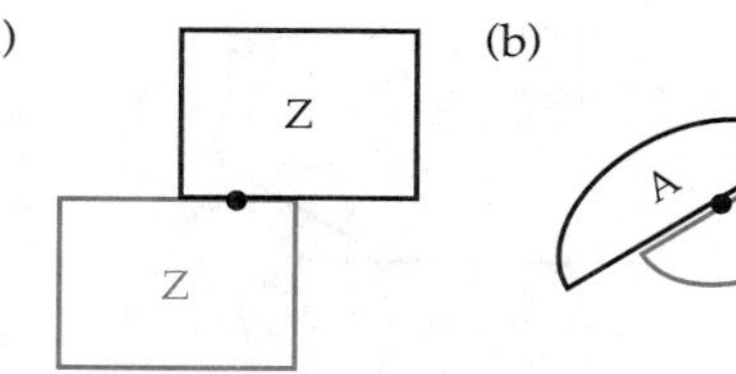

17. (a) C (b) F (c) E (d) E

19. (a) (b)

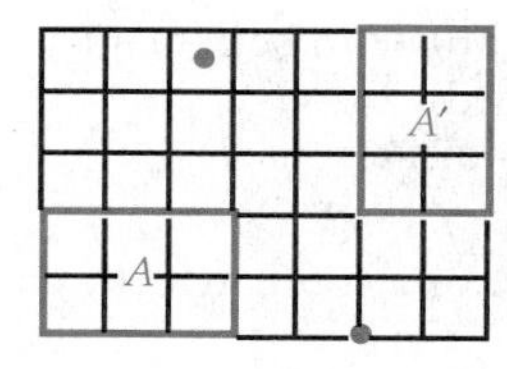

21. When it's reflected in someone's rearview mirror, it says AMBULANCE.

23. (a) Rotation (b) Reflection (c) Translation

27.

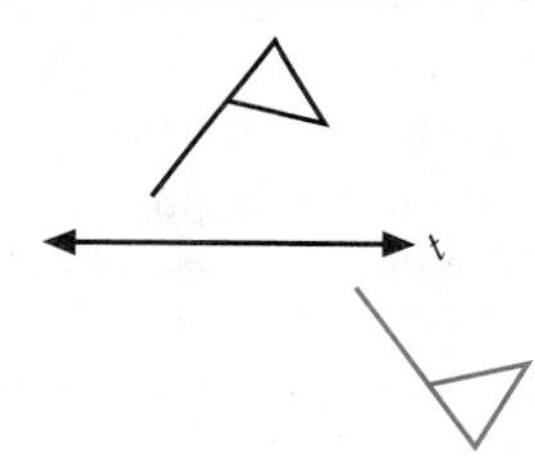

29. (a) False
(b) Reflect the vertical line across any line that is not horizontal or vertical.

31. (b) Only the first pair is congruent.

32. (a) 180° rotation (b) Reflection

33 (a) Translation
(b) 180° rotation around a point midway between the bottom of each F.
(c) Not congruent

35.

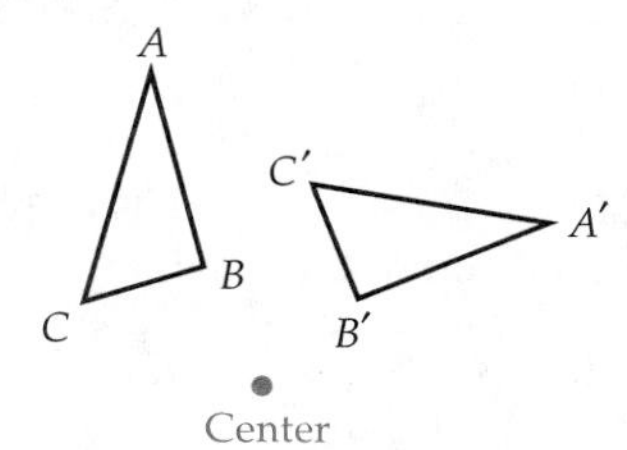

37.

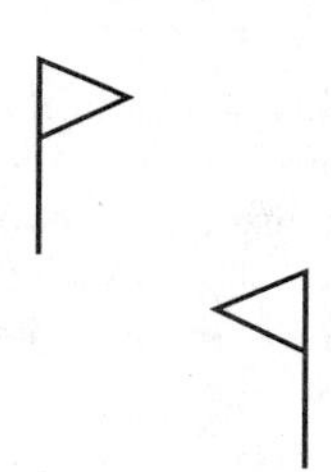

38.

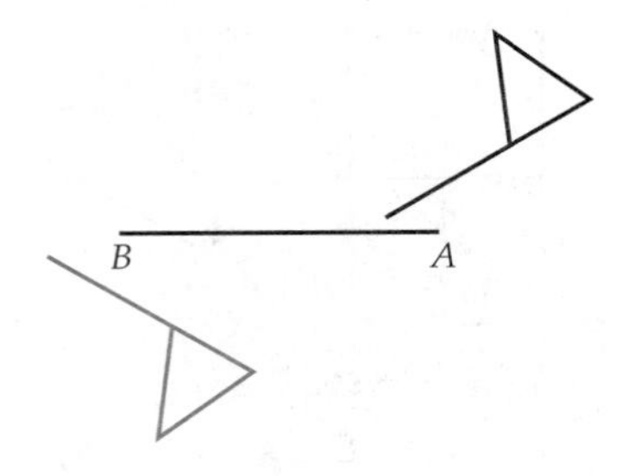

39. (a) 180° turn around E (b) $\cong$ in both blanks
(c) 180° turn around E (d) $\angle ART$
(e) A half turn around E indicates that $\angle BTR \cong \angle RAB$.

40. (a) Reflection through $\overleftrightarrow{AM}$ or a counterclockwise rotation around A of $m\angle CAT$
(c) Reflection through $\overleftrightarrow{AM}$; $\angle C \cong \angle T$

41. (a) $\angle ACB \cong \angle AGF$, $\angle ADE \cong \angle AHI$, $\angle ACD \cong \angle AGH$, $\angle ADC \cong \angle AHG$
(b) $\angle BCG \cong \angle CGH$, $\angle DCG \cong \angle CGF$, $\angle EDH \cong \angle DHG$, $\angle CDH \cong \angle DHI$.

43. (a) The same thing! (b) TIM

45. (a) If they are the same length
(b) If they have one side of equal length
(c) All rays are congruent.

47. (a) Rotate 120° clockwise around a point where their "noses" meet.
(b) Reflect through the body line that passes between the eyes.

49. (a) Yes

53. The boy is photographed in one mirror while holding another.

55. A mirror about half of your height, positioned around the top half of your body

57. Translation

59. (a) $A(2, 4)$ $B(2, 1)$ $C(6, 1)$
(b) $A'(-2, 5)$ $B'(-2, 2)$ $C'(2, 2)$
(d) Translation

61. (a) $A(1, 2)$ $B(3, -1)$ $C(1, -2)$
(b) $A'(-1, 2)$ $B'(-3, -1)$ $C'(-1, -2)$
(d) Reflection through the y-axis

63. (a)  (b) 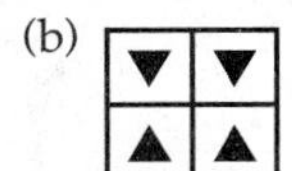

65. (d) A 120° clockwise rotation around a point C, different than A or B
(f) An x° clockwise rotation around a point C followed by a y° clockwise rotation around a point B is the same as a single $(x + y)$° clockwise rotation.

67. (a) 24 (b) 15

69.
```
TO ROTATE
  REPEAT 3 [FD 30 RT 120]
  RT 140
  REPEAT 3 [FD 30 RT 120]
  END
```

71.
```
TO SQUARE
  REPEAT 4 [FD 30 RT 90]
  END
TO PATTERN
  REPEAT 4 [SQUARE RT 90]
  END
```

73. (c) This exercise could lead to a generalization about congruent alternate interior angles or congruent corresponding angles.

75. (a) (1) and (2); Glide reflection
(c)

9.2 *Homework Exercises*

2. (b) By the construction, $AN = CD$, $MA = BC$, and $MN = BD$. Then $\triangle MAN \cong \triangle BCD$ by SSS.

4. (b)

5. Yes, by the SSS property

6. (b) *Hint:* Use SSS.

9. You could construct a right angle at A. Then, mark off the distance AB on the new side and label the endpoint D. Connect D to C.

13. (a) No (b) Yes (c) No (d) Yes (SAS)

15. $AM = AM$. Then $\triangle CAM \cong \triangle TAM$ by SSS. Since the triangles are congruent, $\angle MAC \cong \angle MAT$, which proves that $\overrightarrow{AM}$ bisects $\angle CAT$.

16. (b) *Hint:* Use SSS.

17. (a)

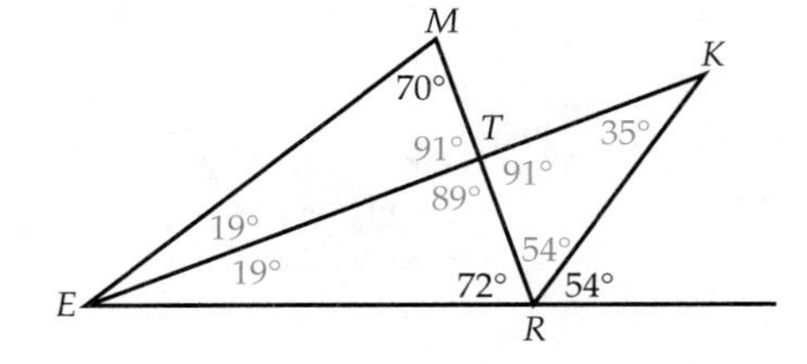

(b) (possible answer) $\angle MTE$ and $\angle ETR$

19. Draw the angle bisector of $\angle A$ and label the point where it intersects side $\overline{BC}$ as point D. Since $AD = AD$, $\angle B \cong \angle C$, and $\angle BAD \cong \angle CAD$, $\triangle ABD \cong \triangle ACD$ by AAS. Since the triangles are congruent, corresponding sides $\overline{AB}$ and $\overline{AC}$ are congruent.

21. (a)

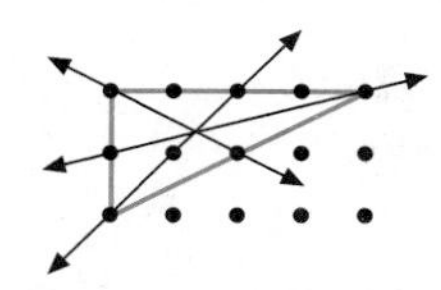

(b) Bisect $\overline{BC}$ to locate the midpoint. Then connect the midpoint to A.
(c) The three medians intersect at one point called the centroid.

23. *Hint:* Use the span of the compass.

25. (b) Fold one side of the angle onto the other.

27. (c)

29. 60°

33. They bisect each other but are not congruent or perpendicular.

37. (a) *Hint:* Construct a 90° angle and bisect it.
(b) Take the 45° angle from the last exercise and construct a perpendicular on one side. The sum of the two adjacent angle measures will be 135°.

39. 4 mm

43.

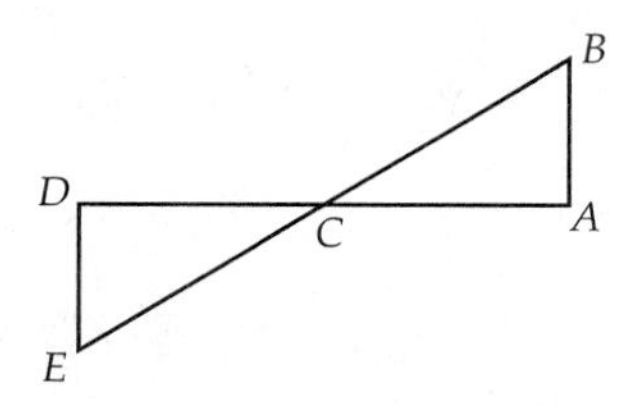

$AC = CD$, $\angle A \cong \angle D$, $\angle DCE \cong \angle ACB$. Therefore $\triangle ACB \cong \triangle DCE$ by AAS. Then $DE = AB$.

45. *Hint:* Let $\angle A = x°$. Answer: 20°

47. $\triangle ABC \cong \triangle EFG$ by SAS. So $AC = EG$ and $\angle BAC \cong \angle FEG$. By subtraction, $\angle DAC \cong \angle GEH$. Then $\triangle ADC \cong \triangle EHG$ by SAS. This means $DC = HG$ and $\angle H \cong \angle D$. Since $\angle ACD \cong \angle EGH$ and $\angle ACB \cong \angle EGF$, we can add to obtain $\angle BCD \cong \angle FGH$. Therefore, $ABCD \cong EFGH$.

49. (a)

(b) It indicates that a quadrilateral is not rigid.

53. A given point on the perpendicular bisector is equidistant from A and B.

9.3 *Homework Exercises*

1. Yes; no

3. (a), (d)

5. (a) 4 (b) 2 (c) 2 (d) 0

7. If the rectangle is reflected through the diagonal, the image rectangle will not be in the same position as the original rectangle.

9. (d) The two lines of symmetry are perpendicular. The two lines of symmetry bisect the angles formed by the intersecting lines.
(e) Induction

11. (a) A butterfly (b) Hexagon

13. (a) 3 (b) 1 (c) 1 (d) 1

15. (a) and (c) have rotational and reflection; (b) has reflection.

17. (a) *U, I, D* (b) *S, I*

19. Usually: (a) all 2's, 4's, 10's, J's, Q's, K's and the A, 3, 5, 6, 8, and 9 of diamonds
(b) None

21. (Number of rotational symmetries) = (number of reflection symmetries) − 1

23. (a)

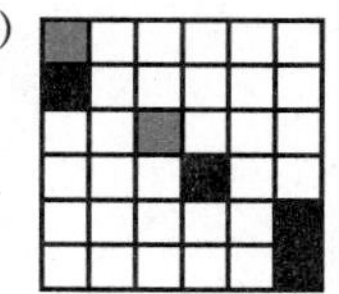

(b)

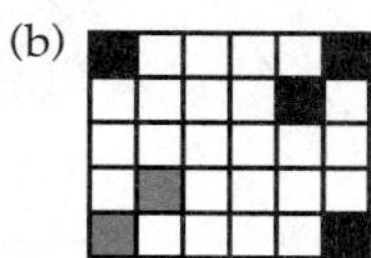

25. Rotational

27. 4

29. 7

31. Right circular cone

33. (a) Usually rotational
(b) Can be laid down in a number of ways

35. (a) A tennis ball has rotational and reflection symmetry if one ignores the ridges; a tennis racket has reflection symmetry; a tennis court has rotational symmetry, and the surface has reflection symmetry.
(b) The tennis ball will bounce well and can be hit at any point without it making a difference; one can strike the ball with either face of the tennis racket; the court on either side of the net is the same, and the court is comparable on the left and right sides on each side of the net.

37. (a), (c), (d), and (e) have rotational and reflection; (b) has none.

39. 4 if both surfaces are the same

41. 3 and 1

43. (a) Reflection (b) Commutative (c) Yes

45. Fold the circle in half from two different positions to create two different diameters. The intersection of the two diameters (folds) is the center.

47. (a) They are congruent.
(b) They are congruent.

49. Impossible

51. (c) 3
(e) *Hint:* Five times the total of all your turns should equal a multiple of 360°.

9.4 *Homework Exercises*

1. (a) 1:2 (b) 1:4 (c) 1:2

2.

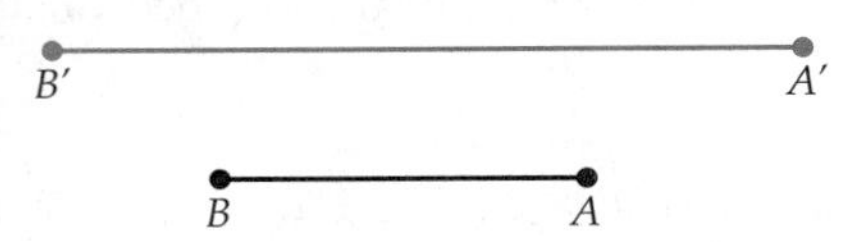

3.

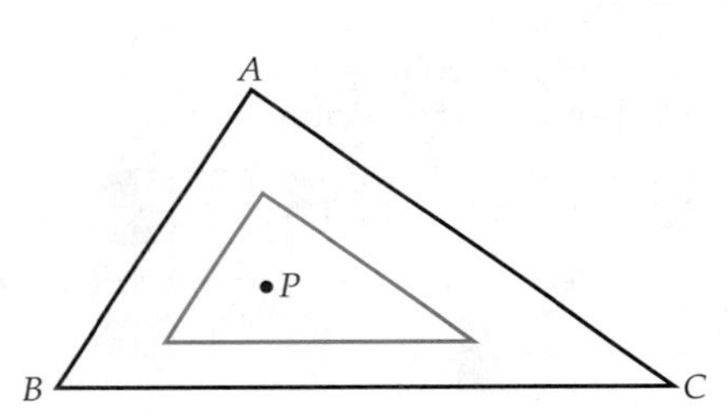

4. Yes

5. (a) $\triangle RMN$ (b) 1/2

7. Corresponding sides are not proportional.
$3/6 \neq 4/7 \neq 5/8$

9. (a) Yes. Corresponding angles are congruent, and corresponding sides are proportional.
(b) No. Corresponding angles may not be congruent.

11. They are similar.

12. (a) $\angle P$ (b) NA; PE

13.

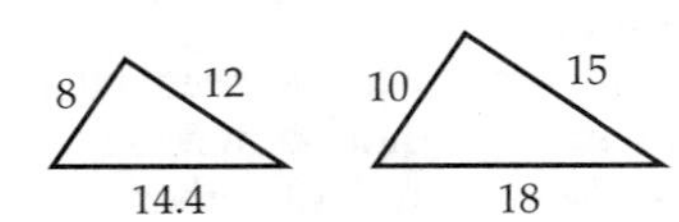

15. $x = 17\frac{1}{7}$ $y = 22\frac{6}{7}$

17. 1.6

19. $x = 5, y = 11\frac{2}{3}$

21. Yes, they are all the same shape.

23. About 297″ or 24′ 9″

25. (a) 6.8 cm by 10.2 cm (b) Yes (c) 46%

27. *Hint:* 8 m would be represented by 2 cm.
10 m would be represented by 2.5 cm.

29. (b) They are similar.
(c) Corresponding sides are proportional.
(d) Similar

31. (c) Yes (e) to be

33. (b) B and C (c) $BC = 2 \cdot DE$
(d) The segment joining the midpoints of two sides of a triangle is half the length of the third side.

35. (b) You obtain a similar figure. The original figure can be mapped to the image figure using a rotation around the origin followed by a size change.

37. Yes

39. 1.10

41. (c) They are similar.
(e) If you reduce a polygon with a scale factor of $\frac{1}{2}$, the result will be a similar polygon.

43. Read Exercise 44.

45. (a) Yes (b) Yes (c) Forever
(d) $r = \dfrac{1 \pm \sqrt{5}}{2}$

Chapter 9 *Review Exercises*

1. Two figures are congruent if one can be mapped onto the other using a rotation, reflection, or translation.

2.

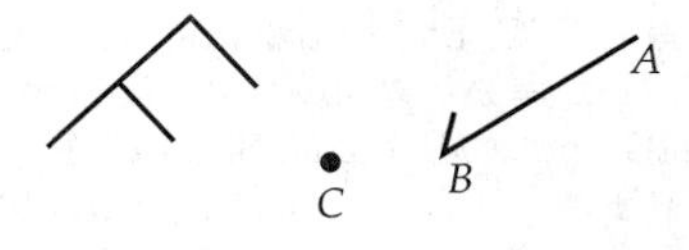

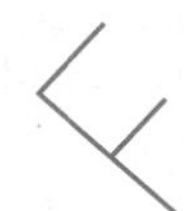

3. (a)

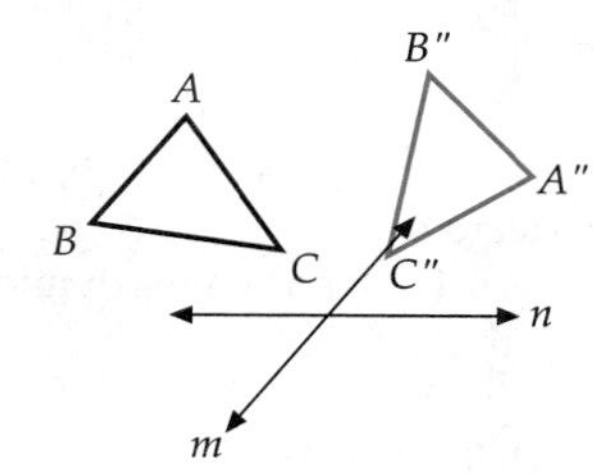

(b) A rotation around the intersection point of the two lines

4. (a) $(-6, -2)$ (b) A translation

5. $\perp$

6. (a) Could say $\angle DAB \cong \angle BCD$, 180° turn around E
(b) $\angle AEB \cong \angle CED$, 180° turn around E

7. (b) From the construction, $AI = MI$, $TI = TI$, and $AT = MT$. So $\triangle AIT \cong \triangle MIT$ by SSS. Then, $\angle AIT \cong \angle MIT$, so $\overrightarrow{IT}$ is the angle bisector.

8. The angles of the hexagon are 120°; smaller angles of the triangle are 30°; the remaining two angles adjacent to the 30° angles are 90°.

9. Two, each passing through the midpoints of two opposite sides

10. Yes, all their angles measure 108°, and corresponding sides are proportional.

11. A square

12.

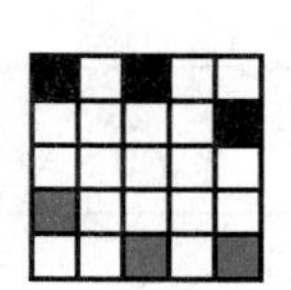

13.

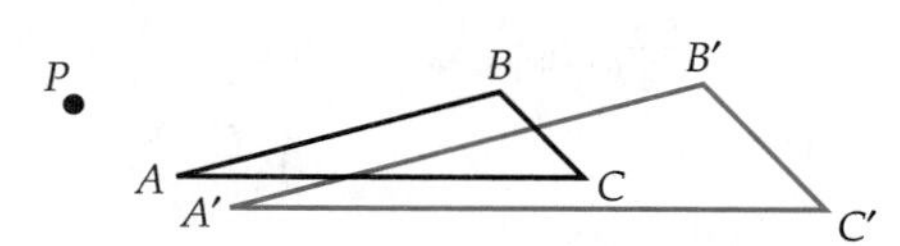

14. 5

15. 26 ft 11 in.

16. $\dfrac{10L + 60}{L}$

17.

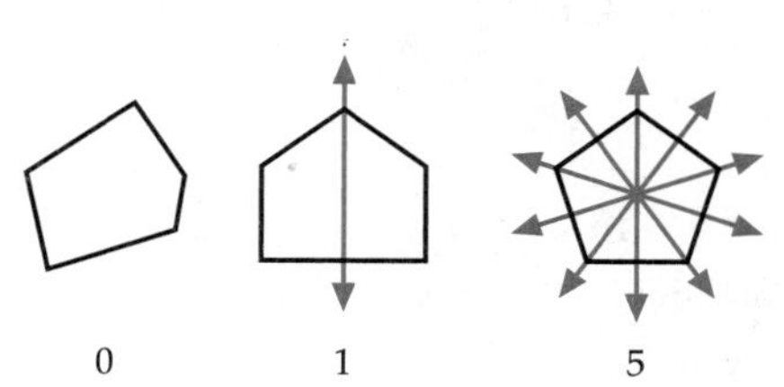

Chapter 10

10.1 *Homework Exercises*

5. (a) The width of a finger
(b) The height of a doorknob

7. (a) 20 (b) 160 (c) 0.082

9. 20

11. 10 cm

13. 70 kg

15. (a) 9400 (b) 0.037 (c) 82

17. 3240

19. 250 mL

21. 5000

23. (a) False (b) False (c) False (d) True

24. (a) g (b) cm (c) mm (d) kg (e) km

25. (a)

27. (b)

29. Most college students can.

30. 46 cm, 871 mm, 137 cm, 37 m, 3 km

31. (a) Approximate (b) Exact (c) Approximate
(d) Exact (e) Approximate

33. 27.5 and 28.5 cm

35. (a) 50 mm (b) 0.5 mm (c) 49.5 and 50.5 mm

36. (a) Nearest g; 0.5 g; 61.5 and 62.5 g
(b) Nearest hundredth of a cm; 0.005 cm; 5.115 and 5.125 cm
(c) Nearest tenth of a km; 0.05 km; 4.55 and 4.65 km

37. (a) 0.0081 (b) 0.00098 (c) 0.011

39. *Hint:* Draw *one* side of the rectangle and the diagonal first.

41. 0.17 km

43. (a) 1,000,000 (b) 0.00006 (c) 0.0821
(d) 4,800,000

45. 1.944

47. $\dfrac{x - 9}{2}$ cm

10.2 *Homework Exercises*

1. (a) 8.4 cm (b) 8.0 cm

3. Draw another square adjacent to the second highest square on its left.

5. $T/4$ m

7. (a) The second has twice the perimeter.
(b) Same as (a)
(c) When the length of each side of a square is doubled, the perimeter also doubles.

9. Because $2r = d$

11. 7958 miles

13. 332

15. (a) 325.68 m (b) 6.284 m

17. The space in its interior

21. 3

23. (a) Show 2 rows of 5 squares. (b) 10 m^2

25. $0.2 \times 0.3 = 0.06$

27. $A =$ about 212,214 km^2, $P = 1930$ km

29. 48 ft^2

31. (a)

33. $\sqrt{48}$ ft

35. 4 square units

37. (a) Draw two vertical line segments, or two horizontal ones, or one of each.
(b) $140,800

39. Package *B*

41. $62.93

42. (a) 100 (b) 0.0001

43. (a) 800 (b) 5,000,000

45. (a) 2.6
(b) Converting square units as if they were units

47. 128

49. 4 rows and 6 columns; $\frac{10}{4}$ by $\frac{14}{6}$ = 2.5 in. by 2.3 in.

51. (c)

53. (a) 5 square units (b) 10, 12

55. (a) 177.17 m (b) 87.60 m

57. (a) 8 m (b) 12 m (c) $4\sqrt{N}$ m

59. (a) 4, 6, 8 (b) ; 10
R4
(c) $2N + 2$ (d) 22

61. Each rectangle could have $L = 42.5$ ft and $W = 60$ ft.

63. (a) A line segment excluding its endpoints that connects (10, 0) and (0, 10).

(b)

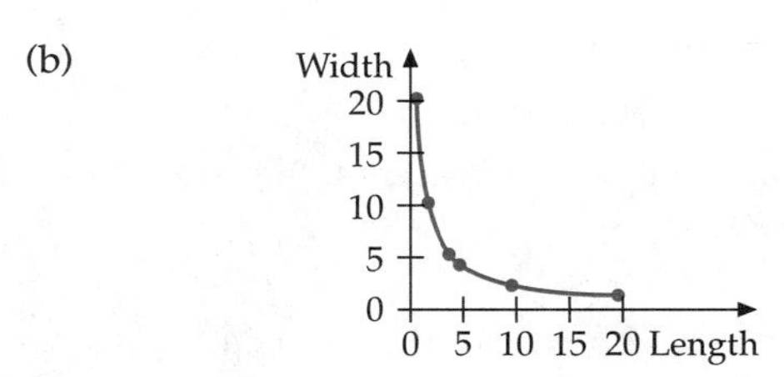

65. (a) $ab + ac = a(b + c)$
(b) $(a + b)^2 = a^2 + 2ab + b^2$

67. $s = 2\pi r\theta/360°$

69. (a) C3 = A3 * B3
(b) C4 = A4 * B4; C5 = A5 * B5; C6 = A6 * B6
(c) They decrease.
(d) The closer in size the length and width are, the greater the area.

10.3 *Homework Exercises*

1. See the answer to Lesson Exercise 10.38.

3. (a) 50 in.^{2}, 32 in. (b) 6 m^2, 12 m

5. Put together two of the triangles as shown to form a parallelogram.

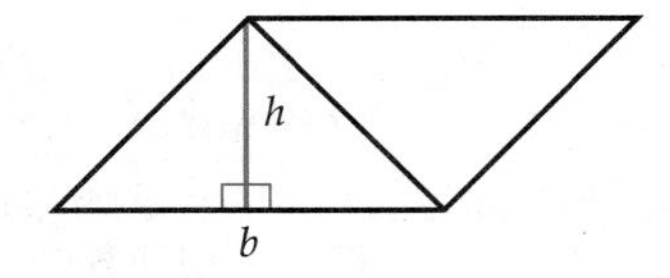

$$A_{\triangle} = \frac{1}{2}A_{\text{parallelogram}} = \frac{1}{2}bh.$$

6. (a) 360 in.^{2} (b) 12 square units

7. 24 ft^2

9. 4000 cm^2

11. (a) 4.5 square units (b) 6 square units

13. Put together two of the trapezoids as shown to form a parallelogram.

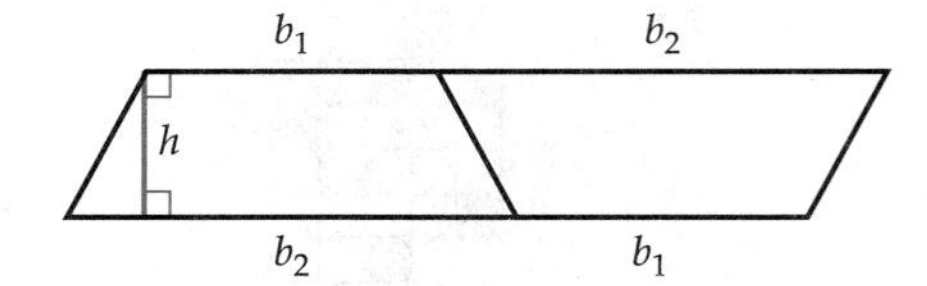

$$A_{\text{trapezoid}} = \frac{1}{2}A_{\text{parallelogram}} = \frac{1}{2}(b_1 + b_2) \cdot h.$$

15. About 270 mm^2

17. B ($28.57/$m^2$)

19. Squares and rectangles

21. (a) 20.25π square units (b) 63.585 square units
(c) $\left(\frac{8}{9}d\right)^2$ or $\left(\frac{16}{9}r\right)^2$ or $\frac{256}{81}r^2$
(d) $\frac{256}{81} \approx 3.16$

23. $\frac{1}{2}Cr = \frac{1}{2}(2\pi r) = \pi r^2$

25. (a) 2 (b) 4

27. Same area either way

29. 2.25π ft^2

31. 100π ft^2

33. $7\pi \approx 22$ m^2

35. (a) $8 + 20\pi/3$ (b) $40\pi/3$ square units

37. (c)

39. (a) 100.3 km (b) No theoretical limit

41. 2.25π square units

43. $9\pi - 18$ square units

45. $1.5\pi + 1$ square units

47. False; circles with radii of 1 and 2 have areas in a ratio of 1:4, not 1:2.

49. 56 square units

51. The 500 miles is $\frac{7.5}{360}$ of the circle. The circumference would be $500 \cdot \frac{360}{7.5} = 24{,}000$ miles.

53. See pp. 148–153 in the 1969 NCTM yearbook, *Historical Topics for the Mathematics Classroom.*

10.4 *Homework Exercises*

1. (a) Right triangles
(b) a and b are the lengths of the legs; c is the length of the hypotenuse.

3. $m\angle ABH + m\angle AHB = 90°$. Since $\angle DBC \cong \angle AHB$, $m\angle ABH + m\angle DBC = 90°$.
$m\angle ABH + m\angle DBC + m\angle 3 = 180°$, so $m\angle 3 = 90°$.

4. (a) $\frac{1}{2}ab, \frac{1}{2}ab, \frac{1}{2}c^2$ (b) $\frac{1}{2}(a + b)(a + b)$
(c) $\frac{1}{2}ab + \frac{1}{2}ab + \frac{1}{2}c^2 = \frac{1}{2}(a + b)(a + b)$

$$ab + \frac{1}{2}c^2 = \frac{1}{2}a^2 + ab + \frac{1}{2}b^2$$
$$\frac{1}{2}c^2 = \frac{1}{2}a^2 + \frac{1}{2}b^2$$
$$c^2 = a^2 + b^2$$

5. Draw a line segment that goes up 2 and over 3.

7. $3 + \sqrt{8} + 3 + \sqrt{8} = 6 + 2\sqrt{8}$ or $6 + 4\sqrt{2}$

9. 3.9 m

11. 71%

13. $4\sqrt{2}$ mph NE

15. 14.1 in.

17. $8\sqrt{20}$ or $16\sqrt{5}$ square units

19. 5 miles

21. (a) 10 (b) Finding the length of the hypotenuse

23. (a) Yes (b) No

25. (a) Obtuse (b) Right (c) Acute

27. (a) Obtuse (b) Right (c) Not (d) Obtuse

29. (a) Yes (b) 6, 8, 10; 5, 12, 13; 9, 12, 15
(c) If $a^2 + b^2 = c^2$ then $k(a^2 + b^2) = kc^2$ or $ka^2 + kb^2 = kc^2$. So ka^2, kb^2, kc^2 is a Pythagorean triple.

31. (a) $x = 3, y = 4, z = 5; x = 6, y = 8, z = 10;$ $x = 9, y = 12, z = 15$
(b) No

33. 169 square units

35. $\frac{25\sqrt{3}}{4}$ square units

37. $30\sqrt{75}$ or $150\sqrt{3}$ square units

39. $12\sqrt{12} \approx 17$ in.

41. 64π square units

43. (a), (b), and (c)

45. $\sqrt{12}$ cm

47. (a) $a^2 + b^2 = c^2$
(b) They both equal $a^2 + b^2$.
(c) $\triangle GFE$; the SSS property

49.
```
TO RECT
  REPEAT 2 [FD 30 RT 90 FD 50 RT 90]
  RT 59 FD 58
  END
```

10.5 *Homework Exercises*

1. 32 square units

3. (a) 1 by 1 by 20 (b) 2 by 2 by 5

5. 2700 cm^2

7. (a) 12,150 cm^2 (b) \$24.30

9. By the square yard

11. $120 + 15\sqrt{24}$ or $120 + 30\sqrt{6}$ square units

12. (a) 72π in.^{2} (b) 180π in.^{2} (c) 252π in.^{2}

13. $r = 2, h = 8$ or $r = 4, h = 1$

15. (a) 916 ft^2 (b) 2.29 gal (c) 3

16. (a) 30 cubic units
(b) The child counts the number of squares that are visible.

17. 38,000 m^2

19. B

21. (a) Surface area (b) Volume (c) Volume

23. (c) Surface area would increase; volume would stay the same.

25. Prisms and cylinders

27. 1544.3 cm^3

29. (a) and (b) 2.7 in. and 3.5 in.
(c) The cylinder with $d = 3.5$ in.

31. $50\sqrt{91}$ m^3

33. 1.30 m^3

35. Cones and pyramids

37. 303,750 cm^3

39. 400 in.^{2}

41. (b) 355 cm^3

43. (a) 115,640,000 ft^3 (b) 15,419

45. 5.8 in.^{3}

47. (a) 180 (b) Omitting π from the answer

49. 47%

51. 1016 lb

53. $L = W = 10$ cm, $H = 20$ cm

55. (a) $4x + 2$; $8x + 2$

(b)

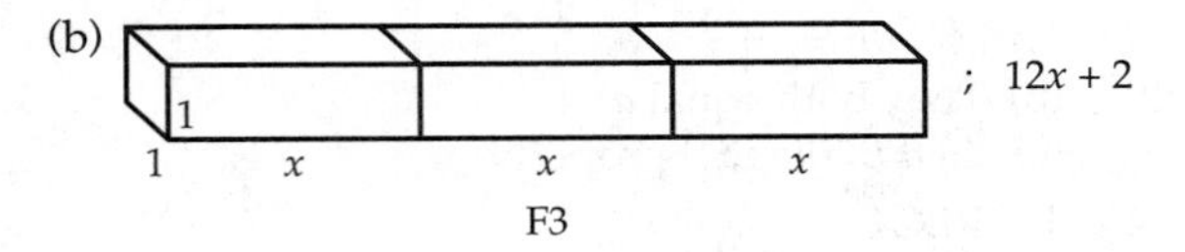

(c) $40x + 2$ (d) $4Nx + 2$

57. (a) 1, 2; 1, 8; 6, 24 (b) Volume

59. 21%

61. Lay it on its side.

63. $16{,}200\pi$ in.^{3}

67. Complete the cone. $432\pi - 54\pi \approx 1188$ in.^{3}
Or compare it to a cylinder. $\pi(4.5)^2 \cdot 18 \approx 1145$ in.^{3}

69. 2,200,000 L

71. (a) 16π and 64π in.^{2}; larger one has 4 times as much.
(b) The material for the larger ball costs about 4 times as much.

73. (a) Least expensive (b) $\frac{V}{\pi r^2}$
(c) $2\pi r^2 + 2\pi rh$ (d) $A = 106.18$ cm^2
(e) $r = 2.37$ cm, $h = 4.75$ cm

75. 100,000 kg

10.6 *Homework Exercises*

1. (a) 20 (b) 1:4 (c) 1:16 (d) 1:4

3. (a) $m:n$ (b) $m^2:n^2$

4. (a) 4:9 (b) 2:3 (c) 2.25

5. (a) 16 in. (b) 9 in.

7. (a) 2:3 (b) 208 and 468 m^2 (c) 4:9
(d) It is the square of the ratio of the edges.
(e) 192 and 648 m^3 (f) 8:27
(g) It is the cube of the ratio of the edges.

9. (a) 4:1 (b) 64:1

11. (a) 3.5 m (b) $3\frac{1}{16}$ (c) 64:343

13. (b)

15. 512

17. (a) $1\frac{1}{4}$ (b) $1\frac{9}{16}$ (c) $1\frac{61}{64}$ (d) The smaller one

19. 1,152,000 ft^3

21. Food consumption is proportional to volume. Cube the 12 from the ratio of lengths to obtain 1728.

22. 144

23. About 0.1 lb

25. (a) Volume (b) \$131.82

27. (a) 3 (b) 2

29. Larger

31. *Hint:* Find the cube root of each number in the first row and the square root of each number in the second row.

Chapter 10 *Review Exercises*

1. (b)

2. $\sqrt{149}$ m

3. 0.0008

4. 72 units^2

5. 10.0 m

6. *A*

7. 7

8. Multiplied by 16

9. Put together two of the triangles as shown to form a parallelogram.

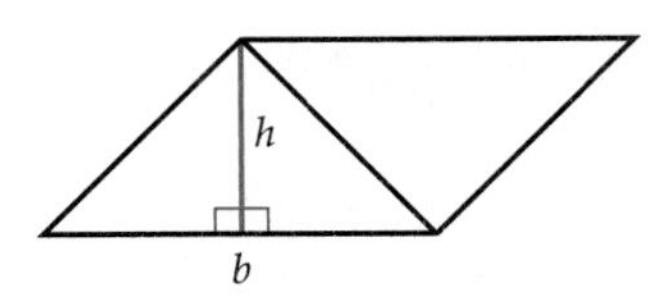

$$A_\triangle = \frac{1}{2}A_{\text{parallelogram}} = \frac{1}{2}bh$$

10. 15.0 m^2

11. $V = 10\sqrt{20}$ or $20\sqrt{5}$ cubic units and
$A = 50 + 9\sqrt{20}$ or $50 + 18\sqrt{5}$ square units

12. $100 - 25\pi\ m^2$

13. $20 + 5\pi$ m

14. $225\pi - 216$ square units

15. $625/\pi\ cm^2$

16. 14 in.

17. $432\pi\ ft^3$ and $216\pi\ ft^2$

18. 4:9

19. 675 m^3

20. (a) 15.625 (b) 6.4 lb

21. (a) No (b) Acute (c) Right

22. The perimeter measures the sum of the lengths of all the sides (the border), and the area measures the size of the interior of the polygon.

Chapter 11

11.1 *Homework Exercises*

1. (a) {(0, 0), (1, 0), (2, 0), (2, 2), (3, 0), (3, 2), (4, 0), (4, 2), (4, 4)}

(b)

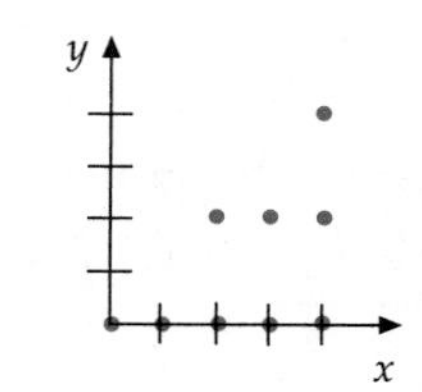

3.

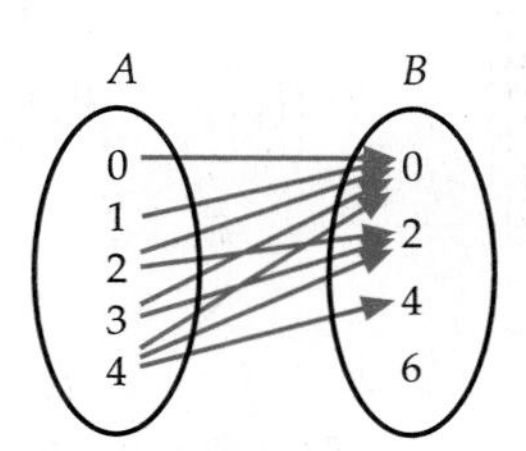

5. (a) $500 \le F \le 800$
(b) $92{,}500 \le C \le 112{,}000$
(c) Each F-value in the domain has exactly one C-value in the range.

7. (a) (1) and (3)
(b) (1) Squaring (2) "is a friend of"
(3) "wants to be a"

9. 30 from set A would be assigned to more than one number in set B.

11. (b) and (c)

13. (a)

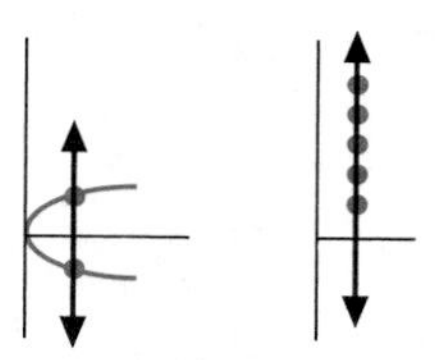

(b) If the line intersects more than one point on the curve, it indicates more than one point with the same x-value and a different y-value.

15. (b) and (c)

17. (a) Add the sums of the squares of the two numbers.
(b) Number of letters in each name
(c) Multiply the first number in the ordered pair by the square of the second number.

19. (a) 13 (b) $y = 2x + 7$

21. (a) 4 (b) 900 (c) $F = S^2/4$
(d) Weight, direction (e) $0 < S \le 100$ mph

23. (a) The set of feet measured
(b) The measurements between 4 and 35 cm

25. "is taller than," "is heavier than"

27. (a)

29. (a) $y \le 8$ (b) $y \ne 0$

31. {(5, 9), (5, 18), (10, 18), (15, 18)}

11.2 *Homework Exercises*

1. (a) Time cannot be negative.
(b) Distance cannot be negative.
(c) and (d) A ray with endpoint at (0, 0) going through (1, 12)
(e) Approximate

3. (a) The average revenue is \$4/customer.
(b) $R/C = 4$

5. (a) Slope > 0 (b) Slope < 0 (c) Slope $= 0$

7. (a) 2/3 (b) $-4/3$ (c) 2/3

9. $\overline{AB}$ has slope -1; $\overline{CD}$ has slope $\frac{1}{3}$; $\overline{EF}$ has slope 2.

11. Ramp length is $\sqrt{(36)^2 + 3^2} \approx 36.1$ ft

13. (a)

Water
360
270
180
90
0
0 3 6 9 12 Time

(b) 30 (c) About 30 quarts
(d) About 30 quarts/min

15. (a)

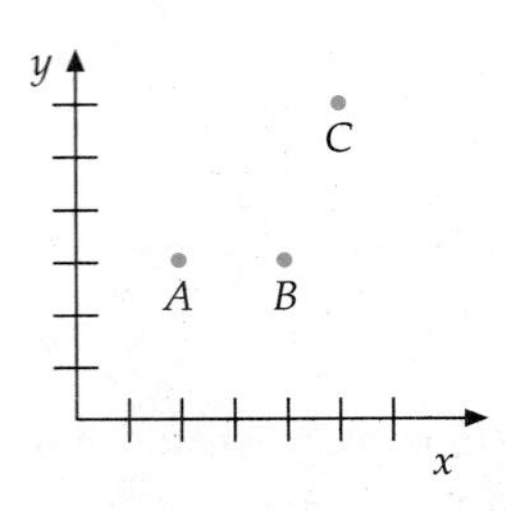

(b) (3, 6) (c) Both 3

17. Up 5 units

19. Parallel lines

21. Both have a slope of 1/2.

23. (a) A line through (0, 1) and (1, 2)
(b) A line through (0, 1) and (1, 3)
(c) A line through (0, 1) and (1, 4)
(d) The y-intercept (0, 1)
(e) $y = 3x + 1$

24. (a) Slope = 2, y-intercept = −5
(b) Slope = −2.5, y-intercept = 5

25. $y = \frac{2}{3}x$

27. $y = -2x + 5$

29. (a) Yes (b) No (c) Yes
(d) When x changes by a fixed amount (such as 1), y also changes by a fixed amount.

31. (a) $T/5$ (b) A line through (0, 0) and (5, 1)
(c) Each change of 1 second is an additional 1/5 mile of distance.
(d) The lightning (light) reaches one in virtually no time, so the formula simply estimates the distance the thunder (sound) travels at the rate of 0.2 mi/sec.

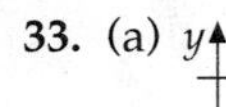

33. (a)

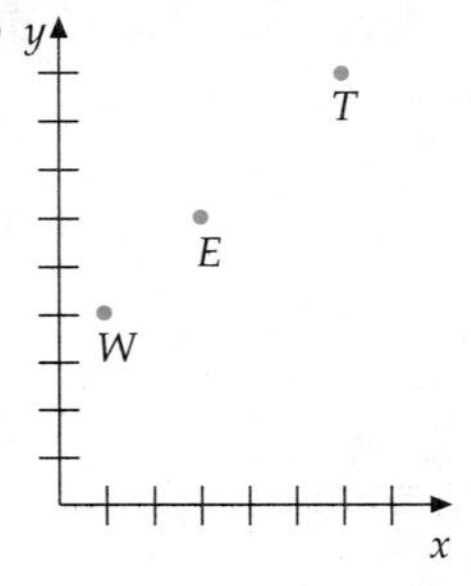

(b) (2, 5) and (4, 7)

(c) y is 3 more than x. (d) $y = x + 3$

35. (a)

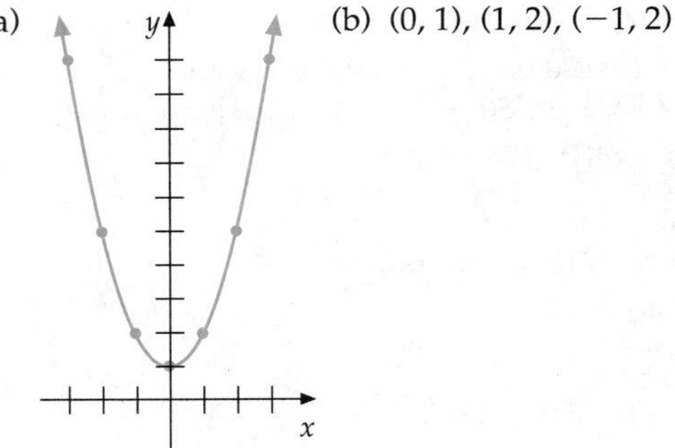

(b) (0, 1), (1, 2), (−1, 2)

(c) For $0 \le x \le 3$ (d) About 2.5

36. It is the graph of $y = x^2$ translated up 3 units.

37. Translate $y = x^2 + 1$ up 4 units to obtain $y = x^2 + 5$.

39. (c)

41. (a)

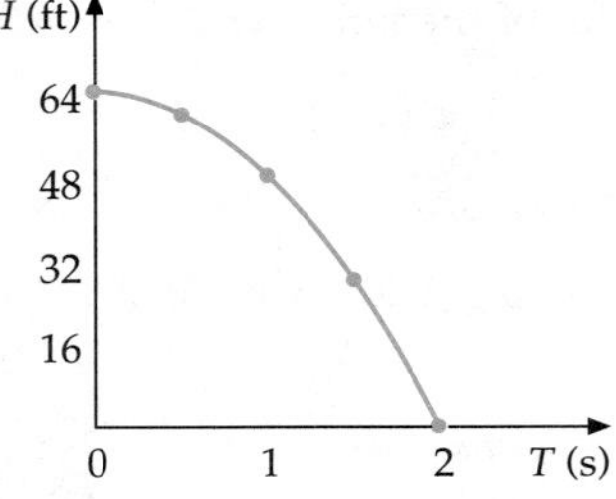

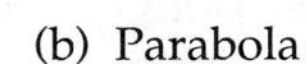

(b) Parabola

(c) The object hits the ground at $t = 2$.

43. (a)

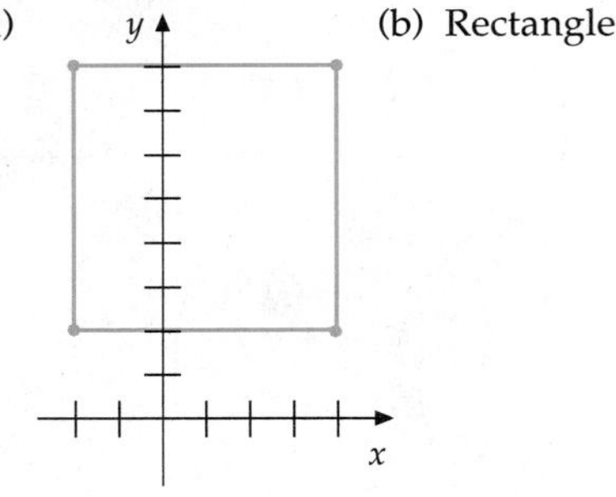

(b) Rectangle

(c) It looks longer and skinnier.

45. (a)

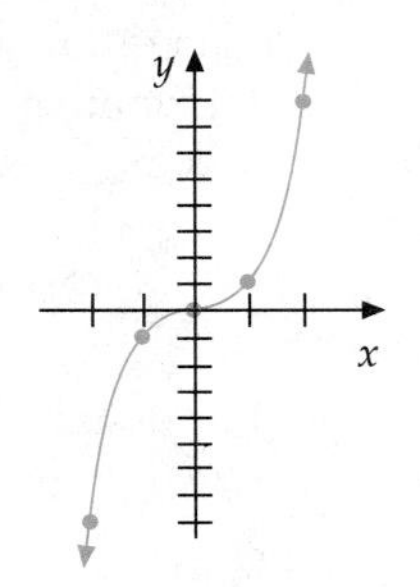

(b) Translate $y = x^3$ up 2 units.

47. (a) -1 and 1; $1/2$ and -2
(b) The product of the slopes is -1.

49. (a) m (b) $-1/m$

51. No

53. (a) A line through $(0, -2)$ and $(1, 1)$
(b) A line through $(0, -2)$ and $(-1, 1)$
(c) $y = -3x - 2$

55. One is the image of the other for a reflection through the x-axis.

57. (a) A rectangle with base 30 and height 20
(b)
```
TO RECT
    PU SETX 10 SETY 10 PD
    SETY 50 SETX -20 SETY 10 SETX 10
    END
```

11.3 *Homework Exercises*

1. $a + 12$, in which a is the average temperature for this day in degrees

3. (a) $20 = 0.85P$, in which P is the list price in dollars
(b) The wholesale price, the quality of the item, what the item is

5. X must represent a quantity.

7. $P = 3L$, in which L is the length and P is the perimeter.

9. $y \geq 2x$, in which x is last year's height and y is this year's height

11. Total cost $C = 5X + 3Y$, in which X is the number of adult tickets and Y is the number of children's tickets

13. (a) $P + 6 = 20$ (b) $N + 8 = G$

15. $x + 0 = x$ where x is any real number.

17. $a(b + c) = ab + ac$, in which a, b, and c are real.

19. (a) $1100 + 0.1M$
(b) A ray with endpoint (0, 1100) through (100, 1110)

21. $W = 8100 - 900Y$

23. (a) $200 - 5W$ lb (b) Possible answer: 10

25. The rental rate is \$10 plus \$22/day.

27. The commission is 2.5% of sales.

29. The car went 120 miles in 3 hours, traveling more quickly at the beginning and the end of the trip.

31. (a) 2 (b) (1)

33. (a) (3) (b) (2) (c) (1)

34. (a) Negative slope (b) Positive slope
(c) Negative slope

35.

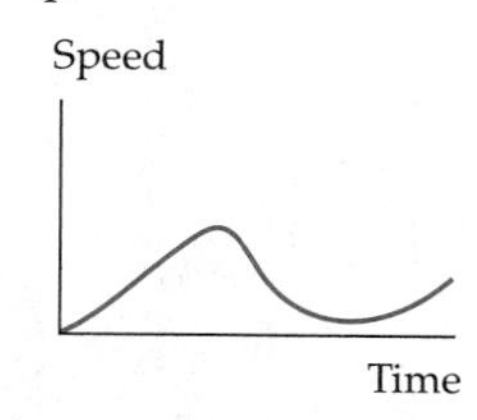

37. $y = 100 + x/10$

39. $y = 24 - x$

41. $W = 3H - 83$

43. $P = 3I - 1$

45. $D = 3M + 2$

47. $x^2 + y^2 = 25$

49. (a)

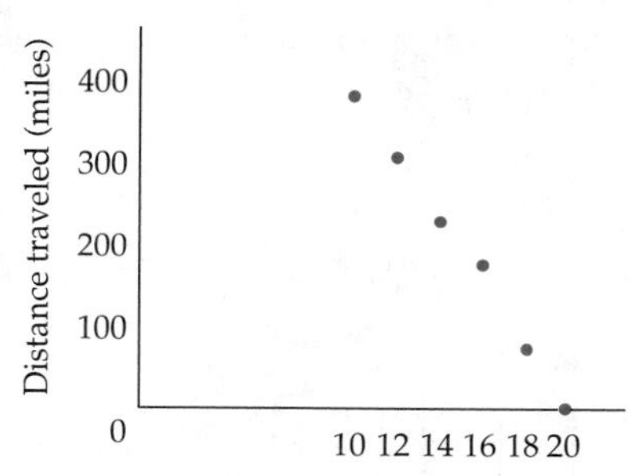

(b) Negative (e) $y = -38x + 761$

51. (a) Positive (b) Positive (c) Negative

53. (a)

L	3	5	7	9	11
W	10	6	30/7	10/3	30/11

(b) $LW = 30$ or $W = 30/L$

(c)

(d) It is all in the first quadrant, and y decreases less and less rapidly (from left to right). The curve gets very close to the x-axis.
(e) Any positive number

55. (a) $WT = 3000$ or $T = 3000/W$ (b) 5 minutes

(c)

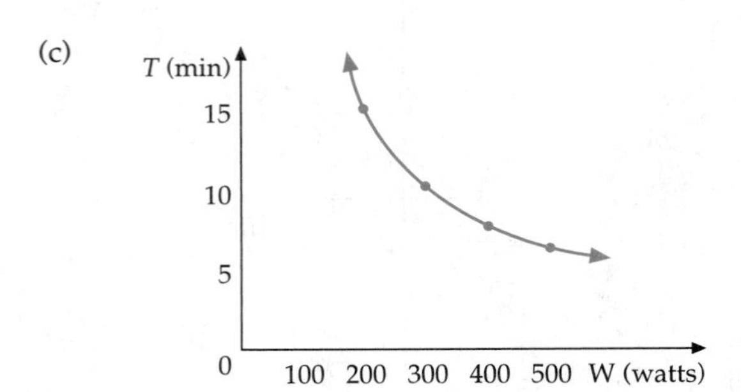

57. Parts (a), (b), and (c) come out the same for both expressions.
(d) Part (c)

59. $S = 1.5x$, in which S is his new salary and x is his old one.

61.

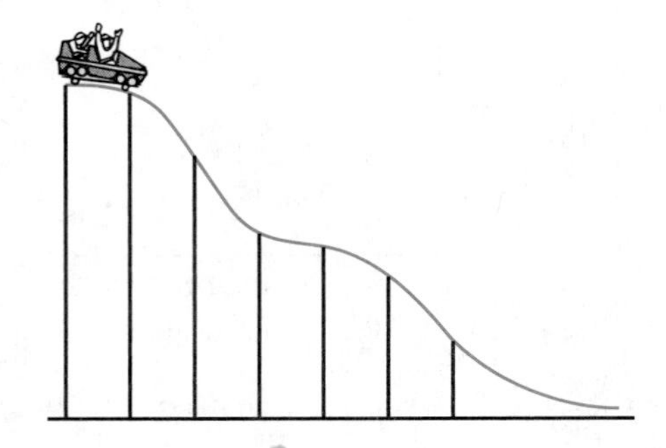

63. (a)

Price ($)
800
700
600
500
400
300
200
100
0 70 80 90 Score

(b) Very slight
(c) A line with positive slope

65. $y = 0.880x + 1.563$

67. (a) $T = \dfrac{300}{S} + 1$ (c) $B3 = \dfrac{300}{A3} + 1$
(d) The total time decreases as the speed increases. The total time decreases more for a 5 mph increase at a lower speed than at a higher speed.

11.4 *Homework Exercises*

1. (a) $I = 100M/A$, in which M = mental age and A = actual age
(b) With tests (c) 125

3. (a) Subtract \$1.80 to get \$1.89 and divide that by \$.27 to get 7 photos
(b) $P = \dfrac{C - 1.80}{.27}$
(c) Formula from part (b), $P = 7$

5.

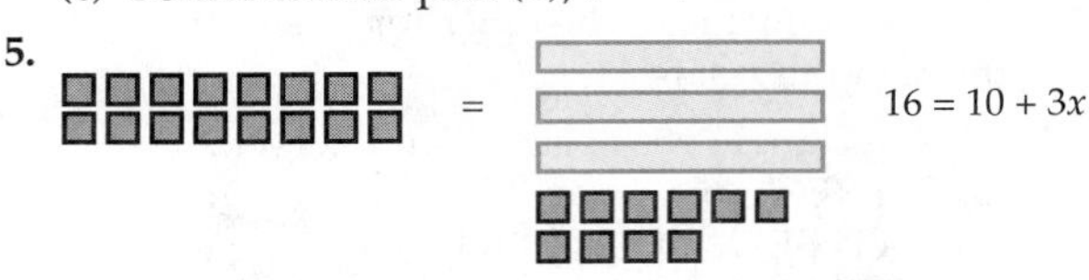

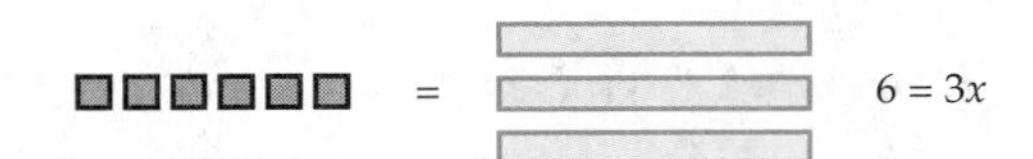

= 2 = x

7. (a) 12 (b) 3 (c) $x = -2$

9. (a) $T = \dfrac{N}{4} + 40$ (b) $N = 4(T - 40)$
(c) 65°F (d) 220

11. (a) Less than 500 miles
(b) $120 + .32x < 280$; $x < 500$ miles

13. \$340

15. (a) 725 (b) 11.45

17. 1870 adults and 590 children

19. 9.6 oz bread, 6.4 oz cheese

21. Between 3 and 4 years

23. A little over 10 years

25. (b) B2 = 39.95 * 3; C2 = 24.95 * 3 + .10 * (A − 100)
(c) Use Drive-U if I plan to drive 700 or more miles. Use Used-a-Bit if I plan to drive 500 or fewer miles. I could investigate more values between 500 and 700 miles.

27. (a) She drove for 5 minutes, then 2 minutes, then 6 minutes, and finally 3 minutes, accelerating at the beginning and decelerating at the end of each time interval. She averaged about 25 mph.
(b) 6 miles

29. (a) 150 miles (b) 150 square units

31. (a) About 100 miles
(b) About $1400 \cdot 3 = 2100$ miles
(c) About 467 miles/h

33. (a)

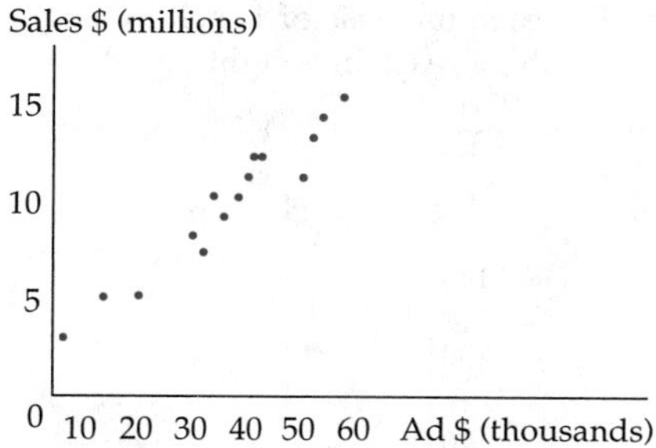

(d) Substitute \$20,000 for x and find y.

35. (a) 96 ft (b)

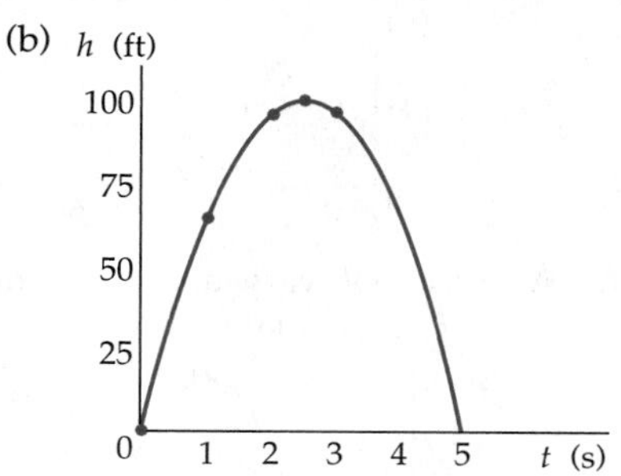

(c) $t = 0.7, 4.3$ s

37. (a) $10 = 2L + 2W$ (b) Exact (c) 2 ft
(d) Between 0 and 5 ft
(e) Line segment excluding its endpoints (0, 5) and (5, 0)

39. (a)

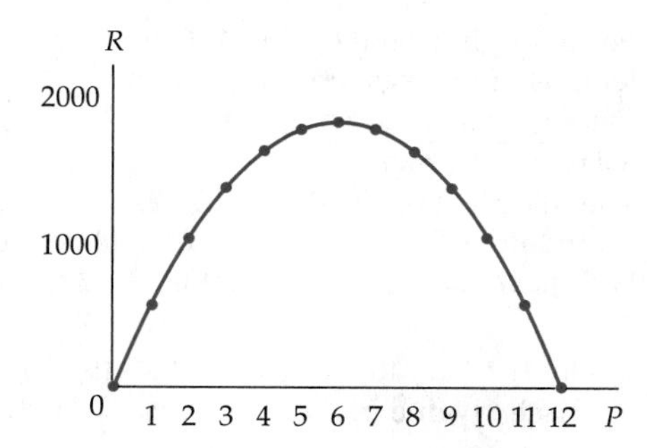

(b) \$6

41. (a) After a little over 6 h (b) $H = 25 - 4T$

43. (a) $A = 400 + 30M$ and $B = 550 + 20M$
(b) A is a ray with endpoint (0, 400) through (10, 700). B is a ray with endpoint (0, 550) through (10, 750).
(c) (15, 850) (d) (15, 850)
(e) For less than 15 months, A is cheaper. For more than 15 months, B is cheaper.
(f) Reliability, design, size

45. (a) $A = 800 + 0.05S$ and $B = 300 + 0.15S$, in which S = sales dollars
(b) A is a ray with endpoint (0, 800) through (1000, 850). B is a ray with endpoint (0, 300) through (1000, 450).
(c) (5000, 1050)
(d) For less than \$5000 in sales, A pays more. For more than \$5000 in sales, B pays more.
(e) Location, co-workers, product being sold

47. 21.3 hours

49. (a) \$.78
(b) Greater than 6 oz, up to and including 7 oz
(c)

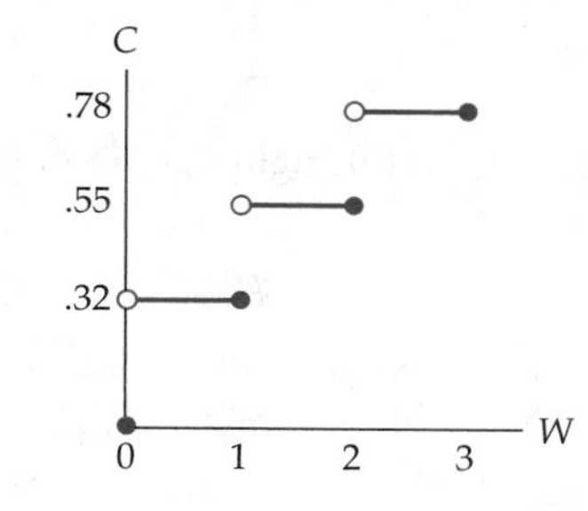

(d) The weights
(e) Counting-number values
(f) Horizontal steps

51. The diesel would be cheaper only after driving about 500,000 miles, so the gas car is a better buy.

53. (a) 0.5, 8.5 (b) 1.6, 7.4

55. The temperature is now 4°F and it is increasing 2°F/h. How long will it take to reach 10°F?

11.5 *Homework Exercises*

1. $\sqrt{17}$

2. Form a right triangle with $C(x_2, y_1)$ as the third vertex. By the Pythagorean Theorem,
$$AB = \sqrt{(AC)^2 + (CB)^2} = \sqrt{(x_2 - x_1)^2 + (y_2 - y_1)^2}$$

3. (a) (a, b)
(b)
$$AC \stackrel{?}{=} BD$$
$$\sqrt{(a-0)^2 + (b-0)^2} \stackrel{?}{=} \sqrt{(a-0)^2 + (0-b)^2}$$
$$\sqrt{a^2 + b^2} = \sqrt{a^2 + b^2}$$

5. $3 + \sqrt{8} + \sqrt{5}$ or $3 + 2\sqrt{2} + \sqrt{5}$

7. $(x - 2)^2 + (y + 1)^2 = 9$

9. $(x - 2)^2 + (y - 3)^2 = 32$

11. (3, 3)

13. (a) $D(a/2, b/2)$, $E((a + c)/2, b/2)$
(b) Slope of $\overleftrightarrow{DE}$ = slope of $\overleftrightarrow{AC} = 0$
(c) $DE = \sqrt{[(a + c)/2 - a/2]^2 + (b/2 - b/2)^2}$
$= \sqrt{c^2/4} = c/2$ and $AC = c$ so $DE = \frac{1}{2}AC$.

15. (5, −5), (5, 1), or (3, −5)

17. 7

19.

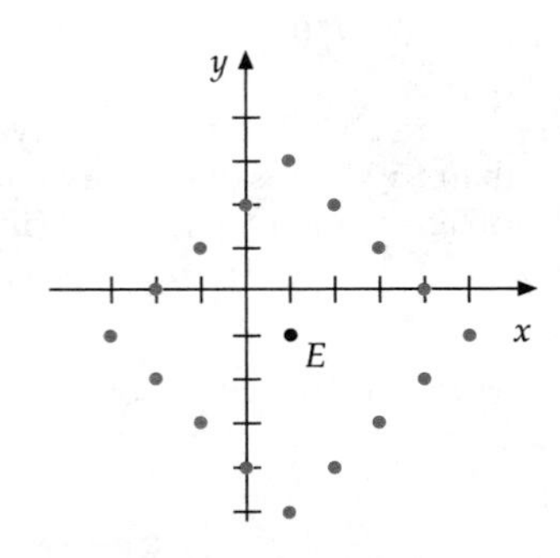

21. Up, right, up, right, right (23 minutes)

23. (a) (a, a)

(b)
$$AC \stackrel{?}{=} BD$$
$$\sqrt{(a-0)^2 + (a-0)^2} \stackrel{?}{=} \sqrt{(a-0)^2 + (0-a)^2}$$
$$\sqrt{2a^2} = \sqrt{2a^2}$$

(c) Slope of $\overleftrightarrow{AC} = 1$ and slope of $\overleftrightarrow{BD} = -1$ so $\overline{AC} \perp \overline{BD}$.

25. 6 points including: B, the two points just below it, and 3 more points that are one to the right of these first 3 points

27. (c) A parabola

(d) Each fold creates a point equidistant from the point and the line, which is what defines the points on a parabola.

29. (3.5, 2)

31. (1, 1) and (2, 0)

33. (a) 1 (b) 2 (c) 3 (d) 4 (e) 11

35. Outside

Chapter 11 *Review Exercises*

1. (a) $J > 2S$, in which J = Joe's weight and S = his sister's weight.

(b) Untranslatable

2. (a) No (b) Yes (c) No

3. $x > 3$

4. (a) Line through (0, 1) and (1, 3)

(b) $y = -2x + 1$

5. 5/2

6. (a) $h = 10 + 0.5n$

(b) A ray with endpoint (0, 10) through (2, 11)

(c) A ray

7. (a) $E = H + 2T$ (b) 16

8. $8445.95

9. It is $y = x^2$ translated up 2 units.

10. (a) For each increase of 1 inch in height, there is a 5.5 lb increase in weight.

(b) $H = \dfrac{W + 220}{5.5}$ (c) The original

11. (a)

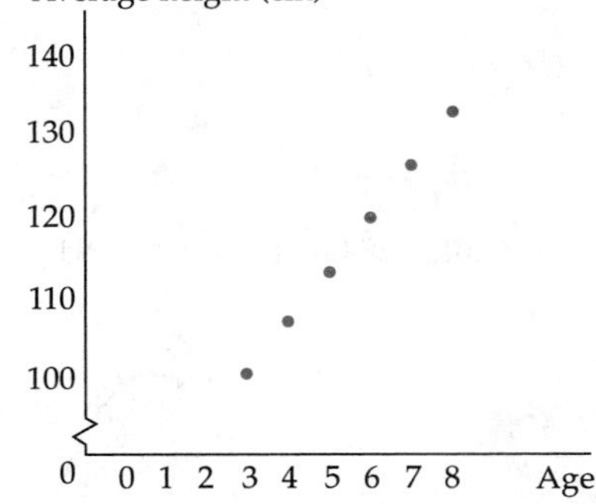

(d) Answers will vary; about 180 cm

12.

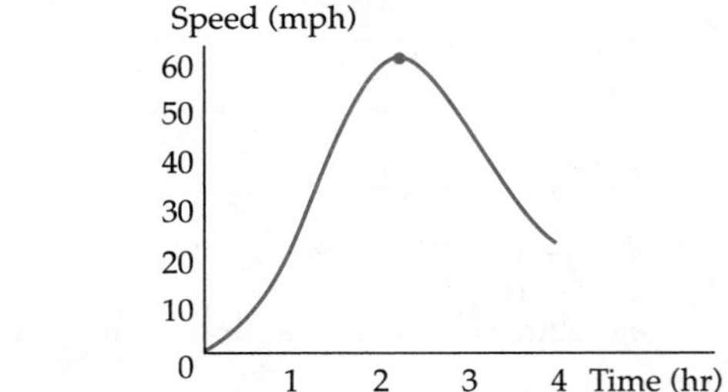

13. Form a right triangle with A, B, and C (x_2, y_1). The lengths of the legs are $(x_2 - x_1)$ and $(y_2 - y_1)$. By the Pythagorean Theorem, $AB = \sqrt{(x_2 - x_1)^2 + (y_2 - y_1)^2}$.

14. The midpoint of $\overline{AC}$ is $((a + 0)/2, (b + 0)/2)$ and the midpoint of $\overline{BD}$ is $((a + 0)/2, (0 + b)/2)$. So they both have the same midpoint $(a/2, b/2)$.

15. 3

16. (a) $\{0, 1, 2, \ldots, 10\}$ (b) $\{-25, -15, -5, \ldots, 75\}$

(c) Each x-value in the domain has exactly one P-value.

17. (a) (1) and (2)

(b) (1): $y = 2x - 1$

(2): The person's birthplace

18. (a) $C = 2.10 + .80d$

(b)

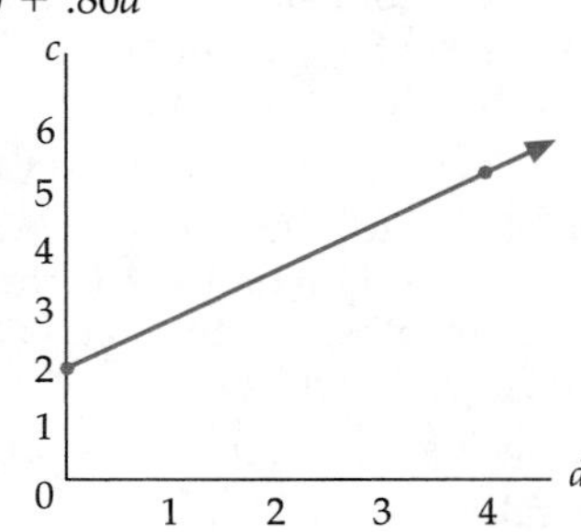

Chapter 12

12.1 *Homework Exercises*

1. (a) Circle (b) Line (c) Bar

3. (a) Education (b) Mechanical Engineering
(c) The salaries range from $22,505 to $34,460, the highest starting salary being about 55% higher than the lowest. Technological fields tend to pay higher salaries.
(f) Average salaries in each field

5. (a) (2) (b) (3) (c) (1)

7.

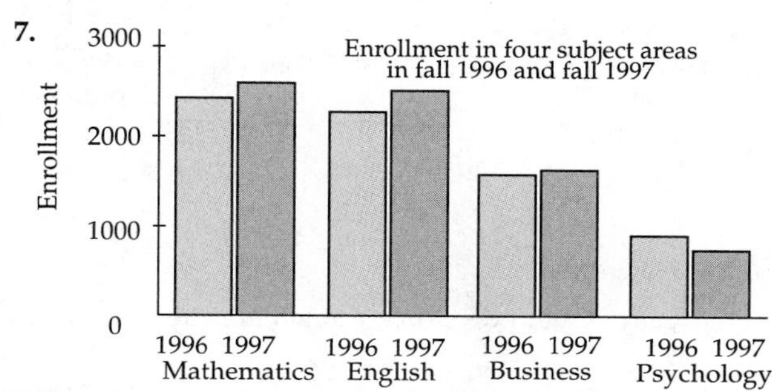

9. (a) Lower from 1977 to 1979, then rose from 1979 to 1983, declined from 1983 to 1989, and rose again from 1989 to 1993
(c) Unemployment data from this time, data on wealthier people

11. (a) 43.5% (b) $27,000 (c) $155,000

13. (a) U.S., Sweden (b) U.S.
(c) Not much (d) Birth control devices
(e) By confidential surveys

15. (a) Class A

9	28
8	015
7	025
6	22
5	8

(b) Class B

9	019
8	02
7	5
6	558
5	
4	2

(c) Class B has more extreme scores than Class A.

17. (a)

7	2
6	4568
5	126889
4	3789

(b) Many people retire in their 50's. Almost as many retire in their 40's or 60's.
(c) The height of each bar corresponds to the length of the row of leaves in part (a).
(d) A stem-and-leaf plot shows each individual's retirement age. A bar graph allows you to choose subintervals of any size.

21. 79.2°

23. (a) Food 25%, clothing 8.3%, rent 50%, entertainment 6.7%, other 10%
(b) Make the central angles: food 90°, clothing 30°, rent 180°, entertainment 24°, other 36°

27. (a) Israel
(b) Start the vertical axis at 0. The bars will be much closer in height.
(c) U.S.

29. The first region is 4 times bigger in area.

31. The university

33. (a) Use 60–69, 70–79, 80–89, 90–99.
(b) Use 60–79 and 80–99.

35. *Hint:* Comment on how various types of workers are affected by downsizing.

39. (a) E, T, A, O
(b) *Hint:* The 11 most frequently occurring letters are A, D, E, H, I, L, N, O, R, S, and T.
(c) C—6, U—4, W—3, A—2, N—2, Q—1, P—1, T—1, I—1, X—1, G—1, Z—1, S—1
(d) C (e) The meeting will be at the pier.

12.2 *Homework Exercises*

1. 56%

3. No. 20% of wages is less than 20% of overall costs.

5. 4%

7. How far each car traveled

9. How many dentists

11. Hard to tell; could be as low as $350 savings

13. (d)

15. 10; 100

17. (a) No one sells them at list price.
(b) The meat protein in steak has little to do with the appeal of steak.

19. (a) AIM with fluoride and AIM without fluoride
(b) Fluoride is effective in reducing dental decay.

21. Worsening external conditions such as water and air pollution, or few people can afford the new treatments, or medical improvements for heart disease are showing an even greater improvement

23. People in hospitals are generally much sicker than people who are not in hospitals.

25. They were using different time periods.

27. $P \rightarrow 1.1P \rightarrow 0.88P$. Overall change is a decline of 12%.

12.3 *Homework Exercises*

1. One method would be to compare their means; Tom's is 141 and Jean's is 133.75.

3. (a) $(V + W + X + Y + Z)/5$ (b) $5M$

5. $9.88

7. Mean = 6, median = 7

9. (a) 4 (b) Mean
(c) Place the stacks in order of increasing size; the median is the height of the middle stack (3).

11. The "median" strip in a roadway—it divides the road down the middle just as the median of a set splits it in the middle.

13. {5, 8, 9, 10}

15. (a) 7 (b) 57 (c) 9 (d) 44

16. (a) 15
(b) Change two 15's into 6's or 8's.

17. In most large groups, close to 50% of the people score below average.

18. (a) 24
(b) Mean = 67.5, median = 70, mode = 70
(c) 15 (d) 6

19. Mean = $3.7 million, median = $3 million, mode = $2 million

21. 2.8

23. 12 Parson and 8 Valencia

25. Probably the mean

27. Mode

29. $8000

31. (a) Mean (b) Median

33. Median

35. (a) Median (b) Mean

37. (a) $1.70 (b) $3

39. New employees in a profession generally earn lower-than-average salaries.

41. 85.4

43. *Hint:* Make up an easy distance for the trip. *Answer:* 120 km/hr

45. (a) Mean or median? (b) Mean

47. {4, 4, 4, 5, 5, 6, 7}

49. The same number of students in each class

51. They computed the average over different time periods.

53. (a) Class 1: 16, Class 2: 11, Class 3: 7.5, Class 4: 7.5
(b) Class 1: 14.25, Class 2: 14.25, Class 4: 8.5, Class 4: 5
(c) Class 4 prefers Trickledown's plan. Classes 2 and 3 prefer Dragdown's plan. Class 1 prefers to leave things as they are.
(d) 1 (e) 3 and 4

12.4 *Homework Exercises*

1. It may be 30 ft deep in a limited area and 2 ft deep everywhere else.

3. Lyle scored better in mathematics than 43% of those in his group.

5. (a) *Hint:* The five-number summary for A is 56, 70, 81, 91, and 97, and the five-number summary for B is 68, 74, 80, 86.5, and 91.
(b) Class A's scores are more spread out.

7. (a) (possible answer) {20, 20, 20, 20, 20}
(b) {20, 20, 20, 10, 30} (c) {20, 20, 10, 25, 25}

9. (b)

11. (a) Mean = 90, $s = 6.3$ (b) 20

13. (a) 53, 56, 60, 64, 67 (b) None

15. (a) The average weight is 140 lb. Most scores are somewhere around 21 lb below average to 21 lb above average.
(b) Weights over 182 lb and weights under 98 lb

19. (a) 50% (b) 81.5% (c) 16%

21. *A* is more likely to last close to 1000 hours. *B* lasts longer than 1000 hours on the average.

23. Skewed to the left

25. Skewed to the right

27. Mean, standard deviation, first quartile, median, third quartile

29. Class with wide ability range → high standard deviation

Honors class → high mean, low standard deviation

Students with very similar abilities → low standard deviation

Remedial class → low standard deviation, low median

31. (a) The mean increases by $2000, and the standard deviation stays the same.
(b) The mean increases by 10% and the standard deviation increases by 10%.

33. The mean is $(A + B + C + D + E)/5$. So the sum of the deviations is $A - (A + B + C + D + E)/5 + B - (A + B + C + D + E)/5 + C - (A + B + C + D + E)/5 + D - (A + B + C + D + E)/5 + E - (A + B + C + D + E)/5 =$
$A + B + C + D + E - A - B - C - D - E = 0$

35. (a) 50% (b) 16% (c) 2.5%

37. (a) $(A + B + C + D)/4$

(b) $$s = \sqrt{\frac{(A - \bar{x})^2 + (B - \bar{x})^2 + (C - \bar{x})^2 + (D - \bar{x})^2}{4}}$$

(c) $$\sqrt{\frac{A^2 - 2\bar{x}A + \bar{x}^2 + B^2 - 2\bar{x}B + \bar{x}^2 + C^2 - 2\bar{x}C + \bar{x}^2 + D^2 - 2\bar{x}D + \bar{x}^2}{4}}$$

$$= \sqrt{\frac{A^2 + B^2 + C^2 + D^2 - 2\bar{x}(A + B + C + D) + 4\bar{x}^2}{4}}$$

Use the hint.

$$= \sqrt{\frac{A^2 + B^2 + C^2 + D^2}{4} - (\bar{x})^2}$$

12.5 *Homework Exercises*

1. (a) Registered voters in Colorado (b) $0.27P$

3. 1776

5. Some people will not answer honestly.

7. No, the first group is more likely to lie to please the interviewer.

8. 29 freshmen, 26 sophomores, 24 juniors, and 21 seniors

9. "Would you favor or oppose building an expensive new missile system?"

11. Fewer people would agree or disagree.

13. 64% OF ALL AMERICANS OPPOSE ABORTIONS UNDER SOME OR ALL CIRCUMSTANCES.

15. The sample is too small.

17. Grades mean different things to different people.

19. (a) Voluntary response (b) Stratified
(c) Cluster

21. Predict that 111 will buy whole milk, 233 will buy low-fat milk, and 156 will buy skim milk.

22. (a) Yes (b) No
(c) The respondents were not at all representative of voters.

25. Look at past sales records of similar clothing by color.

27. (a) The results of a poll can influence what people think, and, consequently, how they respond to subsequent polls.

29. (a) The rating is the percent of *all* TV sets; the share is out of a smaller sample, only those sets that are on.
(b) By showing that the 35% who do not respond are similar to the 65% who do.
(c) As many shows as there are TV sets

31. None of these

33. The most likely kind of death that results from keeping a firearm at home is a suicide. The next most likely result is killing someone you know in a fight. These events are much more likely than using a firearm to defend your home against a burglar.

12.6 *Homework Exercises*

1. (a) 36 out of 60
(b) She got 36 questions right out of 60.

2. (a) 48
(b) She scored better than about 48% of those in her group.

3. Students in this area score below the national average.

5. (a) 7.4
(b) She scored the same as the average child in the fourth month of seventh grade.

7. No

9. 8; 4; 6

11. No, their confidence bands overlap. Their scores are not significantly different.

13. 130 or above

15. 16th

17. Yes

19. (d)

21. (b) and (c)

Chapter 12 *Review Exercises*

1. (a) .99, 1.30, 1.10, 1.05, .84 (b) Whole
(c) Whole (d) Thighs

2. (a) The amount was fairly steady from 1974 to 1978; then it rose from 1978 to 1986 and remained at this higher level from 1986 to 1996.
(b) The U.S. government spends more than it collects in taxes, so it borrows money and pays interest on it.

3. (a) 12%
(b) $P \rightarrow 1.4P \rightarrow 1.12P$; overall change is 12%.

4. The phrase "as much as 50%" could mean anywhere from 0 to 50%.

5. (a) Start the vertical axis at 19,000.
(b) Scale the vertical axis accurately.

6. Mean = 14.1 minutes, median = 15 minutes

7. The mean age will be higher because the older students will pull it up, while the median age will be around 20.

8. (a)

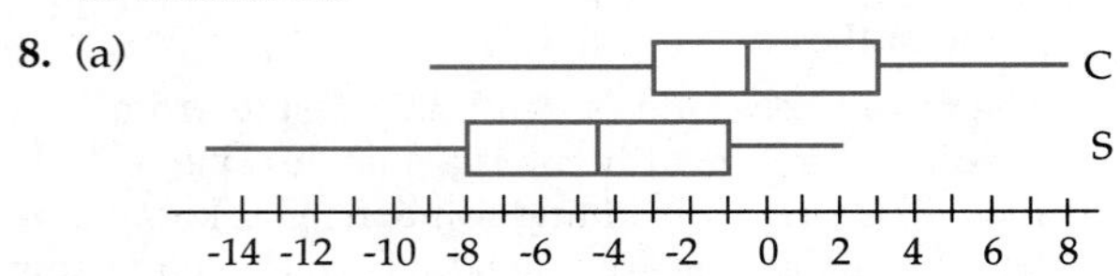

(b) The saturated fat group has more success lowering cholesterol. The other group shows little change on average.

9. (a)

Salaries	52.5%	Utilities	12.5%
Maintenance	15%	Supplies	5%
Insurance	5%	Other	10%

(b)

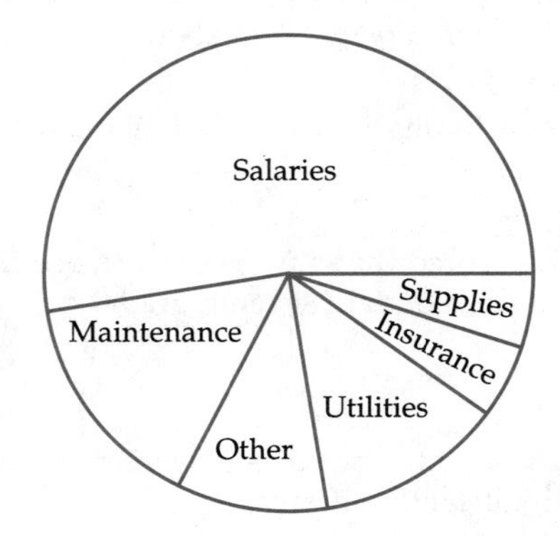

10. (a) Line (b) Bar

11. (a) 95% (b) 16%

12. (a) 84th (b) 97.5th

13. (c)

14. When the grade level is a few grades higher or lower than the general level of the test. Such a score would be based upon test items that are mostly written for a significantly higher or lower grade level.

15. (a) 29 (b) She got 29 questions right. (c) 54
(d) In her region, she scored better than about 54% of those in her age group in reading.

16. $77,920 is the median and $93,422 is the mean. There are more house prices far above these numbers than below them (skewed to the right). These high prices pull up the mean more than the median.
(b)

17. (c)

18. Mean = 7, standard deviation = $\sqrt{14/3} \approx 2.2$

19. Stratified

20. The mean is the sum of all the scores divided by the number of scores. The median is the middle score when all the scores are arranged in increasing order. I would use the mean if I want to count all the scores equally, and the median if I wanted to lessen the impact of a relatively small group of unusually high or low scores.

Chapter 13

13.1 *Homework Exercises*

1. (a) {1, 2, 3, 4, 5, 6} (b) Yes

3. (a) 12/15 (b) 6/25

5. (a) 1/4 (b) 1/6 (c) 1/8 (d) 11/24

7. (c) HHH, HHT, HTH, THH, TTH, THT, HTT, TTT
(d) 3/8

9. (a) 1/36, 1/18, 1/12, 1/9, 5/36, 1/6, 5/36, 1/9, 1/12, 1/18, 1/36
(b) The denominator is always 36, and the numerator counts from 1 up to 6 and then back down to 1.
(c) $\frac{1}{12}$ of 50 $\approx$ 4 times

11. 13/18

13. (a) $P(2) = 1/18, P(3) = 1/9, P(4) = 1/6, P(5) = 1/6,$ $P(6) = 1/6, P(7) = 1/6, P(8) = 1/9, P(9) = 1/18$
(b) $\frac{1}{9}$ of $100 \approx 11$ times

15. (a) 0.75 (b) 0.0001235 (c) Yes

17. After a large number of rolls of a die, we expect 5 to occur about 1/6 of the time.

21. (b)

23. $\frac{1180}{2200} \approx 0.54$

25. Roll a 1, 2, or 3 and Sam does it. Otherwise, Mike does it.

27. (a) $\frac{300}{500} = 0.60$ (b) $\frac{220}{500} = 0.44$ (c) $\frac{280}{480} \approx 0.58$

29. 3/8

31. (a) About 300
(b) Also take into account how the average amount of time traveling abroad per person compares to the average time spent in a car per person.

33. (a) 40% (b) 33

35. One die has 1, 2, 2, 3, 3, and 4. The other has 1, 3, 4, 5, 6, and 8.

37. (a) 1 5 10 10 5 1 (b) 1, 2, 1
(c) It's the second row.
(d) If you flip a coin 3 times, add up the numbers in the third row to get the denominator (8), and the numerators of the probabilities are the numbers in the third row: 1 for 0 heads, 3 for 1 head, 3 for 2 heads, and 1 for 3 heads.

39. (a) From statistical studies
(b) People make mistakes.

41. 10/216 = 5/108

13.2 *Homework Exercises*

1. 1

3. (a)

5. (a) No (b) Yes

7. $A = \{\text{heads}\}, B = \{\text{tails}\}$

9. 0.45

15. (a) The contestants choose door 1 and door 3.
(b) 3/5

17. 2 and 12 are 10°, 3 and 11 are 20°, 4 and 10 are 30°, 5 and 9 are 40°, 6 and 8 are 50°, and 7 is 60°.

21. (a) 0 or 1

23. (b) Should be about 0.17
(c) Change line 80 to:
80 IF R = 10 THEN 5 = S + 1
(d) Should be about 0.08

13.3 *Homework Exercises*

1. (a) 8 (b)

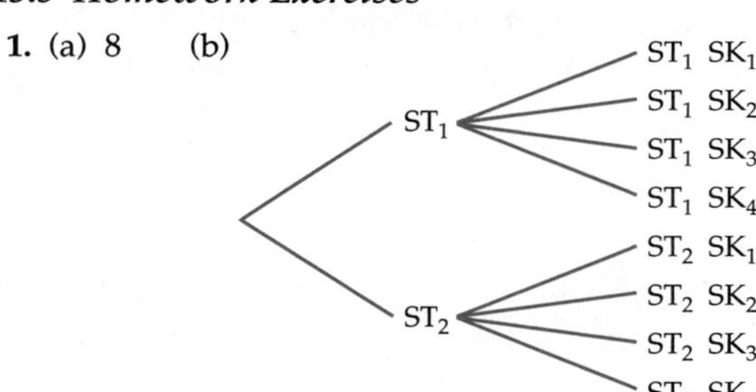

3. 24

5. STRAIGHT, 10 choices for each of 4 digits (10)(10)(10)(10) = 10,000 choices, 1/10,0000 to pick the correct number
4 WAY BOX, 4 ways to match out of 10,000 or 1/2500

7. 210

9. 59,049

11. (a) 2,600,000 (b) 786,240

13. (a) 625 (b) 3 or 4 (c) 17/625

15. (a) 720 (b) 72 (c) 330

17. ${}_{30}P_4 = 30 \cdot 29 \cdot 28 \cdot 27 = 657{,}720$

19. ${}_{52}C_5 = 2{,}598{,}960$

21. When the order matters

23. ${}_8C_2 = 28$

25. ${}_{12}P_2 = 132$

27. ${}_nP_n = \frac{n!}{(n-n)!} = \frac{n!}{1} = n!$

29. (a) 8000 (b) 4 (c) 1 (d) 4
(e) 1/2000, 1/8000, 1/2000 (f) 1692
(g) 148 (h) 423/2000, 37/2000

31. 10

33. (a) 6 (b) 1 (c) 4 (d) 1
(e) 2 (f) 1 (g) 48

35. 72

37. (a) 1 (In all arrangements, person A has person B on one side and person C on the other side.)
(b) 3 (c) 6 (d) $\frac{(N-1)(N-2)}{2}$

13.4 *Homework Exercises*

1. You

3. \$187.50

5. (a) 9/19 (b) \$.05 loss (c) \$.05 loss

7. (a) \$20 (b) \$30
(c) Part (b) protects you from a possible catastrophic loss, while part (a) does not.

9. 1.8

11. 4

13. \$32

15. \$2.32

17. (a)

19. 1:3

23. (a) 199 (b) 49/248 (c) About 1 (1.4)
(d) About $29.4/30.8 \approx 0.95$

25. Discussion question; note that people choosing surgery have a better chance of surviving more than three years.

27. Answers will vary.

13.5 *Homework Exercises*

1. (a) Dependent (for most people)
(b) Independent

3. (a) B = it rains tomorrow
(b) B = the sum of two regular dice is odd

5. (a) Independent (b) {1, 1), (1, 2), (2, 1), (2, 2)}

(c)

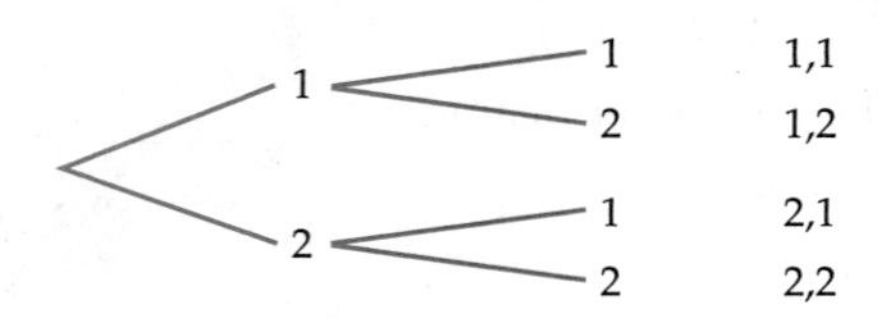

(d) $\frac{1}{4}$ (e) $P(1 \text{ and } 2) = P(1) \cdot P(2) = \frac{1}{2} \cdot \frac{1}{2} = \frac{1}{4}$

7. $\frac{1}{9}$

9. (a) 0.07 (b) 0.27 (c) 0.66

10. (a) Dependent
(b) {(1, 2), (1, 3), (1, 4), (2, 1), (2, 3), (2, 4), (3, 1), (3, 2), (3, 4), (4, 1), (4, 2), (4, 3)}
(c) $P(A) = \frac{1}{4}$, $P(B \text{ given } A) = \frac{1}{3}$, $P(A \text{ and } B) = \frac{1}{12}$
(d) $P(A \text{ and } B) = P(A) \cdot P(B \text{ given } A) = \frac{1}{4} \cdot \frac{1}{3} = \frac{1}{12}$

11. 1/221

13. (a) Independent (b) Dependent

15. $2.4 \cdot 10^{-10}$

17. (a) 0.006125 (b) 0 (c) 0.7695

19. $P(A \text{ and } B) = P(A) \cdot P(B \text{ given } A) = (0.3)(0.7) = 0.21$

21. 0.86

23. (a) Pair 1 has independent events, since the occurrence of one event does not affect the chance that the other event occurs. Pair 2 has dependent events, since the occurrence of one event affects the chance that the other event occurs.
(b) Pair 1 and Pair 2 both have events that are not mutually exclusive, since both events in each group could occur on a given day.

25. 3

27. (a) 0.36 (b) 0.288

29. (a) 0.17 (b) 0.53

31. (a) B = Mike will gain 5 pounds next semester.
(b) B = Mike will go to school next semester.
(c) Impossible

Chapter 13 *Review Exercises*

1. (a) 17/36
(b) Roll two dice 30 or more times and see what fraction of the time a product less than 10 occurs.

2. 1/4

3. 1/2

4. After a large number of repetitions, we would expect 3 heads and 2 tails about 5/16 of the time.

5. {$AB, BA, AC, CA, AD, DA, BC, CB, BD, DB, CD, DC$}

6. 743,600

7. 1 and 2 represent a car; 3, 4, 5, 6, 7, and 8 represent no car passing.

8. (a) 8 (b) 1/11

9. (a) \$1.06 (b) \$.06

10. 0.01

11. \$13.50

12. About 6 (expected value is 5.8)

13. Dependent; independent

14. (a) $1 - 0.2 = 0.8$ (b) $(0.2)(0.1) = 0.02$

15. ${}_{20}C_{12} = 125{,}970$

16. ${}_{25}P_3 = 13{,}800$

17. (a) $\frac{60}{145} = \frac{12}{29}$ (b) $\frac{60}{300} = \frac{1}{5}$

18. An experimental probability is the fraction of times an event occurs in an experiment. A theoretical probability is determined with a sample space or formula. For a large number of trials, the experimental probability is expected to be close to the theoretical probability.

Section A.1

1. (a) 10 PRINT "MAD"
20 PRINT "DOG"
(b) MAD
DOG

3. 10 PRINT "TONY"
20 PRINT "LIKES"
30 PRINT "ROSA"

5. (a) 15 (b) 5 * 3 (c) 5 * 3 = 15

7. PRINT (28.9 + 37.6 + 0.8) ∧ 30

9. (a) Calculator (b) Mentally (c) Computer

11. (a) 4.68 E 05 (b) 1.3 E −03

13. PRINT (4 + 0.1) ∧ 60

15. PRINT "$"; 5000 * (1.08) ∧ 10

17. (c)

19. (a) 20 should say B = 3 and RUN should not be numbered.
(b) 10 should say LET, 14 should say 3 * A, 18 should have quotes around TOTAL PRICE IS and a semicolon before T

21. (a) 6.5 DEGREES (b) THE AREA IS 9

23. (b) It multiplies fractions.
(c) 30 PRINT A * D; "/"; B * C

25. 10 PRINT "COMPUTE A PERCENTAGE TEST SCORE"
20 PRINT "ENTER THE TOTAL NUMBER OF QUESTIONS AND THE NUMBER CORRECT SEPARATED BY A COMMA."
30 INPUT T, C
40 PRINT "YOUR SCORE IS "; 100 * C / T ; "%."

27. (a) A = 2LW + 2LH + 2HW
(b) 10 PRINT "SURFACE AREA OF A RECTANGULAR PRISM"
20 PRINT "ENTER L, W, AND H SEPARATED BY COMMAS"
30 INPUT L, W, H
40 PRINT "THE SURFACE AREA IS"; 2 * L * W + 2 * L * H + 2 * H * W ; "SQUARE UNITS."

29. 10 PRINT "FIND THE LENGTH OF THE HYPOTENUSE."
20 PRINT "ENTER THE LENGTHS OF THE LEGS SEPARATED BY A COMMA."
30 INPUT A, B
40 PRINT "THE HYPOTENUSE IS "; SQR(A ∧ 2 + B ∧ 2)

31. (a) R + N * C (b) $13.44
(c) Change the formula to (R + N * C) * 1.06.

Section A.2

1. (a) NOT QUITE (b) HONORS (c) Deductive

3. IF N >= 5 THEN C = 16 * N

5. (a) 1, 3, 5, 7 (b) 10 FOR N = 2 TO 8 STEP 2
(c) 10 FOR N = 7 TO 23 STEP 4
(d) 10 FOR N = 11 TO −1 STEP −3

7. (a) $T = 230.40 + 8.64(H - 40)$
(b) 10 PRINT "COMPUTE YOUR WEEKLY PAY."
20 PRINT "HOW MANY HOURS DID YOU WORK?"
30 INPUT H
40 IF H > 40 THEN PRINT "YOUR PAY IS $"; 230.40 + 8.64 * (H − 40)
50 IF H <= 40 THEN PRINT "YOUR PAY IS $"; 5.76 * H

9. Add the following lines.
65 IF X2 = X1 THEN PRINT "SLOPE IS UNDEFINED."
66 GOTO 90
90 END

11. (c) Below ground level

13. 10 FOR I = 1 to 20
20 PRINT 9 * I
30 NEXT I

15. (a) $A = 0.6(C - 100)$ (b) $A = 0.8(C - 500)$
(c) 10 FOR I = 1 TO 8
20 LET C = 250 * I
30 PRINT "PLAN 1 PAYS $"; 0.6 * (C − 100)
40 PRINT "PLAN 2 PAYS $"; 0.8 * (C − 500)
50 NEXT I
(d) The first policy pays more for claims under $1700; the second pays more for claims over $1700.

17. An integer

19. 2, 3, 6, 31, 62, 93, 186

21. (b) Whole numbers between 9 and 100 (not including 9 and 100)
(c) Prints all whole numbers

23. 10 PRINT "ROUNDING PROGRAM"
20 PRINT "TYPE IN A DECIMAL NUMBER."
30 INPUT N
40 PRINT INT(N + 0.5)

Index

A

B

C

M

N

O

P

S

T